ASTRONOMY AND
ASTROPHYSICS LIBRARY

More information about this series at http://www.springer.com/series/848

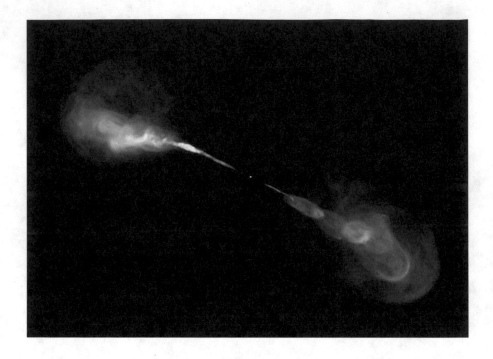

An image from the Karl G. Jansky Very Large Array of the Galaxy Hercules A (also known as 3C348) showing powerful synchrotron jets emerging from its core, the site of a supermassive black hole of 10^9 solar masses. The field center is RA = $16^h51^m8.147^s$, Dec. = $4°$ $59'$ $33.32''$ (2000), and the field of view is 3.3 × 2.4 arcmin. The image has been rotated clockwise by 36 degrees. The data set comprised 70 hours of observations acquired in 2010 and 2011 in bands from 4.2 to 9 GHz in all four array configurations with baselines from 36 m to 36 km. The image resolution is $0.3''$, corresponding to a linear scale of 800 pc at a distance of 730 Mpc, and the image contains about 10.7 Mpixels. The dynamic range is about 1200. The image has been reconstructed with a multiresolution CLEAN algorithm and self-calibration procedures described in Chapter 11. Color coded by intensity. Image from the NRAO, courtesy of B. Saxton, W. Cotton, and R. Perley (NRAO/AUI/NSF). © NRAO.

A. Richard Thompson • James M. Moran •
George W. Swenson Jr.

Interferometry and Synthesis in Radio Astronomy

Third Edition

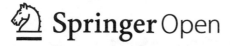

Springer Open

A. Richard Thompson
National Radio Astronomy Observatory
Charlottesville
Virginia, USA

James M. Moran
Harvard Smithsonian Center for
  Astrophysics
Cambridge
Massachusetts, USA

George W. Swenson Jr.
University of Illinois Urbana Champaign
Champaign
Illinois, USA

Previous edition published by John Wiley & Sons 2001, ISBN: 978-0-471-25492-8

ISSN 0941-7834   ISSN 2196-9698   (electronic)
Astronomy and Astrophysics Library
ISBN 978-3-319-83053-7   ISBN 978-3-319-44431-4   (eBook)
DOI 10.1007/978-3-319-44431-4

Cover illustration: The Atacama Large Millimeter/submillimeter Array on the Chajnantor Plateau of the Atacama Desert in northern Chile at 5000 m above sea level.
Credit: ALMA(ESO/NAOJ/NRAO)/W. Garnier (ALMA). ©ESO.

Printed on acid-free paper

This Springer imprint is published by Springer Nature
The registered company is Springer International Publishing AG
The registered company address is: Gewerbestrasse 11, 6330 Cham, Switzerland

To
Sheila, Barbara, Janice,
Sarah, Susan, and Michael

...truste wel that alle the conclusiouns that han been founde, or elles possibly mighten be founde in so noble an instrument as an Astrolabie, ben un-knowe perfitly to any mortal man...

GEOFFREY CHAUCER
A Treatise on the Astrolabe
circa 1391

Preface to the Third Edition

The advances in radio astronomy, especially in instrumentation for interferometry, over the past 15 years since the second edition have been remarkable. With the commissioning of the Atacama Large Millimeter/submillimeter Array (ALMA), high-resolution radio astronomy has reached the high-frequency limit of ground-based observations of about 1 THz. There has been a revitalization of interest at low frequencies, with multiple new instruments such as the LOw Frequency ARray (LOFAR), the Long Wavelength Array (LWA), and the Murchison Widefield Array (MWA). Tremendous advances in signal-processing capabilities have enabled the first instruments with multiple fields of view, the Australian SKA Pathfinder (ASKAP) and APERITIF on the Westerbork array. VLBI has reached submillimeter wavelengths and is being used by the Event Horizon Telescope (EHT) to resolve the structure of the emission surrounding the black hole in the center of our galaxy. VLBI with the elements in Earth orbit, RadioAstron and VSOP, has greatly extended the baselines available.

Much new material has been added to this edition. In Chap. 1, the historical perspective has been brought up to date. An appendix has been added where the radiometer equation, which gives the fundamental limitation in the sensitivity of a radio telescope, has been derived from basic principles. In Chap. 2, a new appendix gives an overview of the Fourier transform theory used throughout the book. Chapter 4 includes a description of the so-called measurement equation, which provides a unified framework for array calibration. Chapter 5 includes a description of the new instruments available, including the fast Fourier Transform Telescope. The discussion of system design has been substantially expanded in Chap. 7. In Chap. 8, which deals with digital signal processing, the coverage of FX-type correlators has been greatly expanded and the operation of polyphase filter banks explained. The analysis of sensitivity loss due to quantization has been generalized. An appendix describing the basic properties of the discrete Fourier transform has been added. Chapter 9 on VLBI has been updated to reflect the conversion from data storage on tape to data storage on disk media. With the prevalence of direct data transmission to correlation facilities, the distinction between VLBI and connected-element interferometry continues to diminish. In Chap. 10, the discussion of model fitting in the (u, v) plane has been greatly expanded to reflect a trend in the field toward fitting the fundamental interferometric data even though image fidelity continues to improve dramatically. The phase and amplitude closure conditions are explored in greater depth because of their underlying importance in data calibration. In Chap. 11, advances in image processing algorithms are described, including the application of compressed sensing techniques. Chapter 12 describes the techniques underlying the tremendous advance in astrometry. Precisions of 10 microarcseconds are now routine as a result of progress in phase-referencing methods. In this edition, discussion of the propagation of the neutral atmosphere and the ionized media from the ionosphere to the interstellar medium has been separated into two chapters, Chaps. 13 and 14, because of the growth in information in these areas. Over the last 15 years, enormous amounts of data have been acquired on site characterization, which are described in Chap. 13. Because of the importance of both two- and three-dimensional turbulence in the troposphere, a detailed analysis of the two regimes is given. Chapter 17, on related techniques, includes new material on the use of radio arrays to track satellites and space debris. It also describes the application of radio interferometry to remote sensing of the Earth. Such application provides important information on soil moisture and ocean salinity.

In the early days of radio interferometry, measurements of the distribution of source intensity were usually referred to as "maps" and the associated technique as "mapping." With the maturity of the field, it seems more appropriate to refer to the results as "images." We have done so, except in a few cases where the term "map" still seems appropriate, as in the determination of the distribution of maser spot positions from fringe rate measurements.

Readers who are new to the field of radio astronomy are strongly encouraged to study the basic principles of the field from other sources. Some of the numerous textbooks are listed under Further Reading at the end of Chap. 1. Of particular usefulness is the book *The Fourer Transform and Its Applications* by Ron Bracewell, a radio astronomer and mathematician, because of its practical approach to the subject. The intellectual roots of this approach can be traced to the lecture notes of J. A. Ratcliffe of Cambridge University, which inspired the book *Fourier Transforms and Convolutions for the Experimentalist* by Roger Jennison.

The authors would be grateful for any feedback from the readers of this book in regard to pedogogical, technical, or grammatical issues or typographical errors.

We have benefited greatly from many of our colleagues who have helped in the preparation of this edition. They include Betsey Adams, Kazunori Akiyama, Subra Ananthakrishnan, Yoshiharu Asaki, Jaap Baars, Denis Barkats, Norbert Bartel, Leo Benkevitch, Mark Birkinshaw, Katie Bouman, Geoff Bower, Michael Bremer, John Bunton, Andrew Chael, Barry Clark, Tim Cornwell, Pierre Cox, Adam Deller, Hélène Dickel, Phil Edwards, Ron Ekers, Pedro Elosegui, Phil Erickson, Hugh Garsden, John Gibson, Lincoln Greenhill, Richard Hills, Mareki Honma, Chat Hull, Michael Johnson, Ken Kellermann, Eric Keto, Robert Kimberk, Jonathon Kocz, Vladimir Kostenko, Yuri Kovalev, Laurent Loinard, Colin Lonsdale, Ryan Loomis, Chopo Ma, Dick Manchester, Satoki Matsushita, John McKean, Russ McWhirter, Arnaud Mialon, George Miley, Eric Murphy, Tara Murphy, Ramesh

Narayan, Scott Paine, Nimesh Patel, Michael Pearlman, Richard Plambeck, Danny Price, Rurik Primiani, Simon Radford, Mark Reid, Maria Rioja, Luis Rodríguez, Nemesio Rodríguez-Fernández, Alan Rogers, Jon Romney, Katherine Rosenfeld, Jean Rüeger, Marion Schmitz, Fred Schwab, Mamoru Sekido, T. K. Sridharan, Anjali Tripathi, Harish Vedantham, Jonathan Weintroub, Alan Whitney, David Wilner, Robert Wilson, and Andre Young.

JM taught a graduate course in radio astronomy at Harvard University biannually for 40 years. He thanks the hundreds of students who took this course for the feedback, stimulation, and challenges they posed.

The publication of this edition under an Open Access license was made possible by grants from the D. H. Menzel Fund at Harvard University and the National Radio Astronomy Observatory. We are particularly grateful to Charles Alcock, director of the Harvard–Smithsonian Center for Astrophysics, and Anthony Beasley, director of the National Radio Astronomy Observatory, for their generous support of all aspects of this project.

We thank John Lewis for much help with the graphics and other creative contributions that improved the presentation of material in this book. We are also grateful to Tania Burchell, Maureen Connors, Christopher Erdmann, Muriel Hodges, Carolyn Hunsinger, Clinton Leite, Robert Reifsnyder, and Larry Selter for their valuable support.

The publication of this edition would not have been possible without the tireless and expert assistance of Carolann Barrett of Harvard University. An experienced editor with a degree in mathematics, she completed both our sentences and our equations. Her capacity to hold every detail of the book in her brain is truly amazing.

Charlottesville, VA, USA	A. Richard Thompson
Cambridge, MA, USA	James M. Moran
Urbana, IL, USA	George W. Swenson Jr.
June 2016	

Preface to the Second Edition

Half a century of remarkable scientific progress has resulted from the application of radio interferometry to astronomy. Advances since 1986, when this book was first published, have resulted in the VLBA (Very Long Baseline Array), the first array fully dedicated to very-long-baseline interferometry (VLBI), the globalization of VLBI networks with the inclusion of antennas in orbit, the increasing importance of spectral line observations, and the improved instrumental performance at both ends of the radio spectrum. At the highest frequencies, millimeter-wavelength arrays of the Berkeley–Illinois–Maryland Association (BIMA), the Institut de Radio Astronomie Millimétrique (IRAM), the Nobeyama Radio Observatory (NRO), and the Owens Valley Radio Observatory (OVRO), which were in their infancy in 1986, have been greatly expanded in their capabilities. The Submillimeter Array (SMA) and the Atacama Large Millimeter/submillimeter Array (ALMA), a major international project at millimeter and submillimeter wavelengths, are under development. At low frequencies, with their special problems involving the ionosphere and wide-field mapping, the frequency coverage of the Very Large Array (VLA) has been extended down to 75 MHz, and the Giant Metrewave Radio Telescope (GMRT), operating down to 38 MHz, has been commissioned. The Australia Telescope and the expanded Multi-Element Radio Linked Interferometer Network (MERLIN) have provided increased capability at centimeter wavelengths.

Such progress has led to this revised edition, the intent of which is not only to bring the material up to date but also to expand its scope and improve its comprehensibility and general usefulness. In a few cases, symbols used in the first edition have been changed to follow the general usage that is becoming established in radio astronomy. Every chapter contains new material, and there are new figures and many new references. Material in the original Chap. 3 that was peripheral to the basic discussion has been condensed and moved to a later chapter. Chapter 3 now contains the essential analysis of the response of an interferometer. The section on polarization in Chap. 4 has been substantially expanded, and a brief introduction to antenna theory has been added to Chap. 5. Chapter 6 contains a discussion of the sensitivity for a wide variety of instrumental configurations. A discussion of spectral line observations is included in Chap. 10. Chapter 13 has been expanded to include a description of the new techniques for atmospheric phase correction, and site-testing data and techniques at millimeter wavelengths. Chapter 14 has been added and contains an examination of the van Cittert–Zernike theorem and discussions of spatial coherence and scattering, some of which is derived from the original Chap. 3.

Special thanks are due to a number of people for reviews or other help during the course of the revision. These include D. C. Backer, J. W. Benson, M. Birkinshaw, G. A. Blake, R. N. Bracewell, B. F. Burke, B. Butler, C. L. Carilli, B. G. Clark, J. M. Cordes, T. J. Cornwell, L. R. D'Addario, T. M. Dame, J. Davis, J. L. Davis, D. T. Emerson, R. P. Escoffier, E. B. Fomalont, L. J. Greenhill, M. A. Gurwell, C. R. Gwinn, K. I. Kellermann, A. R. Kerr, E. R. Keto, S. R. Kulkarni, S. Matsushita, D. Morris, R. Narayan, S.-K. Pan, S. J. E. Radford, R. Rao, M. J. Reid, A. Richichi, A. E. E. Rogers, J. E. Salah, F. R. Schwab, S. R. Spangler, E. C. Sutton, B. E. Turner, R. F. C. Vessot, W. J. Welch, M. C. Wiedner, and J.-H. Zhao. For major contributions to the preparation of the text and diagrams, we thank J. Heidenreich, G. L. Kessler, P. Smiley, S. Watkins, and P. Winn. For extensive help in preparation and editing, we are especially indebted to P. L. Simmons. We are grateful to P. A. Vanden Bout, director of the National Radio Astronomy Observatory, and to I. I. Shapiro, director of the Harvard–Smithsonian Center for Astrophysics, for the encouragement and support. The National Radio Astronomy Observatory is operated by Associated Universities Inc. under contract with the National Science Foundation, and the Harvard–Smithsonian Center for Astrophysics is operated by Harvard University and the Smithsonian Institution.

Charlottesville, VA, USA
Cambridge, MA, USA
Urbana, IL, USA
November 2000

A. Richard Thompson
James M. Moran
George W. Swenson Jr.

Preface to the First Edition

The techniques of radio interferometry as applied to astronomy and astrometry have developed enormously in the past four decades, and the attainable angular resolution has advanced from degrees to milliarcseconds, a range of more than six orders of magnitude. As arrays for synthesis mapping[1] have developed, techniques in the radio domain have overtaken those in optics in providing the finest angular detail in astronomical images. The same general developments have introduced new capabilities in astrometry and in the measurement of the Earth's polar and crustal motions. The theories and techniques that underlie these advances continue to evolve but have reached by now a sufficient state of maturity that it is appropriate to offer a detailed exposition.

The book is intended primarily for graduate students and professionals in astronomy, electrical engineering, physics, or related fields who wish to use interferometric or synthesis-mapping techniques in astronomy, astrometry, or geodesy. It is also written with radio systems engineers in mind and includes discussions of important parameters and tolerances for the types of instruments involved. Our aim is to explain the underlying principles of the relevant interferometric techniques but to limit the discussion of details of implementation. Such details of the hardware and the software are largely specific to particular instruments and are subject to change with developments in electronic engineering and computing techniques. With an understanding of the principles involved, the reader should be able to comprehend the instructions and instrumental details that are encountered in the user-oriented literature of most observatories.

The book does not stem from any course of lectures, but the material included is suitable for a graduate-level course. A teacher with experience in the techniques described should be able to interject easily any necessary guidance to emphasize astronomy, engineering, or other aspects as required.

---

[1] We define synthesis mapping as the reconstruction of images from measurements of the Fourier transforms of their brightness distributions. In this book, the terms map, image, and brightness (intensity) distribution are largely interchangeable.

The first two chapters contain a brief review of radio astronomy basics, a short history of the development of radio interferometry, and a basic discussion of the operation of an interferometer. Chapter 3 discusses the underlying relationships of interferometry from the viewpoint of the theory of partial coherence and may be omitted from a first reading. Chapter 4 introduces coordinate systems and parameters that are required to describe synthesis mapping. It is appropriate then to examine configurations of antennas for multielement synthesis arrays in Chap. 5. Chapters 6–8 deal with various aspects of the design and response of receiving systems, including the effects of quantization in digital correlators. The special requirements of very-long-baseline interferometry (VLBI) are discussed in Chap. 9. The foregoing material covers in detail the measurement of complex visibility and leads to the derivation of radio maps discussed in Chaps. 10 and 11. The former presents the basic Fourier transformation method and the latter the more powerful algorithms that incorporate both calibration and transformation. Precision observations in astrometry and geodesy are the subject of Chap. 12. There follow discussions of factors that can degrade the overall performance, namely, effects of propagation in the atmosphere, the interplanetary medium, and the interstellar medium in Chap. 13 and radio interference in Chap. 14. Propagation effects are discussed at some length since they involve a wide range of complicated phenomena that place fundamental limits on the measurement accuracy. The final chapter describes related techniques including intensity interferometry, speckle interferometry, and lunar occultation observations.

References are included to seminal papers and to many other publications and reviews that are relevant to the topics of the book. Numerous descriptions of instruments and observations are also referenced for purposes of illustration. Details of early procedures are given wherever they are of help in elucidating the principles or origin of current techniques, or because they are of interest in their own right. Because of the diversity of the phenomena described, it has been necessary, in some cases, to use the same mathematical symbol for different quantities. A glossary of principal symbols and usage follows the final chapter.

The material in this book comes only in part from the published literature, and much of it has been accumulated over many years from discussions, seminars, and the unpublished reports and memoranda of various observatories. Thus, we acknowledge our debt to colleagues too numerous to mention individually. Our special thanks are due to a number of people for critical reviews of portions of the book or for other support. These include D. C. Backer, D. S. Bagri, R. H. T. Bates, M. Birkinshaw, R. N. Bracewell, B. G. Clark, J. M. Cordes, T. J. Cornwell, L. R. D'Addario, J. L. Davis, R. D. Ekers, J. V. Evans, M. Faucherre, S. J. Franke, J. Granlund, L. J. Greenhill, C. R. Gwinn, T. A. Herring, R. J. Hill, W. A. Jeffrey, K. I. Kellermann, J. A. Klobuchar, R. S. Lawrence, J. M. Marcaide, N. C. Mathur, L. A. Molnar, P. C. Myers, P. J. Napier, P. Nisenson, H. V. Poor, M. J. Reid, J. T. Roberts, L. F. Rodriguez, A. E. E. Rogers, A. H. Rots, J. E. Salah, F. R. Schwab, I. I. Shapiro, R. A. Sramek, R. Stachnik, J. L. Turner, R. F. C. Vessot, N. Wax, and W. J. Welch. The reproduction of diagrams from other publications is acknowledged in the captions, and we thank the authors and the publishers concerned for the permission to use this material. For major contributions to the preparation of the manuscript, we wish to thank C. C. Barrett, C. F. Burgess, N. J. Diamond, J. M. Gillberg, J. G. Hamwey, E. L. Haynes, G. L. Kessler, K. I. Maldonis, A. Patrick, V. J. Peterson, S. K. Rosenthal, A. W. Shepherd, J. F. Singarella, M. B. Weems, and C. H. Williams. We are grateful to M. S. Roberts and P. A. Vanden Bout, former director and present director of the National Radio Astronomy Observatory, and to G. B. Field and I. I. Shapiro, former director and present director of the Harvard–Smithsonian Center for Astrophysics, for the encouragement and support. Much of the contribution by J. M. Moran was written while on sabbatical leave at the Radio Astronomy Laboratory of the University of California, Berkeley, and he is grateful to W. J. Welch for the hospitality during that period. G. W. Swenson Jr. thanks the Guggenheim Foundation for a fellowship during 1984–1985. Finally, we acknowledge the support of our home institutions: the National Radio Astronomy Observatory, which is operated by Associated Universities Inc. under contract with the National Science Foundation; the Harvard–Smithsonian Center for Astrophysics, which is operated by Harvard University and the Smithsonian Institution; and the University of Illinois.

Charlottesville, VA, USA A. Richard Thompson
Cambridge, MA, USA James M. Moran
Urbana, IL, USA George W. Swenson Jr.
January 1986

Contents

Preface to the Third Edition ix

Preface to the Second Edition xiii

Preface to the First Edition xv

Abbreviations and Acronyms xxxi

Principal Symbols xxxv

1   Introduction and Historical Review 1
    1.1   Applications of Radio Interferometry 1
    1.2   Basic Terms and Definitions 3
            1.2.1   Cosmic Signals 4
            1.2.2   Source Positions and Nomenclature 10
            1.2.3   Reception of Cosmic Signals 11
    1.3   Development of Radio Interferometry 13
            1.3.1   Evolution of Synthesis Techniques 13
            1.3.2   Michelson Interferometer 14
            1.3.3   Early Two-Element Radio Interferometers 18
            1.3.4   Sea Interferometer 20
            1.3.5   Phase-Switching Interferometer 21
            1.3.6   Optical Identifications and Calibration Sources 23
            1.3.7   Early Measurements of Angular Width 24
            1.3.8   Early Survey Interferometers and the Mills Cross 26
            1.3.9   Centimeter-Wavelength Solar Imaging 28
            1.3.10   Measurements of Intensity Profiles 30
            1.3.11   Spectral Line Interferometry 31
            1.3.12   Earth-Rotation Synthesis Imaging 31
            1.3.13   Development of Synthesis Arrays 34

1.3.14   Very-Long-Baseline Interferometry   37
1.3.15   VLBI Using Orbiting Antennas   42
1.4   Quantum Effect   44
Appendix 1.1   Sensitivity of Radio Astronomical Receivers
(the Radiometer Equation)   46
Further Reading   49
References   51

2   Introductory Theory of Interferometry and Synthesis Imaging   59
2.1   Planar Analysis   59
2.2   Effect of Bandwidth   62
2.3   One-Dimensional Source Synthesis   66
2.3.1   Interferometer Response as a Convolution   66
2.3.2   Convolution Theorem and Spatial Frequency   69
2.3.3   Example of One-Dimensional Synthesis   70
2.4   Two-Dimensional Synthesis   73
2.4.1   Projection-Slice Theorem   74
2.4.2   Three-Dimensional Imaging   76
Appendix 2.1   A Practical Fourier Transform Primer   76
A2.1.1   Useful Fourier Transform Pairs   78
A2.1.2   Basic Fourier Transform Properties   80
A2.1.3   Two-Dimensional Fourier Transform   85
A2.1.4   Fourier Series   86
A2.1.5   Truncated Functions   86
References   87

3   Analysis of the Interferometer Response   89
3.1   Fourier Transform Relationship Between Intensity
and Visibility   89
3.1.1   General Case   89
3.1.2   East–West Linear Arrays   95
3.2   Cross-Correlation and the Wiener–Khinchin Relation   98
3.3   Basic Response of the Receiving System   99
3.3.1   Antennas   100
3.3.2   Filters   101
3.3.3   Correlator   102
3.3.4   Response to the Incident Radiation   102
Appendix 3.1   Mathematical Representation of Noiselike Signals ...   104
A3.1.1   Analytic Signal   104
A3.1.2   Truncated Function   107
References   108

**4   Geometrical Relationships, Polarimetry,
and the Interferometer Measurement Equation**   109
4.1   Antenna Spacing Coordinates and (u, v) Loci   109
4.2 (u', v') Plane   113

4.3		Fringe Frequency	115
4.4		Visibility Frequencies	115
4.5		Calibration of the Baseline	117
4.6		Antennas	118
	4.6.1	Antenna Mounts	118
	4.6.2	Beamwidth and Beam-Shape Effects	120
4.7		Polarimetry	121
	4.7.1	Antenna Polarization Ellipse	123
	4.7.2	Stokes Visibilities	126
	4.7.3	Instrumental Polarization	129
	4.7.4	Matrix Formulation	134
	4.7.5	Calibration of Instrumental Polarization	137
4.8		The Interferometer Measurement Equation	142
	4.8.1	Multibaseline Formulation	143
Appendix 4.1		Hour Angle–Declination and Elevation–Azimuth Relationships	146
Appendix 4.2		Leakage Parameters in Terms of the Polarization Ellipse	146
	A4.2.1	Linear Polarization	147
	A4.2.2	Circular Polarization	148
References			149

5 Antennas and Arrays   **153**
5.1		Antennas	153
5.2		Sampling the Visibility Function	157
	5.2.1	Sampling Theorem	157
	5.2.2	Discrete Two-Dimensional Fourier Transform	159
5.3		Introductory Discussion of Arrays	162
	5.3.1	Phased Arrays and Correlator Arrays	162
	5.3.2	Spatial Sensitivity and the Spatial Transfer Function	164
	5.3.3	Meter-Wavelength Cross and T-Shaped Arrays	168
5.4		Spatial Transfer Function of a Tracking Array	169
	5.4.1	Desirable Characteristics of the Spatial Transfer Function	172
	5.4.2	Holes in the Spatial Frequency Coverage	173
5.5		Linear Tracking Arrays	173
5.6		Two-Dimensional Tracking Arrays	179
	5.6.1	Open-Ended Configurations	180
	5.6.2	Closed Configurations	183
	5.6.3	VLBI Configurations	187
	5.6.4	Orbiting VLBI Antennas	188
	5.6.5	Planar Arrays	192
	5.6.6	Some Conclusions on Antenna Configurations	193

5.7   Implementation of Large Arrays 194
      5.7.1   Low-Frequency Range 195
      5.7.2   Midfrequency and Higher Ranges 197
           5.7.2.1   Phased-Array Feeds 197
           5.7.2.2   Optimum Antenna Size 198
      5.7.3   Development of Extremely Large Arrays 199
      5.7.4   The Direct Fourier Transform Telescope 199
Further Reading 201
References 201

6   Response of the Receiving System 207
    6.1   Frequency Conversion, Fringe Rotation,
          and Complex Correlators 207
      6.1.1   Frequency Conversion 207
      6.1.2   Response of a Single-Sideband System 209
      6.1.3   Upper-Sideband Reception 210
      6.1.4   Lower-Sideband Reception 212
      6.1.5   Multiple Frequency Conversions 213
      6.1.6   Delay Tracking and Fringe Rotation 213
      6.1.7   Simple and Complex Correlators 214
      6.1.8   Response of a Double-Sideband System 215
      6.1.9   Double-Sideband System with Multiple
            Frequency Conversions 218
      6.1.10   Fringe Stopping in a Double-Sideband System 220
      6.1.11   Relative Advantages of Double- and
            Single-Sideband Systems 221
      6.1.12   Sideband Separation 221
    6.2   Response to the Noise 223
      6.2.1   Signal and Noise Processing in the Correlator 223
      6.2.2   Noise in the Measurement of Complex Visibility 228
      6.2.3   Signal-to-Noise Ratio in a Synthesized Image 230
      6.2.4   Noise in Visibility Amplitude and Phase 233
      6.2.5   Relative Sensitivities of Different
            Interferometer Systems 235
      6.2.6   System Temperature Parameter α 239
    6.3   Effect of Bandwidth 240
      6.3.1   Imaging in the Continuum Mode 240
      6.3.2   Wide-Field Imaging with a Multichannel System 245
    6.4   Effect of Visibility Averaging 246
      6.4.1   Visibility Averaging Time 246
      6.4.2   Effect of Time Averaging 246
    6.5   Speed of Surveying 249
    Appendix 6.1   Partial Rejection of a Sideband 251
    References 253

7   System Design   255
  7.1   Principal Subsystems of the Receiving Electronics   255
         7.1.1   Low-Noise Input Stages   256
         7.1.2   Noise Temperature Measurement   257
         7.1.3   Local Oscillator   261
         7.1.4   IF and Signal Transmission Subsystems...............   261
         7.1.5   Optical Fiber Transmission............................   262
         7.1.6   Delay and Correlator Subsystems.....................   264
  7.2   Local Oscillator and General Considerations
         of Phase Stability   264
         7.2.1   Round-Trip Phase Measurement Schemes   264
         7.2.2   Swarup and Yang System   265
         7.2.3   Frequency-Offset Round-Trip System   266
         7.2.4   Automatic Correction System.........................   272
         7.2.5   Fiberoptic Transmission of LO Signals................   273
         7.2.6   Phase-Locked Loops and Reference Frequencies   274
         7.2.7   Phase Stability of Filters   276
         7.2.8   Effect of Phase Errors   277
  7.3   Frequency Responses of the Signal Channels   277
         7.3.1   Optimum Response   277
         7.3.2   Tolerances on Variation of the Frequency
                   Response: Degradation of Sensitivity..................   279
         7.3.3   Tolerances on Variation of the Frequency
                   Response: Gain Errors..................................   281
         7.3.4   Delay and Phase Errors in Single- and
                   Double-Sideband Systems   282
         7.3.5   Delay Errors and Tolerances   283
         7.3.6   Phase Errors and Degradation of Sensitivity   284
         7.3.7   Other Methods of Mitigation of Delay Errors   286
         7.3.8   Multichannel (Spectral Line) Correlator Systems.....   287
         7.3.9   Double-Sideband Systems   288
  7.4   Polarization Mismatch Errors.....................................   289
  7.5   Phase Switching   290
         7.5.1   Reduction of Response to Spurious Signals   290
         7.5.2   Implementation of Phase Switching   291
         7.5.3   Timing Accuracy in Phase Switching   296
         7.5.4   Interaction of Phase Switching with Fringe
                   Rotation and Delay Adjustment.......................   298
  7.6   Automatic Level Control and Gain Calibration   299
  7.7   Fringe Rotation   300
  Appendix 7.1   Sideband-Separating Mixers   301
  Appendix 7.2   Dispersion in Optical Fiber...........................   302
  Appendix 7.3   Alias Sampling   303
  References...   304

8   Digital Signal Processing 309
   8.1   Bivariate Gaussian Probability Distribution 311
   8.2   Periodic Sampling 312
       8.2.1   Nyquist Rate 312
       8.2.2   Correlation of Sampled but Unquantized
                Waveforms 313
   8.3   Sampling with Quantization 316
       8.3.1   Two-Level Quantization 316
       8.3.2   Four-Level Quantization 320
       8.3.3   Three-Level Quantization 327
       8.3.4   Quantization Efficiency: Simplified
                Analysis for Four or More Levels 328
       8.3.5   Quantization Efficiency: Full Analysis,
                Three or More Levels 332
       8.3.6   Correlation Estimates for Strong Sources 336
   8.4   Further Effects of Quantization 338
       8.4.1   Correlation Coefficient for Quantized Data 339
       8.4.2   Oversampling 344
       8.4.3   Quantization Levels and Data Processing 346
   8.5   Accuracy in Digital Sampling 347
       8.5.1   Tolerances in Digital Sampling Levels 348
   8.6   Digital Delay Circuits 351
   8.7   Quadrature Phase Shift of a Digital Signal 352
   8.8   Digital Correlators 353
       8.8.1   Correlators for Continuum Observations 353
       8.8.2   Digital Spectral Line Measurements 353
       8.8.3   Lag (XF) Correlator 358
       8.8.4   FX Correlator 360
       8.8.5   Comparison of XF and FX Correlators 360
       8.8.6   Hybrid Correlator 366
       8.8.7   Demultiplexing in Broadband Correlators 367
       8.8.8   Examples of Bandwidths and Bit Data
                Quantization 368
       8.8.9   Polyphase Filter Banks 369
       8.8.10   Software Correlators 373
   Appendix 8.1   Evaluation of $\Sigma_{q=1}^{\infty} R_{\infty}^2 (q\tau_s)$ 375
   Appendix 8.2   Probability Integral for Two-Level Quantization 376
   Appendix 8.3   Optimal Performance for Four-Level Quantization ... 377
   Appendix 8.4   Introduction to the Discrete Fourier Transform 378
       A8.4.1   Response to a Complex Sine Wave 382
       A8.4.2   Padding with Zeros 384
   Further Reading 386
   References 387

9 Very-Long-Baseline Interferometry 391
   9.1 Early Development 391
   9.2 Differences Between VLBI and Conventional Interferometry ... 394
      9.2.1 The Problem of Field of View 396
   9.3 Basic Performance of a VLBI System 399
      9.3.1 Time and Frequency Errors 399
      9.3.2 Retarded Baselines 405
      9.3.3 Noise in VLBI Observations 407
      9.3.4 Probability of Error in the Signal Search 412
      9.3.5 Coherent and Incoherent Averaging 415
   9.4 Fringe Fitting for a Multielement Array 419
      9.4.1 Global Fringe Fitting 419
      9.4.2 Relative Performance of Fringe Detection
               Methods 422
      9.4.3 Triple Product, or Bispectrum......................... 423
      9.4.4 Fringe Searching with a Multielement Array 424
      9.4.5 Multielement Array with Incoherent Averaging 425
   9.5 Phase Stability and Atomic Frequency Standards 425
      9.5.1 Analysis of Phase Fluctuations 426
      9.5.2 Oscillator Coherence Time 434
      9.5.3 Precise Frequency Standards 436
      9.5.4 Rubidium and Cesium Standards 440
      9.5.5 Hydrogen Maser Frequency Standard 442
      9.5.6 Local Oscillator Stability 446
      9.5.7 Phase Calibration System 447
      9.5.8 Time Synchronization 447
   9.6 Data Storage Systems 448
   9.7 Processing Systems and Algorithms 453
      9.7.1 Fringe Rotation Loss (η_R) 454
      9.7.2 Fringe Sideband Rejection Loss (η_S) 457
      9.7.3 Discrete Delay Step Loss (η_D) 459
      9.7.4 Summary of Processing Losses 461
   9.8 Bandwidth Synthesis 462
      9.8.1 Burst Mode Observing 465
   9.9 Phased Arrays as VLBI Elements 466
   9.10 Orbiting VLBI (OVLBI) 470
   9.11 Satellite Positioning 473
   Further Reading 476
   References... 477

10 Calibration and Imaging.. 485
   10.1 Calibration of the Visibility 485
      10.1.1 Corrections for Calculable or Directly
               Monitored Effects.................................... 486
      10.1.2 Use of Calibration Sources 487

10.2   Derivation of Intensity from Visibility   490
        10.2.1   Imaging by Direct Fourier Transformation   490
        10.2.2   Weighting of the Visibility Data   491
                 10.2.2.1   Robust Weighting   494
        10.2.3   Imaging by Discrete Fourier Transformation   495
        10.2.4   Convolving Functions and Aliasing   497
        10.2.5   Aliasing and the Signal-to-Noise Ratio................   501
        10.2.6   Wide-Field Imaging   503
10.3   Closure Relationships   505
10.4   Visibility Model Fitting   510
        10.4.1   Basic Considerations for Simple Models..............   511
        10.4.2   Examples of Parameter Fitting to Models   514
        10.4.3   Modeling Azimuthally Symmetric Sources   518
        10.4.4   Modeling of Very Extended Sources   520
10.5   Spectral Line Observations   522
        10.5.1   VLBI Observations of Spectral Lines   524
        10.5.2   Variation of Spatial Frequency Over the
                 Bandwidth...   527
        10.5.3   Accuracy of Spectral Line Measurements.............   528
        10.5.4   Presentation and Analysis of Spectral Line
                 Observations   528
10.6   Miscellaneous Considerations   530
        10.6.1   Interpretation of Measured Intensity...................   530
        10.6.2   Ghost Images   530
        10.6.3   Errors in Images   532
        10.6.4   Hints on Planning and Reduction of Observations....   534
10.7   Observations of Cosmological Fine Structure   535
        10.7.1   Cosmic Microwave Background   535
        10.7.2   Epoch of Reionization   536
Appendix 10.1   The Edge of the Moon as a Calibration Source   537
Appendix 10.2   Doppler Shift of Spectral Lines   538
Appendix 10.3   Historical Notes   543
        A10.3.1   Images from One-Dimensional Profiles   543
        A10.3.2   Analog Fourier Transformation   544
Further Reading   544
References..   545

11   Further Imaging Techniques   551
11.1   The CLEAN Deconvolution Algorithm   551
        11.1.1   CLEAN Algorithm   552
        11.1.2   Implementation and Performance
                 of the CLEAN Algorithm   553
11.2   Maximum Entropy Method   557
        11.2.1   MEM Algorithm...   557
        11.2.2   Comparison of CLEAN and MEM   559
        11.2.3   Further Deconvolution Procedures....................   562

11.3   Adaptive Calibration and Imaging............................... 563
        11.3.1   Hybrid Imaging 563
        11.3.2   Self-Calibration 565
        11.3.3   Imaging with Visibility Amplitude Data Only 569
11.4   Imaging with High Dynamic Range............................. 569
11.5   Mosaicking 570
        11.5.1   Methods of Producing the Mosaic Image 573
        11.5.2   Short-Baseline Measurements 575
11.6   Multifrequency Synthesis 578
11.7   Noncoplanar Baselines....................................... 579
11.8   Some Special Techniques of Image Analysis 585
        11.8.1   Use of CLEAN and Self-Calibration
                  with Spectral Line Data 585
        11.8.2   A-Projection...................................... 586
        11.8.3   Peeling 587
        11.8.4   Low-Frequency Imaging 587
        11.8.5   Lensclean 589
        11.8.6   Compressed Sensing................................ 590
Further Reading 593
References.. 593

12   Interferometer Techniques for Astrometry and Geodesy 599
12.1   Requirements for Astrometry 600
        12.1.1   Reference Frames.................................. 602
12.2   Solution for Baseline and Source-Position Vectors.............. 603
        12.2.1   Phase Measurements 603
        12.2.2   Measurements with VLBI Systems 606
        12.2.3   Phase Referencing (Position) 610
        12.2.4   Phase Referencing (Frequency) 615
12.3   Time and Motion of the Earth 616
        12.3.1   Precession and Nutation 616
        12.3.2   Polar Motion 617
        12.3.3   Universal Time.................................... 619
        12.3.4   Measurement of Polar Motion and UT1 620
12.4   Geodetic Measurements....................................... 622
12.5   Proper Motion and Parallax Measurements 623
12.6   Solar Gravitational Deflection 628
12.7   Imaging Astronomical Masers................................. 631
Appendix 12.1   Least-Mean-Squares Analysis........................ 636
        A12.1.1   Linear Case 636
        A12.1.2   Nonlinear Case................................... 646
        A12.1.3 (u, v) vs. Image Plane Fitting 646
Appendix 12.2   Second-Order Effects in Phase Referencing........... 648
Further Reading 649
References.. 650

13   Propagation Effects: Neutral Medium 657
   13.1   Theory 658
       13.1.1   Basic Physics 658
       13.1.2   Refraction and Propagation Delay 663
       13.1.3   Absorption 670
       13.1.4   Origin of Refraction 674
       13.1.5   Radio Refractivity 679
       13.1.6   Phase Fluctuations 680
       13.1.7   Kolmogorov Turbulence.............................. 685
       13.1.8   Anomalous Refraction................................ 692
   13.2   Site Evaluation and Data Calibration............................ 693
       13.2.1   Opacity Measurements 693
       13.2.2   Site Testing by Direct Measurement
             of Phase Stability 697
   13.3   Calibration via Atmospheric Emission 701
       13.3.1   Continuum Calibration 701
       13.3.2   22-GHz Water-Vapor Radiometry 703
       13.3.3   183-GHz Water-Vapor Radiometry 705
   13.4   Reduction of Atmospheric Phase Errors by Calibration.......... 708
   Appendix 13.1   Importance of the 22-GHz Line in WWII
             Radar Development..................................... 711
   Appendix 13.2   Derivation of the Tropospheric Phase
             Structure Function 713
   Further Reading 717
   References.. 718

14   Propagation Effects: Ionized Media 725
   14.1   Ionosphere 725
       14.1.1   Basic Physics 726
       14.1.2   Refraction and Propagation Delay 730
       14.1.3   Calibration of Ionospheric Delay 734
       14.1.4   Absorption 735
       14.1.5   Small- and Large-Scale Irregularities.................. 735
   14.2   Scattering Caused by Plasma Irregularities 737
       14.2.1   Gaussian Screen Model 737
       14.2.2   Power-Law Model 742
   14.3   Interplanetary Medium... 744
       14.3.1   Refraction 744
       14.3.2   Interplanetary Scintillation (IPS) 748
   14.4   Interstellar Medium 750
       14.4.1   Dispersion and Faraday Rotation 751
       14.4.2   Diffractive Scattering 753
       14.4.3   Refractive Scattering 756

Appendix 14.1   Refractive Bending in the Ionosphere 758
Further Reading 761
References 761

15   Van Cittert–Zernike Theorem, Spatial Coherence,
and Scattering 767
15.1   Van Cittert–Zernike Theorem 767
    15.1.1   Mutual Coherence of an Incoherent Source 769
    15.1.2   Diffraction at an Aperture and the Response
        of an Antenna 771
    15.1.3   Assumptions in the Derivation and
        Application of the van Cittert–Zernike Theorem 774
15.2   Spatial Coherence 776
    15.2.1   Incident Field 776
    15.2.2   Source Coherence 777
    15.2.3   Completely Coherent Source 780
15.3   Scattering and the Propagation of Coherence 781
References 785

16   Radio Frequency Interference 787
16.1   Detection of Interference.. 788
    16.1.1   Low-Frequency Radio Environment 790
16.2   Removal of Interference 790
    16.2.1   Nulling for Attenuation of Interfering Signals 791
    16.2.2   Further Considerations of Deterministic Nulling 792
    16.2.3   Adaptive Nulling in the Synthesized Beam 793
16.3   Estimation of Harmful Thresholds 793
    16.3.1   Short- and Intermediate-Baseline Arrays 796
    16.3.2   Fringe-Frequency Averaging 796
    16.3.3   Decorrelation of Broadband Signals 800
16.4   Very-Long-Baseline Systems 801
16.5   Interference from Airborne and Space Transmitters.............. 804
16.6   Regulation of the Radio Spectrum................................ 805
Further Reading 806
References 806

17   Related Techniques 809
17.1   Intensity Interferometer 809
17.2   Lunar Occultation Observations 814
17.3   Measurements on Antennas 819
17.4   Detection and Tracking of Space Debris 824
17.5   Earth Remote Sensing by Interferometry 825
17.6   Optical Interferometry 827
    17.6.1   Instruments and Their Usage......................... 829
    17.6.2   Sensitivity of Direct Detection
        and Heterodyne Systems 833

17.6.3   Optical Intensity Interferometer   835
17.6.4   Speckle Imaging   835
Further Reading in Optical Interferometry   838
References..   839

Author Index..   845

Subject Index   855

Abbreviations and Acronyms

3C	Third Cambridge Catalog of Radio Sources
3CR	Revised Cambridge Catalog of Radio Sources
AGN	Active galactic nuclei
AIPS	Astronomical Image Processing System
ALC	Automatic level control
ALMA	Atacama Large Millimeter/submillimeter Array
AM	Atmospheric model (atmospheric modeling code)
APERTIF	Aperture Tile in Focus
ASKAP	Australian Square Kilometre Array Pathfinder
ATM	Atmospheric transmission of microwaves (atmospheric modeling code)
AU	Astronomical unit
AUI	Associated Universities Inc.
B	Besselian
BLH	Bureau International de L'Heure
bpi	Bits per inch
CARMA	Combined Array for Research in Millimeter-Wave Astronomy
CBI	Cosmic Background Imager
CCIR	International Radio Consultative Committee
CIO	Conventional International Origin
CLEAN	Imaging algorithm for removal of unwanted responses due to point spread function
CMB	Cosmic microwave background
COBE	Cosmic Background Explorer
COESA	Committee on the Extension of the Standard Atmosphere
CSIRO	Commonwealth Scientific and Industrial Research Organzation

CS	Compressed sensing
CSO	Caltech Submillimeter Observatory
CVN	Chinese VLBI Network
CW	Continuous wave
DASI	Degree Angular Scale Interferometer
dB	Decibel (formally, one-tenth of a bel)
DD	Direction-dependent
DFT	Discrete Fourier transform
DM	Dispersion measure
DSB	Double sideband
EHT	Event Horizon Telescope
EOR	Epoch of Reionization
EVN	European VLBI Network
FFT	Fast Fourier transform
FIFO	First-in-first-out
FIR	Finite impulse response
FK	Fundamental catalog (stellar positions)
FWHM	Full width at half-maximum
FX	Fourier transform before cross multiplication of data
FXF	Correlator architecture
GBT	Green Bank Telescope
GLONASS	Global Navigation Satellite System
GMRT	Giant Meterwave Radio Telescope
GMSK	Gaussian-filtered minimum shift keying
GPS	Global positioning system
GR	General relativity
HALCA	Highly Advanced Laboratory for Communications and Astronomy (a VLBI satellite of Japan)
IAT	International Atomic Time
IAU	International Astronomical Union
ICRF	International Celestial Reference Frame
ICRS	International Celestial Reference System
IEEE	Institute of Electrical and Electronics Engineers
IF	Intermediate frequency
IPAC	Infrared Processing and Analysis Center (NASA/Caltech)
IRI	International Reference Ionosphere
ITU	International Telecommunication Union
IVS	International VLBI Service
J	Julian
JPL	Jet Propulsion Laboratory

JVN	Japanese VLBI Network
LASSO	Least absolute shrinkage and selection operator
LBA	Long-Baseline Array
LO	Local oscillator
LOD	Length of day
LOFAR	LOw Frequency ARray
LSMF	Least-mean-square fit
LSR	Local standard of rest
LWA	Long Wavelength Array
MCMC	Markov chain Monte Carlo
MeerKAT	Meer and Karoo Array Telescope
MEM	Maximum entropy method
MERLIN	Multi-Element Radio Linked Interferometer Network
MERRA	Modern-Era Retrospective Analysis for Research and Applications (NASA program)
MIT	Massachusetts Institute of Technology
MKS	Meter kilogram second
MMA	Millimeter Array (precursor to ALMA)
MSTID	Midscale traveling ionospheric disturbance
NASA	National Aeronautics and Space Administration
NGC	New General Catalog
NGS	National Geodetic Survey
NNLS	Nonnegative least-squares (algorithm)
NRAO	National Radio Astronomical Observatory (USA)
NRO	Nobeyama Radio Observatory
NVSS	NRAO VLA Sky Survey
OVLBI	Orbiting VLBI
OVRO	Owens Valley Radio Observatory
PAF	Phased-array feed
PFB	Polyphase filter bank
PIM	Parametrized Ionosphere Model
PPN	Parametrized post–Newtonian (formalism of general relativity)
Q-factor	Center frequency divided by bandwidth
QPSK	Quadri-phase-shift keying
RA	Right ascension
RAM	Random access memory
RF	Radio frequency
RFI	Radio frequency interference
RM	Rotation measure

RMS	Root mean square
SAO	Smithsonian Astrophysical Observatory
SEFD	System equivalent flux density
SI	System International (modern MKS units)
SIM	Space Interferometry Mission
SIS	Superconductor–insulator–superconductor
SKA	Square Kilometre Array
SMA	Submillimeter Array
SMOS	Soil Moisture and Ocean Salinity mission
SNR	Signal-to-noise ratio
SSB	Single sideband
STI	Satellite tracking interferometer
TDRSS	Tracking and Data Relay Satellite System
TEC	Total electron content
TID	Traveling ionospheric disturbance
TV	Total variation
USNO	United States Naval Observatory
USSR	Union of Soviet Socialist Republics
UT	Universal time
UT0, UT1, UT2	Modified UT
UTC	Coordinated universal time
UT-R2	Ukrainian Academy of Sciences T-shaped array
VCR	Video cassette recorder
VERA	VLBI Exploration of Radio Astronomy (Japanese-led project)
VLA	Very Large Array
VLBA	Very Long Baseline Array
VLBI	Very-long-baseline interferometry
VLSI	Very-large-scale integrated (circuits)
VSA	Very Small Array
VSOP	VLBI Space Observatory Programme
WIDAR	Wideband Interferometric Digital ARchitecture
WMAP	Wilkinson Microwave Anisotropy Probe
WVR	Water vapor radiometer
XF	Cross-correlation before Fourier transformation
Y-factor	Ratio of receiver power outputs with hot and cold input loads

Principal Symbols

Listed below are the principal symbols used throughout the book. Locally defined symbols with restricted usage are selectively included.

a	Model dimension, scale size, atmospheric model constant (Sect. 13.1), scale size of ionospheric irregularities (Sect. 14.2)
A	Antenna collecting area (reception pattern)
$\mathbf{A}$	Antenna polarization matrix (Chap. 4)
A_1	One-dimensional reception pattern
A_0	Antenna collecting area on axis
A_N	Normalized reception pattern
$\mathcal{A}$	Mirror-image reception pattern, azimuth
b	Galactic latitude (Sect. 14.4)
b_0	Synthesized beam pattern, point-source response
b_N	Normalized synthesized beam pattern
B	Magnetic field magnitude
$\mathbf{B}$	Magnetic field vector
c	Velocity of light
C	Constant (Chap. 1), coherence function (Chap. 9), convolving function (Chap. 10)
C_n^2	Turbulence strength parameters for refractive index (Chap. 13)
C_{ne}^2	Turbulence strength, electron density (Chap. 14)
$\mathcal{C}$	Amplitude of a complex signal (Appendix 3.1)
d	Distance, antenna diameter, baseline declination, projected baseline (Chap. 13)
d_f	Fried length (Chaps. 13, 17)
d_{in}	Inner scale of turbulence

d_{out}	Outer scale of turbulence
d_r	Diffractive limit
d_{tc}	Distance between ray paths to target and calibrator sources in turbulent region
d_0	Distance over which rms phase deviation = 1 rad (Chap. 13)
d_2	Transition from 2-D to 3-D turbulence
D	Baseline (antenna spacing), polarization leakage (Chap. 4)
$\mathbf{D}$	Baseline vector
$D_\lambda, \mathbf{D}_\lambda$	Baseline measured in wavelengths
$D_a, \mathbf{D}_a$	Interaxis distance of antenna mount (Chap. 4)
D_E	Equatorial component of baseline
DM	Dispersion measure (Chap. 13)
D_n	Structure function of refractive index (spatial) (Chap. 13)
D_R	Delay resolution function [Eq. (9.181)]
D_τ	Structure function of phase (temporal) (Chap. 13)
D_ϕ	Structure function of phase (spatial) (Chaps. 12, 13)
$\mathcal{D}$	Dispersion in optical fiber (Sect. 7.1, Appendix 7.2)
e	Magnitude of electronic charge (Chap. 14), emissivity
$E, \mathbf{E}$	Electric field (usually in the measurement plane), spectral components of electric field, energy
E_x, E_y	Components of electric field
$\mathcal{E}$	Electric field at a source or aperture (Chaps. 3, 15, 17), elevation angle
f	Frequency of Fourier components of power spectrum (Chaps. 9, 13)
f_i	Oscillator strength at resonance i (Chap. 13)
f_m, f_n	Phase switching waveforms (Chap. 7)
F	Power flux density (W m^{-2}), fringe function
F_h	Threshold of harmful interference (W m^{-2}) (Chap. 16)
$F(\beta)$	Faraday dispersion function (Chap. 13)
F_1, F_2	See Eq. (9.17)
F_1, F_2, F_3	Entropy measures (Chap. 11)
F_B	Bandwidth pattern (Chap. 2)
$\mathcal{F}$	Sensitivity degradation factor (Chap. 7)
$\mathcal{F}_R, \mathcal{F}_I$	Quantized fringe-rotation functions (Chap. 9)
g	Voltage gain constant for an antenna, gravitational acceleration (Chap. 13)
G	Gravitational constant
G_i	Power gain of receiver for one antenna (Chap. 7)
G_{mn}	Gain factor for a correlated antenna pair

G_0	Gain factor (Chap. 7)
G	Occultation response function (Chap. 17)
h	Planck's constant, impulse response of a filter (Sect. 3.3), hour angle of baseline, height, height above surface
h_0	Atmospheric scale height (Chap. 13)
H	Hour angle, voltage–frequency response, Hadamard matrix (Sect. 7.5)
H_0	Gain constant
i	Electric current
$\mathbf{i}$	Unit vector in direction of polar or azimuth axes (Chap. 4), current vector (Chap. 14)
I	Intensity, Stokes parameter
I^2	Variance of fractional frequency deviation (Chap. 9)
I_s	Speckle intensity (Chap. 17)
I_v	Stokes visibility
I_0	Peak intensity of a point source, derived (synthesized) intensity distribution, modified Bessel function of zero order (Chaps. 6, 9)
I_1	One-dimensional intensity function, modified Bessel function of first order (Chap. 9)
Im	Imaginary part
j	$\sqrt{-1}$
$\mathbf{J}$	Jones Matrix (Chap. 4)
j_v	Volume emissivity of a source (Chap. 13)
J	Mutual intensity (Chap. 15)
J_0	Bessel function of first kind and zero order
J_1	Bessel function of first kind and first order
k	Boltzmann's constant, propagation constant $2\pi/\lambda$ (Chap. 13)
$\mathbf{k}$	Propagation vector with magnitude $2\pi/\lambda$ (Chap. 9)
l	Direction cosine with respect to baseline component u, lapse rate (Chap. 13)
L	Length of a transmission line, loss factor in a transmission line (Chap. 7), probability integral [Eq. (8.109)], path length, likelihood function (Chap. 12), thickness of turbulent atmospheric layer or screen (Chap. 13)
L_{inner}, L_{outer}	Scales of turbulence (Chap. 13)
ℓ	Multipole moment (Chap. 10), length, galactic longitude (Chap. 13)
ℓ_λ	Unit spacing (in wavelengths) in a grating array (Chaps. 1, 5)

$\mathcal{L}$	Latitude, *excess* path length (Chap. 13)
$\mathcal{L}_D, \mathcal{L}_V$	Excess path length of dry air, water vapor
m	Direction cosine with respect to baseline component v, modulation index (Appendix 7.2), measured quantity (Appendix 12.1), electron mass (Chap. 13)
m_ℓ, m_c, m_t	Degree of linear, circular, and total polarization
M	Frequency multiplication factor (Chap. 9), model function (Chap. 10), mass, complex degree of linear polarization (Chap. 13)
$\mathcal{M}, \mathcal{M}_D, \mathcal{M}_V$	Molecular weight; total, dry air, water vapor (Chap. 13)
n	Direction cosine with respect to baseline component w, weighting factor in quantization (Chap. 8), noise component, index of refraction (Chap. 13)
$n = n_R + jn_I$	Complex refractive index
n_a	Number of antennas
n_d	Number of data points
n_e, n_i, n_n, n_m	Density of electrons, ions, neutral particles, and molecules (Chap. 13)
n_p	Number of antenna pairs
n_s	Number of sources
n_r	Number of points in a rectangular array (grid points)
n_0	Refractive index at Earth's surface (Chap. 13)
N	Number of samples (Chap. 8), total refractivity (Chap. 13)
N_b	Number of bits per sample (Chap. 8)
N_D, N_V	Refractivity of; dry air, water vapor (Chap. 13)
N_N	Number of Nyquist rate samples (Chap. 8)
$\mathcal{N}$	$2\mathcal{N}$ and $(2\mathcal{N} + 1)$ are even and odd numbers of quantization levels (Chap. 8)
p	Probability density or probability distribution [i.e. $p(x)\,dx$ is the probability that the random variable lies between x and $x + dx$], bivariate normal probability function (Chap. 8), number of model parameters (Chap. 10), partial pressure (Sect. 13.1), impact parameter (Sects. 12.6, 14.3)
p_D	Partial pressure of dry air (Chap. 13)
p_V	Partial pressure of water vapor (Chap. 13)
P	Power, cumulative probability, total atmospheric pressure (Chap. 13)
P_0	Atmospheric pressure at Earth's surface (Chap. 13)
$\mathbf{P}$	Dipole moment per unit volume
P_3	Triple product (bispectrum)
P_{mnp}	Instrumental polarization factor

P_{ne}	Spectrum of electron density fluctuations
$\mathcal{P}$	Point-source response at Moon's limb (Sect. 17.2), speckle point-spread function (Sect. 17.6.4)
q	Distance in (u, v) plane
q'	Distance in (u', v') plane
q_x, q_y	Components in the spatial frequency (cycles per meter) plane (Chap. 13)
Q	Stokes parameter, quality factor of a line or cavity (Sect. 9.5), number of quantization levels (Sects. 8.3, 9.6)
Q_v	Stokes visibility
r	Correlator output, distance in the (l, m) plane, radial distance
$\mathbf{r}$	Position vector of antenna relative to center of Earth
r_e	Classical electron radius (Chap. 14)
r_ℓ	Correlator output resulting from lower sideband
r_p	Pearson's correlation coefficient
r_u	Correlator output resulting from upper sideband
r_0	Radius of the Earth
R	Autocorrelation function, correlator output, robustness factor (Sect. 10.2.2.1), frequency ratio (Sect. 12.2.4), distance, gas constant (Chap. 13)
$\mathbf{R}$	Correlator output matrix (Chap. 4)
R_a	Response with visibility averaging (Chap. 6)
R_b	Response with finite bandwidth (Chap. 6)
R_e	Radius of electron orbit (Chap. 14)
R_{ff}	Far-field distance (Chap. 15)
RM	Rotation measure (Chap. 14)
R_m	Distance of the Moon's limb (Chap. 17)
R_n	Autocorrelation for n-level quantization (Chap. 8)
R_y	Autocorrelation function of fractional frequency deviation (Chap. 9)
R_0	Distance of Earth to Sun
R_ϕ	Autocorrelation function of phase (Chaps. 9, 13)
$\mathcal{R}e$	Real part
$\mathcal{R}_{sn}$	Signal-to-noise ratio
s	Signal component, smoothness measure (Chap. 11)
$\mathbf{s}$	Unit position vector (Chap. 3)
$\mathbf{s}_0$	Unit position vector of field center (Chap. 3)
S	(spectral) power flux density ($\text{W m}^{-2} \text{Hz}^{-1}$)
S_c	Flux density of a calibrator
SEFD	System equivalent flux density
S_h	Threshold of harmful interference ($\text{W m}^{-2} \text{Hz}^{-1}$) (Chap. 16)

Sq	Square wave functions (Sect. 7.5) (also known as Rademacher functions)
$\mathcal{S}$	Cross power spectrum (Chap. 9)
S_I	Power spectrum of intensity fluctuations (Chap. 14)
S_y, S'_y	Single-sided and double-sided power spectra of fractional frequency deviation (single-sided power spectrum used only in Sect. 9.4)
S_ϕ, S'_ϕ	Single-sided and double-sided power spectra of phase fluctuations (single-sided power spectrum used only in Sect. 9.4)
S_2	Two-dimensional power spectrum of phase (Chap. 13)
t	Time
t_e	Period of the Earth's rotation (Chap. 12)
t_{cyc}	Cycle period for target and calibrator sources
T	Temperature, time interval, transmission factor (Chap. 15)
T_{at}	Atmospheric temperature (Chap. 13)
T_A	Component of antenna temperature resulting from target source
T'_A	Total antenna temperature
T_B	Brightness temperature
T_c	Noise temperature of calibration signal
T_g	Gas temperature (Chap. 9)
T_R	Receiver temperature
T_S	System temperature
$\mathcal{T}$	Time interval
u	Antenna spacing coordinate in units of wavelength (spatial frequency)
u'	Projection of u coordinate onto the equatorial plane
U	Stokes parameter
U_v	Stokes visibility (Chap. 4)
$\mathcal{U}$	Unwanted response (Sect. 7.5)
v	Antenna spacing coordinate in units of wavelength (spatial frequency), phase velocity in a transmission line (Chap. 8)
v'	Projection of v coordinate onto the equatorial plane
v_g	Group velocity (Chap. 14)
v_m	Rate of angular motion of Moon's limb (Chap. 16)
v_p	Phase velocity (Chap. 13)
v_r	Radial velocity
v_s	Velocity of scattering screen (parallel to baseline, if relevant) (Chaps. 12, 13)
v_0	Quantization level (Chap. 8), particle velocity (Chap. 9)

V	Voltage, Stokes parameter
V_A	Voltage response of an antenna
V_v	Stokes visibility (Chap. 4)
$\mathcal{V}, \boldsymbol{\mathcal{V}}$	Complex visibility, vector visibility
$\mathcal{V}_m$	Measured complex visibility
$\mathcal{V}_M$	Michelson's fringe visibility
$\mathcal{V}_N$	Normalized complex visibility
w	Antenna spacing coordinate in units of wavelength (spatial frequency), weighting function, column height of precipitable water (Chap. 13)
w'	w coordinate measured in the polar direction
w_a	Atmospheric weighting function (Chap. 13)
w_{mean}	Mean of weighting factors (Chap. 6)
w_{rms}	Root-mean-square of weighting factors (Chap. 6)
w_t	Visibility tapering function (Chap. 10)
w_u	Function that adjusts visibility amplitude for effective uniform weighting (Chap. 10)
W	Spectral sensitivity function (spatial transfer function); propagator (Chap. 15)
x	General position coordinate, coordinate in antenna aperture, signal voltage
x_λ	x coordinate measured in wavelengths
X	Coordinate of antenna spacing [see Eq. (4.1)], signal waveform measured in units of rms amplitude (Chap. 8), coordinate within a source or an aperture (Chaps. 3, 15), signal spectrum (Sect. 8.7)
X_λ	X coordinate measured in wavelengths
y	General position coordinate, coordinate in antenna aperture, signal voltage, distance along a ray path (Chap. 13)
y_k	Fractional frequency deviation (Chap. 9)
y_λ	y coordinate measured in wavelengths
Y	Coordinate of antenna spacing [Eq. (4.1)], Y-factor (Chap. 7), coordinate within a source or aperture (Chaps. 3, 15), signal waveform measured in units of rms amplitude (Sect. 8.4), signal spectrum (Sect. 8.7)
Y_λ	Y coordinate measured in wavelengths
z	General position coordinate, signal voltage, zenith angle (Chap. 13), redshift
z_λ	z coordinate measured in wavelengths
Z	Coordinate of antenna spacing [Eq. (4.1)], visibility plus noise in correlator output (Chaps. 6, 9)
$\mathcal{Z}_D, \mathcal{Z}_V$	Compressibility factors for dry air and water vapor (Chap. 13)

Z	Visibility-plus-noise vector (Chaps. 6, 9)
Z_λ	Z coordinate measured in wavelengths
α	Right ascension, power attenuation coefficient, quantization threshold in units of σ (Chap. 8), spectral index (Chap. 11), absorption coefficient and power-law exponent in Table 13.2 and related text (Sect. 13.1), exponent in electron density fluctuation (Sect. 13.4)
β	Fractional length change in transmission line (Chap. 7), oversampling factor (Chap. 8), exponent of distance in rms phase fluctuation [Eq. (13.80a)] (Sects. 12.2, 13.1), exponent in solar electron density (Sect. 14.3), Faraday depth (Sect. 14.4)
γ	Instrumental polarization factor (Sect. 4.8), maser relaxation rate (Chap. 9), loop gain in CLEAN (Chap. 11), post-Newtonian GR parameter (Chap. 12), source coherence function (Chap. 15)
Γ	Damping factor (Chap. 13), mutual coherence function (Chap. 15), gamma function
Γ_{12}	Mutual coherence function (Chap. 15)
δ	Declination, increment prefix, (Dirac) delta function, instrumental polarization factor (Sect. 4.8)
$^2\delta$	Delta function in two dimensions
Δ	Small length, increment prefix
$\Delta\nu$	Bandwidth, Doppler shift (Appendix 10.2)
$\Delta\nu_{IF}$	Intermediate frequency bandwidth
$\Delta\nu_{LF}$	Low frequency bandwidth
$\Delta\nu_{LO}$	Frequency difference of local oscillators
$\Delta\tau$	Delay error
$\Delta u, \Delta v$	Increments in (u, v) plane
$\Delta l, \Delta m$	Increments in (l, m) plane
ϵ	Solar elongation (Sect. 12.6)
ϵ	Width of quantization level in units of σ (Chap. 8), noise component in IF signal (Chap. 9), permittivity (Chap. 13)
ϵ_a	Amplitude error (Chap. 11)
ϵ_0	Permittivity of free space (Chap. 13)
ε	Noise component of correlator output (Chaps. 6, 9), residual, error component, dielectric constant (Chap. 13, Sect. 17.5), random surface deviation (Chap. 17)
$\boldsymbol{\varepsilon}$	Noise vector (Chap. 6)
η	Loss factor

η_D	Discrete delay step-loss factor
η_Q	Efficiency (loss) factor for Q-level quantization
η_R	Fringe rotation loss factor
η_S	Fringe sideband rejection loss factor
θ	General angle, angle measured from a plane normal to the baseline, instrumental phase angle, angle between baseline and source direction vector (Chap. 12)
θ_0	Angular position of source or field center
θ_b	Width of synthesized beam, bending angle (Chap. 13)
θ_f	Width of synthesized field (field of view)
θ_F	Width of first Fresnel zone
θ_{LO}	Local oscillator phase
θ_m, θ_n	Local oscillator phase at antennas m and n (Chap. 6)
θ_s	Effective beamwidth resulting from atmospheric fluctuations (Chap. 13), width of source (Chap. 16)
Θ	Variation in Earth-rotation angle (UT1$-$UTC) (Chap. 12)
λ	Wavelength
λ_{opt}	Wavelength of optical carrier (Appendix 7.2)
Λ	Reflected amplitude in a transmission line (Chap. 7)
μ	Power-law exponent in Allan variance (Chap. 9)
ν	Frequency
ν'	Frequency measured with respect to center frequency or local oscillator frequency (Chap. 9)
ν_b	Bit rate
ν_B	Gyrofrequency (Chap. 13)
ν_c	Collision frequency (Chap. 13)
ν_C	Cavity frequency (Chap. 9)
ν_d	Intermediate frequency at which delay is inserted
ν_{ds}	Delay step frequency (Chap. 9)
ν_f	Fringe frequency
ν_{in}	Instrumental component of fringe frequency (Chap. 12)
ν_{IF}	Intermediate frequency
ν_{LO}	Local oscillator frequency
ν_ℓ	Frequency of a correlator channel (Chap. 9)
ν_m	Frequency of modulation on optical carrier (Chap. 7)
ν_{RF}	Radio frequency
ν_{opt}	Frequency of optical carrier (Appendix 7.2)
ν_p	Plasma frequency (Chap. 13)
ν_0	Center frequency of an IF or RF band, frequency of absorption peak (Chap. 13)
Π	Parallax angle (Chap. 12)

ρ	Autocorrelation function, cross-correlation coefficient, reflection coefficient (Chap. 7), gas density (Chap. 13)
ρ_D, ρ_V, ρ_T	Density: dry air, water vapor, total (Chap. 13)
ρ_{mn}	Cross-correlation
ρ_σ	Area density in the (u, v) plane (Chap. 10)
ρ_m, ρ_n	Reflection coefficients in transmission line (Chap. 7)
ρ_w	Density of water (Chap. 13)
σ	Standard deviation, rms noise level; radar cross section (Chap. 17)
$\boldsymbol{\sigma}$	Position vector on the unit sphere
σ_y	Allan standard deviation ($\sigma_y{}^2 =$ Allan variance)
σ_τ	Root-mean-square uncertainty in delay (Chap. 9)
σ_ϕ	Root-mean-square deviation of phase
τ	Time interval
τ_a	Averaging (integration) time
τ_{at}	Atmospheric delay error (Chap. 12)
τ_c	Coherent integration time (Chap. 9)
τ_e	Clock error
τ_g	Geometric delay
τ_i	Instrumental delay
τ_0	Unit increment of instrumental delay, duration of an observation (Chap. 6), zenith optical depth (opacity) of the atmosphere (Chap. 13)
τ_s	Sampling interval in time
τ_{or}	Minimum period of orthogonality (Chap. 7)
τ_{sw}	Interval between switch transitions (Chap. 7)
τ_v	Optical depth (opacity) (Chap. 13)
ϕ	Phase angle
ϕ_m	Phase of signal received by antenna m
ϕ_v	Visibility phase
ϕ_G, ϕ_{in}	Instrumental phase for correlated antenna pair
ϕ_{pp}	Peak-to-peak phase error (Chap. 9)
Φ	Phase of a complex signal (Appendix 3.1), probability integral [Eq. (8.44)] (Chap. 8), phase of a signal (Sect. 13.1)
χ	Arctangent of axial ratio of polarization ellipse
χ^2	Chi-squared statistical parameter
ψ	Position angle, phase angle
ψ_p	Parallactic angle
ω_e	Angular rotation velocity of the Earth
Ω	Solid angle
Ω_s	Solid angle subtended by source
Ω_0	Solid angle of main lobe of synthesized beam

Frequently Used Subscripts

A	Antenna
d	Delay, double sideband
D	Dry component (Chap. 13)
I	Imaginary part
IF	Intermediate frequency
ℓ	Left circular polarization, lower sideband
LO	Local oscillator
0	Center of frequency band or angular field, Earth's surface (Chap. 13)
m, n	Antenna designation
N	Normalized, Nyquist rate (Sects. 8.2, 8.3)
r	Right circular polarization
R	Real part
S	System
u	Upper sideband
V	Water vapor (Chap. 13)
λ	Measured in wavelengths

Other Symbols

Π	Unit rectangle function
$\prod$	Product symbol
III	Shah function in one dimension
^{2}III	Shah function in two dimensions
$\longleftrightarrow$	"is the Fourier transform of"
$*$	Convolution in one dimension
$**$	Convolution in two dimensions
$\star$	Cross correlation in one dimension
$\star\star$	Cross correlation in two dimensions
$\langle \ \rangle$	Expectation (or approximation by a finite average)
dot ($\dot{\ }$)	First derivative with respect to time
double dot ($\ddot{\ }$)	Second derivative with respect to time
overline ($\bar{\ }$)	Average (Chaps. 1, 9, Sect. 14.1); Fourier transform of function (Chaps. 3, 5, 8, 10, 11, 13, Sect. 14.2)
circumflex ($\hat{\ }$)	Quantized variable (Chap. 8)
circumflex ($\hat{\ }$)	Function of frequency (Chap. 3)

Angular Notation

°, $'$, $''$	Degrees, minutes of arc, and seconds of arc
mas	Milliarcseconds
μas	Microarcseconds

Functions

For definitions and descriptions, see, e.g., Abramowitz, M., and Stegun, I.A., *Handbook of Mathematical Functions*, National Bureau of Standards, Washington, DC (1964), reprinted by Dover, New York, (1965).

erf	Error function [Eq. (6.63c)]
J_0	Bessel function of first kind and zero order [Eq. (A2.55)]
J_1	Bessel function of first kind and first order
I_0	Modified Bessel function of zero order [Eq. (9.46)]
I_1	Modified Bessel function of first order [Eq. (9.52)]
Γ	Gamma function [note that $\Gamma(x+1) = x\Gamma(x)$]
δ	Dirac delta function [Eq. (A2.10)]
$\prod$	Unit rectangle function [Eq. (A2.12a)]
$\prod$	Modified unit rectangle function [Table 10.2]
sinc	$\sin \pi x/(\pi x)$ [Eq. (2.4)]

Chapter 1
Introduction and Historical Review

The subject of this book can be broadly described as the principles of radio interferometry applied to the measurement of natural radio signals from cosmic sources. The uses of such measurements lie mainly within the domains of astrophysics, astrometry, and geodesy. As an introduction, we consider in this chapter the applications of the technique, some basic terms and concepts, and the historical development of the instruments and their uses.

The fundamental concept of this book is that the image, or intensity distribution, of a source has a Fourier transform that is the two-point correlation function of the electric field, whose components can be directly measured by an interferometer. This Fourier transform is normally called the fringe visibility function, which in general is a complex quantity. The basic formulation of this principle is called the van Cittert–Zernike theorem (see Chap. 15), derived in the 1930s in the context of optics but not widely appreciated by radio astronomers until the publication of the well-known textbook *Principles of Optics* by Born and Wolf (1959). The techniques of radio interferometry developed from those of the Michelson stellar interferometer without specific knowledge of the van Cittert–Zernike theorem. Many of the principles of interferometry have counterparts in the field of X-ray crystallography (see Beevers and Lipson 1985).

1.1 Applications of Radio Interferometry

Radio interferometers and synthesis arrays, which are basically ensembles of two-element interferometers, are used to make measurements of the fine angular detail in the radio emission from the sky. The angular resolution of a single radio antenna is insufficient for many astronomical purposes. Practical considerations limit the resolution to a few tens of arcseconds. For example, the beamwidth of

© The Author(s) 2017
A.R. Thompson, J.M. Moran, and G.W. Swenson Jr., *Interferometry and Synthesis in Radio Astronomy*, Astronomy and Astrophysics Library,
DOI 10.1007/978-3-319-44431-4_1

a 100-m-diameter antenna at 7-mm wavelength is approximately 17″. In the optical range, the diffraction limit of large telescopes (diameter $\sim$ 8 m) is about 0.015″, but the angular resolution achievable from the ground by conventional techniques (i.e., without adaptive optics) is limited to about 0.5″ by turbulence in the troposphere. For progress in astronomy, it is particularly important to measure the positions of radio sources with sufficient accuracy to allow identification with objects detected in the optical and other parts of the electromagnetic spectrum [see, for example, Kellermann (2013)]. It is also very important to be able to measure parameters such as intensity, polarization, and frequency spectrum with similar angular resolution in both the radio and optical domains. Radio interferometry enables such studies to be made.

Precise measurement of the angular positions of stars and other cosmic objects is the concern of astrometry. This includes the study of the small changes in celestial positions attributable to the parallax introduced by the Earth's orbital motion, as well as those resulting from the intrinsic motions of the objects. Such measurements are an essential step in the establishment of the distance scale of the Universe. Astrometric measurements have also provided a means to test the general theory of relativity and to establish the dynamical parameters of the solar system. In making astrometric measurements, it is essential to establish a reference frame for celestial positions. A frame based on extremely distant high-mass objects as position references is close to ideal. Radio measurements of distant, compact, extragalactic sources presently offer the best prospects for the establishment of such a system. Radio techniques provide an accuracy of the order of 100 μas or less for absolute positions and 10 μas or less for the relative positions of objects closely spaced in angle. Optical measurements of stellar images, as seen through the Earth's atmosphere, allow the positions to be determined with a precision of about 50 mas. However, positions of 10^5 stars have been measured to an accuracy of $\sim$1 mas with the Hipparcos satellite (Perryman et al. 1997). The Gaia[1] mission is expected to provide the positions of 10^9 stars to an accuracy of $\sim$10 μas (de Bruijne et al. 2014).

As part of the measurement process, astrometric observations include a determination of the orientation of the instrument relative to the celestial reference frame. Ground-based observations therefore provide a measure of the variation of the orientation parameters for the Earth. In addition to the well-known precession and nutation of the direction of the axis of rotation, there are irregular shifts of the Earth's axis relative to the surface. These shifts, referred to as polar motion, are attributed to the gravitational effects of the Sun and Moon on the equatorial bulge of the Earth and to dynamic effects in the Earth's mantle, crust, oceans, and atmosphere. The same causes give rise to changes in the angular rotation velocity of the Earth, which are manifest as corrections that must be applied to the system of universal time. Measurements of the orientation parameters are important in the study of the dynamics of the Earth. During the 1970s, it became clear that radio techniques could provide an accurate measure of these effects, and in the late 1970s,

---

[1] An astrometric space observatory of the European Space Agency.

the first radio programs devoted to the monitoring of universal time and polar motion were set up jointly by the U.S. Naval Observatory and the U.S. Naval Research Laboratory, and also by NASA and the National Geodetic Survey. Polar motion can also be studied with satellites, in particular the Global Positioning System, but distant radio sources provide the best standard for measurement of Earth rotation.

In addition to revealing angular changes in the motion and orientation of the Earth, precise interferometer measurements entail an astronomical determination of the vector spacing between the antennas, which for spacings of ~ 100 km or more is usually more precise than can be obtained by conventional surveying techniques. Very-long-baseline interferometry (VLBI) involves antenna spacings of hundreds or thousands of kilometers, and the uncertainty with which these spacings can be determined has decreased from a few meters in 1967, when VLBI measurements were first made, to a few millimeters. Relative motions of widely spaced sites on separate tectonic plates lie in the range 1–10 cm per year and have been tracked extensively with VLBI networks. Interferometric techniques have also been applied to the tracking of vehicles on the lunar surface and the determination of the positions of spacecraft. In this book, however, we limit our concern mainly to measurements of natural signals from astronomical objects.

The attainment of the highest angular resolution in the radio domain of the electromagnetic spectrum results in part from the ease with which radio frequency (RF) signals can be processed electronically with high precision. The use of the heterodyne principle to convert received RF signals to a convenient baseband, by mixing them with a signal from a local oscillator, is essential to this technology. A block diagram of an idealized standard receiving system (also known as a radiometer) is shown in Appendix 1.1. Another advantage in the radio domain is that the phase variations induced by the Earth's neutral atmosphere are less severe than at shorter wavelengths. Future technology will provide even higher resolution at infrared and optical wavelengths from observatories above the Earth's atmosphere. However, radio waves will remain of vital importance in astronomy since they reveal objects that do not radiate in other parts of the spectrum, and they are able to pass through galactic dust clouds that obscure the view in the optical range.

1.2 Basic Terms and Definitions

This section is written for readers who are unfamiliar with the basics of radio astronomy. It presents a brief review of some background information that is useful when approaching the subject of radio interferometry.

1.2.1   Cosmic Signals

The voltages induced in antennas by radiation from cosmic radio sources are generally referred to as *signals*, although they do not contain information in the usual engineering sense. Such signals are generated by natural processes and almost universally have the form of Gaussian random noise. That is to say, the voltage as a function of time at the terminals of a receiving antenna can be described as a series of very short pulses of random occurrence that combine as a waveform with Gaussian amplitude distribution. In a bandwidth $\Delta \nu$, the envelope of the radio frequency waveform has the appearance of random variations with timescale of order $1/\Delta \nu$. For most radio sources (except, for example, pulsars), the characteristics of the signals are invariant with time, at least on the scale of minutes or hours, the duration of a typical radio astronomy observation. Gaussian noise of this type is assumed to be identical in character to the noise voltages generated in resistors and amplifiers and is sometimes called Johnson noise. Such waveforms are usually assumed to be stationary and ergodic, that is, ensemble averages and time averages converge to equal values.

Most of the power is in the form of *continuum radiation*, the power spectrum of which shows gradual variation with frequency. For some wideband instruments, there may be significant variation within the receiver bandwidth. Figure 1.1 shows continuum spectra of eight different types of radio sources. Radio emission from the radio galaxy Cygnus A, the supernova remnant Cassiopeia A, and the quasar 3C48 is generated by the synchrotron mechanism [see, e.g., Rybicki and Lightman (1979), Longair (1992)], in which high-energy electrons in magnetic fields radiate as a result of their orbital motion. The radiating electrons are generally highly relativistic, and under these conditions, the radiation emitted by each one is concentrated in the direction of its instantaneous motion. An observer therefore sees pulses of radiation from those electrons whose orbital motion lies in, or close to, a plane containing the observer. The observed polarization of the radiation is mainly linear, and any circularly polarized component is generally quite small. The overall linear polarization from a source, however, is seldom large, since it is randomized by the variation of the direction of the magnetic field within the source and by Faraday rotation. The power in the electromagnetic pulses from the electrons is concentrated at harmonics of the orbital frequency, and a continuous distribution of electron energies results in a continuum radio spectrum. The individual pulses from the electrons are too numerous to be separable, and the electric field appears as a continuous Gaussian random process with zero mean. The variation of the spectrum as a function of frequency is related to the energy distribution of the electrons. At low frequencies, these spectra turn over due to the effect of self-absorption. M82 is an example of a starburst galaxy. At low frequencies, synchrotron emission dominates, but at high frequencies, emission from dust grains at a temperature of about 45K and emissivity of 1.5 dominates. TW Hydrae is a star with a protoplanetary disk whose emission at radio frequencies is dominated by dust at a temperature of about 30K and emissivity of 0.5.

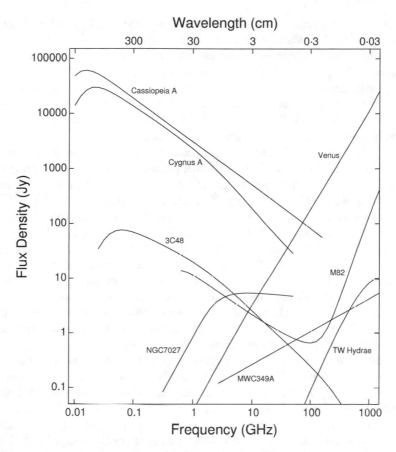

Fig. 1.1 Examples of spectra of eight different types of discrete continuum sources: Cassiopeia A [supernova remnant, Baars et al. (1977)], Cygnus A [radio galaxy, Baars et al. (1977)], 3C48 [quasar, Kellermann and Pauliny-Toth (1969)], M82 [starburst galaxy, Condon (1992)], TW Hydrae [protoplanetary disk, Menu et al. (2014)], NGC7207 [planetary nebula, Thompson (1974)], MWC349A [ionized stellar wind, Harvey et al. (1979)], and Venus [planet, at 9.6″ diameter (opposition), Gurwell et al. (1995)]. For practical purposes, we define the edges of the radio portion of the electromagnetic spectrum to be set by the limits imposed by ionospheric reflection at low frequencies ($\sim$ 10 MHz) and to atmospheric absorption at high frequencies ($\sim$ 1000 GHz). Some of the data for this table were taken from NASA/IPAC Extragalactic Database (2013) [One jansky (Jy) = 10^{-26} W m^{-2} Hz^{-1}].

   NGC7027, the spectrum of which is shown in Fig. 1.1, is a planetary nebula within our Galaxy in which the gas is ionized by radiation from a central star. The radio emission is a thermal process and results from free-free collisions between unbound electrons and ions within the plasma. At the low-frequency end of the spectral curve, the nebula is opaque to its own radiation and emits a blackbody spectrum. As the frequency increases, the absorptivity, and hence the emissivity, decrease approximately as ν^{-2} [see, e.g., Rybicki and Lightman (1979)], where ν is the frequency. This behavior counteracts the ν^2 dependence of the Rayleigh–Jeans law, and thus the spectrum becomes nearly flat when the nebula is no longer opaque to the radiation. Radiation of this type is randomly polarized. MWC349A is an example of an inhomogeneous ionized gas expanding at constant velocity in a stellar envelope, which gives rise to a spectral dependence of $\nu^{0.6}$.

At millimeter wavelengths, opaque thermal sources such as planetary bodies become very strong and often serve as calibrators. Venus has a brightness temperature that varies from 700K (the surface temperature) at low frequencies to 250K (the atmospheric temperature) at high frequencies.

In contrast with continuum radiation, *spectral line radiation* is generated at specific frequencies by atomic and molecular processes. A fundamentally important line is that of neutral atomic hydrogen at 1420.405 MHz, which results from the transition between two energy levels of the atom, the separation of which is related to the spin vector of the electron in the magnetic field of the nucleus. The natural width of the hydrogen line is negligibly small ($\sim 10^{-15}$ Hz), but Doppler shifts caused by thermal motion of the atoms and large-scale motion of gas clouds spread the line radiation. The overall Doppler spread within our Galaxy covers several hundred kilohertz. Information on galactic structure is obtained by comparison of these velocities with those of models incorporating galactic rotation.

Our Galaxy and others like it also contain large molecular clouds at temperatures of 10–100 K in which new stars are continually forming. These clouds give rise to many atomic and molecular transitions in the radio and far-infrared ranges. More than 4,500 molecular lines from approximately 180 molecular species have been observed [see Herbst and van Dishoeck (2009)]. Lists of atomic and molecular lines are given by Jet Propulsion Laboratory (2016), the University of Cologne (2016), and Splatalogue (2016). For earlier lists, see Lovas et al. (1979) and Lovas (1992). A few of the more important lines are given in Table 1.1. Note that this table contains less than 1% of the known lines in the frequency range below 1 THz. Figure 1.2 shows the spectrum of radiation of many molecular lines from the Orion Nebula in the bands from 214 to 246 and from 328 to 360 GHz. Although the radio window in the Earth's atmosphere ends above ~ 1 THz, sensitive submillimeter- and millimeter-wavelength arrays can detect such lines as the $^2P_{3/2} \rightarrow {}^2P_{1/2}$ line of CII at 1.90054 THz (158 μm), which are Doppler shifted into the radio window for redshifts (z) greater than ~ 2. Some of the lines, notably those of OH, H_2O, SiO, and CH_3OH, show very intense emission from sources of very small apparent angular diameter. This emission is generated by a maser process [see, e.g., Reid and Moran (1988), Elitzur (1992), and Gray (2012)].

The strength of the radio signal received from a discrete source is expressed as the *spectral flux density*, or *spectral power flux density*, and is measured in watts per square meter per hertz (W m^{-2} Hz^{-1}). For brevity, astronomers often refer to this quantity as *flux density*. The unit of flux density is the jansky (Jy); 1 Jy = 10^{-26} W m^{-2} Hz^{-1}. It is used for both spectral line and continuum radiation. The measure of radiation integrated in frequency over a spectral band has units of W m^{-2} and is referred to as *power flux density*. In the standard definition of the IEEE (1977), power flux density is equal to the time average of the Poynting vector of the wave. In producing an image of a radio source, the desired quantity is the power flux density

Table 1.1 Some important radio lines

Chemical name	Chemical formula	Transition	Frequency (GHz)
Deuterium	D	$^2S_{1/2}, F = \frac{3}{2} \to \frac{1}{2}$	0.327
Hydrogen	H	$^2S_{1/2}, F = 1 \to 0$	1.420
Hydroxyl radical	OH	$^2\Pi_{3/2}, J = 3/2, F = 1 \to 2$	1.612[a]
Hydroxyl radical	OH	$^2\Pi_{3/2}, J = 3/2, F = 1 \to 1$	1.665[a]
Hydroxyl radical	OH	$^2\Pi_{3/2}, J = 3/2, F = 2 \to 2$	1.667[a]
Hydroxyl radical	OH	$^2\Pi_{3/2}, J = 3/2, F = 2 \to 1$	1.721[a]
Methyladyne	CH	$^2\Pi_{1/2}, J = 1/2, F = 1 \to 1$	3.335
Hydroxyl radical	OH	$^2\Pi_{1/2}, J = 1/2, F = 1 \to 0$	4.766[a]
Formaldehyde	H_2CO	$1_{10} - 1_{11}$, six F transitions	4.830
Hydroxyl radical	OH	$^2\Pi_{3/2}, J = 5/2, F = 3 \to 3$	6.035[a]
Methanol	CH_3OH	$5_1 \to 6_0 A^+$	6.668[a]
Helium	$^3He^+$	$^2S_{1/2}, F = 1 \to 0$	8.665
Methanol	CH_3OH	$2_0 \to 3_{-1} E$	12.179[a]
Formaldehyde	H_2CO	$2_{11} \to 2_{12}$, four F transitions	14.488
Cyclopropenylidene	C_3H_2	$1_{10} \to 1_{01}$	18.343
Water	H_2O	$6_{16} \to 5_{23}$, five F transitions	22.235[a]
Ammonia	NH_3	1, 1 $\to$ 1, 1, eighteen F transitions	23.694
Ammonia	NH_3	2, 2 $\to$ 2, 2, seven F transitions	23.723
Ammonia	NH_3	3, 3 $\to$ 3, 3, seven F transitions	23.870
Methanol	CH_3OH	$6_2 \to 6_1, E$	25.018
Silicon monoxide	SiO	$v = 2, J = 1 \to 0$	42.821[a]
Silicon monoxide	SiO	$v = 1, J = 1 \to 0$	43.122[a]
Carbon monosulfide	CS	$J = 1 \to 0$	48.991
Silicon monoxide	SiO	$v = 1, J = 2 \to 1$	86.243[a]
Hydrogen cyanide	HCN	$J = 1 \to 0$, three F transitions	88.632
Formylium	HCO^+	$J = 1 \to 0$	89.189
Diazenylium	N_2H^+	$J = 1 \to 0$, seven F transitions	93.174
Carbon monosulfide	CS	$J = 2 \to 1$	97.981
Carbon monoxide	$^{12}C^{18}O$	$J = 1 \to 0$	109.782
Carbon monoxide	$^{13}C^{16}O$	$J = 1 \to 0$	110.201
Carbon monoxide	$^{12}C^{17}O$	$J = 1 \to 0$, three F transitions	112.359
Carbon monoxide	$^{12}C^{16}O$	$J = 1 \to 0$	115.271
Carbon monosulfide	CS	$J = 3 \to 2$	146.969
Water	H_2O	$3_{13} \to 2_{20}$	183.310[a]
Carbon monoxide	$^{12}C^{16}O$	$J = 2 \to 1$	230.538
Carbon monosulfide	CS	$J = 5 \to 4$	244.936
Water	H_2O	$5_{15} \to 4_{22}$	325.153[a]
Carbon monosulfide	CS	$J = 7 \to 6$	342.883
Carbon monoxide	$^{12}C^{16}O$	$J = 3 \to 2$	345.796
Water	H_2O	$4_{14} \to 3_{21}$	380.197[b]
Carbon monoxide	$^{12}C^{16}O$	$J = 4 \to 3$	461.041
Heavy water	HDO	$1_{01} \to 0_{00}$	464.925
Carbon	C	$^3P_1 \to {}^3P_0$	492.162
Water	H_2O	$1_{10} \to 1_{01}$	556.936[b]
Ammonia	NH_3	$1_0 \to 0_0$	572.498
Carbon monoxide	$^{12}C^{16}O$	$J = 6 \to 5$	691.473
Carbon monoxide	$^{12}C^{16}O$	$J = 7 \to 6$	806.652
Carbon	C	$^3P_2 \to {}^3P_1$	809.340

[a]Strong maser transition.
[b]High atmospheric opacity (see Fig. 13.14).

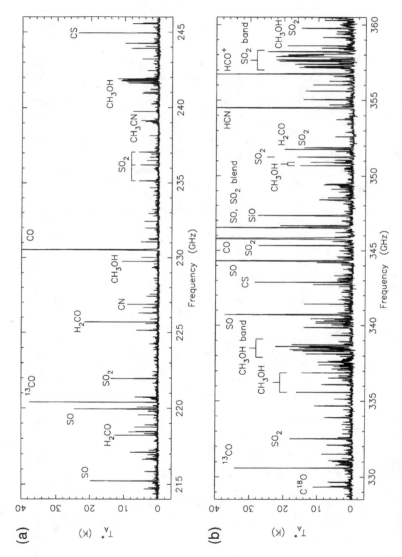

Fig. 1.2 Spectrum of the Orion Nebula for 214–246 and 328–360 GHz. The ordinate is antenna temperature corrected for atmospheric absorption, which is proportional to the power received. The frequency scale has been corrected for motion of the Earth with respect to the local standard of rest. The spectral resolution is 1 MHz, which corresponds to a velocity resolution of 1.3 and 0.87 km s^{-1} at 230 and 345 GHz, respectively. Note the higher density of lines in the higher frequency band. The measurements shown in (**a**) are from Blake et al. (1987), and those in (**b**) are from Schilke et al. (1997).

Fig. 1.3 Elements of solid angle and surface area illustrating the definition of intensity. dA is normal to **s**.

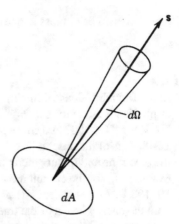

emitted per unit solid angle subtended by the radiating surface, which is measured in units of W m^{-2} Hz^{-1} sr^{-1}. This quantity is variously referred to as the *intensity, specific intensity,* or *brightness* of the radiation. In radio astronomical imaging, we can measure the intensity in only two dimensions on the surface of the celestial sphere, and the measured emission is the component normal to that surface, as seen by the observer.

In radiation theory, the quantity intensity, or specific intensity, often represented by I_ν, is the measure of radiated energy flow per unit area, per unit time, per unit frequency bandwidth, and per unit solid angle. Thus, in Fig. 1.3, the power flowing in direction **s** within solid angle $d\Omega$, frequency band $d\nu$, and area dA is $I_\nu(\mathbf{s})\, d\Omega\, d\nu\, dA$. This can be applied to emission from the surface of a radiating object, to propagation through a surface in space, or to reception on the surface of a transducer or detector. The last case applies to reception in an antenna, and the solid angle then denotes the area of the celestial sphere from which the radiation emanates. Note that in optical astronomy, the specific intensity is usually defined as the intensity per unit bandwidth I_λ, where $I_\lambda = I_\nu \nu^2/c$, and c is the speed of light [see, e.g., Rybicki and Lightman (1979)].

For thermal radiation from a blackbody, the intensity is related to the physical temperature T of the radiating matter by the Planck formula, for which

$$I_\nu = \frac{2kT\nu^2}{c^2}\left[\frac{\frac{h\nu}{kT}}{e^{h\nu/kT}-1}\right], \tag{1.1}$$

where k is Boltzmann's constant, and h is Planck's constant. When $h\nu \ll kT$, we can use the Rayleigh–Jeans approximation, in which case the expression in the square brackets is replaced by unity. The Rayleigh–Jeans approximation requires ν (GHz) $\ll$ 20 T (K) and is violated at high frequencies and low temperatures in many situations of interest to radio astronomers. However, for any radiation mechanism, a brightness temperature T_B can be defined:

$$T_B = \frac{c^2 I_\nu}{2k\nu^2} \, .$$ (1.2)

In the Rayleigh–Jeans domain, the brightness temperature T_B is that of a blackbody at physical temperature $T = T_B$. In the examples in Fig. 1.1, T_B is of the order of 10^4 K for NGC7027 and corresponds to the electron temperature. For Cygnus A and 3C48, T_B is of the order of 10^8 K or greater and is a measure of the energy density of the electrons and the magnetic fields, not a physical temperature. As a spectral line example, T_B for the carbon monoxide (CO) lines from molecular clouds is typically 10–100 K. In this case, T_B is proportional to the excitation temperature associated with the energy levels of the transition and is related to the temperature and density of the gas as well as to the temperature of the radiation field.

1.2.2   Source Positions and Nomenclature

The positions of radio sources are measured in the celestial coordinates *right ascension* and *declination*. On the celestial sphere, these quantities are analogous, respectively, to longitude and latitude on the Earth but tied to the plane of the Earth's orbit around the Sun. The zero of right ascension is arbitrarily chosen as the point at which the Sun crosses the celestial equator (going from negative to positive declination) on the vernal equinox at the first point of Aries at a given epoch. Positions of objects in celestial coordinates vary as a result of precession and nutation of the Earth's axis of rotation, aberration, and proper motion. These positions are usually listed for the standard epoch of the year 2000. Former standard epochs were 1950 and 1900. Methods of naming sources have proceeded haphazardly over the centuries. Important optical catalogs of sources were constructed as numerical lists, often in order of right ascension. Examples include the Messier catalog of nonstellar objects (Messier 1781; now containing 110 objects identified as galaxies, nebulae, and star clusters), the New General Catalog of nonstellar sources (Dreyer 1888; originally with 7,840 objects, mostly galaxies), and the Henry Draper catalog of stars (Cannon and Pickering 1924; now with 359,083 entries). The earliest radio sources were designated by their associated constellation. Hence, Cygnus A is the strongest source in the constellation of Cygnus. As the radio sky was systematically surveyed, catalogs appeared such as the third Cambridge catalog (3C), with 471 entries in the original list [Edge et al. (1959), extragalactic sources, e.g., 3C273] and the Westerhout catalog of 81 sources along the galactic plane [Westerhout (1958); mostly ionized nebula, e.g., W3].

In 1974, the International Astronomical Union adopted a resolution (International Astronomical Union 1974) to standardize the naming of sources based on their coordinates in the epoch of 1950 called the 4 + 4 system, in which the first four characters give the hour and minutes of right ascension (RA); the fifth, the sign of the declination (Dec.); and the remaining three, the degrees and tenths of degrees of declination. For example, the source at RA $01^h34^m49.83^s$, Dec. $32°54'20.5''$ would be designated 0134+329. Note that coordinates were truncated, not rounded. This system no longer has the accuracy needed to distinguish among sources. The current recommendation of the IAU Task Group on Astronomical Designations [International Astronomical Union (2008); see also NASA/IPAC Extragalactic Database (2013)] recommends the following convention. The source name begins with an identification acronym followed by a letter to identify the type of coordinates, followed by the coordinates to requisite accuracy. Examples of identification acronyms are QSO (quasi-stellar object), PSR (pulsar), and PKS (Parkes Radio Source). Coordinate identifiers are usually limited to J for epoch 2000, B for epoch 1950, and G for galactic coordinates. Hence, the radio source at the center of the galaxy M87, also known as NGC4486, contains an active galactic nucleus (AGN) centered at RA $= 12^h30^m49.42338^s$, Dec. $= 12°23'28.0439''$, which might be designated AGN J1230494233+122328043. It is also well known by the designations Virgo A and 3C274. Many catalogs of radio sources have been made, and some of them are described in Sect. 1.3.8. An index of more than 50 catalogs made before 1970, identifying more than 30,000 extragalactic radio sources, was compiled by Kesteven and Bridle (1971).

An example of a more recent survey is the NRAO VLA Sky Survey (NVSS) conducted by Condon et al. (1998) using the Very Large Array (VLA) at 1.4 GHz, which contains approximately 2×10^6 sources (about one source per 100 beam solid angles). Another important catalog derived from VLBI observations is the International Celestial Reference Frame (ICRF), which contains 295 sources with positions accurate to about 40 microarcseconds (Ma et al. 1998; Fey et al. 2015).

1.2.3   Reception of Cosmic Signals

The antennas used most commonly in radio astronomy are of the reflector type mounted to allow tracking over most of the sky. The exceptions are mainly instruments designed for meter or longer wavelengths. The collecting area A of a reflector antenna, for radiation incident in the center of the main beam, is equal to the geometrical area multiplied by an aperture efficiency factor, which is typically within the range 0.3–0.8. The received power P_A delivered by the antenna to a matched load in a bandwidth Δv, from a randomly polarized source of flux density S, assumed to be small compared to the beamwidth, is given by

$$P_A = \tfrac{1}{2}SA\Delta v .\tag{1.3}$$

Note that S is the intensity I_v integrated over the solid angle of the source. The factor $\tfrac{1}{2}$ takes account of the fact that the antenna responds to only one-half the power in the randomly polarized wave. It is often convenient to express random noise power, P, in terms of an effective temperature T, as

$$P = kT\Delta v ,\tag{1.4}$$

where k is Boltzmann's constant. In the Rayleigh–Jeans domain, P is equal to the noise power delivered to a matched load by a resistor at physical temperature T (Nyquist 1928). In the general case, if we use the Planck formula [Eq. (1.1)], we can write $P = kT_{\text{Planck}} \Delta \nu$, where T_{Planck} is an effective radiation temperature, or noise temperature, of a load at physical temperature T, and is given by

$$T_{\text{Planck}} = T \left[\frac{\frac{h\nu}{kT}}{e^{h\nu/kT} - 1} \right] . \tag{1.5}$$

The noise power in a receiving system (see Appendix 1.1) can be specified in terms of the *system temperature T_S* associated with a matched resistive load that would produce an equal power level in an equivalent noise-free receiver when connected to the input terminals. T_S is defined as the power available from this load divided by $k\Delta\nu$. In terms of the Planck formula, the relation between T_S and the physical temperature, T, of such a load is given by replacing T_{Planck} by T_S in Eq. (1.1).

The system temperature consists of two parts: T_R, the *receiver temperature*, which represents the internal noise from the receiver components, plus the unwanted noise incurred from connecting the receiver to the antenna and from the noise components from the antenna produced by ground radiation, atmospheric emission, ohmic losses, and other sources.

We reserve the term *antenna temperature* to refer to the component of the power received by the antenna that results from a cosmic source under study. The power received in an antenna from the source is [see Eq. (1.4)]

$$P_A = kT_A \, \Delta\nu , \tag{1.6}$$

and T_A is related to the flux density by Eqs. (1.3) and (1.6). It is useful to express this relation as T_A (K) $= SA/2k = S$ (Jy) $\times A$ (m^2)/2800. Astronomers sometimes specify the performance of an antenna in terms of *janskys per kelvin*, that is, the flux density (in units of 10^{-26} W m^{-2} Hz^{-1}), of a point source that increases T_A by one kelvin. Thus, this measure is equal to $2800/A$ (m^2) Jy K^{-1}.

Another term that may be encountered is the *system equivalent flux density*, SEFD, which is an indicator of the combined sensitivity of both an antenna and receiving system. It is equal to the flux density of a point source in the main beam of the antenna that would cause the noise power in the receiver to be twice that of the system noise in the absence of a source. Equating P_A in Eq. (1.3) with $kT_S\Delta\nu$, we obtain

$$\text{SEFD} = \frac{2kT_S}{A} . \tag{1.7}$$

The ratio of the signal power from a source to the noise power in the receiving amplifier is T_A/T_S. Because of the random nature of the signal and noise, measurements of the power levels made at time intervals separated by $(2\Delta\nu)^{-1}$ can be considered independent. A measurement in which the signal level is averaged for a time τ contains approximately $2\Delta\nu\tau$ independent samples. The signal-to-noise ratio (SNR), $\mathcal{R}_{sn}$, at the output of a power-measuring device attached to the receiver is increased in proportion to the square root of the number of independent samples and is of the form

$$\mathcal{R}_{sn} = C\frac{T_A}{T_S}\sqrt{\Delta\nu\tau}\,, \tag{1.8}$$

where C is a constant that is greater than or equal to one. This result (derived in Appendix 1.1) appears to have been first obtained by Dicke (1946) for an analog system. $C = 1$ for a simple power-law receiver with a rectangular passband and can be larger by a factor of ~ 2 for more complicated systems. Typical values of $\Delta\nu$ and τ are of order 1 GHz and 6 h, which result in a value of 4×10^6 for the factor $(\Delta\nu\tau)^{1/2}$. As a result, it is possible to detect a signal for which the power level is less than 10^{-6} times the system noise. A particularly effective use of long averaging time is found in the observations with the Cosmic Background Explorer (COBE) satellite, in which it was possible to measure structure at a brightness temperature level less than 10^{-7} of the system temperature (Smoot et al. 1990, 1992).

The following calculation may help to illustrate the low energies involved in radio astronomy. Consider a large radio telescope with a total collecting area of 10^4 m^2 pointed toward a radio source of flux density 1 mJy ($= 10^{-3}$ Jy) and accepting signals over a bandwidth of 50 MHz. In 10^3 years, the total energy accepted is about 10^{-7} J (1 erg), which is comparable to a few percent of the kinetic energy in a single falling snowflake. To detect the source with the same telescope and a system temperature of 50 K would require an observing time of about 5 min, during which time the energy received would be about 10^{-15} J.

1.3 Development of Radio Interferometry

1.3.1 Evolution of Synthesis Techniques

This section presents a brief history of interferometry in radio astronomy. As an introduction, the following list indicates some of the more important steps in the progress from the Michelson stellar interferometer to the development of multi-element, synthesis imaging arrays and VLBI:

1. *Michelson stellar interferometer.* This optical instrument introduced the technique of using two spaced receiving apertures, and the measurement of fringe amplitude to determine angular width (1890–1921).
2. *First astronomical observations with a two-element radio interferometer.* Ryle and Vonberg (1946), solar observations.
3. *Phase-switching interferometer.* First implementation of the voltage-multiplying action of a correlator, which is the device used to combine the signals from two antennas (1952).

4. *Astronomical calibration.* Gradual accumulation during the 1950s and 1960s of accurate positions for small-diameter radio sources from optical identifications and other means. Observations of such sources enabled accurate calibration of interferometer baselines and instrumental phases.

5. *Early measurements of angular dimensions of sources.* Use of variable-baseline interferometers ($\sim$ 1952 onward).

6. *Solar arrays.* Development of multiantenna arrays of centimeter-wavelength tracking antennas that provided detailed maps and profiles of the solar disk (mid-1950s onward).

7. *Arrays of tracking antennas.* General movement from meter-wavelength, non-tracking antennas to centimeter-wavelength, tracking antennas. Development of multielement arrays with a separate correlator for each baseline ($\sim$ 1960s).

8. *Earth-rotation synthesis.* Introduced by Ryle with some precedents from solar imaging. The development of computers to control receiving systems and perform Fourier transforms required in imaging was an essential component (1962).

9. *Spectral line capability.* Introduced into radio interferometry ($\sim$ 1962).

10. *Development of image-processing techniques.* Based on phase and amplitude closure, nonlinear deconvolution and other techniques, as described in Chaps. 10 and 11 ($\sim$ 1974 onward).

11. *Very-long-baseline interferometry* (*VLBI*). First observations (1967). Super-luminal motion in active galactic nuclei discovered (1971). Contemporary plate motion detected (1986). International Celestial Reference Frame adopted (1998).

12. *Millimeter-wavelength instruments* ($\sim$ 100–300 GHz). Major developments mid-1980s onward.

13. *Orbiting VLBI (OVLBI).* U.S. Tracking and Data Relay Satellite System (TDRSS) experiment (1986–88). VLBI Space Observatory Programme (VSOP) (1997). RadioAstron (2011).

14. *Submillimeter-wavelength instruments* (300 GHz–1 THz). James Clerk Maxwell Telescope–Caltech Submillimeter Observatory interferometer (1992). Submillimeter Array of the Smithsonian Astrophysical Observatory (SAO) and Academia Sinica of Taiwan (2004). Atacama Large Millimeter/submillimeter Array (ALMA) (2013).

1.3.2 Michelson Interferometer

Interferometric techniques in astronomy date back to the optical work of Michelson (1890, 1920) and of Michelson and Pease (1921), who were able to obtain sufficiently fine angular resolution to measure the diameters of some of the nearer and larger stars such as Arcturus and Betelgeuse. The basic similarity of the theory of radio and optical radiation fields was recognized early by radio astronomers,

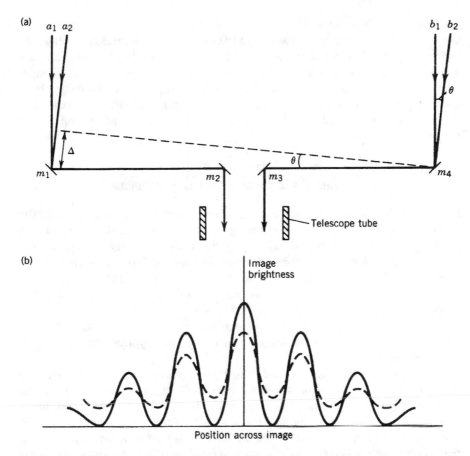

Fig. 1.4 (**a**) Schematic diagram of the Michelson–Pease stellar interferometer. The incoming rays are guided into the telescope aperture by mirrors m_1 to m_4, of which the outer pair define the two apertures of the interferometer. Rays a_1 and b_1 traverse equal paths to the eyepiece at which the image is formed, but rays a_2 and b_2, which approach at an angle θ to the instrumental axis, traverse paths that differ by a distance Δ. (**b**) The intensity of the image as a function of position angle in a direction parallel to the spacing of the interferometer apertures. The solid line shows the fringe profiles for an unresolved star ($\mathcal{V}_M = 1.0$), and the broken line is for a partially resolved star for which $\mathcal{V}_M = 0.5$.

and optical experience has provided valuable precedents to the theory of radio interferometry.

As shown in Fig. 1.4, beams of light from a star fall upon two apertures and are combined in a telescope. The resulting stellar image has a finite width and is shaped by effects that include atmospheric turbulence, diffraction at the mirrors, and the bandwidth of the radiation. Maxima in the light intensity resulting from interference occur at angles θ for which the difference Δ in the path lengths from the star to the point at which the light waves are combined is an integral number of wavelengths at the effective center of the optical passband. If the angular width of the star is small compared with the spacing in θ between adjacent maxima, the image of the star is crossed by alternate dark and light bands, known as interference fringes. If, however, the width of the star is comparable to the spacing between maxima, one can visualize the resulting image as being formed by the superposition of images from a series of points across the star. The maxima and minima of the fringes from different points do not coincide, and the fringe amplitude is attenuated, as shown in Fig. 1.4b. As a measure of the relative amplitude of the fringes, Michelson defined the *fringe visibility*, $\mathcal{V}_M$, as

$$\mathcal{V}_M = \frac{\text{intensity of maxima - intensity of minima}}{\text{intensity of maxima + intensity of minima}} . \tag{1.9}$$

Note that with this definition, the visibility is normalized to unity when the intensity at the minima is zero, that is, when the width of the star is small compared with the fringe width. If the fringe visibility is measurably less than unity, the star is said to be *resolved* by the interferometer. In their 1921 paper, Michelson and Pease explained the apparent paradox that their instrument could be used to detect structure smaller than the seeing limit imposed by atmospheric turbulence. The fringe pattern, as depicted in Fig. 1.4, moves erratically on time scales of 10–100 ms. Over long averaging time, the fringes are smoothed out. However, the "jittering" fringes can be discerned by the human eye, which has a typical response time of tens of milliseconds.

Let $I(l, m)$ be the two-dimensional intensity of the star, or of a source in the case of a radio interferometer. (l, m) are coordinates on the sky, with l measured parallel to the aperture spacing vector and m normal to it. The fringes provide resolution in a direction parallel to the aperture spacing only. In the orthogonal direction, the response is simply proportional to the intensity integrated over solid angle. Thus, the interferometer measures the intensity projected onto the l direction, that is, the one-dimensional profile $I_1(l)$ given by

$$I_1(l) = \int I(l, m) \, dm . \tag{1.10}$$

As will be shown in later chapters, the fringe visibility is proportional to the modulus of the Fourier transform of $I_1(l)$ with respect to the spacing of the apertures measured in wavelengths. Figure 1.5 shows the integrated profile I_1 for three simple models of a star or radio source and the corresponding fringe visibility as a function of u, the spacing of the interferometer apertures in units of the wavelength. At the top of the figure is a rectangular pillbox distribution, in the center a circular pillbox, and at the bottom a circular Gaussian function. The rectangular pillbox represents a uniformly bright rectangle on the sky with sides parallel to the l and m axes and width a in the l direction. The circular pillbox represents a uniformly bright circular disk of diameter a. When projected onto the l axis, the one-dimensional intensity function I_1 has a semicircular profile. The Gaussian model is a circularly symmetric source with Gaussian taper of the intensity from the maximum at the center. The

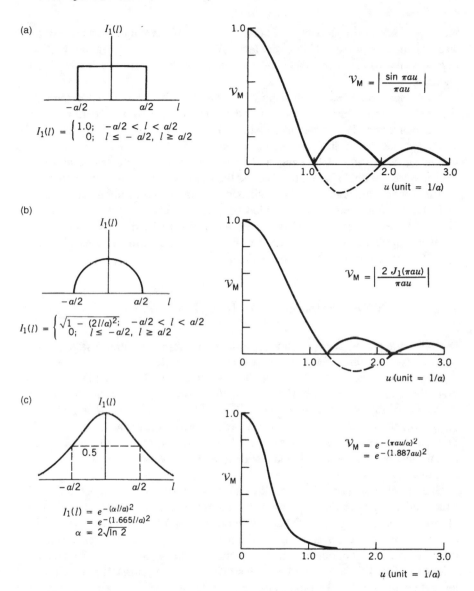

Fig. 1.5 The one-dimensional intensity profiles $I_1(l)$ for three simple intensity models: (**a**) left, a uniform rectangular source; (**b**) left, a uniform circular source; and (**c**) left, a circular Gaussian distribution. The corresponding Michelson visibility functions $\mathcal{V}_M$ are on the right. l is an angular variable on the sky, u is the spacing of the receiving apertures measured in wavelengths, and a is the characteristic angular width of the model. The solid lines in the curves of $\mathcal{V}_M$ indicate the modulus of the Fourier transform of $I_1(l)$, and the broken lines indicate negative values of the transform. See text for further explanation. Models are discussed in more detail in Sect. 10.4.

intensity is proportional to $\exp[-4\ln 2(l^2 + m^2)/a^2]$, resulting in circular contours and a diameter a at the half-intensity level. Any slice through the model in a plane perpendicular to the (l, m) plane has a Gaussian profile with the same half-height width, a.

Michelson and Pease used mainly the circular disk model to interpret their observations and determined the stellar diameter by varying the aperture spacing of the interferometer to locate the first minimum in the visibility function. In the age before electronic instrumentation, the adjustment of such an instrument and the visual estimation of $\mathcal{V}_M$ required great care, since, as described above, the fringes were not stable but vibrated across the image in a random manner as a result of atmospheric fluctuations. The published results on stellar diameters measured with this method were never extended beyond the seven bright stars in Pease's (1931) list; for a detailed review see Hanbury Brown (1968). However, the use of electro-optical techniques now offers much greater instrumental capabilities in optical interferometry, as discussed in Sect. 17.4.

1.3.3   Early Two-Element Radio Interferometers

In 1946, Ryle and Vonberg constructed a radio interferometer to investigate cosmic radio emission, which had been discovered and verified by earlier investigators (Jansky 1933; Reber 1940; Appleton 1945; Southworth 1945). This interferometer used dipole antenna arrays at 175 MHz, with a baseline (i.e., the spacing between the antennas) that was variable between 10 and 140 wavelengths (17 and 240 m). A diagram of such an instrument and the type of record obtained are shown in Fig. 1.6. In this and most other meter-wavelength interferometers of the 1950s and 1960s, the antenna beams were pointed in the meridian, and the rotation of the Earth provided scanning in right ascension.

The receiver in Fig. 1.6 is sensitive to a narrow band of frequencies, and a simplified analysis of the response of the interferometer can be obtained in terms of monochromatic signals at the center frequency ν_0. We consider the signal from a radio source of very small angular diameter that is sufficiently distant that the incoming wavefront effectively lies in a plane. Let the signal voltage from the right antenna in Fig. 1.6 be represented by $V \sin(2\pi\nu_0 t)$. The longer path length to the left antenna (as in Fig. 1.4) introduces a time delay $\tau = (D/c)\sin\theta$, where D is the antenna spacing, θ is the angular position of the source, and c is the velocity of light. Thus, the signal from the left antenna is $V \sin[2\pi\nu_0(t - \tau)]$. The detector of the receiver generates a response proportional to the squared sum of the two signal voltages:

$$\{V \sin(2\pi\nu_0 t) + V \sin[2\pi\nu_0(t - \tau)]\}^2 . \tag{1.11}$$

The output of the detector is averaged in time, i.e., it contains a lowpass filter that removes any frequencies greater than a few hertz or tens of hertz, so in

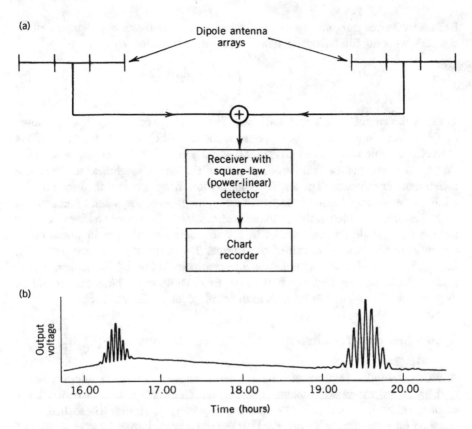

Fig. 1.6 (a) A simple interferometer, also called an adding interferometer, in which the signals are combined additively. (b) Record from such an interferometer with east–west antenna spacing. The ordinate is the total power received, since the voltage from the square-law detector is proportional to power, and the abscissa is time. The source at the left is Cygnus A and the one at the right Cassiopeia A. The increase in level near Cygnus A results from the galactic background radiation, which is concentrated toward the plane of our Galaxy but is completely resolved by the interferometer fringes. The record is from Ryle (1952). Reproduced with permission of the Royal Society, London, and the Master and Fellows of Churchill College, Cambridge. © Royal Society.

expanding (1.11), we can ignore the term in the harmonic of $2\pi \nu_0 t$. The detector output,[2] in terms of the power P_0 generated by either of the antennas alone, is therefore

$$P = P_0[1 + \cos(2\pi \nu_0 \tau)] . \tag{1.12}$$

---

[2]For simplicity, in expression (1.11), we added the signal voltages from each antenna. In practice, such signals must be combined in networks that obey the conservation of power. Thus, if the signal from each antenna is represented as a voltage source V and characteristic impedance R, the power available is V^2/R. Combining two signals in series can be represented by a voltage $2V$ and impedance $2R$, giving a power of $2V^2/R$. In contrast, in free space, the addition of two coherent electric fields of equal strength quadruples the power. This distinction is important in the discussion of the sea interferometer (Sect. 1.3.4).

Because τ varies only slowly as the Earth rotates, the frequency represented by $\cos(2\pi\nu_0\tau)$ is not filtered out. In terms of the source position, θ, we have

$$P = P_0 \left[1 + \cos\left(\frac{2\pi\nu_0 D \sin\theta}{c} \right) \right]. \tag{1.13}$$

Thus, as the source moves across the sky, P varies between 0 and $2P_0$, as shown by the sources in Fig. 1.6b. The response is modulated by the beam pattern of the antennas, of which the maximum is pointed in the meridian. The cosine function in Eq. (1.13) represents the Fourier component of the source brightness to which the interferometer responds. The angular width of the fringes is less than the angular width of the antenna beam by (approximately) the ratio of the width of an antenna to the baseline D, which in this example is about 1/10. The use of an interferometer instead of a single antenna results in a corresponding increase in precision in determining the time of transit of the source. The form of the fringe pattern in Eq. (1.13) also applies to the Michelson interferometer in Fig. 1.4. In the former case (radio), the fringes develop as a function of time, while in the latter case (optical), they appear as a function of position in the pupil plane of the telescope.

1.3.4 Sea Interferometer

A different implementation of interferometry, known as the sea interferometer, or Lloyd's mirror interferometer (Bolton and Slee 1953), was provided by a number of horizon-pointing antennas near Sydney, Australia. These had been installed for radar during World War II at several coastal locations, at elevations of 60–120 m above the sea. Radiation from sources rising over the eastern horizon was received both directly and by reflection from the sea, as shown in Fig. 1.7. The frequencies of the observations were in the range 40–400 MHz, the middle part of the range being the most satisfactory because of the sensitivity of receivers there and because of ionospheric effects at lower frequencies and sea roughness at higher frequencies. The sudden appearance of a rising source was useful in separating individual sources. Because of the reflected wave, the power received at the peak of a fringe was four times that for direct reception with the single antenna, and twice that of an adding interferometer with two of the same antennas (see footnote 2). Observations of the Sun by McCready et al. (1947) using this system provided the first published record of interference fringes in radio astronomy. They recognized that they were measuring a Fourier component of the brightness distribution and used the term "Fourier synthesis" to describe how an image could be produced from fringe visibility measurements on many baselines. Observations of the source Cygnus A by Bolton and Stanley (1948) provided the first positive evidence of the existence of a discrete nonsolar radio source. Thus, the sea interferometer played an important part in early radio astronomy, but the effects of the long atmospheric paths, the roughness of the sea surface, and the difficulty of varying the physical length of the baseline, which was set by the cliff height, precluded further useful development.

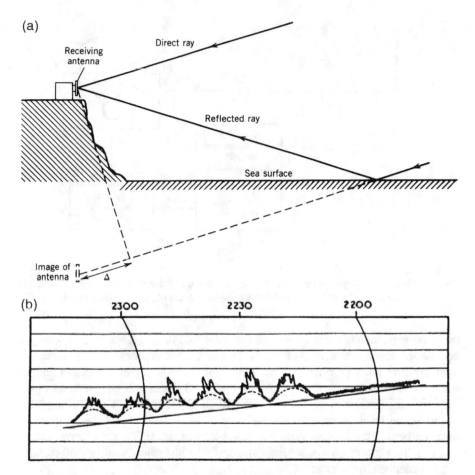

Fig. 1.7 (a) Schematic diagram of a sea interferometer. The fringe pattern is similar to that which would be obtained with the actual receiving antenna and one at the position of its image in the sea. The reflected ray undergoes a phase change of 180° on reflection and travels an extra distance Δ in reaching the receiving antenna. (b) Sea interferometer record of the source Cygnus A at 100 MHz by Bolton and Stanley (1948). The source rose above the horizon at approximately 22^h17^m. The broken line was inserted to show that the record could be interpreted in terms of a steady component and a fluctuating component of the source; the fluctuations were later shown to be of ionospheric origin. The fringe width was approximately 1.0° and the source is unresolved, that is, its angular width is small in comparison with the fringe width. Part (b) is reprinted by permission from MacMillan Publishers Ltd.: *Nature*, **161**, 312–313, © 1948.

1.3.5 Phase-Switching Interferometer

A problem with the interferometer systems in both Figs. 1.6 and 1.7 is that in addition to the signal from the source, the output of the receiver contains components from other sources of noise power such as the galactic background radiation, thermal noise from the ground picked up in the antenna sidelobes, and

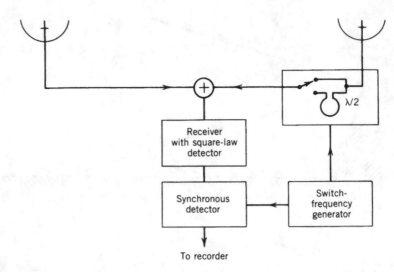

Fig. 1.8 Phase-switching interferometer. The signal from one antenna is periodically reversed in phase, indicated here by switching an additional half-wavelength of path into the transmission line.

the noise generated in the amplifiers of the receiver. For all except the few strongest cosmic sources, the component from the source is several orders of magnitude less than the total noise power in the receiver. Thus, a large offset has been removed from the records shown in Figs. 1.6b and 1.7b. This offset is proportional to the receiver gain, changes in which are difficult to eliminate entirely. The resulting drifts in the output level degrade the detectability of weak sources and the accuracy of measurement of the fringes. With the technology of the 1950s, the receiver output was usually recorded on a paper chart and could be lost when baseline drifts caused the recorder pen to go off scale.

The introduction of *phase switching* by Ryle (1952), which removed the unwanted components of the receiver output, leaving only the fringe oscillations, was the most important technical improvement in early radio interferometry. If V_1 and V_2 represent the signal voltages from the two antennas, the output from the simple adding interferometer is proportional to $(V_1 + V_2)^2$. In the phase-switching system, shown in Fig. 1.8, the phase of one of the signals is periodically reversed, so the output of the detector alternates between $(V_1 + V_2)^2$ and $(V_1 - V_2)^2$. The frequency of the switching is a few tens of hertz, and a synchronous detector takes the difference between the two output terms, which is proportional to $V_1 V_2$. Thus, the output of a phase-switching interferometer is the time average of the product of the signal voltages; that is, it is proportional to the cross-correlation of the two signals. The circuitry that performs the multiplication and averaging of the signals in a modern interferometer is known as a *correlator*: a more general definition of a correlator will be given later. Comparison with the output of the system in Fig. 1.6 shows that if the signals from the antennas are multiplied instead of added and squared, then the constant term within the square brackets in Eq. (1.13)

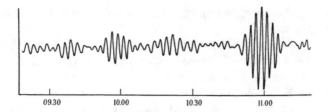

09.30   10.00   10.30   11.00

Fig. 1.9 Output of a phase-switching interferometer as a function of time, showing the response to a number of sources. From Ryle (1952). Reproduced with permission of the Royal Society, London, and the Master and Fellows of Churchill College, Cambridge. © Royal Society.

disappears, and only the cosine term remains. The output consists of the fringe oscillations only, as shown in Fig. 1.9. The removal of the constant term greatly reduces the sensitivity to instrumental gain variation, and it becomes practicable to install amplifiers at the antennas to overcome attenuation in the transmission lines. This advance resulted in the use of longer antenna spacings and larger arrays. Most interferometers from about 1950 onward incorporated phase switching, which provided the earliest means of implementing the multiplying action of a correlator. With more modern instruments, it is no longer necessary to use phase switching to obtain the voltage-multiplying action, but it is often included to help eliminate various instrumental imperfections, as described in Sect. 7.5.

1.3.6   Optical Identifications and Calibration Sources

Interferometer observations by Bolton and Stanley (1948), Ryle and Smith (1948), Ryle et al. (1950), and others provided evidence of numerous discrete sources. Identification of the optical counterparts of these required accurate measurement of radio positions. The principal method then in use for position measurement with interferometers was to determine the time of transit of the central fringe using an east–west baseline, and also the frequency of the fringe oscillations, which is proportional to the cosine of the declination (see Sect. 12.1 for more details). The measurement of position is only as accurate as the knowledge of the interferometer fringe pattern, which is determined by the relative locations of the electrical centers of the antennas. In addition, any inequality in the electrical path lengths in the cables and amplifiers from the antennas to the point where the signals are combined introduces an instrumental phase term, which offsets the fringe pattern. Smith (1952a) obtained positions for four sources with rms errors as small as $\pm 20''$ in right ascension and $\pm 40''$ in declination and gave a detailed analysis of the accuracy that was attainable. The optical identification of Cygnus A and Cassiopeia A by Baade and Minkowski (1954a,b) was a direct result of improved radio positions by Smith (1951) and Mills (1952). Cygnus A proved to be a distant galaxy and Cassiopeia A a supernova remnant, but the interpretation of the optical observations was not fully understood at the time.

The need for absolute calibration of the antennas and receiving system rapidly disappeared after a number of compact radio sources were identified with optical objects. Optical positions accurate to $\sim 1''$ could then be used, and observations of such sources enabled calibration of interferometer baseline parameters and fringe phases. Although it cannot be assumed that the radio and optical positions of a source coincide exactly, the offsets for different sources are randomly oriented. Thus, errors were reduced as more calibration sources became available. Another important way of obtaining accurate radio positions during the 1960s and 1970s was by observation of occultation of sources by the Moon, which is described in Sect. 17.2.

1.3.7   Early Measurements of Angular Width

Comparison of the angular widths of radio sources with the corresponding dimensions of their optical counterparts helped in some cases to confirm identifications as well as to provide important data for physical understanding of the emission processes. In the simplest procedure, measurements of the fringe amplitude are interpreted in terms of intensity models such as those shown in Fig. 1.5. The peak-to-peak fringe amplitude for a given spacing normalized to the same quantity when the source is unresolved provides a measure of the fringe visibility equivalent to the definition in Eq. (1.9).

Some of the earliest measurements were made by Mills (1953), who used an interferometer operating at 101 MHz, in which a small transportable array of Yagi elements could be located at distances up to 10 km from a larger antenna. The signal from this remote antenna was transmitted back over a radio link, and fringes were formed. Smith (1952b,c), at Cambridge, England, also measured the variation of fringe amplitude with antenna spacing but used shorter baselines than Mills and concentrated on precise measurements of small changes in the fringe amplitude. Results by both investigators provided angular sizes of a number of the strongest sources: Cassiopeia A, the Crab Nebula, NGC4486 (Virgo A), and NGC5128 (Centaurus A).

A third early group working on angular widths at the Jodrell Bank Experimental Station,[3] England, used a different technique: *intensity interferometry* (Jennison and Das Gupta 1953, 1956; Jennison 1994). Hanbury Brown and Twiss (1954) had shown that if the signals received by two spaced antennas are passed through square-law detectors, the fluctuations in the intensity that result from the Gaussian fluctuations in the received field strength are correlated. The degree of correlation

---

[3]Later known as the Nuffield Radio Astronomy Laboratories, and since 1999 as the Jodrell Bank Observatory.

varies in proportion to the square of the visibility that would be obtained in a conventional interferometer in which signals are combined before detection. The intensity interferometer has the advantage that it is not necessary to preserve the radio-frequency phase of the signals in bringing them to the location at which they are combined. This simplifies the use of long baselines, which in this case extended up to 10 km. A VHF radio link was used to transmit the detected signal from the remote antenna, for measurement of the correlation. The disadvantage of the intensity interferometer is that it requires a high SNR, and even for Cygnus A and Cassiopeia A, the two highest flux-density sources in the sky, it was necessary to construct large arrays of dipoles, which operated at 125 MHz. The intensity interferometer is discussed further in Sect. 17.1, but it has been of only limited use in radio astronomy because of its lack of sensitivity.

The most important result of these intensity interferometer measurements was the discovery that for Cygnus A, the fringe visibility for the east–west intensity profile falls close to zero and then increases to a secondary maximum as the antenna spacing is increased. Two symmetric source models were consistent with the visibility values derived from the measurements. These were a two-component model in which the phase of the fringes changes by 180° in going through the minimum, and a three-component model in which the phase does not change. The intensity interferometer gives no information on the fringe phase, so a subsequent experiment was made by Jennison and Latham (1959) using conventional interferometry. Because the instrumental phase of the equipment was not stable enough to permit calibration, three antennas were used and three sets of fringes for the three pair combinations were recorded simultaneously. If ϕ_{mn} is the phase of the fringe pattern for antennas m and n, it is easy to show that at any instant, the combination

$$\phi_{123} = \phi_{12} + \phi_{23} + \phi_{31} \tag{1.14}$$

is independent of instrumental and atmospheric phase effects and is a measure of the corresponding combination of fringe phases (Jennison 1958). By moving one antenna at a time, it was found that the phase does indeed change by approximately 180° at the visibility minimum and therefore that the two-component model in Fig. 1.10 is the appropriate one. The use of combinations of simultaneous visibility measurements typified by Eq. (1.14), now referred to as *closure relationships*, became important about 20 years later in image-processing techniques. Closure relationships and the conditions under which they apply are discussed in Sect. 10.3. They are now integral parts of the self-calibration used in image formation (see Sect. 11.3).

The results on Cygnus A demonstrated that the simple models of Fig. 1.5 are not generally satisfactory for representation of radio sources. To determine even the most basic structure, it is necessary to measure the fringe visibility at spacings well beyond the first minimum of the visibility function to detect multiple components, and to make such measurements at a number of position angles across the source.

An early interferometer aimed at achieving high angular resolution with high sensitivity was developed by Hanbury Brown et al. (1955) at the Jodrell Bank

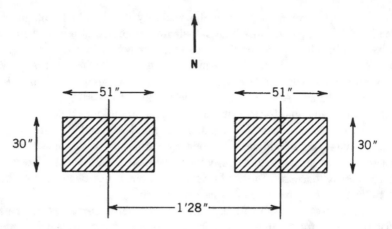

Fig. 1.10 Two-component model of Cygnus A derived by Jennison and Das Gupta (1953) using the intensity interferometer. Reprinted by permission from MacMillan Publishers Ltd.: *Nature*, **172**, 996–997, © 1953.

Experimental Station. This interferometer used an offset local oscillator technique at one antenna that took the place of a phase switch and also enabled the frequency of the fringe pattern to be slowed down to within the response time of the chart recorder used to record the output. A radio link was used to bring the signal from the distant antenna. Three sources were found to have diameters less than 12″ using spacings up to 20 km at 158 MHz observing frequency (Morris et al. 1957). During the 1960s, this instrument was extended to baselines of up to 134 km to achieve resolution of less than 1″ and greater sensitivity (Elgaroy et al. 1962; Adgie et al. 1965). The program later led to the development of a multielement, radio-linked interferometer known as the MERLIN array (Thomasson 1986).

1.3.8 Early Survey Interferometers and the Mills Cross

In the mid-1950s, the thrust of much work was toward cataloging larger numbers of sources with positions of sufficient accuracy to allow optical identification. The instruments operated mainly at meter wavelengths, where the spectrum was then much less heavily crowded with manmade emissions. A large interferometer at Cambridge used four antennas located at the corners of a rectangle 580 m east–west by 49 m north–south (Ryle and Hewish 1955). This arrangement provided both east–west and north–south fringe patterns for measurement of right ascension and declination.

A different type of survey instrument was developed by Mills et al. (1958) at Fleurs, near Sydney, consisting of two long, narrow antenna arrays in the form of a cross, as shown in Fig. 1.11. Each array produced a *fan beam*, that is, a beam that is narrow in a plane containing the long axis of the array and wide in the orthogonal

Fig. 1.11 Simplified diagram
of the Mills cross radio
telescope. The cross-shaped
area represents the apertures
of the two antennas.

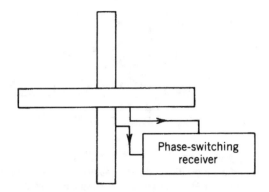

direction. The outputs of these two arrays were combined in a phase-switching receiver, and the voltage-multiplying action produced a power-response pattern equal to the product of the voltage responses of the two arrays. This combined response had the form of a narrow *pencil beam*. The two arrays had a common electrical center, so there were no interferometer fringes. The arrays were 457 m long, and the cross produced a beam of width 49 arcmin and approximately circular cross section at 85.5 MHz. The beam pointed in the meridian and could be steered in elevation by adjusting the phase of the dipoles in the north–south arm. The sky survey made with this instrument provided a list of more than 2,200 sources.

A comparison of the source catalogs from the Mills cross with those from the Cambridge interferometer, which initially operated at 81.5 MHz (Shakeshaft et al. 1955), showed poor agreement between the source lists for a common area of sky (Mills and Slee 1957). The discrepancy was found to result principally from the occurrence of *source confusion* in the Cambridge observations. When two or more sources are simultaneously present within the antenna beams, they produce fringe oscillations with slightly different frequencies, resulting from differences in the source declinations. Maxima in the fringe amplitude, which occur when the fringe components happen to combine in phase, can mimic responses to sources. This was a serious problem because the beams of the interferometer antennas were too wide, a problem that did not arise in the Mills cross, which was designed to provide the required resolution for accurate positions in the single pencil beam. The frequency of the Cambridge interferometer was later increased to 159 MHz, thereby reducing the solid angles of the beams by a factor of four, and a new list of 471 sources was rapidly compiled (Edge et al. 1959). This was the 3C survey (source numbers, listed in order of right ascension, are preceded by 3C, indicating the third Cambridge catalog). The revised version of this survey (Bennett 1962, the 3C catalog) had 328 entries (some additions and deletions) and became a cornerstone of radio astronomy for the following decade. To avoid confusion problems and errors in flux-density distributions determined with these types of instruments as well as single-element telescopes, some astronomers subsequently recommended that the density of sources cataloged should not, on average, exceed 1 in roughly 20 times the solid angle of the resolution element of the measurement instrument (Pawsey

(a)   (b)

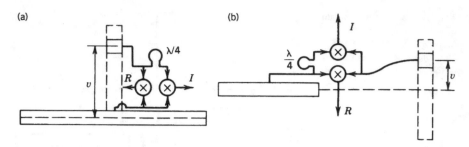

Fig. 1.12 Schematic diagrams of two instruments, in each of which a small antenna is moved to different positions between successive observations to synthesize the response that would be obtained with a full aperture corresponding to the rectangle shown by the broken line. The arrangement of two signal-multiplying correlators producing real (R) and imaginary (I) outputs is explained in Sect. 6.1.7. Instruments of both types, the T-shaped array (**a**), and the two-element interferometer (**b**), were constructed at the Mullard Radio Astronomy Observatory, Cambridge, England.

1958; Hazard and Walsh 1959). This criterion depends on the slope of a source count vs. flux density distribution (Scheuer 1957). For a modern treatment of the effects of source confusion, see Condon (1974) and Condon et al. (2012).

In the 1960s, a generation of new and larger survey instruments began to appear. Two such instruments developed at Cambridge are shown in Fig. 1.12. One was an interferometer with one antenna elongated in the east–west direction and the other north–south, and the other was a large T-shaped array that had characteristics similar to those of a cross, as explained in Sect. 5.3.3. In each of these instruments, the north–south element was not constructed in full, but the response with such an aperture was synthesized by using a small antenna that was moved in steps to cover the required aperture; a different position was used for each 24-h scan in right ascension (Ryle et al. 1959; Ryle and Hewish 1960). The records from the various positions were combined by computer to synthesize the response with the complete north–south aperture. An analysis of these instruments is given by Blythe (1957). The large interferometer produced the 4C (Fourth Cambridge) catalog containing over 4,800 sources (Gower et al. 1967). At Molonglo in Australia, a larger Mills cross (Mills et al. 1963) was constructed with arrays 1 mile long, producing a beam of 2.8-arcmin width at 408 MHz. The development of the Mills cross is described in papers by Mills and Little (1953), Mills (1963), and Mills et al. (1958, 1963). Crosses of comparable dimensions located in the Northern Hemisphere included one at Bologna, Italy (Braccesi et al. 1969), and one at Serpukhov, near Moscow in the former Soviet Union (Vitkevich and Kalachev 1966).

1.3.9   Centimeter-Wavelength Solar Imaging

A number of instruments have been designed specifically for imaging the Sun. The antennas were usually parabolic reflectors mounted to track the Sun, but since the

(a)

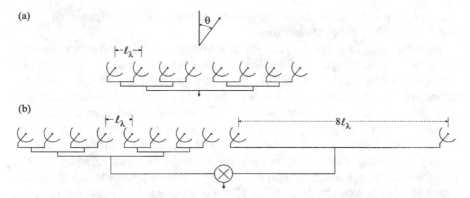

(b)

Fig. 1.13 (a) A linear array of eight equally spaced antennas connected by a branching network in which the electrical path lengths from the antennas to the receiver input are equal. This arrangement is sometimes referred to as a grating array, and in practice, there are usually 16 or more antennas. (b) An eight-element grating array combined with a two-element array to enhance the angular resolution. A phase-switching receiver, indicated by the multiplication symbol, is used to form the product of the signal voltages from the two arrays. The receiver output contains the simultaneous responses of antenna pairs with 16 different spacings. Systems of this general type were known as compound interferometers.

Sun is a strong radio source, the apertures did not have to be very large. Figure 1.13a shows an array of antennas from which the signals at the receiver input are aligned in phase when the angle θ between the direction of the source and a plane normal to the line of the array is such that $\ell_\lambda \sin \theta$ is an integer, where ℓ_λ is the unit antenna spacing measured in wavelengths. This type of array is sometimes referred to as a grating array, since it forms a series of fan-shaped beams, narrow in the θ direction, in a manner analogous to the response of an optical diffraction grating. It is useful only for solar observations in which all but one of the beams falls on "quiet" sky. Christiansen and Warburton (1955) obtained a two-dimensional image of the quiet Sun at 21-cm wavelength using both east–west and north–south grating arrays. These arrays consisted of 32 (east–west) and 16 (north–south) uniformly spaced, parabolic antennas. As the Sun moved through the sky, it was scanned at different angles by the different beams, and a two-dimensional map could be synthesized by Fourier analysis of the scan profiles. To obtain a sufficient range of scan angles, observations extending over eight months were used. In later instruments for solar imaging, it was generally necessary to be able to make a complete image within a day to study the variation of enhanced solar emission associated with active regions. Several instruments used grating arrays, typically containing 16 or 32 antennas and crossed in the manner of a Mills cross. Crossed grating arrays produce a rectangular matrix pattern of beams on the sky, and the rotation of the Earth enables sufficient scans to be obtained to provide daily maps of active regions and other features. Instruments of this type included crosses at 21-cm wavelength at Fleurs, Australia (Christiansen and Mullaly 1963), and at 10-cm wavelength at Stanford, California (Bracewell and Swarup 1961), and a T-shaped array at 1.9-m wavelength at Nançay,

France (Blum et al. 1957, 1961). These were the earliest imaging arrays with large numbers ($\sim$ 16 or more) of antennas.

Figure 1.13b illustrates the principle of a configuration known as a *compound interferometer* (Covington and Broten 1957), which was used to enhance the performance of a grating array or other antenna with high angular resolution in one dimension. The system shown consists of the combination of a grating array with a two-element array. An examination of Fig. 1.13b shows that pairs of antennas, chosen one from the grating array and one from the two-element array, can be found for all spacings from 1 to 16 times the unit spacing ℓ_λ. In comparison, the grating array alone provides only one to seven times the unit spacing, so the number of different spacings simultaneously contributing to the response is increased by a factor of more than two by the addition of two more antennas. Arrangements of this type were used to increase the angular resolution of one-dimensional scans of strong sources (Picken and Swarup 1964; Thompson and Krishnan 1965). By combining a grating array with a single larger antenna, it was also possible to reduce the number of grating responses on the sky (Labrum et al. 1963). Both the crossed grating arrays and the compound interferometers were originally operated with phase-switching receivers to combine the outputs of the two subarrays. In later implementations of similar systems, the signal from each antenna is converted to an intermediate frequency (IF), and a separate voltage-multiplying correlator was used for each spacing. This allows further possibilities in arranging the antennas to maximize the number of different antenna spacings, as discussed in Sect. 5.5.

1.3.10   Measurements of Intensity Profiles

Continuing measurements of the structure of radio sources indicated that in general, the intensity profiles are not symmetrical, so their Fourier transforms, and hence the visibility functions, are complex. This will be explained in detail in later chapters, but at this point, we note that it means that the phase of the fringe pattern (i.e., its position in time with respect to a fiducial reference), as well as the amplitude, varies with antenna spacing and must be measured to allow the intensity profiles to be recovered. To accommodate both fringe amplitude and phase, visibility is expressed as a complex quantity. Measurement of the fringe phase became possible in the 1960s and 1970s, by which time a number of compact sources with well-determined positions, suitable for calibration of the fringe phase, were available. Electronic phase stability had also improved, and computers were available for recording and processing the output data. Improvements in antennas and receivers enabled measurements to be made at wavelengths in the centimeter range (frequencies greater than $\sim$ 1 GHz), using tracking antennas.

An interferometer at the Owens Valley Radio Observatory, California (Read 1961), provides a good example of one of the earliest instruments used extensively for determining radio structure. It consisted of two 27.5-m-diameter parabolic antennas on equatorial mounts with a rail track system that allowed the spacing between them to be varied by up to 490 m in both the east–west and north–south directions. It was used mainly at frequencies from 960 MHz to a few GHz. Studies by Maltby and Moffet (1962) and Fomalont (1968) illustrate the use of this instrument for measurement of intensity distributions, an example of which is shown in Fig. 1.14. Lequeux (1962) studied the structure of about 40 extragalactic sources at 1400 MHz on a reconfigurable two-element interferometer with baselines up to 1460 m (east–west) and 380 m (north–south) at Nançay Observatory in France. These are early examples of model fitting of visibility data, a technique of continuing usefulness (see Sect. 10.4).

1.3.11   Spectral Line Interferometry

The earliest spectral line measurements were made with single narrowband filters. By the early 1960s, the interferometer at Owens Valley and several others had been fitted with spectral line receiving systems. The passband of each receiver was divided into a number of channels by a filter bank, usually in the IF stages, and for each channel, the signals from the two antennas went to a separate correlator. In later systems, the IF signals were digitized and the filtering was performed digitally, as described in Sect. 8.8. The width of the channels should ideally be less than that of the line to be observed so that the line profile can be studied. Spectral line interferometry allows the distribution of the line emission across a radio source to be examined. Roger et al. (1973) describe an array in Canada built specifically for observations in the 1420 MHz (21-cm wavelength) line of neutral hydrogen.

Spectral lines can also be observed in absorption, especially in the case of the neutral hydrogen line. At the line frequency, the gas absorbs the continuum radiation from any more distant source that is observed through it. Comparison of the emission and absorption spectra of neutral hydrogen yields information on its temperature and density. Measurement of absorption spectra of sources can be made using single antennas, but in such cases, the antenna also responds to the broadly distributed emitting gas within the antenna beam. The absorption spectra for weak sources are difficult to separate from the line emission. With an interferometer, the broad emission features on the sky are almost entirely resolved and the narrow absorption spectrum can be observed directly. For early examples of hydrogen line absorption measurements, see Clark et al. (1962) and Hughes et al. (1971).

1.3.12   Earth-Rotation Synthesis Imaging

A very important step in the development of synthesis imaging was the use of the variation of the antenna baseline provided by the rotation of the Earth. Figure 1.15 illustrates this principle, as described by Ryle (1962). For a source at a high declination, the position angle of the baseline projected onto a plane normal to the

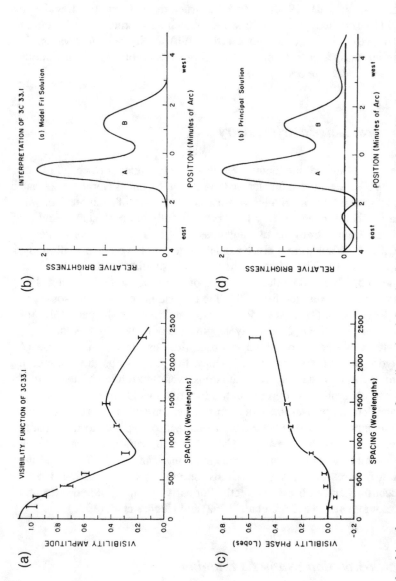

Fig. 1.14 Example of interferometer measurements of one-dimensional intensity (brightness): the east–west profile of source 3C33.1 as determined by Fomalont (1968) using the interferometer at the Owens Valley Radio Observatory at 1425 MHz. (**a,c**) The points show the measured amplitude and phase of the visibility. (**b**) The profile was obtained by fitting Gaussian components to the visibility data, as shown by the curves through the measured visibility points. (**d**) The profile was obtained by Fourier transformation of the observed visibility values. The unit of visibility phase (lobe) is 2π radians. From Fomalont (1968). © AAS. Reproduced by permission.

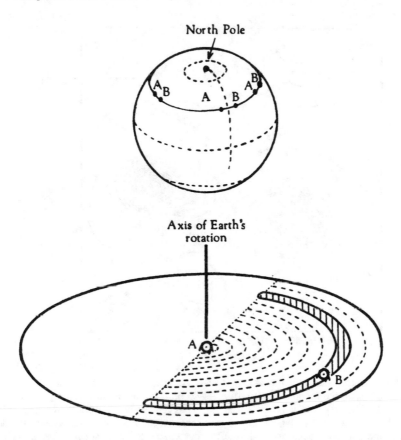

Fig. 1.15 Use of Earth rotation in synthesis imaging, as explained by Ryle (1962). The antennas A and B are spaced on an east–west line. By varying the distance between the antennas from one day to another, and observing for 12 h with each configuration, it is possible to encompass all the spacings from the origin to the elliptical outer boundary of the lower diagram. Only 12 h of observing at each spacing is required, since during the other 12 h, the spacings covered are identical but the positions of the antennas are effectively interchanged. Reprinted by permission from MacMillan Publishers Ltd.: *Nature*, **194**, 517–518, © 1962.

direction of the source rotates through 180° in 12 h. Thus, if the source is tracked across the sky for a series of 12-h periods, each one with a different antenna spacing, the required two-dimensional visibility data can be collected while the antenna spacing is varied in one dimension only. Calculation of two-dimensional Fourier transforms was an arduous task at this time.

The Cambridge One-Mile Radio Telescope was the first instrument designed to exploit fully the Earth-rotation technique and apply it to a large number of radio sources. The use of Earth rotation was not a sudden development in radio astronomy and had been used in solar studies for a number of years. O'Brien (1953) made two-dimensional Fourier synthesis observations with a movable-element interferometer, and, as noted earlier, Christiansen and Warburton (1955) had obtained a two-

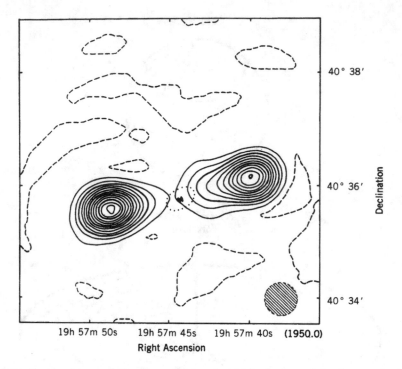

Fig. 1.16 Contour image of the source Cygnus A, which was one of the first results (Ryle et al. 1965) from the Cambridge One-Mile Telescope using the Earth-rotation principle shown in Fig. 1.15. The frequency is 1.4 GHz. The image has been scaled in declination so that the half-power beam contour is circular, as shown by the shaded area in the lower right corner. The dotted ellipse shows the outer boundary of the optical source, and its central structure is also indicated. Reprinted by permission from MacMillan Publishers Ltd.: *Nature*, **205**, 1259–1262, © 1965.

dimensional map of the Sun, using tracking antennas in two grating arrays. At Jodrell Bank, Rowson (1963) had used a two-element interferometer with tracking antennas to map strong nonsolar sources. Also, Ryle and Neville (1962) had imaged the north polar region using Earth rotation to demonstrate the technique. However, the first images published from the Cambridge One-Mile telescope, those of the strong sources Cassiopeia A and Cygnus A (Ryle et al. 1965), exhibited a degree of structural detail unprecedented in earlier studies and heralded the development of synthesis imaging. The image of Cygnus A is shown in Fig. 1.16.

1.3.13   Development of Synthesis Arrays

Following the success of the Cambridge One-Mile Telescope, interferometers such as the NRAO instrument at Green Bank, West Virginia (Hogg et al. 1969), were rapidly adapted for synthesis imaging. Several large arrays designed to provide increased imaging speed, sensitivity, and angular resolution were brought into

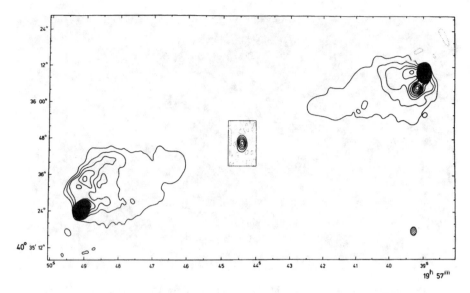

Fig. 1.17 Contour image of the source Cygnus A using the Cambridge Five-Kilometre Radio Telescope at 5 GHz. This showed for the first time the radio nucleus associated with the central galaxy and the high intensity at the outer edges of the radio lobes. From Hargrave and Ryle (1974). © Royal Astronomical Society, used with permission.

operation during the 1970s. Prominent among these were the Five-Kilometre Radio Telescope at Cambridge, England (Ryle 1972), the Westerbork Synthesis Radio Telescope in the Netherlands (Baars et al. 1973), and the Very Large Array (VLA) in New Mexico (Thompson et al. 1980; Napier et al. 1983). With these instruments, imaging of radio sources with a resolution of less than $1''$ at centimeter wavelengths was possible. By using n_a antennas, as many as $n_a(n_a - 1)/2$ simultaneous baselines can be obtained. If the array is designed to avoid redundancy in the antenna spacings, the speed with which the visibility function is measured is approximately proportional to n_a^2. Images of Cygnus A obtained with two of the arrays mentioned above are shown in Figs. 1.17 and 1.18. Resolution of the central source was first achieved with very-long-baseline interferometry (VLBI, see Sect. 1.3.14) (Linfield 1981). A more recent VLBI image is shown in Fig. 1.19. A review of the development of synthesis instruments at Cambridge is given in the Nobel lecture by Ryle (1975). An array with large collecting area, the Giant Metrewave Radio Telescope (GMRT), which operates at frequencies from 38 to 1420 MHz, was completed in 1998 near Pune, India (Swarup et al. 1991). More recently, advances in broadband antenna technology and large-scale integrated circuits have enabled further large increases in performance. For example, the capability of the VLA was greatly improved with an updated electronic system (Perley et al. 2009).[4]

---

[4]The upgraded VLA was formally rededicated as the Karl G. Jansky Very Large Array and is sometimes referred to as the JVLA.

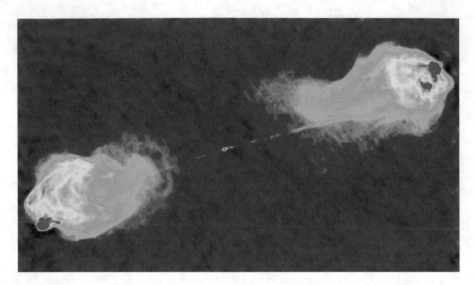

Fig. 1.18 Image of Cygnus A made with the VLA at 4.9 GHz. Observations with four configurations of the array were combined, and the resolution is 0.4″. The display of the image shown here involves a nonlinear process to enhance the contrast of the fine structure. This emphasizes the jet from the central galaxy to the northwestern lobe (top right) and the filamentary structure in the main lobes. Comparison with other records of Cygnus A in this chapter illustrates the technical advances made during three decades. Reproduced by permission of NRAO/AUI. From Perley et al. (1984). © AAS. Reproduced with permission.

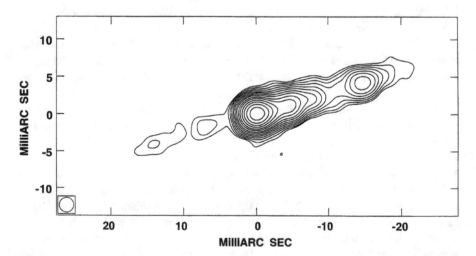

Fig. 1.19 VLBI image of the central part of Cygnus A at 5 GHz, imaged with a ten-station global VLBI array. The resolution is 2 mas, and the rms noise level is 0.4 mJy/beam. The coordinates are centered on the core components. The knots in the jet have apparent expansion speeds of ~ 0.4 c. The counter jet to the left of the core is clearly visible. The jet structure is more clearly defined in an image at 43 GHz with a resolution of 0.15 mas by Boccardi et al. (2016). From Carilli et al. (1994). © AAS. Reproduced with permission.

During the 1980s and 1990s, synthesis arrays operating at short millimeter wavelengths (frequencies of 100 GHz or greater) were developed. Spectral lines are particularly numerous at these frequencies (see Fig. 1.2). Several considerations are more important at millimeter wavelengths than at centimeter wavelengths. Because the wavelengths are much shorter, any irregularity in the atmospheric path length results in a proportionately greater effect on the signal phase. Attenuation in the neutral atmosphere is much more serious at millimeter wavelengths. Also, the beams of the individual antennas become narrower at shorter wavelengths, and maintenance of a sufficiently wide field of view is one reason why the antenna diameter tends to decrease with increasing frequency. Thus, to obtain the necessary sensitivity, larger numbers of antennas are required than at centimeter wavelengths. Arrays for millimeter wavelengths have included those at Hat Creek, California (Welch 1994); Owens Valley, California (Scoville et al. 1994)[5]; Nobeyama, Japan (Morita 1994); the Plateau de Bure, France (Guilloteau 1994); and Mauna Kea, Hawaii (Ho et al. 2004). The largest such array, the Atacama Large Millimeter/submillimeter Array (ALMA) consists of 50 12-m-diameter antennas in one array and 12 7-m-diameter antennas in another. Located in the Atacama Desert of Chile, a dry site at $\sim 5,000$-m elevation, its operating frequency range is 31–950 GHz, and antenna spacings range up to 14 km. The field of view, defined by the beamwidths of the antennas, is only about 8″ at 345 GHz in the primary array. It is an international facility and came into operation in 2013 (Wootten and Thompson 2009).

1.3.14 Very-Long-Baseline Interferometry

Investigation of the angular diameters of quasars and other objects that appear nearly pointlike in structure presented an important challenge throughout the early years of radio astronomy. An advance that led to an immediate increase of an order of magnitude in resolution, and subsequently to several orders more, was the use of independent local oscillators and signal recorders. By using local oscillators at each antenna that are controlled by high-precision frequency standards, it is possible to preserve the coherence of the signals for time intervals long enough to measure interference fringes. In the early years, the received signals were converted to an intermediate frequency low enough that they could be recorded directly on magnetic tape and then brought together and played into a correlator. This technique became known as very-long-baseline interferometry (VLBI), and the early history of its development is discussed by Broten (1988), Kellermann and Cohen (1988), Moran (1998), and Kellermann and Moran (2001). The technical requirements for VLBI were discussed in the USSR in the early 1960s (see, e.g., Matveenko et al. 1965).

A successful early experiment was performed in January 1967 by a group at the University of Florida, who detected fringes from the burst radiation of Jupiter at

---

[5]The arrays at Hat Creek and Owens Valley were combined at Cedar Flats, a high site east of Owens Valley, to form the CARMA array, which operated from 2005 to 2015.

18 MHz (Brown et al. 1968). Because of the strong signals and low frequency, the required recording bandwidth was only 2 kHz and the frequency standards were crystal oscillators. Much more sensitive and precise VLBI systems, which used wider bandwidths and atomic frequency standards, were developed by three other groups. In Canada, an analog recording system was developed with a bandwidth of 1 MHz based on television tape recorders (Broten et al. 1967). Fringes were obtained at a frequency of 448 MHz on baselines of 183 and 3074 km on several quasars in April 1967. In the United States, another group from the National Radio Astronomy Observatory and Cornell University developed a computer-compatible digital recording system with a bandwidth of 360 kHz (Bare et al. 1967). They obtained fringes at 610 MHz on a baseline of 220 km on several quasars in May 1967. A third group from MIT joined in the development of the NRAO–Cornell system in early 1967 and obtained fringes at a frequency of 1665 MHz on a baseline of 845 km on several OH-line masers, with spectroscopic analysis, in June 1967 (Moran et al. 1967).

The initial experiments used signal bandwidths of less than a megahertz, but by the 1980s, systems capable of recording signals with bandwidths greater than 100 MHz were available, with corresponding improvements in sensitivity. Real-time linking of the signals from remote telescopes to the correlator via a geostationary satellite was demonstrated (Yen et al. 1977). Also, experiments were performed in which the local oscillator signal was distributed over a satellite link (Knowles et al. 1982). Neither of these satellite-supported techniques have been used significantly for practical and economic reasons. Most importantly, the accessibility of the world-wide network of fiberoptic transmission lines, which have since become available, allows real-time transmission of the data to the correlator. These developments, as well as the advent of sophisticated data analysis techniques, have lessened the distinction between VLBI and more conventional forms of interferometry. A detailed technical description of issues specific to the VLBI technique is given in Chap. 9.

An early example of the extremely high angular resolution that can be achieved with VLBI is provided by a measurement by Burke et al. (1972), who obtained a resolution of 200 μas using antennas in Westford, Massachusetts, and near Yalta in the Crimea, operating at a wavelength of 1.3 cm. Early measurements, obtained using a few baselines only, were generally interpreted in terms of the simple models in Fig. 1.5. Important results were the discovery and investigation of superluminal (apparently faster-than-light) motions in quasars (Whitney et al. 1971; Cohen et al. 1971), as shown in Fig. 1.20, and the measurement of proper motion in H_2O line masers (Genzel et al. 1981). During the mid-1970s, several groups of astronomers began to combine their facilities to obtain measurements over ten or more baselines simultaneously. In the United States, the Network Users' Group, later called the U.S. VLBI Consortium, included the following observatories: Haystack Observatory in Massachusetts (NEROC); Green Bank, West Virginia (NRAO); Vermilion River Observatory in Illinois (Univ. of Illinois); North Liberty in Iowa (Univ. of Iowa); Fort Davis, Texas (Harvard College Observatory); Hat Creek Observatory, California (Univ. of California); and Owens Valley Radio Observatory, California. Other

Fig. 1.20   VLBI images of
the quasar 3C273 at five
epochs, showing the relative
positions of two components.
From the distance of the
object, deduced from the
optical redshift, the apparent
relative velocity of the
components exceeds the
velocity of light, but this can
be explained by relativistic
and geometric effects. The
observing frequency is
10.65 GHz. An angular scale
of 2 mas is shown in the lower
right corner. From Pearson
et al. (1981). Reprinted by
permission from MacMillan
Publishers Ltd.: *Nature*, **290**,
365–368, © 1981.

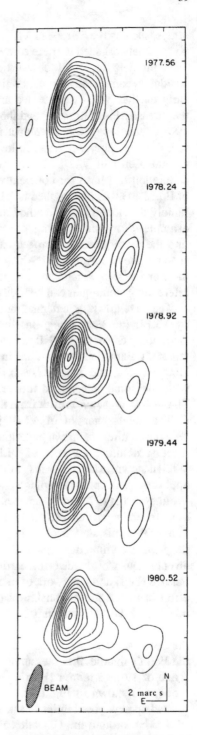

arrays, such as the European VLBI Network (EVN) soon developed. Observations on such networks led to more complex models [see, e.g., Cohen et al. (1975)].

A problem in VLBI observations is that the use of nonsynchronized local oscillators complicates the calibration of the phase of the fringes. It became evident early on that VLBI represented an intermediate form of interferometer between the intensity interferometer and the perfectly stable coherent interferometer (Clark 1968). Techniques were developed to combine coherent averaging on timescales up to a defined coherence time, followed by incoherent averaging. These techniques remain useful in VLBI at very high frequencies. To overcome the problem of calibration of phase for coherently averaged data, the phase closure relationship of Eq. (1.14) was first applied to VLBI data by Rogers et al. (1974). The technique rapidly developed into a method to obtain images known as hybrid mapping. For examples of hybrid mapping, see Figs. 1.19 and 1.20. This method was subsumed into the more general approach called self-calibration (see Chap. 11). For some spectral line observations in which the source consists of spatially isolated masers, the signals from which can be separated by their individual Doppler shifts, phase referencing techniques can be used (e.g., Reid et al. 1980).

The first array of antennas built specifically for astronomical measurements by VLBI, the Very Long Baseline Array (VLBA) of the U.S. National Radio Astronomy Observatory (NRAO), was brought into operation in 1994. It consists of ten 25-m-diameter antennas, one in the U.S. Virgin Islands, eight in the continental United States, and one in Hawaii (Napier et al. 1994). The VLBA is often linked with additional antennas to further improve the baseline coverage and sensitivity. Figure 1.21 presents a result from the combined VLBA and EVN array.

The great potential of VLBI in astrometry and geodesy was immediately recognized after the initial experiments in 1967 [see, e.g., Gold (1967)]. A seminal meeting defining the role of VLBI in Earth dynamics programs was held in Williamstown, Massachusetts, in 1969 (Kaula 1970). The use of VLBI in these applications developed rapidly during the 1970s and 1980s; see, for example, Whitney et al. (1976) and Clark et al. (1985). In the United States, NASA and several other federal agencies set up a cooperative program of geodetic measurements in the mid-1970s. This work evolved in part from the use of the Jet Propulsion Laboratory deep-space communications facilities for VLBI observations. It has expanded into an enormous worldwide effort carried out under the aegis of the International VLBI Service (IVS) and a network of more than 40 antennas. An important result of this effort has been the establishment of the International Celestial Reference Frame adopted by the IAU, which is based on 295 "defining" sources whose positions are known to an accuracy of about 40 μas (Fey et al. 2015). Another striking result of the geodetic VLBI work has been the detection of contemporary plate motions in the Earth's mantle, first measured as a change in the Westford–Onsala baseline at a rate of 17 ± 2 mm/yr (Herring et al. 1986). The VLBI measurements of plate motions is shown in Fig. 1.22. Astrometry with submilliarcsecond accuracy has opened up new possibilities in astronomy, for example, the detection of the motion of the Sun around the Galactic center from the proper motion of Sagittarius A* (Backer and Sramek 1999; Reid and Brunthaler 2004) and measurements of the

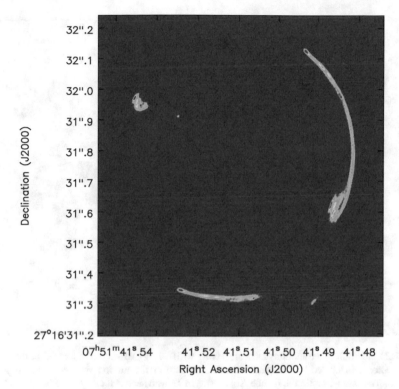

Fig. 1.21 Image of the gravitational lens source MG J0751+2716 made with a 14-h observation on a 21-element global VLBI Array (VLBA and the EVN plus the Green Bank telescope) at a frequency of 1.7 GHz. The rms noise level is 12 μJy, and the resolution is 2.2 × 5.6 mas. This image of an extended background source at redshift 3.2 is highly distorted by an unseen foreground radio-quiet galaxy at a redshift of 0.35 into extended arcs. Image courtesy of and © John McKean.

annual parallaxes of galactic radio sources (Reid and Honma 2014). Astrometric and geodetic methods are described in Chap. 12.

The combination of VLBI with spectral line processing is particularly effective in the study of problems that involve both astrometry and dynamical analysis of astronomical systems. The galaxy NGC4258, which exhibits an active galactic nucleus, has been found to contain a number of small regions that emit strongly in the 22.235-GHz water line as a result of maser processes. VLBI observations have provided an angular resolution of 200 μas, an accuracy of a few microarcseconds in the relative positions of the masers, and measurements of Doppler shifts to an accuracy of 0.1 km s^{-1} in radial velocity (see Fig. 1.23). NGC4258 is fortuitously aligned so that the disk is almost edge-on as viewed from the Earth. The orbital velocities of the masers, which obey Kepler's law, are accurately determined as a function of radius from the center of motion. Hence, the distance can be found by comparing the linear and angular motions. The angular motions are about 30 μas per year. These results provide a value for the central mass of 3.9 × 10^7 times

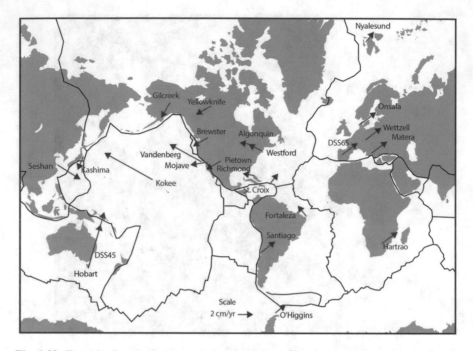

Fig. 1.22 Tectonic plate motions measured with VLBI. A VLBI station is located at the foot of each vector and labeled by the station name. The sum of the motion vectors is constrained to be zero. The largest motion is for the Kokee site in Hawaii, about 8 cm yr^{-1}. Plate boundaries, established by other techniques, are shown by the jagged lines. From Whitney et al. (2014). Reprinted with permission courtesy of and © MIT Lincoln Laboratory, Lexington, MA.

the mass of the Sun, presumably a supermassive black hole (Miyoshi et al. 1995; Herrnstein et al. 1999), and 7.6 ± 0.2 Mpc for the distance (Humphreys et al. 2013). The uncertainty of 3% in the distance of an extragalactic object, measured directly, set a precedent.

1.3.15   VLBI Using Orbiting Antennas

The use of spaceborne antennas in VLBI observations is referred to as the OVLBI (orbiting VLBI) technique. The first observations of this type were made in 1986 using a satellite of the U.S. Tracking and Data Relay Satellite System (TDRSS). These satellites were in geostationary orbit at a height of approximately 36,000 km and were used to relay data from low-Earth-orbit spacecraft to Earth. They carried two 4.9-m antennas used to communicate with other satellites at 2.3 and 15 GHz and a smaller antenna for the space-to-Earth link. In this experiment, one of the 4.9-m antennas was used to observe a radio source, and the other received a reference signal from a hydrogen maser on the ground (Levy et al. 1989). The received signals

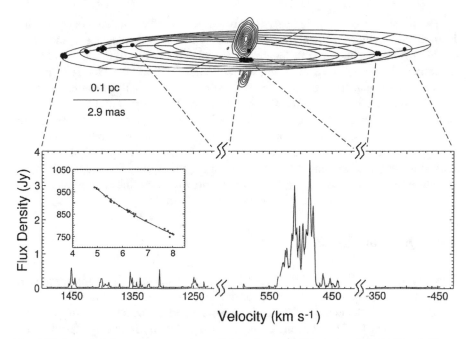

Fig. 1.23 Image of the water vapor maser disk in the core of the galaxy NGC4258 at 1.35 cm made with the VLBA. The spots mark the positions of the unresolved maser components. The elliptical grid lines denote the thin, slightly warped disk that the masers trace. The position of the gravitational center is shown by the black square. The contour plot shows the continuum emission from the central active galactic nucleus. Each maser spot corresponds to a feature in the spectrum in the lower panel. The strongest feature, at 470 km s^{-1}, serves as a phase reference. The inset shows the radial velocity of the masers vs. radial distance from the black hole in milliarcseconds. From Herrnstein et al. (2005). © AAS. Reproduced with permission.

were transmitted to the ground and recorded on a VLBI tape system for correlation with signals from ground-based antennas. The numbers of sources detected were 23 and 11 at 2.3 and 15 GHz, respectively (Linfield et al. 1989, 1990). At 15 GHz, the fringe width was of order 0.3 mas, and interpretation of the results in terms of circular Gaussian models indicated brightness temperatures as high as 2×10^{12} K.

VLBI observations using a satellite in a non-geostationary orbit were first made in 1997 by the VLBI Space Observatory Programme (VSOP) (Hirabayashi et al. 1998), designed specifically for VLBI observations. It was equipped with an antenna of 8-m diameter, and observations were made at 1.6 and 5 GHz. The orbital period was approximately 6.6 h and the apogee height, 21,000 km. VSOP was followed by the RadioAstron satellite, which was launched in 2011 into an orbit with an apogee height of about 300,000 km and a period of 8.3 days (Kardashev et al. 2013). It is equipped with an antenna of 10-m diameter and receivers at 18, 6, and 1.35 cm. Operating with ground-based telescopes, it can attain a resolution of 8 μas at 1.35 cm. More information about satellite VLBI can be found in Sect. 9.10.

The possibility of achieving very long baselines by reflection from the Moon has been discussed by Hagfors et al. (1990). Reflection from the surface of the Moon could provide baselines up to a length approaching the radius of the lunar orbit. An antenna of 100-m aperture, or larger, would be used to track the Moon and receive the reflected signal from the source under study, and a smaller antenna could be used for the direct signal. It is estimated that the sensitivity would be about three orders of magnitude less than would be obtained by observing the source directly with both antennas. Further complications result from the roughness of the lunar surface and from libration. The technique could be useful for special observations requiring very high angular resolution of strong sources, for example, for the burst radiation from Jupiter. However, RadioAstron provides baselines almost as long.

1.4   Quantum Effect

The development of VLBI introduced a new facet into the apparent paradox in the quantum-mechanical description of interferometry (Burke 1969). The radio interferometer is the analog of Young's two-slit interference experiment. It is well known (Loudon 1973) that a single photon creates an interference pattern but that any attempt to determine which slit the photon entered will destroy the interference pattern; otherwise, the uncertainty principle would be violated. Consideration of VLBI suggests that it might be possible to determine at which antenna a particular photon arrived, since its signature is captured in the medium used for transmission to the correlator as well as in the fringe pattern generated during correlation. However, in the radio frequency range, the input stages of receivers used as the measurement devices consist of amplifiers or mixers that conserve the received phase in their outputs. This allows formation of the fringes in subsequent stages. The response of such devices must be consistent with the uncertainty principle, $\Delta E \Delta t \simeq h/2\pi$, where ΔE and Δt are the uncertainties in signal energy and measurement time. This principle can be written in terms of uncertainty in photon number, ΔN_p, and phase, $\Delta \phi$, as

$$\Delta N_p \, \Delta \phi \simeq 1 \, , \tag{1.15}$$

where $\Delta N_p = \Delta E/h\nu$ and $\Delta \phi = 2\pi \nu \Delta t$. To preserve phase, $\Delta \phi$ must be small, so ΔN_p must be correspondingly large, and there must be an uncertainty of at least one photon per unit bandwidth per unit time in the output of the receiving amplifier. Hence, the SNR is less than unity in the single-photon limit, and it is impossible to determine at which antenna a single photon entered. An alternative but equivalent statement is that the output of any receiving system must contain a noise component that is not less than an equivalent input power approximately equal to $h\nu$ per unit bandwidth.

The individual photons that constitute a radio signal arrive at antennas at random times but with an average rate that is proportional to the signal strength. For phenomena of this type, the number of events that occur in a given time interval τ varies statistically in accordance with the Poisson distribution. For a signal power P_{sig}, the average number of photons that arrive within time τ is $\overline{N_p} = P_{\text{sig}}\tau/h\nu$. The rms deviation of the number arriving during a series of intervals τ is, for Poisson statistics, given by $\Delta N_p = \sqrt{\overline{N_p}}$. From Eq. (1.15), the resulting uncertainty in the signal phase is

$$\Delta\phi \simeq \frac{1}{\sqrt{N_p}} = \sqrt{\frac{h\nu}{P_{\text{sig}}\tau}} \ . \tag{1.16}$$

We can also express the uncertainty in the measurement of the signal phase in terms of the noise that is present in the receiving system. The minimum noise power, P_{noise}, is approximately equal to the thermal noise from a matched resistive load at temperature $h\nu/k$, that is, $P_{\text{noise}} = h\nu\Delta\nu$. The uncertainty in the phase, as measured with an averaging time τ, becomes

$$\Delta\phi = \sqrt{\frac{P_{\text{noise}}}{P_{\text{sig}}\tau\Delta\nu}} \ . \tag{1.17}$$

Note that $\Delta\phi$ is the accuracy with which the phase of the amplified signal received from one antenna can be measured: for example, in Doppler tracking of a spacecraft (Cannon 1990). This is not to be confused with the accuracy of measurement of the fringe phase of an interferometer. For a frequency $\nu = 1$ GHz, the effective noise temperature $h\nu/k$ is equal to 0.048 K. Thus, for frequencies up to some tens of gigahertz, the quantum effect noise makes only a small contribution to the receiver noise. At 900 GHz, which is generally considered to be about the high frequency limit for ground-based radio astronomy, $h\nu/k = 43$ K, and the contribution to the system noise is becoming important. In the optical region, $\nu \approx 500$ THz, $h\nu/k \approx 30,000$ K, and heterodyne systems are of limited practicality, as discussed in Sect. 17.6.2. However, in the optical region, it is possible to build "direct detection" devices that detect power without conserving phase, so $\Delta\phi$ in Eq. (1.17) effectively tends to infinity, and there is no constraint on the measurement accuracy of the number of photons. Thus, most optical interferometers form fringes directly from the light received and measure the resulting patterns of light intensity to determine the fringe parameters.

For further reading on the general subject of thermal and quantum noise, see, for example, Oliver (1965) and Kerr et al. (1997). Nityananda (1994) compares quantum issues in the radio and optical domains, and a discussion of basic concepts is given by Radhakrishnan (1999).

Appendix 1.1   Sensitivity of Radio Astronomical Receivers (the Radiometer Equation)

An idealized block diagram of the basic receiver configuration widely used in radio astronomy is shown in Fig. A1.1. We describe its function and analyze its performance in this appendix. The signal from an antenna is first passed through an amplifier. The amplifier is characterized by its power gain factor, G; receiver temperature; and the bandwidth, $\Delta \nu$. The gain factor is assumed to be constant. If the gain is sufficiently high, this amplifier sets the noise performance of the entire system, which we denote as T_S to include the contributions from atmosphere, ground pickup, and ohmic losses. We assume that the passband has a rectangular shape that is flat between a lower cutoff frequency, ν_0, and the upper cutoff frequency, $\nu_0 + \Delta \nu$. The signal then passes through a mixer, where it is multiplied by a sinusoidal local oscillator signal at frequency ν_0 and is converted to a baseband from 0 to $\Delta \nu$. In the next stage, the signal is converted to a digital data stream sampled at the Nyquist rate. According to the Nyquist sampling theorem, a bandlimited signal can be represented by samples taken at intervals of $1/2\Delta \nu$. We assume there is no quantization error in this sampling process. In this case, the original signal can be exactly reconstructed from the sampled sequence by convolution with a sinc function. The sampled signal has the same statistical properties as the corresponding analog signal. The next step is a square-law detector, which squares the amplitudes of the signal samples. This is followed by an averager, which simply averages N samples in a running mean fashion. A system with these features is known as a single-sideband superheterodyne receiver (Armstrong 1921) or, simply, a total-power radiometer. Early interferometer receivers were a variation on this basic design (see Fig. 1.6): Signals from two antennas were added after the mixing stage before entering the square-law detector, and there was no signal digitization.

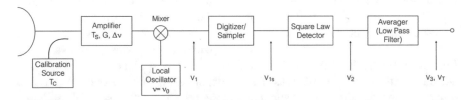

Fig. A1.1 A block diagram of an idealized radiometer used in most radio astronomical systems for measuring total power. The system temperature T_S includes the receiver temperature T_R plus all unwanted additive contributions (e.g., ohmic loses, atmospheric effects, ground pickup). In practice, at very low frequencies (< 100 MHz), downconversion may be omitted, while at high frequencies (more than a few gigahertz), multiple stages of frequency downconversion are required. At very high frequencies (> 100 GHz), where low noise amplifiers are not available, the first stage is usually the mixer. In this case, its losses and those of the amplifiers following it contribute to T_S.

The statistical performance of the idealized system in Fig. A1.1 can be readily evaluated. The power level at any point in the system can be characterized by a temperature T according to the Nyquist relation [e.g., Eq. (1.4)]

$$P = kT \Delta v \, G \,, \tag{A1.1}$$

where we have included the effect of power amplification by the gain factor G. The voltage v_1 is a combination of antenna input, characterized by T_A, and the additive system noise, T_S. We assume the cosmic input signal has a flat spectrum over the baseband frequency range. Hence, v_1 is a zero-mean Gaussian random process with a flat spectrum, i.e., a white noise spectrum. Such a process, described by $p(v)$, has only one parameter, the variance, σ^2. The odd moments of the probability distribution are zero, and the even moments are

$$\langle v^n \rangle = (1 \cdot 3 \cdot 5 \cdot \ldots \cdot n) \, \sigma^n \,. \tag{A1.2}$$

The expectations of v_1 and v_1^2 (the power) are therefore

$$\langle v_1 \rangle = 0 \,, \tag{A1.3}$$

$$\langle v_1^2 \rangle = k(T_S + T_A) \Delta v \, G \,. \tag{A1.4}$$

The statistics of the sampled signal, v_{1s}, and the analog signal v_1 are the same, i.e., $v_{1s} = v_1$, $v_{1s}^2 = v_1^2$, etc. The characteristics of v_2 are

$$\langle v_2 \rangle = \langle v_1^2 \rangle \,, \tag{A1.5}$$

$$\langle v_2^2 \rangle = \langle v_1^4 \rangle = 3 \langle v_1^2 \rangle^2 \,, \tag{A1.6}$$

$$\sigma_2^2 = \langle v_2^2 \rangle - \langle v_2 \rangle^2 = 2 \langle v_1^2 \rangle^2 \,. \tag{A1.7}$$

The averager averages $N = 2 \Delta v \tau$ samples together, where τ is the integration time. Hence,

$$\langle v_3 \rangle = \langle v_1^2 \rangle = k(T_S + T_A) \Delta v \, G \,, \tag{A1.8}$$

$$\sigma_3^2 = \frac{\sigma_2^2}{N} = \frac{2[k(T_S + T_A) \Delta v \, G]^2}{2 \Delta v \tau} \,. \tag{A1.9}$$

v_3 is converted from a power scale to a temperature scale by inserting a thermal noise signal of known temperature T_c in order to remove or calibrate the $k \Delta v \, G$ factor. Formally, the calibrated version of v_3 is

$$v_T = \frac{v_3}{\partial v_3 / \partial T_A} \,, \tag{A1.10}$$

where $(\partial v_3/\partial T_A)^{-1}$ is the conversion factor from power to temperature written as a partial derivative. The mean and rms of the output in temperature units are therefore

$$\langle v_T \rangle = T_S + T_A \, , \tag{A1.11}$$

$$\sigma_T = \frac{T_S + T_A}{\sqrt{\Delta \nu \tau}} \, . \tag{A1.12}$$

It is important to note that the factor of two in the expression for σ_3^2 in Eq. (A1.8) cancels the factor of two in the number of samples averaged. The signal-to-noise ratio (SNR) is therefore

$$\mathcal{R}_{\text{sn}} = \frac{T_A}{T_S + T_A} \sqrt{\Delta \nu \tau} \, . \tag{A1.13}$$

Equation (A1.13) shows that T_A contributes to the fluctuations, and in the limit $T_A \gg T_S$, longer integration does increase the SNR. For $T_A \ll T_S$, the usual case, Eq. (A1.13) becomes Eq. (1.8). Because of the fundamental limitation imposed by the Nyquist sampling theorem, no receiver system can perform better than specified by Eq. (A1.13). The performance of any other system can be written as

$$\sigma_T = C \frac{T_S + T_A}{\sqrt{\Delta \nu \tau}} \, , \tag{A1.14}$$

where C is a factor equal to, or greater than, one. The square-law detector could be replaced by another type of detector. For a linear detector, i.e., $v_2 = |v_1|$, a similar analysis to the one for the square-law detector yields $C = \sqrt{\pi - 2} = 1.07$ when $T_A \ll T_S$. In this calculation, it is necessary to linearize the output by calculation of $\partial v_3/\partial T_A$ in Eq. (A1.10). For a fourth-order detector, $v_2 = v_1^4$, $C = \sqrt{4/3} = 1.15$. More details can be found in Davenport and Root (1958).

G may not be a constant but can vary randomly due to electronic instabilities. In that case, a synchronous detector is added and receiver input is switched between the antenna and a reference signal. This system is known as a Dicke (1946) radiometer. (Note that the phase-switching interferometer [see Fig. 1.8] uses the synchronous detection principle.) The noise performance of a Dicke radiometer is worse by a factor of two, but the effects of gain fluctuations are mitigated. An alternative receiver that reduces the effects of gain fluctuations is called the correlation receiver, in which the signal from the antenna is divided in half and passed through separate amplifiers before being multiplied, where the multiplier replaces the square-law detector.

In older receivers, there was usually no digitization before the final averaging stage. The performance of a comparable analog system is identical to that described above. For analog analysis of radiometers, see Tiuri (1964) or Kraus (1986). A summary of the performance of various receiver types is given in Table A1.1.

Table A1.1 Sensitivity characteristics of various types of receivers

Receiver type	C^a
Total power ($v_2 = v_1^2$)	1
Linear detector ($v_2 = \lvert v_1^2 \rvert$)	1.07[b]
Fourth-order detector ($v_2 = v_1^4$)	1.15[b]
Dicke-switched receiver	2[b]
Correlation receiver	$\sqrt{2}$[b]

[a] C is defined in Eq. (A1.14).
[b] For $T_A \ll T_S$.

There are two major differences between radio and optical systems. Radio systems are characterized by Gaussian noise characteristics of both the signal and the additive receiver noise, whereas optical detectors are limited by Poisson statistics appropriate for counting photons and the SNR, $\mathcal{R}_{sn}$, is $1/\sqrt{N_p}$, where N_p is the number of photons. In terms of quantum mechanics, the Gaussian noise corresponds to photo bunching noise [see Radhakrishnan (1999)].

Further Reading

Textbooks on Radio Astronomy and Radio Interferometry

Burke, B.F., and Graham-Smith, F., *An Introduction to Radio Astronomy*, 3rd ed., Cambridge Univ. Press, Cambridge, UK (2014)

Christiansen, W.N., and Högbom, J.A., *Radiotelescopes*, Cambridge Univ. Press, Cambridge, UK (1969) (2nd ed., 1985)

Condon, J.J., and Ransom, S.M., *Essential Radio Astronomy*, Princeton Univ. Press, Princeton, NJ (2016)

Kraus, J.D., *Radio Astronomy*, 2nd ed., Cygnus-Quasar Books, Powell, OH, 1986. 1st ed., McGraw-Hill, New York (1966)

Lovell, B., and Clegg, J.A., *Radio Astronomy*, Chapman and Hall, London (1952)

Marr, J.M., Snell, R.L., and Kurtz, S.E., *Fundamentals of Radio Astronomy: Observational Methods*, CRC Press, Boca Raton, FL (2016)

Pawsey, J.L., and Bracewell, R.N., *Radio Astronomy*, Oxford Univ. Press, Oxford, UK (1955)

Shklovsky, I.S., *Cosmic Radio Waves*, translated by R. B. Rodman and C. M. Varsavsky, Harvard Univ. Press, Cambridge, MA (1960)

Wilson, T.L., Rohlfs, K., and Hüttemeister, S., *Tools of Radio Astronomy*, 6th ed., Springer-Verlag, Berlin (2013) (See also earlier editions, starting with K. Rohlfs, 1986.)

Wohlleben, R., Mattes, H., and Krichbaum, T., *Interferometry in Radioastronomy and Radar Techniques*, Kluwer, Dordrecht, the Netherlands (1991)

Historical Reviews

Kellermann, K.I., and Moran, J.M., The Development of High-Resolution Imaging in Radio Astronomy, *Ann. Rev. Astron. Astrophys.*, **39**, 457–509 (2001)

Sullivan, W.T., III, Ed., *The Early Years of Radio Astronomy*, Cambridge Univ. Press, Cambridge, UK (1984)

Sullivan, W.T., III, *Cosmic Noise: A History of Early Radio Astronomy*, Cambridge Univ. Press, Cambridge, UK (2009)

Townes, C.H., Michelson and Pease's Interferometric Stellar Diameters, *Astrophys. J.*, **525**, 148–149 (1999)

General Interest

Alder, B., Fernbach, S., Rotenberg, M., Eds., *Methods in Computational Physics*, Vol. 14, *Radio Astronomy*, Academic Press, New York (1975)

Berkner, L.V., Ed., *IRE Trans. Antennas Propag.*, Special Issue on Radio Astronomy, **AP-9**, No. 1 (1961)

Biraud, F., Ed., *Very Long Baseline Interferometry Techniques*, Cépaduès, Tou-louse, France (1983)

Bracewell, R.N., Ed., *Paris Symposium on Radio Astronomy*, IAU Symp. 9, Stanford Univ. Press, Stanford, CA (1959)

Bracewell, R.N., Radio Astronomy Techniques, in *Handbuch der Physik*, Vol. 54, S. Flugge, Ed., Springer-Verlag, Berlin (1962)

Cornwell, T.J., and Perley, R.A., Eds., *Radio Interferometry: Theory, Techniques, and Applications*, IAU Colloq. 131, Astron. Soc. Pacific Conf. Ser., **19** (1991)

Findlay, J.W., Ed., *Proc. IEEE*, Special Issue on Radio and Radar Astronomy, **61**, No. 9 (1973)

Frater, R.H., and Brooks, J.W., Eds., *J. Electric. Electron. Eng. Australia*, Special Issue on the Australia Telescope, **12**, No. 2 (1992)

Goldsmith, P.F., Ed., *Instrumentation and Techniques for Radio Astronomy*, IEEE Press, New York (1988)

Haddock, F.T., Ed., *Proc. IRE*, Special Issue on Radio Astronomy, **46**, No. 1 (1958)

Ishiguro, M., and Welch, W.J., Eds., *Astronomy with Millimeter and Submillimeter Wave Interferometry*, IAU Colloq. 140, Astron. Soc. Pacific Conf. Ser., **59** (1994)

Kraus, J.D., Ed., *IEEE Trans. Mil. Electron.*, Special Issue on Radio and Radar Astronomy, **MIL-8**, Nos. 3 and 4, 1964; also issued by *IEEE Trans. Antennas Propag.*, **AP-12**, No. 7 (1964)

Meeks, M.L., Ed., *Methods of Experimental Physics*, Vol. 12, Parts B, *Astrophysics: Radio Telescopes*, and C, *Astrophysics: Radio Observations*, Academic Press, New York (1976)

Pawsey, J.L., Ed., *Proc. IRE Aust.*, Special Issue on Radio Astronomy, **24**, No. 2 (1963)

Perley, R.A., Schwab, F.R., and Bridle, A.H., Eds., *Synthesis Imaging in Radio Astronomy*, Astron. Soc. Pacific Conf. Ser., **6** (1989)

Raimond, E., and Genee, R., Eds., *The Westerbork Observatory, Continuing Adventure in Radio Astronomy*, Kluwer, Dordrecht, the Netherlands (1996)

Taylor, G.B., Carilli, C.L., and Perley, R.A., Eds., *Synthesis Imaging in Radio Astronomy II*, Astron. Soc. Pacific Conf. Ser., **180** (1999)

Wild, J.P., Ed., *Proc. IREE Aust.*, Special Issue on the Culgoora Radioheliograph, **28**, No. 9 (1967)

Yen, J.L., Image Reconstruction in Synthesis Radio Telescope Arrays, in *Array Signal Processing*, Haykin, S., Ed., Prentice Hall, Englewood Cliffs, NJ (1985), pp. 293–350

References

Adgie, R.I., Gent, H., Slee, O.B., Frost, A.D., Palmer, H.P., and Rowson, B., New Limits to the Angular Sizes of Some Quasars, *Nature*, **208**, 275–276 (1965)

Appleton, E.V., Departure of Long-Wave Solar Radiation from Black-Body Intensity, *Nature*, **156**, 534–535 (1945)

Armstrong, E.H., A New System of Short Wave Amplification, *Proc. IRE*, **9**, 3–11 (1921)

Baade, W., and Minkowski, R., Identification of the Radio Sources in Cassiopeia, Cygnus A, and Puppis A, *Astrophys. J.*, **119**, 206–214 (1954a)

Baade, W., and Minkowski, R., On the Identification of Radio Sources, *Astrophys. J.*, **119**, 215–231 (1954b)

Baars, J.W.M., Genzel, R., Pauliny-Toth, I.I.K., and Witzel, A., The Absolute Spectrum of Cas A: An Accurate Flux Density Scale and a Set of Secondary Calibrators, *Astron. Astrophys*, **61**, 99–106 (1977)

Baars, J.W.M., van der Brugge, J.F., Casse, J.L., Hamaker, J.P., Sondaar, L.H., Visser, J.J., and Wellington, K.J., The Synthesis Radio Telescope at Westerbork, *Proc. IEEE*, **61**, 1258–1266 (1973)

Backer, D.C., and Sramek, R.A., Proper Motion of the Compact, Nonthermal Radio Source in the Galactic Center, Sagittarius A*, *Astrophys. J.*, **524**, 805–815 (1999)

Bare, C., Clark, B.G., Kellermann, K.I., Cohen, M.H., and Jauncey, D.L., Interferometer Experiment with Independent Local Oscillators, *Science*, **157**, 189–191 (1967)

Beevers, C.A., and Lipson, H., A Brief History of Fourier Methods in Crystal-Structure Determination, *Aust. J. Phys.*, **38**, 263–271 (1985)

Bennett, A.S., The Revised 3C Catalog of Radio Sources, *Mem. R. Astron. Soc.*, **68**, 163–172 (1962)

Blake, G.A., Sutton, E.C., Masson, C.R., and Phillips, T.G., Molecular Abundances in OMC-1: The Chemical Composition of Interstellar Molecular Clouds and the Influence of Massive Star Formation, *Astrophys. J.*, **315**, 621–645 (1987)

Blum, E.J., Le Réseau Nord-Sud à Multiples, *Ann. Astrophys.*, **24**, 359–366 (1961)

Blum, E.J., Boischot, A., and Ginat, M., Le Grand Interféromètre de Nançay, *Ann. Astrophys.*, **20**, 155–164 (1957)

Blythe, J.H., A New Type of Pencil Beam Aerial for Radio Astronomy, *Mon. Not. R. Astron. Soc.*, **117**, 644–651 (1957)

Boccardi, B., Krichbaum, T.P., Bach, U., Mertens, F., Ros, E., Alef, W., and Zensus, J.A., The Stratified Two-Sided Jet of Cygnus A, *Astron. Astrophys.*, **585**, A33 (9pp) (2016)

Bolton, J.G., and Slee, O.B., Galactic Radiation at Radio Frequencies. V. The Sea Interferometer, *Aust. J. Phys.*, **6**, 420–433 (1953)

Bolton, J.G., and Stanley, G.J., Variable Source of Radio Frequency Radiation in the Constellation of Cygnus, *Nature*, **161**, 312–313 (1948)

Born, M., and Wolf, E., *Principles of Optics*, 1st ed., Pergamon Press Ltd., London (1959) [many subsequent editions, e.g., 7th ed., Cambridge Univ. Press, Cambridge, UK, 1999]

Braccesi, A., Ceccarelli, M., Colla, G., Fanti, R., Ficarra, A., Gelato, G., Greuff, G., and Sinigaglia, G., The Italian Cross Radio Telescope. III. Operation of the Telescope, *Nuovo Cimento B*, **62**, 13–19 (1969)

Bracewell, R.N., and Swarup, G., The Stanford Microwave Spectroheliograph Antenna, a Microsteradian Pencil Beam Interferometer, *IRE Trans. Antennas Propag.*, **AP-9**, 22–30 (1961)

Broten, N.W., Early Days of Canadian Long-Baseline Interferometry: Reflections and Reminiscences, *J. Roy. Astron. Soc. Can.*, **82**, 233–241 (1988)

Broten, N.W., Legg, T.H., Locke, J.L., McLeish, C.W., Richards, R.S., Chisholm, R.M., Gush, H.P., Yen, J.L., and Galt, J.A., Observations of Quasars Using Interferometer Baselines Up to 3,074 km, *Nature*, **215**, 38 (1967)

Brown, G.W., Carr, T.D., and Block, W.F., Long Baseline Interferometry of S-Bursts from Jupiter, *Astrophys. Lett.*, **1**, 89–94 (1968)

Burke, B.F., Quantum Interference Paradox, *Nature*, **223**, 389–390 (1969)

Burke, B.F., Johnston, K.J., Efanov, V.A., Clark, B.G., Kogan, L.R., Kostenko, V.I., Lo, K.Y., Matveenko, L.I., Moiseev, I.G., Moran, J.M., and five coauthors, Observations of Maser Radio Source with an Angular Resolution of 0.″0002, *Sov. Astron.-AJ*, **16**, 379–382 (1972)

Cannon, A.J., and Pickering, E.C., The Henry Draper Catalogue, 21^h, 22^h, and 23^h, *Annals of the Astronomical Observatory of Harvard College*, **99** (1924)

Cannon, W.H., Quantum Mechanical Uncertainty Limitations on Deep Space Navigation by Doppler Tracking and Very Long Baseline Interferometry, *Radio Sci.*, **25**, 97–100 (1990)

Carilli, C.L., Bartel, N., and Diamond, P., Second Epoch VLBI Observations of the Nuclear Jet in Cygnus A: Subluminal Jet Proper Motion Measured, *Astron. J.*, **108**, 64–75 (1994). doi: 10.1086/117045

Christiansen, W.N., and Mullaly, R.F., Solar Observations at a Wavelength of 20 cm with a Cross-Grating Interferometer, *Proc. IRE Aust.*, **24**, 165–173 (1963)

Christiansen, W.N., and Warburton, J.A., The Distribution of Radio Brightness over the Solar Disk at a Wavelength of 21 cm. III. The Quiet Sun. Two-Dimensional Observations, *Aust. J. Phys.*, **8**, 474–486 (1955)

Clark, B.G., Radio Interferometers of Intermediate Type, *IEEE Trans. Antennas Propag.*, **AP-16**, 143–144 (1968)

Clark, B.G., Radhakrishnan, V., and Wilson, R.W., The Hydrogen Line in Absorption, *Astrophys. J.*, **135**, 151–174 (1962)

Clark, T.A., Corey, A.E., Davis, J.L., Elgered, G., Herring, T.A., Hinteregger, H.F., Knight, C.A., Levine, J.I., Lundqvist, G., Ma, C., and 11 coauthors, Precision Geodesy Using the Mark-III Very-Long-Baseline Interferometer System, *IEEE Trans. Geosci. Remote Sens.*, **GE-23**, 438–449 (1985)

Cohen, M.H., Cannon, W., Purcell, G.H., Shaffer, D.B., Broderick, J.J., Kellermann, K.I., and Jauncy, D.L., The Small-Scale Structure of Radio Galaxies and Quasi-Stellar Sources, at 3.8 Centimeters, *Astrophys. J.*, **170**, 202–217 (1971)

Cohen, M.H., Moffet, A.T., Romney, J.D., Schilizzi, R.T., Shaffer, D.B., Kellermann, K.I., Purcell, G.H., Grove, G., Swenson, G.W., Jr., Yen, J.L., and four coauthors, Observations with a VLB Array. I. Introduction and Procedures, *Astrophys. J.*, **201**, 249–255 (1975)

Condon, J.J., Confusion and Flux-Density Error Distributions, *Astrophys. J.*, **188**, 279–286 (1974)

Condon, J.J., Radio Emission from Normal Galaxies, *Ann. Rev. Astron. Astrophys.*, **30**, 575–611 (1992)

Condon, J.J., Cotton, W.D., Fomalont, E.B., Kellermann, K.I., Miller, N., Perley, R.A., Scott, D., Vernstrom, T., and Wall, J.V., Resolving the Radio Source Background: Deeper Understanding Through Confusion, *Astrophys. J.*, **758**:23 (14pp) (2012)

Condon, J.J., Cotton, W.D., Greisen, E.W., Yin, Q.F., Perley, R.A., Taylor, G.B., and Broderick, J.J., The NRAO VLA Sky Survey, *Astron. J.*, **115**, 1693–1716 (1998)

Covington, A.E., and Broten, N.W., An Interferometer for Radio Astronomy with a Single-Lobed Radiation Pattern, *Proc. IRE Trans. Antennas Propag.*, **AP-5**, 247–255 (1957)

Davenport, W.B., Jr. and Root, W.L., *An Introduction to the Theory of Random Signals and Noise*, McGraw-Hill, New York (1958) de Bruijne, J.H.J., Rygl, K.L.J., and Antoja, T., GAIA Astrometric Science Performance: Post-Launch Predictions, *EAS Publications Ser.*, **67–68**, 23–29 (2014)

Dicke, R.H., The Measurement of Thermal Radiation at Microwave Frequencies, *Rev. Sci. Instrum.*, **17**, 268–275 (1946)

Dreyer, J.L.E., New General Catalog of Nebulae and Clusters of Stars, *Mem. R. Astron. Soc.*, **49**, Part 1 (1888) (reprinted *R. Astron. Soc. London*, 1962)

Edge, D.O., Shakeshaft, J.R., McAdam, W.B., Baldwin, J.E., and Archer, S., A Survey of Radio Sources at a Frequency of 159 Mc/s, *Mem. R. Astron. Soc.*, **68**, 37–60 (1959)

Elgaroy, O., Morris, D., and Rowson, B., A Radio Interferometer for Use with Very Long Baselines, *Mon. Not. R. Astron. Soc.*, **124**, 395–403 (1962)

Elitzur, M., *Astronomical Masers*, Kluwer, Dordrecht, the Netherlands (1992)

Fey, A.L., Gordon, D., Jacobs, C.S., Ma, C., Gamme, R.A., Arias, E.F., Bianco, G., Boboltz, D.A., Böckmann, S., Bolotin, S., and 31 coauthors, The Second Realization of the International Celestial Reference Frame by Very Long Baseline Interferometry, *Astrophys. J.*, **150**:58 (16pp) (2015)

Fomalont, E.B., The East–West Structure of Radio Sources at 1425 MHz, *Astrophys. J. Suppl.*, **15**, 203–274 (1968). doi: 10.1086/190166F

Genzel, R., Reid, M.J., Moran, J.M., and Downes, D., Proper Motions and Distances of H$_2$O Maser Sources. I. The Outflow in Orion-KL, *Astrophys. J.*, **244**, 884–902 (1981)

Gold, T., Radio Method for the Precise Measurement of the Rotation Period of the Earth, *Science*, **157**, 302–304 (1967)

Gower, J.F.R., Scott, P.F., and Wills, D., A Survey of Radio Sources in the Declination Ranges −07° to 20° and 40° to 80°, *Mem. R. Astron. Soc.*, **71**, 49–144 (1967)

Gray, M., *Maser Sources in Astrophysics*, Cambridge Univ. Press, Cambridge, UK (2012)

Guilloteau, S., The IRAM Interferometer on Plateau de Bure, in *Astronomy with Millimeter and Submillimeter Wave Interferometry*, Ishiguro, M., and Welch, W.J., Eds., Astron. Soc. Pacific Conf. Ser., **59**, 27–34 (1994)

Gurwell, M.A., Muhleman, D.O., Shah, K.P., Berge, G.L., Rudy, D.J., and Grossman, A.W., Observations of the CO Bulge on Venus and Implications for Mesospheric Winds, *Icarus*, **115**, 141–158 (1995)

Hagfors, T., Phillips, J.A., and Belcora, L., Radio Interferometry by Lunar Reflections, *Astrophys. J.*, **362**, 308–317 (1990)

Hanbury Brown, R., Measurement of Stellar Diameters, *Ann. Rev. Astron. Astrophys.*, **6**, 13–38 (1968)

Hanbury Brown, R., Palmer, H.P., and Thompson, A.R., A Rotating-Lobe Interferometer and Its Application to Radio Astronomy, *Philos. Mag.*, Ser. 7, **46**, 857–866 (1955)

Hanbury Brown, R., and Twiss, R.Q., A New Type of Interferometer for Use in Radio Astronomy, *Philos. Mag.*, Ser. 7, **45**, 663–682 (1954)

Hargrave, P.J., and Ryle, M., Observations of Cygnus A with the 5-km Radio Telescope, *Mon. Not. R. Astron. Soc.*, **166**, 305–327 (1974)

Harvey, P.M., Thronson, H.A., Jr., and Gatley, I., Far-Infrared Observations of Optical Emission-Line Stars: Evidence for Extensive Cool Dust Clouds, *Astrophys. J.*, **231**, 115–123 (1979)

Hazard, C., and Walsh, D., A Comparison of an Interferometer and Total-Power Survey of Discrete Sources of Radio Frequency Radiation, in *The Paris Symposium on Radio Astronomy*, Bracewell, R.N., Ed., Stanford Univ. Press, Stanford, CA (1959), pp. 477–486

Herbst, E., and van Dishoeck, E.F., Complex Organic Interstellar Molecules, *Ann. Rev. Astron. Astrophyys.*, **47**, 427–480 (2009)

Herring, T.A., Shapiro, I.I., Clark, T.A., Ma, C., Ryan, J.W., Schupler, B.R., Knight, C.A., Lundqvist, G., Shaffer, D.B., Vandenberg, N.R., and nine coauthors, Geodesy by Radio Interferometry: Evidence for Contemporary Plate Motion, *J. Geophys. Res.*, **91**, 8344–8347 (1986)

Herrnstein, J.R., Moran, J.M., Greenhill, L.J., Diamond, P.J., Inoue, M., Nakai, N., Miyoshi, M., Henkel, C., and Riess, A., A Geometric Distance to the Galaxy NGC4258 from Orbital Motions in a Nuclear Gas Disk, *Nature*, **400**, 539–841 (1999)

Herrnstein, J.R., Moran, J.M., Greenhill, L.J., and Trotter, A.S., The Geometry of and Mass Accretion Rate Through the Maser Accretion Disk in NGC4258, *Astrophys. J.*, **629**, 719–738 (2005). doi: 10.1086/431421

Hirabayashi, H., Hirosawa, H., Kobayashi, H., Murata, Y., Edwards, P.G., Fomalont, E.B., Fujisawa, K., Ichikawa, T., Kii, T., Lovell, J.E.J., and 43 coauthors, Overview and Initial Results of the Very Long Baseline Interferometry Space Observatory Programme, *Science*, **281**, 1825–1829 (1998)

Ho, P.T.P., Moran, J.M., and Lo, K.Y., The Submillimeter Array, *Astrophys. J. Lett.*, **616**, L1–L6 (2004)

Hogg, D.E., Macdonald, G.H., Conway, R.G., and Wade, C.M., Synthesis of Brightness Distribution in Radio Sources, *Astron. J.*, **74**, 1206–1213 (1969)

Hughes, M.P., Thompson, A.R., Colvin, R.S., An Absorption-Line Study of Galactic Neutral Hydrogen at 21 cm Wavelength, *Astrophys. J. Suppl.*, **23**, 232–367 (1971)

Humphreys, E.M.L., Reid, M.J., Moran, J.M., Greenhill, L.J., and Argon, A.L., Toward a New Geometric Distance to the Active Galaxy NGC 4258. III. Final Results and the Hubble Constant, *Astrophys. J.*, **775**:13 (10pp) (2013)

International Astronomical Union, *Trans. Int. Astron. Union*, Proc. of the 15th General Assembly Sydney 1974 and Extraordinary General Assembly Poland 1973, Contopoulos, G., and Jappel, A., Eds., **15B**, Reidel, Dordrecht, the Netherlands (1974), p. 142

International Astronomical Union, Specifications Concerning Designations for Astronomical Radiation Sources Outside the Solar System (2008). http://vizier.u-strasbg.fr/vizier/Dic/iau-spec.htx

IEEE, Standard Definitions of Terms for Radio Wave Propagation, Std. 211–1977, Institute of Electrical and Electronics Engineers Inc., New York (1977)

Jansky, K.G., Electrical Disturbances Apparently of Extraterrestrial Origin, *Proc. IRE*, **21**, 1387–1398 (1933)

Jennison, R.C., A Phase Sensitive Interferometer Technique for the Measurement of the Fourier Transforms of Spatial Brightness Distributions of Small Angular Extent, *Mon. Not. R. Astron. Soc.*, **118**, 276–284 (1958)

Jennison, R.C., High-Resolution Imaging Forty Years Ago, in *Very High Angular Resolution Imaging*, IAU Symp. 158, J. G. Robertson and W. J. Tango, Eds., Kluwer, Dordrecht, the Netherlands (1994), pp. 337–341

Jennison, R.C., and Das Gupta, M.K., Fine Structure in the Extra-terrestrial Radio Source Cygnus 1, *Nature*, **172**, 996–997 (1953)

Jennison, R.C., and Das Gupta, M.K., The Measurement of the Angular Diameter of Two Intense Radio Sources, Parts I and II, *Philos. Mag.*, Ser. 8, **1**, 55–75 (1956)

Jennison, R.C., and Latham, V., The Brightness Distribution Within the Radio Sources Cygnus A (19N4A) and Cassiopeia (23N5A), *Mon. Not. R. Astron. Soc.*, **119**, 174–183 (1959)

Jet Propulsion Laboratory, molecular spectroscopy site of JPL (2016). https://spec.jpl.nasa.gov

Kardashev, N.S., Khartov, V.V., Abramov, V.V., Avdeev, V.Yu., Alakoz, A.V., Aleksandrov, Yu.A., Ananthakrishnan, S., Andreyanov, V.V., Andrianov, A.S., Antonov, N.M., and 120 coauthors, "RadioAstron": A Telescope with a Size of 300,000 km: Main Parameters and First Observational Results, *Astron. Reports*, **57**, 153–194 (2013)

Kaula, W., Ed., *The Terrestrial Environment: Solid-Earth and Ocean Physics*, NASA Contractor Report 1579, NASA, Washington, DC (1970). http://ilrs.gsfc.nasa.gov/docs/williamstown_1968.pdf

Kellermann, K.I., The Discovery of Quasars, *Bull. Astr. Soc. India*, **41**, 1–17 (2013)

Kellermann, K.I., and Cohen, M.H., The Origin and Evolution of the N.R.A.O.–Cornell VLBI System, *J. Roy. Astron. Soc. Can.*, **82**, 248–265 (1988)

Kellermann, K.I., and Moran, J.M., The Development of High-Resolution Imaging in Radio Astronomy, *Ann. Rev. Astron. Astrophys.*, **39**, 457–509 (2001)

Kellermann, K.I., and Pauliny-Toth, I.I.K., The Spectra of Opaque Radio Sources, *Astrophys. J. Lett.*, **155**, L71–L78 (1969)

Kerr, A.R., Feldman, M.J., and Pan, S.-K., Receiver Noise Temperature, the Quantum Noise Limit, and the Role of Zero-Point Fluctuations, *Proc. 8th Int. Symp. Space Terahertz Technology*, 1997, pp. 101–111; also available as MMA Memo 161, National Radio Astronomy Observatory (1997)

Kesteven, M.J.L., and Bridle, A.H., Index of Extragalactic Radio-Source Catalogues, *Royal Astron. Soc. Canada J.*, **71**, 21–39 (1971)

Knowles, S.H., Waltman, W.B., Yen, J.L., Galt, J., Fort, D.N., Cannon, W.H., Davidson, D., Petrachenko, W., and Popelar, J., A Phase-Coherent Link via Synchronous Satellite Developed for Very Long Baseline Radio Interferometry, *Radio Sci.*, **17**, 1661–1670 (1982)

Kraus, J.D., *Radio Astronomy*, 2nd ed., Cygnus-Quasar Books, Powell, OH (1986)

Labrum, N.R., Harting, E., Krishnan, T., and Payten, W.J., A Compound Interferometer with a 1.5 Minute of Arc Fan Beam, *Proc. IRE Aust.*, **24**, 148–155 (1963)

Lequeux, J., Mesures Interférométriques à Haute Résolution du Diamètre et de la Structure des Principales Radiosources à 1420 MHz, *Annales d'Astrophysique*, **25**, 221–260 (1962)

Levy, G.S., Linfield, R.P., Edwards, C.D., Ulvestad, J.S., Jordan, J.F., Jr., Dinardo, S.J., Christensen, C.S., Preston, R.A., Skjerve, L.J., Stavert, L.R., and 22 coauthors, VLBI Using a Telescope in Earth Orbit. I. The Observations, *Astrophys. J.*, **336**, 1098–1104 (1989)

Linfield, R., VLBI Observations of Jets in Double Radio Galaxies, *Astrophys. J.*, **244**, 436–446 (1981)

Linfield, R.P., Levy, G.S., Edwards, C.D., Ulvestad, J.S., Ottenhoff, C.H., Hirabayashi, H., Morimoto, M., Inoue, M., Jauncey, D.L., Reynolds, J., and 18 coauthors, 15 GHz Space VLBI Observations Using an Antenna on a TDRSS Satellite, *Astrophys. J.*, **358**, 350–358 (1990)

Linfield, R.P., Levy, G.S., Ulvestad, J.S., Edwards, C.D., DiNardo, S.J., Stavert, L.R., Ottenhoff, C.H., Whitney, A.R., Cappallo, R.J., Rogers, A.E.E., and five coauthors, VLBI Using a Telescope in Earth Orbit. II. Brightness Temperatures Exceeding the Inverse Compton Limit, *Astrophys. J.*, **336**, 1105–1112 (1989)

Longair, M.S., *High Energy Astrophysics*, 2 vols., Cambridge Univ. Press, Cambridge, UK (1992)

Loudon, R., *The Quantum Theory of Light*, Oxford Univ. Press, London (1973), p. 229

Lovas, F.J., Recommended Rest Frequencies for Observed Interstellar Molecular Microwave Transitions—1991 Revision, *J. Phys. and Chem. Ref. Data*, **21**, 181–272 (1992)

Lovas, F.J., Snyder, L.E., and Johnson, D.R., Recommended Rest Frequencies for Observed Interstellar Molecular Transitions, *Astrophys. J. Suppl.*, **41**, 451–480 (1979)

Ma, C., Arias, E.F., Eubanks, T.M., Fey, A.L., Gontier, A.-M., Jacobs, C.S., Sovers, O.J., Archinal, B.A., and Charlot, P., The International Celestial Reference Frame as Realized by Very Long Baseline Interferometry, *Astron. J.*, **116**, 516–546 (1998)

Maltby, P., and Moffet, A.T., Brightness Distribution in Discrete Radio Sources. III. The Structure of the Sources, *Astrophys. J. Suppl.*, **7**, 141–163 (1962)

Matveenko, L.I., Kardashev, N.S., and Sholomitskii, G.B., Large Baseline Radio Interferometers, *Radiofizika*, **8**, 651–654 (1965); Engl. transl. in *Sov. Radiophys.*, **8**, 461–463 (1965)

McCready, L.L., Pawsey, J.L., and Payne-Scott, R., Solar Radiation at Radio Frequencies and Its Relation to Sunspots, *Proc. R. Soc. London A*, **190**, 357–375 (1947)

Menu, J., van Boekel, R., Henning, Th., Chandler, C.J., Linz, H., Benisty, M., Lacour, S., Min, M., Waelkens, C., Andrews, S.M., and 18 coauthors, On the Structure of the Transition Disk Around TW Hydrae, *Astron. Astrophys.*, **564**, A93 (22pp) (2014)

Messier, C., Catalogue des Nébuleuses et des Amas d'Étoiles [Catalogue of Nebulae and Star Clusters], *Connoissance des Temps* for 1784, 227–267, published in 1781. http://messier.seds.org/xtra/history/m-cat81.html

Michelson, A.A., On the Application of Interference Methods to Astronomical Measurements, *Philos. Mag.*, Ser. 5, **30**, 1–21 (1890)

Michelson, A.A., On the Application of Interference Methods to Astronomical Measurements, *Astrophys. J.*, **51**, 257–262 (1920)

Michelson, A.A., and Pease, F.G., Measurement of the Diameter of α Orionis with the Interferometer, *Astrophys. J.*, **53**, 249–259 (1921)

Mills, B.Y., The Positions of the Six Discrete Sources of Cosmic Radio Radiation, *Aust. J. Sci. Res.*, **A5**, 456–463 (1952)

Mills, B.Y., The Radio Brightness Distribution Over Four Discrete Sources of Cosmic Noise, *Aust. J. Phys.*, **6**, 452–470 (1953)

Mills, B.Y., Cross-Type Radio Telescopes, *Proc. IRE Aust.*, **24**, 132–140 (1963)

Mills, B.Y., Aitchison, R.E., Little, A.G., and McAdam, W.B., The Sydney University Cross-Type Radio Telescope, *Proc. IRE Aust.*, **24**, 156–165 (1963)

Mills, B.Y., and Little, A.G., A High Resolution Aerial System of a New Type, *Aust. J. Phys.*, **6**, 272–278 (1953)

Mills, B.Y., Little, A.G., Sheridan, K.V., and Slee, O.B., A High-Resolution Radio Telescope for Use at 3.5 m, *Proc. IRE*, **46**, 67–84 (1958)

Mills, B.Y., and Slee, O.B., A Preliminary Survey of Radio Sources in a Limited Region of the Sky at a Wavelength of 3.5 m, *Aust. J. Phys.*, **10**, 162–194 (1957)

Miyoshi, M., Moran, J., Herrnstein, J., Greenhill, L., Nakal, N., Diamond, P., and Inoue, M., Evidence for a Black Hole from High Rotation Velocities in a Sub-parsec Region of NGC4258, *Nature*, **373**, 127–129 (1995)

Moran, J.M., Thirty Years of VLBI: Early Days, Successes, and Future, in *Radio Emission from Galactic and Extragalactic Compact Sources*, J. A. Zensus, G. B. Taylor, and J. M. Wrobel, Eds., Astron. Soc. Pacific Conf. Ser., **144**, 1–10 (1998)

Moran, J.M., Crowther, P.P., Burke, B.F., Barrett, A.H., Rogers, A.E.E., Ball, J.A., Carter, J.C., and Bare, C.C., Spectral Line Interferometer with Independent Time Standards at Stations Separated by 845 Kilometers, *Science*, **157**, 676–677 (1967)

Morita, K.-I., The Nobeyama Millimeter Array, in *Astronomy with Millimeter and Submillimeter Wave Interferometry*, M. Ishiguro and W. J. Welch, Eds., Astron. Soc. Pacific Conf. Ser., **59**, 18–26 (1994)

Morris, D., Palmer, H.P., and Thompson, A.R., Five Radio Sources of Small Angular Diameter, *Observatory*, **77**, 103–106 (1957)

Napier, P.J., Bagri, D.S., Clark, B.G., Rogers, A.E.E., Romney, J.D., Thompson, A.R., and Walker, R.C., The Very Long Baseline Array, *Proc. IEEE*, **82**, 658–672 (1994)

Napier, P.J., Thompson, A.R., and Ekers, R.D., The Very Large Array: Design and Performance of a Modern Synthesis Radio Telescope, *Proc. IEEE*, **71**, 1295–1320 (1983)

NASA/IPAC Extragalactic Database, Best Practices for Data Publication to Facilitate Integration into NED: A Reference Guide for Authors, version 1.2, Sept. 25 (2013)

Nityananda, R., Comparing Optical and Radio Quantum Issues, in *Very High Resolution Imaging*, IAU Symp. 158, Robertson, J.G., and Tango, W.J., Eds., Kluwer, Dordrecht, the Netherlands (1994), pp. 11–18

Nyquist, H., Thermal Agitation of Electric Charge in Conductors, *Phys. Rev.*, **32**, 110–113 (1928)

O'Brien, P.A., The Distribution of Radiation Across the Solar Disk at Metre Wavelengths, *Mon. Not. R. Astron. Soc.*, **113**, 597–612 (1953)

Oliver, B.M., Thermal and Quantum Noise, *Proc. IEEE*, **53**, 436–454 (1965)

Pawsey, J.L., Sydney Investigations and Very Distant Radio Sources, *Publ. Astron. Soc. Pacific*, **70**, 133–140 (1958)

Pearson, T.J., Unwin, S.C., Cohen, M.H., Linfield, R.P., Readhead, A.C.S., Seielstad, G.A., Simon, R.S., and Walker, R.C., Superluminal Expansion of Quasar 3C273, *Nature*, **290**, 365–368 (1981)

Pease, F.G., Interferometer Methods in Astronomy, *Ergeb. Exakten Naturwiss.*, **10**, 84–96 (1931)

Perley, R.A., Dreher, J.W., and Cowan, J.J., The Jet and Filaments in Cygnus A, *Astrophys. J. Lett.*, **285**, L35–L38 (1984). doi: 10.1086/184360

Perley, R., Napier, P., Jackson, J., Butler, B., Carlson, B., Fort, D., Dewdney, P., Clark, B., Hayward, R., Durand, S., Revnell, M., and McKinnon, M., The Expanded Very Large Array, *Proc. IEEE*, **97**, 1448–1462 (2009)

Perryman, M.A.C., Lindegren, L., Kovalevsky, J., Høg, E., Bastian, U., Bernacca, P.L., Crézé, M., Donati, F., Grenon, M., Grewing, M., and ten coauthors, The Hipparcos Catalogue, *Astron. Astrophys.*, **323**, L49–L52 (1997)

Picken, J.S., and Swarup, G., The Stanford Compound-Grating Interferometer, *Astron. J.*, **69**, 353–356 (1964)

Radhakrishnan, V., Noise and Interferometry, in *Synthesis Imaging in Radio Astronomy II*, Taylor, G.B., Carilli, C.L., and Perley, R.A., Eds., Astron. Soc. Pacific Conf. Ser., **180**, 671–688 (1999)

Read, R.B., Two-Element Interferometer for Accurate Position Determinations at 960 Mc, *IRE Trans. Antennas Propag.*, **AP-9**, 31–35 (1961)

Reber, G., Cosmic Static, *Astrophys. J.*, **91**, 621–624 (1940)

Reid, M.J., and Brunthaler, A., The Proper Motion of Sagittarius A*. II. The Mass of Sagittarius A*, *Astrophys. J.*, **616**, 872–884 (2004)

Reid, M.J., Haschick, A.D., Burke, B.F., Moran, J.M., Johnston, K.J., and Swenson, G.W., Jr., The Structure of Interstellar Hydroxyl Masers: VLBI Synthesis Observations of W3(OH), *Astrophys. J.*, **239**, 89–111 (1980)

Reid, M.J., and Honma, M., Microarcsecond Radio Astrometry, *Ann. Rev. Astron. Astrophys.*, **52**, 339–372 (2014)

Reid, M.J., and Moran, J.M., Astronomical Masers, in *Galactic and Extragalactic Radio Astronomy*, 2nd ed., Verschuur, G.L., and Kellermann, K.I., Eds., Springer-Verlag, Berlin (1988), pp. 255–294

Roger, R.S., Costain, C.H., Lacey, J.D., Landaker, T.L., and Bowers, F.K., A Supersynthesis Radio Telescope for Neutral Hydrogen Spectroscopy at the Dominion Radio Astrophysical Observatory, *Proc. IEEE*, **61**, 1270–1276 (1973)

Rogers, A.E.E., Hinteregger, H.F., Whitney, A.R., Counselman, C.C., Shapiro, I.I., Wittels, J.J., Klemperer, W.K., Warnock, W.W., Clark, T.A., Hutton, L.K., and four coauthors, The Structure of Radio Sources 3C273B and 3C84 Deduced from the "Closure" Phases and Visibility Amplitudes Observed with Three-Element Interferometers, *Astrophys. J.*, **193**, 293–301 (1974)

Rowson, B., High Resolution Observations with a Tracking Radio Interferometer, *Mon. Not. R. Astron. Soc.*, **125**, 177–188 (1963)

Rybicki, G.B., and Lightman, A.P., *Radiative Processes in Astrophysics*, Wiley-Interscience, New York (1979) (reprinted 1985)

Ryle, M., A New Radio Interferometer and Its Application to the Observation of Weak Radio Stars, *Proc. R. Soc. London A*, **211**, 351–375 (1952)

Ryle, M., The New Cambridge Radio Telescope, *Nature*, **194**, 517–518 (1962)

Ryle, M., The 5-km Radio Telescope at Cambridge, *Nature*, **239**, 435–438 (1972)

Ryle, M., Radio Telescopes of Large Resolving Power, *Science*, **188**, 1071–1079 (1975)

Ryle, M., Elsmore, B., and Neville, A.C., High Resolution Observations of Radio Sources in Cygnus and Cassiopeia, *Nature*, **205**, 1259–1262 (1965)

Ryle, M., and Hewish, A., The Cambridge Radio Telescope, *Mem. R. Astron. Soc.*, **67**, 97–105 (1955)

Ryle, M., and Hewish, A., The Synthesis of Large Radio Telescopes, *Mon. Not. R. Astron. Soc.*, **120**, 220–230 (1960)

Ryle, M., Hewish, A., and Shakeshaft, J.R., The Synthesis of Large Radio Telescopes by the Use of Radio Interferometers, *IRE Trans. Antennas Propag.*, **7**, S120–S124 (1959)

Ryle, M., and Neville, A.C., A Radio Survey of the North Polar Region with a 4.5 Minute of Arc Pencil-Beam System, *Mon. Not. R. Astron. Soc.*, **125**, 39–56 (1962)

Ryle, M., and Smith, F.G., A New Intense Source of Radio Frequency Radiation in the Constellation of Cassiopeia, *Nature*, **162**, 462–463 (1948)

Ryle, M., Smith, F.G., and Elsmore, B., A Preliminary Survey of the Radio Stars in the Northern Hemisphere, *Mon. Not. R. Astron. Soc.*, **110**, 508–523 (1950)

Ryle, M., and Vonberg, D.D., Solar Radiation at 175 Mc/s, *Nature*, **158**, 339–340 (1946)

Scheuer, P.A.G., A Statistical Method for Analysing Observations of Faint Radio Stars, *Proc. Cambridge Phil. Soc.*, **53**, 764–773 (1957)

Schilke, P., Groesbeck, T., Blake, G.A., and Phillips, T.G., A Line Survey of Orion KL from 325 to 360 GHz, *Astrophys. J. uppl.*, **108**, 301–337 (1997)

Scoville, N., Carlstrom, J., Padin, S., Sargent, A., Scott, S., and Woody, D., The Owens Valley Millimeter Array, in *Astronomy with Millimeter and Submillimeter Wave Interferometry*, M. Ishiguro and W. J. Welch, Eds., Astron. Soc. Pacific Conf. Ser., **59**, 10–17 (1994)

Shakeshaft, J.R., Ryle, M., Baldwin, J.E., Elsmore, B., and Thomson, J.H., A Survey of Radio Sources Between Declinations -38° and +83°, *Mem. R. Astron. Soc.*, **67**, 106–154 (1955)

Smith, F.G., An Accurate Determination of the Positions of Four Radio Stars, *Nature*, **168**, 555 (1951)

Smith, F.G., The Determination of the Position of a Radio Star, *Mon. Not. R. Astron. Soc.*, **112**, 497–513 (1952a)

Smith, F.G., The Measurement of the Angular Diameter of Radio Stars, *Proc. Phys. Soc. B.*, **65**, 971–980 (1952b)

Smith, F.G., Apparent Angular Sizes of Discrete Radio Sources–Observations at Cambridge, *Nature*, **170**, 1065 (1952c)

Smoot, G.F., Bennett, C.L., Kogut, A., Wright, E.L., Aymon, J., Boggess, N.W., Cheng, E.S., de Amici, G., Gulkis, S., Hauser, M.G., and 18 coauthors, Structure in the COBE Differential Microwave Radiometer First-Year Maps, *Astrophys. J. Lett.*, **396**, L1–L5 (1992)

Smoot, G., Bennett, C., Weber, R., Maruschak, J., Ratliff, R., Janssen, M., Chitwood, J., Hilliard, L., Lecha, M., Mills, R., and 18 coauthors, COBE Differential Microwave Radiometers: Instrument Design and Implementation, *Astrophys. J.*, **360**, 685–695 (1990)

Southworth, G.C., Microwave Radiation from the Sun, *J. Franklin Inst.*, **239**, 285–297 (1945)

Splatalogue, database for astronomical spectroscopy (2016). http://www.cv.nrao.edu/php/splat

Swarup, G., Ananthakrishnan, S., Kapahi, V.K., Rao, A.P., Subrahmanya, C.R., and Kulkarni, V.K., The Giant Metrewave Radio Telescope, *Current Sci.* (Current Science Association and Indian Academy of Sciences), **60**, 95–105 (1991)

Thomasson, P., MERLIN, *Quart. J. R. Astron. Soc.*, **27**, 413–431 (1986)

Thompson, A.R., The Planetary Nebulae as Radio Sources, in *Vistas in Astronomy*, Vol. 16, A. Beer, Ed., Pergamon Press, Oxford, UK (1974), pp. 309–328

Thompson, A.R., Clark, B.G., Wade, C.M., and Napier, P.J., The Very Large Array, *Astrophys. J. Suppl.*, **44**, 151–167 (1980)

Thompson, A.R., and Krishnan, T., Observations of the Six Most Intense Radio Sources with a 1.0′ Fan Beam, *Astrophys. J.*, **141**, 19–33 (1965)

Tiuri, M.E., Radio Astronomy Receivers, *IEEE Trans. Antennas Propag.*, **AP-12**, 930–938 (1964)

University of Cologne, Physics Institute, Molecules in Space (2016). http://www.astro.uni-koeln.de/cdms/molecule

Vitkevich, V.V., and Kalachev, P.D., Design Principles of the FIAN Cross-Type Wide Range Telescope, in *Radio Telescopes*, Proc. P. N. Lebedev Phys. Inst. (Acad. Sci. USSR), Skobel'tsyn, D.V., Ed., Vol. 28, translated by Consultants Bureau, New York (1966)

Welch, W.J., The Berkeley–Illinois–Maryland Association Millimeter Array, in *Astronomy with Millimeter and Submillimeter Wave Interferometry*, Ishiguro, M., and Welch, W.J., Eds., Astron. Soc. Pacific Conf. Ser., **59**, 1–9 (1994)

Westerhout, G., A Survey of the Continuous Radiation from the Galactic System at a Frequency of 1390 Mc/s, *Bull. Astron. Inst. Netherlands*, **14**, 215–260 (1958)

Whitney, A.R., Lonsdale, C.J., and Fish, V.L., Insights into the Universe: Astronomy with Haystack's Radio Telescope, *Lincoln Lab. J.*, **21**, 8–27 (2014)

Whitney, A.R., Rogers, A.E.E., Hinteregger, H.F., Knight, C.A., Levine, J.I., Lippincott, S., Clark, T.A., Shapiro, I.I., and Robertson, D.S., A Very Long Baseline Interferometer System for Geodetic Applications, *Radio Sci.*, **11**, 421–432 (1976)

Whitney, A.R., Shapiro, I.I., Rogers, A.E.E., Robertson, D.S., Knight, C.A., Clark, T.A., Goldstein, R.M., Marandino, G.E., and Vandenberg, N.R., Quasars Revisited: Rapid Time Variations Observed via Very Long Baseline Interferometry, *Science*, **173**, 225–230 (1971)

Wootten, A., and Thompson, A.R., The Atacama Large Millimeter/Submillimeter Array, *Proc. IEEE*, **97**, 1463–1471 (2009)

Yen, J.L., Kellermann, K.I., Rayher, B., Broten, N.W., Fort, D.N., Knowles, S.H., Waltman, W.B., and Swenson, G.W., Jr., Real-Time, Very Long Baseline Interferometry Based on the Use of a Communications Satellite, *Science*, **198**, 289–291 (1977)

Chapter 2
Introductory Theory of Interferometry and Synthesis Imaging

In this chapter, we provide a simplified analysis of interferometry and introduce several important concepts. We first consider an interferometer in one dimension and discuss the effect of finite bandwidth and show how the interferometer response can be interpreted as a convolution. We extend the analysis to two dimensions and discuss circumstances in which three-dimensional imaging can be undertaken. This chapter is intended to provide a broad introduction to the principles of synthesis imaging to facilitate the understanding of more detailed development in later chapters. A brief introduction to the theory of Fourier transforms is given in Appendix 2.1.

2.1 Planar Analysis

The instantaneous response of a radio interferometer to a point source can most simply be analyzed by considering the signal paths in the plane containing the electrical centers of the two interferometer antennas and the source under observation. For an extended observation, it is necessary to take account of the rotation of the Earth and consider the geometric situation in three dimensions, as can be seen from Fig. 1.15. However, the two-dimensional geometry is a good approximation for short-duration observations, and the simplified approach facilitates visualization of the response pattern.

Consider the geometric situation shown in Fig. 2.1, where the antenna spacing is east–west. The two antennas are separated by a distance D, the baseline, and observe the same cosmic source, which is in the *far field* of the interferometer; that is, it is sufficiently distant that the incident wavefront can be considered to be a plane over the distance D. The source will be assumed for the moment to have

© The Author(s) 2017
A.R. Thompson, J.M. Moran, and G.W. Swenson Jr., *Interferometry and Synthesis in Radio Astronomy*, Astronomy and Astrophysics Library,
DOI 10.1007/978-3-319-44431-4_2

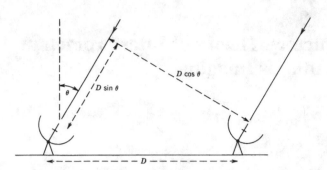

Fig. 2.1 Geometry of an elementary interferometer. D is the interferometer baseline.

infinitesimal angular dimensions. For this discussion, the receivers will be assumed to have narrow bandpass filters that pass only signal components very close to v.

As explained for the phase-switching interferometer in Chap. 1, the signal voltages are multiplied and then time-averaged, which has the effect of filtering out high frequencies. The wavefront from the source in direction θ reaches the right antenna at a time

$$\tau_g = \frac{D}{c} \sin \theta \qquad (2.1)$$

before it reaches the left one. τ_g is called the *geometric delay*, and c is the velocity of light. Thus, in terms of the frequency v, the output of the multiplier is proportional to

$$F = 2 \sin(2\pi v t) \sin 2\pi v (t - \tau_g)$$
$$= 2 \sin^2(2\pi v t) \cos(2\pi v \tau_g) - 2 \sin(2\pi v t) \cos(2\pi v t) \sin(2\pi v \tau_g) . \qquad (2.2)$$

The center frequency of the receivers is generally in the range of tens of megahertz to hundreds of gigahertz. As the Earth rotates, the most rapid rate of variation of θ is equal to the Earth's rotational velocity, which is of the order of 10^{-4} rad s^{-1}. Also, because D cannot be more than, say, 10^7 m for terrestrial baselines, the rate of variation of $v \tau_g$ is smaller than that of $v t$ by at least six orders of magnitude. For an averaging period $T \gg 1/v$, the average value of $\sin^2(2\pi v t) = \frac{1}{2}$ and the average value of $\sin(2\pi v t) \cos(2\pi v t) = 0$, leaving the fringe function

$$F = \cos 2\pi v \tau_g = \cos \left(\frac{2\pi D l}{\lambda} \right) , \qquad (2.3)$$

where $l = \sin \theta$; the definition of the variable l is discussed further in Sect. 2.4. For sidereal sources, the variation of θ with time as the Earth rotates generates quasisinusoidal fringes at the correlator, which are the output of the interferometer. Figure 2.2 shows an example of this function, which can be envisaged as the directional power reception pattern of the interferometer for the case in which the antennas either track the source or have isotropic responses and thus do not affect the shape of the pattern.

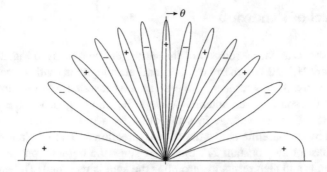

Fig. 2.2 Polar plot to illustrate the fringe function $F = \cos(2\pi Dl/\lambda)$. The radial component is equal to $|F|$, and θ is measured with respect to the vertical axis. Alternate lobes correspond to positive and negative half-cycles of the quasi-sinusoidal fringe pattern, as indicated by the plus and minus signs. To simplify the diagram, a very low value of 3 is used for D/λ. The increase in fringe width due to foreshortening of the baseline as $|\theta|$ increases is clearly shown. The maxima in the horizontal direction ($\theta = \pm 90°$) are a result of the arbitrary choice of an integer value for D/λ.

An alternate and equivalent way of envisaging the formation of the sinusoidal fringes is to note that because of the rotation of the Earth, the two antennas have different components of velocity in the direction of the source. The signals reaching the antennas thus suffer different Doppler shifts. When the signals are combined in the multiplying action of the receiving system, the sinusoidal output arises from the beats between the Doppler-shifted signals.

A development of the simple analysis can be made if we consider two Fourier components of the received signal at frequencies ν_1 and ν_2. These frequency components are statistically independent so that the interferometer output is the linear sum of the responses to each component. Hence, the output has components F_1 and F_2, as in Eq. (2.3). For frequency ν_2, the coefficient $2\pi D/\lambda = 2\pi D\nu_2/c$ will be different from that for ν_1, so F_2 will have a different period from F_1 at any given angle θ. This difference in period gives rise to interference between F_1 and F_2, so that the fringe maxima have superimposed on them a modulation function that also depends on θ. Similar effects occur in the case of a continuous band of frequencies. For example, if the signals at the correlator are of uniform power spectral density over a band of width $\Delta\nu$ and center frequency ν_0, the output becomes

$$F(l) = \frac{1}{\Delta\nu}\int_{\nu_0-\Delta\nu/2}^{\nu_0+\Delta\nu/2} \cos\left(\frac{2\pi Dl\nu}{c}\right) d\nu$$

$$= \cos\left(\frac{2\pi Dl\nu_0}{c}\right) \frac{\sin(\pi Dl\Delta\nu/c)}{\pi Dl\Delta\nu/c} . \tag{2.4}$$

Thus, the fringe pattern has an envelope in the form of a sinc function [$\mathrm{sinc}(x) = (\sin\pi x)/\pi x$]. This is an example of the general result, to be discussed in the following section, that in the case of uniform power spectral density at the antennas, the envelope of the fringe pattern is the Fourier transform of the instrumental frequency response.

2.2   Effect of Bandwidth

Figure 2.3 shows an interferometer of the same general type as in Fig. 2.1 but with the amplifiers H_1 and H_2, the multiplier, and an integrator (with respect to time) shown explicitly. An instrumental time delay τ_i is inserted into one arm. Assume that for a point source, each antenna delivers the same signal voltage $V(t)$ to the correlator, and that one voltage lags the other by a time delay $\tau = \tau_g - \tau_i$, as determined by the baseline D and the source direction θ. The integrator within the correlator has a time constant $2T$; that is, it sums the output from the multiplier for $2T$ seconds and then resets to zero after the sum is recorded. The output of the correlator may be a voltage, a current, or a coded set of logic levels, but in any case, it represents a physical quantity with the dimensions of voltage squared.

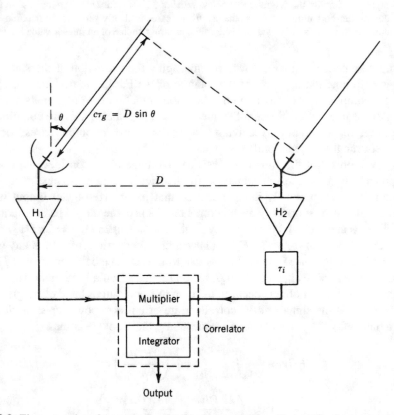

Fig. 2.3 Elementary interferometer showing bandpass amplifiers H_1 and H_2, the geometric time delay τ_g, the instrumental time delay τ_i, and the correlator consisting of a multiplier and an integrator.

The output from the correlator resulting from a point source[1] is

$$r = \frac{1}{2T} \int_{-T}^{T} V(t)V(t - \iota)\, dt \,.$$ (2.5)

We have ignored system noise and assumed that the two amplifiers have identical bandpass characteristics, including finite bandwidths Δv outside which no frequencies are admitted. The integration time $2T$ is typically milliseconds to seconds, that is, very much larger than Δv^{-1}. Thus, Eq. (2.5) can be written as

$$r(\tau) = \lim_{T \to \infty} \frac{1}{2T} \int_{-T}^{T} V(t)V(t - \tau)\, dt \,,$$ (2.6)

which is an (unnormalized) autocorrelation function. The condition $T \to \infty$ is satisfied if a large number of variations of the signal amplitude, which have a duration $\sim \Delta v^{-1}$, occur in time $2T$. The integration time used in practice must clearly be finite and much less than the fringe period.

As described in Chap. 1, the signal from a natural cosmic source can be considered as a continuous random process that results in a broad spectrum, of which the phases are a random function of frequency. It will be assumed for our immediate purpose that the time-averaged amplitude of the cosmic signal in any finite band is constant with frequency over the passband of the receiver.

The squared amplitude of a frequency spectrum is known as the power density spectrum, or power spectrum. The power spectrum of a signal is the Fourier transform of the autocorrelation function of that signal. This statement is known as the Wiener–Khinchin relation (see Appendix A2.1.5) and is discussed further in Sect. 3.2. It applies to signals that are either deterministic or statistical in nature and can be written

$$|H(v)|^2 = \int_{-\infty}^{\infty} r(\tau)e^{-j2\pi v\tau}\, d\tau \,,$$ (2.7)

and

$$r(\tau) = \int_{-\infty}^{\infty} |H(v)|^2 e^{j2\pi v\tau}\, dv \,,$$ (2.8)

where $H(v)$ is the amplitude (voltage) response, and hence $|H(v)|^2$ is the power spectrum of the signal input to the correlator. In this case, because the cosmic signal is assumed to have a spectrum of constant amplitude, the spectrum $H(v)$ is determined solely by the passband characteristics (frequency response) of the receiving system from the outputs of the antennas to the output of the integrator. Thus, the output of the interferometer as a function of the time delay τ is the

---

[1]For simplicity, we consider only the signals from a point source, which are identical except for a time delay. In practical systems, the input waveforms at the correlator may contain the partially correlated signals from a partially resolved source as well as instrumental noise.

Fourier transform of the power spectrum of the cosmic signal as bandlimited by the
receiving system. Assume, as a simple example, a Gaussian passband centered at ν_0:

$$|H(\nu)|^2 = \frac{1}{2\sigma\sqrt{2\pi}} \left\{ \exp\left[-\frac{(\nu - \nu_0)^2}{2\sigma^2}\right] + \exp\left[-\frac{(\nu + \nu_0)^2}{2\sigma^2}\right] \right\} , \qquad (2.9)$$

where σ is the bandwidth factor (the full bandwidth at half-maximum level is
$\sqrt{8\ln 2}\,\sigma$). Note that to perform the Fourier transforms in Eqs. (2.7) and (2.8),
we include a negative frequency response centered on $-\nu_0$. The spectrum is then
symmetrical with respect to zero frequency, which is consistent with the fact that
the autocorrelation function (which is the Fourier transform of the power spectrum)
is real. The negative frequencies have no physical meaning but arise mathematically
from the use of the exponential function. The interferometer response is

$$r(\tau) = e^{-2\pi^2\tau^2\sigma^2} \cos(2\pi\nu_0\tau) , \qquad (2.10)$$

which is illustrated in Fig. 2.4a. Note that $r(\tau)$ is a cosinusoidal function multiplied
by an envelope function, in this case a Gaussian, whose shape and width depend on
the amplifier passband. This envelope function is referred to as the *delay pattern* or
bandwidth pattern.

By setting the instrumental delay τ_i to zero and substituting for the geometric
delay $\tau_g = (D/c)\sin\theta$ in Eq. (2.10), we obtain the response

$$r(\tau_g) = \exp\left[-2\left(\frac{\pi D\sigma}{c}\sin\theta\right)^2\right] \cos\left(\frac{2\pi\nu_0 D}{c}\sin\theta\right) . \qquad (2.11)$$

The period of the fringes (the cosine term) varies inversely as the quantity
$\nu_0 D/c = D/\lambda$ and does not depend on the bandwidth parameter σ. The width
of the bandwidth pattern (the exponential term), however, is a function of both
σ and D; wide bandwidths and long baselines result in narrow fringe envelopes.
This result is quite general. For example, a rectangular amplifier passband of
width $\Delta\nu$, as considered in Eq. (2.4), results in an envelope pattern of the form
$[\sin(\pi\Delta\nu\tau)]/(\pi\Delta\nu\tau)$, as shown in Fig. 2.4b.

In imaging applications, it is usually desirable to observe the fringes in the
vicinity of the maximum of the pattern, where the fringe amplitude is greatest. This
condition can be achieved by changing the instrumental delay τ_i continuously or
periodically so as to keep $\tau = \tau_g - \tau_i$ suitably small. If τ_i is adjusted in steps of
the reciprocal of the center frequency[2] ν_0, the response remains cosinusoidal with
τ_g. Note that for wide bandwidths, as $\Delta\nu$ approaches ν, the width of the envelope
function becomes so narrow that only the central fringe remains. This occurs mainly
in optics, where a central fringe of this type is often called the "white light" fringe.

---

[2]This adjustment method is useful to consider here, but more commonly used methods are
described in Sects. 7.3.5 and 7.3.6.

Fig. 2.4 Point-source response of an interferometer with (**a**) Gaussian and (**b**) rectangular passbands. The abscissa is the geometric delay τ_g. The bandwidth pattern determines the envelope of the fringe term.

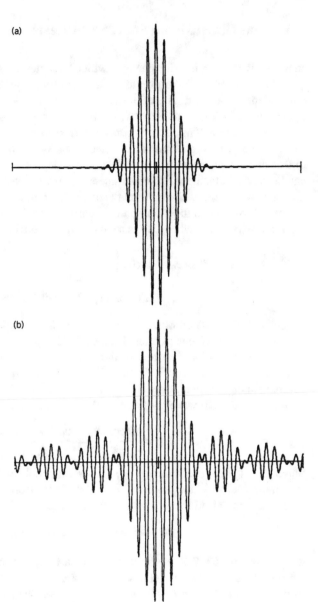

2.3   One-Dimensional Source Synthesis

In the analysis of an interferometer in which the antennas and the instrumental delay track the position of the source, as is the norm for frequencies above ~ 1 GHz, it is convenient to specify angles of the antenna beam and other variables with respect to a reference position on the sky, usually the center or nominal position of the source under observation. This is commonly referred to as the *phase reference position*. Since the range of angles required to specify the source intensity distribution relative to this point is generally no more than a few degrees, small-angle approximations can be used to advantage. The instrumental delay is constantly adjusted to equal the geometric delay for radiation from the phase reference position. If we designate this reference position as the direction θ_0, then $\tau_i = (D/c) \sin \theta_0$. For radiation from a direction $(\theta_0 - \Delta\theta)$, where $\Delta\theta$ is a small angle, the fringe response term is

$$
\cos(2\pi \nu_0 \tau) = \cos\left\{ 2\pi \nu_0 \left[\frac{D}{c} \sin(\theta_0 - \Delta\theta) - \tau_i \right] \right\}
$$
$$
\simeq \cos[2\pi \nu_0 (D/c) \sin \Delta\theta \cos \theta_0] \tag{2.12}
$$

for $\cos \Delta\theta \simeq 1$. When observing a source at any position in the sky, the angular resolution of the fringes is determined by the length of the baseline projected onto a plane normal to the direction of the source. In Fig. 2.1, for example, this is the distance designated $D \cos \theta$. We therefore introduce a quantity u that is equal to the component of the antenna spacing normal to the direction of the reference position θ_0. u is measured in wavelengths, λ, at the center frequency ν_0, that is,

$$
u = \frac{D \cos \theta_0}{\lambda} = \frac{\nu_0 D \cos \theta_0}{c} . \tag{2.13}
$$

Since $\Delta\theta$ in Eq. (2.12) is small, we can assume that the bandwidth pattern is near maximum (unity) in the direction $\theta_0 - \Delta\theta$. Then, from Eqs. (2.12) and (2.13), the response to radiation from that direction is proportional to

$$
F(l) = \cos(2\pi \nu_0 \tau) = \cos(2\pi u l) , \tag{2.14}
$$

where $l = \sin \Delta\theta$. This is the response to a point source at $\theta = \theta_0 - \Delta\theta$ of an interferometer whose net delay $\tau_g - \tau_i$ is zero at $\theta = \theta_0$. As we shall show, the quantity u is interpreted as *spatial frequency*. It can be measured in cycles per radian, since the spatial variable l, being small, can be expressed in radians.

2.3.1   Interferometer Response as a Convolution

The response of a single antenna or an interferometer to a source can be expressed in terms of a convolution. Consider first the response of a single antenna and a receiver that measures the power received. Figure 2.5 shows the power reception

Fig. 2.5 The power pattern $A(\theta)$ of an antenna pointed in the direction OC, and the intensity profile of a source $I_1(\theta')$, used to illustrate the convolution relationship. The angle θ is measured with respect to the beam center OC. The profile of the source is a function of θ', measured with respect to the direction of the nominal position of the source OB.

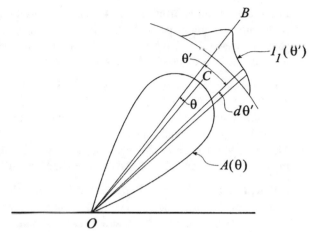

pattern of the antenna $A(\theta)$, which is a polar plot of the effective area of the antenna as a function of angle from the center of the antenna beam. Also shown is the one-dimensional intensity profile of a source $I_1(\theta')$, as defined in Eq. (1.9), in which θ' is measured with respect to the center, or nominal position, of the source. The component of the output power in bandwidth $\Delta\nu$ contributed by each element $d\theta'$ of the source is $\frac{1}{2}\Delta\nu A(\theta' - \theta)I_1(\theta')d\theta'$, where the factor $\frac{1}{2}$ takes account of the ability of the antenna to respond to only one component of randomly polarized radiation. The total output power from the antenna, omitting the constant factor $\frac{1}{2}\Delta\nu$, is proportional to

$$\int_{\text{source}} A(\theta' - \theta)I_1(\theta')d\theta' . \qquad (2.15)$$

This integral is equal to the cross-correlation of the antenna reception pattern and the intensity distribution of the source. It is convenient to define $\mathcal{A}(\theta) = A(-\theta)$, where $\mathcal{A}$ is the mirror image of A with respect to θ. Then expression (2.14) becomes

$$\int_{\text{source}} \mathcal{A}(\theta - \theta')I_1(\theta')d\theta' . \qquad (2.16)$$

The integral in expression (2.15) is an example of the *convolution integral*; see Appendix 2.1, Eq. (A2.33). We can say that the output power of the antenna is given by the convolution of the source with the mirror image of the power reception pattern of the antenna. The mirror-image[3] reception pattern can be described as the response of the antenna to a point source.

---

[3] In many cases, the beam is symmetrical, and the mirror image is identical to the beam.

In the case of an interferometer, we can express the response as a convolution by replacing the antenna power pattern in Eq. (2.16) by the overall power pattern of the interferometer. From the results presented earlier, we find that the response of an interferometer is determined by three functions:

- The reception pattern of the antennas, which we represent as $A(l)$,
- The fringe pattern, $F(l)$, as in the example of Fig. 2.2 and given by Eq. (2.14). Note that the fringe term in the interferometer output, being the product of two voltages, is proportional to power.
- The bandwidth pattern, for example, as given by the sinc-function factor in Eq. (2.4). In the general case, we can represent this by $F_B(l)$.

Note that the antenna beam is often symmetrical, in which case, if the interferometer fringes are aligned with the beam center, we can disregard the distinction between the interferometer power pattern and its mirror image in using the convolution relationship.

Next, consider an interferometer with tracking antennas and an instrumental delay that is adjusted so the bandwidth pattern also tracks the source across the sky. In effect, the intensity distribution is modified by the antenna and bandwidth patterns. We can therefore envisage the output of the interferometer as the convolution of (the mirror image of) the fringe pattern with the modified intensity. In terms of the convolution integral, the response can be written as

$$R(l) = \int_{source} \cos\left[2\pi u(l - l')\right] A(l')F_B(l')I_1(l')dl' . \tag{2.17}$$

or, more concisely,

$$R(l) = \cos(2\pi u l) * [A(l)F_B(l)I_1(l)] , \tag{2.18}$$

where the in-line asterisk symbol ($*$) denotes convolution. The intensity distribution measured with the interferometer is modified by $A(l)$ and $F_B(l)$, but since these are measurable instrumental characteristics, $I_1(l)$ can generally be recovered from the product $A(l)F_B(l)I_1(l)$. In many cases, the angular size of the source is small compared with the antenna beams and the bandwidth pattern, so these two functions introduce only a constant in the expression for the response. To simplify the discussion, we shall consider this case, and omitting constant factors, we can write the essential response of the interferometer as

$$R(l) = \cos(2\pi u l) * I_1(l) . \tag{2.19}$$

In the case of the early interferometer shown in Fig. 1.6, in which the antennas are fixed in the meridian and do not track the source, the delays in the signal paths between the antennas and the point at which the signals are multiplied are equal, and there is no variable instrumental delay. Thus, the three functions that determine the interferometer power pattern are all fixed with respect to the interferometer baseline. The interferometer power pattern is of the form $A(l) \cos(2\pi u l) F_B(l)$, and the response of the interferometer to the source is $[A(l) \cos(2\pi u l) F_B(l)] * I_1(l)$.

Most interferometers for operation at meter wavelengths, that is, at frequencies below about 300 MHz, use antennas that are arrays of fixed dipoles. At such long wavelengths, it is possible to obtain large collecting areas and still have wide enough beams that some minutes of observing time are obtained as a source passes through in sidereal motion. Often the bandwidth of such low-frequency instruments is small, so that the bandwidth pattern, $F_B(l)$, is wide and this factor can be omitted. Also, the antenna beams are usually wider than the source and sufficiently wide that several cycles of the fringe pattern can be measured as the source transits the beam. So in the nontracking case, the essential form of the response is also represented by Eq. (2.19). However, fixed antennas with nontracking beams are mainly a feature of the early years of radio astronomy, and in more recent meter-wavelength arrays, the phases of individual dipoles, or small clusters of dipoles, can be adjusted to provide steerable beams.

2.3.2   Convolution Theorem and Spatial Frequency

We now examine the interferometer response, as given in Eq. (2.19), using the *convolution theorem* of Fourier transforms (see the derivation in Appendix A2.1.2), which can be expressed as:

$$f * g \longleftrightarrow FG , \tag{2.20}$$

where $f \longleftrightarrow F$, $g \longleftrightarrow G$, and $\longleftrightarrow$ indicates Fourier transformation. Consider the Fourier transforms with respect to l and u of the three functions in Eq. (2.19). For the interferometer response, we have $r(u) \longleftrightarrow R(l)$. For a particular value $u = u_0$, the Fourier transform of the fringe term is given by [see Fourier transform example in Eq. (A2.15)]

$$\cos(2\pi u_0 l) \longleftrightarrow \tfrac{1}{2} [\delta(u + u_0) + \delta(u - u_0)] , \tag{2.21}$$

where δ is the delta function defined in Appendix 2.1. The Fourier transform of $I_1(l)$ is the visibility function $\mathcal{V}(u)$. Thus, from Eqs. (2.19), (2.20), and (2.21), we obtain

$$
\begin{aligned}
r(u) &= \tfrac{1}{2} [\delta(u + u_0) + \delta(u - u_0)] \, \mathcal{V}(u) \\
&= \tfrac{1}{2} [\mathcal{V}(-u_0)\delta(u + u_0) + \mathcal{V}(u_0)\delta(u - u_0)] .
\end{aligned} \tag{2.22}
$$

This result shows that the instantaneous output of the interferometer as a function of spatial frequency consists of two delta functions situated at plus and minus u_0 on the u axis. Now, $\mathcal{V}(u)$, the Fourier transform of $I_1(l)$, represents the amplitude and phase of the sinusoidal component of the intensity profile with spatial frequency u

cycles per radian. The interferometer acts as a filter that responds only to spatial frequencies $\pm u_0$. The negative spatial frequency $-u_0$ has no physical meaning. It arises from the use, for mathematical convenience, of the exponential Fourier transform rather than the sine and cosine transforms, which correspond more directly to the physical situation. As a result, the spatial frequency spectra are symmetrical about the origin in the Hermitian sense, that is, with even real parts and odd imaginary parts, which is appropriate since the intensity is a real, not complex, quantity.

Fringe visibility, as originally defined by Michelson [$\mathcal{V}_M$, see Eq. (1.9)], is a real quantity and is normalized to unity for an unresolved source. Complex visibility (Bracewell 1958) was defined to take account of the phase of the visibility, measured as the fringe phase, to allow imaging of asymmetric and complicated sources. The normalization is convenient when comparing measurements with simple models, as shown in Fig. 1.5. However, in images, it is desirable to display the magnitude of the intensity or brightness temperature, so the general practice is to retain the measured value of visibility, without normalization, since this incorporates the required information. Thus, visibility $\mathcal{V}$ as used here is an unnormalized complex quantity with units of flux density (W m^{-2} Hz^{-1}). The quantity u, which was introduced as the projected baseline in wavelengths, is seen also to represent the spatial frequency of the Fourier components of the intensity. The concepts of spatial frequency and spatial frequency spectra are fundamental to the Fourier synthesis of astronomical images, and this general subject is discussed in a seminal paper by Bracewell and Roberts (1954).

2.3.3 Example of One-Dimensional Synthesis

To illustrate the observing process outlined in this chapter, we present a rudimentary simulation of measurements of the complex visibility of a source using arbitrary parameters. The source consists of two components separated by 0.34° of angle, the flux densities of which are in the ratio 2 : 1. The measurements are made with pairs of antennas placed along a line parallel to the direction of separation of the two components. Measurements are made for antenna spacings that are integral multiples of a unit spacing of 30 wavelengths. All spacings from 1 to 23 times the unit spacing are measured. These results could be obtained using two antennas and a single correlator, observing the source as it transits the meridian on 23 different days and moving the antennas to provide a new spacing each day. Alternately, the 23 measurements could be made simultaneously using 23 correlators and a number of antennas that could be as small as 8 (if they were set out with minimum redundancy in the spacings, as discussed in Sect. 5.5). The angular sizes of the two components of the source are too small to be resolved by the interferometer, so they can be regarded as point radiators. The two components radiate noise, and their two outputs are uncorrelated. The source is at a sufficient distance that incoming wavefronts can be considered to be plane over the measurement baselines.

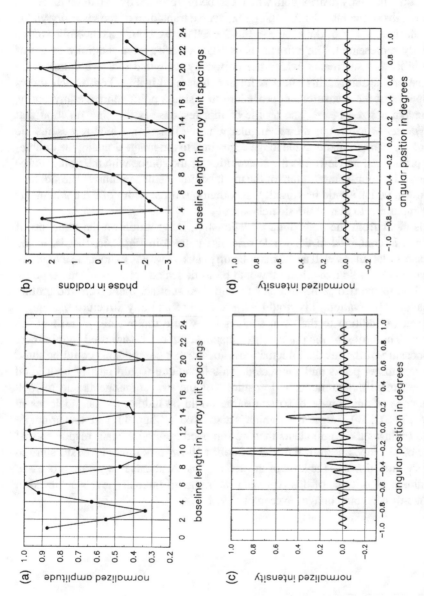

Fig. 2.6 Simulated measurements of visibility of a double source: (**a**) visibility amplitude and (**b**) visibility phase, each plotted as a function of antenna spacing as a multiple of the unit spacing; (**c**) the profile derived from the measurements and (**d**) the response to a point source.

Figure 2.6a and b show, respectively, the amplitude[4] and phase of the visibility function as it would be measured. Since the data are derived from a model, there are no measurement errors, so the points indicate samples of the Fourier transform of the source intensity distribution, which can be represented by two delta functions with strengths in the ratio 1 : 2. Taking the inverse transform of the visibility yields the synthesized image of the source in Fig. 2.6c. The two components of the source are clearly represented. The extraneous oscillations arise from the finite extent of the visibility measurements, which are uniformly weighted out to a cutoff at 23 times the unit spacing. This effect is further illustrated in Fig. 2.6d, which shows the response of the measurement procedure to a single point source; equivalently, it is the synthesized beam. The profile of this response is the sinc function that is the Fourier transform of the rectangular window function, which represents the cutoff of the measurements at the longest spacing. In the image domain, the double-source profile can be viewed as the convolution of the source with the point-source response. The point-source nature of the model components maximizes the sidelobe oscillations, which would be partially smoothed out if the width of the components were comparable to that of the sidelobes.

As is clear from the convolution relationship, information on the structure of the source is contained in the whole response pattern in Fig. 2.6c, that is, in the sidelobe oscillations as well as the main-beam peaks. A way to extract the maximum information on the source structure would be to fit scaled versions of the response in Fig. 2.6d to the two peaks in Fig. 2.6c and then subtract them from the profile. In an actual observation, this would leave the noise and any structure that might be present in addition to the point sources but would remove all or most of the sidelobes. The fitting of the point-source responses could be adjusted to minimize some measure of the residual fluctuations, and further components could be fitted to any remaining peaks and subtracted. This technique would clearly be a good way to estimate the strengths and positions of the two components and to look for evidence of any low-level structure that could be hidden by the sidelobes in Fig. 2.6c. The CLEAN algorithm, which is discussed in Chap. 11, uses this principle but also replaces the components that are removed by model beam responses that are free of sidelobes. Removal of the sidelobes allows any lower-level structure to be investigated, down to the level of the noise. Most synthesis images are processed by nonlinear algorithms of this type, and the range of intensity levels achieved in some two-dimensional images exceeds 10^5 to 1.

---

[4]It is arguable that the modulus of the complex visibility should be referred to as *magnitude* rather than *amplitude* since the dimensions of visibility include power rather than voltage. However, the term *visibility amplitude* is widely used in radio astronomy, probably resulting from the early practice of recording the fringe pattern as a quasi-sinusoidal waveform, and subsequently analyzing the amplitude and phase of the oscillations.

2.4   Two-Dimensional Synthesis

Synthesis of an image of a source in two dimensions on the sky requires measurement of the two-dimensional spatial frequency spectrum in the (u, v) plane, where v is the north–south component as shown in Fig. 2.7a. Similarly, it is necessary to define a two-dimensional coordinate system (l, m) on the sky. The (l, m) origin is the reference position, or phase reference position, introduced in the last section. In considering functions in one dimension in the earlier part of this chapter, it was possible to define l in Eq. (2.3) as the sine of an angle. In two-dimensional analysis, l and m are defined as the cosines of the angles between the direction (l, m) and the u and v axes, respectively, as shown in Fig. 2.7c. If the angle between the direction (l, m) and the w axis is small, l and m can be considered as the components of this angle measured in radians in the east–west and north–south directions, respectively.

For a source near the celestial equator, measuring the visibility as a function of u and v requires observing with a two-dimensional array of interferometers, that is, an array in which the baselines between pairs of antennas contain components in the north–south as well as the east–west directions. Although we have considered only east–west baselines, the results derived in terms of angles measured with respect to a plane that is normal to the baseline hold for any baseline direction.

A source at a high declination (near the celestial pole) can be imaged in two dimensions with either one- or two-dimensional arrays, as shown in Fig. 1.15 and

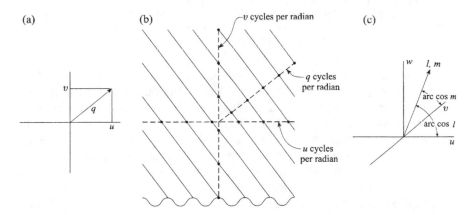

Fig. 2.7 (**a**) The (u, v) plane in which the arrow point indicates the spatial frequency, q cycles per radian, of one Fourier component of an image of the intensity of a radio source. The components u and v of the spatial frequency are measured along axes in the east–west and north–south directions, respectively. (**b**) The (l, m) plane in which a single component of spatial frequency in the intensity domain has the form of sinusoidal corrugations on the sky. The figure shows corrugations that represent one such component. The diagonal lines indicate the ridges of maximum intensity. The dots indicate the positions of these maxima along lines in three directions. In a direction normal to the ridges, the frequency of the oscillations is q cycles per radian, and in directions parallel to the u and v axes, it is u and v cycles per radian, respectively. (**c**) The u and v coordinates define a plane, and the w coordinate is perpendicular to it. The coordinates (l, m) are used to specify a direction on the sky in two dimensions. l and m are defined as the cosines of the angles made with the u and v axes, respectively.

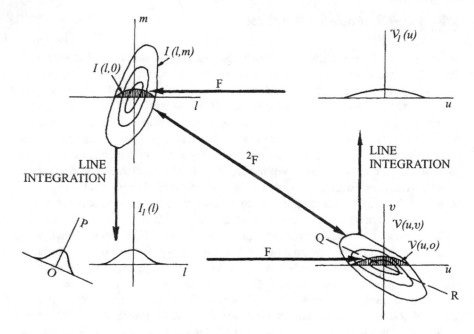

Fig. 2.8 Illustration of the projection-slice theorem, which explains the relationships between one-dimensional projections and cross sections of intensity and visibility functions. One-dimensional Fourier transforms are organized horizontally and projections vertically. The symbols F and ^{2}F indicate one-dimensional and two-dimensional Fourier transforms, respectively. See the text for further explanation. From Bracewell (1956). © CSIRO 1956. Published by CSIRO Publishing, Melbourne, Victoria, Australia. Reproduced with permission.

further explained in Sect. 4.1. As the Earth rotates, the baseline projection on the celestial sphere rotates and foreshortens. A plot of the variation of the length and direction of the projected baseline as the antennas track the source across the sky is an arc of an ellipse in the (u, v) plane. The parameters of the ellipse depend on the declination of the source, the length and orientation of the baseline, and the latitude of the center of the baseline. In the design of a synthesis array, the relative positions of the antennas are chosen to provide a distribution of measurements in u and v consistent with the angular resolution, field of view, declination range, and sidelobe level required, as discussed in Chap. 5. The two-dimensional intensity distribution is then obtained by taking a two-dimensional Fourier transform of the observed visibility, $\mathcal{V}(u, v)$.

2.4.1   Projection-Slice Theorem

Some important relationships between one-dimensional and two-dimensional functions of intensity and visibility are summarized in Fig. 2.8, which illustrates the projection-slice theorem of Fourier transforms (Bracewell 1956, 1995, 2000). At the top left is the two-dimensional intensity distribution of a source $I(l, m)$, and at the bottom right is the corresponding visibility function $\mathcal{V}(u, v)$. These two functions are related by a two-dimensional Fourier transform, as indicated on the arrows shown between them. Note the general property of Fourier transforms that the width in one domain is inversely related to the width in the other domain. At the lower left is the projection of $I(l, m)$ on the l axis, which is equal to the one-dimensional intensity distribution $I_1(l)$. This projection is obtained by line integration along lines parallel to the m axis, as defined in Eq. (1.10). I_1 is related by a one-dimensional Fourier transform to the visibility measured along the u axis at the lower right, that is, the profile of a slice $\mathcal{V}(u, 0)$ through the visibility function $\mathcal{V}(u, v)$, indicated by the shaded area in the diagram. $\mathcal{V}(u, 0)$ could be measured, for example, by observations of a source made at meridian transit with a series of interferometer baselines in an east–west direction. This relationship was encountered in Chap. 1 in the description of the Michelson interferometer, and examples of such pairs of functions are shown in Fig. 1.5. At the upper right is a projection of $\mathcal{V}(u, v)$ on the u axis, $\mathcal{V}_1(u) = \int \mathcal{V}(u, v)dv$, and this is related by a one-dimensional Fourier transform to a slice profile of the source intensity $I(l, 0)$ along the l axis at the upper left, indicated by the shaded area. The relationships between the projections and slices are not confined to the u and l axes but apply to any sets of axes that are parallel in the two domains. For example, integration of $I(l, m)$ along lines parallel to OP results in a curve, the Fourier transform of which is the profile of a slice through $\mathcal{V}(u, v)$ along the line QR.

The relationships in Fig. 2.8 apply to Fourier transforms in general, and their application to radio astronomy was recognized during the early development of the subject. For example, in determining the two-dimensional intensity of a source from a series of fan-beam scans at different angles, one can perform one-dimensional transforms of the scans to obtain values of $\mathcal{V}$ along a series of lines through the origin of the (u, v) plane, thus obtaining the two-dimensional visibility $\mathcal{V}(u, v)$. Then, $I(l, m)$ can be obtained by two-dimensional Fourier transformation. In the early years of radio astronomy, before computers were widely available, such computation was a very laborious task, and various alternative procedures for image formation from fan-beam scans were devised (Bracewell 1956; Bracewell and Riddle 1967).

As this introductory chapter has shown, much of the theory of interferometry is concerned with data in two forms or domains. Within the literature, there is some variation in the associated terminology. The observations provide data in the *visibility* domain, also variously referred to as the *spatial frequency*, (u, v), or *correlation* domain. The astronomical results are shown in the *image* domain, also variously referred to as the *brightness*, *intensity*, *sky*, or *map* domain. "Map" was appropriate in earlier years when the image was sometimes in the form of contours of intensity.

2.4.2   Three-Dimensional Imaging

Three-dimensional images can be made of objects that are optically thin and rotating. An image taken at a particular time is the projected image along the line of sight. A series of images taken at different projection angles can be combined to obtained an estimate of the three-dimensional distribution of emitters in the source. This can be done in a straightforward fashion by use of the three-dimensional generalization of the projection-slice theorem, described in Sect. 2.4.1, to build up a three-dimensional visibility function. Such a technique was developed and first used to image the radiation belts of Jupiter by Sault et al. (1997). A somewhat different tomographic technique was developed by de Pater et al. (1997). The techniques were compared by de Pater and Sault (1998). These techniques might be applicable to extended stellar atmospheres observed with VLBI arrays.

Appendix 2.1   A Practical Fourier Transform Primer

This appendix is intended to provide a brief introduction to the principles of Fourier transform theory most relevant to radio interferometry. For more comprehensive treatment, see Bracewell (1995, 2000), Champeney (1973), and Papoulis (1962).

The Fourier transform of a function $f(x)$ can be written as

$$F(s) = \int_{-\infty}^{\infty} f(x)\, e^{-j2\pi sx} dx \; . \tag{A2.1}$$

The inverse transform is

$$f(x) = \int_{-\infty}^{\infty} F(s)\, e^{j2\pi sx} ds \; . \tag{A2.2}$$

The transform pair is written symbolically as

$$f(x) \longleftrightarrow F(s) \; . \tag{A2.3}$$

If x has units of meters, then s has units of cycles/meter; if x has units of time, then s has units of cycles/second, i.e., hertz. The Fourier transform pair can also be written in the form normally used in the time-frequency domains as

$$F(\omega) = \int_{-\infty}^{\infty} f(t)\, e^{-j\omega t} dt \; , \tag{A2.4}$$

$$f(t) = \frac{1}{2\pi} \int_{-\infty}^{\infty} F(\omega)\, e^{j\omega t} d\omega \; . \tag{A2.5}$$

In this case, the frequency is an angular frequency in radians/sec. We use the formulation in Eqs. (A2.1) and (A2.2) for three reasons: It is widely used in image analysis, it allows for easier tracking of 2π factors, and it provides a more natural segue to the discussion of the discrete Fourier transform (see Appendix 8.4).

We can check that $f(x)$ can be recovered from $F(s)$ by the substitution of Eq. (A2.1) into Eq. (A2.2),

$$f(x) = \int_{-\infty}^{\infty} \left[\int_{-\infty}^{\infty} f(x') e^{-j2\pi sx'} \, dx' \right] e^{j2\pi sx} ds ,$$

(A2.6)

where we switched the variable x to x' to allow us to interchange the order of integration, thereby obtaining

$$f(x) = \int_{-\infty}^{\infty} f(x') \left[\int_{-\infty}^{\infty} e^{-j2\pi s(x'-x)} ds \right] dx' .$$

(A2.7)

The integral in brackets can be evaluated by a limit process, i.e.,

$$\int_{-\infty}^{\infty} e^{-j2\pi s(x'-x)} ds = \lim_{s_0 \to \infty} \int_{-s_0}^{s_0} e^{j2\pi s(x'-x)} ds$$

$$= \lim_{s_0 \to \infty} 2s_0 \left[\frac{\sin 2\pi s_0(x'-x)}{2\pi s_0(x'-x)} \right] .$$

(A2.8)

The function in the brackets is a sinc function (see Fig. A2.1) centered at $x' = x$, having a width between first nulls of $2/s_0$ and an integral, which happens to equal the area of the triangle formed by the peak and the first nulls, of unity. The limit of this function can be used as a definition of the Dirac delta function (often called the impulse function in much of engineering literature),

$$\delta(x'-x) \equiv \lim_{s_0 \to \infty} 2s_0 \left[\frac{\sin 2\pi s_0(x'-x)}{2\pi s_0(x'-x)} \right] ,$$

(A2.9)

which is undefined at $x' = x$ and has the properties

$$\delta(x'-x) = 0 , \qquad\qquad x' \neq x$$

(A2.10a)

and

$$\int_{-\infty}^{\infty} \delta(x'-x) \, dx' = 1 .$$

(A2.10b)

Substitution of Eqs. (A2.9) and (A2.8) into Eq. (A2.7) gives

$$f(x) = \int_{-\infty}^{\infty} f(x') \, \delta(x'-x) \, dx' .$$

(A2.11)

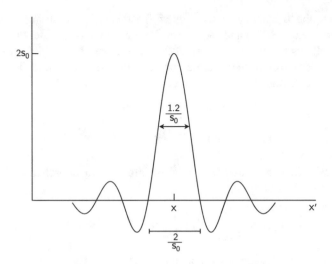

Fig. A2.1 The sinc function in Eq. (A2.9), whose limiting form is a delta function, $\delta(x' - x)$.

Since $\delta(x' - x)$ is nonzero only at $x' = x$, it is clear from Eq. (A2.10b) that we can factor $f(x)$ out of the integral in Eq. (A2.11), which gives the desired result, $f(x) = f(x)$, and proves that $f(x)$ can be recovered from its transform, $F(s)$. Equation (A2.11) is called the sifting property of $\delta(x)$.

A2.1.1   Useful Fourier Transform Pairs

We mention five Fourier transform pairs of particular interest to readers of this book. The first pair is

$$f(x) = 1 , \qquad\qquad |x| \leq \frac{x_0}{2},$$

$$= 0 , \qquad\qquad \text{otherwise,} \qquad\qquad \text{(A2.12a)}$$

$$F(s) = x_0 \frac{\sin \pi s x_0}{\pi s x_0} = x_0 \text{sinc}(s x_0) . \qquad\qquad \text{(A2.12b)}$$

$f(x)$ is called a boxcar or unit rectangular function and denoted as $\prod(x)$.

The second Fourier transform is of a Gaussian function

$$f(x) = e^{-\frac{x^2}{2a^2}} , \qquad\qquad \text{(A2.13a)}$$

$$F(s) = \sqrt{2\pi} \, a \, e^{-2\pi^2 a^2 s^2} . \qquad\qquad \text{(A2.13b)}$$

$F(s)$ can be calculated by a procedure called "completing the square":

$$F(s) = \int_{-\infty}^{\infty} e^{-\frac{x^2}{2a^2}} e^{-j2\pi sx} dx .$$
(A2.14)

The term in the exponent is $(x^2 + j4\pi a^2 sx)/2a^2 = [(x - j2\pi a^2 s)^2 + 4\pi^2 a^4 s^2]/2a^2$. The term involving $4\pi^2 a^4 s^2$ can be factored out of the integral, which leads to Eq. (A2.13b).

The third useful Fourier transform pair is

$$f(x) = \cos 2\pi s_0 x ,$$
(A2.15a)

$$F(s) = \frac{1}{2}[\delta(s - s_0) + \delta(s + s_0)] .$$
(A2.15b)

$F(s)$ is calculated by writing $f(x)$ in terms of exponentials and by use of the same limiting process used in deriving Eq. (A2.9).

The fourth Fourier transform pair is for an infinite train of delta functions, which is also an infinite train of delta functions, i.e.,

$$\sum_{k=-\infty}^{\infty} \delta(x - kx_0) \longleftrightarrow \sum_{m=-\infty}^{\infty} \delta\left(s - \frac{m}{x_0}\right) .$$
(A2.16)

This relation can be proved by starting with a finite train of impulses and applying the shift property [Eq. (A2.22)]. The Fourier transform is an infinite series of sinc functions at intervals of x_0^{-1}. Then, by the same process used in Eq. (A2.9), the sinc functions become Dirac delta functions in the limit as $k \to \infty$.

The fifth Fourier transform pair is for the Heaviside step function

$$f(x) = 1 , \qquad\qquad x \geq 0 ,$$
$$f(x) = 0 , \qquad\qquad x < 0 ,$$
(A2.17a)

$$F(s) = \frac{1}{2}\delta(s) + \frac{1}{j2\pi s} .$$
(A2.17b)

The calculation of $F(s)$ requires some care. Decompose $f(x)$ into $f_e(x) = \frac{1}{2}$ and $f_o(x) = \frac{1}{2}\text{sgn}(x) \equiv \frac{1}{2}$ for $x \geq 0$ and $-\frac{1}{2}$ for $x < 0$. The Fourier transform of $f_e(x)$ is $F_e(s) = \frac{1}{2}\delta(s)$. We replace $f_o(x)$ with the functions $\frac{1}{2}e^{-ax}$, $x \geq 0$, and $-\frac{1}{2}e^{ax}$, $x < 0$, and evaluate $F_o(s)$ in a limit as $a \to 0$. Hence

$$F_o(s) = \lim_{a \to 0}\left[-\int_{-\infty}^{0} e^{ax} e^{-j2\pi sx} dx + \int_{0}^{\infty} e^{-ax} e^{-j2\pi sx} dx \right]$$

$$= \lim_{a \to 0} -\frac{j2\pi s}{a^2 + (2\pi s)^2} = \frac{1}{2\pi js} .$$
(A2.18)

Combining these results gives $F(s) = F_e(s) + F_o(s)$, which proves Eq. (A2.17b).

A2.1.2 Basic Fourier Transform Properties

We list several important properties that are readily provable.

- **Integral property**

$$F(0) = \int_{-\infty}^{\infty} f(x)\, dx \,, \tag{A2.19a}$$

$$f(0) = \int_{-\infty}^{\infty} F(s)\, ds \,. \tag{A2.19b}$$

The application of Eq. (A2.19) to example five above [Eq. (A2.17)] gives the interesting result that $f(0) = \frac{1}{2}$ [see Bracewell (2000) for a discussion of this point].

- **Linearity property**. If $f(x)$ and $g(x)$ have transforms $F(s)$ and $G(s)$, then

$$af(x) \longleftrightarrow aF(s) \,, \tag{A2.20}$$

and

$$f(x) + g(x) \longleftrightarrow F(s) + G(s) \,. \tag{A2.21}$$

Equation (A2.21) is fundamental and particularly useful. In terms of interferometry, it means that the visibility function is the sum of the visibility functions of all the components in the image.

- **Shift property**

$$f(x - x_0) \longleftrightarrow e^{-j2\pi s x_0} F(s) \,, \tag{A2.22a}$$

and

$$F(s - s_0) \longleftrightarrow e^{j2\pi s_0 x} f(x) \,. \tag{A2.22b}$$

- **Modulation property**. From the shift property, it follows that

$$f(x) \cos s_0 x \longleftrightarrow \tfrac{1}{2} \left[F(s - s_0) + F(s + s_0) \right] \,. \tag{A2.23}$$

- **Similarity property**

$$f(ax) \longleftrightarrow \frac{1}{|a|} F\left(\frac{s}{a}\right) \,. \tag{A2.24}$$

This important relation shows that if a function $f(x)$ narrows, then $F(s)$ broadens proportionally and vice versa, so that the product of the widths of functions in the x and s domains, Δx and Δs, respectively, satisfies the relation

$$\Delta x \, \Delta s \sim 1 \,. \tag{A2.25}$$

This result is the basis of the uncertainty principle in quantum mechanics, a wave theory. It is called the time-bandwidth product in signal-processing applications and the ambiguity function in radar astronomy. If Δx and Δs are defined as the full width at half-maximum (FWHM), then for the boxcar–sinc function pair [Eq. (A2.12)], $\Delta x \Delta s = 1.21$, and for the Gaussian function pair [Eq. (A2.13)], $\Delta x \Delta s = 4 \ln 2 / \pi = 0.88$.

- **Derivative property**

$$\frac{d^n f}{dx^n} \longleftrightarrow (j2\pi s)^n F(s) , \tag{A2.26}$$

and

$$\frac{d^n F}{ds^n} \longleftrightarrow (-j2\pi x)^n f(x) . \tag{A2.27}$$

- **Symmetry properties**. Symmetry properties are very useful in calculating and visualizing Fourier transforms. Any function can be divided into even and odd components, $f_e(x)$ and $f_o(x)$, respectively, which are defined as

$$f_e(x) = \tfrac{1}{2} [f(x) + f(-x)] , \tag{A2.28a}$$

$$f_o(x) = \tfrac{1}{2} [f(x) - f(-x)] . \tag{A2.28b}$$

Hence, if $f(x)$ is real and even, then $F(s)$ is also real and even. If $f(x)$ is real and odd, then $F(s)$ is imaginary and odd. The Fourier transform pair in Eq. (A2.17) is a nice example of these symmetry properties.

- **Moment property**. The moments of $f(x)$ are

$$m_n = \int_{-\infty}^{\infty} x^n f(x) \, dx . \tag{A2.29}$$

Hence, from the derivative and the integral properties,

$$\frac{d^n F(0)}{ds^n} \longleftrightarrow (-j2\pi)^n m_n . \tag{A2.30}$$

If these moments exist, then the Taylor expansion of $F(s)$ is

$$F(s) - \sum_{n=0}^{\infty} \frac{(-j2\pi)^n}{n!} m_n s^n . \tag{A2.31}$$

Hence, if $f(x)$ is an even function and its moments exist, the lead terms of $F(s)$ are

$$F(s) = m_0 - 2\pi^2 m_2 s^2 . \tag{A2.32}$$

- **Convolution property**. The convolution of two functions, $f(x)$ and $g(x)$, which have Fourier transforms $F(s)$ and $G(s)$, respectively, is defined as

$$h(y) = \int_{-\infty}^{\infty} f(x)g(y-x)\,dx \,,$$ (A2.33)

which can be written with the convolution operator, $*$, as

$$h(y) = f(y) * g(y) \,.$$ (A2.34)

Note that $f * g = g * f$. The convolution property is

$$f(y) * g(y) \longleftrightarrow F(s)G(s) \,.$$ (A2.35)

This property can be demonstrated as follows. The Fourier transform of $h(y)$ is

$$H(s) = \int_{-\infty}^{\infty} \left[\int_{-\infty}^{\infty} f(x)g(y-x)\,dx \right] e^{-j2\pi sy} dy \,,$$ (A2.36)

or, interchanging the order of integration,

$$H(s) = \int_{-\infty}^{\infty} f(x) \left[\int_{-\infty}^{\infty} g(y-x)\,e^{-j2\pi sy} dy \right] dx \,.$$ (A2.37)

We make the variable substitution, $z = y - x$, to obtain

$$H(s) = \int_{-\infty}^{\infty} f(x) \left[\int_{-\infty}^{\infty} g(z)\,e^{-j2\pi sz} dz \right] e^{-j2\pi sx}\,dx \,.$$ (A2.38)

The term in brackets is $G(s)$, which can be factored out of the remaining integral, which is $F(s)$, so

$$H(s) = F(s)\,G(s) \,.$$ (A2.39)

Hence, the Fourier transform of the convolution of two functions is the product of their Fourier transforms. This relationship, known as the convolution theorem, is shown diagrammatically in Fig. A2.2. It follows that the convolution of two functions in the frequency domain corresponds to multiplication in the time domain.

- **Correlation property**. The correlation function is defined as

$$r(y) = \int_{-\infty}^{\infty} f(x)\,g(x-y)\,dx \,,$$ (A2.40)

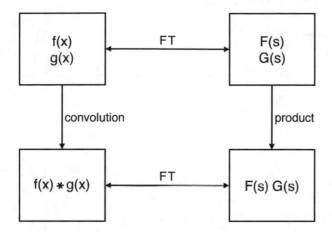

Fig. A2.2 Relationships involving Fourier transforms and convolution. As elsewhere in this book, the in-line asterisk indicates convolution.

which can be written with the correlation operator, $\star$, as

$$r(y) = f(x) \star g(x) \,. \tag{A2.41}$$

The correlation property is

$$f(x) \star g(x) \longleftrightarrow F(s)G^*(s) \,. \tag{A2.42}$$

The Fourier transform of Eq. (A2.40) is

$$R(s) = \int_{-\infty}^{\infty} \left[\int_{-\infty}^{\infty} f(x)\, g(x-y)\, dx \right] e^{-j2\pi sy} dy \,. \tag{A2.43}$$

Interchanging the order of integration and making the substitution $z = x-y$ gives

$$R(s) = \int_{-\infty}^{\infty} f(x) \left[\int_{-\infty}^{\infty} g(z)\, e^{j2\pi z} dz \right] e^{-j2\pi sx} dx \,, \tag{A2.44}$$

which results in

$$R(s) - F(s)\, G^*(s) \,. \tag{A2.45}$$

This relationship is shown in Fig. 8.1. An example where $f(x) = g(x) =$ boxcar is shown in Fig. A2.3. Since $f(x)$ is an even function, convolution and correlation are the same, both producing even functions. Hence, $F(s)$ is real and even, and $F(s)F(s) = F(s)F^*(s)$.

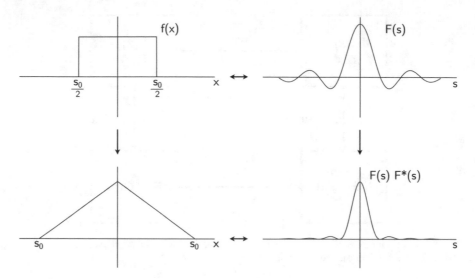

Fig. A2.3 Example of the correlation and convolution theorems for an even function $f(x)$. The vertical arrow on the left indicates $f * f$ for the case of convolution and $f \star f$ for correlation. The vertical arrow on the right indicates $F(s)F(s)$ for convolution and $F(s)F^*(s)$ for correlation.

- **Parseval's theorem**. The relationship

$$\int_{-\infty}^{\infty} |f(x)|^2 dx = \int_{-\infty}^{\infty} |F(s)|^2 ds \qquad (A2.46)$$

is known generally as Parseval's theorem.[5] To prove it, we write

$$\int_{-\infty}^{\infty} f(x)f^*(x)\, dx = \int_{-\infty}^{\infty} \left[\int_{-\infty}^{\infty} F(s)e^{j2\pi sx} ds \right] \left[\int_{-\infty}^{\infty} F^*(s')e^{-j2\pi s'x} ds' \right] dx \qquad (A2.47)$$

or

$$\int_{-\infty}^{\infty} f(x)f^*(x)\, dx = \int_{-\infty}^{\infty} \int_{-\infty}^{\infty} F(s)F^*(s') \left[\int_{-\infty}^{\infty} e^{j2\pi(s-s')x} dx \right] ds\, ds' . \qquad (A2.48)$$

The integral in brackets is $\delta(s - s')$, so that

$$\int_{-\infty}^{\infty} f(x)f^*(x)\, dx = \int_{-\infty}^{\infty} F(s)F^*(s)\, ds . \qquad (A2.49)$$

---

[5]Parseval's theorem originally applied to Fourier series (see Appendix A2.1.4). Rayleigh generalized it for application to Fourier transforms. Mathematicians often refer to it as Plancherel's theorem. As is common practice, we use only the name "Parseval's theorem" in this book.

A useful theorem in interferometry is the projection–slice theorem, which is proved in Sect. 2.4.1.

A2.1.3   Two-Dimensional Fourier Transform

The two-dimensional Fourier transform between $f(x, y)$ and $F(u, v)$ can be written

$$F(u, v) = \int_{-\infty}^{\infty} \int_{-\infty}^{\infty} f(x, y) \, e^{-j2\pi(ux+vy)} \, dx \, dy \, ,$$

$$f(x, y) = \int_{-\infty}^{\infty} \int_{-\infty}^{\infty} F(u, v) \, e^{j2\pi(ux+vy)} \, du \, dv \, .$$

$$(A2.50)$$

If x and y are in radians, then u and v are in units of cycles/radian. We write symbolically

$$f(x, y) \longleftrightarrow F(u, v) \, . \qquad (A2.51)$$

All of the properties in Appendix A2.1.2 have analogs in the two-dimensional Fourier transform. For example, the shift theorem is

$$f(x - x_0, y - y_0) \longleftrightarrow e^{-j2\pi(ux_0 + vy_0)} F(u, v) \, . \qquad (A2.52)$$

The two-dimensional Fourier transform can be converted to polar coordinates by defining $x = r\cos\theta$, $y = r\sin\theta$, $u = q\cos\phi$, and $v = q\sin\phi$, which leads to

$$F(q, \phi) = \int_{0}^{2\pi} \int_{0}^{\infty} f(r, \theta) \, e^{-j2\pi rq(\theta-\phi)} r \, dr \, d\theta \, . \qquad (A2.53)$$

If $f(r, \theta) = f(r)$, i.e., f is azimuthally symmetric, then

$$F(q, \phi) = \int_{0}^{\infty} f(r) \, r \, dr \int_{0}^{2\pi} e^{-j2\pi rq(\theta-\phi)} d\theta \, . \qquad (A2.54)$$

Since the zeroth-order Bessel function is defined as

$$J_0(z) = \frac{1}{2\pi} \int_{0}^{2\pi} e^{-jz\cos\theta} d\theta \, , \qquad (A2.55)$$

$F(q, \phi) = F(q)$ and

$$F(q) = 2\pi \int_{0}^{\infty} f(r) J_0(2\pi qr) r \, dr \, . \qquad (A2.56a)$$

By symmetry,

$$f(r) = 2\pi \int_0^\infty F(q) J_0(2\pi qr) q \, dq \ . \tag{A2.56b}$$

Equations (A2.56a) and (A2.56b) are called the Hankel transform pair.

A2.1.4   Fourier Series

The Fourier series is a special case of the Fourier transform. A periodic function $f(x)$, which repeats over the interval $-x_0/2, x_0/2$, has the complex Fourier series representation

$$f(x) = \sum_{-\infty}^{\infty} \alpha_k \, e^{\frac{j2\pi kx}{x_0}} \ , \tag{A2.57}$$

where

$$\alpha_k = \int_{-\frac{x_0}{2}}^{\frac{x_0}{2}} f(x) \, e^{-\frac{j2\pi kx}{x_0}} \, dx \ . \tag{A2.58}$$

If we define $f_0(x)$ as $f(x)$ over the interval $-x_0/2, x_0/2$, then its Fourier transform, $F(s)$, is given by

$$F(s) = \sum_{k=0}^{\infty} F_0(ks_0) \, \delta(s - ks_0) \ , \tag{A2.59}$$

where $s_0 = 1/x_0$ and $F_0(ks_0) = \alpha_k$. This is called a line spectrum: $F(s)$ consists of delta functions spaced at intervals $s = 1/x_0$ with amplitudes corresponding to the Fourier coefficients. Parseval's theorem for the Fourier series can be found by substituting Eqs. (A2.57) and (A2.59) into Eq. (A2.49), yielding

$$\sum_{-\infty}^{\infty} \alpha_k^2 = \int_{-\frac{x_0}{2}}^{\frac{x_0}{2}} f(x) f^*(x) \, dx \ . \tag{A2.60}$$

A2.1.5   Truncated Functions

The Fourier transform theory described above can be applied to functions that are random processes. If an ergodic random process has an associated temporal function $f(x)$, that function generally extends to infinity, and $\int |f(x)|^2 = \infty$, which presents certain theoretical difficulties. These difficulties are mitigated by choosing a truncated version of the function

$$f_T(x) = f(x)\Pi(x/x_0) , \tag{A2.61}$$

where $\Pi(x)$ is the boxcar function defined after Eq. (A2.12). By the convolution property [Eq. (A2.35)],

$$F_T(s) = F(s) * \text{sinc}(sx_0) . \tag{A2.62}$$

Truncation has the effect of smoothing, or limiting the resolution of, $F(s)$.

The power spectrum of a truncated function is usually defined as

$$P_T(s) = \frac{1}{T}F(s)F^*(s) , \tag{A2.63}$$

which has units of power and does not depend on T. Note that the Fourier transform as defined for deterministic functions in previous sections is actually an energy density spectrum. The conditions under which the Fourier transform of an autocorrelation function and its power spectrum exist for random processes were first explored and clarified by Wiener and Khinchin. Hence, the Fourier transform between the autocorrelation function of a random process and its power spectrum is formally called the Wiener–Khinchin theorem (or relation).

References

Bracewell, R.N., Strip Integration in Radio Astronomy, *Aust. J. Phys.*, **9**, 198–217 (1956)
Bracewell, R.N., Radio Interferometry of Discrete Sources, *Proc. IRE*, **46**, 97–105 (1958)
Bracewell, R.N., *Two-Dimensional Imaging*, Prentice-Hall, Englewood Cliffs, NJ (1995)
Bracewell, R.N., *The Fourier Transform and Its Applications*, McGraw-Hill, New York (2000) (earlier eds. 1965, 1978)
Bracewell, R.N., and Riddle, A.C., Inversion of Fan Beam Scans in Radio Astronomy, *Astrophys. J.*, **150**, 427–434 (1967)

Bracewell, R.N., and Roberts, J.A., Aerial Smoothing in Radio Astronomy, *Aust. J. Phys.*, **7**, 615–640 (1954)

Champeney, D.C., *Fourier Transforms and Their Physical Applications*, Academic Press, London (1973)

de Pater, I., and Sault, R.J., An Intercomparison of Three-Dimensional Reconstruction Techniques Using Data and Models of Jupiter's Synchrotron Radiation, *J. Geophys. Res.*, **103**, 19973–19984 (1998)

de Pater, I., van der Tak, F., Strom, R.G., and Brecht, S.H., The Evolution of Jupiter's Radiation Belts after the Impact of Comet D/Shoemaker–Levy 9, *Icarus*, **129**, 21–47 (1997)

Papoulis, A., *The Fourier Integral and Its Applications*, McGraw-Hill, New York (1962)

Sault, R.J., Oosterloo, T., Dulk, G.A., and Leblanc, Y., The First Three-Dimensional Reconstruction of a Celestial Object at Radio Wavelengths: Jupiter's Radiation Belts, *Astron. Astrophys.*, **324**, 1190–1196 (1997)

Chapter 3
Analysis of the Interferometer Response

In this chapter, we introduce the full two-dimensional analysis of the interferometer response, without small-angle assumptions, and then investigate the small-field approximations that simplify the transformation from the measured visibility to the intensity distribution. There is a discussion of the relationship between the cross-correlation of the received signals and the cross power spectrum, which results from the Wiener–Khinchin relation and is fundamental to spectral line interferometry. An analysis of the basic response of the receiving system is also given. The appendix considers some approaches to the representation of noiselike signals, including the analytic signal, and truncation of the range of integration.

3.1 Fourier Transform Relationship Between Intensity and Visibility

3.1.1 General Case

We begin by deriving the relationship between intensity and visibility in a coordinate-free form and then show how the choice of a coordinate system results in an expression in the familiar form of the Fourier transform. Suppose that the antennas track the source under observation, which is the most common situation, and let the unit vector $\mathbf{s}_0$ in Fig. 3.1 indicate the phase reference position introduced in Sect. 2.3. This position, sometimes also known as the phase-tracking center, becomes the center of the field to be imaged. For one polarization, an element of the source of solid angle $d\Omega$ at position $\mathbf{s} = \mathbf{s}_0 + \boldsymbol{\sigma}$ contributes a component of power $\frac{1}{2}A(\boldsymbol{\sigma})I(\boldsymbol{\sigma})\Delta\nu d\Omega$ at each of the two antennas. Here, $A(\boldsymbol{\sigma})$ is the effective collecting area of each antenna, $I(\boldsymbol{\sigma})$ is the source intensity distribution as observed

© The Author(s) 2017

89

A.R. Thompson, J.M. Moran, and G.W. Swenson Jr., *Interferometry and Synthesis in Radio Astronomy*, Astronomy and Astrophysics Library, DOI 10.1007/978-3-319-44431-4_3

Fig. 3.1 Baseline and
position vectors that specify
the interferometer and the
source. The source is
represented by the outline on
the celestial sphere.

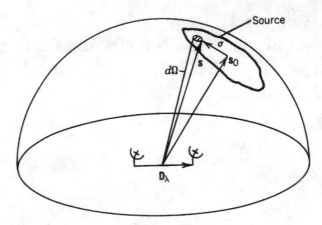

from the distance of the antennas, and $\Delta\nu$ is the bandwidth of the receiving system.
It is easily seen that this expression has the dimensions of power since the units
of I are W m^{-2} Hz^{-1} sr^{-1}. From the considerations outlined in the derivation of
Eqs. (2.1) and (2.2), including the far-field condition for the source, the resulting
component of the correlator output is proportional to the received power and to the
fringe term $\cos(2\pi\nu\tau_g)$, where τ_g is the geometric delay. The vector $\mathbf{D}_\lambda$ will specify
the baseline measured in wavelengths, and then $\nu\tau_g = \mathbf{D}_\lambda \cdot \mathbf{s} = \mathbf{D}_\lambda \cdot (\mathbf{s}_0 + \boldsymbol{\sigma})$. Thus,
the output from the correlator is represented by

$$r(\mathbf{D}_\lambda, \mathbf{s}_0) = \Delta\nu \int_{4\pi} A(\boldsymbol{\sigma})I(\boldsymbol{\sigma}) \cos\left[2\pi\,\mathbf{D}_\lambda \cdot (\mathbf{s}_0 + \boldsymbol{\sigma})\right] d\Omega$$

$$= \Delta\nu \cos(2\pi\mathbf{D}_\lambda \cdot \mathbf{s}_0) \int_{4\pi} A(\boldsymbol{\sigma})I(\boldsymbol{\sigma}) \cos(2\pi\mathbf{D}_\lambda \cdot \boldsymbol{\sigma})\, d\Omega$$

$$- \Delta\nu \sin(2\pi\mathbf{D}_\lambda \cdot \mathbf{s}_0) \int_{4\pi} A(\boldsymbol{\sigma})I(\boldsymbol{\sigma}) \sin(2\pi\mathbf{D}_\lambda \cdot \boldsymbol{\sigma})\, d\Omega \;. \qquad (3.1)$$

Note that the integration of the response to the element $d\Omega$ over the source in
Eq. (3.1) requires the assumption that the source is spatially incoherent, that is,
that the radiated waveforms from different elements $d\Omega$ are uncorrelated. This
assumption is justified for essentially all cosmic radio sources. Spatial coherence is
discussed further in Sect. 15.2. Let A_0 be the antenna collecting area in direction
$\mathbf{s}_0$ in which the beam is pointed. We introduce a normalized reception pattern
$A_N(\boldsymbol{\sigma}) = A(\boldsymbol{\sigma})/A_0$ and consider the modified intensity distribution $A_N(\boldsymbol{\sigma})I(\boldsymbol{\sigma})$. Now

we define the complex visibility[1] as

$$\mathcal{V} = |\mathcal{V}|e^{j\phi_v} = \int_{4\pi} A_N(\boldsymbol{\sigma})I(\boldsymbol{\sigma})\, e^{-j2\pi \mathbf{D}_\lambda \cdot \boldsymbol{\sigma}}\, d\Omega \; . \tag{3.2}$$

Then by separating the real and imaginary parts, we obtain

$$\int_{4\pi} A_N(\boldsymbol{\sigma})I(\boldsymbol{\sigma})\, \cos(2\pi\mathbf{D}_\lambda \cdot \boldsymbol{\sigma})\, d\Omega = |\mathcal{V}|\cos\phi_v \; , \tag{3.3}$$

$$\int_{4\pi} A_N(\boldsymbol{\sigma})I(\boldsymbol{\sigma})\, \sin(2\pi\mathbf{D}_\lambda \cdot \boldsymbol{\sigma})\, d\Omega = -|\mathcal{V}|\sin\phi_v \; , \tag{3.4}$$

and from Eq. (3.1)

$$r(\mathbf{D}_\lambda, \mathbf{s}_0) = A_0 \Delta v |\mathcal{V}|\cos(2\pi\mathbf{D}_\lambda \cdot \mathbf{s}_0 - \phi_v) \; . \tag{3.5}$$

Thus, the output of the correlator can be expressed in terms of a fringe pattern corresponding to that for a hypothetical point source in the direction $\mathbf{s}_0$, which is the phase reference position. As noted earlier, this is usually the center or nominal position of the source to be imaged. The modulus and phase of $\mathcal{V}$ are equal to the amplitude and phase of the fringes; the phase is measured relative to the fringe phase for the hypothetical source. As defined above, $\mathcal{V}$ has the dimensions of flux density (W m^{-2} Hz^{-1}), which is consistent with its Fourier transform relationship with I. Some authors have defined visibility as a normalized, dimensionless quantity, in which case it is necessary to reintroduce the intensity scale in the resulting image. Note that the bandwidth has been assumed to be small compared with the center frequency in deriving Eq. (3.5).

In introducing a coordinate system, the geometry we now consider is illustrated in Fig. 3.2. The two antennas track the center of the field to be imaged. They are assumed to be identical, but if they differ, $A_N(\boldsymbol{\sigma})$ is the geometric mean of the beam patterns of the two antennas. The magnitude of the baseline vector is measured in wavelengths at the center frequency of the observing band, and the baseline has components (u, v, w) in a right-handed coordinate system, where u and v are measured in a plane normal to the direction of the phase reference position. The spacing component v is measured toward the north as defined by the plane through the origin, the source, and the pole, and u is measured toward the east.

---

[1]In formulating the fundamental Fourier transform relationship in synthesis imaging, which follows from Eq. (3.2), we use the negative exponent to derive the complex visibility function (or mutual coherence function) from the intensity distribution, and the positive exponent for the inverse operation. From a physical viewpoint, the choice is purely arbitrary, and the literature contains examples of both this and the reverse convention. Our choice follows Born and Wolf (1999) and Bracewell (1958).

Fig. 3.2 Geometric relationship between a source under observation $I(l, m)$ and an interferometer or one antenna pair of an array. The antenna baseline vector, measured in wavelengths, has length D_λ and components (u, v, w).

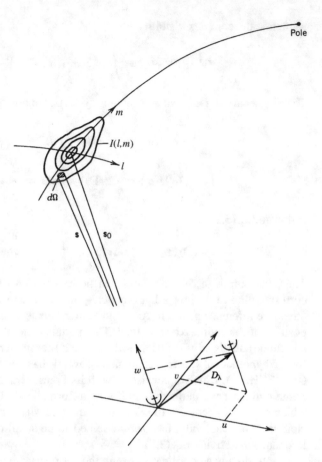

The component w is measured in the direction $\mathbf{s}_0$, which is the phase reference position. On Fourier transformation, the phase reference position becomes the origin of the derived intensity distribution $I(l, m)$, where l and m are direction cosines measured with respect to the axes u and v. In terms of these coordinates, we find

$$\mathbf{D}_\lambda \cdot \mathbf{s}_0 = w$$

$$\mathbf{D}_\lambda \cdot \mathbf{s} = \left(ul + vm + w\sqrt{1 - l^2 - m^2} \right)$$

$$d\Omega = \frac{dl\,dm}{\sqrt{1 - l^2 - m^2}} , \tag{3.6}$$

where $\sqrt{1 - l^2 - m^2}$ is equal to the third direction cosine n measured with respect to the w axis.[2] Note also that $\mathbf{D}_\lambda \cdot \boldsymbol{\sigma} = \mathbf{D}_\lambda \cdot \mathbf{s} - \mathbf{D}_\lambda \cdot \mathbf{s}_0$. Thus, from Eq. (3.2):

$$
\mathcal{V}(u, v, w) = \int_{-\infty}^{\infty} \int_{-\infty}^{\infty} A_N(l, m) I(l, m)
$$
$$
\times \exp\left\{-j2\pi \left[ul + vm + w\left(\sqrt{1 - l^2 - m^2} - 1\right)\right]\right\} \frac{dl\,dm}{\sqrt{1 - l^2 - m^2}}.
$$
$$(3.7)$$

A factor $e^{j2\pi w}$ on the right side in Eq. (3.7) results from the measurement of angular position with respect to the w axis. For a source on the w axis, $l = m = 0$, and the argument of the exponential term in Eq. (3.7) is zero. For any other source, the fringe phase is measured relative to that for a source on the w axis, which is the phase reference position, $\mathbf{s}_0$ in Fig. 3.2. The function $A_N I$ in Eq. (3.7) is zero for $l^2 + m^2 \geq 1$, and in practice, it usually falls to very low values for directions outside the field to be imaged, as a result of the antenna beam pattern, the bandwidth pattern, or the finite size of the source. Thus, we can extend the limits of integration to $\pm\infty$. Note, however, that Eq. (3.7) requires no small-angle assumptions. The reason why we use direction cosines rather than a linear measure of angle in interferometer theory is that they occur in the exponential term of this relationship.

The coordinate system (l, m) defined above is a convenient one in which to present an intensity distribution. It corresponds to the projection of the celestial sphere onto a plane that is a tangent at the field center, as shown in Fig. 3.3. The distance of any point in the image from the (l, m) origin is proportional to the sine of the corresponding angle on the sky, so for small fields, distances on the image are closely proportional to the corresponding angles. The same relationship usually applies to the field of an optical telescope. For a detailed discussion of relationships on the celestial sphere and tangent planes, see König (1962).

If all the measurements could be made with the antennas in a plane normal to the w direction so that $w = 0$, Eq. (3.7) would reduce to an exact two-dimensional Fourier transform. In general, this is not possible, and we now consider ways in which the transform relationship can be applied. Recall first that the basis of the

---

[2]The expression for $d\Omega$ is obtained by considering the unit sphere centered on the (u, v, w) origin. A point P on the sphere with coordinates (u, v, w) is projected onto the (u, v) plane at $u = l, v = m$, and the increments dl, dm define a column of square cross section running through $(u, v, 0)$ parallel to the w axis. The column makes an angle $\cos^{-1} n$ with the normal to the spherical surface at P, and $d\Omega$ is equal to the surface area intersected by the column, which is $dl\,dm/n$, or $dl\,dm/\sqrt{1 - l^2 - m^2}$. Alternately, the solid angle can be expressed in polar coordinates as $d\Omega = \sin\theta\,d\theta\,d\phi$, where θ and ϕ are the polar and azimuthal angles in the (u, v, w) plane, that is, $\theta = \sin^{-1}\sqrt{l^2 + m^2}$ and $\phi = \tan^{-1} m/l$. Calculation of the Jacobian of the transformation from (θ, ϕ) coordinates to (l, m) coordinates gives the result $d\Omega = dl\,dm/\sqrt{1 - l^2 - m^2}$ (Apostol 1962).

Fig. 3.3 Mapping of the celestial sphere onto an image plane, shown in one dimension. The position of the point P is measured in terms of the direction cosine m with respect to the v axis. When projected onto a plane surface with a scale linear in m, P appears at P' at a distance from the field center C proportional to $\sin \psi$.

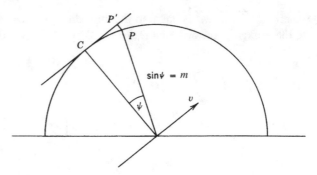

synthesis imaging process is the measurement of $\mathcal{V}$ over a wide range of u and v. For a ground-based array, this can be achieved by varying the length and direction of the antenna spacing and also by tracking the field-center position as the Earth rotates. The rotation causes the projection of $\mathbf{D}_\lambda$ to move across the (u, v) plane, and an observation may last for 6–12 h. As the Earth's rotation carries the antennas through space, the baseline vector remains in a plane only if $\mathbf{D}_\lambda$ has no component parallel to the rotation axis, that is, the baseline is an east–west line on the Earth's surface. In the general case, there is a three-dimensional distribution of the measurements of $\mathcal{V}$. The simplest form of the transform relationship that can then be used is based on an approximation that is valid so long as the synthesized field is not too large. If l and m are small enough that the term

$$\left(\sqrt{(1 - l^2 - m^2)} - 1 \right) w \simeq -\tfrac{1}{2}(l^2 + m^2)w \tag{3.8}$$

can be neglected, Eq. (3.7) becomes

$$\mathcal{V}(u, v, w) \simeq \mathcal{V}(u, v, 0) = \int_{-\infty}^{\infty} \int_{-\infty}^{\infty} \frac{A_N(l, m)I(l, m)}{\sqrt{1 - l^2 - m^2}} e^{-j2\pi(ul+vm)} \, dl \, dm \ . \tag{3.9}$$

Thus, for a restricted range of l and m, $\mathcal{V}(u, v, w)$ is approximately independent of w, and for the inverse transform, we can write

$$\frac{A_N(l, m)I(l, m)}{\sqrt{1 - l^2 - m^2}} = \int_{-\infty}^{\infty} \int_{-\infty}^{\infty} \mathcal{V}(u, v) \, e^{j2\pi(ul+vm)} \, du \, dv \ . \tag{3.10}$$

With this approximation, it is usual to omit the w dependence and write the visibility as the two-dimensional function $\mathcal{V}(u, v)$. Note that the factor $\sqrt{1 - l^2 - m^2}$ in Eqs. (3.9) and (3.10) can be subsumed into the function $A_N(l, m)$. Equation (3.10) is a form of the van Cittert–Zernike theorem, which originated in optics and is discussed in Sect. 15.1.1.

The approximation in Eq. (3.9) introduces a phase error equal to 2π times the neglected term, that is, $\pi(l^2 + m^2)w$. Limitation of this error to some tolerable value places a restriction on the size of the synthesized field, which can be estimated

Fig. 3.4 When observations are made at a low angle of elevation and at an azimuth close to that of the baseline, the spacing component w becomes comparable to the baseline length D_λ, which is measured in wavelengths.

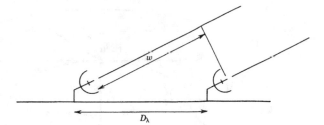

approximately as follows. If the antennas track the source under observation down to low elevation angles, the values of w can approach the maximum spacings $(D_\lambda)_{max}$ in the array, as shown in Fig. 3.4. Also, if the spatial frequencies measured are evenly distributed out to the maximum spacing, the synthesized beamwidth θ_b is approximately equal to $(D_\lambda)_{max}^{-1}$. Thus, the maximum phase error is approximately

$$\pi \left(\frac{\theta_f}{2}\right)^2 \theta_b^{-1} , \qquad (3.11)$$

where θ_f is the width of the synthesized field. The condition that no phase errors can exceed, say, 0.1 rad then requires that

$$\theta_f < \tfrac{1}{3}\sqrt{\theta_b} , \qquad (3.12)$$

where the angles are measured in radians. For example, if $\theta_b = 1''$, $\theta_f < 2.5$ arcmin. Much synthesis imaging in astronomy is performed within this restriction, and ways of imaging larger fields will be discussed later.

3.1.2 East–West Linear Arrays

We now turn to the case of arrays with east–west spacings only and discuss further the conditions for which we can put $w = 0$, and the resulting effects. Let us first rotate the (u, v, w) coordinate system about the u axis until the w axis points toward the pole, as shown in Fig. 3.5. We indicate by primes the quantities measured in the rotated system. The (u', v') axes lie in a plane parallel to the Earth's equator. The east–west antenna spacings contain components in this plane only (i.e., $w' = 0$), and as the Earth rotates, the spacing vectors sweep out circles concentric with the (u', v') origin. From Eq. (3.7), we can write

$$\mathcal{V}(u', v') = \int_{-\infty}^{\infty} \int_{-\infty}^{\infty} A_N(l', m')I(l', m') \, e^{-j2\pi(u'l'+v'm')} \frac{dl'\,dm'}{\sqrt{1 - l'^2 - m'^2}} , \qquad (3.13)$$

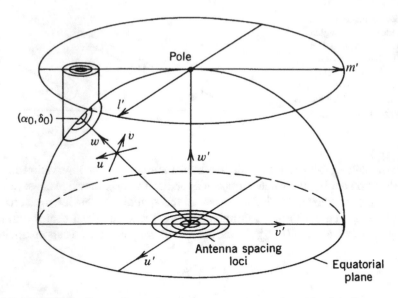

Fig. 3.5 The (u', v', w') coordinate system for an east–west array. The (u', v') plane is the equatorial plane and the antenna spacing vectors trace out arcs of concentric circles as the Earth rotates. Note that the directions of the u' and v' axes are chosen so that the v' axis lies in the plane containing the pole, the observer, and the point under observation (α_0, δ_0). In Fourier transformation from the (u', v') to the (l', m') planes, the celestial hemisphere is imaged as a projection onto the tangent plane at the pole. The (u, v, w) coordinates for observation in the direction (α_0, δ_0) are also shown.

where (l', m') are direction cosines measured with respect to (u', v'). Equation (3.13) holds for the whole hemisphere above the equatorial plane. The inverse transformation yields

$$\frac{A_N(l', m')I(l', m')}{\sqrt{1 - l'^2 - m'^2}} = \int_{-\infty}^{\infty} \int_{-\infty}^{\infty} \mathcal{V}(u', v') \, e^{j2\pi(u'l' + v'm')} du' dv' . \tag{3.14}$$

In this imaging, the hemisphere is projected onto the tangent plane at the pole, as shown in Fig. 3.5. In practice, however, an image may be confined to a small area within the antenna beams. In the vicinity of such an area, centered at right ascension and declination (α_0, δ_0), angular distances in the image are compressed by a factor $\sin \delta_0$ in the m' dimension. Also, in imaging the (α_0, δ_0) vicinity, it is convenient if the origin of the angular position variables is shifted to (α_0, δ_0). Expansion of the scale and shift of the origin can be accomplished by the coordinate transformation

$$l = l', \qquad m'' = (m' - \cos \delta_0) \operatorname{cosec} \delta_0 . \tag{3.15}$$

If we write $F(l', m')$ for the left side of Eq. (3.14), then

$$F(l', m') \longleftrightarrow \mathcal{V}(u', v') , \tag{3.16}$$

and

$$F\left[l', (m' - \cos \delta_0) \operatorname{cosec} \delta_0\right] \longleftrightarrow |\sin \delta_0| \mathcal{V}(u', v' \sin \delta_0) e^{-j2\pi v' \cos \delta_0} , \tag{3.17}$$

where $\longleftrightarrow$ indicates Fourier transformation. Equation (3.17) follows from the behavior of Fourier pairs with change of variable and involves the shift and similarity properties of Fourier transforms (see Appendix 2.1). The coordinates $(u', v' \sin \delta_0)$ on the right side of Eq. (3.17) represent the projection of the equatorial plane onto the (u, v) plane, which is normal to the direction (α_0, δ_0). In the (u, v, w) system, $u = u'$ and $v = v' \sin \delta_0$. The coordinate w shown in Fig. 3.5 is equal to $-v' \cos \delta_0$. Thus, $e^{-j2\pi v' \cos \delta_0}$ in Eq. (3.17) is the same factor that occurs in Eq. (3.7) as a result of the measurement of visibility phase relative to that for a point source in the w direction. Equation (3.14) now becomes

$$\frac{A_N(l, m'')I(l, m'')}{\sqrt{1 - l^2 - m''^2}} = \int_{-\infty}^{\infty} \int_{-\infty}^{\infty} \mathcal{V}(u', v' \sin \delta_0) |\sin \delta_0| e^{-j2\pi v' \cos \delta_0}$$
$$\times e^{j2\pi(u'l' + v'm')} du' dv'$$
$$= \int_{-\infty}^{\infty} \int_{-\infty}^{\infty} \mathcal{V}(u, v) e^{j2\pi(ul + vm'')} du \, dv . \tag{3.18}$$

A similar analysis is given by Brouw (1971).

The derivation of Eq. (3.18) from Eq. (3.14) involves a redefinition of the m coordinate but no approximations. Equation (3.18) is of the same form as Eq. (3.10), in which the term in Eq. (3.8) was neglected. Thus, if we apply the imaging scheme of Eq. (3.10), which is based on omitting this term, to observations made with an east–west array, the phase errors introduced distort the image in a way that corresponds exactly to the change of definition of the m variable to m''. Since m'' is derived from a direction cosine measured from the v' axis in the equatorial plane, there is a progressive change in the north–south angular scale over the image. The factor cosec δ_0 in Eq. (3.15) establishes the correct angular scale at the center of the image, but this simple correction is acceptable only for small fields. The crucial point to note here is that when visibility data measured in a plane are projected into (u, v, w) coordinates, w is a linear function of u and v (and a linear function of v alone for east–west baselines). Hence, the phase error $\pi(l^2 + m^2)w$ is linear in u and v. Phase errors of this kind have the effect of introducing position shifts in the resulting image, but there remains a one-to-one correspondence between points in the image and on the sky. The effect is simply to produce a predictable, and hence correctable, distortion of the coordinates.

It is clear from Fig. 3.5 that if all the measurements lie in the (u', v') plane, then the values of v in the (u, v) plane become seriously foreshortened for directions close to the celestial equator. Obtaining two-dimensional resolution in such directions requires components of antenna spacing parallel to the Earth's axis. The design of such arrays is discussed in Chap. 5. The effect of the Earth's rotation is then to distribute the measurements in (u, v, w) space so that they no longer lie in a plane, unless the observation is of short time duration. In some cases, the restriction of the synthesized field in Eq. (3.12) is acceptable. In other cases, it may be necessary to image the entire beam to avoid source confusion, and several techniques are possible based on the following approaches:

1. Equation (3.7) can be written in the form of a three-dimensional Fourier transform. The resulting intensity distribution is then taken from the surface of a unit sphere in (l, m, n) space.
2. Large images can be constructed as mosaics of smaller ones that individually comply with the field restriction for two-dimensional transformation. The centers of the individual images must be taken at tangent points on the same unit sphere referred to in 1.
3. Since in most terrestrial arrays the antennas are mounted on an approximately plane area of ground, measurements taken over a short time interval lie close to a plane in (u, v, w) space. It is therefore possible to analyze an observation lasting several hours as a series of short duration images, which are subsequently combined after adjustment of the coordinate scales.

Practical implementation of the three approaches outlined above requires the nonlinear deconvolution techniques described in Chap. 11. A more detailed discussion of the resulting methods is given in Sect. 11.7.

3.2   Cross-Correlation and the Wiener–Khinchin Relation

The Fourier transform relationship between the power spectrum of a waveform and its autocorrelation function, the Wiener–Khinchin relation, is expressed in Eqs. (2.6) and (2.7). It is also useful to examine the corresponding relation for the cross-correlation function of two different waveforms. The response of a correlator, as used in a radio interferometer, can be written as

$$r(\tau) = \lim_{T \to \infty} \frac{1}{2T} \int_{-T}^{T} V_1(t) V_2^*(t - \tau) \, dt \, , \tag{3.19}$$

where the superscript asterisk indicates the complex conjugate. In practice, the correlation is measured for a finite time period $2T$, which is usually a few seconds or minutes but is long compared with both the period and the reciprocal bandwidth of the waveforms. The factor $1/2T$ is sometimes omitted, but for the waveforms considered here, it is required to obtain convergence. Cross-correlation is represented by the pentagram symbol ($\star$):

$$V_1(t) \star V_2(t) = \lim_{T \to \infty} \frac{1}{2T} \int_{-T}^{T} V_1(t) V_2^*(t - \tau) \, dt \ . \tag{3.20}$$

This integral can be expressed as a convolution in the following way:

$$V_1(t) \star V_2(t) = \lim_{T \to \infty} \frac{1}{2T} \int_{-\infty}^{\infty} V_1(t) V_{2-}^*(\tau - t) \, dt = V_1(t) * V_{2-}^*(t) \ , \tag{3.21}$$

where $V_{2-}(t) = V_2(-t)$. Now the ν, t Fourier transforms are as follows[3]: $V_1(t) \longleftrightarrow \widehat{V}_1(\nu)$, $V_2(t) \longleftrightarrow \widehat{V}_2(\nu)$, and $V_{2-}^*(t) \longleftrightarrow \widehat{V}_2^*(\nu)$. Then from the convolution theorem,

$$V_1(t) \star V_2(t) \longleftrightarrow \widehat{V}_1(\nu) \widehat{V}_2^*(\nu) \ . \tag{3.22}$$

The right side of Eq. (3.22) is known as the cross power spectrum of $V_1(t)$ and $V_2(t)$. The cross power spectrum is a function of frequency, and we see that it is the Fourier transform of the cross-correlation, which is a function of τ. This is a useful result, and in the case where $V_1 = V_2$, it becomes the Wiener–Khinchin relation. The relationship expressed in Eq. (3.22) is the basis of cross-correlation spectrometry, described in Sect. 8.8.2.

3.3   Basic Response of the Receiving System

From a mathematical viewpoint, the basic components of the interferometer receiving system are the antennas that transform the incident electric fields into voltage waveforms, the filters that select the frequency components to be processed, and the correlator that forms the averaged product of the signals. In the filter and the correlator, the signals may be in either analog or digital form. These components are shown in Fig. 3.6. Most other effects can be represented by multiplicative gain constants, which we shall ignore here, or as variations of the frequency response that can be subsumed into the expressions for the filters. Thus, we assume that the frequency response of the antennas and the strength of the received signal are effectively constant over the filter passband, which is realistic for many continuum observations.

---

[3]In this chapter, in cases where the same letter is used for functions of both time and frequency, the circumflex (hat) accent is used to indicate functions of frequency.

Fig. 3.6 Basic components
of the receiving system of a
two-element interferometer.

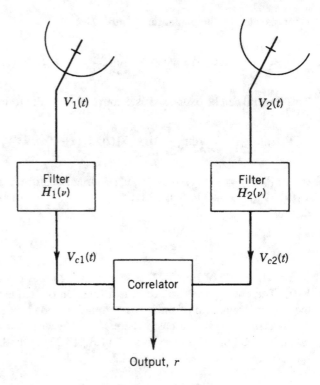

3.3.1 Antennas

In order to consider the responses of the two antennas independently, we should introduce their voltage reception patterns, since the correlator responds to the product of the signal voltages. The voltage reception pattern of an antenna $V_A(l, m)$ has the dimension *length* and responds to the electric field specified in volts per meter. $V_A(l, m)$ is the Fourier transform of the field distribution in the aperture $\overline{\mathcal{E}}(X, Y)$, as shown in Sect. 15.1.2. X and Y are coordinates of position within the antenna aperture. Omitting constant factors, we can write

$$V_A(l, m) \propto \int \int_{-\infty}^{\infty} \overline{\mathcal{E}}(X, Y)\, e^{j2\pi[(X/\lambda)l + (Y/\lambda)m]} dX\, dY \,, \qquad (3.23)$$

where λ is the wavelength. In applying Eq. (3.23), X and Y are measured from the center of each antenna aperture. The power reception pattern is proportional to the squared modulus of the voltage reception pattern. $V_A(l, m)$ is a complex quantity, and it represents the phase of the radio frequency voltage at the antenna terminals as well as the amplitude. For an interferometer (with antennas denoted by subscripts 1 and 2), the response is proportional to $V_{A1}V_{A2}^*$, which is purely real if the antennas are identical. For each antenna, the collecting area $A(l, m)$ is a real quantity. In practice, it is usual to specify the antenna response in terms of $A(l, m)$ and to replace

$V_A(l, m)$ by $\sqrt{A(l, m)}$, which is proportional to the modulus of $V_A(l, m)$. Any phase introduced by differences between the antennas is ignored in this analysis but in effect is combined with the phase responses of the amplifiers, filters, transmission lines, and other elements that make up the signal path to the correlator input. The overall instrumental response of the interferometer in both phase and amplitude is calibrated by observing an unresolved source of known position and flux density.

For the case in which the antennas track the source, both the antenna beam center and the center of the source are at the (l, m) origin. If $E(l, m)$ is the incident field, the output voltage of an antenna can be written (omitting constant gain factors) as

$$\widehat{V} = \int\int_{-\infty}^{\infty} E(l, m) \sqrt{A(l, m)} \, dl \, dm .\tag{3.24}$$

If the antennas do not track the source, a convolution relationship of the form shown in Eq. (2.15) applies.

3.3.2   Filters

The filters in Fig. 3.6 will be regarded as a representation of the overall effect of components that determine the frequency response of the receiving channels, including amplifiers, cables, filters, and other components. The frequency response of a filter will be represented by $H(\nu)$, which can also be called the bandpass function. The output of the filter $\widehat{V}_c(\nu)$ is related to the input $\widehat{V}(\nu)$ by

$$\widehat{V}_c(\nu) = H(\nu)\widehat{V}(\nu) .\tag{3.25}$$

The Fourier transform of $H(\nu)$ with respect to time and frequency is the impulse response of the filter $h(t)$, which is the response to a voltage impulse $\delta(t)$ at the input. Thus, in the time domain, the corresponding expression to Eq. (3.25) is

$$V_c(t) = \int_{-\infty}^{\infty} h(t')V(t - t') \, dt' = h(t) * V(t) ,\tag{3.26}$$

where the centerline asterisk represents convolution. In specifying filters, it is usual to use the frequency response rather than the impulse response because the former is more directly related to the properties of interest in a receiving system and is usually easier to measure.

3.3.3 Correlator

The correlator[4] produces the cross-correlation of the two voltages fed to it. If $V_1(t)$ and $V_2(t)$ are the input voltages, the correlator output is

$$r(\tau) = \lim_{T \to \infty} \frac{1}{2T} \int_{-T}^{T} V_1(t) V_2^*(t - \tau) \, dt \, , \tag{3.27}$$

where τ is the time by which voltage V_2 is delayed with respect to voltage V_1. For continuum observations, τ is maintained small or zero. The functions V_1 and V_2 that represent the signals may be complex. The output of a single multiplying device is a real voltage or number. To obtain the complex cross-correlation, which represents both the amplitude and the phase of the visibility, one can record the fringe oscillations and measure their phase, or use a *complex correlator* that contains two multiplying circuits, as described in Sect. 6.1.7. As follows from Eqs. (3.20) and (3.22), the Fourier transform of $r(\tau)$ is the cross power spectrum, which is required in observations of spectral lines. This can be obtained by inserting a series of instrumental delays in the signal to determine the cross-correlation as a function of τ, as described in Sect. 8.8.3.

3.3.4 Response to the Incident Radiation

We use subscripts 1 and 2 to indicate the two antennas and receiving channels as in Fig. 3.6. The response of antenna 1 to the signal field $E(l, m)$ given by Eq. (3.24) is the voltage spectrum $\widehat{V}(\nu)$. We multiply this by $H(\nu)$ to obtain the signal at the output of the filter, and then take the Fourier transform to go from the frequency to the time domain. Thus

$$V_{c1}(t) = \int_{-\infty}^{\infty} \int_{-\infty}^{\infty} \int_{-\infty}^{\infty} E(l, m) \sqrt{A_1(l, m)} H_1(\nu) e^{j2\pi \nu t} \, dl \, dm \, d\nu \, . \tag{3.28}$$

A similar expression can be written for the signal $V_{c2}(t)$ from antenna 2, and the output of the correlator is obtained from Eq. (3.27). Note also that if the radiation were to have some degree of spatial coherence, we should integrate over (l, m) independently for each antenna (Swenson and Mathur 1968), but here we make

---

[4]The term *correlator* basically refers to a device that measures the complex cross-correlation function $r(\tau)$, as given in Eq. (3.27). It is also used to denote simpler systems in which the time delay τ is zero or where both signals are represented by real functions. Large systems that cross-correlate the signal pairs of multielement arrays may contain 10^7 or more correlator circuits to accommodate many antennas and many spectral channels. Complete systems of this type are also commonly referred to as correlators.

the usual assumption of incoherence. Thus, the correlator output is

$$
r(\tau) = \lim_{T \to \infty} \frac{1}{2T} \int_{-\infty}^{\infty} \int_{-\infty}^{\infty} \int_{-\infty}^{\infty} \int_{-\infty}^{\infty} E(l,m)E^*(l,m)\sqrt{A_1(l,m)A_2(l,m)}
$$

$$
\times \, H_1(v)H_2^*(v)e^{j2\pi vt}\, e^{-j2\pi v(t-\tau)}\, dl\, dm\, dt\, dv
$$

$$
= \int_{-\infty}^{\infty} \int_{-\infty}^{\infty} \int_{-\infty}^{\infty} I(l,m)\sqrt{A_1(l,m)A_2(l,m)}H_1(v)H_2^*(v)\, e^{j2\pi v\tau}\, dl\, dm\, dv \ .
$$

$$
(3.29)
$$

Here, we have replaced the squared field amplitude by the intensity I. The result is a very general one since the use of separate response functions A_1 and A_2 for the two antennas can accommodate different antenna designs, or different pointing offset errors, or both. Also, different frequency responses H_1 and H_2 are used. In the case in which the antennas and filters are identical, Eq. (3.29) becomes

$$
r(\tau) = \int_{-\infty}^{\infty} \int_{-\infty}^{\infty} \int_{-\infty}^{\infty} I(l,m)A(l,m)|H(v)|^2 e^{j2\pi v\tau}\, dl\, dm\, dv \ . \qquad (3.30)
$$

The result is a function of the delay τ of the signal $V_{c2}(t)$ with respect to $V_{c1}(t)$. The geometric component of the delay is generally compensated by an adjustable instrumental delay (discussed in Chaps. 6 and 7), so that $\tau = 0$ for radiation from the direction of the (l, m) origin. For a wavefront incident from the direction (l, m), the difference in propagation times through the two antennas to the correlator results from a difference in path lengths of $(ul + vm)$ wavelengths, for the conditions indicated in Eqs. (3.8) and (3.9). The corresponding time difference is $(ul + vm)/v$. If we take as V_1 the signal from the antenna for which the path length is the greater (for positive l and m), then from Eq. (3.30), the correlator output becomes

$$
r = \int_{-\infty}^{\infty} \int_{-\infty}^{\infty} \int_{-\infty}^{\infty} I(l,m)A(l,m)|H(v)|^2 e^{-j2\pi(lu+mv)}\, dl\, dm\, dv \ . \qquad (3.31)
$$

Equation (3.31) indicates that the correlator output measures the Fourier transform of the intensity distribution modified by the antenna pattern. Let us assume that, as is often the case, the intensity and the antenna pattern are constant over the bandpass range of the filters, and the width of the source is small compared with the antenna beam. The correlator output then becomes

$$
r = \int_{-\infty}^{\infty} \int_{-\infty}^{\infty} I(l,m)A(l,m)e^{-j2\pi(lu+mv)}\, dl\, dm \int_{-\infty}^{\infty} |H(v)|^2 dv
$$

$$
= A_0 \mathcal{V}(u,v) \int_{-\infty}^{\infty} |H(v)|^2 dv \ , \qquad (3.32)
$$

where A_0 is the collecting area of the antennas in the direction of the maximum beam response and $\mathcal{V}$ is the visibility as in Eq. (3.2). The filter response $H(\nu)$ is a dimensionless (gain) quantity. If the filter response is essentially constant over a bandwidth $\Delta\nu$, Eq. (3.32) becomes

$$r = A_0 \mathcal{V}(u, v)\Delta\nu \; . \tag{3.33}$$

$\mathcal{V}(u, v)$ has units of W m^{-2} Hz^{-1}, A_0 has units of m^2, and $\Delta\nu$ has units of Hz. This is consistent with r, the output of the correlator, which is proportional to the correlated component of the received power.

Appendix 3.1   Mathematical Representation of Noiselike Signals

Electromagnetic fields and voltage waveforms that result from the emissions of astronomical objects are generally characterized by variations of a random nature. The received waveforms are usually described as ergodic (time averages and ensemble averages converge to equal values), which implies strict stationarity. For a detailed discussion, see, for example, Goodman (1985). Although such fields and voltages are entirely real, it is often convenient to represent them mathematically as complex functions. These complex functions can be manipulated in exponential form, and it is then necessary to take the real part as a final step in a calculation.

A3.1.1   Analytic Signal

A formulation that is often used in optical and radio signal analysis to represent a function of time is known as the *analytic signal*, which was introduced by Gabor (1946): see, for example, Born and Wolf (1999), Bracewell (2000), or Goodman (1985). Let $V_R(t)$ represent a real function of which the Fourier (voltage) spectrum is

$$\widehat{V}(\nu) = \int_{-\infty}^{\infty} V_R(t)\, e^{-j2\pi\nu t} dt \; . \tag{A3.1}$$

The inverse transform is

$$V_R(t) = \int_{-\infty}^{\infty} \widehat{V}(\nu)\, e^{j2\pi\nu t} d\nu \; . \tag{A3.2}$$

To form the analytic signal, the imaginary part that is added to produce a complex function is the Hilbert transform [see, e.g., Bracewell (2000)] of $V_R(t)$. One way of forming the Hilbert transform is to multiply the Fourier spectrum of the original function by $j\,\mathrm{sgn}(\nu)$.[5] In forming the Hilbert transform of a function, the amplitudes of the Fourier spectral components are unchanged, but the phases are shifted by $\pi/2$, with the sign of the shift reversed for negative and positive frequencies. The Hilbert transform of $V_R(t)$, which becomes the imaginary part $V_I(t)$, is obtained as the inverse Fourier transform of the modified spectrum, as follows:

$$V_I(t) = -j \int_{-\infty}^{\infty} \mathrm{sgn}(\nu)\widehat{V}(\nu)\, e^{j2\pi\nu t} d\nu$$

$$= j \int_{-\infty}^{0} \widehat{V}(\nu)\, e^{j2\pi\nu t} d\nu - j \int_{0}^{\infty} \widehat{V}(\nu)\, e^{j2\pi\nu t} d\nu \ . \qquad (A3.3)$$

The analytic signal is the complex function that represents $V_R(t)$, and is

$$V(t) = V_R(t) + jV_I(t)$$

$$= \int_{-\infty}^{0} (1 + j^2)\widehat{V}(\nu)\, e^{j2\pi\nu t} d\nu + \int_{0}^{\infty} (1 - j^2)\widehat{V}(\nu)\, e^{j2\pi\nu t} d\nu$$

$$= 2 \int_{0}^{\infty} \widehat{V}(\nu)\, e^{j2\pi\nu t} d\nu \ . \qquad (A3.4)$$

It can be seen that the analytic signal contains no negative-frequency components. From Eq. (A3.4), another way of obtaining the analytic signal for a real function $V_R(t)$ is to suppress the negative-frequency components of the spectrum and double the amplitudes of the positive ones. It can also be shown [see, e.g., Born and Wolf (1999)] that

$$\langle [V_R(t)]^2 \rangle = \langle [V_I(t)]^2 \rangle = \tfrac{1}{2}\langle V(t)V^*(t) \rangle \ , \qquad (A3.5)$$

where angle brackets $\langle\ \rangle$ indicate the expectation. The analytic signal is so called because, considered as a function of a complex variable, it is analytic in the lower half of the complex plane.

From Eqs. (A3.2) and (A3.4), we obtain

$$\int_{-\infty}^{\infty} \widehat{V}(\nu)\, e^{j2\pi\nu t} dt = 2\,\mathcal{Re}\left[\int_{0}^{\infty} \widehat{V}(\nu)\, e^{j2\pi\nu t} dt\right] \ . \qquad (A3.6)$$

This is a useful equality that can be used with any *Hermitian*[6] function and its conjugate variable.

---

[5]The function $\mathrm{sgn}(\nu)$ is equal to 1 for $\nu \geq 0$ and -1 for $\nu < 0$. The Fourier transform of $\mathrm{sgn}(\nu)$ is $-j/\pi t$ (see Appendix 2.1).

[6]A Hermitian function is one in which the real part of the Fourier transform is an even function and the imaginary part is an odd function.

In many cases of interest in radio astronomy and optics, the bandwidth of a signal is small compared with the mean frequency ν_0, which in many instrumental situations is the center frequency of a filter. Such a waveform resembles a sinusoid with amplitude and phase that vary with time on a scale that is slow compared with the period $1/\nu_0$. The analytic signal can then be written as

$$V(t) = C(t)\, e^{j[2\pi \nu_0 t - \Phi(t)]} \,, \tag{A3.7}$$

where C and Φ are real. The spectral components of the function under consideration are appreciable only for small values of $|\nu - \nu_0|$. Thus, $C(t)$ and $\Phi(t)$ consist of low-frequency components, and the period of the time variation of C and Φ is characteristically the reciprocal of the bandwidth. The real and imaginary parts of the analytic signal can be written as

$$V_R(t) = C(t) \cos[2\pi \nu_0 t - \Phi(t)] \,, \tag{A3.8}$$

$$V_I(t) = C(t) \sin[2\pi \nu_0 t - \Phi(t)] \,. \tag{A3.9}$$

The modulus $C(t)$ of the complex analytic signal can be regarded as a modulation envelope, and $\Phi(t)$ represents the phase. In cases where the width of the signal band and the effect of the modulation are not important, it is clearly possible to consider C and Φ as constants, that is, to represent the signals as monochromatic waveforms of frequency ν_0, as in the introductory discussion. The case in which the bandwidth is small compared with the center frequency, as represented by Eq. (A3.7), is referred to as the quasi-monochromatic case.

As a simple example, $e^{j2\pi \nu t}$ is the analytic signal corresponding to the real function of time $\cos(2\pi \nu t)$. The Fourier spectrum of $e^{j2\pi \nu t}$ has a component at frequency ν only, but the Fourier spectrum of $\cos(2\pi \nu t)$ has components at the two frequencies $\pm \nu$. In general, it is necessary to consider the negative-frequency components in the analysis of waveforms, unless they are represented by the analytic signal formulation, for which negative-frequency components are zero. For example, in Eq. (2.8), we included negative-frequency components. If we had omitted the negative frequencies and doubled the amplitude of the positive ones, the cosine term in Eq. (2.9) would have been replaced by $e^{j2\pi \nu_0 \tau}$. We would then have taken the real part to arrive at the correct result. In the approach used in Chap. 2, it is necessary to include the negative frequencies since the autocorrelation function is purely real, and thus its Fourier transform is Hermitian. In this book, we have generally included the negative frequencies rather than using the analytic signal and have made use of the relationship in Eq. (A3.6) when it is advantageous to do so.

It is interesting to note another property of functions of which the real and imaginary parts are a Hilbert transform pair. If the real and imaginary parts of a waveform (i.e., a function of time) are a Hilbert transform pair, then its spectral components are zero for negative frequencies. If we consider the inverse Fourier transforms, it is seen that if the waveform amplitude is zero for $t < 0$, the real and imaginary parts of the spectrum are a Hilbert transform pair. The response of any electrical system to an impulse function applied at time $t = 0$ is zero for $t < 0$, since an effect cannot precede its cause. A function representing such a response is referred to as a *causal function*, and the Hilbert transform relationship applies to its spectrum.

A3.1.2   Truncated Function

Another consideration in the representation of waveforms concerns the existence of the Fourier transform. A condition of the existence of the transform is that the Fourier integral over the range $\pm\infty$ be finite. Although this is not always the case, it is possible to form a function for which the Fourier transform exists and that approaches the original function as the value of some parameter tends toward a limit. For example, the original function can be multiplied by a Gaussian so that the product falls to zero at large values, and the Fourier integral exists. The Fourier transform of the product approaches that of the original function as the width of the Gaussian tends to infinity. Such transforms in the limit are applicable to periodic functions such as $\cos(2\pi\nu t)$, as shown by Bracewell (2000). In the case of noiselike waveforms, the frequency spectrum of a time function can always be determined with satisfactory accuracy by analyzing a sufficiently long (but finite) time interval. In practice, the time interval needs to be long compared with the physically significant timescales that are associated with the waveform, such as the reciprocals of the mean frequency and of the bandwidth. Thus, if the function $V(t)$ is truncated at $\pm T$, the Fourier transform with respect to frequency becomes

$$\widehat{V}(\nu) = \lim_{T\to\infty} \frac{1}{2T} \int_{-T}^{T} V(t)\, e^{-j2\pi\nu t} dt \ . \qquad (A3.10)$$

It is sometimes useful to define the truncated function as $V_T(t)$, where

$$V_T(t) = V(t) \ , \qquad\qquad |t| \leq T \ ,$$
$$V_T(t) = 0 \ , \qquad\qquad |t| > T \ , \qquad\qquad (A3.11)$$

and to write the Fourier transform as

$$\widehat{V}(\nu) = \lim_{T\to\infty} \frac{1}{2T} \int_{-\infty}^{\infty} V_T(t)\, e^{-j2\pi\nu t} dt \ . \qquad (A3.12)$$

In the case of the analytic signal, truncation of the real part does not necessarily result in truncation of its Hilbert transform. It may therefore be necessary that the limits of the integral over time be $\pm\infty$, as in Eq. (A3.12), rather than $\pm T$.

References

Apostol, T.M., *Calculus*, Vol. II, Blaisdel, Waltham, MA (1962), p. 82

Born, M., and Wolf, E., *Principles of Optics*, 7th ed., Cambridge Univ. Press, Cambridge, UK (1999)

Bracewell, R.N., Radio Interferometry of Discrete Sources, *Proc. IRE*, **46**, 97–105 (1958)

Bracewell, R.N., *The Fourier Transform and Its Applications*, McGraw-Hill, New York (2000) (earlier eds. 1965, 1978)

Brouw, W.N., "Data Processing for the Westerbork Synthesis Radio Telescope," Ph.D. thesis, Univ. of Leiden (1971)

Gabor, D., Theory of Communication, *J. Inst. Elect. Eng.*, **93**, Part III, 429–457 (1946)

Goodman, J.W., *Statistical Optics*, Wiley, New York (1985)

König, A., Astrometry with Astrographs, in *Astronomical Techniques, Stars, and Stellar Systems*, Vol. 2, Hiltner, W.A., Ed., Univ. Chicago Press, Chicago (1962), pp. 461–486

Swenson, G.W., Jr., and Mathur, N.C., The Interferometer in Radio Astronomy, *Proc. IEEE*, **56**, 2114–2130 (1968)

Chapter 4
Geometrical Relationships, Polarimetry, and the Interferometer Measurement Equation

In this chapter, we start to examine some of the practical aspects of interferometry. These include baselines, antenna mounts and beam shapes, and the response to polarized radiation, all of which involve geometric considerations and coordinate systems. The discussion is concentrated on Earth-based arrays with tracking antennas, which illustrate the principles involved, although the same principles apply to other systems such as those that include one or more antennas in Earth orbit.

4.1 Antenna Spacing Coordinates and (u, v) Loci

Various coordinate systems are used to specify the relative positions of the antennas in an array, and of these, one of the more convenient for terrestrial arrays is shown in Fig. 4.1. A right-handed Cartesian coordinate system is used, where X and Y are measured in a plane parallel to the Earth's equator, X in the meridian plane[1] (defined as the plane through the poles of the Earth and the reference point in the array), Y toward the east, and Z toward the north pole. In terms of hour angle H and declination δ, coordinates (X, Y, Z) are measured toward $(H = 0, \delta = 0)$, $(H = -6^{\mathrm{h}}, \delta = 0)$, and $(\delta = 90°)$, respectively. If $(X_\lambda, Y_\lambda, Z_\lambda)$ are the components

---

[1] In VLBI observations, it is customary to set the X axis in the Greenwich meridian, in which case H is measured with respect to that meridian rather than a local one.

© The Author(s) 2017
A.R. Thompson, J.M. Moran, and G.W. Swenson Jr., *Interferometry and Synthesis in Radio Astronomy*, Astronomy and Astrophysics Library, DOI 10.1007/978-3-319-44431-4_4

Fig. 4.1 The (X, Y, Z) coordinate system for specification of relative positions of antennas. Directions of the axes specified are in terms of hour angle H and declination δ.

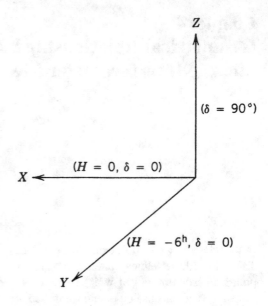

of $\mathbf{D}_\lambda$ in the (X, Y, Z) system, the components (u, v, w) are given by

$$
\begin{bmatrix} u \\ v \\ w \end{bmatrix} = \begin{bmatrix} \sin H & \cos H & 0 \\ -\sin\delta\cos H & \sin\delta\sin H & \cos\delta \\ \cos\delta\cos H & -\cos\delta\sin H & \sin\delta \end{bmatrix} \begin{bmatrix} X_\lambda \\ Y_\lambda \\ Z_\lambda \end{bmatrix}.
\tag{4.1}
$$

Here (H, δ) are usually the hour angle and declination of the phase reference position. The elements of the transformation matrix given above are the direction cosines of the (u, v, w) axes with respect to the (X, Y, Z) axes and can easily be derived from the relationships in Fig. 4.2. Another method of specifying the baseline vector is in terms of its length, D, and the hour angle and declination, (h, d), of the intersection of the baseline direction with the Northern Celestial Hemisphere. The coordinates in the (X, Y, Z) system are then given by

$$
\begin{bmatrix} X \\ Y \\ Z \end{bmatrix} = D \begin{bmatrix} \cos d\cos h \\ -\cos d\sin h \\ \sin d \end{bmatrix}.
\tag{4.2}
$$

The coordinates in the (u, v, w) system are, from Eqs. 4.1 and 4.2,

$$
\begin{bmatrix} u \\ v \\ w \end{bmatrix} = D_\lambda \begin{bmatrix} \cos d\sin(H - h) \\ \sin d\cos\delta - \cos d\sin\delta\cos(H - h) \\ \sin d\sin\delta + \cos d\cos\delta\cos(H - h) \end{bmatrix}.
\tag{4.3}
$$

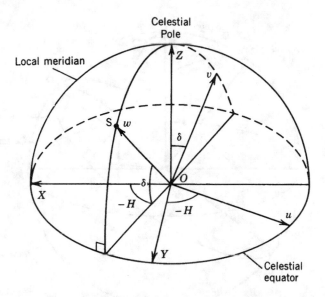

Fig. 4.2 Relationships between the (X, Y, Z) and (u, v, w) coordinate systems. The (u, v, w) system is defined for observation in the direction of the point S, which has hour angle and declination H and δ. As shown, S is in the eastern half of the hemisphere and H is therefore negative. The direction cosines in the transformation matrix in Eq. (4.1) follow from the relationships in this diagram. The relationship in Eq. (4.2) can also be derived if we let S represent the direction of the baseline and put the baseline coordinates (h, d) for (H, δ).

The (D, h, d) system was used more widely in the earlier literature, particularly for instruments involving only two antennas; see, for example, Rowson (1963).

When the (X, Y, Z) components of a new baseline are first established, the usual practice is to determine the elevation $\mathcal{E}$, azimuth $\mathcal{A}$, and length of the baseline by field surveying techniques. Figure 4.3 shows the relationship between $(\mathcal{E}, \mathcal{A})$ and other coordinate systems; see also Appendix 4.1. For latitude $\mathcal{L}$, using Eqs. (4.2) and (A4.2), we obtain

$$\begin{bmatrix} X \\ Y \\ Z \end{bmatrix} = D \begin{bmatrix} \cos \mathcal{L} \sin \mathcal{E} - \sin \mathcal{L} \cos \mathcal{E} \cos \mathcal{A} \\ \cos \mathcal{E} \sin \mathcal{A} \\ \sin \mathcal{L} \sin \mathcal{E} + \cos \mathcal{L} \cos \mathcal{E} \cos \mathcal{A} \end{bmatrix}. \tag{4.4}$$

Examination of Eq. (4.1) or (4.3) shows that the locus of the projected antenna spacing components u and v defines an ellipse with hour angle as the variable. Let

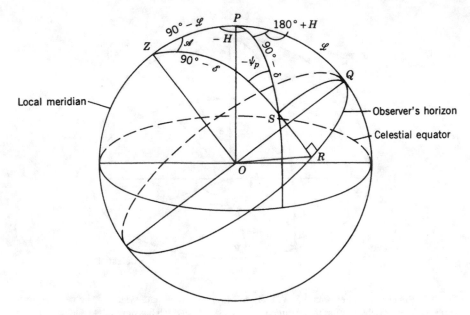

Fig. 4.3 Relationship between the celestial coordinates (H, δ) and the elevation and azimuth $(\mathcal{E}, \mathcal{A})$ of a point S as seen by an observer at latitude $\mathcal{L}$. P is the celestial pole and Z the observer's zenith. The parallactic angle ψ_p is the position angle of the observer's vertical on the sky measured from north toward east. The lengths of the arcs measured in terms of angles subtended at the center of the sphere O are as follows:

$$ZP = 90° - \mathcal{L} \qquad PQ = \mathcal{L} \qquad SR = \mathcal{E} \qquad RQ = \mathcal{A}$$
$$SZ = 90° - \mathcal{E} \qquad SP = 90° - \delta \qquad SQ = \cos^{-1}(\cos \mathcal{E} \cos \mathcal{A}).$$

The required relationships can be obtained by application of the sine and cosine rules for spherical triangles to ZPS and PQS and are given in Appendix 4.1. Note that with S in the eastern half of the observer's sky, as shown, H and ψ_p are negative.

(H_0, δ_0) be the phase reference position. Then from Eq. (4.1), we have

$$u^2 + \left(\frac{v - Z_\lambda \cos \delta_0}{\sin \delta_0} \right)^2 = X_\lambda^2 + Y_\lambda^2 . \tag{4.5}$$

In the (u, v) plane, Eq. (4.5) defines an ellipse[2] with the semimajor axis equal to $\sqrt{X_\lambda^2 + Y_\lambda^2}$, and the semiminor axis equal to $\sin \delta_0 \sqrt{X_\lambda^2 + Y_\lambda^2}$, as in Fig. 4.4a. The ellipse is centered on the v axis at $(u, v) = (0, Z_\lambda \cos \delta_0)$. The arc of the ellipse that is traced out during any observation depends on the azimuth, elevation, and latitude of the baseline; the declination of the source; and the range of hour angle covered,

---

[2]The first mention of elliptical loci appears to have been by Rowson (1963).

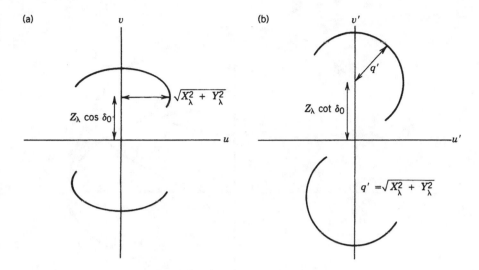

Fig. 4.4 (**a**) Spacing vector locus in the (u, v) plane from Eq. (4.5). (**b**) Spacing vector locus in the (u', v') plane from Eq. (4.8). The lower arc in each diagram represents the locus of conjugate values of visibility. Unless the source is circumpolar, the cutoff at the horizon limits the lengths of the arcs.

as illustrated in Fig. 4.5. Since $\mathcal{V}(-u, -v) = \mathcal{V}^*(u, v)$, any observation supplies simultaneous measurements on two arcs, which are part of the same ellipse only if $Z_\lambda = 0$.

4.2 (u', v') Plane

The (u', v') plane, which was introduced in Sect. 3.1.2 with regard to east–west baselines, is also useful in discussing certain aspects of the behavior of arrays in general. This plane is normal to the direction of the pole and can be envisaged as the equatorial plane of the Earth. For non-east–west baselines, we can also consider the projection of the spacing vectors onto the (u', v') plane. All such projected vectors sweep out circular loci as the Earth rotates. The spacing components in the (u', v') plane are derived from those in the (u, v) plane by the transformation $u' = u$, $v' = v \operatorname{cosec} \delta_0$. In terms of the components of the baseline $(X_\lambda, Y_\lambda, Z_\lambda)$ for two antennas, we obtain from Eq. (4.1)

$$u' = X_\lambda \sin H_0 + Y_\lambda \cos H_0 \tag{4.6}$$

$$v' = -X_\lambda \cos H_0 + Y_\lambda \sin H_0 + Z_\lambda \cot \delta_0 . \tag{4.7}$$

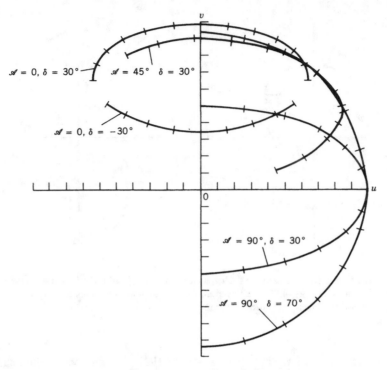

Fig. 4.5 Examples of (u, v) loci to show the variation with baseline azimuth $\mathcal{A}$ and observing declination δ (the baseline elevation $\mathcal{E}$ is zero). The baseline length in all cases is equal to the length of the axes measured from the origin. The tracking range is -4 to $+4$ h for $\delta = -30°$, and -6 to $+6$ h in all other cases. Marks along the loci indicate 1-h intervals in tracking. Note the change in ellipticity for east–west baselines $(\mathcal{A} = 90°)$ with $\delta = 30°$ and with $\delta = 70°$. The loci are calculated for latitude $40°$.

The loci are circles centered on $(0, Z_\lambda \cot \delta_0)$, with radii q' given by

$$q'^2 = u'^2 + (v' - Z_\lambda \cot \delta_0)^2 = X_\lambda^2 + Y_\lambda^2 , \qquad (4.8)$$

as shown in Fig. 4.4b. The projected spacing vectors that generate the loci rotate with constant angular velocity ω_e, the rotation velocity of the Earth, which is easier to visualize than the elliptic motion in the (u, v) plane. In particular, problems involving the effect of time, such as the averaging of visibility data, are conveniently dealt with in the (u', v') plane. Examples of its use will be found in Sects. 4.4, 6.4.2, and 16.3.2. In Fourier transformation, the conjugate variables of (u', v') are (l', m'), where $l' = l$ and $m' = m \sin \delta_0$, that is, the image plane is compressed by a factor $\sin \delta_0$ in the m direction.

4.3 Fringe Frequency

The component w of the baseline represents the path difference to the two antennas for a plane wave incident from the phase reference position. The corresponding time delay is w/v_0, where v_0 is the center frequency of the observing band. The relative phase of the signals at the two antennas changes by 2π radians when w changes by unity. Thus, the frequency of the oscillations at the output of the correlator that combines the signals is

$$\frac{dw}{dt} = \frac{dw}{dH}\frac{dH}{dt} = -\omega_e \left[X_\lambda \cos\delta \sin H + Y_\lambda \cos\delta \cos H \right] = -\omega_e u \cos\delta, \quad (4.9)$$

where $\omega_e = dH/dt = 7.29115 \times 10^{-15}$ rad s$^{-1} = \omega_e$ is the rotation velocity of the Earth with respect to the fixed stars: for greater accuracy, see Seidelmann (1992). The sign of dw/dt indicates whether the phase is increasing or decreasing with time. The result shown above applies to the case in which the signals suffer no time-varying instrumental phase changes between the antennas and the correlator inputs. In an array in which the antennas track a source, time delays to compensate for the space path differences w are applied to maintain correlation of the signals. If an exact compensating delay were introduced in the radio frequency section of the receivers, the relative phases of the signals at the correlator input would remain constant, and the correlator output would show no fringes. However, except in some low-frequency systems like LOFAR (de Vos et al. 2009), the compensating delays are usually introduced at an intermediate frequency, of which the band center v_d is much less than the observing frequency v_0. The adjustment of the compensating delay introduces a rate of phase change $2\pi v_d(dw/dt)/v_0 = -\omega_e u(\cos\delta)v_d/v_0$. The resulting fringe frequency at the correlator output is

$$v_f - \frac{dw}{dt}\left(1 \mp \frac{v_d}{v_0}\right) = -\omega_e u \cos\delta \left(1 \mp \frac{v_d}{v_0}\right), \quad (4.10)$$

where the negative sign refers to upper-sideband reception and the positive sign to lower-sideband reception; these distinctions and the double-sideband case are explained in Sect. 6.1.8. From Eq. (4.3), the right side of Eq. (4.10) is equal to $-\omega_e D \cos d \cos\delta \sin(H - h)(v_0 \mp v_d)/c$. Note that $(v_0 \mp v_d)$ is usually determined by one or more local oscillator frequencies.

4.4 Visibility Frequencies

As explained in Sect. 3.1, the phase of the complex visibility is measured with respect to that of a hypothetical point source at the phase reference position. The fringe-frequency variations do not appear in the visibility function, but slower

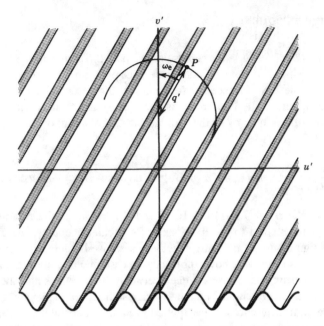

Fig. 4.6 The (u', v') plane showing sinusoidal corrugations that represent the visibility of a point source. For simplicity, only the real part of the visibility is included. The most rapid variation in the visibility is encountered at the point P, where the direction of the spacing locus is normal to the ridges in the visibility. ω_e is the rotation velocity of the Earth.

variations occur that depend on the position of the radiating sources within the field. We now examine the maximum temporal frequency of the visibility variations. Consider a point source represented by the delta function $\delta(l_1, m_1)$. The visibility function is the Fourier transform of $\delta(l_1, m_1)$, which is

$$e^{-j2\pi(ul_1+vm_1)} = \cos 2\pi(ul_1 + vm_1) - j\sin 2\pi(ul_1 + vm_1) . \qquad (4.11)$$

This expression represents two sets of sinusoidal corrugations, one real and one imaginary. The corrugations represented by the real part of Eq. (4.11) are shown in (u', v') coordinates in Fig. 4.6, where the arguments of the trigonometric functions in Eq. (4.11) become $2\pi(u'l_1 + v'm_1 \sin \delta_0)$. The frequency of the corrugations in terms of cycles per unit distance in the (u', v') plane is l_1 in the u' direction, $m_1 \sin \delta_0$ in the v' direction, and

$$r'_1 = \sqrt{l_1^2 + m_1^2 \sin^2 \delta_0} \qquad (4.12)$$

in the direction of most rapid variations. Expression (4.12) is maximized at the pole and then becomes equal to r_1, which is the angular distance of the source from the (l, m) origin. For any antenna pair, the spatial frequency locus in the (u', v') plane is a circle of radius q' generated by a vector rotating with angular velocity ω_e, where

q' is as defined in Eq. (4.8). From Fig. 4.6, it is clear that the temporal variation of the measured visibility is greatest at the point P and is equal to $\omega_e r'_1 q'$. This is a useful result, since if r_1 represents a position at the edge of the field to be imaged, it indicates that to follow the most rapid variations, the visibility must be sampled at time intervals sufficiently small compared with $(\omega_e r'_1 q')^{-1}$. Also, we may wish to alternate between two frequencies or polarizations during an observation, and these changes must be made on a similarly short timescale. Note that this requirement is also covered by the sampling theorem in Sect. 5.2.1.

4.5 Calibration of the Baseline

The position parameters (X, Y, Z) for each antenna relative to a common reference point can usually be established to a few centimeters or millimeters by a conventional engineering survey. Except at long wavelengths, the accuracy required is greater than this. We must be able to compute the phase at any hour angle for a point source at the phase reference position to an accuracy of, say, $1°$ and subtract it from the observed phase. This reference phase is represented by the factor $e^{j2\pi w}$ in Eq. (3.7), and it is therefore necessary to calculate w to 1/360 of the observing wavelength. The baseline parameters can be obtained to the required accuracy from observations of calibration sources for which the positions are accurately known. The phase of such a calibrator observed at the phase reference position (H_0, δ_0) should ideally be zero. However, if practical uncertainties are taken into account, the measured phase is, from Eq. (4.1),

$$2\pi \Delta w + \phi_{in} = 2\pi (\cos \delta_0 \cos H_0 \Delta X_\lambda - \cos \delta_0 \sin H_0 \Delta Y_\lambda + \sin \delta_0 \Delta Z_\lambda) + \phi_{in} , \tag{4.13}$$

where the prefix Δ indicates the uncertainty in the associated quantity, and ϕ_{in} is an instrumental phase term for the two antennas involved. If a calibrator is observed over a wide range of hour angle, ΔX_λ and ΔY_λ can be obtained from the even and odd components, respectively, of the phase variation with H_0. To measure ΔZ_λ, calibrators at more than one declination must be included. A possible procedure is to observe several calibrators at different declinations, repeating a cycle of observations for several hours. For the kth observation, we can write, from Eq. (4.13),

$$a_k \Delta X_\lambda + b_k \Delta Y_\lambda + c_k \Delta Z_\lambda + \phi_{in} = \phi_k , \tag{4.14}$$

where a_k, b_k, and c_k are known source parameters, and ϕ_k is the measured phase. The calibrator source position need not be accurately known since the phase measurements can be used to estimate both the source positions and the baselines. Techniques for this analysis are discussed in Sect. 12.2. In practice, the instrumental phase ϕ_{in} will vary slowly with time: instrumental stability is discussed in Chap. 7.

Also, there will be atmospheric phase variations, which are discussed in Chap. 13. These effects set the final limit on the attainable accuracy in observing both calibrators and sources under investigation.

Measurement of baseline parameters to an accuracy of order 1 part in 10^7 (e.g., 3 mm in 30 km) implies timing accuracy of order $10^{-7}\omega_e^{-1} \simeq 1$ ms. Timekeeping is discussed in Sects. 9.5.8 and 12.3.3.

4.6   Antennas

4.6.1   Antenna Mounts

In discussing the dependence of the measured phase on the baseline components, we have ignored any effects introduced by the antennas, which is tantamount to assuming that the antennas are identical and their effects on the signals cancel out. This, however, is only approximately true. In most synthesis arrays, the antennas must have collecting areas of tens or hundreds of square meters for reasons of sensitivity. Except for dipole arrays at meter wavelengths, the antennas required are large structures that must be capable of accurately tracking a radio source across the sky. Tracking antennas are almost always constructed either on equatorial mounts (also called polar mounts) or on altazimuth mounts, as illustrated in Fig. 4.7. In an equatorial mount, the polar axis is parallel to the Earth's axis of rotation, and tracking a source requires only that the antenna be turned about the polar axis at the

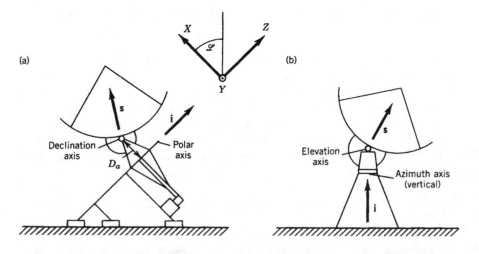

Fig. 4.7 Schematic diagrams of antennas on (**a**) equatorial (polar) and (**b**) altazimuth mounts. In the positions shown, the declination and elevation axes are normal to the plane of the page. In the equatorial mount, there is a distance D_a between the two rotational axes, but in the altazimuth mount, the axes often intersect, as shown.

sidereal rate. Equatorial mounts are mechanically more difficult to construct than altazimuth ones and are found mainly on antennas built prior to the introduction of computers for control and coordinate conversion.

In most tracking arrays used in radio astronomy, the antennas are circularly symmetrical reflectors. A desirable feature is that the axis of symmetry of the reflecting surface intersect both the rotation axes of the mount. If this is not the case, pointing motions will cause the antenna to have a component of motion along the direction of the beam. It is then necessary to take account of phase changes associated with small pointing corrections, which may differ from one antenna to another. In most antenna mounts, however, whether of equatorial or altazimuth type, the reflector axis intersects the rotation axes with sufficient precision that phase errors of this type are negligible.

It is convenient but not essential that the two rotation axes of the mount intersect. The intersection point then provides an appropriate reference point for defining the baseline between antennas, since whatever direction in which the antenna points, its aperture plane is always the same distance from that point as measured along the axis of the beam. In most large equatorially mounted antennas, the polar and declination axes do not intersect. In many cases, there is an offset of several meters between the polar and declination axes. Wade (1970) considered the implication of this offset for high-accuracy phase measurements and showed that it is necessary to take account of variations in the offset distance and in the accuracy of alignment of the polar axis. These results can be obtained as follows. Let $\mathbf{i}$ and $\mathbf{s}$ be unit vectors in the direction of the polar axis and the direction of the source under observation, respectively, and let $\mathbf{D}_a$ be the spacing vector between the two axes measured perpendicular to $\mathbf{i}$ (see Fig. 4.7a). The quantity that we need to compute is the projection of $\mathbf{D}_a$ in the direction of observation, $\mathbf{D}_a \cdot \mathbf{s}$. Since $\mathbf{D}_a$ is perpendicular to $\mathbf{i}$, the cosine of the angle between $\mathbf{D}_a$ and $\mathbf{s}$ is $\sqrt{1 - (\mathbf{i} \cdot \mathbf{s})^2}$. Thus,

$$\mathbf{D}_a \cdot \mathbf{s} = D_a \sqrt{1 - (\mathbf{i} \cdot \mathbf{s})^2} \,, \tag{4.15}$$

where D_a is the magnitude of $\mathbf{D}_a$. In the (X, Y, Z) coordinate system in which the baseline components are measured, $\mathbf{i}$ has direction cosines (i_X, i_Y, i_Z), and $\mathbf{s}$ has direction cosines given by the transformation matrix on the right side of Eq. (4.2), but with h and d replaced by H and δ, which refer to the direction of observation. If the polar axis is correctly aligned to within about 1 arcmin, i_X and i_Y are of order 10^{-3} and $i_Z \simeq 1$. Thus, we can use the direction cosines to evaluate Eq. (4.15), and ignoring second-order terms in i_X and i_Y, we obtain

$$\mathbf{D}_a \cdot \mathbf{s} = D_a(\cos \delta - i_X \sin \delta \cos H + i_Y \sin \delta \sin H) \,. \tag{4.16}$$

If the magnitude of $\mathbf{D}_a$ is expressed in wavelengths, the difference in the values of $\mathbf{D}_a \cdot \mathbf{s}$ for the two antennas must be added to the w component of the baseline given by Eq. (4.1) when calculating the reference phase at the field center. To do this, it is first necessary to determine the unknown constants in Eq. (4.16), which can be done by adding a term of the form $2\pi(\alpha\cos\delta_0 + \beta\sin\delta_0\cos H_0 + \gamma\sin\delta_0\sin H_0)$ to the right side of Eq. (4.13) and extending the solution to include α, β, and γ. The result then represents the differences in the corresponding mechanical dimensions of the two antennas. Note that the terms in i_X and i_Y in Eq. (4.16) are important only when D_a is large. If D_a is no more than one wavelength, it should be possible to ignore them.

The preceding analysis can be extended to the case of an altazimuth mount by letting $\mathbf{i}$ represent the direction of the azimuth axis, as in Fig. 4.7b. Then $i_X = \cos(\mathcal{L}+\varepsilon)$, $i_Y = \sin\varepsilon'$, and $i_Z = \sin(\mathcal{L}+\varepsilon)$, where $\mathcal{L}$ is the latitude and ε and ε' are, respectively, the tilt errors in the XZ plane and in the plane containing the Y axis and the local vertical. The errors again should be quantities of order 10^{-3}. In many altazimuth mounts, the axes are designed to intersect, and D_a represents only a structural tolerance. Thus, we assume that D_a is small enough to allow terms in $i_Y D_a$ and εD_a to be ignored, and evaluation of Eq. (4.15) gives

$$\mathbf{D}_a \cdot \mathbf{s} = D_a \left[1 - (\sin\mathcal{L}\sin\delta + \cos\mathcal{L}\cos\delta\cos H)^2\right] = D_a\cos\mathcal{E}, \qquad (4.17)$$

where $\mathcal{E}$ is the elevation of direction $\mathbf{s}$: see Eq. (A4.1) of Appendix 4.1. Correction terms of this form can be added to the expressions for the baseline calibration and for w.

4.6.2 Beamwidth and Beam-Shape Effects

The interpretation of data taken with arrays containing antennas with nonidentical beamwidths is not always a straightforward matter. Each antenna pair responds to an effective intensity distribution that is the product of the actual intensity of the sky and the geometric mean of the normalized beam profiles. If different pairs of antennas respond to different effective distributions, then, in principle, the Fourier transform relationship between $I(l, m)$ and $\mathcal{V}(u, v)$ cannot be applied to the ensemble of observations. Mixed arrays are sometimes used in VLBI when it is necessary to make use of antennas that have different designs. However, in VLBI studies, the source structure under investigation is very small compared with the widths of the antenna beams, so the differences in the beams can usually be ignored. If cases arise in which different beams are used and the source is not small compared with beamwidths, it is possible to restrict the measurements to the field defined by the narrowest beam by convolution of the visibility data with an appropriate function in the (u, v) plane.

A problem similar to that of unmatched beams occurs if the antennas have altazimuth mounts and the beam contours are not circularly symmetrical about the nominal beam axis. As a point in the sky is tracked using an altazimuth mount, the beam rotates with respect to the sky about this nominal axis. This rotation does not occur for equatorial mounts. The angle between the vertical at the antenna and the direction of north at the point being observed (defined by the great circle through the point and the North Pole) is the parallactic angle ψ_p in Fig. 4.3. Application of the sine rule to the spherical triangle ZPS gives

$$\frac{-\sin \psi_p}{\cos \mathcal{L}} = \frac{-\sin H}{\cos \mathcal{E}} = \frac{\sin \mathcal{A}}{\cos \delta} , \tag{4.18}$$

which can be combined with Eq. (A4.1) or (A4.2) to express ψ_p as a function of $(\mathcal{A}, \mathcal{E})$ or (H, δ). If the beam has elongated contours and width comparable to the source under observation, rotation of the beam causes the effective intensity distribution to vary with hour angle. This is particularly serious in the case of observations to reveal the structure of the most distant Universe, for which foreground sources need to be accurately removed. For the Australia Pathfinder Array (DeBoer et al. 2009), the 12-m-diameter antennas have altazimuth mounts, with a third axis that allows the reflector, feed supports, and feeds to be rotated about the reflector axis so the beam pattern and the angle of polarization remain fixed relative to the sky.

4.7 Polarimetry

Polarization measurements are very important in radio astronomy. Most synchrotron radiation shows a small degree of polarization that indicates the distribution of the magnetic fields within the source. As noted in Chap. 1, this polarization is generally linear (plane) and can vary in magnitude and position angle over the source. As frequency is increased, the percentage polarization often increases because the depolarizing action of Faraday rotation is reduced. Polarization of radio emission also results from the Zeeman effect in atoms and molecules, cyclotron radiation and plasma oscillations in the solar atmosphere, and Brewster angle effects at planetary surfaces. The measure of polarization that is almost universally used in astronomy is the set of four parameters introduced by Sir George Stokes in 1852. It is assumed here that readers have some familiarity with the concept of Stokes parameters or can refer to one of numerous texts that describe them [e.g., Born and Wolf (1999); Kraus and Carver (1973); Wilson et al. (2013)].

Stokes parameters are related to the amplitudes of the components of the electric field, E_x and E_y, resolved in two perpendicular directions normal to the direction of propagation. Thus, if E_x and E_y are represented by $\mathcal{E}_x(t) \cos[2\pi \nu t + \delta_x(t)]$ and $\mathcal{E}_y(t) \cos[2\pi \nu t + \delta_y(t)]$, respectively, Stokes parameters are defined as follows:

$$I = \langle \mathcal{E}_x^2(t) \rangle + \langle \mathcal{E}_y^2(t) \rangle$$
$$Q = \langle \mathcal{E}_x^2(t) \rangle - \langle \mathcal{E}_y^2(t) \rangle$$
$$U = 2 \langle \mathcal{E}_x(t) \, \mathcal{E}_y(t) \cos \left[\delta_x(t) - \delta_y(t) \right] \rangle$$
$$V = 2 \langle \mathcal{E}_x(t) \, \mathcal{E}_y(t) \sin \left[\delta_x(t) - \delta_y(t) \right] \rangle , \tag{4.19}$$

where the angular brackets denote the expectation or time average. This averaging is necessary because in radio astronomy, we are dealing with fields that vary with time in a random manner. Of the four parameters, I is a measure of the total intensity of the wave, Q and U represent the linearly polarized component, and V represents the circularly polarized component. Stokes parameters can be converted to a measure of polarization with a more direct physical interpretation as follows:

$$m_\ell = \frac{\sqrt{Q^2 + U^2}}{I} \tag{4.20}$$

$$m_c = \frac{V}{I} \tag{4.21}$$

$$m_t = \frac{\sqrt{Q^2 + U^2 + V^2}}{I} \tag{4.22}$$

$$\theta = \frac{1}{2}\tan^{-1}\left(\frac{U}{Q}\right), \qquad 0 \le \theta \le \pi, \tag{4.23}$$

where m_ℓ, m_c, and m_t are the degrees of linear, circular, and total polarization, respectively, and θ is the position angle of the plane of linear polarization. For monochromatic signals, $m_t = 1$ and the polarization can be fully specified by just three parameters. For random signals such as those of cosmic origin, $m_t \le 1$, and all four parameters are required. The Stokes parameters all have the dimensions of flux density or intensity, and they propagate in the same manner as the electromagnetic field. Thus, they can be determined by measurement or calculation at any point along a wave path, and their relative magnitudes define the state of polarization at that point. Stokes parameters combine additively for independent waves. When they are used to specify the total radiation from any point on a source, I, which measures the total intensity, is always positive, but Q, U, and V can take both positive and negative values depending on the position angle or sense of rotation of the polarization. The corresponding visibility values measured with an interferometer are complex quantities, as will be discussed later.

In considering the response of interferometers and arrays, up to this point we have ignored the question of polarization. This simplification can be justified by the assumption that we have been dealing with completely unpolarized radiation for which only the parameter I is nonzero. In that case, the response of an interferometer with identically polarized antennas is proportional to the total flux density of the radiation. As will be shown below, in the more general case, the response is proportional to a linear combination of two or more Stokes parameters, where the combination is determined by the polarizations of the antennas. By observing with different states of polarization of the antennas, it is possible to separate the responses to the four parameters and determine the corresponding components of the visibility. The variation of each parameter over the source can thus be imaged individually, and the polarization of the radiation emitted at any point can be determined. There are alternative methods of describing the polarization state of a wave, of which the coherency matrix is perhaps the most important (Ko 1967a,b). However, the classical treatment in terms of Stokes parameters remains widely used by astronomers, and we therefore follow it here.

4.7.1 Antenna Polarization Ellipse

The polarization of an antenna in either transmission or reception can be described in general by stating that the electric vector of a transmitted signal traces out an elliptical locus in the wavefront plane. Most antennas are designed so that the ellipse approximates a line or circle, corresponding to linear or circular polarization, in the central part of the main beam. However, precisely linear or circular responses are hardly achievable in practice. As shown in Fig. 4.8, the essential characteristics of the polarization ellipse are given by the position angle ψ of the major axis, and by

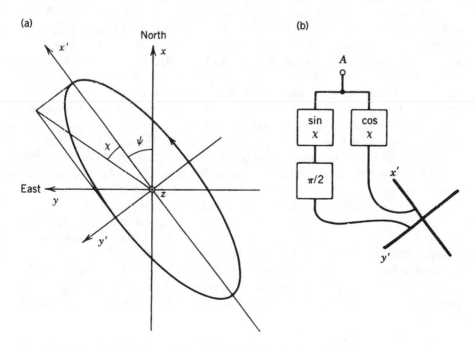

Fig. 4.8 (a) Description of the general state of polarization of an antenna in terms of the characteristics of the ellipse generated by the electric vector in the transmission of a sinusoidal signal. The position angle ψ of the major axis is measured with respect to the x axis, which points toward the direction of north on the sky. A wave approaching from the sky is traveling toward the reader, in the direction of the positive z axis. For such a wave, the arrow on the ellipse indicates the direction of right-handed polarization. (b) Model antenna that radiates the electric field represented by the ellipse in (a) when a signal is applied to the terminal A. Cos χ and sin χ indicate the amplitudes of the voltage responses of the units shown, and $\pi/2$ indicates a phase lag.

the axial ratio, which it is convenient to express as the tangent of an angle χ, where $-\pi/4 \leq \chi \leq \pi/4$.

An antenna of arbitrary polarization can be modeled in terms of two idealized dipoles as shown in Fig. 4.8b. Consider *transmitting* with this antenna by applying a signal waveform to the terminal A. The signals to the dipoles pass through networks with voltage responses proportional to $\cos \chi$ and $\sin \chi$, and the signal to the y' dipole also passes through a network that introduces a $\pi/2$ phase lag. Thus, the antenna produces field components of amplitude $\mathcal{E}_{x'}$ and $\mathcal{E}_{y'}$ in phase quadrature along the directions of the major and minor axes of the ellipse. If the antenna input is a radio frequency sine wave $V_0 \cos 2\pi \nu t$, then the field components are

$$
\begin{aligned}
\mathcal{E}_{x'} \cos(2\pi \nu t) &\propto V_0 \cos \chi \cos(2\pi \nu t) \\
\mathcal{E}_{y'} \sin(2\pi \nu t) &\propto V_0 \sin \chi \sin(2\pi \nu t) \ .
\end{aligned}
\tag{4.24}
$$

In these equations, the y' component lags the x' component by $\pi/2$. If $\chi = \pi/4$, the radiated electric vector traces a circular locus with the sense of rotation from the x' axis to the y' axis (i.e., *counterclockwise* in Fig. 4.8a). This is consistent with the quarter-cycle delay in the signal to the y' dipole. Then a wave propagating in the positive z' direction of a right-handed coordinate system (i.e., toward the reader in Fig. 4.8a) is right circularly polarized in the IEEE (1977) definition. (This definition is now widely adopted, but in some of the older literature, such a wave would be defined as left circularly polarized.) The International Astronomical Union (IAU 1974) has adopted the IEEE definition and states that the position angle of the electric vector on the sky should be measured from north through east with reference to the system of right ascension and declination. The IAU also states that "the polarization of incoming radiation, for which the position angle, θ, of the electric vector, measured at a fixed point in space, increases with time, is described as right-handed and positive." Note that Stokes parameters in Eq. (4.19) specify only the field in the (x, y) plane, and to determine whether a circularly polarized wave is left- or right-handed, the direction of propagation must be given. From Eq. (4.19) and the definitions of E_x and E_y that precede them, a wave traveling in the positive z direction in right-handed coordinates is right circularly polarized for positive V.

In reception, an electric vector that rotates in a *clockwise* direction in Fig. 4.8 produces a voltage in the y' dipole that leads the voltage in the x' dipole by $\pi/2$ in phase, and the two signals therefore combine in phase at A. For counterclockwise rotation, the signals at A are in antiphase and cancel one another. Thus, the antenna in Fig. 4.8 receives right-handed waves incident from the positive z direction (that is, traveling toward negative z), and it transmits right-handed polarization in the direction toward positive z. To receive a right-handed wave propagating down from the sky (in the positive z direction), the polarity of one of the dipoles must be reversed, which requires that $\chi = -\pi/4$.

To determine the interferometer response, we begin by considering the output of the antenna modeled in Fig. 4.8b. We define the field components in complex form:

$$E_x(t) = \mathcal{E}_x(t)\, e^{j[2\pi\nu t + \delta_x(t)]}\ ,$$
$$E_y(t) = \mathcal{E}_y(t)\, e^{j[2\pi\nu t + \delta_y(t)]}\ . \tag{4.25}$$

The signal voltage received at A in Fig. 4.8b, expressed in complex form, is

$$V' = E_{x'} \cos\chi - jE_{y'}\sin\chi\ , \tag{4.26}$$

where the factor $-j$ represents the $\pi/2$ phase lag applied to the y' signal, for the fields represented by Eq. (4.25). Now we need to specify the polarization of the incident wave in terms of Stokes parameters. In accordance with IAU (1974), the axes used are in the directions of north and east on the sky, which are represented by x and y in Fig. 4.8a. In terms of the field in the x and y directions, the components of the field in the x' and y' directions are

$$E_{x'}(t) = \left[\mathcal{E}_x(t)\, e^{j\delta_x(t)}\cos\psi + \mathcal{E}_y(t)\, e^{j\delta_y(t)}\sin\psi\right] e^{j2\pi\nu t}$$
$$E_{y'}(t) = \left[-\mathcal{E}_x(t)\, e^{j\delta_x(t)}\sin\psi + \mathcal{E}_y(t)\, e^{j\delta_y(t)}\cos\psi\right] e^{j2\pi\nu t}\ . \tag{4.27}$$

Derivation of the response at the output of the correlator for antennas m and n of an array involves straightforward manipulation of some rather lengthy expressions that are not reproduced here. The steps are as follows:

1. Substitute $E_{x'}$ and $E_{y'}$ from Eq. (4.27) into Eq. (4.26) to obtain the output of each antenna.
2. Indicate values of ψ, χ, and V' for the two antennas by subscripts m and n and calculate the correlator output, $R_{mn} = G_{mn}\langle V'_m V'^*_n\rangle$, where G_{mn} is an instrumental gain factor.
3. Substitute Stokes parameters for $\mathcal{E}_x$, $\mathcal{E}_y$, δ_x, δ_y using Eq. (4.19) as follows:

$$\langle(\mathcal{E}_x e^{j\delta_x})(\mathcal{E}_x e^{j\delta_x})^*\rangle = \langle\mathcal{E}_x^2\rangle = \tfrac{1}{2}(I + Q)$$
$$\langle(\mathcal{E}_y e^{j\delta_y})(\mathcal{E}_y e^{j\delta_y})^*\rangle = \langle\mathcal{E}_y^2\rangle = \tfrac{1}{2}(I - Q)$$
$$\langle(\mathcal{E}_x e^{j\delta_x})(\mathcal{E}_y e^{j\delta_y})^*\rangle = \langle\mathcal{E}_x\mathcal{E}_y e^{j(\delta_x-\delta_y)}\rangle = \tfrac{1}{2}(U + jV)$$
$$\langle(\mathcal{E}_x e^{j\delta_x})^*(\mathcal{E}_y e^{j\delta_y})\rangle = \langle\mathcal{E}_x\mathcal{E}_y e^{-j(\delta_x-\delta_y)}\rangle = \tfrac{1}{2}(U - jV)\ . \tag{4.28}$$

The result is

$$
\begin{aligned}
R_{mn} = \tfrac{1}{2}G_{mn} \{ &I_v \left[\cos(\psi_m - \psi_n)\cos(\chi_m - \chi_n) + j\sin(\psi_m - \psi_n)\sin(\chi_m + \chi_n)\right] \\
&+ Q_v \left[\cos(\psi_m + \psi_n)\cos(\chi_m + \chi_n) + j\sin(\psi_m + \psi_n)\sin(\chi_m - \chi_n)\right] \\
&+ U_v \left[\sin(\psi_m + \psi_n)\cos(\chi_m + \chi_n) - j\cos(\psi_m + \psi_n)\sin(\chi_m - \chi_n)\right] \\
&- V_v \left[\cos(\psi_m - \psi_n)\sin(\chi_m + \chi_n) + j\sin(\psi_m - \psi_n)\cos(\chi_m - \chi_n)\right] \} \; .
\end{aligned}
$$

(4.29)

In this equation, a subscript v has been added to Stokes parameter symbols to indicate that they represent the complex visibility for the distribution of the corresponding parameter over the source, not simply the intensity or brightness of the radiation. Equation (4.29) is a useful general formula that applies to all cases. It was originally derived by Morris et al. (1964) and later by Weiler (1973). In the derivation by Morris et al., the sign of V_v is opposite to that given by Weiler and in Eq. (4.29). This difference results from the convention for the sense of rotation for circular polarization. In the convention we have followed in Fig. 4.8, two identical antennas both adjusted to receive right circularly polarized radiation would have parameters $\psi_m = \psi_n$ and $\chi_m = \chi_n = -\pi/4$. In Eq. (4.29), these values correspond to a positive sign for V_v. Thus, in Eq. (4.29), positive V_v represents right circular polarization incident from the sky, which is in agreement with the IAU definition in 1973 (IAU 1974). The derivation by Morris et al. predates the IAU definition and follows the commonly used convention at that time, in which the sign for V was the reverse of that in the IAU definition. Note that in what follows, the factor 1/2 in Eq. (4.29) is omitted and considered to be subsumed within the overall gain factor. Equation (4.29) was the main basis for polarization measurements in radio interferometry for at least three decades until an alternative formulation was developed by Hamaker et al. (1996). This later formulation is introduced in Sect. 4.8.

4.7.2 Stokes Visibilities

As noted above, the symbols I_v, Q_v, U_v, and V_v in Eq. (4.29) refer to the corresponding visibility values as measured by the spaced antennas. We shall therefore refer to these quantities as *Stokes visibilities*, following the nomenclature of Hamaker et al. (1996). Stokes visibilities are the quantities required in imaging polarized emission, and they can be derived from the correlator output values by using Eq. (4.29). This equation is considerably simplified when the nominal polarization characteristics of practical antennas are inserted. First, consider the case in which both antennas are identically polarized. Then $\chi_m = \chi_n$, $\psi_m = \psi_n$, and Eq. (4.29) becomes

$$
R_{mn} = G_{mn}[I_v + Q_v \cos 2\psi_m \cos 2\chi_m + U_v \sin 2\psi_m \cos 2\chi_m - V_v \sin 2\chi_m] \; .
$$

(4.30)

In considering linearly polarized antennas, it is convenient to use subscripts x and y to indicate two orthogonal planes of polarization. For example, R_{xy} represents the correlator output for antenna m with polarization x and antenna n with polarization y. For linearly polarized antennas, $\chi_m = \chi_n = 0$. Consider two antennas, each with separate outputs for linear polarizations x and y. Then for parallel polarizations, omitting gain constants, we obtain from Eq. (4.30)

$$R_{xx} = I_v + Q_v \cos 2\psi_m + U_v \sin 2\psi_m \ . \tag{4.31}$$

Here, ψ_m is the position angle of the antenna polarization measured from celestial north in the direction of east. The y polarization angle is equal to the x polarization angle plus $\pi/2$. For ψ_m equal to $0°, 45°, 90°$, and $135°$, the output R_{xx} is proportional to $(I_v + Q_v)$, $(I_v + U_v)$, $(I_v - Q_v)$, and $(I_v - U_v)$, respectively. By using antennas with these polarization angles, I_v, Q_v, and U_v, but not V_n, can be measured. In many cases, circular polarization is negligibly small, and the inability to measure V_v is not a serious problem. However, Q_v and U_v are often only a few percent of I_v, and in attempting to measure them with identical feeds, one faces the usual problems of measuring a small difference in two much larger quantities. The same is true if one attempts to measure V_v using identical circular feeds for which $\chi = \pm\pi/4$ and the response is proportional to $(I_v \mp V_v)$. These problems are reduced by using oppositely polarized feeds to measure Q_v, U_v, or V_v. For an example of measurement of V_v, see Weiler and Raimond (1976).

With oppositely polarized feeds, we insert in Eq. (4.29) $\psi_n = \psi_m + \pi/2$, and $\chi_m = -\chi_n$. For linear polarization, the χ terms are zero and the planes of polarization orthogonal. The antennas are then described as cross-polarized, as typified by crossed dipoles. Omitting constant gain factors and using the x and y subscripts defined above, we obtain for the correlator output

$$\begin{aligned} R_{xy} &= -Q_v \sin 2\psi_m + U_v \cos 2\psi_m + jV_v \\ R_{yx} &= -Q_v \sin 2\psi_m + U_v \cos 2\psi_m - jV_v \ , \end{aligned} \tag{4.32}$$

where ψ_m refers to the angle of the plane of polarization in the direction (x or y) indicated by the first subscript of the R term in the same equation. Then for ψ_m equal to $0°$ and $45°$, the R_{xy} response is proportional to $(U_v + jV_v)$ and $(-Q_v + jV_v)$. If V_v is assumed to be zero, this suffices to measure the polarized component. If both antennas provide outputs for cross-polarized signals, the outputs of which go to two separate receiving channels at each antenna, four correlators can be used for each antenna pair. These provide responses for both crossed and parallel pairs, as listed in Table 4.1. Thus, if the planes of polarization can be periodically rotated through $45°$ as indicated by position angles I and II in Table 4.1, for example, by rotating antenna feeds, then Q_v, U_v, and V_v can be measured without taking differences between responses involving I_v. The use of rotating feeds has, however, proved to be of limited practicality. Rotating the feed relative to the main reflector is likely to have a small but significant effect on the beam shape and polarization properties. This is

Table 4.1 Stokes visibilities vs. position angles

| Position angles | | Stokes visibilities | |
m	n	measured	
0°	0°	$I_v + Q_v$	Position angle I
0°	90°	$U_v + jV_v$	"
90°	0°	$U_v - jV_v$	"
90°	90°	$I_v - Q_v$	"
45°	45°	$I_v + U_v$	Position angle II
45°	135°	$-Q_v + jV_v$	"
135°	45°	$-Q_v - jV_v$	"
135°	135°	$I_v - U_v$	"

because the rotation will cause deviations from circular symmetry in the radiation pattern of the feeds to interact differently with the shadowing effects of the focal support structure and any departures from circular symmetry in the main reflector. Furthermore, in radio astronomy systems designed for the greatest sensitivity, the feed together with the low-noise amplifiers and a cryogenically refrigerated Dewar are often built as one monolithic unit that cannot easily be rotated. However, for antennas on altazimuth mounts, the variation of the parallactic angle with hour angle causes the antenna response pattern to rotate on the sky as a source is tracked in hour angle. Conway and Kronberg (1969) pointed out this advantage of altazimuth mounts, which enables instrumental effects to be distinguished from the true polarization of the source if observations continue for a period of several hours.

An example of a different arrangement of linearly polarized feeds, which has been used at the Westerbork Synthesis Radio Telescope, is described by Weiler (1973). The antennas are equatorially mounted and the parallactic angle of the polarization remains fixed as a source is tracked. The outputs of the antennas that are movable on rail track are correlated with those from the antennas in fixed locations. Table 4.2 shows the measurements when the position angles of the planes of polarization for the movable antennas are 45° and 135° and those of the fixed antennas 0° and 90°. Although the responses are reduced by a factor of $\sqrt{2}$ relative to those in Table 4.1, there is no loss in sensitivity since each Stokes visibility appears at all four correlator outputs. Note that since only signals from antennas with different polarization configurations are cross-correlated, this scheme does not make use of all possible polarization products.

Opposite circularly polarized feeds offer certain advantages for measurements of linear polarization. In determining the responses, an arbitrary position angle ψ_m for antenna m is included to represent the effect of rotation caused, for example, by an altazimuth antenna mount. If the antennas provide simultaneous outputs for opposite

Table 4.2 Stokes visibilities vs. position angles

Position angles		Stokes visibilities measured
m	n	
$0°$	$45°$	$(I_v + Q_v + U_v + jV_v)/\sqrt{2}$
$0°$	$135°$	$(-I_v - Q_v + U_v + jV_v)/\sqrt{2}$
$90°$	$45°$	$(I_v - Q_v + U_v - jV_v)/\sqrt{2}$
$90°$	$135°$	$(I_v - Q_v - U_v + jV_v)/\sqrt{2}$

senses of rotation (denoted by r and ℓ) and four correlation products are generated for each antenna pair, the outputs are proportional to the quantities in Table 4.3.

Here, we have made $\psi_\ell = \psi_r + \pi/2$, and $\chi = -\pi/4$ for right circular polarization and $\chi = \pi/4$ for left circular. The feeds need not be rotated during an observation, and the responses to Q_v and U_v are separated from those to I_v. The expressions in Table 4.3 can be simplified by choosing values of ψ_r such as $\pi/2$, $\pi/4$, or 0. For example, if $\psi_r = 0$, the sum of the $r\ell$ and ℓr responses is a measure of Stokes visibility U_v. Again, the effects of the rotation of the position angle with altazimuth mounts must be taken into account. Conway and Kronberg (1969) appear to have been the first to use an interferometer with circularly polarized antennas to measure linear polarization in weakly polarized sources. Circularly polarized antennas have since been commonly used in radio astronomy.

4.7.3 Instrumental Polarization

The responses with the various combinations of linearly and circularly polarized antennas discussed above are derived on the assumption that the polarization is exactly linear or circular and that the position angles of the linear feeds are exactly determined. This is not the case in practice, and the polarization ellipse can never be maintained as a perfect circle or straight line. The nonideal characteristics of the antennas cause an unpolarized source to appear polarized and are therefore referred to as *instrumental polarization*. The effect of these deviations from ideal behavior

Table 4.3 Stokes visibilities vs. sense of rotation

Sense of rotation		Stokes visibilities measured
m	n	
r	r	$I_v + V_v$
r	ℓ	$(-jQ_v + U_v)e^{-j2\psi_m}$
ℓ	r	$(-jQ_v - U_v)e^{j2\psi_m}$
ℓ	ℓ	$I_v - V_v$

can be calculated from Eq. (4.29) if the deviations are known. In the expressions in Tables 4.1–4.3, the responses given are only the major terms, and if the instrumental terms are included, all four Stokes visibilities are, in general, involved. For example, consider the case of crossed linear feeds with nominal position angles 0° and 90°. Let the actual values of ψ and χ be such that $(\psi_x + \psi_y) = \pi/2 + \Delta\psi^+$, $(\psi_x - \psi_y) = -\pi/2 + \Delta\psi^-$, $\chi_x + \chi_y = \Delta\chi^+$, and $\chi_x - \chi_y = \Delta\chi^-$. Then from Eq. (4.29),

$$R_{xy} \simeq I_v(\Delta\psi^- - j\Delta\chi^+) - Q_v(\Delta\psi^+ - j\Delta\chi^-) + U_v + jV_v . \qquad (4.33)$$

Generally, antennas can be adjusted so that the Δ terms are no more than $\sim 1°$, and here we have assumed that they are small enough that their cosines can be approximated by unity, their sines by the angles, and products of two sines by zero. Instrumental polarization is often different for the antennas even if they are structurally similar, and corrections must be made to the visibility data before they are combined into an image.

   Although we have derived expressions for deviations of the antenna polarizations from the ideal in terms of the ellipticity and orientation of the polarization ellipse in Eq. (4.29), it is not necessary to know these parameters for the antennas so long as it is possible to remove the instrumental effects from the measurements, so that they do not appear in the final image. In calibrating the antenna responses, an approach that is widely preferred is to specify the instrumental polarization in terms of the response of the antenna to a wave of polarization that is orthogonal or opposite-handed with respect to the nominal antenna response. Thus, for linearly polarized antennas, following the analysis of Sault et al. (1991), we can write

$$v'_x = v_x + D_x v_y \quad \text{and} \quad v'_y = v_y + D_y v_x , \qquad (4.34)$$

where subscripts x and y indicate two orthogonal planes of polarization, the v' terms indicate the signal received, the v terms indicate the signal that would be received with an ideally polarized antenna, and the D terms indicate the response of the real antenna to the polarization orthogonal to the nominal polarization. The D terms are often described as the *leakage* of the orthogonal polarization into the antenna (Bignell 1982) and represent the instrumental polarization. For each polarization state, the leakage is specified by one complex number, that is, the same number of terms as the two real numbers required to specify the ellipticity and orientation of the polarization ellipse. In Appendix 4.2, expressions for D_x and D_y are derived in terms of the parameters of the polarization ellipse:

$$D_x \simeq \psi_x - j\chi_x \quad \text{and} \quad D_y \simeq -\psi_y + j\chi_y , \qquad (4.35)$$

where the approximations are valid for small values of the χ and ψ parameters. Note that in Eq. (4.35), ψ_y is measured with respect to the y direction. For an ideal linearly polarized antenna, χ_x and χ_y are both zero, and the polarization in the x and y planes is precisely aligned with, and orthogonal to, the x direction with respect to the antenna. Thus, for an ideal antenna, ψ_x and ψ_y are also zero. For a practical antenna, the terms in Eq. (4.35) represent limits of accuracy in the hardware, and we see that the real and imaginary parts of the leakage terms can be related to the misalignment and ellipticity, respectively.

For a pair of antennas m and n, the leakage terms allow us to express the measured correlator outputs R'_{xx}, R'_{yy}, R'_{xy}, and R'_{yx} in terms of the unprimed quantities that represent the corresponding correlations as they would be measured with ideally polarized antennas:

$$
\begin{aligned}
R'_{xx}/(g_{xm}g^*_{xn}) &= R_{xx} + D_{xm}R_{yx} + D^*_{xn}R_{xy} + D_{xm}D^*_{xn}R_{yy} \\
R'_{xy}/(g_{xm}g^*_{yn}) &= R_{xy} + D_{xm}R_{yy} + D^*_{yn}R_{xx} + D_{xm}D^*_{yn}R_{yx} \\
R'_{yx}/(g_{ym}g^*_{xn}) &= R_{yx} + D_{ym}R_{xx} + D^*_{xn}R_{yy} + D_{ym}D^*_{xn}R_{xy} \\
R'_{yy}/(g_{ym}g^*_{yn}) &= R_{yy} + D_{ym}R_{xy} + D^*_{yn}R_{yx} + D_{ym}D^*_{yn}R_{xx} \, .
\end{aligned}
\tag{4.36}
$$

The g terms represent the voltage gains of the corresponding signal channels. They are complex quantities representing amplitude and phase, and the equations can be normalized so that the values of the individual g terms do not differ greatly from unity. Note that Eq. (4.36) contain no small-term approximations. However, the leakage terms are typically no more than a few percent, and products of two such terms will be omitted at this point. Then, from Eqs. (4.31) and (4.32), the responses can be written in terms of the Stokes visibilities as follows:

$$
\begin{aligned}
R'_{xx}/(g_{xm}g^*_{xn}) &= I_v + Q_v[\cos 2\psi_m - (D_{xm} + D^*_{xn})\sin 2\psi_m] \\
&\quad + U_v[\sin 2\psi_m + (D_{xm} + D^*_{xn})\cos 2\psi_m] - jV_v(D_{xm} - D^*_{xn}) \\
R'_{xy}/(g_{xm}g^*_{yn}) &= I_v(D_{xm} + D^*_{yn}) - Q_v[\sin 2\psi_m + (D_{xm} - D^*_{yn})\cos 2\psi_m] \\
&\quad + U_v[\cos 2\psi_m - (D_{xm} - D^*_{yn})\sin 2\psi_m] + jV_v \\
R'_{yx}/(g_{ym}g^*_{xn}) &= I_v(D_{ym} + D^*_{xn}) - Q_v[\sin 2\psi_m - (D_{ym} - D^*_{xn})\cos 2\psi_m] \\
&\quad + U_v[\cos 2\psi_m + (D_{ym} - D^*_{xn})\sin 2\psi_m] - jV_v \\
R'_{yy}/(g_{ym}g^*_{yn}) &= I_v - Q_v[\cos 2\psi_m + (D_{ym} + D^*_{yn})\sin 2\psi_m] \\
&\quad - U_v[\sin 2\psi_m - (D_{ym} + D^*_{yn})\cos 2\psi_m] + jV_v(D_{ym} - D^*_{yn}) \, .
\end{aligned}
\tag{4.37}
$$

Note that ψ_m refers to the polarization (x or y) indicated by the first of the two subscripts of the R' term in the same equation. Sault et al. (1991) describe Eq. (4.37) as representing the strongly polarized case. In deriving them, no restriction was placed on the magnitudes of the Stokes visibility terms, but the leakage terms of the antennas are assumed to be small. In the case where the source is only weakly polarized, the products of Q_v, U_v, and V_v with leakage terms can be omitted.

Equation (4.37) then become

$$R'_{xx}/(g_{xm}g^*_{xn}) = I_v + Q_v \cos 2\psi_m + U_v \sin 2\psi_m$$

$$R'_{xy}/(g_{xm}g^*_{yn}) = I_v(D_{xm} + D^*_{yn}) - Q_v \sin 2\psi_m + U_v \cos 2\psi_m + jV_v$$

$$R'_{yx}/(g_{ym}g^*_{xn}) = I_v(D_{ym} + D^*_{xn}) - Q_v \sin 2\psi_m + U_v \cos 2\psi_m - jV_v$$

$$R'_{yy}/(g_{ym}g^*_{yn}) = I_v - Q_v \cos 2\psi_m - U_v \sin 2\psi_m \ .$$

$$(4.38)$$

If the antennas are operating well within the upper frequency limit of their performance, the polarization terms can be expected to remain largely constant with time since gravitational deflections that vary with pointing should be small. The instrumental gain terms can contain components due to the atmosphere, which may vary on time scales of seconds or minutes, and they also include any effects of the receiver electronics.

In the case of circularly polarized antennas, leakage terms can also be defined and similar expressions for the instrumental response derived. The leakage terms are given by the following equations:

$$v'_r = v_r + D_r v_\ell \quad \text{and} \quad v'_\ell = v_\ell + D_\ell v_r \ , \tag{4.39}$$

where, as before, the v' terms are the measured signal voltages, the unprimed v terms are the signals that would be observed with an ideally polarized antenna, and the D terms are the leakages. The subscripts r and ℓ indicate the right and left senses of rotation. Again, the relationship between the leakage terms and the orientation and ellipticity of the antenna responses is derived in Appendix 4.2. The results, which in this case require no small-angle approximations, are

$$D_r = e^{j2\psi_r} \tan \Delta\chi_r \quad \text{and} \quad D_\ell = e^{-j2\psi_\ell} \tan \Delta\chi_\ell \ , \tag{4.40}$$

where the Δ terms are defined by $\chi_r = -45° + \Delta\chi_r$ and $\chi_\ell = 45° + \Delta\chi_\ell$. To derive expressions for the outputs of an interferometer in terms of the leakage terms and Stokes visibilities, the four measured correlator outputs are represented by $R'_{rr}, R'_{\ell\ell}, R'_{r\ell}$, and $R'_{\ell r}$. These are related to the corresponding (unprimed) quantities that would be observed with ideally polarized antennas as follows:

$$R'_{rr}/(g_{rm}g^*_{rn}) = R_{rr} + D_{rm}R_{\ell r} + D^*_{rn}R_{r\ell} + D_{rm}D^*_{rn}R_{\ell\ell}$$

$$R'_{r\ell}/(g_{rm}g^*_{\ell n}) = R_{r\ell} + D_{rm}R_{\ell\ell} + D^*_{\ell n}R_{rr} + D_{rm}D^*_{\ell n}R_{\ell r}$$

$$R'_{\ell r}/(g_{\ell m}g^*_{rn}) = R_{\ell r} + D_{\ell m}R_{rr} + D^*_{rn}R_{\ell\ell} + D_{\ell m}D^*_{rn}R_{r\ell}$$

$$R'_{\ell\ell}/(g_{\ell m}g^*_{\ell n}) = R_{\ell\ell} + D_{\ell m}R_{r\ell} + D^*_{\ell n}R_{\ell r} + D_{\ell m}D^*_{\ell n}R_{rr} \ .$$

$$(4.41)$$

Now, from the expressions in Table 4.3, the outputs in terms of the Stokes visibilities are

$$R'_{rr}/(g_{rm}g^*_{rn}) = I_v(1 + D_{rm}D^*_{rn}) - jQ_v(D_{rm}e^{j2\psi_m} + D^*_{rn}e^{-j2\psi_m})$$
$$- U_v(D_{rm}e^{j2\psi_m} - D^*_{rn}e^{-j2\psi_m}) + V_v(1 - D_{rm}D^*_{rn})$$
$$R'_{r\ell}/(g_{rm}g^*_{\ell n}) = I_v(D_{rm} + D^*_{\ell n}) - jQ_v(e^{-j2\psi_m} + D_{rm}D^*_{\ell n}e^{j2\psi_m})$$
$$+ U_v(e^{-j2\psi_m} - D_{rm}D^*_{\ell n}e^{j2\psi_m}) - V_v(D_{rm} - D^*_{\ell n})$$
$$R'_{\ell r}/(g_{\ell m}g^*_{rn}) = I_v(D_{\ell m} + D^*_{rn}) - jQ_v(e^{j2\psi_m} + D_{\ell m}D^*_{rn}e^{-j2\psi_m})$$
$$- U_v(e^{j2\psi_m} - D_{\ell m}D^*_{rn}e^{-j2\psi_m}) + V_v(D_{\ell m} - D^*_{rn})$$
$$R'_{\ell\ell}/(g_{\ell m}g^*_{\ell n}) = I_v(1 + D_{\ell m}D^*_{\ell n}) - jQ_v(D_{\ell m}e^{-j2\psi_m} + D^*_{\ell n}e^{j2\psi_m})$$
$$+ U_v(D_{\ell m}e^{-j2\psi_m} - D^*_{\ell n}e^{j2\psi_m}) - V_v(1 - D_{\ell m}D^*_{\ell n}) .$$

$$(4.42)$$

Here again, ψ_m refers to the polarization (r or ℓ) indicated by the first of the two subscripts of the R' term in the same equation. The angle ψ_m represents the parallactic angle plus any instrumental offset. We have made no approximations in deriving Eq. (4.42) [in the similar Eq. (4.37), products of two D terms were omitted]. If the leakage terms are small, then any product of two of them can be omitted, as in the strongly polarized case for linearly polarized antennas in Eq. (4.37). The weakly polarized case is derived from the strongly polarized case by further omitting products of Q_v, U_v, and V_v with the leakage terms and is as follows:

$$R'_{rr}/(g_{rm}g^*_{rn}) = I_v + V_v$$
$$R'_{r\ell}/(g_{rm}g^*_{\ell n}) = I_v(D_{rm} + D^*_{\ell n}) - (jQ_v - U_v)e^{-j2\psi_m}$$
$$R'_{\ell r}/(g_{\ell m}g^*_{rn}) = I_v(D_{\ell m} + D^*_{rn}) - (jQ_v + U_v)e^{j2\psi_m}$$
$$R'_{\ell\ell}/(g_{\ell m}g^*_{\ell n}) = I_v - V_v .$$

$$(4.43)$$

Similar expressions[3] are given by Fomalont and Perley (1989). To make use of the expressions that have been derived for the response in terms of the leakage and gain factors, we need to consider how such quantities can be calibrated, and this is discussed later.

---

[3]In comparing expressions for polarimetry by different authors, note that differences of signs or of the factor j can result from differences in the way the parallactic angle is defined with respect to the antenna, and similar arbitrary factors.

4.7.4 Matrix Formulation

The description of polarimetry given above, using the ellipticity and orientation of the antenna response, is based on a physical model of the antenna and the electromagnetic wave, as in Eq. (4.29). Historically, studies of *optical* polarization have developed over a much longer period. A description of radio polarimetry following an approach originally developed in optics is given in Hamaker et al. (1996) and in more detail in four papers: Hamaker et al. (1996), Sault et al. (1996), Hamaker (2000), and Hamaker (2006). The mathematical analysis is largely in terms of matrix algebra, and in particular, it allows the responses of different elements of the signal path such as the atmosphere, the antennas, and the electronic system to be represented independently and then combined in the final solution. This approach is convenient for detailed analysis including effects of the atmosphere, ionosphere, etc.

In the matrix formulation, the electric fields of the polarized wave are represented by a two-component column vector. The effect of any linear system on the wave, or on the voltage waveforms of the signal after reception, can be represented by a 2×2 matrix of the form shown below:

$$\begin{bmatrix} E'_p \\ E'_q \end{bmatrix} = \begin{bmatrix} a_1 & a_2 \\ a_3 & a_4 \end{bmatrix} \begin{bmatrix} E_p \\ E_q \end{bmatrix} , \tag{4.44}$$

where E_p and E_q represent the input polarization state (orthogonal linear or opposite circular) and E'_p and E'_q represent the outputs. The 2×2 matrix in Eq. (4.44) is referred to as a Jones matrix (Jones 1941), and any simple linear operation on the wave can be represented by such a matrix. Jones matrices can represent a rotation of the wave relative to the antenna; the response of the antenna, including polarization leakage effects; or the amplification of the signals in the receiving system up to the correlator input. The combined effect of these operations is represented by the product of the corresponding Jones matrices, just as the effect on a scalar voltage can be represented by the product of gains and response factors for different stages of the receiving system. For a wave specified in terms of opposite circularly polarized components, Jones matrices for these operations can take the following forms:

$$\mathbf{J}_{\text{rotation}} = \begin{bmatrix} \exp(j\theta) & 0 \\ 0 & \exp(-j\theta) \end{bmatrix} \tag{4.45}$$

$$\mathbf{J}_{\text{leakage}} = \begin{bmatrix} 1 & D_r \\ D_\ell & 1 \end{bmatrix} \tag{4.46}$$

$$\mathbf{J}_{\text{gain}} = \begin{bmatrix} G_r & 0 \\ 0 & G_\ell \end{bmatrix} . \tag{4.47}$$

Here, θ represents a rotation relative to the antenna, and the cross polarization in the antenna is represented by the off-diagonal[4] leakage terms D_r and D_ℓ. For a nonideal antenna, the diagonal terms will be slightly different from unity, but in this case, the difference is subsumed into the gain matrix of the two channels. The gain of both the antenna and the electronics can be represented by a single matrix, and since any cross coupling of the signals in the amplifiers can be made negligibly small, only the diagonal terms are significant in the gain matrix.

Let $\mathbf{J}_m$ represent the product of the Jones matrices required to represent the linear operations on the signal of antenna m up to the point where it reaches the correlator input. Let $\mathbf{J}_n$ be the same matrix for antenna n. The signals at the inputs to the correlator are $\mathbf{J}_m\mathbf{E}_m$ and $\mathbf{J}_n\mathbf{E}_n$, where $\mathbf{E}_m$ and $\mathbf{E}_n$ are the vectors representing the signals at the antenna. The correlator output is the *outer product* (also known as the Kronecker, or tensor, product) of the signals at the input:

$$\mathbf{F}'_m \otimes \mathbf{E}'^{*}_n = (\mathbf{J}_m\mathbf{E}_m) \otimes (\mathbf{J}^*_n\mathbf{E}^*_n) \, , \qquad (4.48)$$

where $\otimes$ represents the outer product. The outer product $\mathbf{A} \otimes \mathbf{B}$ is formed by replacing each element a_{ik} of $\mathbf{A}$ by $a_{ik}\mathbf{B}$. Thus, the outer product of two $n \times n$ matrices is a matrix of order $n^2 \times n^2$. It is also a property of the outer product that

$$(\mathbf{A}_i\mathbf{B}_i) \otimes (\mathbf{A}_k\mathbf{B}_k) = (\mathbf{A}_i \otimes \mathbf{A}_k)(\mathbf{B}_i \otimes \mathbf{B}_k) \, . \qquad (4.49)$$

Thus, we can write Eq. (4.48) as

$$\mathbf{E}'_m \otimes \mathbf{E}'^{*}_n = (\mathbf{J}_m \otimes \mathbf{J}^*_n)(\mathbf{E}_m \otimes \mathbf{E}^*_n) \, . \qquad (4.50)$$

The time average of Eq. (4.50) represents the correlator output, which is

$$\mathbf{R}_{mn} = \langle \mathbf{E}'_m \otimes \mathbf{E}'^{*}_n \rangle = \begin{bmatrix} R^{pp}_{mn} \\ R^{pq}_{mn} \\ R^{qp}_{mn} \\ R^{qq}_{mn} \end{bmatrix} \, , \qquad (4.51)$$

where p and q indicate opposite polarization states. The column vector in Eq. (4.51) is known as the *coherency vector* and represents the four cross products from the correlator outputs for antennas m and n. From Eq. (4.50), it is evident that the measured coherency vector $\mathbf{R}'_{mn}$, which includes the effects of instrumental responses, and the true coherency vector $\mathbf{R}_{mn}$, which is free from such effects, are related by the outer product of the Jones matrices that represent the instrumental effects:

$$\mathbf{R}'_{mn} = (\mathbf{J}_m \otimes \mathbf{J}^*_n)\mathbf{R}_{mn} \, . \qquad (4.52)$$

---

[4]The diagonal terms are those that move downward from left to right, and the off-diagonal terms slope in the opposite direction.

To determine the response of an interferometer in terms of the Stokes visibilities of the input radiation, which are complex quantities, we introduce the Stokes visibility vector

$$\mathcal{V}_{Smn} = \begin{bmatrix} I_v \\ Q_v \\ U_v \\ V_v \end{bmatrix}. \tag{4.53}$$

The Stokes visibilities can be regarded as an alternate coordinate system for the coherency vector. Let $\mathbf{S}$ be a 4×4 transformation matrix from Stokes parameters to the polarization coordinates of the antennas. Then we have

$$\mathbf{R}'_{mn} = (\mathbf{J}_m \otimes \mathbf{J}_n^*)\mathbf{S}\mathcal{V}_{Smn}. \tag{4.54}$$

For ideal antennas with crossed (orthogonal) linear polarization, the response in terms of Stokes visibilities is given by the expressions in Table 4.1. We can write this result in matrix form as

$$\begin{bmatrix} R_{xx} \\ R_{xy} \\ R_{yx} \\ R_{yy} \end{bmatrix} = \begin{bmatrix} 1 & 1 & 0 & 0 \\ 0 & 0 & 1 & j \\ 0 & 0 & 1 & -j \\ 1 & -1 & 0 & 0 \end{bmatrix} \begin{bmatrix} I_v \\ Q_v \\ U_v \\ V_v \end{bmatrix}, \tag{4.55}$$

where the subscripts x and y here refer to polarization position angles $0°$ and $90°$, respectively. Similarly for opposite-hand circular polarization, we can write the expressions in Table 4.3 as

$$\begin{bmatrix} R_{rr} \\ R_{r\ell} \\ R_{\ell r} \\ R_{\ell\ell} \end{bmatrix} = \begin{bmatrix} 1 & 0 & 0 & 1 \\ 0 & -je^{-j2\psi_m} & e^{-j2\psi_m} & 0 \\ 0 & -je^{j2\psi_m} & -e^{j2\psi_m} & 0 \\ 1 & 0 & 0 & -1 \end{bmatrix} \begin{bmatrix} I_v \\ Q_v \\ U_v \\ V_v \end{bmatrix}. \tag{4.56}$$

The 4×4 matrices in Eqs. (4.55) and (4.56) are transformation matrices from Stokes visibilities to the coherency vector for crossed linear and opposite circular polarizations, respectively. These 4×4 matrices are known as Mueller matrices following the terminology established in optics.[5] Note that these matrices depend on the particular formulation we have used to specify the angles ψ and χ, and other factors in Fig. 4.8, which may not be identical to corresponding parameters used by other authors.

---

[5]Further explanation of Jones and Mueller matrices can be found in textbooks on optics [e.g., O'Neill (1963)].

The expression $\mathbf{S}^{-1}(\mathbf{J}_m \otimes \mathbf{J}_n^*)\mathbf{S}$ is a matrix that relates the input and output coherency vectors of a system where these quantities are in Stokes coordinate form. As an example of the matrix usage, we can derive the effect of the leakage and gain factors in the case of opposite circular polarizations. For antenna m, the Jones matrix $\mathbf{J}_m$ is the product of the Jones matrices for leakage and gain as follows:

$$\mathbf{J}_m = \begin{bmatrix} g_{rm} & 0 \\ 0 & g_{\ell m} \end{bmatrix} \begin{bmatrix} 1 & D_{rm} \\ D_{\ell m} & 1 \end{bmatrix} = \begin{bmatrix} g_{rm} & g_{rm}D_{rm} \\ g_{\ell m}D_{\ell m} & g_{\ell m} \end{bmatrix}. \tag{4.57}$$

Here, the g terms represent voltage gain, the D terms represent leakage, and the subscripts r and ℓ indicate polarization. A corresponding matrix $\mathbf{J}_n$ is required for antenna n. Then if we use primes to indicate the components of the coherency vector (i.e., the correlator outputs) for antennas m and n, we can write

$$\begin{bmatrix} R'_{rr} \\ R'_{r\ell} \\ R'_{\ell r} \\ R'_{\ell\ell} \end{bmatrix} = \mathbf{J}_m \otimes \mathbf{J}_n^* \begin{bmatrix} 1 & 0 & 0 & 1 \\ 0 & -je^{-j2\psi_m} & e^{-j2\psi_m} & 0 \\ 0 & -je^{j2\psi_m} & -e^{j2\psi_m} & 0 \\ 1 & 0 & 0 & -1 \end{bmatrix} \begin{bmatrix} I_v \\ Q_v \\ U_v \\ V_v \end{bmatrix}, \tag{4.58}$$

where the 4×4 matrix is the one relating Stokes visibilities to the coherency vector in Eq. (4.56). Also, we have

$$\mathbf{J}_m \otimes \mathbf{J}_n^* =$$

$$\begin{bmatrix} g_{rm}g_{rn}^* & g_{rm}g_{rn}^*D_{rn}^* & g_{rm}g_{rn}^*D_{rm} & g_{rm}g_{rn}^*D_{rm}D_{rn}^* \\ g_{rm}g_{\ell n}^*D_{\ell n}^* & g_{rm}g_{\ell n}^* & g_{rm}g_{\ell n}^*D_{rm}D_{\ell n}^* & g_{rm}g_{\ell n}^*D_{rm} \\ g_{\ell m}g_{rn}^*D_{\ell m} & g_{\ell m}g_{rn}^*D_{\ell m}D_{rn}^* & g_{\ell m}g_{rn}^* & g_{\ell m}g_{rn}^*D_{rn}^* \\ g_{\ell m}g_{\ell n}^*D_{\ell m}D_{\ell n}^* & g_{\ell m}g_{\ell n}^*D_{\ell m} & g_{\ell m}g_{\ell n}^*D_{\ell n}^* & g_{\ell m}g_{\ell n}^* \end{bmatrix}. \tag{4.59}$$

Insertion of Eq. (4.59) into Eq. (4.58) and reduction of the matrix products results in Eq. (4.42) for the response with circularly polarized feeds. The use of matrices is convenient since they provide a format for expressions representing different effects, which can then be combined as required.

4.7.5 Calibration of Instrumental Polarization

The fractional polarization of many astronomical sources is of magnitude comparable to that of the leakage and gain terms that are used above to define the instrumental polarization. Thus, to obtain an accurate measure of the polarization of a source, the leakage and gain terms must be accurately calibrated. It may be necessary to determine the calibration independently for each set of observations since the gain terms may be functions of the temperature and state of adjustment of the electronics and cannot be assumed to remain constant from one observing session to another. Making observations (i.e., measuring the coherency vector) of sources for which the polarization parameters are already known is clearly a way of determining the leakage and gain terms. The number of unknown parameters to be calibrated is proportional to the number of antennas, n_a, but the number of measurements is proportional to the number of baselines, $n_a(n_a - 1)/2$. The unknown parameters are therefore usually overdetermined, and a least-mean-squares solution may be the best procedure.

For any antenna with orthogonally polarized receiving channels, there are seven degrees of freedom, that is, seven unknown quantities, that must be calibrated to allow full interpretation of the measured Stokes visibilities. This applies to the general case, and the number can be reduced if approximations are made for weak polarization or small instrumental polarization. In terms of the polarization ellipses, these unknowns can be regarded as the orientations and ellipticities of the two orthogonal feeds and the complex gains (amplitudes and phases) of the two receiving channels. When the outputs of two antennas are combined, only the differences in the instrumental phases are required, leaving seven degrees of freedom per antenna. Sault et al. (1996) make the same point from the consideration of the Jones matrix of an antenna, which contains four complex quantities. They also give a general result that illustrates the seven degrees of freedom or unknown terms. This expresses the relationship between the uncorrected (measured) Stokes visibilities (indicated by primes) and the true values of the Stokes visibilities, in terms of seven γ and δ terms:

$$
\begin{bmatrix} I'_v - I_v \\ Q'_v - Q_v \\ U'_v - U_v \\ V'_v - V_v \end{bmatrix} = -\frac{1}{2} \begin{bmatrix} \gamma_{++} & \gamma_{+-} & \delta_{+-} & -j\delta_{-+} \\ \gamma_{+-} & \gamma_{++} & \delta_{++} & -j\delta_{--} \\ \delta_{+-} & -\delta_{++} & \gamma_{++} & j\gamma_{--} \\ -j\delta_{-+} & -j\delta_{--} & j\gamma_{--} & \gamma_{++} \end{bmatrix} \begin{bmatrix} I_v \\ Q_v \\ U_v \\ V_v \end{bmatrix} . \tag{4.60}
$$

The seven γ and δ terms are defined as follows:

$$
\begin{aligned}
\gamma_{++} &= (\Delta g_{xm} + \Delta g_{ym}) + (\Delta g^*_{xn} + \Delta g^*_{yn}) \\
\gamma_{+-} &= (\Delta g_{xm} - \Delta g_{ym}) + (\Delta g^*_{xn} - \Delta g^*_{yn}) \\
\gamma_{--} &= (\Delta g_{xm} - \Delta g_{ym}) - (\Delta g^*_{xn} - \Delta g^*_{yn}) \\
\delta_{++} &= (D_{xm} + D_{ym}) + (D^*_{xn} + D^*_{yn}) \\
\delta_{+-} &= (D_{xm} - D_{ym}) + (D^*_{xn} - D^*_{yn}) \\
\delta_{-+} &= (D_{xm} + D_{ym}) - (D^*_{xn} + D^*_{yn}) \\
\delta_{--} &= (D_{xm} - D_{ym}) - (D^*_{xn} - D^*_{yn}) .
\end{aligned} \tag{4.61}
$$

Here, it is assumed that Eqs. (4.36) are normalized so that the gain terms are close to unity, and the Δg terms are defined by $g_{ik} = 1 + \Delta g_{ik}$. The D (leakage) terms and the Δg terms are often small enough that products of two such terms can be neglected. The results, as shown in Eqs. (4.60) and (4.61), apply to antennas that are linearly polarized in directions x and y. The same results apply to circularly polarized antennas if the subscripts x and y are replaced by r and ℓ, respectively, and, in the column matrices on the left and right sides of Eq. (4.60), terms in Q_v, U_v, and V_v are replaced by corresponding terms in V_v, Q_v, and U_v, respectively. A similar result is given by Sault et al. (1991). The seven γ and δ terms defined above are subject to errors in the calibration process, so there are seven degrees of freedom in the error mechanisms.

An observation of a single calibration source for which the four Stokes parameters are known enables four of the degrees of freedom to be determined. However, because of the relationships of the quantities involved, it takes at least three calibration observations to solve for all seven unknown parameters (Sault et al. 1996). In the calibration observations, it is useful to observe one unpolarized source, but observing a second unpolarized one would add no further solutions. At least one observation of a linearly polarized source is required to determine the relative phases of the two oppositely polarized channels, that is, the relative phases of the complex gain terms $g_{xm}g_{yn}^*$ and $g_{ym}g_{xn}^*$, or $g_{rm}g_{\ell n}^*$ and $g_{\ell m}g_{rn}^*$. Note that with antennas on altazimuth mounts, observations of a calibrator with linear polarization, taken at intervals between which large rotations of the parallactic angle occur, can essentially be regarded as observations of independent calibrators. Under these circumstances, three observations of the same calibrator will suffice for the full solution. Furthermore, the polarization of the calibrator need not be known in advance but can be determined from the observations.

In cases in which only an unpolarized calibrator can be observed, it may be possible to estimate two more degrees of freedom by introducing the constraint that the sum of the leakage factors over all antennas should be small. As shown by the expressions for the leakage terms in Appendix 4.2, this is a reasonable assumption for a homogeneous array, that is, one in which the antennas are of nominally identical design. However, the phase difference between the signal paths from the feeds to the correlator for the two orthogonal polarizations of each antenna remains unknown. This requires an observation of a calibrator with a component of linear polarization, or a scheme to measure the instrumental component of the phase. For example, on the compact array of the Australia Telescope (Frater and Brooks 1992), noise sources are provided at each antenna to inject a common signal into the two polarization channels (Sault et al. 1996). With such a system, it is necessary to provide an additional correlator for each antenna, or to be able to rearrange correlator inputs, to measure the relative phase of the injected signals in the two polarizations.

In the case of the approximations for weak polarization, Eqs. (4.38) and (4.43) show that if the gain terms are known, the leakage terms can be calibrated by observing an unpolarized source. For opposite circular polarizations, Eq. (4.43) shows that if V_v is small, it is possible to obtain solutions for the gain terms from the outputs for the $\ell\ell$ and rr combinations only, provided also that the number of baselines is several times larger than the number of antennas. The leakage terms can then be solved for separately. For crossed linear polarizations, Eq. (4.38) shows that this is possible only if the linear polarization (Q_v and U_v parameters) for the calibrator have been determined independently.

Optimum strategies for calibration of polarization observations is a subject that leads to highly detailed discussions involving the characteristics of particular synthesis arrays, the hour angle range of the observations, the availability of calibration sources (which can depend on the observing frequency), and other factors, especially if the solutions for strong polarization are used. Such discussions can be found, for example, in Conway and Kronberg (1969), Weiler (1973), Bignell (1982), Sault et al. (1991), Sault et al. (1996), and Smegal et al. (1997). Polarization measurements with VLBI involve some special considerations: see, for example, Roberts et al. (1991), Cotton (1993), Roberts et al. (1994), and Kemball et al. (1995).

For most large synthesis arrays, effective calibration techniques have been devised and the software to implement them has been developed. Thus, a prospective observer need not be discouraged if the necessary calibration procedures appear complicated. Some general considerations relevant to observations of polarization are given below.

- Since the polarization of many sources varies on a timescale of months, it is usually advisable to regard the polarization of the calibration source as one of the variables to be solved for.
- Two sources with relatively strong linear polarization at position angles that do not appear to vary are 3C286 and 3C138. These are useful for checking the phase difference for oppositely polarized channels.
- For most sources, the circular polarization parameter V_v is very small, $\sim 0.2\%$ or less, and can be neglected. Measurements with circularly polarized antennas of the same sense therefore generally give an accurate measure of I_v. However, circular polarization is important in the measurement of magnetic fields by Zeeman splitting. As an example of positive detection at a very low level, Fiebig and Güsten (1989) describe measurements for which $V/I \simeq 5 \times 10^{-5}$. Zeeman splitting of several components of the OH line at 22.235 GHz was observed using a single antenna, the 100-m paraboloid of the Max Planck Institute for Radio Astronomy, with a receiving system that switched between opposite circular polarizations at 10 Hz. Rotation of the feed and receiver unit was used to identify spurious instrumental responses to linearly polarized radiation, and calibration of the relative pointing of the two beams to $1''$ accuracy was required.
- Although the polarized emission from most sources is small compared with the total emission, it is possible for Stokes visibilities Q_v and U_v to be comparable to I_v in cases in which there is a broad unpolarized component that is highly resolved and a narrower polarized component that is not resolved. In such cases, errors may occur if the approximations for weak polarization [Eqs. (4.38) and (4.43)] are used in the data analysis.
- For most antennas, the instrumental polarization varies over the main beam and increases toward the beam edges. Sidelobes that are cross polarized relative to the main beam tend to peak near the beam edges. Thus, polarization measurements

are usually made for cases in which the source is small compared with the width of the main beam, and for such measurements, the beam should be centered on the source.

- Faraday rotation of the plane of polarization of incoming radiation occurs in the ionosphere and becomes important for frequencies below a few gigahertz; see Table 14.1. During polarization measurements, periodic observations of a strongly polarized source are useful for monitoring changes in the rotation, which varies with the total column density of electrons in the ionosphere. If not accounted for, Faraday rotation can cause errors in calibration; see, for example, Sakurai and Spangler (1994).

- In some antennas, the feed is displaced from the axis of the main reflector, for example, when the Cassegrain focus is used and the feeds for different bands are located in a circle around the vertex. For circularly polarized feeds, this departure from circular symmetry results in pointing offsets of the beams for the two opposite hands. The pointing directions of the two beams are typically separated by ~ 0.1 beamwidths, which makes measurements of circular polarization difficult because V_v is proportional to $(R_{rr} - R_{\ell\ell})$. For linearly polarized feeds, the corresponding effect is an increase in the cross-polarized sidelobes near the beam edges.

- In VLBI, the large distances between antennas result in different parallactic angles at different sites, which must be taken into account.

- The quantities m_ℓ and m_t, of Eqs. (4.20) and (4.22), have Rice distributions of the form of Eq. (6.63a), and the position angle has a distribution of the form of Eq. (6.63b). The percentage polarization can be overestimated, and a correction should be applied (Wardle and Kronberg 1974).

The following points concern choices in designing an array for polarization measurements.

- The rotation of an antenna on an altazimuth mount, relative to the sky, can sometimes be used to advantage in polarimetry. However, the rotation could be a disadvantage in cases in which polarization imaging over a large part of the antenna beam is being attempted. Correction for the variation of instrumental polarization over the beam may be more complicated if the beam rotates on the sky.

- With linearly polarized antennas, errors in calibration are likely to cause I_v to corrupt the linear parameters Q_v and U_v, so for measurements of linear polarization, circularly polarized antennas offer an advantage. Similarly, with circularly polarized antennas, calibration errors are likely to cause I_v to corrupt V_v, so for measurements of circular polarization, linearly polarized antennas may be preferred.

- Linearly polarized feeds for reflector antennas can be made with relative bandwidths of at least $2:1$, whereas for circularly polarized feeds, the maximum relative bandwidth is commonly about $1.4:1$. In many designs of circularly polarized feeds, orthogonal linear components of the field are combined with

±90° relative phase shifts, and the phase-shifting element limits the bandwidth. For this reason, linear polarization is sometimes the choice for synthesis arrays [see, e.g., James (1992)], and with careful calibration, good polarization performance is obtainable.

- The stability of the instrumental polarization, which greatly facilitates accurate calibration over a wide range of hour angle, is perhaps the most important feature to be desired. Caution should therefore be used if feeds are rotated relative to the main reflector or if antennas are used near the high end of their frequency range.

4.8   The Interferometer Measurement Equation

The set of equations for the visibility values that would be measured for a given brightness distribution—taking account of all details of the locations and characteristics of the individual antennas, the path of the incoming radiation through the Earth's atmosphere including the ionosphere, the atmospheric transmission, etc.—is commonly referred to as the *measurement equation* or the *interferometer measurement equation*. For any specified brightness distribution and any system of antennas, the measurement equation provides accurate values of the visibility that would be observed. The reverse operation, i.e., the calculation of the optimum estimate of the brightness distribution from the measured visibility values, is more complicated. Taking the Fourier transform of the observed visibility function usually produces a brightness function with physically distorted features such as negative brightness values in some places. However, starting with a physically realistic model for the brightness, the measurement equation can accurately provide the corresponding visibility values that would be observed. This provides a basis for derivation of realistic brightness distributions that represent the observed visibilities, using an iterative procedure.

The formulation of the interferometer measurement equation is based on the analysis of Hamaker et al. (1996) and further developed by Rau et al. (2009), Smirnov (2011a,b,c,d), and others. It traces the variations of the signals from a source to the output of the receiving system. Direction-dependent effects include the direction of propagation of the signals, the primary beams of the antennas, polarization effects that vary with the alignment of the polarization of the source relative to that of the antennas, and also the effects of the ionosphere and troposphere. Direction-independent effects include the gains of the signal paths from the outputs of the antennas to the correlator. It is necessary to take account of all these various effects to calculate accurately the visibility values corresponding to the source model. Several of these effects are dependent upon the types of the interferometer antennas and the observing frequencies, so the details of the measurement equation are to some extent specific to each particular instrument to which it is applied.

The variations in the signal characteristics can generally be expressed as the effects of Faraday rotation, parallactic rotation, tilting of the wavefront by propagation effects, and variations in feed responses. These are linear effects on the signal and, as noted in Sect. 4.7.4, each of them can be represented by a 2×2 (Jones) matrix. Their effect on the signal matrix is given by a series of outer products as explained with respect to Eq. (4.48). If the original signal is represented by the vector I and the series of effects along the signal path by Jones matrices $\mathbf{J}_1$ to $\mathbf{J}_n$ for antenna p and $\mathbf{J}_1$ to $\mathbf{J}_m$ for antenna q, then the voltage at the correlator output from the pair of antennas m and n is represented by

$$\mathcal{V} = \mathbf{J}_{pn}(\dots (\mathbf{J}_{p2}(\mathbf{J}_{p1} I \, \mathbf{J}_{q1}^H) \mathbf{J}_{q2}^H) \dots) \mathbf{J}_{qm}^H \,, \qquad (4.62)$$

where the superscript H indicates the Hermitian (complex) conjugate. Each of the $\mathbf{J}_p$ terms represents a 2×2 (Jones) matrix. This analysis is from Smirnov (2011a,b,c,d). The combination of the various corrections into a single equation is helpful in ensuring that no significant effects have been overlooked.

An alternative formulation takes each product $\mathbf{J}_{pn} \otimes \mathbf{J}_{pn}^H$, which results in a 4×4 (Mueller) matrix for each of the effects to be corrected along the signal path. If the resulting matrices are represented by $[\mathbf{J}_{pn} \otimes \mathbf{J}_{pn}^H]$, where n indicates the physical order in which the effects are encountered in the propagation path, then the correction for the effects is obtained as a series of products:

$$\mathcal{V} = [\mathbf{J}_{pn} \otimes \mathbf{J}_{pn}^H] \dots [\mathbf{J}_{p2} \otimes \mathbf{J}_{p2}^H][\mathbf{J}_{p1} \otimes \mathbf{J}_{p1}^H] S I \,, \qquad (4.63)$$

where S is a Fourier transform matrix that converts the Stokes visibility to brightness. Each of the $\mathbf{J}_p \otimes \mathbf{J}_p^H$ terms represents a 4×4 matrix. This is basically the form used by Rau et al. (2009). The details of the interferometer equation will vary for different instruments, depending upon which factors need to be included. Here, the intention is to give a general outline of how the calibration factors can be applied. Further details can be found in papers by Hamaker et al. (1996), Hamaker (2000), Rau et al. (2009), and Smirnov (2011a,b,c,d).

4.8.1 Multibaseline Formulation

In this chapter thus far, we have mainly considered the response of a single pair of antennas. The data gathered from a multielement array can conveniently be expressed in the form of a covariance matrix. The discussion here largely follows Leshem et al. (2000) and Boonstra and van der Veen (2003). We start from the expression for the two-element interferometer response and, for simplicity, consider the small-angle case in which the w component can be omitted, as in Eq. (3.9),

$$\mathcal{V}(u, v) = \int_{-\infty}^{\infty} \int_{-\infty}^{\infty} \frac{A_N(l, m) I(l, m)}{\sqrt{1 - l^2 - m^2}} e^{-j 2\pi (ul + vm)} \, dl \, dm \,. \qquad (4.64)$$

Here, $\mathcal{V}$ is the complex visibility, and u and v represent the projected baseline coordinates measured in wavelengths in a plane normal to the phase reference direction. We make four adjustments to the equation. (1) We assume that both the astronomical brightness function and the visibility function can each be represented by a point-source model with a number of points p. For a point k, the direction is specified by direction cosines (l_k, m_k). We replace the integrals in Eq. (4.64) with summations over the points. (2) We replace A_N by the product of the corresponding complex voltage gain factors $g_i(l, m)g_j^*(l, m)$, where i and j indicate antennas. Constants representing conversion of aperture to gain, etc., can be ignored since, in practice, the intensity scale is determined by calibration. (3) We allow the factor $\sqrt{1 - l^2 - m^2})$ to be subsumed within the intensity function $I(l, m)$. (4) For each antenna, we specify the components in the (u, v) plane relative to a reference point that can be chosen, for example, to be the center of the array. The (u, v) values for a pair of antennas i and j then become $(u_i - u_j, v_i - v_j)$. The second and fourth modifications allow the parameters involved to be specified in terms of individual antennas rather than antenna pairs. Equation (4.64) can now be written as:

$$\mathcal{V}(u_i - u_j, v_i - v_j) = \sum_{k=1}^{p} I_k \, g_i(l_k, m_k) \, e^{-j2\pi(u_i l_k + v_i m_k)} g_j^*(l_k, m_k) \, e^{j2\pi(u_j l_k + v_j m_k)} \,,$$

(4.65)

where $I_k = I(l_k, m_k)$. Note that u and v do not vary with the source positions within the field of view but are defined for the phase reference position (field center). Equations (4.64) and (4.65) represent the visibility as measured by a single pair of antennas.

It is useful to put Eq. (4.65) in matrix form. For an array of n antennas, we define an $n \times p$ matrix containing terms corresponding to the first antenna gain and exponential terms of Eq. (4.65) (i.e., the terms associated with antenna i):

$$\mathbf{A} =$$

$$\cdot \begin{bmatrix} g_1(l_1, m_1)e^{-j2\pi(u_1 l_1 + v_1 m_1)} & g_1(l_2, m_2)e^{-j2\pi(u_1 l_2 + v_1 m_2)} & \ldots & g_1(l_p, m_p)e^{-j2\pi(u_1 l_p + v_1 m_p)} \\ g_2(l_1, m_1)e^{-j2\pi(u_2 l_1 + v_2 m_1)} & \ldots & \ldots & \ldots \\ \vdots & \vdots & \vdots & \vdots \\ g_n(l_1, m_1)e^{-j2\pi(u_n l_1 + v_n m_1)} & \ldots & \ldots & g_n(l_p, m_p)e^{-j2\pi(u_n l_p + v_n m_p)} \end{bmatrix} .$$

(4.66)

The antenna index increases downward across the n rows, and the point-source index increases toward the right across the p columns.

To generate the covariance matrix, we first define a $p \times p$ diagonal matrix containing the intensity values of the p source-model points:

$$\mathbf{B} = \begin{bmatrix} I_1 & & & \\ & I_2 & & \\ & & \ddots & \\ & & & I_p \end{bmatrix}. \tag{4.67}$$

Then we can write

$$\mathbf{R} = \mathbf{ABA}^H, \tag{4.68}$$

where the superscript H indicates the Hermitian transpose (transposition of the matrix plus complex conjugation). $\mathbf{R}$ is the covariance matrix, which is Hermitian with dimensions $n \times n$. Each element of $\mathbf{R}$ is of the form of the right side of Eq. (4.65), that is, the sum of responses to the p intensity points for a specific pair of antennas. For row i and column j, the element is $r_{i,j}$, which is equal to the right side of Eq. (4.65). The elements $r_{i,j}$ represent the cross-correlation of signals from antennas i and j. When the gain factors g are equal to unity, the elements represent the source visibility $\mathcal{V}$. The diagonal elements are the n self-products ($i = j$), which represent the total power responses of the antennas. Note that $\mathbf{R}$ is Hermitian: $r_{i,j} = r_{j,i}^*$. $\mathbf{R}$ contains the full set of correlator output terms for an array of n antennas for a single averaging period and a single frequency channel. These data, when calibrated as visibility, can provide a snapshot image. In cases in which the w component is important, a term of the form $w(\sqrt{1 - l^2 - m^2} - 1)$ [as in Eq. (3.7)] with appropriate subscripts, can be included within each exponent. If the response patterns of the antennas are identical, i.e., $g_i = g_j$ for all (i, j), then $g_i g_j^* = |g|^2$, and this (real) gain factor can be taken outside the matrix $\mathbf{R}$. Thus, to determine the angle of incidence (l, m) of a signal from the covariance measurements [the (u, v) values being known], the gain factors need not be known if they are identical from one antenna to another but otherwise must be known.

The covariance matrix can also be formulated in terms of the complex signal voltages from the antennas of an array. Let the signal from antenna k be x_k, which is a function of time. For the array, the signals can be represented by a (column) vector $\mathbf{x}$ of dimensions $n \times 1$, each term of which corresponds to the sum of the terms in the corresponding row of the matrix in Eq. (4.66). The outer (or Kronecker) product $\mathbf{x} \otimes \mathbf{x}^H$ leads to a covariance matrix:

$$\mathbf{R}' = \begin{bmatrix} x_1 \\ x_2 \\ \vdots \\ x_n \end{bmatrix} \otimes \begin{bmatrix} x_1^* & x_2^* & \cdots & x_n^* \end{bmatrix} = \begin{bmatrix} x_1 x_1^* & x_1 x_2^* & \cdots & x_1 x_n^* \\ x_2 x_1^* & \cdots & \cdots & \cdots \\ \vdots & \vdots & \vdots & \vdots \\ x_n x_1^* & \cdots & \cdots & x_n x_n^* \end{bmatrix}. \tag{4.69}$$

The elements $r_{i,j}$ of the matrix $\mathbf{R}$ represent the correlator outputs, which involve a time average of the signal products. If the signal products in the elements of $\mathbf{R}'$ are similarly understood to represent time-averaged products, then $\mathbf{R}'$ is equivalent to the covariance matrix $\mathbf{R}$.

An example of the application of matrix formulation in radio astronomy is provided by the discussion of gain calibration by Boonstra and van der Veen (2003). Also, the eigenvectors of the matrix can be used to identify interfering signals that are strong enough to be distinguished in the presence of the noise. Such signals can then be removed from the data, as discussed, for example, by Leshem et al. (2000).

Appendix 4.1   Hour Angle–Declination and Elevation–Azimuth Relationships

Although the positions of cosmic sources are almost always specified in celestial coordinates, for purposes of observation, it is generally necessary to convert to elevation and azimuth. The conversion formulas between hour angle and declination (H, δ) and elevation and azimuth $(\mathcal{E}, \mathcal{A})$ can be derived by applying the sine and cosine rules for spherical triangles to the system in Fig. 4.3. For an observer at latitude $\mathcal{L}$, they are, for (H, δ) to $(\mathcal{A}, \mathcal{E})$,

$$\sin \mathcal{E} = \sin \mathcal{L} \sin \delta + \cos \mathcal{L} \cos \delta \cos H$$
$$\cos \mathcal{E} \cos \mathcal{A} = \cos \mathcal{L} \sin \delta - \sin \mathcal{L} \cos \delta \cos H \qquad (A4.1)$$
$$\cos \mathcal{E} \sin \mathcal{A} = - \cos \delta \sin H \ ,$$

Similarly, for $(\mathcal{A}, \mathcal{E})$ to (H, δ),

$$\sin \delta = \sin \mathcal{L} \sin \mathcal{E} + \cos \mathcal{L} \cos \mathcal{E} \cos \mathcal{A}$$
$$\cos \delta \cos H = \cos \mathcal{L} \sin \mathcal{E} - \sin \mathcal{L} \cos \mathcal{E} \cos \mathcal{A} \qquad (A4.2)$$
$$\cos \delta \sin H = - \cos \mathcal{E} \sin \mathcal{A} \ .$$

Here, azimuth is measured from north through east.

Appendix 4.2   Leakage Parameters in Terms of the Polarization Ellipse

The polarization leakage terms used to express the instrumental polarization are related to the ellipticity and orientation of the polarization ellipses of each antenna, as shown below.

A4.2.1   Linear Polarization

Consider the antenna in Fig. 4.8, and suppose that it is nominally linearly polarized in the x direction, in which case χ and ψ are small angles that represent engineering tolerances. A field E aligned with the x axis in Fig. 4.8a produces components $E_{x'}$ and $E_{y'}$ along the (x', y') axes with which the dipoles in Fig. 4.8b are aligned. Then from Eq. (4.26), we obtain the voltage at the output of the antenna (point A in Fig. 4.8b), which is

$$V'_x = E(\cos \psi \cos \chi + j \sin \psi \sin \chi) . \tag{A4.3}$$

The response to the same field, but aligned with the y axis, is

$$V'_y = E(\sin \psi \cos \chi - j \cos \psi \sin \chi) . \tag{A4.4}$$

V'_x represents the wanted response to the field along the x axis, and V'_y represents the unwanted response to a cross-polarized field. The leakage term is equal to the cross-polarized response expressed as a fraction of the wanted x-polarization response, that is,

$$D_x = \frac{V'_y}{V'_x} = \frac{(\sin \psi_x \cos \chi_x - j \cos \psi_x \sin \chi_x)}{(\cos \psi_x \cos \chi_x + j \sin \psi_x \sin \chi_x)} \simeq \psi_x - j\chi_x , \tag{A4.5}$$

where the subscript x indicates the x-polarization case. The corresponding term D_y, for the condition in which Fig. 4.8 represents the nominal y polarization of the antenna, is obtained as V'_x/V'_y by inverting Eq. (A4.5), replacing ψ_x by $\psi_y + \pi/2$, and replacing χ_x by χ_y. Then ψ_y is measured from the y axis in the same sense as ψ_x is measured from the x axis, that is, increasing in a counterclockwise direction in Fig. 4.8. Thus, we obtain

$$D_y = \frac{V'_x}{V'_y} = \frac{[\cos (\psi_y + \pi/2) \cos \chi_y + j \sin (\psi_y + \pi/2) \sin \chi_y]}{[\sin (\psi_y + \pi/2) \cos \chi_y - j \cos (\psi_y + \pi/2) \sin \chi_y]}$$

$$= \frac{(-\sin \psi_y \cos \chi_y + j \cos \psi_y \sin \chi_y)}{(\cos \psi_y \cos \chi_y + j \sin \psi_y \sin \chi_y)} \simeq -\psi_y + j\chi_y . \tag{A4.6}$$

Similar expressions for D_x and D_y have also been derived by Sault et al. (1991). Note that D_x and D_y are of comparable magnitude and opposite sign, so one would expect the average of all the D terms for an array of antennas to be very small. As used earlier in this chapter, subscripts m and n are added to the D terms to indicate individual antennas.

A4.2.2 Circular Polarization

To receive right circular polarization from the sky, the antenna in Fig. 4.8b must respond to a field with counterclockwise rotation in the plane of the diagram, as explained earlier. This requires $\chi = -45°$. In terms of fields in the x and y directions, counterclockwise rotation requires that E_x leads E_y in phase by $\pi/2$; that is, $E_x = jE_y$ for the fields as defined in Eq. (4.25). For fields E_x and E_y, we determine the components in the x' and y' directions and then obtain expressions for the output of the antenna for both counterclockwise and clockwise rotation of the incident field. For counterclockwise rotation:

$$E'_x = E_x \cos \psi + E_y \sin \psi = E_x(\cos \psi - j \sin \psi) \,, \tag{A4.7}$$

$$E'_y = -E_x \sin \psi + E_y \cos \psi = -E_x(\sin \psi + j \cos \psi) \,. \tag{A4.8}$$

For nominal right-circular polarization, $\chi_r = -\pi/4 + \Delta\chi_r$, where $\Delta\chi_r$ is a measure of the departure of the polarization from circularity. Then from Eq. (4.26), we obtain

$$V'_r = E_x e^{-j\psi_r}(\cos \chi_r - \sin \chi_r) = \sqrt{2} E_x e^{-j\psi_r} \cos \Delta\chi_r \,. \tag{A4.9}$$

The next step is to repeat the procedure for left circular polarization from the sky, for which we have clockwise rotation of the electric vector and $E_y = jE_x$. The result is

$$V'_\ell = E_x e^{j\psi_r}(\cos \chi_r + \sin \chi_r) = \sqrt{2} E_x e^{j\psi_r} \sin \Delta\chi_r \,. \tag{A4.10}$$

The relative magnitude of the opposite-hand response of the nominally right-handed polarization state, that is, the leakage term, is

$$D_r = \frac{V'_\ell}{V'_r} = e^{j2\psi_r} \tan \Delta\chi_r \simeq e^{j2\psi_r} \Delta\chi_r \,. \tag{A4.11}$$

For nominal left-handed polarization, the relative magnitude of the opposite-hand response is obtained by inverting the right side of Eq. (A4.11) and also substituting $\Delta\chi_\ell + \pi/2$ for $\Delta\chi_r$ and $\psi_\ell - \pi/2$ for ψ_r. For the corresponding leakage term D_ℓ, which represents the right circular leakage of the nominally left circularly polarized antenna, we then obtain

$$D_\ell = e^{-j2\psi_\ell} \tan \Delta\chi_\ell \simeq e^{-j2\psi_\ell} \Delta\chi_\ell \,. \tag{A4.12}$$

Since $-\pi/4 \le \chi \le \pi/4$, $\Delta\chi_r$ and $\Delta\chi_\ell$ take opposite signs. Thus, as in the case of the leakage terms for linear polarization, D_r and D_ℓ are of comparable magnitude and opposite sign.

References

Bignell, R.C., Polarization, in *Synthesis Mapping, Proceedings of NRAO Workshop No. 5*, Socorro, NM, June 21–25, 1982, Thompson, A.R., and D'Addario, L.R., Eds., National Radio Astronomy Observatory, Green Bank, WV (1982)

Boonstra, A.-J., and van der Veen, A.-J., Gain Calibration Methods for Radio Telescope Arrays, *IEEE Trans. Signal Proc.*, **51**, 25–38 (2003)

Born, M., and Wolf, E., *Principles of Optics*, 7th ed., Cambridge Univ. Press, Cambridge, UK (1999)

Conway, R.G., and Kronberg, P.P., Interferometer Measurement of Polarization Distribution in Radio Sources, *Mon. Not. R. Astron. Soc.*, **142**, 11–32 (1969)

Cotton, W.D., Calibration and Imaging of Polarization Sensitive Very Long Baseline Interferometer Observations, *Astron. J.*, **106**, 1241–1248 (1993)

DeBoer, D.R., Gough, R.G., Bunton, J.D., Cornwell, T.J., Beresford, R.J., Johnston, S., Feain, I.J., Schinckel, A.E., Jackson, C.A., Kesteven, M.J., and nine coauthors, Australian SKA Pathfinder: A High-Dynamic Range Wide-Field of View Survey Telescope, *Proc. IEEE*, **97**, 1507–1521 (2009)

de Vos, M., Gunst, A.W., and Nijboer, R., The LOFAR Telescope: System Architecture and Signal Processing, *Proc. IEEE*, **97**, 1431–1437 (2009)

Fiebig, D., and Güsten, R., Strong Magnetic Fields in Interstellar Maser Clumps, *Astron. Astrophys.*, **214**, 333–338 (1989)

Fomalont, E.B., and Perley, R.A., Calibration and Editing, in *Synthesis Imaging in Radio Astronomy*, Perley, R.A., Schwab, F.R., and Bridle, A.H., Eds., Astron. Soc. Pacific Conf. Ser., **6**, 83–115 (1989)

Frater, R.H., and Brooks, J.W., Eds., *J. Electric. Electron. Eng. Aust.*, Special Issue on the Australia Telescope, **12**, No. 2 (1992)

Hamaker, J.P., A New Theory of Radio Polarimetry with an Application to the Westerbork Synthesis Radio Telescope (WSRT), in *Workshop on Large Antennas in Radio Astronomy*, ESTEC, Noordwijk, the Netherlands (1996)

Hamaker, J.P., Understanding Radio Polarimetry. IV. The Full-Coherency Analogue of Scalar Self-Calibration: Self-Alignment, Dynamic Range, and Polarimetric Fidelity, *Astron. Astrophys. Suppl.*, **143**, 515–534 (2000)

Hamaker, J.P., Understanding Radio Polarimetry. V. Making Matrix Self-Calibration Work: Processing of a Simulated Observation, *Astron. Astrophys.*, **456**, 395–404 (2006)

Hamaker, J.P., Bregman, J.D., and Sault, R.J., Understanding Radio Polarimetry. I. Mathematical Foundations, *Astron. Astrophys. Suppl.*, **117**, 137–147 (1996)

IAU, *Trans. Int. Astron. Union*, Proc. of the 15th General Assembly Sydney 1973 and Extraordinary General Assembly Poland 1973, G. Contopoulos and A. Jappel, Eds., **15B**, Reidel, Dordrecht, the Netherlands (1974), p. 166

IEEE, Standard Definitions of Terms for Radio Wave Propagation, Std. 211–1977, Institute of Electrical and Electronics Engineers Inc., New York (1977)

James, G.L., The Feed System, *J. Electric. Electron. Eng. Aust.*, Special Issue on the Australia Telescope, **12**, No. 2, 137–145 (1992)

Jones, R.C., A New Calculus for the Treatment of Optical Systems. I. Description and Discussion of the Calculus, *J. Opt. Soc. Am.*, **31**, 488–493 (1941)

Kemball, A.J., Diamond, P.J., and Cotton, W.D., Data Reduction Techniques for Spectral Line Polarization VLBI Observations, *Astron. Astrophys. Suppl.*, **110**, 383–394 (1995)

Ko, H.C., Coherence Theory of Radio-Astronomical Measurements, *IEEE Trans. Antennas Propag.*, **AP-15**, 10–20 (1967a)

Ko, H.C., Theory of Tensor Aperture Synthesis, *IEEE Trans. Antennas Propag.*, **AP-15**, 188–190 (1967b)

Kraus, J.D., and Carver, K.R., *Electromagnetics*, 2nd ed., McGraw-Hill, New York (1973) p. 435

Leshem, A., van der Veen, A.-J., and Boonstra, A.-J., Multichannel Interference Mitigation Techniques in Radio Astronomy, *Astrophys. J. Suppl.*, **131**, 355–373 (2000)

Morris, D., Radhakrishnan, V., and Seielstad, G.A., On the Measurement of Polarization Distributions Over Radio Sources, *Astrophys. J.*, **139**, 551–559 (1964)

O'Neill, E.L., *Introduction to Statistical Optics*, Addison-Wesley, Reading, MA (1963)

Rau, U., Bhatnagar, S., Voronkov, M.A., and Cornwell, T.J., Advances in Calibration and Imaging Techniques in Radio Interferometry, *Proc. IEEE*, **97**, 1472–1481 (2009)

Roberts, D.H., Brown, L.F., and Wardle, J.F.C., Linear Polarization Sensitive VLBI, in *Radio Interferometry: Theory, Techniques, and Applications*, Cornwell, T.J., and Perley, R.A., Eds., Astron. Soc. Pacific Conf. Ser., **19**, 281–288 (1991)

Roberts, D.H., Wardle, J.F.C., and Brown, L.F., Linear Polarization Radio Imaging at Milliarcsecond Resolution, *Astrophys. J.*, **427**, 718–744 (1994)

Rowson, B., High Resolution Observations with a Tracking Radio Interferometer, *Mon. Not. R. Astron. Soc.*, **125**, 177–188 (1963)

Sakurai, T., and Spangler, S.R., Use of the Very Large Array for Measurement of Time Variable Faraday Rotation, *Radio Sci.*, **29**, 635–662 (1994)

Sault, R.J., Hamaker, J.P., and Bregman, J.D., Understanding Radio Polarimetry. II. Instrumental Calibration of an Interferometer Array, *Astron. Astrophys. Suppl.*, **117**, 149–159 (1996)

Sault, R.J., Killeen, N.E.B., and Kesteven, M.J., *AT Polarization Calibration*, Aust. Tel. Tech. Doc. Ser. 39.3/015, CSIRO, Epping, New South Wales (1991)

Seidelmann, P.K., Ed., *Explanatory Supplement to the Astronomical Almanac*, Univ. Science Books, Mill Valley, CA (1992)

Smegal, R.J., Landecker, T.L., Vaneldik, J.F., Routledge, D., and Dewdney, P.E., Aperture Synthesis Polarimetry: Application to the Dominion Astrophysical Observatory Synthesis Telescope, *Radio Sci.*, **32**, 643–656 (1997)

Smirnov, O.M., Revisiting the Radio Interferometer Measurement Equation. 1. A Full-Sky Jones Formalism, *Astron. Astrophys.*, **527**, A106 (11pp) (2011a)

Smirnov, O.M., Revisiting the Radio Interferometer Measurement Equation. 2. Calibration and Direction-Dependent Effects, *Astron. Astrophys.*, **527**, A107 (10pp) (2011b)

Smirnov, O.M., Revisiting the Radio Interferometer Measurement Equation. 3. Addressing Direction-Dependent Effects in 21-cm WSRT Observations of 3C147, *Astron. Astrophys.*, **527**, A108 (12pp) (2011c)

Smirnov, O.M., Revisiting the Radio Interferometer Measurement Equation. 4. A Generalized Tensor Formalism, *Astron. Astrophys.*, **531**, A159 (16pp) (2011d)

Wade, C.M., Precise Positions of Radio Sources. I. Radio Measurements, *Astrophys. J.*, **162**, 381–390 (1970)

Wardle, J.F.C., and Kronberg, P.P., The Linear Polarization of Quasi-Stellar Radio Sources at 3.71 and 11.1 Centimeters, *Astrophys. J.*, **194**, 249–255 (1974)

Weiler, K.W., The Synthesis Radio Telescope at Westerbork: Methods of Polarization Measurement, *Astron. Astrophys.*, **26**, 403–407 (1973)

Weiler, K.W., and Raimond, E., Aperture Synthesis Observations of Circular Polarization, *Astron. Astrophys.*, **52**, 397–402 (1976)

Wilson, T.L., Rohlfs, K., and Hüttemeister, S., *Tools of Radio Astronomy*, 6th ed., Springer-Verlag, Berlin (2013)

Chapter 5
Antennas and Arrays

This chapter opens with a brief review of some basic considerations of antennas. The main part of the chapter is concerned with the configurations of antennas in interferometers and synthesis arrays. It is convenient to classify array designs as follows:

1. Arrays with nontracking antennas
2. Interferometers and arrays with antennas that track the sidereal motion of a source:

   - Linear arrays
   - Arrays with open-ended arms (crosses, T-shaped arrays, and Y-shaped arrays)
   - Arrays with closed configurations (circles, ellipses, and Reuleaux triangles)
   - VLBI arrays
   - Planar arrays.

Examples of these types of arrays are described, and their spatial transfer functions (i.e., spatial sensitivities) are compared. Other concerns include the size and number of antennas needed in an array. Also discussed is the technique of forming images from direct Fourier transformation of the electric field on an aperture.

5.1 Antennas

The subject of antennas is well covered in numerous books; see Further Reading at the end of this chapter. Baars (2007) gives an informative review of parabolic antennas, including details of testing and surface adjustment. Here, we are concerned with the special requirements of antennas for radio astronomy. As discussed in Chap. 1, early radio astronomy antennas operated mainly at meter wavelengths and often consisted of arrays of dipoles or parabolic-cylinder reflectors. These had large areas,

© The Author(s) 2017
A.R. Thompson, J.M. Moran, and G.W. Swenson Jr., *Interferometry and Synthesis in Radio Astronomy*, Astronomy and Astrophysics Library, DOI 10.1007/978-3-319-44431-4_5

but the operating wavelengths were long enough that beamwidths were usually of order 1° or more. For detection and cataloging of sources, satisfactory observations could be obtained during the passage of a source through a stationary beam or interferometer fringe pattern. Thus, it was not always necessary for such antennas to track the sidereal motion of a source. More recent meter-wavelength systems use dipole arrays with computer-controlled phasing to provide tracking beams [see, e.g., Koles et al. (1994) and Lonsdale et al. (2009)]. For higher frequencies, synthesis arrays use tracking antennas that incorporate equatorial or altazimuth mounts.

The requirement for high sensitivity and angular resolution has resulted in the development of large arrays of antennas. Such instruments are usually designed to cover a range of frequencies. For centimeter-wavelength instruments, the coverage typically includes bands extending from a few hundred megahertz to some tens of gigahertz. For such frequency ranges, the antennas are most often parabolic or similar-type reflectors, with separate feeds for the different frequency bands. In addition to wide frequency coverage, another advantage of the parabolic reflector is that all of the power collected is brought, essentially without loss, to a single focus, which allows full advantage to be taken of low-loss feeds and cryogenically cooled input stages to provide the maximum sensitivity.

Figure 5.1 shows several focal arrangements for parabolic antennas, of which the Cassegrain is perhaps the most often used. The Cassegrain focus offers a number of advantages. A convex hyperbolic reflector intercepts the radiation just before it reaches the prime focus and directs it to the Cassegrain focus near the vertex of the main reflector. Sidelobes resulting from spillover of the beam of the feed around the edges of the subreflector point toward the sky, for which the noise temperature is generally low. With a prime-focus feed, the sidelobes resulting from spillover around the main reflector point toward the ground and thus result in a higher level of unwanted noise pickup. The Cassegrain focus also has the advantage that in all but the smallest antennas, an enclosure can be provided behind the main reflector to

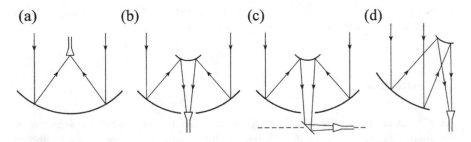

Fig. 5.1 Focus arrangements of reflector antennas: (**a**) prime focus; (**b**) Cassegrain focus; (**c**) Naysmith focus; (**d**) offset Cassegrain. With the Naysmith focus, the feed horn is mounted on the alidade structure below the elevation axis (indicated by the dashed line), and for a linearly polarized signal, the angle of polarization relative to the feed varies with the elevation angle. In some other arrangements, for example, beam-waveguide antennas (not shown), there are several reflectors, including one on the azimuth axis, which allows the feed horn to remain fixed relative to the ground. The polarization then rotates relative to the feed for both azimuth and elevation motions.

accommodate the low-noise input stages of the electronics. However, the aperture of the feed for a prime-focus location is less than that for a feed at the Cassegrain focus, and as a result, the feeds for the longer wavelengths are often at the prime focus.

The Cassegrain design also allows the possibility of improving the aperture efficiency by shaping the two reflectors of the antenna (Williams 1965). The principle involved can best be envisioned by considering the antenna in transmission. With a conventional hyperboloid subreflector and parabolic main reflector, the radiation from the feed is concentrated toward the center of the antenna aperture, whereas for maximum efficiency, the electric field should be uniformly distributed. If the profile of the subreflector is slightly adjusted, more power can be directed toward the outer part of the main reflector, thus improving the uniformity. The main reflector must then be shaped to depart slightly from the parabolic profile to regain uniform phase across the wavefront after it leaves the main reflector. This type of shaping is used, for example, in the antennas of the VLA in New Mexico, for which the main reflector is 25 m in diameter. For the VLA, the rms difference between the reflector surfaces and the best fit paraboloid is ~ 1 cm, so the antennas can be used with prime-focus feeds for wavelengths longer than ~ 16 cm. Shaping is not always to be preferred since it introduces some restriction in off-axis performance, which is detrimental for multibeam applications. Multiple beams for a large parabolic antenna can greatly increase sky coverage, which is particularly useful for survey observations. A beamformer feed system in which beams are formed using phased arrays of feed elements is described by Elmer et al. (2012), who consider various designs (see discussion in Sect. 5.7.2.1).

For tracking parabolic reflectors, there are numerous differences in the detailed design. For example, when a number of feeds for different frequency bands are required at the Cassegrain focus, these are sometimes mounted on a turntable structure, and the feed that is in use is brought to a position on the axis of the main reflector. Alternately, the feeds may be in fixed positions on a circle centered on the vertex, and by using a rotatable subreflector of slightly asymmetric design, the incoming radiation can be focused onto the required feed.

Parabolic reflector antennas with asymmetrical feed geometry can exhibit undesirable instrumental polarization effects that would largely cancel out in a circularly symmetrical antenna. This may occur in an unblocked aperture design, as in Fig. 5.1d, or in a design in which a cluster of feeds is used for operation on a number of frequency bands, where the feeds are close to, but not exactly on, the axis of the paraboloid. With crossed linearly polarized feeds, the asymmetry results in cross-polarization sidelobes within the main beam. With opposite circularly polarized feeds, the two beams are offset in opposite directions in a plane that is normal to the plane containing the axis of symmetry of the reflector and the center of the feed. This offset can be a serious problem in measurements of circular polarization, since the result is obtained by taking the difference between measurements with opposite circularly polarized responses (see Table 4.3). For measurements of linear polarization, the offset is less serious since this involves taking the product of two opposite-hand outputs, and the resulting response is symmetrical about the axis of the parabola. The effects can be largely canceled by inserting a compensating offset in a secondary reflector. For further details, see Chu and Turrin (1973) and Rudge and Adatia (1978).

A basic point concerns the accuracy of the reflector surface. Deviations of the surface from the ideal profile result in variations in the phase of the electromagnetic field as it reaches the focus. We can think of the reflector surface as consisting of many small sections that deviate from the ideal surface by ϵ, a Gaussian random variable with probability distribution

$$p(\epsilon) = \frac{1}{\sqrt{2\pi}\,\sigma} e^{-\epsilon^2/2\sigma^2} , \tag{5.1}$$

where $\langle \epsilon \rangle = 0$, $\langle \epsilon^2 \rangle = \sigma^2$, and $\langle\ \rangle$ indicates the expectation. A relation of general importance in probabilistic calculations is $\langle e^{j\epsilon} \rangle$, which is

$$\langle e^{j\epsilon} \rangle = \int_{-\infty}^{\infty} p(\epsilon) e^{j\epsilon} d\epsilon = \frac{1}{\sqrt{2\pi}\sigma} \int_{-\infty}^{\infty} e^{-\frac{\epsilon^2}{2\sigma^2} - j\epsilon} d\epsilon = e^{-\sigma^2/2} . \tag{5.2}$$

The rightmost integral is accomplished by the method of completing the square in the argument of the exponential, i.e., $-\left(\frac{\epsilon^2}{2\sigma^2} + j\epsilon\right) = -\frac{1}{2\sigma^2}\left(\epsilon + j\sigma^2\right)^2 - \frac{\sigma^2}{2}$. The $e^{-\sigma^2/2}$ factor can be moved outside the integral, the rest of which is unity.

A surface deviation ϵ produces a deviation of approximately 2ϵ in the path length of a reflected ray; this approximation improves as the focal ratio is increased. Thus, a deviation ϵ causes a phase shift $\phi \simeq 4\pi\epsilon/\lambda$, where λ is the wavelength. As a result, the electric field components at the focus have a Gaussian phase distribution with $\sigma_\phi = 4\pi\sigma/\lambda$. If there are N independent sections of the surface, then the collecting area, which is proportional to the square of the electric field, is given by

$$A = A_0 \left\langle \left| \frac{1}{N} \sum_i e^{j\phi_i} \right|^2 \right\rangle = \frac{A_0}{N^2} \sum_{i,k} \langle e^{j(\phi_i - \phi_k)} \rangle \simeq A_0 e^{\frac{-(\sqrt{2}\sigma_\phi)^2}{2}} , \tag{5.3}$$

where A_0 is the collecting area for a perfect surface, and it has been assumed that N is large enough that terms for which $i = k$ can be ignored. The $\sqrt{2}$ factor comes from differencing two random variables. Then from Eqs. (5.2) and (5.3), we obtain

$$A = A_0 e^{-(4\pi\sigma/\lambda)^2} . \tag{5.4}$$

This equation is known in radio engineering as the Ruze formula (Ruze 1966) and in some other branches of astronomy as the Strehl ratio. As an example, if $\sigma/\lambda = 1/20$, the aperture efficiency, A/A_0, is 0.67. In the case of antennas with multiple reflecting surfaces, the rms deviations can be combined in the usual root-sum-squared manner. Secondary reflectors, such as a Cassegrain subreflector, are smaller than the main reflector, and for smaller surfaces, the rms deviation is usually correspondingly smaller. The surface adjustment of the 12-m-diameter antennas of the Atacama Large Millimeter/submillimeter Array (ALMA) array, which are capable of operation up to $\sim$ 900 GHz, is a good example of the accuracy that can be achieved (Mangum et al. 2006). A study of the dynamics of the surface of the antennas is described by Snel et al. (2007).

Several techniques have been developed for improving the performance of parabolic antennas. An example is the adjustment of the subreflector shape to compensate for errors in the main reflector [see, e.g., Ingalls et al. (1994), Mayer et al. (1994)]. Another improvement is in the design of the focal support structure to minimize blockage of the aperture and reduce sidelobes in the direction of the ground (Lawrence et al. 1994; Welch et al. 1996). A common method of supporting equipment near the reflector focus is the use of a tripod or quadrupod structure. If the legs of the structure are connected to the edge of the main reflector rather than to points within the reflector aperture, they interrupt only the plane wave incident on the aperture, not the spherical wavefront between the reflector and the focus. Use of an offset-feed reflector avoids any blockage of the incident wavefront in reaching the focus. However, both of these methods of reducing blockage increase the complexity and cost of the structure.

5.2   Sampling the Visibility Function

5.2.1   Sampling Theorem

The choice of configuration of the antennas of a synthesis array is largely based on optimizing the sampling of the visibility function in (u, v) space. Thus, in considering array design, it is logical to start by examining the sampling requirements. These are governed by the sampling theorem of Fourier transforms (Bracewell 1958). Consider first the measurement of the one-dimensional intensity distribution of a source, $I_1(l)$. It is necessary to measure the complex visibility $\mathcal{V}$ in the corresponding direction on the ground at a series of values of the projected antenna spacing. For example, to measure an east–west profile, a possible method is to make observations near meridian transit of the source using an east–west baseline and to vary the length of the baseline from day to day.

Figure 5.2a–c illustrates the sampling of the one-dimensional visibility function $\mathcal{V}(u)$. The sampling operation can be represented as multiplication of $\mathcal{V}(u)$ by the series of delta functions in Fig. 5.2b, which can be written

$$\left[\frac{1}{\Delta u}\right] \mathrm{III}\left(\frac{u}{\Delta u}\right) = \sum_{i=-\infty}^{\infty} \delta(u - i\Delta u) , \qquad (5.5)$$

where the left side is included to show how the series can be expressed in terms of the *shah function*, III, introduced by Bracewell and Roberts (1954). The series extends to infinity in both positive and negative directions, and the delta functions

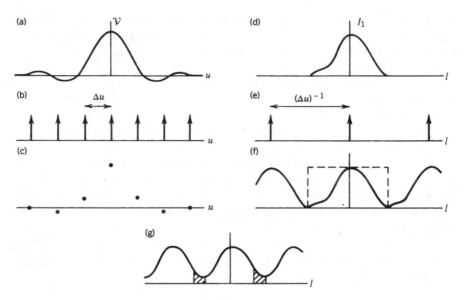

Fig. 5.2 Illustration of the sampling theorem: (**a**) visibility function $\mathcal{V}(u)$, real part only; (**b**) sampling function in which the arrows represent delta functions; (**c**) sampled visibility function; (**d**) intensity function $I_1(l)$; (**e**) replication function; (**f**) replicated intensity function. Functions in (**d**), (**e**), and (**f**) are the Fourier transforms of those in (**a**), (**b**), and (**c**), respectively. (**g**) is the replicated intensity function showing aliasing in the shaded areas resulting from using too large a sampling interval.

are uniformly spaced with an interval Δu. The Fourier transform of Eq. (5.5) is the series of delta functions shown in Fig. 5.2e:

$$\mathrm{III}(l\Delta u) = \frac{1}{\Delta u} \sum_{p=-\infty}^{\infty} \delta\left(l - \frac{p}{\Delta u}\right). \qquad (5.6)$$

In the l domain, the Fourier transform of the sampled visibility is the convolution of the Fourier transform of $\mathcal{V}(u)$, which is the one-dimensional intensity function $I_1(l)$, with Eq. (5.6). The result is the replication of $I_1(l)$ at intervals $(\Delta u)^{-1}$ shown in Fig. 5.2f. If $I_1(l)$ represents a source of finite dimensions, the replications of $I_1(l)$ will not overlap as long as $I_1(l)$ is nonzero only within a range of l that is no greater than $(\Delta u)^{-1}$. Hence, if l_w is the range over which $I_1(l)$ is nonzero or, more generally, the field of view of an observation, then the avoidance of aliasing requires $\Delta u \leq 1/l_w$. An example of overlapping replications is shown in Fig. 5.2g. The loss of information resulting from such overlapping is commonly referred to as *aliasing*, because the components of the function within the overlapping region lose their identity with respect to which end of the replicated function they properly belong. The distortion in the replicated intensity function is said to be caused by "leakage" [see Bracewell (2000)].

The requirement for the restoration of a function from a set of samples, for example, deriving the function in Fig. 5.2a from the samples in Fig. 5.2c, is easily understood by considering the Fourier transforms in Fig. 5.2d and f. Interpolation in the u domain corresponds to removing the replications in the l domain, which can be achieved by multiplying the function in Fig. 5.2f by the rectangular function indicated by the broken line. In the u domain, this multiplication corresponds to convolution of the sampled values with the Fourier transform of the rectangular function, which is the unit area sinc function,

$$\frac{\sin \pi u/\Delta u}{\pi u}. \tag{5.7}$$

If aliasing is avoided, convolution with (5.7) provides exact interpolation of the original function from the samples. Note that perfect restoration requires a sum over all samples except when the sinc function is centered on a specific sample. Thus, we can state, as the sampling theorem for the visibility, that *if the intensity distribution is nonzero only within an interval of width* l_w, $I_1(l)$ *is fully specified by sampling the visibility function at points spaced* $\Delta u = l_w^{-1}$ *in u.* The interval $\Delta u = l_w^{-1}$ is called the critical sampling interval. Sampling at a finer interval in u is called oversampling and usually does no harm nor does it provide any benefit. Sampling at a coarser interval is called undersampling, which leads to aliasing.

Aliasing can lead to serious misinterpretation of source structure. For example, suppose the intensity function $I_1(l)$ consists of a number of compact separated components. A component that lies outside the proper sampling window, i.e., $|l| > l_w/2$, at negative l will be aliased to a position on the positive side of the replicated intensity function. Thus, its appears at the wrong position. This error can be discovered by regridding the data at a finer interval Δu. An aliased component will move in an unexpected way in the image plane.

The spatial sampling theorem described here is just a formulation of the standard Shannon–Nyquist theorem normally written in the time (t)-frequency (v) domain. Here, the critical sampling frequency for a temporal waveform of bandwidth Δv is $1/(2\Delta v)$. The factor of two appears because the spectrum in Fourier space extends from $-\Delta v$ to $+\Delta v$.

In two dimensions, it is simply necessary to apply the theorem separately to the source in the l and m directions. A compact source that is just beyond the sampling limit at the lower left of the image will be aliased into the sampling interval in the upper right. For further discussion of the sampling theorem, see, for example, Unser (2000).

5.2.2 Discrete Two-Dimensional Fourier Transform

The derivation of an image (or map) from the visibility measurements is the subject of Chap. 10, but it is important at this point to understand the form in which the visibility data are required for this transformation. The discrete Fourier transform (DFT) is very widely used in synthesis imaging because of the computational advantages of the fast Fourier transform (FFT) algorithm [see, e.g., Brigham (1988)]. The basic properties of the DFT in one dimension are described in Appendix 8.4. In two dimensions, the functions $\mathcal{V}(u, v)$ and $I(l, m)$ are expressed as rectangular matrices of sampled values at uniform increments in the two variables involved. The rectangular grid points on which the intensity is obtained provide a convenient form for further data processing.

The two-dimensional form of the discrete transform for a Fourier pair f and g is defined by

$$f(p,q) = \sum_{i=0}^{M-1} \sum_{k=0}^{N-1} g(i,k)\, e^{-j2\pi ip/M} e^{-j2\pi kq/N} \; , \tag{5.8}$$

and the inverse is

$$g(i,k) = \sum_{p=0}^{M-1} \sum_{q=0}^{N-1} f(p,q)\, e^{j2\pi ip/M} e^{j2\pi kq/N} \; . \tag{5.9}$$

The functions are periodic with periods of M samples in the i and p dimensions and N samples in the k and q dimensions. Evaluation of Eqs. (5.8) or (5.9) by direct computation requires approximately $(MN)^2$ complex multiplications. In contrast, if M and N are powers of 2, the FFT algorithm requires only $\frac{1}{2}MN\log_2(MN)$ complex multiplications.

The transformation between $\mathcal{V}(u, v)$ and $I(l, m)$, where I is the source intensity in two dimensions, is obtained by substituting $g(i, k) = I(i\Delta l, k\Delta m)$ and $f(p, q) = \mathcal{V}(p\Delta u, q\Delta v)$ in Eqs. (5.8) and (5.9). The relationship between the integral and discrete forms of the Fourier transform is found in several texts; see, for example, Rabiner and Gold (1975) or Papoulis (1977). The dimensions of the (u, v) plane that contain these data are $M\Delta u$ by $N\Delta v$. In the (l, m) plane, the points are spaced Δl in l and Δm in m, and the image dimensions are $M\Delta l$ by $N\Delta m$. The dimensions in the two domains are related by

$$\Delta u = (M\Delta l)^{-1}, \qquad \Delta v = (N\Delta m)^{-1} \; ,$$
$$\Delta l = (M\Delta u)^{-1}, \qquad \Delta m = (N\Delta v)^{-1} \; . \tag{5.10}$$

The spacing between points in one domain is the reciprocal of the total dimension in the other domain. Thus, if the size of the array in the intensity domain is chosen to be large enough that the intensity function is nonzero only within the area $M\Delta l \times N\Delta m$, then the spacings Δu and Δv in Eq. (5.10) satisfy the sampling theorem.

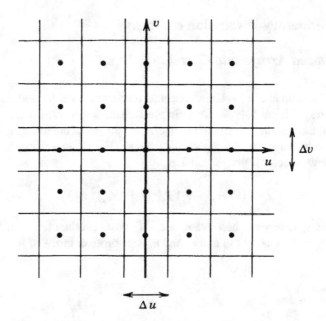

Fig. 5.3 Points on a rectangular grid in the (u, v) plane at which the visibility is sampled for use with the discrete Fourier transform. As shown, the spacings Δu and Δv are equal. The division of the plane into grid cells of size $\Delta u \times \Delta v$ is also shown.

To apply the discrete transform to synthesis imaging, it is necessary to obtain values of $\mathcal{V}(u, v)$ at points separated by Δu in u and by Δv in v, as shown in Fig. 5.3. However, the measurements are generally not made at (u, v) points on a grid since for tracking interferometers, they fall on elliptical loci in the (u, v) plane, as explained in Sect. 4.1. Thus, it is necessary to obtain the values at the grid points by interpolation or similar processes. In Fig. 5.3, the plane is divided into cells of size $\Delta u \times \Delta v$ centered on the grid points. A very simple method of determining a visibility value to assign at each grid point is to take the mean of all values that fall within the same cell. This procedure has been termed *cell averaging* (Thompson and Bracewell 1974). Better procedures are generally used; see Sect. 10.2.2. However, the cell averaging concept helps one to visualize the required distribution of the measurements; ideally there should be at least one measurement, or a small number of measurements, within each cell. Thus, the baselines should be chosen so that the spacings between the (u, v) loci are no greater than the cell size, to maximize the number of cells that are intersected by a locus. Cells that contain no measurements result in holes in the (u, v) coverage, and minimization of such holes is an important criterion in array design. Lobanov (2003) and Lal et al. (2009) discuss the performance of arrays based on uniformity of (u, v) coverage (see Sect. 5.4.2).

5.3   Introductory Discussion of Arrays

5.3.1   Phased Arrays and Correlator Arrays

An array of antennas can be interconnected to operate as a phased array or as a correlator array. Figure 5.4a shows a simple schematic diagram of a phased array connected to a square-law detector, in which the number of antennas, n_a, is equal to four. If the voltages at the antenna outputs are V_1, V_2, V_3, and so on, the output of the square-law detector is proportional to

$$(V_1 + V_2 + V_3 + \cdots + V_{n_a})^2 . \tag{5.11}$$

Note that for n_a antennas, there are $n_a(n_a - 1)$ cross-product terms of form $V_m V_n$ involving different antennas m and n, and n_a self-product terms of form V_m^2. If the

Fig. 5.4 Simple four-element linear array. ℓ_λ is the unit antenna spacing measured in wavelengths, and θ indicates the angle of incidence of a signal. (**a**) Connected as a phased array with an adjustable phase shifter in the output of each antenna, and the combined signal applied to a square-law detector. The voltage combiner is a matching network in which the output is proportional to the sum of the radio-frequency input voltages. (**b**) The same antennas connected as a correlator array. (**c**) The ordinate is the response of the array: the scale at the left applies to the phased array, and at the right to the correlator array. The abscissa is proportional to θ in units of ℓ_λ^{-1} rad. The equal spacing between antennas in this simple grating array gives rise to sidelobes in the form of replications of the central beam.

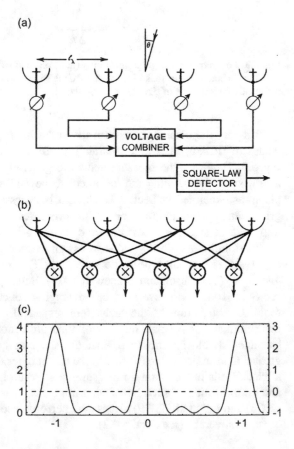

signal path (including the phase shifter) from each antenna to the detector is of the same electrical length, the signals combine in phase when the direction of the incoming radiation is given by

$$\theta = \sin^{-1}\left(\frac{N}{\ell_\lambda}\right) , \tag{5.12}$$

where N is an integer, including zero, and ℓ_λ is the spacing interval measured in wavelengths. The position angles of the maxima, which represent the beam pattern of the array, can be varied by adjusting the phase shifters at the antenna outputs. Thus, the beam pattern can be controlled and, for example, scanned to form an image of an area of sky.

In correlator arrays, a correlator generates the cross product of the signal voltages $V_m V_n$ for every antenna pair, as in Fig. 5.4b. These outputs take the form of fringe patterns and can be combined to produce maxima similar to those of the phased array. If a phase shift is introduced at the output of one of the correlator array antennas, the result appears as a corresponding change in the phase of the fringes measured with the correlator connected to that antenna. Conversely, the effect of an antenna phase shift can be simulated by changing the measured phases when combining the correlator outputs. Thus, a beam-scanning action can be accomplished by combining measured cross-correlations in a computer with appropriate variations in the phase. This is what happens in computing the Fourier transform of the visibility function, that is, the Fourier transform of the correlator outputs as a function of spacing. The loss of the self-product terms reduces the instantaneous sensitivity of the correlator array by a factor $(n_a - 1)/n_a$ in power, which is close to unity if n_a is large. However, at any instant, the correlator array responds to the whole field of the individual antennas, whereas the response of the phased array is determined by the narrow beam that it forms, unless it is equipped with a more complex signal-combining network that allows many beams to be formed simultaneously. Thus, in imaging, the correlator array gathers data more efficiently than the phased array.

The response pattern of the correlator array to a point source is the same as that of the phased array, except for the self-product terms. The response of the phased array consists of one or more beams in the direction in which the antenna responses combine with equal phase. These are surrounded by sidelobes, the pattern and magnitude of which depend on the number and configuration of antennas. Between individual sidelobe peaks, there will be nulls that can be as low as zero, but the response is positive because the output of the square-law detector cannot go negative. Now consider subtracting the self-product terms, to simulate the response of the correlator array. Over a field of view small compared with the beamwidth of an individual antenna, each self-product term represents a constant level, and each cross product represents a fringe oscillation. In the response to a point source, all of these terms are of equal magnitude. Subtracting the self-products from the phased-array response causes the zero level to be shifted in the positive direction by an amount equal to $1/n_a$ of the peak level, as indicated by the broken line in Fig. 5.4c. The points that represent zeros in the phased-array response become the peaks of negative sidelobes. Thus, in the response of the correlator array, the positive values are decreased by a factor $(n_a - 1)/n_a$ relative to those of the phased array. In the negative direction, the response extends to a level of $-1/(n_a - 1)$ of the positive peak but no further since this level corresponds to the zero level of the phased array. Kogan (1999) pointed out this limitation on the magnitude of the negative sidelobes of a correlator array and also noted that this limit depends not on the configuration of the individual antennas but only on their number. Neither of these conclusions applies to the positive sidelobes. This result is strictly true only for snapshot observations [i.e., those in which the (u, v) coverage is not significantly increased by Earth rotation] and for uniform weighting of the correlator outputs.

Finally, consider some characteristics of a phased array as in Fig. 5.4a. The power combiner is a passive network, for example, the branched transmission line in Fig. 1.13a. If a *correlated* waveform of power P is applied to each combiner input, then the output power is $n_a P$. In terms of the voltage V at each input, a fraction $1/\sqrt{n_a}$ of each *voltage* combines additively to produce an output of $\sqrt{n_a} V$, or $n_a P$ in power. Now if the input waveforms are *uncorrelated*, again each contributes $V/\sqrt{n_a}$ in voltage but the resulting *powers* combine additively (i.e., as the sum of the squared voltages), so in this case, the power at the output is equal to the power P at one input. Each input then contributes only $1/n_a$ of its power to the output, and the remaining power is dissipated in the terminating impedances of the combiner inputs (i.e., radiated from the antennas if they are directly connected to the combiner). The signals from an unresolved source received in the main beam of the array are fully correlated, but the noise contributions from amplifiers at the antennas are uncorrelated. Thus, if there are no losses in the transmission lines or the combiner, the same signal-to-noise ratio at the detector is obtained by inserting an amplifier at the output of each antenna, or a single amplifier at the output of the combiner. However, such losses are often significant, so generally it is advantageous to use amplifiers at the antennas. Note that if half of the antennas in a phased array are pointed at a radio source and the others at blank sky, the signal power at the combiner output is one-quarter of that with all antennas pointed at the source.

5.3.2 Spatial Sensitivity and the Spatial Transfer Function

We now consider the sensitivity of an antenna or array to the spatial frequencies on the sky. The angular response pattern of an antenna is the same in reception or transmission, and at this point it may be easier to consider the antenna in transmission. Then power applied to the terminals produces a field at the antenna aperture. A function $W(u, v)$ is equal to the autocorrelation function of the distribution of the electric field across the aperture, $\mathcal{E}(x_\lambda, y_\lambda)$. Here x_λ and y_λ are coordinates in the aperture plane of the antenna and are measured in wavelengths. Thus,

$$W(u, v) = \mathcal{E}(x_\lambda, y_\lambda) \star \star \mathcal{E}^*(x_\lambda, y_\lambda)$$

$$= \int_{-\infty}^{\infty} \int_{-\infty}^{\infty} \mathcal{E}(x_\lambda, y_\lambda) \, \mathcal{E}^*(x_\lambda - u, y_\lambda - v) \, dx_\lambda \, dy_\lambda \,. \tag{5.13}$$

The double-pentagram symbol represents two-dimensional autocorrelation. The integral in Eq. (5.13) is proportional to the number of ways, suitably weighted by the field intensity, in which a specific spacing vector (u, v) can be found within the antenna aperture. In reception, $W(u, v)$ is a measure of the sensitivity of the antenna to different spatial frequencies. In effect, the antenna or array acts as a spatial frequency filter, and $W(u, v)$ is widely referred to as the *transfer function* by analogy with the usage of this term in filter theory. $W(u, v)$ has also been called the spectral sensitivity function (Bracewell 1961, 1962), which refers to the spectrum of spatial frequencies (not the radio frequencies) to which the array responds. We use the terms *spatial transfer function* and *spatial sensitivity* when discussing $W(u, v)$. The area of the (u, v) plane over which measurements can be made [i.e., the support of $W(u, v)$, defined as the closure of the domain within which $W(u, v)$ is nonzero] is referred to as the *spatial frequency coverage*, or the (u, v) coverage.

Consider the response of the antenna or array to a point source. Since the visibility of a point source is constant over the (u, v) plane, the measured spatial frequencies are proportional to $W(u, v)$. Thus, the point-source response $\mathcal{A}(l, m)$ is the Fourier transform of $W(u, v)$. This result is formally derived by Bracewell and Roberts (1954). [Recall from the discussion preceding Eq. (2.15) that the point-source response is the mirror image of the antenna power pattern: $\mathcal{A}(l, m) = A(-l, -m)$.] The spatial transfer function $W(u, v)$ is an important feature in this chapter, and Fig. 5.5 further illustrates its place in the interrelationships between functions involved in radio imaging.

Figure 5.6a shows an interferometer in which the antennas do not track and are represented by two rectangular areas. We shall assume that $\mathcal{E}(x_\lambda, y_\lambda)$ is uniformly distributed over the apertures, such as in the case of arrays of uniformly excited dipoles. First suppose that the output voltages from the two apertures are summed and fed to a power-measuring receiver, as in some early instruments. The three rectangular areas in Fig. 5.6b represent the autocorrelation function of the aperture distributions, that is, the spatial transfer function. Note that the autocorrelation of the two apertures contains the autocorrelation of the individual apertures (the central rectangle in Fig. 5.6b) plus the cross-correlation of the two apertures (the shaded rectangles). If the two antennas are combined using a correlator instead of a receiver that responds to the total received power, the spatial sensitivity is represented by only the shaded rectangles since the correlator forms only the cross products of signals from the two apertures.

The interpretation of the spatial transfer function as the Fourier transform of the point-source response can be applied to both the adding and correlator cases. For example, for the correlator implementation of the interferometer in Fig. 5.6a, the

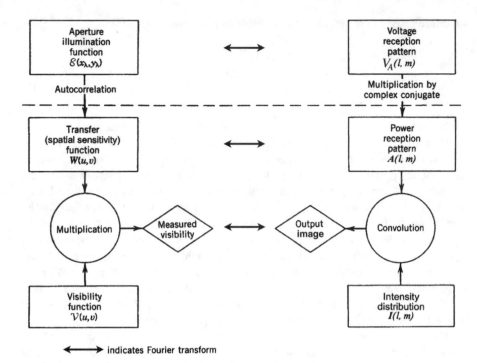

← → indicates Fourier transform

Fig. 5.5 Relationships between functions involved in imaging a source. Starting at the top left, the autocorrelation of the aperture distribution of the electric field over an antenna $\mathcal{E}(x_\lambda, y_\lambda)$ gives the spatial transfer function $W(u, v)$. The measured visibility in the observation of a source is the product of the source visibility $\mathcal{V}(u, v)$ and the spatial transfer function. At the top right, the multiplication of the voltage reception pattern $V_A(l, m)$ with its complex conjugate produces the power reception pattern $A(l, m)$. Imaging of the source intensity distribution $I(l, m)$ results in convolution of this function with the antenna power pattern. The Fourier transform relationships between the quantities in the (x_λ, y_λ) and (u, v) domains, and those in the (l, m) domain, are indicated by the bidirectional arrows. When the spatial sensitivity is built up by Earth rotation, as in tracking arrays, it cannot, in general, be described as the autocorrelation function of any field distribution. Only the part of the diagram below the broken line applies in such cases.

response to a point source is the Fourier transform of the function represented by the shaded areas. This Fourier transform is

$$\left[\frac{\sin \pi x_{\lambda 1} l}{\pi x_{\lambda 1} l}\right]^2 \left[\frac{\sin \pi y_{\lambda 1} m}{\pi y_{\lambda 1} m}\right]^2 \cos 2\pi D_\lambda l, \tag{5.14}$$

where $x_{\lambda 1}$ and $y_{\lambda 1}$ are the aperture dimensions, and D_λ is the aperture separation, all measured in wavelengths. The sinc-squared functions in (5.14) represent the power pattern of the uniformly illuminated rectangular apertures, and the cosine term represents the fringe pattern. In early instruments, the relative magnitude of the spatial sensitivity was controlled only by the field distribution over the antennas, but image processing by computer enables the magnitude to be adjusted after an observation has been made.

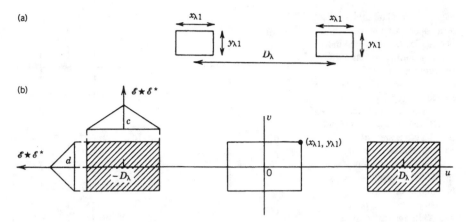

Fig. 5.6 The two apertures in (**a**) represent a two-element interferometer, the spatial transfer function of which is shown in (**b**). The shaded areas contain the spatial sensitivity components that result from the cross-correlation of the signals from the two antennas. If the field distribution is uniform over the apertures, the magnitude of the spatial sensitivity is linearly tapered. This is indicated by c and d, which represent cross sections of the spatial transfer function.

Some commonly used configurations of antenna arrays, and the boundaries of their autocorrelation functions, are shown in Fig. 5.7. The autocorrelation functions indicate the instantaneous spatial sensitivity for a continuous aperture in the form of the corresponding figure. Equation (5.13) shows that the autocorrelation function is the integral of the product of the field distribution with its complex conjugate displaced by u and v. By investigating the values of u and v for which the two aperture figures overlap, it is easy to determine the boundary within which the spatial transfer function is nonzero, using graphical procedures described by Bracewell (1961, 1995). It is also possible to identify ridges of high autocorrelation that occur for displacements at which the arms of figures such as those in Fig. 5.7a, b, or c are aligned. In the case of the ring, Fig. 5.7g, the autocorrelation function is proportional to the area of overlap at the two points where the ring intersects with its displaced replication. This area decreases monotonically for a ring of unit diameter until $q = \sqrt{u^2 + v^2} = 1/\sqrt{2}$, where the tangents to the two rings at the intersection points are $\pi/2$. For $q > 1/\sqrt{2}$, the autocorrelation function increases as the tangents realign. The analytic form of the autocorrelation function, shown in Fig. 5.7j, is the Fourier transform of a J_0^2 Bessel function, which is proportional to $1/(q\sqrt{1-q^2})$, for $0 \leq q \leq 1$. Another interesting aperture is a filled circle, for which the autocorrelation function decreases monotonically from $q = 0$ to 1 with the form $\cos^{-1}(q) - q\sqrt{1-q^2}$, which Bracewell (2000) calls the Chinese hat function. When the aperture is not completely filled, that is, when the figure represents an array of discrete antennas, the spatial sensitivity takes the form of samples of the autocorrelation function. For example, for a cross of uniformly spaced antennas, the square in Fig. 5.7b would be represented by a matrix pattern within the square boundary.

Fig. 5.7 Configurations for array apertures and the boundaries within which the corresponding autocorrelation functions are nonzero. The configurations represent the aperture (x_λ, y_λ) plane and the autocorrelations, the spatial frequency (u, v) plane. (**a**) The cross and (**b**) its autocorrelation boundary. (**c**) The T-shaped array and (**d**) its autocorrelation boundary. (**e**) The equiangular Y-shaped array and (**f**) its autocorrelation boundary. The broken lines in (**b**), (**d**), and (**f**) indicate ridges of high autocorrelation value. (**g**) The ring and (**h**) its autocorrelation boundary. The autocorrelation function of the ring is circularly symmetrical and (**j**) shows the radial profile of the function from the center to the edge of the circle in (**h**). (**i**) The Reuleaux triangle. The broken lines indicate an equilateral triangle, and the circular arcs that form the Reuleaux triangle have radii centered on the vertices of the triangle. The autocorrelation of the Reuleaux triangle is bounded by the same circle shown in (**h**) but does not have the same autocorrelation function as the ring.

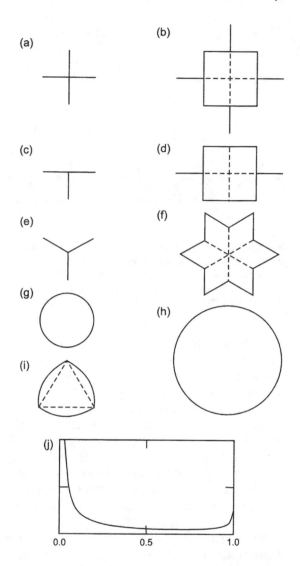

5.3.3   Meter-Wavelength Cross and T-Shaped Arrays

A cross and its autocorrelation function are shown in Fig. 5.7a and b. It is assumed that the width of the arms is finite but small compared with the length of the arms. In the case of the Mills cross (Mills 1963) described briefly in Chap. 1, the outputs of the two arms go to a single cross-correlating receiver, so the spatial sensitivity is represented by the square in Fig. 5.7b. The narrow extensions on the centers of the sides of the square represent parts of the autocorrelation functions of the individual arms, which are not formed in the cross-correlation of the arms. However, they are formed if the arms consist of lines of individual antennas, for which the cross-correlation is formed for pairs on the same arm as well as those on crossed arms. The case for a T-shaped array is similar and is shown in Fig. 5.7c and d.

If the sensitivity (i.e., the collecting area per unit length) is uniform along the arms for a cross or a corresponding T, then the weighting of the spatial sensitivity is uniform over the square (u, v) area; note that it does not taper linearly from the center as in the situation in Fig. 5.6. At the edge of the square area, the spatial sensitivity falls to zero in a distance equal to the width of the arms. Such a sharp edge, resulting from the uniform sensitivity, results in strong sidelobes. Therefore, an important feature of the Mills cross design was a Gaussian taper of the coupling of the elements along the arms to reduce the sensitivity to about 10% at the ends. This greatly reduced local maxima in the response resulting from sidelobes outside the main beam, at the expense of some broadening of the beam.

Figure 1.12a shows an implementation of a T-shaped array that is an example of a nontracking correlator interferometer. Here, a small antenna is moved in steps, with continuous coverage, to simulate a larger aperture; see Blythe (1957), Ryle et al. (1959), and Ryle and Hewish (1960). The spatial frequency coverage is the same as would be obtained in a single observation with an antenna of aperture equal to that simulated by the movement of the small antenna, although the magnitude of the spatial sensitivity is not exactly the same. The term *aperture synthesis* was introduced to describe such observations, but to be precise, it is the *autocorrelation* of the aperture that is synthesized (see Sect. 5.4).

5.4   Spatial Transfer Function of a Tracking Array

The range of spatial frequencies that contribute to the output of an interferometer with tracking antennas is illustrated in Fig. 5.8b. The two shaded areas represent the cross-correlation of the two apertures of an east–west interferometer for a source on the meridian. As the source moves in hour angle, the changing (u, v) coverage is represented by a band centered on the spacing locus of the two antennas. Recall from Sect. 4.1 that the locus for an Earth-based interferometer is an arc of an ellipse, and that since $\mathcal{V}(-u, -v) = \mathcal{V}^*(u, v)$, any pair of antennas measures visibility along two arcs symmetric about the (u, v) origin, both of which are included in the spatial transfer function.

Because the antennas track the source, the antenna beams remain centered on the same point in the source under investigation, and the array measures the product of the source intensity distribution and the antenna pattern. Another view of this effect is obtained by considering the radiation received by small areas of the apertures of two antennas, the centers of which are A_1 and A_2 in Fig. 5.9. The antenna apertures encompass a range of spacings from $u - d_\lambda$ to $u + d_\lambda$ wavelengths, where d_λ is the antenna diameter measured in wavelengths. If the antenna beams remain

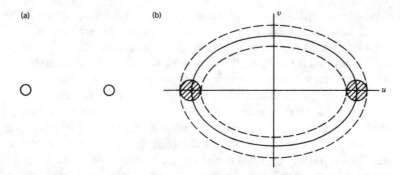

Fig. 5.8 (**a**) The aperture of an east–west, two-element interferometer. The corresponding spatial frequency coverage for cross-correlated signals is shown by the shaded areas in (**b**). If the antennas track the source, the spacing vector traces out an elliptical locus (the solid line) in the (u, v) plane. The area between the broken lines in (**b**) indicates the spatial frequencies that contribute to the measured values. The spacing between the broken lines is determined by the cross-correlation of the antenna aperture.

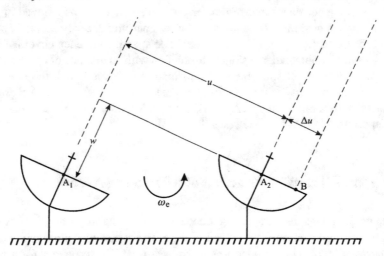

Fig. 5.9 Illustration of the effect of tracking on the fringe frequency at the correlator output. The u component of the baseline is shown, and the v component is omitted since it does not affect the fringe frequency. The curved arrow indicates the tracking motion of the antennas.

fixed in position as a source moves through them, then the correlator output is a combination of fringe components with frequencies from $\omega_e(u - d_\lambda)\cos\delta$ to $\omega_e(u+d_\lambda)\cos\delta$, where ω_e is the angular velocity of the Earth and δ is the declination of the source. To examine the effect when the antennas track the source, consider the point B, which, because of the tracking, has a component of motion toward the source equal to $\omega_e\Delta u\cos\delta$ wavelengths per second. This causes a corresponding Doppler shift in the signal received at B. To obtain the fringe frequency for waves arriving at A_1 and B, we subtract the Doppler shift from the nontracking fringe frequency and obtain $[\omega_e(u + \Delta u)\cos\delta] - (\omega_e\Delta u\cos\delta) = (\omega_e u\cos\delta)$. The fringe frequency when tracking is thus the same as for the central points A_1 and A_2 of the apertures. (This is true for any pair of points; choosing one point at an antenna center in the example above slightly simplifies the discussion.) Thus, if the antennas track, the contributions from all pairs of points within the apertures appear at the same fringe frequency at the correlator output. As a result, such contributions cannot be separated by Fourier analysis of the correlator output waveform, and information on how the visibility varies over the range $u - d_\lambda$ to $u + d_\lambda$ is lost. However, if the antenna motion differs from a purely tracking one, the information is, in principle, recoverable. In imaging sources wider than the antenna beams, an additional scanning motion to cover the source is added to the tracking motion. In effect, this scanning allows the visibility to be sampled at intervals in u and v that are fine enough for the extended width of the source. This technique, known as *mosaicking*, is described in Sect. 11.5.

To accommodate the effects that result when the antennas track the source position, the normalized antenna pattern is treated as a modification to the intensity distribution, which then becomes $A_N(l,m)I(l,m)$. The spatial transfer function $W(u,v)$ for a pair of tracking antennas is represented at any instant by a pair of two-dimensional delta functions $^2\delta(u,v)$ and $^2\delta(-u,-v)$. For an array of antennas, the resulting spatial transfer function is represented by a series of delta functions weighted in proportion to the magnitude of the instrumental response. As the Earth rotates, these delta functions generate the ensemble of elliptical spacing loci. The loci represent the spatial transfer function of a tracking array.

Consider observation of a source $I(l,m)$, for which the visibility function is $\mathcal{V}(u,v)$, with normalized antenna patterns $A_N(l,m)$. Then if $W(u,v)$ is the spatial transfer function, the measured visibility is

$$\left[\mathcal{V}(u,v) ** \overline{A}_N(u,v)\right]W(u,v) , \qquad (5.15)$$

where the double asterisk indicates two-dimensional convolution and the bar denotes the Fourier transform. The Fourier transform of (5.15) gives the measured intensity:

$$[I(l,m)A_N(l,m)] ** \overline{W}(l,m) . \qquad (5.16)$$

If we observe a point source at the (l,m) origin, where $A_N = 1$, expression (5.16) becomes the point-source response $b_0(l,m)$. We then obtain

$$b_0(l,m) = \left[^2\delta(l,m)A_N(l,m)\right] ** \overline{W}(l,m) = \overline{W}(l,m) , \qquad (5.17)$$

where the two-dimensional delta function, $^2\delta(l,m)$, represents the point source. Here again, the point-source response is the Fourier transform of the spatial transfer function. In the tracking case, the spatial frequencies that contribute to the measurement are represented by $W(u,v) ** \overline{A}_N(u,v)$. Note that $\overline{A}_N(u,v)$ is twice as wide as the corresponding antenna aperture in the (x,y) domain.

The term *aperture synthesis* is sometimes extended to include observations that involve hour-angle tracking. However, it is not possible to define an exactly equivalent antenna aperture for a tracking array. For example, consider the case of two antennas with an east–west baseline tracking a source for a period of 12 h. The spatial transfer function is an ellipse centered on the origin of the (u, v) plane, with zero sensitivity within the ellipse (except for a point at the origin that could be supplied by a measurement of total power received in the antennas). The equivalent aperture would be a function, the autocorrelation of which is the same elliptical ring as the spatial transfer function. No such aperture function exists, and thus the term "aperture synthesis" can only loosely be applied to describe most observations that include hour-angle tracking.

5.4.1 Desirable Characteristics of the Spatial Transfer Function

As a first step in considering the layout of the antennas, it is useful to consider the desired spatial (u, v) coverage [see, e.g., Keto (1997)]. For any specific observation, the optimum (u, v) coverage clearly depends on the expected intensity distribution of the source under study, since one would prefer to concentrate the capacity of the instrument in (u, v) regions where the visibility is nonzero. However, most large arrays are used for a wide range of astronomical objects, so some compromise approach is required. Since, in general, astronomical objects are aligned at random in the sky, there is no preferred direction for the highest resolution. Thus, it is logical to aim for visibility measurements that extend over a circular area centered on the (u, v) origin.

As described in Sect. 5.2.2, the visibility data may be interpolated onto a rectangular grid for convenience in Fourier transformation, and if approximately equal numbers of measurements are used for each grid point, they can be given equal weights in the transformation. Uneven weighting results in loss of sensitivity, since some values then contain a larger component of noise than others. From this viewpoint, one would like the natural weighting (i.e., the weighting of the measurements that results from the array configuration without further adjustment) to be as uniform as possible within the circular area.

For a general-purpose array, it is difficult to improve on the circularity of the measurement area. However, there are exceptions to the uniformity of the measurements within the circle. As mentioned above, in the Mills cross, uniform coupling of the radiating elements along the arms would result in uniform spatial sensitivity. To reduce sidelobes, a Gaussian taper of the coupling was introduced, resulting in a similar taper in the spatial sensitivity. This was particularly important because at the frequencies for which this type of instrument was constructed, typically in the range 85–408 MHz, source confusion can be a serious problem. Sidelobe responses can be mistaken for sources and can also mask genuine sources.

For a spatial sensitivity function of uniform rectangular character, the beam has a sinc function $(\sin \pi x / \pi x)$ profile, for which the first sidelobe has a relative strength of 0.217. For a uniform, circular, spatial transfer function, the beam has a profile of the form $J_1(\pi x) / \pi x$ for which the first sidelobe has a relative strength of 0.132. Sidelobes for a uniform circular (u, v) coverage are less than for a rectangular one but would still be a problem in conditions of source confusion. Tapering of the antenna illumination reduces the sidelobe responses. Thus, the uniform weighting may not be optimum for conditions of high source density.

5.4.2 Holes in the Spatial Frequency Coverage

Consider a circular (u, v) area of diameter a_λ wavelengths in which there are no holes in the data; that is, the visibility data interpolated onto a rectangular grid for Fourier transformation has no missing values. Then for uniform weighting, the synthesized beam, which is obtained from the Fourier transform of the gridded transfer function, has the form $J_1(\pi a_\lambda \theta) / \pi a_\lambda \theta$, where θ is the angle measured from the beam center. If centrally concentrated weighting is used, the beam is a smoothed form of this function. Let us refer to the (u, v) area described above as the complete (u, v) coverage and the resulting beam as the complete response. Now if some data are missing, the actual (u, v) coverage is equal to the complete coverage minus the (u, v) hole distribution. By the additive property of Fourier transforms, the corresponding synthesized beam is equal to the complete response minus the Fourier transform of the hole distribution. The holes result in an unwanted component to the complete response, in effect adding sidelobes to the synthesized beam. From Parseval's theorem, the rms amplitude of the hole-induced sidelobes is proportional to the rms value of the missing spatial sensitivity represented by the holes. Other sidelobes also occur as a result of the oscillations in the $J_1(\pi a_\lambda \theta) / \pi a_\lambda \theta$ profile of the complete response, but there is clearly a sidelobe component from the holes.

5.5 Linear Tracking Arrays

We now consider interferometers or arrays in which the locations of the antennas are confined to a straight line. We have seen that for pairs of antennas with east–west spacings, the tracking loci in the (u, v) plane are a series of ellipses centered on the (u, v) origin. To obtain complete ellipses, it is necessary that the tracking covers a range of 12 h in hour angle. If the antenna spacings of an east–west array increase in uniform increments, the spatial sensitivity is represented by a series of concentric ellipses with uniform increments in their axes. The angular resolution obtained is inversely proportional to the width of the (u, v) coverage in the corresponding

Fig. 5.10 Two linear array configurations in which the antennas are represented by filled circles. (**a**) Arsac's (1955) configuration containing all spacings up to six times the unit spacing, with no redundancy. (**b**) Bracewell's (1966) configuration containing all spacings up to nine times the unit spacing, with the unit spacing occurring twice.

direction; the width in the v direction is equal to that in the u direction times the sine of the declination, δ. East–west linear arrays containing spacings at multiples of a basic interval have found wide use, especially in earlier radio astronomy, for observations at $|\delta|$ greater than $\sim 30°$.

In the simplest type of linear array, the antennas are spaced at uniform intervals ℓ_λ (see Fig. 1.13a). This type of array is sometimes known as a grating array, by analogy with an optical diffraction grating. If there are n_a antennas, such an array output contains $(n_a - 1)$ combinations with the unit spacing, $(n_a - 2)$ with twice the unit spacing, and so on. Thus, short spacings are highly redundant, and one is led to seek other ways to configure the antennas to provide larger numbers of different spacings for a given n_a. Note, however, that redundant observations can be used as an aid in calibration of the instrumental response and atmospheric effects, so some degree of redundancy is arguably beneficial (Hamaker et al. 1977).

Early examples of antenna configurations include one in Fig. 5.10a, used by Arsac (1955), with no redundant spacings. The six possible pair combinations all have different spacings. With more than four antennas, there is always either some redundancy or some missing spacings. A five-element, *minimum-redundancy*[1] configuration devised by Bracewell (1966) is shown in Fig. 5.10b. Moffet (1968) listed examples of minimum-redundancy arrays of up to 11 elements, and solutions for larger arrays are discussed by Ishiguro (1980). Moffet defined two classes. These are restricted arrays in which all spacings up to the maximum spacing, $n_{\max}\ell_\lambda$ (that is, the total length of the array), are present; general arrays in which all spacings up to some particular value are present; and also some longer ones. Examples for eight elements are shown in Fig. 5.11. A measure of redundancy for a linear array is given by the expression

$$\frac{1}{2}n_a(n_a - 1)/n_{\max} , \tag{5.18}$$

which is the number of antenna pairs divided by the number of unit spacings in the longest spacing. This is equal to 1.0 and 1.11 for the configurations in Fig. 5.10a

---

[1]The mathematical theory of minimum redundancy is known as the optimal Golomb ruler (Golomb 1972), which has roots in the mathematical literature going back to the 1930s.

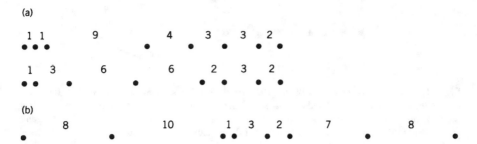

Fig. 5.11 Eight-element, minimum-redundancy, linear arrays: the numbers indicate spacings in multiples of the unit spacing. (**a**) Two arrays that uniformly cover the range of 1 to 23 times the unit spacing. (**b**) An array that uniformly covers 1 to 24 times the unit spacing but has a length of 39 times the unit spacing. The extra spacings are 8, 31 (twice), and 39 times the unit spacing. © 1968 IEEE. Reprinted with permission, from A. T. Moffet (1968).

and 5.10b, respectively. A study in number theory by Leech (1956) indicates that for large numbers of elements, this redundancy factor approaches 4/3. A linear minimum-redundancy array that uses the configuration in Fig. 5.10b is described by Bracewell et al. (1973). For arrays with such small numbers of antennas, the choice of the configuration is particularly important.

The ability to move a small number of elements adds greatly to the range of performance of an array. Figure 5.12 shows the arrangement of three antennas in an early synthesis instrument, the Cambridge One-Mile Radio Telescope (Ryle 1962). Antennas 1 and 2 are fixed, and their outputs are correlated with that from antenna 3, which can be moved on a rail track. In each position of antenna 3, the source under observation is tracked for 12 h, and visibility data are obtained over two elliptical loci in the (u, v) plane. The observation is repeated as antenna 3 is moved progressively along the track, and the increments in the position of this antenna determine the spacing of the elliptical loci in the (u, v) plane. From the sampling theorem (Sect. 5.2.1), the required (u, v) spacing is the reciprocal of the angular width (in radians) of the source under investigation. The ability to vary the incremental spacing adds versatility to the array and reduces the number of antennas required. The configuration of a larger instrument of this type, the Westerbork Synthesis Radio Telescope (Baars and Hooghoudt 1974; Högbom and Brouw 1974; Raimond and Genee 1996), is shown in Fig. 5.13. Here, ten fixed antennas are

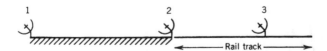

Fig. 5.12 The Cambridge One-Mile Radio Telescope. Antennas 1 and 2 are at fixed locations, and the signals they receive are each correlated with the signal from antenna 3, which can be located at various positions along a rail track. The fixed antennas are 762 m apart, and the rail track is a further 762 m long. The unit spacing is equal to the increment of the position of antenna 3, and all multiples up to 1524 m can be obtained.

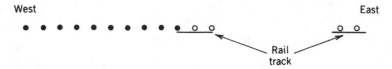

Fig. 5.13 Antenna configuration of the Westerbork Synthesis Radio Telescope. The ten filled circles represent antennas at fixed locations, and the four open circles represent antennas that are movable on rail tracks. The signals from each of the fixed antennas are combined with the signals from each of the movable ones. The diameter of the antennas is 25 m, and the spacing of the fixed antennas is 144 m.

combined with four movable ones, and the rate of gathering data is approximately 20 times greater than with the three-element array.

The sampling of the visibility function at points on concentric, equally spaced ellipses results in the introduction of ringlobe responses. These may be understood by noting that for a linear array, the instantaneous spacings are represented in one dimension by a series of δ functions, as shown in Fig. 5.14a. If the array contains all multiples of the unit spacings up to $N\ell_\lambda$, and if the corresponding visibility measurements are combined with equal weights, the instantaneous response is a series of fan beams, each with a profile of sinc-function form, as in Fig. 5.14b. This follows from the Fourier transform relationship for a truncated series of delta functions (see Appendix 2.1):

$$\sum_{i=-N}^{N} \delta(u - i\ell_\lambda) \longleftrightarrow \frac{\sin\left[(2N + 1)\pi\ell_\lambda l\right]}{\pi\ell_\lambda l} * \sum_{k=-\infty}^{\infty} \delta\left(l - \frac{k}{\ell_\lambda}\right). \tag{5.19}$$

The delta functions on the left side represent the spacings in the u domain. The series on the left is truncated and can be envisaged as selected from an infinite series by multiplication with a rectangular window function. The right side represents the

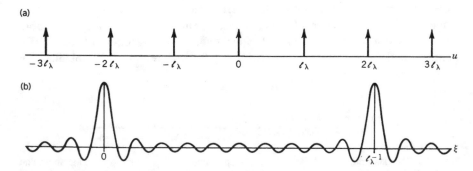

Fig. 5.14 Part of a series of δ functions representing the instantaneous distribution of spacings for a uniformly spaced linear array with equal weight for each spacing. (**b**) Part of the corresponding series of fan beams that constitute the instantaneous response. Parts (**a**) and (**b**) represent the left and right sides of Eq. (5.19), respectively.

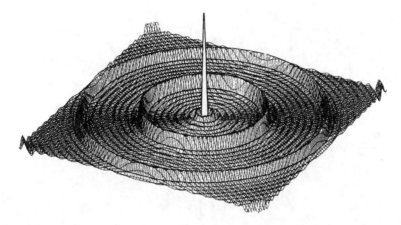

Fig. 5.15 Example of ringlobes. The response of an array for which the spatial transfer function is a series of nine circles concentric with the (u, v) origin, resulting, for example, from observations with an east–west linear array with 12-h tracking at a high declination. The radii of these circles are consecutive integral multiples of the unit antenna spacing. The weighting corresponds to the principal response discussed in Sect. 10.2. From Bracewell and Thompson (1973). © AAS. Reproduced with permission.

beam pattern in which the Fourier transform of the window function is replicated by convolution with delta functions. As the Earth's rotation causes the spacing vectors to sweep out ellipses in the (u, v) plane, the corresponding rotation of the array relative to the sky can be visualized as causing a central fan beam to rotate into a narrow pencil beam, while its neighbors give rise to lower-level, ring-shaped responses concentric with the central beam, as shown in Fig. 5.15. This general argument gives the correct spacing of the ringlobes, the profile of which is modified from the sinc-function form.

If the spatial sensitivity in the (u, v) plane is a series of circular delta functions of radius $q, 2q, \ldots, Nq$, the profile of the kth ringlobe is of the form

$$\text{sinc}^{1/2} \left[2(N + \frac{1}{2})(qr - k) \right] , \qquad (5.20)$$

where $r = \sqrt{l^2 + m^2}$. The function $\text{sinc}^{1/2}(\chi)$ is plotted in Fig. 5.16 and is the half-order derivative of $(\sin \pi \chi)/\pi \chi$. It can be computed using Fresnel integrals (Bracewell and Thompson 1973).

The application of the sampling theorem (Sect. 5.2.1) to the choice of incremental spacing requires that the increment be no greater than the reciprocal of the source width. In terms of ringlobes, this condition ensures that the minimum ringlobe spacing is no less than the source width. Thus, if the sampling theorem is followed, the main-beam response to a source just avoids being overlapped by a ringlobe response to the same source. In arrays such as those in Figs. 5.12 and 5.13, ringlobes can be effectively suppressed if the movable antennas are positioned in steps slightly

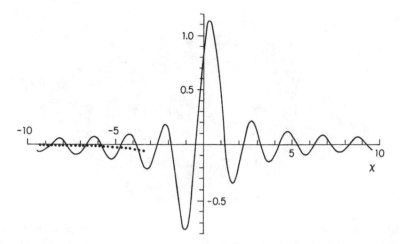

Fig. 5.16 Cross section of a ringlobe in the principal response to a point source of an east–west array with uniform increments in antenna spacing. The left side is the inside of the ring, and the right is the outside. The dotted line indicates a negative mean level of the oscillations on the inner side. From Bracewell and Thompson (1973). © AAS. Reproduced with permission.

less than the antenna diameter, in which case the ringlobe lies outside the primary antenna beam. Note, however, that the first spacing cannot be less than the antenna diameter, and the missing low-spacing measurements may have to be obtained by other means (see the discussion of mosaicking in Sect. 11.5). Ringlobes can also be greatly reduced by image-processing techniques such as the CLEAN algorithm, which is described in Sect. 11.1.

Although the elliptical loci in the (u, v) plane are spaced at equal intervals, the natural weighting of the data for an east–west linear array is not uniform, because in any interval of time, the antenna-spacing vectors move a distance proportional to their length. In the projection of the (u, v) plane onto the equatorial plane of the Earth, which is discussed in Sect. 4.2 as the (u', v') plane, the spacing vectors rotate at constant angular velocity, and the density of measured points is proportional to

$$q'^{-1} = (u'^2 + v'^2)^{-1/2} = (u^2 + v^2\mathrm{cosec}^2\delta)^{-1/2} . \tag{5.21}$$

In the (u, v) plane, the density of measurements, averaged over an area of dimensions comparable to the unit spacing of the antennas, is inversely proportional to $\sqrt{u^2 + v^2\mathrm{cosec}^2\delta}$. Along a straight line through the (u, v) origin, the density is inversely proportional to $\sqrt{u^2 + v^2}$.

5.6   Two-Dimensional Tracking Arrays

As noted previously, the spatial frequency coverage for an east–west linear array becomes severely foreshortened in the v dimension for observations near the celestial equator. For such observations, a configuration of antennas is required in which the Z component of the antenna spacing, as defined in Sect. 4.1, is comparable to the X and Y components. This is achieved by including spacings with azimuths other than east–west. The configuration is then two-dimensional. An array located at an intermediate latitude and designed to operate at low declinations can cover the sky from the pole to declinations of about 30° into the opposite celestial hemisphere. This range includes about 70% of the total sky, that is, almost three times as much as that of an east–west array. Since the Z component is not zero, the elliptical (u, v) loci are broken into two parts, as shown in Fig. 4.4. As a result, the pattern of the (u, v) coverage is more complex than is the case for an east–west linear array, and the ringlobes that result from uniform spacing of the loci are replaced by more complex sidelobe structure. In two dimensions, the choice of a minimum-redundancy configuration of antennas is not as simple as for a linear array. A first step is to consider the desired spatial transfer function $W(u, v)$. There is no direct analytical way to go from $W(u, v)$ to the antenna configuration, but iterative methods of finding an optimum, or near-optimum, solution can be used.

First, consider the effect of tracking a source across the sky, and suppose that for a source near the zenith, the *instantaneous* spatial frequency coverage results in approximately uniform sampling within a circle centered on the (u, v) origin. At any time during the period of tracking of the source, the (u, v) coverage is the zenith coverage projected onto the plane of the sky, with some degree of rotation that depends on the hour angle and declination of the source. The projection results in foreshortening of the coverage from a circular to an elliptical area, still centered on the (u, v) origin, and this foreshortening is least at meridian transit. The effect of observing over a range of hour angle can be envisaged as averaging a range of elliptical (u, v) areas that suffer some rotation of the major axis. At the center of the (u, v) plane will be an area that remained within the foreshortened coverage over the whole observation, and if the instantaneous coverage is uniform, then it will remain uniform within this area. Outside the area, the foreshortening will cause the coverage to taper off smoothly. These effects depend on the declination of the source and the range of hour-angle tracking. Practical experience indicates that some tapering of the visibility measurements is seldom a serious problem. Thus, it can generally be expected that two-dimensional arrays in which the number of antennas is large enough to provide good instantaneous (u, v) coverage will also provide good performance when used with hour-angle tracking.

5.6.1   Open-Ended Configurations

For configurations with open-ended arms such as the cross, T, and Y, the spatial frequency coverage is shown in Fig. 5.7. The spatial frequency coverage of the cross and T has fourfold symmetry in both cases; we ignore the effect of the missing small extensions on the top and bottom sides of the square for the T. The spatial frequency coverage of the equiangular Y-shaped array (120° between adjacent arms) has sixfold symmetry. (n-fold symmetry denotes a figure that is unchanged by rotation through $2\pi/n$. For a circle, n becomes infinite, and other figures approach circular symmetry as n increases.) The autocorrelation function of the equiangular Y-shaped array is closer to circular symmetry than that of a cross or T-shaped array. In this respect, a five-armed array, as suggested by Hjellming (1989), would be better still, but more expensive.

As an example of the open-ended configuration, we examine some details of the design of the VLA (Thompson et al. 1980; Napier et al. 1983; Perley et al. 2009). This array is located at latitude 34° N in New Mexico and is able to track objects as far south as $-30°$ for almost 7 h without going below 10° in elevation. Performance specifications called for imaging with full resolution down to at least $-20°$ declination and for obtaining an image in no more than 8 h of observation without moving antennas to new locations. In designing the array, comparison of the performance of various antenna configurations was accomplished by computing the spatial transfer function with tracking over an hour-angle range ±4 h at various declinations. In judging the merit of any configuration, the basic concern was to minimize sidelobes in the synthesized beam. It was found that the percentage of holes in the (u, v) coverage was a consistent indication of the sidelobe levels of the synthesized beam, and to judge between different configurations, it was not always necessary to calculate the detailed response (National Radio Astronomy Observatory 1967, 1969). For a given number of antennas, the equiangular Y-shaped array was found to be superior to the cross and T-shaped array; see Fig. 5.17.

Inverting the Y has no effect on the beam, but if the antennas have the same radial disposition on each arm, the performance near zero declination is improved by rotating the array so that the nominal north or south arm makes an angle of about 5° with the north–south direction. Without this rotation, the baselines between corresponding antennas on the other two arms are exactly east–west, and for $\delta = 0°$, the spacing loci degenerate to straight lines that are coincident with the u axis and become highly redundant. The total number of antennas, 27, was chosen from a consideration of (u, v) coverage and sidelobe levels and resulted in peak sidelobes at least 16 dB below the main-beam response, except at $\delta = 0°$, where Earth rotation is least effective. The 27 antennas provide 351 pair combinations.

The positions of the antennas along the arms provide another set of variables that can be adjusted to optimize the spatial transfer function. Figure 5.17 shows two approaches to the problem. Configuration (a) was obtained by using a pseudo-dynamic computation technique (Mathur 1969), in which arbitrarily chosen initial conditions were adjusted by computer until a near-optimum (u, v) coverage was

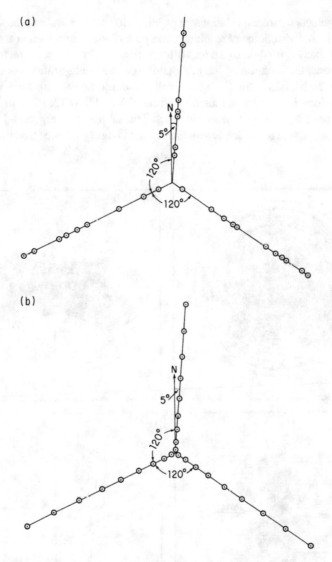

Fig. 5.17 (a) Proposed antenna configuration for the VLA that resulted from Mathur's (1969) computer-optimized design. (b) Power-law design (Chow 1972) adopted for the VLA. © 1983 IEEE. Reprinted, with permission, from P. J. Napier et al. (1983).

reached. Configuration (b) shows a power-law configuration derived by Chow (1972). This analysis led to the conclusion that a spacing in which the distance of the nth antenna on an arm is proportional to n^α would provide good (u, v) coverage. Comparison of the empirically optimized configuration with the power-law spacing with $\alpha \simeq 1.7$ showed the two to be essentially equal in performance. The power-law result was chosen largely for reasons of economy. A requirement of the design was that four sets of antenna stations be provided to vary the scale of the spacings in four steps, to allow a choice of resolution and field of view for different astronomical objects. By making α equal to the logarithm to the base 2 of the scale factor between configurations, the location of the nth station for one configuration coincides with that of the $2n$th station for the next-smaller configuration. The total number of antenna stations required was thereby reduced from 108 to 72. Plots of the spatial frequency coverage are shown in Fig. 5.18. The snapshot in Fig. 5.18d shows the instantaneous coverage, which is satisfactory for imaging simple structure in strong sources.

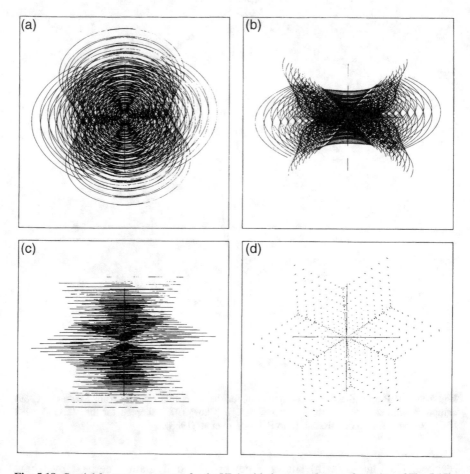

Fig. 5.18 Spatial frequency coverage for the VLA with the power-law configuration of Fig. 5.17b: (**a**) $\delta = 45°$; (**b**) $\delta = 30°$; (**c**) $\delta = 0°$; (**d**) snapshot at zenith. The range of hour angle is ± 4 h or as limited by a minimum pointing elevation of $9°$, and ± 5 min for the snapshot. The lengths of the (u, v) axes from the origin represent the maximum distance of an antenna from the array center, that is, 21 km for the largest configuration. © 1983 IEEE. Reprinted, with permission, from P. J. Napier et al. (1983).

5.6.2 Closed Configurations

The discussion here largely follows that of Keto (1997). Returning to the proposed criterion of uniform distribution of measurements within a circle in the (u, v) plane, we note that a configuration of antennas around a circle (a ring array) provides a useful starting point since the distribution of antenna spacings cuts off sharply in all directions at the circle diameter. This is shown in Fig. 5.7g and h. We begin by considering the instantaneous (u, v) coverage for a source at the zenith. This is shown in Fig. 5.19a for 21 equally spaced antenna locations indicated by triangles. There are 21 antenna pairs at the unit spacing, uniformly distributed in azimuth, and each of these is represented by two points in the (u, v) plane. The same statement can be made for any other paired spacings around the circle. As a result, the spatial transfer function consists of points that lie on a pattern of circles and radial lines. Note also that as the spacings approach the full diameter of the circle, the distance between antennas increases only very slowly. For example, the direct distance between antennas spaced 10 intervals around the circle is very little more than that for antennas at 9 intervals. Thus, there is an increase in the density of measurements at the longest spacings (the points along any radial line become more closely spaced) as well as a marked increase toward the center. Note that the density of points closely follows the radial profile of the autocorrelation function in Fig. 5.7j, except close to the origin, since Fig. 5.19 includes only cross-correlations between antennas.

One way of obtaining a more uniform distribution is to randomize the spacings of the antennas around the circle. The (u, v) points are then no longer constrained to lie on the pattern of circles and lines, and Fig. 5.19b shows an example in which a partial optimization has been obtained by computation using a neural-net algorithm. Keto (1997) discussed various algorithms for optimizing the uniformity of the spatial sensitivity. An earlier investigation of circular arrays by Cornwell (1988) also resulted in good uniformity within a circular (u, v) area. In this case, an optimizing program based on simulated annealing was used, and the spacing of the antennas around the circle shows various degrees of symmetry that result in patterns resembling crystalline structure in the (u, v) spacings.

Optimizing the antenna configurations can also be considered more broadly, and Keto (1997) noted that the cutoff in spacings at the same value for all directions is not unique to the circular configuration. There are other figures, such as the Reuleaux triangle, for which the width is constant in all directions. The Reuleaux triangle is shown in Fig. 5.7i and consists of three equal circular arcs indicated by the solid lines. The total perimeter is equal to that of a circle with diameter equal to one of the sides of the equilateral triangle shown by the broken lines. Similar figures can be constructed for any regular polygon with an odd number of sides, and a circle represents such a figure for which the number tends to infinity. The Reuleaux triangle is the least symmetrical of this family of figures. Other facts about the Reuleaux triangle and similar figures can be found in Rademacher and Toeplitz (1957).

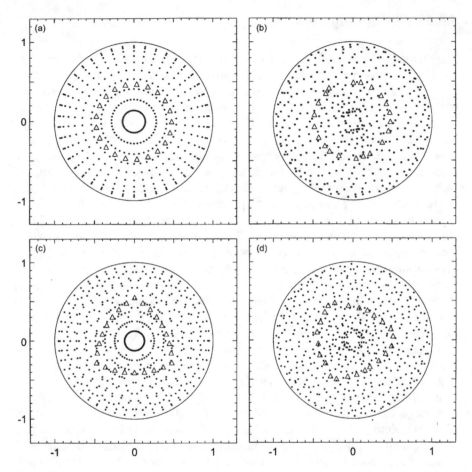

Fig. 5.19 (a) A circular array with 21 uniformly spaced antennas indicated by the triangles, and the instantaneous spatial frequency coverage indicated by the points. The scale of the diagrams is the same for both the antenna positions and the spatial frequency coordinates u and v. (b) The array and spatial frequency coverage as in (a) but after adjustment of the antenna positions around the circle to improve the uniformity of the coverage. (c) An array of 24 antennas equally spaced around a Reuleaux triangle, and the corresponding spatial frequency coverage. (d) The array and spatial sensitivity as in (c) with adjustment of the antenna spacing to optimize the uniformity of the coverage. From Keto (1997). © AAS. Reproduced with permission.

Since the optimization of the circular array in Fig. 5.19b results in a reduction in the symmetry, it may be expected that an array based on the Reuleaux triangle would provide better uniformity in the spatial frequency coverage than the circular array. This is indeed the case, as can be seen by comparing Figs. 5.19a and c, where the spacing between adjacent antennas for both is uniform. The circular array with irregular antenna spacings in Fig. 5.19b was obtained by starting with a circular array and allowing antenna positions to be moved small distances. In this case, the program was not allowed to reach a fully optimized solution. Allowing the optimization to run to convergence results in antennas at irregular spacings around a Reuleaux triangle, as shown in Fig. 5.19d. This result does not depend on the starting configuration. Comparison of Figs. 5.19b and d shows that the difference between the circle and the Reuleaux triangle is much less marked when they have both been subjected to some randomization of the antenna positions around the figure, although a careful comparison shows the uniformity in Fig. 5.19d to be a little better than in b.

Figure 5.20 shows the spatial frequency coverage for an array in an optimized Reuleaux triangle configuration. The tracking range is $\sim \pm 3$ h of hour angle, and the latitude is equal to that of the VLA. Comparison of these figures with corresponding ones for the VLA in Fig. 5.18 shows that the Reuleaux triangle produces spatial frequency coverage that is closer to the uniformly sampled circular area than does the equiangular Y configuration. As indicated in Fig. 5.7, the autocorrelation function of a figure with linear arms contains high values in directions where the

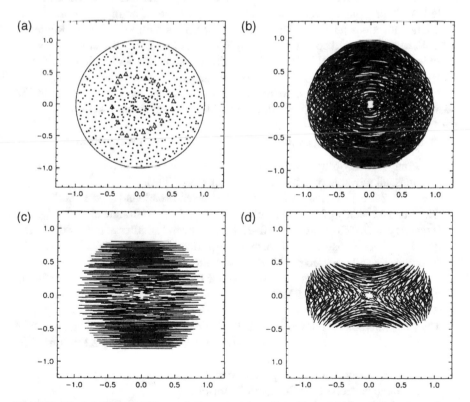

Fig. 5.20 Spatial frequency coverage for a closed configuration of 24 antennas optimized for uniformity of measurements in the snapshot mode: (**a**) snapshot at zenith; (**b**) $\delta = +30°$; (**c**) $\delta = 0°$; (**b**) $\delta = -28°$. The triangles in (**a**) indicate the positions of the antennas. The tracking is calculated for an array at $34°$ latitude to simplify comparison with the VLA (Fig. 5.18). For each declination shown, the tracking range is the range of hour angle for which the source elevation is greater than $25°$. From Keto (1997). © AAS. Reproduced with permission.

arms of overlapping figures line up. This effect contributes to the lack of uniformity in the spatial sensitivity of the Y-shaped array. Curvature of the arms or quasi-random lateral deviations of the antennas from the arms helps to smear the sharp structure in the spatial transfer function. The high values along radial lines do not occur in the autocorrelation function of a circle or similar closed figure, which is one reason why configurations of this type provide more uniform spatial frequency coverage.

Despite some less-than-ideal features of the equiangular Y-shaped array, the VLA produces astronomical images of very high quality. Thus, although the circularity and uniformity of the spatial frequency coverage are useful criteria, they are not highly critical factors. As long as the measurements cover the range of u and v for which the visibility is high enough to be measurable, and the source is strong enough that any loss in sensitivity resulting from nonuniform weighting can be tolerated, excellent results can be obtained. The Y-shaped array has a number of practical advantages over a closed configuration. When several scaled configurations are required to allow for a range of angular resolution, the alternative locations lie along the same arms, whereas with the circle or Reuleaux triangle, separate scaled configurations are required. The flexibility of the Y-shaped array is particularly useful in VLA observations at southern declinations for which the projected spacings are seriously foreshortened in the north–south direction. For such cases, it is possible to move the antennas on the north arm onto the positions for the next-larger configuration and thereby substantially compensate for the foreshortening.

Some further interesting examples of configurations are given below.

- The compact array of the Australia Telescope is an east–west linear array of six antennas, all movable on a rail track (Frater et al. 1992).
- The UTR-2 is a T-shaped array of large-diameter, broadband dipoles built by the Ukrainian Academy of Sciences near Grakovo, Ukraine (Braude et al. 1978). The frequency range of operation is 10–25 MHz. Several smaller antennas of similar type have been constructed at distances up to approximately 900 km from the Grakovo site and are used for VLBI observations.
- An array of 720 conical spiral antennas in a T-shaped configuration operating in the frequency range 15–125 MHz was constructed at Borrego Springs, California (Erickson et al. 1982).
- The Mauritius Radio Telescope, near Bras d'eau, Mauritius, is a T-shaped array of helix antennas operating at 150 MHz. The east–west arm is 2 km long. The south arm is 880 m long and is synthesized by moving a group of antennas on trolleys. The array is similar in principle to the one in Fig. 1.12a. It is intended to cover a large portion of the Southern Hemisphere.
- The GMRT (Giant Metrewave Radio Telescope) near Pune, India, consists of 30 antennas, 16 of which are in a Y-shaped array with curved arms approximately 15 km long. The remaining 14 are in a quasi-random cluster in the central 2 km (Swarup et al. 1991). The antennas are 45 m in diameter and are at fixed locations. The highest operating frequency is approximately 1.6 GHz.

- A circular array with 96 uniformly spaced antennas was constructed at Culgoora, Australia, for observations of the Sun (Wild 1967). This was a multibeam, scanning, phased array rather than a correlator array, consisting of 96 antennas uniformly spaced around a circle of diameter 3 km and operating at 80 and 160 MHz. To suppress unwanted sidelobes of the beam, Wild (1965) devised an ingenious phase-switching scheme called J^2 synthesis. The spatial sensitivity of this ring array was analyzed by Swenson and Mathur (1967).
- The Multielement Radio-Linked Interferometer Network (MERLIN) of the Jodrell Bank Observatory, England, consists of six antennas with baselines up to 233 km (Thomasson 1986).
- The Submillimeter Array (SMA) of the Smithsonian Astrophysical Observatory and Academia Sinica of Taiwan, located on Mauna Kea, Hawaii, is the first array to be built using a Reuleaux triangle configuration (Ho et al. 2004).
- In large arrays in which the antennas cover areas extending over several kilometers, there is usually a central area with relatively dense antenna coverage, surrounded by extensive areas with sparser coverage. These outer parts may be in the form of extended arms, but the placement of the individual antennas is often irregular as a result of details of the landscape. Examples include ALMA (Wootten and Thompson 2009), the Murchison Widefield Array (Lonsdale et al. 2009), the Australian SKA Pathfinder (DeBoer et al. 2009), and the Low-Frequency Array (LOFAR) (de Vos et al. 2009). For discussion of projects for large arrays, see Carilli and Rawlings (2004).

5.6.3   VLBI Configurations

In VLBI (very-long-baseline interferometry) arrays, which are discussed in more detail in Chap. 9, the layout of antennas results from considerations of both (u, v) coverage and practical operating requirements. During the early years of VLBI, the signals were recorded on magnetic tapes that were then sent to the correlator location for playback. The use of tape has been superseded by magnetic disks and in some cases by direct transmission of the signals to the correlator using fiberoptic or other transmission media. Observing periods are limited by the ranges of hour angle and declination that are simultaneously observable from widely spaced locations. Although these locations usually deviate significantly from a plane, the angular widths of the sources under observation are generally sufficiently small that the small-field approximation (i.e., l and m small) can be used in deriving the radio image, as in Eq. (3.9).

For the first two decades after the inception of the VLBI technique, observations were mainly joint ventures among different observatories. Consideration of arrays dedicated solely to VLBI occurred as early as 1975 (Swenson and Kellermann 1975), but construction of such arrays did not begin for another decade. A study of antenna locations for a VLBI array has been discussed by Seielstad et al. (1979).

To obtain a single index as a measure of the performance of any configuration, the spatial transfer function was computed for a number of declinations. The fraction of appropriately sized (u, v) cells containing measurements was then weighted in proportion to the area of sky at each declination and averaged. Maximizing the index, in effect, minimizes the number of holes (unfilled cells). Other studies have involved computing the response to a model source, synthesizing an image, and improving the model as necessary.

The design of an array dedicated to VLBI, the Very Long Baseline Array (VLBA) of the United States, is described by Napier et al. (1994). The antenna locations [and associated (u, v) loci] are shown in Fig. 5.21 and listed in Table 5.1. A discussion of the choice of sites is given by Walker (1984). Antennas in Hawaii and St. Croix provide long east–west baselines. New Hampshire to St. Croix is the longest north–south spacing. A site in Alaska would be farther north but would be of limited benefit because it would provide only restricted accessibility for sources at southern declinations. An additional site within the Southern Hemisphere would enhance the (u, v) coverage at southern declinations. The southeastern region of the United States is avoided because of the higher levels of water vapor in the atmosphere. Intermediate north–south baselines are provided by the drier West Coast area. The Iowa site fills in a gap between New Hampshire and the southwestern sites. The short spacings are centered on the VLA, and as a result, the spatial frequency coverage shows a degree of central concentration. This enables the array to make measurements on a wider range of source sizes than would be possible with the same number of antennas and more uniform coverage. However, this results in some sacrifice in capability for imaging complex sources.

5.6.4   Orbiting VLBI Antennas

The discussion of placing a VLBI station in Earth orbit to work with ground-based arrays started as early as 1969 (Preston et al. 1983; Burke 1984; Kardashev et al. 2013). The combination of orbiting VLBI (OVLBI) and ground-based antennas has several obvious advantages. Higher angular resolution can be achieved, and the ultimate limit may be set by interstellar scintillation (see Sect. 14.4). The orbital motion of the spacecraft helps to fill in the coverage in the (u, v) plane and has the potential to improve the detail and dynamic range in the resulting images. Furthermore, a satellite in low Earth orbit provides rapid (u, v) plane variation, which can be valuable for obtaining information on time variability of source structure.

Figure 5.22 shows an example of the (u, v) coverage for observations with the VSOP project spacecraft known as HALCA (Hirabayashi et al. 1998) and a series of terrestrial antennas: one at Usuda, Japan, one at the VLA site, and the ten VLBA antennas. The spacecraft orbit is inclined at an angle of 31° to the Earth's equator, and the height above the Earth's surface is 21,400 km at apogee and 560 km at perigee. The mission of this spacecraft was to extend the resolution by a factor of

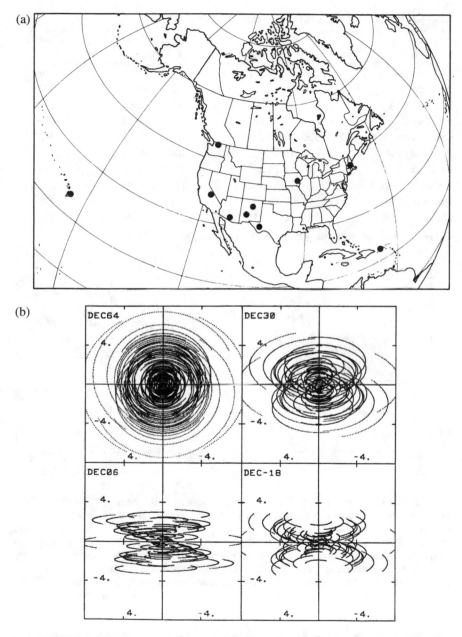

Fig. 5.21 Very Long Baseline Array in the United States: (**a**) locations of the ten antennas, and (**b**) spatial frequency coverage (spacings in thousands of kilometers) for declinations of 64°, 30°, 6°, and −18°, in which the observing time at each antenna is determined by an elevation limit of 10°. From Walker (1984). Reprinted with the permission of and © Cambridge University Press.

Table 5.1 Locations of antennas in the VLBA[a]

Location	N. Latitude (deg min sec)	W. Longitude (deg min sec)	Elevation (m)
St. Croix, VI	17 45 30.57	64 35 02.61	16
Hancock, NH	42 56 00.96	71 59 11.69	309
N. Liberty, IA	41 46 17.03	91 34 26.35	241
Fort Davis, TX	30 38 05.63	103 56 39.13	1615
Los Alamos, NM	35 46 30.33	106 14 42.01	1967
Pie Town, NM	34 18 03.61	108 07 07.24	2371
Kitt Peak, AZ	31 57 22.39	111 36 42.26	1916
Owens Valley, CA	37 13 54.19	118 16 33.98	1207
Brewster, WA	48 07 52.80	119 40 55.34	255
Mauna Kea, HI	19 48 15.85	155 27 28.95	3720

[a]© 1994 IEEE. Reprinted, with permission, from P. J. Napier et al. (1994).

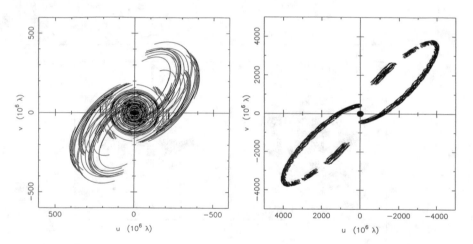

Fig. 5.22 (u, v) plane tracks for arrays with a satellite station for the source 1622+633 at 5 GHz. (**left**) Coverage with VSOP and 12 ground-based antennas. The roughly circular tracks within $2 \times 10^6 \lambda$ are the baselines among the ground-based antennas. Produced with the FAKESAT software developed by D. W. Murphy, D. L. Meier, and T. J. Pearson. (**right**) Coverage with RadioAstron and six ground-based antennas. The gaps in the coverage correspond to actual satellite constraints for hypothetical observations in February 2016. The satellite period is 8.3 days, and the "wobbly" appearance of the tracks is caused by the Earth's diurnal motion. Produced with the FAKERAT software, a derivative of FAKESAT (http://www.asc.rssi.ru/radioastron/software/fakerat).

three over ground-based arrays and to retain good imaging capability. The spacings shown are for a frequency of 5 GHz, and the units of u and v are 10^6 wavelengths; the maximum spacing is 5×10^8 wavelengths, which corresponds to a fringe width of 0.4 mas. The approximately circular loci at the center of the figure represent baselines between terrestrial antennas. The orbital period is 6.3 h, and the data shown correspond to an observation of duration about four orbital periods. The spacecraft orbit precesses at a rate of order $1°$ per day, and over the course of one to two years, the coverage of any particular source can be improved by combining observations.

Figure 5.22 also shows examples of the (u, v) coverage for observations with the RadioAstron project spacecraft known as Spektr-R (Kardashev et al. 2013) and a set of ground-based antennas. The spacecraft orbit is inclined at an angle of $80°$ to the Earth's equator, and, for the case shown here, the ellipticity is 0.86, and the height above the Earth's surface is 289,000 km at apogee and 47,000 km at perigee (orbit on April 14, 2012). The mission of RadioAstron is to provide ultrahigh resolution to explore new astrophysical phenomena while sacrificing imaging quality because of the gap between satellite–Earth and Earth-only baselines. The orbital period is 8.3 days. The orbit evolves substantially with time because of the influences of the Sun and Moon. Occasions when the orbit eccentricity reaches its maximum of 0.95 offer opportunities for better imaging capability.

Figure 5.23 shows an example of the (u, v) coverage that could be obtained between two spacecraft in circular orbits of radius about ten Earth radii, with orthogonal planes that have periods differing by 10%. Multispacecraft operation offers satellite-to-satellite baselines, which are free from the effects of atmospheric delay. In practice, there are likely to be restrictions on coverage resulting from the limited steerability of the astronomy and communication antennas relative to the spacecraft. It is necessary for the spacecraft to maintain an attitude in which the solar power panels remain illuminated and the communications antenna can be pointed toward the Earth. Further discussion of orbiting VLBI is given in Sect. 9.10.

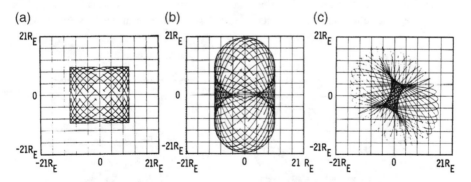

Fig. 5.23 Spatial frequency coverage for two antennas on satellites with circular orbits of radius approximately ten times the Earth's radius R_E: (**a**) source along the X axis; (**b**) source along Y or Z axes; (**c**) source centered between X, Y, and Z axes. The orbits lie in the XY and XZ planes of a rectangular coordinate system. The satellite periods differ by 10%, and the observing period is approximately 20 days. From R. A. Preston et al. (1983), © Cépaduès Éditions, 1983.

5.6.5   Planar Arrays

Studies of cosmic background radiation and the Sunyaev–Zel'dovich effect require observations with very high brightness sensitivity at wavelengths of order 1 cm and shorter: see also Sect. 10.7. Unlike the sensitivity to point sources, the sensitivity to a broad feature that largely fills the antenna beam does not increase with increasing collecting area of the antenna. Thus, for cosmic background measurements, large antennas are not required. Extremely good stability is necessary to allow significant measurements at the level of a few tens of microkelvins per beam, that is, of order $10~\mu$Jy arcmin^{-2}. Special arrays have been designed for this purpose. A number of antennas are mounted on a platform, with their apertures in a common plane. The whole structure is then supported on an altazimuth mount so the antennas can be pointed to track any position on the sky. An example of such an instrument, the Cosmic Background Imager (CBI), was developed by Readhead and colleagues at Caltech (Padin et al. 2001). Thirteen Cassegrain focus paraboloids, each of diameter 90 cm, were operated in the 26- to 36-GHz range. In this instrument, the antenna mounting frame had the shape of an irregular hexagon with threefold symmetry and maximum dimensions of approximately 6.5 m, as shown in Fig. 5.24. For the particular type of measurements required, the planar array has a number of desirable properties compared with a single antenna of similar aperture, or a number of individually mounted antennas, as outlined below:

- The use of a number of individual antennas allows the output to be measured in the form of cross-correlations between antenna pairs. Thus, the output is not sensitive to the total power of the receiver noise but only to correlated signals entering the antennas. The effects of gain variations are much less severe than

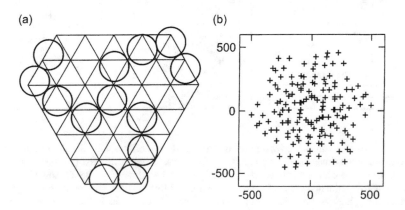

Fig. 5.24 (a) Face view of the antenna platform of the Cosmic Background Imager, showing a configuration of the 13 antennas. (b) The corresponding antenna spacings in (u, v) coordinates for a wavelength of approximately 1 cm.

in the case of a total-power receiver. Thermal noise from ground pickup in the sidelobes is substantially resolved.

- The antennas can be mounted with the closest spacing physically possible. There are then no serious gaps in the spatial frequencies measured, and structure can be imaged over the width of the primary antenna beams. The apertures cannot block one another because the antenna platform tracks, as can occur for individually mounted antennas in closely spaced arrays.

- In the array in Fig. 5.24, the whole antenna mounting platform can be rotated about an axis normal to the plane of the apertures. Thus, rotation of the baselines can be controlled as desired and is independent of Earth rotation. For a constant pointing direction and rotation angle relative to the sky, the pattern of (u, v) coverage remains constant as the instrument tracks. Variations in the correlator outputs with time can result from ground radiation in the sidelobes, which varies with azimuth and elevation as the array tracks. This variation can help to separate out the unwanted response.

- The close spacing of the antennas results in some cross coupling by which spurious correlated noise is introduced into the receiving channels of adjacent antennas. However, because the antennas are rigidly mounted, the coupling does not vary as the system tracks a point on the sky, as is the case for individually mounted antennas. The effects of the coupling are therefore more easily calibrated out. In the CBI design, the coupling is reduced to -110 to -120 dB by the use of a cylindrical shield around each antenna and by designing the subreflector supports to minimize scattering.

At a frequency of 30 GHz, a pointing error of $1''$ in a 6-m baseline produces a visibility phase error of $1°$. Pointing accuracy is critical, and the CBI antenna is mounted in a retractable dome to shield it from wind, which can be strong at the 5000-m-elevation site at Llano de Chajnantor, Chile. Observations of the cosmic microwave background with this system are briefly described in Sect. 10.7.

5.6.6 Some Conclusions on Antenna Configurations

The most accurate prediction of the performance of an array is obtained by computation of the response of the particular design to models of sources to be observed. However, here we are more concerned with broad comparisons of various configurations to illustrate the general considerations in array design. Some conclusions are summarized below.

- A circle centered on the (u, v) origin can be considered an optimum boundary for the distribution of measurements of visibility. Uniformity of the distribution within the circle is a further useful criterion in many circumstances. An exception is the condition in which sidelobes of the synthesized beam are a serious problem, for example, in low-frequency arrays operating in conditions of source confusion. In arrays in which the scale of the configuration cannot be varied to accommodate a wide range of source dimensions, a centrally concentrated distribution allows a greater range of angular sizes to be measured with a limited number of antennas. If sensitivity to broad, low-brightness objects is important, it is preferable to have more antenna pairs with short spacings at which such sources are not highly resolved. Note that two of the largest arrays in which the antennas are not movable, the GMRT (in India) and the VLBA (North America), each have a cluster of antennas at relatively short spacings, as well as other antennas at longer spacings, in order to cover a wide range of source dimensions.

- Although the effect of sidelobes on the synthesized beam can be greatly reduced by CLEAN and other image-processing algorithms described in Chap. 11, obtaining the highest dynamic range in radio images (that is, a range of reliable intensity measurements of order 10^6 or more) requires both good spatial frequency coverage and effective image processing. Reducing holes (unsampled cells), which are found to be a consistent indicator of sidelobe levels in this coverage, is a primary objective in array design.
- The east–west linear array has been used for both large and small instruments and requires tracking over ± 6 h to obtain full two-dimensional coverage. It is most useful for regions of the sky within about 60° of the celestial poles and is the most economical configuration with respect to land use for road or rail track.
- The equiangular Y-shaped array gives the best spatial frequency coverage of the existing configurations with linear, open-ended arms. Autocorrelation functions of configurations with odd numbers of arms have higher-order symmetry than those with even numbers in which opposite arms are aligned. Curvature of the arms or random displacement of the antennas helps to smooth out the linear ridges in the (u, v) coverage (e.g., in the snapshot in Fig. 5.18). Such features are also smoothed out by hour-angle tracking and are most serious for snapshot observations.
- The circle and Reuleaux triangle provide the most uniform distributions of measurements. With uniformly spaced antennas, the Reuleaux triangle provides more uniform (u, v) coverage than the circle, but varying the spacing in a quasi-random manner greatly improves both cases and reduces the difference between them; see Fig. 5.19. However, if higher resolution is needed, these configurations are not so easily extended as ones with open-ended arms.

5.7   Implementation of Large Arrays

Of the large arrays that have contributed prominently to progress in radio astronomy, those that developed first have largely been in the range of roughly 500 MHz to 30 GHz, i.e., approximately the wavelength range of 1–60 cm. Examples are the VLA and the arrays at Westerbork (the Netherlands) and the Australia Telescope at Narrabri (Australia). This wavelength range is most conducive for construction of large parabolic reflectors with surface accuracy better than $\sim 1/16$ of a wavelength. Arrays for millimeter-wavelength observations such as the SMA on Mauna Kea followed a decade or two later, as technology for more accurate surfaces developed, leading to ALMA on the Atacama plateau in Chile, which came into operation in 2013 (Wootten and Thompson 2009). For the 12-m-diameter antennas of ALMA, the specified surface accuracy is less than 25 μm, allowing useful operation up to a frequency of almost 1 THz. For details of measuring and adjusting the surface, see Mangum et al. (2006), Snel et al. (2007), and papers in Baars et al. (2009). The main ALMA array consists of 50 12-m-diameter antennas movable between foundation pads that allow a wide range of spacings up to $\sim$ 15 km. A second, compact, array uses 12 7-m-diameter antennas, and 4 other antennas are available for total power measurements.

At the long-wavelength end of the spectrum, radio astronomy was, for the first few decades, largely limited to measurements of relatively small numbers of the stronger sources, for example, Erickson et al. (1982). A major problem is presented by the ionosphere, calibration of the effects of which requires that the antenna elements be arranged in phased clusters, or subarrays, the beams of which are no wider than the aplanatic structure of the ionosphere. The outputs of these clusters are cross correlated to provide the visibility values. These long-wavelength observations are important for the study of the most distant Universe including redshifted neutral hydrogen just prior to the Epoch of Reionization. In LOFAR [de Vos et al. (2009) and van Haarlem et al. (2013)], the clusters of dipoles have diameters of $\sim$ 81 m for 10–90 MHz and $\sim$ 40 m for 115–240 MHz. LOFAR is based in the Netherlands, and baselines between the clusters extend up to 1200 km in a generally eastward direction. The dipoles take the form of an inverted V configuration, in which four conductors run outward and downward at an angle of 45° from a point roughly 2 m above the ground, forming two orthogonal dipoles over a ground plane. Note that since the need to calibrate the effect of the ionosphere places a lower limit on the size of the dipole clusters that are used, in this long-wavelength range, large-scale arrays are generally the most successful.

5.7.1 Low-Frequency Range

At frequencies up to about 300 MHz, arrays of broadband dipoles mounted over a ground-plane reflecting screen provide a very practical antenna system. Dipoles are robust, and crossed dipoles provide full polarization coverage. Low-noise transistor front ends can operate at ambient temperature at these frequencies, where the system noise level is set largely by radiation from the sky. Signals from groups of dipoles are combined and the phases adjusted to form beams that can be pointed as required without the need for moving parts. If the spacing between the centers of the dipoles is greater than $\lambda/2$, the array is described as sparse. The collecting area is maximized at $\lambda^2/4$ per element, but because of the spacing, the grating sidelobes begin to be significant as $\lambda/2$ is exceeded. If the spacing is less than $\lambda/2$, the array is described as compact. The effective area is then less than $\lambda^2/4$ per element, but grating lobes are avoided. The variation of the path length through the ionosphere is a serious problem in imaging at these low frequencies, but it is possible to calibrate the ionosphere over a wide angular range by forming beams in the directions of calibration sources for which the positions are accurately known. LOFAR and the Murchison Widefield Array (Lonsdale et al. 2009) and the Allen Telescope Array (Welch et al. 2009) are examples of this type.

Ellingson (2005) describes a system using dipoles below 100 MHz. To achieve the maximum sensitivity, it is necessary only to match the antennas to the receivers sufficiently well that the total noise is dominated by the background component received by the antennas. This is an advantageous situation since it allows the dipoles to be used over a much wider frequency range than is possible when the impedance must be well matched. To investigate the performance of an inverted-V dipole under these conditions, let γ be the power ratio of the background noise received from the sky to the noise contributed by the receiver. Then we have

$$\gamma \simeq e_r \frac{T_{\text{sky}}}{T_{\text{rec}}} (1 - |\Gamma^2|) , \tag{5.22}$$

where e_r (< 1) is an efficiency factor that results largely from the ohmic losses in the ground and in the dipole, T_{sky} is the noise brightness temperature of the sky, T_{rec} is the noise temperature of the receiver, and Γ is the voltage reflection coefficient at the antenna looking toward the receiver. Γ is given by

$$\Gamma = \frac{Z_{\text{rec}} - Z_{\text{ant}}}{Z_{\text{rec}} + Z_{\text{ant}}} , \tag{5.23}$$

where Z_{rec} and Z_{ant} are the impedances at the receiver and antenna terminals, respectively. For dominance of the sky noise, one can take γ greater than ~ 10. T_{sky} is related to the intensity of the background radiation I_ν (W m^{-2} Hz^{-1} sr^{-1}), by $T_{\text{sky}} = c^2 I_\nu / 2k\nu^2$, where c is the speed of light and k is Boltzmann's constant. An expression for the sky background intensity I_ν as a function of frequency is given by Dulk et al. (2001) based on measurements by Cane (1979):

$$I_\nu = I_g \nu^{-0.52} \frac{1 - e^{-\tau(\nu)}}{\tau(\nu)} + I_{eg} \nu^{-0.80} e^{-\tau(\nu)} , \tag{5.24}$$

where $I_g = 2.48 \times 10^{-20}$ W m^{-2} Hz^{-1} sr^{-1} is the galactic component of the intensity, $I_{eg} = 1.06 \times 10^{-20}$ W m^{-2} Hz^{-1} sr^{-1} is the extragalactic component, and $\tau(\nu) = 5.0\nu^{-2.1}$. This model applies broadly over the sky except near the galactic plane where higher intensities are encountered. In the system described by Ellingson (2005), a wide frequency response for the dipoles is obtained with Z_{rec} in the range 200–800 ohms. Computed responses indicate usable beamwidths in the range 120–140°. Stewart et al. (2004) describe design of an inverted-V dipole in which the effective width of the conducting arms is increased in one dimension, which reduces the impedance variation with frequency compared with that of a dipole with single-wire elements.

5.7.2 *Midfrequency and Higher Ranges*

In the midfrequency range, approximately 0.3–2 GHz, there are two main possibilities. For the frequencies up to about 1 GHz, aperture arrays (van Ardenne et al. 2009) can take the form of half-wave dipoles over a ground screen or, especially at the shorter wavelengths, arrays of Vivaldi antennas (Schaubert and Chio 1999) are used. The Vivaldi elements are formed on strips of aluminum or of copper-clad insulating board. By using two sets of Vivaldi elements running in orthogonal directions, full polarization is obtained. The approximate spacing between adjacent Vivaldi elements is $\lambda/2$, and approximately four amplifiers are required for each square wavelength of collecting area, e.g., $\sim$ 44 amplifiers per square meter at 1 GHz. Aperture arrays provide multiple beams with rapid and flexible pointing.

5.7.2.1 Phased-Array Feeds

For the range from $\sim$ 700 MHz and above, parabolic dish-type antennas with single or multiple beams become more practicable than aperture arrays since, for a given collecting area, they do not require such large numbers of low-noise amplifiers and phasing components. With feeds in the form of a focal-plane array, i.e., an array of individual feed elements in the focal plane of an antenna, it is usually not possible to get the feeds close enough together to avoid gaps between the individual beams. Thus, it is often preferable to use phased-array feeds in which an array of closely spaced receiving elements is arranged in the focal plane. Any one antenna beam is formed as a phased combination of the signals from a number of the feed elements, and such combinations can be designed to provide optimum beam spacings for efficient sky coverage. It is the beamformer that distinguishes the phased-array feed from the focal-plane array. The elements are individually terminated with matched amplifiers, but mutual coupling between the elements cannot be avoided, so the design and adjustment of phased-array feeds is generally more critical than for focal-plane arrays. A general analysis of a phased-array feed can be found in van Ardenne et al. (2009) and Roshi and Fisher (2016).

Designs of phased-array feeds include ones using the Vivaldi system mentioned above and others using a "checkerboard" conductor pattern (Hay et al. 2007). The checkerboard scheme can be envisaged as a series of conducting elements on a circuit board that are arranged like the black squares of a checkerboard. At each point where two corners of conducting squares meet, the corners do not touch, but each feeds one input of a balanced amplifier. The patterns of conducting and nonconducting surfaces are identical and thus self-complimentary. A screen of this form in free space is well matched with load impedances of 377 ohms between the corner pairs of conducting squares where the amplifiers are connected.[2] For use

---

[2] This follows from a formula by Booker: see, e.g., *Antennas*, J. D. Kraus (1950 or later edition).

as a feed array, the checkerboard screen is mounted over a ground plane, which introduces some frequency variation in the impedance. In this frequency range, the input stages of amplifiers at the feeds may be cryogenically cooled to minimize the system temperature.

The use of phased-array feeds in interferometric arrays presents a huge challenge in signal processing because separate correlators are required for each beam. The first interferometer to be designed specifically for phased-array feed technology is ASKAP at the Murchison Radio Observatory. The system has 36 dual-polarized beams operating in the 0.7–1.8 GHz band (Hay et al. 2007; Hotan et al. 2014). A 52-element phased array called APERTIF at 21-cm wavelength has been implemented on the Westerbork telescope (van Cappellen and Bakker 2010; van Cappellen et al. 2011; and Ivashina et al. 2011).

5.7.2.2 Optimum Antenna Size

An array with fixed collecting area can be built with a large number of small antennas (called the "large N, small d solution") or a small number of large antennas (the "small n, large D solution"). Determining the right antenna size is a complex problem. With smaller antennas, the field of view is larger, which enhances survey speed, but with larger antennas, phase calibration sources can be found closer to the target.

A cost analysis is an important element in the determination of antenna size. The critical fact in cost optimization is that the cost of parabolic antenna elements of diameter D scales approximately as $D^{2.7}$ (Meinel 1979). Because the exponent on D is greater than two, the total cost of the antennas in an array increases with diameter for a fixed array area. On the other hand, a larger array of smaller antennas requires more receivers and a larger correlator. A crude cost model can be written

$$C + f_1 n_a D^{2.7} + f_2 n_a + f_3 n_a^2 , \qquad (5.25)$$

where n_a is the number of antennas, f_1 is the antenna cost factor, f_2 is the receiver cost factor, and f_3 is the correlator cost factor, where we assume the correlator cost scales as n_a^2. For a fixed array collecting area, A,

$$n_a = \frac{A}{(\pi \eta D^2 / 4)} , \qquad (5.26)$$

where η is the aperture efficiency. We can substitute Eq. (5.26) into Eq. (5.25) and find the value of D that minimizes C. These values of D are typically in the range of 4 to 20 m. The proposals for the antenna sizes for ALMA ranged in diameter from 6 to 15 m before the decision was made for 12-m-diameter elements, based on cost and many other factors.

5.7.3 Development of Extremely Large Arrays

The concept of an array with a collecting area of ~ 1 square kilometer arose in the late 1990s after the Westerbork Synthesis Radio Telescope, the VLA, and similar instruments had demonstrated the power of the synthesis technique in high-resolution imaging and in cataloging and studying large numbers of sources. Such an array would have a collecting area of about two orders of magnitude greater than existing arrays at that time but would require significant technological development to be financially feasible. An initial objective was to extend the redshift range at which HI in galaxies can be studied by an order of magnitude to $z \sim 2$. The concept has been developed into a plan to build multiple arrays spanning the frequency interval of 70 MHz to greater than 25 GHz, with baselines up to about 5000 km. This instrument, collectively called the Square Kilometre Array (SKA)[3] would have an enormous impact on a broad range of astronomical problems from planet formation to cosmology. The science case for the instrument has been presented by Carilli and Rawlings (2004) and Bourke et al. (2015). Technical details are given in Hall (2004) and Dewdney et al. (2009). The concept of such an array has led to the development of several smaller arrays to test the practicality and performance of possible technologies, including antenna and correlator designs. These include ASKAP, with 12-m-diameter antennas with a checkerboard phased-array feed system providing multiple beams (see Sect. 5.7.2.1), located in Western Australia (DeBoer et al. 2009), and MeerKAT, an array of low-cost 12-m-diameter dish antennas with single-pixel feeds to cover 0.7–10 GHz, located in the Karoo region of South Africa (Jonas 2009).

5.7.4 The Direct Fourier Transform Telescope

The normal practice in radio astronomy is to measure the correlation function of the incident electric field and then take its Fourier transform to obtain the image of the source intensity distribution. An alternative approach is to measure the Fourier transform of the incident electric field with a uniform array of antennas and take its square modulus to obtain the image. Either the correlation function or the direct Fourier transform approach must be implemented at the Nyquist rate appropriate for the bandwidth. The latter approach is simply an implementation of the Fraunhofer diffraction equation, which relates the aperture field distribution to the far field distribution (see Chap. 15). For this reason, instruments based on this method are sometimes called digital lenses. The Fraunhofer equation is also the basis of the holographic method of measuring the surface accuracy of parabolic antennas, as described in Sect. 17.3.

---

[3]The SKA Memo Series can be found at http://www.skatelescope.org/publications.

Daishido et al. (1984) described the operation and prototype of a direct Fourier transform telescope operating at 11 GHz. They called the instrument a "phased array telescope" because its operation was equivalent to forming phased array beams pointed at a grid of positions on the sky. The Fourier transform was affected though the use of Butler matrices. A 64-element array (8×8 elements on a uniform grid) was built at Waseda University and used for wide-field searches of transient sources (Nakajima et al. 1992, 1993; Otobe et al. 1994). The signal processing was further improved in another instrument aimed at pulsar observations (Daishido et al. 2000; Takeuchi et al. 2005).

Interest has been renewed in the direct Fourier transform telescope because of the advent of arrays with very large numbers of antennas. In this case, the direct Fourier transform configuration can take advantage of the computational speed of the fast Fourier transform, which scales as $n_a \log_2 n_a$, where n_a is the number of antennas. A detailed analysis of the direct Fourier transform telescope was developed by Tegmark and Zaldarriaga (2009, 2010). They were motived by the challenges of measuring the wide-field distribution of redshifted HI emission, the signature of the Epoch of Reionization (see Sect. 10.7.2), and called their instrument the Fast Fourier Transform Telescope. Zheng et al. (2014) built a prototype 8×8 array at 150 MHz to develop techniques for such measurements.

One characteristic of the direct Fourier transform telescope based on the FFT with a uniform-grid antenna layout is the high redundancy of short baselines. The situation is similar to that encountered in the design of the digital FFT spectrometers described in Sect. 8.8.5, wherein the number of equivalent baselines at large spacings is underrepresented. Methods of relaxing the requirement of uniform spacings have been explored by Tegmark and Zaldarriaga (2010) and Morales (2011).

A disadvantage of the direct Fourier transform telescope relates to calibration. Since no baseline-based measurements are made, the traditional techniques of self-calibration based on amplitude and phase closure cannot be directly applied. There are several approaches to the calibration problem. The most straightforward approach is to transform the images back to the visibility domain on the time scale of instrumental and atmospheric variability, and apply the techniques described in Chap. 11. Auxiliary measurements can also be made to supply calibration information. More sophisticated methods are under development (e.g., Foster et al. 2014; Beardsley et al. 2016).

Further Reading

Baars, J.W.M., *The Paraboloidal Reflector Antenna in Radio Astronomy and Communication*, Springer, New York (2007)

Balanis, C.A., *Antenna Theory Analysis and Design*, Wiley, New York, 1982 (1997)

Collin, R.E., *Antennas and Radiowave Propagation*, McGraw-Hill, New York (1985)

Imbriale, W.A., and Thorburn, M., Eds., *Proc. IEEE*, Special Issue on Radio Telescopes, **82**, 633–823 (1994)

Johnson, R.C., and Jasik, H., Eds., *Antenna Engineering Handbook*, McGraw Hill, New York (1984)

Kraus, J.D., *Antennas*, McGraw-Hill, New York, 1950, and 2nd ed., McGraw-Hill, New York, 1988. The 3rd ed. is Kraus, J.D., and Marhefka, R.J., *Antennas for All Applications*, McGraw-Hill, New York (2002)

Love, AW, Ed., *Reflector Antennas*, IEEE Press, Institute of Electrical and Electronics Engineers, New York (1978)

Milligan, T.A., *Modern Antenna Design*, McGraw-Hill, New York (1985)

Stutzman, W.L., and Thiele, G.A., *Antenna Theory and Design*, 2nd ed., Wiley, New York (1998)

References

Arsac, J., Nouveau réseau pour l'observation radioastronomique de la brillance sur le soleil à 9530 Mc/s, *Compt. Rend. Acad. Sci.*, **240**, 942–945 (1955)

Baars, J.W.M., *The Paraboloidal Reflector Antenna in Radio Astronomy and Communication*, Springer, New York (2007)

Baars, J.W.M., D'Addario, L.R., Thompson, A.R., Eds., *Proc. IEEE*, Special Issue on Advances in Radio Telescopes, **97**, 1369–1548 (2009)

Baars, J.W.M., and Hooghoudt, B.G., The Synthesis Radio Telescope at Westerbork: General Layout and Mechanical Aspects, *Astron. Astrophys.*, **31**, 323–331 (1974)

Beardsley, A.P., Thyagarajan, N., Bowman, J.D., and Morales, M.F., An Efficient Feedback Calibration Algorithm for Direct Imaging Radio Telescopes, *Mon. Not. R. Astron. Soc.*, in press (2016), arXiv:1603.02126

Blythe, J.H., A New Type of Pencil Beam Aerial for Radio Astronomy, *Mon. Not. R. Astron. Soc.* **117**, 644–651 (1957)

Bourke, T.L., Braun, R., Fender, R., Govoni, F., Green, J., Hoare, M., Jarvis, M., Johnston-Hollitt, M., Keane, E., Koopmans, L., and 14 coauthors, *Advancing Astrophysics with the Square Kilometre Array*, 2 vols., Dolman Scott Ltd., Thatcham, UK (2015) (available at http://www.skatelescope.org/books)

Bracewell, R.N., Interferometry of Discrete Sources, *Proc. IRE*, **46**, 97–105 (1958)

Bracewell, R.N., Interferometry and the Spectral Sensitivity Island Diagram, *IRE Trans. Antennas Propag.*, **AP-9**, 59–67 (1961)

Bracewell, R.N., Radio Astronomy Techniques, in *Handbuch der Physik*, Vol. 54, S. Flugge, Ed., Springer-Verlag, Berlin (1962), pp. 42–129

Bracewell, R.N., Optimum Spacings for Radio Telescopes with Unfilled Apertures, in *Progress in Scientific Radio*, Report on the 15th General Assembly of URSI, Publication 1468 of the National Academy of Sciences, Washington, DC (1966), pp. 243–244

Bracewell, R.N., *Two-Dimensional Imaging*, Prentice-Hall, Englewood Cliffs, NJ (1995)

Bracewell, R.N., *The Fourier Transform and Its Applications*, McGraw-Hill, New York (2000) (earlier eds. 1965, 1978).

Bracewell, R.N., Colvin, R.S., D'Addario, L.R., Grebenkemper, C.J., Price, K.M., and Thompson, A.R., The Stanford Five-Element Radio Telescope, *Proc. IEEE*, **61**, 1249–1257 (1973)

Bracewell, R.N., and Roberts, J.A., Aerial Smoothing in Radio Astronomy, *Aust. J. Phys.*, **7**, 615–640 (1954)

Bracewell, R.N., and Thompson, A.R., The Main Beam and Ringlobes of an East–West Rotation-Synthesis Array, *Astrophys. J.*, **182**, 77–94 (1973).

Braude, S. Ya., Megn, A.V., Ryabov, B.P., Sharykin, N.K., and Zhouck, I.N., Decametric Survey of Discrete Sources in the Northern Sky, *Astrophys. Space Sci.*, **54**, 3–36 (1978)

Brigham, E.O., *The Fast Fourier Transform and Its Applications*, Prentice Hall, Englewood Cliffs, NJ (1988)

Burke, B.F., Orbiting VLBI: A Survey, in *VLBI and Compact Radio Sources*, Fanti, R., Kellermann, K., and Setti, G., Eds., Reidel, Dordrecht, the Netherlands (1984)

Cane, H.V., Spectra of the Nonthermal Radio Radiation from the Galactic Polar Regions, *Mon. Not. R. Astron. Soc.*, **149**, 465–478 (1979)

Carilli, C., and Rawlings, S., Eds., Science with the Square Kilometre Array, *New Astron. Rev.*, **48**, 979–1605 (2004)

Chow, Y.L., On Designing a Supersynthesis Antenna Array, *IEEE Trans. Antennas Propag.*, **AP-20**, 30–35 (1972)

Chu, T.-S., and Turrin, R.H., Depolarization Effects of Offset Reflector Antennas, *IEEE Trans. Antennas Propag.*, **AP-21**, 339–345 (1973)

Cornwell, T.J., A Novel Principle for Optimization of the Instantaneous Fourier Plane Coverage of Correlation Arrays, *IEEE Trans. Antennas Propag.*, **36**, 1165–1167 (1988)

Daishido, T., Ohkawa, T., Yokoyama, T., Asuma, K., Kikuchi, H., Nagane, K., Hirabayashi, H., and Komatsu, S., Phased Array Telescope with Large Field of View to Detect Transient Radio Sources, in *Indirect Imaging: Measurement and Processing for Indirect Imaging*, Roberts, J.A., Ed., Cambridge Univ. Press, Cambridge, UK (1984), pp. 81–87

Daishido, T., Tanaka, N., Takeuchi, H., Akamine, Y., Fujii, F., Kuniyoshi, M., Suemitsu, T., Gotoh, K., Mizuki, S., Mizuno, K., Suziki, T., and Asuma, K., Pulsar Huge Array with Nyquist Rate Digital Lens and Prism, in *Radio Telescopes*, Butcher, H.R., Ed., Proc. SPIE, **4015**, 73–85 (2000)

DeBoer, D.R., Gough, R.G., Bunton, J.D., Cornwell, T.J., Beresford, R.J., Johnston, S., Feain, I.J., Schinckel, A.E., Jackson, C.A., Kesteven, M.J., and nine coauthors, Australian SKA Pathfinder: A High-Dynamic Range Wide-Field of View Survey Telescope, *Proc. IEEE*, **97**, 1507–1521 (2009)

de Vos, M., Gunst, A.W., and Nijboer, R., The LOFAR Telescope: System Architecture and Signal Processing, *Proc. IEEE*, **97**, 1431–1437 (2009)

Dewdney, P.E., Hall, P.J., Schilizzi, R.T., and Lazio, T.J.L.W., The Square Kilometre Array, *Proc. IEEE*, **97**, 1482–1496 (2009)

Dulk, G.A., Erickson, W.C., Manning, R., and Bougeret, J.-L., Calibration of Low-Frequency Radio Telescopes Using Galactic Background Radiation, *Astron. Astrophys.*, **365**, 294–300 (2001)

Ellingson, S.W., Antennas for the Next Generation of Low-Frequency Radio Telescopes, *IEEE Trans. Antennas Propag.*, **53**, 2480–2489 (2005)

Elmer, M., Jeffs, B.J., Warnick, K.F., Fisher, J.R., and Norrod, R.D., Beamformer Design Methods for Radio Astronomical Phased Array Feeds, *IEEE Trans. Antennas Propag.*, **60**, 903–914 (2012)

Erickson, W.C., Mahoney, M.J., and Erb, K., The Clark Lake Teepee-Tee Telescope, *Astrophys. J. Suppl.*, **50**, 403–420 (1982)

Foster, G., Hickish, J., Magro, A., Price, D., and Zarb Adami, K., Implementation of a Direct-Imaging and FX Correlator for the BEST-2 Array, *Mon. Not. R. Astron. Soc.*, **439**, 3180–3188 (2014)

Frater, R.H., Brooks, J.W., and Whiteoak, J.B., The Australia Telescope—Overview, in *J. Electric. Electron. Eng. Australia*, Special Issue on the Australia Telescope, **12**, 103–112 (1992)

Golomb, S.W., How to Number a Graph, in *Graph Theory and Computing*, Read, R.C., Ed., Academic Press, New York (1972), pp. 23–27

Hall, P.J., Ed., The Square Kilometre Array: An Engineering Perspective, *Experimental Astron.*, **17(1–3)** (2004) (also as a single volume, Springer, Dordrecht, the Netherlands, 2005)

Hamaker, J.P., O'Sullivan, J.D., and Noordam, J.E., Image Sharpness, Fourier Optics, and Redundant Spacing Interferometry, *J. Opt. Soc. Am.*, **67**, 1122–1123 (1977)

Hay, S.G., O'Sullivan, J.D., Kot, J.S., Granet, C., Grancea, A., Forsythe, A.R., and Hayman, D.H., Focal Plane Array Development for ASKAP, in *Antennas and Propagation*, Proc. European Conf. on Ant. and Prop. (2007)

Hirabayashi, H., Hirosawa, H., Kobayashi, H., Murata, Y., Edwards, P.G., Fomalont, E.B., Fujisawa, K., Ichikawa, T., Kii, T., Lovell, J.E.J., and 43 coauthors, Overview and Initial Results of the Very Long Baseline Interferometry Space Observatory Program, *Science*, **281**, 1825–1829 (1998)

Hjellming, R.M., The Design of Aperture Synthesis Arrays, *Synthesis Imaging in Radio Astronomy*, Perley, R.A., Schwab, F.R., and Bridle, A.H., Eds., Astron. Soc. Pacific. Conf. Ser., **6**, 477–500 (1989)

Ho, P.T.P., Moran, J.M., and Lo, K.-Y., The Submillimeter Array, *Astrophys. J. Lett.*, **616**, L1–L6 (2004)

Högbom, J.A., and Brouw, W.N., The Synthesis Radio Telescope at Westerbork, Principles of Operation, Performance, and Data Reduction, *Astron. Astrophys.*, **33**, 289–301 (1974)

Hotan, A.W., Bunton, J.D., Harvey-Smith, L., Humphreys, B., Jeffs, B.D., Shimwell, T., Tuthill, J., Voronkov, M., Allen, G., Amy, S., and 91 coauthors, The Australian Square Kilometre Array Pathfinder: System Architecture and Specifications of the Boolardy Engineering Test Array, *Publ. Astron. Soc. Aust.*, **31**, c041 (15pp) (2014)

Ingalls, R.P., Antebi, J., Ball, J.A., Barvainis, R., Cannon, J.F., Carter, J.C., Charpentier, P.J., Corey, B.E., Crowley, J.W., Dudevoir, K.A., and six coauthors, Upgrading of the Haystack Radio Telescope for Operation at 115 GHz, *Proc. IEEE*, **82**, 742–755 (1994)

Ishiguro, M., Minimum Redundancy Linear Arrays for a Large Number of Antennas, *Radio Sci.*, **15**, 1163–1170 (1980)

Ivashina, M.V., Iupikov, O., Maaskant, R., van Cappellen, W.A., and Oosterloo, T., An Optimal Beamforming Strategy for Wide-Field Survey with Phased-Array-Fed Reflector Antennas, *IEEE Trans. Antennas Propag.*, **59**, 1864–1875 (2011)

Jonas, J.L., MeerKAT—The South African Array with Composite Dishes and Wide-Band Single Pixel Feeds, *Proc. IEEE*, **97**, 1522–1530 (2009)

Kardashev, N.S., Khartov, V.V., Abramov, V.V., Avdeev, V.Yu., Alakoz, A.V., Aleksandrov, Yu.A., Ananthakrishnan, S., Andreyanov, V.V., Andrianov, A.S., Antonov, N.M., and 120 coauthors, "RadioAstron": A Telescope with a Size of 300,000 km: Main Parameters and First Observational Results, *Astron. Reports*, **57**, 153–194 (2013)

Keto, E., The Shapes of Cross-Correlation Interferometers, *Astrophys. J.*, **475**, 843–852 (1997).

Kogan, L., Level of Negative Sidelobes in an Array Beam, *Publ. Astron. Soc. Pacific*, **111**, 510–511 (1999)

Koles, W.A., Frehlich, R.G., and Kojima, M., Design of a 74-MHz Antenna for Radio Astronomy, *Proc. IEEE*, **82**, 697–704 (1994)

Lal, D.V., Lobanov, A.P., and Jiménez-Monferrer, S., Array Configuration Studies for the Square Kilometre Array: Implementation of Figures of Merit Based on Spatial Dynamic Range, Square Kilometre Array Memo 107 (2009)

Lawrence, C.R., Herbig, T., and Readhead, A.C.S., Reduction of Ground Spillover in the Owens Valley 5.5-m Telescope, *Proc. IEEE*, **82**, 763–767 (1994)

Leech, J., On Representation of 1, 2, ... , *n* by Differences, *J. London Math. Soc.*, **31**, 160–169 (1956)

Lobanov, A.P., Imaging with the SKA: Comparison to Other Future Major Instruments, Square Kilometre Array Memo 38 (2003)

Lonsdale, C.J., Cappallo, R.J., Morales, M.F., Briggs, F.H., Benkevitch, L., Bowman, J.D., Bunton, J.D., Burns, S., Corey, B.E., deSouza, L., and 38 coauthors, The Murchison Widefield Array: Design Overview, *Proc. IEEE*, **97**, 1497–1506 (2009)

Mangum, J.G., Baars, J.W.M., Greve, A., Lucas, R., Snel, R.C., Wallace, P., and Holdaway, M., Evaluation of the ALMA Prototype Antennas, *Publ. Astron. Soc. Pacific*, **118**, 1257–1301 (2006)

Mathur, N.C., A Pseudodynamic Programming Technique for the Design of Correlator Supersynthesis Arrays, *Radio Sci.*, **4**, 235–244 (1969)

Mayer, C.E., Emerson, D.T., and Davis, J.H., Design and Implementation of an Error-Compensating Subreflector for the NRAO 12-m Radio Telescope, *Proc. IEEE*, **82**, 756–762 (1994)

Meinel, A.B., Multiple Mirror Telescopes of the Future, in *The MMT and the Future of Ground-Based Astronomy*, Weeks, T.C., Ed., SAO Special Report 385 (1979), pp. 9–22

Mills, B.Y., Cross-Type Radio Telescopes, *Proc. IRE Aust.*, **24**, 132–140 (1963)

Moffet, A.T., Minimum-Redundancy Linear Arrays, *IEEE Trans. Antennas Propag.*, **AP-16**, 172–175 (1968)

Morales, M.F., Enabling Next-Generation Dark Energy and Epoch of Reionization Radio Observatories with the MOFF Correlator, *Pub. Astron. Soc. Pacific*, **123**, 1265–1272 (2011)

Nakajima, J., Otobe, E., Nishibori, K., Watanabe, N., Asuma, K., and Daishido, T., First Fringe with the Waseda FFT Radio Telescope, *Pub. Astron. Soc. Japan*, **44**, L35–L38 (1992)

Nakajima, J., Otobe, E., Nishibori, K., Kobayashi, H., Tanaka, N., Saitoh, T., Watanabe, N., Aramaki, Y., Hoshikawa, T., Asuma, K., and Daishido, T., One-Dimensional Imaging with the Waseda FFT Radio Telescope, *Pub. Astron. Soc. Japan*, **45**, 477–485 (1993)

Napier, P.J., Bagri, D.S., Clark, B.G., Rogers, A.E.E., Romney, J.D., Thompson, A.R., and Walker, R.C., The Very Long Baseline Array, *Proc. IEEE*, **82**, 658–672 (1994)

Napier, P.J., Thompson, A.R., and Ekers, R.D., The Very Large Array: Design and Performance of a Modern Synthesis Radio Telescope, *Proc. IEEE*, **71**, 1295–1320 (1983)

National Radio Astronomy Observatory, *A Proposal for a Very Large Array Radio Telescope*, National Radio Astronomy Observatory, Green Bank, WV, Vol. 1 (1967); Vol. 3, Jan. 1969.

Otobe, E., Nakajima, J., Nishibori, K., Saito, T., Kobayashi, H., Tanaka, N., Watanabe, N., Aramaki, Y., Hoshikawa, T., Asuma, K., and Daishido, T., Two-Dimensional Direct Images with a Spatial FFT Interferometer, *Pub. Astron. Soc. Japan*, **46**, 503–510 (1994)

Padin, S., Cartwright, J.K., Mason, B.S., Pearson, T.J., Readhead, A.C.S., Shepherd, M.C., Sievers, J., Udomprasert, P.S., Holzapfel, W.L., Myers, S.T., and five coauthors, First Intrinsic Anisotropy Observations with the Cosmic Background Imager, *Astrophys. J. Lett.*, **549**, L1–L5 (2001)

Papoulis, A., *Signal Analysis*, McGraw-Hill, New York (1977), p. 74

Perley, R., Napier, P., Jackson, J., Butler, B., Carlson, B., Fort, D., Dewdney, P., Clark, B., Hayward, R., Durand, S., Revnell, M., and McKinnon, M., The Expanded Very Large Array, *Proc. IEEE*, **97**, 1448–1462 (2009)

Preston, R.A., Burke, B.F., Doxsey, R., Jordan, J.F., Morgan, S.H., Roberts, D.H., and Shapiro, I.I., The Future of VLBI Observations in Space, in *Very-Long-Baseline Interferometry Techniques*, Biraud, F., Ed., Cépaduès Éditions, Toulouse, France (1983), pp. 417–431

Rabiner, L.R., and Gold, B., *Theory and Application of Digital Signal Processing*, Prentice-Hall, Englewood Cliffs, NJ (1975), p. 50

Rademacher, H., and Toeplitz, O., *The Enjoyment of Mathematics*, Princeton Univ. Press, Princeton, NJ (1957)

Raimond, E., and Genee, R., Eds., *The Westerbork Observatory, Continuing Adventure in Radio Astronomy*, Kluwer, Dordrecht, the Netherlands (1996)

Roshi, D.A., and Fisher, J.R., A Model for Phased Array Feed, Electronics Div. Internal Report 330, National Radio Astronomy Observatory, Charlottesville, VA (2016)

Rudge, A.W., and Adatia, N.A., Offset-Parabolic-Reflector Antennas: A Review, *Proc. IEEE*, **66**, 1592–1618 (1978)

Ruze, J., Antenna Tolerance Theory—A Review, *Proc. IEEE*, **54**, 633–640 (1966)

Ryle, M., The New Cambridge Radio Telescope, *Nature*, **194**, 517–518 (1962)

Ryle, M., and Hewish, A., The Synthesis of Large Radio Telescopes, *Mon. Not. R. Astron. Soc.*, **120**, 220–230 (1960)

Ryle, M., Hewish, A., and Shakeshaft, J.R., The Synthesis of Large Radio Telescopes by the Use of Radio Interferometers, *IRE Trans. Antennas Propag.*, **7**, S120–S124 (1959)

Schaubert, D.H., and Chio, T.H., Wideband Vivaldi Arrays for Large Aperture Arrays, in *Perspectives on Radio Astronomy: Technologies for Large Arrays*, Smolders, A.B., and van Haarlem, M.P., Eds., ASTRON, Dwingeloo, the Netherlands, pp. 49–57 (1999)

Seielstad, G.A., Swenson, G.W., Jr., and Webber, J.C., A New Method of Array Evaluation Applied to Very Long Baseline Interferometry, *Radio Sci.*, **14**, 509–517 (1979)

Snel, R.C., Mangum, J.G., and Baars, J.W.M., Study of the Dynamics of Large Reflector Antennas with Accelerometers, *IEEE Antennas Propag. Mag.*, **49**, 84–101 (2007)

Stewart, K.P., Hicks, B.C., Ray, P.S., Crane, P.C., Kassim, N.E., Bradley, R.F., and Erickson, W.C., LOFAR Antenna Development and Initial Observations of Solar Bursts, *Planetary Space Sci.*, **52**, 1351–1355 (2004)

Swarup, G., Ananthakrishnan, S., Kapahi, V.K., Rao, A.P., Subrahmanya, C.R., and Kulkarni, V.K., The Giant Metrewave Radio Telescope, *Current Sci.* (Current Science Association and Indian Academy of Sciences), **60**, 95–105 (1991)

Swenson, G.W., Jr. and Kellermann, K.I., An Intercontinental Array—A Next-Generation Radio Telescope, *Science*, **188**, 1263–1268 (1975)

Swenson, G.W., Jr., and Mathur, N.C., The Circular Array in the Correlator Mode, *Proc. IREE Aust.*, **28**, 370–374 (1967)

Takeuchi, H., Kuniyoshi, M., Daishido, T., Asuma, K., Matsumura, N., Takefuji, K., Niinuma, K., Ichikawa, H., Okubo, R., Sawano, A., and four coauthors, Asymmetric Sub-Reflectors for Spherical Antennas and Interferometric Observations with an FPGA-Based Correlator, *Pub. Astron. Soc. Japan*, **57**, 815–820 (2005)

Tegmark, M., and Zaldarriaga, M., The Fast Fourier Transform Telescope, *Phys. Rev. D*, **79**, 08530 (2009)

Tegmark, M., and Zaldarriaga, M., Omniscopes: Large Area Telescope Arrays with Only $N \log N$ Computational Cost, *Phys. Rev. D*, **82**, 103501(10 pp) (2010)

Thomasson, P., MERLIN, *Quart. J. R. Astron. Soc.*, **27**, 413–431 (1986)

Thompson, A.R., and Bracewell, R.N., Interpolation and Fourier Transformation of Fringe Visibilities, *Astron. J.*, **79**, 11–24 (1974)

Thompson, A.R., Clark, B.G., Wade, C.M., and Napier, P.J., The Very Large Array, *Astrophys. J. Suppl.*, **44**, 151–167 (1980)

Unser, M., Sampling—50 Years After Shannon, *Proc. IEEE*, **88**, 569–587 (2000)

van Ardenne, A, Bregman, J.D., van Cappellen, W.A., Kant, G.W., and Bij de Vaate, J.G., Extending the Field of View with Phased Array Techniques: Results of European SKA Research, *Proc. IEEE*, **97**, 1531–1542 (2009)

van Cappellen, W.A., and Bakker, L., APERTIF: Phased Array Feeds for the Westerbork Synthesis Radio Telescope, in Proc. IEEE International Symposium on Phased Array Systems and Technology (ARRAY), Boston, MA, Oct. 12–15 (2010), pp. 640–647

van Cappellen, W.A., Bakker, L., and Oosterloo, T.A., Experimental Results of the APERTIF Phased Array Feed, in Proc. 30th URSI General Assembly and Scientific Symposium, Istanbul, Turkey, Aug. 13–20 (2011), 4 pp van Haarlem, M.P., Wise, M.W., Gunst, A.W., Heald, G., McKean, J.P., Hessels, J.W.T., de Bruyn, A.G., Nijboer, R., Swinbank, J., Fallows, R., and 191 coauthors, LOFAR: The LOw-Frequency ARray, *Astron. Astrophys.*, **556**, A2 (53pp) (2013)

Walker, R.C., VLBI Array Design, in *Indirect Imaging*, J. A. Roberts, Ed., Cambridge Univ. Press, Cambridge, UK (1984), pp. 53–65

Welch, J., Backer, D., Blitz, L., Bock, D., Bower, G.C., Cheng, C., Croft, S., Dexter, M., Engargiola, G., Fields, E., and 36 coauthors, The Allen Telescope Array: The First Widefield, Panchromatic, Snapshot Radio Camera for Radio Astronomy and SETI, *Proc. IEEE*, **97**, 1438–1447 (2009)

Welch, W.J., Thornton, D.D., Plambeck, R.L., Wright, M.C.H., Lugten, J., Urry, L., Fleming, M., Hoffman, W., Hudson, J., Lum, W.T., and 27 coauthors, The Berkeley–Illinois–Maryland Association Millimeter Array, *Publ. Astron. Soc. Pacific*, **108**, 93–103 (1996)

Wild, J.P., A New Method of Image Formation with Annular Apertures and an Application in Radio Astronomy, *Proc. R. Soc. Lond. A*, **286**, 499–509 (1965)

Wild, J.P., Ed., *Proc. IREE Aust.*, Special Issue on the Culgoora Radioheliograph, **28**, No. 9 (1967)

Williams, W.F., High Efficiency Antenna Reflector, *Microwave J.*, **8**, 79–82 (1965) [reprinted in Love (1978); see Further Reading]

Wootten, A., and Thompson, A.R., The Atacama Large Millimeter/Submillimeter Array, *Proc. IEEE*, **97**, 1463–1471 (2009)

Zheng, H., Tegmark, M., Buza, V., Dillon, J., Gharibyan, H., Hickish, J., Kunz, E., Liu, A., Losh, J., Lutomirski, A., and 28 coauthors, Mapping Our Universe in 3D with MITEoR, in Proc. IEEE International Symposium on Phased Array Systems and Technology, Waltham, MA (2013), pp. 784–791

Zheng, H., Tegmark, M., Buza, V., Dillon, J.S., Gharibyan, H., Hickish, J., Kunz, E., Liu, A., Losh, J., Lutomirski, A., and 27 coauthors, MITEoR: A Scalable Interferometer for Precision 21 cm Cosmology, *Mon. Not. R. Astron. Soc.*, **445**, 1084–1103 (2014)

Chapter 6
Response of the Receiving System

This chapter is concerned with the response of the receiving system that accepts the signals from the antennas, amplifies and filters them, and measures the cross-correlations for the various antenna pairs. We show how the basic parameters of the system affect the output. Some of the effects were introduced in earlier chapters and are here presented in a more detailed development that leads to consideration of system design in Chaps. 7 and 8. At some point in the processing chain between the antenna and the correlator output, the form of the signals is changed from an analog voltage to a digital format, and the resulting data are thereafter processed by computer-type hardware. This does not affect the mathematical analysis of the processing and is not considered in this chapter. However, the digitization introduces a component of quantization noise, which is analyzed in Chap. 8.

6.1 Frequency Conversion, Fringe Rotation, and Complex Correlators

6.1.1 Frequency Conversion

With the exception of some systems operating below ~ 100 MHz, in most radio astronomy instruments, the frequencies of the signals received at the antennas are changed by mixing with a local oscillator (LO) signal. This feature, referred to as *frequency conversion* or (heterodyne frequency conversion), enables the major part of the signal processing to be performed at intermediate frequencies that are most appropriate for amplification, transmission, filtering, delaying, recording, and similar processes. For observations at frequencies up to roughly 50 GHz, the best sensitivity is generally obtained by using a low-noise amplifying stage before the frequency conversion.

© The Author(s) 2017   207
A.R. Thompson, J.M. Moran, and G.W. Swenson Jr., *Interferometry and Synthesis in Radio Astronomy*, Astronomy and Astrophysics Library,
DOI 10.1007/978-3-319-44431-4_6

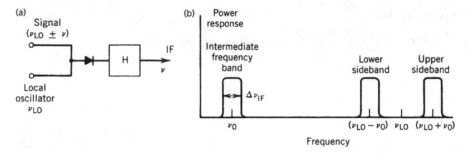

Fig. 6.1 Frequency conversion in a radio receiving system. (**a**) Simplified diagram of a mixer and a filter H that defines the intermediate-frequency (IF) band. The nonlinear element shown is a diode. (**b**) Signal spectrum showing upper and lower sidebands that are converted to the IF. Frequency v_0 is the center of the IF band.

Frequency conversion takes place in a mixer in which the signal to be converted, plus an LO waveform, are applied to a circuit element with a nonlinear voltage–current response. This element may be a diode, as shown in Fig. 6.1a. The current i through the diode can be expressed as a power series in the applied voltage V:

$$i = a_0 + a_1 V + a_2 V^2 + a_3 V^3 + \cdots . \tag{6.1}$$

Now let V consist of the sum of an LO voltage $b_1 \cos(2\pi v_{LO}t + \theta_{LO})$ and a signal, of which one Fourier component is $b_2 \cos(2\pi v_s t + \phi_s)$. The second-order term in V then gives rise to a product in the mixer output of the form

$$b_1 \cos(2\pi v_{LO}t + \theta_{LO})$$

$$\times b_2 \cos(2\pi v_s t + \phi_s) = \frac{1}{2}b_1 b_2 \cos\left[2\pi(v_s + v_{LO})t + \phi_s + \theta_{LO}\right] \tag{6.2}$$

$$+ \frac{1}{2}b_1 b_2 \cos\left[2\pi(v_s - v_{LO})t + \phi_s - \theta_{LO}\right] .$$

Thus, the current through the diode contains components at the sum and difference of v_s and v_{LO}. Other terms in Eq. (6.1) lead to other components, such as $3v_{LO} \pm v_s$, but the filter H shown in Fig. 6.1 passes only the wanted output components, and with proper design, unwanted combinations can be prevented from falling within the filter passband. Usually the signal voltage is much smaller than the LO voltage, so harmonics and intermodulation products (i.e., spurious signals that arise as a result of cross products of different frequency components within the input signal band) are small compared with the wanted terms containing v_{LO}.

In most cases of frequency conversion, the signal frequency is being reduced, and the second term on the right side in Eq. (6.2) is the important one. The filter H then defines an intermediate-frequency (IF) band centered on v_0, as shown in Fig. 6.1b. Signals from within the bands centered on $v_{LO} - v_0$ and $v_{LO} + v_0$ are converted to the

IF band and admitted by the filter. These bands are known as the *lower* and *upper* *sidebands*, as shown, and if only a single sideband is wanted, the other can often be removed by a suitable filter inserted before the mixer. In some cases, both sidebands are accepted, resulting in a double-sideband response.

6.1.2   Response of a Single-Sideband System

Figure 6.2 shows a basic receiving system for two antennas, m and n, of a synthesis array. Here, we are interested in further effects of frequency conversion. The time difference τ_g between the arrival at the antennas of the signals from a radio source varies continuously as the Earth rotates and the antennas track the source across the sky. A variable instrumental delay τ_i is continuously adjusted to compensate for the

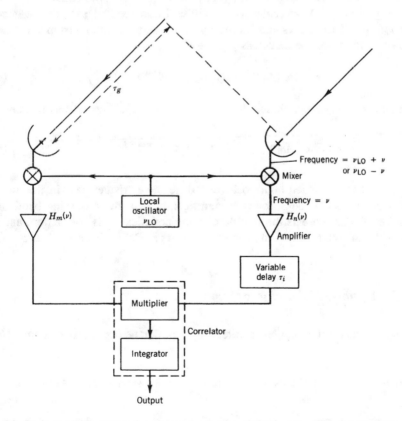

Fig. 6.2 Basic receiving system for two antennas of a synthesis array. The variable delay τ_i is continuously adjusted under computer control to compensate for the geometric delay τ_g. The frequency response functions $H_m(\nu)$ and $H_n(\nu)$ represent the overall bandpass characteristics of the amplifiers and filters in the signal channels.

geometric delay τ_g, so that the signals arrive simultaneously at the correlator. The receiving channels through which the signals pass contain amplifiers and filters, the overall amplitude (voltage) responses of which are $H_m(\nu)$ and $H_n(\nu)$ for antennas m and n. Here, ν represents a frequency at the correlator input; the corresponding frequency at the antenna is $\nu_{LO} \pm \nu$. The voltage waveforms that are processed by the receiving system result from cosmic noise and system noise; we consider the usual case in which these processes are approximately constant across the receiver passband. The spectra at the correlator inputs are thus determined mainly by the response of the receiving system. Let ϕ_m be the phase change in the signal path through antenna m resulting from τ_g and the LO phase, and let ϕ_n be the corresponding phase change in the signal for the path through antenna n, including τ_i. ϕ_m and ϕ_n, together with the instrumental phase resulting from the amplifiers and filters, represent the phases of the cosmic signal at the correlator inputs. Negative values of these parameters indicate phase lag (signal delay). The response to a source for which the visibility is $\mathcal{V}(u, v) = |\mathcal{V}|e^{j\phi_v}$ is most easily obtained by returning to Eq. (3.5) and replacing the phase difference $2\pi \mathbf{D}_\lambda \cdot \mathbf{s}_0$ by the general term $\phi_n - \phi_m$. Then the response at the correlator output resulting from a frequency band of width $d\nu$ can be written as

$$dr = \mathcal{R}e \left\{ A_0 |\mathcal{V}| H_m(\nu) H_n^*(\nu) \, e^{j(\phi_n - \phi_m - \phi_v)} d\nu \right\} , \tag{6.3}$$

where ϕ_v is the visibility phase. The response from the full system passband is

$$r = \mathcal{R}e \left\{ A_0 |\mathcal{V}| \int_{-\infty}^{\infty} H_m(\nu) H_n^*(\nu) \, e^{j(\phi_n - \phi_m - \phi_v)} d\nu \right\} , \tag{6.4}$$

where we have included both positive and negative frequencies[1] in the integral and assumed that $\mathcal{V}$ does not vary significantly over the observing bandwidth. Equation (6.4) represents the real part of the complex cross-correlation, and the way to obtain both the real and imaginary parts is explained later in this section.

6.1.3   Upper-Sideband Reception

For upper-sideband reception, a filter or amplifier at the receiver input selects frequencies in a band defined by the correlator input spectrum (frequency ν) plus ν_{LO}. In Fig. 6.2, the signal entering antenna m traverses the geometric delay τ_g at a frequency $\nu_{LO} + \nu$ and thus suffers a phase shift $2\pi(\nu_{LO} + \nu)\tau_g$. At the mixer, its

---

[1]The negative frequencies have no physical meaning but arise as part of the mathematical representation of the frequency conversion.

phase is also decreased by the LO phase θ_m. Thus, we obtain

$$\phi_m(v) = -2\pi(v_{LO} + v)\tau_g - \theta_m . \tag{6.5}$$

The phase of the signal entering antenna n is decreased by the LO phase θ_n, and the signal then traverses the instrumental delay τ_i at a frequency v, thus suffering a shift $2\pi v \tau_i$. The total phase shift for antenna n is

$$\phi_n(v) = -2\pi v \tau_i - \theta_n . \tag{6.6}$$

From Eqs. (6.4), (6.5), and (6.6), the correlator output is

$$r_u = \mathcal{Re}\left\{A_0|\mathcal{V}|e^{j[2\pi v_{LO}\tau_g + (\theta_m - \theta_n) - \phi_v]} \int_{-\infty}^{\infty} H_m(v)H_n^*(v)\, e^{j2\pi v \Delta \tau} dv\right\} . \tag{6.7}$$

The real part of the integral in Eq. (6.7) is one-half the Fourier transform of the (Hermitian) cross power spectrum $H_m(v)H_n^*(v)$ with respect to the delay compensation error, $\Delta\tau = \tau_g - \tau_i$, which introduces a linear phase slope across the band.[2] We assume that $\mathcal{V}$ does not vary significantly over the observing bandwidth. For example, if the IF passbands are rectangular with center frequency v_0, width Δv_{IF}, and identical phase responses, then for positive frequencies,

$$|H_m(v)| = |H_n(v)| = \begin{cases} H_0 , & |v - v_0| < \dfrac{\Delta v_{IF}}{2} , \\[2mm] 0 , & |v - v_0| > \dfrac{\Delta v_{IF}}{2} . \end{cases} \tag{6.8}$$

Using the equality in Eq. (A3.6) of Appendix 3.1 for the Hermitian[3] function $H_m H_n$, we can write

$$\int_{-\infty}^{\infty} H_m(v)H_n^*(v)\, e^{j2\pi v \Delta \tau} dv = 2\mathcal{Re}\left\{\int_{v_0 - (\Delta v_{IF}/2)}^{v_0 + (\Delta v_{IF}/2)} H_0^2\, e^{j2\pi v \Delta \tau} dv\right\}$$

$$= 2H_0^2 \Delta v_{IF} \left[\frac{\sin(\pi \Delta v_{IF} \Delta \tau)}{\pi \Delta v_{IF} \Delta \tau}\right] \cos 2\pi v_0 \Delta \tau . \tag{6.9}$$

---

[2] Here, we assume that the source is sufficiently close to the center of the field being imaged that the condition $\Delta\tau = 0$ maintains zero delay error. The effect of the variation of the delay error across a wider field of view is considered in Sect. 6.3.

[3] The term "Hermitian" indicates a function in which the real part is even and the imaginary part is odd.

In the general case, we define an instrumental gain factor $G_{mn} = |G_{mn}|e^{j\phi_G}$ as follows:

$$A_0 \int_{-\infty}^{\infty} H_m(\nu)H_n^*(\nu)\, e^{j2\pi\nu\Delta\tau}d\nu = G_{mn}(\Delta\tau)\, e^{j2\pi\nu_0\Delta\tau}$$

$$= |G_{mn}(\Delta\tau)|e^{j(2\pi\nu_0\Delta\tau+\phi_G)}, \qquad (6.10)$$

where the $\Delta\tau$ dependence in $|G_{mn}(\Delta\tau)|$ is the sinc function in Eq. (6.9). The phase ϕ_G results from the difference in the phase responses of the amplifiers and filters in the two channels. The LO phases θ_m and θ_n are not included within the general instrumental phase term ϕ_G because they enter into the upper and lower sidebands with different signs.

Substituting Eq. (6.10) into Eq. (6.7), we obtain for upper-sideband reception

$$r_u = |\mathcal{V}||G_{mn}(\Delta\tau)|\cos\left[2\pi(\nu_{LO}\tau_g + \nu_0\Delta\tau) + (\theta_m - \theta_n) - \phi_\nu + \phi_G\right]. \qquad (6.11)$$

The term $2\pi\nu_{LO}\tau_g$ in the cosine function results in a quasi-sinusoidal oscillation as the source moves through the fringe pattern. The phase of this oscillation depends on the delay error $\Delta\tau$, the relative phases of the LO signals, the phase responses of the signal channels, and the phase of the visibility function. The frequency of the output oscillation $\nu_{LO}d\tau_g/dt$ is often referred to as the *natural fringe frequency*. The oscillations result because the signals traverse the delays τ_g and τ_i at different frequencies, that is, at the input radio frequency for τ_g and at the intermediate frequency for τ_i, and these two frequencies differ by ν_{LO}. Thus, even if these two delays are identical, they introduce different phase shifts, and they increase or decrease progressively as the Earth rotates.

6.1.4 Lower-Sideband Reception

Consider now the situation where the frequencies accepted from the antenna are those in the lower sideband, at ν_{LO} minus the correlator input frequencies. The phases are

$$\phi_m = 2\pi(\nu_{LO} - \nu)\tau_g + \theta_m \qquad (6.12)$$

and

$$\phi_n = -2\pi\nu\tau_i + \theta_n. \qquad (6.13)$$

The signs of these terms and of ϕ_ν differ from those in the upper-sideband case because increasing the phase of the signal at the antenna here decreases the phase at the correlator. The expression for the correlator output is

$$r_\ell = \mathcal{Re} \left\{ A_0 |\mathcal{V}| e^{-j[2\pi \nu_{LO} \tau_g + (\theta_m - \theta_n) - \phi_v]} \int_{-\infty}^{\infty} H_m(\nu) H_n^*(\nu) \, e^{j2\pi \Delta \tau} dv \right\} . \qquad (6.14)$$

Proceeding as in the upper-sideband case, we obtain

$$r_\ell = |\mathcal{V}||G_{mn}(\Delta \tau)| \cos \left[2\pi (\nu_{LO} \tau_g - \nu_0 \Delta \tau) + (\theta_m - \theta_n) - \phi_v - \phi_G \right] . \qquad (6.15)$$

6.1.5 Multiple Frequency Conversions

In an operational system, the signals may undergo several frequency conversions between the antennas and the correlators. A frequency conversion in which the output is at the lower sideband (i.e., the LO frequency minus the input frequency) results in a reversal of the signal spectrum in which frequencies at the high end at the input appear at the low end at the output, and vice versa. If there is no net reversal (that is, an even number of lower-sideband conversions), Eq. (6.11) applies, except that ν_{LO} must be replaced by a corresponding combination of LO frequencies. Similarly, the oscillator phase terms θ_m and θ_n are replaced by corresponding combinations of oscillator phases.

6.1.6 Delay Tracking and Fringe Rotation

Adjustment of the compensating delay τ_i of Fig. 6.2 is usually accomplished under computer control, the required delay being a function of the antenna positions and the position of the phase center of the field under observation. This can be achieved by designating one antenna of the array as the delay reference and adjusting the instrumental delays of other antennas so that, for an incoming wavefront from the phase reference direction, the signals intercepted by the different antennas all arrive at the correlator simultaneously.

To control the frequency of the sinusoidal fringe variations in the correlator output, a continuous phase change can be inserted into one of the LO signals. Equations (6.11) and (6.15) show that the fringe frequency can be reduced to zero by causing $\theta_m - \theta_n$ to vary at a rate that maintains constant, modulo 2π, the term $[2\pi \nu_{LO} \tau_g + (\theta_m - \theta_n)]$. This requires adding a frequency $2\pi \nu_{LO}(d\tau_g/dt)$ to θ_n or subtracting it from θ_m. Note that $d\tau_g/dt$ can be evaluated from Eq. (4.9) in which w, the third component of the interferometer baseline, is equal to $c\tau_g$ measured in wavelengths; for example, for an east–west antenna spacing of 1 km, the maximum value of $d\tau_g/dt$ is 2.42×10^{-10}, so the fringe frequencies are generally small compared with the radio frequencies involved. Reduction of the output frequency reduces the quantity of data to be processed, since each correlator output must be sampled at least twice per cycle of the output frequency (the Nyquist rate) to preserve the information, as is discussed in Sect. 8.2.1. With antenna spacings required for angular resolution of milliarcsecond order, which occur in VLBI, the natural fringe frequency, $\nu_{LO}d\tau_g/dt$, can exceed 10 kHz. For an array with more than one antenna pair, it is possible to reduce each output frequency to the same fraction of its natural frequency, or to zero. Reduction to zero frequency (fringe stopping) is generally the preferred practice. Some special technique, such as the use of a complex correlator, described in the following section, is then required to extract the amplitude and phase of the output.

6.1.7 Simple and Complex Correlators

A method of measuring the amplitude and phase of the correlator output signal when the fringe frequency is reduced to zero is shown in Fig. 6.3. Two correlators are used, one of which has a quadrature phase shift network in one input. For signals of finite bandwidth, this phase shift is not equivalent to a delay. The phase shift can also be effected by feeding the signal into two separate mixers and converting it with two LOs in phase quadrature. The output of the second correlator can be represented by replacing $H_m(\nu)$ by $H_m(\nu)e^{-j\pi/2}$. From Eq. (6.10), the result is to add $-\pi/2$ to ϕ_G, and thus in Eq. (6.11) and Eq. (6.15), the cosine function is replaced by $\pm$sine. Another way of comparing the two correlator outputs in Fig. 6.3 is to note that the real output of the complex correlator, omitting constant factors, is

$$r_{\text{real}} = \mathcal{R}e\left\{\mathcal{V}\int_{-\infty}^{\infty} H_m(\nu)H_n^*(\nu)\,d\nu\right\} = \mathcal{R}e\{\mathcal{V}\}\int_{-\infty}^{\infty} H_m(\nu)H_n^*(\nu)\,d\nu \;, \qquad (6.16)$$

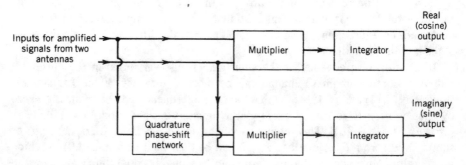

Fig. 6.3 Use of two correlators to measure the real and imaginary parts of the visibility. This system is called a complex correlator.

where the integral is real since $H_m(\nu)$ and $H_n^*(\nu)$ are Hermitian and thus $H_m(\nu)H_n^*(\nu)$ is Hermitian. The imaginary output of the correlator is proportional to

$$r_{\text{imag}} = \mathcal{R}e\left\{\mathcal{V}\int_{-\infty}^{\infty} H_m(\nu)H_n^*(\nu)e^{-j\pi/2}\,d\nu\right\} = \mathcal{I}m\{\mathcal{V}\}\int_{-\infty}^{\infty} H_m(\nu)H_n^*(\nu)\,d\nu\ .$$

(6.17)

Thus, the two outputs respond to the real and imaginary parts of the visibility $\mathcal{V}$.

The combination of two correlators and the quadrature network is usually referred to as a *complex correlator*, and the two outputs as the cosine and sine, or real and imaginary, outputs. For continuum observations, the compensating delay is adjusted so that $\Delta\tau = 0$ and the fringe rotation maintains the condition $2\pi\nu_{\text{LO}}\tau_i + (\theta_m - \theta_n) = 0$. Thus, the cosine and sine outputs represent the real and imaginary parts of $G_{mn}\mathcal{V}(u, v)$. With the use of the complex correlator, the rotation of the Earth, which sweeps the fringe pattern across the source, is no longer a necessary feature in the measurement of visibility. An important feature of the complex correlator is that the noise fluctuations in the cosine and sine outputs are independent, as discussed in Sect. 6.2.2.

Spectral correlator systems, in which a number of correlators are used to measure the correlation as a function of time offset or "lag" [i.e., τ in Eq. (3.27)], are discussed in Sect. 8.8. The correlation as a function of τ measured using a correlator with a quadrature phase shift in one input is the Hilbert transform of the same quantity measured without the quadrature phase shift (Lo et al. 1984).

6.1.8 Response of a Double-Sideband System

A double-sideband (DSB) receiving system is one in which both the upper- and lower-sideband responses are accepted. From Eqs. (6.11) and (6.15), the output is

$$r_d = r_u + r_\ell = 2|\mathcal{V}||G_{mn}(\Delta\tau)|\cos(2\pi\nu_0\Delta\tau + \phi_G)$$
$$\times \cos\left[2\pi\nu_{\text{LO}}\tau_g + (\theta_m - \theta_n) - \phi_v\right]\ .$$ (6.18)

There is a significant difference from the single-sideband (SSB) cases. The phase of the fringe-frequency term, which is the cosine function containing the term $2\pi\nu_{\text{LO}}\tau_g$, is no longer dependent on $\Delta\tau$ or ϕ_G, but instead these quantities appear in the term that controls the fringe amplitude.

$$|G_{mn}(\Delta\tau)|\cos(2\pi\nu_0\Delta\tau + \phi_G)\ .$$ (6.19)

If the delay τ_i is held constant, $\Delta\tau$ varies continuously, resulting in cosinusoidal modulation of the fringe oscillations through the cosine term in (6.19). Also, as

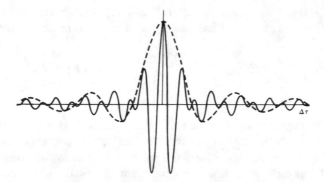

Fig. 6.4 Example of the variation of the fringe amplitude as a function of $\Delta\tau$ for a DSB system (solid line). In this case, the centers of the two sidebands are separated by three times the IF bandwidth, that is, $\nu_0 = 1.5\Delta\nu_{IF}$, and the IF response is rectangular. The broken line shows the equivalent function for an SSB system with the same IF response.

shown in Fig. 6.4, the cross-correlation (fringe amplitude) falls off more rapidly because of the cosine term in (6.19) than it does in the SSB case, in which it depends only on $G_{mn}(\Delta\tau)$. The required precision in matching the geometric and instrumental delays is correspondingly increased. The lack of dependence of the fringe phase on the phase response of the signal channel occurs because the latter has equal and opposite effects on the signals from the two sidebands.

The response of a DSB system with a complex correlator is given by Eq. (6.18) for the cosine output, and the sine output is obtained by replacing ϕ_G by $\phi_G - \pi/2$:

$$(r_d)_{\text{sine}} = 2|\mathcal{V}||G_{mn}(\Delta\tau)| \sin(2\pi\nu_0\Delta\tau + \phi_G)$$

$$\times \cos\left[2\pi\nu_{LO}\tau_g + (\theta_m - \theta_n) - \phi_v\right] . \tag{6.20}$$

If the term $2\pi\nu_0\Delta\tau + \phi_G$ is adjusted to maximize either the real output [Eq. (6.18)] or the imaginary output [Eq. (6.20)], the other will be zero. Thus, for continuum observations in which the signal is of equal strength in both sidebands, the complex correlator offers no increase in sensitivity. However, it can be useful for observations in the sideband-separation mode described later.

To help visualize the difference between SSB and DSB interferometer systems, Fig. 6.5 illustrates the correlator outputs in the complex plane. The SSB case is shown in Fig. 6.5a. The output of the complex correlator is represented by the vector **r**. If the fringes are not stopped, the vector **r** rotates through 2π each time the geometric delay τ_g changes by one wavelength (that is, one wavelength at the LO frequency if the instrumental delay is tracking the geometric delay). The projections of the radial vector on the real and imaginary axes indicate the real and imaginary outputs of the complex correlator, which are two fringe-frequency sinusoids in phase quadrature. If the fringes are stopped, **r** remains fixed in position angle. Figure 6.5b represents the DSB case. Vectors $\mathbf{r}_u$ and $\mathbf{r}_\ell$ represent the output components from

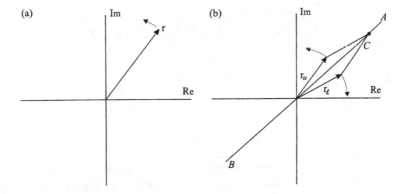

Fig. 6.5 Representation in the complex plane of the output of a correlator with (**a**) an SSB and (**b**) a DSB receiving system. The point C in (**b**) represents the sum of the upper- and lower-sideband outputs of the correlator.

the upper and lower sidebands. Here the variation of τ_g causes $\mathbf{r}_u$ and $\mathbf{r}_\ell$ to rotate in opposite directions. To verify this statement, note that the real parts of the correlator output are given in Eqs. (6.11) and (6.15), and the corresponding imaginary parts are obtained by replacing ϕ_G by $\phi_G - \pi/2$. Then with $(\theta_m - \theta_n) = 0$ (no fringe rotation), consider the effect of a small change in τ_g.

The contra-rotating vectors representing the two sidebands at the correlator output coincide at an angle determined by instrumental phase, which we represent by the line AB in Fig. 6.5b. Thus, the vector sum oscillates along this line, and the fringe-frequency sinusoids at the real and imaginary outputs of the correlator are in phase. Now suppose we adjust the phase term $(2\pi\nu_0\Delta\tau + \phi_G)$ in Eq. (6.18) to maximize the fringe amplitude at the real output. This action has the effect of rotating the line AB to coincide with the real axis. The imaginary output of the complex correlator then contains no signal, only noise. From Eq. (6.18), it can be seen that the visibility phase ϕ_v is represented by the phase of the vector that oscillates in amplitude along the real axis. The phase can be recovered by letting the fringes run and fitting a sinusoid to the waveform at the real output. If the fringes are stopped, it is possible to determine the amplitude and phase of the fringes by $\pi/2$ switching of the LO phase at one antenna. In Eq. (6.18), this phase switch action can be represented by $\theta_m \rightarrow (\theta_m - \pi/2)$, which results in a change of the second cosine function to a sine, thus enabling the argument in square brackets to be determined. However, in such a case, the data representing the cosine and sine components of the output are not measured simultaneously, so the effective data-averaging time is half that for the SSB, complex-correlator case. In Fig. 6.5b, a $\pi/2$ switch of the LO phase results in a rotation of $\mathbf{r}_u$ and $\mathbf{r}_\ell$ by $\pi/2$ in opposite directions, so the vector sum of the two sideband outputs remains on the line AB. Relative sensitivities of different systems are discussed in Sect. 6.2.5.

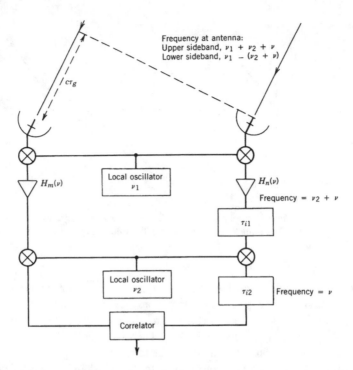

Fig. 6.6 Receiving system for two antennas that incorporates two frequency conversions, the first being DSB and the second upper-sideband. Two compensating delays, τ_{i1} and τ_{i2}, are included so that in deriving the response for a DSB system, the effect of the position of the delay relative to the first mixer can be investigated. In practice, only one compensating delay is required. The overall frequency responses H_m and H_n are specified as functions of v, which is the corresponding frequency at the correlator input.

6.1.9 Double-Sideband System with Multiple Frequency Conversions

The response with multiple frequency conversions is more complicated for a DSB interferometer than for an SSB one and is illustrated by considering the system in Fig. 6.6. Note that for the case in which the IF signal undergoes a number of SSB frequency conversions after the first mixer, the second mixer of each antenna in Fig. 6.6 can be considered to represent several mixers in series, and v_2 is equal to the sum of the LO frequencies with appropriate signs to take account of upper- or lower-sideband conversions. The signal phase terms are determined by considerations similar to those described in the derivation of Eqs. (6.5) and (6.6). Thus, we obtain

$$\phi_m = \mp 2\pi(v_1 \pm v_2 \pm v)\tau_g \mp \theta_{m1} - \theta_{m2} \qquad (6.21)$$

and

$$\phi_n = -2\pi(\nu_2 + \nu)\tau_{i1} - 2\pi\nu\tau_{i2} \mp \theta_{n1} - \theta_{n2} , \tag{6.22}$$

where the upper signs correspond to upper-sideband conversion at both the first and second mixers for each antenna, and the lower signs to lower-sideband conversion at the first mixer for each antenna and upper-sideband conversion at the second. We then proceed as in the previous examples; that is, use Eqs. (6.21) and (6.22) to substitute for ϕ_m and ϕ_n in Eq. (6.4), separate out the integral of $H_m H_n^*$ with respect to frequency, ν, as in Eq. (6.7), and substitute for the integral using Eq. (6.10). The results are

$$r_u = |\mathcal{V}||G_{mn}(\Delta\tau)| \cos[2\pi\nu_1\tau_g + 2\pi\nu_2(\tau_g - \tau_{i1}) + 2\pi\nu_0\Delta\tau$$
$$+ (\theta_{m1} - \theta_{n1}) + (\theta_{m2} - \theta_{n2}) - \phi_v + \phi_G] \tag{6.23}$$

and

$$r_\ell = |\mathcal{V}||G_{mn}(\Delta\tau)| \cos[2\pi\nu_1\tau_g - 2\pi\nu_2(\tau_g - \tau_{i1}) - 2\pi\nu_0\Delta\tau$$
$$+ (\theta_{m1} - \theta_{n1}) - (\theta_{m2} - \theta_{n2}) - \phi_v - \phi_G] . \tag{6.24}$$

The DSB response is

$$r_d = r_u + r_\ell$$
$$= 2|\mathcal{V}||G_{mn}(\Delta\tau)| \cos\left\{2\pi\left[\nu_2(\tau_{i1} \quad \tau_g) - \nu_0\Delta\tau\right] - (\theta_{m2} - \theta_{n2}) - \phi_G\right\}$$
$$\times \cos\left[\nu_1\tau_g + (\theta_{m1} - \theta_{n1}) - \phi_v\right] , \tag{6.25}$$

where $\Delta\tau = \tau_g - \tau_{i1} - \tau_{i2}$. Note that the phase of the output fringe pattern, given by the second cosine term, depends only on the phase of the first LO. Thus, in the implementation of fringe rotation, the phase shift must be applied to this oscillator. The first cosine term in Eq. (6.25) affects the fringe amplitude, and two cases should be considered:

1. The delay τ_{i1}, at the IF immediately following the DSB mixer, is used as the compensating delay, and $\tau_{i2} = 0$. Then in the first cosine function in Eq. (6.25), $\tau_{i1} - \tau_g \simeq 0$, and ϕ_G should be small if the frequency responses of the two channels are similar. It is necessary only to equalize θ_{m2} and θ_{n2} to maximize the amplitude of the fringe-frequency term. This is similar to the single conversion case in Eq. (6.18).
2. The delay τ_{i2}, located after the last mixer, is used as the compensating delay, and $\tau_{i1} = 0$. (This is the case in any array in which the compensating delays are implemented digitally, which includes almost all currently operational systems.)

Then a continuously varying phase shift is required in θ_{m2} or θ_{n2} of Eq. (6.25) to keep the value of the first cosine function close to unity as τ_g varies. This phase shift does not affect the phase of the output fringe oscillations, only the amplitude [see, e.g., Wright et al. (1973)].

6.1.10 Fringe Stopping in a Double-Sideband System

Consider two antennas of an array as shown in Fig. 6.6 and the case in which the instrumental delay that compensates for τ_g is the one immediately preceding the correlator, so that $\tau_{i1} = 0$. One can think of interferometer fringes as being caused by a Doppler shift in the signal at one antenna, which results in a beat frequency when the signals are combined in the correlator. Suppose that the geometric delay, τ_g, in the signal path to antenna m (on the left side of the diagram) is increasing with time, that is, antenna m is moving away from the source relative to antenna n. Then a signal at frequency ν_{RF} at the wavefront from a source appears at frequency $\nu_{RF}(1 - d\tau_g/dt)$ when received at antenna m. If the signal is in the upper sideband, its frequency at the correlator input will be

$$\nu_{RF}\left(1 - \frac{d\tau_g}{dt}\right) - \nu_1 - \nu_2 \ . \tag{6.26}$$

To stop the fringes, we need to apply a corresponding decrease to the frequency of the signal from antenna n so that the signals arrive at the correlator at the same frequency. To do this, we increase the frequencies of the two LOs for antenna n by the factor $(1+d\tau_g/dt)$. Note that this is equivalent to adding $2\pi(d\tau_g/dt)\nu_1$ to θ_{n1} and $2\pi(d\tau_g/dt)\nu_2$ to θ_{n2}, which are the rates of change of the oscillator phases required to maintain each of the two cosine functions in Eq. (6.25) at constant value. The corresponding signal from antenna n traverses the delay τ_{i2} at a frequency $\nu_{RF} - (\nu_1 + \nu_2)(1 + d\tau_g/dt)$, and since the delay is continuously adjusted to equal τ_g, the signal suffers a reduction in frequency by a factor $(1 - d\tau_g/dt)$. Thus, at the correlator input, the frequency of the antenna-n signal is

$$\left[\nu_{RF} - (\nu_1 + \nu_2)\left(1 + \frac{d\tau_g}{dt}\right)\right]\left(1 - \frac{d\tau_g}{dt}\right) \ , \tag{6.27}$$

which is equal to (6.26) when second-order terms in $d\tau_g/dt$ are neglected. (Recall that for, e.g., a 1-km baseline, the highest possible value of $d\tau_g/dt$ is 2.42×10^{-10}.) For the lower sideband, (6.26) and (6.27) apply if the signs of both ν_{RF} and ν_1 are reversed, and again the frequencies at the correlator input are equal. Thus, the overall effect is that the fringes are stopped for *both* sidebands.

6.1.11 Relative Advantages of Double- and Single-Sideband Systems

The principal reason for using DSB reception in interferometry is that in certain cases, the lowest receiver noise temperatures are obtained by using input stages that are inherently DSB devices. As frequency increases above $\sim$ 100 GHz, it becomes increasingly difficult to make low-noise amplifiers, and receiving systems often use a mixer of the superconductor–insulator–superconductor (SIS) type [see, e.g., Tucker and Feldman (1985)] as the input stage followed by a low-noise IF amplifier. Both the mixer and the IF amplifier are cryogenically cooled to obtain superconductivity in the mixer and to minimize the amplifier noise. If a filter is placed between the antenna and the mixer to cut out one sideband, the received signal power is halved, but there is no reduction in the receiver noise generated in the mixer and IF stages. Thus, the signal-to-noise ratio (SNR) in the IF stages is reduced, and in this case, the best continuum sensitivity may be obtained if both sidebands are retained. As a historical note, DSB systems were used at centimeter wavelengths during the 1960s and early 1970s [see, e.g., Read (1961)], sometimes with a degenerate type of parametric amplifier as the low-noise input stage. These amplifiers were inherently DSB devices, and their use in interferometry is discussed by Vander Vorst and Colvin (1966).

DSB systems have a number of disadvantages. Increased accuracy of delay setting is required, frequency and phase adjustment on more than one LO is likely to be required, interpretation of spectral line data is complicated if there are lines in both sidebands, and the width of the interference-free spectrum required is doubled. Also, the smearing effect of a finite bandwidth, to be discussed in Sect. 6.3, is increased. These problems have stimulated the development of schemes by which the responses for upper and lower sidebands can be separated.

6.1.12 Sideband Separation

To illustrate the method by which the responses for the two sidebands can be separated at the correlator output of a DSB receiving system, we examine the sum of the upper- and lower-sideband responses from Eqs. (6.11) and (6.15). This is

$$r_d = r_u + r_\ell = |\mathcal{V}||G_{mn}(\Delta\tau)| \left\{ \cos\left[2\pi(\nu_{\mathrm{LO}}\tau_g + \nu_0\Delta\tau) + \theta_{mn} - \phi_\nu + \phi_G \right] \right.$$
$$\left. + \cos\left[2\pi(\nu_{\mathrm{LO}}\tau_g - \nu_0\Delta\tau) + \theta_{mn} - \phi_\nu - \phi_G \right] \right\} , \qquad (6.28)$$

where $\theta_{mn} = \theta_m - \theta_n$. Equation (6.28) represents the real output of a complex correlator. We rewrite Eq. (6.28) as

$$r_d = |\mathcal{V}||G_{mn}|(\cos\Psi_u + \cos\Psi_\ell) , \qquad (6.29)$$

where Ψ_u and Ψ_ℓ represent the corresponding expressions in square brackets in Eq. (6.28). The responses considered above represent the normal output of the interferometer, which we call condition 1. The expression for the imaginary output of the correlator is obtained by replacing ϕ_G by $\phi_G - \pi/2$. Consider a second condition in which a $\pi/2$ phase shift is introduced into the first LO signal of antenna m, so that θ_{mn} becomes $\theta_{mn} - \pi/2$. The correlator outputs for the two conditions are obtained from Eqs. (6.28) and (6.29):

$$\text{condition 1} \tag{6.30}$$

$$r_1 = |\mathcal{V}||G_{mn}|(\cos\Psi_u + \cos\Psi_\ell)$$

$$r_2 = |\mathcal{V}||G_{mn}|(\sin\Psi_u - \sin\Psi_\ell)$$

$$\underline{\text{condition 2}} \quad (\theta_{mn} \to \theta_{mn} - \pi/2) \tag{6.31}$$

$$r_3 = |\mathcal{V}||G_{mn}|(\sin\Psi_u + \sin\Psi_\ell)$$

$$r_4 = |\mathcal{V}||G_{mn}|(-\cos\Psi_u + \cos\Psi_\ell)$$

where r_1 and r_3 represent the real outputs of the correlator and r_2 and r_4 the imaginary outputs. Thus, the upper-sideband response, expressed in complex form, is

$$|\mathcal{V}||G_{mn}|(\cos\Psi_u + j\sin\Psi_u) = \frac{1}{2}[(r_1 - r_4) + j(r_2 + r_3)] \ . \tag{6.32}$$

Similarly, the lower-sideband response is

$$|\mathcal{V}||G_{mn}|(\cos\Psi_\ell + j\sin\Psi_\ell) = \frac{1}{2}[(r_1 + r_4) - j(r_2 - r_3)] \ . \tag{6.33}$$

If the $\pi/2$ phase shift is periodically switched into and out of the LO signal, the upper- and lower-sideband responses can be obtained as indicated by Eqs. (6.32) and (6.33).

A similar implementation of sideband separation that makes use of fringe frequencies is attributable to B. G. Clark. This method is based on the fact that a small frequency shift in the first LO adds the same frequency shift to the fringes at the correlator for both sidebands, but a similar shift in a later LO adds to the fringe frequency for one sideband but subtracts from it for the other. Consider two antennas of an array in which the fringes have been stopped as in the discussion associated with expressions (6.26) and (6.27). Now suppose that we increase the frequency of the first LO at antenna n by a frequency $\delta\nu$ and decrease the frequency of the second LO by the same amount. The fringe frequency for the upper-sideband signal will be unchanged; that is, the fringes will remain stopped. For the lower sideband, the signal frequencies after the second mixer will be decreased by $2\delta\nu$. The lower-sideband output will consist of fringes at frequency $2\delta\nu(1 - d\tau_g/dt) \approx 2\delta\nu$ and will be averaged to a small residual if $(2\delta\nu)^{-1}$ is small compared with the integration period at the correlator output, or if an integral number of fringe cycles fall within such an integration period. If the frequency of the second LO is increased by $\delta\nu$ instead of decreased, the lower sideband will be stopped and the upper one averaged out. To apply this scheme to an array of n_a antennas, the offset must be different for each antenna, and this can be achieved by using an offset $n\delta\nu$ for antenna n, where n runs from 0 to $n_a - 1$. An advantage of this sideband-separating scheme is that it can be implemented using the variable LOs required for fringe stopping, and no other special hardware is needed. Unlike the $\pi/2$ phase-switching scheme, one sideband is lost in this method. However, as mentioned above, sideband separation schemes of this type separate only the correlated component of the signal and not the noise. To separate the noise, the SIS mixers at the receiver inputs can be mounted in a sideband-separating circuit of the type described in Appendix 7.1. In such cases, the isolation of the sidebands achieved in the mixer circuit may be only ~ 15 dB, which is sufficient to remove most of the noise contributed by an unwanted sideband, but not sufficient to remove strong spectral lines. The Clark technique described above is nicely suited to increasing the suppression of an unwanted sideband that has already suffered limited rejection at the mixer.

Fringe-frequency effects can also be used for sideband separation in VLBI observations. In VLBI systems, the fringe rotation is usually applied during playback. Fringe rotation then has the effect of reducing the fringe frequency for one sideband and increasing it for the other. If the fringe rotation is set to stop the fringes in one sideband, then since the baselines are so long, fringes resulting from the other sideband will often have a sufficiently high frequency that they will be reduced to a negligible level by the time averaging at the correlator output. The data are played back to the correlator twice, once for each sideband, with appropriate fringe rotation.

6.2 Response to the Noise

The ultimate sensitivity of a receiving system is determined principally by the system noise. We now consider the response to the noise and the resulting threshold of sensitivity, beginning with the effect at the correlator output and the resulting uncertainty in the real and imaginary parts of the visibility, $\mathcal{V}$. This leads to calculation of the rms noise level in a synthesized image in terms of the peak response to a source of given flux density. Finally, we consider the effect of noise in terms of the rms fluctuations in the amplitude and phase of $\mathcal{V}$.

6.2.1 Signal and Noise Processing in the Correlator

Consider an observation in which the field to be imaged contains only a point source located at the phase reference position. Let $V_m(t)$ and $V_n(t)$ be the waveforms at the correlator input from the signal channels of antennas m and n. The output is

$$r = \langle V_m(t)V_n(t) \rangle \,, \tag{6.34}$$

where all three functions are real, and the expectation denoted by the angular brackets is approximated in practice by a finite time average. To determine the relative power levels of the signal and noise components of r, we determine their power spectra by first calculating the autocorrelation functions. The autocorrelation of the signal product in Eq. (6.34) is

$$\rho_r(\tau) = \langle V_m(t)V_n(t)V_m(t-\tau)V_n(t-\tau) \rangle \,. \tag{6.35}$$

This expression can be evaluated using the following fourth-order moment relation[4]:

$$\langle z_1 z_2 z_3 z_4 \rangle = \langle z_1 z_2 \rangle \langle z_3 z_4 \rangle + \langle z_1 z_3 \rangle \langle z_2 z_4 \rangle + \langle z_1 z_4 \rangle \langle z_2 z_3 \rangle \,, \tag{6.36}$$

where z_1, z_2, z_3, and z_4 are joint Gaussian random variables with zero mean. Thus,

$$\begin{aligned}
\rho_r(\tau) &= \langle V_m(t)V_n(t) \rangle \langle V_m(t-\tau)V_n(t-\tau) \rangle \\
&+ \langle V_m(t)V_m(t-\tau) \rangle \langle V_n(t)V_n(t-\tau) \rangle \\
&+ \langle V_m(t)V_n(t-\tau) \rangle \langle V_m(t-\tau)V_n(t) \rangle \\
&= \rho_{mn}^2(0) + \rho_m(\tau)\rho_n(\tau) + \rho_{mn}(\tau)\rho_{mn}(-\tau) \,,
\end{aligned} \tag{6.37}$$

where ρ_m and ρ_n are the unnormalized autocorrelation functions of the two signals V_m and V_n, respectively, and ρ_{mn} is their cross-correlation function. Each V term is the sum of a signal component s and a noise component n, and to examine how these components contribute to the correlator output, we substitute them in Eq. (6.37). Products of uncorrelated terms, that is, products of signal and noise voltages, or noise voltages from different antennas, have an expectation of zero, and omitting them, we obtain

$$\begin{aligned}
\rho_r(\tau) &= \langle s_m(t)s_n(t) \rangle \langle s_m(t-\tau)s_n(t-\tau) \rangle \\
&+ \langle s_m(t)s_m(t-\tau) + n_m(t)n_m(t-\tau) \rangle \langle s_n(t)s_n(t-\tau) + n_n(t)n_n(t-\tau) \rangle \\
&+ \langle s_m(t)s_n(t-\tau) \rangle \langle s_m(t-\tau)s_n(t) \rangle \,,
\end{aligned} \tag{6.38}$$

where the three lines on the right side correspond to the three terms on the last line of Eq. (6.37). To determine the effect of the frequency response of the receiving

---

[4]This relation is a special case of a more general expression for the expectation of the product of N such variables, which is zero if N is odd and a sum of pair products if N is even. A form of Eq. (6.36) can be found in Lawson and Uhlenbeck (1950), Middleton (1960), and Wozencraft and Jacobs (1965).

system on the various terms of $\rho(\tau)$, we need to convert them to power spectra. By the Wiener–Khinchin relation, we should therefore examine the Fourier transforms of each term on the right sides of Eqs. (6.37) and (6.38).

The first term from Eq. (6.37), $\rho^2_{mn}(0)$, is a constant, and its Fourier transform is a delta function at the origin in the frequency domain, multiplied by $\rho^2_{mn}(0)$. From Eq. (6.38), we see that $\rho^2_{mn}(0)$ involves only the signal terms, which it is convenient to express as antenna temperatures. By the integral theorem of Fourier transforms, $\rho_{mn}(0)$ is the infinite integral of the Fourier transform of $\rho_{mn}(\tau)$, and thus the Fourier transform of $\rho^2_{mn}(0)$ is

$$ k^2 T_{Am} T_{An} \left[\int_{-\infty}^{\infty} H_m(v) H_n^*(v)\, dv \right]^2 \Delta(v) , \tag{6.39} $$

where k is Boltzmann's constant, T_{Am} and T_{An} are the components of antenna temperature resulting from the source, $H_m(v)$ and $H_n(v)$ are the frequency responses of the signal channels, and $\Delta(v)$ is the bandwidth.

The Fourier transform of the second term of Eq. (6.37), $\rho_m(\tau)\rho_n(\tau)$, is the convolution of the transforms of ρ_m and ρ_n, that is

$$ k^2 (T_{Sm} + T_{Am})(T_{Sn} + T_{An}) \int_{-\infty}^{\infty} H_m(v) H_m^*(v) H_n(v' - v) H_n^*(v' - v)\, dv , \tag{6.40} $$

where T_{Sm} and T_{Sn} are the system temperatures. Note that the magnitude of this term is proportional to the product of the total noise temperatures.

The Fourier transform of the third term of Eq. (6.37), $\rho_{mn}(\tau)\rho_{mn}(-\tau)$, is the convolution of the transforms of $\rho_{mn}(\tau)$ and $\rho_{mn}(-\tau)$, and the latter is the complex conjugate of the former, since ρ_{mn} is real. Thus, the Fourier transform of $\rho_{mn}(\tau)\rho_{mn}(-\tau)$ is

$$ k^2 T_{Am} T_{An} \int_{-\infty}^{\infty} H_m(v) H_n^*(v) H_m^*(v' - v) H_n(v' - v)\, dv . \tag{6.41} $$

In expression (6.39), as in Eq. (6.37), only the antenna temperatures appear, because the receiver noise for different antennas makes no contribution to the cross-correlation.

Expression (6.39) represents the signal power in the correlator output, and (6.40) and (6.41) represent the noise. The effect of the time averaging at the correlator output can be modeled in terms of a filter that passes frequencies from 0 to Δv_{LF}. The output bandwidth Δv_{LF} is less than the correlator input bandwidth by several or many orders of magnitude. Therefore, the spectral density of the output noise can be assumed to be equal to its value at zero frequency, that is, for $v' = 0$ in (6.40) and (6.41). From these considerations, and because $H_m(v)$ and $H_n(v)$ are Hermitian, the ratio of the signal voltage to the rms noise voltage after averaging at the correlator output is

$$\mathcal{R}_{sn} =$$

$$\frac{\sqrt{T_{Am}T_{An}} \int_{-\infty}^{\infty} H_m(v)H_n^*(v)\,dv}{\sqrt{(T_{Am}+T_{Sm})(T_{An}+T_{Sn})+T_{Am}T_{An}}\sqrt{2\Delta v_{LF} \int_{-\infty}^{\infty} |H_m(v)|^2 |H_n(v)|^2 dv}},$$

(6.42)

where $2\Delta v_{LF}$ is the equivalent bandwidth after averaging, with negative frequencies included. It is unusual for $\mathcal{R}_{sn}$, the estimate of the SNR at the output of a simple correlator, to be required to an accuracy better than a few percent. Indeed, it is usually difficult to specify T_S to any greater accuracy since the effects of ground radiation and atmospheric absorption on T_S vary as the antennas track. Thus, it is usually satisfactory to approximate $H_m(v)$ and $H_n(v)$ by identical rectangular functions of width Δv_{IF}. Also, in sensitivity calculations, one is concerned most often with sources near the threshold of detectability, for which $T_A \ll T_S$. With these simplifications, Eq. (6.42) becomes

$$\mathcal{R}_{sn} = \sqrt{\frac{T_{Am}T_{An}}{T_{Sm}T_{Sn}}} \sqrt{\frac{\Delta v_{IF}}{\Delta v_{LF}}}.$$

(6.43)

Figure 6.7 shows the signal and noise spectra for the rectangular bandpass approximation. Note that the input spectra $|H_m(v)|^2$ and $|H_n(v)|^2$ contain both positive and negative frequencies and are symmetric about the origin in v. Thus, the output noise spectrum can be described as proportional to either the convolution or the cross-correlation function of $|H_m(v)|^2$ and $|H_n(v)|^2$.

The output bandwidth is related to the data averaging time τ_a since the averaging can be described as convolution in the time domain with a rectangular function of unit area and width τ_a. The power response of the averaging circuit as a function of frequency is the square of the Fourier transform of the rectangular function, that is, $\sin^2(\pi\tau_a v)/(\pi\tau_a v)^2$. The equivalent bandwidth, including both positive and negative frequencies, is

$$2\Delta v_{LF} = \int_{-\infty}^{\infty} \frac{\sin^2(\pi\tau_a v)}{(\pi\tau_a v)^2}dv = \frac{1}{\tau_a}.$$

(6.44)

Then from Eq. (6.43), we obtain

$$\mathcal{R}_{sn} = \sqrt{\left(\frac{T_{Am}T_{An}}{T_{Sm}T_{Sn}}\right)2\Delta v_{IF}\tau_a}.$$

(6.45)

Note that $2\Delta v_{IF}\tau_a$ is the number of independent samples of the signal in time τ_a.

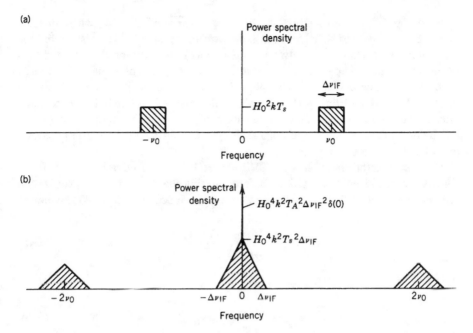

Fig. 6.7 Spectra of (a) the input and (b) the output waveforms of a correlator. The input passbands are rectangular of width $\Delta\nu_{IF}$. Shown in (b) is the complete spectrum of signals generated in the multiplication process, including noise bands at twice the input frequency. Only frequencies very close to zero are passed by the averaging circuit at the correlator output. These include the wanted signal, the spectrum of which has the form of a delta function and is represented by the arrow. It is assumed that $T_A \ll T_S$.

If the source is unpolarized, each antenna responds to half the total flux density S, and the received power density is

$$kT_A = \frac{1}{2}AS , \qquad (6.46)$$

where A is the effective collecting area of the antenna. For identical antennas and system temperatures, we obtain, from Eqs. (6.45) and (6.46),

$$\mathcal{R}_{sn} = \frac{AS}{kT_S}\sqrt{\frac{\Delta\nu_{IF}\tau_a}{2}} . \qquad (6.47)$$

Similar derivations of this result can be found in Blum (1959), Colvin (1961), and Tiuri (1964). Usually the result in Eq. (6.47), in which we have assumed $T_A \ll T_S$, is the one needed. At the other extreme, which may be encountered in observations of very strong, unresolved sources for which $T_A \gg T_S$, we have $\mathcal{R}_{sn} = \sqrt{\Delta\nu_{IF}\tau_a}$. The SNR is then determined by the fluctuations in signal level and is independent of the areas of the antennas. Anantharamaiah et al. (1989) give a discussion of noise levels in the observation of very strong sources.

From Fig. 6.7, we can see how the factor $\sqrt{\Delta\nu_{\mathrm{IF}}\tau_a}$ in Eq. (6.47), which enables very high sensitivity to be achieved in radio astronomy, arises. The noise within the correlator results from beats between components in the two input bands and thus extends in frequency up to $\Delta\nu_{\mathrm{IF}}$. The triangular noise spectrum in Fig. 6.7 is simply proportional to the number of beats per unit frequency interval. However, only the very small fraction of this noise that falls within the output bandwidth is retained after the averaging. Note that the signal bandwidth $\Delta\nu_{\mathrm{IF}}$ that is important here is the bandwidth at the correlator input. In a DSB system, this is only one-half of the total input bandwidth at the antenna.

One other factor that affects the SNR should be introduced at this point. If the signals are quantized and digitized before entering the correlators, a quantization efficiency η_Q related to the quantization must be included, and Eq. (6.47) becomes

$$\mathcal{R}_{\mathrm{sn}} = \frac{AS\eta_Q}{kT_S}\sqrt{\frac{\Delta\nu_{\mathrm{IF}}\tau_a}{2}}, \tag{6.48}$$

or in terms of antenna temperature,

$$\mathcal{R}_{\mathrm{sn}} = \frac{T_A\eta_Q}{T_S}\sqrt{2\Delta\nu_{\mathrm{IF}}\tau_a}. \tag{6.49}$$

Values of η_Q vary between 0.637 and 1.0 and are discussed in Chap. 8 (see Table 8.1). In VLBI observing, other losses affect the SNR, as discussed in Sect. 9.7.

6.2.2 Noise in the Measurement of Complex Visibility

To understand precisely what $\mathcal{R}_{\mathrm{sn}}$ represents, note that in deriving Eqs. (6.48) and (6.49), no delay was introduced between the signal components at the correlator, and the phase responses of the signal channels were assumed to be identical. Thus, the source is in the central fringe of the interferometer pattern, and the response is the peak fringe amplitude, which represents the modulus of the visibility. To express the rms noise level at the correlator output in terms of the flux density σ of an unresolved source for which the peak fringe amplitude produces an equal output, we put $\mathcal{R}_{\mathrm{sn}} = 1$ in Eq. (6.48) and replace S by σ:

$$\sigma = \frac{\sqrt{2}\,kT_S}{A\eta_Q}\bigg/\sqrt{\Delta\nu_{\mathrm{IF}}\tau_a}, \tag{6.50}$$

where σ is in units of W m^{-2} Hz^{-1}. Consider the case of an instrument with a complex correlator in which the output oscillations are slowed to zero frequency as described earlier. The noise fluctuations in the real and imaginary outputs are uncorrelated, as we now show. Suppose the antennas are pointed at blank sky so

Fig. 6.8 Complex quantity
Z, which is the sum of the
modulus of the true complex
visibility $\mathcal{V}$ and the noise $\boldsymbol{\varepsilon}$.
The noise has real and
imaginary components of rms
amplitude σ, and ϕ is the
phase deviation resulting
from the noise.

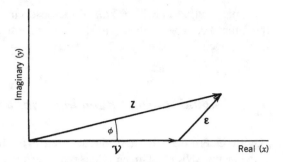

that the only inputs to the correlators in Fig. 6.3 are the noise waveforms n_m, n_n, and n_m^H, where the last is the Hilbert transform of n_m produced by the quadrature phase shift. The expectation of the product of the real and imaginary outputs is $\langle n_m n_n n_m^H n_n \rangle$, which can be shown to be zero by using Eq. (6.36) and noting that the expectations $\langle n_m n_n \rangle$, $\langle n_m n_m^H \rangle$, and $\langle n_m^H n_n \rangle$ must all be zero. Thus, the noise from the real and imaginary outputs is uncorrelated.[5]

The signal and noise components in the measurement of the complex visibility are shown in Fig. 6.8 as vectors in the complex plane. Here $\mathcal{V}$ represents the visibility as it would be measured in the absence of noise, which is assumed to be along the x, or real, axis; and **Z** represents the sum of the visibility and noise, $\mathcal{V} + \boldsymbol{\varepsilon}$. We consider **Z** and $\boldsymbol{\varepsilon}$ to be vectors whose components correspond to the real and imaginary parts of the corresponding quantities. The components of $\boldsymbol{\varepsilon}$ are independent Gaussian random variables with zero mean and variance σ^2. Hence, the noise in both components of **Z** has an rms amplitude σ, and

$$\langle |\mathbf{Z}|^2 \rangle = |\mathcal{V}|^2 + 2\sigma^2 . \tag{6.51}$$

The factor of two arises because of the contributions of the real and imaginary parts of $\boldsymbol{\varepsilon}$. If the measurement is made using only a single-multiplier correlator, one can periodically introduce a quadrature phase shift at one input, thus obtaining real and

---

[5]The noise in the correlator outputs is composed of an ensemble of components of frequency $|\nu_m - \nu_n|$, where ν_m and ν_n are frequency components of the correlator inputs n_m and n_n. Components of the imaginary output are shifted in phase by $\pm \pi/2$ relative to the corresponding components of the real output. Note that for any pair of input components, the sign of the shift in the imaginary output takes opposite values depending on whether $\nu_m > \nu_n$ or $\nu_m < \nu_n$. As a result, the noise waveforms at the correlator outputs are not a Hilbert transform pair, and one cannot be derived from the other.

imaginary outputs, each for half of the observing time. Then the data are half of those that would be obtained with a complex correlator, and the noise in the visibility measurement is greater by $\sqrt{2}$.

6.2.3   Signal-to-Noise Ratio in a Synthesized Image

Having determined the noise-induced error in the visibility, the next step is to consider the SNR in an image. Consider an array with n_p antenna pairs, and suppose that the visibility data are averaged for time τ_a and that the whole observation covers a time interval τ_0. The total number of independent data points in the (u, v) plane is therefore

$$n_d = n_p \frac{\tau_0}{\tau_a} . \tag{6.52}$$

In imaging an unresolved source at the field center for which the visibility data combine in phase, we should thus expect the SNR in the image to be greater than that in Eqs. (6.48) and (6.49) by a factor $\sqrt{n_p \tau_0/\tau_a}$. This simple consideration gives the correct result for the case in which the data are combined with equal weights. We now derive the result for the more general case of arbitrarily weighted data.

The ensemble of measured data can be represented by

$$\sum_{i=1}^{n_d} \left[{}^2\delta(u - u_i, v - v_i)(\mathcal{V}_i + \varepsilon_i) + {}^2\delta(u + u_i, v + v_i)(\mathcal{V}_i^* + \varepsilon_i^*) \right] , \tag{6.53}$$

where ${}^2\delta$ is the two-dimensional delta function and ε_i is the complex noise contribution to the ith measurement. Each such data point appears at two (u, v) locations, reflected through the origin of the (u, v) plane. Before taking the Fourier transform of the data in Eq. (6.53), each data point is assigned a weight w_i (the choice of weighting factors is discussed in Sect. 10.2.2). To simplify the calculation, we assume that the source is unresolved and located at the phase reference point of the image and therefore produces a constant real visibility $\mathcal{V}$ equal to its flux density S. The intensity at the center of the image is then

$$I_0 = \frac{\displaystyle\sum_{i=1}^{n_d} w_i(\mathcal{V} + \varepsilon_{\mathcal{R}i})}{\displaystyle\sum w_i} , \tag{6.54}$$

where $\varepsilon_{\mathcal{R}i}$ is the real part of the noise, ε_i. Note that the imaginary part of ε_i vanishes at the origin of the image when the conjugate components are summed. For neighboring points in the image, the same rms level of noise is distributed between the real and imaginary parts of ε. The expectation of I_0 is

$$\langle I_0 \rangle = \mathcal{V} = S \,, \tag{6.55}$$

since $\langle \varepsilon_{\mathcal{R}i} \rangle = 0$. The variance of the estimate of the intensity, σ_m^2, is

$$\sigma_m^2 = \langle I_0^2 \rangle - \langle I_0 \rangle^2 = \frac{\sum w_i^2 \langle \varepsilon_{\mathcal{R}i}^2 \rangle}{\left(\sum w_i \right)^2} \,. \tag{6.56}$$

Equation (6.56) is derived directly from Eq. (6.54) using the fact that the noise terms from different (u, v) locations are uncorrelated, that is, $\langle \varepsilon_{\mathcal{R}i} \varepsilon_{\mathcal{R}j} \rangle = 0$, for $i \neq j$. We define the mean weighting factor w_{mean} and rms weighting factor w_{rms} by the equations

$$w_{\mathrm{mean}} = \frac{1}{n_d} \sum w_i \tag{6.57}$$

and

$$w_{\mathrm{rms}}^2 = \frac{1}{n_d} \sum w_i^2 \,. \tag{6.58}$$

The rms noise contribution [see Eq. (6.51)] is the same for each (u, v) point and is equal to $\langle \varepsilon_{\mathcal{R}i}^2 \rangle = \sigma^2$, where σ is given by Eq. (6.50). Thus, the SNR can be calculated from Eqs. (6.55), (6.56), (6.57), and (6.58) as

$$\frac{\langle I_0 \rangle}{\sigma_m} = \frac{S \sqrt{n_d}}{\sigma} \frac{w_{\mathrm{mean}}}{w_{\mathrm{rms}}} \,. \tag{6.59}$$

For an array with complex correlators, we have, from Eq. (6.50),

$$\frac{\langle I_0 \rangle}{\sigma_m} = \frac{A S \eta_Q \sqrt{n_d \Delta \nu_{\mathrm{IF}} \tau_a}}{\sqrt{2} k T_S} \frac{w_{\mathrm{mean}}}{w_{\mathrm{rms}}} \,. \tag{6.60}$$

If combinations of all pairs of antennas are used, $n_p = \frac{1}{2} n_a (n_a - 1)$, where n_a is the number of antennas. Since, from Eq. (6.52), $n_d = n_p \tau_0 / \tau_a$, we obtain

$$\frac{\langle I_0 \rangle}{\sigma_m} = \frac{A S \eta_Q \sqrt{n_a (n_a - 1) \Delta \nu_{\mathrm{IF}} \tau_0}}{2 k T_S} \frac{w_{\mathrm{mean}}}{w_{\mathrm{rms}}} \,. \tag{6.61}$$

To express the rms noise level in terms of flux density, we put $I_0/\sigma_m = 1$ in Eq. (6.61). S then represents the flux density of a point source for which the peak response is equal to the rms noise level. If we represent this particular value of S by S_{rms}, then

$$S_{rms} = \frac{2kT_S}{A\eta_Q \sqrt{n_a(n_a - 1)\Delta\nu_{IF}\tau_0}} \frac{w_{rms}}{w_{mean}} . \tag{6.62}$$

If all the weighting factors w_i are equal, $w_{mean}/w_{rms} = 1$, and this situation is referred to as the use of *natural weighting*. In such a case, the SNR given by Eq. (6.61) is equal to the corresponding sensitivity for a total-power receiver combined with an antenna of aperture $\sqrt{n_a(n_a - 1)}A$, which approaches n_aA as n_a becomes large. For an analysis of the sensitivity of single-antenna systems, see Appendix 1.1.

We have considered the *point-source sensitivity* in Eq. (6.62). In the case of a source that is wider than the synthesized beam, it is useful to know the *brightness sensitivity*. The flux density (in W m^{-2} Hz^{-1}) received from a broad source of mean intensity I (W m^{-2} Hz^{-1} sr^{-1}) across the synthesized beam is $I\Omega$, where Ω is the effective solid angle of the synthesized beam. Thus, the intensity level that is equal to the rms noise is S_{rms}/Ω. Note that the brightness sensitivity decreases as the synthesized beam becomes smaller, so compact arrays are best for detecting broad, faint sources. However, to measure the intensity of a uniform background, a measurement of the total power received in an antenna is required because a correlation interferometer does not respond to such a background.

The ratio w_{mean}/w_{rms} is less than unity except when the weighting is uniform. Although the SNR depends on the choice of weighting, in practice, this dependence is not highly critical. The use of natural weighting maximizes the sensitivity for detection of a point source in a largely blank field but can also substantially broaden the synthesized beam. The advantage in sensitivity is usually small. For example, if the density of data points is inversely proportional to the distance from the (u, v) origin, as is the case for an east–west array with uniform increments in antenna spacing, the weighting factors required to obtain effective uniform density of data result in $w_{mean}/w_{rms} = 2\sqrt{2}/3 = 0.94$. In this case, the natural weighting results in an undesirable beam profile in which the response remains positive for large angular distances from the beam axis and dies away only slowly.

Methods of Fourier transformation of visibility data are reviewed in Chap. 10, and the results derived in Eqs. (6.61) and (6.62) can be applied to these by using the appropriate values of w_{mean} and w_{rms}. Convolution of the visibility data in the (u, v) plane to obtain values at points on a rectangular grid is a widely used process. In general, the data at adjacent grid points are then not independent, and a tapering of the signal and noise is introduced into the image. Aliasing can also cause the SNR to vary across the image. (These effects are explained in Fig. 10.5 and the associated discussion.) In such cases, the results derived here apply near the origin of the image, where the effects of tapering and aliasing are unimportant. The rms noise level over the image can be obtained by the application of Parseval's theorem to the noise in the visibility data (see Appendix 2.1).

In practice, a number of factors that affect the SNR are difficult to determine precisely. For example, T_S varies somewhat with antenna elevation. There are also a number of effects that can reduce the response to a source without reducing the noise, but these are important only for sources not near the (l, m) origin of an image. These include the smearing resulting from the receiving bandwidth and from visibility averaging, discussed later in this chapter, and the effect of non-coplanar baselines, discussed in Sect. 11.7.

Note also that in many instruments, two oppositely polarized signals (with crossed linear or opposite circular polarizations) are received and processed using separate IF amplifiers and correlators. For unpolarized sources, the overall SNR is then $\sqrt{2}$ greater than the values derived above, which include only one signal from each antenna.

6.2.4   Noise in Visibility Amplitude and Phase

In synthesis imaging, we are usually concerned with data in the form of the real and imaginary parts of the complex visibility $\mathcal{V}$, but sometimes it is necessary to work with the amplitude and phase. The sum of the visibility and noise is represented by $\mathbf{Z} = Ze^{j\phi}$, where we choose the real axis so that the phase ϕ is measured with respect to the phase of $\mathcal{V}$, as in Fig. 6.8. Then for $T_A \ll T_S$ (the antenna temperature resulting from the source is much less than the system temperature), the probability distributions of the resulting amplitude and phase are

$$p(Z) = \frac{Z}{\sigma^2} \exp\left(-\frac{Z^2 + |\mathcal{V}|^2}{2\sigma^2}\right) I_0 \left(\frac{Z|\mathcal{V}|}{\sigma^2}\right), \qquad Z > 0 \qquad (6.63a)$$

$$p(\phi) = \frac{1}{2\pi} \exp\left(-\frac{|\mathcal{V}|^2}{2\sigma^2}\right) \left\{1 + \sqrt{\frac{\pi}{2}} \frac{|\mathcal{V}|\cos\phi}{\sigma} \exp\left(\frac{|\mathcal{V}|^2 \cos^2\phi}{2\sigma^2}\right)\right.$$
$$\left. \times \left[1 + \mathrm{erf}\left(\frac{|\mathcal{V}|\cos\phi}{\sqrt{2}\sigma}\right)\right]\right\}, \qquad (6.63b)$$

and erf is the error function (Abramowitz and Stegun 1968).

$$\mathrm{erf}\left(\frac{x}{\sqrt{2}}\right) = \frac{2}{\sqrt{\pi}} \int_0^x e^{-t^2/2}\, dt, \qquad (6.63c)$$

where I_0 is the modified Bessel function of zero order and σ is as given by Eq. (6.50). The amplitude distribution is identical to that for a sine wave in noise, and the derivation is given by Rice (1944, 1954), Vinokur (1965), and Papoulis (1965), of which the last two also derive the result for the phase. $p(Z)$ is sometimes referred to as the Rice distribution, and for $\mathcal{V} = 0$, it reduces to the Rayleigh distribution. Curves of $p(Z)$ and $p(\phi)$ are given in Fig. 6.9. Comparison of the curves

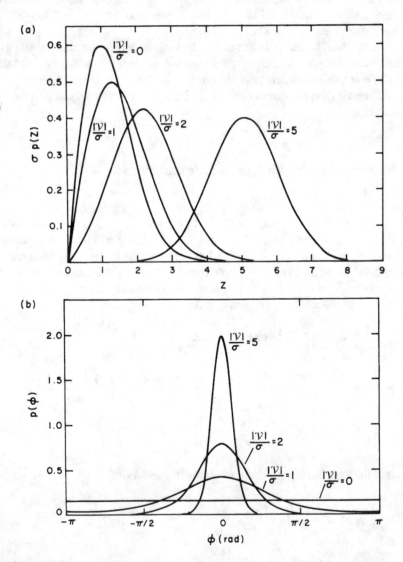

Fig. 6.9 Probability distributions of (**a**) the amplitude, and (**b**) the phase, of the measured complex visibility as functions of the SNR. $|\mathcal{V}|$ is the modulus of the signal component. Reprinted from Moran (1976), © 1976, with permission from Elsevier.

for $|\mathcal{V}|/\sigma = 0$ and 1 indicates that the presence of a weak signal is more easily detected by examining the visibility phase than by examining the amplitude.

Approximation for $p(Z)$ and $p(\phi)$ for the cases in which $|\mathcal{V}|/\sigma \ll 1$ and $|\mathcal{V}|/\sigma \gg 1$ are given in Sect. 9.3. Expressions for the moments of Z and ϕ and their rms deviations are also given in that section. The rms phase deviation σ_ϕ is a particularly useful quantity, especially for astrometric and diagnostic work. The expression for σ_ϕ, valid for the case in which $|\mathcal{V}|/\sigma \gg 1$, is $\sigma_\phi \simeq \sigma/|\mathcal{V}|$ [Eq. (9.67)]. This result also follows intuitively from an examination of Fig. 6.8. By substituting Eq. (6.50) into the expression for σ_ϕ, setting $|\mathcal{V}|$ equal to the flux density S of the source, which is appropriate if the source is unresolved, and using Eq. (6.46) to relate the flux density and antenna temperature, we obtain

$$\sigma_\phi = \frac{T_S}{\eta_Q T_A \sqrt{2\Delta\nu_{\mathrm{IF}}\tau_a}} . \tag{6.64}$$

This equation is valid for the conditions $T_S/\sqrt{2\Delta\nu_{\mathrm{IF}}\tau_a} \ll T_A \ll T_S$, which are the conditions most frequently encountered, and is useful for determining whether the noise in the phase measurements of an interferometer is due exclusively to receiver noise. Excess phase noise can be contributed by the atmosphere, by system instabilities, and, in the case of VLBI, by the frequency standards.

6.2.5 Relative Sensitivities of Different Interferometer Systems

Next we compare the sensitivity of several different interferometer systems, using as a measure of sensitivity the modulus of the signal divided by the rms noise, that is, $\mathcal{V}/\varepsilon$ in terms of the quantities at the correlator output in Fig. 6.8. Parameters such as averaging times and IF bandwidths are the same for all cases considered. To compare DSB and SSB cases, it is convenient to introduce a factor

$$\alpha = \frac{\text{double-sideband system temperature of double-sideband system}}{\text{system temperature of single-sideband system}} . \tag{6.65}$$

Recall that the system temperature of a receiver can be defined as the noise temperature of a thermal source at the input of a hypothetical noise-free (but otherwise identical) receiver that would produce the same noise level at the receiver output. [Equation (1.4) can be used for the equivalent noise temperature if the Rayleigh–Jeans approximation does not apply.] For a DSB receiver, the system temperature is described as DSB or SSB depending on whether the thermal noise source emits noise in both sidebands or only one. With these definitions, the SSB noise temperature is twice the DSB noise temperature.

For an SSB system, the rms noise from one output of a correlator (either the real or imaginary output in the case of a complex correlator) is σ after averaging for a time τ_a, as given by Eq. (6.50). The corresponding noise power is σ^2. For a DSB system, the rms output noise at a correlator output is $2\alpha\sigma$. In all cases, the signal results from an unresolved source. For an SSB system, we take the signal voltage from the correlator output to be $\mathcal{V}$, as in Fig. 6.8. For a DSB system with the input signal in one sideband only, the signal at the correlator output is $\mathcal{V}$, and for a DSB system with input in both sidebands, the correlator output is $2\mathcal{V}$.

Values of the relative sensitivity for various systems are discussed below and summarized in Table 6.1. Similar results are given by Rogers (1976).

1. *SSB system with complex correlator.* The output signal is $\mathcal{V}$, and the rms noise from each correlator output is σ. As shown by Fig. 6.9 and Eq. (6.51), the ratio of the signal amplitude to rms noise is $\mathcal{V}/(\sqrt{2}\sigma)$. We shall take this as the standard with respect to which the relative sensitivities of other systems are defined.

2. *SSB system and simple correlator with fringe fitting.* To measure both the real and imaginary parts of the complex visibility, the fringes are not stopped but appear as a sinusoid of amplitude $\mathcal{V}$ at the fringe frequency ν_f. The signal is accompanied by noise of rms amplitude σ. The amplitude and phase are measured by "fringe fitting," that is, performing a least-mean-squares fit of a sinusoid to the correlator output. This procedure involves multiplying the correlator output waveform by $\cos(2\pi\nu_f t)$ and $\sin(2\pi\nu_f t)$ and integrating over the period τ_a. The results

Table 6.1 Relative signal-to-noise ratios for several types of systems

System type	Relative SNR
1. SSB with complex correlator	1
2. SSB with simple correlator	$\frac{1}{\sqrt{2}}$
3. SSB, simple correlator, fringe stopping, $\pi/2$ phase switching	$\frac{1}{\sqrt{2}}$
4. DSB, simple correlator,[a] fringe fitting, continuum signal	$\frac{1}{\sqrt{2}\alpha}$
5. DSB, simple correlator, fringe stopping, $\pi/2$ phase switching, continuum signal	$\frac{1}{\sqrt{2}\alpha}$
6a. DSB, fringe stopping, sideband-separation [Eqs. (6.30) to (6.33)], signal in one sideband only	$\frac{1}{2\alpha}$
6b. As (6a), but for continuum signal and visibilities in both sidebands combined	$\frac{1}{\sqrt{2}\alpha}$
7a. VLBI, DSB, complex correlator, one sideband removed by averaging of fast fringes	$\frac{1}{2\alpha}$
7b. As (7a), but for continuum signal, correlated separately for each sideband and results combined	$\frac{1}{\sqrt{2}\alpha}$
8. SSB, digital spectral correlator with simple correlator elements and correlation measured as a function of time offsets (see Sect. 8.8)	1

[a]For DSB with complex correlator, see text pertaining to Fig. 6.5

represent the real and imaginary parts, respectively, of the cross-correlation. We calculate the effects of fringe fitting on the signal and noise separately and assume, with no loss of generality, that the fringes are in phase with the cosine component in the fringe fitting, in which case the sine component of the signal is zero. The correlator output has a bandwidth $\Delta \nu_c$ which is sufficient to pass the fringe-frequency waveform, and it is sampled at time intervals $\tau_s = 1/(2\nu_c)$ and digitized. Within the period τ_a, there are $N = 2\Delta\nu_c\tau_a$ samples. Thus, for the cosine component of the signal, the amplitude is

$$\frac{1}{N} \sum_{i=1}^{N} \mathcal{V} \cos^2(2\pi i \nu_f \tau_s) = \frac{\mathcal{V}}{2} + \frac{\mathcal{V}}{2N} \sum_{i=1}^{N} \cos(4\pi i \nu_f \tau_s) \,. \tag{6.66}$$

The second term on the right side represents the end effects and is approximately zero if there are an integral number of half-cycles of the fringe frequency within the period τ_a. It also becomes relatively small as $\nu_f \tau_a$ increases, and we assume here that there are enough fringe cycles (say, ten or more) within time τ_a that end effects can be neglected. To determine the effect of fringe fitting on the noise, we represent the sampled noise by $n(i\tau_s)$, multiply by the cosine function, and determine the variance (mean squared value). Averaged over time τ_a, the result is

$$\frac{1}{N} \left[\sum_{i=1}^{N} n(i\tau_s) \cos(2\pi i \nu_f \tau_s) \right]^2$$

$$= \frac{1}{N} \sum_{i=1}^{N} \sum_{k=1}^{N} n(i\tau_s) \cos(2\pi i \nu_f \tau_s) \, n(k\tau_s) \cos(2\pi k \nu_f \tau_s) \,. \tag{6.67}$$

We need to determine the expectation value of this expression, denoted by angle brackets. Only terms for which $i = k$ contribute to the expectation. Thus, the noise variance becomes

$$\left\langle \frac{1}{2N} \sum_{i=1}^{N} n^2(i\tau_s)[1 + \cos(4\pi i \nu_f \tau_s)] \right\rangle = \frac{\sigma^2}{2} \,. \tag{6.68}$$

This result shows that half of the noise power, σ^2, that is available at the correlator output appears in the cosine component of the fringe fitting. Similarly, the other half appears in the sine component. The combined rms noise of the two components is σ, and the SNR after fringe fitting is $\mathcal{V}/(2\sigma)$. The relative sensitivity is $1/\sqrt{2}$.

3. *SSB system with simple correlator and $\pi/2$ phase switching of LO.* In this case, the fringes have been stopped, and to determine the complex visibility, a phase change of $\pi/2$ is periodically inserted into one oscillator [e.g., $\theta_n \to \theta_n + \pi/2$ in Eq. (6.11) or (6.15)] so that the correlator is effectively time-shared between the real and imaginary parts of the cross-correlation function, which are averaged

separately. The visibility phase can thereby be determined. The signal in the two phase conditions is $\mathcal{V}\cos(\phi_v)$ and $\mathcal{V}\sin(\phi_v)$, and the rms noise associated with each of these terms is $\sqrt{2}\sigma$ (the $\sqrt{2}$ factor enters because the noise in each output is averaged over time $\tau/2$ only). Thus, the modulus of the signal is $\mathcal{V}$ and the rms noise from the two components is 2σ. The SNR is $\mathcal{V}/(2\sigma)$ and the relative sensitivity is $1/\sqrt{2}$.

4. *DSB system with simple correlator and fringe fitting.* We consider the case of a continuum source with signal in both sidebands and assume that the instrumental delay is adjusted so that the signal appears entirely in the (real) output of a simple correlator, as a fringe-frequency cosine wave of amplitude $\mathcal{V}$. In terms of Eq. (6.18), the factor $\cos(2\pi\nu_0\Delta\tau_a + \phi_G)$ is unity. Then for the DSB system, the signal amplitude is $2\mathcal{V}$, and the rms noise is $2\alpha\sigma$. The fringe-fitting procedure follows that of case 2, but in this case, the signal amplitude is greater by a factor of two and is equal to $\mathcal{V}$. The rms noise is greater by a factor of 2α. Thus, the SNR is $\mathcal{V}/(2\alpha\sigma)$, and the relative sensitivity is $1/(\sqrt{2}\alpha)$.

5. *DSB system with simple correlator and $\pi/2$ phase switching of LO.* Here, the fringes have been stopped, and to determine the visibility phase, it is necessary to perform $\pi/2$ phase switching as in case 3 above. (For a DSB system, the phase switching must be on the first LO.) The amplitude of the signal is $2\mathcal{V}$ because the system is DSB, and the rms noise level from the correlator output is increased to $2\sqrt{2}\alpha\sigma$ because the averaging time for each component is reduced to $\tau_a/2$ by the time sharing of the correlator between the two phase conditions. This rms level is associated with both the cosine and sine components of the signal, so the SNR is $\mathcal{V}/(2\alpha\sigma)$. The relative sensitivity is $1/(\sqrt{2}\alpha)$.

6. *One sideband of a DSB system with $\pi/2$ phase switching of the LO and sideband separation after correlation.* A complex correlator is used, and the procedure corresponding to Eqs. (6.30) to (6.33) is followed. We consider the upper sideband and ignore lower-sideband signal terms. The components r_1, r_2, r_3, r_4 have amplitudes $\mathcal{V}$ multiplied by the cosine or sine of Ψ_u. Thus, from Eqs. (6.30) and (6.31), ignoring lower-sideband terms, the right side of Eq. (6.32) becomes $\frac{1}{2}(2\mathcal{V}\cos\Psi_u + j2\mathcal{V}\sin\Psi_u)$, the modulus of which is $\mathcal{V}$. The rms noise associated with each term r_1, r_2, r_3, and r_4 is $2\sqrt{2}\alpha\sigma$ since the system is DSB, and because of the LO switching, the effective averaging time is $\tau_a/2$. Thus, the rms noise associated with the right side of Eq. (6.32) is $2\sqrt{2}\alpha\sigma$, as in case 5. The SNR is $\mathcal{V}/(2\sqrt{2}\alpha\sigma)$, and the relative sensitivity is $1/(2\alpha)$. This applies to a signal in one sideband such as a spectral line. For a continuum source, the cross-correlation can be measured for each of the two sidebands, and if the results are then averaged, the relative sensitivity becomes $1/(\sqrt{2}\alpha)$. The terms r_2 and r_4 are eliminated in averaging the right sides of Eqs. (6.32) and (6.33), and the result is the same as for a simple correlator with LO phase switching described under case 5 above.

7. *VLBI observations with a DSB system and complex correlator.* In VLBI observations, a DSB system is sometimes used, and fringe rotation is inserted after playback of the recorded signal, as mentioned in Sect. 6.1. For one sideband,

the fringes are stopped, but for the other, they are lost in the averaging at the correlator output because the fringe frequencies are high. Thus, for one playback, we have the signal of an SSB system and the noise of a DSB system in each of the real and imaginary outputs, that is, an SNR of $\mathcal{V}/(2\sqrt{2}\alpha\sigma)$ and a relative sensitivity of $1/(2\alpha)$ for each individual sideband.

8. *Measurement of cross-correlation as a function of time offset.* Digital spectral correlators that measure cross-correlation as a function of time delay are described in Sect. 8.8. In a lag-type correlator, the cross-correlation is measured as a function of time offset, implemented by introducing instrumental delays. The Fourier transform of the cross-correlation as a function of relative time delay between the signals is the cross-correlation as a function of frequency, as required in spectral line measurements. As mentioned in Sect. 6.1.7, it is necessary to use only simple correlators for this measurement. The range of time offsets of the two signals covers both positive and negative values, and the resulting measurements of cross-correlation contain both even and odd components. Fourier transformation then provides both the real and imaginary components of the cross-correlation as a function of frequency. The full sensitivity is obtained so long as the range of time offsets is comparable to the reciprocal signal bandwidth or greater; see Table 9.7. Note that in Table 6.1, we have not included the quantization loss discussed in Sect. 8.3.3. A demonstration of the sensitivity using a simple correlator when the measurements are made as a function of time delay is given by Mickelson and Swenson (1991).

Of the cases included in Table 6.1, the SSB with complex correlator is the one generally used where possible, because of the sensitivity and avoidance of the complications of DSB operation. Cases 2 and 3 in the table are included mainly for completeness of the discussion. As mentioned earlier, for frequencies of several hundred gigahertz, the most sensitive type of receiver input stage may be an SIS mixer. This has an inherently double-sided response, and if necessary, a sideband can be removed by filtering or using a sideband-separating arrangement (Appendix 7.1). For DSB operation, the most important cases in Table 6.1 are 6a and 6b. The case in which the unwanted sideband is only partially rejected is discussed in Appendix 6.1.

6.2.6 System Temperature Parameter α

As already noted, DSB systems are mainly used at millimeter and submillimeter wavelengths, at which the receiver input stage is commonly a cooled SIS mixer. Such a system can be converted to SSB operation by filtering out the unwanted sideband and terminating the corresponding input in a cold load. If the atmospheric losses are high and the receiver temperature is low, most of the system noise will come from the antenna, and terminating one sideband in a cold load will approximately halve the level of noise within the receiver. The system temperature of the SSB system will then be approximately equal to the *double*-sideband system temperature of the DSB system, and the value of α [defined in Eq. (6.65)] tends toward 1. On the other hand, if atmospheric and antenna losses are low and most of the system noise comes from the mixer and IF stages, then terminating one sideband input in a cold load rather than the cold sky makes little difference to the noise level in the receiver. The system temperature of the SSB system will be close to the *single*-sideband system temperature of the DSB system, which is twice the DSB value. The value of α then tends toward 1/2. To recapitulate: If the atmospheric noise dominates the receiver noise, then α tends toward 1, but if the receiver noise dominates, then α tends toward 1/2. Note, however, that α is not confined to the range $1/2 < \alpha < 1$. For example, if noise from the antenna is low but the termination of the image sideband in the SSB system is uncooled and injects a high noise level, then α can be $< 1/2$. If the front end is tuned close to an atmospheric absorption line in such a way that the additional sideband of the DSB system falls in a frequency range of enhanced atmospheric noise, then α can be > 1.

6.3   Effect of Bandwidth

As seen in the preceding section, the sensitivity of a receiving system to a broadband cosmic signal increases with the system bandwidth. Here we are concerned with the effect of bandwidth on the angular range over which fringes are detected, and on the fringe amplitude. These effects result from the variation of fringe frequency, in cycles per radian on the sky, with the received radio frequency. If the monochromatic response is integrated over the bandwidth, the fringes are reinforced for directions close to that for which the time delays from the source to the correlator inputs are equal, but for other directions, the fringes vary in phase across the bandwidth. This effect, when measured in a plane containing the interferometer baseline, causes the fringe amplitude to decrease with angle in a manner similar to that caused by the antenna beams (Swenson and Mathur 1969) and is sometimes referred to as the *delay beam*. It can be used to confine the response of an interferometer to a limited area of the sky and thereby reduce the possibility of source confusion, which can occur when the fringe patterns of two or more sources are recorded simultaneously. Examples of such usage can be found in some early interferometers built for operation at frequencies below 100 MHz (Goldstein 1959; Douglas et al. 1973).

6.3.1   Imaging in the Continuum Mode

The effect of bandwidth on the fringe amplitude was discussed in Sect. 2.2. Equation (2.3) gives an expression for the fringes observed for a point source with an east–west baseline of length D and a rectangular signal passband of width $\Delta \nu$.

The fringe amplitude is proportional to a factor

$$R'_b = \frac{\sin(\pi Dl\Delta\nu/c)}{\pi Dl\Delta\nu/c} . \tag{6.69}$$

Consider an array for which D is typical of the longest baselines. The synthesized beamwidth of the array, θ_b, is approximately equal to $\lambda_0/D = c/\nu_0 D$, where ν_0 is the observing frequency and λ_0 the corresponding wavelength. (Note that in this section, ν_0 is the center frequency of the RF input band, not an IF band.) Thus, Eq. (6.69) becomes

$$R'_b \simeq \frac{\sin(\pi\Delta\nu l/\nu_0\theta_b)}{\pi\Delta\nu l/\nu_0\theta_b} . \tag{6.70}$$

The parameter $\Delta\nu l/\nu_0\theta_b$ is equal to the fractional bandwidth multiplied by the angular distance of the source from the (l, m) origin measured in beamwidths. If this parameter is equal to unity, $R'_b = 0$ and the measured visibility is reduced to zero. To keep R'_b close to unity, we require $\Delta\nu l/\nu_0\theta_b \ll 1$. Thus, to avoid underestimation of the visibility at long baselines, there is a limit on the angular size of the image that is inversely proportional to the fractional bandwidth.

We now examine the same effect in more detail by considering the distortion in the synthesized image. First, recall that the response of an array can be written as

$$'V(u, v)W(u, v) \longleftrightarrow I(l, m) ** b_0(l, m) , \tag{6.71}$$

where $\longleftrightarrow$ represents Fourier transformation, and the double asterisk represents convolution in two dimensions. The fringe visibility is multiplied by $W(u, v)$, the spatial sensitivity function of the array for a particular observation. The Fourier transform of the left side of Eq. (6.71) gives the intensity distribution $I(l, m)$ convolved with the synthesized beam function $b_0(l, m)$. For simplicity, we have omitted the primary antenna beam and minor effects related to use of the discrete Fourier transform. The synthesized beam is defined here as the Fourier transform of $W(u, v)$.

In operation in the continuum mode, the visibility data measured with bandwidth $\Delta\nu$ are treated as though they were measured with a single-frequency receiving system tuned to the center frequency, ν_0. Thus, for all frequencies within the bandwidth, the assigned values of u and v are those appropriate to frequency ν_0. At another frequency ν within the passband, the true spatial frequency coordinates u_ν and v_ν are related to the assigned values u and v by

$$(u, v) = \left(\frac{u_\nu \nu_0}{\nu}, \frac{v_\nu \nu_0}{\nu}\right) . \tag{6.72}$$

The contribution to the measured visibility from a narrow band of frequencies centered on ν is

$$\mathcal{V}_\nu(u, v) = \mathcal{V}_\nu\left(\frac{u_\nu \nu_0}{\nu}, \frac{v_\nu \nu_0}{\nu}\right) \longleftrightarrow \left(\frac{\nu}{\nu_0}\right)^2 I\left(\frac{l\nu}{\nu_0}, \frac{m\nu}{\nu_0}\right) , \tag{6.73}$$

where we have used the similarity theorem of Fourier transforms.[6] Thus, the contribution to the measured intensity is the true intensity distribution scaled in (l, m) by a factor ν/ν_0 and in intensity by $(\nu/\nu_0)^2$. The derived intensity distribution is convolved with $b_0(l, m)$, the synthesized beam corresponding to frequency ν_0. We have assumed that the beam does not vary significantly with frequency and have used the same spacial sensitivity function $W(u, v)$ to represent the whole frequency passband. The overall response is obtained by integrating over the passband with appropriate weighting and is

$$I_b(l, m) = \left[\frac{\displaystyle\int_0^\infty \left(\frac{\nu}{\nu_0}\right)^2 |H_{RF}(\nu)|^2 I\left(\frac{l\nu}{\nu_0}, \frac{m\nu}{\nu_0}\right) d\nu}{\displaystyle\int_0^\infty |H_{RF}(\nu)|^2 d\nu}\right] ** b_0(l, m) . \tag{6.74}$$

Note that the integrals must be taken over the whole radio-frequency passband, denoted by the subscript RF, which includes both sidebands in the case of a DSB system. We assume that the passband function $H_{RF}(\nu)$ is identical for all antennas. The values of l and m in the intensity function in Eq. (6.74) are multiplied by the factor ν/ν_0, which varies as we integrate over the passband, being equal to unity at the band center. Thus, one can envisage the integrals in the square brackets in Eq. (6.74) as resulting in a process of averaging a large number of images, each with a different scale factor. The scale factors are equal to ν/ν_0, and the range of values of ν is determined by the observing passband. The images are aligned at the origin, and thus the effect of the integration over frequency is to produce a radial smearing of the intensity distribution before it is convolved with the beam. The response to a point source at position (l, m) is radially elongated by a factor equal to $\sqrt{l^2 + m^2}\Delta\nu/\nu_0$. For distances from the origin at which the elongation is large compared with the synthesized beamwidth, features on the sky become attenuated by the smearing, so there is an effective limitation of the useful field of view. The measured intensity is the smeared distribution convolved with the synthesized beam.

Details of the behavior of the derived intensity distribution can be deduced from Eq. (6.74). For example, suppose that the beam contains a circularly symmetrical sidelobe at a large angular distance from the beam axis and that in an image, the response to a distant source causes the sidelobe to fall near the origin. Is the sidelobe broadened near the origin? Since the distant source is elongated, the sidelobe will be smeared in a direction parallel to that of a line joining the source and the origin, as

---

[6]If $f(x)$ has a Fourier transform $F(s)$, then $f(ax)$ has a Fourier transform $|a|^{-1}F(s/a)$ (Bracewell 2000).

Fig. 6.10 Radial smearing
resulting from the bandwidth
effect for a point source at
(l_1, m_1). The effects on the
responses of the main beam
and a ringlobe (i.e., a sidelobe
of the form in Fig. 5.15) are
shown.

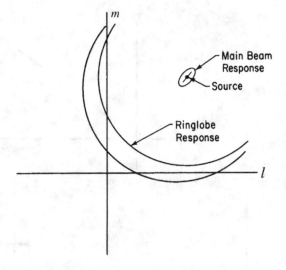

shown in Fig. 6.10. It will be broadened near the origin but not at a point 90° around
the sidelobe as measured from the source.

To estimate the magnitude of the suppression of distant sources, it is useful to
calculate R_b, the peak response to a point source at a distance r_1 from the origin
of the (l, m) plane, as a fraction of the response to the same source at the origin.
Because the effect we are considering is a radial smearing, we need consider only
the intensity along a radial line through the (l, m) origin, as shown in Fig. 6.11a. We
use idealized parameters; the passband is represented by a rectangular function of
width Δv and the synthesized beam by a circularly symmetrical Gaussian function
of standard deviation $\sigma_b = \theta_b/\sqrt{8 \ln 2}$, where θ_b is the half-power beamwidth.
For simplicity, the factor $(v/v_0)^2$ in the integral in the numerator of Eq. (6.74) is
omitted. The convolution becomes a one-dimensional (radial) process, as shown in
Fig. 6.11b. The radially elongated source is represented by a rectangular function
from $r_1(1 - \Delta v/2v_0)$ to $r_1(1 + \Delta v/2v_0)$, normalized to unit area. The beam is
represented by the function $e^{-r^2/2\sigma_b^2}$, which is normalized to unity on the beam axis.
When the beam is centered on the source, as shown in Fig. 6.11, R_b is given by

$$R_b = \frac{v_0}{r_1 \Delta v} \int_{r_1(1-\Delta v/2v_0)}^{r_1(1+\Delta v/2v_0)} e^{-(r-r_1)^2/2\sigma_b^2} \, dr$$

$$= \sqrt{2\pi} \frac{\sigma_b v_0}{r_1 \Delta v} \operatorname{erf}\left(\frac{r_1 \Delta v}{2\sqrt{2}\sigma_b v_0}\right)$$

$$= 1.0645 \frac{\theta_b v_0}{r_1 \Delta v} \operatorname{erf}\left(0.8326 \frac{r_1 \Delta v}{\theta_b v_0}\right). \tag{6.75}$$

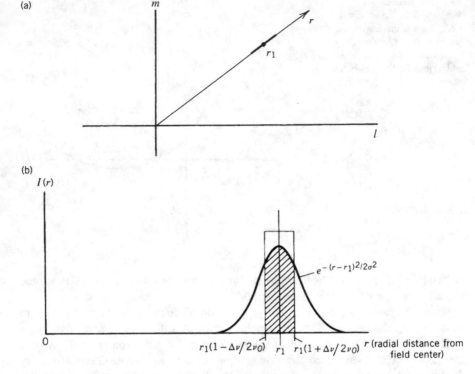

Fig. 6.11 Response of an array with a broadband receiving system to a point source at distance r_1 from the origin of the (l, m) plane. (**a**) The point source (delta function) at r_1 becomes radially broadened into a rectangular function of unit area indicated by the heavy line. (**b**) Cross section of the intensity distribution in the r direction. The synthesized beam is represented by the Gaussian function. The peak intensity of the response to the source is proportional to the shaded area.

A curve of R_b as a function of the parameter $r_1 \Delta\nu / \theta_b \nu_0$, which is the distance of the source from the origin measured in beamwidths, multiplied by the fractional bandwidth, is shown in Fig. 6.12. Values of 0.2 and 0.5 for this parameter reduce the response by 0.9% and 5.5%, respectively.

If the receiving passband is represented by a Gaussian function of equivalent width $\Delta\nu$ (i.e., standard deviation = $\Delta\nu/2.5066$), the reduction factor becomes

$$R_b = \frac{1}{\sqrt{1 + (0.939 r_1 \Delta\nu / \theta_b \nu_0)^2}} . \qquad (6.76)$$

A curve of this function is also included in Fig. 6.12. Comparison of the two curves illustrates the dependence on the passband shape.

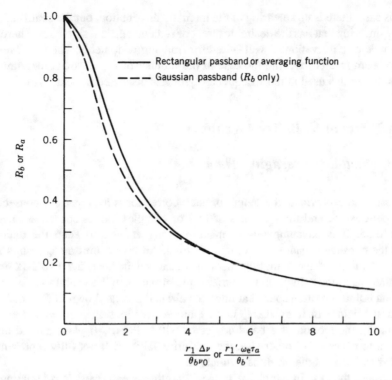

Fig. 6.12 Relative amplitude of the peak response to a point source as a function of the distance from the field center and either the fractional bandwidth or the averaging time.

6.3.2   Wide-Field Imaging with a Multichannel System

Broadband images can also be obtained by observing with a multichannel system (i.e., a spectral line system as described in Sect. 8.8.2). In this case, the passband is divided into a number of channels by using either a bank of narrowband filters or a multichannel digital correlator. The visibility is measured independently for each channel, so the values of u and v can be scaled correctly and an independent image obtained for each channel. This scaling causes the spatial sensitivity function to vary over the band, and at frequency v, the synthesized beam is $(v/v_0)^2 b_0(lv/v_0, mv/v_0)$, where $b_0(l, m)$ is the monochromatic beam at frequency v_0. The images can be combined by summation, and if given equal weights, the result for N channels is represented by

$$I(l, m) ** \left[\frac{1}{N} \sum_{i=1}^{N} \left(\frac{v_i}{v_0} \right)^2 b_0 \left(\frac{lv_i}{v_0}, \frac{mv_i}{v_0} \right) \right].$$ (6.77)

In this case, there is no smearing of the intensity distribution, but the beam suffers a radial smearing that has the desirable effect of reducing distant sidelobes. Therefore, this mode of observation is well suited for imaging wide fields. The improvement in the beam results from the increase in the number of (u, v) points measured, an effect that is also used in multifrequency synthesis discussed in Sect. 11.6.

6.4   Effect of Visibility Averaging

6.4.1   Visibility Averaging Time

In most synthesis arrays, the output of each correlator is averaged for consecutive time periods, τ_a, and thus consists of real or complex values spaced at intervals τ_a in time. It is advantageous to make τ_a long enough to keep the data rate from the correlator readout conveniently small. An upper limit on τ_a results from a consideration of the sampling theorem discussed in Sect. 5.2.1 and is briefly explained as follows. In discrete Fourier transformation of the visibility to intensity, the data points are often spaced at intervals Δu and Δv, as shown in Fig. 5.3. If the size of the field to be imaged is θ_f in the l and m directions, then $\Delta u = \Delta v = 1/\theta_f$. In time τ_a, the motion of a baseline vector within the (u, v) plane should not be allowed to exceed Δu; otherwise, the visibility values will not fully represent the angular variation of the brightness function.

Consider the case in which the longest baseline is east–west in orientation and the source under observation is at a high declination, which results in the fastest motion of the baseline vector. If the baseline length is D_λ wavelengths, the vector in the (u, v) plane traces out an approximately circular locus, the tip of which moves at a speed of $\omega_e D_\lambda$ wavelengths per unit time, where ω_e is the angular velocity of rotation of the Earth. Thus, we require that $\tau_a \omega_e D_\lambda < 1/\theta_f$, which results, in practice, in $\tau_a \approx C/(\omega_e D_\lambda \theta_f)$, where C is a factor likely to be in the range 0.1–0.5. Note that $D_\lambda \theta_f$ is approximately the number of synthesized beamwidths across the field, and thus τ_a must be somewhat smaller than the time taken for the Earth to rotate through one radian, divided by this number. Although shorter baselines could be averaged for longer times, in most synthesis arrays, all correlator outputs are read at the same time, at a rate appropriate for the longest baselines. Another consideration is that sporadic interference can be edited out of the data with minimal information loss if τ_a is not too long. For large arrays, τ_a is generally in the range of tens of milliseconds to tens of seconds. Determining the visibility at the $(\Delta u, \Delta v)$ grid points from the sampled data on the (u, v) loci is discussed in Sect. 10.2.3.

6.4.2   Effect of Time Averaging

We now examine in more detail the effect of the averaging on the synthesized intensity distribution. In reducing the data, all visibility values within each interval

τ_a are treated as though they applied to the time at the center of the averaging period. Thus, for example, the measurements at the beginning of each averaging period enter into the visibility data with assigned values of u and v that apply to times $\tau_a/2$ later than the true values. In effect, the resulting image consists of the average of a large number of images, each with a different timing offset distributed progressively throughout the range $-\tau_a/2$ to $\tau_a/2$. These timing offsets apply only to the assignment of (u, v) values and do not resemble a clock error that would affect the whole receiving system.

Consider an unresolved source, represented by a delta function. To simplify the situation, we consider observations with east–west baselines and examine the effects in the (u', v') plane and the corresponding (l', m') sky plane (see Sect. 4.2). The spacing loci are circular arcs generated by vectors rotating at angular velocity ω_e, as shown in Fig. 6.13a. Consider first the case of an east–west linear array; then, of the antenna spacing components (X, Y, Z) defined in Fig. 4.1, only Y is nonzero. The circular arcs of the spacing loci are centered on the (u', v') origin as in Fig. 6.13b, and a timing offset δt is equivalent to a rotation of the (u', v') axes through an angle $\omega_e \delta t$. The visibility of the source is the combination of two sets of sinusoidal corrugations, one real and one imaginary:

$$\delta(l'_1, m'_1) \longleftrightarrow \cos 2\pi(u'l'_1 + v'm'_1) - j \sin 2\pi(u'l'_1 + v'm'_1) . \qquad (6.78)$$

The angle of the corrugations is related to the position angle $\psi' = \tan^{-1}(m'_1/l'_1)$ of the point source, as shown in Fig. 6.14. A change in ψ' causes an equivalent rotation of the corrugations and vice versa. For an east–west array, time offsets therefore correspond to proportional rotations of the intensity in the (l', m') plane. It

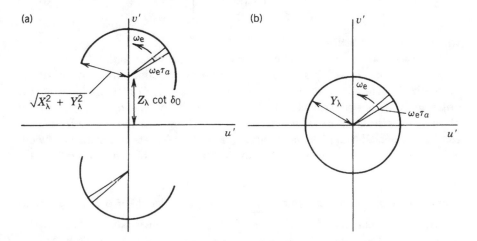

Fig. 6.13 Spacing loci in the (u', v') plane, (**a**) for the general case and (**b**) for an east–west baseline. The angle $\omega_e \tau_a$ over which the averaging takes place is enlarged for clarity: for example, with an averaging time of 30 s, the angle would be 7.5 arcmin.

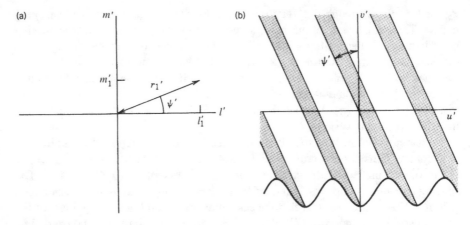

Fig. 6.14 (a) Point source at (l_1', m_1') and (b) the real part of the corresponding visibility function. The ridges of the sinusoidal corrugations that represent the visibility in the (u', v') plane are orthogonal to the radius vector r_1' at the position of the source in the (l', m') plane.

follows that the effect of the time averaging is to produce a circumferential smearing similar to that resulting from the receiving bandwidth but orthogonal to it. If we express positions in the (l', m') plane in terms of the radial coordinates (r', ψ') shown in Fig. 6.14a, the image obtained from the averaged data can be expressed in terms of the sky brightness $I(r', \psi')$ by

$$I_a(r', \psi') = \left[\frac{1}{\omega_e \tau_a} \int_{-\omega_e \tau_a/2}^{\omega_e \tau_a/2} I(r', \psi') \, d\psi' \right] ** b_0(r', \psi') , \qquad (6.79)$$

where b_0 is the synthesized beam.

The fractional decrease in the peak response to the point source is most easily considered in the (l', m') plane. With an east–west baseline, the contours of the synthesized beam are approximately circular in the (l', m') plane, as long as the observing time is approximately 12 h, which results in spacing loci in the form of complete circles in the (u', v') plane. If we assume that the synthesized beam can be represented by a Gaussian function, as in the calculations for the bandwidth effect, the curve for the rectangular bandwidth in Fig. 6.12 can also be used for the averaging effect. In one case, the spreading function is radial and of width $r_1 \Delta \nu / \nu_0$, and in the other, it is circumferential and of width $r_1' \omega_e \tau_a$. Thus, for the averaging effect, we can replace $r_1 \Delta \nu / \theta_b \nu_0$ in Eq. (6.75) and Fig. 6.12 (solid curve) by $r_1' \omega_e \tau_a / \theta_b'$, noting that $r_1' = \sqrt{l_1^2 + m_1^2 \sin^2 \delta_0}$ and θ_b', the synthesized beamwidth in the (l', m') plane, is equal to the east–west beamwidth in the (l, m) plane. Hence, for the decrease in the response to a point source resulting from averaging, using

Eq. (6.75), we can write

$$R_a = 1.0645 \frac{\theta'_b}{r'_1 \omega_e \tau_a} \mathrm{erf} \left(0.8326 \frac{r'_1 \omega_e \tau_a}{\theta'_b} \right) . \tag{6.80}$$

Generally, one chooses τ_a so that R_a is only slightly less than unity at any point in the image, in which case we can approximate the error function by the integral of the first two terms in the power series for a Gaussian function:

$$R_a \simeq 1 - \frac{1}{3} \left(\frac{0.8326 \omega_e \tau_a}{\theta'_b} \right)^2 (l_1^2 + m_1^2 \sin^2 \delta_0) . \tag{6.81}$$

This formula can be used for checking that τ_a is not too large.

Two aspects of the behavior predicted by Eq. (6.81) should be mentioned. First, if the source is near the m' axis and at a low declination, the averaging has very little effect. This is because the ridges of the sinusoidal corrugations of the visibility function then run approximately parallel to the u' axis, and in the transformation $u' = u \csc \delta_0$, the period of the variations in the v direction is expanded by a large factor. In comparison, the arc through which any baseline vector moves in time τ_a is small, and hence, the averaging has only a small effect on the visibility amplitude. Second, for a source on the l' axis, R_a is independent of δ_0. In this case, the ridges of the corrugations run parallel to the v axis, and the expansion of the scale in the v direction has no effect on the sinusoidal period.

For arrays that contain baselines other than east–west, the centers of the corresponding loci in the (u', v') plane are offset from the origin, as in Fig. 6.13a, and a time offset is no longer equivalent to a simple rotation of axes. However, this may not increase the smearing of the visibility, so the effect may be no worse than for an east–west array with baselines of similar lengths.

6.5   Speed of Surveying

The requirement to maximize the efficiency of use of large instruments requires consideration of the best procedures for surveying, i.e., searching large areas of the sky for radio sources of various types including transient sources. In the frequency range below about 2 GHz, four of the key science applications of the proposed Square Kilometre Array that require imaging of a significant fraction of the sky are as follows.

1. Searching for pulsars in binary combinations with neutron stars or black holes, for gravitational studies.
2. Measurement of Faraday rotation in very large numbers of radio galaxies to determine the structure of galactic and intergalactic magnetic fields.

3. Imaging of very large numbers of HI galaxies out to redshifts of $z \simeq 1.5$ to study galactic evolution and provide further constraints on the nature of dark energy.
4. Detection of transient events such as afterglows of gamma-ray bursts.

The choice of parameters for optimization of speed in survey observations is not the same as for optimization of sensitivity in targeted studies (Bregman 2005). Consider first the case of targeted observations of individual continuum sources that have angular dimensions small compared with a station[7] beam. We can adapt the expression for the rms noise [Eq. (6.62)] as a measure of the minimum detectable flux density S_{min} observed with two stations in a time τ:

$$ S_{min} = \frac{2kT_s}{A\sqrt{(\Delta\nu\tau)}} , \tag{6.82} $$

where A is here the collecting area of a station, equal to $A\eta_Q$ of Eq. (6.62). For an array with n_s stations, there are $n_s(n_s - 1)/2 \approx n_s^2/2$ correlated pairs of signals, so the right side of Eq. (6.82) is multiplied by a factor $\sqrt{2}/n_s$. The observing speed, i.e., the number of observations per unit time, is

$$ 1/\tau = \frac{A^2 \Delta\nu S_{min}^2 n_s^2}{8k^2 T_s^2} . \tag{6.83} $$

Next, consider the case of a survey in which we are concerned with the speed of coverage of a specified solid angle of sky down to a sensitivity level S_{min}. Since A is the area of a station, the solid angle of a station beam is λ^2/A sr, where λ is the wavelength. If each station forms n_{sb} simultaneous beams, the instantaneous field of view is

$$ F_v = \lambda^2 n_{sb}/A . \tag{6.84} $$

The reciprocal of the time τ required to cover solid angle F_v down to flux density level S_{min} is given by Eq. (6.83), and the corresponding survey speed is

$$ F_v/\tau = \frac{\lambda^2 A \Delta\nu S_{min}^2 n_s^2 n_{sb}}{8k^2 T_s^2} \quad \text{sr per unit time .} \tag{6.85} $$

For surveys to detect spectral line features, $\Delta\nu$ in Eqs. (6.83) and (6.85) represents the bandwidth of the line. Then, if it is necessary to search in frequency, the

---

[7]For an array with a large total collecting area, it may be more practical to use a large number of small antennas rather than a smaller number of large antennas. To limit the number of cross-correlated signal pairs, the small antennas are located in groups. These groups are commonly referred to as *stations*, and the signals from the antennas at a station are combined to provide a number of beams within the main beam of the individual antennas. For pairs of stations, the signals from corresponding beams are cross-correlated to provide the visibility data.

bandwidth of the receiving system can be included as an additional factor in the expression for the speed in Eq. (6.85).

Comparison of Eqs. (6.83) and (6.85) shows the effect of the field-of-view dependence in the survey case. The survey speed is proportional to the number of simultaneous beams and is less strongly dependent on the station aperture area A. The wavelength-squared factor results from the increased beamwidth with decreasing frequency, but the benefit of lower frequency (increasing λ) on the survey speed applies only so long as the effect of the galactic background radiation on the system temperature is small. From the galactic background model of Dulk et al. (2001), the brightness temperature in the range 10–1,000 MHz is approximately proportional to $\nu^{-2.5}$, so for frequencies at which this background is the dominant contributor to T_s, the frequency dependence of the survey speed is approximately proportional to ν^3. For directions that are not close to the galactic plane, the background temperature is about 20 K at 500 MHz and 2 K at 1 GHz, so if the receiver contribution to T_s is $\sim$ 20 K, there is a broad maximum in survey speed between these two frequencies.

Note that the discussion above involves the assumption that the sensitivity is limited only by the system noise. If dynamic range is the limiting factor, then the density of the (u, v) coverage, which improves with increasing n_s and τ, may become the most important consideration. In either case, performance improves with increasing number of stations.

Survey speed can be increased by increasing the number of stations as well as the number of station beams. However, the size of the correlator system for the full array is proportional to n_s^2 and to n_{sb}, so increasing the number of stations, or the number of station beams, requires an increase in the size of the correlator. Increasing the station aperture A is likely to require adding more antennas to the station subarray and thus increases the station beamforming hardware. The only way of increasing the observing speed that does not increase the signal-processing requirements is reducing the system temperature T_s. However, the complexity of phased-array feeds for the formation of multiple beams from a single parabolic antenna can degrade the system temperature. If cryogenic cooling in necessary, the required cooling capacity is considerably greater for multiple-beam systems than for single-beam ones. Thus, optimization of the array performance for a given overall cost requires a broad consideration of the performance of various parts of the receiving system.

Appendix 6.1   Partial Rejection of a Sideband

In an SSB system using a mixer as the input stage, the unwanted (image) sideband may be rejected by one of several schemes. These include use of a waveguide filter, a Martin–Puplett interferometer (Martin and Puplett 1969; Payne 1989), a tuned backshort, or a sideband-separating configuration of two mixers (as in Appendix 7.1). Practical considerations, particularly at millimeter wavelengths, can limit the effectiveness of the rejection of the image sideband. Let the response to the

Fig. A6.1 Vectors in the complex plane representing the parameters in Eqs. (A6.1) and (A6.2). The constant gain factor G_{mn} is omitted. If ρ is known and C_0 and $C_{\pi/2}$ are measured, Eq. (A6.3) gives the optimum estimate of $\mathcal{V}$.

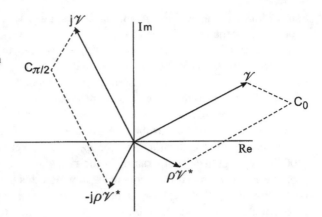

image sideband, in terms of the power gain of the receiver, be ρ times the response to the wanted (signal) sideband, where $0 < \rho < 1$. In practice, ρ could be as large as $\sim 1/10$.

In the case of spectral line observation, where the wanted line occurs only in the signal sideband, the effect of the noise introduced by the image sideband is to increase the rms noise at the correlator output by a factor $(1 + \rho)$. Thus, the sensitivity is correspondingly reduced (Fig. A6.1).

In the case of a continuum observation, the image sideband also introduces a component of signal and noise at the correlator. Assume that the visibility is the same in both sidebands, the fringes are stopped, and $\pi/2$ phase switching of the first LO allows measurement of the complex visibility. A complex correlator is used, and for simplicity, we consider that the instrumental phase is adjusted so that the line AB in Fig. 6.5b is coincident with the real axis. We can represent the complex correlator output with zero phase shift of the LO as

$$C_0 = G_{mn}(\mathcal{V} + \rho \mathcal{V}^*) , \tag{A6.1}$$

and with the $\pi/2$ phase switch as

$$C_{\pi/2} = G_{mn}(j\mathcal{V} - j\rho \mathcal{V}^*) . \tag{A6.2}$$

Here, G_{mn} is the gain in the signal sideband, so ρG_{mn} is the gain in the image sideband. Note that in the expression for $C_{\pi/2}$, the j factors have opposite signs for the two sidebands, because the $\pi/2$ phase shift causes the corresponding vectors in the complex plane to rotate through $\pi/2$ in opposite directions, as in Fig. A6.1. The optimum estimate of the visibility is then found to be

$$\mathcal{V} = \frac{1}{2G_{mn}} \left[\frac{1}{1+\rho^2}(C_0 - jC_{\pi/2}) + \frac{\rho}{1+\rho^2}(C_0 + jC_{\pi/2})^* \right] . \tag{A6.3}$$

The first term within the square brackets represents the response of the signal sideband, and the second term represents the response of the image. The total noise power delivered to the correlator input [i.e., the sum of the two terms in the square brackets in Eq. (A6.3)] is proportional to $(1 + \rho)$, so the noise associated with the first term in the square brackets is proportional to $(1 + \rho)/(1 + \rho^2)$. Similarly, for the second term, the associated noise is proportional to $\rho(1 + \rho)/(1 + \rho^2)$. Thus, the total noise in the estimate of $\mathcal{V}$ from Eq. (A6.3) is proportional to $(1 + \rho)^2/(1 + \rho^2)$. In terms of the rms noise, the sensitivity is proportional to the square root of the reciprocal of the last term, i.e., $\sqrt{(1 + \rho^2)}/(1 + \rho)$. For $\rho \approx 1/10$ or less, the ρ^2 term is very small, and the sensitivity degradation factor is approximately $(1 + \rho)^{-1}$ (Thompson and D'Addario 2000).

References

Abramowitz, M., and Stegun, I.A., Eds., *Handbook of Mathematical Functions*, National Bureau of Standards, Washington, DC (1968)

Anantharamaiah, K.R., Ekers, R.D., Radhakrishnan, V., Cornwell, T.J., and Goss, W.M., Noise in Images of Very Bright Sources, in *Synthesis Imaging In Radio Astronomy*, Perley, R.A., Schwab, F.R., and Bridle, A.H., Eds., Astron. Soc. Pacific Conf. Ser., **6**, 431–442 (1989)

Blum, E.J., Sensibilité des Radiotélescopes et Récepteurs à Corrélation, *Ann. Astrophys.*, **22**, 140–163 (1959)

Bracewell, R.N., *The Fourier Transform and Its Applications*, McGraw-Hill, New York (2000) (also see earlier eds. 1965, 1978)

Bregman, J.D., System Optimization of Multibeam Aperture Synthesis Arrays for Survey Performance, in *The Square Kilometre Array: An Engineering Perspective*, Hall, P.J., Ed., Springer, Dordrecht (2005)

Colvin, R.S., "A Study of Radio Astronomy Receivers," Ph.D. thesis, Stanford Univ., Stanford, CA (1961)

Douglas, J.N., Bash, F.N., Ghigo, F.D., Moseley, G.F., and Torrence, G.W., First Results from the Texas Interferometer: Positions of 605 Discrete Sources, *Astron. J.*, **78**, 1–17 (1973)

Dulk, G.A., Erickson, W.C., Manning, R., and Bougeret, J.-L., Calibration of Low-Frequency Radio Telescopes Using Galactic Background Radiation, *Astron. Astrophys.*, **365**, 294–300 (2001)

Goldstein, S.J., Jr., The Angular Size of Short-Lived Solar Radio Disturbances, *Astrophys. J.*, **130**, 393–399 (1959)

Lawson, J.L., and Uhlenbeck, G.E., *Threshold Signals*, Radiation Laboratory Series, Vol. 24, McGraw-Hill, New York (1950), p. 68

Lo, W.F., Dewdney, P.E., Landecker, T.L., Routledge, D., and Vaneldik, J.F., A Cross-Correlation Receiver for Radio Astronomy Employing Quadrature Channel Generation by Computed Hilbert Transform, *Radio Sci.*, **19**, 1413–1421 (1984)

Martin, D.H., and Puplett, E., Polarized Interferometric Spectrometry for the Millimeter and Submillimeter Spectrum, *Infrared Phys.*, **10**, 105–109 (1969)

Mickelson, R.L., and Swenson, G.W., Jr., A Comparison of Two Correlation Schemes, *IEEE Trans. Instrum. Meas.*, **IM-40**, 816–819 (1991)

Middleton, D., *An Introduction to Statistical Communication Theory*, McGraw-Hill, New York (1960), p. 343

Moran, J.M., Very Long Baseline Interferometric Observations and Data Reduction, in *Methods of Experimental Physics*, Vol. 12, Part C *(Astrophysics: Radio Observations)*, Meeks, M. L., Ed., Academic Press, New York (1976), pp. 228–260

Papoulis, A., *Probability, Random Variables and Stochastic Processes*, McGraw-Hill, New York (1965)

Payne, J.M., Millimeter and Submillimeter Wavelength Radio Astronomy, *Proc. IEEE*, **77**, 993–1017 (1989)

Read, R.B., Two-Element Interferometer for Accurate Position Determinations at 960 Mc, *IRE Trans. Antennas Propag.*, **AP-9**, 31–35 (1961)

Rice, S.O., Mathematical Analysis of Random Noise, *Bell Syst. Tech. J.*, **23**, 282–332 (1944); **24**, 46–156, 1945 (repr. in *Noise and Stochastic Processes*, Wax, N., Ed., Dover, NY, 1954)

Rogers, A.E.E., Theory of Two-Element Interferometers, in *Methods of Experimental Physics*, Vol. 12, Part C *(Astrophysics: Radio Observations)*, Meeks, M.L., Ed., Academic Press, New York (1976), pp. 139–157 (see Table 1)

Swenson, G.W., Jr., and Mathur, N.C., On the Space-Frequency Equivalence of a Correlator Interferometer, *Radio Sci.*, **4**, 69–71 (1969)

Thompson, A.R., and D'Addario, L.R., Relative Sensitivity of Double- and Single-Sideband Systems for Both Total Power and Interferometry, ALMA Memo 304, National Radio Astronomy Observatory (2000)

Tiuri, M.E., Radio Astronomy Receivers, *IEEE Trans. Antennas Propag.*, **AP-12**, 930–938 (1964)

Tucker, J.R., and Feldman, M.J., Quantum Detection at Millimeter Wavelengths, *Rev. Mod. Phys.*, **57**, 1055–1113 (1985)

Vander Vorst, A.S., and Colvin, R.S., The Use of Degenerate Parametric Amplifiers in Interferometry, *IEEE Trans. Antennas Propag.*, **AP-14**, 667–668 (1966)

M. Vinokur, Optimisation dans la Recherche d'une Sinusoide de Période Connue en Présence de Bruit, *Ann. d'Astrophys.*, **28**, 412–445 (1965)

Wozencraft, J.M., and Jacobs, I.M., *Principles of Communication Engineering*, Wiley, New York (1965) p. 205

Wright, M.C.H., Clark, B.G., Moore, C.H., and Coe, J., Hydrogen-Line Aperture Synthesis at the National Radio Astronomy Observatory: Techniques and Data Reduction, *Radio Sci.*, **8**, 763–773 (1973)

Chapter 7
System Design

In this chapter, we consider certain aspects of the design of the interferometric system in more detail. This discussion primarily involves parts of the system where the signals are in analog form. The trend in technology has been to convert signals as early as possible in the signal chain, following the antennas, into digital form to facilitate data handling, avoid low-level distortions, and generally take advantage of the rapid progress in the development of digital equipment and computers. Three key items are discussed: (1) low noise amplification of signals at the antenna output to minimize the effect of additive noise, (2) phase-stable transmission systems that allow the transfer of reference timing and phase signals from the central communications hub of the instrument to the antennas, and (3) the synchronous phase switching systems needed to eliminate spurious responses in the correlator output. The analysis here leads to specification of tolerances on system parameters that are necessary to achieve the goals of sensitivity and accuracy.

7.1 Principal Subsystems of the Receiving Electronics

Optimum techniques and components for implementation of the electronic hardware vary continuously as the state of the art advances, and descriptions in the literature provide examples of the practical techniques current at various times: see, for example, Read (1961), Elsmore et al. (1966), Baars et al. (1973), Bracewell et al. (1973), Wright et al. (1973), Welch et al. (1977, 1996), Thompson et al. (1980), Batty et al. (1982), Erickson et al. (1982), Napier et al. (1983), Sinclair et al. (1992), Young et al. (1992), Napier et al. (1994), de Vos et al. (2009), Perley et al. (2009), Wootten and Thompson (2009), and Prabu et al. (2015). The earlier papers in this list are mainly of interest from the viewpoint of the development of the technology.

© The Author(s) 2017
A.R. Thompson, J.M. Moran, and G.W. Swenson Jr., *Interferometry and Synthesis in Radio Astronomy*, Astronomy and Astrophysics Library,
DOI 10.1007/978-3-319-44431-4_7

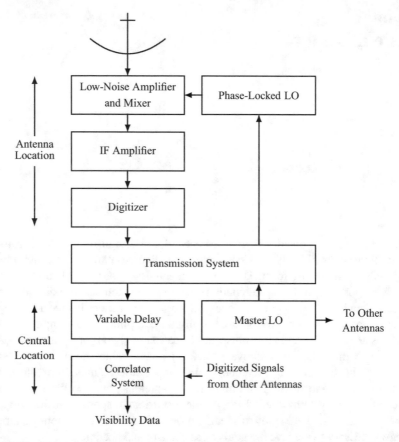

Fig. 7.1 Basic elements of the receiving system of a synthesis array. Here, the received signals are converted to an intermediate frequency (IF), digitized, and then transmitted by optical fiber to the central location for the derivation of visibility data. In systems of earlier design, and some smaller systems, the IF signal is transmitted to the central location in analog form and then digitized. LO indicates a local oscillator (i.e., usually one within the receiving system).

Figure 7.1 shows a simplified schematic diagram of the receiving system associated with one antenna of a synthesis array. Note that digitization of the signals is introduced as early as possible in the system, thus allowing most of the signal processing to be implemented digitally. In very early interferometers, there was no digitization, and the output was displayed on a chart recorder. In the original VLA system, the digitization occurred at the central location just before the delay and correlator processing. In the later VLA system (Perley et al. 2009), the signals are digitized at the antenna stations.

7.1.1 Low-Noise Input Stages

In radio astronomy receivers, minimizing the noise temperature usually involves cryogenic cooling of the amplifier or mixer stages from the input up to a point at which noise from succeeding stages is unimportant. The low-noise input stages are often packaged with a cooling system, and sometimes also a feed horn, in a single package often referred to as the *front end*. The active components are usually transistor amplifiers or, for millimeter wavelengths, SIS (superconductor–insulator–superconductor) mixers followed by transistor amplifiers. For descriptions, see, for example, Reid et al. (1973), Weinreb et al. (1977a), Weinreb et al. (1982), Casse et al. (1982), Phillips and Woody (1982), Tiuri and Räisänen (1986), Payne (1989), Phillips (1994), Payne et al. (1994), Webber and Pospieszalski (2002), and Pospieszalski (2005).

In discussing the level of noise associated with a receiver, we begin by considering the case in which the Rayleigh–Jeans approximation suffices. This is the domain in which $h\nu/kT \ll 1$, where h is Planck's constant and T is the temperature of the thermal noise source involved. As noted in the discussion following Eq. (1.1), this condition can be written as ν (GHz) $\ll 20T$, where T is the system noise temperature in kelvins. It is convenient to specify noise power in terms of the temperature of a resistive load matched to the receiver input. In the Rayleigh–Jeans approximation, noise power available at the terminals of a resistor at temperature T is $kT\Delta\nu$, where k is Boltzmann's constant and $\Delta\nu$ is the bandwidth within which the noise is measured (Nyquist 1928). One kelvin of temperature represents a power spectral density of $(1/k)$ W Hz^{-1}. The receiver temperature T_R is a measure of the internally generated noise power within the system and is equal to the temperature of a matched resistor at the input of a hypothetical noise-free (but otherwise identical) receiver that would produce the same noise power at the output. The system temperature, T_S, is a measure of the total noise level and includes, in addition to T_R, the noise power from the antenna and any lossy components between the antenna and the receiver:

$$T_S = T'_A + (L - 1)T_L + LT_R , \qquad (7.1)$$

where T'_A is the antenna temperature resulting from the atmosphere and other unwanted sources of noise, L is the power loss factor of the transmission line from the antenna to the receiver [defined as (power in)/(power out)], and T_L is the temperature of the line. In defining the noise temperature of the receiver, we should note that in practice, a receiver is always used with the input attached to some source impedance that is itself a source of noise. The noise at the receiver output thus consists of two components, the noise from the source at the input, which is the antenna and transmission line in Eq. (7.1), and the noise generated within the receiver.

7.1.2 Noise Temperature Measurement

The noise temperature of a receiver is often measured by the Y-factor method. The thermal noise sources used in this measurement are usually impedance-matched

resistive loads connected to the receiver input by waveguide or coaxial line. The
receiver input is connected sequentially to two loads at temperatures T_{hot} and T_{cold}.
The measured ratio of the receiver output powers in these two conditions is the
factor Y:

$$Y = \frac{T_R + T_{hot}}{T_R + T_{cold}}, \tag{7.2}$$

and thus,

$$T_R = \frac{T_{hot} - Y T_{cold}}{Y - 1}. \tag{7.3}$$

Commonly used values are $T_{hot} = 290$ K (ambient temperature) and $T_{cold} \simeq 77$ K
(liquid nitrogen temperature). For very precise measurements of T_R, it is important
to note that the boiling point of liquid nitrogen depends on the ambient pressure.

The receiver temperature can be expressed in terms of the noise temperatures of
successive stages through which the signal flows [see, e.g., Kraus (1986)]:

$$T_R = T_{R1} + T_{R2} G_1^{-1} + T_{R3} (G_1 G_2)^{-1} + \cdots . \tag{7.4}$$

Here T_{Ri} is the noise temperature of the ith receiver stage, and G_i is its power gain.
If the first stage is a mixer instead of an amplifier, G_1 may be less than unity, and
the second-stage noise temperature then becomes very important.

For cryogenically cooled receivers for millimeter and shorter wavelengths, the
Rayleigh–Jeans approximation can introduce significant errors. The power spectral
density (power per unit bandwidth) of the noise is no longer linearly proportional
to the temperature of the radiator or source. The ratio h/k is equal to 0.048 K
per gigahertz, so if, for example, $T = 4$ K (liquid helium temperature), then
$h\nu/kT = 1$ for $\nu = 83$ GHz. Thus, quantum effects become important as frequency
is increased and temperature decreased. Under these conditions, the noise power
per unit bandwidth divided by k provides an effective noise temperature that can be
used in noise calculations, instead of the physical temperature. Two formulas are in
use that give the effective temperature for a thermal source when quantum effects
become important. One is the Planck formula and the other the Callen and Welton
formula (Callen and Welton 1951). The effective noise temperatures for a waveguide
carrying a single mode and terminated in a thermal load, or for a transmission line
terminated in a resistive load, given by the two formulas are as follows:

$$T_{Planck} = T \left[\frac{\frac{h\nu}{kT}}{e^{h\nu/kT} - 1} \right] \tag{7.5}$$

$$T_{C\&W} = T \left[\frac{\frac{h\nu}{kT}}{e^{h\nu/kT} - 1} \right] + \frac{h\nu}{2k}, \tag{7.6}$$

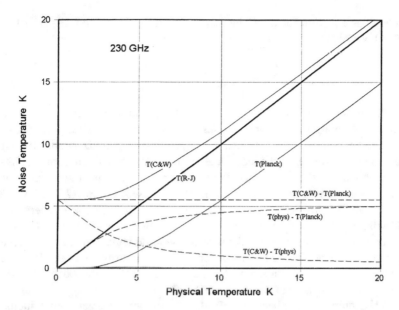

Fig. 7.2 Noise temperature vs. physical temperature for blackbody radiators at 230 GHz, according to the Rayleigh–Jeans, Planck, and Callen and Welton formulas. Also shown (broken lines) are the differences between the three radiation curves. The Rayleigh–Jeans curve converges with the Callen and Welton curve at high temperature, while the Planck curve is always $h\nu/2k$ below the Callen and Welton curve. From Kerr et al. (1997).

where T is the physical temperature. From Eqs. (7.5) and (7.6), we have

$$T_{C\&W} = T_{Planck} + \frac{h\nu}{2k}. \tag{7.7}$$

The Callen and Welton formula is equal to the Planck formula with an additional term, $h\nu/2k$, which represents an additional half-photon. This half-photon is the noise level from a body at absolute zero temperature and is referred to as the zero-point fluctuation noise. Figure 7.2 shows the relationships between physical temperature and noise temperature corresponding to the Rayleigh–Jeans, Planck, and Callen and Welton formulas, for a frequency of 230 GHz. Note that for the case of $h\nu/kT \ll 1$, we can put $\exp(h\nu/kT) - 1 \simeq (h\nu/kT) + \frac{1}{2}(h\nu/kT)^2$, in which case the Callen and Welton formula reduces to the Rayleigh–Jeans formula, but the result from the Planck formula is lower by $h\nu/2k$.

When using Eq. (7.3) to derive the noise temperature of a receiver, the values of T_{hot} and T_{cold} should be the noise temperatures derived from the Planck or Callen and Welton formulas, not the physical temperatures of the loads (except in the Rayleigh–Jeans domain). Thus, for the Planck formula, we can write

$$T_{R(Planck)} = \frac{T_{hot(Planck)} - YT_{cold(Planck)}}{Y - 1} \tag{7.8}$$

and a similar equation for the Callen and Welton formula. From Eqs. (7.4), (7.5), and (7.6), we obtain

$$T_{R(\text{Planck})} = T_{R(\text{C\&W})} + \frac{h\nu}{2k} \, . \tag{7.9}$$

In using any measurement of receiver noise temperature, it is important to know whether, in deriving it, the Planck formula, the Callen and Welton formula, or the physical temperature of the loads (i.e., the Rayleigh–Jeans approximation) was used. If the noise temperatures of the individual components are derived from the physical temperatures using the Callen and Welton formula, the temperature sum will be greater by $h\nu/2k$ than if the Planck formula were used; see Eq. (7.7). However, if the Callen and Welton formula is used to derive the receiver noise temperature, the result will be less by $h/2k$ than if the Planck formula were used; see Eq. (7.9). Thus the system temperature, which is the sum of the input temperature and the receiver temperature, will be the same whichever of the two formulas is used. However, to avoid confusion, it is important to use one formula or the other consistently throughout the derivation of the noise temperatures.

Differing opinions have been expressed on the nature of the zero-point fluctuation noise, and whether it should be considered as originating in the load connected to the receiver or in the receiver input stages; see, for example, Tucker and Feldman (1985), Zorin (1985), and Wengler and Woody (1987). At frequencies at which quantum effects become most important, the usual type of input stage in radio astronomy receivers is the SIS mixer, for which the quantum theory of operation is given by Tucker (1979). For a summary from various authors of some conclusions relevant to noise temperature considerations, see Kerr et al. (1997) and Kerr (1999).

To recapitulate: The radiation level predicted by the Callen and Welton formula is equal to the Planck radiation level plus the zero-point fluctuation component $h\nu/2$. The latter component is attributable to the power from a blackbody or matched resistive load at absolute zero temperature. An amplifier noise temperature derived using the Callen and Welton formula to interpret the measured Y factor is lower than that derived using the Planck formula by $h\nu/2k$. However, an antenna temperature obtained using the Callen and Welton formula is higher by $h\nu/2k$ than the corresponding Planck formula value. The system temperature, which is the sum of the noise temperature and the antenna temperature, is the same in either case. Since the system temperature determines the sensitivity of a radio telescope, these details may seem unimportant. However, in procuring an amplifier or mixer for a receiver input stage, it is important to know how the noise temperature is specified.

In addition to the noise generated in the electronics, the noise in a receiving system contains components that enter from the antenna. These components arise from cosmic sources, the cosmic background radiation, the Earth's atmosphere, the ground, and other objects in the sidelobes of the antenna. The opacity of the atmosphere, from which the atmospheric contribution to the system noise arises, is discussed in Chap. 13.

7.1.3   Local Oscillator

As explained in the previous chapter, local oscillator (LO) signals are required at the antenna locations and sometimes at other points along the signal paths to the correlators. The corresponding oscillator frequencies for different antennas must be maintained in phase synchronism to preserve the coherence of the signals. The phases of the oscillators at corresponding points on different antennas need not be identical, but the differences should be stable enough to permit calibration. Maintaining synchronism at different antennas requires transmitting one or more reference frequencies from a central master oscillator to the required points, where they may be used to phase-lock other oscillators. The frequencies required at the mixers can then be synthesized.

Special phase shifts are required at certain mixers to implement fringe rotation (fringe stopping), as described in Sect. 6.1, and to implement phase switching, as described in Sect. 7.5. Often these can best be synthesized by digital techniques, which can provide a signal at a frequency of, say, a few megahertz that contains the required frequency offsets and phase changes. These can be transferred to the LO frequency by using the synthesized signal as a reference frequency in a phase-locked loop.

7.1.4   IF and Signal Transmission Subsystems

After amplification in the low-noise front-end stages, the signals pass through various IF amplifiers and a transmission system before reaching the correlators. Transmission between the antennas and a central location can be effected by means of coaxial or parallel-wire lines, waveguide, optical fibers, or direct radiation by microwave radio link. Cables are often used for small distances, but for long distances, the cable attenuation may require the use of too many line amplifiers, and optical fiber, for which the transmission loss is much lower, is generally preferred. Low-loss TE_{01}-mode waveguide (Weinreb 1977b; Archer et al. 1980) was used in the construction of the original VLA system, which preceded the development of optical fiber by a few years. Optical fiber is now used for the Very Large Array (VLA)[1] (Perley et al. 2009). Cable or optical fiber can be buried at depths of 1–2 m to reduce temperature variations. Bandwidths of signals transmitted by cables are usually limited to some tens or hundreds of megahertz by attenuation, and radio links are similarly limited by available frequency allocations. For very wide bandwidths, optical fibers offer the greatest possibilities.

In the (mostly earlier) systems in which the signals are transmitted from the antennas to the central location in analog form, phase errors resulting from

---

[1] With the upgraded receiving system.

temperature effects in filters, and delay-setting errors, can be minimized by using the lowest possible intermediate frequency (IF) at this point. Accordingly, the final IF amplifiers may have a *baseband* response defined by a lowpass filter.[2] The response at the low-frequency end falls off at a frequency that is a few percent of the upper cutoff frequency.

7.1.5   Optical Fiber Transmission

The introduction of optical fiber systems provided a very great advance in transmission capability for broadband signals over long distances. Signals are modulated onto optical carriers, commonly in the wavelength range 1300–1550 nm, and transmitted along glass fiber. The fiber attenuation is a minimum of approximately 0.2 dB km^{-1} near 1550 nm and is about 0.4 dB km^{-1} at 1300 nm. These values are much lower than can be obtained in radio frequency transmission lines. In the fiber, a glass core is surrounded by a glass cladding of lower refractive index, so light waves launched into the core at a small enough angle with respect to the axis of the fiber can propagate by total internal reflection. If the inner-core diameter is approximately 50 μm, a number of different modes can be supported. These modes travel with slightly different velocities, which results in a limitation in performance of this multimode fiber. If the core is reduced to approximately 10 μm in diameter, only the HE_{11} mode propagates. Single-mode fiber of this type is required for the longest distances and/or the highest frequencies and bandwidths. At 1550 nm, an interval of 1 nm in wavelength corresponds to a bandwidth of approximately 125 GHz. The low attenuation and the bandwidth capacity facilitate the use of wide bandwidths and long baselines in linked-element arrays. Signals can be transmitted in analog form or digitized and transmitted as pulse trains. Design of a fiber transmission system involves the characteristics of the lasers that generate the optical carriers and the detectors that recover the modulation, as well as the characteristics of the fiber. For further information, see, for example, Agrawal (1992), Borella et al. (1997), and Perley et al. (2009).

In practice, the bandwidth and distance of the transmission are limited by the noise in the laser that generates the optical signal at the transmitting end of the fiber, and the noise in the diode demodulator and the amplifier at the receiving end. To avoid degradation of the sensitivity in analog transmission, the power spectral density of the signal (measured in W Hz^{-1}) must be greater than the power spectral density of the noise generated in the transmission system by $\sim$ 20 dB for most radio astronomy applications. However, the total signal power is limited by the need to avoid nonlinearity of the response of the modulator or demodulator. The result is a limit on the bandwidth of the signal, since for signals with a flat spectrum, the power

---

[2]In some cases, an image rejection mixer (see Appendix 7.1) is used for the conversion to baseband, but the suppression of the unwanted sideband may then be no greater than 20–30 dB.

is proportional to the bandwidth. In practice, a single transmitter and receiver pair can operate with a bandwidth of 10–20 GHz for transmission distances of some tens of kilometers. Optical amplifiers, which most commonly operate at wavelengths near 1550 nm, can be used to increase the range of transmission.

In the modulation process, the *power* of the carrier is varied in proportion to the *voltage* of the signal. Because of this, the effect of small unwanted components in fiber transmission systems is greatly reduced. Consider, for example, a small component of the optical signal resulting from a reflection within the fiber. If the optical power of the reflected component is x dB less than that of the main component, then after demodulation at the photodetector, the signal power contributed by the reflected component is $2x$ dB less than that from the main optical component. This also applies to small unwanted effects resulting from finite isolation of couplers, isolators, and other elements. Variations in the frequency response resulting from standing waves in microwave transmission lines are significantly less in optical fiber than in cable.

A feature that must be taken into account in applications of optical fiber is the dispersion in velocity, $\mathcal{D}$, usually specified in $ps(nm \cdot km)^{-1}$. The difference in the time of propagation for two optical wavelengths that differ by $\Delta\lambda$ traveling a distance ℓ in the fiber is $\mathcal{D}\Delta\lambda\ell$. Figure 7.3 shows the dispersion for two types of fiber. Curve 1 is for a type of fiber widely used in early applications, and curve 2 represents a design in which the zero-dispersion wavelength is shifted to coincide approximately with the minimum-attenuation wavelength of 1550 nm. This optimization of the performance at 1550 nm is achieved by designing the fiber so that the dispersion of the cylindrical waveguide formed by the core of the fiber cancels the intrinsic dispersion of the glass at that wavelength.

Consider a spectral component, at frequency v_m, of a broadband signal that is modulated onto an optical carrier. Amplitude modulation of the signal results in sidebands spaced $\pm v_m$ in frequency with respect to the carrier. Because of the velocity dispersion, the two sidebands and the carrier each propagate down the fiber with slightly different velocities and thus exhibit relative offsets in time at the receiving end. Such time offsets result in attenuation of the amplitude of the high-frequency components of analog signals and in broadening of the pulses used to represent digital data. Thus, for both analog and digital transmission, dispersion

Fig. 7.3 Dispersion $\mathcal{D}$ in single-mode optical fiber of two different designs, as a function of the optical wavelength.

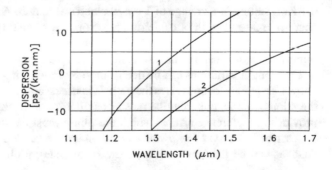

as well as noise can limit the bandwidth × distance product. An analysis of the effect of dispersion on analog signals is given in Appendix 7.2.

7.1.6   Delay and Correlator Subsystems

The compensating delays and correlators can be implemented by either analog or digital techniques. An analog delay system may consist of a series of switchable delay units with a binary sequence of values in which the delay of the nth unit is $2^{n-1}\tau_0$, where τ_0 is the delay of the smallest unit. Such an arrangement, with N units, provides a range of delay from zero to $(2^N - 1)\tau_0$ in steps of τ_0. For delays up to about 1 μs, lengths of coaxial cable or optical fiber have been used. The design of analog multiplying circuits for correlators has been discussed by Allen and Frater (1970). An example of a broadband analog correlator is described by Padin (1994). However, the development of digital circuitry capable of operating at high clock frequencies has led to the general practice of digitizing the IF signal so that the delay and correlators are generally implemented digitally, as discussed in Chap. 8.

7.2   Local Oscillator and General Considerations of Phase Stability

7.2.1   Round-Trip Phase Measurement Schemes

Synchronizing of the oscillators at the antennas can be accomplished by phase-locking them to a reference frequency that is transmitted out from a central master oscillator. Buried cables or fibers offer the advantage of the greatest stability of the transmission path. At a depth of 1–2 m, the diurnal temperature variation is almost entirely eliminated, but the annual variation is typically attenuated by a factor of 2–10 only. For a discussion of temperature variation in soil as a function of depth, see Valley (1965). As an example, a 10-km-long buried cable with a temperature coefficient of length of 10^{-5} K^{-1} might suffer a diurnal temperature variation of 0.1 K, resulting in a change of 1 cm in electrical length. A similar variation would occur in a 50-m length of cable running from the ground to the receiver enclosure on an antenna and subjected to a diurnal temperature variation of 20 K. Rotating joints and flexible cables can also contribute to phase variations.

Path length variations can be determined by monitoring the phase of a signal of known frequency that traverses the path. It is necessary for the signal to travel in two directions, that is, out from the master oscillator and back again, since the master provides the reference against which the phase must be measured. This technique is described as *round-trip phase measurement*. Correction for the measured phase changes can be implemented in hardware by using a phase shifter driven by the measurement system, or in software by inserting corrections in the data from the correlator, either in real time or during the later stages of data analysis. It is also possible to generate a signal in which the phase changes are greatly reduced by combining signals that travel in opposite directions in the transmission line. As an illustration of the last procedure, consider a signal applied to the near end of a loss-free transmission line that results in a voltage $V_0 \cos(2\pi \nu t)$ at the far end. At a distance ℓ, measured back from the far end, the outgoing signal is $V_1 = V_0 \cos 2\pi \nu (t + \ell/v)$, where v is the phase velocity along the line. Suppose that the signal is reflected from the far end without change in phase. At the same point, distant ℓ from the far end, the returned signal is $V_2 = V_0 \cos 2\pi \nu (t - \ell/v)$, and the total signal voltage is

$$V_1 + V_2 = 2V_0 \cos (2\pi \nu t) \cos \left(\frac{2\pi \nu \ell}{v} \right) . \qquad (7.10)$$

The first cosine function in Eq. (7.10) represents the radio frequency signal, the phase of which (modulo π) is independent of ℓ and of line length variations. The second cosine function is a standing-wave amplitude term. Such a system cannot easily be implemented in practice because of attenuation and unwanted reflections, and thus more complicated schemes have evolved. In what follows, we consider cable transmission, although the basic principles are applicable to other systems. Some general considerations, including the use of microwave links, are given by Thompson et al. (1968).

7.2.2   Swarup and Yang System

Several different round-trip schemes have been devised as instruments have developed, and one of the earliest of these was by Swarup and Yang (1961). A system based on this scheme is shown in Fig. 7.4. Part of the outgoing signal is reflected from a known reflection point at an antenna, and variation in the path length to the reflector is monitored by measuring the relative phase of the reflected component at the detector. The phase of the reflected signal is compared with that of a reference signal. The phase of the latter is variable by means of a movable probe that samples the outgoing signal. Since many other reflections may occur in the transmission line, it is necessary to identify the desired component. To do this, a modulated reflector, for example, a diode loosely coupled to the line, is used. This is switched between conducting and nonconducting states by a square wave voltage, and a synchronous detector is used to separate the modulated component of the reflected signal.

An increase $\Delta\ell$ in the length of the transmission line is detected as a corresponding movement of $2\Delta\ell$ in the probe position for the null. It results in an increase of $2\pi \Delta\ell \nu_1/v$ in the phase of the frequency ν_1 at the antenna, where v is the phase velocity in the line. The corresponding changes in LO phases and IF

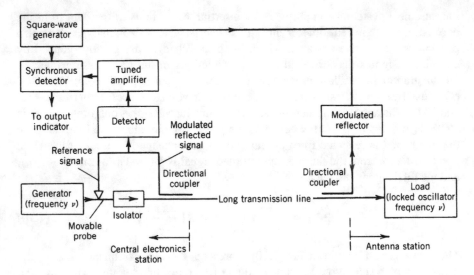

Fig. 7.4 System for measuring variations in the electrical path length in a transmission line, based on the technique of Swarup and Yang (1961). The output of the synchronous detector is a sinusoidal function of the difference between the phases of the reference (outgoing) and reflected components at the detector. A null output is obtained when these signal phases are in quadrature, and the position of the probe for a null is thus a measure of the phase of the reflected signal. Because of the isolator in the line, the probe samples only the outgoing component of the signal.

phases transmitted over the same path can be calculated and applied as a correction to the visibility phases.

7.2.3   *Frequency-Offset Round-Trip System*

A second scheme, shown in Fig. 7.5, is one in which the round-trip phase is measured directly. The signals traveling in opposite directions are at frequencies ν_1 and ν_2 that differ by only a small amount, but enough to enable them to be separated easily. This type of system is widely used, and we examine its performance in some detail. Note that although directional couplers or circulators allow the signals at the same frequency but going in opposite directions in the line to be separated, the signal from the unwanted direction is suppressed by only 20–30 dB relative to the wanted one. An unwanted component at a level of −30 dB can cause a phase error of 1.8°. However, the frequency offset enables the signals to be separated with much higher isolation.

An oscillator at frequency ν_2 at an antenna is phase-locked to the difference frequency of signals at ν_1 and $\nu_1 - \nu_2$, which travel to the antenna via a transmission line. The difference frequency $\nu_1 - \nu_2$ is small compared with ν_1 and ν_2. The

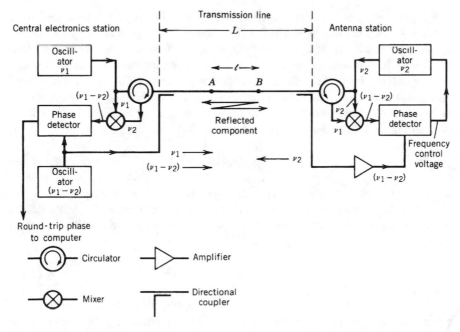

Fig. 7.5 Phase-lock scheme for the oscillator ν_2 at the antenna. Frequencies ν_1 and $\nu_1 - \nu_2$ are transmitted to the antenna station where they provide the phase reference to lock the oscillator. ν_1 and ν_2 are almost equal, so $\nu_1 - \nu_2$ is small. A signal at frequency ν_2 is returned to the central station for the round-trip phase measurement.

frequency ν_2 is returned to the master oscillator location for the round-trip phase comparison.

At the antenna, the phases of the signals at frequencies ν_1 and $\nu_1 - \nu_2$ relative to their phases at the central location are $2\pi \nu_1 L/v$ and $2\pi(\nu_1 - \nu_2)L/v$, where L is the length of the cable. The phase of the ν_2 oscillator at the antenna is constrained by a phase-locked loop to equal the difference of these phases, that is, $2\pi \nu_2 L/v$. The phase change in the ν_2 signal in traveling back to the central location is $2\pi \nu_2 L/v$, and thus the measured round-trip phase (modulo 2π) is $4\pi \nu_2 L/v$. Now suppose that the length of the line changes by a small fraction, β. The phase of the oscillator ν_2 at the antenna relative to the master oscillator changes to $2\pi \nu_2 L(1 + \beta)/v$. The required correction to the ν_2 oscillator is just half the change in the measured round-trip phase. The problem that arises is that several effects, including reflections and velocity dispersion in the transmission line, can cause errors in the round-trip phase. Such errors result in phase offsets of the oscillator at the antenna, which is not serious if the offsets remains constant. However, in practice, it is likely to vary with ambient temperature. The largest error usually results from reflections, and control of this error places an upper limit on the difference frequency $\nu_1 - \nu_2$. We now examine this limit.

Consider what happens if reflections occur at points A and B separated by a distance ℓ along the line as in Fig. 7.5. The complex voltage reflection coefficients at these points are ρ_A and ρ_B, and their values will be assumed to be the same at frequencies ν_1 and ν_2. Signals ν_1 and ν_2, after traversing the cable, include components that have been reflected once at A and once at B. The coefficients ρ_A and ρ_B are sufficiently small that components suffering more than one reflection at each point can be neglected. For the frequency ν_1 arriving at the antenna, the amplitude (voltage) of the reflected component relative to the unreflected one is

$$\Lambda = |\rho_A||\rho_B|10^{-\ell\alpha/10} , \tag{7.11}$$

where α is the (power) attenuation coefficient of the cable in decibels per unit length. Note that the attenuation in voltage is equal to the square root of the attenuation in power. The phase of the reflected component relative to the unreflected one is (modulo 2π)

$$\theta_1 = 4\pi\ell\nu_1 v^{-1} + \phi_A + \phi_B , \tag{7.12}$$

where ϕ_A and ϕ_B are the phase angles of ρ_A and ρ_B (that is, $\rho_A = |\rho_A|e^{j\phi_A}$, etc.), and v is the phase velocity in the line. Figure 7.6 shows a phasor representation of the reflected and unreflected components and their phase θ_1. The reflected component causes the resultant phase to be deflected through an angle ϕ_1 given by

$$\phi_1 \simeq \tan\phi_1 = \frac{\Lambda \sin\theta_1}{1 + \Lambda \cos\theta_1} . \tag{7.13}$$

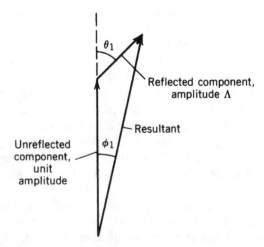

Fig. 7.6 Phasor diagram of components at frequency ν_1 transmitted by the cable.

Similarly, the phase of the frequency v_2 is deflected through an angle ϕ_2, given by equations equivalent to Eqs. (7.12) and (7.13) with subscript 1 replaced by 2.

With the reflection effects represented by ϕ_1 and ϕ_2, the round-trip phase for a line of length L is

$$4\pi v_2 L v^{-1} + \phi_1 + \phi_2 \,. \tag{7.14}$$

If the line length increases uniformly to $L(1 + \beta)$, the angles ϕ_1 and ϕ_2 vary in a nonlinear manner with ℓ and become $\phi_1 + \delta\phi_1$ and $\phi_2 + \delta\phi_2$, respectively. The round-trip phase then becomes

$$4\pi v_2 L v^{-1}(1 + \beta) + \phi_1 + \delta\phi_1 + \phi_2 + \delta\phi_2 \,. \tag{7.15}$$

(The effect of the reflection on the phase of the signal at frequency $v_1 - v_2$ has been omitted since $v_1 - v_2$ is much smaller than v_1 or v_2, and reflections for the relatively low frequency may be very small. Also, the rate of change of phase of $v_1 - v_2$ with line length is correspondingly small.) The applied correction for the increase in line length is half the measured change in round-trip phase:

$$2\pi v_2 \beta L v^{-1} + \frac{1}{2}(\delta\phi_1 + \delta\phi_2) \,. \tag{7.16}$$

However, the exact correction would be equal to the change in the phase of v_2 at the antenna, which is

$$2\pi v_2 \beta L v^{-1} + \delta\phi_2 \,. \tag{7.17}$$

Consequently, the phase correction is in error by

$$\frac{1}{2}(\delta\phi_1 + \delta\phi_2) - \delta\phi_2 = \frac{1}{2}(\delta\phi_1 - \delta\phi_2) \,. \tag{7.18}$$

If v_1 and v_2 were equal, the phase error would be zero. It is possible therefore to specify a maximum allowable difference frequency in terms of the maximum tolerable error.

The difference between the phase angles ϕ_1 and ϕ_2 is obtained from Eq. (7.13) as follows:

$$
\begin{aligned}
\phi_1 - \phi_2 &= \frac{\partial \phi_1}{\partial v_1}(v_1 - v_2) \\
&= \frac{4\pi \ell v^{-1} \Lambda \cos\theta_1 (1 + \Lambda \cos\theta_1) + 4\pi \ell v^{-1} \Lambda^2 \sin^2\theta_1}{(1 + \Lambda \cos\theta_1)^2}(v_1 - v_2) \,.
\end{aligned}
\tag{7.19}
$$

The reflected amplitude Λ must be much less than unity if phase errors are to be tolerable, so terms in Λ^2 can be omitted from the numerator in Eq. (7.19), and the denominator is approximately unity. Thus,

$$\phi_1 - \phi_2 \simeq 4\pi \ell v^{-1} \Lambda (v_1 - v_2) \cos \theta_1 . \tag{7.20}$$

The variation of $\phi_1 - \phi_2$ with line length is given by

$$
\begin{aligned}
\delta\phi_1 - \delta\phi_2 &= \beta\ell \frac{\partial}{\partial \ell}(\phi_1 - \phi_2) \\
&= 4\pi v^{-1} \Lambda \left[\cos \theta_1 - 0.1\ell\alpha(\ln 10)\cos \theta_1 - 4\pi v^{-1}\ell v_1 \sin \theta_1\right] \\
&\quad \times (v_1 - v_2)\beta\ell .
\end{aligned}
\tag{7.21}
$$

The maximum values of the terms in square brackets in Eq. (7.21) are dominated by the third term, which is of the order of the number of wavelengths in the line. If the two smaller terms are neglected, we obtain the magnitude of the phase error as follows:

$$\frac{1}{2}(\delta\phi_1 - \delta\phi_2) \simeq 8\pi^2 v^{-2} |\rho_A||\rho_B|\beta\ell^2 10^{-\alpha\ell/10} v_1 (v_1 - v_2) \sin \theta_1 . \tag{7.22}$$

The factor $\ell^2 10^{-\alpha\ell/10}$ has a maximum value at

$$\ell = 20(\alpha \ln 10)^{-1} . \tag{7.23}$$

This maximum occurs because for small values of ℓ, the change in the angle θ with frequency or cable expansion is small, and for large values of ℓ, the reflected component is greatly attenuated. The maximum value is equal to

$$\left[\ell^2 10^{-\alpha\ell/10}\right]_{\text{max}} = 10.21\alpha^{-2} . \tag{7.24}$$

Curves of $\ell^2 10^{-\alpha\ell/10}$ are plotted in Fig. 7.7 for various values of α that correspond to good-quality cables. It is evident that reducing the attenuation in a cable increases the error in the round-trip phase correction in Eq. (7.22).

The type of reflections that may be encountered depends on the type of transmission line and how it is used. For example, consider a buried coaxial cable that runs along a set of stations used for a movable antenna. The principal cause of reflections in such a cable is the connectors that are inserted at the antenna stations. Unless the antenna is at the closest station, there are one or more interconnecting loops, where unused stations are bypassed, between the antenna and the master oscillator. If there are n connectors in the cable, there are $N = n(n-1)/2$ pairs between which reflections can occur. Also, if the phasors of the corresponding reflected components combine randomly, the overall rms error in the phase correction is, from Eq. (7.22),

$$\delta\phi_{\text{rms}} = \sqrt{32}\pi^2 v^{-2} |\rho|^2 \beta v_1 (v_1 - v_2) F(\alpha, \ell) , \tag{7.25}$$

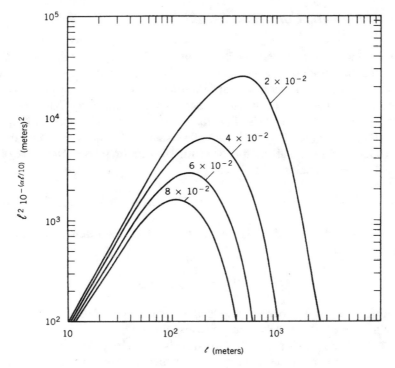

Fig. 7.7 The function $\ell^2 10^{-\alpha\ell/10}$ plotted against ℓ for four values of the transmission-line attenuation, α dB m^{-1}. This function is a factor in the round-trip phase error given by Eq. (7.22).

where

$$F(\alpha, \ell) - \sqrt{\sum_{i=1}^{n}\sum_{k<i} \ell_{ik}^4 10^{-2\alpha\ell_{ik}/10}}, \tag{7.26}$$

the rms value has been used for $\sin \theta_1$, and the reflection coefficients are all approximated by an average magnitude $|\rho|$.

As an example, suppose that an interferometer is designed for observations near 100 GHz and that it incorporates ten antenna stations in a linear configuration at approximately equal increments in distance up to 1 km from the master oscillator. The interconnecting oscillator cable carries a reference signal at $\nu_1 - 2$ GHz, and for this cable $|\rho| = 0.1$, $\alpha = 0.06$ dB m^{-1}, $v = 2.4 \times 10^8$ ms^{-1}, and the temperature coefficient of electrical length is 10^{-5} K^{-1}. From Eq. (7.26), we find that $F(\alpha, \ell) = 1.1 \times 10^4$. For a temperature variation of 0.1 K in the cable, $\beta = 10^{-6}$. If phase errors at 100 GHz are required to be less than 1°, $\delta\phi_{rms}$ must not exceed 0.02°, and from Eq. (7.25), ν_1 and ν_2 must not differ by more than 1.6 MHz.

7.2.4   Automatic Correction System

An interesting variation on the round-trip scheme, shown in Fig. 7.8, was suggested by J. Granlund (National Radio Astronomy Observatory 1967). It is particularly suitable for providing a stable reference frequency at a number of points along a linear array of antennas. Frequencies ν_1 and ν_2 are generated by stable oscillators and are injected at opposite ends of the transmission line. The difference frequency $\nu_1 - \nu_2$ is again very small. At an intermediate station, the two signals are extracted by directional couplers and multiplied to form the sum frequency. The phase of this sum at the antenna station in Fig. 7.8 is

$$2\pi\nu_1\ell_1 v^{-1} + 2\pi\nu_2(L-\ell_1)v^{-1} = 2\pi\nu_1 L v^{-1} - 2\pi(\nu_1-\nu_2)(L-\ell_1)v^{-1} . \qquad (7.27)$$

For two points at positions ℓ_1 and ℓ_2 on the line, the difference in the sum-frequency phases is

$$\Delta\phi = 2\pi(\nu_1-\nu_2)(\ell_1-\ell_2)v^{-1} . \qquad (7.28)$$

This difference would be zero if ν_1 and ν_2 were equal, but it is necessary to maintain a finite difference frequency because the directivity of the couplers alone is seldom sufficient to separate the two signals adequately. The effect of the line length variation is not measured explicitly in this case, but the correction occurs automatically, except for the small term in Eq. (7.28). Reflections in the cable can

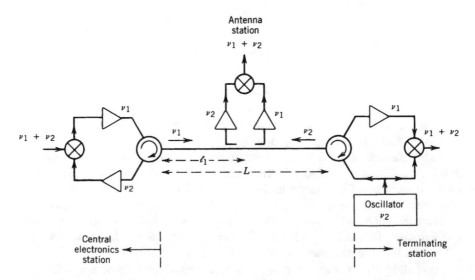

Fig. 7.8 Scheme proposed by J. Granlund (National Radio Astronomy Observatory 1967) for establishing a reference signal at frequency $\nu_1 + \nu_2$ at various stations along a transmission line. One such antenna station is shown.

produce errors, as described for the previous scheme, and may be the limiting consideration for the frequency offset. A practical implementation of the scheme of Fig. 7.8 is described by Little (1969).

7.2.5 Fiberoptic Transmission of LO Signals

Optical fiber can replace cables and transmission lines in most of the LO schemes discussed above. Some features of optical fiber transmission that should be taken into account are outlined below.

- Different optical wavelengths can be used in the two directions of a round-trip system to help separate the signals. At the antenna, the frequency of the laser signal from the master LO can be offset by a few tens of megahertz by using a special modulating device, and injected into the line in the return direction. Alternately, a different laser can be used for the return signal. It is important to take into account the effects of the fiber dispersion and temperature-induced changes in the laser wavelengths, particularly in the case in which two different lasers are used. However, if the laser wavelengths are chosen to be very close to the zero-dispersion wavelength of the fiber, the resulting errors can be minimized.
- As mentioned in Sect. 7.1, the performance of optical components such as isolators and directional couplers is much better than that of corresponding microwave components. With careful design, it is possible to use such components to separate signals at the same laser wavelength traveling in opposite directions in a fiber. Round-trip phase systems have been made in which a radio frequency signal is transmitted on an optical carrier, and at the receiving end, a half-silvered mirror is used to return a component of the signal back along the fiber for a round-trip measurement. It may be necessary to use an optical isolator at the transmitting end to ensure that any of the returned signal that reaches the laser is very small. Reflection of a laser signal back into the output can disturb the operation of the laser.
- In general, when a multifiber cable is flexed, the effective lengths of the individual fibers vary smoothly and remain matched to a much greater degree than is the case for bundled coaxial cables. As a result, it may be possible to use two separate fibers for the two different directions in a round-trip scheme, depending on the accuracy required.
- Twisting of a straight fiber that is held under constant tension has been found to cause less change in the electrical length than bending of a fiber. Twisting, however, can result in small changes in the amplitude of the transmitted signal, resulting from the residual sensitivity of the optical receiver to the angle of the linear polarization of the light.
- It is possible to stabilize the length of the path through a fiber by use of round-trip phase measurement at the optical wavelength. In practice, this requires the use of an automatic correction loop in which a length adjustment device is

controlled by the round-trip phase, since length variations comparable to the optical wavelength can occur on timescales of much less than one second.

- An LO frequency can be transmitted as the difference frequency of two optical laser signals that travel in the same fiber. The radio frequency is generated by combining the optical signals in a photo-optic diode. Radio power of several microwatts can be obtained, which is sufficient to provide LO power for an SIS mixer. This scheme is particularly attractive for receivers at millimeter and submillimeter wavelengths (Payne et al. 1998).
- For standard optical fiber, the temperature coefficient of length is approximately 7×10^{-6} K^{-1}. High-stability fiber, developed by Sumitomo for special applications, has a temperature coefficient that is about an order of magnitude less and was used in the Submillimeter Array without a round-trip correction system (Moran 1998).

7.2.6   Phase-Locked Loops and Reference Frequencies

Some practical points in the implementation of LO systems should be briefly mentioned. In two of the schemes described above, an oscillator at the antenna is controlled by a phase-locked loop. Details of the design of phase-locked loops are given, for example, by Gardner (1979), and here we mention only the choice of the natural frequency of the loop. Unless the natural frequency is about an order of magnitude less than the frequency at the inputs of the phase detector, the loop response may be fast enough to introduce undesirable phase modulation at the phase detector frequency. In the system in Fig. 7.5, the frequency of the input signals to the phase detector is the offset frequency $v_1 - v_2$, an upper limit on which has been placed by consideration of the reflections in the line. Also, the bandwidth of the noise to which the loop responds is proportional to the natural frequency. These considerations place an upper limit on the natural frequency of the loop, which in turn limits the choice of the oscillator to be locked. An oscillator with inherently poor phase stability (when unlocked) requires a loop with a higher natural frequency than does a more stable oscillator. Crystal-controlled oscillators are highly stable and require loop natural frequencies of only a few hertz. They are especially suitable for long transmission lines because the noise bandwidth of the loop is correspondingly small. With crystal-controlled oscillators at the antennas, it is possible to send out the reference frequency in bursts, rather than continuously. Signals traveling in opposite directions can then be separated by time multiplexing, and no frequency offset is required. However, the change in impedance of the circuits at the ends of the cable when the direction of the signal is reversed could become a limiting factor in the accuracy of the round-trip phase measurement. Systems of this type have been designed for several large arrays (Thompson et al. 1980; Davies et al. 1980).

In addition to the establishment of a phase-locked oscillator at each antenna at a reference frequency (equal to v in Fig. 7.4, v_2 in Fig. 7.5, and $v_1 + v_2$ in

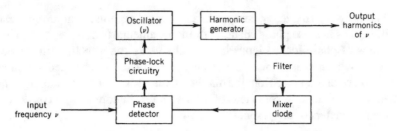

Fig. 7.9 Scheme for generating a comb spectrum of harmonics of a frequency v, in which phase changes in the harmonic generator are eliminated by enclosing it within a phase-locked loop. The filter passes two harmonics that combine in the mixer diode to generate a signal at frequency v.

Fig. 7.8), it is necessary to generate the multiples or submultiples of this frequency that are required for frequency conversions of the received signal. In frequency multiplication, phase variations increase in proportion to the frequency. Within the multiplier chain from the frequency standard to the first LO frequency, the choice of frequency that is transmitted from the central location to the antenna is generally not critical. However, if significant noise is added in the transmission process, it may be better to transmit a high frequency to minimize multiplication of phase errors resulting from the added noise.

Minimization of phase variations in the frequency-multiplication circuit is largely a matter of reducing temperature-related effects, and in this regard, the scheme depicted in Fig. 7.9 is worthy of mention. It may be useful to generate a "comb" spectrum consisting of many harmonics that can be used, for example, for tuning in discrete frequency intervals. This can be done by applying the fundamental frequency to a varactor diode, but the voltage at which the varactor goes into conduction varies with temperature, so the phase of the waveform at which it starts to conduct during each cycle varies. This causes variation in the phases of the harmonics that are generated. In the circuit in Fig. 7.9, the effect of this variation is eliminated. The input fundamental waveform at frequency v is not applied directly to the harmonic generator but is used to lock an oscillator at frequency v. This oscillator drives the harmonic generator. The waveform at the oscillator frequency that is compared with the input frequency is taken after the varactor by selecting two adjacent harmonics and combining them in a mixer diode. The phase-locked loop holds constant the phase of this output waveform relative to the input frequency v and adjusts the phase of the oscillator to compensate for a change in time of switch-on of the varactor.

In the case of a connected-element array, low-frequency components of the phase noise of the master oscillator cause similar effects in the LO phase at each antenna, and therefore their contributions to the relative phase of the signals at the correlator input tend to cancel. However, the frequency components of the phase noise suffer phase changes as a result of the time delay in the path of the reference signal from the master oscillator to each antenna, and also as a result of the time delay of the IF signal from the corresponding mixer to the correlator input (including the variable delay that compensates for the geometric delay). Thus, the cancellation is important only for frequency components of the phase noise that are low enough that differences in these phase changes, from one antenna to another, are small. The bandwidths of phase-locked loops in the LO signals can also limit the frequency range over which phase noise in the master oscillator is canceled. In practice, cancellation of phase noise from the master oscillator is likely to be effective up to a frequency in the range of some tens of hertz to a few hundred kilohertz, depending upon the parameters of the particular system.

7.2.7 Phase Stability of Filters

Tuned filters used for selecting LO frequencies are also a source of temperature-related phase variations. The phase response ϕ of a filter changes by approximately $n\pi/2$ across the 3-dB bandwidth $\Delta\nu$, where n is the number of sections (poles). Thus, the rate of change of phase with frequency, measured at the center frequency ν_0, is

$$\left.\frac{\partial\phi}{\partial\nu}\right|_{\nu_0} = \frac{n\pi k_1}{2\Delta\nu} , \tag{7.29}$$

where k_1 is a constant of order unity that depends on the design of the filter. The center frequency varies with physical temperature T by

$$\frac{\partial\nu_0}{\partial T} = k_2\nu_0 , \tag{7.30}$$

where k_2 is a constant related to the coefficients of expansion and variation of the dielectric constant of the filter. Thus, the rate of variation of phase with temperature is given by

$$\frac{\partial\phi}{\partial T} = \left.\frac{\partial\phi}{\partial\nu}\right|_{\nu_0} \frac{\partial\nu_0}{\partial T} = nk_1k_2\left(\frac{\pi}{2}\right)\left(\frac{\nu_0}{\Delta\nu}\right) . \tag{7.31}$$

The factor $\nu_0/\Delta\nu$ is the Q-factor of the filter. The combined constant k_1k_2 can be determined empirically and is typically of order 10^{-5} K^{-1} for tubular bandpass filters with center frequencies in the range 1 MHz to 1 GHz. Thus, for example, if one allows a 1-K temperature variation for such a filter and places an upper limit of 0.1° on its contribution to the phase variation, the fractional bandwidth must not be less than $n/100$, or 5.4%, for a six-pole filter. Filters of narrow fractional bandwidth should be used with caution. To pick out a particular frequency from a series of closely spaced harmonics, it may be preferable to use a phase-locked oscillator rather than a filter.

7.2.8 Effect of Phase Errors

Rapidly varying phase errors, such as those resulting from noise in LO circuits, cause a loss in signal amplitude and hence in sensitivity. They may also cause errors in the visibility phase, but the effect is small, since fast variations in the visibility phase are substantially reduced by the visibility averaging. To determine the loss in sensitivity, the signals from two antennas can be represented by $V_m e^{\phi_m(t)}$ and $V_n e^{\phi_n(t)}$ at the correlator inputs, where the ϕ terms are the phase errors for antennas m and n. The correlator output is

$$r = \langle V_m e^{\phi_m(t)} V_n^* e^{\phi_n(t)} \rangle \,, \tag{7.32}$$

where the angle brackets represent the expectation. Then if $\Delta\phi = [\phi_m(t) - \phi_n(t)]$ is the phase error, we have

$$r = V_m V_n^* [\langle \cos \Delta\phi \rangle + j \langle \sin \Delta\phi \rangle] \,. \tag{7.33}$$

If the probability distribution of $\Delta\phi$ is an even function with zero mean, which is frequently the case, the time average of the sine term has an expectation of zero. Then, by using the first two terms of the series expression for a cosine, we obtain a result in terms of the rms phase error, $\Delta\phi_{rms}$:

$$r \simeq [1 - \frac{1}{2}\Delta\phi_{rms}^2] \,. \tag{7.34}$$

The cosine approximation is accurate to 1% for values of $\Delta\phi_{rms}$ less than $\sim 37°$. A reduction in sensitivity of 1% occurs for $\Delta\phi_{rms} = 8.1°$.

7.3 Frequency Responses of the Signal Channels

7.3.1 Optimum Response

The signals in a synthesis array may pass through amplifiers, filters, and mixers before being converted to digital form. The characteristics of these components can vary with temperature, etc. However, the resulting problems have become less serious as improvements in the technology allow digitization to take place at earlier stages in the receiving system. Also, for systems with multichannel correlators, as used for spectral line observations, gains of the individual channels can be adjusted to provide a uniform response across the full receiving band. However, it is important to consider the effect of gain variations in analog components since low-noise input stages are generally followed by further amplification to increase the signal levels to something of order 20 dbm or more before they are digitized.

Except in cases in which the astronomical signals cover a wide relative band-width, the signal and the receiver noise both have largely flat spectra over the width of an IF band, and the broad spectral limits of the signal delivered to the digitizer, or, in earlier systems, to the analog correlator, are determined by the frequency response of the receiving equipment. If $H(v) = |H(v)|e^{j\phi(v)}$ is the voltage–frequency response function, the output from the correlator for antennas m and n, resulting from cosmic signals, is proportional to

$$\frac{1}{2}\int_{-\infty}^{\infty} H_m(v)H_n^*(v)\,dv = \mathcal{R}e\left[\int_0^{\infty} H_m(v)H_n^*(v)\,dv\right]$$

$$= \mathcal{R}e\left[\int_0^{\infty} |H_m(v)||H_n(v)|e^{j(\phi_m-\phi_n)}dv\right], \qquad (7.35)$$

where we have used the relation in Eq. (A3.6) of Appendix 3.1, $H_m H_n^*$ being Hermitian, and the subscripts denote the antennas. We are concerned here with the dependence of the signal-to-noise ratio (SNR) of an observation on the frequency responses of the signal channels. In practice, the frequency responses are nonzero only within a limited frequency band of width Δv. From Eq. (6.42), we can define a factor $\mathcal{F}$ equal to the SNR relative to that with identical rectangular responses of width Δv:

$$\mathcal{F} = \frac{\mathcal{R}e\left[\int_0^{\infty} H_m(v)H_n^*(v)\,dv\right]}{\sqrt{\Delta v \int_0^{\infty} |H_m(v)|^2\,|H_n(v)|^2\,dv}}. \qquad (7.36)$$

This equation has a maximum value if $|H_m(v)|$ and $|H_n(v)|$ are constant across the band Δv, that is, if the amplitude response is a rectangular function. If, in addition, $\phi(v)$ is identical for both antennas, $\mathcal{F}$ is equal to unity. Thus, a rectangular passband yields the greatest sensitivity within a limited bandwidth. Note that the same integral of $H_m H_n^*$ applies to both the real and imaginary parts of a complex correlator, and hence it also applies to the modulus of the visibility.

Of the other ways in which the receiving passband modifies the response of a synthesis array, the most important is the smearing of detail in the synthesized response, which limits the field of view that can usefully be imaged. This effect has been described in Sect. 6.3. For a given sensitivity, a rectangular passband results in the least smearing, since it is the most compact in the frequency dimension.

An exact rectangular passband is only an ideal concept. In practice, the steepness of the sides of the passband must be determined by the particular design and the number of poles in the response. The response can be made to approximate a rect-angular shape more closely as the number of poles increases, with a proportionate increase in $\partial\phi/\partial T$ as shown by Eq. (7.31). To examine the tolerable deviations of the actual passband responses, two effects must be considered: the decrease in the SNR, and the introduction of errors in determining gain factors for individual antennas, as will be described.

7.3.2   Tolerances on Variation of the Frequency Response: Degradation of Sensitivity

We first consider the effects on the sensitivity. Equation (7.36) provides a degradation factor $\mathcal{F}$, which is the SNR with frequency responses $H_m(\nu)$ and $H_n(\nu)$, expressed as a fraction of that which would be obtained with rectangular passbands of width $\Delta(\nu)$. In constructing a receiving system, the usual goal is to keep the passband flat with steep edges, but in practice, effects such as differential attenuation and reflections in cables introduce slopes and ripples in the frequency response that are not identical from one antenna to another. To examine these effects, $\mathcal{F}$ can be calculated for an initially rectangular passband with various distortions imposed. The distortions considered are the following:

1. Amplitude slope across the passband, with the logarithm of the amplitude varying linearly with frequency.
2. Sinusoidal amplitude ripple; this could result from a reflection in a transmission line.
3. Displacement of the center frequency of the passband.
4. Variation in phase response as a function of frequency.
5. Delay-setting error, which introduces a component of phase linear with frequency.

The first four of these effects apply mainly to signals in analog form and so are of most importance in systems of earlier design in which digitization of the signals occurred only in the later stages. Expressions for the frequency response involving the above effects are given in the first column of Table 7.1. The second column of the table gives the signal-to-noise degradation factor $\mathcal{F}$, and subscripts m and n indicate parameter values for particular antennas. The expressions in Table 7.1 have been used to derive the maximum tolerable passband distortion for each of the effects, allowing a loss in sensitivity of no more than 2.5% ($\mathcal{F} = 0.975$). The resulting limits on the passband distortion are shown in Table 7.2. A discussion of limits with more stringent tolerances associated with the ALMA array is given by D'Addario (2003).

Table 7.1 Deviation of the frequency characteristic from an ideal rectangular response, and corresponding expressions for $\mathcal{F}$ and G_{mn}

Frequency response	Signal-to-noise degradation, $\mathcal{F}$	Antenna-pair gain, G_{mn}
Amplitude slope[a] $$|H(\nu)|^2 = H_0^2 e^{\sigma(\nu-\nu_0)}\prod\left(\frac{\nu-\nu_0}{\Delta\nu}\right)$$	$$\sqrt{\frac{4\left[e^{(\sigma_m+\sigma_n)\Delta\nu/2}-1\right]}{\Delta\nu(\sigma_m+\sigma_n)\left[e^{(\sigma_m+\sigma_n)\Delta\nu/2}+1\right]}}$$	$$\frac{2G_0}{\Delta\nu(\sigma_m+\sigma_n)}\left[e^{(\sigma_m+\sigma_n)\Delta\nu/4}-e^{-(\sigma_m+\sigma_n)\Delta\nu/4}\right]$$
Sinusoidal ripple $$H(\nu)=H_0\left[1+\gamma e^{j2\pi(\nu-\nu_0)\tau}\right]\prod\left(\frac{\nu-\nu_0}{\Delta\nu}\right)$$	$$\left[\frac{1+2\mathcal{Re}\{\gamma_m\gamma_n^*\}+|\gamma_m\gamma_n|^2}{1+|\gamma_m|^2+|\gamma_n|^2+2\mathcal{Re}\{\gamma_m\gamma_n^*\}+|\gamma_m\gamma_n|^2}\right]^{1/2}$$ (see footnote b)	$$G_0\left[1+\frac{2}{\pi}\left(\gamma_m+\gamma_n^*\right)+\gamma_m\gamma_n^*\right]$$ (see footnote c)
Center-frequency displacement[d] $$H(\nu)=H_0\prod\left(\frac{\nu-\delta\nu-\nu_0}{\Delta\nu}\right)e^{jN\pi(\nu-\delta\nu-\nu_0)/\Delta\nu}$$	$$\sqrt{1-\frac{\delta\nu_m-\delta\nu_n}{\Delta\nu}}$$	$$G_0\left[1-\frac{\delta\nu_m-\delta\nu_n}{\Delta\nu}\right]e^{jN\pi(\delta\nu_m-\delta\nu_n)/\Delta\nu}$$
Phase variation $$H(\nu)=H_0\prod\left(\frac{\nu-\nu_0}{\Delta\nu}\right)e^{j\phi(\nu)}$$	$$1-\frac{1}{2}\langle\phi_{mn}^2\rangle$$ $$\phi_{mn}(\nu)=\phi_m(\nu)-\phi_n(\nu)-\langle\phi_m(\nu)-\phi_n(\nu)\rangle$$ (see footnote e)	$$G_0\left[1-\frac{1}{2}\langle\phi_{mn}^2\rangle\right]$$
Delay-setting error $$H(\nu)=H_0 e^{j2\pi\nu\tau}\prod\left(\frac{\nu-\nu_0}{\Delta\nu}\right)$$	$$\frac{\sin[\pi\Delta\nu(\tau_m-\tau_n)]}{\pi\Delta\nu(\tau_m-\tau_n)}$$	$$G_0\left[\frac{\sin\pi\Delta\nu(\tau_m-\tau_n)}{\pi\Delta\nu(\tau_m-\tau_n)}\right]e^{j\pi\Delta\nu(\tau_m-\tau_n)}$$ (see footnote f)

[a] The unit rectangle function $\prod(x)$ is equal to 1 for $|x|\leq\frac{1}{2}$ and zero for $|x|>\frac{1}{2}$. Parameters are as follows: H_0 and G_0, gain constants; σ, slope parameter; ν_0, passband center frequency; γ, relative amplitude of sinusoidal component; $\delta\nu$, frequency offset; τ, delay error.

[b] For integral value of $\Delta\nu\tau$ (integral number of cycles across passband).

[c] $\Delta\nu\tau=\frac{1}{2}$ (half cycle of sinusoidal ripple across passband).

[d] Linear phase response with difference $N\pi$ between passband edges.

[e] The brackets $\langle\ \rangle$ indicate a mean over the passband.

[f] Phase term corresponds to baseband response (center frequency = $\Delta\nu/2$).

Table 7.2 Examples of frequency response tolerances

	Criterion	
	2.5% Degradation in	1% Maximum
Type of variation	Signal-to-noise ratio	Gain error
Amplitude slope	3.5 dB edge-to-edge	2.7 dB edge-to-edge
Sinusoidal ripple	2.9 dB peak-to-peak	2.0 dB peak-to-peak
Center-frequency displacement	$0.05\Delta v$	$0.007\Delta v$
Phase variation	$\phi_{mn} = 12.8°$ rms	$\phi_{mn} = 9.1°$ rms
Delay-setting error	$0.12/\Delta v$	$0.05/\Delta v$

7.3.3 Tolerances on Variation of the Frequency Response: Gain Errors

A second effect that sets limits on the deviations of the frequency responses results from errors that can be introduced in the calibration procedure. If we omit the noise terms, the output of the correlator for an antenna pair can be expressed as

$$r_{mn} = G_{mn}\mathcal{V}_{mn} , \qquad (7.37)$$

where $\mathcal{V}_{mn}$ is the source-dependent complex visibility from which the intensity map can be computed, and G_{mn} is a gain factor related to the frequency responses of the signal channels. We suppose that these responses incorporate the characteristics of the antennas and electronics in such a way that G_{mn} is proportional to the correlator output for antenna pair (m, n) when a point source of unit flux density at the field center is observed. In practice, the G_{mn} values may be determined from observations of calibration sources for which the visibilities are known. The measured antenna-pair gains can be used to correct the correlator output data directly, but there are advantages if, instead, they are used to determine (voltage) gain factors $g = |g|e^{j\phi}$ for the individual antennas such that

$$G_{mn} = g_m g_n^* . \qquad (7.38)$$

Since, in a large array, there are many more correlated antenna pairs than antennas [up to $n_a(n_a - 1)/2$ pairs for n_a antennas], not all the calibration data need be used. This adds important flexibility to the calibration procedure; for example, a source resolved at the longest spacings of an array can be used to determine the antenna gains from measurements made only at the shorter spacings. The same principle leads to adaptive calibration described in Sect. 11.3.

In general, the factoring in Eq. (7.38) requires that the frequency responses be identical for all antennas or differ only by constant multiplicative factors. If this requirement is fulfilled, we can assign gain factors

$$g = \sqrt{\int_0^\infty |H(v)|^2 dv} . \qquad (7.39)$$

In practice, the frequency responses differ, and an approximate solution to Eq. (7.38) can be obtained by choosing the g values to minimize

$$\sum |G_{mn} - g_m g_n^*|^2 , \qquad (7.40)$$

where the summation is taken over all antenna pairs (m, n) for which G_{mn} can be measured by observation of a calibration source. In calibrating subsequent observations of unknown sources, $g_m g_n^*$ is used in place of G_{mn} in Eq. (7.37) for all antenna pairs, whether they are directly calibrated or not. To avoid introducing errors with this scheme, the residuals

$$\varepsilon_{mn} = G_{mn} - g_m g_n^* \qquad (7.41)$$

must be small, which requires that the frequency responses be sufficiently similar. Thus, we are concerned here with the deviations of the frequency responses from one another rather than from an ideal response.

By using model responses for groups of antennas—calculating the pair gains, the best-fit antenna gains, and the residuals—tolerances on the bandpass distortion can be assigned. Pair gains for the various distortions discussed earlier are given in the third column of Table 7.1. Table 7.2 shows examples of tolerances. The results depend to some extent on the distribution of distortions in the model responses, which for the results shown were chosen with the intention of maximizing the residuals. The criteria of 2.5% loss in sensitivity and 1% maximum gain error shown in Table 7.2 were used during the early operation of the VLA (Thompson and D'Addario 1982). More stringent criteria may be appropriate, depending on the sensitivity and dynamic range to be achieved. The acceptable level of gain error for any instrument can be determined by making calculations of the response to source models with simulated errors of various levels introduced into the model visibility data. Bagri and Thompson (1991) give a discussion of the sources and effects of gain errors in the VLA.

7.3.4   Delay and Phase Errors in Single- and Double-Sideband Systems

For an incoming wavefront from a source, the path lengths to different antennas of an array are generally unequal. The relative time differences in the wavefront arrival at the antennas are referred to as the geometric delays, τ_g. To compensate for the different geometric delays, the signal received at each antenna is subjected to an instrumental delay τ_i that is continuously adjusted so that $\tau_g + \tau_i$ is the same for all antennas. Thus, the signals at the correlator inputs are aligned in time with respect to a common wavefront incident from the phase reference position. The fringes at the correlator output result from the fact that the signals traverse the geometric and the instrumental delays at different frequencies and that the phase shifts resulting from the delays vary as the delays themselves change. In an ideal situation, the instrumental delays would be continuously adjusted, and if there were no frequency changes within the receivers, no fringe oscillations would occur. In practice, the situation is rather more complicated. The instrumental delays are inserted after the signals have been digitized, and the sample interval τ_s provides a convenient unit for coarse adjustment. For Nyquist sampling, $\tau_s = 1/2\Delta\nu$, where $\Delta\nu$ is the signal bandwidth.

7.3.5   Delay Errors and Tolerances

In an array with a number of antennas, the delays are adjusted relative to the delay for a designated reference antenna. We consider the reference antenna to be the one that is the last one encountered by the approaching wavefront, and its instrumental delay remains fixed. The delay error for any antenna is the difference between the sum of the geometric and instrumental delays for that antenna and for the reference antenna. When the delay error becomes as large as $\pm\tau_s/2$, the delay is adjusted by an increment $\pm\tau_s$. Thus, the delay error for a single antenna is uniformly distributed over $\pm\tau_s/2$. Coarse delay adjustments in units of the digital sampling interval are implemented in a FIFO (first-in-first-out) memory. These provide the major part of the instrumental delay, but the residual delay errors are large enough that they can cause serious loss of sensitivity if not mitigated. In the original VLA system, for example, finer steps were provided by an adjustment in the timing of the sampler action in steps of $\tau_0 = \tau_s/16$. The spacing of adjacent samples remains τ_s except when a delay adjustment is made and the sample occurs earlier or later by τ_0. When the delay error becomes equal to $\tau_0/2$, the instrumental delay is adjusted by τ_0, as represented by the staircase function in Fig. 7.10. One can see from Fig. 7.10 that the probability distribution of the delay error is uniform within a range $\pm\tau_0/2$. For a pair of antennas, it can usually be assumed that the times of delay adjustment are unrelated (in general, the rates of change of the geometric delay will be different for each antenna), so the probability distribution of their combined delay errors is a triangular function with extreme values of $\pm\tau_0$, as in Fig. 7.11. The rms value of this delay error is:

$$\left[\frac{\int_0^{\tau_0} p(\Delta\tau)\Delta\tau^2 d\Delta\tau}{\int_0^{\tau_0} p(\Delta\tau)\, d\Delta\tau} \right]^{1/2} = \frac{\tau_0}{\sqrt{6}}, \qquad (7.42)$$

where $p(\Delta\tau)$ is the expression for the probability distribution of $\Delta\tau$ in Fig. 7.11.

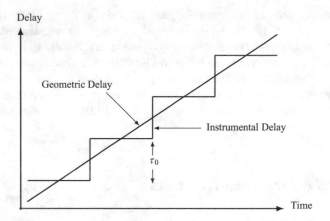

Fig. 7.10 Adjustment of instrumental delay in steps τ_0 to compensate for geometric delay. The vertical sections of the staircase function indicate change of instrumental delay, and the horizontal sections are the time intervals during which the signal is sampled. Over small time intervals, the geometric delay can be represented as a linear function of time. Both axes have the dimensions of time but, for example, a baseline of 1 km east–west, the maximum slope of the line representing space delay is 0.24 ns per second, and the timescales of the two axes differ by a factor of order 10^{10}.

7.3.6   *Phase Errors and Degradation of Sensitivity*

A delay error $\Delta\tau$ results in a phase error in a signal equal to $2\pi\Delta\tau\nu$ where, for systems using analog delays, ν represents the frequency in the IF band in which the delays are inserted. For systems in which the signal passband is Nyquist sampled, which is the most usual case, ν is a baseband frequency in the range 0 to $\Delta\nu$. With a spectral correlator, the highest frequency channel within such a band has a center frequency that is approximately equal to the high-frequency edge, $\Delta\nu$. For frequencies in this top channel, the maximum delay error τ_0 for an antenna pair results in a phase error of $2\pi\Delta\nu\tau_0 = (\tau_0/\tau_s)\pi$. Thus, the probability distribution of this phase error is a triangular function as in Fig. 7.11 with extreme values $\pm(\tau_0/\tau_s)\pi$. As shown for the delay error in Eq. (7.42), the rms phase error is the maximum value divided by $\sqrt{6}$.

To determine the effect of delay errors on sensitivity, note that for frequency ν, a delay error $\Delta\tau$ results in a phase error $2\pi\nu\Delta\tau$. Let α be the size of the fine step as a fraction of the coarse step τ_s. The sensitivity (i.e., the relative response) is determined by averaging the cosine of the phase error, weighted by the triangular distribution of delay error in Fig. 7.11,

$$\frac{2}{\alpha\tau_s} \int_0^{\alpha\tau_s} \left(1 - \frac{\Delta\tau_s}{\alpha\tau_s}\right) \cos(2\pi\nu\Delta\tau)\, d\Delta\tau = \left[\frac{\sin(\pi\nu\alpha\tau_s)}{\pi\nu\alpha\tau_s}\right]^2. \tag{7.43}$$

This is the sensitivity for a very small bandwidth centered on frequency ν, as in the case for spectral line observations. For continuum observations, the sensitivity for a band extending from $N\Delta\nu$ to $(N+1)\Delta\nu$, where N is any integer (including zero),

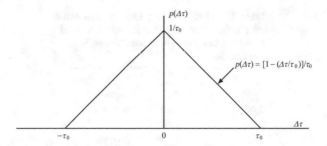

Fig. 7.11 Probability distribution $p(\Delta\tau)$ of the delay error $\Delta\tau$ for a pair of antennas. τ_0 is the minimum increment of the instrumental compensating delay. The expression shown for $p(\Delta\tau)$ applies to the part of the probability function for which $\Delta\tau \geq 0$.

is obtained by averaging over the baseband response,

$$\frac{1}{\Delta\nu}\int_0^{\Delta\nu\alpha\tau_s}\left[\frac{\sin(\pi\nu\alpha\tau_s)}{\pi\nu\alpha\tau_s}\right]^2 d\nu = \frac{2}{\alpha}\int_0^{\frac{\alpha}{2}}\left[\frac{\sin(\pi x)}{\pi x}\right]^2 dx . \tag{7.44}$$

Here, we use $\Delta\nu\tau_s = \frac{1}{2}$ and have put $\nu\alpha\tau_s = x$ for convenience in numerical evaluation of the integral.[3] For the case in which we use only the coarse delay steps ($\alpha = 1$), Eq. (7.44) is equal to 0.774, so, as noted earlier, the performance with coarse delay steps without further mitigation is not acceptable. Some values of rms phase error and sensitivity loss averaged across the bandwidth are given in Table 7.3.

In Table 7.3, sensitivity loss for $\alpha = 1/4$ is approaching an acceptable level. However, the maximum phase errors are $\sqrt{6} = 2.45$ times the rms value, i.e., $2.45\times 10.6° = 26°$ in this case. Depending upon how fast the delay error is changing with time, the maximum error will be decreased somewhat in the data averaging after cross-correlation. In the (u, v) plane, the rate of change of delay error goes through zero as the u-component of the baseline crosses the v axis. Thus, averaged data in which the phase error is close to the maximum are to be expected, especially for short-baseline configurations. Hence, in considering the acceptable delay errors, phase errors should be considered as well as sensitivity loss. The original VLA system used $\tau_s = 16\tau_0$ and a baseband IF response.

---

[3]In the case in which the phase errors are small, it may be convenient to use $\langle\cos(\phi)\rangle \approx (1 - \langle\phi^2\rangle/2)$, where $\phi = 2\pi\nu\alpha\Delta\tau$ and $\langle\rangle$ indicates the mean value. Then noting that $\Delta\tau$ and ν vary independently, $\langle\phi^2\rangle = (2\pi)^2\langle\Delta\tau^2\rangle\langle\nu^2\rangle$. From Eq. (7.42), $\langle\Delta\tau^2\rangle = \tau_0^2/6$, and for a baseband response, $\langle\nu^2\rangle = \Delta\nu^2/3$.

Table 7.3 Values of the loss in signal-to-noise ratio (sensitivity) for the full baseband response from 0 to Δv, as used in continuum observations

$\alpha = \tau_0/\tau_s$	ϕ_{rms}	SNR loss
1/4	10.6°	1.7%
1/8	5.30°	0.43%
1/16	2.65°	0.11%
1/32	1.33°	0.027%

7.3.7 Other Methods of Mitigation of Delay Errors

Conceptually, the most straightforward way of keeping the loss in sensitivity resulting from delay errors within a tolerable limit[4] (say, $\sim 1\%$) is to use a small enough value for the minimum delay increment. This may not always be easy in systems with wide bandwidths, which require correspondingly high sample rates in the digitization. A possible scheme to reduce phase errors (D'Addario 2003) is one in which whenever a delay increment is inserted or removed, a phase jump of magnitude $2\pi v_0 \tau_0$, and opposite sign to the delay-induced phase jump, is inserted in the corresponding signal through an LO.[5] Here, v_0 is the IF center frequency, for which the phase error is exactly canceled. The overall effect for the full bandwidth can be found by determining the value of $\langle (v - v_0)^2 \rangle$, that is, the mean squared value of frequency measured with respect to the band center:

$$\langle (v - v_0)^2 \rangle = \frac{1}{\Delta v} \int_{v_0 - \Delta v/2}^{v_0 + \Delta v/2} (v - v_0)^2 \, dv = \frac{\Delta v^2}{12} . \quad (7.45)$$

This result applies to any IF band of width Δv. Since the phase changes resulting from the changes in the instrumental delay provide a component of the frequency offset used to stop the interferometer fringes, it is necessary to account for this effect by inserting a smooth component in the form of a frequency offset, $2\pi v_0 \, d\tau_g/dt$, where τ_g is the geometric delay. This could be combined with the fringe-rotation offset in an LO.[6] The combination of the inserted phase jumps and the frequency

---

[4]Various effects in an interferometer system limit the sensitivity. There are some large effects, such as aperture efficiency and quantization efficiency, and more numerous smaller ones, such as phase irregularities in frequency responses, LO noise, timing errors, delay errors, etc. The combined effect of the smaller losses can become serious, so for each one, it is reasonable to aim at a fairly stringent limit such as the 1% figure suggested here.

[5]This method of mitigation of the delay errors was considered but not implemented during the early development of the VLA. The original idea is attributed to B. G. Clark.

[6]In the case of a double-sideband system, the fringe rotation must be applied to the first LO, but the frequency offset required by the phase error reduction scheme must be applied to the second or a later LO so that, like the delay-induced phase errors, the offsets are applied with the same sign to each sideband component within the IF band.

offset provides a sawtooth phase component that, at the band center, exactly cancels the phase sawtooth induced by the delay error. If this method were used with no fine delay steps, i.e., $\tau_0 = \tau_s$, the loss in sensitivity would be ~ 13%, so a combination with some finer steps would be necessary.

7.3.8 Multichannel (Spectral Line) Correlator Systems

In multichannel correlators, the input band is divided into many channels, and the signals for corresponding channels are cross-correlated. The number of channels is usually an integral power of two and commonly 1024 or more. Within any channel, the relative variation of frequency is very small. Thus, at any instant, the effect of a delay error $\Delta\tau$ is to introduce a phase error $2\pi\Delta\tau\nu_c$, where ν_c is the center frequency of the channel. Since the frequency variation is small across a single channel, the loss in signal amplitude that occurs in a wide (continuum) band, resulting from the frequency variation of the phase error, is avoided. The time variation of the delay error results in a varying phase error that can be corrected by inserting a phase correction for each channel at the correlator. Thus, with a multichannel correlator, it is possible to avoid the need for delay increments finer than τ_s, so long as the extra processing steps to correct the phase can be incorporated. For an individual antenna, the maximum delay error is $\tau_s/2 = 1/(2\Delta\nu)$, and for the highest channel, centered very close to frequency $\Delta\nu$, the maximum phase error is π. The time for the delay error variation to complete one cycle is $\frac{d\tau}{dt}/\tau_s = \frac{d\tau}{dt}\Delta\nu/2$. (Note that the rate of change of delay, $\frac{d\tau}{dt}$, is different for each antenna.) This is greater than the time for one fringe cycle by a factor of $2 \times$ (signal frequency at the antenna)/(signal frequency in the baseband $0 - \Delta\nu$). At any instant, the phase error for the signal from an antenna is equal to $2\pi \times$ (delay error) $\times$ (channel frequency in the baseband $0 - \Delta\nu$). If the correction is applied at the correlator output, the corrections for both antennas of each pair must be included.

Carlson and Dewdney (2000) describe a multichannel correlator designed to handle wide bandwidth signals (the WIDAR system). The signals from the antennas are Nyquist sampled, and then the band is divided into a number of channels, N_c. The Nyquist sample rate appropriate for each channel is equal to the original sample rate divided by N_c, and the sample rates are adjusted to this value at the filter outputs. The outputs of the filters then go to separate cross-correlators. In this way, the total bandwidth that can be processed is not limited by the capacity of a single correlator. A value of $N_c = 32$ would be sufficient to reduce the loss in sensitivity resulting from the delay errors to an acceptably small value. Adjusting the phases of the signals at the correlator inputs removes the phase errors resulting from delay errors and also provides fringe stopping. Phase adjustment at this point is possible because the samples are in complex form, having been through the filtering process. Since multichannel correlators give a means of removing channels that are contaminated by interference, they are widely used for continuum as well as spectral line observations.

7.3.9   Double-Sideband Systems

The considerations up to this point have applied to single-sideband (SSB) systems. For double-sideband (DSB) systems, some differences must be considered (Thompson and D'Addario 2000). For an SSB system, the main effect of a phase error is to cause a rotation of the correlation vector, as indicated in Fig. 6.5a, resulting in an error in the correlator output phase, as considered above.[7] For a DSB system, the delay error causes the components of the correlation vector resulting from the two sidebands to rotate in opposite directions in the complex plane, as shown in Fig. 6.5b, where the line AB represents the phase angle when the delay error is zero.[8] The amplitude of the vector sum of the two components is proportional to $\cos(2\pi\nu_0\Delta\tau)$, where ν_0 is the IF center frequency, but the phase of the correlation is not changed by a variation in the instrumental delay.

Consider a case in which the geometric delay is varying rapidly enough that the delay error changes sign several times during the minimum averaging time at the correlator output. For an SSB system (Fig. 6.5a), the phase of the correlation vector swings back and forth, following the difference of the error patterns for the two antennas (the small arrows indicating variation of the vector phase in Fig. 6.5 reverse direction when the sign of the phase error changes). For a DSB system (Fig. 6.5b), the phase angles of the vectors representing the two sideband responses move in opposite senses. In both the SSB and DSB cases, components of the correlation that are normal to the vector time-average (in Fig. 6.5b, the line AB) cancel, and the magnitude of the correlation is proportional to the time average of the cosine of the phase measured with respect to the mean phase. Over an averaging period in which the SSB phase error changes sign, the loss in sensitivity is effectively the same for the SSB and DSB systems. Note, however, that in the SSB case, the loss in sensitivity occurs in the averaging, whereas in the DSB case, the loss occurs immediately in the correlation process. Thus, in the SSB case, there is an opportunity to correct for phase errors after cross-correlation, but in the DSB case, this is possible only if the sideband responses can be separated.

If we are considering delay errors that are quasi-constant, or vary only slowly with time, the tolerance on the errors is more stringent in the DSB case. Such errors were more important in early interferometers with analog delay systems using coaxial cable or ultrasonic elements (see, e.g., Coe 1973), which could be temperature sensitive and difficult to calibrate accurately. In digital systems, the delays are controlled by a highly accurate master clock, and the only significant

---

[7]There is also a relatively small decrease in the amplitude, which results from the variation of the phase error with frequency across the IF band and is proportional to $\mathrm{sinc}(\Delta\nu\Delta\tau)$, where $\Delta\nu$ is the IF bandwidth. This results from the averaging of the varying phase over time. The same sinc function appears as part of Eq. (6.9) and is shown by the broken line in Fig. 6.4.

[8]To measure the phase of the cross-correlation of both sidebands in combination, it is necessary to periodically insert a $\pi/2$ phase shift into the IF signal of one antenna, or not to stop the fringes and fit a sine wave to the fringe function.

errors result from the incremental nature of the adjustment. With digital delays, the effects are in most respects the same for SSB and DSB systems, except that for DSB systems, again, post-correlation corrections are possible only if the sidebands can be separated.

In addition to causing a loss in sensitivity, delay-induced phase errors contribute to errors in the phase of the measured visibility. In this case, the values after time averaging, not the instantaneous values, are critical. The effective averaging time is of the order of the time taken for the baseline vector to cross a cell in the simple case of cell averaging, discussed in Sect. 5.2.2. In a synthesis array, the compensating delay for each antenna is adjusted to equalize the delay relative to some celestial reference point as the source moves across the sky. If the antenna spacings are large, the delay may change by several increments during most cell crossings, and the resulting phase errors are reduced by the data averaging. However, for any pair of antennas, the rate of change of the geometric delay, which is proportional to u, goes through zero when the baseline vector crosses the v axis.

In conclusion, the tolerances in Table 7.2 apply to the overall system from the antennas to the correlator inputs. Specifications of filters that define the passband should include consideration of the temperature effects discussed in Sect. 7.2.7. The frequency selectivity of elements in the earlier stages can then be held to the minimum required for rejection of interfering signals. There are advantages to implementing the filtering that defines the passband digitally, after the sampling, instead of in the analog stages [see, e.g., Prabu et al. (2015)].

7.4 Polarization Mismatch Errors

The response of two antennas to an unpolarized source is greatest when the antennas are identically polarized. Small variations in the polarization characteristics of one antenna relative to another occur as a result of mechanical tolerances. These variations lead to errors in the assignment of antenna gains in a manner similar to the variations in frequency responses. To examine this effect, we calculate the response of two arbitrarily polarized antennas to a randomly polarized source, which is given by the term for the Stokes parameter I_v in Eq. (4.29). Definitions of symbols are in terms of the polarization ellipse (see Fig. 4.8 and related text). The position angle of the major axis is ψ, the axial ratio is $\tan \chi$, and subscripts m and n indicate two antennas of an array. As an example, we consider antennas with nominally identical circular polarization for which we can write $\chi_m = \pi/4 + \Delta\chi_m$ and $\chi_n = \pi/4 + \Delta\chi_n$, where the Δ terms represent the deviations of the corresponding parameter from the ideal value. The required response is

$$G_{mn} = G_0 \left[\cos(\psi_m - \psi_n)\cos(\Delta\chi_m - \Delta\chi_n) \right.$$
$$\left. + j\sin(\psi_m - \psi_n)\cos(\Delta\chi_m + \Delta\chi_n)\right] . \tag{7.46}$$

Now $\psi_m - \psi_n$ and the Δ terms represent construction tolerances and are all small. Thus, we can expand the trigonometric functions and retain only the first- and second-order terms. Equation (7.46) then becomes

$$G_{mn} = G_0 \left\{ 1 - \frac{1}{2} \left[(\psi_m - \psi_n)^2 + (\Delta\chi_m - \Delta\chi_n)^2 \right] + j(\psi_m - \psi_n) \right\} . \qquad (7.47)$$

An analysis similar to the procedure for frequency responses in Sect. 7.3 can be made by assigning polarization characteristics to a model group of antennas and determining pair gains, best-fit antenna gains, and gain residuals. For simplicity, it is assumed that the spread of values is of similar magnitude for the parameters χ and ψ. A 1% maximum gain residual then results from a spread of $\pm 3.6°$ in χ and ψ. A value of $\Delta\chi = 3.6°$ corresponds to an axial ratio of 1.13 for the polarization ellipse, and it is not difficult to obtain feeds for which the deviation from circularity is within this value near the beam center. A similar analysis for linearly polarized antennas gives tolerances of the same order (Thompson 1984).

7.5   Phase Switching

7.5.1   Reduction of Response to Spurious Signals

The technique of phase switching for a two-element interferometer has been described in Chap. 1, where it was explained as an early method of obtaining analog multiplication of signals. The principle is as indicated in Fig. 1.8. However, in later instruments, the power-law detector is replaced by a correlator. Although more direct methods of signal multiplication are now used, phase switching is still useful to eliminate small offsets in correlator outputs that can result from imperfections in circuit operation or from spurious signals. The latter are difficult to eliminate entirely in any complicated receiving system, since combinations of harmonics of oscillator frequencies that fall within the observing frequency band or any IF band may infiltrate the electronics. Such signals, at levels too low to detect by simple test procedures, can be strong enough to produce unwanted components in the output. For an array of n_a antennas, a receiving bandwidth $\Delta\nu$, and an observing duration τ, signals at the limit of detectability are at a power level of order $(n_a\sqrt{\Delta\nu\tau})^{-1}$ relative to the noise; for example, this gives 75 dB below the noise for $n_a = 27$, $\Delta\nu = 50$ MHz, and $\tau = 8$ h. Similar effects can also be produced by cross coupling of small amounts of noise from one IF system to another. Because such spurious signals produce components of the visibility that change only slowly with time, they show up as spurious detail near the origin of the image. If they enter the signal channel at a point that comes after the phase switch, so that they produce a component with no switch-frequency variation at the synchronous detector, they can generally be reduced by several orders of magnitude by the phase switching.

7.5.2 Implementation of Phase Switching

Consider the problem of phase switching in a multielement array in which the products of the signals from all possible pairs of antennas are formed. Phase switching can be represented by multiplication of the received signals by periodic functions that alternate in time between values of +1 and −1. For the mth and nth antennas, let these functions be $f_m(t)$ and $f_n(t)$. Synchronous detection of the correlator output for these two antennas requires a reference waveform $f_m(t)f_n(t)$, and any nonvarying, unswitched components from the multiplier are reduced by a factor

$$\frac{1}{\tau}\int_0^{\tau} f_m(t)f_n(t)\,dt \tag{7.48}$$

after averaging for a time τ. For the periodic waveforms that we are concerned with, this factor will be zero if τ is a multiple of the minimum period of orthogonality τ_{or} for $f_m(t)$ and $f_n(t)$. In fact, unwanted output components may not be exactly constant, because the tracking of the compensating delays introduces slow changes in the phases with which the spurious signals are combined. However, the unwanted outputs will be strongly reduced by the synchronous detection as long as their variation is small over the period τ_{or}. If the orthogonality of the phase-switching functions depends on the relative timing of transitions, the timing should be adjusted so that the functions are orthogonal at the correlator inputs. Thus, it may be necessary to adjust the timing of the switching waveforms at the antennas to compensate for the varying instrumental delays inserted as a source moves across the sky.

Implementation of phase switching on an array of n_a antennas calls for n_a mutually orthogonal, two-state waveforms. Square waves whose frequencies are proportional to integral powers of two are orthogonal, with τ_{or} equal to the period of the lowest nonzero frequency.[9] In phase switching, τ_{or} is equal to the data averaging time, which is typically a few seconds but for special cases may be as low as 10 ms. The shortest interval between switching transitions τ_{sw} is equal to the half-period of the fastest square wave. Technically, it is convenient if τ_{or}/τ_{sw} does not greatly exceed about two orders of magnitude. If one antenna remains unswitched, then $\tau_{or}/\tau_{sw} = 2^{n_a-1}$. Square waves of the *same* frequency are orthogonal if their phases differ by a quarter of a cycle in time. When this condition for orthogonality is also included, $\tau_{or}/\tau_{sw} = 2^{n+1}$, where n is the smallest integer greater than or equal to $(n_a - 3)/2$. This reduces the value of τ_{or}/τ_{sw}, but the orthogonality then depends upon the relative timing of the transitions at the correlator, which is not the case for square waves of different frequencies. In either case, τ_{or}/τ_{sw} is inconveniently large for a large array and, for example, for $n_a = 27$, it is of order 10^8 in the first case and 10^4 in the second.

---

[9] Such waveforms are sometimes referred to as Rademacher functions.

It is useful to note that a condition for a pair of square waves of different frequency to be orthogonal, for arbitrary time shifts, is that they do not contain Fourier components of the same frequency. A property of square waves is that all even-numbered Fourier components (i.e., even harmonics of the fundamental frequency) have zero coefficients, but odd-numbered components have nonzero coefficients. Thus, although sinusoids with frequencies proportional to 1, 2, 3, ... are mutually orthogonal, square waves with such frequencies, in general, are not. For example, square waves of frequencies 1, 2, and 4 have no common Fourier components and are mutually orthogonal, but 1, 3, and 5 have common components and are not mutually orthogonal. D'Addario (2001) shows by generalization of this analysis that the lowest frequency sets of N mutually orthogonal square waves consist of those with frequencies proportional to 2^n for $n = 0, 1, \ldots, (N - 1)$, that is, the square-wave sets discussed above. Since the different square waves of a set that we are considering contain no common Fourier components, their orthogonality is not affected by relative time shifts. Note, also, that exact orthogonality is not essential for phase switching. Unwanted responses can be reduced by a factor of 10^4 or more by using square waves with k cycles per averaging period for values of k that are prime numbers greater than 100.

For arrays with large numbers of antennas, Walsh functions are generally the preferred waveforms for phase switching. Walsh functions are rectangular waveforms in which transitions between $+1$ and -1 occur at intervals that are a varying integral submultiple of a basic time cycle, as in Fig. 7.12. For a description of Walsh functions (Walsh 1923; also Fowle 1904) see, for example, Harmuth (1969, 1972) or Beauchamp (1975). Various systems of designating and ordering Walsh functions are in use. In one system (Harmuth 1972), those with even symmetry are designated as $\text{cal}(k, t)$ and those with odd symmetry as $\text{sal}(k, t)$. Here, t is time expressed as a fraction of the time base T, which is the interval at which the waveform repeats, and k is the *sequency*, which is equal to half the number

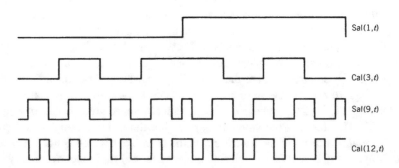

Fig. 7.12 Four examples of Walsh functions, each of which repeats after the one cycle of the time base interval plotted above. Within this interval, the sal functions are odd, and the cal functions are even. The value of each function alternates between 1 and -1. The first number in parentheses in the name of each function is the sequency, which is equal to half the number of zero crossings in the time base interval. Time t is measured as a fraction of the time base.

of zero crossings within the time base. Walsh functions with different sequencies are orthogonal, and cal and sal functions of the same sequency are orthogonal but differ only by a time offset. The orthogonality requires that the time bases of the individual Walsh functions be aligned in time, so time offsets are not permitted. Walsh functions with sequencies that are integral powers of two are square waves. If one antenna is unswitched, and if only the cal or only the sal functions are used, the highest sequency required is $n_a - 1$. Then $\tau_{or}/\tau_{sw} = 2n$, where n is the smallest power-of-two integer greater than or equal to $n_a - 1$. If both cal and sal functions are used, then n is the smallest power-of-two integer greater than or equal to $(n_a - 1)/2$. For example, for $n_a = 64$, τ_{or}/τ_{sw} is 128 in the first case and 64 in the second. Another designation for Walsh functions, wal(n, t), includes both cal and sal functions, cal(n, t) = wal($2n, t$) and sal(n, t) = wal($2n - 1, t$).

One method of generating Walsh functions makes use of Hadamard matrices, of which the one of lowest order is

$$H_2 = \begin{bmatrix} 1 & 1 \\ 1 & -1 \end{bmatrix}. \tag{7.49}$$

Higher-order Hadamard matrices can be obtained by replacing each element of H_2 by the matrix H_2 multiplied by the element replaced [which is equivalent to forming an outer product; see Eq. (4.48)]. If this is performed twice, for example, we obtain

$$H_8 = \begin{bmatrix} 1 & 1 & 1 & 1 & 1 & 1 & 1 & 1 \\ 1 & -1 & 1 & -1 & 1 & -1 & 1 & -1 \\ 1 & 1 & -1 & -1 & 1 & 1 & -1 & -1 \\ 1 & -1 & -1 & 1 & 1 & -1 & -1 & 1 \\ 1 & 1 & 1 & 1 & -1 & -1 & -1 & -1 \\ 1 & -1 & 1 & -1 & -1 & 1 & -1 & 1 \\ 1 & 1 & -1 & -1 & -1 & -1 & 1 & 1 \\ 1 & -1 & -1 & 1 & -1 & 1 & 1 & -1 \end{bmatrix} \quad \begin{matrix} \text{cal}(0,t), \ \text{pal}(0,t) \\ \text{sal}(4,t), \ \text{pal}(4,t) \\ \text{sal}(2,t), \ \text{pal}(2,t) \\ \text{cal}(2,t), \ \text{pal}(6,t) \\ \text{sal}(1,t), \ \text{pal}(1,t) \\ \text{cal}(3,t), \ \text{pal}(5,t) \\ \text{cal}(1,t), \ \text{pal}(3,t) \\ \text{sal}(3,t), \ \text{pal}(7,t) \ . \end{matrix} \tag{7.50}$$

The rows of the matrices correspond to the Walsh functions indicated, the signs being reversed for odd sequencies in this particular generation process. The waveform required at the phase detector is the product of the phase-switching functions at the two antennas involved. The product of two such Walsh functions is a Walsh function, the sequency of which is greater than, or equal to, the difference between the sequencies of the two original functions.

Walsh functions can also be generated as products of square-wave functions. Square-wave functions are here designated Sq(n, t), where n is an integer and the half-period of the square wave is $T/2^n$; that is, there are 2^{n-1} complete cycles within the time base, T. The function Sq($0, t$) has a constant value of unity. In the examples in Fig. 7.12, sal($1, t$) is a square-wave function, and cal($3, t$) and sal($9, t$) are each products of sal($1, t$) and one other square-wave function. When considering Walsh functions as products of square-wave functions, it is convenient to use the Paley designation of Walsh functions, pal(n, t) (Paley 1932). The integer n is called the *natural order* of the Walsh function. A Walsh function pal(n, t), which is the product of square-wave functions Sq(i, t), Sq(j, t), ..., Sq(m, t), has a natural order number $n = 2^{i-1} + 2^{j-1}+, \ldots, +2^{m-1}$. The product of two Walsh functions is another Walsh function, of which the natural order number is given by modulo-2 addition (that is, no-carry addition) of the binary natural order numbers of the component Walsh functions.

Table 7.4 shows the relationship between the natural order numbers for a series of Walsh functions and the square-wave functions of which they are composed. The product of two Walsh functions can be expressed as the product of the component square-wave functions, for example,

$$
\begin{aligned}
\text{pal}(7, t) \times \text{pal}(10, t) &= [\text{Sq}(1, t) \times \text{Sq}(2, t) \times \text{Sq}(3, t)] \times [\text{Sq}(2, t) \times \text{Sq}(4, t)] \\
&= \text{Sq}(1, t) \times \text{Sq}(2, t) \times \text{Sq}(2, t) \times \text{Sq}(3, t) \times \text{Sq}(4, t) \\
&= \text{Sq}(1, t) \times \text{Sq}(3, t) \times \text{Sq}(4, t) \\
&= \text{pal}(13, t) ,
\end{aligned}
\tag{7.51}
$$

where we have used the fact that the product of a Walsh or square-wave function with itself is equal to unity. The natural orders of the two Walsh functions, 7 and 10, in binary form are 0111 and 1010. The modulo-2 addition of these binary numbers is 1101, which is equal to 13, the natural order of the Walsh function product.

Table 7.4 Square-wave components of some Walsh functions

Natural order designation	Sq($0, t$)	Sq($1, t$)	Sq($2, t$)	Sq($3, t$)	Sq($4, t$)	Sequency designation
pal($0,t$)	1					cal($0,t$)
pal($1,t$)		1				sal($1,t$)
pal($2,t$)			1			sal($2,t$)
pal($3,t$)		1	1			cal($1,t$)
pal($4,t$)				1		sal($4,t$)
pal($5,t$)		1		1		cal($3,t$)
pal($6,t$)			1	1		cal($2,t$)
pal($7,t$)		1	1	1		sal($3,t$)
pal($8,t$)					1	sal($8,t$)
pal($9,t$)		1			1	cal($7,t$)
pal($10,t$)			1		1	cal($6,t$)
pal($11,t$)		1	1		1	sal($7,t$)
pal($12,t$)				1	1	cal($4,t$)
pal($13,t$)		1		1	1	sal($5,t$)
pal($14,t$)			1	1	1	sal($6,t$)
pal($15,t$)		1	1	1	1	cal($5,t$)

The examination of Walsh functions as products of square-wave functions leads to a useful insight into the efficiency of Walsh function phase switching in eliminating unwanted components (Emerson 1983, 2009). Let $\mathcal{U}(t)$ be an unwanted response within the receiving system, for example, resulting from cross talk in IF signals or from an error in the sampling level of a digitizer. $\mathcal{U}(t)$ arises after the initial phase switching, so when synchronous detection with the phase-switching waveform is performed at a later stage, $\mathcal{U}(t)$ becomes $\mathcal{U}(t)\mathrm{pal}(n, t)$, and this product is significantly reduced in the subsequent averaging. Suppose that $\mathrm{pal}(n, t)$ is the product of m square-wave functions, $\mathrm{Sq}(1, t), \mathrm{Sq}(2, t), \ldots, \mathrm{Sq}(m, t)$. We can consider multiplying $\mathcal{U}(t)$ by $\mathrm{pal}(n, t)$ as equivalent to multiplying by each of the square-wave components in turn. Also, we assume that the period of the square-wave functions is small compared with the timescale of variations of $\mathcal{U}(t)$. Then, after the first multiplication and averaging, the mean residual spurious voltage is

$$\mathcal{U}_1(t) = \frac{[\mathcal{U}(t) + \delta t \frac{d\mathcal{U}}{dt}] - \mathcal{U}(t)}{2} = \frac{\delta t}{2} \frac{d\mathcal{U}}{dt} = \frac{T}{2^{i+1}} \frac{d\mathcal{U}}{dt} , \tag{7.52}$$

where δt is equal to the half-period of the square-wave function, $T/2^i$. $\mathcal{U}_1$ is calculated for one cycle of $\mathrm{Sq}(i, t)$, but within the assumption that $\mathcal{U}(t)$ is slowly varying, $\mathcal{U}_1$ can be taken as equal to the average over the Walsh time base T. Multiplication by the second square-wave function is obtained by replacing $\mathcal{U}$ in Eq. (7.52) by $\mathcal{U}_1$, which yields

$$\mathcal{U}_2(t) = \frac{T}{2^{j+1}} \frac{d\mathcal{U}_1}{dt} = \frac{T^2}{2^{i+j+2}} \frac{d^2\mathcal{U}}{dt^2} . \tag{7.53}$$

For the m square-wave components, we obtain

$$\mathcal{U}_m(t) = \frac{T^m}{2^{(1+2+\cdots+m)}} \frac{d^m\mathcal{U}}{dt^m} , \tag{7.54}$$

so only the higher derivatives of $\mathcal{U}$ remain.

Walsh functions $\mathrm{pal}(n, t)$ for which n is an integral power of two are the least effective in eliminating unwanted responses, since they are each just a single square-wave function. As shown by examination of Table 7.4, those for which $n = 2^k - 1$, where k is an integer, contain the largest number of square-wave components. In arrays with a small number of antennas, for which a large number of different switching functions is not required, it is possible to select Walsh functions that are the most effective in reducing unwanted components. Similarly, Walsh functions can be more effective than square waves in some applications to single antennas, such as beam switching between a source and a reference position on the sky (Emerson 1983). Another set of possible phase-switching functions are m-sequences, considered by Keto (2000) for cases in which both $90°$ and $180°$ phase changes are required.

7.5.3   Timing Accuracy in Phase Switching

In designing a phase-switching system, timing tolerances should be considered. In general, the accuracy should be much smaller than the minimum interval between transitions in any function. For example, in ALMA [the Atacama Large Millimeter/submillimeter Array, Wootten (2003), Wootten and Thompson (2009)], two phase-switching actions are used, one nested within the other (Emerson 2007). For sideband separation, $\frac{\pi}{2}$ phase-switching is used, with Walsh functions from a 128-element set with time base 2.048 s and minimum interval between transitions of 16 ms. A second 128-element set with time base 16 ms and minimum interval 125 μs is used for π-shift switching. Thus, timing errors must be very small compared with 125 μs.

In general, the orthogonality of Walsh functions requires that there be no relative time shifts between the functions. The first switching occurs at the antenna location, that is, as early in the signal path as possible. Digitization of the received signal may occur at the antenna or after transmission of the signal in analog form to a central processing location. The major system delays are shown in Fig. 7.13, in which τ_g is the geometric delay, τ_{tr} is the transmission delay (antenna to processing location), and τ_i is the instrumental compensating delay. Delays in the analog or digital circuitry are generally small enough to be neglected with regard to the timing of phase switching. There are three main timing requirements in the receiving system.

1. The total delay from the incident wavefront to the correlator input, $\tau_g + \tau_{tr} + \tau_i$, must be the same for all antennas to preserve the correlation of the wanted signals. This is implemented through adjustment of τ_i.
2. The corresponding transitions in the first and second phase switchings should be aligned in time so that the phase switching of the wanted signals is precisely

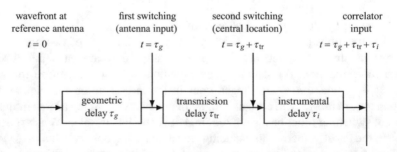

Fig. 7.13 Delays in an array that are large enough to affect the timing of the Walsh functions used in phase switching. Here, t is time relative to a signal wavefront at the point where it intercepts the delay reference antenna. The second switching is shown after the transmission delay, which applies when the signals are transmitted in analog form to the central processing location. When digitization occurs at the antenna location, both the first and second switchings can be applied before the transmission delay.

canceled. For example, if the second phase switching is done at the central location, the timing of the second switching should be delayed relative to the first one by τ_{tr}.

3. Both the first and second switchings in any signal path should be delayed by the geometric delay of the corresponding antenna τ_g so that, at the correlator input, the switching transitions in the unwanted components are aligned in time from one antenna to another. The delay τ_g varies with time as the antennas track a source.

Requirement 2 above is concerned only with the relative accuracy of switchings within the same signal path from one antenna to the correlator. This is the simplest case because it is concerned only with offsets in two switchings of the same Walsh function. Consider the effect of a small time offset δ in the relative timing of the first and second switchings. For each transition, the timing difference causes the correlator output voltage to be reversed for a period δ and thereby cancels an equivalent interval of the unreversed output. Hence, for each transition, there is an effective loss of signal for a period 2δ. The average fractional loss of sensitivity is $2n_t\delta/\tau_{tb}$, where n_t is the number of transitions within the time base τ_{tb} (i.e., twice the Walsh sequency). Thus, for a tolerable limit of, e.g., 1% correlation loss, the tolerable value of δ can be determined for any given time base and maximum sequency used. Since the correlation loss is proportional to n_t, use of the lowest sequencies within the Walsh set helps to minimize loss in sensitivity. For arrays in which the numbers of antennas and the baseline lengths are not too long, the delaying of the switchings by τ_g (as noted in the third requirement above) can often be neglected. This introduces a timing error τ_g that is greatest for the longest baselines. The effect of this error can be minimized by using the lowest values of n_t for the antennas for which the geometric delay is greatest.

Requirements 1 and 3 are concerned with the relative timing of transitions at different antennas, i.e., between different Walsh functions. The effect of a timing offset on the rejection of the unwanted components depends on the loss in orthogonality of the Walsh functions used for different antennas. This is more complicated than the effect of an offset on two identical Walsh functions discussed above. The loss in orthogonality depends upon the sequencies of the two functions involved and is greatest for sequencies in the middle range of the Walsh set, as shown by Emerson (2005). Pairs consisting of a function with an even sequency and one with an odd sequency remain orthogonal in the presence of time shifts, but such combinations are possible for no more than half of the pairs in a complete Walsh set. Of the other pairs, some remain orthogonal with time offsets, as can be shown by numerical trials, and some do not [as shown in Fig. 3 of Emerson (2005)]. It is clearly beneficial to use equal numbers of odd and even sequencies in an array so that for approximately half of the antenna pairs, the orthogonality is independent of time offsets.

7.5.4   Interaction of Phase Switching with Fringe Rotation and Delay Adjustment

The effectiveness of phase switching in reducing the response to spurious signals depends on the point in the signal channel at which these unwanted signals are introduced. The three following cases illustrate the most important possibilities.

1. The unwanted signal enters the antennas or some point in the signal channels that is ahead of the phase switching, the fringe rotation, and the compensating delays. The unwanted signal then suffers phase switching like the wanted signals and is not suppressed in the synchronous detection (although it may be reduced by the fringe rotation if the fringe frequency is high, as in the case of VLBI). Externally generated interference behaves in this manner, and its effect is discussed in Chap. 16.
2. The unwanted signal enters after the phase switching but before the fringe rotation and delay compensation. The fringe-rotation phase shifts, designed to reduce to zero the fringe frequencies of the desired signals at the correlator output, act on the spurious signal and cause it to appear at the correlator output as a component at the natural fringe frequency for a point source at the phase reference position. This component then undergoes synchronous detection with a Walsh function. If the natural fringe frequency transiently matches the frequency of a Fourier component of this Walsh function, a spurious response can occur.
3. The spurious signal enters after the phase switching and the fringe rotation but before the delay compensation. The signal then suffers phase shifts resulting from the changing of the compensating delay. The resulting component at the correlator output has a frequency equal to the natural fringe frequency that would occur if the observing frequency were equal to the IF at which the compensating delays are introduced. Thus, the oscillations are one to three orders of magnitude lower in frequency than the natural fringe frequency, and it is consequently easier to avoid coincidence with the frequency of a component of the Walsh function.

From these considerations, it is usually advantageous to perform both the phase switching and the fringe rotation as early in the signal channel as possible. Figure 7.14 shows, as an example, the phase-switching scheme that was used in the original VLA system, from a description by Granlund et al. (1978). The phase switching at the antenna was performed on an LO, rather than on the full signal band, so that a broadband phase switch was not needed. The signals were digitized at the output of the final IF amplifier and thereafter were delayed and multiplied digitally. In such a system, slow phase drifts that may occur after the phase switching are removed by the synchronous detection at the digitizing sampler. The synchronous detection could be performed by reversing the sign bit in the digitized signal data and needed to be applied only to n_a signal channels rather than $n_a(n_a - 1)/2$ correlator outputs.

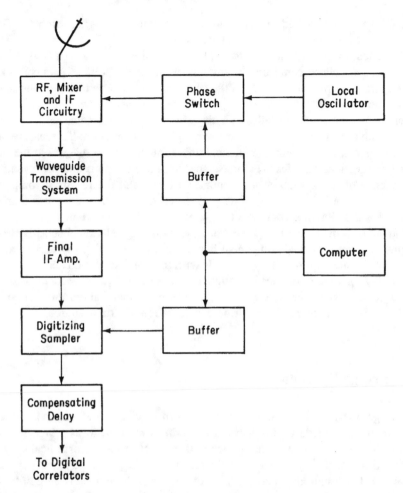

Fig. 7.14 Simplified schematic diagram of the receiving channel for one antenna of the original VLA system. Walsh functions generated by the computer were periodically fed to digital buffers, from which they were clocked out to the phase switch and to sign-reversal circuitry at the digitizing sampler. © 1978 IEEE. Reprinted, with permission, from J. Granlund et al. (1978).

7.6   Automatic Level Control and Gain Calibration

In most synthesis arrays, automatic level control (ALC) circuits are used to hold constant the level of the total signal, that is, the cosmic signal plus the system noise, at certain critical points. A fraction of the total signal level is detected, and the resulting voltage is compared with a preset value to generate a control signal that is fed back to a variable-gain element of the signal chain. Points at which the signal level is critical include modulators for transmission of IF signals on optical or microwave carriers and inputs to analog correlators or digital samplers. For a discussion of level tolerances in samplers, see Sect. 8.5.1.

The effect of an ALC loop is to hold constant the quantity $|g|^2(T_S + T_A)\Delta v$, where g is the voltage gain from the antenna output to the point of gain control, T_S is the system temperature, and T_A is the component of antenna temperature due to the source under observation. Thus, $|g|^2$ is made to vary inversely as $(T_S + T_A)$, which can change substantially with the antenna pointing angle as a result of ground radiation in the sidelobes and atmospheric attenuation. To measure such gain changes, a signal from a broadband noise source can be injected at the input of the receiving electronics. This noise source is switched on and off, usually at a frequency of a few hertz to a few hundred hertz, and the resulting component is sampled and monitored using a synchronous detector. When the noise source is on, it adds a calibrating component T_C to the overall system temperature, which should not be more than a few percent of T_S to avoid degradation of sensitivity. The amplitude of the switched component is a direct measure of the system gain, and for $T_A \ll T_S$, the ratio of the signal levels with the noise source on and off is equal to $1 + T_C/T_S$, which provides a continuous measure of T_S. This scheme does not correct for changes in antenna gain resulting from mechanical deformation, which must be calibrated separately by periodic observation of a radio source.

7.7 Fringe Rotation

The fringe oscillations in the data from the correlator must be removed before an image can be formed. This process is sometimes referred to as fringe stopping (i.e., stopping the motion of the fringes with respect to the astronomical sky). As described in Chap. 6, this can be achieved by inserting a fringe-frequency offset on an LO. For a multiantenna array, the offset for each antenna is chosen to stop the fringes for that antenna when combined with a common reference antenna. It is also possible to stop the fringes by inserting corrections in the phase of the signals at the correlator. If the corrections are inserted before the cross multiplication that occurs in the correlation, they can be applied to each of the n_a antennas of the array (see, e.g., Carlson and Dewdney 2000), whereas after cross multiplication, the corrections must be applied to all of the $n_a^2/2$ antenna pairs. However, corrections inserted before cross multiplication must be applied to each signal sample that goes to the cross-correlator, whereas corrections applied to the cross products can be performed after some time-limited averaging of the products. (The averaging must not be so long that the fringe oscillations are attenuated.) The effect of time averaging on the fringes is to convolve the sinusoidal fringe function of frequency v_f with a rectangular function of width equal to the averaging time τ_{av}. A 1% loss in sensitivity occurs for $v_f\tau_{av} = 0.078$ and 2% loss for $v_f\tau_{av} = 0.111$. As an approximate criterion, the averaging time should be no more than $\sim 1/10$ of a fringe period. For a DSB system, the LO offset must be applied to the first LO, or if the fringe stopping is applied at the correlator, the sidebands must first be separated. For sideband separation, see Sect. 6.1.12.

Appendix 7.1   Sideband-Separating Mixers

The principle of the sideband-separating mixer, or image-rejection mixer, is shown in Fig. A7.1. The terms $\cos(2\pi\nu_u t)$ and $\cos(2\pi\nu_\ell t)$ represent frequency components of the input waveform at the upper- and lower-sideband frequencies, respectively. The input is applied to two mixers, for which the LO waveforms at frequency ν_{LO} are in phase quadrature. The mixers generate products of the signal and LO waveforms, and the filters pass only the terms of frequency equal to the difference of ν_{LO} and ν_u or ν_ℓ. The output from the lower mixer also passes through a $\pi/2$ phase lag network. From the resulting terms at points A and B, one can see that by applying the waveforms at these points to a summing network, the upper-sideband response is obtained. Similarly, by using a differencing network, the lower-sideband response is obtained. In either case, the accuracy of the suppression of the response to the unwanted sideband depends on the accuracy of the quadrature phase relationships, the matching of the frequency responses of the mixers and filters, and the insertion loss of the phase lag network. In practice, for conversion from a few gigahertz to baseband, suppression to a level of -20 dB is routinely achievable. With careful design, suppression to a level approaching -30 dB can be obtained (Archer et al. 1981).

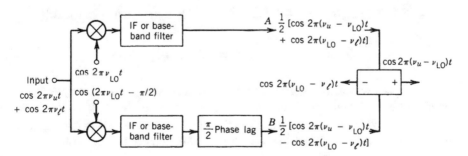

Fig. A7.1 Schematic illustration of the principle of the sideband-separating (image-rejection) mixer. The upper-sideband response is obtained from the sum of the outputs A and B, and the lower-sideband response from the difference of these outputs.

Appendix 7.2   Dispersion in Optical Fiber

For a frequency component, ν_m, of a signal modulated onto an optical carrier, $A\sin(2\pi\nu_{\text{opt}} + \phi)$, and transmitted down a fiber, the resulting signal intensity at the output of the fiber can be represented by

$$A^2[1 + m\cos(2\pi\nu_m t)]\sin^2(2\pi\nu_{\text{opt}}t + \phi)$$

$$= A^2\sin^2(2\pi\nu_{\text{opt}}t + \phi)$$

$$+ \frac{A^2 m}{2}\sin[2\pi(\nu_{\text{opt}} + \nu_m)(t - \Delta t) + \phi]\sin(2\pi\nu_{\text{opt}}t + \phi)$$

$$+ \frac{A^2 m}{2}\sin[2\pi(\nu_{\text{opt}} - \nu_m)(t + \Delta t) + \phi]\sin(2\pi\nu_{\text{opt}}t + \phi) , \qquad (A7.1)$$

where m is the modulation index. This equation resembles the usual representation for amplitude modulation in communications, except that here, the carrier *power* varies linearly with the modulation. Thus, on the left side, the square of the carrier expression is used. For the terms of frequency $\nu_{\text{opt}} \pm \nu_m$, the time has been offset by $\pm\Delta t$ to represent the effects of the variation of propagation velocity with frequency. Δt can take both positive and negative values depending on the sign of the dispersion $\mathcal{D}$ shown in Fig. 7.3. Each term in Eq. (A7.1) is proportional to optical power and, thus, also to the modulation amplitude. By applying the identity for the product of two sines to each term on the right side of Eq. (A7.1), and ignoring DC and optical frequency terms, we obtain for the amplitude at the output of the optical receiver,

$$\frac{A^2 m}{4}\left\{\cos[2\pi\nu_m(t + \Delta t) - 2\pi\nu_{\text{opt}}\Delta t] + \cos[2\pi\nu_m(t - \Delta t) - 2\pi\nu_{\text{opt}}\Delta t]\right\}$$

$$= \frac{A^2 m}{2}\left\{\cos[2\pi(\nu_m t - \nu_{\text{opt}}\Delta t)]\cos(2\pi\nu_m\Delta t)\right\} . \qquad (A7.2)$$

The free-space wavelength corresponding to frequency ν_{opt} is λ_{opt}, and the wavelength difference between frequencies ν_{opt} and $\nu_{\text{opt}} + \nu_m$ is $\lambda_{\text{opt}}^2\nu_m/c$ (since $\nu_m \ll \nu_{\text{opt}}$). If $\mathcal{D}$ is the dispersion and ℓ is the length of the fiber, $\Delta t = \mathcal{D}\ell\lambda_{\text{opt}}^2\nu_m/c$, and $\nu_{\text{opt}}\Delta t = \mathcal{D}\ell\lambda_{\text{opt}}\nu_m$. Thus, the recovered modulation can be written as

$$\frac{A^2 m}{2}\left\{\cos[2\pi\nu_m(t - \mathcal{D}\ell\lambda_{\text{opt}})]\cos(2\pi\nu_m^2\mathcal{D}\ell\lambda_{\text{opt}}^2/c)\right\} . \qquad (A7.3)$$

The phase change induced by Δt at the carrier frequency ν_{opt} appears in the phase of the modulation frequency in the first cosine function in Eq. (A7.2). At frequency ν_m, this phase term is equivalent to a time delay $\mathcal{D}\ell\lambda_{\text{opt}}$, as seen in Eq. (A7.3). This delay is much larger than Δt and represents the difference between the phase and the group velocities in the fiber. The second cosine modifies the amplitude of the modulation component ν_m. For example, with dispersion $\mathcal{D} = 2$ ps/(km·nm) (note that this is equal to 2×10^{-6}s m^{-2}), $\ell = 50$ km, $\lambda_{opt} = 1550$ nm, and $\nu_m = 10$ GHz, we obtain $\Delta t = 8$ ps, $\mathcal{D}\ell\lambda_{opt} = 155$ ns, and the response at frequency ν_m is reduced by 1.1 dB relative to the low-frequency end of the modulation spectrum. Note that we have assumed above that the frequency spread of the laser results entirely from the modulation spectrum, which is justifiable for a high-quality laser with an external modulator. Modulation of a diode laser by varying the voltage across it can result in unwanted frequency modulation, further spreading the optical spectrum.

Appendix 7.3   Alias Sampling

After Nyquist sampling of a signal band $n\Delta\nu$ to $(n + 1)\Delta\nu$, where n is an integer, the frequency band of the sampled data is 0 to $\Delta\nu$ and does not depend on the frequency of the band at the sampler input.[10] This effect is known as alias sampling. To illustrate this situation, consider a Fourier component $A\sin(2\pi\nu t + \phi)$, with arbitrary amplitude and phase, within a band 0 to $\Delta\nu$. The band is sampled at the Nyquist rate, the sample times being $t = m/(2\Delta\nu)$ where $m = 0, 1, 2, \ldots$. The sampled values of the component are $A\sin(\phi), A\sin(\frac{\pi\nu}{\Delta\nu} + \phi), A\sin(\frac{2\pi\nu}{\Delta\nu} + \phi), \ldots$. Now consider the case in which the same input band has been converted to the range $\Delta\nu$ to $2\Delta\nu$. The frequency is higher by $\Delta\nu$, so the original component becomes

$$A\sin[2\pi(\nu + \Delta\nu)t + \phi] =$$

$$A\sin(2\pi\nu t + \phi)\cos(2\pi\Delta\nu t) + A\cos(2\pi\nu t + \phi)\sin(2\pi\Delta\nu t) .$$

$$(A7.4)$$

Again, sampling at times $m/(2\Delta\nu)$, we obtain for the components: $A\sin(\phi), -A\sin(\frac{\pi\nu}{\Delta\nu} + \phi), A\sin(\frac{2\pi\nu}{\Delta\nu} + \phi), \ldots$. The result is the same as before except that the sign is reversed for odd values of m. Further investigation shows that this sign reversal occurs when n has an odd value. Since the sign reversal occurs for both signals of a cross-correlated pair, it has no effect on the product. Thus, for any value of n, the result at the correlator output is the same as for a baseband input to the sampler. Thus, sampling of the band $n\Delta\nu$ to $(n + 1)\Delta\nu$ has the effect of converting the band downward by $n\Delta\nu$, sometimes referred to as alias sampling.

---

[10]This is the case, for example, in both the VLA and ALMA, where a 1 : 2 frequency ratio between the lower and upper edges of the final analog IF response is used because it is easier to maintain uniform gain than with a baseband response.

References

Agrawal, G.P., *Fiber-Optic Communication Systems*, Wiley, New York (1992)

Allen, L.R., and Frater, R.H., Wideband Multiplier Correlator, *Proc. IEEE*, **117**, 1603–1608 (1970)

Archer, J.W., Caloccia, E.M., and Serna, R., An Evaluation of the Performance of the VLA Circular Waveguide System, *IEEE Trans. Microwave Theory Tech.*, **MTT-28**, 786–791 (1980)

Archer, J.W., Granlund, J., and Mauzy, R.E., A Broadband UHF Mixer Exhibiting High Image Rejection Over a Multidecade Baseband Frequency Range, *IEEE J. Solid-State Circuits*, **SC-16**, 385–392 (1981)

Baars, J.W.M., van der Brugge, J.F., Casse, J.L., Hamaker, J.P., Sondaar, L.H., Visser, J.J., and Wellington, K.J., The Synthesis Radio Telescope at Westerbork, *Proc. IEEE*, **61**, 1258–1266 (1973)

Bagri, D.S., and Thompson, A.R., Hardware Considerations for High Dynamic Range Imaging, *Radio Interferometry: Theory, Techniques and Applications*, IAU Colloq. 131, T. J. Cornwell and R. A. Perley, Eds., Astron. Soc. Pacific Conf. Ser., **19**, 47–54 (1991)

Batty, M.J., Jauncey, D.L., Rayner, P.T., and Gulkis, S., Tidbinbilla Two-Element Interferometer, *Astron. J.*, **87**, 938–944 (1982)

Beauchamp, K.G., *Walsh Functions and Their Applications*, Academic Press, London (1975)

Borella, M.S., Jue, J.P., Banerjee, D., Ramamurthy, B., and Mukherjee, B., Optical Components for WDM Lightwave Networks, *Proc. IEEE*, **85**, 1274–1307 (1997)

Bracewell, R.N., Colvin, R.S., D'Addario, L.R., Grebenkemper, C.J., Price, K.M., and Thompson, A.R., The Stanford Five-Element Radio Telescope, *Proc. IEEE*, **61**, 1249–1257 (1973)

Callen, H.B., and Welton, T.A., Irreversibility and Generalized Noise, *Phys. Rev.*, **83**, 34–40 (1951)

Carlson, B.R., and Dewdney, P.E., Efficient Wideband Digital Correlation, *Electronics Lett.*, **36**, 987–988 (2000)

Casse, J.L., Woestenburg, E.E.M., and Visser, J.J., Multifrequency Cryogenically Cooled Front-End Receivers for the Westerbork Synthesis Radio Telescope, *IEEE Trans. Microwave Theory Tech.*, **MTT-30**, 201–209 (1982)

Coe, J.R., Interferometer Electronics, *Proc. IEEE*, **61**, 1335–1339 (1973)

D'Addario, L.R., Orthogonal Functions for Phase Switching and a Correction to ALMA Memo 287, ALMA Memo 385, National Radio Astronomy Observatory (2001)

D'Addario, L.R., Passband Shape Deviation Limits, ALMA Memo 452, National Radio Astronomy Observatory (2003)

Davies, J.G., Anderson, B., and Morison, I., The Jodrell Bank Radio-Linked Interferometer Network, *Nature*, **288**, 64–66 (1980)

de Vos, M., Gunst, A.W., and Nijboer, R., The LOFAR Telescope: System Architecture and Signal Processing, *Proc. IEEE*, **97**, 1431–1437 (2009)

Elsmore, B., Kenderdine, S., and Ryle, M., Operation of the Cambridge One-Mile Diameter Radio Telescope, *Mon. Not. R. Astron. Soc.*, **134**, 87–95 (1966)

Emerson, D.T., *The Optimum Choice of Walsh Functions to Minimize Drift and Cross-Talk*, Working Report 127, IRAM, Grenoble, July 18 (1983)

Emerson, D.T., Walsh Function Demodulation in the Presence of Timing Errors, Leading to Signal Loss and Crosstalk, ALMA Memo 537, National Radio Astronomy Observatory (2005)

Emerson, D.T., Walsh Function Definition for ALMA, ALMA Memo 565, National Radio Astronomy Observatory (2007)

Emerson, D.T., Walsh Function Choices for 64 Antennas, ALMA Memo 586, National Radio Astronomy Observatory (2009)

Erickson, W.C., Mahoney, M.J., and Erb, K., The Clark Lake Teepee-Tee Telescope, *Astrophys. J. Suppl.*, **50**, 403–420 (1982)

Fowle, F.F., The Transposition of Electrical Conductors, *Trans. Am. Inst. Elect. Eng.*, **23**, 659–689 (1904)

Gardner, F.M., *Phaselock Techniques*, 2nd ed., Wiley, New York (1979)

Granlund, J., Thompson, A.R., and Clark, B.G., An Application of Walsh Functions in Radio Astronomy Instrumentation, *IEEE Trans. Electromag. Compat.*, **EMC-20**, 451–453 (1978)

Harmuth, H.F., Applications of Walsh Functions in Communications, *IEEE Spectrum*, **6(11)**, 82–91 (1969)

Harmuth, H.F., *Transmission of Information by Orthogonal Functions*, 2nd ed., Springer-Verlag, Berlin (1972)

Kerr, A.R., Suggestions for Revised Definitions of Noise Quantities, Including Quantum Effects, *IEEE Trans. Microwave Theory Tech.*, **47**, 325–329 (1999)

Kerr, A.R., Feldman, M.J., and Pan, S.-K., Receiver Noise Temperature, the Quantum Noise Limit, and the Role of Zero-Point Fluctuations, *Proc. 8th Int. Symp. Space Terahertz Technology* (1997), pp. 101–111; also available as MMA Memo 161, National Radio Astronomy Observatory (1997)

Keto, E., Three-Phase Switching with m-Sequences for Sideband Separation in Radio Interferometry, *Publ. Astron. Soc. Pacific*, **112**, 711–715 (2000)

Kraus, J.D., *Radio Astronomy*, 2nd ed., Cygnus-Quasar Books, Powell, OH (1986)

Little, A.G., A Phase-Measuring Scheme for a Large Radiotelescope, *IEEE Trans. Antennas Propag.*, **AP-17**, 547–550 (1969)

Moran, J.M., The Submillimeter Array, in *Advanced Technology MMW, Radio, and Terahertz Telescopes*, Phillips, T.G., Ed., Proc. SPIE, **3357**, 208–219 (1998)

Napier, P.J., Bagri, D.S., Clark, B.G., Rogers, A.E.E., Romney, J.D., Thompson, A.R., and Walker, R.C., The Very Long Baseline Array, *Proc. IEEE*, **82**, 658–672 (1994)

Napier, P.J., Thompson, A.R., and Ekers, R.D., The Very Large Array: Design and Performance of a Modern Synthesis Radio Telescope, *Proc. IEEE*, **71**, 1295–1320 (1983)

National Radio Astronomy Observatory, *A Proposal for a Very Large Array Radio Telescope*, Vol. II, National Radio Astronomy Observatory, Green Bank, WV (1967), ch. 14

Nyquist, H., Thermal Agitation of Electric Charge in Conductors, *Phys. Rev.*, **32**, 110–113 (1928)

Padin, S., A Wideband Analog Continuum Correlator for Radio Astronomy, *IEEE Trans. Instrum. Meas.*, **IM-43**, 782–784 (1994)

Paley, R.E.A.C., A Remarkable Set of Orthogonal Functions, *Proc. London Math. Soc.*, **34**, 241–279 (1932)

Payne, J.M., Millimeter and Submillimeter Wavelength Radio Astronomy, *Proc. IEEE*, **77**, 993–1071 (1989)

Payne, J.M., D'Addario, L., Emerson, D.T., Kerr, A.R., and Shillue, B., Photonic Local Oscillator for the Millimeter Array, in *Advanced Technology MMW, Radio, and Terahertz Telescopes*, Phillips, T.G., Ed., Proc. SPIE, **3357**, 143–151 (1998)

Payne, J.M., Lamb, J.W., Cochran, J.G., and Bailey, N.J., A New Generation of SIS Receivers for Millimeter-Wave Radio Astronomy, *Proc. IEEE*, **82**, 811–823 (1994)

Perley, R., Napier, P., Jackson, J., Butler, B., Carlson, B., Fort, D., Dewdney, P., Clark, B., Hayward, R., Durand, S., Revnell, M., and McKinnon, M., The Expanded Very Large Array, *Proc. IEEE*, **97**, 1448–1462 (2009)

Phillips, T.G., Millimeter and Submillimeterwave Receivers, in *Astronomy with Millimeter and Submillimeter Wave Interferometry*, Ishiguro, M., and Welch, W.J., Eds., Astron. Soc. Pacific Conf. Ser., **59**, 68–77 (1994)

Phillips, T.G., and Woody, D.P., Millimeter- and Submillimeter-Wave Receivers, *Ann. Rev. Astron. Astrophys.*, **20**, 285–321 (1982)

Pospieszalski, M.W., Extremely Low-Noise Amplification with Cryogenic FET's and HFET's: 1970–2004, *Microw. Mag.*, **6**, 62–75 (2005)

Prabu, T., Srivani, K.S., Roshi, D.A., Kamini, P.A., Madhavi, S., Emrich, D., Crosse, B., Williams, A.J., Waterson, M., Deshpande, A.A., and 48 coauthors, A Digital Receiver for the Murchison Widefield Array, *Experimental Astron.*, **39**, 73–93 (2015)

Read, R.B., Two-Element Interferometer for Accurate Position Determinations at 960 Mc, *IRE Trans. Antennas Propag.*, **AP-9**, 31–35 (1961)

Reid, M.S., Clauss, R.C., Bathker, D.A., and Stelzried, C.T., Low Noise Microwave Receiving Systems in a Worldwide Network of Large Antennas, *Proc. IEEE*, **61**, 1330–1335 (1973)

Sinclair, M.W., Graves, G.R., Gough, R.G., and Moorey, G.G., The Receiver System, *J. Elect. Electronics Eng. Aust.*, **12**, 147–160 (1992)

Swarup, G., and Yang, K.S., Phase Adjustment of Large Antennas, *IEEE Trans. Antennas Propag.*, **AP-9**, 75–81 (1961)

Thompson, A.R., *Tolerances on Polarization Mismatch*, VLB Array Memo 346, National Radio Astronomy Observatory (1984)

Thompson, A.R., Clark, B.G., Wade, C.M., and Napier, P.J., The Very Large Array, *Astrophys. J. Suppl.*, **44**, 151–167 (1980)

Thompson, A.R., and D'Addario, L.R., Frequency Response of a Synthesis Array: Performance Limitations and Design Tolerances, *Radio Sci.*, **17**, 357–369 (1982)

Thompson, A.R., and D'Addario, L.R., Relative Sensitivity of Double- and Single- Sideband Systems for Both Total Power and Interferometry, ALMA Memo 304, National Radio Astronomy Observatory (2000)

Thompson, M.C., Wood, L.E., Smith, D., and Grant, W.B., Phase Stabilization of Widely Separated Oscillators, *IEEE Trans. Antennas Propag.*, **AP-16**, 683–688 (1968)

Tiuri, M.E., and Räisänen, A.V., Radio-Telescope Receivers, in *Radio Astronomy*, 2nd ed., J. D. Kraus, Cygnus-Quasar Books, Powell, OH (1986), ch. 7

Tucker, J.R., Quantum Limited Detection in Tunnel Junction Mixers, *IEEE J. Quantum Elect.*, **QE-15**, 1234–1258 (1979)

Tucker, J.R., and Feldman, M.J., Quantum Detection at Millimeter Wavelengths, *Rev. Mod. Phys.*, **57**, 1055–1113 (1985)

Valley, S.L., Ed., *Handbook of Geophysics and Space Environments*, Air Force Cambridge Research Laboratories, Bedford, MA (1965), pp. 3-20–3-22

Walsh, J.L., A Closed Set of Orthogonal Functions, *Ann. J. Math.*, **55**, 5–24 (1923)

Webber, J.C., and Pospieszalski, M.W., Microwave Instrumentation for Radio Astronomy, *IEEE Trans. Microwave Theory Tech.*, **MTT-50**, 986–995 (2002)

Weinreb, S., Balister, M., Maas, S., and Napier, P.J., Multiband Low-Noise Receivers for a Very Large Array, *IEEE Trans. Microwave Theory Tech.*, **MTT-25**, 243–248 (1977a)

Weinreb, S., Fenstermacher, D.L., and Harris, R.W., Ultra-Low-Noise 1.2- to 1.7-GHz Cooled GaSaFET Amplifiers, *IEEE Trans. Microwave Theory Tech.*, **MTT-30**, 849–853 (1982)

Weinreb, S., Predmore, R., Ogai, M., and A. Parrish, Waveguide System for a Very Large Antenna Array, *Microwave J.*, **20**, 49–52 (1977b)

Welch, W.J., Forster, J.R., Dreher, J., Hoffman, W., Thornton, D.D., and Wright, M.C.H., An Interferometer for Millimeter Wavelengths, *Astron. Astrophys.*, **59**, 379–385 (1977)

Welch, W.J., Thornton, D.D., Plambeck, R.L., Wright, M.C.H., Lugten, J., Urry, L., Fleming, M., Hoffman, W., Hudson, J., Lum, W.T., and 27 coauthors, The Berkeley–Illinois–Maryland–Association Millimeter Array, *Publ. Astron. Soc. Pacific*, **108**, 93–103 (1996)

Wengler, M.J., and Woody, D.P., Quantum Noise in Heterodyne Detection, *IEEE J. Quantum Electr.*, **QE-23**, 613–622 (1987)

Wootten, A., Atacama Large Millimeter Array (ALMA), in *Large Ground-Based Telescopes*, Oschmann, J.M., and Stepp, L.M., Eds., Proc. SPIE, **4837**, 110–118 (2003)

Wootten, A., and Thompson, A.R., The Atacama Large Millimeter/Submillimeter Array, *Proc. IEEE*, **97**, 1463–1471 (2009)

Wright, M.C.H., Clark, B.G., Moore, C.H., and Coe, J., Hydrogen-Line Aperture Synthesis at the National Radio Astronomy Observatory: Techniques and Data Reduction, *Radio Sci.*, **8**, 763–773 (1973)

Young, A.C., McCulloch, M.G., Ables, S.T., Anderson, M.J., and Percival, T.M., The Local Oscillator System, *Proc. IREE Aust.*, **12**, 161–172 (1992)

Zorin, A.B., Quantum Noise in SIS Mixers, *IEEE Trans. Magn.*, **MAG-21**, 939–942 (1985)

Chapter 8
Digital Signal Processing

The use of digital rather than analog instrumentation offers important practical advantages in the transmission of signals over long baselines, the implementation of compensating time delays, and the measurement of cross-correlation of signals. In digital delay circuits, the accuracy of the delay depends on the accuracy of the timing pulses in the system, and long delays accurate to tens of picoseconds are more easily achieved digitally than by using analog delay lines. Furthermore, there is no distortion of the signal by the digital units other than the calculable effects of quantization. In contrast, with an analog system, it is difficult to keep the shape of the frequency response within tolerances while delay elements are switched into and out of the signal channels. Correlators with wide dynamic range are readily implemented digitally, including those with multichannel output, as needed for spectral line observations. Analog multichannel correlators employ filter banks to divide the signal passband into many narrow channels. Such filters, when subject to temperature variations, can be a source of phase instability. Finally, except at the highest bit rates (frequencies), digital circuits need less adjustment than analog ones and are better suited to replication in large numbers for large arrays.

Digitization of the signal waveforms requires sampling of the voltages at periodic intervals and quantizing the sampled values so that each can be represented by a finite number of bits. The number of bits per sample is usually not large, especially in cases in which the signal bandwidth is large, requiring high sampling rates. However, coarse quantization results in a loss in sensitivity, since modification of the signal levels to the quantized values effectively results in the addition of a component of "quantization noise." In most cases, this loss is small and is outweighed by the other advantages. In designing digital correlators, there are compromises to be made between sensitivity and complexity, and the number of quantization levels to use is an important consideration.

There are two ways to determine the spectrum of a random noise signal, as shown in Fig. 8.1. The autocorrelation function of the signal can be measured and

© The Author(s) 2017

A.R. Thompson, J.M. Moran, and G.W. Swenson Jr., *Interferometry and Synthesis in Radio Astronomy*, Astronomy and Astrophysics Library, DOI 10.1007/978-3-319-44431-4_8

Fig. 8.1 The relationship between two random processes, $x(t)$ and $y(t)$, of duration T, and their cross-correlation function, $R_{xy}(\tau)$ and the cross spectrum, $S_{xy}(\nu)$. If $x(t)$ and $y(t)$ are the same, $R_{xx}(\tau)$ is the autocorrelation function, and $S_{xx}(\nu)$ is the power spectrum. The spatial counterpart to this diagram is shown in Fig. 5.5.

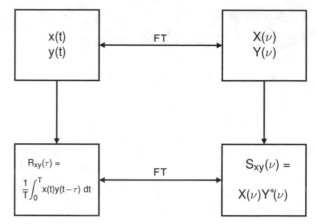

then Fourier transformed into a power spectrum after a specified integration period. Alternately, the signal can be Fourier transformed first and the square modulus taken. In the first case, the resolution of the spectral estimate is approximately the reciprocal of the number of lags of the autocorrelation function calculated. In the direct Fourier transform route, the data stream must be segmented to control the spectral resolution, i.e., the resolution is approximately the reciprocal of the data segment length. The power spectra from all of the segments are summed over the integration period. To compare results between these methods, the number of lags in the correlator is set equal to the number of segment samples. For interferometry, the same two methods can be applied. The cross-correlation function can be calculated and Fourier transformed into a cross spectrum (called the XF technique), or the direct Fourier transform of one can be multiplied by the conjugate of the other to form the cross spectrum (the FX technique). These two methods are explored in detail in this chapter.

Digital signal processing in radio astronomy began in the early 1960s when Weinreb (1963) built a digital 64-channel autocorrelator that operated on the signal sampled at the Nyquist rate and quantized with one bit per sample.[1] At that time, the modern fast Fourier transform (FFT) algorithm (Cooley and Tukey 1965) was not known, although there are historical precedents in the mathematical literature going back to Gauss in the early nineteenth century. For the next two decades, virtually all spectrometers for single-dish and interferometric applications were based on the auto- or cross-correlation approach. By the 1990s and the advent of very large spectral processing systems (in terms of frequency channels and baselines), the advantages of the FX approach became apparent. All modern interferometers have spectral analysis capabilities, not only for observations of spectral lines but also for mitigation of the effects of radio frequency interference (RFI) and of instrumental bandwidth smearing.

---

[1] A similar device was used by Goldstein (1962) to detect radar echoes from Venus.

8.1   Bivariate Gaussian Probability Distribution

The bivariate normal probability function is central to all signal analysis. If x and y are joint Gaussian random variables with zero mean and variance σ^2, the probability that one variable is between x and $x + dx$ and, simultaneously, the other is between y and $y + dy$ is $p(x, y)\,dx\,dy$, where

$$p(x, y) = \frac{1}{2\pi\sigma^2\sqrt{1-\rho^2}} \exp\left[\frac{-(x^2 + y^2 - 2\rho xy)}{2\sigma^2(1-\rho^2)}\right], \qquad (8.1)$$

and ρ is the correlation coefficient equal to $\langle xy\rangle/\sqrt{\langle x^2\rangle\langle y^2\rangle}$, where $\langle\ \rangle$ denotes the expectation, which, with the usual assumption of ergodicity, is approximated by the average over many samples. The form of this function is shown in Fig. 8.2. Note that $-1 \le |\rho| \le 1$. For $|\rho| \ll 1$, the exponential can be expanded, giving

$$p(x, y) \simeq \left[\frac{1}{\sigma\sqrt{2\pi}}\exp\left(\frac{-x^2}{2\sigma^2}\right)\right]\left[\frac{1}{\sigma\sqrt{2\pi}}\exp\left(\frac{-y^2}{2\sigma^2}\right)\right]\left(1 + \frac{\rho xy}{\sigma^2}\right), \qquad (8.2)$$

which for $\rho = 0$ is simply the product of two Gaussian functions. Equation (8.1) can also be written as

$$p(x, y) = \frac{1}{\sigma\sqrt{2\pi}}\exp\left(\frac{-x^2}{2\sigma^2}\right)\frac{1}{\sigma\sqrt{2\pi(1-\rho^2)}}\exp\left[\frac{-(y-\rho x)^2}{2\sigma^2(1-\rho^2)}\right]. \qquad (8.3)$$

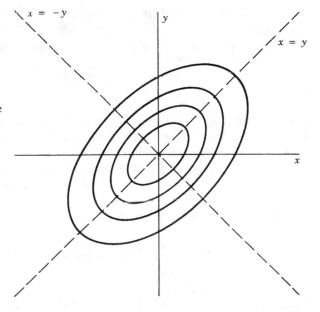

Fig. 8.2 Contours of equal probability density from the bivariate Gaussian distribution in Eq. (8.1). The contours are given by $x^2 + y^2 - 2\rho xy = $ const. For $\rho = 0$, they become circles; for $\rho = 1$, they merge into the line $x = y$; and for $\rho = -1$, they merge into $x = -y$.

If this expression is integrated with respect to y from $-\infty$ to $+\infty$, it reduces to a Gaussian function in x. As ρ approaches unity, Eq. (8.3) becomes the product of a Gaussian in x and a Gaussian in $y - x$; the latter has a standard deviation $\sigma \sqrt{1 - \rho^2}$, which tends to zero as ρ approaches 1. Equations (8.1) and (8.2) will be used in examining the response of various types of samplers and correlators. For autocorrelators used with single antennas, the quantity to be measured is the autocorrelation function $R(\tau) = \langle v(t)v(t - \tau) \rangle$, where v is the received signal. This case can be treated with $x = v(t)$ and $y = v(t - \tau)$.

8.2 Periodic Sampling

8.2.1 Nyquist Rate

If the signal is bandlimited, that is, its power spectrum is nonzero only within a finite band of frequencies, no information is lost in the sampling process as long as the sampling rate is high enough. This follows from the sampling theorem discussed in Sect. 5.2.1. Here, we sample a function of time and must avoid aliasing in the frequency domain. For a baseband (lowpass) rectangular spectrum with an upper cutoff frequency Δv, the width of the frequency spectrum, including negative frequencies, is $2\Delta v$. The function is fully specified by samples spaced in time with an interval no greater than $1/(2\Delta v)$, that is, a sampling frequency of $2\Delta v$ or greater. This critical sampling frequency, $2\Delta v$, is called the *Nyquist rate*[2] for the waveform. For further discussion, see, for example, Bracewell (2000) or Oppenheim and Schafer (2009). In some digital systems in radio astronomy, the waveform that is digitized has a baseband spectrum and is sampled at the Nyquist rate. For a rectangular passband of this type, the autocorrelation function, which by the Wiener–Khinchin relation is the Fourier transform of the power spectrum, is

$$R_\infty(\tau) = \frac{\sin(2\pi \Delta v \, \tau)}{2\pi \Delta v \, \tau} , \tag{8.4}$$

where the subscript ∞ indicates unquantized sampling (that is, the accuracy is not limited by a finite number of quantization levels). Nyquist sampling can also be applied to bandpass spectra, and if the spectrum is nonzero only within a range of $n\Delta v$ to $(n + 1)\Delta v$, where n is an integer, the Nyquist rate is again $2\Delta v$. Thus, for sampling at the Nyquist rate, the lower and upper bounds of the spectral band must be integral multiples of the bandwidth. The autocorrelation function of a signal that has a flat spectrum over such a band is

$$R_\infty(\tau) = \frac{\sin(\pi \Delta v \, \tau)}{\pi \Delta v \, \tau} \cos \left[2\pi \left(n + \tfrac{1}{2} \right) \Delta v \, \tau \right] . \tag{8.5}$$

---

[2]Shannon (1949) cites several references relevant to the development of this result, of which the earliest is Nyquist (1928).

Zeros in this function occur at time intervals τ that are integral multiples of $1/(2\Delta v)$. Therefore, for a rectangular passband, successive samples at the Nyquist rate are uncorrelated. Sampling at frequencies greater or less than the Nyquist rate is referred to as oversampling or undersampling, respectively. For any signal, adjusting the center frequency so that the spectrum conforms to the bandpass sampling requirement described above minimizes the sampling rate required to avoid aliasing.

8.2.2 Correlation of Sampled but Unquantized Waveforms

We now investigate the response of a hypothetical correlator for which the input signals are sampled at the Nyquist rate but are not quantized. It is necessary to consider only single-multiplier correlators since complex correlators can be implemented as combinations of them, as indicated in Fig. 6.3. The system under discussion can be visualized as one in which the samples either remain as analog voltages or are encoded with a sufficiently large number of bits that quantization errors are negligible. Since no information is lost in sampling, the signal-to-noise ratio of the correlation measurement may be expected to be the same as would be obtained by applying the waveforms without sampling to an analog correlator. There is probably no reason, in practice, to build a correlator for inputs with unquantized sampling. However, by comparing the results with those for quantized sampling, which we discuss later, the effects of quantization are more easily understood.

Two bandlimited waveforms, $x(t)$ and $y(t)$, are sampled at the Nyquist rate, and for each pair of samples, the multiplier within the correlator produces an output proportional to the product of the input amplitudes. The integrator allows the output to be averaged for any required time interval. Now the (normalized) cross-correlation coefficient of $x(t)$ and $y(t)$ for zero time delay between the two waveforms is

$$\rho = \frac{\langle x(t)y(t) \rangle}{\sqrt{\langle [x(t)]^2 \rangle \langle [y(t)]^2 \rangle}} . \tag{8.6}$$

(The cross-correlation coefficient ρ should not be confused with the autocorrelation function of x or y, R_∞.) Since x and y have equal variance σ^2,

$$\langle x(t)y(t) \rangle = \rho\sigma^2 . \tag{8.7}$$

The left side is the averaged product of the two waveforms and thus represents the correlator output. The output of the digital correlator after N_N samples is

$$r_\infty = N_N^{-1} \sum_{i=1}^{N_N} x_i y_i , \tag{8.8}$$

where the subscript N denotes the Nyquist rate. Since the samples x_i and y_i obey the same Gaussian statistics as the continuous waveforms $x(t)$ and $y(t)$, we can clearly write

$$\langle r_\infty \rangle = \rho \sigma^2 \, . \tag{8.9}$$

Thus, the output of the correlator is a linear measure of the correlation ρ. The variance of the correlator output is

$$\sigma_\infty^2 = \langle r_\infty^2 \rangle - \langle r_\infty \rangle^2 \, , \tag{8.10}$$

and

$$
\begin{aligned}
\langle r_\infty^2 \rangle &= N_N^{-2} \sum_{i=1}^{N_N} \sum_{k=1}^{N_N} \langle x_i y_i x_k y_k \rangle \\
&= N_N^{-2} \sum_{i=1}^{N_N} \langle x_i y_i \rangle^2 + N_N^{-2} \sum_{i=1}^{N_N} \sum_{k \neq i} \langle x_i y_i x_k y_k \rangle \, ,
\end{aligned}
\tag{8.11}
$$

where we have separated the terms for which $i = k$ and $i \neq k$. The first summation on the right side of Eq. (8.11) has a value of $\sigma^4(1+2\rho^2)N_N^{-1}$: from Eq. (8.3), it can be shown that

$$\int_{-\infty}^{\infty} \int_{-\infty}^{\infty} x^2 y^2 p(x, y) dx \, dy = \sigma^4(1 + 2\rho^2) \, . \tag{8.12}$$

The second summation term in Eq. (8.11) is readily evaluated by using the fourth-order moment relation in Eq. (6.36). Because successive samples of each signal are uncorrelated (a rectangular passband is assumed), $\langle x_i y_i x_k y_k \rangle = \langle x_i y_i \rangle \langle x_k y_k \rangle$, and the second summation term has a value of $(1 - N_N^{-1})\rho^2\sigma^4$. Returning to Eq. (8.10), we can write

$$
\begin{aligned}
\sigma_\infty^2 &= (1 + 2\rho^2)\sigma^4 N_N^{-1} + (1 - N_N^{-1})\rho^2\sigma^4 - \rho^2\sigma^4 \\
&= \sigma^4 N_N^{-1}(1 + \rho^2) \, .
\end{aligned}
\tag{8.13}
$$

The signal-to-noise ratio with unquantized sampling is

$$\mathcal{R}_{sn\infty} = \frac{\langle r_\infty \rangle}{\sigma_\infty} = \frac{\rho \sqrt{N_N}}{\sqrt{(1 + \rho^2)}} \simeq \rho \sqrt{N_N} \, , \tag{8.14}$$

where the approximation applies for $\rho \ll 1$. Note that the condition $\rho \ll 1$ is satisfactory in many practical circumstances. For the case in which $\rho \gtrsim 0.2$, see Sect. 8.3.6. (The signal-to-noise ratio at the correlator output, which we are calculating here, is of interest mainly for weak signals.) For a measurement period τ, $N_N = 2\Delta\nu\tau$, which is commonly 10^6–10^{12}. From Eq. (8.14), the threshold of detectability of a signal is given by $\rho\sqrt{N_N} \simeq 1$, that is, $\rho \simeq 10^{-3}$–10^{-6}. In terms of the signal bandwidth and measurement duration, $\mathcal{R}_{sn\infty} = \rho\sqrt{2\Delta\nu\,\tau}$. Now for observations of a point source with identical antennas and receivers, ρ is equal to the ratio of the resulting antenna temperature to the system temperature, T_A/T_S. Thus, the present result is equal to that given by Eq. (6.45) for an analog correlator with continuous unsampled inputs and $T_A \ll T_S$.

Before leaving the subject of unquantized sampling, we should consider the effect of sampling at rates other than the Nyquist rate. Successive sample values from any one signal are then no longer independent. We consider a sampling frequency that is β times the Nyquist rate[3] and a number of samples $N = \beta N_N$. The sample interval is $\tau_s = (2\beta\Delta\nu)^{-1}$. Samples spaced by $q\tau_s$, where q is an integer, have a correlation coefficient that, from Eq. (8.4), is equal to

$$R_\infty(q\tau_s) = \frac{\sin(\pi q/\beta)}{\pi q/\beta} \tag{8.15}$$

for a rectangular baseband response. Since the samples are not independent, we must reconsider the evaluation of the second summation term on the right side of Eq. (8.11). For those terms for which $q = |i - k|$ is small enough that $R_\infty(q\tau_s)$ is significant, there will be an additional contribution given by

$$\left[\sigma^2 R_\infty(q\tau_s)\right]^2 . \tag{8.16}$$

Now R_∞^2 is very small for all but a very small fraction of the $N(N-1)$ terms in the second summation in Eq. (8.11). From Eq. (8.15), R_∞^2, at its maxima, is equal to $(\beta/\pi q)^2$ and for $q = 10^3$ is of order 10^{-6}. However, as shown above, N is likely to be as high as 10^6–10^{12}. Thus, in the second summation in Eq. (8.11), the contribution made by the terms for which the i and k samples are effectively independent remains essentially unchanged. The products for which R_∞^2 is significant make an additional contribution equal to

$$2\sigma^4 N^{-2} \sum_{q=1}^{N-1}(N-q)R_\infty^2(q\tau_s) \simeq 2\sigma^4 N^{-1} \sum_{q=1}^{\infty} R_\infty^2(q\tau_s) . \tag{8.17}$$

The variance of the correlator output now becomes

$$\sigma_\infty^2 = \sigma^4 N^{-1}\left[1 + 2\sum_{q=1}^{\infty} R_\infty^2(q\tau_s)\right] , \tag{8.18}$$

---

[3] β is referred to as the oversampling factor.

and the signal-to-noise ratio of the correlation measurement is (see Appendix 8.1)

$$\mathcal{R}_{\mathrm{sn}\infty} = \frac{\rho\sqrt{\beta N_N}}{\sqrt{1 + 2\sum_{q=1}^{\infty} R_{\infty}^2(q\tau_s)}} \, . \tag{8.19}$$

Compare this result with Eq. (8.14) for Nyquist sampling. For values of β of $\frac{1}{2}$, $\frac{1}{3}$, $\frac{1}{4}$, and so on, which correspond to undersampling, $R_{\infty} = 0$ and the denominator in Eq. (8.19) is unity. The sensitivity thus drops as one would expect from the decrease in the data. For oversampling, $\beta > 1$, and the summation of $R_{\infty}^2(q\tau_s)$ in Eq. (8.19) is shown in Appendix 8.1 to be equal to $(\beta - 1)/2$. The denominator in Eq. (8.19) is then equal to $\sqrt{\beta}$, so the sensitivity is the same as that for sampling at the Nyquist rate. This is as expected, since in Nyquist sampling, no information is lost, and thus there is none to be gained by increased sampling. The result is different for quantized sampling, as will appear in the following sections.

8.3   Sampling with Quantization

In some sampling schemes, the signal is first quantized and then sampled, and in others, it is sampled and then quantized. Ideally, the end result is the same in either case, and in analyzing the process, we can choose the order that is most convenient. Suppose that a bandlimited signal is first quantized and then sampled. Quantization generates new frequency components in the signal waveform, so it is no longer bandlimited. If it is sampled at the Nyquist rate corresponding to the unquantized waveform, as is the usual practice, some information will be lost, and the sensitivity will be less than for unquantized sampling. Also, because quantization is a nonlinear operation, we cannot assume that the measured correlation of the quantized waveforms will be a linear function of ρ, which is what we want to measure. Thus, to utilize digital signal processing, there are three main points that should be investigated: (1) the relation between ρ and the measured correlation, (2) the loss in sensitivity, and (3) the extent to which oversampling can restore the lost sensitivity. Investigations of these points can be found in the work of Weinreb (1963), Cole (1968), Burns and Yao (1969), Cooper (1970), Hagen and Farley (1973), Bowers and Klingler (1974), Jenet and Anderson (1998), and Gwinn (2004).

Note that in discussing sampling with quantization, it is common practice to refer to Nyquist sampling when what is meant is sampling at the Nyquist rate for the *unquantized* waveform. We also follow this usage.

8.3.1   Two-Level Quantization

Sampling with two-level (one bit) quantization provided the earliest digital form of radio astronomy signals (Weinreb 1963). Although larger numbers of levels are

Fig. 8.3 Characteristic curve
for two-level quantization.
The abscissa is the input
voltage x and the ordinate is
the quantized output $\hat{x}$.

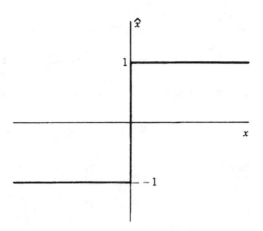

now routinely used, this subsection is included as an introduction to the subject.
The quantization characteristic for two-level sampling is shown in Fig. 8.3. The
quantizing action senses only the sign of the instantaneous signal voltage. In
many samplers, the signal voltage is first amplified and strongly clipped. The zero
crossings are more sharply defined in the resulting waveform, and errors that might
occur if the sampling time coincides with a sign reversal are thereby minimized.

The correlator for two-level signals consists of a multiplying circuit followed by
a counter that sums the products of the input samples. The input signals are assigned
values of $+1$ or -1 to indicate positive or negative signal voltages, and the products
at the multiplier output thus take values of $+1$ or -1 for identical or different input
values, respectively. We consider sampling both at the Nyquist rate and at multiples
of it and represent by N the number of sample pairs fed to the correlator. The two-
level correlation coefficient is

$$\rho_2 = \frac{(N_{11} + N_{\bar{1}\bar{1}}) - (N_{\bar{1}1} + N_{1\bar{1}})}{N} , \tag{8.20}$$

where N_{11} is the number of products for which both samples have the value $+1$,
$N_{1\bar{1}}$ is the number of products in which the x sample has the value $+1$ and the y
sample -1, and so on. The denominator in Eq. (8.20) is equal to the output that
would occur if, for each sample pair, the signs of the signals were identical. ρ_2 can
be related to the correlation coefficient ρ of the unquantized signals through the
bivariate probability distribution Eq. (8.1), from which

$$P_{11} = \frac{N_{11}}{N} = \frac{1}{2\pi\sigma^2\sqrt{1 - \rho^2}} \int_0^\infty \int_0^\infty \exp\left[\frac{-(x^2 + y^2 - 2\rho xy)}{2\sigma^2(1 - \rho^2)}\right] dx\, dy , \tag{8.21}$$

where P_{11} is the probability of the two unquantized signals being simultaneously
greater than zero. The other required probabilities are obtained by changing the
limits of the integrals in Eq. (8.21) as follows: $\int_{-\infty}^0 \int_{-\infty}^0$ for $P_{\bar{1}\bar{1}}$; $\int_{-\infty}^0 \int_0^\infty$ for $P_{\bar{1}1}$;

and $\int_0^\infty \int_{-\infty}^0$ for $P_{1\bar{1}}$. Note that $P_{11} = P_{\bar{1}\bar{1}}$ and $P_{1\bar{1}} = P_{\bar{1}1}$. Thus,

$$\rho_2 = 2(P_{11} - P_{1\bar{1}}) \ . \tag{8.22}$$

The integral in Eq. (8.21) is evaluated in Appendix 8.2, from which we obtain

$$P_{11} = \frac{1}{4} + \frac{1}{2\pi} \sin^{-1} \rho \ . \tag{8.23}$$

Similarly,

$$P_{1\bar{1}} = \frac{1}{4} - \frac{1}{2\pi} \sin^{-1} \rho \ , \tag{8.24}$$

so

$$\rho_2 = \frac{2}{\pi} \sin^{-1} \rho \ . \tag{8.25}$$

Equation (8.25), known as the Van Vleck relationship,[4] allows ρ to be obtained from the measured correlation ρ_2. For small values, ρ is proportional to ρ_2.

To determine the signal-to-noise ratio of the correlation measurement, we now calculate σ_2^2, the variance of the correlator output r_2:

$$\sigma_2^2 = \langle r_2^2 \rangle - \langle r_2 \rangle^2 \ , \tag{8.26}$$

where

$$r_2 = N^{-1} \sum_{i=1}^N \hat{x}_i \hat{y}_i \ . \tag{8.27}$$

In this chapter, the circumflex ($\hat{\ }$) is used to denote quantized signal waveforms. Since $\rho_2 = \langle \hat{x}\hat{y} \rangle$, then from Eq. (8.27), $\langle r_2 \rangle = \rho_2$. Thus, r_2 is an unbiased estimator of ρ_2. The expression for $\langle r_2^2 \rangle$ is equivalent to Eq. (8.11) for unquantized waveforms:

$$\langle r_2^2 \rangle = N^{-2} \sum_{i=1}^N \langle \hat{x}_i^2 \hat{y}_i^2 \rangle + N^{-2} \sum_{i=1}^N \sum_{k \neq i} \langle \hat{x}_i \hat{y}_i \hat{x}_k \hat{y}_k \rangle \ . \tag{8.28}$$

---

[4]This result was first derived by J. H. Van Vleck during World War II in a classified report, when studying the power spectrum of strongly clipped noise, which was used for electromagnetic jamming (Van Vleck 1943). The work was later declassified, and a brief summary of it appeared in Vol. 24 of MIT's Radiation Laboratory Series (Lawson and Uhlenbeck 1950). A fuller account was given by Van Vleck and Middleton (1966).

The first summation term on the right side of Eq. (8.28) is equal to N^{-1} since the products $\hat{x}_i \hat{y}_i$ take values of ± 1 for two-level sampling. In evaluating the second summation term, the situation is similar to that for unquantized sampling. The factor σ^4 in Eq. (8.17) is here replaced by the square of the variance of the quantized waveform, which is unity for two-level quantization. For all except a small fraction of the terms, $q = |i-k|$ is large enough that samples i and k from the same waveform are uncorrelated. These terms make a total contribution closely equal to ρ_2^2. Those terms for which samples i and k are correlated make an additional contribution closely equal to

$$2N^{-1} \sum_{q=1}^{\infty} R_2^2(q\tau_s) \, , \tag{8.29}$$

where $R_2(\tau)$ is the autocorrelation coefficient for a signal after two-level quantization. Thus,

$$\sigma_2^2 = N^{-1} + (1 - N^{-1})\rho_2^2 + 2N^{-1} \sum_{q=1}^{\infty} R_2^2(q\tau_s) - \rho_2^2 \tag{8.30a}$$

$$\simeq N^{-1} \left[1 + 2 \sum_{q=1}^{\infty} R_2^2(q\tau_s) \right] , \tag{8.30b}$$

where we have assumed that $\rho_2 \ll 1$ and also that the term $-N^{-1}\rho_2^2$ can be neglected, since here we are mostly interested in signals near the threshold of detectability. Then the signal-to-noise ratio is

$$\mathcal{R}_{sn2} = \frac{\langle r_2 \rangle}{\sigma_2} = \frac{2\rho\sqrt{N}}{\pi \sqrt{1 + 2\sum_{q=1}^{\infty} R_2^2(q\tau_s)}} . \tag{8.31}$$

This ratio, relative to that for unquantized sampling at the Nyquist rate given by Eq. (8.14), defines an efficiency factor for the quantized correlation process:

$$\eta_2 = \frac{\mathcal{R}_{sn2}}{\mathcal{R}_{sn\infty}} = \frac{2\sqrt{\beta}}{\pi \sqrt{1 + 2\sum_{q=1}^{\infty} R_2^2(q\tau_s)}} . \tag{8.32}$$

Here, we have used $N = \beta N_N$, so we are considering the same observing time as in the Nyquist-sampled case but sampling β times as rapidly. Note that τ_s is correspondingly reduced. η_2 is one case of the general quantization efficiency factor, η_Q (introduced in Sect. 6.2), where Q is the number of quantization levels.

Equation (8.25) gives the relationship between the correlation coefficients for a pair of signals before and after two-level quantization. This result includes the case of autocorrelation in which the two signals differ only because of a delay. Thus, we may write

$$R_2(q\tau_s) = \frac{2}{\pi} \sin^{-1}[R_\infty(q\tau_s)] .$$ (8.33)

Equation (8.15) gives $R_\infty(q\tau_s)$ for a rectangular baseband signal spectrum sampled at β times the Nyquist rate, and Eq. (8.33) becomes

$$R_2(q\tau_s) = \frac{2}{\pi} \sin^{-1}\left[\frac{\beta \sin(\pi q/\beta)}{\pi q}\right] .$$ (8.34)

$R_2(q\tau_s)$ thus has zeros at the same values of $q\tau_s$ that $R_\infty(q\tau_s)$ does (the principal value is taken for the inverse sine function), and for $\beta = 1, \frac{1}{2}, \frac{1}{3}$, and so on, we obtain

$$\sum_{q=1}^{\infty} R_2^2(q\tau_s) = 0 .$$ (8.35)

In these cases, the signal-to-noise ratio is a factor of $2/\pi$ ($= 0.637$) times that for unquantized sampling at the same rate given in Eq. (8.15). For oversampling with $\beta = 2$ and $\beta = 3$, the corresponding signal-to-noise factors from Eqs. (8.32) and (8.34) are 0.744 and 0.773, respectively. Note, however, that the increased bit rate used in oversampling could produce a bigger increase in the signal-to-noise ratio if used to increase the number of quantization levels. Doubling the bit rate could be used to increase the number of levels to four, for which the signal-to-noise factor is 0.881 (as derived in Sect. 8.3.3). For a bit rate increase of three, the number of levels could be increased to eight, for which the signal-to-noise factor is 0.963. Note also that in the calculations given above, there is an implicit dependence on the bandpass shape of the signal through the assumption that $\rho_2 \ll 1$ for samples for which i is not equal to k in Eq. (8.28). For $\beta \geq 2$, a further dependence on the bandpass shape enters through the autocorrelation function $R_2(q\tau_s)$.

It has been mentioned that quantization generates additional spectral components. We can compare the power spectra of a signal before and after quantization since these spectra are the Fourier transforms of autocorrelation functions that are related by Eq. (8.25). Figure 8.4 shows the spectrum, after two-level quantization, of noise with an originally rectangular spectrum. A fraction of the original bandlimited spectrum is converted into a broad, low-level skirt that dies away very slowly with frequency.

8.3.2   Four-Level Quantization

The use of two digital bits to represent the amplitude of each sample results in less degradation of the signal-to-noise ratio than is obtained with one-bit quantization.

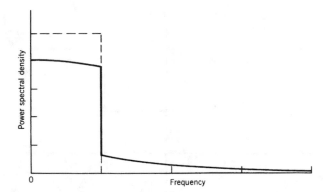

Fig. 8.4 Spectra of rectangular bandpass noise before and after two-level quantization. The unquantized spectrum is of lowpass form, as shown by the broken line. The spectrum after quantization is shown by the solid curve. The power levels of the two waveforms (represented by the areas under the curves) are equal, and the Fourier transforms of their spectra are related by Eq. (8.25).

Consideration of two-bit sampling leads naturally to four-level quantization, the performance of which has been investigated by several authors, notably Cooper (1970) and Hagen and Farley (1973). The quantization characteristic is shown in Fig. 8.5, where the quantization thresholds are $-v_0$, 0, and v_0. The four quantization states have designated values $-n$, -1, $+1$, and $+n$, where n, which is not necessarily an integer, can be chosen to optimize the performance. Products of two samples can take the value ± 1, $\pm n$, or $\pm n^2$. The four-level correlation coefficient ρ_4 can be specified by an expression similar to Eq. (8.20) for the two-level case, that is,

$$\rho_4 = \frac{2n^2 N_{nn} - 2n^2 N_{n\bar{n}} + 4n N_{1n} - 4n N_{1\bar{n}} + 2N_{11} - 2N_{1\bar{1}}}{(2n^2 N_{nn} + 2N_{11})_{\rho=1}} , \tag{8.36}$$

where a bar on the subscript indicates a negative sign. The numerator is proportional to the correlator output and reduces to the form in the denominator for $\rho = 1$, that is, when the two input waveforms are identical. The numbers of the various level combinations can be derived from the corresponding joint probabilities. Thus, for example,

$$N_{nn} = N P_{nn}$$

$$= \frac{N}{2\pi\sigma^2 \sqrt{1-\rho^2}} \int_{v_0}^{\infty} \int_{v_0}^{\infty} \exp\left[\frac{-(x^2 + y^2 - 2\rho xy)}{2\sigma^2(1-\rho^2)}\right] dx\, dy , \tag{8.37}$$

and, as in the two-level case, the other probabilities are obtained by using the appropriate limits for the integrals. For the case of $\rho \ll 1$, the approximate form of the probability distribution in Eq. (8.2) simplifies the calculation.

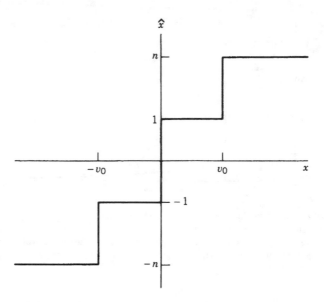

Fig. 8.5 Characteristic curve for four-level quantization, with weighting factor n for outer levels. The abscissa is the unquantized voltage x, and the ordinate is the quantized output $\hat{x}$. v_0 is the threshold voltage.

Although ρ_4 can be evaluated from Eq. (8.36) in the above manner, an alternative derivation that provides a more rapid approach to the desired result is used here. This approach follows the treatment of Hagen and Farley (1973) and is based on a theorem by Price (1958). The form of the theorem that we require is

$$\frac{d\langle r_4 \rangle}{d\rho} = \sigma^2 \left\langle \frac{\partial \hat{x}}{\partial x} \frac{\partial \hat{y}}{\partial y} \right\rangle , \tag{8.38}$$

where r_4 is the unnormalized correlator output, and $\hat{x}$ and $\hat{y}$ are again the quantized versions of the input signals. For four-level sampling,

$$\frac{\partial \hat{x}}{\partial x} = (n-1)\delta(x+v_0) + 2\delta(x) + (n-1)\delta(x-v_0) , \tag{8.39}$$

where δ is the delta function, and a similar expression can be written for $\partial \hat{y}/\partial y$. Equation (8.39) is the derivative of the function in Fig. 8.5. To determine the expectation of the product of the two derivatives on the right side of Eq. (8.38), the magnitudes of each of the nine terms in the product of the derivatives must be multiplied by the probability of occurrence. Thus, for example, the term $(n-1)^2\delta(x+v_0)\delta(y+v_0)$ has a magnitude of $(n-1)^2$ and probability

$$\frac{1}{2\pi\sigma^2\sqrt{1-\rho^2}} \exp\left[\frac{-2v_0^2}{2\sigma^2(1+\rho)}\right] . \tag{8.40}$$

By consolidating terms with equal probabilities, we obtain

$$\frac{d\langle r_4 \rangle}{d\rho} = \frac{1}{\pi\sqrt{1-\rho^2}}\left\{(n-1)^2\left[\exp\left(\frac{-v_0^2}{\sigma^2(1+\rho)}\right) + \exp\left(\frac{-v_0^2}{\sigma^2(1-\rho)}\right)\right]\right.$$
$$\left. + 4(n-1)\exp\left(\frac{-v_0^2}{2\sigma^2(1-\rho^2)}\right) + 2\right\} , \qquad (8.41)$$

and

$$\langle r_4 \rangle = \frac{1}{\pi}\int_0^\rho \frac{1}{\sqrt{1-\xi^2}}\left\{(n-1)^2\left[\exp\left(\frac{-v_0^2}{\sigma^2(1+\xi)}\right) + \exp\left(\frac{-v_0^2}{\sigma^2(1-\xi)}\right)\right]\right.$$
$$\left. + 4(n-1)\exp\left(\frac{-v_0^2}{2\sigma^2(1-\xi^2)}\right) + 2\right\} d\xi ,$$
$$\qquad (8.42)$$

where ξ is a dummy variable of integration. To obtain the correlation coefficient ρ_4, $\langle r_4 \rangle$ must be divided by the expectation of the correlator output when the inputs are identical four-level waveforms, as in Eq. (8.36):

$$\rho_4 = \frac{\langle r_4 \rangle}{\Phi + n^2(1-\Phi)} , \qquad (8.43)$$

where Φ is the probability that the unquantized level lies between $\pm v_0$, that is,

$$\Phi = \frac{1}{\sigma\sqrt{2\pi}}\int_{-v_0}^{v_0}\exp\left(\frac{-x^2}{2\sigma^2}\right)dx = \text{erf}\left(\frac{v_0}{\sigma\sqrt{2}}\right) . \qquad (8.44)$$

Equations (8.42)–(8.44) provide a relationship between ρ_4 and ρ that is equivalent to the Van Vleck relationship for two-level quantization.

The choice of values for n and v_0 is usually made to maximize the signal-to-noise ratio for weak signals, which we now derive. For $\rho \ll 1$, Eqs. (8.42) and (8.43) reduce to

$$(\rho_4)_{\rho\ll1} = \rho\frac{2[(n-1)E+1]^2}{\pi[\Phi + n^2(1-\Phi)]} , \qquad (8.45)$$

where $E = \exp(-v_0^2/2\sigma^2)$. The variance in the measurement of r_4 is

$$\sigma_4^2 = \langle r_4^2 \rangle - \langle r_4 \rangle^2 = \langle r_4^2 \rangle - \rho_4^2\left[\Phi + n^2(1-\Phi)\right]^2 . \qquad (8.46)$$

The factor $[\Phi + n^2(1-\Phi)]$ is the variance of the quantized waveform and here takes the place of σ^2 in the corresponding equations for unquantized sampling. Again, we follow the procedure explained for the unquantized case and write

$$\langle r_4^2 \rangle = N^{-2} \sum_{i=1}^{N} \langle \hat{x}_i^2 \hat{y}_i^2 \rangle + N^{-2} \sum_{i=1}^{N} \sum_{i \neq k} \langle \hat{x}_i \hat{y}_i \hat{x}_k \hat{y}_k \rangle \ . \tag{8.47}$$

To evaluate the first summation, note that $(\hat{x}_i \hat{y}_i)^2$ can take values of 1, n^2, or n^4, and the sum of these values multiplied by their probabilities is equal to $[\Phi + n^2(1-\Phi)]^2$. The contribution of the second summation is

$$(1 - N^{-1})\rho_4^2 \left[\Phi + n^2(1 - \Phi)\right]^2 + 2N^{-1}[\Phi + n^2(1 - \Phi)]^2 \sum_{q=1}^{\infty} R_4^2(q\tau_s) \ , \tag{8.48}$$

where the second term represents the effect of oversampling and is similar to Eq. (8.17), and R_4 is the autocorrelation function after four-level quantization. Thus, from Eq. (8.46), we have

$$\sigma_4^2 = N^{-1} \left[\Phi + n^2(1 - \Phi)\right]^2 \left[1 + 2\sum_{q=1}^{\infty} R_4^2(q\tau_s) - \rho_4^2\right] \ . \tag{8.49}$$

Since we have assumed $\rho \ll 1$, the ρ_4^2 term can be neglected, and the signal-to-noise ratio for the four-level correlation measurement is

$$\mathcal{R}_{sn4} = \frac{\langle r_4 \rangle}{\sigma_4} = \frac{2\rho[(n-1)E + 1]^2 \sqrt{N}}{\pi \left[\Phi + n^2(1 - \Phi)\right] \sqrt{1 + 2\sum_{q=1}^{\infty} R_4^2(q\tau_s)}} \ . \tag{8.50}$$

The signal-to-noise ratio relative to that for unquantized Nyquist sampling is obtained from Eq. (8.14) for $N = \beta N_N$ and is

$$\eta_4 = \frac{\mathcal{R}_{sn4}}{\mathcal{R}_{sn\infty}} = \frac{2[(n-1)E + 1]^2 \sqrt{\beta}}{\pi[\Phi + n^2(1 - \Phi)] \sqrt{1 + 2\sum_{q=1}^{\infty} R_4^2(q\tau_s)}} \ . \tag{8.51}$$

For sampling at the Nyquist rate, $\beta = 1$ and

$$\eta_4 = \frac{\mathcal{R}_{sn4}}{\mathcal{R}_{sn\infty}} = \frac{2[(n-1)E + 1]^2}{\pi[\Phi + n^2(1 - \Phi)]} \ . \tag{8.52}$$

Values of η_4 very close to optimum sensitivity are obtained for $n = 3$ with $v_0 = 0.996\sigma$, and for $n = 4$, with $v_0 = 0.942\sigma$: see Table A8.1 in Appendix 8.3. Note that the choice of an integer for the value of n simplifies the correlator. For these two cases, η_4, the signal-to-noise ratio relative to that for unquantized sampling, is equal to 0.881 and 0.880, respectively. Curves of the relative sensitivity as a function of

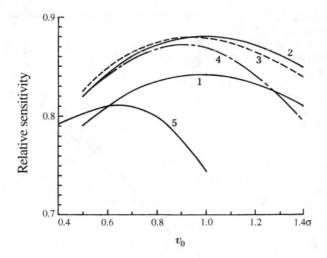

Fig. 8.6 Signal-to-noise ratio relative to that for unquantized correlation for the four-level system and several modifications of it. The abscissa is the quantization threshold v_0 in units of the rms level of the waveforms at the quantizer input. The ordinate is sensitivity (signal-to-noise ratio) relative to an unquantized system. The curves are for: (1) full four-level system with $n = 2$; (2) full four-level system with $n = 3$; (3) full four-level system with $n = 4$; (4) four-level system with $n = 3$ and low-level products omitted; (5) three-level system. From Cooper (1970). © CSIRO 1970. Published by CSIRO Publishing, Melbourne, Victoria, Australia. Reproduced with permission.

v_0/σ for $n = 2$, 3, and 4 are shown in Fig. 8.6. Similar conclusions are derived by Hagen and Farley (1973) and Bowers and Klingler (1974).

Having chosen values for n and v_0, we can now return to Eqs. (8.42) and (8.43) to examine the relationship of ρ and ρ_4. Curve 1 of Fig. 8.7 shows a plot of ρ and ρ_4. Extrapolation of a linear relationship with slope chosen to fit low values of ρ results in errors of 1% at $\rho = 0.5$, 2% at $\rho = 0.7$, and 2.8% at $\rho = 0.8$, where the error is a percentage of the true value of ρ. Thus, for many purposes, a linear approximation is satisfactory for values of ρ up to ~ 0.6. This linearity assumption simplifies the final step that we require in discussing four-level sampling, namely, calculation of the improvement in sensitivity resulting from oversampling.

The relationship between the autocorrelation function for unquantized noise R_∞ and that for the same waveform after four-level quantization is the same as for the corresponding cross-correlation functions in Eq. (8.45), so we can write

$$R_4 = \frac{2[(n-1)E + 1]^2 R_\infty}{\pi[\Phi + n^2(1 - \Phi)]}, \tag{8.53}$$

provided that $R_\infty \lesssim 0.6$. Now R_∞ as given by Eq. (8.15) fulfills this condition for $q = 1$ with an oversampling factor $\beta = 2$. For $n = 3$ and the corresponding optimum value of v_0, $E = 0.6091$, $\Phi = 0.6806$, and $R_4 = 0.881 R_\infty$. For $\beta = 2$, we use Eqs. (8.15) and (8.53) and Eq. (A8.5) of Appendix 8.1 to evaluate the summation

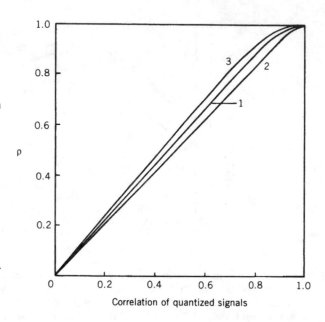

Fig. 8.7 Correlation coefficient ρ for unquantized signals plotted as a function of the correlation that would be measured after quantization. The curves are for: (1) full four-level system with $n = 3$ and $v_0 = \sigma$, or $n = 4$ and $v_0 = 0.95\sigma$; (2) four-level system with low-level products omitted, $n = 4$ and $v_0 = 0.9\sigma$; (3) three-level system with $v_0 = 0.6\sigma$. From Cooper (1970). © CSIRO 1970. Published by CSIRO Publishing, Melbourne, Victoria, Australia. Reproduced with permission.

Correlation of quantized signals

in the denominator of Eq. (8.51), and obtain $\eta_4 = 0.935$, which is a factor of 1.06 greater than for $\beta = 1$. Bowers and Klingler (1974) have pointed out that the optimum value of the quantization level v_0 changes slightly with the oversampling factor. However, the optimum values are rather broad (see Fig. 8.6), and the effect on the sensitivity is very small.

In a discussion of two-bit quantization, Cooper (1970) considered the effect of omitting certain products in the multiplication process. For example, if all products of the two low-level bits are counted as zero instead of ± 1, the loss in signal-to-noise ratio is approximately 1%, as shown in curve 4 of Fig. 8.6. The products to be accumulated are then only those counted as $\pm n$ and $\pm n^2$ in the full four-level system described above, and in the modified system, they can be assigned values of ± 1 and $\pm n$, respectively, thereby simplifying the counter circuitry of the integrator. An even greater simplification can be accomplished by omitting the intermediate-level products also and assigning values ± 1 to the high-level products. This last type of modification yields 92% of the sensitivity of a full four-level correlator. We shall not analyze the case where only the low-level products are omitted, but we note that to derive the correlation coefficient as a function of ρ, one can express the action of the correlator in terms of two different quantization characteristics (Hagen and Farley 1973) or else return to Eq. (8.36) and omit the appropriate terms. If both the low- and intermediate-level products are omitted, however, the action can be described more simply in terms of a new quantization characteristic, known as three-level quantization, without arbitrary omission of product terms.

8.3.3   Three-Level Quantization

Three-level quantization has proved to be an important practical technique, and the quantization characteristic is shown in Fig. 8.8. In this case, the approach using Price's theorem will again be followed.

The expressions for the operating characteristics of a three-level correlator can be obtained from those in the preceding section by omitting the terms that refer to low- and intermediate-level products and adjusting the weighting factors as appropriate. Thus, the equivalent derivative needed in Price's theorem is,

$$\frac{\partial \hat{x}}{\partial x} = \delta(x - v_0) + \delta(x + v_0) , \tag{8.54}$$

and the expectation of the correlator output $\langle r_3 \rangle$ is, from Price's theorem,

$$\langle r_3 \rangle = \frac{1}{\pi} \int_0^\rho \frac{1}{\sqrt{1 - \xi^2}} \left[\exp\left(\frac{-v_0^2}{\sigma^2(1 + \xi)} \right) + \exp\left(\frac{-v_0^2}{\sigma^2(1 - \xi)} \right) \right] d\xi , \tag{8.55}$$

where ξ is a dummy variable of integration. The normalized correlation coefficient is

$$\rho_3 = \frac{\langle r_3 \rangle}{1 - \Phi} , \tag{8.56}$$

Fig. 8.8 Characteristic curve for three-level quantization. The abscissa is the unquantized voltage x, and the ordinate is the quantized output $\hat{x}$. v_0 is the threshold voltage. Since the magnitude of $\hat{x}$ takes only one nonzero value, it is perfectly general to set this value to unity.

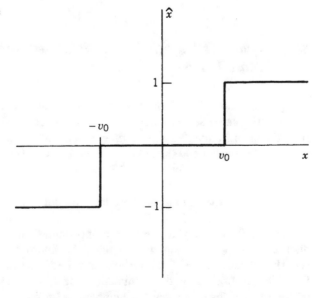

where Φ is given by Eq. (8.44). For $\rho \ll 1$, Eqs. (8.55) and (8.56) yield

$$(\rho_3)_{\rho \ll 1} = \rho \frac{2E^2}{\pi(1 - \Phi)} \, , \tag{8.57}$$

where E is defined following Eq. (8.45). The variance of r_3 is

$$\sigma_3^2 = \langle r_3^2 \rangle - \langle r_3 \rangle^2 = N^{-1}(1 - \Phi)^2 \left[1 + 2\sum_{q=1}^{\infty} R_3^2(q\tau_s) - \rho_3^2 \right] , \tag{8.58}$$

where R_3 is the autocorrelation coefficient after three-level quantization. If ρ_3^2 in Eq. (8.58) can be neglected, the signal-to-noise ratio relative to a nonquantizing correlator is

$$\eta_3 = \frac{\mathcal{R}_{sn3}}{\mathcal{R}_{sn\infty}} = \frac{\langle r_3 \rangle}{\sigma_3 \mathcal{R}_{sn\infty}} = \frac{2\sqrt{\beta}E^2}{\pi(1 - \Phi)\sqrt{1 + 2\sum_{q=1}^{\infty} R_3^2(q\tau_s)}} \, . \tag{8.59}$$

For Nyquist sampling, the maximum sensitivity relative to the nonquantizing case is obtained with $v_0 = 0.6120\sigma$, for which η_3 is equal to 0.810 (see curve 5 of Fig. 8.6). With this optimized threshold value, $\Phi = 0.4595$, $E = 0.8292$, and we can write $R_3(q\tau_s) = 0.810R_\infty(q\tau_s)$, assuming that ρ is an approximately linear function of r_3. Then from Eqs. (8.15), (8.59), and Eq. (A8.5), we find that for a rectangular baseband spectrum with the oversampling factor $\beta = 2$, η_3 becomes 0.890, which is a factor of 1.10 greater than for $\beta = 1$.

8.3.4   Quantization Efficiency: Simplified Analysis for Four or More Levels

For quantization into two, three, or four levels, the quantization efficiency, η_Q, is 0.636, 0.810, and 0.881. For more quantization levels, the loss in efficiency resulting from the quantization decreases further, and an approximate method of calculating the loss (Thompson 1998) can be used, as follows. This is simpler than the more accurate method given in Sect. 8.3.3. In either case, the principle is to calculate the fractional increase in the variance of a signal that results from the quantization. The signal-to-noise ratio at the correlator output is inversely proportional to this variance.

Figure 8.9 shows a piecewise linear approximation of the Gaussian probability distribution of a signal from one antenna. This approximation simplifies the analysis. The intersections with the vertical lines indicate exact values of the Gaussian. For eight-level sampling, the quantization thresholds are indicated by the positions of the vertical lines between the numbers ± 3.5 on the abscissa. The horizontal

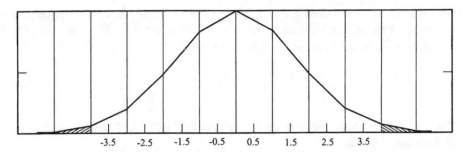

Fig. 8.9 Piecewise linear representation of the Gaussian probability distribution of the amplitude of a signal within the receiver. The intersections of the curve with the vertical lines denote exact values of the Gaussian. The abscissa is the signal amplitude (voltage) in units of $\epsilon\sigma$, and the numbers indicate the values assigned to the levels after quantization. For eight-level sampling the quantization thresholds are indicated by the seven vertical lines that lie between $-3.5\epsilon\sigma$ and $3.5\epsilon\sigma$ on the abscissa. For signal levels outside the range $\pm4\epsilon\sigma$, indicated by the shaded areas, the assigned values are $\pm3.5\epsilon\sigma$.

spacing between adjacent levels is represented by ϵ, in units of the (unquantized) rms voltage, σ, i.e., $\epsilon\sigma$ is the spacing between the levels in volts. We consider first the case in which the number of levels is even, as in Fig. 8.9. Any one sample that falls between the two consecutive thresholds at $m\epsilon\sigma$ and $(m + 1)\epsilon\sigma$ will be assigned a value $(m+\frac{1}{2})\epsilon\sigma$. The normalized trapezoidal probability distribution for the voltage in this segment of the overall probability distribution in Fig. 8.9 can be written as

$$p(v) = \frac{1}{\epsilon\sigma} + \left[v - \left(m + \frac{1}{2} \right) \epsilon\sigma \right] \Delta_m \qquad m\epsilon\sigma < v < (m+1)\epsilon\sigma , \qquad (8.60)$$

where Δ_m is the change in probability, over the voltage range $m\epsilon\sigma$ to $(m + 1)\epsilon\sigma$. The extra variance that is incurred by quantizing the voltage is

$$\left\langle \left[v - (m + \frac{1}{2})\epsilon\sigma \right]^2 \right\rangle = \int_{m\epsilon\sigma}^{(m+1)\epsilon\sigma} \left[v - \left(m + \frac{1}{2} \right) \epsilon\sigma \right]^2 p(v) \, dv . \qquad (8.61)$$

If we make the substitution $x = v - (m + \frac{1}{2})\epsilon\sigma$, the excess variance becomes

$$\int_{-\epsilon\sigma/2}^{\epsilon\sigma/2} x^2 \left[\frac{1}{\epsilon\sigma} + x\Delta_m \right] dx , \qquad (8.62)$$

or

$$\frac{2}{\epsilon\sigma} \int_0^{\epsilon\sigma/2} x^2 dx = \frac{1}{3} \left(\frac{\epsilon\sigma}{2} \right)^2 . \qquad (8.63)$$

Note that the Δ_m factor does not appear in Eq. (8.63). Hence, the excess variance is the same for all voltage bins from $-4\epsilon\sigma$ to $4\epsilon\sigma$. The fraction of the area under the Gaussian probability curve that lies between these levels is

$$\frac{1}{\sqrt{2\pi}\sigma} \int_{-4\epsilon\sigma}^{4\epsilon\sigma} e^{-x^2/2\sigma^2} dx = \mathrm{erf}\left(\frac{4\epsilon}{\sqrt{2}}\right) . \tag{8.64}$$

Thus, the variance resulting from quantization of the signal samples with amplitudes in the range $\pm4\epsilon\sigma$ is

$$\frac{1}{3}\left(\frac{\epsilon\sigma}{2}\right)^2 \mathrm{erf}\left(\frac{4\epsilon}{\sqrt{2}}\right) . \tag{8.65}$$

We shall assume that the quantization error is essentially uncorrelated with the unquantized signal. In the extreme case of two-level sampling, the quantization error is highly correlated with the unquantized signal, so the treatment used here would not apply. Consider, however, the case of multilevel quantization, as in Fig. 8.10. If the signal voltage is increased steadily, the quantization error decreases from a maximum at each quantization threshold to zero when the voltage is equal to the midpoint of two thresholds. At each threshold, the quantization error changes sign, and the cycle repeats. This behavior greatly reduces any correlation between the quantization error and the signal waveform.

It is also necessary to take account of the effect of counting all signals below $-4\epsilon\sigma$ as level $-3.5\epsilon\sigma$, and those above $+4\epsilon\sigma$ as $+3.5\epsilon\sigma$. To make an approximate

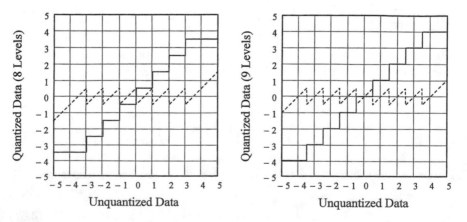

Fig. 8.10 Examples of quantization characteristics for (left diagram) an even number of levels (eight), and (right diagram) an odd number of levels (nine). Units on both axes are equal to ϵ. The abscissa is the analog (unquantized) voltage, and the ordinate is the quantized output. The dotted curves show the analog level minus the quantized level. Note that for even numbers of levels, the thresholds occur at integral values on the abscissa, whereas for odd numbers of levels, they occur at values that are an integer plus one-half.

estimate of this effect, we divide the range of signal level outside of $\pm 4\epsilon\sigma$ into intervals of width $\epsilon\sigma$. Consider, for example, the interval centered on $6.5\epsilon\sigma$. The probability of the signal falling within this level is equal to the corresponding area under the curve, which for the piecewise linear approximation is

$$\frac{1}{2}\frac{\epsilon}{\sqrt{2\pi}}\left[e^{-(6\epsilon)^2/2}+e^{-(7\epsilon)^2/2}\right].\tag{8.66}$$

The variance resulting from quantization of the signal within this range is closely approximated by $[(6.5-3.5)\epsilon\sigma]^2$, so the total variance of the quantization error for signals outside the range $\pm 4\epsilon\sigma$ is

$$\frac{\epsilon^3\sigma^2}{\sqrt{2\pi}}\sum_{m=4}^{\infty}(m-3)^2\left[e^{-m^2\epsilon^2/2}+e^{-(m+1)^2\epsilon^2/2}\right].\tag{8.67}$$

In practice, the summation in (8.67) converges rapidly, and only a few terms are needed (i.e., those for $m\epsilon \lesssim 3$). The quantization error resulting from the truncation of the signal values outside the range $\pm 4\epsilon\sigma$ clearly has some degree of correlation with the unquantized signal level. However, this is a small effect because the fraction of samples for which the signal lies outside $\pm 4\epsilon\sigma$ is less than 1.6% for eight-level quantization, with ϵ optimized for sensitivity. The percentage decreases as the number of quantization levels increases. We shall therefore treat the quantization error resulting from the truncation of the signal peaks as uncorrelated with the signal, but bear in mind that this assumption may introduce a small uncertainty into the calculation.

The variance of the quantized signal is equal to the variance of the unquantized signal (σ^2) plus the variance of the quantization errors in (8.65) and (8.67), that is,

$$\sigma^2+\frac{1}{3}\left(\frac{\epsilon\sigma}{2}\right)^2\text{erf}\left(\frac{4\epsilon}{\sqrt{2}}\right)+\frac{\epsilon^3\sigma^2}{\sqrt{2\pi}}\sum_{m=4}^{\infty}(m-3)^2\left[e^{-m^2\epsilon^2/2}+e^{-(m+1)^2\epsilon^2/2}\right].\tag{8.68}$$

If the variance is the same for both signals at the correlator input, and if the correlation of the signals is small (i.e., $\rho \ll 1$), then the signal-to-noise ratio at the correlator output is inversely proportional to the variance. Thus, the quantization efficiency is

$$\eta_{(2N)}=\left\{1+\frac{1}{3}\left(\frac{\epsilon}{2}\right)^2\text{erf}\left(\frac{N\epsilon}{\sqrt{2}}\right)\right.$$

$$\left.+\frac{\epsilon^3}{\sqrt{2\pi}}\sum_{m=N}^{\infty}(m-N+1)^2\left[e^{-m^2\epsilon^2/2}+e^{-(m+1)^2\epsilon^2/2}\right]\right\}^{-1}.\tag{8.69}$$

Table 8.1 Quantization efficiency and other factors for four or more levels

Number of levels (Q)	N	ϵ	P	η_Q
4	2	1.08	0.03	0.86
8	4	0.60	0.016	0.960
9	4	0.55	0.013	0.968
16	8	0.34	0.006	0.988
32	16	0.19	0.002	0.996
256	16	0.03	<0.001	1.000

Here, the equation has been generalized for $2N$ levels. For an odd number of levels, $2N + 1$, one of which is centered on zero signal level, the equivalent equation for the quantization efficiency is

$$\eta_{(2N+1)} = \left\{ 1 + \frac{1}{3} \left(\frac{\epsilon}{2} \right)^2 \mathrm{erf} \left(\frac{(N+\frac{1}{2})\epsilon}{\sqrt{2}} \right) \right.$$

$$\left. + \frac{\epsilon^3}{\sqrt{2\pi}} \sum_{m=N+1}^{\infty} (m-N)^2 \left[e^{-(m-\frac{1}{2})^2 \epsilon^2 / 2} + e^{-(m+\frac{1}{2})^2 \epsilon^2 / 2} \right] \right\}^{-1}. \quad (8.70)$$

Results from Eqs. (8.69) and (8.70) are given in Table 8.1. The values of ϵ are those that maximize η_Q. The fourth column of the table gives P, which is the fraction of samples for which the signal amplitude is greater than $\pm N\epsilon\sigma$ for an even number of levels or greater than $\pm \left(N+\frac{1}{2}\right)\epsilon\sigma$ for an odd number of levels. For eight levels, P is the fraction of signal samples that contribute to the variance in (8.67). The values of η_Q calculated here are accurate to about 2% for $Q = 4$ and to 0.1% for $Q = 8$ and higher.

8.3.5   Quantization Efficiency: Full Analysis, Three or More Levels

This section presents a general analysis of quantized systems for three or more levels, [e.g., Thompson et al. (2007)]. Let x represent the voltage of the unquantized signal samples, which have a Gaussian probability distribution with variance σ^2. Let $\hat{x}$ represent the quantized values of x. The difference $x - \hat{x}$ represents an inequality introduced by the quantization. The inequality contains a component that is correlated with x, and an uncorrelated component that behaves much like random noise. Consider the correlation coefficient between x and $x' = x - \alpha\hat{x}$, where α is a scaling factor. The correlation coefficient is

$$\frac{\langle xx' \rangle}{x_{\mathrm{rms}} x'_{\mathrm{rms}}} = \frac{\langle x^2 \rangle - \alpha \langle x\hat{x} \rangle}{x_{\mathrm{rms}} x'_{\mathrm{rms}}}. \quad (8.71)$$

Here, the angle brackets $\langle\rangle$ indicate the mean value. If $\alpha = \langle x^2\rangle/\langle x\hat{x}\rangle$, then the correlation coefficient is zero, and x' represents purely random noise. We refer to this random component as the quantization noise, equal to $x - \alpha_1\hat{x}$, where $\alpha_1 = \langle x^2\rangle/\langle x\hat{x}\rangle$. Without loss of generality, we take $\sigma^2 = \langle x^2\rangle = 1$ in this analysis and use $\alpha_1 = 1/\langle x\hat{x}\rangle$. The variance of the quantization noise is

$$\langle q^2\rangle = \langle(x - \alpha_1\hat{x})^2\rangle = \langle x^2\rangle - 2\alpha_1\langle x\hat{x}\rangle + \alpha_1^2\langle\hat{x}^2\rangle = \alpha_1^2\langle\hat{x}^2\rangle - 1 . \tag{8.72}$$

The total variance of the digitized signal is $1 + \langle q^2\rangle$, and the quantization efficiency η_Q is equal to the variance of the unquantized signal expressed as a fraction of the total variance. Thus,

$$\eta_Q = \frac{1}{(1 + \langle q^2\rangle)} = \frac{1}{\alpha_1^2\langle\hat{x}^2\rangle} = \frac{\langle x\hat{x}\rangle^2}{\langle\hat{x}^2\rangle} . \tag{8.73}$$

Consider the case for an even number of equally spaced levels, as in the eight-level case in Fig. 8.10. When the number of levels is even, it is convenient to define N as half the number of levels. We first determine $\langle x\hat{x}\rangle$. Note that for each sample value, x and $\hat{x}$ have the same sign, so $x\hat{x}$ is always positive. Let ϵ represent the spacing between adjacent quantization levels. The values of x that fall within the quantization level between $m\epsilon$ and $(m + 1)\epsilon$ are assigned values $\hat{x} = (m + \frac{1}{2})\epsilon$, and their contribution to $\langle x\hat{x}\rangle$ is

$$\frac{1}{\sqrt{2\pi}}\int_{m\epsilon}^{(m+1)\epsilon} (m + \frac{1}{2})\epsilon\, x\, e^{-x^2/2}\, dx . \tag{8.74}$$

The contribution from the level between $-m\epsilon$ and $-(m + 1)\epsilon$ is the same as the expression above, so to obtain $\langle x\hat{x}\rangle$, we sum the integrals for the positive levels and include a factor of two:

$$\langle x\hat{x}\rangle = \sqrt{\frac{2}{\pi}} \left[\left(\sum_{m=0}^{N-2}\int_{m\epsilon}^{(m+1)\epsilon} \left(m + \frac{1}{2}\right)\epsilon\, x\, e^{-x^2/2}\, dx\right) \right.$$
$$\left. + \int_{(N-1)\epsilon}^{\infty} \left(N - \frac{1}{2}\right)\epsilon\, x\, e^{-x^2/2}\, dx \right] . \tag{8.75}$$

The summation term contains one integral for each positive quantization level except the highest one. The integral on the lower line covers the highest level and the range of x above it, for both of which the assigned value is $\hat{x} = (N - \frac{1}{2})\epsilon$. Then, performing the integration, Eq. (8.75) reduces to

$$\langle x\hat{x}\rangle = \sqrt{\frac{2}{\pi}}\,\epsilon \left(\frac{1}{2} + \sum_{m=1}^{N-1} e^{-m^2\epsilon^2/2} \right) . \tag{8.76}$$

To evaluate the variance of $\hat{x}$, again consider first the contribution from values of x that fall between $m\epsilon$ and $(m + 1)\epsilon$. For this level, the quantized data $\hat{x}$ all have the value $(m + \frac{1}{2})\epsilon$. The variance of $\hat{x}$ for all values of x within this level is

$$\left(m + \frac{1}{2}\right)^2 \epsilon^2 \frac{1}{\sqrt{2\pi}} \int_{m\epsilon}^{(m+1)\epsilon} e^{-x^2/2}\, dx \,. \tag{8.77}$$

For negative x, we again include a factor of 2, sum over all positive quantization levels except the highest, and add a term for the highest level and the range of x above it. Thus, the total variance of $\hat{x}$ is:

$$\langle \hat{x}^2 \rangle = \sqrt{\frac{2}{\pi}} \left[\left(\sum_{m=0}^{N-2} (m + \frac{1}{2})^2 \epsilon^2 \int_{m\epsilon}^{(m+1)\epsilon} e^{-x^2/2}\, dx \right) \right.$$
$$\left. + \left(N - \frac{1}{2}\right)^2 \epsilon^2 \int_{(N-1)\epsilon}^{\infty} e^{-x^2/2}\, dx \right] \,. \tag{8.78}$$

The integrals in Eq. (8.78) can be represented by error functions. Then, using Eqs. (8.73), (8.76), and (8.78), we obtain

$$\eta_{(2N)} = \frac{\frac{2}{\pi} \left(\frac{1}{2} + \sum_{m=1}^{N-1} e^{-m^2\epsilon^2/2} \right)^2}{\left(N - \frac{1}{2}\right)^2 - 2 \sum_{m=1}^{N-1} m \operatorname{erf}\left(\frac{m\epsilon}{\sqrt{2}} \right)} \,. \tag{8.79}$$

For the cases in which the number of levels is odd, the thresholds of the levels occur at values that are an integer plus $\frac{1}{2}$, as in the nine-level case in Fig. 8.10. We represent the odd level number by $2N + 1$. Consider the values of x that fall within the quantization level between $(m - \frac{1}{2})\epsilon$ and $(m + \frac{1}{2})\epsilon$. These are assigned the value $m\epsilon$, i.e., zero for the level centered on $x = 0$. For this level, the contribution to $\langle x\hat{x} \rangle$ is

$$\frac{1}{\sqrt{2\pi}} \int_{(m-\frac{1}{2})\epsilon}^{(m+\frac{1}{2})\epsilon} m\epsilon\, x\, e^{-x^2/2}\, dx \,. \tag{8.80}$$

Summing over all levels, as in Eq. (8.75), we obtain

$$\langle x\hat{x} \rangle = \sqrt{\frac{2}{\pi}} \left[\left(\sum_{m=1}^{N-1} \int_{(m-\frac{1}{2})\epsilon}^{(m+\frac{1}{2})\epsilon} m\epsilon\, x\, e^{-x^2/2}\, dx \right) + \int_{(N-\frac{1}{2})\epsilon}^{\infty} N\epsilon\, x\, e^{-x^2/2}\, dx \right] \,. \tag{8.81}$$

Then, as in Eq. (8.78), we determine $\langle \hat{x}^2 \rangle$:

$$\langle \hat{x}^2 \rangle = \sqrt{\frac{2}{\pi}} \left[\sum_{m=1}^{(N-1)} \left(\int_{(m-\frac{1}{2})\epsilon}^{(m+\frac{1}{2})\epsilon} (m\epsilon)^2 \, e^{-x^2/2} \, dx \right) + \int_{(N-\frac{1}{2})\epsilon}^{\infty} (N\epsilon)^2 \, e^{-x^2/2} \, dx \right] .$$

(8.82)

Performing the integration in Eqs. (8.81) and (8.82), from (8.73), we obtain

$$\eta_{(2N+1)} = \frac{\dfrac{2}{\pi} \left(\displaystyle\sum_{m=1}^{N} e^{-\left(m-\frac{1}{2}\right)^2 \epsilon^2/2} \right)^2}{N^2 - 2 \displaystyle\sum_{m=1}^{N} \left(m - \frac{1}{2} \right) \mathrm{erf}\left(\frac{\left(m - \frac{1}{2}\right)\epsilon}{\sqrt{2}} \right)} .$$

(8.83)

Equations (8.79) and (8.83) can easily be evaluated numerically and provide values of quantization efficiency for any number of equally spaced levels.

Since no significant approximations were made, the same method can be used for cases in which the number of quantization levels is small and consequently the quantization noise is relatively large. Values of η_Q for two, three, and four levels can be obtained by considering the effect of the quantization noise at a correlator input, following the method used above. In cases such as that in Appendix 8.3, for which the assigned values for the levels are chosen to optimize η_Q, or for which the spacing between the level thresholds is not uniform, the formulas derived here cannot be applied directly. However, the same general approach of considering the spacings between levels can be used. For three-level quantization, the levels for maximum quantization efficiency are $\pm 0.612\sigma$ ($\epsilon = 1.224$). Then we have

$$\langle x\hat{x} \rangle = \sqrt{\frac{2}{\pi}} \, \epsilon \int_{\epsilon/2}^{\infty} x e^{-x^2/2} \, dx ,$$

(8.84)

$$\langle \hat{x}^2 \rangle = \sqrt{\frac{2}{\pi}} \, \epsilon^2 \int_{\epsilon/2}^{\infty} e^{-x^2/2} \, dx ,$$

(8.85)

and

$$\eta_3 = \frac{\langle x\hat{x} \rangle^2}{\langle \hat{x}^2 \rangle} = \frac{\sqrt{\frac{2}{\pi}} e^{-0.612^2/2}}{1 - \mathrm{erf}\left(\frac{0.612}{\sqrt{2}} \right)} .$$

(8.86)

Examples of results derived using Eqs. (8.79), (8.83), and (8.86) are shown in Table 8.2. In each case, the value of ϵ is chosen to maximize η_Q. The values of η_Q are given to five decimal places to show how they approach 1.0 as the number

Table 8.2 Examples of quantization efficiency, η_Q, for sampling at the Nyquist rate

Number of levels (Q)	N	ϵ	η_Q
3	1	1.224	0.80983
4	2	0.995	0.88115
8	4	0.586	0.96256
9	4	0.534	0.96930
16	8	0.335	0.98846
32	16	0.188	0.99651
256	128	0.0312	0.99991

of levels increases. However, this is for the case of an ideal rectangular passband, which in a practical receiving system may be closely approximated. Figure 8.11 shows the quantization efficiency η_Q as a function of the threshold spacing ϵ.

If the constant voltage spacing between adjacent thresholds for both input and output values is not maintained, the individual levels can sometimes be adjusted to obtain an improvement in η_Q of a few tenths of a percent, decreasing with increasing number of levels. The values of η_Q in Table 8.2 are in agreement with results by Jenet and Anderson (1998), who give detailed calculations of performance for two- to eight-bit quantization, for both uniform and nonuniform threshold spacing. See also Appendix 8.3 for optimization in the case of four-level quantization.

In recent designs of radio telescopes, the level increment ϵ is frequently chosen so that signals at levels much higher than the rms system noise can be accommodated within the range of levels of the quantizer. This preserves an essentially linear response to interfering signals so that they can be eliminated or mitigated by further processing. For example, with 256 levels (8-bit representation) and $\epsilon = 0.5$, we find that $\eta_Q = 0.9796$. The range of ± 128 levels then corresponds to $\pm 64\sigma$, i.e., ± 36 dB above the system noise, for a $\sim 2\%$ sacrifice in signal-to-noise ratio.

8.3.6   *Correlation Estimates for Strong Sources*

The efficiency calculations of the previous sections are based on estimates of the correlation from the averaged signal products before or after quantization, $\langle x_i y_i \rangle$ or $\langle \hat{x}_i \hat{y}_i \rangle$, in the limit of small correlation, $|\rho| \ll 1$. Johnson et al. (2013) show that when the correlation is small ($|\rho| \ll 1$) and the signal variances are known (as is assumed when setting sampler thresholds), averaged products $\langle \hat{x}_i \hat{y}_i \rangle$ do provide optimal estimates of the correlation. That is, when the correlation is small, no combination of the quantized signals will produce an unbiased estimate of correlation that has smaller variance than that of the correlator output, if suitable weights are chosen. This result arises from the form of the bivariate Gaussian distribution in this limit, which can be written such that the factor including the correlation coefficient ρ includes only terms of the form xy [see, e.g., Eq. (8.2)].

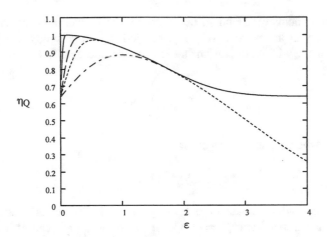

Fig. 8.11 Quantization efficiency as a function of the threshold spacing, ϵ, in units equal to the rms amplitude, σ. The curves are for 64-level (solid line), 16-level (long-dashed line), 9-level (short-dashed line), and 4-level (long-and-short-dashed line). As ϵ becomes very small, the output of the quantizer depends mainly on the sign of the input, so the curves meet the ordinate axis at the two-level value of $\eta_Q = 2/\pi$. As ϵ increases, more of the higher (positive and negative) levels contain only values in the extended tails of the Gaussian distribution, so the number of levels that make a significant contribution to the output decreases, and the curves merge together. The curves for even-level numbers move asymptotically to the two-level value, and curves for odd-level numbers move toward zero. The working point in each case is chosen to be near the maximum of the curve.

However, when the correlation is large, alternative estimates of the correlation will have lower noise. Thus, in the high correlation regime, it is necessary to revise our expression for quantization efficiency. For instance, when the signals are not quantized, the optimal estimate of correlation for two zero-mean signals is Pearson's correlation coefficient [e.g., Wall and Jenkins (2012)],

$$
r_p = \frac{\displaystyle\sum_{i=1}^{N_N}(x_i - \bar{x})(y_i - \bar{y})}{\sqrt{\displaystyle\sum_{i=1}^{N_N}(x_i - \bar{x})^2}\sqrt{\displaystyle\sum_{i=1}^{N_N}(y_i - \bar{y})^2}}, \tag{8.87}
$$

where, $\bar{x} = \frac{1}{N_N}\sum_{i=1}^{N_N} x_i$ is the sample mean, and the sums in the denominator are proportional to the sample variances. The standard error in the estimate of r_p, i.e., σ_p, is

$$
\sigma_p = N_N^{-1/2}(1 - \rho^2) . \tag{8.88}
$$

As ρ approaches unity, σ_p goes to zero. In this limit, the probability function $p(x, y)$ given in Eq. (8.1) collapses to a one-dimensional Gaussian distribution along the line $x = y$ (see Fig. 8.2). When $\rho = 1$, that line is perfectly defined by a set of measurements of x_i and y_i, i.e., there are no deviations from the line, and the uncertainty in the estimate of ρ is zero. Perhaps, surprisingly, the estimate of correlation made without the sample means and variances, as in Eq. (8.8), i.e.,

$$r_\infty = \frac{1}{N_N} \sum_{i=1}^{N_N} x_i y_i \, , \tag{8.89}$$

has an error when normalized by σ^2 of

$$\sigma_\infty = N_N^{-1/2} (1 + \rho^2)^{1/2} \tag{8.90}$$

[see Eq. (8.13)], which equals σ_p only for $\rho = 0$. For two-level quantization, for a rectangular passband and $\beta = 1$, the error on the correlation estimate is [see Eq. (8.30a) and Eq. (8.35)]

$$\sigma_2 = N_N^{-1/2} (1 - \rho^2)^{1/2} \, . \tag{8.91}$$

In the case of large correlation, the Van Vleck relation [Eq. (8.25)]

$$\rho = \sin \frac{\pi}{2} \rho_2, \tag{8.92}$$

will require a nonlinear scaling of the error in ρ_2, denoted σ_{2V}, which can be written

$$\sigma_{2V} = N_N^{-1/2} \left[\left(\frac{\pi}{2} \right)^2 - (\sin^{-1} \rho)^2 \right] (1 - \rho^2)^{1/2} \, . \tag{8.93}$$

These errors in correlation for the various cases described above [σ_p, σ_∞, and σ_{2V} as well as the error for the case of four-level sampling, derived from formulas by Gwinn (2004)] are shown in Fig. 8.12. The interesting result is that the performance of the two-level correlator is better than that of the unquantized correlator for $\rho > 0.6$ and approaches that of the Pearson estimator as ρ approaches unity. The peculiarity in the two-level scenario was noted by Cole (1968) and is related to the fact that the sample variance is irrelevant in the two-level quantization estimate.

Johnson et al. (2013) derive maximum likelihood estimators (MLEs) for the unsampled case in which the signal variance is known. Its standard deviation, σ_q, falls slightly below σ_p. The authors also derive MLEs for various quantization levels and show that their performance $\sigma_q(Q)$ approaches σ_q for large values of Q.

8.4 Further Effects of Quantization

Various forms of analysis in radio astronomy involve cross-correlation of signals from different antennas or autocorrelation of a signal as a function of time. The values of the correlation of quantized signals deviate from the true correlation of

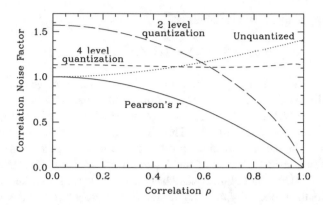

Fig. 8.12 The correlation noise factor, $N_N^{1/2}$ times the standard deviation, vs. correlation ρ. The correlation factors for $\rho = 0$ are equal to n_Q^{-1}. The curves labeled unquantized and two-level quantization give correlation factors based on the standard signal-product estimate of ρ [see Eqs. (8.90) and (8.93) for unquantized and two-level quantization, respectively]. The factor for the Pearson's r curve, given in Eq. (8.88), is based on the estimator r_p, which involves the sample mean and variance. Adapted from Johnson et al. (2013).

the unquantized signals to an extent that is most serious for two-level sampling, and the deviation decreases as the number of levels is increased. Correction for this effect requires determination of how the cross-correlation of the quantized data, here designated R, is related to the true cross-correlation, ρ. To examine the effect of quantization, we consider the effect of a time offset τ on two Gaussian waveforms that are otherwise identical. In the case of two-level sampling, the required relationship is given by the Van Vleck equation [Eq. (8.25)] and is

$$R_2(\tau) = \frac{2}{\pi} \sin^{-1} \rho(\tau) . \tag{8.94}$$

For more than two quantization levels, the relationship is more complicated, and although the nonlinearity of the quantized correlation becomes less serious with an increasing number of levels, correction may still be necessary. As very large instruments come into operation, it becomes increasingly important to remove the responses to strong radio sources in order to study the fainter emission from the most distant regions of the Universe. This requires very accurate calibration of the received signal strengths.

8.4.1   Correlation Coefficient for Quantized Data

Let x and y represent two Gaussianly distributed data streams that differ only by a time offset, τ. The correlation coefficient, $\rho(\tau)$, is equal to $\langle xy \rangle / \langle x^2 \rangle$. The quantized values of x and y are identified by circumflex accents, i.e., $\hat{x}$ and $\hat{y}$. The correlation coefficient of the quantized variables is

$$R(\tau) = \langle \hat{x}\hat{y} \rangle / \langle \hat{x}^2 \rangle \; . \tag{8.95}$$

To determine $\rho(\tau)$ as a function of the correlation coefficient ρ, we need to consider the probabilities of occurrence of the unquantized variables x and y within each quantization interval. First, consider the case in which the number of quantization intervals is even and equal to $2N$. Thus, there are N positive intervals plus N negative ones. The mean value of the products of pairs of the quantized values, $\langle \hat{x}\hat{y} \rangle$, is obtained by considering each of the $2N \times 2N = 4N^2$ possible pairings of the levels of $\hat{x}$ and $\hat{y}$. Only half of these need be calculated, since if the x and y values are interchanged, the probability remains the same. The probability of the unquantized variables x and y falling within any pair of intervals is given by integration of the Gaussian bivariate probability distribution, Eq. (8.1), over the corresponding range of x and y. In Eq. (8.1), x and y have variance σ and cross-correlation coefficient ρ. Here, we are concerned with samples of x and y taken at the Nyquist interval τ_s, and n is the number of Nyquist intervals between the pairs of samples considered. For a rectangular passband of width $\Delta \nu$, the correlation coefficient is given by

$$\rho(n\tau_s) = \frac{\sin(\pi n \tau_s)}{\pi n \tau_s} \; . \tag{8.96}$$

To calculate $\langle \hat{x}\hat{y} \rangle$ for each combination of two quantization intervals, the joint probability of the required unquantized variables falling within these intervals is multiplied by the product of the corresponding values assigned in the quantization process. These results are then summed for all the pairs of intervals. Since the probability distributions of $\hat{x}$ and $\hat{y}$ are both symmetrical about zero, first consider the case in which both of these variables are positive and run from zero to N. As noted above, we take the step size to be unity. Let $L(i)$ be the series of $N + 1$ values that define the positive quantization steps, i.e., $0, 1, 2, \ldots, (N - 1), \ldots, \infty$. Thus, for $i = 1$ to N, $L(i) = i - 1$, and $L(N + 1) = \infty$. For y, there is an identical series of levels represented as $L(j)$. Then the component of $\langle \hat{x}\hat{y} \rangle$ that results from the positive ranges of x and y is

$$\sum_{i=1}^{N}(i - 1/2) \left[\sum_{j=1}^{N}(j - 1/2) \int_{L(j)}^{L(j+1)} \int_{L(i)}^{L(i+1)} p(x, y) \, dx \, dy \right], \tag{8.97}$$

where $(i - 1/2)$ and $(j - 1/2)$ are the values of the digital data assigned to the corresponding quantization intervals, and $p(x, y)$ is the Gaussian bivariate probability distribution, Eq. (8.1). The case in which both x and y are negative provides an equal component of $\langle \hat{x}\hat{y} \rangle$. Thus, the component of $\langle \hat{x}\hat{y} \rangle$ for cases in which x and y have the same sign is

$$\langle \hat{x}\hat{y} \rangle = \frac{1}{\pi\sigma^2\sqrt{1-\rho^2}} \sum_{i=1}^{N}(i-1/2)$$

$$\left\{ \sum_{j=1}^{N}(j-1/2) \int_{L(j)}^{L(j+1)} \int_{L(i)}^{L(i+1)} \left[\exp\left(\frac{-(x^2+y^2-2\rho xy)}{2\sigma^2(1-\rho^2)}\right) \right] dx\,dy \right\} . \qquad (8.98)$$

For the cases in which x and y have opposite signs, one of either $(i-1/2)$ or $(j-1/2)$ is negative, and the sign of either x or y within the exponential function in Eq. (8.98) is negative. When the corresponding expression is included (with negative sign since the component of $\langle \hat{x}\hat{y} \rangle$ is negative), we obtain

$$\langle \hat{x}\hat{y} \rangle = \frac{1}{\pi\sigma^2\sqrt{1-\rho^2}} \sum_{i=1}^{N}(i-1/2)$$

$$\left\{ \sum_{j=1}^{N}(j-1/2) \int_{L(j)}^{L(j+1)} \int_{L(i)}^{L(i+1)} \left[\exp\left(\frac{-(x^2+y^2-2\rho xy)}{2\sigma^2(1-\rho^2)}\right) \right. \right.$$

$$\left. \left. - \exp\left(\frac{-(x^2+y^2+2\rho xy)}{2\sigma^2(1-\rho^2)}\right) \right] dx\,dy \right\} . \qquad (8.99)$$

Equation (8.99) shows how $\langle \hat{x}\hat{y} \rangle$ is derived using the usual form of the bivariate distribution in Eq. (8.1). An equivalent form of the probability distribution of x and y in Eq. (8.1) is given by Abramowitz and Stegun (1968, see Eqs. 26.2.1 and 26.3.2), which avoids the explicit use of the double integrals. Equation (8.99) can then be written as follows:

$$\langle \hat{x}\hat{y} \rangle = \frac{1}{\sqrt{2\pi}\sigma}$$

$$\left\{ \sum_{i=1}^{N}\left[(i-1/2)\sum_{j=1}^{N}\left[(j-1/2)\int_{L(i)}^{L(i+1)} \mathrm{erfc}\left(\frac{L(j)-\rho x}{\sigma\sqrt{2(1-\rho^2)}}\right)\exp\left(\frac{-x^2}{2\sigma^2}\right)dx \right] \right] \right.$$

$$- \sum_{i=1}^{N}\left[(i-1/2)\sum_{j=1}^{N}\left[(j-1/2)\int_{L(i)}^{L(i+1)} \mathrm{erfc}\left(\frac{(L(j+1)-\rho x}{\sigma\sqrt{2(1-\rho^2)}}\right)\exp\left(\frac{-x^2}{2\sigma^2}\right)dx \right] \right]$$

$$- \sum_{i=1}^{N}\left[(i-1/2)\sum_{j=1}^{N}\left[(j-1/2)\int_{L(i)}^{L(i+1)} \mathrm{erfc}\left(\frac{L(j)+\rho x}{\sigma\sqrt{2(1-\rho^2)}}\right)\exp\left(\frac{-x^2}{2\sigma^2}\right)dx \right] \right]$$

$$\left. + \sum_{i=1}^{N}\left[(i-1/2)\sum_{j=1}^{N}\left[(j-1/2)\int_{L(i)}^{L(i+1)} \mathrm{erfc}\left(\frac{(L(j+1)+\rho x}{\sigma\sqrt{2(1-\rho^2)}}\right)\exp\left(\frac{-x^2}{2\sigma^2}\right)dx \right] \right] \right\} .$$

$$(8.100)$$

where erfc is the complementary error function $(1-\mathrm{erf})$.

Fig. 8.13 Curves of the correlation coefficient of quantized data as a function of the true correlation (i.e., the correlation of the unquantized data). The lowest (solid) curve is for 2-level quantization, and moving upward, the curves are for 3 levels (long dashes), 4 levels (long and short dashes), 8 levels (small dashes) and 16 levels (solid line). Similar curves for three and four quantization levels are given in Fig. 8.7.

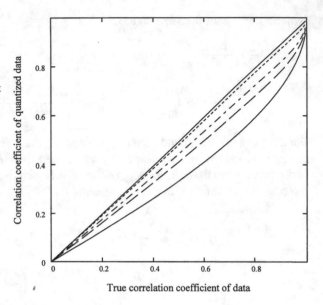

To calculate $\hat{x}^2$, the Gaussian probability function for a single variable is used, taking double the expression for the positive range of x:

$$\hat{x}^2 = \frac{\sqrt{2}}{\sqrt{\pi}\sigma} \sum_{i=1}^{N} (i - 1/2)^2 \int_{L(i)}^{L(i+1)} \exp\left(\frac{-x^2}{2\sigma^2}\right) dx$$

$$= \sum_{i=1}^{N} (i - 1/2)^2 \left[\mathrm{erfc}\left(\frac{L(i)}{\sqrt{2}\sigma}\right) - \mathrm{erfc}\left(\frac{L(i+1)}{\sqrt{2}\sigma}\right) \right]. \tag{8.101}$$

Thus, for a given value of the time interval between samples, the correlation coefficient for the quantized data is as given in Eqs. (8.95), (8.99) or (8.100), and (8.101). Note that the ratio $\langle xy \rangle / \langle x^2 \rangle$ is independent of the frequency response of the system considered and is based on a Gaussian distribution of the amplitude.

Figures 8.13 and 8.14 show examples of the relationship between the correlation of the quantized signals and the true signal correlation. Both of the figures result from the same analysis, but the presentation in Fig. 8.14, in which the correlation of the quantized data is shown as a fraction of the true correlation, helps to emphasize the nonlinearity in the response. A linear response would appear as a horizontal line in Fig. 8.14, and the curves approach this condition as the number of quantization

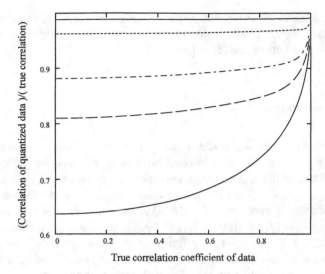

Fig. 8.14 Curves of the correlation coefficient of quantized data, expressed as a fraction of the true correlation. These are from the same data as used in Fig. 8.13, but here they are plotted as fractions of the true (unquantized) correlation, in which the nonlinearity appears as a deviation from a horizontal line. The lowest curve is for 2-level quantization, and moving upward, the curves are for 3, 4, 8, and 16 levels. The points at which the curves meet the left vertical axis indicate the reduction in correlation resulting from quantization when the signal-to-noise ratio is low, as given in Table 8.2. The signal-to-noise ratio increases as the curves move from left to right, and the correlation coefficients of both the quantized and unquantized data move toward 1.0 for the theoretical case of complete correlation between the two signals.

levels increases. Except for observations of the strongest sources, the signal-to-noise ratio from an individual element of a synthesis array is small. Thus, the working point on the curves in Figs. 8.13 and 8.14 is generally near the left side, where the linearity for signals from cross-correlated pairs is best. As the number of quantization levels increases, the accuracy of the correlation increases. The curves provide an indication of the extent to which the quantization affects the measurement of cross-correlation of signals with Gaussian amplitude distribution. A detailed discussion of the effects of quantization of the signal amplitude is given by Benkevitch et al. (2016). This includes the case in which the cross-correlated signals have different amplitudes, and the effects of quantization as the cross-correlation of the analog waveforms approach unity.

For ease of computation, the correlation can be expressed as a rational function, or similar approximation, of the correlator output: See Appendix 8.3 for four-level quantization. For three-level quantization, procedures for determination of the cross-correlation, ρ, from the correlator output are given by Kulkarni and Heiles (1980) and D'Addario et al. (1984). However, with the continuing increase in computer power, larger numbers of levels are generally used.

8.4.2   Oversampling

Sampling of signals at the Nyquist rate results in no loss of information, but quantization causes a reduction in sensitivity as represented by the quantization efficiency. Some of the loss due to quantization can be recovered by oversampling, that is, sampling faster than the Nyquist rate. For sampling of random noise with an ideal rectangular spectrum of width Δv, the time interval between adjacent Nyquist samples is $1/(2\Delta v)$. With Nyquist sampling, the noise within each sample is uncorrelated with respect to the noise in any other sample, and when such data are combined, the noise combines additively in power. Consider the case of oversampling in which the number of samples per second is β times the Nyquist rate. When the sample rate exceeds the Nyquist rate, the samples are no longer independent, and for any particular sample, there are components of the noise within other samples that are correlated with the noise in the sample considered. [Note, however, that for any two samples spaced by β times the sample interval (i.e., spaced at the Nyquist interval), or by an integral multiple of the Nyquist interval, the noise is uncorrelated.[5]] The correlated components of the noise in different samples combine additively in voltage, rather than additively in power, as is the case for uncorrelated noise.

To illustrate how the components of noise combine, consider one pair of antennas and, for example, just four consecutive samples at the correlator output. Let a_1, a_2, a_3, and a_4 be these voltages, which are proportional to the product of the voltages at the correlator inputs. Then we have for the squared sum of these correlated noise voltages, i.e., the total noise power,

$$[a_1 + a_2 + a_3 + a_4]^2 =$$
$$a_1^2 + a_2^2 + a_3^2 + a_4^2 + 2(a_1 a_2 + a_1 a_3 + a_1 a_4 + a_2 a_3 + a_2 a_4 + a_3 a_4) . \quad (8.102)$$

The autocorrelation coefficient of the quantized signals at the correlator input is $R(n\tau_s)$, where n is an integer and τ_s is the spacing in time between adjacent samples. The output of the correlator consists of values that are the product of two input samples, so the autocorrelation coefficient of the samples at the correlator output is

---

[5]It can be assumed that the noise components of the signals from any two antennas are uncorrelated, because noise from the sky background that is received in separate antennas is resolved, and generally the antennas are sufficiently far apart that cross talk of instrumental noise can be ignored.

$R^2(n\tau_s)$. The mean noise power is given by the mean of the terms in the right side of Eq. (8.102), in which each of the a_n^2 terms can be replaced by the mean squared noise amplitude $\langle a^2 \rangle$, and each of the $a_m a_n$ terms by $\langle a^2 \rangle R^2(|n - m|\tau_s)$. Thus, the squared sum of the four noise voltages becomes

$$4\langle a^2 \rangle + 2\langle a^2 \rangle [3R^2(\tau_s) + 2R^2(2\tau_s) + R^2(3\tau_s)] . \tag{8.103}$$

If the four noise terms were uncorrelated, i.e., if the R^2 terms were zero, the noise power would be the sum of the individual noise powers, $4\langle a^2 \rangle$. The effect of the correlation of the noise is to increase the averaged noise power by a factor equal to (8.103) divided by $\langle 4a^2 \rangle$:

$$1 + 2[(3/4)R^2(\tau_s) + (1/2)R^2(2\tau_s) + (1/4)R^2(3\tau_s)] . \tag{8.104}$$

In the general case, averaging a total of N samples at the correlator output, this factor becomes

$$1 + 2\left[\left(\frac{N-1}{N}\right) R^2(\tau_s) + \left(\frac{N-2}{N}\right) R^2(2\tau_s) + \left(\frac{N-3}{N}\right) R^2(3\tau_s) + \ldots \right.$$
$$\left. + \left(\frac{1}{N}\right) R^2[(N-1)\tau_s] \right] . \tag{8.105}$$

In practice, in radio astronomy, the rate at which the data are sampled is in the range of MHz to GHz. The averaging times are in the range milliseconds to seconds, so N is likely to be within the range 10^3 to 10^9. The autocorrelation coefficient decreases as the time interval between samples increases, and in practice, $R^2(n\tau_s)$ becomes very small for $n\tau_s \gtrsim 200$ times the Nyquist sample interval. Thus, for the terms within the square brackets in Eq. (8.105), those after about the first $\sim 200\beta$ can be neglected. Since, in most cases, $N \gg 200\beta$, the squared sum of the noise voltages simplifies to

$$1 + 2[R^2(\tau_s) + R^2(2\tau_s) + R^2(3\tau_s) + \ldots] = 1 + 2\sum_{n=1}^{\infty} R^2(n\tau_s) . \tag{8.106}$$

Equation (8.106) is the fractional increase in the squared noise voltage (i.e., the noise power) that results from the fact that the noise in the samples is no longer independent when the data are oversampled. The quantization efficiency η_Q is equal to the quantization efficiency for Nyquist sampling, η_{QN}, multiplied by $\sqrt{\beta}$ to take account of the increase in the number of samples, but divided by the square root of Eq. (8.106) because the noise in different samples is no longer independent. Thus,

Table 8.3 Variation of quantization efficiency, η_Q, with oversampling factor β

No. of levels	ϵ	$\beta = 1$	$\beta = 2$	$\beta = 4$	$\beta = 8$	$\beta = 16$	$\beta = 32$
2		0.6366	0.744	0.784	0.795	0.798	0.799
3	1.224	0.8098	0.882	0.912	0.920	0.922	0.923
4	0.995	0.8812	0.930	0.951	0.958	0.960	0.960
8	0.586	0.9626	0.980	0.987	0.991	0.991	0.992
16	0.335	0.9885	0.994	0.996	0.998	0.998	0.998

noting that $\tau_s = 1/(2\beta\Delta\nu)$, we obtain

$$\eta_Q = \frac{\eta_{QN}\sqrt{\beta}}{\sqrt{1 + 2\sum_{n=1}^{\infty} R^2\left(\frac{n}{2\beta\Delta\nu}\right)}}. \tag{8.107}$$

To illustrate the effect of oversampling, examples of the quantization efficiency η_Q, derived using Eqs. (8.95), (8.100), (8.101), and (8.107), are shown in Table 8.3. These are for 2-, 3-, 4-, 8-, and 16-level sampling and values of β equal to 1, 2, 4, 8, 16, and 32. In each case, the value of ϵ used is the one that maximizes η_Q for Nyquist sampling, as given in Thompson et al. (2007).[6] Note that as β is increased, the improvement gained by each further increase declines, because the correlation between adjacent samples increases, and thus, the new information provided by finer sampling becomes progressively smaller.

8.4.3   Quantization Levels and Data Processing

At this point, it is useful to put into perspective the characteristics of quantization schemes, which are summarized in Tables 8.2 and 8.3. It should be remembered that the assumption $\rho \ll 1$ was used in determining these values. In considering the relative advantages of different quantization schemes, we note first that both the quantization efficiency η_Q and the receiving bandwidth $\Delta\nu$ may be limited by the size and speed of the correlator system. The overall sensitivity is proportional to $\eta_Q\sqrt{\Delta\nu}$. Consider two conditions. In the first, the observing bandwidth is limited by factors other than the capacity of the digital system. This can occur in spectral line observing or when the interference-free band is of limited width. The sensitivity limitation imposed by the correlator system then involves only the quantization efficiency η_Q in Table 8.2, and the choice of quantization scheme is one between

---

[6]In this reference, σ is taken to be unity and ϵ, the size of the quantization steps, is chosen to maximize the quantization efficiency.

Table 8.4 Sensitivity factor $\frac{\eta_Q}{\sqrt{\beta N_b}}$ for a correlator-limited system

Number of quantization levels (Q)	$\beta = 1$	$\beta = 2$
2 (1 bit)	0.64	0.53
3 (2 bits)	0.57	0.44
4 (2 bits)	0.62	0.47
8 (3 bits)	0.56	0.40
16 (4 bits)	0.49	0.35

simplicity and sensitivity. In the second case, the observing bandwidth is set by the maximum bit rate that the digital system can handle, as may occur in continuum observation in the higher-frequency bands. For a fixed bit rate v_b, the sample rate is v_b/N_b, where N_b is the number of bits per sample, and the maximum signal bandwidth Δv is $v_b/(2\beta N_b)$, where β is the oversampling factor. Thus, the sensitivity is proportional to $\eta_Q/\sqrt{\beta N_b}$, and this factor is listed for various systems in Table 8.4, in which $N_b = 1$ for $Q = 2$ and $N_b = 2$ for $Q = 3$ or 4. Note that oversampling always reduces the performance under these conditions. For those situations in which the capacity of the correlator is limited by the maximum bit rate, the value of 0.64 for Nyquist sampling with two-level quantization results in the highest overall performance. Four-level sampling is almost as good, and four or more levels would be preferred if the bandwidth is limited, as in spectral line observations.

A three-level × five-level correlator, for which the quantization efficiency η_Q is 0.86, was constructed by Bowers et al. (1973) for spectral line imaging with a two-element interferometer.

A further point to be noted is that with an analog correlator, the sin × sin and cos × cos products for signals from two antennas provide, in principle, exactly the same information. However, with a digital correlator, the quantization noise is largely uncorrelated between the sine and cosine components of the signal, so the quantization loss can be reduced by generating both products and averaging them.

8.5 Accuracy in Digital Sampling

Deviations from ideal performance in practical samplers result in errors that, if not corrected for, can limit the accuracy of images synthesized from the data. Once the signal is in digital form, however, the rate at which errors are introduced is usually negligibly small.

Two-level samplers, which sense only the sign of the signal voltages, are the simplest to construct. The most serious error that is likely to occur is in the definition of the zero level, in which a small voltage offset may occur. The effect of offsets in the samplers is to produce small offsets of positive or negative polarity in the correlator outputs, which can be largely eliminated by phase switching, as described in Sect. 7.5. Alternately, the offsets in the samplers can be measured by incorporating counters to compare the numbers of positive and negative samples produced. Correction for the offsets can then be applied to the correlator output data [see, e.g., Davis (1974)].

In samplers with three or more quantization levels, the performance depends on the specification of the levels with respect to the rms signal level, σ. An automatic level control (ALC) circuit is therefore sometimes used at the sampler input. Errors resulting from incorrect signal amplitude become less important as the number of quantization levels is increased; with many levels, the signal amplitude becomes simply a linear factor in the correlator output. In systems using complex correlators, two samplers are usually required for each signal, one at each output of a quadrature network. The accuracy of the quadrature network and the relative timing of the two sample pulses are also important considerations.

8.5.1   Tolerances in Digital Sampling Levels

This section provides an example of the accuracy required in sampling. It is based on a study of errors in three-level sampling thresholds by D'Addario et al. (1984). We start by considering the diagram in Fig. 8.15, which shows the sampling thresholds for a pair of signals to be correlated. Thresholds v_1 and $-v_2$ apply to the signal waveform $x(t)$ and v_3 and $-v_4$ to $y(t)$. The Gaussian probability distribution of x and y is given by Eq. (8.1), and the correlator output is proportional to this probability integrated over the (x, y) plane with the weighting factors ± 1 and zero indicated in the figure. This approach enables one to investigate the effect of deviations of the sampler thresholds from the optimum, $v_0 = 0.612\sigma$. For three-level sampling, the

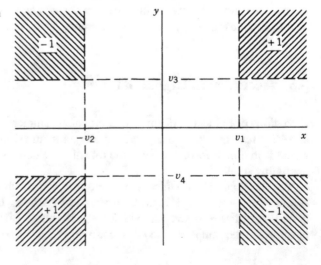

Fig. 8.15 Threshold diagram for a correlator, the inputs of which are three-level quantized signals. x and y represent the unquantized signals, and the shaded areas show the combinations of input levels for which the output is nonzero.

correlator output can be written

$$\langle r_3(\boldsymbol{\alpha}, \rho) \rangle = [L(\alpha_1, \alpha_3, \rho) + L(\alpha_2, \alpha_4, \rho) - L(\alpha_1, \alpha_4, -\rho) - L(\alpha_2, \alpha_3, -\rho)] ,$$

$$(8.108)$$

where $\alpha_i = v_i/\sigma$, and

$$L(\alpha_i, \alpha_k, \rho) = \int_{\alpha_i}^{\infty} \int_{\alpha_k}^{\infty} \frac{1}{2\pi \sqrt{1 - \rho^2}} \exp\left[\frac{-(X^2 + Y^2 - 2\rho XY)}{2(1 - \rho^2)}\right] dX \, dY .$$

$$(8.109)$$

Here, $X = x/\sigma$, $Y = y/\sigma$, and the integrand in Eq. (8.109) is equivalent to the expression in Eq. (8.1) but with the variables measured in units of σ.

D'Addario et al. (1984) point out that since less than 5% loss in signal-to-noise ratio occurs for threshold departures of $+40\%$ from optimum, the required accuracy of the threshold settings, in practice, depends mainly on the algorithm used to correct the result. Suppose that the thresholds are kept close to, but not exactly equal to, the optimum value. For the x sampler in Fig. 8.15, the deviations from the ideal threshold value α_0 can be expressed in terms of an even part

$$\Delta_{\mathrm{gx}} = \frac{1}{2}(\alpha_1 + \alpha_2) - \alpha_0 , \qquad (8.110)$$

and an odd part

$$\Delta_{\mathrm{ox}} = \frac{1}{2}(\alpha_1 - \alpha_2) . \qquad (8.111)$$

For the y sampler, Δ_{gy} and Δ_{oy} are similarly defined. The Δ_{g} terms produce gain errors. They are equivalent to an error in the level of the signal at the sampler, and they have the effect of introducing a multiplicative error in the measured cross-correlation. The Δ_{o} terms produce offset errors in the correlator output and are potentially more damaging since such errors can be large compared with the low levels of cross-correlation resulting from weak sources. The offset errors, however, can be removed with high precision by phase switching. The cancellation of the offset results from the sign reversal of the digital samples, or of the correlator output, as described in Sect. 7.5. The correlator output of a phase-switched system is of the form

$$r_{3s}(\boldsymbol{\alpha}, \rho) = \frac{1}{2}[r_3(\boldsymbol{\alpha}, \rho) - r_3(\boldsymbol{\alpha}, -\rho)] . \qquad (8.112)$$

If all α values are within $\pm 10\%$ of α_0, the output is always within 10^{-3} (relative error) of the output of a correlator with the same gain errors, but no offset errors, in the samplers. Thus, with phase switching, errors of up to $\sim 10\%$ in the thresholds may be tolerable. Also, corrections can be made for gain errors if the actual threshold levels are known. Since the probability density distribution of the signal amplitudes can be assumed to be Gaussian, the threshold levels can be determined by counting the relative numbers of $+1$, 0, and -1 outputs from each sampler. When ρ is small (a few percent), a simple correction for the gain error can be obtained by dividing the correlator output by the arithmetic mean of the numbers of high-level (± 1) samples for the two signals. Then 10% errors in the threshold settings result in errors of less than 1% in ρ.

Another nonideal aspect of the behavior of the sampler and quantizer is that the threshold level may not be precisely defined but may be influenced by effects such as the direction and rate of change of the signal voltage, the previous sample value (hysteresis), and noise in the sampling circuitry. The result can be modeled by including an indecision region in the sampler response extending from $\alpha_k - \Delta$ to $\alpha_k + \Delta$. It is assumed that a signal that falls within this region results in an output that takes either of the two values associated with the threshold randomly and with equal probability. The three-level threshold diagram with indecision regions included is shown in Fig. 8.16.

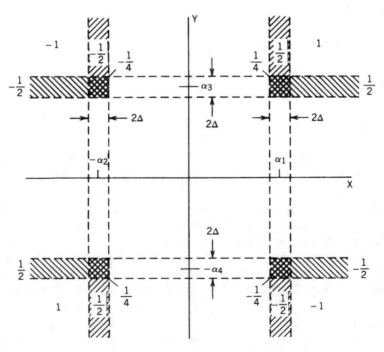

Fig. 8.16 Threshold diagram for a three-level correlator showing indecision regions and the shaded areas within them for which the response is nonzero. The figures ± 1, $\pm\frac{1}{2}$, and $\pm\frac{1}{4}$ indicate the correlator response. The diagram shows the (X, Y) plane in which the signals are normalized to the rms value σ.

Fig. 8.17 Effect of indecision regions on the output of a three-level correlator. The thresholds are assumed to be set to the optimum value 0.612σ, and the widths of the indecision regions are $2\sigma\Delta$. The output is given as a fraction of the output for $\Delta = 0$.

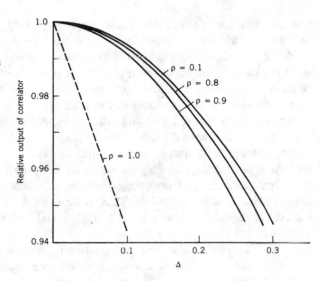

The weighting in the indecision regions depends on the probability of the random sample values and is 1/4 when both signals fall within indecision regions, and 1/2 when one signal is within an indecision region and the other produces a nonzero output. As before, the correlator output can be obtained by integrating the weighted probability of the signal values over the (X, Y) plane. Figure 8.17 shows the decrease in the correlator output as a function of Δ for several values of ρ, computed by expressing the output decrease as a Maclaurin series in Δ (D'Addario et al. 1984). For all cases except those in which ρ approaches unity, the relatively small decrease in output results from the fact that when one input waveform falls within an indecision region, the other generally does not. For the particular case of $\rho = 1$, the input waveforms are identical and fall within these regions simultaneously. The output decrease is then proportional to Δ, as shown by the broken line in Fig. 8.17: However, this case is only of limited practical importance. For a 1% maximum error, Δ must not exceed 0.11σ, so the indecision region can be as large as $\pm18\%$ of the threshold value. For a maximum error of 0.1%, the above limits must be divided by $\sqrt{10}$. Thus, the indecision regions have large enough tolerances that their effect may be negligible.

8.6   Digital Delay Circuits

Time delays that are multiples of the sample interval can be applied to streams of digital bits by passing them through shift registers that are clocked at the sampling frequency. Shift registers with different numbers of stages thus provide different fixed delays. A method of using two shift registers to obtain a delay that is variable in increments of the clock pulse interval is described by Napier et al. (1983). However, integrated circuits for random access memory (RAM), developed for computer applications, provide an economical solution for large digital delays.

Another useful technique is serial-to-parallel conversion, that is, the division of a bit stream at frequency v into n parallel streams at frequency v/n, where n is a power-of-two integer. This allows the use of slower and more economical types of digital circuits for delay, correlation, and other processes.

The precision required in setting a delay has been discussed in Sect. 7.3.5 and is usually some fraction of the reciprocal of the signal bandwidth. In any form of delay that operates at the frequency of the sampler clock, the basic delay increment is the reciprocal of the sampling frequency. A finer delay step can be obtained digitally by varying the timing of the sample pulse in a number of steps, for example, 16, between the basic timing pulses. Thus, if an extra delay of, say, 5/16 of a clock interval is required, the sampler is activated 11/16 of a clock interval after the previous clock pulse, and the data are held for 5/16 of an interval to bring them into phase with the clock-pulse timing. Correction for delay steps equal to the sampling interval can also be made after the signals have been cross-correlated, by applying a phase correction to the cross power spectrum.

8.7   Quadrature Phase Shift of a Digital Signal

We have mentioned that complex correlators for digital signals can be implemented by introducing the quadrature phase shift in the analog signal, as in Fig. 6.3, and then using separate samplers for the signal and its phase-shifted version. The Hilbert transformation that the phase shift represents can also be performed on the digital signal, thus eliminating the quadrature network and saving samplers and delay lines, but the accuracy is limited. Hilbert transformation is mathematically equivalent to convolution with the function $(-\pi\tau)^{-1}$, which extends to infinity in both directions [see, e.g., Bracewell (2000), p. 364]. A truncated sequence of the same form, for example, $\frac{1}{3}, 0, 1, 0, -1, 0, -\frac{1}{3}$, provides a convolving function for the digital data that introduces the required phase shift. However, the truncation results in convolution of the resulting signal spectrum with the Fourier transform of the truncation function, that is, a sinc function. This introduces ripples and degrades the signal-to-noise ratio by a few percent. Also, the summation process in the digital convolution increases the number of bits in the data samples, but the low-order bits can be discarded to avoid a major increase in the complexity of the correlator. This results in a further quantization loss. The overall result is that the imaginary output of the correlator suffers spectral distortion and some loss in signal-to-noise ratio relative to the real output. These effects are most serious in broad-bandwidth systems, in which the high data rate permits only simple processing. Lo et al. (1984) have described a system in which the real part of the correlation is measured as a function of time offset, as described below for the spectral correlator, and the imaginary part is then computed by Hilbert transformation.

8.8   Digital Correlators

8.8.1   Correlators for Continuum Observations

In continuum observations, the average correlation over the signal bandwidth is measured, and data on a finer frequency scale may not be required. In such cases, the correlation of the signals is measured usually for zero time-delay offset. Digital correlators can be designed to run at the sampling frequency of the signals or at a submultiple resulting from dividing the bit stream from the sampler into a number of parallel streams. In the latter case, the number of correlator units must be proportionally increased, and their outputs can subsequently be additively combined. Two-level and three-level correlators, for which the products are represented by values of -1, 0, and $+1$, are the simplest. Correlators in which one of the inputs is a two-level or three-level signal and the other input is more highly quantized also have a degree of simplicity. In this case, the correlator is essentially an accumulating register into which the higher-quantization value is entered. The two-level or three-level value is used to specify whether the other number is to be added, subtracted, or ignored. In correlators in which both inputs have more than three levels of quantization, the multiplier output for any single product can be one of a range of numbers. One method of implementing such a multiplier is to use a read-only memory unit as a lookup table in which the possible product values are stored. The input bits to be multiplied are used to specify the address of the required product in the memory.

The output of a multiplier can take both positive and negative values, and, ideally, an up–down counter is required as an integrator. Since such counters are usually slower than simple adding counters, two of the latter are sometimes used to accumulate the positive and negative counts independently. Another technique is to count, for example, -1, 0, and $+1$ as 0, 1, and 2, and then subtract the excess values, in this case equal to the number of products, in the subsequent processing.

Spectral line (multichannel) correlators are used with most large general-purpose arrays. For continuum observations, they offer advantages such as the ability to reject narrowband interfering signals or to divide a band into narrower sub-bands to reduce the smearing of spectral details.

8.8.2   Digital Spectral Line Measurements

In spectral line observations, measurements at different frequencies across the signal band are required. These measurements can be obtained by digital techniques using a spectral correlator system, which is commonly implemented by measuring the correlation of the signals as a function of time offset. The Fourier transform of this quantity is the cross power spectrum, which can be regarded as the complex visibility as a function of frequency. (This Fourier transform relationship is a form of the Wiener–Khinchin relation discussed in Sect. 3.2.) In the case of an autocorrelator (for use with a single antenna), the two input signals are the same waveform with a time offset. Thus, the autocorrelation function is symmetric, and the power spectrum is entirely real and even. However, the cross power spectrum of the signals from two different antennas is complex, and the cross-correlation function has odd as well as even parts.

The output of a spectral correlator system provides values of the visibility at N frequency intervals across the signal band. These intervals are sometimes spoken of as frequency channels and their spacing as the channel bandwidth. To explain the action of a digital spectral correlator, we consider the cross power spectrum $S(\nu)$ of the signals from two antennas, as shown in idealized form in Fig. 8.18. Here it is assumed that the source under observation has a flat spectrum with no line features, and the final IF amplifier before the sampler has a rectangular baseband response. In Fig. 8.18, we have included the negative frequencies since they are necessary in the Fourier transform relationships. For $-\Delta\nu \le \nu \le \Delta\nu$, the real and imaginary parts of $S(\nu)$ have magnitudes a and b, respectively, and the corresponding visibility phase is $\tan^{-1}(b/a)$. The cross-correlation function $\rho(\tau)$ is the Fourier transform of $S(\nu)$, where τ is the time offset:

$$\rho(\tau) = (a - jb) \int_{-\Delta\nu}^{0} e^{j2\pi\nu\tau} d\nu + (a + jb) \int_{0}^{\Delta\nu} e^{j2\pi\nu\tau} d\nu$$

$$= 2\Delta\nu \left[a \frac{\sin(2\pi\Delta\nu\,\tau)}{2\pi\Delta\nu\,\tau} - b \frac{1 - \cos(2\pi\Delta\nu\,\tau)}{2\pi\Delta\nu\,\tau} \right]. \qquad (8.113)$$

Thus, $\rho(\tau)$ has an even component of the form $(\sin x)/x$, which is related to the real part of $S(\nu)$, and an odd component of the form $(1 - \cos x)/x$, which is related to the imaginary part. The spectral correlator measures $\rho(\tau)$ for integral values

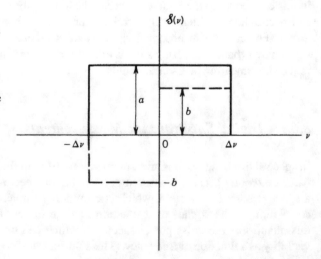

Fig. 8.18 Cross power spectrum $S(\nu)$ of two signals for which the power spectra are rectangular bands extending in frequency from zero to $\Delta\nu$. Negative frequencies are included. The solid line represents the real part of $S(\nu)$ and the dashed line the imaginary part. The corresponding correlation function is derived in Eq. (8.113).

of the sampling interval τ_s. We consider the case of Nyquist sampling, for which $\tau_s = 1/(2\Delta v)$. The measured cross-correlation refers to the quantized waveforms, and the analysis in Sect. 8.4.1 shows how this is related to the cross-correlation of the unquantized waveforms. For correlation levels that are not too large, the two quantities are closely proportional, so for simplicity, we assume that Eq. (8.113) represents the behavior of the measured cross-correlation. The measurements are made with $2N$ time offsets from $-N\tau_s$ to $(N-1)\tau_s$ between the signals, and Fourier transformation of these discrete values yields the cross power spectrum at frequency intervals of $(2N\tau_s)^{-1} = \Delta v/N$ for Nyquist sampling. The N complex values of the positive frequency spectrum are the data required. Of these, the imaginary part comes from the odd component of the correlator output $r(\tau)$. Thus, in the correlation measurement, it suffices to use single-multiplier correlators to measure $2N$ real values of $r(\tau)$ over both positive and negative values of τ for one antenna with respect to the other. As an alternative to measuring only the real part of the correlation, complex correlators could be used to measure both the real and imaginary parts for a range of time offsets from zero to $(N-1)\tau_s$. However, complex correlators require broadband quadrature networks.

Measurement of the cross-correlation over the limited time offset range is equivalent to measuring $r(\tau)$ multiplied by a rectangular function of width $2N\tau_s$. The cross power spectrum derived from the limited measurements is therefore equal to the true cross power spectrum convolved with the Fourier transform of the rectangular function, that is, with the sinc function

$$\frac{\sin(\pi v N/\Delta v)}{\pi v} , \tag{8.114}$$

which is normalized to unit area with respect to v. Any line feature within the spectrum is broadened by the sinc function (8.114) and, depending on its frequency profile, may show the characteristic oscillating skirts. The width of the sinc function at the half-maximum level is $1.2\Delta v/N$, that is, 1.2 times the channel separation, and this width defines the effective frequency resolution.

The oscillations of the sinc function introduce structure in the frequency spectrum similar to the sidelobe responses of an antenna beam. They result from the sharp edges of the rectangular function that multiplies the correlation function. Such sidelobes are undesirable and can be reduced by choosing weighting functions, other than rectangular truncation, that are constrained to be zero outside the measurement range. It is desirable that weighting functions should taper smoothly to zero at $|\tau| = N\tau_s$, thereby reducing unwanted ripples in the smoothing (convolving) function, but also to be as wide as possible in order to keep the width of the smoothing function as narrow as possible. These requirements are not generally compatible, so weighting functions that produce smoothing functions with very low sidelobes have poor frequency resolution. Some commonly used weighting functions are listed in Table 8.5.

Hann weighting, also known as raised cosine weighting, reduces the first sidelobe by a factor of 9 but degrades the resolution by 1.67, compared with uniform

Table 8.5 Commonly used weighting functions

Weighting function $w(\tau)$	$[w(\tau) = 0, \|\tau\| > \tau_1 = N\tau_s]$	Half-amplitude width (Unit=$\Delta\nu/N$)	Peak sidelobe
Uniform	$w(\tau) = 1$	1.21	0.22
Bartlett	$w(\tau) = 1 - (\|\tau\|/\tau_1)$	1.77	0.047
Hann[a]	$w(\tau) = 0.5 + 0.5\cos(\pi\tau/\tau_1)$	2.00	0.027
Hamming	$w(\tau) = 0.54 + 0.46\cos(\pi\tau/\tau_1)$	1.82	0.0073
Blackman	$w(\tau) = 0.42 + 0.50\cos(\pi\tau/\tau_1)$ $\quad + 0.08\cos(2\pi\tau/\tau_1)$	2.30	0.0011
Blackman-Harris	$w(\tau) = 0.35875 + 0.48829\cos(\pi\tau/\tau_1)$ $\quad + 0.14128\cos(2\pi\tau/\tau_1)$ $\quad + 0.0106411\cos(3\pi\tau/\tau_1)$	2.67	0.000025

[a]Hann weighting is named after the nineteenth-century meteorologist Julius von Hann and is sometimes colloquially referred to as "Hanning weighting." Hamming weighting is named after R. W. Hamming, an engineer at Bell Telephone Laboratories.

weighting. The Fourier transform of the Hann weighting function is the sum of three sinc functions of relative amplitudes $0.25, 0.5$, and 0.25. This is the smoothing function in the spectral domain shown in Fig. 8.19b, which corresponds to Hann weighting. For the usual case in which the number of points in the discretely sampled spectrum equals the number of points in the correlation function (i.e., no zero padding, as in the FX correlator, Sect. 8.8.4), the smoothing or convolution can be implemented as a three-point running mean with relative weights of 0.25, 0.5, and 0.25. Thus, the smoothed value of the cross power spectrum at frequency channel n is given by

$$S'\left(\frac{n\Delta\nu}{N}\right) = \frac{1}{4}S\left[\frac{(n-1)\Delta\nu}{N}\right] + \frac{1}{2}S\left(\frac{n\Delta\nu}{N}\right) + \frac{1}{4}S\left[\frac{(n+1)\Delta\nu}{N}\right]. \quad (8.115)$$

The Hamming weighting function is very similar to the Hann function and would appear to be superior because it produces a better resolution and a lower peak sidelobe level. However, the sidelobes of the Hamming smoothing function do not decrease in amplitude as rapidly as those of the Hann smoothing function. Weighting functions are discussed in detail by Blackman and Tukey (1959) and Harris (1978).

A further effect of the finite time-offset range complicates the calibration of the instrumental frequency response in the following way (Willis and Bregman 1981). The frequency responses of the amplifiers associated with the different antennas may not be exactly identical, as discussed in Sect. 7.3. To calibrate the response of each antenna pair over the spectral channels, it is usual to measure the cross power spectrum of an unresolved source for which the actual radiated spectrum is known to be flat across the receiving passband. We can consider the result in terms of the idealized power spectra in Fig. 8.18. If no special weighting function is used, the real and imaginary parts are both convolved with the sinc function (8.114).

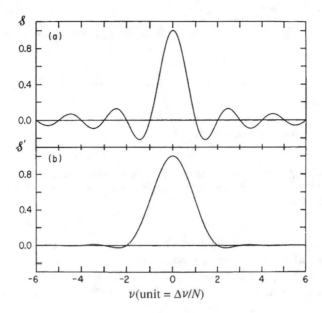

Fig. 8.19 (a) The ordinate is the sinc function $\sin(\pi v N/\Delta v)/(\pi v N/\Delta v)$, which represents the frequency response of a spectral correlator with channels of width $\Delta v/N$ to a narrow line at $v = 0$. The abscissa is frequency v measured with respect to the center of the received signal band. (b) The same curve after the application of Hann smoothing, as in Eq. (8.115).

When a function with a sharp edge is convolved with a sinc function, the result is the appearance of oscillations (the Gibbs phenomenon) near the edge, as shown in Fig. 8.20. The point here is that the real component of $S(v)$ in Fig. 8.18 is continuous through zero frequency, but the imaginary part shows a sharp sign reversal. Thus, near zero frequency, the observed imaginary part of $S(v)$ will show oscillations that may be as high as 18% in peak amplitude, whereas the real component will show relatively small oscillations at that point (see also Fig. 10.14b and associated text). As a result, the magnitude and phase measured for $S(v)$ will show oscillations or ripples, the amplitude of which will depend on the relative amplitudes of the real and imaginary parts, that is, on the phase of the uncalibrated visibility. The uncalibrated phase measured for any source depends on instrumental factors such as the lengths of cables as well as the source position, which may not be known. In general, the phase will not be the same for the source under investigation and the calibrator. Hence, near zero frequency, some precautions must be taken in applying the calibration. Possible solutions to the problem include (1) calibrating the real and imaginary parts separately, (2) observing over a wide enough band that the end channels in which the ripples are strongest can be discarded, or (3) applying smoothing in frequency to reduce the ripples.

Another problem encountered when observing a spectral line in the presence of a continuum background is caused by reflections in the antenna structure. These reflections cause a sinusoidal gain variation across the passband, the period of which

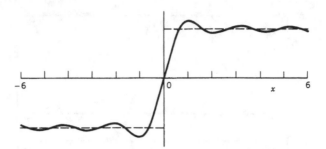

Fig. 8.20 Convolution of a step function at the origin (broken line) with the sinc function $(\sin \pi x)/\pi x$. Here, $x = \nu N/\Delta \nu$, and the half-cycle period of the ripple is approximately equal to the width of a spectral channel.

is equal to the reciprocal of the delay of the signal caused by the reflection. In a correlation interferometer, the magnitude of the ripple is a nearly constant fraction of the correlated continuum flux density, and the ripple is removed when the spectrum of the source under investigation is divided by the spectrum of the calibration source.

8.8.3   Lag (XF) Correlator

Correlators can be classified as two general types. In a lag (or XF) correlator, cross-correlation is followed by Fourier transformation, and in an FX correlator, Fourier transformation is followed by cross-correlation. A simplified schematic diagram of a lag correlator is shown in Fig. 8.21. Practical systems are often more complicated and are designed to take full advantage of the flexibility of digital processing techniques. The bandwidths of channels required for spectral line studies vary greatly, from a few tens of hertz to hundreds of megahertz. This versatility is necessary because the widths of spectral features are influenced by Doppler shifts, which are proportional to the rest frequencies of the lines and the velocities of the emitting atoms and molecules. The correlator of the upgraded VLA system (Perley et al. 2009) is fundamentally an XF design, as is the ALMA system, following its digital filter (Escoffier et al. 2000).

A recirculating correlator is one that can store blocks of data and process them multiple times through the correlator. This can be done only when the correlator is capable of running faster than the incoming data rate. These multiple passes allow the number of correlator channels to be increased. For example, if data samples are processed by the correlator twice, the range of delays can be doubled, so the spectral resolution is improved by a factor of two.

To implement the above scheme, recirculator units are required, which are basically memories that store blocks of input samples and allow them to be read out

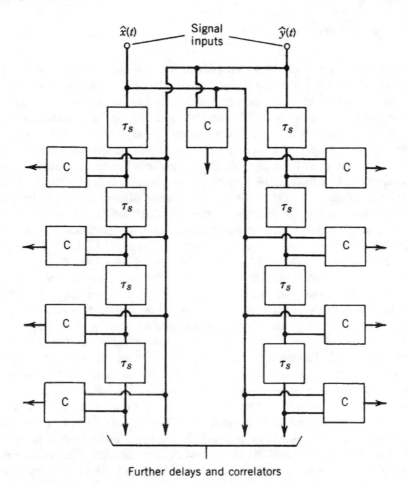

Fig. 8.21 Simplified schematic diagram of a lag (XF) spectral correlator for two sampled signals. τ_s indicates a time delay equal to the sampling interval and C indicates a correlator. The correlation is measured for zero delay, for the $\hat{x}$ input delayed with respect to the $\hat{y}$ input (left correlator bank), and for $\hat{y}$ delayed with respect to $\hat{x}$ (right correlator bank). The delays are integral multiples of τ_s.

at the correlator input rate. These memory units are required in pairs, so that one is filled with data at the Nyquist rate appropriate to the chosen signal bandwidth, while the other is being read at the maximum data rate. One memory becomes filled in the time that the other is read for the required number of times, and the two are then interchanged. Examples of recirculating lag correlators are described by Ball (1973) and Okumura et al. (2000). The WIDAR correlator on the VLA uses recirculation (Perley et al. 2009).

8.8.4 FX Correlator

The designation FX indicates a correlator in which Fourier transformation to the frequency domain is performed before cross multiplication of data from different antennas. In such a correlator, the input bit stream from each antenna is converted to a frequency spectrum by a real-time FFT, and then for each antenna pair, the complex amplitudes for each frequency are multiplied to produce the cross power spectrum. A major part of the computation occurs in the Fourier transformation, for which the total number of operations is proportional to the number of antennas. In comparison, in a lag correlator, the total computation is largely proportional to the number of antenna pairs. Thus, the FX scheme offers economy in hardware, especially if the number of antennas is large (see Sect. 8.8.5). The principle of the FX correlator, based on the use of the FFT algorithm, was discussed by Yen (1974) and first used in a large practical system by Chikada et al. (1984, 1987). Description of system designed for the VLBA are given by Benson (1995) and Romney (1999).

Two slightly different implementations of the FX correlator have been used. In one, both in-phase and quadrature components of the signal are sampled to provide a sequence of N complex samples, which is then Fourier-transformed to provide N values of complex amplitude, distributed in frequency over positive and negative frequencies. In the other, N real samples are transformed to provide N values of complex amplitude. However, the negative frequencies are redundant, and only $N/2$ spectral points need be retained. We follow the second scheme in the discussion below.

Figure 8.22 is a schematic diagram of the basic operations of an FX correlator. The input sample stream from an antenna is Fourier transformed in contiguous sequences of length-N samples, where N is usually a power-of-two integer for efficiency in the FFT algorithm. The output of each transformation is a series of N complex signal amplitudes as a function of frequency. The frequency spacing of the data after transformation is $1/(Nt_s)$, where t_s is the time interval between samples of the signals. In the cross-multiplication process that follows the FFT stage, the complex amplitude from one antenna of each pair is multiplied by the complex conjugate of the amplitude of the other. These multiplications occur in the correlator elements in Fig. 8.22. Note that the data in any one input sequence are combined only with data from other antennas for the same time sequence. This leads to some differences in the effective weighting of the data in the FX and XF designs.

8.8.5 Comparison of XF and FX Correlators

Spectral Response. In the FX configuration, the F engine (DFT processor) operates on short segmented blocks of data in order to control the spectral resolution. The equivalent correlation function constructed from a block of data has N ways, or N possible multiplications for the zero lag component. There are progressively fewer

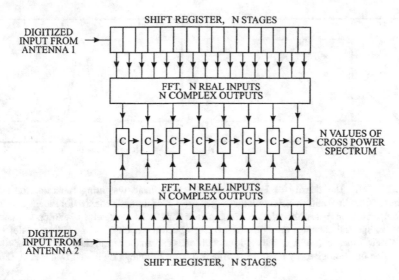

Fig. 8.22 Simplified schematic diagram of an FX correlator for two antennas. The digitized signals are read into the shift registers and an FFT performed at intervals of N sample periods. The correlator elements, indicated by C, form products of one signal with the complex conjugate of the other. In an array with n_a antennas, the outputs of each FFT are split $(n_a - 1)$ ways for combination with the complex amplitudes from all other antennas.

multiplications available for increasing lags because of the data block boundaries. The correlator function at the maximum lags of $\pm(N - 1)t_s$ can be obtained in only one way. Hence, the density of lag multiplications has a triangular shape as a function of lag over the range $\pm Nt_s$, as shown in Fig. 8.23 (see also Moran 1976). Hence, the spectral response, the Fourier transform of this triangular function, is $\text{sinc}^2(Nt_s\nu) = \text{sinc}^2(n)$, where $\nu = n/Nt_s$ and n is the spectral channel number. An alternate derivation of this result is given in Sect. A8.4.1, where it is shown that the spectral response to a sine wave is a sinc^2 function.

For the XF configuration, the spectral resolution depends on the length of the correlation function, calculated as described in Sect. 8.8.2 Since the correlation function is calculated on a segment of data that is much longer than the block length, the density of lag multiplications is essentially uniform, except for a very small end effect. Hence, the spectral response is $\text{sinc}(Nt_s\nu)$ or $\text{sinc}(n)$. The spectral responses for the FX and XF correlators are shown in Fig. 8.23.

Note that the integral over frequency of both of these spectral responses is unity. Therefore, the flux or the area under a spectral line profile, the convolution of the source spectrum with the spectral response function, is conserved. The peak amplitude of a spectral feature narrower than the resolution will depend on where it falls with respect to the spectral channels. A line that falls midway between two channels will have its peak amplitude reduced by $\text{sinc}^2(1/2) = 0.41$ for the FX processor compared with $\text{sinc}(1/2) = 0.81$ for the XF processor. This is the well-known effect called scalloping. It can be mitigated by the technique of padding with zeros to obtain an interpolated spectrum (see Sect. A8.4.2).

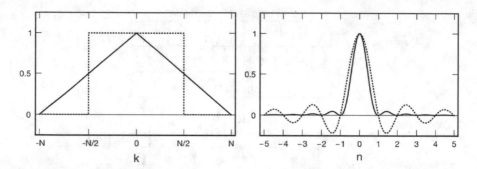

Fig. 8.23 (left) The density of lag calculations, or intrinsic weighting function, for an FX correlator (solid line) and an XF correlator (dotted line). N is the segment size for the FX correlator. For comparison purposes, the width of the function for the XF correlator is chosen to make the number of spectral channels the same. (right) The spectral response for the FX correlator (solid line), a sinc^2 function (see Appendix 8.4 for additional explanation) and the XF correlator (dotted line), a sinc function given by Eq. (8.114). Adapted from Romney (1999) and Deller et al. (2016).

In some cases, it may be desirable to actually calculate the correlation functions from the output of the F engine, such as for application of a full nonlinear quantization correction. It is well known (Press et al. 1992) that it is necessary to pad the spectrum with N zeros in order to obtain the correct result [see discussion after Eq. (A8.40)]. The implementation of this calculation is discussed by O'Sullivan (1982) and Granlund (1986).

Signal-to-Noise Ratio. The fundamental difference between the FX and XF processors is the density weighting in the lag domain. Both systems have the same number of equivalent multiplications, as can be seen in Fig. 8.23. The FX covers twice the number of lags as the XF processor but has lower density as the lag number increases. In particular, the FX provides half the lag density for $k = N/2$. For a continuum source, the signal-to-noise ratio of the FX and XF systems is the same. This can be appreciated by the fact that only the zero lag multiplications are important, and they are equal in both systems. Similarly, the signal-to-noise ratio for a very-narrow-bandwidth source, less than the resolution, is also the same because the total number of equivalent multiplications is the same.

There is a small difference in response for signals that have line widths about equal to the resolution. In particular, for this case, the amplitude of a spectral line is reduced by a factor of about 0.82 (Okumura et al. 2001; Bunton 2005). This is a problem only for slightly resolved spectral features. In any event, most spectrometers are designed to produce several channels per resolution element in order to properly analyze the lines. This perceived deficiency in the FX correlator is due to the distribution of lags. The FX correlator has a larger range but fewer multiplications at lag(k) $\pm N/2$ (see Fig. 8.23). There are several approaches to recovering this loss of information. The classic method (Welch 1967; Percival and Walden 1993) is to overlap the segments in the block processing in the F engine.

A 50% overlap recovers most of the lost signal-to-noise ratio but at a cost of doubling the processing time in the F engine. This overlap feature was available in the original FX VLBA processors but rarely, if ever, used (Romney 1995). Another approach is to simply channel-average the spectrum, but this wastes the resolution capability of the F engine. Note that with polyphase filter banks, scalloping for narrow spectral lines and signal-to-noise ratio loss are very small.

Number of Operations. We can make an approximate comparison of the workload requirements of XF and FX signal processors by comparing the number of multiplications needed in each system. For this rather simplistic analysis, we assume that the data are streams of real numbers at the Nyquist interval appropriate for bandwidth Δv, i.e., $t_s = 1/(2\Delta v)$. To make this comparison, we further assume that the number of lags computed in the X engine (lag correlator), N, is equal to the data segment length into the F engine. This makes the spectral resolution of both systems approximately equal (see Fig. 8.23 for exact responses).

Consider the analysis of one second of data, i.e., $2\Delta v$ samples. For the XF system, a lag correlator is required for each baseline. Thus, $2N\Delta v$ multiplications are required for each baseline. Since $Nt_s \ll 2\Delta v$, the edge effects in calculating the correlation function are negligible (i.e., all lags have almost the same number of multiplications approaching N), and the workload of the single Fourier transform at the end of the integration period is negligible compared with the workload of calculating the correlation function. Thus, the rate of multiplications (multiplies per second), r_{XF}, is

$$r_{XF} = 2\Delta v N n_b \, , \tag{8.116}$$

where n_b is the number of baselines. For the FX processor, one DFT engine is required for each antenna. We assume that the number of multiplications for the FFT implementation of the N-point DFT is $N \log_2 N$. (Some variation exists, depending on the FFT implementation, e.g., an FFT with N being a power of four would run somewhat faster.) The cross power spectrum calculation requires the pairwise cross multiplication of the outputs of DFT engines for all baselines. These multiplications are complex, requiring four real multiplications each. In addition, only the $N/2$ spectral points at positive frequencies need to be calculated and retained. The number of multiplications is therefore $[n_a N \log_2 N + 4Nn_b/2]M$, where M is the number of segments processed, $2\Delta v/N$. Since $MN = 2\Delta v$, the aggregate multiplication rate is

$$r_{FX} = 2\Delta v [n_a \log_2 N + 2n_b] \, . \tag{8.117}$$

The workload ratio, $\mathcal{R} = r_{XF}/r_{FX}$ is therefore

$$\mathcal{R} = \frac{n_b N}{n_a \log_2 N + n_b} \, . \tag{8.118}$$

The $n_a \log_2 N$ factor reflects the $\log_2 N$ advantage and the antenna-based processing of the DFT engine. The $n_b N$ factor reflects the baseline processing of the X engine. Since $n_b = n_a(n_a - 1)/2$, we can rewrite Eq. (8.118) as

$$R = \frac{N}{\frac{2\log_2 N}{n_a-1} + 2} \; . \tag{8.119}$$

Note that this relation holds for $n_a \geq 2$ because no single antenna spectra are calculated. [Analysis for a spectrometer on a single antenna would yield $R = N/(\log_2 N + 1)$.] The limiting forms of Eq. (8.119) are

$$R = N/2 \, , \qquad\qquad n_a \gg 1 + \log_2 N \, ,$$
$$R = \frac{N(n_a - 1)}{2\log_2 N} \qquad\qquad n_a \ll 1 + \log_2 N \, . \tag{8.120}$$

In general, the larger the values of N or n_a, the more the FX design is favored. For example, with $n_a = 10$ and $N = 1024$, $R \sim 240$. Perhaps the most important limitation in Eq. (8.119) is that the X engine operates on one or a few-bit representation of the signal and multiplication can be achieved by simple table lookup, whereas the F engine needs more bits per sample and there is additional bit growth in its internal operations. Furthermore, the detailed architecture of chips has a major influence on calculation speed. Hence, Eq. (8.119) is a useful guide for the general dependence of R on N and n_a but does not accurately specify a crossover point favoring one design over the other. The advantage clearly shifts to the FX design for very large n_a or N.

Digital Fringe Rotation. In early systems, fringe rotation was often applied to the signal as an analog process, but generally it is advantageous to implement it after digitization. For example, in VLBI observations in which the data are recorded as digital samples, it is useful to be able to repeat the analysis with different fringe rates if the position of the source on the sky is not known with sufficient accuracy before the observation. Digital fringe rotation is usually applied to the digitized IF waveform before it goes to the correlator and involves multiplication with a digitized fringe rotation waveform. It is desirable to use a multibit representation for the rotated data to maintain the required accuracy, and thus, the number of bits in the input data to the correlator may be increased. Increasing the number of bits per sample in a lag correlator results in a proportional increase in complexity. Thus, it may be necessary to truncate the data before input to the correlator, which effectively introduces the quantization loss a second time. In contrast, in the FX design, multibit data representation is required in the FFT processing, so the bit increase that fringe rotation presents is more easily accommodated. See Sect. 9.7.1 for more details.

Fractional Sample Delay Correction. In digital implementation of the compensating delays, one way of adjusting the delay in steps smaller than the sampling interval is to adjust the timing of the sampler pulses, as described in Sect. 8.6. Another way of introducing a fractional sample period delay is done after transformation to the frequency domain by incrementing the phase values by an amount that varies in proportion to the frequency across the IF band. In the FX correlator, this is easily done because the signals appear as an amplitude spectrum every FFT cycle, and the correction can be applied as required for each antenna before the data are combined in antenna pairs. With a lag correlator, there are two problems in this process. First, the transformation to a spectrum occurs after the data are combined for antenna pairs, so many more values require correction. Second, for long baselines, the corrections required may occur more rapidly than the rate at which cross-correlation values are transformed to cross power spectra. Thus, it may be possible to apply only a statistical correction rather than an exact one. See Sect. 9.7.3 for a description of the statistical corrections.

Quantization Correction. The nonlinearity of the amplitude of the cross-correlation measured using coarsely quantized samples is seen in the Van Vleck relationship [Eq. (8.25)]. Application of a correction for the nonlinearity in quantization in the lag (XF) correlator is a relatively straightforward process because the cross-correlation values are directly calculated. To obtain the cross-correlation values in the FX correlator, the cross power spectrum at the correlator output must be Fourier transformed from the frequency domain to the lag domain. After applying the correction, the data must then be transformed back to a frequency spectrum. The correction is necessary only if the correlation of the total waveform (signal plus noise) is large for any pair of antennas. This condition implies observation of a source that is largely unresolved and sufficiently strong that the signal power in the receiver is comparable to the noise or greater. In the case of a spectral line observation, it is the power averaged over the receiver bandwidth that is important.

Adaptability. The FX design is somewhat more easily expanded or adapted to special requirements because more of the system is modularized per antenna rather than per baseline, as in the lag correlator. Addition of an extra antenna to an FX correlator requires less modification of the reduction procedure than is necessary for a lag correlator. Thus, the FX design is convenient for projects in which the number of antennas is planned to increase over time and is more efficient for larger arrays (Parsons et al. 2008).

Pulsar Observations. For pulsar observations, a gating system at the correlator output is required to separate data received during the pulsar-on period, so that the sensitivity is not degraded by noise received when the pulsar is off. For many pulsars, which have periods ≥ 0.1 s, time resolution of order 1 ms is adequate in the gating.[7] With an FX correlator, it is necessary to collect data in complete sequences of N samples, so the gating process has to accommodate data that arrive

---

[7]Many arrays can also be used in a phased-array mode (e.g., for VLBI, see Sect. 9.9), which provides one signal output per polarization. A specially designed pulsar processor can then provide measurements with high time resolution for study of the pulse profile and timing. In such cases, the array is used only to provide a large collecting area for high sensitivity. See Sect. 9.9.

at time intervals of $\sim N$ times the sample interval t_s. For example, with $N = 1,024$ and a total bandwidth of 10 MHz, $Nt_s \simeq 500 \ \mu s$. Again, this might restrict flexibility for the fastest pulsars. However, a nice feature of the FX correlator is that complete spectra are obtained during each Nt_s interval in time. In the subsequent time averaging, it is possible to process the frequency channels individually and to vary the time of the gating pulse for each one so as to match the variation in pulse timing that results from dispersion in the interstellar medium.

Choice of Correlator Design. Because the relative advantages of the lag and FX schemes discussed above involve a number of different features, the best choice of architecture for any particular application may not be immediately obvious. Detailed design studies for different approaches, taking account of the precise requirements and the implementation of the very-large-scale integrated (VLSI) circuits, are required. For discussions of lag and FX correlators, see D'Addario (1989), Romney (1995, 1999), and Bunton (2003). The widespread use of polyphase filter banks for precise channel definitions and radio frequency interference (RFI) excision favors the FX approach (see Sect. 8.8.9).

8.8.6   Hybrid Correlator

In designing a broadband correlator, it may be advantageous or necessary to divide the analog signal of total bandwidth, Δv, from each antenna into n_f contiguous narrow sub-bands. A separate digital sampler is used for each such sub-band, and the correlator is designed as n_f sections operating in parallel to cover the full signal band. A system of this type that incorporates both analog filtering and digital frequency analysis is referred to as a *hybrid correlator*. If the digital part uses a lag design, then the rate of digital operations is reduced by a factor n_f relative to the rate for a lag correlator that processes the whole bandwidth without subdivision. This can be seen from Eq. (8.116), where for one sub-band, the bandwidth is $\Delta v_s = \Delta v/n_f$, the number of channels required is N/n_f, but n_f such sections of digital processing are required. We can write a cost equation for a hybrid correlator (Weinreb 1984), as

$$C = A_1 \frac{\Delta v n_a (n_a - 1) N}{n_f} + A_2 n_f n_a + A_3 , \qquad (8.121)$$

where A_1 and A_2 are coefficients for the digital and analog hardware, respectively, and A_3 is another constant.

In this equation, the cost can be minimized with respect to n_f, with the result that

$$n_f = \left[\frac{A_1}{A_2} \Delta v (n_a - 1) N \right]^{1/2} . \qquad (8.122)$$

Table 8.6 Hybrid channelization

Instrument	Year Commissioned	$\Delta\nu$ (GHz)	$\Delta\nu_s$ (MHz)	n_f	Reference
SMA	2003	2.0^b	104	24^a	Ho et al. (2004)
SMA	2015	8.0	2000	4	–
Plateau de Bure	1992	0.5	50	10	Guilloteau et al. (1992)
Plateau de Bure	2013	8.0	2000	4	Jan (2008)
ALMA	2013	4.0^b	2000^c	2	Escoffier et al. (2007)

[a]n_f is only approximately $\Delta\nu/\Delta\nu_s$ because of sub-band overlap.
[b]Two polarizations or two bands.
[c]These sub-bands are subsequently reduced to 128 channels of 62.5 MHz each by digital filtering. Each 2000-MHz band can be positioned independently.

Equation (8.122) is useful only if the digital electronics are fast enough to handle a bandwidth of $\Delta\nu/n_f$. Over the last decades, the sampling rates have steadily risen and the costs have dropped for digital hardware, while the cost of analog electronics has remained relatively flat. The evolution of design in hybrid correlators can be seen in Table 8.6. A general disadvantage of the hybrid correlator is that very careful calibration of the frequency responses of the sub-bands is required to avoid discontinuities in gain at the sub-band edges. In general, it is advantageous to use the fastest samplers to minimize the analog filtering required. However, at millimeter wavelengths, where very wide bandwidths are needed and can be accommodated by receivers, the restriction on digital sampling speed requires some channelization. If an FX implementation is used for the digital section, a similar cost equation can be written, but there is less reduction in the number of operations since in Eq. (8.117), N enters logarithmically.

8.8.7   Demultiplexing in Broadband Correlators

The bit rate for the VLSI circuits used in large correlator systems is generally slower than that of the digital samplers that are used with broadband correlators. Serial-to-parallel conversion at the sampler output, that is, demultiplexing in the time domain, allows use of optimum bit rates for the correlator. Consider a system in which each sampler output is demultiplexed into n streams, and assume for simplicity that there is one bit per sample; parallel architecture accommodates multiple bits. Any n contiguous samples all go to different streams. To obtain all the products required in a lag correlator for a pair of IF signals with this configuration of the data, it would be necessary to include cross-correlations between each stream of one signal with every stream of the other signal. To simplify the system, Escoffier (1997) developed a scheme in which the n demultiplexed bit streams from each signal are fed into a large random access memory (RAM) and read out in reordered form. Each demultiplexed stream then contains a series of discontinuous blocks of

$\sim 10^5$ samples. Each block contains data contiguous in time, as sampled. Cross-correlations are performed between data in corresponding blocks only. Thus, for any pair of input signals, n cross-correlators running at the demultiplexed rate are required for each value of lag. Also, each signal requires two RAM units so that one is filled as the other is read out. In Escoffier's system, the sample rate is 4 Gbit s^{-1}, $n = 32$, and the length of a block of the demultiplexed data is approximately 1 ms. Since cross-correlations do not extend across the boundaries of any given block, there is a very small loss of efficiency, which in this case is about 0.2%. Another possible approach is based on demultiplexing in the frequency domain, as in the case of the hybrid correlator. It is then necessary only to cross-correlate corresponding frequency channels between each antenna, so the number of cross-correlators per signal pair is again equal to n for each lag. Carlson and Dewdney (2000) have described an all-digital development of the frequency demultiplexing principle used in the hybrid correlator. This is used with the expanded VLA (Perley et al. 2009), and the system is described as a WIDAR correlator. Broadband signals are digitized at full bandwidth, divided into frequency channels using digital filters, and resampled at the appropriate lower rate before cross-correlation between all antenna pairs. (The use of digital filters avoids the small differences in the responses of analog filters, which in some systems provide the initial channelization.) As a final step, the cross-correlated data are Fourier transformed to the frequency domain. This scheme is sometimes referred to as an FXF system. Both Escoffier's reordering scheme and the WIDAR system of demultiplexing provide approaches to the design of large broadband correlators. The latter requires fewer lags because the digital filters provide part of the spectral resolution.

For filtering sampled signals, digital filters of the FIR (finite impulse response) type can be used, in which the incoming sample stream is convolved with a series of numbers, referred to as tap weights, the Fourier transform of which represents the filter response (Escoffier et al. 2000). The tap weights can be stored in a RAM and readily changed as required. An important advantage of digital filters is the freedom from individual variations of the characteristics. However, it may be necessary to truncate the output data samples to match the number of bits per sample that can be handled by the correlator, and thus a further quantization loss may be incurred.

8.8.8   Examples of Bandwidths and Bit Data Quantization

The initial observing bandwidth of the 27-antenna VLA, when it came into operation in the early 1980s, was 100 MHz per polarization with three-level (2-bit) sampling. The expanded system that came into operation around 2010, covering a frequency range of 1–50 GHz, has a maximum observing bandwidth of 8 GHz per polarization with 3-bit sampling, or 8-bit sampling with a reduced bandwidth (Perley et al. 2009). This large increase in data capacity is possible as a result of the increase in computing speed and in signal transmission capacity using optical fiber.

The Atacama Large Millimeter/submillimeter Array, which came into operation in 2012, covers bandwidths of 8 GHz per polarization with 3-bit (8-level) quantization (Wootten and Thompson 2009). The number of antennas is 64 and the correlator is FX with initial digital filtering, sometimes referred to as an FXF system.

In the meter-wavelength range, observing bandwidths are generally narrower than at shorter wavelengths, but the spectrum is often more heavily used by transmitting services, so the requirement for avoiding or removing interfering signals is important. Larger numbers of bits allow for greater dynamic range in the system response, which helps to reduce the probability that interfering signals will cause overloading. The LWA (Long Wavelength Array) covers 20–80 MHz using 8-bit (256-level) sampling with the option of 12-bit (4096-level) sampling. The sample frequency is 196 mega-samples/sec (Ellingson et al. 2009). The LOFAR system covers 15–80 MHz and 110–240 MHz using 12-bit digitization (de Vos et al. 2009).

8.8.9   Polyphase Filter Banks

Polyphase filtering is a digital signal-processing technique that was developed for applications such as the separation of signals in multichannel communication systems with high interchannel rejection (Bellanger et al. 1976). The disadvantages of the nonoverlapping-segment discrete Fourier transform (DFT) processing, which we will call the single-block Fourier transform (SBFT) method, have been noted in earlier sections of this chapter and are also described in Appendix 8.4. Namely, this approach has high spectral leakage since the spectral response is a sinc-squared function that has sidelobe levels as high as -13.5 db. In addition, the amplitude of a monochromatic signal, or unresolved cosmic line, depends on its relative location with respect to the channel boundaries, going from 1 at channel center to $(2/\pi)^2 = 0.41$ at the edge. This effect is called scalloping. There is a slight loss in sensitivity for signals whose line widths are close to the spectral resolution, which is related to the effective lag distribution of the DFT (see Sect. 8.8.5).

Polyphase filtering and polyphase filter banks (PFBs) correct these deficiencies at a modest computational overhead. PFBs have become an important tool in radio astronomy as a way of excising radio frequency interference since they make possible the elimination of only the specific channels in which the interference occurs. It is also helpful in spectroscopic observations of some cosmic sources such as masers, where a very strong and narrow line in the passband makes it difficult to study other nearby lines because of the effect of spectral leakage. For detailed treatments of PFBs, see Crochiere and Rabiner (1981), Vaidyanathan (1990), and Harris et al. (2003). Useful tutorials are available by Harris (1999) and Chennamangalam (2014). For applications to radio interferometry, see Bunton (2000, 2003).

Before describing the PFB, consider an elementary design of a digital filter bank based on a conventional analog filter bank with M equally spaced filters spanning the frequency range 0 to Δv. Suppose the input voltage is $x(t)$, which is a bandlimited Gaussian process in the frequency range 0 to Δv. $x(t)$ can be represented by a digital sequence $x(n)$, sampled at the Nyquist interval $1/2\Delta v$. A crude lowpass filter can be constructed by taking a running mean of M samples in the time domain. The spectral response to this "boxcar" averaging is a sinc function with its first null at $2\Delta v/M$. Obtaining a perfect lowpass filter response in the frequency domain with a cutoff at $v_c = \Delta v/M$ would require the convolution in the time domain with a sinc function $[\mathrm{sinc}(x) = \sin(\pi x)/(\pi x)]$,

$$h(t) = \mathrm{sinc}(2v_c t) = \mathrm{sinc}(n/M) . \tag{8.123}$$

Note that for $M = 1$, $h(t) = 1$ for $n = 0$ and otherwise is zero, so $x(n)$ remains unchanged. However, for $M > 1$, perfect lowpass filtering action requires a convolution over infinite time. As an approximation, we can use N-point smoothing. The filter shape will be

$$H(v) = \sum_{n=0}^{N-1} \mathrm{sinc}(n/M)\, e^{j(\pi vn/\Delta v)} , \tag{8.124}$$

which has a fairly sharp cutoff at $v = \Delta v/M$. $y(n)$, the smoothed version of $x(n)$, will be oversampled by a factor of about M. The normal process at this point is to resample $y(n)$, taking every Mth sample. This process is called decimation,[8] or downsampling. To make the rest of the filter bank, multiply $x(n)$ by $e^{j2\pi vt}$, where $v = m/\Delta v$ and $m = 1$ to $M - 1$, filter each stream by $h(n)$, and downsample. This process is inefficient since the downsampling discards most of the arithmetic computations. The PFB provides a more efficient processing structure to obtain a filter shape with a sharp cutoff.

We now describe the PFB, following the analysis of Bunton (2000, 2003). Consider a sample sequence of data, $x(n)$, of length N, which is multiplied by a window function $h(n)$. Its DFT is

$$X(k) = \sum_{n=0}^{N-1} h(n)x(n)\, e^{-j(2\pi/N)nk} , \tag{8.125}$$

where k ranges from 0 to $N - 1$. The frequency steps are $2\Delta v/N$, i.e., covering both positive and negative frequencies. If $H(k)$, the DFT of $h(n)$, has a width of approximately $2\Delta v/M$, then $X(k)$ will be oversampled, and only $r = N/M$ samples need be retained. N and M are chosen so that r is an integer. If $H(k)$ is desired to be

---

[8]"Decimation" formally means a reduction to 1/10; however, the broader definition is to "reduce drastically, especially in number."

the discrete idealization of a perfect lowpass filter, i.e.,

$$H(k) = 1 , \qquad N - M < k \leq M ,$$
$$= 0 , \qquad \text{otherwise} , \qquad (8.126)$$

then

$$h(n) \simeq \text{sinc}\left[\left(\frac{n - \frac{N}{2}}{N}\right) r\right] . \qquad (8.127)$$

The decimated spectrum, i.e., taking every rth point of $X(k)$ in Eq. (8.125), is

$$X(k') = \sum_{n=0}^{N-1} h(n) x(n) \, e^{-j(2\pi/N)nrk'} , \qquad (8.128)$$

where k' goes from 0 to $M - 1$. We can rewrite Eq. (8.128) as a double summation over r subsegments, each of length M, as

$$X(k') = \sum_{m=0}^{r-1} \sum_{n=0}^{M-1} h(n + mM) \, x(n + mM) \, e^{-j(2\pi/N)(n+mM)rk'} . \qquad (8.129)$$

Notice that

$$e^{-j(2\pi/N)(n+mM)rk'} = e^{-j(2\pi nk'/M)} e^{-j2\pi mk'} . \qquad (8.130)$$

The rightmost exponential factor is unity. Hence,

$$X(k') = \sum_{m=0}^{r-1} \sum_{n=0}^{M-1} h(n + mM) \, x(n + mM) \, e^{-j(2\pi/M)nk'} . \qquad (8.131)$$

In Eq. (8.131), there are r DFTs of length M, and in Eq. (8.128), there is one DFT of length $N = rM$, so there is only a slight reduction in the workload, approximated by the number of multiplications required. Note that the FFT algorithm, which has a workload proportional to $M \log_2 M$, is used for the DFT calculation.

The kernel of the exponential in Eq. (8.131) does not contain r, so we can interchange the order of summation and rewrite it as

$$X(k') = \sum_{n=0}^{M-1} \left[\sum_{m=0}^{r-1} h(n + mM) \, x(n + mM)\right] e^{-j(2\pi/M)nk'} . \qquad (8.132)$$

This step reduces the calculation from r DFTs of length M to one DFT of length M. The workload for applying the window function $h(n)$ remains proportional to N. Hence, the workload for Eq. (8.128) is $N + N \log_2 N$, while the workload for Eq. (8.131) is $N + M \log_2 M$. The workload is thus reduced by a factor of $\mathcal{R}$, given by

$$\mathcal{R} = \frac{N + N \log_2 N}{N + M \log_2 M} = \frac{1 + \log_2 N}{1 + \frac{1}{r}(\log_2 N - \log_2 r)} \simeq r , \tag{8.133}$$

where the approximation holds for $N \gg 1$.

After the calculation in Eq. (8.132) is performed, the N-point window is moved by M steps, and the process is repeated. Each segment of M points is thus processed r times. Therefore, the input and output data rates are the same, except when spectral values at negative frequencies are discarded.

The calculation in Eq. (8.132) is expressed diagramatically in Fig. 8.24. This process may seem counterintuitive in the following sense. The data stream is severely decimated by the action of the commutator, which distributes the time samples among the branches, or "partitions," with a cycling period M. That is, the data samples into each of the M partitions are

$$\begin{array}{lllll}
x(0), & x(M), & x(2M), & \cdots, & x(rM - M) \\
x(1), & x(M + 1), & x(M + 2), & \cdots, & x(rM - M + 1) \\
x(2), & x(M + 2), & x(M + 3), & \cdots, & x(rM - M + 2) \\
\vdots & & & & \\
x(M - 1), & x(M), & x(M + 1), & \cdots, & x(rM - 1) .
\end{array} \tag{8.134}$$

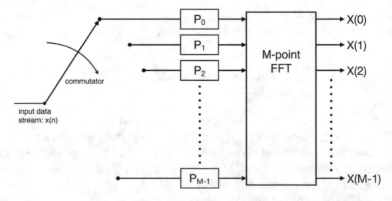

Fig. 8.24 A diagram of a polyphase filter bank, which converts a set of N data samples into an M-point spectrum. The input data stream is distributed among the M filter partitions by a commutator. Each partition receives a data stream that has been downsampled by a factor M. In each partition, P_M represents the action of the decimated version of $h(n)$, as described by the term in brackets in Eq. (8.132). The nonaliased M-point spectrum is assembled by the action of the FFT. Note that if the data samples are real numbers, then only $M/2$ values of the spectrum, corresponding to the positive frequencies, need be retained.

Each of these decimated data streams is undersampled by a factor of M, and its corresponding spectrum is heavily aliased. The action of the PFB undoes this aliasing.

Consider an example where $N = 1024$, $M = 256$, and $r = 4$ (a four-tap polyphase filter), as shown in Fig. 8.25. The first polyphase partition, P_0, calculates only the four-term sum $x(0)h(0) + x(256)h(256) + x(512)h(512) + x(768)h(768)$, and P_1 calculates $x(1)h(1) + x(257)h(257) + x(513)h(513) + x(769)h(769)$.

We can now compare the performance and requirements of the SBFT and the PFB. The SBFT produces an M-point spectrum for each M data samples. It moves successively from block to block, so the data rate remains the same. The PFB takes in N data samples and produces an M-point spectrum and then steps by M samples for the next spectral calculation. Hence, its data rate also remains the same. The overhead in the PFB is due to the windowing. Hence, the workload ratio $\mathcal{R}$ needed for the PFB with respect to the SBFT is

$$\mathcal{R} = \frac{N + M \log_2 M}{M \log_2 M} = 1 + \frac{r}{\log_2 M} \,. \tag{8.135}$$

For $M = 1024$ and $r = 4$, there is a 40% overhead incurred with the PFB structure. The flat response on low leakage from the PFB is made possible because there are N samples available to provide the filter action rather than M. Note that in hardware implementation, the buffering requirement increases with r.

It is advantageous to apply additional weighting to $h(n)$ such as Hann, Hamming, or Blackman weighting to further reduce spectral leakage. This does not reduce the resolution significantly as long as the weighting function remains at a level of ~ 1 over M samples. Examples of PFB and SBFT filter shapes are shown in Fig. 8.26. If the weighting is applied in the SBFT mode over M samples, the leakage is reduced, but the resolution is also reduced.

Note that PFBs can be concatenated. The output of any subset or all of the channels of a PFB can be fed into an additional PFB to obtain finer resolution. The Murchison Widefield Array uses such a scheme. Another application is to use a PFB only for course channelization. Its output can then be fed to an XF or FX correlator.

8.8.10   Software Correlators

Since, in practice, the signals for which cross-correlations are formed are in digital form, having also been subjected to a digital delay system, the cross multiplication and averaging processes can be carried out in a computer system. This is useful in

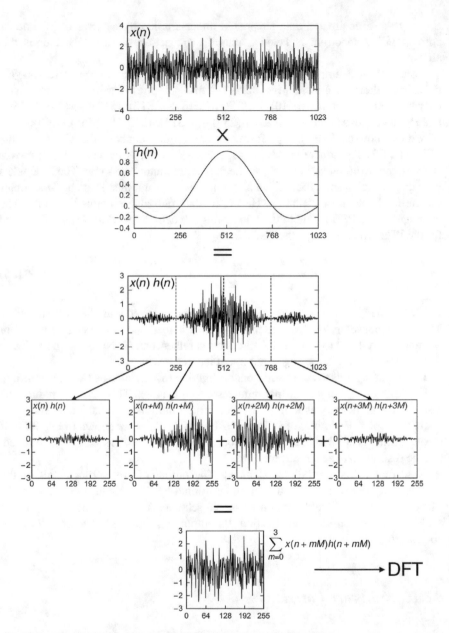

Fig. 8.25 A graphical representation of the action of a polyphase filter bank with $r = 4$, or four taps. A random noise data stream represented by a set of N independent Gaussianly distributed random noise (i.e., white noise) is shown in the top panel. It is multiplied by a window function $h(n)$, the envelope of which is shown in the next panel. Here, $h(n)$ is chosen to be a sinc functions with exactly four zero crossings, equal to the number of taps. The result is separated into four segments, which are coadded to form the M-term time series shown in the lowest panel, which is then Fourier transformed into an M-point spectrum, $-\Delta \nu$ to $\Delta \nu$, as formulated by Eq. (8.132). After this calculation, the window is moved by M samples, and the process repeated. Adapted from Gary (2014).

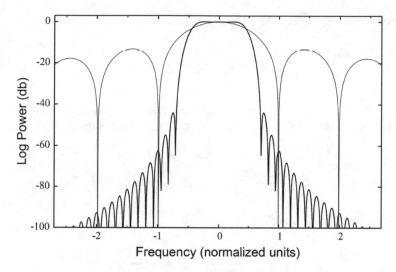

Fig. 8.26 The thick line shows the response of a filter element in a PFB having $r = 8$ and Hann weighting applied. The thin line shows the response for a SBF, a sinc² function. Both filters have a response of about $(2/\pi)^2$ at the filter edge of ± 0.5 in normalized frequency units of $2\Delta\nu/M$.

small systems for which the development of special correlator hardware is avoided. Also, in the case of large systems in which antennas are brought into operation over a period of years, changes in the correlation requirements are more easily accommodated. An example of a software correlator and the advantages of the design are described by Deller et al. (2007). Most VLBI processing is done in software correlators.

Appendix 8.1   Evaluation of $\Sigma_{q=1}^{\infty} R_{\infty}^2 (q\tau_s)$

The periodic function $f(t)$ can be expressed as a Fourier series as follows:

$$f(t) = \frac{a_0}{2} + \sum_{q=1}^{\infty} \left[a_q \cos\left(\frac{2\pi q t}{\beta}\right) + b_q \sin\left(\frac{2\pi q t}{\beta}\right) \right], \tag{A8.1}$$

where β is the period and

$$a_q = \frac{2}{\beta} \int_0^\beta f(t) \cos\left(\frac{2\pi q t}{\beta}\right) dt, \tag{A8.2a}$$

$$b_q = \frac{2}{\beta} \int_0^\beta f(t) \sin\left(\frac{2\pi q t}{\beta}\right) dt. \tag{A8.2b}$$

Parseval's theorem for Eq. (A8.1) takes the form

$$\frac{2}{\beta} \int_0^\beta f^2(t) \, dt = \frac{a_0^2}{2} + \sum_{q=1}^\infty (a_q^2 + b_q^2) \, . \tag{A8.3}$$

Now let $f(t)$ be a series of rectangular functions of unit height and width, one centered on $t = 0$ and the others centered on integral multiples of $\pm \beta$. Then, one obtains

$$a_0 = \frac{2}{\beta} \, , \qquad a_q = \frac{2}{\beta} \frac{\sin(\pi q/\beta)}{\pi q/\beta} \, , \qquad b_q = 0 \, , \qquad \int_0^\beta f^2(t) \, dt = 1 \, . \tag{A8.4}$$

From Eqs. (A8.3) and (A8.4),

$$\sum_{q=1}^\infty \left[\frac{\sin(\pi q/\beta)}{\pi q/\beta} \right]^2 = \frac{\beta - 1}{2} \, , \tag{A8.5}$$

which, from Eq. (8.15), is the summation needed to evaluate Eq. (8.19).

Appendix 8.2   Probability Integral for Two-Level Quantization

The probability integration required in Eq. (8.21) can be performed as follows. The integral is

$$P_{11} = \frac{1}{2\pi\sigma^2 \sqrt{1 - \rho^2}} \int_0^\infty \int_0^\infty \exp \left[\frac{-(x^2 + y^2 - 2\rho xy)}{2\sigma^2(1 - \rho^2)} \right] dx \, dy \, . \tag{A8.6}$$

Restore circular symmetry in the integral by the substitutions

$$z = \frac{y - \rho x}{\sqrt{1 - \rho^2}} \, , \qquad dy = \sqrt{1 - \rho^2} \, dz \, . \tag{A8.7}$$

Then

$$P_{11} = \frac{1}{2\pi\sigma^2} \int_0^\infty dx \int_{\frac{-\rho x}{\sqrt{1-\rho^2}}}^\infty \exp \left[\frac{-(x^2 + z^2)}{2\sigma^2} \right] dz \, . \tag{A8.8}$$

Next, substitute $x = r \cos \theta$ and $z = r \sin \theta$. The lower limit of the z integral in Eq. (A8.8) represents the line $z = -\rho x/\sqrt{1 - \rho^2}$, which makes an angle θ with the x axis given by $\theta = -\sin^{-1} \rho$. The integral covers an area of the (x, z) plane between this line and the z axis ($\theta = \pi/2$). Thus,

$$P_{11} = \frac{1}{2\pi\sigma^2} \int_0^\infty dr \int_{-\sin^{-1}\rho}^{\pi/2} r \, \exp\left(\frac{-r^2}{2\sigma^2}\right) d\theta \ . \tag{A8.9}$$

Finally, substitute $u = r^2/2\sigma^2$:

$$P_{11} = \frac{1}{2\pi} \int_0^\infty du \int_{-\sin^{-1}\rho}^{\pi/2} e^{-u} d\theta \ . \tag{A8.10}$$

Equation (A8.10) can be integrated directly to give

$$P_{11} = \frac{1}{4} + \frac{1}{2\pi} \sin^{-1}\rho \ . \tag{A8.11}$$

Appendix 8.3   Optimal Performance for Four-Level Quantization

Schwab (1986) has investigated various aspects of the performance of correlators with four-level quantization. These include precise values for optimal thresholds and quantization efficiencies, and expressions for computation of the cross-correlation as a function of the correlator output. The threshold values and the efficiencies are given Table A8.1.

The values of quantization efficiency η_4 for $n = 3$ and 4 are within 0.3% of the highest value and are useful because nonintegral values of the weighting factor n would require more complicated implementation in a lag-type correlator. Rational approximations for the cross-correlation $\tilde{\rho}$ are minimax solutions; that is, they minimize the maximum relative error. The variable r_N is the normalized correlator output, that is, the measured output divided by the corresponding output for $\rho = 1$. The first three approximations given below are valid for all $|r_N| \leq 1$.

For $n = 3$ and the corresponding value of v_0/σ in Table A8.1, the following approximation yields a maximum relative error of 1.51×10^{-4}:

$$\tilde{\rho}(r_N) = \frac{1.1347043 - 3.0971312r_N^2 + 2.9163894r_N^4 - 0.89047693r_N^6}{1 - 2.6892104r_N^2 + 2.4736683r_N^4 - 0.72098190r_N^6} r_N \ . \tag{A8.12}$$

For $n \approx 3.3359$ and the corresponding value of v_0/σ in Table A8.1, the following approximation yields a maximum relative error of 1.46×10^{-4}:

$$\tilde{\rho}(r_N) = \frac{1.1329552 - 3.1056902r_N^2 + 2.9296994r_N^4 - 0.90122460r_N^6}{1 - 2.7056559r_N^2 + 2.5012473r_N^4 - 0.73985978r_N^6} r_N \ . \tag{A8.13}$$

Table A8.1 Optimal thresholds and efficiencies for four-level quantization

n	v_0/σ	η_4
3	0.99568668	0.8811539496
3.3358750	0.98159883	0.8825181522
4	0.94232840	0.8795104597

For $n = 4$ and the corresponding value of v_0/σ in Table A8.1, the following approximation yields a maximum relative error of 1.50×10^{-4}:

$$\tilde{\rho}(r_N) = \frac{1.1368256 - 3.0533973r_N^2 + 2.8171512r_N^4 - 0.85148929r_N^6}{1 - 2.6529114r_N^2 + 2.4027335r_N^4 - 0.70073934r_N^6} \, r_N \; .$$

(A8.14)

The following approximation also applies for $n = 4$ and the corresponding value of v_0/σ in Table A8.1 but is valid for only $|r_N| \leq 0.95$. It yields a maximum relative error of 2.77×10^{-5}:

$$\tilde{\rho}(r_N) =$$

$$\frac{1.1369813 - 1.2487891r_N^2 + 4.5380174 \times 10^{-2}r_N^4 - 9.1448344 \times 10^{-3}r_N^6}{1 - 1.0617975r_N^2} \, r_N \; .$$

(A8.15)

Appendix 8.4   Introduction to the Discrete Fourier Transform

This appendix provides a brief introduction to the discrete Fourier transform (DFT), with specific emphasis on applications important to topics covered in this book. For more comprehensive discussion, see Bracewell (2000) or Oppenheim and Schafer (2009).

Consider the Fourier transform integral of a function $x(t)$, a bandlimited signal (0 to Δv), which has finite duration T.

$$X(v) = \int_0^T x(t) \, e^{-j2\pi vt} dt \; .$$

(A8.16)

We can approximate this integral as

$$X(v) \simeq \Delta \sum_{n=0}^{N-1} x(t_n) \, e^{-j2\pi vt_n} \; ,$$

(A8.17)

where $x(t_n)$ is a sampled version of $x(t)$ at the Nyquist interval $\Delta = 1/2\Delta v$ so that $t_n = n\Delta$. For simplicity, we assume $x(t)$ to be a real function. We calculate $X(v)$ at a set of N frequencies, $v_k = 2k\Delta v/N$, where $k = 0$ to $N - 1$, as

$$X(v_k) \simeq \Delta \sum_{n=0}^{N-1} x(t_n)\, e^{-j2\pi kn/N} \ . \tag{A8.18}$$

The important next step is to use Eq. (A8.18) as the basis for a definition of the DFT by writing

$$X_k \equiv \sum_{n=0}^{N-1} x_n\, e^{-j2\pi kn/N} \ , \tag{A8.19}$$

where x_n are the samples $x(t_n)$, and X_k are the corresponding spectral components, $X(v_k)$. For $k = 0$,

$$X_0 = \sum_{n=1}^{N-1} x_n \ , \tag{A8.20}$$

which corresponds to the component of X at $v = 0$. For $k = N/2$,

$$X_{N/2} = \sum_{n=1}^{N-1} x_n\, e^{-j\pi n} = \sum_{n=1}^{N-1} x_n(-1)^n \ , \tag{A8.21}$$

which corresponds to X at $v = \Delta v$. The negative frequency components lie between $k = N/2$ and $N - 1$, and $X_N = X_0$. The inverse DFT is

$$x_n = \frac{1}{N} \sum_{k=0}^{N-1} X_k\, e^{j2\pi kn/N} \ . \tag{A8.22}$$

We can show that Eq. (A8.22) is indeed the inverse discrete transform of Eq. (A8.19) by substituting Eq. (A8.19) into Eq. (A8.22), that is,

$$x_n = \frac{1}{N} \sum_{k=0}^{N-1} \left(\sum_{\ell=0}^{N-1} x_\ell\, e^{-j2\pi k\ell/N} \right) e^{j2\pi kn/N} \ . \tag{A8.23}$$

We introduced time index ℓ to distinguish it from n. Interchanging the order of summation gives

$$x_n = \frac{1}{N} \sum_{\ell=0}^{N-1} x_\ell \left(\sum_{k=0}^{N-1} e^{j2\pi k(n-\ell)/N} \right) \ . \tag{A8.24}$$

In the summation in parentheses in Eq. (A8.24), the phasor steps uniformly around the complex plane exactly $n - \ell$ times. Hence,

$$\frac{1}{N} \sum_{k=0}^{N-1} e^{j2\pi k(n-\ell)/N} = \delta_{n\ell} , \qquad (A8.25)$$

where $\delta_{n\ell}$ is called the Kronecker delta function, which has the properties

$$\delta_{n\ell} = 0 , \qquad n \neq \ell ,$$
$$= 1 , \qquad n = \ell . \qquad (A8.26)$$

The Kronecker delta function is nonzero only for $\ell = n$, so Eq. (A8.24) yields $x_n = x_n$ and demonstrates that x_n is recovered from the original data after a DFT followed by an inverse DFT. Note that x_n and X_k are both periodic with period N. Thus,

$$X_{k+mN} = X_k \qquad (A8.27)$$

and

$$x_{n+mN} = x_n , \qquad (A8.28)$$

where m is the period number. Thus, for example, $X_N = X_0 = \sum_{n=0}^{N-1} x_n$.

A very useful concept is to think of x_n and X_k as lying on a circle instead of on a line. Most of the well-known theorems in Fourier transform theory have counterparts in DFT theory where the data lie on a circle. The shift theorem of Fourier transforms illustrates the circular nature of the DFT. The DFT of a circularly shifted, by one step, sequence of x_n, i.e., $y_n = x_{n-1}$, or

$$y = \{x_{N-1}, x_0, x_1, x_2, \ldots, x_{N-2}\} , \qquad (A8.29)$$

is

$$Y_k = \sum_{n=0}^{N-1} y_n\, e^{-j2\pi kn/N} , \qquad (A8.30)$$

$$= \sum_{n=1}^{N-1} x_{n-1} e^{-j2\pi kn/N} + x_{N-1} , \qquad (A8.31)$$

$$= \sum_{n=0}^{N-2} x_n\, e^{-j2\pi k(n+1)/N} + x_{N-1} , \qquad (A8.32)$$

$$= e^{-j2\pi k/N} \sum_{n=0}^{N-2} x_n e^{-j2\pi kn/N} + x_{N-1} , \tag{A8.33}$$

$$= e^{-j2\pi k/N} \sum_{n=0}^{N-1} x_n e^{-j2\pi kn/N} . \tag{A8.34}$$

The last step of absorbing x_{N-1} into the summation was accomplished by recognizing that the x_{N-1} term of the summation in Eq. (A8.34) is

$$e^{-j2\pi k/N} x_{N-1} e^{-j2\pi k(N-1)/N} = x_{N-1} . \tag{A8.35}$$

Thus,

$$Y_k = e^{-j2\pi k/N} X_k . \tag{A8.36}$$

In general, for a shift of ℓ steps,

$$y_n = x_{n-\ell} , \tag{A8.37}$$

the DFT becomes

$$Y_k = e^{-j2\pi \ell k/N} X_k . \tag{A8.38}$$

Hence, the shift theorem for DFT is clearly based on a circular shift of x_n. It is straightforward to prove the cyclic convolution and correlation theorems:

$$x_n * y_n \xrightarrow{\text{DFT}} \frac{1}{N} X_k Y_k , \tag{A8.39}$$

$$x_n \star y_n \xrightarrow{\text{DFT}} \frac{1}{N} X_k Y_k . \tag{A8.40}$$

It is important to understand that to use expressions (A8.39) and (A8.40) to calculate either convolution or correlation function, it is necessary to pad the spectral array with N zeros to avoid unwanted products in the circular correlation (see Sect. A8.4.2).

Parseval's theorem for the DFT can be easily proved, by using Eq. (A8.19) and writing

$$\sum_{k=0}^{N-1} X_k X_k^* = \sum_{k=0}^{N-1} \left[\sum_{n=0}^{N-1} x_n e^{-j2\pi kn/N} \right] \left[\sum_{\ell=0}^{N-1} x_\ell e^{j2\pi k\ell/N} \right] . \tag{A8.41}$$

Interchanging the order of summations gives

$$\sum_{k=0}^{N-1} X_k X_k^* = \sum_{n=0}^{N-1} \sum_{\ell=0}^{N-1} x_n x_\ell \sum_{k=0}^{N-1} e^{-j2\pi k(n-\ell)/N} \ . \tag{A8.42}$$

The rightmost sum is proportional to the Kronecker delta function [Eq. (A8.25)], so

$$\sum_{k=0}^{N-1} X_k X_k^* = N \sum_{n=0}^{N-1} x_n^2 \ . \tag{A8.43}$$

If x_n is complex, then the general form of Eq. (A8.43) becomes[9]

$$\sum_{n=0}^{N-1} |x_n|^2 = \frac{1}{N} \sum_{k=0}^{N-1} |X_k|^2 \ . \tag{A8.44}$$

A8.4.1   Response to a Complex Sine Wave

We now calculate the DFT response to a complex sine wave,

$$x_n = e^{j2\pi v t_n} \ , \qquad t_n = \frac{n}{2\Delta v} \ . \tag{A8.45}$$

We introduce a normalizing frequency $v' = \dfrac{v}{\Delta v} \dfrac{N}{2}$. For the positive frequency range of 0 to Δv, v' ranges from 0 to $N/2$. Note that v' does not have to be an integer. The DFT of x_n is

$$X_k = \sum_{n=0}^{N-1} e^{-j2\pi n(k-v')/N} \ . \tag{A8.46}$$

We use the formula for the sum of a geometric series,

$$\sum_{n=0}^{N-1} y^n = \frac{1-y^N}{1-y} \ , \tag{A8.47}$$

---

[9]In some DFT formulations, the N factor of Eq. (A8.22) appears in Eq. (A8.19). In that case, the N factor in Eq. (A8.44) moves to the numerator on the left side.

where $y = e^{-j2\pi(k-v')/N}$, to write Eq. (A8.46) as

$$X_k = \frac{1 - e^{j2\pi(k-v')}}{1 - e^{j2\pi(k-v')/N}} \cdot \qquad\qquad (A8.48)$$

By factoring out $e^{j\pi(k-v')}$ from the numerator and $e^{j\pi(k-v')/N}$ from the denominator of Eq. (A8.48), we can write

$$X_k = \left[\frac{e^{j\pi(k-v')}}{e^{j\pi(k-v')/N}}\right] \left[\frac{\sin \pi(k-v')}{\sin(\pi(k-v')/N)}\right] \cdot \qquad\qquad (A8.49)$$

We are interested in the power response

$$S_k = |X_k|^2 = \left[\frac{\sin \pi(k-v')}{\sin(\pi(k-v')/N)}\right]^2 \cdot \qquad\qquad (A8.50)$$

This is the circular form of the sinc function, which repeats on the interval N, that is, $X_{k+N} = X_k$, or $S_{k+N} = S_k$.

We approximate the denominator of Eq. (A8.50) as $\pi(k-v')/N$, so that

$$S_k = |X|^2 \simeq N \operatorname{sinc}^2(k-v') . \qquad\qquad (A8.51)$$

If $v' = m$, an integer, then

$$S_k = N , \qquad k = m ,$$
$$= 0 , \qquad k \neq m . \qquad\qquad (A8.52)$$

In Fig. A8.1, we show the response to the complex sine wave for $v' = m$ and $v' = m + 1/2$. Unless v' corresponds exactly to a DFT channel, S_k will be nonzero in every channel, demonstrating the problem of spectral leakage.

The DFT of Eq. (A8.50) gives the corresponding response for the correlation function, which is a triangle function. This reflects the fact that the number of ways the correlation function, given by Eq. (A8.39), can be computed from a segment of data decreases linearly from N ways for lag zero to one way for lag $N-1$, as shown in Fig. 8.23. If it is desired to calculate the correlation function from the power spectrum, i.e., Eq. (A8.39), it is important to note that the spectrum must be padded with zeros to length $2N$.

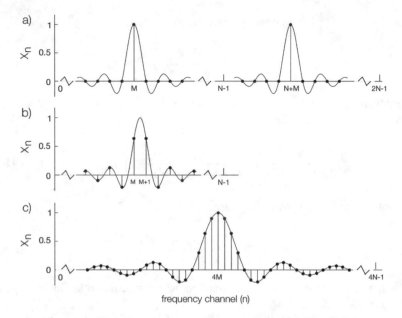

Fig. A8.1 (**a**) The response of the DFT of length N to a complex sine wave of frequency $v' = m$. This is a plot of Eq. (A8.49) without the phase factor. The continuous envelope [e.g., as calculated from Eq. (A8.54)] is shown along with the function values at the sample points. One repetition of this periodic function is shown. (**b**) Response where $v' = m + 1/2$ (the frequency falls midway between two DFT channels). (**c**) The data array size has been increased from N to $4N$ by padding with zeros, which results in a more finely defined spectrum.

A8.4.2   *Padding with Zeros*

Padding with zeros is a very important concept in DFT theory. Padding with zeros means adding a block of zeros to the data sequence, usually at the end, to increase its length from N to N'. The three main reasons to pad with zeros are:

1. It provides a way to interpolate X_k and define it at more finely spaced frequency intervals.
2. If N is not a power of two, x_n can be padded with zeros to N', a power of two. This makes the FFT used to calculate the DFT more computationally efficient. The N' spectrum is a properly interpolated spectrum in the Nyquist sense.
3. If a linear correlation function of two functions is to be properly calculated from $X_n Y_n$ via the circular convolution theorem [Eq. (A8.39)], then X_n and Y_n must be padded with zeros to $2N$ to avoid unwanted multiplications.

   To understand how interpolation is achieved, consider a data sequence of length N to which M zeros are added, giving a sequence of length $N' = N + M$. The DFT of the new data set is

$$X_k = \sum_{n=0}^{N-1} x_n \, e^{-j2\pi kn/N'} + \sum_{n=N}^{N'-1} 0 \cdot e^{-j2\pi kn/N'}$$

$$= \sum_{n=0}^{N-1} x_n \, e^{j2\pi kn/Nr} , \qquad\qquad 0 \le k \le N' - 1 , \qquad (A8.53)$$

where $r = N'/N$. The frequency spacing interval is now $2\Delta v/Nr$. Hence, if $r = 2$, the spectrum X_k gives a halfway interpolation of the unpadded version of X_k.

   Padding with zeros can provide arbitrarily fine definition of X_k. However, it is often helpful to define a continuous spectrum associated with the discrete series x_n as

$$X(v) = \sum_{n=0}^{N-1} x_n \, e^{-j\pi nv'} , \qquad\qquad (A8.54)$$

where $v' = v/\Delta v$. This is sometimes called the discrete time Fourier transform (DTFT). It can be calculated for arbitrary values of v'.

   It is often useful to load the FFT in such a way as to avoid unwanted phase factors from appearing in the transform. For example, suppose we want to calculate the spectrum from an autocorrelation function $R(\tau) = R_n$, where $n = 0, N/2$ ranging over positive delays only. Load R_n into the array

$$R'_n = R_n , \qquad\qquad n = 0, N/2 ,$$

$$R'_{N-1-n} = R_n . \qquad\qquad (A8.55)$$

   The DFT of R'_n will have a real valued spectrum S_k in the positive frequencies $k = 0$ to $k = N/2$, with the negative frequencies being in $k = N$ to $N - N/2 - 1$, as shown in Fig. 8.2.

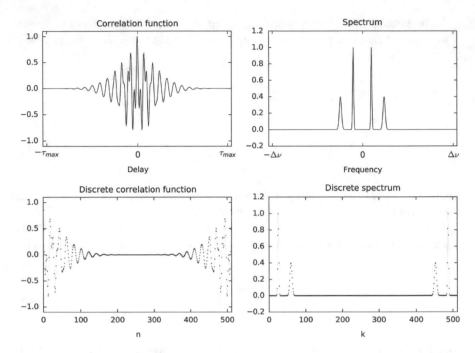

Fig. 8.2 An example of the loading of an autocorrelation function into a DFT array. (**top**) A continuously defined autocorrelation function (left) and its power spectrum (right) via Fourier transform. (**bottom**) Positive lags loaded into 0 to $N/2$ indices, and negative lags loaded into $N/2$ to $N-1$ indices and its spectrum (right) via DFT. Loading the data in this manner gives a real power spectrum. Zero padding should be done in the middle of the delay values to keep the spectrum real valued.

Further Reading

Deller, A.T., Romney, J.D., Gunaratne, T., and Carlson, B., Digital Signal Processing and Cross Correlators, in *Synthesis Imaging in Radio Astronomy III*, Mioduszewski, A., Ed., *Publ. Astron. Soc. Pacific* (2016), in press

Harris, F.J., *Multirate Signal Processing for Communication Systems*, Prentice Hall Professional Technical Reference, Upper Saddle River, NJ (2008)

Lyons, R.G., *Understanding Digital Signal Processing*, Addison-Wesley Longman Inc., Reading, MA (1997)

Oppenheim, A.V., and Schafer, R.W., *Discrete-Time Signal Processing*, Pearson Prentice Hall, Upper Saddle River, NJ (2009)

References

Abramowitz, M., and Stegun, I.A., *Handbook of Mathematical Functions*, National Bureau of Standards, Washington, DC (1968)

Ball, J.A., The Harvard Minicorrelator, *IEEE Trans. Instrum. Meas.*, **IM-22**, 193 (1973)

Bellanger, M.G., Bonnerot, G., and Coudreuse, M., Digital Filtering by Polyphase Network: Application to Sample-Rate Alteration and Filter Banks, *IEEE Trans. Acoust., Speech, and Signal Proc.*, **ASSP-24**, 109–114 (1976)

Benkevitch, L.V., Rogers, A.E.E., Lonsdale, C.J., Cappallo, R.J., Oberoi, D., Erickson, P.J., and Baker, K.D., Van Vleck Correction Generalization for Correlators with Multilevel Quantization, arXiv: 1608.04367v1

Benson, J.M., The VLBA Correlator, in *Very Long Baseline Interferometry and the VLA*, Zensus, J.A., Diamond, P.J., and Napier, P.J., Eds., Astron. Soc. Pacific Conf. Ser., **82**, 117–131 (1995)

Blackman, R.B., and Tukey, J.W., *The Measurement of Power Spectra*, Dover, New York (1959)

Bowers, F.K., and Klingler, R.J., Quantization Noise of Correlation Spectrometers, *Astron. Astrophys. Suppl.*, **15**, 373–380 (1974)

Bowers, F.K., Whyte, D.A., Landecker, T.L., and Klingler, R.J., A Digital Correlation Spectrometer Employing Multiple-Level Quantization, *Proc. IEEE*, **61**, 1339–1343 (1973)

Bracewell, R.N., *The Fourier Transform and Its Applications*, McGraw-Hill, New York (2000) (earlier eds. 1965, 1978)

Bunton, J.D., An Improved FX Correlator, ALMA Memo 342 (2000)

Bunton, J.D., Multi-Resolution FX Correlator, ALMA Memo 447 (2003)

Bunton, J.D., New Generation Correlators, in Proceedings of the 28th URSI General Assembly in New Delhi (2005). http://www.ursi.org/Proceedings/Proc\GA05/pdf/J06.6(0311).pdf

Burns, W.R., and Yao, S.S., Clipping Noise Loss in the One-Bit Autocorrelation Spectral Line Receiver, *Radio Sci.*, **4**, 431–436 (1969)

Carlson, B.R., and Dewdney, P.E., Efficient Wideband Digital Correlation, *Electronics Lett.*, **36**, 987–988 (2000)

Chennamangalam, J., The Polyphase Filter Bank Technique, CASPER Memo 41 (2014). https://casper.berkeley.edu/wiki/Memos (no. 42 in list but labeled no. 41 in the memo)

Chikada, Y., Ishiguro, M., Hirabayashi, H., Morimoto, M., Morita, K., Miyazawa, K., Nagane, K., Murata, K., Tojo, A., Inoue, S., Kanzawa, T., and Iwashita, H., A Digital FFT Spectro-Correlator for Radio Astronomy, in *Indirect Imaging*, Roberts, J.A., Ed., Cambridge Univ. Press, Cambridge, UK (1984), pp. 387–404

Chikada, Y., Ishiguro, M., Hirabayashi, H., Morimoto, M., Morita, K.I., Kanzawa, T., Iwashita, H., Nakazimi, K., Ishiwaka, S.I., Takashi, T., and seven coauthors, A 6 × 320-MHz 1024-Channel FFT Cross Spectrum Analyzer for Radio Astronomy, *Proc. IEEE*, **75**, 1203–1210 (1987)

Cole, T., Finite Sample Correlations of Quantized Gaussians, *Aust. J. Phys.*, **21**, 273–282 (1968)

Cooley, J.W., and Tukey, J.W., An Algorithm for the Machine Calculation of Complex Fourier Series, *Math. Comp.*, **19**, 297–301 (1965)

Cooper, B.F.C., Correlators with Two-Bit Quantization, *Aust. J. Phys.*, **23**, 521–527 (1970)

Crochiere, R.E., and Rabiner, L.R., Interpolation and Decimation of Digital Signals—A Tutorial Review, *Proc. IEEE*, **69**, 300–331 (1981)

D'Addario, L.R., Cross Correlators, in *Synthesis Imaging in Radio Astronomy*, Perley, R.A., Schwab, F.R., and Bridle, A.H., Eds., Astron. Soc. Pacific Conf. Ser., **6**, 59–82 (1989)

D'Addario, L.R., Thompson, A.R., Schwab, F.R., and Granlund, J., Complex Cross Correlators with Three-Level Quantization: Design Tolerances, *Radio Sci.*, **19**, 931–945 (1984)

Davis, W.F., Real-Time Compensation for Autocorrelation Clipper Bias, *Astron. Astrophys. Suppl.* **15**, 381–382 (1974)

Deller, A.T., Tingay, S.J., Bailes, M., and West, C., DiFX: A Software Correlator for Very Long Baseline Interferometry Using Multiprocessor Computing Environments, *Publ. Astron. Soc. Pacific*, **119**, 318–336 (2007)

Deller, A.T., Romney, J.D., Gunaratne, T., and Carlson, B., Digital Signal Processing and Cross Correlators, in *Synthesis Imaging in Radio Astronomy III*, Mioduszewski, A., Ed., *Publ. Astron. Soc. Pacific* (2016), in press

de Vos, M., Gunst, A.W., and Nijboer, R., The LOFAR Telescope: System Architecture and Signal Processing, *Proc. IEEE*, **97**, 1431–1437 (2009)

Ellingson, S.W., Clarke, T.E., Cohen, A., Craig, J., Kassim, N.E., Pihlström, Y., Rickard, L.J., and Taylor, G.B., The Long Wavelength Array, *Proc. IEEE*, **97**, 1421–1430 (2009)

Escoffier, R.P., The MMA Correlator, MMA Memo 166, National Radio Astronomy Observatory (1997)

Escoffier, R.P., Webber, J.C., D'Addario, L.R., and Broadwell, C.M., A Wideband Digital Filter Using FPGAs, in *Radio Telescopes*, Proc. SPIE, **4015**, 106–113 (2000)

Escoffier, R.P., Comoretto, G., Webber, J.C., Baudry, A., Broadwell, C.M., Greenberg, J.H., Treacy, R.R., Cais, P., Quertier, B., Camino, P., Bos, A., and Gunst, A.W., The ALMA Correlator, *Astron. Astrophys.*, **462**, 801–810 (2007)

Gary, D.E., Digital Cross-Correlators, Radio Astronomy Lecture 8, New Jersey's Science and Technology University, fall (2014). http://web.njit.edu/~gary/728/Lecture8.html

Goldstein, R.M., A Technique for the Measurement of the Power Spectra of Very Weak Signals, *IRE Trans. Space Electron. Telem.*, **8**, 170–173 (1962)

Granlund, J., O'Sullivan's Zero-Padding, VLBA Correlator Memo 66, National Radio Astronomy Observatory (1986)

Guilloteau, S., Delannoy, J., Downes, D., Greve, A., Guélin, M., Lucas, R., Morris, D., Radford, S.J.E., Wink, J., Cernicharo, J., and seven coauthors, The IRAM Interferometer on Plateau de Bure, *Astron. Astrophys.*, **262**, 624–633 (1992)

Gwinn, C.R., Correlation Statistics of Quantized Noiselike Signals, *Publ. Astron. Soc. Pacific*, **116**, 84–96 (2004)

Hagen, J.B., and Farley, D.T., Digital Correlation Techniques in Radio Science, *Radio Sci.*, **8**, 775–784 (1973)

Harris, F., Tutorial on Polyphase Transforms (1999). http://ubm.io/1XzFUbY

Harris, F., Dick, C., and Rice, M., Digital Receivers and Transmitters Using Polyphase Filter Banks for Wireless Communications, *IEEE Trans. Microwave Theory Tech.*, **51**, 1395–1412 (2003)

Harris, F.J., The Use of Windows for Harmonic Analysis with the Discrete Fourier Transform, *Proc. IEEE*, **66**, 51–83 (1978)

Ho, P.T.P., Moran, J.M., and Lo, K.-Y., The Submillimeter Array, *Astrophys. J. Lett.*, **616**, L1–L6 (2004)

Jan, M.T., Main Technical Features of the WideX Correlator, 2008. http://www.iram.fr/IRAMFR/TA/backend/WideX

Jenet, F.A., and Anderson, S.B., The Effects of Digitization on Nonstationary Stochastic Signals with Applications to Pulsar Signal Baseband Recording, *Publ. Astron. Soc. Pacific*, **110**, 1467–1478 (1998)

Johnson, M.D., Chou, H.H., and Gwinn, C.R., Optimal Correlation Estimators for Quantized Signals, *Astrophys. J.*, **765**:135 (7pp) (2013)

Kulkarni, S.R., and Heiles, C., How to Obtain the True Correlation from a Three-Level Digital Correlator, *Astron. J.*, **85**, 1413–1420 (1980)

Lawson, J.L., and Uhlenbeck, G.E., *Threshold Signals*, McGraw-Hill, New York (1950). Reprinted by Dover Publications (p. 58), New York, 1965

Lo, W.F., Dewdney, P.E., Landecker, T.L., Routledge, D., and Vaneldik, J.F., A Cross-Correlation Receiver for Radio Astronomy Employing Quadrature Channel Generation by Computed Hilbert Transform, *Radio Sci.*, **19**, 1413–1421 (1984)

Moran, J.M., Very Long Baseline Interferometer Systems, in *Methods of Experimental Physics*, Vol. 12, Part C *(Astrophysics: Radio Observations)*, Meeks, M.L., Ed., Academic Press, New York (1976), pp. 174–197

Napier, P.J., Thompson, A.R., and Ekers, R.D., The Very Large Array; Design and Performance of a Modern Synthesis Radio Telescope, *Proc. IEEE*, **71**, 1295–1320 (1983)

Nyquist, H., Certain Topics in Telegraph Transmission Theory, *Trans. Am. Inst. Elect. Eng.*, **47**, 617–644 (1928)

Okumura, S.K., Momose, M., Kawaguchi, N., Kanzawa, T., Tsutsumi, T., Tanaka, A., Ichikawa, T., Suzuki, T., Ozeki, K., Natori, K., and Hashimoto, T., 1-GHz Bandwidth Digital Spectro-Correlator System for the Nobeyama Millimeter Array, *Publ. Astron. Soc. Japan*, **52**, 393–400 (2000)

Okumura, S.K., Chikada, Y., Momose, M., and Iguchi, S., Feasibility Study of the Enhanced Correlator for Three-Way ALMA, ALMA Memo 350 (2001)

Oppenheim, A.V., and Schafer, R.W., *Digital-Time Signal Processing*, 3rd ed., Pearson, Prentice Hall, Upper Saddle River, NJ (2009)

O'Sullivan, J.D., *Efficient Digital Spectrometers—A Survey of Possibilities*, Note 375, Netherlands Foundation for Radio Astronomy, Dwingeloo (1982)

Parsons, A., Backer, D., Siemion, A., Chen, H., Werthimer, D., Droz, P., Filiba, T., Manley, J., McMahon, P., Paarsa, A., MacMahon, D., and Wright, M., A Stable Correlator Architecture Based on Modular FPGA Hardware, Reusable Gateware, and Data Packetization, *Publ. Astron. Soc. Pacific*, **120**, 1207–1221 (2008)

Percival, B.D., and Walden, A.T., *Spectral Analysis for Physical Applications*, Cambridge Univ. Press, Cambridge, UK (1993), p. 289

Perley, R., Napier, P., Jackson, J., Butler, B., Carlson, B., Fort, D., Dewdney, P., Clark, B., Hayward, R., Durand, S., Revnell, M., and McKinnon, M., The Expanded Very Large Array, *Proc. IEEE*, **97**, 1448–1462 (2009)

Press, W.H., Teukolsky, S.A., Vetterling, W.T., and Flannery, B.P., *Numerical Recipes*, 2nd ed., Cambridge Univ. Press, Cambridge, UK (1992)

Price, R., A Useful Theorem for Nonlinear Devices Having Gaussian Inputs, *IRE Trans. Inf. Theory*, **IT-4**, 69–72 (1958)

Romney, J.D., Theory of Correlation in VLBI, in *Very Long Baseline Interferometry and the VLA*, Zensus, J.A., Diamond, P.J., and Napier, P.J., Eds., Astron. Soc. Pacific Conf. Ser., **82**, 17–37 (1995)

Romney, J.D., Cross Correlators, in *Synthesis Imaging in Radio Astronomy II*, Taylor, G.B., Carilli, C.L., and Perley, R.A., Eds., Astron. Soc. Pacific Conf. Ser., **180**, 57–78 (1999)

Schwab, F.R., Two-Bit Correlators: Miscellaneous Results, VLBA Correlator Memo 75, National Radio Astronomy Observatory (1986)

Shannon, C.E., Communication in the Presence of Noise, *Proc. IRE*, **37**, 10–21 (1949)

Thompson, A.R., Quantization Efficiency for Eight or More Sampling Levels, MMA Memo 220, National Radio Astronomy Observatory (1998)

Thompson, A.R., Emerson, D.T., and Schwab, F.R., Convenient Formulas for Quantization Efficiency, *Radio Sci.*, **42**, RS3022 (2007)

Vaidyanathan, P.P., Multirate Digital Filters, Filter Banks, Polyphase Networks, and Applications: A Tutorial, *Proc. IEEE*, **78**, 56–93 (1990)

Van Vleck, J.H., The Spectrum of Clipped Noise, Radio Research Lab., Harvard Univ., Report RRL-51 (July 21, 1943)

Van Vleck, J.H., and Middleton, D., The Spectrum of Clipped Noise, *Proc. IEEE*, **54**, 2–19 (1966)

Wall, J.V., and Jenkins, C.R., *Practical Statistics for Astronomers*, 2nd ed., Cambridge Univ. Press, Cambridge, UK (2012)

Weinreb, S., *A Digital Spectral Analysis Technique and Its Application to Radio Astronomy*, Technical Report 412, Research Lab. for Electronics, MIT, Cambridge, MA (1963)

Weinreb, S., Analog-Filter Digital-Correlator Hybrid Spectrometer, *IEEE Trans. Instrum. Meas.*, **IM-34**, 670–675 (1984)

Welch, P.D., The Use of Fast Fourier Transform for the Estimation of Power Spectra: A Method Based on Time Averaging over Short, Modified Periodograms, *IEEE Trans. Audio Electroacoust.*, **AU-15**, 70–73 (1967)

Willis, A.G., and Bregman, J.D., Effects in Fourier Transformed Spectra, in *User's Manual for Westerbork Synthesis Radio Telescope*, Netherlands Foundation for Radio Astronomy, Westerbork, the Netherlands (1981), ch. 2, app. 2

Wootten, A., and Thompson, A.R., The Atacama Large Millimeter/Submillimeter Array, *Proc. IEEE*, **97**, 1463–1471 (2009)

Yen, J.L., The Role of Fast Fourier Transform Computers in Astronomy, *Astron. Astrophys. Suppl.*, **15**, 483–484 (1974)

Chapter 9
Very-Long-Baseline Interferometry

In 1967, a new technique of interferometry was developed in which the receiving elements were separated by such a large distance that it was expedient to operate them independently with no real-time communication link. This was accomplished by recording the data on magnetic tape for later cross-correlation at a central processing station. The technique was called very-long-baseline interferometry (VLBI), a term recalling the earlier long-baseline interferometers at Jodrell Bank Observatory, in which the elements were connected by microwave links that had reached 127 km in length. The principles involved in VLBI are fundamentally the same as those involved in interferometers with connected elements. The tape recorder and its successor, disk storage, can be considered as an IF delay line of limited capacity with an unusually long propagation time, weeks instead of microseconds. The use of tape and disk recording media is motivated entirely by economics and places substantial limitations on the system. Satellite links have been demonstrated (Yen et al. 1977), but their high cost discourages their use.

Tape recorders have been entirely replaced by compact disks. Data can also be transmitted to correlation facilities via the Internet in quasi real time. However, latency and throughput are significant issues, and data buffering is usually required.

9.1 Early Development

The motivation to develop VLBI came from the realization that many radio sources have structures that cannot be resolved by interferometers with baselines of a few hundred kilometers. By the mid-1960s, it was well known that scintillation (discussed in Chap. 14) and time variability of the radiation from quasars implied angular sizes of $< 0.01''$. Maser emission from OH molecules at 18-cm wavelength was unresolved at $0.1''$. Low-frequency burst radiation from Jupiter was believed to

© The Author(s) 2017
A.R. Thompson, J.M. Moran, and G.W. Swenson Jr., *Interferometry and Synthesis in Radio Astronomy*, Astronomy and Astrophysics Library,
DOI 10.1007/978-3-319-44431-4_9

emanate from regions of small angular size. The aim of the first VLBI experiments was to measure the angular sizes of these radio sources. It is instructive to consider the operation of these early VLBI experiments in their most primitive form. Consider two telescopes with system temperatures T_{S1} and T_{S2}, which are pointed at a compact source giving antenna temperatures T_{A1} and T_{A2}. Each station records N data samples within the *coherence time*, that is, the interval during which the independent oscillators remain sufficiently stable that fringes can be averaged. In the subsequent processing, these data streams are aligned, cross-correlated, and time-averaged after removing the quasi-sinusoidal fringes. The expected correlation for a point source is

$$\rho_0 \simeq \eta \sqrt{\frac{T_{A1}T_{A2}}{(T_{S1} + T_{A1})(T_{S2} + T_{A2})}} \,, \tag{9.1}$$

where η is a factor of value ~ 0.5 to account for losses due to quantization and processing (see Sect. 9.7). Here, it is convenient to consider a normalized form of the visibility:

$$\mathcal{V}_N = \frac{\rho}{\rho_0} = \frac{\rho}{\eta} \sqrt{\frac{T_{S1}T_{S2}}{T_{A1}T_{A2}}} \,, \tag{9.2}$$

where ρ is the measured correlation, and we assume $T_A \ll T_S$. The rms noise level is

$$\Delta\rho \simeq \frac{1}{\sqrt{N}} \simeq \frac{1}{\sqrt{2\Delta\nu\tau_c}} \,, \tag{9.3}$$

where $\Delta\nu$ is the IF bandwidth, and τ_c is the coherent integration time. Hence, from Eqs. (9.1)–(9.3), the signal-to-noise ratio is

$$\frac{\rho}{\Delta\rho} = \eta\mathcal{V}_N \sqrt{\frac{T_{A1}T_{A2}}{T_{S1}T_{S2}}(2\Delta\nu\tau_c)} \,. \tag{9.4}$$

If the minimum useful signal-to-noise ratio is 4, the smallest detectable flux density is as follows, from Eqs. (1.3), (1.5), and (9.4):

$$S_{\min} \simeq \frac{8k}{\mathcal{V}_N\eta} \sqrt{\frac{T_{S1}T_{S2}}{A_1 A_2}} \frac{1}{\sqrt{2\Delta\nu\tau_c}} \,, \tag{9.5}$$

where k is Boltzmann's constant, and A_1 and A_2 are the antenna collecting areas. Typical parameters in 1967 were $A \simeq 250$ m^2 (25-m-diameter telescope), $T_S \simeq 100$ K, $\eta \simeq 0.5$, and $N = 1.4 \times 10^8$ bits (one bit per sample), the capacity of a tape at a density of 800 bpi (bits per inch) used in the NRAO Mark I system, which was based on standard IBM compatible technology. For an unresolved source, $S_{min} \simeq 2$ Jy. The development after three decades is indicated by the following parameter values: $A \sim 1600$ m^2 (64-m-diameter telescope), $T_S \simeq 30$ K, and $N = 5 \times 10^{12}$ bits, the capacity of an instrumentation tape operated at 64 MHz bandwidth. For $\mathcal{V}_N = 1$, Eq. (9.5) gives $S_{min} \simeq 0.6$ mJy. In both examples, the coherence time is assumed to be greater than the running time of the tape. The source size can be estimated from a single measurement of $\mathcal{V}_N$ by comparison with the visibility expected for a symmetric Gaussian model. Hence, as in Fig. 1.5, the full width at half-maximum, a, is given by

$$a = \frac{2\sqrt{\ln 2}}{\pi u} \sqrt{-\ln \mathcal{V}_N} , \qquad (9.6)$$

where u is the projected baseline (in wavelengths).

VLBI can be used only to study objects of exceedingly high intensity. Thus, the emission processes must normally be of nonthermal origin. To be detected on a baseline of length D, the source must be smaller than the fringe spacing. Since the flux density S is $2kT_B\Omega/\lambda^2$, where T_B is the brightness temperature, λ is the wavelength, and Ω is the source solid angle, the minimum detectable brightness temperature is

$$(T_B)_{min} \simeq \frac{2}{\pi k} D^2 S_{min} , \qquad (9.7)$$

since $\Omega \simeq \pi(\lambda/2D)^2$. If $D = 10^3$ km and $S_{min} = 2$ mJy, then $(T_B)_{min} \simeq 10^6$ K. Therefore, observations of thermal phenomena occurring in molecular clouds, compact HII regions, and most stars are generally not possible. On the other hand, synchrotron sources such as supernova remnants, radio galaxies, and quasars, which are limited to 10^{12} K by Compton losses; masers in which $T_B \simeq 10^{15}$ K; and pulsars can be readily studied.

Three things were accomplished by early VLBI measurements:

1. Simple intensity distributions were derived by comparing measured visibilities with source models.
2. The distribution of the various spectral components of masers was mapped by comparing fringe frequencies for different spectral features.
3. Source positions were measured to an accuracy of $\sim 1''$ and baselines to an accuracy of a few meters.

For a review of early techniques, see Klemperer (1972). Since then, the technique has moved steadily toward the mainstream of interferometry in terms of being able to produce reliable images of complex radio sources. The principal reason for this is the use of phase closure (see Sect. 10.3), which provides most of the phase information when a large enough number of antennas is available in the VLBI network. A list of various VLBI networks is shown in Table 9.1.

It is interesting to note that the correlation of data in the earliest systems was accomplished in software on general-purpose computers. After about 30 years,

Table 9.1 Examples of VLBI arrays[a]

Name	Inception	Center	Stations	Antenna size (m)	Maximum baseline (km)	Frequency (GHz)	Weeks/ Year	Ref.
EVN[b]	1980	Europe	18	10–100	8000	0.3–86	12	1
VLBA[c]	1993	USA	10	25	8610	1.3–86	52	2
LBA[d]	1997	Australia	6	22–70	3100	1.3–22	3	3
CVN[e]	2000	China	5	25–65	3250	1.4–22	2	4
VERA[f]	2005	Japan	4	20	2273	6–43	52	5
KVN[g]	2011	Korea	3	21	476	22–129	52	6
LOFAR[h]	2012	Europe	8	∼ 50	1300	0.15–0.24	52	7
EHT[i]	2012	USA	3	10–30	4700	230	1	8
IVS[j]	1980	Global	32	10–100	10000	2 and 8	26	9

[a]There are also networks of networks, such as the KaVA, which is a combination of VERA (VLBI Exploration of Radio Astronomy) and KVN (Korean VLBI Network).
[b]European VLBI Network.
[c]Very Long Baseline Array.
[d]Long Baseline Array, which often operates with Hartebeesthoek in South Africa and the Warkworth Telescope in New Zealand.
[e]Chinese VLBI Network. First antenna commissioned at Sheshan in 1987.
[f]VLBI Exploration of Radio Astrometry.
[g]Korean VLBI Network, dedicated to astrometry and geodesy. Dual-beam capability part of larger Japanese VLBI Network (JVN) [see Doi et al. (2007)].
[h]LOw Frequency ARray.
[i]Event Horizon Telescope.
[j]International VLBI Service.
References: 1, Porcas (2010); 2, Napier et al. (1994); 3, Edwards (2012); 4, Zhang et al. (2012); 5, Kobayashi et al. (2003), VERA (2015); 6, Lee et al. (2014); 7, van Haarlem et al. (2013); 8, Doeleman (2010); 9, Behrend and Baver (2012).

during which correlation was done with custom-built hardware, this task has largely reverted back to general-purpose computers because of the rapid growth of their capabilities (Deller et al. 2007, 2011).

9.2   Differences Between VLBI and Conventional Interferometry

In this section, we briefly discuss the differences between VLBI and connected-element interferometry. Later sections in this chapter elaborate on these differences. Before beginning, we emphasize the theoretical unity of interferometry. The fundamental aim of all interferometry is to measure the coherence properties of the electromagnetic field. Thus, the principles of connected-element interferometry and VLBI are basically identical. However, certain special techniques used in VLBI are needed because of the particular observational constraints. As the continuity of (u, v) coverage is improved, from a few meters to more than 10^5 km (with the largest spacing achieved by elements on distant satellites), and fiberoptic or other advanced communication systems make recording unnecessary, the concept of VLBI as a distinct technique will become a matter of history. Here, we deal with certain limitations that make classical VLBI practices somewhat distinct from those of connected-element interferometry.

Early VLBI experiments were conducted by organizing a diverse group of observatories that had been constructed for general radio astronomical research. Each telescope had its own limitations, calibration procedures, and management personnel. Various networks were formed to standardize procedures and automate the execution of VLBI experiments. Such ad hoc VLBI networks operated on an intermittent basis, and during observations, the communication between elements to verify proper operation was limited. Small amounts of data from strong sources could be transmitted from the antennas to the correlator over telephone lines and cross-correlated to determine the instrumental delays and to check that the equipment was working properly. Later, arrays dedicated to VLBI were brought into operation [see, e.g., Napier et al. (1994)].

In VLBI, one has less control over the system stability because independent frequency standards are used at each element. Frequency offsets in the standards can cause instrumental timing errors. These errors usually include an epoch error of a few microseconds and a drift of a few tenths of a microsecond per day (Sect. 9.5). Therefore, the correlation function of the received signals [with respect to time offset, τ, as defined in Eq. (3.27)] must be measured to determine and track the instrumental delay. In contrast, delay errors in connected-element interferometers, due mainly to baseline errors and atmospheric propagation delays, are usually less than 30 ps, corresponding to 1 cm of path length. These errors are negligible for bandwidths less than 1 GHz. Thus, the response in connected-element, delay-tracking interferometers is always centered on the white light fringe. Delay becomes important only when the field of view becomes too large for the bandwidth (see Sects. 2.2 and 6.3) or when spectral line measurements are made by introducing time offsets. In VLBI, it is necessary to search a range of delay values to find the correct time relationship that maximizes the correlation. Correlations for a number of delay offsets are usually formed simultaneously, so a VLBI correlator may resemble a digital spectral correlator, although the number of frequency channels may be less than generally used for spectral line observations. The frequency offsets in the standards, which cause drifts with time in the instrumental delay, also introduce offsets in the fringe frequency. Thus, analysis of a VLBI experiment must begin with a two-dimensional search in delay and fringe frequency (delay rate) to find the peak of the correlation function. This process is referred to as fringe finding (see Sect. 9.3.4).

The concept of coherence has different implications in VLBI and connected-element interferometry. In connected-element interferometry, there is generally a suitable calibration source within a few degrees of the source of interest that can be observed every few minutes. Even if the instrumental phase drifts, there is no fundamental limit on integration time, and the concept of coherence time is replaced by that of the interval between calibrations. In VLBI, the use of calibrators to extend coherence time is more difficult because the short-term phase stability ($t < 10^3$ s) is worse. Atmospheric fluctuations above the stations are generally completely uncorrelated, and the frequency standards and frequency multipliers introduce phase noise in the fringes. Furthermore, a fundamental difference between connected-element interferometry and VLBI comes from the fact that there are many fewer sources that are unresolved at VLBI spacings and that can be used as calibrators. It is not always possible to find a calibrator close enough to the source under investigation to use as a phase reference. The time required to repoint the antennas and the decorrelation introduced by the atmosphere both increase with angular spacing. Thus, VLBI is subject to a fundamental coherence time that limits its sensitivity. For integration beyond the coherence time, it is necessary to average the fringe *amplitudes*, for which sensitivity improves only as the fourth root of the integration time (Sect. 9.3.5). It is also more difficult to calibrate phase in VLBI systems, although the situation has steadily improved as enhanced sensitivity has increased the number of sources that can be used as calibrators. Improved instrumental phase stability and more accurate modeling of the baselines, atmosphere, and similar factors have allowed the phase to be related to that of a calibrator several degrees away. Phase referencing in this manner is discussed in Sect. 12.2.3, and an example is shown in Fig. 12.1. Phase information can also be derived from phase closure analysis. In measuring positions, fringe frequency and group delay (the delay pattern effect discussed in Sects. 2.2 and 6.3.1) have also proved useful as measurement quantities.

Storage of the undetected signals before correlation presents VLBI with several problems. The average IF bandwidth is limited by the recording medium, which therefore limits the sensitivity of VLBI. The data must be stored as efficiently as possible, which requires a coarsely quantized representation of the signal, sampled at the Nyquist rate. With such a representation, the basic operations of fringe rotation and delay tracking, when performed on the recorded data, introduce significant effects that must be allowed for in deriving the visibility (Sect. 9.7).

9.2.1   The Problem of Field of View

In most VLBI applications, the ratio of the extent of the source under study to the resolution is typically less than about 10^2 (see Figs. 1.19–1.21). It is interesting to consider the challenge of imaging the entire primary beam of the antennas used in a VLBI observation. Consider an array of the following parameters:

D (longest baseline) = 4000 km

d (antenna diameter) = 25 m

N (number of array elements) = 10

$\nu = 10$ GHz ($\lambda = 3$ cm)

$\Delta\nu$ = bandwidth = 1 GHz

T_{obs} (observation time) = 12 hrs.

The nominal resolution is λ/D, or 1.5 mas, and the field of view is $\Delta\theta \sim \lambda/d$, or 250″. Hence, the number of pixels required for an image (at 2 pixels per resolution element) of the entire primary beam is

$$N_p \simeq \pi \left(\frac{D}{d}\right)^2 \simeq 5 \times 10^{11} . \tag{9.8}$$

Note that N_p is independent of wavelength because the resolution and field of view both scale as wavelength.

The processing and data storage requirements are considerable because of the large range of geometric delay and fringe rate that must be covered. The geometric delay is $\tau_g = D\cos\theta/c$, where θ is the angle between the baseline vector and the source direction. Thus, the range of delay over the primary beam is $D\sin\theta\,\Delta\theta/c$, and the maximum delay range requirement is

$$\Delta\tau_{g,\text{max}} = \frac{D}{\nu d} . \tag{9.9}$$

At the Nyquist sampling interval of $(2\Delta\nu)^{-1}$, the number of lags in the correlation function needed to cover this range will be

$$N_c = 2 \left(\frac{D}{d}\right)\left(\frac{\Delta\nu}{\nu}\right) , \tag{9.10}$$

which is about 30,000 for our example.

The fringe rate, in Hz, $\omega(d\tau_g/dt)/2\pi$, is $D\omega_e \sin\theta/\lambda$, where $\omega_e = 1/T_e$ and T_e is the Earth's sidereal period. This leads to a range of fringe rates of

$$\Delta\nu_{f,\text{max}} = \left(\frac{D}{d}\right)\left(\frac{2\pi}{T_e}\right) , \tag{9.11}$$

which requires a minimum sampling time of $(\Delta\nu_{f,\text{max}})^{-1}$, or about 34 ms. Thus, the number of fringe rate samples in time $T_{\text{obs}} = T_e/2$ is about 2.9×10^6. The total amount of data in the delay–fringe rate domain on $N(N-1)/2 \sim N^2$ baselines is

$$N_T \simeq \pi N^2 \left(\frac{D}{d}\right)^2\left(\frac{\Delta\nu}{\nu}\right) . \tag{9.12}$$

For our case, $N_T \sim 5 \times 10^{12}$ samples. With 2 bytes/sample and complex numbers, the minimum storage requirement would be about 160 Tbytes.

Because of the high brightness requirement of VLBI, most of the primary beam field will be largely empty but may contain a significant number of compact sources. A simple approach would be to image these sources with separate passes through the data processing system with separate field centers for each source. The advent of software correlators has provided a more efficient approach. The data from the correlation step (correlation functions of 30,000 lags at interval cadence of 34 ms, in our case) can be shifted to various phase centers and the resulting data streams reduced in volume substantially before imaging. The details of the phase center shifting, called "(u, v) shifting," are described by Morgan et al. (2011). This process can be embedded in the software architecture without the need for intermediate storage of the entire delay–fringe rate data set. An implementation is described by Deller et al. (2011), and an example is shown in Fig. 9.1.

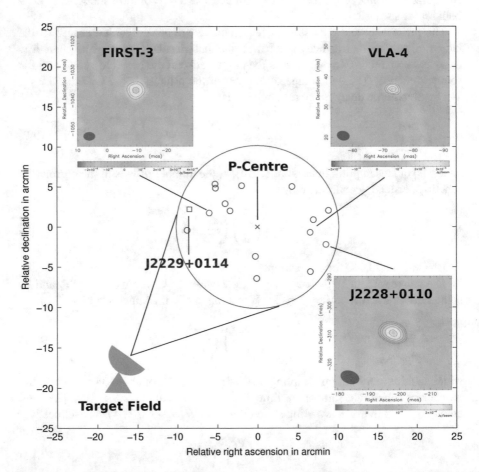

Fig. 9.1 An example of the multiple field center imaging technique with data from the EVN at 1.6 GHz. "P-Centre" is the pointing center of the individual antennas, and the circle shows the primary beam size (FWHM) of a 32-m-diameter antenna. The phase calibrator is J2229+0114. Fifteen other sources were detected in the field, and the images of three of them are shown in the inset panels. The contour levels start at the 3σ level and increase by factors of $\sqrt{2}$. From H.-M. Cao et al. (2014), reproduced with permission. © ESO.

9.3 Basic Performance of a VLBI System

9.3.1 Time and Frequency Errors

A block diagram of a basic VLBI system and a generic processor configuration is shown in Fig. 9.2. The atomic frequency standards control the phases of the local oscillators and the clock pulses for sampling the data. In many VLBI applications, such as spectral line observations or astrometric programs, frequency-dependent effects must be accounted for precisely. To understand the spectral response of the system, we consider the phase shifts encountered by a single frequency component. The signals received from a plane wave are $e^{j2\pi\nu t}$ at antenna 1, which we designate as the time-reference antenna, and $e^{j2\pi\nu(t-\tau_g)}$ at antenna 2, where τ_g is the geometric delay. The local oscillators have phases $2\pi\nu_{LO}t + \theta_1$ and $2\pi\nu_{LO}t + \theta_2$, where ν_{LO} is the local oscillator frequency, and θ_1 and θ_2 are the slowly varying terms that represent the phase noise due to the frequency standards. To start, we consider the upper-sideband response in Fig. 9.2, for which the local oscillator frequency is below the signal frequency. Thus, the phases after mixing are

$$
\begin{aligned}
\phi_1^{(1)} &= 2\pi(\nu - \nu_{LO})t - \theta_1 , \\
\phi_2^{(1)} &= 2\pi(\nu - \nu_{LO})t - 2\pi\nu\tau_g - \theta_2 .
\end{aligned}
\tag{9.13}
$$

The recorded signals each have clock errors τ_1 and τ_2, so the phases of the recorded signals are

$$
\begin{aligned}
\phi_1^{(2)} &= 2\pi(\nu - \nu_{LO})(t - \tau_1) - \theta_1 , \\
\phi_2^{(2)} &= 2\pi(\nu - \nu_{LO})(t - \tau_2) - 2\pi\nu\tau_g - \theta_2 .
\end{aligned}
\tag{9.14}
$$

During processing, the time series of signal samples from antenna 2 is advanced by τ_g', the estimate of τ_g, so

$$
\phi_2^{(3)} = 2\pi(\nu - \nu_{LO})(t - \tau_2 + \tau_g') - 2\pi\nu\tau_g - \theta_2 .
\tag{9.15}
$$

The output of the multidelay correlator and Fourier transform processor is the cross power spectrum. The phase at the output of the processor for the signal component at frequency ν is

$$
\begin{aligned}
\phi_{12} &= \phi_1^{(2)} - \phi_2^{(3)} \\
&= 2\pi(\nu - \nu_{LO})(\tau_2 - \tau_1) + 2\pi(\nu\Delta\tau_g + \nu_{LO}\tau_g') + \theta_{21} \\
&= 2\pi(\nu - \nu_{LO})(\tau_e + \Delta\tau_g) + 2\pi\nu_{LO}\tau_g + \theta_{21} ,
\end{aligned}
\tag{9.16}
$$

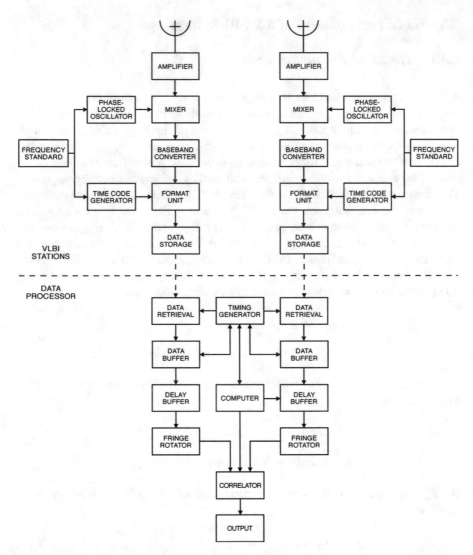

Fig. 9.2 Block diagram of the essential elements of a VLBI system, including data acquisition and processing. The system may pass the upper, lower, or both sidebands at the mixer inputs, depending on the passband of the amplifiers. For millimeter-wavelength observations, there may be no amplifier preceding the mixer, in which case both sidebands may be accepted. Quantization and sampling of the signals occur in the format units. The processor system shown illustrates the configuration described analytically by Eqs. (9.21)–(9.26). Major variations in the processing system relate to the relative positions of the correlator, fringe rotator (see Fig. 9.21), and FFT operation in the correlator.

where $\Delta\tau_g = \tau_g - \tau_g'$ is the delay error, $\tau_e = \tau_2 - \tau_1$ is the clock error, and $\theta_{21} = \theta_2 - \theta_1$. Equation (9.16) applies to the upper-sideband frequency conversion in the mixers in Fig. 9.2, for which the intermediate frequency (IF), $(\nu - \nu_{LO})$, is positive. For generality, we also give the lower-sideband response, for which the IF is $(\nu_{LO} - \nu)$. For the lower sideband,

$$\phi_{12} = 2\pi(\nu_{LO} - \nu)(\tau_e + \Delta\tau_g) - 2\pi\nu_{LO}\tau_g - \theta_{21} . \tag{9.17}$$

Note that in the ideal case where $\tau_1 = \tau_2$, $\theta_1 = \theta_2$, and $\tau_g = \tau'_g$, Eqs. (9.16) and (9.17) reduce to $\phi_{12} = 2\pi\nu_{LO}\tau_g$ for the upper sideband, and $\phi_{12} = -2\pi\nu_{LO}\tau_g$ for the lower sideband.

The correlation function at the correlator output is real, but not even; thus, the cross power spectrum S_{12} for a source of continuum radiation has the property

$$S_{12}(\nu') = S_{12}^*(-\nu') , \tag{9.18}$$

where ν' is the intermediate frequency $(\nu - \nu_{LO})$. We assume that the filters in the electronics have identical responses and therefore do not introduce any net phase shifts. The power response function of the instrumental filters is therefore real, and in terms of the voltage response, $H(\nu)$, of the filters for the two antennas, $S(\nu') = H_1(\nu')H_2^*(\nu')$. By combining the phase from Eq. (9.16) and the magnitude of the power response, the cross power spectrum for the upper sideband can be written

$$S_{12}(\nu') = S(\nu') \exp\left\{j\left[2\pi\nu'(\tau_e + \Delta\tau_g) + 2\pi\nu_{LO}\tau_g + \theta_{21}\right]\right\} . \tag{9.19}$$

The corresponding equation for the lower sideband can be obtained from Eq. (9.17). For the upper sideband, the cross-correlation function can be calculated from Eqs. (9.18) and (9.19) as

$$\rho_{12}(\tau) = \int_{-\infty}^{\infty} S_{12}(\nu')e^{j2\pi\nu'\tau}d\nu' . \tag{9.20}$$

For either sideband, integration includes both positive and negative frequencies, and since S_{12} is Hermitian and S is purely real, we obtain

$$\rho_{12}(\tau) = 2F_1(\tau')\cos(2\pi\nu_{LO}\tau_g + \theta_{21}) - 2F_2(\tau')\sin(2\pi\nu_{LO}\tau_g + \theta_{21}) , \tag{9.21}$$

where $\tau' = \tau + \tau_e + \Delta\tau_g$ and

$$F_1(\tau) = \int_0^{\infty} S(\nu')\cos(2\pi\nu'\tau)d\nu' ,$$
$$F_2(\tau) = \int_0^{\infty} S(\nu')\sin(2\pi\nu'\tau)d\nu' . \tag{9.22}$$

If $\mathcal{S}(\nu')$ is a rectangular lowpass spectrum with bandwidth $\Delta\nu$, then

$$F_1(\tau) = \Delta\nu \frac{\sin 2\pi \Delta\nu\tau}{2\pi \Delta\nu\tau} \ ,$$

$$F_2(\tau) = \Delta\nu \frac{\sin^2 \pi \Delta\nu\tau}{\pi \Delta\nu\tau} \ . \tag{9.23}$$

These functions are shown in Fig. 9.3. By substituting Eq. (9.23) into Eq. (9.21), the cross-correlation function can be written

$$\rho_{12}(\tau) = 2\Delta\nu \cos(2\pi \nu_{\mathrm{LO}}\tau_g + \theta_{21} + \pi \Delta\nu\tau') \frac{\sin \pi \Delta\nu\tau'}{\pi \Delta\nu\tau'} \ . \tag{9.24}$$

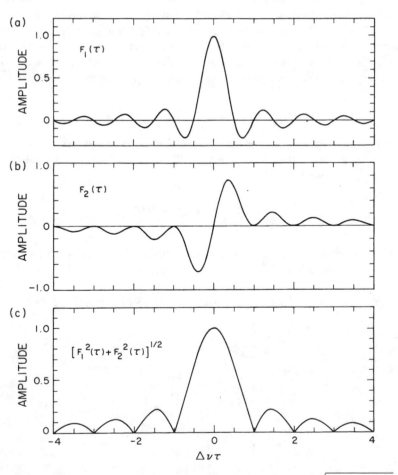

Fig. 9.3 Functions $F_1(\tau)$ and $F_2(\tau)$, defined in Eq. (9.23), and the quantity $\sqrt{F_1^2(\tau) + F_2^2(\tau)}$.

A similar analysis is given by Rogers (1976).

The variation of τ_g with time results in fringe oscillations at the correlator output. The fringe frequency, $(1/2\pi)d\phi_{12}/dt$, is constant across the receiver bandwidth because the (instrumental) delay tracking removes the (geometric) delay-induced phase variation across the band. For the upper and lower sidebands, the rate of change of phase has opposite signs; note the term $2\pi \nu_{LO}\tau_g$ in Eqs. (9.16) and (9.17). See also Fig. 6.5 and the related discussion. In VLBI, the natural fringe frequency is fast enough that the fringes would be lost in the final averaging of the correlated data, so rotation of the phase to stop the fringes is applied before the correlator in Fig. 9.2. In a double-sideband system, if the fringes are stopped for one sideband, the fringe frequency is doubled for the other sideband. However, it is possible to obtain the data from each sideband by processing the data twice with appropriate fringe offsets each time. In VLBI, the source position and other parameters are not always known with sufficient accuracy when the observation is made, so in Fig. 9.2, the fringes are stopped after recovery of the data streams to permit trial of different fringe rotation rates. This involves applying a phase shift to the quantized signals at the correlator input or output (see Sect. 9.7.1). The effect on the cross-correlation function or the cross power spectrum can be described as multiplication by $e^{-j2\pi \nu_{LO}\tau_g'}$ for the upper sideband and filtering to select the low-frequency term. This process results in a complex correlation function:

$$\rho_{12}'(\tau) = \Delta\nu \exp\left[j\left(2\pi \nu_{LO}\Delta\tau_g + \theta_{21} + \pi\Delta\nu\tau'\right)\right] \frac{\sin \pi\Delta\nu\tau'}{\pi\Delta\nu\tau'} \,. \tag{9.25}$$

Note that the principal fringe term, $2\pi \nu_{LO}\tau_g$, has been eliminated, but residual fringes can result from terms in $\Delta\tau_g$ and $\Delta\nu$. The resulting cross power spectrum is

$$S_{12}'(\nu') = S(\nu') \exp\left\{j\left[2\pi\nu'(\tau_e + \Delta\tau_g)2\pi \nu_{LO}\Delta\tau_g + \theta_{21}\right]\right\} \,. \tag{9.26}$$

This applies to the upper sideband, for which the fringes have been stopped, and the correlator output for the other sideband averages to zero.

An example of $\rho_{12}'(\tau)$ for eight values of τ is shown in Fig. 9.4. The waveforms represent the correlator output as a function of time for eight different delay offsets (lags) that differ sequentially by one Nyquist sample interval. Note that there is a phase shift of $\pi/2$ between adjacent delay steps. The fringe phase can be recovered by a proper interpolation (see Sect. 9.7.3) to the peak of the correlation function, or from the phase of the cross power spectrum at $\nu' = 0$. The group delay can be derived from the position of the correlation peak or the slope of the phase of the cross power spectrum. Note that the measured delay is $(1/2\pi)d\phi_{12}/d\nu$ and is therefore a group delay, not a phase delay.

The actual local oscillator frequencies may differ from the nominal value ν_{LO} due to an intentional offset from the nominal frequency or due to an offset error in the frequency standard. We can expand the phase terms θ_1 and θ_2 to include these

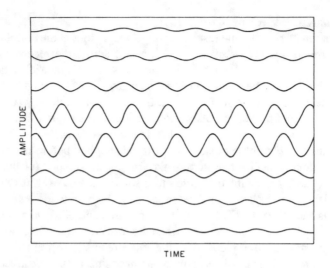

Fig. 9.4 Each sinusoid represents the correlation function [the real part of Eq. (9.25)] vs. time for a particular delay offset (from the top: $\frac{7}{2}$, $\frac{5}{2}$, $\frac{3}{2}$, $\frac{1}{2}$, $-\frac{1}{2}$, $-\frac{3}{2}$, $-\frac{5}{2}$, $-\frac{7}{2}$ times the Nyquist interval). The oscillations result from the residual fringe frequency, which includes any offsets in the frequency standards at the two antennas. Note the progressive phase shift of 90° between values of the correlation function at successive delay offsets.

frequency offsets, Δv_1 and Δv_2, and zero-mean phase components, θ'_1 and θ'_2:

$$\theta_1 = 2\pi \Delta v_1 t + \theta'_1 ,$$
$$\theta_2 = 2\pi \Delta v_2 t + \theta'_2 . \tag{9.27}$$

Thus, the fringe phase from Eq. (9.26) becomes

$$\phi_{12}(v') = 2\pi \left[v'(\tau_e + \Delta\tau_g) + v_{LO}\Delta\tau_g + \Delta v_{LO}t \right] + \theta'_{21} , \tag{9.28}$$

where $\Delta v_{LO} = \Delta v_2 - \Delta v_1$, the difference in the local oscillator frequencies, and $\theta'_{21} = \theta'_2 - \theta'_1$. The fringe frequency $(1/2\pi)d\phi_{12}/dt$ contains this local oscillator difference term. If Δv_1 is due to an offset in a frequency standard and is not zero, the measured fringe phase is actually more complicated than shown in Eq. (9.28). The clock error changes with time because of the frequency standard offset and is

$$\tau_1 = (\tau_1)_{t=0} + \frac{\Delta v_1}{v_{LO}}t . \tag{9.29}$$

The recovered time in the processor, based on the time of station 1, is related to the "true" time t by

$$t_1 = (\tau_1)_{t=0} + \left(1 + \frac{\Delta v_1}{v_{LO}} \right) t , \tag{9.30}$$

so that there is a slight shift in all measured frequencies and phases. Thus, there is a fundamental asymmetry in the processing between the reference station from which time is derived and the other stations (Whitney et al. 1976).

For spectral line observations, the quantity $S(v')$ in Eq. (9.26) is the (temporal frequency) spectrum of the visibility of the source multiplied by the bandpass response of the interferometer. The bandpass response can be obtained by observation of the cross power spectrum of a continuum source with a flat spectrum. Alternately, if the phase responses of the interferometer elements are identical, the bandpass response can be obtained from the geometric mean of the power spectra from the individual elements. These power spectra are obtained by observing a continuum source or blank sky and measuring the autocorrelation of the waveform from each individual antenna. The frequency spectrum of the normalized visibility can be obtained by dividing the visibility spectrum by the geometric mean of the power spectra of the source as measured with each antenna. To correct for nonidentical phase responses, it is necessary to measure the complex power spectrum on a strong continuum source. Details of calibration procedures in VLBI spectral line observations are given by Moran (1973), Reid et al. (1980), Moran and Dhawan (1995), and Reid (1995, 1999).

9.3.2   Retarded Baselines

The estimate of delay τ_g must be accurate enough to ensure that the signal is within the delay and fringe-frequency ranges of the processor. The simplest approximation is

$$\tau_g = \frac{1}{c} \mathbf{D} \cdot \mathbf{s}_0 , \tag{9.31}$$

where $\mathbf{D} = \mathbf{r}_1 - \mathbf{r}_2$, $\mathbf{r}_1$, and $\mathbf{r}_2$ are vectors from the center of the Earth to each station, and $\mathbf{s}_0$ is the unit vector to the center of the field. Account must be taken of the fact that the Earth moves in the time between the arrival of a wave crest at one station and at another, since the Earth is not an inertial reference. Therefore, in calculating the delay, we should use not the instantaneous baseline but the "retarded" baseline (Cohen and Shaffer 1971). A plane wave reaches the first station at time t_1 and the second station at a time t_2, which satisfies the equation

$$\mathbf{k} \cdot \mathbf{r}_1(t_1) - 2\pi v t_1 = \mathbf{k} \cdot \mathbf{r}_2(t_2) - 2\pi v t_2 , \tag{9.32}$$

where $\mathbf{k} = (2\pi/\lambda)\mathbf{s}_0$. Now $t_2 - t_1 = \tau_g$, so

$$2\pi v \tau_g = \mathbf{k} \cdot [\mathbf{r}_2(t_1 + \tau_g) - \mathbf{r}_1(t_1)] . \tag{9.33}$$

Expansion of $\mathbf{r}_2$ in a Taylor series gives

$$\mathbf{r}_2(t_1 + \tau_g) \simeq \mathbf{r}_2(t_1) + \dot{\mathbf{r}}_2(t_1)\tau_g + \cdots , \tag{9.34}$$

where the dot over $\mathbf{r}_2$ indicates the derivative and

$$2\pi \nu \tau_g \simeq \mathbf{k} \cdot [\mathbf{D}(t_1) + \dot{\mathbf{r}}_2(t_1)\tau_g] . \tag{9.35}$$

Solving for τ_g yields

$$\tau_g = \frac{\mathbf{D} \cdot \mathbf{s}_0}{c}\left[1 - \frac{\mathbf{s}_0 \cdot \dot{\mathbf{r}}_2}{c}\right]^{-1} , \tag{9.36}$$

where all quantities are evaluated at t_1. Since $\dot{\mathbf{r}} = \boldsymbol{\omega}_e \times \mathbf{r}$, where $\boldsymbol{\omega}_e$ is the angular velocity vector of the Earth and $\times$ indicates the vector cross product, we can rewrite Eq. (9.36) as

$$\tau_g \simeq \frac{\mathbf{D} \cdot \mathbf{s}_0}{c}\left[1 - \frac{\mathbf{s}_0 \cdot (\boldsymbol{\omega}_e \times \mathbf{r}_2)}{c}\right]^{-1} , \tag{9.37}$$

or

$$\tau_g \simeq \tau_{g0}(1 + \Delta) , \tag{9.38}$$

where $1 + \Delta$ is the term in brackets on the right side of Eq. (9.37). From the w term in Eq. (4.3),

$$\tau_{g0} = \frac{D}{c}\left[\sin d \sin \delta + \cos d \cos \delta \cos(H - h)\right] . \tag{9.39}$$

Here (H, δ) and (h, d) are the hour angle and declination coordinates of the source and baseline, respectively, the hour angles usually being specified with respect to the Greenwich meridian in VLBI practice. Also, we have

$$\Delta = \frac{\omega_e r_2}{c}\cos \mathcal{L}_2 \cos \delta \sin(h_2 - H) , \tag{9.40}$$

where $\mathcal{L}_2$, h_2, and r_2 are the latitude, hour angle, and magnitude of $\mathbf{r}_2$, where ω_e is the magnitude of $\boldsymbol{\omega}_e$. The function Δ has a maximum value of 1.5×10^{-6}, and τ_g can differ from τ_{g0} by a maximum of about 0.05 μs. Note that the appropriate coordinates in Eq. (9.39) are those that are uncorrected for refraction or diurnal aberration. An equivalent way of accounting for the retarded baseline is to use Eq. (9.31) for the delay but correct h and δ for the diurnal aberration at the remote site. We introduced the concept of retarded baseline mainly for pedagogical purposes. It does not appear explicitly when interferometry variables are calculated in a heliocentric frame.

There are different ways to formulate VLBI observables. One system that may be described as station-oriented is to refer the measurements to the center of the Earth, so that if recordings from two antennas are processed once and then interchanged and reprocessed, the phase obtained on the second pass will be the negative of that obtained on the first pass. This method presupposes an Earth model, since the radius vectors must be known. For applications to astrometry or geodesy, a baseline-oriented system is usually preferred, in which the observables have no dependence on a priori values of Earth parameters. A more precise discussion of VLBI observables can be found in Shapiro (1976) and Cannon (1978). For a full barycentric formulation, see Sovers et al. (1998).

9.3.3   Noise in VLBI Observations

In VLBI, it is often necessary to identify and calibrate the fringe visibility in situations of low signal-to-noise ratio and short coherence time. In such cases, a thorough understanding of the noise properties of interferometers can be very useful. The properties of the fringe amplitude and phase were briefly introduced in Sect. 6.2.4. We now develop this discussion further [see Moran (1976) and Hjellming (1992)]. The measured visibility is represented by a vector $\mathbf{Z} = \boldsymbol{\mathcal{V}} + \boldsymbol{\varepsilon}$], where $\boldsymbol{\mathcal{V}}$ and $\boldsymbol{\varepsilon}$ represent the true visibility (the signal) and noise components, respectively. We select coordinates with x (real) and y (imaginary) so that $\boldsymbol{\mathcal{V}}$ lies along the x axis, as shown in Fig. 6.8. There is no loss in generality by having $\boldsymbol{\mathcal{V}}$ lie along the x axis. The phase of the measured visibility resulting from the noise is a random variable denoted by ϕ. The components of $\boldsymbol{\varepsilon}$ have independent zero-mean Gaussian probability distributions in their x and y coordinates, with an rms deviation σ given by Eq. (6.50). In polar coordinates, the amplitude of $\boldsymbol{\varepsilon}$ has a Rayleigh probability distribution, and the phase of $\boldsymbol{\varepsilon}$ has a uniform probability distribution. $\mathbf{Z}$ is therefore a random variable whose x and y components, Z_x and Z_y, have a probability distribution given by

$$p(Z_x, Z_y) = \frac{1}{2\pi\sigma^2} \exp\left[-\frac{(Z_x - |\mathcal{V}|)^2 + Z_y^2}{2\sigma^2} \right] . \tag{9.41}$$

We convert this probability distribution to polar coordinates,

$$Z_x = Z \cos\phi \tag{9.42a}$$

$$Z_y = Z \sin\phi , \tag{9.42b}$$

by noting that the Jacobian of the transformation is simply $|\mathcal{V}|$ [see, e.g., Sivia (2006)] and obtain the result

$$p(Z, \phi) = \frac{|\mathcal{V}|}{2\pi\sigma^2} \exp\left[-\frac{(Z\cos\phi + |\mathcal{V}|)^2 + Z^2 \sin^2\phi}{2\sigma^2} \right] , \tag{9.43}$$

where $Z = \sqrt{Z_x^2 + Z_y^2}$.

The marginal distribution of Z is given by

$$p(Z) = \int_{-\pi}^{\pi} p(Z, \phi) \, d\phi \, , \qquad (9.44)$$

which, as in Eq. (6.63a), is

$$p(Z) = \frac{Z}{\sigma^2} \exp\left(-\frac{Z^2 + |\mathcal{V}|^2}{2\sigma^2}\right) I_0\left(\frac{Z|\mathcal{V}|}{\sigma^2}\right) \, , \qquad Z > 0 \, , \qquad (9.45)$$

where I_0 is a modified Bessel function of order zero, which is defined by

$$I_0(x) = \frac{1}{\pi} \int_0^{\pi} e^{x \cos \theta} \, d\theta \, . \qquad (9.46)$$

$p(Z)$ is known as the Rice distribution.

The marginal distribution ϕ is

$$p(\phi) = \int_0^{\infty} p(Z, \phi) \, dZ \, , \qquad (9.47)$$

which becomes

$$p(\phi) = \frac{1}{2\pi} \exp\left(-\frac{|\mathcal{V}|^2}{2\sigma^2}\right) + \left\{ \frac{1}{\sqrt{8\pi}} \frac{|\mathcal{V}| \cos \phi}{\sigma} \exp\left(-\frac{|\mathcal{V}|^2 \sin^2 \phi}{2\sigma^2}\right) \right.$$
$$\left. \times \left[1 + \mathrm{erf}\left(\frac{|\mathcal{V}| \cos \phi}{\sqrt{2}\sigma}\right) \right] \right\} \, , \qquad (9.48)$$

where erf is the error function defined in Eq. (6.63c). Note that $p(\phi)$ is an even function of ϕ, as expected, since the phase of $\mathcal{V}$ was set to zero. Hence, $\langle \phi \rangle = 0$. $p(\phi)$ was first derived in the interferometry literature by Vinokur (1965). Equations (9.45) and (9.48) correspond to Eqs. (6.63a) and (6.63b). However, here we have written $p(\phi)$ in a slightly different but equivalent form to make its asymptotic behavior more obvious. These probability distributions are plotted in Fig. 6.9.

The expectations of Z, Z^2, and Z^4 are

$$\langle Z \rangle = \sqrt{\frac{\pi}{2}} \sigma \exp\left(-\frac{|\mathcal{V}|^2}{4\sigma^2}\right) \left[\left(1 + \frac{|\mathcal{V}|^2}{2\sigma^2}\right) I_0\left(\frac{|\mathcal{V}|^2}{4\sigma^2}\right) + \frac{|\mathcal{V}|^2}{2\sigma^2} I_1\left(\frac{|\mathcal{V}|^2}{4\sigma^2}\right) \right] \, , \qquad (9.49)$$

$$\langle Z^2 \rangle = |\mathcal{V}|^2 + 2\sigma^2 \, , \qquad (9.50)$$

and

$$\langle Z^4 \rangle = |\mathcal{V}|^4 + 8\sigma^2 |\mathcal{V}|^2 + 8\sigma^4 , \tag{9.51}$$

where I_1 is the modified Bessel function of order one, defined by

$$I_1(x) = \frac{1}{\pi} \int_0^\pi e^{x\cos\theta} \cos\theta \, d\theta . \tag{9.52}$$

Higher even-order moments of Z can be readily calculated using the moment theorem for a Gaussian random distribution. When no signal is present, i.e., when $|\mathcal{V}| = 0$, $I_0(0) = 1$, and the probability distributions of Z and ϕ are those of the noise, which are Rayleigh and uniform distributions, respectively:

$$p(Z) = \frac{Z}{\sigma^2} \exp\left(-\frac{Z^2}{2\sigma^2} \right) , \qquad Z > 0 , \tag{9.53}$$

and

$$p(\phi) = \frac{1}{2\pi} , \qquad 0 \le \phi < 2\pi . \tag{9.54}$$

For the no-signal case,

$$\langle Z \rangle = \sqrt{\pi/2}\,\sigma , \tag{9.55}$$

$$\sigma_Z = \sqrt{\langle Z^2 \rangle - \langle Z \rangle^2} = \sigma \sqrt{2 - \pi/2} , \tag{9.56}$$

and

$$\sigma_\phi = \frac{\pi}{\sqrt{3}} . \tag{9.57}$$

For the weak-signal case, defined as $|\mathcal{V}| \ll \sigma$, we use the approximations $I_0(x) \simeq 1 + x^2/4$ and $I_1(x) \simeq x/2$. The probability distributions of Z and ϕ are

$$p(Z) \simeq \frac{Z}{\sigma^2} \exp\left(-\frac{Z^2}{2\sigma^2} \right) \left[1 - \frac{1}{2}\frac{|\mathcal{V}|^2}{\sigma^2} + \frac{1}{4}\left(\frac{Z|\mathcal{V}|}{\sigma^2} \right)^2 \right] \tag{9.58}$$

and

$$p(\phi) \simeq \frac{1}{2\pi} + \frac{1}{\sqrt{8\pi}}\frac{|\mathcal{V}|}{\sigma} \cos\phi , \tag{9.59}$$

to first order in $|\mathcal{V}|/\sigma$. Thus,

$$\langle Z \rangle \simeq \sigma \sqrt{\frac{\pi}{2}} \left(1 + \frac{|\mathcal{V}|^2}{4\sigma^2} \right) , \tag{9.60}$$

$$\sigma_Z \simeq \sigma \sqrt{2 - \frac{\pi}{2}} \left(1 + \frac{|\mathcal{V}|^2}{4\sigma^2} \right) , \tag{9.61}$$

and

$$\sigma_\phi \simeq \frac{\pi}{\sqrt{3}} \left(1 - \sqrt{\frac{9}{2\pi^3}} \frac{|\mathcal{V}|}{\sigma} \right) . \tag{9.62}$$

Note that Z departs from a Rayleigh distribution slowly as $|\mathcal{V}|/\sigma$ increases, whereas the probability distribution of ϕ is confined to a spread (full width at half-maximum) of only $110°$ and $70°$ for $|\mathcal{V}|/\sigma$ equal to 1 and 2, respectively (see Fig. 6.9). Hence, as a practical matter, it is often easier to identify a weak signal by its phase rather than by its amplitude, as shown in Fig. 9.5.

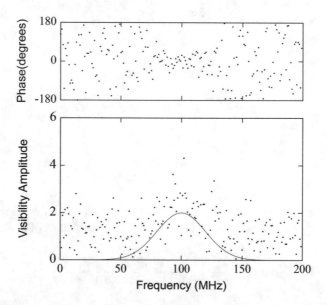

Fig. 9.5 A simulated visibility spectrum of a source with a single spectral line with a Gaussian profile of amplitude equal to 2 and centered at 100 MHz (solid line). The spectral resolution is 1 MHz, and $\sigma = 1$ (hence $|\mathcal{V}|/\sigma = 2$) at line center. This demonstrates that weak signals can be more easily identified (by eye) in phase than in amplitude.

For the strong-signal case, $|\mathcal{V}| \gg \sigma$, $I_0(x) \simeq e^x/\sqrt{2\pi x}$. The probability functions for Z and ϕ are approximately Gaussian distributions and are given by

$$p(Z) \simeq \frac{1}{\sqrt{2\pi}\,\sigma} \sqrt{\frac{Z}{|\mathcal{V}|}} \exp\left[-\frac{(Z - |\mathcal{V}|)^2}{2\sigma^2}\right] \tag{9.63}$$

and

$$p(\phi) \simeq \frac{1}{\sqrt{2\pi}} \frac{|\mathcal{V}|}{\sigma} \exp\left(-\frac{|\mathcal{V}|^2\phi^2}{2\sigma^2}\right) . \tag{9.64}$$

For this case,

$$\langle Z \rangle \simeq |\mathcal{V}| \left(1 + \frac{\sigma^2}{2|\mathcal{V}|^2}\right) , \tag{9.65}$$

$$\sigma_Z \simeq \sigma \left(1 - \frac{\sigma^2}{8|\mathcal{V}|^2}\right) , \tag{9.66}$$

and

$$\sigma_\phi \simeq \frac{\sigma}{|\mathcal{V}|} . \tag{9.67}$$

The quantities σ_Z and σ_ϕ for the full range of $|\mathcal{V}|/\sigma$ are shown in Fig. 9.6. Hence, in the strong-signal case, the statistics of Z are approximately Gaussian (see Fig. 6.9), and $\langle Z \rangle$ approaches $|\mathcal{V}|$. In this case, N samples of Z can be averaged, and the signal-to-noise ratio improves with $\sqrt{N}$. In the weak-signal case, the perturbation

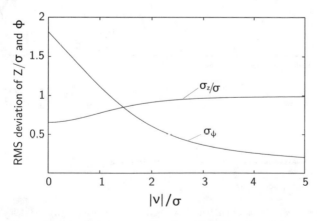

Fig. 9.6 The values of σ_Z/σ and σ_ϕ as a function of $|\mathcal{V}|/\sigma$. Approximate expressions for $|\mathcal{V}|/\sigma \ll 1$ are given in Eqs. (9.61) and (9.62) and for $|\mathcal{V}|/\sigma \gg 1$ in Eqs. (9.66) and (9.67).

of the Rayleigh noise distribution by the signal is small, and as we shall discuss in Sect. 9.5, it is difficult to improve the signal-to-noise ratio by averaging beyond the coherence time of the system.

9.3.4 Probability of Error in the Signal Search

When starting a new session of VLBI observations with an ad hoc array, the first task in the processing is to search for fringes, i.e., fringe finding. This is necessary because of the uncertainties in the station clocks and their drift rates and means that the instrumental delay and fringe frequency must be found. This step is often unnecessary with a dedicated VLBI array, for which the values of fringe rate and delay are continuously updated from successive observations. A fringe search must be carried out on a large two-dimensional grid, as shown in Fig. 9.7. For example,

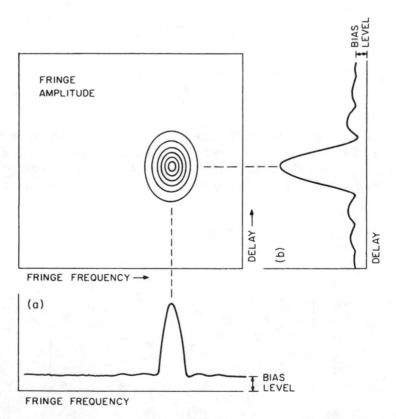

Fig. 9.7 Fringe amplitude as a function of residual fringe frequency (**a**) and delay (**b**). The one-dimensional plots are the peak fringe amplitude vs. delay and fringe frequency. The probability distribution of the noise in these plots is given by Eq. (9.71) and the bias level by Eq. (9.72).

consider an experiment in which $\Delta \nu = 50$ MHz at an observing frequency of 10^{11} Hz. The delay increments are equal to the sampling interval of 0.01 μs. An instrumental delay uncertainty of ± 1 μs requires a search of 200 delay intervals. If the coherent integration time is 200 s and the frequency standards are set only to a fractional accuracy of 10^{-11}, then ± 1 Hz must be searched, which at an interval size of 0.005 Hz is 400 discrete frequencies. The total number of cells to be searched is 80,000. If there is no signal present, then $p(Z)$ will be given by Eq. (9.53). The cumulative probability distribution (that is, the probability that Z is less than Z_0) in this case is the integral of Eq. (9.53) from zero to Z_0, or

$$ P(Z_0) = 1 - \exp\left(-\frac{Z_0^2}{2\sigma^2}\right) . \tag{9.68} $$

The cumulative probability distribution for the maximum of n independent samples $Z_m = \max \{Z_1, Z_2, \ldots, Z_n\}$ is

$$ P(Z_m) = \left[1 - \exp\left(-\frac{Z_m^2}{2\sigma^2}\right)\right]^n . \tag{9.69} $$

Thus, the probability of one or more samples exceeding Z_m, which we call the probability of error, p_e, is

$$ p_e = 1 - \left[1 - \exp\left(-\frac{Z_m^2}{2\sigma^2}\right)\right]^n . \tag{9.70} $$

This function is shown in Fig. 9.8. The probability distribution of Z_m is obtained by differentiating Eq. (9.69),

$$ p(Z_m) = \frac{nZ_m}{\sigma^2} \exp\left(-\frac{Z_m^2}{2\sigma^2}\right) \left[1 - \exp\left(-\frac{Z_m^2}{2\sigma^2}\right)\right]^{n-1} . \tag{9.71} $$

For large n, this probability distribution is nearly Gaussian, with mean value and standard deviation given by

$$ \langle Z_m \rangle \simeq \sigma\sqrt{2\ln n} , \tag{9.72} $$

$$ \sigma_m \simeq \frac{0.77\sigma}{\sqrt{\ln n}} . \tag{9.73} $$

Examples of $p(Z_m)$ for various values of n are shown in Fig. 9.9. It is frequently useful to reduce a two-dimensional function, such as the one shown in Fig. 9.7 of fringe amplitude vs. fringe frequency and delay, to a one-dimensional function by searching for the maximum value of the function over one variable. This search

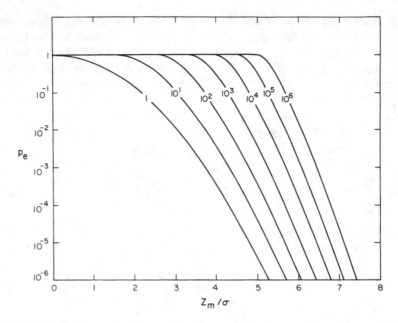

Fig. 9.8 Probability that one or more samples of the fringe amplitude will exceed the value Z_m/σ in the absence of a signal, as given by Eq. (9.70). The curves are labeled by the number of samples measured.

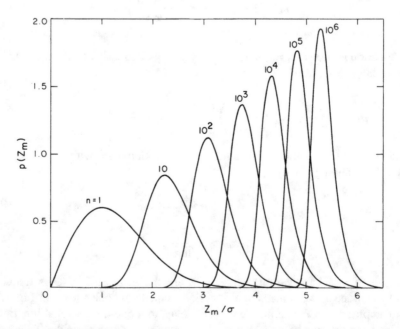

Fig. 9.9 Probability distribution of the maximum of n random variables that have Rayleigh distributions, as given Eq. (9.71).

process introduces a bias, equal to $\langle Z_m \rangle$, into the one-dimensional function. This bias increases with the number of samples and obscures weak signals.

We can also calculate the probability of misidentifying a signal. Suppose that we have measurements of fringe amplitude at two values of delay or fringe frequency with the signal present at one value. The probability that the amplitude in the channel with the signal (Z_1) is larger than the amplitude in the channel with only the noise (Z_2) is

$$p(Z_1 > Z_2) = \int_0^\infty p(Z_1) \left[\int_0^{Z_1} p(Z_2)dZ_2 \right] dZ_1 . \tag{9.74}$$

$p(Z_1)$ is given by Eq. (9.45), and $p(Z_2)$ is given by Eq. (9.53). We can generalize this result for a search over n channels where the signal channel amplitude is Z_s. The probability that Z_s will exceed the values of Z in the other channels is, from Eqs. (9.68) and (9.74),

$$p(Z_s > Z_1, \ldots, Z_n) = \int_0^\infty p(Z) \left[1 - \exp\left(-\frac{Z^2}{2\sigma^2} \right) \right]^{n-1} dZ , \tag{9.75}$$

where $p(Z)$ is given by Eq. (9.45). Thus, the probability of one or more samples exceeding the amplitude of the signal is

$$p'_e = 1 - \int_0^\infty p(Z) \left[1 - \exp\left(-\frac{Z^2}{2\sigma^2} \right) \right]^{n-1} dZ . \tag{9.76}$$

p'_e is plotted in Fig. 9.10. For example, if the search is over 100 channels, a probability of misidentification of less than 0.1% requires $|\mathcal{V}|/\sigma > 6.5$.

9.3.5 Coherent and Incoherent Averaging

We wish to estimate the amplitude of a barely detectable signal. We examine a time series of correlator output values in which the phase, $\phi(t)$, represents the effects of receiver noise, fluctuations in the frequency standards, or fluctuations in the atmospheric path. An example of phase vs. time from a VLBI measurement is shown in Fig. 9.11. The correlator output is

$$r(t) = Z(t)e^{j\phi(t)} . \tag{9.77}$$

How do we estimate $|\mathcal{V}|$ when the time range of the data exceeds the coherence time? There are two useful procedures, the first in the spectral domain and the second in the time domain. Suppose that $r(t)$ is sampled at intervals short with respect to the coherence time, τ_c, thus generating a time series of samples r_n. The

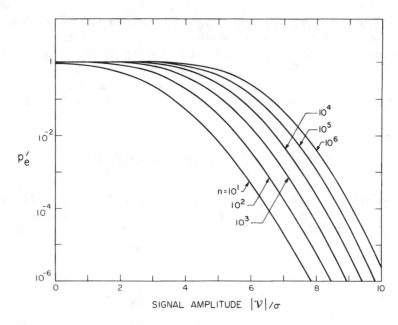

Fig. 9.10 Probability that one or more samples of fringe amplitude among the samples with no signal will exceed the fringe amplitude of the sample with the signal, vs. the signal amplitude, $|\mathcal{V}|$, as given in Eq. (9.76). The curves are labeled according to the total number of samples n. The asymptotic value of p'_e as $|\mathcal{V}|/\sigma$ goes to zero is $1 - 1/n$.

discrete Fourier transform (see Appendix 8.4) of r_n is

$$R_k = \sum_{n=0}^{N-1} r_n \, e^{-j2\pi kn/N} \; , \tag{9.78}$$

where R_k is the N-point discrete fringe rate spectrum ranging in frequency from $-N/2\tau_c$ to $N/2\tau_c$. Hence, from Parseval's theorem [Eq. (8.179)],

$$\sum_{n=0}^{N-1} |r_n|^2 = \frac{1}{N} \sum_{k=0}^{N-1} |R_k|^2 \; . \tag{9.79}$$

Using Eq. (9.50), we can write an unbiased estimator of $|\mathcal{V}|^2$, valid for large N, as

$$|\mathcal{V}|_e^2 = \left(\frac{1}{N^2} \sum_{k=1}^{N-1} |R_k|^2 \right) - 2\sigma^2 \; . \tag{9.80}$$

When the total span of the data exceeds the coherence time of the interferometer, the fringe rate spectrum becomes complicated, but Eq. (9.80) provides a prescription

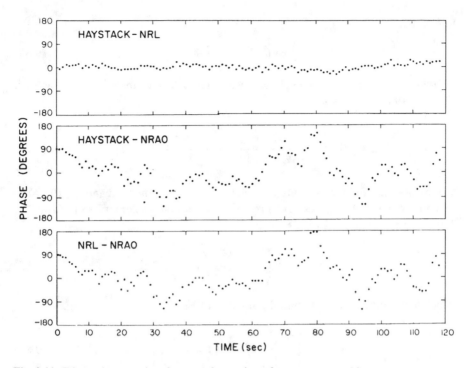

Fig. 9.11 Fringe phase vs. time from an observation of a strong source [the water vapor maser in W3 (OH)] on a three-baseline VLBI experiment at 22 GHz. Two of the stations, Haystack Observatory and the Naval Research Laboratory (Maryland Point Observatory), were equipped with hydrogen maser frequency standards, while the National Radio Astronomy Observatory used a rubidium (vapor frequency standard). The phase noise in the top plot is dominated by contributions from the receivers and the atmosphere, while the phase noise in the bottom two plots is dominated by the phase noise in the rubidium frequency standard. These data were obtained in 1971 with the Mark I VLBI system.

for gathering all of its frequency components into an unbiased estimate of $|\mathcal{V}|^2$. See Clark (1968) and Clark et al. (1968) for applications of this method.

The second method for estimating $|\mathcal{V}|^2$, based on the time series, comes directly from Eq. (9.50),

$$|\mathcal{V}|_e^2 = \left(\frac{1}{N} \sum_{i=1}^{N} Z_i^2 \right) - 2\sigma^2 . \qquad (9.81)$$

Imaging or model analysis is usually based on estimates of $|\mathcal{V}|$, not $|\mathcal{V}|^2$. To obtain an unbiased estimate of $|\mathcal{V}|$, we first examine the properties of the quantity

$$|\mathcal{V}|_b = \left[\frac{1}{N} \sum_{i=1}^{N} Z_i^2 \right]^{1/2} . \qquad (9.82)$$

Recall that

$$Z_i^2 = (|\mathcal{V}| + \epsilon_{x_i})^2 + \epsilon_{y_i}^2 \ , \tag{9.83}$$

where ϵ_{x_i} and ϵ_{y_i} are Gaussian random variables with zero mean and variance σ^2. Equation (9.82) becomes

$$|\mathcal{V}|_b = |\mathcal{V}| \left\{ 1 + \frac{1}{N} \sum_{i=1}^{N} \left[\frac{2\epsilon_{x_i}}{|\mathcal{V}|} + \frac{\epsilon_{x_i}^2 + \epsilon_{y_i}^2}{|\mathcal{V}|^2} \right] \right\}^{1/2} . \tag{9.84}$$

We assume that the terms in the brackets are $\ll 1$ and then expand Eq. (9.84) to second order, which is necessary to retain all the second-order terms involving ϵ_{x_i}. Then the expectation of $|\mathcal{V}|_b$ becomes

$$|\mathcal{V}|_b \simeq |\mathcal{V}| \left[1 + \frac{\sigma^2}{|\mathcal{V}|^2} \left(1 - \frac{1}{2N} \right) \right] , \tag{9.85}$$

which leads directly to an unbiased estimate of $|\mathcal{V}|$ of

$$|\mathcal{V}|_e \simeq \left[\frac{1}{N} \sum_{i=1}^{N} Z_i^2 - \sigma^2 \left(2 - \frac{1}{N} \right) \right]^{1/2} . \tag{9.86}$$

Equation (9.86) is accurate to $< 5\%$ for $\mathcal{V}/\sigma > 2$ and $N = 1$, and $\mathcal{V}/\sigma > 0.3$ and $N = 100$. This estimator has several interesting properties. For $N \gg 1$, it leads to the result suggested by Eq. (9.81). However, for $N = 1$ and $Z_i = Z$, it leads to the result

$$|\mathcal{V}|_e = \left[Z^2 - \sigma^2 \right]^{1/2} . \tag{9.87}$$

Equation (9.87) is used to determine the polarized flux from single measurements of Stokes Q and U [see Wardle and Kronberg (1974)]. For one measurement of Z, $|\mathcal{V}|_e$ in Eq. (9.87) is a good approximation for the most likely value of $|\mathcal{V}|$ given $p(Z)$ defined in Eq. (9.45). See Johnson et al. (2015) for further discussion and applications.

From Eqs. (9.50), (9.51), and (9.81), we have $\langle |\mathcal{V}|_e^2 \rangle = |\mathcal{V}|^2$ and $\langle |\mathcal{V}|_e^4 \rangle = |\mathcal{V}|^4 + 4\sigma^2(|\mathcal{V}|^2 + \sigma^2)/N$, so that the signal-to-noise ratio is

$$\mathcal{R}_{\text{sn}} = \frac{\langle |\mathcal{V}|_e^2 \rangle}{\sqrt{\langle |\mathcal{V}|_e^4 \rangle - \langle |\mathcal{V}|_e^2 \rangle^2}} = \frac{\sqrt{N}}{2\sigma^2} |\mathcal{V}|^2 \frac{1}{\sqrt{1 + |\mathcal{V}|^2/\sigma^2}} . \tag{9.88}$$

$|\mathcal{V}|/\sigma$ is equal to the signal-to-noise ratio at the output of a single-multiplier correlator, as given by Eqs. (6.49) and (6.50). For VLBI observations, the quantization efficiency described in Sect. 8.3, η_Q, is replaced by the general loss factor η, described in Sect. 9.7, and from Eq. (6.64), we obtain $|\mathcal{V}|/\sigma = (T_A \eta / T_S) \sqrt{2 \Delta \nu \tau_c}$.

Equation (9.88) then becomes

$$\mathcal{R}_{\text{sn}} = \frac{T_A^2 \eta^2}{T_S^2} \sqrt{\frac{\Delta v^2 \tau \tau_c}{(1 + 2T_A^2 \eta^2 \Delta v \tau_c / T_S^2)}} , \tag{9.89}$$

where $\tau = N\tau_c$ is the total integrating time. The two limiting cases of Eq. (9.89) are

$$\mathcal{R}_{\text{sn}} \simeq \frac{\eta}{\sqrt{2}} \frac{T_A}{T_S} \sqrt{\Delta v \tau} , \qquad\qquad T_A \gg \frac{T_S}{\sqrt{2\Delta v \tau_c}} , \tag{9.90}$$

$$\mathcal{R}_{\text{sn}} \simeq \left(\frac{T_A \eta}{T_S}\right)^2 \Delta v \sqrt{\tau \tau_c} , \qquad\qquad T_A \ll \frac{T_S}{\sqrt{2\Delta v \tau_c}} . \tag{9.91}$$

Note that in the strong-signal case, incoherent averaging is not needed. When incoherent averaging is used, the coherent averaging time should be as long as possible without decreasing the fringe amplitude. If we assume that $\mathcal{R}_{\text{sn}} = 4$ for detection, and recall that $\tau = N\tau_c$, then for the weak-signal case, the minimum detectable antenna temperature can be found from Eq. (9.91) to be

$$(T_A)_{\text{min}} = \frac{2T_S}{\eta N^{1/4} \sqrt{\Delta v \tau_c}} . \tag{9.92}$$

Thus, because of the $N^{1/4}$ dependence in Eq. (9.92), incoherent averaging is effective only if N is very large. If the coherence time is of the order of $1/\Delta v$, then the observing system reduces to a form of incoherent, or intensity, interferometer [see Sect. 17.1 and Clark (1968)]. For the weak-signal case, Eq. (9.91) then becomes

$$\mathcal{R}_{\text{sn}} \simeq \left(\frac{T_A \eta}{T_S}\right)^2 \sqrt{\Delta v \tau} . \tag{9.93}$$

9.4 Fringe Fitting for a Multielement Array

9.4.1 Global Fringe Fitting

In Sect. 9.3, we considered the problem of searching for fringes in the output from a single baseline. For VLBI, the basic requirement in fringe fitting is to determine the fringe phase (i.e., the phase of the visibility) and the rate of change of the fringe phase, with time and with frequency or delay. Fringe rate offsets result from errors in the positions of the source or antennas as well as antenna-related effects such as frequency offsets in local oscillators. Most of these can be specified as factors that relate to individual antennas, rather than to baselines. Because of this, data from all baselines can be used simultaneously to determine the fringe rate parameters. By simultaneously using all of the data from a multielement VLBI array, it is possible to detect fringes that are too weak to be seen on a single baseline. This is particularly important for VLBI arrays with similar antennas and receivers; with an ad hoc array, a possible alternative is to use the data from the two most sensitive antennas to find the fringes and let this result constrain the solutions for other baselines.

A method of analysis that is based on simultaneous use of the complete data set from a multiantenna observation was developed by Schwab and Cotton (1983) and is referred to as *global fringe fitting*. Let $Z_{mn}(t)$ be the correlator output, that is, the measured visibility, from the baseline for antennas m and n. The complex (voltage) gain for antenna n and the associated receiving system is $g_n(t_k, \nu_\ell)$, where t_k represents a (coherently) time-integrated sample of the correlator output for frequency channel ν_ℓ. Thus,

$$Z_{mn}(t_k, \nu_\ell) = g_m(t_k, \nu_\ell)g_n^*(t_k, \nu_\ell)\mathcal{V}_{mn}(t_k, \nu_\ell) + \epsilon_{mnk\ell} , \tag{9.94}$$

where $\mathcal{V}_{mn}$ is the true visibility for baseline mn, and $\epsilon_{mnk\ell}$ represents the observational errors that result principally from noise. It should be remembered that the noise terms are present in all the measurements, but beyond this point, they will usually be omitted from the equations. The gain terms can be written as

$$g_n(t_k, \nu_\ell) = |g_n|e^{j\psi_n(t_k, \nu_\ell)} . \tag{9.95}$$

To simplify the situation in Eq. (9.95), we assume that the gain terms and the amplitude of the source visibility are constant over the range of (t, ν) space covered by the observation. To first order, we can then write

$$Z_{mn}(t_k, \nu_\ell) = |g_m||g_n||\mathcal{V}| \exp\left[j(\psi_m - \psi_n)(t_0, \nu_0)\right]$$

$$\times \exp\left[j\left(\left.\frac{\partial(\psi_m - \psi_n + \phi_{mn})}{\partial t}\right|_{(t_0,\nu_0)}(t_k - t_0)\right.\right. \tag{9.96}$$

$$\left.\left. + \left.\frac{\partial(\psi_m - \psi_n + \phi_{mn})}{\partial \nu}\right|_{(t_0,\nu_0)}(\nu_\ell - \nu_0)\right)\right] ,$$

where ϕ_{mn} is the phase of the (true) visibility $\mathcal{V}_{mn}$. The rates of change of the phase of the measured visibility with respect to time and frequency are the fringe rate

$$r_{mn} = \left.\frac{\partial(\psi_m - \psi_n + \phi_{mn})}{\partial t}\right|_{(t_0,\nu_0)} , \tag{9.97}$$

and the delay

$$\tau_{mn} = \left.\frac{\partial(\psi_m - \psi_n + \phi_{mn})}{\partial \nu}\right|_{(t_0,\nu_0)} , \tag{9.98}$$

for the baseline mn at time and frequency (t_0, ν_0). In terms of these quantities, we can relate the measured visibility (correlator output) to the true visibility as follows:

$$
\begin{aligned}
Z_{mn}(t_k, \nu_\ell) = |g_m||g_n|\mathcal{V}_{mn}(t_k, \nu_\ell) \exp \Big\{ j \Big[(\psi_m - \psi_n)|_{t=t_0} \\
+ (r_m - r_n)(t_k - t_0) + (\tau_m - \tau_n)(\nu_\ell - \nu_0) \Big] \Big\} .
\end{aligned}
\tag{9.99}
$$

For each antenna, there are four unknown parameters: the modulus of the gain, the phase of the gain, the fringe rate, and the delay. Since all of the data are in the form of relative phases of two antennas, it is necessary to designate one antenna as the reference. For this antenna, the phase, fringe rate, and delay are usually taken to be zero, leaving $4n_a - 3$ parameters to be determined. However, it is possible to simplify further and consider only the phase terms in the fringe fitting. The amplitudes of the antenna gains are subsequently calibrated separately. The number of parameters to be determined is thereby reduced to $3(n_a - 1)$. Then to obtain the global fringe solution, the source visibility $\mathcal{V}_{mn}$ is represented by a model of the source, and a least-mean-squares fit of the parameters in Eq. (9.99) to the visibility measurements is made. For details on a method for the least-mean-squares solution, see Schwab and Cotton (1983). The source model, which is a "first guess" of the true structure, could in some cases be as simple as a point source.

Another method of using the data for several baselines simultaneously in fringe fitting is an extension of the method described earlier for single baselines. The measured visibility data are required to be specified in terms of fringe frequency and delay, which can be obtained, for example, by a time-to-frequency Fourier transformation of the data from a lag correlator. Then for each antenna pair, there is a matrix of values of the interferometer response at incremental steps in the delay and fringe rate. The maximum amplitude indicates the solution for delay and fringe rate for the corresponding baseline, as illustrated in Fig. 9.7. However, the method can be extended to include the responses from a number of baselines by using the closure phase principle, which is discussed in more detail in Sect. 10.3. Because we are considering fringe fitting in phase only, the measured data are represented by ϕ_{mn}. Since ψ_{mn}, the instrumental phase for baseline mn, is equal to the difference between the measured and true visibility phases, we can write

$$
\psi_{mn} = \psi_m - \psi_n = \tilde{\phi}_{mn} - \phi_{mn} ,
\tag{9.100}
$$

where the ψ terms represent the instrumental phases, the ϕ terms represent the visibility phases, and the tilde ($\tilde{\ }$) indicates measured visibility phases. Now consider including a third antenna, designated p. For this combination, we can write

$$
\psi_{mpn} = \psi_{mp} + \psi_{pn} = (\psi_m - \psi_p) + (\psi_p - \psi_n) = \psi_m - \psi_n .
\tag{9.101}
$$

Thus, ψ_{mpn} provides another measured value of ψ_{mn}, equal to

$$
\psi_{mp} + \psi_{pn} = (\tilde{\phi}_{mp} - \phi_{mp}) + (\tilde{\phi}_{pn} - \phi_{pn}) .
\tag{9.102}
$$

Similarly, for four antennas

$$\psi_{mpqn} = \psi_m - \psi_n = (\tilde{\phi}_{mp} + \tilde{\phi}_{pq} + \tilde{\phi}_{qn}) - (\phi_{mp} + \phi_{pq} + \phi_{qn}) \,. \tag{9.103}$$

Thus, estimated values of ψ_{mn} can be obtained from the measurements from loops of antenna pairs, starting with antenna m and ending with antenna n. Combinations of more than three baselines (four antennas) can be expressed as combinations of smaller numbers of antennas, and the noise in such larger combinations is not independent. Loops of three and four antennas provide additional information that contributes to the sensitivity and accuracy of the fringe fitting for antennas m and n. Note, however, that the model visibilities are also required.

Of the two techniques, the least-mean-squares fitting is better with respect to uniform combination of the data, but it requires a good starting estimate if it is to converge efficiently. Schwab and Cotton (1983) used the second of the two methods to provide a starting point for the full least-mean-squares solution. This procedure has subsequently become the basis of standard reduction programs for VLBI data (Walker 1989a,b).

Although global fringe fitting provides sensitivity superior to that of baseline-based fitting, in practice, some experience is needed to determine when use of the global method is appropriate. If the source under study has complicated structure, with large variations in the visibility amplitude, it will probably not be well represented by the model visibility required in the global fitting method. In such a case, it may be better to start with a smaller number of antennas in the fringe fitting or, if the source is sufficiently strong, to consider baselines separately. On the other hand, if the source contains a strong unresolved component, it may be adequate to consider smaller groups of antennas separately and thus reduce the overall computing load.

9.4.2 Relative Performance of Fringe Detection Methods

In the regime in which the phase noise limits the sensitivity, careful investigation of detection techniques is warranted. The most important of these have been examined by Rogers et al. (1995) to determine their relative performance. We assume in all cases that the visibility data from the correlator outputs have been averaged for a time equal to the coherence time, τ_c, discussed earlier. We have seen in Eq. (9.92) that incoherent averaging of N time segments of data reduces the level at which a signal is detectable by an amount proportional to $N^{-1/4}$. Rogers et al. show that for a detection threshold for which the probability of a false detection is $< 0.01\%$ in a search of 10^6 values, the threshold of detection is lower than that without incoherent averaging (in effect, $N = 1$) by a factor $0.53N^{-1/4}$. This result is accurate only for large N, and they find empirically that for smaller N, the detection threshold decreases in proportion to $N^{-0.36}$; that is, the improvement with increasing N is greater when N is small. Table 9.2 includes the improvement factor $0.53N^{-1/4}$, together with other results that are discussed below. The fourth column of Table 9.2

Table 9.2 Relative thresholds for various detection methods[a]

Method		Threshold (relative flux density)	
1	One baseline, coherent averaging	1	1
2	One baseline, incoherent averaging	$0.53N^{-1/4}$	$0.14\ (N = 200)$
3	3-baseline triple product	$\left(\frac{4}{N}\right)^{1/6}$	$0.52\ (N = 200)$
4	Array of n_a elements, coherent global search	$\left(\frac{2}{n_a}\right)^{1/2}$	$0.45\ (n_a = 10)$
5	Global search with incoherent averaging	$0.53\left(\frac{4}{Nn_a{}^2}\right)^{1/4}$	$0.05\ (N = 200, n_a = 10)$
6	Incoherent averaging over time segments and baselines	$0.53\left(\frac{2}{Nn_a(n_a-1)}\right)^{1/4}$	$0.05\ (N = 200, n_a = 10)$

From Rogers et al. (1995). © AAS. Used by permission.
[a]See text for detection criterion.

gives numerical examples of relative sensitivity for $N = 200$ time segments and $n_a = 10$ antennas. Note that for lines 1–5 of Table 9.2, the criterion for detection is a probability of error of less than 1% in a search of 10^6 values of delay and fringe rate for each of n_a-1 elements of the array, the values for the reference antenna taken to be zero. For line 6, the search spans only the two dimensions of right ascension and declination.

9.4.3   Triple Product, or Bispectrum

Another form of the output of a multielement array that can be considered is the triple product, or bispectrum, which is the product of the complex outputs for three baselines that form a triangle. The triple product is given by the product of measured visibilities

$$P_3 = |Z_{12}||Z_{23}||Z_{31}|e^{j(\tilde{\phi}_{12}+\tilde{\phi}_{23}+\tilde{\phi}_{31})} = |Z_{12}||Z_{23}||Z_{31}|e^{j\phi_c} , \tag{9.104}$$

where ϕ_c represents the closure phase (Sect. 10.3), which is zero if the source is unresolved. We assume here that the amplitude of the measured visibility, Z, is calibrated separately, so that the moduli of the gain factors g_m and g_n in Eq. (9.94) are unity. Each of the measured visibility terms includes noise of power $2\sigma^2$, that is, the noise power in the output of a complex correlator. For the weak-signal case, the noise determines the variance of the triple product, which is

$$\langle |P_3|^2 \rangle = \langle |Z_{12}|^2|Z_{23}|^2|Z_{31}|^2 \rangle = 8\sigma^6 . \tag{9.105}$$

For a point source, the signal is real and is equal to $\langle (\mathcal{Re}P_3)^2 \rangle = \langle |P_3|^2 \rangle/2$, where $\mathcal{Re}$ indicates the real part. The ratio of this triple product signal term to the noise in the real output of the correlator is $\mathcal{V}^3/2\sigma^3$. Rogers et al. (1995) also give an expression for the signal-to-noise ratio that is not restricted to the weak-signal case, and Kulkarni (1989) gives a general expression in a detailed analysis of the subject.

Now consider the incoherent average of N values of the triple product for three antennas, each of which represents an average of the correlator output over the coherence interval, τ_c. We represent this average of triple products by

$$\overline{P}_3 = \frac{1}{N} \sum_N |Z_{12}||Z_{23}||Z_{31}|e^{j\phi_c} . \tag{9.106}$$

If the signal amplitudes are equal, the expectation of the real part of $\overline{P}_3$ is

$$\langle \mathcal{R}e\overline{P}_3 \rangle = \mathcal{V}^3 , \tag{9.107}$$

and the second moment of $\mathcal{R}e\overline{P}_3$ is

$$\langle (\mathcal{R}e\overline{P}_3)^2 \rangle = \frac{1}{N} \langle |P_3|^2 \rangle \langle \cos^2 \phi_c \rangle . \tag{9.108}$$

In the weak-signal case, in which the value of $\langle |P_3|^2 \rangle$ results mainly from noise, the expectation of the second moment is, from Eq. (9.105), $4\sigma^6/N$. The signal-to-noise ratio is equal to the expectation of $\overline{P}_3$ divided by the square root of the expectation of the second moment,

$$\mathcal{R}_{\text{sn}} = \frac{\sqrt{N}\mathcal{V}^3}{2\sigma^3} , \tag{9.109}$$

from which

$$\mathcal{V} = (2\mathcal{R}_{\text{sn}})^{1/3}\sigma N^{-1/6} . \tag{9.110}$$

Line 3 of Table 9.2 gives the signal strength for a value of $\mathcal{R}_{\text{sn}}$ that allows detection at a level corresponding to the specified error criterion.

9.4.4   Fringe Searching with a Multielement Array

With an array of n_a VLBI antennas, the amount of information gathered in a given time is greater than that with a single antenna pair by a factor $n_a(n_a - 1)/2$. One might thus expect that the array would offer an increase in sensitivity $\simeq [n_a(n_a - 1)/2]^{1/2}$. However, the larger number of antennas also introduces a very large increase in the parameter space to be searched. Thus, the probability of encountering high noise amplitudes within this parameter space is correspondingly greater. It is therefore necessary to increase the signal level used as a detection threshold in order to avoid increasing the probability of false detection.

Consider a two-element array in which the number of data points to be searched in the parameter space (frequency × delay) is n_d. If a third antenna is then introduced, and correlation is measured for all baselines, the number of data points to be searched becomes n_d^2. For n_a antennas, it becomes $n_d^{(n_a-1)}$. The probability distribution of the maximum of n Rayleigh-distributed values of the signal plus noise, Z_m, is given in Eq. (9.71) and for large n has a mean value of $\sigma(2\ln n)^{1/2}$; see Eq. (9.72). Thus, for a given probability of occurrence, increasing the number of points to be searched from n_d to $n_d^{(n_a-1)}$ increases the level Z_m from $\sigma(2\ln n_d)^{1/2}$ to $\sigma[2(n_a-1)\ln n_d]^{1/2}$; that is, the probability of finding a level $(n_a-1)^{1/2}Z_m$ in a search of $n_d^{(n_a-1)}$ points is the same as that of finding a level Z_m in a search of n_d points. By increasing the number of antennas from 2 to n_a, the overall rms uncertainty in the signal level is reduced by a factor $[n_a(n_a-1)/2]^{1/2}$, but since the detection threshold has increased by $(n_a-1)^{1/2}$, the effective gain in sensitivity for detection of sources is increased by only $(n_a/2)^{1/2}$. Rogers (1991) and Rogers et al. (1995) consider other factors in deriving this result and show that the sensitivity increase $(n_a/2)^{1/2}$ should be multiplied by a factor that lies between 0.94 and 1. This factor is not included in Table 9.2.

9.4.5 Multielement Array with Incoherent Averaging

In Table 9.2, the last two lines are concerned with incoherent averaging of data taken with a multielement array. The method on line 5 involves data that have been averaged over the coherence time and subsequently averaged incoherently before the application of a global fringe search. The relative threshold value is the product of the threshold on line 4 for a multielement global search with that on line 2 for incoherent averaging over a single baseline. The method in line 6 involves incoherent averaging over both time segments (equal to the coherence time) and baselines. The relative threshold is obtained from that in line 2 by increasing the number of data from N (the number of time segments per baseline) to N multiplied by the number of baselines.

9.5 Phase Stability and Atomic Frequency Standards

Precision oscillators have been steadily improved since the 1920s, when the invention of the crystal-controlled (quartz) oscillator had immediate application to the problem of precise timekeeping. In the early 1950s, cesium-beam clocks allowed better timekeeping than could be obtained from astronomical observations. This development led to an atomic definition of time that differs from the astronomical one, and to the establishment of the definition of the second of time based on a particular transition frequency of cesium.

The mathematical theory of the interpretation of measurements of oscillator phase was systematized by an IEEE committee (Barnes et al. 1971). This paper helped standardize the approach to handling low-frequency divergence in the noise of oscillators. The physical theory of noise in oscillators was treated by Edson (1960). In this section, we develop relevant aspects of the theory and describe the operation of atomic frequency standards, with particular emphasis on the hydrogen maser. The theory and analysis of phase fluctuations are discussed in more detail by Blair (1974) and Rutman (1978).

9.5.1 Analysis of Phase Fluctuations

The desired signal from an oscillator is a pure sine wave:

$$V(t) = V_0 \cos 2\pi \nu_0 t . \tag{9.111}$$

This is unobtainable since all devices have some phase noise. A more realistic model is given by

$$V(t) = V_0 \cos [2\pi \nu_0 t + \phi(t)] , \tag{9.112}$$

where $\phi(t)$ is a random process characterizing the phase departure from a pure sine wave. We ignore amplitude fluctuations since they do not directly affect performance in VLBI applications. The instantaneous frequency $\nu(t)$ is the derivative of the argument of Eq. (9.112) divided by 2π, that is,

$$\nu(t) = \nu_0 + \delta\nu(t) , \tag{9.113}$$

where

$$\delta\nu(t) = \frac{1}{2\pi} \frac{d\phi(t)}{dt} . \tag{9.114}$$

The instantaneous fractional frequency deviation is defined as

$$y(t) = \frac{\delta\nu(t)}{\nu_0} = \frac{1}{2\pi\nu_0} \frac{d\phi(t)}{dt} . \tag{9.115}$$

This definition allows the performance of oscillators at different frequencies to be compared. We assume that the random processes $\phi(t)$ and $y(t)$ are statistically stationary, so that correlation functions can be defined. This assumption is not always valid and can cause difficulty (Rutman 1978). The autocorrelation function of $y(t)$ is

$$R_y(\tau) = \langle y(t)\, y(t+\tau) \rangle \,. \tag{9.116}$$

$R_y(\tau)$ is a real and even function, so $S'_y(f)$, the power spectrum of $y(t)$, is a real and even function of frequency f. In order to prevent confusion between $v(t)$ and its frequency components, we use the symbol f for the frequency variable in the following spectral analysis. Following the somewhat nonstandard convention that is used in most of the literature on phase stability (Barnes et al. 1971), we replace the double-sided spectrum $S'_y(f)$ with a single-sided spectrum $S_y(f)$, where $S_y(f) = 2S'_y(f)$ for $f \geq 0$, and $S_y(f) = 0$ for $f < 0$. Since $S'_y(f)$ is even, no information is lost in this procedure. Thus, the Fourier transform relation $R_y(\tau) \longleftrightarrow S'_y(f)$, can also be written as

$$S_y(f) = 4\int_0^\infty R_y(\tau) \cos(2\pi f\tau)\, d\tau \,,$$
$$R_y(\tau) = \int_0^\infty S_y(f) \cos(2\pi f\tau)\, df \,. \tag{9.117}$$

Similarly, the autocorrelation function of the phase is

$$R_\phi(\tau) = \langle \phi(t)\, \phi(t+\tau) \rangle \,. \tag{9.118}$$

$S_\phi(f)$, the power spectrum of ϕ, and $R_\phi(\tau)$ are related by a Fourier transform. From the derivative property of Fourier transforms, the relationship between $S_y(f)$ and $S_\phi(f)$ can be shown to be

$$S_y(f) = \frac{f^2}{v_0^2} S_\phi(f) \,. \tag{9.119}$$

$S_y(f)$ and $S_\phi(f)$ serve as primary measures of frequency stability. They both have the dimensions of Hz^{-1}. Another commonly used specification of oscillator performance is $\mathcal{L}(f)$, which is defined as the power in 1-Hz bandwidth at frequency f in one sideband of a double-sided spectrum, expressed as a fraction of the total power of the oscillator. When the phase deviation is small compared with one radian, $\mathcal{L}(f) \simeq S_\phi(f)/2$.

A second approach to frequency stability is based on time-domain measurements. The average fractional frequency deviation is

$$\bar{y}_k = \frac{1}{\tau}\int_{t_k}^{t_k+\tau} y(t)\, dt \,, \tag{9.120}$$

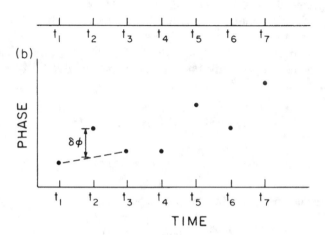

Fig. 9.12 (a) Time intervals involved in the measurement of $\bar{y}_k$ as defined in Eq. (9.121). (b) Plot of a series of phase samples vs. time. The Allan variance, defined in Eq. (9.123), is the average of the square of the deviation, $(\delta\phi)^2$, of each sample from the mean of its two adjacent samples.

which, from Eq. (9.115), becomes

$$\bar{y}_k = \frac{\phi(t_k + \tau) - \phi(t_k)}{2\pi \nu_0 \tau} , \tag{9.121}$$

where the measurements of $\bar{y}_k$ are made with a repetition interval $T(T \geq \tau)$ such that $t_{k+1} = t_k + T$ (see Fig. 9.12a). Measurements of $\bar{y}_k$ are directly obtainable with conventional frequency counters. The measure of frequency stability is the sample variance of $\bar{y}_k$, given by

$$\langle \sigma_y^2(N, T, \tau) \rangle = \frac{1}{N-1} \left\langle \sum_{n=1}^{N} \left(\bar{y}_n - \frac{1}{N} \sum_{k=1}^{N} \bar{y}_k \right)^2 \right\rangle , \tag{9.122}$$

where N is the number of samples in a single estimate of σ_y^2. In the limit as $N \to \infty$, the quantity presented above is the true variance, which we represent as $I^2(\tau)$. However, in many cases, Eq. (9.122) does not converge because of the low-frequency behavior of $S_y(f)$, and $I^2(\tau)$ is then not defined. To avoid some of the convergence problems, a particular case of Eq. (9.122), the two-sample or Allan variance, $\sigma_y^2(\tau)$, has gained wide acceptance (Allan 1966). The Allan variance, for which $T = \tau$ (no dead time between measurements) and $N = 2$, is defined as follows:

$$\sigma_y^2(\tau) = \frac{\langle (\bar{y}_{k+1} - \bar{y}_k)^2 \rangle}{2} , \tag{9.123}$$

or, from Eq. (9.121):

$$\sigma_y^2(\tau) = \frac{\langle [\phi(t + 2\tau) - 2\phi(t + \tau) + \phi(t)]^2 \rangle}{8\pi^2 \nu_0^2 \tau^2} . \tag{9.124}$$

The procedure for estimating the Allan variance can be understood as follows. Take a series of phase measurements at interval T, as shown in Fig. 9.12b. For each set of three independent points, draw a straight line between the outer two and determine the deviation of the center point from the line. With m samples of $\bar{y}$, the average of the squared deviations divided by $(2\pi\nu_0\tau)^2$ is an estimate of $\sigma_y^2(\tau)$, denoted $\sigma_{ye}^2(\tau)$, where

$$\sigma_{ye}^2(\tau) = \frac{1}{2(m-1)} \sum_{k=1}^{m-1} (\bar{y}_{k+1} - \bar{y}_k)^2 . \tag{9.125}$$

The accuracy of this estimate is (Lesage and Audoin 1979)

$$\sigma(\sigma_{ye}) \simeq \frac{K}{\sqrt{m}} \sigma_y , \tag{9.126}$$

where K is a constant of order unity, whose exact value depends on the power spectrum of y.

We can now relate the true variance and the Allan variance to the power spectrum of y or ϕ. From Eq. (9.121), the true variance is $I^2(\tau) = \langle \bar{y}_k^2 \rangle$, given by

$$I^2(\tau) = \frac{1}{(2\pi\nu_0\tau)^2} \left[\langle \phi^2(t + \tau) \rangle - 2\langle \phi(t + \tau)\phi(t) \rangle + \langle \phi^2(t) \rangle \right] , \tag{9.127}$$

which, from Eq. (9.118), is

$$I^2(\tau) = \frac{1}{2(\pi\nu_0\tau)^2} \left[R_\phi(0) - R_\phi(\tau) \right] . \tag{9.128}$$

Then, since $R_\phi(\tau)$ is the Fourier transform of $S_\phi(f)$, by using Eq. (9.119), we obtain from Eq. (9.128) the result

$$I^2(\tau) = \int_0^\infty S_y(f) \left(\frac{\sin \pi f \tau}{\pi f \tau} \right)^2 df . \tag{9.129}$$

Similarly, from Eq. (9.124), we obtain

$$\sigma_y^2(\tau) = \frac{1}{(2\pi \nu_0 \tau)^2} \left[3R_\phi(0) - 4R_\phi(\tau) + R_\phi(2\tau) \right] , \tag{9.130}$$

and therefore,

$$\sigma_y^2(\tau) = 2 \int_0^\infty S_y(f) \left[\frac{\sin^4 \pi f \tau}{(\pi f \tau)^2} \right] df . \tag{9.131}$$

$I^2(\tau)$ and $\sigma_y^2(\tau)$ are dimensionless quantities, measured in rad^2, but we can think of them as the power obtained after filtering $y(t)$ with two different frequency responses, $H_I^2(f)$ and $H_A^2(f)$, respectively. These are

$$H_I^2(f) = \left(\frac{\sin \pi f \tau}{\pi f \tau} \right)^2 \tag{9.132}$$

and

$$H_A^2(f) = \frac{2 \sin^4 \pi f \tau}{(\pi f \tau)^2} . \tag{9.133}$$

The functions $H_I^2(f)$ and $H_A^2(f)$ and the corresponding impulse responses $h_I(t)$ and $h_A(t)$ are shown in Fig. 9.13. Note that $I^2(\tau)$ can be estimated from a series of measurements $\bar{y}_k$ as the average of the square of $h_I(t_k) * \bar{y}_k$, where the asterisk indicates convolution. Similarly, $\sigma_y^2(\tau)$ can be estimated as the average of the square of $h_A(t_k) * \bar{y}_k$. Other transfer functions could be chosen. In time-domain measurements, additional filtering with high- and low-frequency cutoffs can be performed. For example, removing a long-term trend from the frequency data is a form of highpass filtering. Clearly, measurements of $S_y(f)$ are preferable to those of $\sigma_y^2(\tau)$, because σ_y^2 can be calculated from S_y using Eq. (9.131), but S_y cannot be calculated from σ_y^2. However, in many cases of interest, as in the power-law spectra discussed below, the form of σ_y^2 is indicative of the behavior of S_y. Traditionally, it has been easier to make time-domain measurements, and most published results are given in terms of the Allan variance σ_y^2.

The effect of local oscillator noise on the measured coherence of signals received at two antennas is given by Eq. (7.34) in terms of the rms deviation of the phase of the oscillator at one antenna relative to that at the other. For VLBI, this rms deviation is equal to the square root of the sum of the true variances of the local oscillators at the two antennas. In the case of a connected-element array, low-frequency components of the phase noise of the master oscillator cause similar effects in the local oscillator phase at each antenna, and therefore their contributions to the relative phase at different antennas tend to cancel. For exact cancellation, the time delay in the path of the reference signal from the master oscillator to each

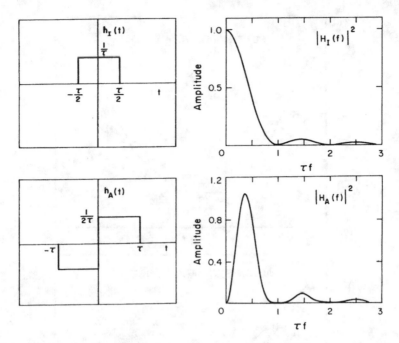

Fig. 9.13 (**top**) The impulse function $h_I(t)$ and the square of its Fourier transform $|H_I(f)|^2$, given by Eq. (9.132), which is used to relate the power spectrum $S_y(f)$ to the true variance $I^2(\tau)$, as defined in Eq. (9.129). (**bottom**) The impulse response $h_A(t)$ and the square of its Fourier transform, $|H_A(f)|^2$, given by Eq. (9.133), which is used to relate the power spectrum $S_y(f)$ to the Allan variance $\sigma_y^2(\tau)$, as defined in Eq. (9.131). Note that the sensitivity of the Allan variance decreases rapidly with decreasing frequency for $f < 0.3/\tau$.

antenna, plus the time delay of the IF signal from the corresponding mixer to the correlator input (including the variable delay that compensates for the geometric delay), should be equal for each antenna. It is generally impractical to preserve this equality. The bandwidths of phase-locked loops in the local oscillator signals at the antennas can also limit the frequency range over which phase noise in the master oscillator is canceled. In practice, cancellation of phase noise resulting from the master oscillator should generally be effective up to a frequency f in the range of a few hundred hertz to a few hundred kilohertz, depending on the parameters of the particular system.

Laboratory measurements show that $S_y(f)$ is often a combination of power-law components. A useful model, shown in Fig. 9.14, is

$$S_y(f) = \sum_{\alpha=-2}^{2} h_\alpha f^\alpha , \qquad 0 < f < f_h , \qquad (9.134)$$

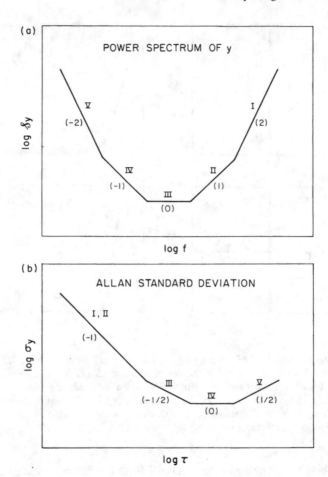

Fig. 9.14 (**a**) The idealized power spectrum $S_y(f)$ of the fractional frequency deviation $y(t)$ [see Eq. (9.134)]. The various spectral regimes are marked by Roman numerals, and the power-law coefficients are given in parentheses. The regimes are I, white-phase noise; II, flicker-phase noise; III, white-frequency noise; IV, flicker-frequency noise; and V, random-walk-of-frequency noise. (**b**) Two-point rms deviation, or Allan standard deviation, vs. the time between samples. The spectral regimes are marked by the Roman numerals, and the power-law coefficients are given in parentheses.

where α is a power-law exponent with integer values between –2 and 2, and f_h is the cutoff frequency of a lowpass filter. An equation similar to Eq. (9.134) can be written for $S_\phi(f)$ using Eq. (9.119). Each term in Eq. (9.134) or the equivalent equation for $S_\phi(f)$ has a name based on traditional terminology (see Table 9.3). Noise with a power-law dependence f^0, independent of frequency, is called "white-phase noise"; f^{-1} is called "flicker-phase noise," or colloquially, "one-over-f noise"; and f^{-2} is called "random-walk noise." There are well-known origins for some of these processes, which we discuss briefly [see also Vessot (1976)]. The frequency dependence given in parentheses below is for S_y.

Table 9.3 Characteristics of noise in oscillators[a]

Noise type	$S_y(f)$	$S_\phi(f)$	$\sigma_y^2(\tau)$	μ[b]	$I^2(\tau)$
White phase[c]	$h_2 f^2$	$v_0^2 h_2$	$\dfrac{3h_2 f_h}{4\pi^2 \tau^2}$	-2	$\dfrac{h_2 f_h}{2\pi^2 \tau^2}$
Flicker phase	$h_1 f$	$v_0^2 h_1 f^{-1}$	$\dfrac{3h_1}{4\pi^2 \tau^2}\ln(2\pi f_h \tau)$	~ -2	–
White frequency or random walk of phase	h_0	$v_0^2 h_0 f^{-2}$	$\dfrac{h_0}{2\tau}$	-1	$\dfrac{h_0}{2\tau}$
Flicker frequency	$h_{-1} f^{-1}$	$v_0^2 h_{-1} f^{-3}$	$(2\ln 2)h_{-1}$	0	–
Random walk of frequency	$h_{-2} f^{-2}$	$v_0^2 h_{-2} f^{-4}$	$\dfrac{2\pi^2 \tau}{3} h_{-2}$	1	–

[a] Adapted from Barnes et al. (1971).
[b] Power-law exponent of Allan variance: $\sigma_y^2(\tau) \propto \tau^\mu$.
[c] For $\sigma_y^2(\tau)$, $2\pi f_h \tau \gg 1$.

1. *White-phase noise* (f^2) is usually due to additive noise outside the oscillator, for example, noise introduced by amplifiers. This process dominates at large values of f, corresponding to short averaging times.
2. *Flicker-phase noise* (f^1) is seen in transistors and may be due to diffusion processes across junctions.
3. *White-frequency* or *random-walk-of-phase noise* (f^0) is due to internal additive noise within the oscillator, such as the thermal noise inside the resonant cavity. Shot noise also has this spectral dependence.
4. *Flicker-frequency noise* (f^{-1}) and *random-walk-of-frequency noise* (f^{-2}) are the processes that limit the long-term stability of oscillators. They are due to random changes in temperature, pressure, and magnetic field in the oscillator environment. This noise is associated with long-term drift. There is a large body of literature on flicker-frequency noise, which is encountered in many situations [see Keshner (1982) for a general discussion, Dutta and Horn (1981) for applications in solid-state physics, and Press (1978) for applications in astrophysics].

The variances $I^2(\tau)$ and $\sigma_y^2(\tau)$ can be calculated for the various types of noise described above. For $\alpha = 1$ and 2, the variances converge only if a high-frequency cutoff f_h is specified. With this restriction, σ_y^2 converges for all cases. $I^2(\tau)$ converges only for $\alpha \geq 0$. These functions are listed in Table 9.3. Except for the logarithmic dependence in flicker-phase noise, each noise component maps into a component of Allan variance of the form τ^μ. From Table 9.3, we can write the total Allan variance as

$$\sigma_y^2(\tau) = [K_2^2 + K_1^2 \ln(2\pi f_h \tau)]\tau^{-2} + K_0^2 \tau^{-1} + K_{-1}^2 + K_{-2}^2 \tau, \qquad (9.135)$$

where the K values are constants. The subscripts correspond to the subscripts of h (see Table 9.3). White-phase and flicker-phase noise both result in $\mu \simeq -2$, but these two processes can be distinguished by varying f_h. Note that for white-phase and white-frequency noise, the following relations hold [see Eqs. (9.129) and (9.131)]:

$$\sigma_y^2(\tau) = \frac{3}{2} I^2(\tau) , \qquad \alpha = 2 , \qquad (9.136)$$

$$\sigma_y^2(\tau) = I^2(\tau) , \qquad \alpha = 0 . \qquad (9.137)$$

In general, when $I^2(\tau)$ is defined, we see from Eqs. (9.128) and (9.130) that

$$\sigma_y^2(\tau) = 2[I^2(\tau) - I^2(2\tau)] . \qquad (9.138)$$

9.5.2   Oscillator Coherence Time

A quantity of special interest in VLBI is the coherence time. The approximate coherence time is that time τ_c for which the rms phase error is 1 radian:

$$2\pi \nu_0 \tau_c \sigma_y(\tau_c) \simeq 1 . \qquad (9.139)$$

Rogers and Moran (1981) calculated a more exact expression for the coherence time that they defined in terms of the coherence function

$$C(T) = \left| \frac{1}{T} \int_0^T e^{j\phi(t)} dt \right| , \qquad (9.140)$$

where $\phi(t)$ is the component of fringe phase of instrumental origin, and T is an arbitrary integration time. $\phi(t)$ includes effects that cause the fringe phase to wander, such as atmospheric irregularities and noise in frequency standards. The rms value of $C(T)$ is a monotonically decreasing function of time with the range 1–0. The coherence time is defined as the value of T for which $\langle C^2(T) \rangle$ drops to some specified value, say, 0.5. The mean-squared value of C is

$$\langle C^2(T) \rangle = \frac{1}{T^2} \int_0^T \int_0^T \langle \exp \{ j [\phi(t) - \phi(t')] \} \rangle \, dt \, dt' . \qquad (9.141)$$

If ϕ is a Gaussian random variable, then

$$\langle C^2(T) \rangle = \frac{1}{T^2} \int_0^T \int_0^T \exp \left[-\frac{\sigma^2(t, t')}{2} \right] dt \, dt' , \qquad (9.142)$$

where $\sigma^2(t, t')$ is the variance $\langle[\phi(t) - \phi(t')]^2\rangle$, which we assume depends only on $\tau = t' - t$. Then from Eq. (9.118),

$$\sigma^2(t, t') = \sigma^2(\tau)$$
$$= \langle[\phi(t) - \phi(t')]^2\rangle = 2[R_\phi(0) - R_\phi(\tau)] . \tag{9.143}$$

Note that $\sigma^2(\tau)$ is the structure function of phase and is related to $I^2(\tau)$ by Eq. (9.128):

$$\sigma^2(\tau) = 4\pi^2\tau^2 v_0^2 I^2(\tau) . \tag{9.144}$$

The integral in Eq. (9.142) can be simplified by noting that the integrand is constant along diagonal lines in (t, t') space for which $t' - t = \tau$. These lines have length $\sqrt{2}(T - \tau)$ so that

$$\langle C^2(T)\rangle = \frac{2}{T}\int_0^T \left(1 - \frac{\tau}{T}\right)\exp\left[-\frac{\sigma^2(\tau)}{2}\right]d\tau . \tag{9.145}$$

Thus, from Eqs. (9.129) and (9.144),

$$\langle C^2(T)\rangle = \frac{2}{T}\int_0^T \left(1 - \frac{\tau}{T}\right)\exp\left[-2(\pi v_0 \tau)^2\int_0^\infty S_y(f)H_I^2(f)df\right]d\tau , \tag{9.146}$$

where $H_I^2(f)$ is defined in Eq. (9.132). Since $S_y(f)$ is often not available, it is useful to relate $\langle C^2(T)\rangle$ to $\sigma_y^2(\tau)$. We can solve Eq. (9.138) for $I^2(\tau)$ by series expansion, obtaining

$$2I^2(\tau) = \sigma_y^2(\tau) + \sigma_y^2(2\tau) + \sigma_y^2(4\tau) + \sigma_y^2(8\tau) + \cdots , \tag{9.147}$$

provided that the series converges. Therefore, from Eqs. (9.144), (9.145), and (9.147),

$$\langle C^2(T)\rangle = \frac{2}{T}\int_0^T \left(1 - \frac{\tau}{T}\right)\exp\left\{-\pi^2 v_0^2\tau^2\left[\sigma_y^2(\tau) + \sigma_y^2(2\tau) + \cdots\right]\right\}d\tau . \tag{9.148}$$

This integral is readily calculable for the cases where $I^2(\tau)$ is defined.

We now consider white-phase noise and white-frequency noise, which are important processes in frequency standards on short time scales. For the case of white-phase noise, $\sigma_y^2 = K_2^2\tau^{-2}$, where $K_2^2 = 3h_2 f_h/4\pi^2$ is the Allan variance in 1 s (Table 9.3), and the coherence function can be evaluated from Eq. (9.146) or Eq. (9.148):

$$\langle C^2(T)\rangle = \exp\left(-\frac{4\pi^2 v_0^2 K_2^2}{3}\right) = \exp(-h_2 f_h v_0^2) . \tag{9.149}$$

For white-frequency noise, $\sigma_y^2 = K_0^2 \tau^{-1}$, where $K_0^2 = h_0/2$, and we obtain

$$\langle C^2(T) \rangle = \frac{2(e^{-aT} + aT - 1)}{a^2 T^2} . \tag{9.150}$$

Here, $a = 2\pi^2 v_0^2 K_0^2 = \pi^2 h_0 v_0^2$. The limiting cases for white-frequency noise are

$$\langle C^2(T) \rangle = 1 - \frac{2\pi^2 v_0^2 K_0^2 T}{3} , \qquad 2\pi^2 v_0^2 K_0^2 T \ll 1 ,$$

$$= \frac{1}{\pi^2 v_0^2 K_0^2 T} , \qquad 2\pi^2 v_0^2 K_0^2 T \gg 1 . \tag{9.151}$$

The approximate relation for coherence time in Eq. (9.139) corresponds to rms values of the coherence function of 0.85 and 0.92 for white-phase noise and white-frequency noise, respectively. These calculations assume that one station has a perfect frequency standard. In practice, the effective Allan variance is the sum of the Allan variances of the two oscillators:

$$\sigma_y^2 = \sigma_{y1}^2 + \sigma_{y2}^2 . \tag{9.152}$$

Thus, if two stations have similar standards, the coherence loss is doubled if the loss is small. If the short-term stability is dominated by white-phase noise, which is usually the case for hydrogen masers, the coherence function is independent of time. This means there is a maximum frequency above which a particular standard will not be usable for VLBI, regardless of the integration time. This frequency is approximately $1/(2\pi K_2)$ Hz, which for a hydrogen maser is about 1000 GHz.

In practice, the coherence $C(T)$ is measured at the peak amplitude of the correlator output, which varies as a function of fringe frequency. This operation is equivalent to removing a constant frequency drift from the phase data and can be considered as highpass filtering of the data with a cutoff frequency of $1/T$. Modeling this operation as the response of a single-pole, highpass filter, one can show that it ensures the convergence of Eq. (9.148) for all processes for which the Allan variance exponent $\mu < 1$. To compare the various representations of frequency stability, we show in Figs. 9.15 and 9.16 examples of the performance of a hydrogen maser given by the functions σ_y^2, $S_y(f)$, and $\langle C^2(T) \rangle^{1/2}$.

9.5.3 Precise Frequency Standards

Precise frequency standards of interest for VLBI include crystal oscillators and atomic frequency standards such as rubidium vapor cells, cesium-beam resonators, and hydrogen masers (Lewis 1991). Atomic frequency standards incorporate crystal oscillators that are phase-locked or frequency-locked to the atomic process, using

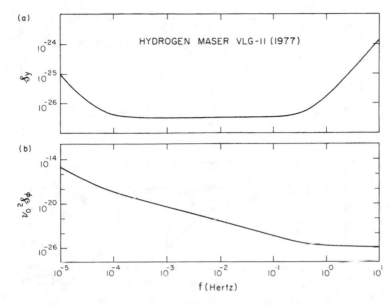

Fig. 9.15 (**a**) Power spectrum of the fractional frequency deviation $S_y(f)$ for a hydrogen maser frequency standard, and (**b**) the normalized power spectrum of the phase noise $v_0^2 S_\phi(f)$. $S_y(f)$ and $S_\phi(f)$ are related by Eq. (9.119). For frequencies above 10 Hz, $S_\phi(f)$ approaches the spectrum of the crystal oscillator to which the maser is locked, which declines as f^{-3}. Adapted from Vessot (1979).

loops with time constants in the range 0.1–1 s, so that short-term performance becomes that of the crystal oscillator. Details of how these loops are implemented are given by Vanier et al. (1979). The performance of the crystal oscillator is very important because unless it has high spectral purity, the phase-locked loops involved in generating the local oscillator signal from the frequency standard will not operate properly (Vessot 1976).

We first consider a frequency standard as a "black box" that puts out a stable sinusoid at a convenient frequency such as 5 MHz, or some higher frequency, at which the crystal oscillator is locked to the atomic process. The performance of various devices is shown in Fig. 9.17. These somewhat idealized plots show that the Allan variances of the standards have three regions: short-term noise dominated by either white-phase or white-frequency noise; flicker-frequency noise, which gives the lowest value of Allan variance and is therefore referred to as the "flicker floor"; and finally, for long periods, random-walk-of-frequency noise. Two other parameters can be specified, a drift rate and an accuracy. The drift rate is the linear change in frequency per unit time interval. Note that if the standard drives a clock, then a constant drift rate results in a clock error that accumulates as time squared. The accuracy refers to how well the standard can be set to its nominal frequency. The performance parameters are summarized in Table 9.4.

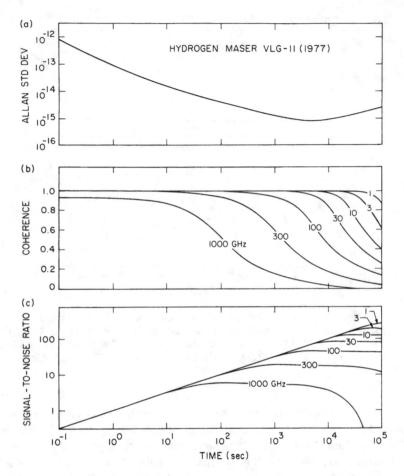

Fig. 9.16 (a) Allan standard deviation vs. sample time for a hydrogen maser frequency standard. Data from Vessot (1979). (b) Coherence $\sqrt{\langle C^2(T)\rangle}$, defined by Eq. (9.145), for various radio frequencies based on two frequency standards with Allan standard deviations given in (a). (c) Signal-to-noise ratio, normalized to unity at one second, of the measured visibility vs. integration time for various frequencies. In a VLBI system, the coherence and signal-to-noise ratios will be further reduced by atmospheric fluctuations.

Atomic frequency standards are based on the detection of an atomic or molecular resonance. There are three parts to any frequency standard [e.g., Kartashoff and Barnes (1972)]: particle preparation, particle confinement, and particle interrogation. Particle preparation involves enhancing the population difference in the desired transition. This is necessary for radio transitions in a gas with temperature T_g for which $h\nu/kT_g \ll 1$, so that the level populations are nearly equal. Preparation is usually done either by state selection in a beam passing through a magnetic or electric field, or by optical pumping. Particle confinement makes it possible to obtain narrow resonance lines from long interaction times, since according to the

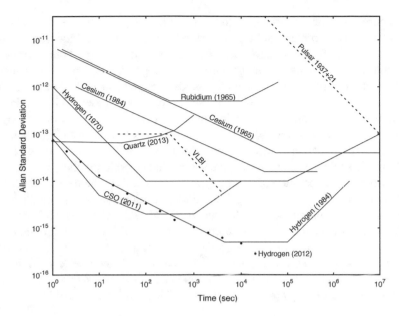

Fig. 9.17 Idealized performance of various frequency standards and other systems. Rubidium (1965) = Hewlett-Packard (HP) 5065; cesium (1965) = HP 5061-004; cesium (1984) = NBS Laboratory device no. 4; hydrogen (1970) = early Varian/HP hydrogen maser oscillator; hydrogen (1984) = hydrogen maser SAO VLG-11; quartz (2013) = crystal oscillator, Oscilloquartz 8607. Dots represent performance of the hydrogen maser oscillator by T4 Science, iMaser 3000; CSO (2011) = cryogenic sapphire oscillator stabilizer by GPS (Doeleman et al. 2011). Millisecond pulsars are very stable clocks, and the data on one of them from Davis et al. (1985) are shown. The stability of some of them, i.e., those with small amounts of "red noise," reaches 10^{-15} on a time scale of ten years (3×10^8 s) [see Verbiest et al. (2009) and Hobbs et al. (2012)]. VLBI data, which show the effect of path length stability through the atmosphere in approximately average conditions at low elevation sites, are from Rogers and Moran (1981).

Table 9.4 Typical performance[a] data on available frequency standards[b]

Type	K_2 (10^{-12}s)	K_0 (10^{-12} s$^{1/2}$)	K_{-1} (10^{-15})	K_{-2} (10^{-17} s$^{-1/2}$)	Drift rate[c] (10^{-15})	Fractional accuracy (10^{-12})
H (active)	0.1	0.03	0.4	0.1	< 1	1
Cs	–	50	100	3	1	5
Cs[d]	–	7	40	3	1	2
Rb	–	7	500	300	10^2	10^2
Crystal	1	–	500	300	10^3	–

[a]Two-point Allan standard deviation; coefficient defined by Eq. (9.135).
[b]Updated from Hellwig (1979).
[c]Fractional frequency change per day.
[d]High-performance Cs.

Heisenberg (uncertainty) principle, the line width is equal to the reciprocal of the interaction time. Particles can be confined in beams or storage cells. Storage cells either contain a buffer gas or have specially coated walls so that particle collisions do not result in phase changes. Finally, particle interrogation is the process of sensing the interaction of particles and radiation fields. Frequency standards can be either active or passive. An example of an active standard is a maser oscillator. Passive standards require an external radiation field, and transitions are observed by (1) absorption, (2) re-emission, (3) detection of particles having made the transition, or (4) indirect detection of a quantity such as a variation in the rate of optical pumping. To show how some principles are implemented in practice, we give brief descriptions of the operation of several types of standards in the next two sections.

Other types of frequency standards are under development. For a general review of types of technology, see Drullinger et al. (1996). The cryogenic sapphire oscillator has excellent short-term stability (better than that of the hydrogen maser) and may be useful for VLBI at frequencies approaching 1 THz (Doeleman et al. 2011; Rioja et al. 2012). Other laboratory devices include the laser-cooled mercury ion frequency standard (Berkeland et al. 1998) and the ultracold atomic ytterbium oscillator, whose stability approaches 10^{-18} in 7 h (Hinkley et al. 2013).

9.5.4 Rubidium and Cesium Standards

Rubidium is an alkali metal with a single valence electron and thus a hydrogenlike spectrum. The electronic ground state is split into two levels, with a transition frequency of 6835 MHz. These levels correspond to the spin of the unpaired electron being parallel or antiparallel to the nuclear spin vector. A schematic diagram of the oscillator system is shown in Fig. 9.18. An RF plasma discharge in a tube containing ^{87}Rb excites the gas to an electronic level about 0.8 μm above the ground state. The light from this discharge passes through a filter that removes the components involving the $F = 2$ level and passes the light at 0.7948 μm. This filter consists of a cell of ^{85}Rb atoms whose energy levels are slightly shifted from those of the ^{87}Rb atoms, such that both gases have transitions near 0.7800 μm. The filtered light passes through another cell of ^{87}Rb gas inside a microwave cavity resonant at the transition frequency between the $F = 2$ and $F = 1$ levels. With no RF signal applied to the cavity, the gas is nearly transparent, and the discharge beam is unattenuated as it reaches the photodetector. The application of an RF signal at 6835 MHz stimulates transitions from the $F = 2$ to $F = 1$ level. The atoms reaching the lower level are then pumped to the excited state by the light from the filtered ^{87}Rb lamp. The ^{87}Rb light therefore suffers absorption. A buffer gas, consisting of inert atoms that collide elastically with the ^{87}Rb atoms in the resonance cell, extends the interaction time to about 10^{-2} s, the mean collision time with the cell walls, and gives an absorption resonance with a line width of about 10^2 Hz. The cavity is magnetically shielded to minimize external fields. A weak homogeneous field is applied so that only $\Delta M_F = 0$ transitions, which have zero first-order Doppler shift,

RUBIDIUM VAPOR FREQUENCY STANDARD

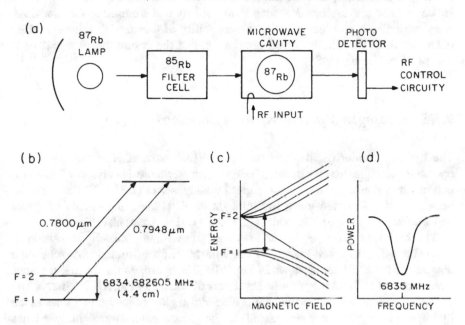

Fig. 9.18 (**a**) Schematic diagram of a rubidium gas-cell frequency standard; (**b**) pump and microwave transitions; (**c**) magnetic sublevels of microwave transition vs. magnetic field; (**d**) absorption of ^{87}Rb light vs. microwave frequency. Adapted from Vessot (1976).

are obtained. The absorption resonance has a width of 10^2–10^3 Hz. The shot noise of individual arriving photons leads to white-frequency noise.

The radio frequency signal is frequency- or phase-modulated so that the resonance line is continuously scanned. A control voltage is generated by comparing the modulation signal and the detector signal and is fed back to the slave oscillator driving the cavity to correct its frequency to the peak of the resonance.

Rubidium standards have the advantage of being small, inexpensive, and readily transportable. They are sometimes used in VLBI below 1 GHz, where the ionosphere dominates system stability. At higher frequencies, the use of rubidium standards results in degraded performance. They are useful as a backup for a primary standard and can also be used in OVLBI spacecraft to reduce the uncertainty in the timing when the radio link from the ground station is interrupted.

Cesium, like rubidium, is an alkali metal with a single valence electron. The cesium standard is important because it is used to define the standard of atomic time. The frequency of the ground-state, spin-flip transition is exactly 9192.631770 MHz, by definition of the second of atomic time. A ribbon-shaped beam of cesium gas is passed through a state-selector magnet that passes the atoms in the $F = 3$ level into a resonator. Cesium frequency standards are larger and substantially more expensive than rubidium standards. Because of their low signal-to-noise ratio, their short-term stability is poor. Thus, they are not used in VLBI for controlling local oscillators. However, they provide excellent long-term stability and are used to monitor time. They have also been used to verify the capability of transferring time via VLBI (Clark et al. 1979). The historical development of the cesium-beam resonator is described by Forman (1985).

9.5.5   *Hydrogen Maser Frequency Standard*

The hydrogen maser oscillator is the usual VLBI standard, and we discuss its operating principles in some detail.[1] The quantum mechanical analysis of the hydrogen maser is presented in a classic paper by Kleppner et al. (1962). Fundamental principles of masers are given by Shimoda et al. (1956), and details of maser construction are given by Kleppner et al. (1965) and Vessot et al. (1976).

The hydrogen maser oscillator uses the ground-state, spin-flip transition at 1420.405 MHz, the well-known 21-cm line in radio astronomy. A schematic diagram of the oscillator is shown in Fig. 9.19. The hydrogen for the maser comes from a tank of molecular hydrogen gas that is dissociated in an RF discharge. The gas in the discharge is ionized and emits the reddish glow of the Balmer lines as the hydrogen atoms recombine and cascade to the ground state. The atomic gas flows out of the dissociator through a hexapole-magnet state selector. The inhomogeneous magnetic field separates the two upper states, $F = 1, M_F = 1$ and $F = 1, M_F = 0$, from the lower states, $F = 1, M_F = -1$ and $F = 0, M_F = 0$. The beam of atoms in the two upper states is directed into the storage bulb that is located inside a microwave cavity resonant in the TE_{011} or TE_{111} mode at 1420.405 MHz. The atoms bounce around the inside of the bulb about 10^5 times before escaping through the entrance hole. The spent atoms are evacuated from the system, which operates at low pressure, by an ion pump. The cavity is surrounded by several layers of material with high magnetic permeability that shield it from ambient magnetic fields. Inside the shield is a solenoid that creates a weak homogenous field. This field allows the $(F = 1, M_F = 0)$-to-$(F = 0, M_F = 0)$ transition to radiate and minimizes transitions from the $F = 1, M_F = 1$ level. There is no first-order Zeeman effect for the $\Delta M_F = 0$ transition (see Fig. 9.19). The maser will oscillate if the cavity is tuned close to the transition frequency and the losses are small enough. In the active maser, the 1420-MHz signal is picked up by a cavity probe and used to phase-lock a crystal oscillator from which a signal at the hydrogen line frequency has been synthesized.

---

[1]This section describes the operation of the active hydrogen maser. There is another device called a passive hydrogen maser frequency standard, in which the hydrogen in the cavity does not reach self-oscillation. This type of standard has about one order of magnitude poorer performance than the active hydrogen maser.

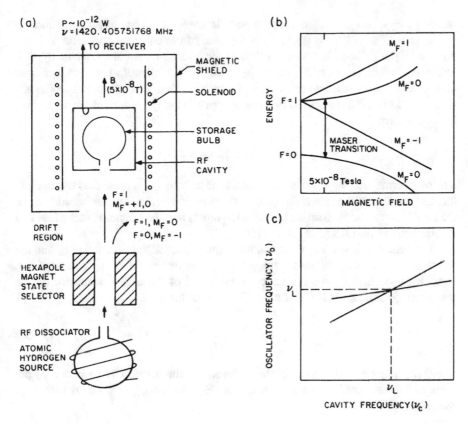

Fig. 9.19 (a) Schematic diagram of a hydrogen maser frequency standard. The line frequency shown is the rest frequency of the transition in free space from Hellwig et al. (1970). The actual frequency will differ typically by ~ 0.1 Hz because of cavity pulling, second-order Doppler, and the wall shift. (b) Energies of magnetic sublevels vs. magnetic field for the 21-cm transition. Adapted from Vessot (1976). (c) Curves of resonance frequency v_0 vs. cavity frequency v_C for two values of line width [see Eq. (9.158)]. The intersection of the curves, which can be found empirically, gives the best operating frequency.

The interaction lifetime of an atom in the bulb can be described by an exponential probability function

$$f(t) = \gamma e^{-\gamma t}, \tag{9.153}$$

where γ is the total relaxation rate. The line has an approximately Lorentzian profile with a line width (full width at half-maximum) Δv_0 of γ/π. The most important contribution to γ is the rate at which atoms escape through the entrance hole. This rate is

$$\gamma_e = \frac{v_0 A_h}{6V}, \tag{9.154}$$

where $v_0 = \sqrt{8kT_g/m}$ is the average particle speed, T_g is the gas temperature, m is the mass of a hydrogen atom, A_h is the area of the entrance hole, and V is the volume of the bulb. γ_e is about $1 \ \mathrm{s^{-1}}$. The atoms lose coherence after many wall collisions, and this leads to a loss rate $\gamma_w \simeq 10^{-4} \ \mathrm{s^{-1}}$. Collisions between hydrogen atoms cause spin-exchange relaxation at a rate γ_{se} that is proportional to the gas density and to v_0. The net relaxation rate is approximately the sum of the three most important terms:

$$\gamma = \gamma_e + \gamma_w + \gamma_{se} = \pi \Delta v_0 \ . \tag{9.155}$$

All three terms are proportional to v_0 and thus also to $\sqrt{T_g}$. Note that the random thermal motions of the atoms do not give rise to a first-order Doppler broadening of the line, because the interaction between the atoms and the RF field takes place in a resonant cavity [see Kleppner et al. (1962)].

The maser oscillator has two resonant frequencies, the line frequency v_L and the electromagnetic cavity resonance frequency v_C, defined by the cavity's dimensions. In classical oscillators, the frequency is the mean of these two, weighted by the respective Q factors, Q_L for the line and Q_C for the cavity:

$$v_0 = \frac{v_L Q_L + v_C Q_C}{Q_L + Q_C} \ . \tag{9.156}$$

The Q factor is defined as π times the reciprocal of the fractional loss in energy per cycle of the resonant frequency. Hence, from Eq. (9.153), Q_L is given by [see, e.g., Siegman (1971)]

$$Q_L \simeq \frac{\pi v_0}{\gamma} = \frac{v_0}{\Delta v_0} \ . \tag{9.157}$$

A typical value of Q_L is about 10^9. The practical value of Q_C for a silver-plated cavity is about 5×10^4. Since $Q_L \gg Q_C$, the resonance frequency is approximately

$$v_0 \simeq v_L + \frac{Q_C}{Q_L}(v_C - v_L) \ . \tag{9.158}$$

Equation (9.158) describes the effect of "cavity pulling" on the resonance frequency. Temperature changes cause the size, and thus the resonant frequency, of the cavity to change. Hence, a fractional frequency stability of 10^{-15} for the maser requires a fractional mechanical stability of about 5×10^{-10} for the cavity. The cavity dimensions therefore must be stable to about 10^{-8} cm. The cavity must be made from material with a small thermal expansion coefficient or the temperature must be carefully controlled. Extreme mechanical stability is also required so that atmospheric pressure changes do not affect the frequency. The TE_{011} cavity is a cylinder about 27 cm in length and diameter, appreciably larger than the free-space wavelength because of the loading by the storage bulb. Coarse tuning is accomplished by moving the end plate of the cavity and fine tuning by a varactor diode. From Eq. (9.158), it is clear that the maser frequency is most stable when v_C is set to v_L so that v_0 equals v_L regardless of the values of Q_C and Q_L. This optimal tuning point of the maser can be found by making a plot of v_0 vs. v_C, which is a straight line with slope Q_C/Q_L, according to Eq. (9.158). By varying Q_L (for example, by varying the gas pressure and thereby changing γ), a family of straight lines can be generated that intersect at the desired frequency $v_0 = v_L = v_C$ (see Fig. 9.19c). Servomechanisms are used in some systems to keep the maser cavity continuously tuned.

The performance of hydrogen masers is shown in Figs. 9.16 and 9.17. For periods less than 10^3 s, the performance is limited by two fundamental processes: white-frequency noise due to thermal noise generated inside the cavity and white-phase noise due to thermal noise in the external amplifier. The thermal noise generated inside the cavity produces a fractional frequency variance (Allan variance) of

$$\sigma_{yf}^2 = \frac{1}{Q_L^2} \frac{kT_g}{P_0 \tau}, \qquad (9.159)$$

where P_0 is the power delivered by the atoms (Edson 1960; Kleppner et al. 1962). There is also shot noise in the cavity due to the discrete radiation of photons. However, this process, described by the Allan variance σ_{ys}^2, is smaller than σ_{yf}^2 by the ratio hv/kT_g, which is 2×10^{-4} at room temperature. Spontaneous emission also contributes a small amount of noise, equivalent to increasing T_g by $hv/k \simeq 0.07$ K. Finally, the maser receiver adds a noise power, $kT_R \Delta v$, to the signal coupled out of the cavity, where T_R is the receiver noise temperature and Δv is the receiver bandwidth. This noise causes an Allan variance of (Cutler and Searle 1966)

$$\sigma_{yR}^2 = \frac{1}{(2\pi v_0 \tau)^2} \frac{kT_R \Delta v}{P_0}. \qquad (9.160)$$

These two processes are independent, so the net Allan variance is $\sigma_y^2 = \sigma_{yf}^2 + \sigma_{yR}^2$. The effects of both processes are clearly evident in the data in Fig. 9.17. Note that a flicker floor is not reached because of long-term drifts. The short-term performance can be improved by increasing the atomic flux level, which increases P_0. However, increasing the flux increases the spin-exchange rate, which decreases Q_L, thereby making the oscillator more susceptible to the long-term effects of cavity pulling.

The frequency of a maser is not exactly equal to the atomic transition frequency because of several effects. These effects limit the accuracy to which the frequency can be set, and because most of them are temperature dependent, they probably contribute to flicker-frequency and random-walk-of-frequency noise. Cavity pulling, which has been described already, is an important effect, and to minimize it, the cavity must be tuned carefully. The collision-induced spin-exchange process gives a frequency shift that varies with Q_L in the same way as the cavity pulling. Thus, the cavity-tuning procedure also eliminates this shift. Collisions with the cavity walls produce an effect called the "wall shift," which is difficult to predict and may be the ultimate limiting factor in the absolute precision of the maser frequency (Vessot and Levine 1970). This shift depends on the temperature and wall coating material. Its fractional value is about 10^{-11}. The first-order Doppler effect cancels, but the second-order Doppler effect does not, because of its v^2/c^2 dependence [see Kleppner et al. (1962)]. The fractional frequency shift is about equal to $-1.4 \times 10^{-13} T_g$. Finally, there is no first-order Zeeman effect in the $(F = 1, M_F = 0)$-to-$(F = 0, M_F = 0)$ transition. However, the second-order Zeeman fractional-frequency shift is $2.0 \times 10^2 B^2$, where B is the magnetic field in tesla.

9.5.6 Local Oscillator Stability

Local oscillator signals are generated by multiplying a signal from the locked oscillator of the frequency standard. The multipliers must have exceptional stability, as discussed in Sect. 7.2, to avoid the introduction of additional noise and drift. Imperfect multipliers are sensitive to vibration and temperature and may have modulation at harmonics of the power line frequency. In an ideal multiplier, a signal of the form of Eq. (9.112) is converted to

$$V(t) = \cos[2\pi M\nu_0 t + M\phi(t)] \,, \qquad (9.161)$$

where M is the multiplication factor, ν_0 is the fundamental frequency, and ϕ is the random phase noise of the frequency standard. If the phase noise is small, $M\phi(t) \ll 1$, then the single-sided power spectrum of $V(t)$ is given by

$$S_v(\nu) = \delta(\nu - M\nu_0) + M^2 S_\phi(\nu - M\nu_0) \,, \qquad (9.162)$$

where δ is a delta function representing the desired signal, and S_ϕ is the power spectrum of the phase noise. Thus, the noise power increases as the square of the multiplication factor. In the general case, S_v can be written (Lindsey and Chie 1978)

$$S_v(\nu) = \delta(\nu - M\nu_0) + \sum_{n=1}^{\infty} \frac{M^{2n}}{n!} \left[S_\phi(\nu - M\nu_0) * S_\phi(\nu - M\nu_0) * \cdots \right], \qquad (9.163)$$

where the term in brackets contains n replications of the same function convolved together. When only the leading term in the summation is retained, Eq. (9.163) reduces to Eq. (9.162). The higher-order terms in Eq. (9.163) represent a series of approximately Gaussian components because of the repeated convolutions. The rms phase deviation of the multiplier output frequency, $M\nu_0$, is proportional to the rms voltage of the noise in the output bandwidth, that is, to the square root of the noise power. Thus, for the case represented by Eq. (9.162), the rms phase fluctuation is proportional to M.

9.5.7 Phase Calibration System

One way to check the integrity of an entire VLBI system is to inject into the front end of the receiver an RF signal that is independently derived from the frequency standard. The RF test signal can be derived by driving a step-recovery diode with, say, a 1-MHz signal from the frequency standard so as to generate a pulse train with 1-μs period. Such a signal has harmonics at 1-MHz intervals throughout the microwave region, all of which have the same phase at the reference intervals. When the RF band is mixed down to baseband, one of the injected harmonics can be made to appear at a convenient frequency of order 10 kHz. This is then compared with a reference signal from the frequency standard. The phase calibration signal can be continuously injected during VLBI recording since a low enough level can be used that it can be detected only by very narrowband filtering in the processor ($\sim$ 10-Hz bandwidth). The calibration allows one to compensate for variations such as those caused by thermal effects in cables (Whitney et al. 1976; Thompson and Bagri 1991; Thompson 1995). Similar methods are used in some connected-element interferometers.

9.5.8 Time Synchronization

The clocks at VLBI stations must be synchronized accurately enough to avoid time-consuming searches for interference fringes. Until around 1980, Loran C was widely used to monitor time at VLBI stations. Loran, an acronym for *Long Range Navigation*, is a system originally developed during World War II for ocean navigation (Pierce et al. 1948). The transmission frequency is 100 kHz. The relative time of arrival of signals from three stations defines the observer's location on the Earth's surface. For a detailed discussion of Loran C, see Frank (1983). Accuracies from a few hundred nanoseconds to a few tens of microseconds are possible, depending on the accuracy of the estimate of propagation time.

The Global Positioning System (GPS) provides higher accuracy than Loran and has been used in almost all VLBI systems since the early 1980s. In the GPS system, the user receives signals at 1.23 or 1.57 GHz from a number of satellites whose positions are known and whose clocks are synchronized to Coordinated Universal Time (UTC; see Sect. 12.3). If timing measurements from four satellites are made, and corrected for propagation effects in the atmosphere, users can determine their positions in three coordinates and their clock errors. The accuracies available to civilian users have improved over about a decade from 100 ns in time (Parkinson and Gilbert 1983; Lewandowski et al. 1999) to $\sim$ 1 ns (Rose et al. 2014), and further improvement down to 100 ps is expected (Ray and Senior 2005). An analysis of the time-transfer problem, including relativistic effects, is given by Ashby and Allan (1979). For general information on GPS usage, see, for example, Leick (1995).

For time scales of a year, the accuracy of timing from pulsar observations approaches 1 part in 10^{14} (Davis et al. 1985). Ultimately, the best time transfer may be obtainable from the processed VLBI data (Counselman et al. 1977; Clark et al. 1979).

9.6 Data Storage Systems

The basic consideration for any storage system is the representation of the signal and the method of incorporating the time information. Recording can be either analog or digital, and various data storage technologies are available. Here, we discuss only digital recording since the technologies involved are well suited to VLBI and are widely used.

A basic parameter of a recording system is its data rate, v_b (bits s^{-1}). This parameter limits the number of bits that can be recorded in a given time and, thus, also the sensitivity of continuum observations in which the potential IF bandwidth is larger than $v_b/2N_b$, where N_b is the number of bits per sample. The signal is represented by samples having Q quantization levels taken at β times the Nyquist rate. For N samples, there are Q^N possible data configurations, which require a minimum of $N \log_2 Q$ bits. Therefore, as noted in Sect. 8.4.3, the maximum RF bandwidth is

$$\Delta v = \frac{v_b}{2\beta N_b} = \frac{v_b}{2\beta \log_2 Q} \, . \tag{9.164}$$

The signal-to-noise ratio obtained in time τ is proportional to $\eta_Q \sqrt{\Delta v \tau}$, where η_Q is the quantization efficiency (see Table 8.3). From Eq. (9.164),

$$\eta_Q \sqrt{\Delta v \tau} = \eta_Q \sqrt{\frac{v_b \tau}{2\beta N_b}} \, . \tag{9.165}$$

If τ is the recording time, $v_b \tau$ is equal to the number of recorded bits. The quantity $\eta_Q/\sqrt{\beta N_b}$ thus provides an indication of the performance per bit, which it is desirable to maximize. For two- and four-level sampling, the obvious encoding schemes are one bit and two bits per sample, respectively. For three-level sampling, a problem arises since encoding one sample (one of three possible states) in two data bits (representing four possible states) is inefficient. Putting three samples into five bits or five samples into eight bits gives data rates of 1.67 and 1.60 bits per sample, respectively, compared with the theoretical optimum value of $\log_2 3 = 1.585$. The values of $\eta_Q/\sqrt{\beta N_b}$ for various values of Q and β, and several encoding schemes, are listed in Table 9.5. The highest signal-to-noise ratio is achieved with three-level sampling at the Nyquist rate, although two- and four-level sampling give almost the same performance.

Table 9.5 Performance of various signal representations as a function of number of quantization levels, sampling rate, and encoding format[a]

Signal representation		η_Q	N_b	$\dfrac{\eta_Q}{\sqrt{\beta N_b}}$
Sampling at Nyquist rate ($\beta = 1$)				
Two-level		0.637	1.0	0.637
Three-level	"Ideal" encoding[b]	0.810	1.585	0.643
	5 samples/8 bit	0.810	1.60	0.640
	3 samples/5 bit	0.810	1.667	0.627
	1 sample/2 bit	0.810	2.0	0.573
Four-level	All products	0.881	2.0	0.623
	Low-level omitted	0.87	2.0	0.61
Sampling at 2 × Nyquist rate ($\beta = 2$)				
Two-level		0.74	1.0	0.52
Three-level	"Ideal" encoding[b]	0.89	1.585	0.50
	5 samples/8 bit	0.89	1.60	0.50
	3 samples/5 bit	0.89	1.667	0.49
	1 sample/2 bit	0.89	2.0	0.45
Four-level	All products	0.94	2.0	0.47

[a] η_Q = quantization efficiency; N_b = number of bits per sample; β = oversampling factor.
[b] N samples encoded in $N \log_2 3$ bits.

In addition to the encoding schemes discussed above, in which the number of bits required for a given number of samples is constant, one can also envisage a scheme in which the number of bits depends on the sample values, that is, a variable-length code. For example, D'Addario (1984) has suggested encoding the +1, 0, and −1 values in three-level quantization as the binary numbers 11, 0, and 10, respectively. It is possible to decode such a data string uniquely, since all one-bit representations begin with 0 and all two-bit representations with 1. The average number of bits per sample depends on the amplitude probability distribution of the signal waveform and the threshold level settings. For a given number of bits, the threshold settings that maximize the signal-to-noise ratio are generally not the same as those derived in Sect. 8.3, which are optimum for a given number of bits per sample. With D'Addario's encoding scheme, the best performance is achieved with the threshold set such that $\eta_Q = 0.769$ and $N_b = 1.370$ bits per sample, giving a performance factor $\eta_Q/\sqrt{\beta N_b}$ equal to 0.657. Thus, an increase in sensitivity of about 3% compared with the use of the scheme with 1.6 bits per sample could be achieved. However, the effects of bit errors or interfering signals that change the amplitude distribution could be more serious. Finally, the data could be encoded statistically in large blocks that would allow a theoretically optimal value of N_b of 1.317 bits per sample, which, with η_Q of 0.769, would give a performance factor of 0.670 (D'Addario 1984).

In practice, the desirability of a simple encoding scheme and other design considerations have usually resulted in the choice of two-level quantization. All five

VLBI systems developed in the United States during the period 1968–1997 (Mark I, Mark II, Mark III, VLBA, and Mark IV) use two-level sampling, but for the last two of these, four-level sampling is also an option. For spectral line observations, where the bandwidth of the signal is small with respect to the bandwidth of the recording system, multilevel sampling is advantageous. Note that multilevel sampling is a more effective way of using recording capacity than sampling faster than the Nyquist rate (Table 9.5).

Each data sample must have either an implicit or explicit time tag. Although an error rate of 10^{-3} in decoding the data bits is acceptable, a one-bit shift in the time axis can be a serious defect and is not acceptable. In virtually all recording systems, the data are blocked into records. Each new record begins at a precise time so that the temporal registration of the data stream can be recovered if it is lost during the previous record. These record lengths are: Mark I, 0.2 s (144,000 bits); Mark II, 16.7 ms (66,600 bits); and Mark III, 5 ms (20,000 bits). In the Mark I system, which used standard computer tape format, the accuracy of recording was very high, and the time of any bit was obtained by counting bits from the beginning of the record and counting records from the beginning of the tape. In the Mark II system, which uses video cassette recorders (VCRs), the data are recorded with a self-clocking code, while in the Mark III system, which uses instrumentation recorders, the data transitions themselves serve as the clock. The characteristics of several systems are given in Fig. 9.20 and Table 9.6. In all of these, the recording is in digital form, except for the Canadian system used during 1971–83. Wietfeldt and D'Addario (1991) discuss the compatibility of some of these systems.

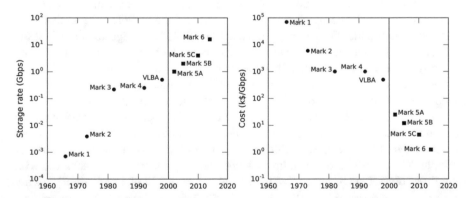

Fig. 9.20 Trends in VLBI recording system data rates (circles) and storage costs (squares). (left) Data rate in Gbits s^{-1} (Gbps) vs. time for various systems. (right) Cost of data storage system in K\$/Gbps. Note that before 2000, data storage was on magnetic tape and after 2000 on disk. From Whitney et al. (2013), courtesy of and © the Astronomical Society of the Pacific.

Table 9.6 Characteristics of some VLBI data storage systems

System	Year of inception	Storage medium[a]	Basic description	Storage unit[b]	Analog channels (no.×MHz)	Quantization (bits/sample)	Sample rate (Mbits/s)	Storage time[c] (min)	References
Canadian	1967	T	Analog recording on TV recorder	Ampex VR660C/ IVC 830	1 × 4	–	8	64	Broten et al. (1967), Moran (1976)
NRAO Mark I[d]	1967	T	IBM computer-compatible format	Ampex TM-12	1 × 0.36	1	0.72	3	Bare et al. (1967)
NRAO Mark II(A)	1971	T	Digital recording on TV recorder	Ampex VR660C	1 × 2	1	4	190	Clark (1973)
NRAO Mark II(B)	1976	T	Digital recording on TV recorder	IVC 800	1 × 2	1	4	64	
NRAO Mark II(C)	1979	T	Digital recording on TV recorder	RCA VCT 500	1 × 2	1	4	246	
MIT/NASA Mark III	1977	T	Instrumentation recorder	Honeywell 96	28 × 2	1	112	13	Rogers et al. (1983)
MIT/NASA Mark III(A)	1984	T	Instrumentation recorder	Honeywell 96[e]	28 × 2	1 or 2	112	164	Clark et al. (1985)
NRAO VLBA[f]	1990	T	Instrumentation recorder	Honeywell 96[e]	8 × 8	1 or 2	128	720	Hinteregger et al. (1991), Rogers (1995)
NICT K-4 (Japan)	1990	T	Video cassette recorder	Sony DIR-1000	1 × 128	1	256	63	Kawaguchi (1991)
S2 (Canada)	1992	T	8 Video cassette recorders (VHS)	Panasonic 2530	8 × 8	1 or 2	128	256	Wietfeldt et al. (1996), Cannon et al. (1997)

(continued)

Table 9.6 Characteristics of some VLBI data storage systems (continued)

System	Year of inception	Storage medium[a]	Basic description	Storage unit[b]	Analog channels (no.×MHz)	Quantization (bits/sample)	Sample rate (Mbits/s)	Storage time[c] (min)	References
MIT/NASA Mark IV	1997	T	Instrumentation recorder	Honeywell 96[e]	16 × 8	1 or 2	512	90	Whitney (1993), Rogers (1995)
MIT/NASA Mark V(A)	2002	D	Disk replacement for Mark III and IV	COTS[g]	16 × 8	1 or 2	512		Whitney (2002)
MIT/NASA Mark V(B)	2005	D	VSI-H interface	COTS[g]	flexible	1 or 2	2048	400	Whitney (2004)
MIT/NASA Mark V(C)	2010	D	10 Gig E interface	COTS[g]	flexible	1 or 2	4094		Whitney (2008)
MIT/NASA Mark VI	2012	D	10 Gig E interface	COTS	8 × 512[h]	1 or 2	16384	1800[i]	Whitney et al. (2013)
NICT K-6 (Japan)[j]	2014	D	10 Gig E interface	COTS	4 × 1000	1 or 2	16384		Sekido (2015)

[a]T = magnetic tape, D = disk.
[b]COTS = commercial off-the-shelf equipment.
[c]Without mechanical intervention.
[d]A similar system was developed in the former Soviet Union (Kogan and Chesalin 1981).
[e]Track width reduced to 40 μm.
[f]VLBA transitioned to Mark V(A) in 2007.
[g]Plus custom interface boards.
[h]Additional digital backend with 2 × 2048 MHz channels (Vertatschitsch et al. 2015).
[i]With rack configuration of four packs of eight disks, each with a capacity of 7 Tbyte/disk (1.8 × 10^{15} bits of storage).
[j]For parameters of K-3 and K-5 systems, see Koyama (2013).

9.7 Processing Systems and Algorithms

A VLBI processor has two main functions: (1) reproduction of smooth data streams and (2) cross-correlation analysis of the data streams. Before 2000, VLBI data were stored on magnetic tape. During that period, the data stream from a tape recorder could have time-base irregularities of up to 100 μs, caused by jitter in the mechanical playback system, and could be subject to dropouts because of tape imperfections. The processor had to derive the true time base either from the encoded clock transitions, in the case of a self-clocking code, or from the data transitions themselves, when a bit synchronizer was used. There had to be enough buffer storage to handle at least the mechanical jitter. The geometric delay was corrected with minimal buffer space by shifting the playback time, thereby retaining the data on the tape until they were needed by the correlator. If the data were read in synchronism from the tapes, a buffer memory of sample capacity about 5×10^4 times the clock rate in megahertz would be needed for geometric delay compensation. Even today, with disk storage or transmission of data over fiber optic networks, some buffer storage is required.

The major differences between the design of the correlation part of the processor for VLBI and for a conventional interferometer are related to the fact that in VLBI, fringe rotation and delay compensation are usually performed on the quantized and sampled signal. This leads to special problems, which we discuss here. Digitization of the signals introduces several signal-to-noise loss factors: η_Q, the loss factor associated with amplitude quantization of the recorded signals, discussed in Sect. 8.3; η_R, the loss factor incurred by quantizing the phase of the fringe rotation waveform; η_S, the loss factor incurred by inadequate sideband rejection as a result of the limited number of delays in the correlator; and η_D, the loss caused by compensating the geometric delay in discrete steps.

Fringe rotation and delay compensation can be done on the analog signals at the telescope before recording. For example, the fringe rotation can be done at the telescopes by offsetting the local oscillators, as described in Sect. 6.1.6 for a connected-element array. The advantage of this arrangement is that only a real correlation function (with both positive and negative delays) needs to be calculated (see Sects. 8.8 and 9.1). Hence, only half the correlator circuits are required. Also, the sensitivity loss from a digital fringe rotator is not incurred. A disadvantage is that the output of the correlator must be averaged over a short enough interval to accommodate the residual fringe frequency of a source anywhere in the primary beams of the antennas. The maximum residual fringe frequency of a source at the half-power point of the primary beam is $\Delta \nu_f \simeq D\omega_e/d$ [see Eq. (9.11)], where D is the baseline length, d is the antenna diameter, and ω_e is the angular velocity of the Earth in radians per second. Hence, the averaging time of the correlator output must be less than $1/(2\Delta \nu_f)$; for example, it should not exceed 30 ms for a baseline equal to the Earth's diameter and $d = 25$ m. The correlation functions can be averaged further after they have been passed through a fringe rotator, which removes the residual fringe frequency. Also, the unit at the telescope that continually changes the local oscillator frequency must be carefully designed so that full phase accountability is provided for astrometric work. Further information on VLBI systems and processing algorithms can be found in Thomas (1981); Herring (1983), and Deller et al. (2007, 2011).

9.7.1   Fringe Rotation Loss (η_R)

Fringe rotation is used to reduce to near zero the frequency of the fringe component of the correlated signals (see Sect. 6.1.6). Here we consider the fringe frequency to include the effect of offsets in the frequency standards. Fringe rotation in the processor can be implemented in a number of ways, as shown in Fig. 9.21. If the fringe rotator is placed after the correlator (Fig. 9.21a), then the correlation function from the correlator must be averaged over an interval short with respect to the fringe period. If the local oscillators at the antennas are offset to slow the fringes, so that only a little further adjustment is required after the correlator, then this scheme is convenient. Otherwise, the short averaging time required and the resulting high data rate from the correlator make this arrangement unattractive. Alternately, before correlation, one of the data streams can be passed through a digital single-sideband mixer that shifts the Fourier components of the signal by the appropriate fringe frequency, as shown in Fig. 9.21b. The 90° phase shift in this mixer is difficult to implement without introducing spectral distortion, so this type of fringe rotator is rarely used (see also Sect. 8.7). The fringe rotation scheme shown in Fig. 9.21c is commonly used, but application of fringe rotation to the quantized signal introduces two complications. First, the fringe function with which the signal is multiplied must be coarsely quantized so as not to increase the number of bits per sample going to the correlator: this also applies to scheme (b). Second, the multiplication introduces an unwanted noise sideband, which is described below in Sect. 9.7.2. We now consider the first of these effects.

The data stream is multiplied by a complex function $\mathcal{F}$ whose real and imaginary parts, $\mathcal{F}_R$ and $\mathcal{F}_I$, approximate $\cos\phi$ and $\sin\phi$, where ϕ is the desired phase function. In the simplest approximation, these functions are square waves with the appropriate frequency and phases. Thus, as shown in Fig. 9.22, the quantized signal is multiplied by a fringe rotation function whose amplitude is constant but whose phase steps by 90° every quarter cycle instead of smoothly progressing. The resulting visibility function then has a phase component with a 90° sawtooth modulation at the fringe frequency. This resembles phase noise in which the phase is uniformly distributed between ±45°. Therefore, the average signal amplitude is degraded by $\sin(\pi/4)/(\pi/4) = 0.900$. Another approach to calculating the loss in signal-to-noise ratio is to calculate the harmonics in the fringe rotation function. The first harmonic of $\mathcal{F}_R$ or $\mathcal{F}_I$ has an amplitude of $4/\pi = 1.273$. Only the signal mixed with the first harmonic appears in the processor output, since the other harmonics are removed by time averaging. Thus, part of the signal is scattered out of the fringe passband. The fraction retained is the square root of the ratio of the power in the first harmonic to the total power of the fringe rotation function, which is $\sqrt{8}/\pi = 0.900$. This represents the loss in signal-to-noise ratio. There is also a scale-factor change

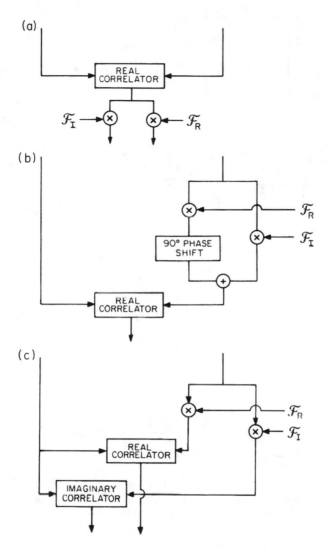

Fig. 9.21 Various processor configurations showing possible locations of fringe rotator. $\mathcal{F}_R$ and $\mathcal{F}_I$ are cosine and sine representations of the fringe function. See text for discussion of relative merits.

since the fringe amplitudes are increased by the action of the fringe rotator. Thus, the fringe amplitudes must be divided by $4/\pi$, the relative amplitude of the first harmonic of $\mathcal{F}_R$.

A better fringe rotation function is the three-level approximation of a sine wave (Clark et al. 1972) shown in Fig. 9.22b. When the fringe rotation function is zero, the correlator is inhibited. Since the real and imaginary parts of $\mathcal{F}$ are never zero simultaneously, all data bits are used at least once. This fringe rotation function can

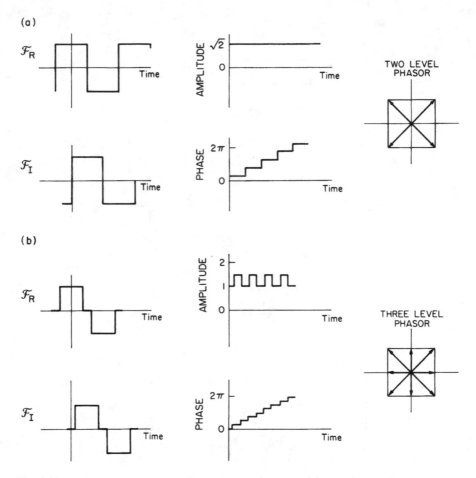

Fig. 9.22 (a) Mathematical model of two-level fringe rotator showing $\mathcal{F}_R$ and $\mathcal{F}_I$, functions that approximate $\cos\phi$ and $\sin\phi$ (left); the amplitude and phase representation of $\mathcal{F}$ (center); and the phasor plot of $\mathcal{F}$ (right). (b) Same plots for a three-level fringe rotator.

be thought of as a phasor whose tip traces out a square such that it has phase jumps in $45°$ increments and its amplitude alternates between $\sqrt{2}$ and 1. The resulting jitter in phase is uniformly distributed between $\pm 22.5°$ and results in a loss of signal amplitude of $\sin(\pi/8)/(\pi/8) = 0.974$. Also, the variation in the amplitude of the phasor introduces a nonuniform weighting of the signal samples. This reduces the signal-to-noise ratio by a further factor equal to $(1 + \sqrt{2})/\sqrt{6} = 0.986$. The net loss in signal-to-noise ratio is 0.960. The reduction in signal-to-noise ratio is also equal to the square root of the ratio of the power in the first harmonic to the total power in $\mathcal{F}_R$. The first harmonic of $\mathcal{F}_R$ is $(4/\pi)\cos(\pi/8) = 1.18$, which is the scale factor correction for the visibility. The three-level fringe function considered here is used in many VLBI processors. The fringe period is divided into 16 parts to generate $\mathcal{F}$. The transitions in $\mathcal{F}$, which then occur at integral multiples of 1/16 of the fringe period, are not optimally located, but this approximation results in no more than 0.1% additional loss. Note that an FX correlator can be made to accept input data with more than one or two bits per sample rather more easily than a lag correlator. With more data bits per signal sample, more accurate representations of sine and cosine functions can be used.

9.7.2 Fringe Sideband Rejection Loss (η_S)

The digital fringe rotator shown in Fig. 9.21c is not a single-sideband mixer. Thus, as well as the wanted output, shifted in frequency by the fringe frequency, an unwanted component of noise corresponding to the image response of a mixer also appears. To understand the effect of this noise, consider the cross power spectrum of the correlator output. Recall that v' is the intermediate frequency defined following Eq. (9.18), and note that in the output of a spectral correlator, $v' > 0$ and $v' < 0$ refer to the upper and lower sidebands, respectively. For upper-sideband operation, the cross power spectrum of the signal is given by Eq. (9.26), which is nonzero only for the upper sideband. However, there will be noise at both positive and negative frequencies. Thus, the cross power spectrum of the correlator output is

$$S'_{12}(v') = \begin{cases} S(v')e^{j\Phi(v')} + n_u(v'), & v' > 0, \\ n_\ell(v'), & v' < 0, \end{cases} \quad (9.166)$$

where $S(v')$ is the instrumental response defined in Eq. (9.19), $j\Phi(v')$ is the exponent in Eq. (9.26), and n_u and n_ℓ are the noise spectra for the upper- and lower-sideband responses. For observations in which a spectral line correlator is used, $S'_{12}(v')$ is computed and the noise at $v' < 0$ is simply ignored. For continuum observations using a correlator with only a small number of channels (lags), the noise at $v' < 0$ contributes excess noise in the correlation function and must be removed. A straightforward way to remove the noise at $v' < 0$ is to compute $S'_{12}(v')$ and multiply it by the filtering function

$$H_F(v') = \begin{cases} 1, & 0 < v' < \Delta v \\ 0, & \text{elsewhere}. \end{cases} \quad (9.167)$$

The resulting function, $S'_{12}(v')H_F(v')$, can be Fourier transformed back into a correlation function. Alternately, the filtering can be applied by convolving the correlation function at the output of the correlator with the Fourier transform of $H_F(v')$, which is

$$h_F(\tau) = \Delta v e^{j\pi\Delta v\tau}\left(\frac{\sin \pi\Delta v\tau}{\pi\Delta v\tau}\right), \quad (9.168)$$

or

$$h_F(\tau) = F_1(\tau) + jF_2(\tau) ,\tag{9.169}$$

where F_1 and F_2 are as defined in Eq. (9.23). The convolution leaves the desired signal unchanged but removes the negative (lower) sideband noise. Thus, the resulting correlation function still has the form of Eq. (9.25), plus the positive (upper) sideband noise that cannot be removed.

The role of $h_F(\tau)$ can be understood in a different way. The correlation function at the output of the correlator is computed at discrete delays at intervals of $(2\Delta \nu)^{-1}$. Therefore, the correlation function in Eq. (9.25) has a full width at half-maximum of about three delay steps. In order to estimate the amplitude and phase of the correlation function, one would like to do more than just take these values from the peak of $\rho'_{12}(\tau)$. Rather, one would like to use all the information provided by the correlation function at various delays. $h_F(\tau)$ is the appropriate interpolation function that properly weights the correlation function, gathering up the power at different delays to provide an optimal estimate of the fringe amplitude, phase, and delay. Note that $h_F(\tau)$ and $\rho'_{12}(\tau)$ are identical forms except for the unknown amplitude, phase, and delay. These unknown quantities can be estimated by the usual procedure of matched filtering or, equivalently, least-mean-squares analysis in which the correlation function is convolved with $h_F(\tau)$. However, $\rho'_{12}(\tau)$ is measured only over a finite number of delay steps, and some information is lost, so the signal-to-noise ratio is reduced. Assume that the system lowpass response is rectangular and the delay errors $\Delta \tau_g$ and τ_e are zero, so that the correlation function is centered in the delay range of the correlator. Let M be the number of delay steps (lags) in the correlator. The loss factor η_S is the signal-to-noise ratio when M values of the correlation function are available, divided by the signal-to noise ratio when the entire function is available:

$$\eta_S = \sqrt{\frac{\displaystyle\sum_{k=-M'}^{M'} |h_F(\tau_k)|^2}{\displaystyle\sum_{k=-\infty}^{\infty} |h_F(\tau_k)|^2}} ,\tag{9.170}$$

where $\tau_k = k/2\Delta \nu$, $M' = (M-1)/2$, and M is an odd integer. The denominator in Eq. (9.170) equals $2\Delta \nu^2$, so

$$\eta_S = \sqrt{\frac{1}{2} + \sum_{k=1}^{M'} \left[\frac{\sin(\frac{\pi k}{2})}{\frac{\pi k}{2}} \right]^2} .\tag{9.171}$$

For $M = 1$, $\eta_S = 1/\sqrt{2}$, which corresponds to the case of no image rejection. M must be at least 3 to ensure that the peak of the correlation function can be determined; $M \simeq 7$, for which $\eta_S = 0.975$, is adequate for most purposes. For large M, η_S approaches unity [see Eq. (A8.5)]. Note that because we assumed the correlation function was exactly centered, its value will be zero at delay steps 2, 4, 6, 8, ... , and so on. This suggests, for example, that a nine-delay correlator ($M' = 4$) is no better than a seven-delay correlator ($M' = 3$). In practice, the nine-delay correlator is better because the correlation function is rarely aligned perfectly in the correlator. In general, η_S is slightly smaller than given in Eq. (9.171) if the correlation function is not perfectly aligned (Herring 1983).

9.7.3 Discrete Delay Step Loss (η_D)

The delay introduced to align the bit streams is quantized at the sampling rate, which we assume to be the Nyquist rate. Thus, there is a periodic sawtooth delay error with a peak-to-peak amplitude equal to the sampling period. This effect is also known as the *fractional bit-shift error*. The delay error gives rise to a periodic phase shift that is a function of the baseband frequency, as shown in Fig. 9.23. The phase error has a peak-to-peak value of

$$\phi_{pp} = \frac{\pi \nu'}{\Delta \nu} , \tag{9.172}$$

and the sawtooth frequency is proportional to the fringe frequency and has a maximum value of

$$\nu_{ds(max)} = \frac{2\Delta\nu D\omega_e}{c} \text{ (delay steps per second) ,} \tag{9.173}$$

where D is the baseline length and ω_e is the angular velocity of the Earth's rotation in radians per second. If nothing is done to correct for this effect and the fringe amplitude is averaged over many times $1/\nu_{ds}$, then the phase at any frequency ν' is uniformly distributed over ϕ_{pp}. The amplitude loss as a function of baseband frequency is

$$L(\nu') = \frac{\int_0^{\phi_{pp}/2} \cos(\phi_{pp}/2)d\phi}{\int_0^{\phi_{pp}/2} d\phi} = \frac{\sin(\phi_{pp}/2)}{\phi_{pp}/2} , \tag{9.174}$$

and the net signal-to-noise reduction over a baseband response of width $\Delta\nu$ is, using Eqs. (9.172) and (9.174),

$$\eta_D = \frac{1}{\Delta\nu} \int_0^{\Delta\nu} \frac{\sin(\pi\nu'/2\Delta\nu)}{\pi\nu'/2\Delta\nu} d\nu' = 0.873 . \tag{9.175}$$

Unless the fringe amplitude averaging is done over an integral number of fringe periods, there is also a residual phase error, the amplitude of which decreases with

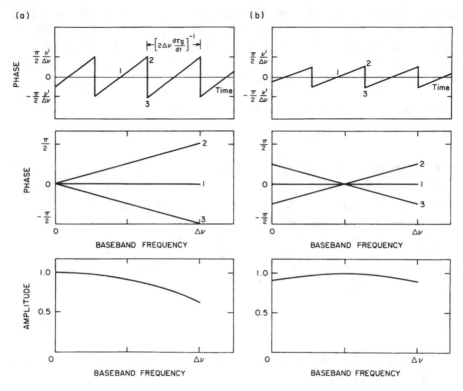

Fig. 9.23 Discrete delay step effect. Case (**a**) applies when the fringe rotator corrects the phase for zero baseband frequency, and case (**b**) applies when the fringe rotator also inserts a $\pi/2$ phase shift when the delay changes by one Nyquist sample. The top plots show the phase vs. time at baseband frequency ν'. The middle plots show the phase across the baseband at three different times denoted by 1, 2, and 3. The bottom plots show the average amplitude across the baseband.

the number of periods. When the fringe frequency is near zero, this phase error can be significant.

The effect of the discrete delay step can be compensated for, and no sensitivity loss need occur. The delay error caused by delay quantization is a known quantity that introduces a phase slope in the cross power spectrum. Therefore, if the cross power spectra are calculated on a period short with respect to $1/\nu_{ds}$, which can be as small as 20 ms on a 5000-km baseline with $\Delta\nu = 20$ MHz [see Eq. (9.173)], then the effect of the discrete delay step can be removed by adjusting the slope of the phase of the cross power spectrum. This correction is easily done in spectral line work where spectra are calculated anyway. Note that if this correction is not made, the sensitivity loss factor is 0.64 at the high-frequency edge of the band, as given by Eq. (9.174). In this case, the amplitude response should be compensated for by dividing the cross power spectra by $L(\nu')$. In continuum work, the correction is sometimes omitted because of the need to Fourier transform to the frequency domain and then back to cross-correlation.

A way to compensate partially for the effect of discrete delay steps is to move the frequency at which the phase is unperturbed from zero to $\Delta v/2$, the baseband center. The phase of the fringe rotator is increased by $\pi \Delta v \Delta \tau_s$, where $\Delta \tau_s$ is the delay error. Thus, when the delay changes by one sampling interval, a phase jump of $\pi/2$ is inserted in the fringe rotator. The resulting loss at the band edges is then only 0.90. The average loss over the band is given by an equation similar to Eq. (9.175), but with the upper limit of integration changed to $\Delta v/2$, and equals 0.966. Also, for a symmetrical bandpass response, the residual phase error is zero because the net phase shift over the band at any instant is zero.

9.7.4 Summary of Processing Losses

The loss factors we have considered are all multiplicative, so the total loss is given by the equation

$$\eta = \eta_Q \, \eta_R \, \eta_S \, \eta_D \, , \tag{9.176}$$

where η_Q = quantization loss, η_R = fringe rotation loss, η_S = fringe sideband rejection loss, and η_D = discrete delay step loss.

If there are fringe rotators in each signal path to the correlator, the fringe rotation loss will be η_R^2 because the fringe rotator phases will be uncorrelated. A summary of the loss factors is given in Table 9.7. As an example, a processor might have two-level sampling ($\eta_Q = 0.637$), three-level fringe rotators in each signal path ($\eta_R = 0.922$), 11-channel correlation function ($\eta_S = 0.983$), and band-center delay compensation ($\eta_D = 0.966$), giving a net loss of 0.558. Thus, the sensitivity is worse than that of an ideal analog system with the same bandwidth by a factor of about 2.

There are other loss factors we have not discussed here. The passband will not in reality be perfectly flat, or the response zero, for frequencies above half the Nyquist sampling frequency. These imperfections introduce loss, which for an ideal nine-pole Butterworth filter amounts to 2% (Rogers 1980). The frequency responses will not be perfectly matched for different antennas (see Sect. 7.3). The phase settings of the fringe rotator may be calculated exactly at convenient intervals and extrapolated by Taylor series; this approximation will introduce periodic phase jumps. The local oscillators may have power-line harmonic and noise sidebands that put some fringe power outside the usual fringe filter passband. Empirical values of η typical of the first decade of VLBI development were about 0.4 (Cohen 1973).

The η values refer to loss in signal-to-noise ratio. The fringe amplitudes must also be corrected for scale changes due to signal quantization and fringe rotation. We summarize the multiplicative normalization factors to be applied to the fringe amplitudes in Table 9.8.

Table 9.7 Signal-to-noise loss factors

Quantization loss (η_Q)[a]	
(a) Two-level	0.637
(b) Three-level	0.810
(c) Four-level, all products	0.881
Fringe rotation loss (η_R)	
(a) Two-level, one path	0.900
(b) Three-level, one path	0.960
(c) Two-level, both paths	0.810
(d) Three-level, both paths	0.922
Fringe sideband rejection loss (η_S)	
(a) 1 channel	0.707
(b) 3 channels	0.952
(c) 7 channels	0.975
(d) 11 channels	0.983
Discrete delay step loss (η_D)	
(a) Spectral correction	1.000
(b) Baseband center correction	0.966
(c) No correction	0.873

[a] See Sect. 8.3.

Table 9.8 Normalization factors[a]

Quantization[b]	
(a) Two-level	1.57
(b) Three-level	1.23
(c) Four-level	1.13
Fringe rotation	
(a) Two-level, one path	0.786
(b) Three-level, one path	0.850
(c) Two-level, both paths	0.617
(d) Three-level, both paths	0.723

[a]Multiply correlator output by listed value to obtain normalized correlation function.
[b]See Sect. 8.3.

9.8   Bandwidth Synthesis

For geodetic and astrometric purposes, it is useful to measure the geometric group delay

$$\tau_g = \frac{1}{2\pi} \frac{\partial \phi}{\partial \nu} \tag{9.177}$$

as accurately as possible. With a single RF band, the delay can be found by fitting a straight line to the phase vs. frequency of the cross power spectrum. The uncertainty in this delay, from the usual application of least-mean-squares analysis, is

$$\sigma_\tau = \frac{\sigma_\phi}{2\pi \Delta v_{\mathrm{rms}}} , \tag{9.178}$$

where σ_ϕ is the rms phase noise for a bandwidth Δv and Δv_{rms} is the rms bandwidth, which for a single band of width Δv is equal to $\Delta v/(2\sqrt{3})$ [see discussion following Eq.(A12.28) in Appendix 12.1]. σ_ϕ can be obtained from Eq. (6.64), and if processing losses are neglected, Eq. (9.178) becomes

$$\sigma_\tau = \frac{T_S}{\zeta \, T_A \sqrt{\Delta v_{\mathrm{rms}}^3 \tau}} , \tag{9.179}$$

where ζ is a constant equal to $\pi(768)^{1/4} \simeq 16.5$ [see derivation of Eq. (A12.33)], and T_S and T_A are the geometric mean system and antenna temperatures. A much higher value of Δv_{rms} can be realized by observing at several different radio frequencies. This can be accomplished by switching the local oscillator of a single-band system sequentially in time among N frequencies, or by dividing up the recorded signal into N simultaneous RF bands (channels), which are spread over a wide frequency interval. The temporal switching method has the disadvantage that phase changes during the switching cycle degrade or bias the delay estimate. These methods are commonly referred to as bandwidth synthesis (Rogers 1970, 1976).

In a practical system, signals from a small number of RF bands (about ten) are recorded. The problem of determining the optimum distribution of these bands in frequency is similar to the problem of finding a minimum-redundancy distribution of antenna spacings in a linear array, as discussed in Sect. 5.5. However, here we do not need to have all multiples of the unit (frequency) spacing up to the maximum value, and some gaps are not necessarily detrimental. From the spectral point of view, we wish to have the bands placed in some geometric sequence of increasing separation so that phase can be extrapolated from one band to the next, as shown in Fig. 9.24, without having any 2π ambiguities in the phase connection process. The rms bandwidth depends critically on the unit spacing, which depends on the minimum signal-to-noise ratio. The delay accuracy for a multiband system is obtained from Eq. (9.178) in the same way as for Eq. (9.179) but without the condition $\Delta v_{\mathrm{rms}} = \Delta v/(2\sqrt{3})$. Thus, we obtain

$$\sigma_\tau = \frac{T_S}{2\sqrt{2}\pi \, T_A \sqrt{\Delta v \, \tau} \, \Delta v_{\mathrm{rms}}} , \tag{9.180}$$

where Δv_{rms} for a typical bandwidth synthesis system is approximately 40% of the total frequency interval spanned, Δv is the total bandwidth, and τ is the integration time for each band. To avoid explicitly the problem of phase connection, we can form an equivalent delay function from the cross power spectra [see Eq. (9.26)] of

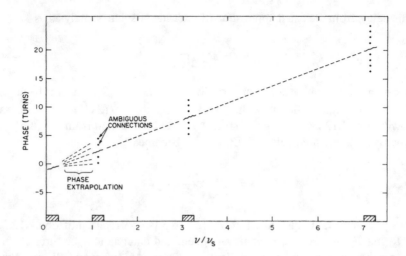

Fig. 9.24 Fringe phase vs. frequency for a bandwidth synthesis system. The phase is measured over discrete bands (crosshatched) spaced at multiples of the fundamental band separation frequency ν_s. The turn ambiguities give rise to sidelobes in the delay resolution function defined in Eq. (9.181) and shown in Fig. 9.25.

the various bands observed:

$$D_R(\tau) = \sum_{i=1}^{N} \int_0^{\Delta\nu} \mathcal{S}_{12i}(\nu - \nu_i) e^{j2\pi\nu\tau} d\nu \;, \tag{9.181}$$

where the ν_i are the local oscillator frequencies relative to the lowest one, and $\nu - \nu_i$ is the baseband frequency. The maximum of $|D_R(\tau)|$ gives the maximum-likelihood estimate of the interferometer delay (Rogers 1970). The a priori normalized delay resolution function, obtained from Eq. (9.181) by setting $\mathcal{S}_{12} = 1$ at frequencies where it is measured and $\mathcal{S}_{12} = 0$ otherwise, is

$$|D_R(\tau)| = \Delta\nu \frac{\sin \pi \Delta\nu\tau}{\pi \Delta\nu\tau} \left| \sum_{i=1}^{N} e^{j2\pi\nu_i\tau} \right| \;. \tag{9.182}$$

The sinc-function envelope is the delay resolution function for a single channel. The frequencies ν_i should be chosen to minimize the width of $D_R(\tau)$ while not allowing any subsidiary maximum to rise above a level such that it could be confused with the principal peak. In situations with low signal-to-noise ratio, the minimum unit spacing should be about four times the bandwidth of a single channel. The delay resolution function for a five-channel system is shown in Fig. 9.25.

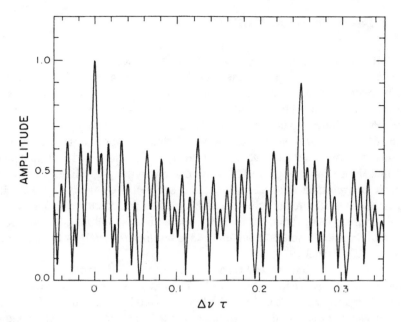

Fig. 9.25 Delay resolution function for a five-channel system with a unit spacing $\nu_s = 4\Delta\nu$ and spacing of 0, 1, 3, 7, and 15 ν_s, as shown in part in Fig. 9.24. The "grating" lobe at $\tau\Delta\nu = 0.25$ need only be reduced sufficiently below unity to avoid delay ambiguity.

9.8.1 Burst Mode Observing

For certain observations, there are advantages in limiting the observing time to short bursts, during which the bit rate can be much higher than the mean data acquisition rate as limited by the recording technology [see, e.g., Wietfeldt and Frail (1991)]. In pulsar observations, the duration of the pulsed emission is typically $\sim$ 3% of the total time, so by recording data taken only during pulse-on time, the bandwidth can be increased by a factor of $\sim$ 33 over the maximum bandwidth for continuous observation. This technique requires the use of a high-speed sampler, high-speed memory, and pulse-timing circuitry at each antenna. During the pulse, the data are stored in the memory at the high rate and then read out continuously at a lower rate. If the ratio of these two rates is a factor w, then the bandwidth can be increased by the same factor over constant-rate observing. For pulsars, this results in an increase in sensitivity by a factor w, of which $\sqrt{w}$ can be attributed to the increased bandwidth, and $\sqrt{w}$ to the fact that noise is not being recorded during the pulse-off time. The second of these $\sqrt{w}$ factors can be obtained without an increase in the data rate by simply deleting data during the pulse-off periods. Burst mode observing is also useful for astrometry and geodesy because it increases the accuracy of measurement of the geometric delay, and it has been used for this purpose in observations of continuum sources at millimeter wavelengths.

9.9   Phased Arrays as VLBI Elements

A phased array is a series of antennas for which the received signals are combined, as indicated in Fig. 5.4. Such systems could be used to form multiple beams, but we describe only the case of forming a single beam. The phase and delay of the signal from each antenna can be adjusted so that the signals from a particular direction in the sky combine in phase, thereby maximizing the sensitivity. It is important to consider the use of phased arrays as VLBI elements for two reasons. First, the elements of a connected-element synthesis array can be combined to form a phased array, thus improving the signal-to-noise ratio of a very-long-baseline interferometer in which they participate as a single station. Second, if elements with very large collecting area are desired to achieve a high signal-to-noise ratio on each baseline, it may be advantageous to build phased arrays rather than monolithic antennas because the cost of a parabolic reflector antenna increases approximately as the diameter to the power 2.7 (Meinel 1979).

Synthesis arrays such as the Westerbork Array, the VLA, the SMA, the Plateau de Bure interferometer, and ALMA are also used as phased arrays to provide a large collecting area for one element in a VLBI system or other applications. Phasing the array consists of adjusting the phase and delay of the signal from each antenna so as to compensate for the different geometric paths for a wavefront from the desired direction. These corrections are easily made through the delay and fringe rotation systems that are used for synthesis imaging. The signals are then summed and go to a VLBI recorder.

We can analyze the performance of a phased array that is used to simulate a single large antenna. Consider an array of n_a identical antennas for which the system temperature is T_S and the antenna temperature for a given source that is unresolved by the longest spacings in the array is T_A. The output of the summing port is

$$V_{\text{sum}} = \sum_i (s_i + \epsilon_i) , \qquad (9.183)$$

where s_i and ϵ_i represent the random signal and random noise voltages, respectively, from antenna i. Now $\langle s_i \rangle = \langle \epsilon_i \rangle = 0$ and, omitting constant gain factors, we can write $\langle s_i^2 \rangle = T_A$ and $\langle \epsilon_i^2 \rangle = T_S$. The power level of the combined signals is represented as the average squared value of Eq. (9.183),

$$\langle V_{\text{sum}}^2 \rangle = \sum_{i,j} [\langle s_i s_j \rangle + \langle s_i \epsilon_j \rangle + \langle s_j \epsilon_i \rangle + \langle \epsilon_i \epsilon_j \rangle] . \qquad (9.184)$$

If the array is accurately phased, $s_i = s_j$. Also, since we are considering an unresolved source, $\langle s_i s_j \rangle = T_A$. If the array is unphased, that is, if the signal phases at the combination point are random, then $\langle s_i s_j \rangle = T_A$ only for $i = j$ and is otherwise zero. In either case, $\langle s_i \epsilon_i \rangle = 0$ and $\langle \epsilon_i \epsilon_j \rangle = 0$. Thus, Eq. (9.184) can be reduced to

$$\langle V_{\text{sum}}^2 \rangle = n_a^2 T_A + n_a T_S \qquad \text{(array phased)} \qquad (9.185)$$

$$\langle V_{\text{sum}}^2 \rangle = n_a T_A + n_a T_S \qquad \text{(array unphased)} , \qquad (9.186)$$

where the first term on the right side represents the signal and the second term represents the noise. When the array is phased, the signal-to-noise (power) ratio is $n_a T_A / T_S$, and when it is unphased, it is T_A / T_S. Thus, the collecting area of the phased array is equal to the sum of the collecting areas of the individual antennas, but when it is unphased, it is, on average, equal to that of a single antenna.

A question of interest concerns the case in which the antennas have different sensitivities resulting from different effective collecting areas and/or system temperatures. This is a matter of practical importance even for nominally uniform arrays, since maintenance or upgrading programs can result in differences in sensitivity. Consider a phased array in which the individual system temperatures and antenna temperatures are represented by T_{Si} and T_{Ai}, respectively. Here, T_{Ai} is defined as the signal from a point source of *unit* flux density,[2] so T_{Ai} is a characteristic of the antenna alone and is proportional to the collecting area. We consider only the weak-signal case for which $T_A \ll T_S$. For antenna i, the output voltage from a source of flux density S is $V_i = s_i + \epsilon_i$, and we can write $\langle s_i^2 \rangle = S T_{Ai}$ and $\langle \epsilon_i^2 \rangle = T_{Si}$.

It is convenient to think of the output of each antenna as providing a measure of the flux density of the source, which is equal to V_i^2 / T_{Ai}. The expectation of the measured value of S should be the same for each antenna. The corresponding voltages are $\sqrt{S} = V_i / \sqrt{T_{Ai}}$ for the signal and $\epsilon_i / \sqrt{T_{Ai}}$ for the noise. In the cross-correlation of the array output with another VLBI antenna, the signal-to-noise ratio at the correlator output is proportional to the signal-to-noise *voltage* ratio of the signal from the array. Thus, in combining the signal voltages in the array, we are, in effect, interested in maximizing the signal-to-noise ratio in an estimate of $\sqrt{S}$. Because the array antennas are not identical, we should use weighting factors w_i in combining their signals. The weights should be chosen to maximize the signal-to-noise ratio of the combined array signals which, in voltage, is

$$\mathcal{R}_{\text{sn}} = \sum_i \frac{w_i V_i}{\sqrt{T_{Ai}}} \Bigg/ \sqrt{\sum_i \frac{w_i^2 T_{Si}}{T_{Ai}}} . \qquad (9.187)$$

Note that we add the signal voltages and the squares of the rms noise voltages. Selecting the weights to provide the best signal-to-noise ratio for $V_i / \sqrt{T_{Ai}}$ is mathematically equivalent to the general problem of obtaining the best estimate of

---

[2]Since it is only the relative values of the weighting factors that matter, T_{Ai} could be defined with respect to any source that is common to all antennas, but consideration of unit flux density simplifies the explanation.

a measured quantity from a series of measurements for which the rms error levels are different but are known. The optimum procedure is to take a mean in which the weight of each measurement is inversely proportional to the variance of the error of that measurement [see Eq. (A12.6)]. The variance of V_i is proportional to T_{Si}, and thus the variance of $V_i/\sqrt{T_{Ai}}$ is T_{Si}/T_{Ai}. Thus, we insert $w_i = T_{Ai}/T_{Si}$ in Eq. (9.187) and obtain

$$
\mathcal{R}_{\text{sn1}} = \sum_i \frac{V_i}{\sqrt{T_{Ai}}} \frac{T_{Ai}}{T_{Si}} \Bigg/ \sqrt{\sum_i \frac{T_{Si}}{T_{Ai}} \left(\frac{T_{Ai}}{T_{Si}}\right)^2}
$$

$$
= \sum_i \frac{V_i\sqrt{T_{Ai}}}{T_{Si}} \Bigg/ \sqrt{\sum_i \frac{T_{Ai}}{T_{Si}}} . \tag{9.188}
$$

Note that in the numerator, V_i is multiplied by $\sqrt{T_{Ai}}/T_{Si}$, which is therefore the (voltage) weighting factor for optimum sensitivity in the signal combination. This conclusion is in agreement with an analysis by Dewey (1994). (Note that the weighting factors for the signal voltages at the combination point are not w_i but $w_i/\sqrt{T_{Ai}}$.) The corresponding weighting of the signal *power* at the combination point is proportional to T_{Ai}/T_{Si}^2.

In synthesis arrays such as the VLA, the IF signals from the antennas are each delivered to a digital sampler at the same power level (of signal plus noise), and the signals are combined after that point so that the time delays required can be inserted digitally. Thus, to avoid modifying the receiving system (which is designed for synthesis imaging), the signals are combined with equal powers when the array is used in the phased mode. For the case of $T_A \ll T_S$ that we are considering, the corresponding weighting is $w_i = 1/\sqrt{T_{Si}}$, and the signal-to-noise ratio becomes

$$
\mathcal{R}_{\text{sn2}} = \sum_i \frac{V_i}{\sqrt{T_{Ai}T_{Si}}} \Bigg/ \sqrt{\sum_i \frac{1}{T_{Ai}}} . \tag{9.189}
$$

Equal-power weighting usually provides sensitivity within a few percent of optimum weighting.

With optimum weighting in the signal combination, all antennas make some contribution to increasing the signal-to-noise ratio. With other weighting, the overall sensitivity may be improved by omitting antennas with poor performance. Moran (1989) has investigated this effect for equal-power weighting. To simplify the situation, it was assumed that T_A is the same for all antennas and only T_S varies. Consider an array undergoing an upgrade of the receiver input stages, in which a fraction n_1/n_a have been refitted with new input stages that reduce the system temperature from T_S to T_S/ξ. After a certain fraction of the antennas have been refitted, the array sensitivity is improved by omitting the unimproved antennas because their input stages are noisier. When T_A does not vary, we can represent the signal voltage received by each antenna by V, and Eq. (9.189) for equal-power weighting becomes

$$\mathcal{R}_{sn2} = \frac{V}{\sqrt{N}} \sum_i \frac{1}{\sqrt{T_{Si}}} \, . \tag{9.190}$$

Thus, we can write

$$\frac{\mathcal{R}_{sn2}(n_1 \text{ refits only})}{\mathcal{R}_{sn2}(\text{all } n_a \text{ antennas})} = \frac{1}{\sqrt{n_1}} \left(\frac{n_1 \sqrt{\xi}}{\sqrt{T_S}} \right) \bigg/ \frac{1}{\sqrt{n_a}} \left(\frac{n_1 \sqrt{\xi}}{\sqrt{T_S}} + \frac{n_a - n_1}{\sqrt{T_S}} \right) . \tag{9.191}$$

The unimproved antennas should be omitted if the expression above is greater than unity, which occurs for

$$\frac{n_1}{n_a} > \left(\frac{\sqrt{\xi}}{2} + \sqrt{1 - \sqrt{\xi} + \frac{\xi}{4}} \right)^{-2} . \tag{9.192}$$

Figure 9.26 shows n_1/n_a as a function of ξ. Thus, for example, if the refitting reduces T_S by a factor of six, then when about half the antennas have been refitted, the others should be omitted. However, unless $\xi > 4$, all antennas should be retained.

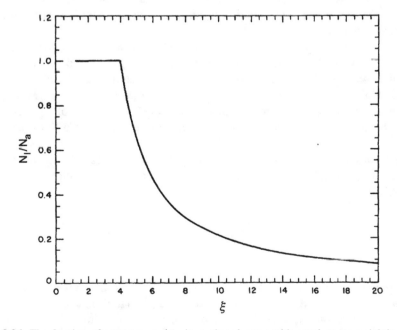

Fig. 9.26 The fraction of antennas, n_1/n_a, in a phased array with equal-power weighting, for which the system temperature must be reduced by a factor ξ before the remaining antennas should be omitted. From Moran (1989), © Kluwer Academic Publishers. With kind permission from Springer Science and Business Media.

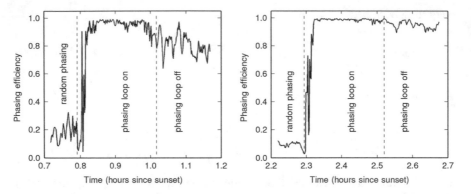

Fig. 9.27 Examples of the phased-array operation of the SMA at 280 GHz on the source 3C354.3, which had a flux density of 10 Jy. The phasing efficiency is the ratio of the sum of the pairwise visibilities divided by their scaler sum. Seven antennas of the SMA were used in the extended configuration with baselines between 44 and 226 m. The weather conditions were: clear sky, 1.3 mm of precipitable water vapor, and wind speed of 2 m/s. The elevation angle range was 44–50° in the left panel and 65–71° in the right panel. The improving atmospheric stability with increasing elevation angle and time since sunset is evident. Adapted from Young et al. (2016).

In practice, a factor of four would be an unusually big improvement, so it can be concluded that omitting antennas is rarely useful. A similar analysis based on Eq. (9.188) shows that with optimum weighting, the sensitivity is never improved by omitting antennas.

For VLBI, the output of a phased array is usually requantized to fit the recording format. The first quantization of the signals, before they are combined, introduces quantization noise that, after combination, has a probability distribution that tends to Gaussian as the number of antennas becomes large. Thus, for such arrays, the additional loss in sensitivity in requantizing is close to the values of η_Q derived in Chap. 8, for which Gaussian noise is assumed. For other cases, see Kokkeler et al. (2001).

The phasing of the SMA is described by Young et al. (2016) and of ALMA by Baudry et al. (2012). An example of a phased array in operation is shown in Fig. 9.27.

9.10   Orbiting VLBI (OVLBI)

The basic requirements for a VLBI station, whether orbiting or terrestrial, include a timing system so that the time associated with each digital sample of the received signal is recoverable, and a position for the antenna known with sufficient accuracy that the fringe frequency (but not necessarily the fringe phase) can be determined. The timing system must be stable to a fraction of the period of the received signal frequency over a coherence time of tens or hundreds of seconds. If it is not possible to put a precise frequency standard on a satellite, then a timing link of equivalent stability must be implemented. Establishing this timing system, which provides the local oscillators and the sampling clock at the satellite, is a major technical challenge in OVLBI. The radial motion of the satellite introduces Doppler shifts, and the tangential motion causes the link path to move relative to the atmospheric irregularities. One or more reference frequencies are transmitted to the satellite over a radio link. The position of the satellite at any time is known from standard orbit-tracking procedures to an accuracy of some tens of meters. This is sufficient to determine the (u, v) coordinates of the baseline but not sufficient for the timing accuracy required. To solve the timing problem, a round-trip phase system implemented by radio link is required. This is identical in principle to the round-trip systems for cables discussed in Sect. 7.2. A discussion of the basic requirements of the timing system is given by D'Addario (1991).

Figure 9.28 shows a simplified example of a system at the satellite and Earth station, which illustrates the essential functions. In this case, a frequency standard is not included in the satellite. A frequency standard in the Earth station provides a reference frequency to synthesizer S_x, from which a signal is transmitted to the satellite. This signal provides a reference for synthesizers S_y, S_L, and S_s that produce signals for the round-trip phase measurement, the local oscillator (LO) of the radio astronomy receiver, and the sampling clock, respectively. The signal from S_y is radiated to the Earth station, where its phase is compared in a correlator with a

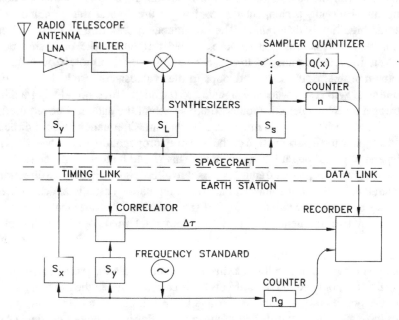

Fig. 9.28 Simplified block diagram of the basic signal transmission and processing required on an OVLBI spacecraft and at the Earth station. See text for further explanation. © 1991 IEEE. Reprinted, with permission, from L. R. D'Addario (1991).

locally generated signal at the same frequency. The correlator output is a measure of $\Delta\tau$, the change in the time delay of the round-trip path. The signal from the radio telescope on the spacecraft goes to a low-noise amplifier, a filter, and a mixer in which it is converted to intermediate frequency (IF) by the LO signal from S_L. The IF signal then goes to an IF amplifier, a sampler (represented by a switch), and a quantizer, $Q(x)$. The counter n is driven by the sampler clock signal from synthesizer S_s and provides timing signals. These provide a record of when each data point was taken, information for formatting the data, and other timing functions required on the satellite. The counter n_g provides timing at the ground location. Some complications with the operation of the scheme just outlined are:

1. The round-trip phase measures the length of the round-trip path with an ambiguity of an integral number of wavelengths. It provides a measure of changes in path length that are continuous.
2. Unless the frequencies generated by the three synthesizers at the satellite are harmonics of one or more reference frequencies supplied (so that no frequency division is necessary in the synthesizers), then the phases of the frequencies will be ambiguous.
3. The transmission times for the reference frequencies and the data may differ because of dispersion in the path or differences in the electronics.

These limitations cause problems when there are discontinuities in the link contact between the satellite and the Earth station. If there is continuous contact during an observing period, then once fringes are found, the combined effect of the ambiguities is determined. The continuous monitoring of the variation of the path enables the solution to be extended throughout the observing period. However, if signal contact is lost due to interference, atmospheric effects, or equipment problems, phase-locked loops in the synthesizers lose lock, and a phase discontinuity will result when the signals are regained. If the round-trip tracking is interrupted for a long period, another fringe search of the data may be required.

For any round-trip measurement, use of the same frequency in both directions would simplify the determination of the one-way propagation time, since the effects of dispersion would be largely eliminated. This would be technically feasible with time sharing or a very small frequency offset to allow signals in the two directions to be separated. However, the international radio regulations usually allocate different frequency bands for the two directions of transmission. Measurement of the round-trip path at two frequencies is therefore important in determining the relative contributions of the neutral and ionized media to the propagation time. If a high-stability frequency standard is included on a satellite, it could serve as the primary clock or as a backup to a radio-link timing system to help keep time at the satellite during link dropouts. Relativistic effects are a complication in the use of an onboard clock, causing its time to vary with respect to Earth-station clocks as the satellite moves through regions of differing strength of the Earth's gravitational field (Ashby and Allan 1979; Vessot 1991).

The first OVLBI experiment was carried out with a satellite in the NASA Tracking and Data Relay Satellite System (TDRSS), which was adapted for VLBI use (Levy et al. 1986, 1989). The purpose of this geostationary satellite was to relay signals from low Earth orbit to ground stations. It was equipped with two 4.9-m-diameter antennas, both with receivers at 2.3 and 15 GHz, and an up–down link communication system at 15.0 and 13.7 GHz. One of the 4.9-m-diameter antennas was used to receive the astronomical signals. The experiment provided limited astronomical data [see Linfield et al. (1989, 1990)] but proved to be an invaluable test bed for time and phase transfer techniques as well as data recovery and processing methods. It was necessary to time-tag the data at the ground station, and hence, the satellite range was part of the interferometer delay. The onboard oscillators were phase-locked via the timing link, much as described in the previous paragraph. However, the coherence of the interferometer was greatly improved by using the second 4.9-m-diameter antenna as part of a separate two-way link at 2.278 GHz. The coherence of the interferometer at 2.3 GHz was found to be 0.98, 0.95, and 0.94 for integration times of 100, 200, and 700 s, respectively. This shows the effective Allan variance of the whole interferometer system of better than 10^{-13} (see discussion in Sect. 9.5.2).

The first satellite specifically designed for use as an orbiting element in a VLBI array was the HALCA satellite (VSOP project), launched in 1997, followed by the Spektr-R satellite (RadioAstron project), launched in 2011. Some of the key specifications of these satellites are listed in Table 9.9. Typical (u, v) plane tracks are shown in Fig. 5.22. RadioAstron has an onboard hydrogen maser frequency standard so that the timing transfer link is not required to synchronize the local oscillator. However, the search for fringes can be a significant task. With orbit position and velocity uncertainty of ± 500 m and 20 mm/s, the delay uncertainty is about 30 ns, or the equivalent of about ± 2000 delay steps, and the fringe rate uncertainty is ± 3 Hz at 6-cm wavelength. The processing must also include a fringe acceleration term. A description of a lag-type correlator designed specifically to include OVLBI stations is given by Carlson et al. (1999).

9.11 Satellite Positioning

The three-dimensional locations of geostationary satellites can be determined with a VLBI array because they lie within its near field (see Section 15.1.3). To understand the sensitivity of a VLBI array to the range, or altitude, of a satellite, consider the simplified geometry shown in Fig. 9.29. In this exercise, the satellite is directly overhead at station 1 of a three-station linear array whose baseline is normal to the direction to the satellite. As a result of being in the near field, the curvature of the spherical wavefront of a broadband-transmitted signal from the satellite can be measured. Note that at least three stations are required, because with only two stations, wavefront tilt cannot be distinguished from wavefront curvature. For the purpose of this exercise, we assume that the bandwidth of the transmitted signal is broad enough that delays can be measured accurately since the signal-to-noise ratio can be expected to be very high. The accuracy of the measurement of R will be

Table 9.9 Parameters of orbiting VLBI stations

	HALCA (VSOP)[a]	Spektr-R (RadioAstron)[b]
Lead institution	Institute of Space and Astronautical Science (Japan)	Astro Space Center (IKI) of the Lebedev Physics Institute (Russia)
Launch date	February 12, 1997[c]	July 18, 2011
Orbital parameters[d]		
Semimajor axis	17,350 km	174,714 km
Eccentricity	0.60	0.69
Inclination	31°	80°
Period	6.3 h	8.3 days
Apogee height	21,400 km	289,246 km
Perigee height	560 km	47,442 km
Maximum resolution	580 μas ($\lambda = 6$ cm)	8 μas ($\lambda = 1.3$ cm)
Orbital determination[e]	±15 m, 6 mm/s	± 500 m, 20 mm/s
Antenna diameter	8 m	10 m
Slew rate	2.25°/min	2°/min
Pointing accuracy	1′	1.5′
Polarization	LCP	RCP/LCP
Operating bands	6, 18 cm[f]	1.3, 6, 18, 92 cm
T_{sys}	95, 75 K	127, 147, 41, 145 K
Aperture efficiency	0.35, 0.24	0.10/0.45/0.52/0.38
Channels/bandwidth	2 × 16 MHz	4 × 16 MHz
Sampling	2 bit	1 bit
Total data rate	128 Mbits/s	128 Mbits/s
Onboard frequency std.	Crystal oscillation[g]	Hydrogen maser
Timing transfer	15.3/14.2 GHz	8.4/7.2 GHz[h]
Ground stations	Usuda (Japan)	Pushchino (Russia)
	Goldstone (USA)	Green Bank (USA)[i]
	Green Bank (USA)	
	Robledo (Spain)	
	Tidbinbilla (Australia)	
Satellite control	Kagoshima (Japan)	Bear Lake (Russia)

[a]Information from Hirabayashi et al. (1998, 2000) and Kobayashi et al. (2000).
[b]Information from RadioAstron Science and Technical Operations Group (2015) and Kardashev et al. (2013).
[c]Operational until 2003.
[d]HALCA: The argument of perigee and longitude of ascending node have periods of 1 and 1.6 years, respectively. RadioAstron: Orbit on April 14, 2012, after repositioning to an orbit of lower eccentricity. Orbit subject to perturbations by the Sun and Moon; eccentricity varies between 0.58 and 0.96.
[e]Accuracy of reconstructed orbit available ∼ 2 weeks after observation determined from Doppler tracking and orbital analysis.
[f]1.35-cm channel not used because of poor sensitivity.
[g]Phase-locked to uplink signal.
[h]Available as backup to onboard hydrogen maser.
[i]See Ford et al. (2014).

Fig. 9.29 Simplified
geometry for tracking a
geostationary satellite with a
three-station VLBI array.
Because the satellite will be
in the near field of the array,
i.e., $R \ll D^2/\lambda$ for typical
values of D and λ, the
wavefront curvature can be
measured.

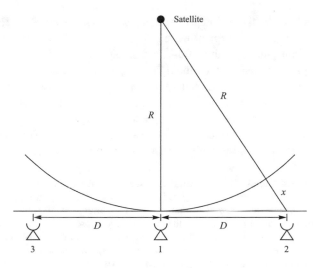

limited by the effects of the atmosphere and ionosphere. Phase measurement could
also be used but might be subject to phase ambiguities.

The excess geometric path length, x, to station 2 or 3 with respect to station 1 is
determined by the relation $(R + x)^2 = R^2 + D^2$. To first order, $x \simeq D^2/2R$, and the
delay is $\tau = x/c = D^2/2Rc$.

Taking the differential of this expression gives the result that the sensitivity of
the delay, $\Delta\tau$, to the sensitivity in range, ΔR, is

$$\Delta\tau = \frac{1}{2c}\left(\frac{D}{R}\right)^2 \Delta R . \tag{9.193}$$

This expression can be used to determine the range accuracy, given the accuracy of
the delay measurement.

Now consider the limitations imposed by the atmosphere. Normal astrometric
positioning can be done to an accuracy of σ_θ, which implies an uncertainty in delay
of $\sigma_\tau \simeq D\sigma_\theta/c$. Hence, the uncertainty in range σ_R is given by

$$\sigma_R = \frac{2R^2}{D}\sigma_\theta , \tag{9.194}$$

while the uncertainty in the transverse direction, σ_{R_T}, is

$$\sigma_{R_T} = R\sigma_\theta . \tag{9.195}$$

Hence, the relative accuracy of the longitudinal and transverse position is

$$\frac{\sigma_R}{\sigma_{R_T}} \simeq 2\left(\frac{R}{D}\right) . \tag{9.196}$$

Consider the following example, with parameters $R = 36,000$ km, $D = 3600$ km, $\lambda = 3$ cm (a typical wavelength for geostationary satellites), and $\sigma_\theta = 100$ μas. From the above equations, we obtain $\sigma_\tau = 0.2$ ns (which corresponds to an rms phase uncertainty of about 20°), $\sigma_R = 40$ cm, $\sigma_{R_T} = 2$ cm, and $\sigma_R/\sigma_{R_T} = 40$. The position can be determined from a single short observation without reliance on Earth rotation. The velocity of the satellite can be determined from the rate of change of position parameters. For an operational system, at least four systems are required since three are needed to define a reference plane. The earliest attempt to measure a satellite position with VLBI was done by Preston et al. (1972).

Further Reading

Biraud, F., Ed., *Very Long Baseline Interferometry Techniques*, Cépaduès, Toulouse, France (1983)

Chi, A.R., Ed., *Proc. IEEE*, Special Issue on Frequency Stability, **54**, No. 2 (1966)

Deller, A.T., and Walker, R.C., Very Long Baseline Interferometry, in *Synthesis Imaging in Radio Astronomy III*, Mioduszewski, A., Ed., *Publ. Astron. Soc. Pacific* (2016), in press

Enge, P., and Misra, P., Eds., *Proc. IEEE*, Special Issue on Global Positioning System, **87**, No. 1, 16–172 (1999)

Felli, M., and Spencer, R.E., Eds., *Very Long Baseline Interferometry: Techniques and Applications*, NATO Science Series C, Kluwer, Dordrecht, the Netherlands (1989)

Hirabayashi, H., Inoue, M., and Kobayashi, H., Eds., *Frontiers of VLBI*, Universal Academy Press, Tokyo (1991)

Jespersen, J., and Hanson, D.W., Eds., *Proc. IEEE*, Special Issue on Time and Frequency, **79**, No. 7 (1991)

Kroupa, V.F., Ed., *Frequency Stability: Fundamentals and Measurement*, IEEE Press, New York (1983)

Morris, D., Ed., *Radio Sci.*, Special Issue Devoted to the Open Symposium on Time and Frequency, **14**, No. 4 (1979)

Zensus, J.A., Diamond, P.J., and Napier, P.J., Eds., *Very Long Baseline Interferometry and the VLBA*, Astron. Soc. Pacific Conf. Ser., **82** (1995)

References

Allan, D.W., Statistics of Atomic Frequency Standards, *Proc. IEEE*, **54**, 221–230 (1966)

Ashby, N., and Allan, D.W., Practical Implications of Relativity for a Global Coordinate Time Scale, *Radio Sci.*, **14**, 649–669 (1979)

Bare, C., Clark, B.G., Kellermann, K.I., Cohen, M.H., and Jauncey, D.L., Interferometer Experiments with Independent Local Oscillators, *Science*, **157**, 189–191 (1967)

Barnes, J.A., Chi, A.R., Cutler, L.S., Healey, D.J., Leeson, D.B., McGunigal, T.E., Mullen, J.A., Smith, W.L., Sydnor, R.L., Vessot, R.F.C., and Winkler, G.M.R., Characterization of Frequency Stability, *IEEE Trans. Instrum. Meas.*, **IM-20**, 105–120 (1971)

Baudry, A., Lacasse, R., Escoffier, R., Greenberg, J., Treacy, R., and Saez, A., Phasing ALMA with the 64-Antenna Correlator, Proc. of the 11th European VLBI Network Symposium and Users Meeting (2012), Bordeaux, France, Session 9: VLBI at Extremely High Angular Resolution, PoS(11th EVN Symposium)054

Behrend, D., and Baver, K.D., Eds., *Launching the Next-Generation IVS Network*, IVS 2012 General Meeting Proceedings, NASA/CP-2012-217504 (2012)

Berkeland, D.J., Miller, J.D., Bergquist, J.C., Itano, W.M., and Wineland, D.J., Laser-Cooled Mercury Ion Frequency Standard, *Phys. Rev. Lett.*, **80**, 2089–2092 (1998)

Blair, B.E., *Time and Frequency: Theory and Fundamentals*, National Bureau of Standards Monograph 140, U.S. Government Printing Office, Washington, DC (1974), pp. 223–313

Broten, N.W., Legg, T.H., Locke, J.L., McLeish, C.W., Richards, R.S., Chisholm, R.M., Gush, H.P., Yen, J.L., and Galt, J.A., Long Baseline Interferometry: A New Technique, *Science*, **156**, 1592–1593 (1967)

Cannon, W.H., The Classical Analysis of the Response of a Long Baseline Radio Interferometer, *Geophys. J. R. Astron. Soc.*, **53**, 503–530 (1978)

Cannon, W.H., Baer, D., Feil, G., Feir, B., Newby, P., Novikov, A., Dewdney, P., Carlson, B., Petrachenko, W.P., Popelar, J., Mathieu, P., and Wietfeldt, R.D., The S2 VLBI System, *Vistas in Astronomy*, **41**, 297–302 (1997)

Cao, H.-M., Frey, S., Gurvits, L.I., Yang, J., Hong, X.-Y., Paragi, Z., Deller, Z.T., and Ivezić, Z., VLBI Observations of the Radio Quasar J2228+0110 at $z = 5.95$ and Other Field Sources in Multiple-Phase-Centre Mode, *Astron. Astrophys.*, **563**, A111 (8 pp) (2014)

Carlson, B.R., Dewdney, P.E., Burgess, T.A., Casoro, R.V., Petrachenko, W.T., and Cannon, W.H., The S2 VLBI Correlator: A Correlator for Space VLBI and Geodetic Signal Processing, *Publ. Astron. Soc. Pacific*, **111**, 1025–1047 (1999)

Clark, B.G., Radio Interferometers of Intermediate Type, *IEEE Trans. Antennas Propag.*, **AP-16**, 143–144 (1968)

Clark, B.G., The NRAO Tape-Recorder Interferometer System, *Proc. IEEE*, **61**, 1242–1248 (1973)

Clark, B.G., Kellermann, K.I., Bare, C.C., Cohen, M.H., and Jauncey, D.L., High-Resolution Observations of Small-Diameter Radio Sources at 18-Centimeter Wavelength, *Astrophys. J.*, **153**, 705–714 (1968)

Clark, B.G., Weimer, R., and Weinreb, S., *The Mark II VLB System*, NRAO Electronics Division Internal Report 118, National Radio Astronomy Observatory, Green Bank, WV (1972)

Clark, T.A., Corey, B.E., Davis, J.L., Elgered, G., Herring, T.A., Hinteregger, H.F., Knight, C.A., Levine, J.I., Lundqvist, G., Ma, C., and 11 coauthors, Precision Geodesy Using the Mark III Very-Long-Baseline Interferometer System, *IEEE Trans. Geosci. Remote Sens.*, **GE-23**, 438–449 (1985)

Clark, T.A., Counselman, C.C., Ford, P.G., Hanson, L.B., Hinteregger, H.F., Klepczynski, W.J., Knight, C.A., Robertson, D.S., Rogers, A.E.E., Ryan, J.W., Shapiro, I.I., and Whitney, A.R., Synchronization of Clocks by Very Long Baseline Interferometry, *IEEE Trans. Instrum. Meas.*, **IM-28**, 184–187 (1979)

Cohen, M.H., Introduction to Very-Long-Baseline Interferometry, *Proc. IEEE*, **61**, 1192–1197 (1973)

Cohen, M.H., and Shaffer, D.B., Positions of Radio Sources from Long-Baseline Interferometry, *Astron. J.*, **76**, 91–100 (1971)

Counselman, C.C., III, Shapiro, I.I., Rogers, A.E.E., Hinteregger, H.F., Knight, C.A., Whitney, A.R., and Clark, T.A., VLBI Clock Synchronization, *Proc. IEEE*, **65**, 1622–1623 (1977)

Cutler, L.S., and Searle, C.L., Some Aspects of the Theory and Measurement of Frequency Fluctuations in Frequency Standards, *Proc. IEEE*, **54**, 136–154 (1966)

D'Addario, L.R., Minimizing Storage Requirements for Quantized Noise, VLBA Memo 332, National Radio Astronomy Observatory (1984)

D'Addario, L.R., Time Synchronization in Orbiting VLBI, *IEEE Trans. Instrum. Meas.*, **IM-40**, 584–590 (1991)

Davis, M.M., Taylor, J.H., Weisberg, J.M., and Backer, D.C., High-Precision Timing of the Millisecond Pulsar PSR 1937+21, *Nature*, **315**, 547–550 (1985)

Deller, A.T., Brisken, W.F., Phillips, C.J., Morgan, J., Alef, W., Cappallo, R., Middelberg, E., Romney, J., Rottmann, H., Tingay, S.J., and Wayth, R., DiFX-2: A More Flexible, Efficient, Robust, and Powerful Software Correlator, *Publ. Astron. Soc. Pacific*, **123**, 275–287 (2011)

Deller, A.T., Tingay, S.J., Bailes, M., and West, C., DiFX: A Software Correlator for Very-Long-Baseline Interferometry Using Multiprocessor Computing Environments, *Publ. Astron. Soc. Pacific*, **119**, 318–336 (2007)

Dewey, R.J., The Effects of Correlated Noise in Phased-Array Observations of Radio Sources, *Astron. J.*, **108**, 337–345 (1994)

Doeleman, S., Building an Event Horizon Telescope: (Sub)mm VLBI in the ALMA Era, Proc. of the Tenth European VLBI Network Symposium and EVN Users Meeting: VLBI and the New Generation of Radio Arrays, Proc. Science, PoS(10th EVN Symposium)053 (2010)

Doeleman, S., Mai, T., Rogers, A.E.E., Hartnett, J.G., Tobar, M.E., and Nand, N., Adapting a Cryogenic Sapphire Oscillator for Very Long Baseline Interferometry, *Publ. Astron. Soc. Pacific*, **123**, 582–595 (2011)

Doi, A., Fujisawa, K., Honma, M., Sugiyama, K., Murata, Y., Mochizuki, N., and Isono, Y., Japanese VLBI Network Observations of 6.7-GHz Methanol Masers. I. Array, in *Astrophysical Masers and Their Environments*, Proc. IAU Symp. 242, Chapman, J.M., and Baan, W.A., Eds. (2007), pp. 148–149

Drullinger, R.E., Rolston, S.L., and Itano, W.M., Primary Atomic Frequency Standards: New Developments, in *Review of Radio Science 1993–1996*, Stone, W.R., Ed., Oxford Univ. Press, Oxford, UK (1996), pp. 11–41

Dutta, P., and Horn, P.M., Low-Frequency Fluctuations in Solids: $1/f$ Noise, *Rev. Mod. Phys.*, **53**, 497–516 (1981)

Edson, W.A., Noise in Oscillators, *Proc. IRE*, **48**, 1454–1466 (1960)

Edwards, P., Novice's Guide to Using the LBA, version 1.5 (2012). http://www.atnf.csiro.au/vlbi/LBA-Novices-Manual-v1.5.pdf

Ford, H.A., Anderson, R., Belousov, K., Brandt, J.J., Ford, J.M., Kanevsky, B., Kovalenko, A., Kovalev, Y.Y., Maddalena, R.J., Sergeev, S., and three coauthors, The RadioAstron Green Bank Earth Station, in *Ground-Based and Airborne Telescopes V*, Proc. SPIE, **9145**, 91450B-1 (11pp) (2014)

Forman, P., Atomichron: The Atomic Clock from Concept to Commercial Product, *Proc. IEEE*, **73**, 1181–1204 (1985)

Frank, R.L., Current Developments in Loran-C, *Proc. IEEE*, **71**, 1127–1139 (1983)

Hellwig, H., Microwave Time and Frequency Standards, *Radio Sci.*, **14**, 561–572 (1979)

Hellwig, H., Vessot, R.F.C., Levine, M.W., Zitzewitz, P.W., Allen, D.W., and Glaze, D.J., Measurement of the Unperturbed Hydrogen Hyperfine Transition Frequency, *IEEE Trans. Instrum. Meas.*, **IM-19**, 200–209 (1970)

Herring, T.A., *Precision and Accuracy of Intercontinental Distance Determinations Using Radio Interferometry*, Air Force Geophysics Laboratory, Hanscom Field, MA, AFGL-TR-84-0182 (1983)

Hinkley, N., Sherman, J.A., Phillips, N.B., Schioppo, M., Lemke, N.D., Beloy, K., Pizzocaro, M., Oates, C.W., and Ludlow, A.D., An Atomic Clock with 10^{-18} Instability, *Science*, **341**, 1215–1218 (2013)

Hinteregger, H.F., Rogers, A.E.E., Capallo, R.J., Webber, J.C., Petrachenko, W.T., and Allen, H., A High Data Rate Recorder for Astronomy, *IEEE Trans. Magn.*, **MAG-27**, 3455–3465 (1991)

Hirabayashi, H., Hirosawa, H., Kobayashi, H., Murata, Y., Asaki, Y., Avruch, I.M., Edwards, P.G., Fomalont, E.B., Ichikawa, T., Kii, T., and 45 coauthors, The VLBI Space Observatory Programme and the Radio-Astronomical Satellite HALCA, *Pub. Astron. Soc. Japan*, **52**, 955–965 (2000)

Hirabayashi, H., Hirosawa, H., Kobayashi, H., Murata, Y., Edwards, P.G., Fomalont, E.B., Fujisawa, K., Ichikawa, T., Kii, T., Lovell, J.E.J., and 44 coauthors, Overview and Initial Results of the Very Long Baseline Interferometry Space Observatory Programme, *Science*, **281**, 1825–1829 (1998)

Hjellming, R.M., An Introduction to the NRAO Very Large Array, National Radio Astronomy Observatory, Socorro, NM (1992), p. 43

Hobbs, G., Coles, W., Manchester, R.N., Keith, M.J., Shannon, R.M., Chen, D., Bailes, M., Bhat, N.D.R., Burke-Spolaor, S., Champion, D., and 14 coauthors, Development of a Pulsar-Based Time Scale, *Mon. Not. R. Astron. Soc.*, **427**, 2780–2787 (2012)

Johnson, M.D., Fish, V.L., Doeleman, S.S., Marrone, D.P., Plambeck, R.L., Wardle, J.F.C., Akiyama, K., Asada, K., Beaudoin, C., Blackburn, L., and 38 coauthors, Resolved Magnetic-Field Structure and Variability near the Event Horizon of Sagittarius A*, *Science*, **350**, 1242–1245 (2015)

Kardashev, N.S., Khartov, V.V., Abramov, V.V., Avdeev, V.Yu., Alakoz, A.V., Aleksandrov, Yu.A., Ananthakrishnan, S., Andreyanov, V.V., Andrianov, A.S., Antonov, N.M., and 120 coauthors, "RadioAstron": A Telescope with a Size of 300,000 km: Main Parameters and First Observational Results, *Astron. Reports*, **57**, 153–194 (2013)

Kartashoff, P., and Barnes, J.A., Standard Time and Frequency Generation, *Proc. IEEE*, **60**, 493–501 (1972)

Kawaguchi, N., VLBI Recording System in Japan, in *Frontiers of VLBI*, Hirabayashi, H., Inoue, M., and Kobayashi, H., Eds., Universal Academy Press, Tokyo (1991), pp. 75–77

Keshner, M.S., 1/f Noise, *Proc. IEEE*, **70**, 212–218 (1982)

Klemperer, W.K., Long Baseline Radio Interferometry with Independent Frequency Standards, *Proc. IEEE*, **60**, 602–609 (1972)

Kleppner, D., Berg, H.C., Crampton, S.B., Ramsey, N.F., Vessot, R.F.C., Peters, H.E., and Vanier, J., Hydrogen-Maser Principles and Techniques, *Phys. Rev. A*, **138**, 972–983 (1965)

Kleppner, D., Goldenberg, H.M., and Ramsey, N.F., Theory of the Hydrogen Maser, *Phys. Rev.*, **126**, 603–615 (1962)

Kobayashi, H., Sasao, T., Kawaguchi, N., Manabe, S., Omodaka, T., Kameya, O., Shibata, K.M., Miyaji, T., Honma, M., Tamura, Y., and 16 coauthors, VERA: A New VLBI Instrument Free from the Atmosphere, in *New Technologies in VLBI*, Minh, Y.C., Ed., Astron. Soc. Pacific Conf. Ser., **306**, 367–370 (2003)

Kobayashi, H., Wajima, K., Hirabayashi, H., Murata, Y., Kawaguchi, N., Kameno, S., Shibata, K.M., Fujisawa, K., Inoue, M., and Hirosawa, H., HALCA's Onboard VLBI Observing System, *Publ. Astron. Soc. Japan*, **52**, 967–973 (2000)

Kogan, L.R., and Chesalin, L.S., Software for VLBI Experiments for CS-Type Computers, *Sov. Astron.*, **25**, 510–513 (1982), transl. from *Astron. Zh.*, **58**, 898–903 (1981)

Kokkeler, A.B.J., Fridman, P., and van Ardenne, A., Degradation Due to Quantization Noise in Radio Astronomy Phased Arrays, *Experimental Astron.*, **11**, 33–56 (2001)

Koyama, Y., Developments of K3, K4, and K5 VLBI Systems and Considerations for the New K6 VLBI System, Technology Development Center News, National Institute of Information and Communications Technology, **33**, 39–45 (2013)

Kulkarni, S.R., Self-Noise in Interferometers: Radio and Infrared, *Astron. J.*, **98**, 1112–1130 (1989)

Lee, S.-S., Petrov, L., Byun, D.-Y., Kim, J., Jung, T., Song, M.-G., Oh, C.S., Roy, D.-G., Je, D.-H., Wi, S.-O., and 14 coauthors, Early Science with the Korean VLBI Network: Evaluation of System Performance, *Astron. J.*, **147**, 77 (14pp) (2014)

Leick, A., *GPS Satellite Surveying*, 2nd ed., Wiley, New York (1995)

Lesage, P., and Audoin, C., Characterization and Measurement of Time and Frequency Stability, *Radio Sci.*, **14**, 521–539 (1979)

Levy, G.S., Linfield, R.P., Edwards, C.D., Ulvestad, J.S., Jordan, J.F., Jr., Di Nardo, S.J., Christensen, C.S., Preston, R.A., Skjerve, L.J., Stavert, L.R., and 22 coauthors, VLBI Using a Telescope in Earth Orbit. I. The Observations, *Astrophys. J.*, **336**, 1098–1104 (1989)

Levy, G.S., Linfield, R.P., Ulvestad, J.S., Edwards, C.D., Jordan, J.F., Jr., Di Nardo, S.J., Christensen, C.S., Preston, R.A., Skjerve, L.J., Stavert, L.R., and 19 coauthors, Very Long Baseline Interferometric Observations Made with an Orbiting Radio Telescope, *Science*, **234**, 187–189 (1986)

Lewandowski, W., Azoubib, J., and Klepczynski, W.J., GPS: Primary Tool for Time Transfer, *Proc. IEEE*, Special Issue on Global Positioning System, **87**, No. 1, 163–172 (1999)

Lewis, L.L., An Introduction to Frequency Standards, *Proc. IEEE*, **79**, 927–935 (1991)

Lindsey, W.C., and Chie, C.M., Frequency Multiplication Effects on Oscillator Instability, *IEEE Trans. Instrum. Meas.*, **IM-27**, 26–28 (1978)

Linfield, R.P., Levy, G.S., Edwards, C.D., Ulvestad, J.S., Ottenhoff, C.H., Hirabayashi, H., Morimoto, M., Inoue, M., Jauncey, D.L., Reynolds, J., and 18 coauthors, 15 GHz Space VLBI Observations Using an Antenna on a TDRSS Satellite, *Astrophys. J.*, **358**, 350–358 (1990)

Linfield, R.P., Levy, G.S., Ulvestad, J.S., Edwards, C.D., DiNardo, S.J., Stavert, L.R., Ottenhoff, C.H., Whitney, A.R., Cappallo, R.J., Rogers, A.E.E., and five coauthors, VLBI Using a Telescope in Earth Orbit. II. Brightness Temperatures Exceeding the Inverse Compton Limit, *Astrophys. J.*, **336**, 1105–1112 (1989)

Meinel, A.B., Multiple Mirror Telescopes of the Future, in *MMT and the Future of Ground-Based Astronomy*, Weeks, T.C., Ed., SAO Special Report 385, Smithsonian Astrophysical Obs., Cambridge, MA (1979), pp. 9–22

Moran, J.M., Spectral-Line Analysis of Very-Long-Baseline Interferometric Data, *Proc. IEEE*, **61**, 1236–1242 (1973)

Moran, J.M., Very Long Baseline Interferometric Observations and Data Reduction, in *Methods of Experimental Physics*, Vol. 12, Part C *(Astrophysics: Radio Observations)*, Meeks, M.L., Ed., Academic Press, New York (1976), pp. 228–260

Moran, J.M., Introduction to VLBI, in *Very Long Baseline Interferometry: Techniques and Applications*, Felli, M., and Spencer, R.E., Eds., Kluwer, Dordrecht, the Netherlands (1989), pp. 27–45

Moran, J.M., and Dhawan, V., An Introduction to Calibration Techniques for VLBI, in *Very Long Baseline Interferometry and the VLBA*, Zensus, J.A., Diamond, P.J., and Napier, P.J., Eds., Astron. Soc. Pacific Conf. Ser., **82**, 161–188 (1995)

Morgan, J.S., Mantovani, F., Deller, A.T., Brisken, W., Alef, W., Middelberg, E., Nanni, M., and Tingay, S.J., VLBI Imaging Throughout the Primary Beam Using Accurate UV Shifting, *Astron. Astrophys.*, **526**, A140–A148 (2011)

Napier, P.J., Bagri, D.S., Clark, B.G., Rogers, A.E.E., Romney, J.D., Thompson, A.R., and Walker, R.C., The Very Long Baseline Array, *Proc. IEEE*, **82**, 658–672 (1994)

Parkinson, B.W., and Gilbert, S.W., NAVSTAR: Global Positioning System—Ten Years Later, *Proc. IEEE*, **71**, 1177–1186 (1983)

Pierce, J.A., McKenzie, A.A., and Woodward, R.H., *Loran*, Radiation Laboratory Ser., Vol. 4, McGraw-Hill, New York (1948)

Porcas, R.W., A History of the EVN: Thirty Years of Fringes, Proc. of the Tenth European VLBI Network Symposium and EVN Users Meeting: VLBI and the New Generation of Radio Arrays, Proc. Science, PoS(10th EVN Symposium)011 (2010)

Press, W.H., Flicker Noises in Astronomy and Elsewhere, *Comments Astrophys.*, **7**, 103–119 (1978)

Preston, R.A., Ergas, R., Hinteregger, H.F., Knight, C.A., Robertson, D.S., Shapiro, I.I., Whitney, A.R., Rogers, A.E.E., and Clark, T.A., Interferometric Observations of an Artificial Satellite, *Science*, **178**, 407–409 (1972)

RadioAstron Science and Technical Operations Group, *RadioAstron User Handbook*, version 2.7 (Dec. 2015). http://www.asc.rssi.ru/radioastron

Ray, J., and Senior, K., Geodetic Techniques for Time and Frequency Comparisons Using GPS Phase and Code Measurements, *Metrologia*, **42**, 215–232 (2005)

Reid, M.J., Spectral-Line VLBI, in *Very Long Baseline Interferometry and the VLBA*, Zensus, J.A., Diamond, P.J., and Napier, P.J., Eds., Astron. Soc. Pacific Conf. Ser., **82**, 209–225 (1995)

Reid, M.J., Spectral-Line VLBI, in *Synthesis Imaging in Radio Astronomy II*, Taylor, G.B., Carilli, C.L., and Perley, R.A., Eds., Astron. Soc. Pacific Conf. Ser., **180**, 481–497 (1999)

Reid, M.J., Haschick, A.D., Burke, B.F., Moran, J.M., Johnston, K.J., and Swenson, G.W., Jr., The Structure of Interstellar Hydroxyl Masers: VLBI Synthesis Observations of W3(OH), *Astrophys. J.*, **239**, 89–111 (1980)

Rioja, M., Dodson, R., Asaki, Y., Hartnett, J., and Tingay, S., The Impact of Frequency Standards on Coherence in VLBI at the Highest Frequencies, *Astron. J.*, **144**, 121 (11pp) (2012)

Rogers, A.E.E., Very Long Baseline Interferometry with Large Effective Bandwidth for Phase Delay Measurements, *Radio Sci.*, **5**, 1239–1247 (1970)

Rogers, A.E.E., Theory of Two-Element Interferometers, in *Methods of Experimental Physics*, Vol. 12, Part C *(Astrophysics: Radio Observations)*, Meeks, M.L., Ed., Academic Press, New York (1976), pp. 139–157

Rogers, A.E.E., The Sensitivity of a Very Long Baseline Interferometer, *Radio Interferometry Techniques for Geodesy*, NASA Conf. Pub. 2115, National Aeronautics and Space Administration, Washington, DC (1980), pp. 275–281

Rogers, A.E.E., Very Long Baseline Fringe Detection Thresholds for Single Baselines and Arrays, in *Frontiers of VLBI*, Hirabayashi, H., Inoue, M., and Kobayashi, H., Eds., Universal Academy Press, Tokyo (1991), pp. 341–349

Rogers, A.E.E., VLBA Data Flow: Formatter to Tape, in *Very Long Baseline Interferometry and the VLBA*, Zensus, J.A., Diamond, P.J., and Napier, P.J., Eds., Astron. Soc. Pacific Conf. Ser., **82**, 93–115 (1995)

Rogers, A.E.E., Cappallo, R.J., Hinteregger, H.F., Levine, J.I., Nesman, E.F., Webber, J.C., Whitney, A.R., Clark, T.A., Ma, C., Ryan, J., and 12 coauthors, Very-Long-Baseline Interferometry: The Mark III System for Geodesy, Astrometry, and Aperture Synthesis, *Science*, **219**, 51–54 (1983)

Rogers, A.E.E., Doeleman, S.S., and Moran, J.M., Fringe Detection Methods for Very-Long-Baseline Arrays, *Astron. J.*, **109**, 1391 1401 (1995)

Rogers, A.E.E., and Moran, J.M., Coherence Limits for Very-Long-Baseline Interferometry, *IEEE Trans. Instrum. Meas.*, **IM-30**, 283–286 (1981)

Rose, J.A.R., Watson, R.J., Allain, D.J., and Mitchell, C.N., Ionospheric Corrections for GPS Time Transfer, *Radio Sci.*, **49**, 196–206 (2014)

Rutman, J., Characterization of Phase and Frequency Instability in Precision Frequency Sources: Fifteen Years of Progress, *Proc. IEEE*, **66**, 1048–1075 (1978)

Schwab, F.R., and Cotton, W.D., Global Fringe Search Techniques for VLBI, *Astron. J.*, **88**, 688–694 (1983)

Sekido, M., VGOS-Related Developments in Japan, Notebook of the Eighth IVS Technical Operations Workshop, May 4–7, 2015, Haystack Observatory, Westford, MA (2015). ftp://ivscc.gsfc.nasa.gov/pub/TOW/tow2015/notebook/Sekido.Lec.pdf

Shapiro, I.I., Estimation of Astrometric and Geodetic Parameters, in *Methods of Experimental Physics*, Vol. 12, Part C *(Astrophysics: Radio Observations)*, Meeks, M.L., Ed., Academic Press, New York (1976), pp. 261–276

Shimoda, K., Wang, T.C., and Townes, C.H., Further Aspects of the Theory of the Maser, *Phys. Rev.*, **102**, 1308–1321 (1956)

Siegman, A.E., *An Introduction to Lasers and Masers*, McGraw-Hill, New York (1971), p. 404

Sivia, D.S. with Skilling, J., *Data Analysis: A Bayesian Tutorial*, 2nd. ed., Oxford Univ. Press, Oxford, UK (2006)

Sovers, O.J., Fanselow, J.L., and Jacobs, C.S., Astrometry and Geodesy with Radio Interferometry: Experiments, Models, Results, *Rev. Mod. Phys.*, **70**, 1393–1454 (1998)

Thomas, J.B., *An Analysis of Radio Interferometry with the Block O System*, JPL Pub. 81–49, Jet Propulsion Laboratory, Pasadena, CA (1981)

Thompson, A.R., The VLBA Receiving System: Antenna to Data Formatter, in *Very Long Baseline Interferometry and the VLBA*, Zensus, J.A., Diamond, P.J., and Napier, P.J., Eds., Astron. Soc. Pacific Conf. Ser., **82**, 73–92 (1995)

Thompson, A.R., and Bagri, D.S., A Pulse Calibration System for the VLBA, in *Radio Interferometry: Theory, Techniques and Applications*, Cornwell, T.J., and Perley, R.A., Eds., Astron. Soc. Pacific Conf. Ser., **19**, 55–59 (1991)

van Haarlem, M.P., Wise, M.W., Gunst, A.W., Heald, G., McKean, J.P., Hessels, J.W.T., de Bruyn, A.G., Nijboer, R., Swinbank, J., Fallows, R., and 191 coauthors, LOFAR: The LOw-Frequency ARray, *Astron. & Astrophys.*, **556**, A2 (53pp) (2013)

Vanier, J., Têtu, M., and Bernier, L.G., Transfer of Frequency Stability from an Atomic Frequency Reference to a Quartz-Crystal Oscillator, *IEEE Trans. Instrum. Meas.*, **IM-28**, 188–193 (1979)

VERA Status Report, Mizusawa VLBI Observatory, National Astronomical Observatory of Japan (2015), 28 pp

Verbiest, J.P.W., Bailes, M., Coles, W.A., Hobbs, G.B., van Straten, W., Champion, D.J., Jenet, F.A., Manchester, R.N., Bhat, N.D.R., Sarkissian, J.M., and four coauthors, Time Stability of Millisecond Pulsars and Prospects for Gravitational-Wave Detection, *Mon. Not. R. Astron. Soc.*, **400**, 951–968 (2009)

Vertatschitsch, L., Primiani, R., Weintroub, J., Young, A., and Blackburn, L., R2DBE: A Wideband Digital Backend for the Event Horizon Telescope, *Publ. Astron. Soc. Pacific*, **127**, 1226–1239 (2015)

Vessot, R.F.C., Frequency and Time Standards, in *Methods of Experimental Physics*, Vol. 12, Part C (*Astrophysics: Radio Observations*), Meeks, M.L., Ed., Academic Press, New York (1976), pp. 198–227

Vessot, R.F.C., Relativity Experiments with Clocks, *Radio Sci.*, **14**, 629–647 (1979)

Vessot, R.F.C., Applications of Highly Stable Oscillators to Scientific Measurements, *Proc. IEEE*, **79**, 1040–1053 (1991)

Vessot, R.F.C., and Levine, M.W., A Method for Eliminating the Wall Shift in the Atomic Hydrogen Maser, *Metrologia*, **6**, 116–117 (1970)

Vessot, R.F.C., Levine, M.W., Mattison, E.M., Hoffman, T.E., Imbier, E.A., Têtu, M., Nystrom, G., Kelt, J.J., Trucks, H.F., and Vaniman, J.L., Space-Borne Hydrogen Maser Design, in *Proc. 8th Annual Precise Time and Interval Meeting*, U.S. Naval Research Laboratory, X-814-77-149 (1976), pp. 277–333

Vinokur, M., Optimisation dans la Recherche d'une Sinusoide de Période Connue en Présence de Bruit, *Ann. d'Astrophys.*, **28**, 412–445 (1965)

Walker, R.C., Very Long Baseline Interferometry I: Principles and Practice, in *Synthesis Imaging in Radio Astronomy*, Perley, R.A., Schwab, F.R., and Bridle, A.H., Eds., Astron Soc. Pacific Conf. Ser., **6**, 355–378 (1989a)

Walker, R.C., Calibration Methods, in *Very Long Baseline Interferometry: Techniques and Applications*, Felli, M., and Spencer, R.E., Eds., Kluwer, Dordrecht, the Netherlands (1989b), pp. 141–162

Wardle, J.F.C., and Kronberg, P.P., The Linear Polarization of Quasi-Stellar Radio Sources at 3.71 and 11.1 Centimeters, *Astrophys. J.*, **194**, 249–255 (1974)

Whitney, A.R., The Mark IV Data-Acquisition and Correlation System, in *Developments in Astrometry and Their Impact on Astrophysics and Geodynamics*, IAU Symp. 156, Mueller, I.I., and Kolaczek, B., Eds., Kluwer, Dordrecht, the Netherlands (1993), pp. 151–157

Whitney, A.R., Mark 5 Disc-Based Gbps VLBI Data System, in Proc. General Meeting of the International VLBI Service for Geodesy and Astrometry, Vandenberg, N.R., and Baver, K.D., Eds., National Technical Information Service, Alexandria, VA (2002), pp. 132–136

Whitney, A.R., The Mark 5B VLBI Data System, in Proc. 7th European VLBI Network Symposium on New Developments in VLBI Science and Technology, Bachiller, R., Colomer, F., Desmurs, J.F., and de Vicente, P., Eds., Observatorio Astronómico Nacional, Madrid (2004), pp. 251–252

Whitney, A.R., The Mark 5C VLBI Data System, in *Measuring the Future*, Proc. 5th General Meeting of the International VLBI Service for Geodesy and Astrometry, Finkelstein, A., and Behrend, D., Eds., Nauka, St. Petersburg, Russia (2008), pp. 390–394

Whitney, A.R., Beaudoin, C.R., Cappallo, R.J., Corey, B.E., Crew, G.B., Doeleman, S.S., Lapsley, D.E., Hinton, A.A., McWhirter, S.R., Niell, A.E., and five coauthors, Demonstration of a 16-Gbps Station^{-1} Broadband-RF VLBI System, *Publ. Astron. Soc. Pacific*, **125**, 196–203 (2013)

Whitney, A.R., Rogers, A.E.E., Hinteregger, H.F., Knight, C.A., Levine, J.I., Lippincott, S., Clark, T.A., Shapiro, I.I., and Robertson, D.S., A Very Long Baseline Interferometer System for Geodetic Applications, *Radio Sci.*, **11**, 421–432 (1976)

Wietfeldt, R.D., and D'Addario, L.R., Compatibility Issues in VLBI, in *Radio Interferometry: Theory, Techniques, and Applications*, Cornwell, T.J., and Perley, R.A., Eds., Astron. Soc. Pacific Conf. Scr., **19**, 98–101 (1991)

Wietfeldt, R.D., Baer, D., Cannon, W.H., Feil, G., Jakovina, R., Leone, P., Newby, P.S., and Tan, H., The S2 Very Long Baseline Interferometry Tape Recorder, *IEEE Trans. Instrum. Meas.*, **IM-45**, 923–929 (1996)

Wietfeldt, R.D., and Frail, D.A., Burst Mode VLBI and Pulsar Applications, in *Radio Interferometry: Theory, Techniques, and Applications*, Cornwell, T.J., and Perley, R.A., Eds., Astron. Soc. Pacific Conf. Ser., **19**, 76–80 (1991)

Yen, J.L., Kellermann, K.I., Rayhrer, B., Broten, N.W., Fort, D.N., Knowles, S.H., Waltman, W.B., and Swenson, G.W., Jr., Real-Time, Very-Long-Baseline Interferometry Based on the Use of a Communcations Satellite, *Science*, **198**, 289–291 (1977)

Young, A., Primiani, R., Weintroub, J., Moran, J., Young, K., Blackburn, L., Johnson, M., and Wilson, R., Performance Assessment of a Beamformer for the Submillimeter Array, SMA Technical Memo 163 (2016)

Zhang, X., Qian, Z., Hong, X., Shen, Z., and Team of CVN, Technology Development in Chinese VLBI Network, First International VLBI Technology Workshop, Haystack Observatory (2012). http://www.haystack.mit.edu/workshop/ivtw/program.html

Chapter 10
Calibration and Imaging

This chapter is concerned with the calibration and Fourier transformation of visibility data, mainly as applied to Earth-rotation synthesis. Methods for the evaluation of the visibility measurements on a rectangular grid of points, necessary for the use of the discrete Fourier transform as implemented with the fast Fourier transform (FFT) algorithm, are discussed. Phase and amplitude closure conditions, which are valuable calibration tools, are also described. Analysis of the causes of certain types of image defects is given. Special consideration is given for certain observing modes, such as spectral line, and conversion of frequency to velocity is described. In addition, methods of extracting astronomical information directly from visibility data by model fitting are described. These techniques are important even with arrays having excellent (u, v) coverage. Some methods of calculating Fourier transforms before the advent of the FFT are discussed in Appendix 10.3.

10.1 Calibration of the Visibility

The purpose of calibration is to remove, insofar as possible, the effects of instrumental and atmospheric factors in the measurements. Such factors depend largely on the individual antennas or antenna pairs and their associated electronics, so correction must be applied to the visibility data before they are combined into an image. Editing the visibility data to delete any that show evidence of radio interference or equipment malfunction is usually performed before the full calibration process. This largely entails examining samples of data for unexpected amplitude or phase variations. Data taken on unresolved calibration sources are particularly useful here, since the response to such a source is predictable and should vary only slowly and smoothly with time.

© The Author(s) 2017
A.R. Thompson, J.M. Moran, and G.W. Swenson Jr., *Interferometry and Synthesis in Radio Astronomy*, Astronomy and Astrophysics Library,
DOI 10.1007/978-3-319-44431-4_10

In the calibration procedure, we first consider instrumental factors that are stable with time over periods of weeks or more. These include:

1. antenna position coordinates that specify the baselines,
2. antenna pointing corrections resulting from axis misalignments or other mechanical tolerances,
3. zero-point settings of the instrumental delays, that is, the settings for which the delays from the antennas to the correlator inputs are equal.

These parameters vary only as a result of major changes such as the relocation of an antenna. They can be calibrated by observing unresolved sources with known positions (see Sect. 12.2). We assume here that they have been determined in advance of the imaging observations. We also assume that correction for the nonlinearity of signal quantization, which is discussed in Sect. 8.4, has been applied if required.

10.1.1   Corrections for Calculable or Directly Monitored Effects

Calibration of the visibility measurements for effects that vary during an observation principally involves correction of the complex gains of the antenna pairs. Such factors can be divided into those for which the behavior can be predicted or directly measured and those for which it must be determined by observing a calibration source during the observation period. Examples of effects that can be corrected for by calculation of their effects include:

1. the constant component of atmospheric attenuation as a function of zenith angle (see Sect. 13.1.3),
2. variation of antenna gain as a function of elevation caused by elastic deformation of the structure under gravity. This may be based on pointing observations as well as structural calculations.

Shadowing, in which one antenna partially blocks the aperture of another, can occur at close spacings and low elevation angles. In principle, it is a problem that should be calibratable, since the positions and structures of the antennas are known. However, the effect of the geometrical blockage is complicated by diffraction, the shape of the primary beam is modified, and the position of the phase center of the aperture is shifted, thus affecting the baseline. Overall, these effects are often too complicated to be analyzed, and data from shadowed antennas are often discarded.

Effects within the receiving system, or external to it, that can be continuously monitored during an observation include:

1. variation of system noise temperature, which can result from changes in the ground radiation picked up in the sidelobes as the antenna tracks or from changes in atmospheric opacity. This effect may also cause variation in the gain as a result

of automatic level control (ALC) action that is used in some instruments to adjust the signal levels at the sampler or correlator (see Sect. 7.6). Monitoring can be performed by injection of a low-level, switched, noise signal at the receiver input and detection of it later in the system.

2. phase variations in the local oscillator system monitored by round-trip phase measurement (see Sect. 7.2),
3. the variable component of atmospheric delay monitored by using water vapor radiometers mounted at the antennas (see Sect. 13.3).

Corrections for these effects are usually performed at an early stage of the calibration procedure.

10.1.2   Use of Calibration Sources

Further steps in the calibration involve parameters that may vary on timescales of minutes or hours and require the observation of one or more calibration sources. Note that the source that is the subject of the astronomical investigation will be referred to as the *target source* to distinguish it from the calibration source, or *calibrator*. From Eq. (3.9), we can write the small-field expression for the interferometer response as follows:

$$[\mathcal{V}(u, v)]_{\text{uncal}} = G_{mn}(t) \int_{\infty}^{\infty} \int_{-\infty}^{\infty} \frac{A_N(l, m)I(l, m)}{\sqrt{1 - l^2 - m^2}} e^{-j2\pi(ul+vm)} dl \, dm , \qquad (10.1)$$

where $[\mathcal{V}(u, v)]_{\text{uncal}}$ is the uncalibrated visibility, and $I(l, m)$ is the source intensity. The complex gain factor $G_{mn}(t)$ is a function of the antenna pair (m, n) and, as a result of unwanted effects, may vary with time. A_N is the antenna aperture normalized to unity for the direction of the main beam. It can be removed from the source image as a final step in the image processing. The factor $A_N(l, m)/\sqrt{1 - l^2 - m^2}$ is close to unity, and from here on, we generally omit it, except in the case of wide-field imaging. To calibrate $G_{mn}(t)$, an unresolved calibrator can be observed, for which the measured response is

$$\mathcal{V}_c(u, v) = G_{mn}(t)S_c , \qquad (10.2)$$

where the subscript c indicates a calibrator, and S_c is the flux density of the calibrator. In calibrating the gain, it is best to consider the amplitude and phase separately, since the errors in these two quantities generally arise through different mechanisms. For example, atmospheric fluctuations due to tropospheric inhomogeneity cause phase fluctuations but have little effect on the amplitudes. To calibrate the visibility of the target source, we can write

$$\mathcal{V}(u, v) = \frac{[\mathcal{V}(u, v)]_{\text{uncal}}}{G_{mn}(t)} = [\mathcal{V}(u, v)]_{\text{uncal}} \left[\frac{S_c}{\mathcal{V}_c} \right] . \qquad (10.3)$$

To observe the calibration source, it is usually placed at the phase center of its field. Then assuming that the calibrator is unresolved, the phase is a direct measure of the instrumental phase. Thus, phase calibration for the target source requires subtracting the calibrator phase from the observed phase. The visibility amplitude can be calibrated by using the moduli of the visibility terms in Eq. (10.3). The response to the calibrator should be corrected for the calculable and/or directly monitored effects before the gain calibration is performed. Where there are separate receiving channels for two opposite polarizations at each antenna, the calibration should be performed separately for each one. For measurements of source polarization, further calibration procedures are necessary, as described in Sect. 4.7.5.

Calibration observations require periodic interruption of observations of the target source. At centimeter wavelengths, the interval between calibration observations depends on the stability of the instrument and typically falls within the range of 15 min to 1 h. At meter and centimeter wavelengths, the ionosphere and the neutral atmosphere introduce gain and phase changes, and elimination of these may require observation of a calibrator at time intervals as short as a few minutes. At millimeter and submillimeter wavelengths, calibration at time intervals less than a minute is usually required.

As indicated by Eq. (7.38), $G_{mn} = g_m g_n^*$, so the measured gains for antenna pairs can be used to determine gain factors for the individual antennas. Using the individual antenna gain factors rather than the baseline gain factors reduces the calibration data to be stored and helps in monitoring the performance of individual antennas. Also, with this technique, some of the spacings can be omitted from the calibration observation so long as each of the antennas is included. In practice, gain tables including both amplitude and phase are generated for the antennas as a function of time, and the values are interpolated to the times at which data from the target source were taken. The interpolation should be done separately for the amplitude and phase, not for the real and imaginary parts of the gain; otherwise, the phase errors can degrade the amplitude, and vice versa. The desirable characteristics of a calibration source are the following.

Flux density.   The calibrator should be strong, so that a good signal-to-noise ratio is obtained in a short time, to reduce the (u, v) coverage lost from the target source. The gaps in the (u, v) coverage are more serious for a linear array, in which complete sectors are lost, than for a two-dimensional array, in which the instantaneous coverage is more widely distributed in u and v.

Angular width.   The calibrator should, if possible, be unresolved so that precise details of its visibility are not required.

Position.   The position of the calibrator should be close to that of the target source. Effects in the atmosphere or antennas that cause the gain to vary with pointing angle are then more effectively removed, and time lost in driving the antennas between the target source and calibrator positions is kept small. At millimeter wavelengths, where the atmospheric phase path is the main factor being calibrated, the calibrator distance must be within the angular scale of the irregularities. This usually means a distance of no more than a few degrees on the sky (see Sect. 13.4).

It is not always possible to find a calibrator that satisfies all of the above requirements. In such cases, it may be necessary to find a source that is largely unresolved and close to the target source and then calibrate it against one of the more commonly used flux density references such as 3C48, 3C147, 3C286, and 3C295. The last of these is the most reliable with regard to nonvariability. Thermal sources such as the compact planetary nebula NGC7027 may be useful as amplitude calibrators for short baselines. At millimeter wavelengths, it may be more difficult to find a source that provides a strong signal for test purposes or calibration. Disks of planets become resolved at rather short baselines, but the limb of the Moon or a planet can be useful: see Appendix 10.1.

The use of clusters of small sources as calibrators has been investigated by Kazemi et al. (2013). Such clusters might typically consist of two to ten sources of small angular diameter, and flux densities are correspondingly lower than required for single calibration sources. This approach allows calibrators to be found closer to the object under investigation and thus potentially increases the number available as well as reducing errors related to angular distance.

For VLBI observations with milliarcsecond resolution, there are fewer suitable calibrators. Angular structure on this scale is sometimes variable over periods of months, and caution is necessary if a previously measured and partially resolved source is to be used as a calibrator. An alternative approach to amplitude calibration of VLBI data involves use of the system temperatures and collecting areas of the individual antennas, as follows. The cross-correlation data should first be normalized to unity for the case in which the two input data streams are fully correlated. To obtain this normalization, the data are divided by the product of the rms values of the data streams at the two correlator inputs. (For two-level sampling, this rms value is unity, and for other types of sampling, the rms depends on the setting of the sampler thresholds with respect to the level of the analog signal.) Then, to convert the normalized correlation to visibility $\mathcal{V}$ with units of flux density (janskys), the amplitude is multiplied by the geometric mean of the system equivalent flux density (SEFD) values for the two antennas involved. The system equivalent flux density, SEFD $= 2kT_S/A$, is defined in Eq. (1.7). If the value of T_S corresponds to a signal plane above the atmosphere, then the resulting visibility values will be corrected for atmospheric losses. For VLBI data in which the phase may sometimes not be calibrated, the closure relationships in Sect. 10.3 allow images to be formed if absolute position is not required.

10.2   Derivation of Intensity from Visibility

10.2.1   Imaging by Direct Fourier Transformation

A straightforward method of obtaining an estimate of the intensity distribution from measured visibility data is by *direct* Fourier transformation, that is, by performing the transformation without putting the visibility into any special form such as interpolating it onto a uniform grid. The measured visibility $\mathcal{V}_{\text{meas}}(u, v)$ can be written

$$\mathcal{V}_{\text{meas}}(u, v) = W(u, v)w(u, v)\mathcal{V}(u, v) \,, \tag{10.4}$$

where $W(u, v)$ is the transfer function or spatial sensitivity function introduced in Sect. 5.3, and $w(u, v)$ represents any applied weighting. The Fourier transform of Eq. (10.4) is the measured intensity distribution (i.e., the image), which is

$$I_{\text{meas}}(l, m) = I(l, m) * * b_0(l, m) \,, \tag{10.5}$$

where the double asterisk indicates two-dimensional convolution, and b_0 is the synthesized beam, which is the Fourier transform of the weighted transfer function:

$$b_0(l, m) \longleftrightarrow W(u, v)w(u, v) \,, \tag{10.6}$$

where $\longleftrightarrow$ indicates the Fourier transform relationship. Effects such as those of noncoplanar baselines, signal bandwidth, and visibility averaging are not included here. $b_0(l, m)$ is also known as the point-source response function or the dirty beam, in the context of the CLEAN deconvolution algorithm, which is discussed in Sect. 11.1.

The visibility is measured at an ensemble of n_d points in the (u, v) plane. If the antennas are identically polarized and the source is unpolarized, the direct Fourier transform of these data is represented by

$$I_{\text{meas}}(l, m) = \sum_{i=1}^{n_d} w_i \left[\mathcal{V}_{\text{meas}}(u_i, v_i)e^{j2\pi(u_i l + v_i m)} + \mathcal{V}_{\text{meas}}(-u_i, -v_i)e^{-j2\pi(u_i l + v_i m)} \right] \,. \tag{10.7}$$

The fundamental issue in image synthesis is whether we can recover $I(l, m)$ from $I_{\text{meas}}(l, m)$. In principle, Eq. (10.4) can be used to determine $\mathcal{V}(u, v)$ as $\mathcal{V}_{\text{meas}}(u, v)/W(u, v)w(u, v)$. The image can be calculated exactly if $W(u, v)w(u, v)$ is everywhere nonzero.

Bracewell and Roberts (1954) pointed out that, in principle, there are an infinite number of solutions to the convolution in Eq. (10.5), since one can add any arbitrary visibility values in the unsampled areas of the (u, v) plane. The Fourier transform of these added values constitutes an invisible distribution that cannot be detected by any instrument with corresponding zero areas in the transfer function. It may be argued that in interpreting observations from any radio telescope, one should maintain only zeros in the unmeasured regions of spectral sensitivity, to avoid arbitrarily generating information. On the other hand, the zeros are themselves arbitrary values, some of which are certainly wrong. What is wanted is a procedure that allows the visibility at the unmeasured points to take values consistent with the most reasonable or likely intensity distribution, while minimizing the addition of arbitrary detail. Positivity of intensity and limitation of size of the angular structure of a source are expected characteristics that can be introduced into the imaging process. Image restoration techniques that implicitly generate nonzero visibility values at unmeasured (u, v) points include CLEAN, maximum entropy, and compressed sensing, which are discussed in Chap. 11.

10.2.2   Weighting of the Visibility Data

To obtain the best signal-to-noise ratio in the summation of measurements that contain Gaussian noise, the individual data values should be weighted inversely as their variances. The same is true for the combination of sinusoidal components in an image of a source, the amplitudes of which are proportional to the corresponding visibility points. Thus, for the best signal-to-noise ratio, the weights w_i in Eq (10.7) should be inversely proportional to the variances. If the data are obtained with a uniform array of antennas and receivers, and the averaging time is the same for all data points, then the variances should all be the same, and maximum signal-to-noise ratio is obtained by including all measurements with the same weight. This is known as *natural weighting*. For many arrays, natural weighting results in a poor beam shape with wide skirts because the shorter spacings are overemphasized. Thus, the usual approach is to include in the weighting a factor that is inversely related to the area density of the data in the (u, v) plane. The area density $\rho_\sigma(u, v)$ can be defined such that the number of points in the range $u \pm \frac{1}{2}du$, $v \pm \frac{1}{2}dv$ is $\rho_\sigma(u, v)du\,dv$ (Thompson and Bracewell 1974). Although ρ_σ at any given point depends on the size of the increments du and dv, it is usually possible to specify the variation of relative density and correct for it satisfactorily. As a simple example, in the observation of a high-declination source with an east–west array in which the antenna spacings are nonredundant integral multiples of a unit value, the visibility points lie on concentric circles, as in Fig. 10.1. Then, if the visibility is measured at uniform increments in hour angle, the area density at any ring is inversely proportional to the radius of the ring. With $w(u, v)$ proportional to $1/\rho_\sigma(u, v)$, the effective density of the data is uniform within a circle of radius u_{max} determined by the maximum spacing. The beam then closely approximates the Fourier transform of a circular disk function, which, normalized to unity at the maximum, is given by

$$\frac{J_1(2\pi l u_{max})}{\pi l u_{max}}, \tag{10.8}$$

Fig. 10.1 Transfer function (spacing loci) in the (u, v) plane for observations of a high-declination source using an east–west array with uniform increments in antenna spacing. The points indicate visibility measurements, and their (u, v) positions reflected through the origin, for uniform intervals of time. The angle ϕ indicates data for a specific hour angle. If the visibility values are weighted in proportion to the radii of the loci, the density of the visibility data is effectively uniform out to a radius u_{max}.

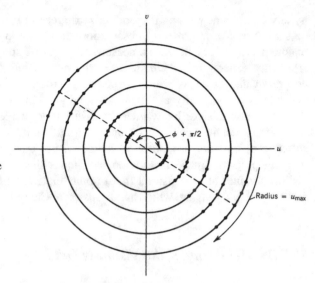

where J_1 is the Bessel function of the first kind and first order. $2J_1(x)/x$ is called a jinc function, by analogy to a sinc function. The full width of the beam at half-maximum (FWHM) is $0.705\,u_{max}^{-1}$, and the first sidelobe response is 13.2% of the main beam.[1] Similarly, if the effective density of measurements is uniform within a rectangular area of dimensions $2u_{max} \times v_{max}$, the synthesized beam is closely approximated by

$$\frac{\sin(2\pi u_{max} l)}{2\pi u_{max} l} \times \frac{\sin(2\pi v_{max} m)}{2\pi v_{max} m}. \tag{10.9}$$

This beam is not circularly symmetrical, and the first sidelobe has a maximum value of 22% in the east–west and north–south directions through the beam center.

With uniform weighting, the strong, near-in sidelobes (close to the main beam) in Fig. 10.2 obscure low-level detail and thereby reduce the range of intensity levels that can be reliably measured. The near-in sidelobes of the functions in expressions (10.8) and (10.9) can be reduced at the expense of some increase in the width of the synthesized beam by introducing a Gaussian or similar taper into the weighting function. The effect of such tapering of the visibility is shown in Fig. 10.2. The taper can be specified in terms of the amplitude of the tapering function at a distance u_{max} from the (u, v) origin; a taper to ~ -13 dB of the central value

---

[1] This synthesized response should not be confused with the power pattern of a uniformly illuminated antenna with circular aperture of radius r, which is proportional to $[J_1(2\pi rl/\lambda)/(\pi rl/\lambda)]^2$ and has a full width at half-maximum of $0.514\lambda/r$, first null at $0.610\lambda/r$, and first sidelobe of 1.7%. The antenna pattern is proportional to the Fourier transform of the autocorrelation function of a uniform circular aperture.

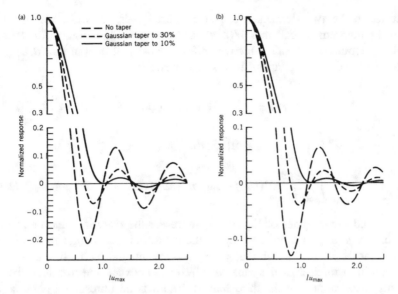

Fig. 10.2 Examples of synthesized beam profiles. Curves for no taper correspond to a visibility distribution that is uniform within (**a**) a rectangular area of width $2u_{max}$, and (**b**) a circular area of diameter $2u_{max}$. For no taper, the responses correspond to expression (10.9) for (**a**) and (10.8) for (**b**). The effects of Gaussian tapers that reduce the visibility at the edge of the distribution to 30% and to 10% are also shown. Note the difference in the ordinate scales.

is commonly used. With such a taper, the weighting $w(u, v)$ is the product of two functions: $w_u(u, v)$, the weighting required to obtain uniform effective density, and $w_t(u, v)$, the tapering function. Thus, the synthesized beam is the Fourier transform of $W(u, v)w_u(u, v)w_t(u, v)$:

$$b_0(l, m) = \overline{W}(l, m) * * \overline{w}_u(l, m) * * \overline{w}_t(l, m) , \qquad (10.10)$$

where the bar denotes a Fourier transform. The Fourier transform of $W(u, v)w_u(u, v)$ is simply the beam obtained with uniform effective density, for example, as in expressions (10.8) or (10.9). If $w_t(u, v)$ is a two-dimensional Gaussian function, its Fourier transform is also a Gaussian. Thus, the sidelobe reduction results from convolution with a Gaussian in the (l, m) domain. The variances of functions are additive under convolution [see, e.g., Bracewell (2000)], so the beam obtained by convolution with $\overline{w}_t$ is broader than that with no tapering, as is evident in Fig. 10.2.

An interesting property of the uniform weighting is that it minimizes the mean-squared deviation of the resulting intensity from the true intensity, within the constraint that unmeasured visibility values remain zero. This can be understood as follows. Since the true intensity distribution $I(l, m)$ and the true visibility function $\mathcal{V}(u, v)$ are a Fourier pair, and the weighted measured visibility and the derived intensity $I_0(l, m)$ are a Fourier pair, it follows that the differences between these quantities in the two domains are also a Fourier pair, to which we can apply Parseval's theorem. Recall that $W(u, v)$ is the transfer function, $w_u(u, v)$ is the weighting required to obtain effective uniform density of data in the (u, v) plane, and $w_t(u, v)$ is an applied taper. Thus, we can write

$$
\int\int_{\text{meas}} |\mathcal{V}(u, v) - \mathcal{V}(u, v)W(u, v)w_u(u, v)w_t(u, v)|^2 \, du \, dv
$$

$$
+ \int\int_{\text{unmeas}} |\mathcal{V}(u, v)|^2 \, du \, dv
$$

$$
= \int_{-\infty}^{\infty} \int_{-\infty}^{\infty} |I(l, m) - I_0(l, m)|^2 \, dl \, dm . \tag{10.11}
$$

The first and second lines of Eq. (10.11) represent the measured and unmeasured areas of the (u, v) plane, respectively. In the measured area, $W(u, v)w_u(u, v) = 1$. For the case of uniform weighting, $w_t = 1$, so the integral on the first line is zero. This condition minimizes the squared difference between the true and observed intensity distributions on the third line. If $I(l, m)$ is an unresolved point source, then $I_0(l, m)$ is equal to the synthesized beam. The uniform weighting minimizes the squared difference, over 4π steradians, between the synthesized beam and the response to a point source as it would be observed with unlimited (u, v) coverage. In this sense, it is sometimes said that uniform weighting minimizes the sidelobes of the synthesized beam. However, as shown in Fig. 10.2, a Gaussian taper reduces the sidelobes outside of the main beam at the expense of widening the beam. Images derived from visibility data that are uniformly weighted within the measured area of the (u, v) plane have been referred to as the *principal solution* or *principal response* (Bracewell and Roberts 1954). The related process of reducing the sidelobe response in optical imaging is called apodization, for which there is an extensive literature; see, for example, Jacquinot and Roizen-Dossier (1964) and Slepian (1965).

10.2.2.1   Robust Weighting

With large arrays, the visibility data must be interpolated onto a uniform grid as described in Sect. 5.2 in order to make computations tractable. The simplest approach is called cell averaging, where each data point is associated with the (u, v) grid point nearest to it. The number of points averaged in a cell will decrease with increasing (u, v) distance, and many cells will have zero entries. Thus, the variance of the visibility estimates will vary considerably over the (u, v) plane. A conflict arises between the goal of forming a synthesized beam that is narrow and has low sidelobes and achieving the optimum sensitivity for the detection of weak sources. The best strategy for detecting a weak point source in the field is to use natural weighting, i.e., performing the image transform with variance weighting. On the other hand, if the signal-to-noise ratio is high, an image with better resolution and lower sidelobes can be obtained with uniform weighting.

Briggs (1995) introduced a logarithmic parametrized scheme that allows a continuous variation in weighting between uniform and variance weighting. The process is called robust weighting. The weighting of cell (i, k) in the (u, v) plane whose visibility has an rms error of σ_{ik} is specified as

$$w_{ik} = \frac{1}{S^2 + \sigma_{ik}^2}, \qquad (10.12)$$

where S is a parameter defined by

$$S^2 = \frac{(5 \times 10^{-R})^2}{\overline{w}}. \qquad (10.13)$$

R is the robustness factor, and $\overline{w}$ is the average variance weighting factor over the n_c cells in the image,

$$\overline{w} = \frac{1}{n_c} \sum \frac{1}{\sigma_{ik}^2}. \qquad (10.14)$$

The nominal range of R is –2 to 2. $R = 2$ makes S very small with respect to w so that the weighting approaches natural weighting, whereas $R = -2$ makes S large with respect to w so that the weighting approaches the uniform weighting. $R = 0$ produces an rms that is midway between the values for $R = -2$ and 2. R is called the robustness factor because as it increases, the image is more immune to errors in calibration or errors due to radio frequency interference, because the effect of a bad point in a cell with few data points is deemphasized as R increases. An example of how the synthesized beamwidth and rms noise vary with R is shown in Fig. 10.3. In the vicinity of $R = 0$, which is the normal default value, the beamwidth and rms noise are most sensitive to changes in R. For the example shown in Fig. 10.3, the beamwidth increases by 5%, and the rms noise decreases by 45% as R increases from –0.5 to 0.5. For inhomogeneous arrays such as those used in VLBI, the gain in sensitivity can increase markedly for little increase in beamwidth.

10.2.3 Imaging by Discrete Fourier Transformation

The speed of the fast algorithm for the discrete Fourier transform (FFT), briefly discussed in Sect. 5.2, is a major advantage in computing large images. However, the use of the FFT introduces two complications in addition to those discussed for the direct transform: (1) the necessity to evaluate the visibility at points on a rectangular grid and (2) the resulting possibility of aliasing of parts of the image from outside the

Fig. 10.3 Synthesized beamwidth vs. normalized rms noise level in an image for robustness factor R ranging from -2 to 2. The calculations are for the source 1987A (Dec. $= -69°$) observed with two tracks of the Australia Telescope (configurations 6A and 6C) of about 7-h duration each. Adapted from Briggs (1995).

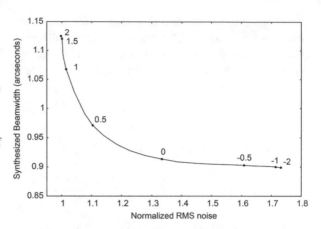

synthesized field. The evaluation at the grid points is often referred to as *gridding*. The output of such a process can be represented by the following expression:

$$\frac{w(u, v)}{\Delta u \Delta v} \, {}^{2}\mathrm{III}\left(\frac{u}{\Delta u}, \frac{v}{\Delta v}\right) \{C(u, v) * * [W(u, v)\mathcal{V}(u, v)]\} \ . \tag{10.15}$$

Here the visibility $\mathcal{V}(u, v)$, measured at the points denoted by the transfer function $W(u, v)$, is convolved with a function $C(u, v)$ to produce a continuous visibility distribution. This is then resampled at points in a rectangular grid with incremental spacings Δu and Δv. This process is sometimes referred to as *convolutional gridding*. The resampling is here represented by the two-dimensional shah function ${}^{2}\mathrm{III}$ (Bracewell 1956b), defined by

$$\,{}^{2}\mathrm{III}\left(\frac{u}{\Delta u}, \frac{v}{\Delta v}\right) = \Delta u \Delta v \sum_{i=-\infty}^{\infty} \sum_{k=-\infty}^{\infty} {}^{2}\delta(u - i\Delta u, v - k\Delta v) \,, \tag{10.16}$$

where ${}^{2}\delta$ is the two-dimensional delta function. The weighting to optimize the beam is applied to the resampled data. Although this process is described mathematically in terms of convolution and resampling, in practice the convolution is evaluated only at the grid points. The Fourier transform of (10.15) represents the measured intensity:

$$I_{\mathrm{meas}}(l, m) = \, {}^{2}\mathrm{III}(l\Delta u, m\Delta v) * * \overline{w}(l, m) * * \{\overline{C}(l, m)\left[\overline{W}(l, m) * * I(l, m)\right]\} \ . \tag{10.17}$$

As a result of the Fourier transformation, the intensity function $I(l, m)$ is convolved with the Fourier transform of the transfer function; multiplied by $\overline{C}(l, m)$, which is the Fourier transform of the convolving function; and then convolved with the Fourier transforms of the weighting and resampling functions. This last convolution causes the whole image to be replicated at intervals Δu^{-1} in l and Δv^{-1} in m.

These intervals are equal to the dimensions of the image in the (l, m) plane; that is, $\Delta u^{-1} = M\Delta l$ and $\Delta v^{-1} = N\Delta m$, for an $M \times N$ point array. The function $\overline{C}(l, m)$ takes the form of a taper applied to the image, and if this function does not vary greatly on the scale of the width of $\overline{w}(l, m)$, which is usually the case for large images, then $\overline{w}(l, m)$ in Eq. (10.17) can be convolved directly with $\overline{W}(l, m) * * I(l, m)$, and Eq. (10.17) becomes

$$I_{\text{meas}}(l, m) \simeq {}^{2}\text{III}(l\,\Delta u, m\,\Delta v) * * \left\{\overline{C}(l, m)\,[I(l, m) * * b_0(l, m)]\right\} , \qquad (10.18)$$

where the synthesized beam $b_0(l, m)$ enters through the relationship in Eq. (10.6). Comparison with Eq. (10.5) shows that the effect of the gridding and resampling is to multiply the image by $\overline{C}(l, m)$ and replicate it. This replication introduces the aliasing.

Returning to the estimation of the visibility at the grid points, we might perhaps expect the best technique to be some form of exact interpolation so that the resulting values are equal to those that would be obtained by measurement at the grid points. A method of this type has been described by Thompson and Bracewell (1974). However, the problem of aliasing remains, and the most effective way to deal with this is to convolve the data in the (u, v) plane with the Fourier transform of a function that, in the (l, m) plane, varies very little over the image and then falls off rapidly at the image edges. We therefore look for a convolving function $C(u, v)$ for which the Fourier transform $\overline{C}(l, m)$ has these properties. An ideal function with infinitely sharp cutoff at the field edges would completely eliminate the aliasing since there would be no overlap of the replicated images. Unfortunately, this ideal is not practical because the required convolving function is not bounded in the (u, v) plane. Nevertheless, a very worthwhile degree of suppression of the aliasing is possible with a careful choice of functions. A common and convenient practice is to combine both the gridding, and the convolution to minimize aliasing, into a single operation. Note, however, that at the (u, v) points at which the measurements are made, the function $C(u, v) * * [W(u, v)\mathcal{V}(u, v)]$, in general, is not equal to the measured visibility $\mathcal{V}(u, v)$. Thus, the gridding process cannot precisely be described as interpolation. Also, because of the convolution, the sampled points represent averages of the visibility local to the grid points, rather than samples of the visibility function. Finally, note also that although convolution is effective in suppressing artifacts that result from gridding of the data, it does not reduce sidelobe or ringlobe responses to sources located outside the area of the image.

10.2.4   Convolving Functions and Aliasing

From the foregoing discussion, we can conclude that the point of principal concern in the use of the FFT is the choice of convolving function. A detailed discussion of convolving functions is given by Schwab (1984). It is convenient to consider those that are separable into one-dimensional functions of the same form for u and v, that is,

$$C(u, v) = C_1(u)C_1(v) .$$

(10.19)

We therefore discuss some examples of the function C_1.

Rectangular Function. This function is the one used in cell averaging discussed in Sect. 5.2.2. It can be written

$$C_1(u) = (\Delta u)^{-1} \Pi \left(\frac{u}{\Delta u} \right) ,$$

(10.20)

where Π is the unit rectangle function defined by

$$\Pi(x) = \begin{cases} 1, & |x| \leq \frac{1}{2} \\ 0, & |x| > \frac{1}{2} . \end{cases}$$

(10.21)

The Fourier transform of $C_1(u)$ is

$$\overline{C}_1(l) = \frac{\sin(\pi \Delta u l)}{\pi \Delta u l} .$$

(10.22)

At the edge of the synthesized field, $l = (2\Delta u)^{-1}$ and $\overline{C}_1(1/2\Delta u) = 2/\pi$. The image is tapered by a sinc-function profile in the l and m directions and a sinc-squared profile along the diagonals. Equation (10.22) is plotted in Fig. 10.4, and the value at the first maximum outside the edge of the image is 0.22 of the value at the image center. The effect of aliasing is shown more directly in Fig. 10.5a, which is a plot of $\overline{C}_1(l)/\overline{C}_1[f(l)]$, where $f(l)$ is the value of l within the image [i.e., $|f(l)| < (2\Delta u)^{-1}$] at which the alias of a feature of l would appear. This quantity gives the relative response to an aliased feature in an image that has been corrected for the taper imposed by $\overline{C}_1(l)$. It is clear that simple averaging of points within a rectangular cell performs poorly in suppressing aliasing.

Gaussian Function. Here we have

$$C_1(u) = \frac{1}{\alpha \Delta u \sqrt{\pi}} e^{-(u/\alpha \Delta u)^2}$$

(10.23)

and

$$\overline{C}_1(l) = e^{-(\pi \alpha \Delta u l)^2} .$$

(10.24)

The value of the constant α can be chosen to vary the widths of the functions as desired. If α is too small, $C_1(u)$ will be too narrow, and only visibility measurements that are close to grid points will be used effectively in the imaging. If α is too large, the function $\overline{C}_1(u)$ will taper the resulting image too severely. The Gaussian

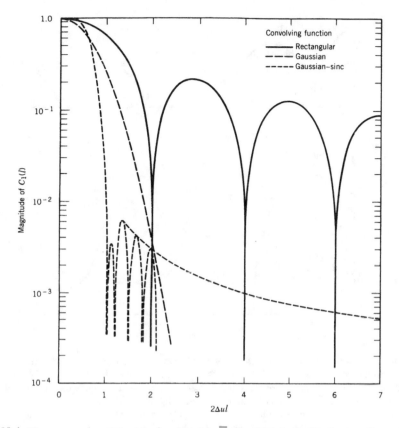

Fig. 10.4 Three examples of the tapering function $\overline{C}_1(l)$, which is the Fourier transform of the convolving function $C_1(u)$. For the Gaussian convolving function, $\alpha = 0.75$. For the Gaussian-sinc convolving function, $\alpha_1 = 1.55$, $\alpha_2 = 2.52$, and beyond the fourth subsidiary maximum, only the envelope of the maxima is shown. On the abscissa scale, the center of the image is at zero and the edge at 1.0. The data for the Gaussian-sinc function were computed by F. R. Schwab.

convolving function was used in the early years of the Westerbork array with $\alpha = 2\sqrt{\ln 4}/\pi = 0.750$ (Brouw 1971). The value of the factor $e^{-(u/\alpha\Delta u)^2}$ in $C_1(u)$ is then equal to 0.41 for a point on a diagonal in the (u, v) plane midway between two grid points. Thus, all measured points enter into the image with significant weights, and at the edge of the image, the tapering factor $\overline{C}_1 = \frac{1}{4}$. A curve for the Gaussian function is shown in Fig. 10.4.

Gaussian-Sinc Function. The ideal form for the image tapering function $\overline{C}_1(l)$ would be a rectangle, which corresponds to convolution with a sinc function, as in Eq. (10.22). However, the envelope of a sinc function falls to zero slowly as its argument increases, and the computation required for the convolution becomes large. Truncation of the sinc function is undesirable because in the l domain, the desired rectangular function is convolved with the Fourier transform of the

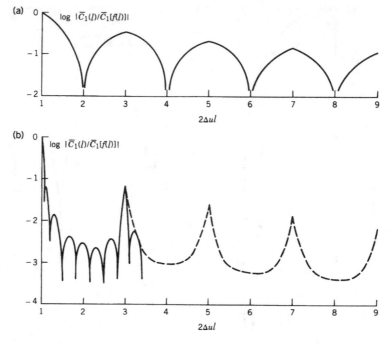

Fig. 10.5 Logarithmic plot of the factor by which the amplitudes of structures outside the image are multiplied when aliased into the image. On the abscissa scale, 1.0 is the edge of the image and 2, 4, 6, ..., are the centers of the adjacent replications. (**a**) Aliasing factor for a rectangular convolving function of width equal to Δu (cell averaging). (**b**) Aliasing factor for a Gaussian-sinc convolving function with the optimized parameters given in the text. The broken line indicates the envelope of the maxima. Data computed by F. R. Schwab.

truncation function, and this destroys the sharp cutoff at the edges of the image. A better procedure is to multiply the sinc function with a Gaussian, which gives

$$C_1(u) = \frac{\sin(\pi u/\alpha_1 \Delta u)}{\pi u} e^{-(u/\alpha_2 \Delta u)^2} \tag{10.25}$$

and

$$\overline{C}_1(l) = \Pi(\alpha_1 \Delta u l) * \left[\sqrt{\pi}\alpha_2 \Delta u e^{-(\pi \alpha_2 \Delta u l)^2} \right]. \tag{10.26}$$

Good performance is obtained with $\alpha_1 = 1.55$ and $\alpha_2 = 2.52$, with the convolution extending over an area about $6\Delta u$ in width. Corresponding curves for $\overline{C}_1(l)$ and the resulting aliasing are given in Figs. 10.4 and 10.5b. This convolving function is much better than either of the two previous examples.

Spheroidal Functions. Various other functions can be found that have the features desirable for convolution. As a measure of the effectiveness of the suppression of aliasing, (Brouw 1975) has suggested the following quantity:

$$\frac{\int \int_{\text{image}} \left[\overline{C}(l, m) \right]^2 dl \, dm}{\int_{-\infty}^{\infty} \int_{-\infty}^{\infty} \left[\overline{C}(l, m) \right]^2 dl \, dm}, \tag{10.27}$$

which shows the fraction of the integrated squared amplitude of the tapering function that falls within the image. Maximization of (10.27) provides a criterion for choosing a convolving function. This approach led to consideration of the prolate spheroidal wave functions [see, e.g., Slepian and Pollak (1961)] and the spheroidal functions (Rhodes 1970). Schwab (1984) found that among functions investigated, the latter provide the best approach to an optimum convolving function. The spheroidal functions are solutions to certain differential equations and are not expressible in simple analytic form. In applying such functions for convolution of visibility data, they are computed in advance to provide a look-up table. Comparison of some functions of this type with the Gaussian-sinc function shows that the aliasing factor $\overline{C}_1(l)/\overline{C}_1 [f(l)]$ falls off about as rapidly from the center to the edge of the image, but as l increases beyond the edge of the image, it reaches values an order of magnitude or more lower than those for the Gaussian-sinc function Briggs et al. (1999). Computational capacity complicates the choice of the optimal function, since it limits the area of the (u, v) plane over which the convolution can be performed. Commonly, this area is six to eight grid cells wide and centered on the point to be interpolated. Roundoff errors in the Fourier transform are amplified in the removal of the tapering function and may limit the allowable taper at the edges of the image.

10.2.5   Aliasing and the Signal-to-Noise Ratio

Features aliased into an image from outside the boundary include not only the images of features on the sky but also the random variations resulting from the system noise. If we consider a direct Fourier transform of the noise component of the measured visibility, it is clear from Eq. (10.7) that for any point (l, m), the visibility data are weighted by complex exponential factors, all of which have the same modulus. Since the noise is independent at each data point in the (u, v) plane, the variance of the noise in the (l, m) plane is statistically constant in all parts of the image. If the FFT is used, however, the rms noise level across the image is multiplied by the function $\overline{C}(l, m)$, and details beyond the image edge are aliased into the image. Note that the noise contributions combine additively in the variance. Thus, in one dimension, the noise variance as a function of l is proportional to

$$\text{III}(l\Delta u) * |C_1(l)|^2 . \tag{10.28}$$

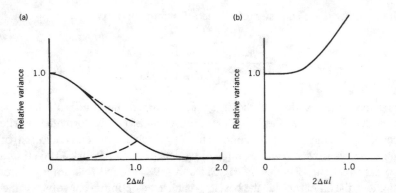

Fig. 10.6 Effect of aliasing on the variance of the noise across an image. The abscissa in each case is l in units of half the image width; the image center is at 0, the edge at 1.0, and the center of the adjacent replication at 2.0. (**a**) Solid curve shows the taper for a Gaussian convolving function C_1, and dashed curves show the effect of aliasing. (**b**) Variance of the noise including aliased component after correction for taper C_1. Adapted from Napier and Crane (1982) [see also Crane and Napier (1989)].

The replication resulting from the FFT can also be written in terms of a summation, and the variance of the noise at a point l within the image is then proportional to

$$\sum_{i=-\infty}^{\infty} |\overline{C}_1(l + i\Delta u^{-1})|^2 . \tag{10.29}$$

Usually $\overline{C}_1(l)$ decreases sufficiently with l that only the noise from the adjacent replication of the image makes a serious contribution through aliasing. This contribution is greatest near the edge of the image, as shown in Fig. 10.6.

If the convolving function is the Gaussian-sinc type, we see from Fig. 10.5b that, except for values of $2\Delta ul$ between 1.0 and 1.1, aliased features are reduced in amplitude by a factor $< 10^{-1}$, and in the square of the amplitude by $< 10^{-2}$. Thus, there is no significant increase in the noise level as a result of aliasing, except in a narrow zone at the edge of the image.

At the other extreme, the aliasing is most serious in the case of cell averaging, for which $C_1(u)$ is the sinc function given by Eq. (10.22). Expression (10.29) then becomes

$$\sum_{i=-\infty}^{\infty} \frac{\sin^2 [\pi(\Delta ul + i)]}{[\pi(\Delta ul + i)]^2} = 1 , \tag{10.30}$$

which indicates that the aliasing exactly cancels the taper, and the variance of the noise is constant with l, that is, before any correction for tapering of the astronomical features in the image is applied. (This result could also be deduced from the fact that in cell averaging, each visibility measurement contributes to one grid point only, and the noise components of the visibility at the grid points are therefore independent.) However, the intensity distribution of the sky within the field being imaged is tapered by the function $\overline{C}_1(l)$, and correction for this taper then causes the noise to increase toward the edges of the image. For the sinc-function taper, the noise is increased by a factor of $\pi/2$ at the edge of the image on the l and m axes and by $(\pi/2)^2$ at the corners. At the center of the image, the aliased contribution originates at points for which $2\Delta ul$ is an even integer in the plots in Fig. 10.5, and in both cases shown, the aliasing factor $\overline{C}_1(l)/\overline{C}_1[f(l)]$ drops to a very low value. With any of the convolving functions that we have considered, there is no significant increase in the noise at the center of the image, and the signal-to-noise ratio for a source at that point is determined by the factors discussed in Sect. 6.2.

10.2.6   Wide-Field Imaging

To take full advantage of large new instruments with wide bandwidths, high sensitivity, and full polarization responses, it is necessary to measure the radio sky down to the level of the background radiation from the Epoch of Reionization (EoR) and to be able to separate out components from individual radio sources that overlie the background. The width of the synthesized field may be much greater than a few degrees, so the image is no longer the Fourier transform of the visibility function. The basic requirement for such an analysis is an equation for the visibility values that would be measured for a given brightness distribution, taking account of all details of the locations and characteristics of the individual antennas, the path of the incoming radiation through the Earth's atmosphere including the ionosphere, the atmospheric transmission, etc. This is the *interferometer measurement equation* introduced in Sect. 4.8. In its basic form, it describes the response of a single pair of antennas and is thus applicable to any specified system of antennas and any brightness distribution, to provide values of the visibility for each antenna pair. It includes direction-dependent effects such as the primary beam patterns of the antennas, polarization effects that vary with the alignment of the polarization of the source relative to that of the antennas, and the baselines of the antenna pairs. These must be accounted for without small-field or other approximations. Direction-independent effects such as large-scale propagation in the atmosphere and the ionosphere, and the response of the receiving system, can also be included.

The reverse operation, i.e., the calculation of the optimum estimate of the image from the measured visibility values, is less simple. Taking the Fourier transform of the observed visibility function usually produces a brightness function with physically distorted features such as negative brightness values in some places. However, starting with a simple but physically realistic model for the brightness, the measurement equation can accurately provide the corresponding visibility values that would be observed. By comparing these with the observed values, it is possible to adjust the brightness model toward the observed distribution and, by iterative repetition of this process, to arrive at an image that agrees with the visibility measurements to within the uncertainties resulting from the noise. An example of this process of making an image of a radio source is described by Rau et al. (2009), who use an iterative Newton–Raphson approach, as follows.

1. Calibrate the interferometer responses by making observations of sources with known position and structure. This includes measurement of both parallel and cross polarizations (for circular or linear polarization, whichever is used).
2. Make observations of the area of sky under investigation and, using the calibration data from (1), determine the (complex) visibility function for points in a rectangular grid in the (u, v) plane.
3. Using the measurement equation, calculate visibility values for a model source centered in the area in (2), for the (u, v) values of the gridded visibility measurements in (2). The model can make use of any prior information on the source under observation, but otherwise a point source model will generally suffice.
4. Subtract the calculated visibilities for the model source from the corresponding observed values in (2), and take the Fourier transform of the difference to provide a brightness function that represents the difference between the sky and the model.
5. Use the brightness function from (4) to improve the model brightness function, i.e., to make it closer to the visibilities measured in (2). To do this, add a fraction γ of the brightness function from (4), to the model, to provide a new model source. γ is the loop gain in the process.
6. Calculate the visibility values (Vm_j) for the improved source model from (5), and if they are sufficiently close to the observed visibilities (Vo_j), go to (7). Otherwise, return to (4) with the improved model from (5). Comparison of the observed and model visibilities involves computation of $\chi^2 = \sum_j [(Vo_j - Vm_j) (Vo_j - Vm_j)^*]$, which is minimized by the iterative process.
7. Take the residual differences between the observed and model visibility values in (6), Fourier transform them to brightness, and add them to the model values from (6). This step ensures that the Fourier transform of the final model is equal to the observed visibilities.

The number of iterations (from step 6 back to step 4) required varies inversely with the value of γ in step 5. A value of $\gamma = 0.5$ or less allows the optimum solution to be approached more accurately by using smaller steps. The choice of the model source in step 3 is not critical. For example, if the source is actually a wide one and a point source is used as the model, then in step 3, the model visibility values will have significant values over a much wider range of (u, v) spacings than that of the measured visibilities. However, in step 4, the fraction γ of the excess visibilities is subtracted, and the model sequentially moves toward the measured visibility, within the limits of the noise. Obtaining an image that is a realistic model of the sky, and is in agreement with the measured visibility, is the essential goal in synthesis imaging. This iterative procedure with χ^2 minimization illustrates the basic approach to a number of the processes used in imaging.

10.3   Closure Relationships

Closure effects are relationships between visibility values for baselines that form a closed figure, for example, a triangle or quadrilateral with the antennas at the vertices. As shown by Eqs. (7.37) and (7.38), the correlator output for antenna pair (m, n) can be written as

$$r_{mn} = G_{mn}\mathcal{V}_{mn} = g_m g_n^* \mathcal{V}_{mn} , \qquad (10.31)$$

where G_{mn} is the complex gain for the antenna pair, and g_m and g_n are gain factors for the individual antennas. We ignore any gain terms that do not factor into the terms for individual antennas (see Sect. 7.3.3), i.e., those that are baseline dependent.

Considering first the phase relationships, we represent the arguments of the exponential terms of r_{mn}, g_m, g_n, and $\mathcal{V}_{mn}$ by ϕ_{mn}, ϕ_m, ϕ_n, and ϕ_{vmn}, respectively. Thus, we can write

$$\phi_{mn} = \phi_m - \phi_n + \phi_{vmn} . \qquad (10.32)$$

For three antennas m, n, and p, the phase closure relationship is

$$\begin{aligned} \phi_{c_{mnp}} &= \phi_{mn} + \phi_{np} + \phi_{pm} \\ &= \phi_m - \phi_n + \phi_{vmn} \\ &\quad + \phi_n - \phi_p + \phi_{vnp} \\ &\quad + \phi_p - \phi_m + \phi_{vpm} \end{aligned} \qquad (10.33)$$

or

$$\phi_{c_{mnp}} = \phi_{vmn} + \phi_{vnp} + \phi_{vpm} . \qquad (10.34)$$

The antenna gain terms, g_m and so on, contain the effects of the atmospheric paths to the antennas as well as instrumental effects, and since these terms do not appear in Eq. (10.34), it is evident that the combination of the three correlator output phases constitutes an observable quantity that depends only on the phase of the visibility. This property of the phase closure relationships was first recognized and used by Jennison (1958).

If a point source is observed, then the visibility phases are all zero, and, in the absence of receiver noise, the closure phase is also zero. Note that if the rms phase noise on each baseline is σ, the rms noise in the closure phase is $\sqrt{3}\sigma$.

To help visualize the phase closure concept, consider three stations of an array observing a point source, as shown in Fig. 10.7. We depict the origin of the instrumental phase terms associated with each station as being caused by atmospheric delay along each line of sight. The total visibility phase on each

Fig. 10.7 A three-baseline triangle for antennas m, n, and p. s is the unit vector in the direction of the source. The phases of the antenna-based gain factors are represented by atmospheric cloudlets that cause excess phase shifts of ϕ_m, ϕ_n, and ϕ_p, respectively.

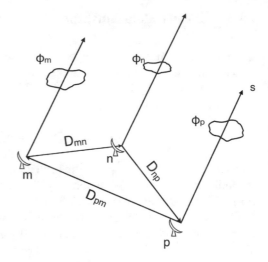

baseline is $\phi_v = \frac{2\pi}{\lambda} \mathbf{D} \cdot \mathbf{s}$; hence, the closure phase is

$$\phi_{c_{mnp}} = \frac{2\pi}{\lambda} \left(\mathbf{D}_{mn} + \mathbf{D}_{np} + \mathbf{D}_{pm} \right) \cdot \mathbf{s} = 0 \qquad (10.35)$$

because the sum of the baselines around a triangle is identically zero. This shows that the closure phase for a point source is zero, even if it is not at the phase-tracking center or if the station coordinates have errors. A corollary of this result is that the position of a source cannot be deduced from closure phase measurements alone.

If we have n_a antennas and we measure the correlation of all pairs, the number of independent phase closure relationships is equal to the number of correlator output phases less the number of unknown instrumental phases, one of which can be arbitrarily chosen. If there are no redundant spacings, then each closure relationship provides different information on the source structure. The number of phase closure relationships is

$$\frac{1}{2}n_a(n_a - 1) - (n_a - 1) = \frac{1}{2}(n_a - 1)(n_a - 2) . \qquad (10.36)$$

It is often important to be able to identify which set of closure triangles can be considered to be independent. This is necessary if closure phases are to be used directly in model fits. Combinatorial mathematics is useful in this regard. The question of how many triangles can be formed among n_a antennas can be rephrased as: Among n_a objects, how many unique ways can three of them be chosen without replacement or regard to order? The answer is the binomial coefficient

$$n_{PT} = \binom{n_a}{3} = \frac{n_a!}{(n_a - 3)!3!} = \frac{n_a(n_a - 1)(n_a - 2)}{6} . \qquad (10.37)$$

Fig. 10.8 The four closure triangles among four antennas. The three triangles involving the reference antenna, denoted by 1, are independent. The phase closure on the fourth triangle linking antennas m, n, and p can be derived from the three independent phase closures.

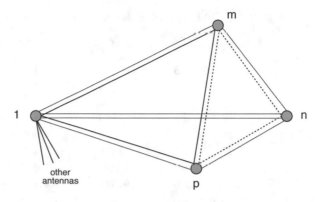

Similarly, the number of baselines, n_b, is

$$\binom{n_a}{2} = \frac{n_a(n_a - 1)}{2} . \tag{10.38}$$

A set of independent triangles can be found by the following process. Select one antenna as a reference, as shown in Fig. 10.8. The set of independent triangles is all of those that include the reference antenna. The nonindependent triangles are the ones that do not involve the reference antenna, taken to be antenna 1, i.e.,

$$\phi_{c_{mnp}} = \phi_{mn} + \phi_{np} + \phi_{pm} , \tag{10.39}$$

where none of n, m, and p are not equal to one. The sum of closure phases

$$\phi_{c_{1nm}} = \phi_{1n} + \phi_{nm} + \phi_{m1}$$
$$\phi_{c_{1mp}} = \phi_{1m} + \phi_{mp} + \phi_{p1} \tag{10.40}$$
$$\phi_{c_{1pn}} = \phi_{1p} + \phi_{pn} + \phi_{n1}$$

is

$$\phi_{nm} + \phi_{mp} + \phi_{pn} , \tag{10.41}$$

since $\phi_{1n} = -\phi_{n1}$, $\phi_{1m} = -\phi_{m1}$, and $\phi_{1p} = -\phi_{p1}$. The number of independent closure triangles is thus given by

$$n_{P\,\text{indep}} = \binom{n_a - 1}{2} = \frac{(n_a - 1)(n_a - 2)}{2} , \tag{10.42}$$

Table 10.1 Baselines and phase closures for an array of n_a elements[a]

n_a	n_b	n_{PT}	$n_{P\,indep}$	f_P [b]
2	1	0	0	0
3	3	1	1	0.33
4	6	4	3	0.50
5	10	10	6	0.60
8	28	56	21	0.75
10	45	120	36	0.80
27	351	2,925	325	0.93
50	1,225	19,600	1,176	0.96
100	4,950	161,700	4,851	0.98

[a] $n_b = n_a(n_a - 1)/2.\ n_{PT} = n_a(n_a - 1)\,(n_a - 2)/6.$
$n_{P\,indep} = (n_a - 1)(n_a - 2)/2.\ f_p = n_{P\,indep}/n_b = 1 - \frac{2}{n_a}.$
[b] See Fig. 11.4.

in agreement with Eq. (10.36). The fraction of the phase information that can recovered from phase closures in an array is

$$f_P = n_{P\,indep}/n_b = \frac{(n_a - 1)(n_a - 2)}{2} \Bigg/ \frac{n_a(n_a - 1)}{2} = 1 - \frac{2}{n_a} . \tag{10.43}$$

Representative numbers are given in Table 10.1.

We now discuss the amplitude closure relations. An amplitude closure relationship involves four antenna pairs, for which four antennas $m, n, p,$ and q are required:

$$\frac{|r_{mn}||r_{pq}|}{|r_{mp}||r_{nq}|} = \frac{|\mathcal{V}_{mn}||\mathcal{V}_{pq}|}{|\mathcal{V}_{mp}||\mathcal{V}_{nq}|} . \tag{10.44}$$

The proof of Eq. (10.44) is obtained by substituting terms of the form $g_m g_n^* \mathcal{V}_{mn}$ into the left side of Eq. (10.44), using Eq. (10.31). The moduli of the g terms then cancel out because the numerator and denominator both contain the product of the moduli of all four g terms. A total of six closure amplitudes can be formed. Three will be reciprocals of the other three and ignored. The basic three configurations are shown in Fig. 10.9. The product of these three closure amplitudes—$|r_{mn}||r_{pq}|/|r_{mp}||r_{nq}|$, $|r_{mp}||r_{nq}|/|r_{mq}||r_{np}|$, and $|r_{mn}||r_{pq}|/|r_{mq}||r_{np}|$—is unity, so only two of them are independent. The number of independent amplitude closure relationships for n_a antennas with no redundant baselines is equal to the number of measured amplitudes, $\frac{1}{2}n_a(n_a - 1)$, less the number of unknown antenna gain factors n_a, that is,

$$n_{A\,indep} = \frac{1}{2}n_a(n_a - 1) - n_a = \frac{1}{2}n_a(n_a - 3) . \tag{10.45}$$

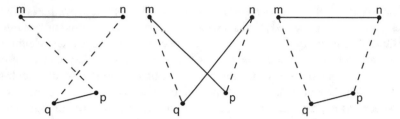

Fig. 10.9 The three closure amplitudes that can be formed among four antennas [see Eq. (10.34)]. (We have not included the trivially redundant reciprocal cases, i.e., solid band dotted lines interchanged.) In each case, the two visibility moduli that go in the numerator of the closure amplitude are shown by the solid lines, and the two that go in the denominator are shown by the dashed lines. The product of the three closure amplitudes is unity, so only two of the closure amplitudes are independent.

The fraction of amplitude information that can be recovered from amplitude closures is

$$f_A = \frac{n-3}{n-1}.$$
(10.46)

For early usage of the principle of taking ratios of observed visibility amplitudes to eliminate instrumental gains, see Smith (1952) and Twiss et al. (1960). The total number of closure quadrangles is

$$n_{AT} = 6 \binom{n_a}{4},$$
(10.47)

which is on the order of n_a^4. Systematic procedures can be devised to select an independent set. For a detailed analysis of amplitude closure structures, see Lannes (1991).

Note that a fundamental requirement for the validity of the closure relationships is that at any instant, it must be possible to represent the effect of any signal path from the source to the correlator by a single complex gain factor. Thus, the effects of the atmosphere must be constant over the source under observation, that is, the angular width of the source should be no greater than the isoplanatic patch size for the atmosphere. The isoplanatic patch is the area of sky within which the path length for an incident wave remains constant to within a small fraction of a wavelength; see also Sect. 11.8.4. The size of the isoplanatic patch varies with frequency. At a few hundred megahertz or less, it is common to have more than one source within an antenna beam, and these may be separated sufficiently in angle that ionospheric conditions may be different for each one. The closure conditions will then be different for each source, and use of the closure principle then becomes more complicated than in the single-source case discussed above.

The closure relationships have proved to be very important in synthesis imaging. When applied to unresolved point sources, the phase closure should be zero and the amplitude closure unity. Thus, they are useful in checking the accuracy of calibration and examining instrumental effects. For resolved sources, they can be used as observables in situations in which direct calibration by observation of a calibration source is not practicable, as is sometimes the case in VLBI. Most importantly, they can be used to improve calibration accuracy for observations where high dynamic range is required. The amplitude closure relationships are less frequently used because it is generally easier to calibrate the visibility amplitudes than the phases. However, they provide a useful check in cases in which the amplitude is required with especially high accuracy [for examples, see Trotter et al. (1998); Bower et al. (2014), and Ortiz-León et al. (2016)].

10.4 Visibility Model Fitting

The fitting of simple intensity models to visibility data was practiced extensively in early radio interferometry, especially when the visibility phase was poorly calibrated or the data were not sufficiently complete to allow Fourier transformation. Examples of simple models are shown in Figs. 1.5, 1.10, and the Gaussian components in Fig. 1.14.

Model fitting continues to be the only recourse for data interpretation in sparse VLBI arrays such as those used at short millimeter wavelengths [see, e.g., Doeleman et al. (2008)]. However, model fitting is very important even in large, well-sampled arrays that can generate high-quality images. These images are produced by a complex process that includes Fourier transformation of visibility data that have been interpolated onto a grid, followed by self-calibration and application of nonlinear deconvolution algorithms such as CLEAN, as described in Chap. 11. The noise in these images is correlated among pixels and can have poorly understood characteristics. Such images are not unique and can be considered to be models of the true brightness distributions. Extractions of source parameters in the image plane can therefore be characterized as "modeling the model."

In contrast, the fundamental data product of an array, the visibilities, has well-characterized noise properties, i.e., it is uncorrelated Gaussian noise with known variance. If the characteristics of the source emission structure are to be interpreted with a specific model in mind, the parameters of such a model can often best be obtained from direct analysis of the visibility data. Important examples of the application of model fitting include the cases of sources whose intensity decreases as a power law, as described in Sect. 10.4.4. In these cases, the proper estimate of the total flux density and other parameters from image plane analysis is difficult. Another application of visibility model fitting is in the determination of the changes in parameters of a source in which time-separated observations may not have identical (u, v) coverage. Fitting the same model to both data sets, but allowing the parameters of interest to vary, is likely to give the best evidence of change. An interesting example is provided by Masson (1986) in a measurement of angular expansion of a compact planetary nebula. From several data sets obtained at different epochs, the image from the one with the best (u, v) coverage was used as a model to fit to the others, thereby avoiding direct comparison of images made with different synthesized beams.

A useful discussion of the general principles of model fitting can be found in Pearson (1999). For the estimate of large numbers of parameters in a Bayesian framework, see Lochner et al. (2015). There are advantages for searching for transient sources in the (u, v) data (Trott et al. 2011).

10.4.1   Basic Considerations for Simple Models

We consider the case of the small field of view $(l, m \ll 1, A(l, m) \simeq 1)$, where the transform between image intensity and visibility given in Eqs. (3.7) and (3.10) can be written as

$$\mathcal{V}(u, v) = \int_{-\infty}^{\infty} I(l, m) \, e^{-j2\pi(ul+vm)} \, dl \, dm \tag{10.48}$$

$$I(l, m) = \int_{-\infty}^{\infty} \mathcal{V}(u, v) \, e^{j2\pi(ul+vm)} \, du \, dv \ . \tag{10.49}$$

A simple common source model is a Gaussian intensity distribution centered at position (l_1, m_1) with peak intensity I_0 and width parameter a:

$$I(l, m) = I_0 \exp\left[\frac{-(l - l_1)^2 - (m - m_1)^2}{2a^2} \right], \tag{10.50}$$

which has FWHM, θ_G, of $\sqrt{8 \ln 2} a$. The corresponding model visibility distribution is

$$\mathcal{V}_m(u, v) = S_0 e^{-2\pi^2 a^2(u^2+v^2) - j2\pi(ul_1+vm_1)} , \tag{10.51}$$

where $S_0 = 2\pi I_0 a^2$, the total flux density. The visibility has real and imaginary components that are sinusoidal corrugations, the ridges of which are normal to the radius vector to the point (l_1, m_1) in the image domain. These visibility components are modulated in amplitude by a Gaussian function centered on the (u, v) origin and of width inversely proportional to σ. Examination of the visibility distribution can thus indicate the form and position of the main intensity components. For early discussions and examples of this type of model fitting, see, for example, Maltby and Moffet (1962); Fomalont (1968), and Fomalont and Wright (1974). Fitting the four parameters (I_0, a, l_1, m_1) in the image plane or (S_0, a, l_1, m_1) in the visibility plane is a nonlinear process. It requires an initial guess for the parameters. The choice of these initial parameters is more obvious in the image plane than in the visibility plane, but final analysis is best done in the visibility plane.

To fit model parameters, it is necessary to choose a criterion for the goodness of fit. Since the real and imaginary components of the visibility usually have Gaussian noise, the optimum criterion from a maximum likelihood point of view (see Appendix 12.1) is the χ^2 criterion, which minimizes the weighted mean-squared difference between the model and the data set of n_d points, i.e.,

$$\chi^2 = \sum_{i=1}^{n_d} \frac{[\mathcal{V}(u_i, v_i) - \mathcal{V}_m(u_i, v_i, \mathbf{p})][\mathcal{V}(u_i, v_i) - \mathcal{V}_m(u_i, v_i, \mathbf{p})]^*}{\sigma_i^2} \, , \tag{10.52}$$

where $\mathcal{V}(u_i, v_i)$ are the measured visibilities, $\mathcal{V}_m(u_i, v_i, \mathbf{p})$ are the model visibilities with n_p parameter $\mathbf{p}$, and the σ_i's are the measurement errors. For a perfect fit, the expected minimum value of χ^2 is $n_d - n_p$, and the standard deviation of χ^2 is $\sqrt{2(n_d - n_p)}$. The reduced chi square, χ_r^2, which is $\chi^2/(n_d - n_p)$, should be close to unity for a good fit. $\chi_r^2 > 1$ indicates that the model is not correctly parametrized or that the estimates of errors are not correct. In any fitting procedures, the residuals, i.e., $[\mathcal{V}(u_i, v_i) - \mathcal{V}_m(u_i, v_i, \mathbf{p})]/\sigma_i$, should be examined for any systematic deviations from a Gaussian probability distribution. Such deviations suggest that more or different parameters are required. If the deviations follow a Gaussian probability distribution, then the problem may be that values of σ_i are misestimated by a constant factor that can be chosen to make $\chi_r^2 = 1$. Another common defect is that the data have a noise floor. In this case, the σ_i^2 terms can be replaced by $\sigma_i^2 + \sigma_f^2$, where σ_f represents a noise floor and σ_f^2 is chosen so that $\chi_r^2 = 1$. A $\sigma_f^2 > 0$ has the effect of reducing the importance of measurements with low σ_i, and $\sigma_f \gg \sigma_i$ tends toward a solution that gives equal weight for all data regardless of σ_i.

Note that Eq. (10.52) can be written as

$$\chi^2 = \sum_{i=1}^{n_d} \frac{(\mathcal{V}_{R_i} - \mathcal{V}_{mR_i})^2 + (\mathcal{V}_{I_i} - \mathcal{V}_{mI_i})^2}{\sigma_i^2} \, , \tag{10.53}$$

where $\mathcal{V}_R$ and $\mathcal{V}_I$ are the real and imaginary parts of $\mathcal{V}$, and $\mathcal{V}_{mR}$ and $\mathcal{V}_{mI}$ are the real and imaginary parts of $\mathcal{V}_m$. The data to be fitted may consist of visibility amplitudes and closure phases. In this case, the χ^2 can be written as

$$\chi^2 = \sum_{i=1}^{n_d} \frac{[|\mathcal{V}| - |\mathcal{V}_m|]^2}{\sigma_{A_i}^2} + \sum_{i=1}^{n_c} \frac{(\phi_{c_i} - \phi_{mc_i})^2}{\sigma_{c_i}^2} \, , \tag{10.54}$$

where $\sigma_{A_i}^2$ and $\sigma_{c_i}^2$ are the measurement variances on the closure amplitudes and closure phases, respectively. In the strong signal case (see Sect. 9.3.3),

$$\sigma_{A_i}^2 = \sigma_i^2 \quad \text{and} \quad \sigma_{c_i}^2 = \left(\frac{\sigma_1}{\mathcal{V}_1}\right)^2 + \left(\frac{\sigma_2}{\mathcal{V}_2}\right)^2 + \left(\frac{\sigma_3}{\mathcal{V}_3}\right)^2 \, . \tag{10.55}$$

In cases of weaker signal, the application of Eq. (10.54) may not yield an optimum solution because the probability distributions for the closure and amplitude become non-Gaussian. In particular, the probability distribution of the closure amplitude becomes progressively more skewed as the signal-to-noise ratio (SNR) decreases.

Examples of models fitted to visibility data sets with limited amounts of closure data can be found in Akiyama et al. (2015); Fish et al. (2016), and Lu et al. (2013).

The computation challenge of finding the minimum value of χ^2 can be daunting. A popular method that is straightforward to implement but that can require larg e computation resources is the Markov chain Monte Carlo (MCMC) algorithm based on Bayesian theory. It provides a way to systematically vary the parameters in search of a χ^2 minimum. It also produces posterior probability functions for the parameters [see, e.g., Sivia (2006)].

There is an important relationship between the moments of the intensity distribution and the visibility. The zero-order moment is equal to the flux density S, the odd-order moments contribute to the imaginary components of the visibility, and the even-order moments contribute to the real part. If the source is symmetrical in l, the odd-order terms are zero. If, in addition, the source is only slightly resolved, the decrease in $\mathcal{V}$ results mainly from the second-moment term. Then the source can be represented by a symmetrical model with an appropriate second moment.

For simplicity, consider the one-dimensional problem

$$\mathcal{V}_1(u) = \int_{-\infty}^{\infty} I_1(l) \, e^{-j2\pi ul} dl \,, \tag{10.56}$$

where $\mathcal{V}_1(u) = \mathcal{V}(u, 0)$ and

$$I_1(l) = \int_{-\infty}^{\infty} I(l, m) \, dm \,. \tag{10.57}$$

Each derivative of $\mathcal{V}_1$ with respect to u introduces a factor of $-j2\pi l$, so that the nth derivative can be written as

$$\mathcal{V}_1^{(n)}(u) = \int_{-\infty}^{\infty} (-j2\pi l)^n I_1 \, e^{-j2\pi ul} dl \tag{10.58}$$

or

$$\mathcal{V}_1^{(n)}(0) = (-j2\pi)^n \int_{-\infty}^{\infty} l^n I_1(l) \, dl \,. \tag{10.59}$$

The Taylor expansion of $\mathcal{V}_1(u)$ is

$$\mathcal{V}_1(u) = \mathcal{V}_1(0) + \mathcal{V}_1'(0)u + \mathcal{V}_1''(0)\frac{u^2}{2} + \ldots + \mathcal{V}_1^{(n)}(0)\frac{u^n}{n!} + \ldots \tag{10.60}$$

or

$$\mathcal{V}_1(u) = M_0 + \sum_{n=1}^{\infty} \frac{(-j2\pi)^n}{n!} M_n u^n \, , \tag{10.61}$$

where

$$M_n = \int_{-\infty}^{\infty} l^n I_1(l) \, dl \, . \tag{10.62}$$

The Taylor expansion requires that the moments be finite.

10.4.2 Examples of Parameter Fitting to Models

The model most commonly encountered in interferometry is a simple Gaussian distribution with unknown flux density, size, and position, as described by Eq. (10.51). The four model parameters, S_0, a, l_1, and m_1 can be estimated from standard procedures for nonlinear least-mean-squares analysis (Appendix 12.1). This analysis requires initial guesses for the parameters. The model can be generalized to an elliptical Gaussian source described by major and minor axis diameters and a position angle (six-parameter fit).

To gain an understanding of the accuracy to which parameters of a simple model can be deduced, consider a slightly resolved source having an azimuthally symmetric distribution of unknown position observed at a set of n_d points with noise σ. In the case of high SNR, we can analyze the visibility amplitude and phase separately. The model for the visibility phase and amplitude can be written

$$\phi = 2\pi(u_1 l_1 + v_1 m_1) \tag{10.63}$$

$$|\mathcal{V}| = S_0 - bq^2 \, , \tag{10.64}$$

where $q^2 = u^2 + v^2$ and l_1, m_1, and b are parameters to be determined. We further assume that m_1 is zero.

A simulated data set is shown in Fig. 10.10. The models are linear in the parameters l_1, S_0, and b. These parameters can be estimated via the usual linear solutions to the χ^2 minimization equations for phase and amplitude [see Appendix 12.1 or Bevington and Robinson (1992)]. The estimate of l_1 is

$$l_1 = \frac{\dfrac{1}{2\pi} \sum_{i=1}^{n_d} \phi_i u_i / \sigma_{\phi_i}^2}{\sum_{i=1}^{n_d} u_i^2 / \sigma_{\phi_i}^2} \, , \tag{10.65}$$

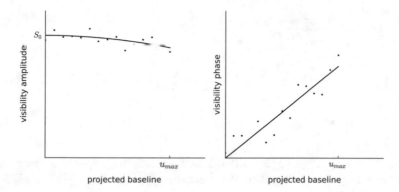

Fig. 10.10 Fringe visibility model and data for a slightly resolved, azimuthally symmetric source. (left) Visibility amplitude quadratically declining, indicative of the source being resolved; (right) visibility phase, indicative of a position offset.

where $\sigma_{\phi_i} \simeq \sigma_i/|\mathcal{V}|_i$ and σ_i is defined in Eq. (6.50). We assume that all antennas have the same sensitivity, so $\sigma_i = \sigma$, and that $|\mathcal{V}| \sim S_0$, so that σ_{ϕ_i} is approximately constant. In this case,

$$\sigma_{l_1} = \frac{\sigma/S_0}{2\pi \left[\sum_{i=1}^{n_d} u_i^2 \right]^{1/2}} . \tag{10.66}$$

If the data are uniformly spaced at intervals Δu, i.e., $u_i = i\Delta u$, then $\sum u_i^2 = (\Delta u)^2 \sum i^2 = (\Delta u)^2 n_d(n_d + 1)(2n_d + 1)/6 \simeq (\Delta u)^2 n_d^3/3$ for $n_d \gg 1$. Hence,

$$\sigma_{l_1} \simeq \frac{1}{2\pi} \sqrt{\frac{3}{n_d}} \frac{\sigma}{S_0} \frac{1}{u_{\max}} , \tag{10.67}$$

where $u_{\max} = n_d \Delta u$, or

$$\sigma_{l_1} \simeq \frac{0.3}{\sqrt{n_d}} \frac{\sigma}{S_0} \frac{\lambda}{D_{\max}} , \tag{10.68}$$

where $D_{\max} = \lambda u_{\max}$. This formula is close to the one used in astrometry for direct image fitting [see Eq. (12.16)].

The estimates of S_0 and b, along with their errors, σ_{S_0} and σ_b, are

$$S_0 = \frac{1}{\Delta} \left[\sum_{i=1}^{n_d} q_i^4 \sum_{i=1}^{n_d} |\mathcal{V}|_i - \sum_{i=1}^{n_d} q_i^2 \sum_{i=1}^{n_d} |\mathcal{V}|_i q_i^2 \right] \tag{10.69}$$

$$\sigma_{S_0}^2 = \frac{\sigma^2}{\Delta} \sum_{i=1}^{n_d} q_i^4 \tag{10.70}$$

$$b = \frac{1}{\Delta} \left[n_d \sum_{i=1}^{n_d} |\mathcal{V}|_i q_i^2 - \sum_{i=1}^{n_d} q_i^2 \sum_{i=1}^{n_d} |\mathcal{V}|_i \right] \tag{10.71}$$

$$\sigma_b^2 = \frac{n_d}{\Delta} , \tag{10.72}$$

where $\Delta = n_d \sum q_i^4 - \left(\sum q_i^2 \right)^2$. If the data are uniformly spaced at intervals of Δq from 0 to $q_{max} = n_d \Delta q$, then, if we use the approximations $\sum q_i^4 \simeq n_d^5/5$ and $\sum q_i^2 \simeq n_d^3/3$, for $n_d \gg 1$,

$$\sigma_{S_0} \simeq \frac{\sigma}{\sqrt{n_d}} , \tag{10.73}$$

and

$$\sigma_b \simeq \sqrt{\frac{5}{n_d}} \frac{1}{q_{max}^2} . \tag{10.74}$$

For a Gaussian source distribution, the Taylor expansion of the visibility function in Eq. (10.51) (see Table 10.2) gives $b = 2\pi^2 a^2 S_0$. Since θ_G, the FWHM angular diameter, is $\sqrt{8 \ln 2} a$, we obtain

$$\theta_G = \left[\frac{4 \ln 2}{\pi^2} \frac{b}{S_0} \right]^{1/2} . \tag{10.75}$$

The uncertainty in θ_G, σ_{θ_G}, for the case $\sigma_{\theta_G} \ll \theta_G$, will be

$$\sigma_{\theta_G} \simeq \frac{4 \ln 2}{2\pi^2} \sqrt{\frac{5}{n_d}} \frac{\sigma}{S_0} \frac{1}{\theta_G q_{max}^2} . \tag{10.76}$$

The minimum source size that can actually be measured at the 1-sigma error level is about $\theta_{min} \sim \sigma_{\theta_G} \sim \theta_G$, which is

$$\theta_{min} \simeq \frac{0.6}{\sqrt{\mathcal{R}_{sn}}} \frac{\lambda}{D_{max}} , \tag{10.77}$$

where the signal-to-noise ratio $\mathcal{R}_{sn} = S_0 \sqrt{n_d}/\sigma$ and $D_{max} = \lambda q_{max}$. A more precise and general analysis for various levels of statistical significance is given by Martí-Vidal et al. (2012).

Note that position and angular parameters can be estimated to an accuracy limited only by the SNR and the confidence in the model. When the SNR is very high, the

Table 10.2 Visibility functions for azimuthally symmetric source distributions[a]

Model	$I(r)/I_0$	FWHM	$\mathcal{V}(q)/I_0$	$\mathcal{V}(0)/I_0$	A [b]
Delta function	$\delta(r)$	–	1	–	–
Ring	$\delta(r-a)$	$2a$	$J_0(2\pi aq)$	1	$\pi^2 a^2$
Disk[c]	$\Pi\left(\frac{r}{a}\right)$	$2a$	$\pi a^2 \dfrac{J_1(2\pi aq)}{\pi aq}$	πa^2	$\dfrac{\pi^2}{2}a^2$
Annulus[d]	$\Pi\left(\frac{r}{a_2}\right) - \Pi\left(\frac{r}{a_1}\right)$	$2a_2$	$\pi a_2^2\left[\dfrac{J_1(2\pi a_2 q)}{\pi a_2 q}\right] - \pi a_1^2\left[\dfrac{J_1(2\pi a_1 q)}{\pi a_1 q}\right]$	$\pi(a_2^2 - a_1^2)$	$\dfrac{\pi^2}{2}\dfrac{a_2^4-a_1^4}{a_2^2-a_1^2}$
Gaussian	$e^{-r^2/2a^2}$	$\sqrt{8\ln 2}\,a$	$2\pi a^2 e^{-2\pi^2 a^2 q^2}$	$2\pi a^2$	$2\pi^2 a^2$
Uniform sphere[c]	$\sqrt{1-\left(\frac{r}{a}\right)^2}\,\Pi\left(\frac{r}{a}\right)$	$\dfrac{\sqrt{3}}{2}a$	$\sqrt{\dfrac{\pi}{2}}(2\pi a^2)\dfrac{J_{3/2}(2\pi aq)}{(2\pi aq)^{3/2}}$ $=\dfrac{2\pi a^2}{(2\pi aq)^3}\left[\sin(2\pi aq) - 2\pi aq\cos(2\pi aq)\right]$	$\dfrac{2\pi}{3}a^2$	$\dfrac{\pi^2 a^2}{4}$

[a] For additional models and fitting algorithms, see Lobanov (2015), Ng et al. (2008), and Martí-Vidal et al. (2014).

[b] Taylor expansion: $\mathcal{V}(q) = \mathcal{V}(0)[1 - Aq^2]$.

[c] Π, modified unit rectangle function: $\Pi(x) = 1$, $0 < x \le 1$; $\Pi(x) = 0$ otherwise.

[d] a_2 = outer radius, a_1 = inner radius.

size can be determined even though it is much less than the nominal beam size. Model fitting should not be confused with super-resolution deconvolution.[2]

10.4.3   Modeling Azimuthally Symmetric Sources

A very important class of models is those that have azimuthal symmetry, i.e., $I(l, m) = I(r)$, where $r = \sqrt{l^2 + m^2}$. For the following analysis, the position of the source is assumed to be known. In this case, the Fourier transform between the image and visibility becomes a Hankel transform [see Bracewell (1995, 2000), Baddour (2009)], i.e.,

$$\mathcal{V}(q) = 2\pi \int_0^\infty I(r) J_0 (2\pi r q) r \, dr \tag{10.78}$$

$$I(r) = 2\pi \int_0^\infty \mathcal{V}(q) J_0 (2\pi r q) q \, dq , \tag{10.79}$$

where $q = \sqrt{u^2 + v^2}$. $\mathcal{V}(q)$ is a real function, i.e., the visibility phase is zero.

A useful model is one of a uniform bright circular source of intensity I_0 and radius a. Since $\int J_0(x) x \, dx = x J_1(x)$,

$$\mathcal{V}(q) = \pi a^2 I_0 \frac{J_1(2\pi a q)}{\pi a q} , \tag{10.80}$$

where $J_1(2\pi a q)/\pi a q = 1$ for $q = 0$ and $\pi a^2 I_0 = S_0$, the total flux density. The visibility of an annulus of inner and outer radii a_1 and a_2 can be represented as the difference of two disk visibility functions

$$\mathcal{V}(q) = \pi a_2^2 I_0 \frac{J_1(2\pi a_2 q)}{\pi a_2 q} - \pi a_1^2 I_0 \frac{J_1(2\pi a_1 q)}{\pi a_1 q} , \tag{10.81}$$

i.e., the difference of two area-normalized jinc functions. The visibility functions for these and a number of other models are listed in Table 10.2 and shown in Fig. 10.11. An important lesson is that circularly symmetric models are very hard to distinguish for short baselines where the visibility decreases quadratically according to a size parameter. It is interesting to compare the visibility functions for a ring and thin annular disk, as shown in Fig. 10.12. The visibilities become significantly different only when q reaches about 1/ring thickness.

A useful model for the analysis of an azimuthally symmetric source might be a superposition of annuli in image space with intensities I_i and outer and inner radii

---

[2]In some fields, such model fitting is called "breaking the diffraction barrier" [e.g., Betzig et al. (1991)].

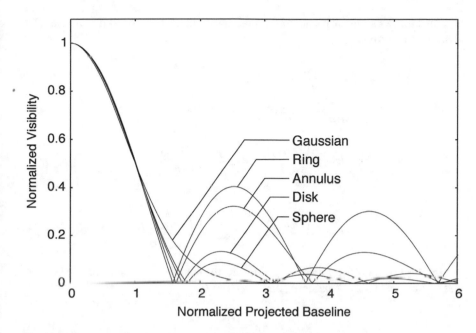

Fig. 10.11 Normalized visibility models, $|\mathcal{V}|/\mathcal{V}_0$, vs. projected baseline length, q, for azimuthally symmetric source models described in Table 10.2.

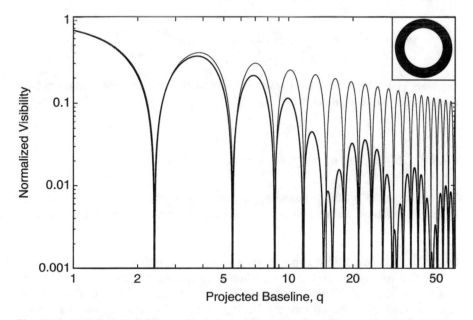

Fig. 10.12 (thin line) Visibility amplitude for a ring source with radius 1. (thick line) Visibility amplitude for an annular source with inner and outer radii of 0.8 and 1.2, respectively. Adapted from Bracewell (2000).

of a_i and a_{i-1}. The inner radius of the innermost annuli is zero, so it is a disk. The visibility function is

$$\mathcal{V}(q) = \pi I_0 a_0^2 \frac{J_1(2\pi a_0 q)}{\pi a_0 q}$$

$$+ \pi I_1 a_1^2 \frac{J_1(2\pi a_1 q)}{\pi a_1 q} - \pi I_1 a_0^2 \frac{J_1(2\pi a_0 q)}{\pi a_0 q}$$

$$+ \pi I_2 a_2^2 \frac{J_1(2\pi a_2 q)}{\pi a_2 q} - \pi I_2 a_1^2 \frac{J_1(2\pi a_1 q)}{\pi a_1 q}$$

$$\vdots$$

$$+ \pi I_n a_n^2 \frac{J_1(2\pi a_n q)}{\pi a_n q} - \pi I_n a_{n-1}^2 \frac{J_1(2\pi a_{n-1} q)}{\pi a_{n-1} q} \ . \tag{10.82}$$

For the case of a uniform disk, all the I_is are the same, i.e., I_0, and the visibility is that of a uniform disk of radius a_n and intensity I_0,

$$\mathcal{V}(q) = \pi a_n^2 \frac{I_0 J_1(2\pi a_n q)}{\pi a_n q} \ , \tag{10.83}$$

as expected. Equation (10.82) can be rearranged as

$$\mathcal{V}(q) = \pi \sum_{i=0}^{n-1} (I_i - I_{i+1}) a_i^2 \frac{J_1(2\pi a_i q)}{\pi a_i q} + \pi I_n a_n^2 \frac{J_1(2\pi a_n q)}{\pi a_n q} \ . \tag{10.84}$$

Equation (10.84) can be fitted to data from sources with elliptical symmetry by a simple change in coordinates.

10.4.4   Modeling of Very Extended Sources

The technique of visibility modeling can be of particular importance for diffuse symmetric sources. The models for these sources often do not have finite moments, although they can have well-defined visibility functions. However, the Taylor expansion of visibility function around $q = 0$ described in the previous section cannot be used. We discuss two important practical examples.

The first example is that of a radio source created by a fully ionized wind, i.e., thermal plasma at constant temperature T_e, surrounding a star. If the wind has a constant velocity of expansion, the electron density will decrease as the inverse square of the distance from the star. It can be shown (Wright and Barlow 1975)

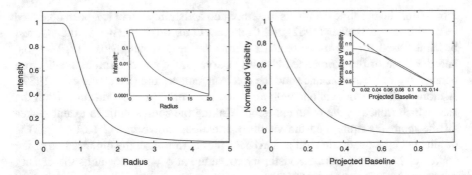

Fig. 10.13 (**left**) The intensity distribution defined by Eq. (10.85) for a stellar wind source where the radius is in units of a. The inset shows the intensity on a logarithmic scale. (**right**) Visibility function for the intensity distribution. The inset shows the visibility function near $q = 0$ and also for the case in which the intensity distribution is truncated at $r = 5a$. Note that the visibility function departs from the untruncated distribution for $q \lesssim 1/$truncation radius and approaches $q = 0$ quadratically.

that the intensity distribution for such a source can be written as

$$I(r) = I_0[1 - e^{-(r/a)^3}] ,$$
$$\simeq I_0 , \qquad\qquad\qquad r \ll a , \qquad (10.85)$$
$$\simeq I_0(r/a)^3 , \qquad\qquad r \gg a ,$$

where $I_0 = 2kT_e(\nu/c)^2$ (the Planck function), and a is the angular radius where the optical depth is unity. The rather benign-looking intensity profile, shown in Fig. 10.13, has an FWHM of about $1.25a$, and the intensity falls off as r^{-3}. The flux density is given by

$$S_0 = \frac{2\pi^2 a^2 I_0}{\sqrt{3}\Gamma(2/3)} , \qquad (10.86)$$

where Γ is the gamma function. S_0 is 1.3 times the flux density of a uniformly bright source of radius a. This source has the interesting characteristic that its angular size varies as $\nu^{0.7}$ (because a scales as $\nu^{-2.1}$), and the flux density varies as $\nu^{0.6}$ (see the example of MWC349A in Fig. 1.1). However, the second and higher moments of the intensity distribution are infinite. Nonetheless, the visibility function can be calculated from Eq. (10.78). It is shown in Fig. 10.13. It has the interesting characteristic that it decreases linearly (rather than quadratically) with q, that is,

$$\mathcal{V}(q) \simeq S_0(1 - bq) , \qquad (10.87)$$

where $b = 2\pi/S_0$. This behavior can be understood intuitively from the fact that the source extends smoothly to infinity. Hence, the correlated flux density continues to increase as the baseline decreases to zero. Such a visibility curve has been observed [e.g., White and Becker (1982) and Contreras et al. (2000)] down to the shortest baselines used for the measurements. From the zero spacing flux, $\mathcal{V}(0) = S_0$, and the slope of the normalized visibility curve, b, we can determine the electron density at a reference distance and the electron temperature (Escalante et al. 1989). A more realistic model is one with an ionization cutoff at some distance from the star, which truncates the radio emission. Making the source finite in extent makes all the moments finite, and the visibility function, shown in Fig. 10.13 (right) is dominated by a quadratic term at zero baseline. In this case, the outer radius of the source can be found from the visibility curvature at $q = 0$ as well as the density parameter and electron temperature.

The second example is useful in modeling the Sunyaev–Zeldovich effect. An isothermal spherical distribution of ionized gas in a cluster of galaxies causes a decrement in the cosmic microwave background. For many clusters, the profile of this decrement can be modeled as

$$I(r) = \frac{I_0}{\sqrt{1 + \left(\frac{r}{a}\right)^2}} \, , \tag{10.88}$$

where I_0 is the decrement at the cluster center, and a is the cluster core angular radius. The visibility function for this distribution has the analytic form (Bracewell 2000)

$$\mathcal{V}(q) = 2\pi a I_0 \frac{e^{-2\pi aq}}{2\pi aq} \, . \tag{10.89}$$

The visibility increases very rapidly as q decreases, and synthesis images made with missing short spacings are likely to underestimate I_0. However, the parameters I_0 and a can be readily estimated by fitting Eq. (10.89) to the visibility data (Hasler et al. 2012; Carlstrom et al. 1996). As with the wind case of the stellar wind source, an actual cluster source will be truncated at some radius, r_c, which will keep the flux density finite and will give the visibility function a parabolic shape for baselines less than $1/r_c$.

10.5   Spectral Line Observations

A basic requirement for observation of spectral lines is a receiving system that provides measurements of the signal intensity in a bandwidth less than, or comparable to, that of the expected spectral feature. Thus, a spectral line correlator produces separate visibility measurements at many points across the receiver passband, and the intensity distribution of the line features can be obtained. The data reduction involved is in principle the same as used in continuum imaging but differs in some practical details. The number of channels into which the received signal is divided is typically in the range 100–10,000. The discussion in this section is largely based on Ekers and van Gorkom (1984) and van Gorkom and Ekers (1989).

Calibration of the instrumental bandpass response is perhaps the most important step in obtaining accurate spectral line data. Generally, the channel-to-channel differences are relatively stable with time and need not be calibrated as frequently as the time-variable effects of the overall receiver gain. Except in very early systems, the channel filtering (see Sect. 8.8) is performed digitally and is not susceptible to ambient variations in temperature or voltage. The overall gain variations require periodic observation of a calibration source, as described for continuum observations. For this purpose, the summed response of the individual channels is often used, since a much longer observing time would be required to obtain a sufficient SNR in each narrow channel. For the bandpass calibration, a longer observation of a calibrator can be made to determine the relative gains of the spectral channels. Since the relative gains of the different channels into which the passband is divided change very little with time, the bandpass calibration need only be performed once or twice during, say, an 8-h observation. The bandpass calibration source should be unresolved and strong enough to provide good SNR in the spectral channels and should have a sufficiently flat spectrum. However, it need not be close in position to the source being observed.

Bandpass ripples resulting from standing waves between the antenna feed and the reflector, which pose a serious problem for single-antenna total-power systems, are much less important for interferometers. This is because the instrumental noise, including thermal noise picked up in the antenna sidelobes, is not correlated between antennas. On the other hand, for digital correlators, the Gibbs phenomenon ripples in the passband, which arise in Fourier transformation from the delay to the frequency domains, introduce a problem not found in autocorrelators. Because the cross-correlation of the signals from two antennas is real but not symmetrical as a function of delay, the cross power spectrum as a function of frequency is complex. (The autocorrelation function of the signal from a single antenna is real and symmetrical, and the power spectrum is real.) As explained in Sect. 8.8.8 (see Fig. 8.18), the imaginary part of the cross power spectrum changes sign at the origin, but the real part does not. Because of this large discontinuity at the frequency origin, ripples in the imaginary part of the frequency spectrum are of larger relative amplitude than those in the real part. The peak overshoot in the imaginary part is 18% (9% of the full step size); see also Bos (1984, 1985). Figure 10.14 shows a calculated example. The ratio of the real and imaginary parts depends on the instrumental phase (which is not calibrated out at this stage of the analysis) and on the position of the source of the radiation relative to the phase center of the field.

Increasing the number of lags of a lag correlator, or the size of the FFT in an FX correlator, improves the spectral resolution and confines the Gibbs phenomenon ripples more closely to the bandpass edges. The data from the channels at the band edges are sometimes discarded because of the ripples and the roll-off of the frequency response. However, variations in the passband are less important in later systems in which the signals are in digital form and the passband is defined by

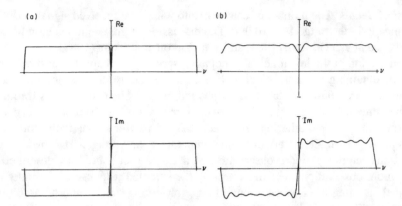

Fig. 10.14 (**a**) The cross power spectrum resulting from a continuum source in which the phase is arbitrarily chosen such that the amplitudes of the real and imaginary parts are equal. (**b**) Computed response of a cross-correlator with 16 channels to the spectrum in (**a**). Note the difference in amplitude of the ripples in the real and imaginary parts. From D'Addario (1989), courtesy of and © the Astronomical Society of the Pacific.

digital filtering. An effective way to reduce the amplitude of the ripples is to taper the cross-correlation function and thus introduce smoothing into the cross power spectrum. For this smoothing, the Hann function (see Table 8.5) is often used. van Gorkom and Ekers (1989) draw attention to the following examples:

1. If the field contains a line source but no continuum, and the line is confined to the central part of the passband, then the spectrum has no discontinuity at the passband edges. This is the only case in which it is advisable to use different tapering of the cross-correlation function for the source and the continuum calibrator.
2. If, in addition to the line source, the field contains one continuum point source, and if both this source and the bandpass calibrator are at the centers of their respective fields, then an accurate calibration of the bandpass ripples is possible. The same weighting must be used for the source and calibrator.
3. In more complicated cases—for example, when there is both a line source and an extended continuum source within the field—the ripples will be different in the two cases, and exact calibration is not possible. Hann smoothing of the spectra of both the source and the calibrator is recommended.

10.5.1   VLBI Observations of Spectral Lines

Since VLBI observations are limited to sources of very high brightness temperature, spectral line measurements in VLBI are used mainly for the study of masers and absorption of emission from bright extragalactic sources by molecular clouds.

Frequently observed maser lines include those arising from OH, H_2O, CH_3OH, and SiO. For absorption studies, many atomic and molecular species can be observed since the brightness temperature requirement is fulfilled by the background source. The formalism of spectral line signal processing is described in Sect. 9.3. Special considerations for astrometric measurements are given in Sect. 12.7. Here we discuss several practical issues related to the handling of spectroscopic data. The use of independent frequency standards at the antennas results in time-dependent timing errors, which introduce linear phase slopes across the basebands. The difference in Doppler shifts among the antennas can be large, and hence the residual fringe rates can also be large, which may necessitate short integration times for calibration. For masers, the phase calibration can usually be obtained from the use of the phase of a particular spectral feature as a reference. The amplitude calibration can be obtained from the measurement of the spectra derived from the data recorded at individual antennas. More details of procedures for handling spectral line data can be found in Reid (1995, 1999).

In spectral line VLBI, it is usual to observe a compact continuum calibrator several times an hour, preferably one strong enough to give an accurate fringe measurement in 1 or 2 min of integration. If a lag-type correlator is used to cross-correlate the signals, the output is a function of time and delay. Equation (9.21), in which $\Delta\tau_g$ and θ_{21} are functions of time, shows cross-correlation as a function of time and delay. By Fourier transformation, the arguments t and τ can be changed to the corresponding conjugate variables, which are fringe frequency, ν_f, and the frequency of the spectral feature, ν, respectively. Thus, the correlator output can be expressed as a function of (t, τ), (ν_f, τ), (t, ν), or (ν_f, ν) and can be interchanged between these domains by Fourier transformation. This is important because some steps in the calibration are best performed in particular domains. Note that the fringe frequency in VLBI observations results mainly from the difference between the true fringe frequency and the model fringe frequency used to stop the fringes. Consider first the data from the continuum calibrator. In fringe fitting for a continuum source, it is advantageous to use visibility data as a function of fringe frequency and delay, (ν_f, τ), as shown in Fig. 9.7. In that domain, the visibility data are most compactly concentrated and therefore most easily identified in the presence of the noise. In the absence of errors, the visibility will be concentrated at the origin in the (ν_f, τ) domain. A shift from the origin in the τ coordinate indicates timing errors resulting from clock offsets or baseline errors. The shift $\Delta\tau$ represents the difference in the errors for the two antennas. Values of $\Delta\tau$ determined from the continuum calibrator are used to apply corrections to the spectral line data. Variation of the $\Delta\tau$ values over time requires interpolation to the times of the spectral line data. The continuum data can also be used for bandpass calibration, to determine the relative amplitude and phase characteristics of the spectral channels.

For fringe fitting to spectral line data, it is advantageous to transform to the (t, ν) domain since, in contrast with the continuum case, the spectral line data contain features that are narrow in frequency. The cross-correlation function is therefore correspondingly broad in the delay dimension and generally more compact in frequency. Note that in the τ-to-ν transformation, ν is not the frequency of the radiation as received at the antenna, since the frequency of a local oscillator (or a combination of more than one local oscillators), ν_{LO}, has been subtracted. Thus, ν here represents the frequency within the intermediate-frequency (IF) band that is sampled and recorded for transmission to the correlator. The (t, ν) domain is also appropriate for inserting corrections for the timing errors, $\Delta\tau$, determined from the continuum data. These corrections are made by inserting phase offsets that are proportional to frequency. Thus, the data as a function of (t, ν) are multiplied by[3] $\exp(j2\pi\Delta\tau\nu)$. If the variation in the $\Delta\tau$ values over time results from a clock rate error at one or both of the antennas, correction should be made for the associated error in the frequency ν_{LO} at the antennas. The resulting phase error is corrected by multiplying the correlator output data by $\exp(j2\pi\Delta\tau\nu_{LO})$.

Since Doppler shift corrections (see Appendix 10.2) are rarely made as local oscillator offsets at the antennas, these corrections must be made at the correlator or subsequently in the post-processing analysis. The diurnal Doppler shift is normally removed at the station level in the precorrelation fringe rotation, where the signals are delayed and frequency-shifted to a reference point at the center of the Earth. Correction for the Doppler shift due to the Earth's orbital motion and the local standard of rest, as well as any other frequency offset, can conveniently be made on the post-correlation data by use of the shift theorem, that is, multiplication of the correlation functions by $\exp(j2\pi\Delta\nu\tau)$, where $\Delta\nu$ is the total frequency shift desired.

The visibility spectra can be calibrated in units of flux density by multiplication of the normalized visibility spectra by the geometric mean of the system equivalent flux densities (SEFDs) of the two antennas concerned, as discussed in Sect. 10.1.2. The SEFD is defined in Eq. (1.7). It can be determined from occasional supplemental measurements at the antennas, and the results interpolated in time. A better method for strong sources is to calculate the total-power spectrum of the source from the autocorrelation functions of the data from each antenna. These must be corrected for the bandpass response, which can be obtained from the autocorrelation functions on a continuum fringe calibrator. The amplitude of a specific spectral feature is proportional to the reciprocal of the SEFD. If greater sensitivity is required, then each measured spectrum can be matched to a spectral template obtained from a global average of all the single-antenna data or from a spectrum taken with the most sensitive antenna in the array. The difficulty with this method is that it is seldom convenient to acquire bandpass spectra often enough to ensure sufficiently accurate baseline subtraction on weak sources.

If the total frequency bandwidth in the measurements is covered by using two or more IF bands of the receiving system, it is necessary to correct for differences in their instrumental phase responses. This can be done using the continuum calibrator measurements, by averaging the phase values for the different channels in each IF

---

[3]Note that the required sign of the exponent in this and similar expressions used in this subsection may be positive or negative, depending on the sign conventions used.

band and subtracting these averages from the corresponding spectral line visibility data.

Finally, it is necessary to correct for remaining instrumental phases and for the different atmospheric and ionospheric phase shifts, which may be large for widely separated sites. In imaging strong continuum sources, this can be achieved by using phase closure, as described in Sect. 10.3. A similar approach can be used in imaging a distribution of maser point sources, by selecting a strong spectral component that is seen at all baselines and assuming that it represents a single point source. Then if the phase for this component at one arbitrarily chosen antenna is assumed to be zero, the relative phases for the other antennas can be deduced from the fringe phases. Since these phases are attributed to the atmosphere over each antenna, the correction can be applied to all frequency components within the measured spectrum. This method of using one maser component to provide a phase reference is discussed in more detail in Sect. 12.7, together with fringe frequency mapping, a technique that is useful in determining the positions of major components in a large field of masers.

10.5.2   Variation of Spatial Frequency Over the Bandwidth

The effect of using the center frequency of the receiver passband in calculating the values of u and v for all frequencies within the passband is discussed in Sect. 6.3.1. Consider, for example, a single discrete source for which the visibility function has a maximum centered on the (u, v) origin and decreases monotonically for a range of increasing u and v. If we use the frequency at the band center ν_0 to calculate u and v for a frequency at the high end of the band, that is, $\nu > \nu_0$, then the values of u and v will be underestimated. The measured visibility will fall off too quickly with u and v, and the central peak of the visibility function will be too narrow. Hence, the width of the image in l and m will be too wide. Thus, if the source radiates a spectral line at the blueshifted side of the bandwidth, the angular dimensions may be overestimated and similarly underestimated at the redshifted side. This effect can be described as chromatic aberration.

As discussed in Sect. 6.3, for observations with a spectral line (multichannel) correlator, the visibility measured for each channel can be expressed as a function of the (u, v) values appropriate for the frequency of the channel. This corrects the chromatic aberration but causes the (u, v) range over which the visibility is measured to increase over the bandwidth in proportion to the frequency. Thus, the width of the synthesized beam (i.e., the angular resolution) and the angular scale of the sidelobes vary over the bandwidth. The variation of the resolution can, if necessary, be corrected by truncation or tapering of the visibility data to reduce the resolution to that of the lowest frequency within the passband.

10.5.3   Accuracy of Spectral Line Measurements

The *spectral dynamic range* of an image after final calibration is an estimate of the accuracy of the measurement of spectral features expressed as a fraction of the maximum signal amplitude. It can be defined as the variation in the response of different channels to a continuum signal divided by the maximum response, the variation being a result of noise and instrumental errors. When the amplitude of a spectral line is only a few percent of the continuum that is present, as in the case of a recombination line or a weak absorption line, the accuracy of spectral line features depends on the accuracy with which the response to the continuum can be separated from that to the line. In such a case, a dynamic range of order 10^3 is required to measure a line profile to an accuracy of 10%. Hence, we see the importance of accurate bandpass calibration and of correction for chromatic aberration.

Various techniques have been used to help subtract the continuum response from an image. It is necessary to choose the receiver bandwidth so as to include some channels that contain continuum only, at frequencies on either side of the line features. A straightforward method is to use an average of the line-free channel data to make a continuum image and subtract this image from each of the images derived for a channel with line emission. Unless the receiver bandwidth is sufficiently small compared with the center frequency, it is likely that a correction for chromatic aberration should be used in making the continuum image. If the continuum emanates from point sources, the positions and flux densities of the sources provide a convenient model. For the most precise subtraction, the continuum response should be calculated separately for each line channel, using the individual channel frequencies in determining the (u, v) values. The subtraction should be performed in the visibility data. Use of deconvolution algorithms in the continuum subtraction is briefly discussed in Sect. 11.8.1.

10.5.4   Presentation and Analysis of Spectral Line Observations

Spectral line data can be presented as three-dimensional distributions of pixels in (l, m, v). For physical interpretation, the Doppler shift in the frequency dimension is often converted to radial velocity v_r with respect to the rest frequency of the line. The relationship between frequency and velocity is given in Appendix 10.2. A model of such a three-dimensional distribution is shown in Fig. 10.15. Continuum sources are represented by cylindrical functions of constant cross section in l and m.

The three-dimensional data cube that contains the images for the individual channels can be thought of as representing a line profile for each pixel in two-dimensional (l, m) space. To simplify the ensemble of images, it is often useful

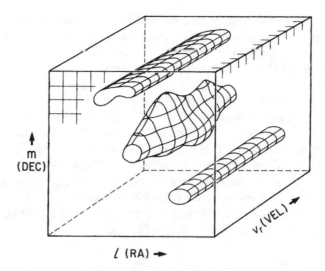

Fig. 10.15 Three-dimensional representation of spectral line data in right ascension, declination, and frequency. The frequency axis is calibrated in velocity corresponding to the Doppler shift of the rest frequency of the line. The flux density or intensity of the radiation is not shown but could be represented by color or shading. The indicated velocity has no physical meaning for continuum sources, which are represented by cylindrical forms of constant cross section normal to the velocity dimension. Spectral line emission is indicated by the variation of position or intensity with velocity. From Roelfsema (1989), courtesy of and © the Astronomical Society of the Pacific.

to plot a single (l, m) image of some feature of the line profile. This feature might be the integrated intensity

$$\Delta v \sum_i I_i(l, m) \,, \tag{10.90}$$

where i indicates the range of spectral channels, which are spaced at intervals Δv in frequency. For an optically thin radiating medium such as neutral hydrogen, this is proportional to the column density of radiating atoms or molecules. The intensity-weighted mean velocity is an indicator of large-scale motion,

$$\langle v_r(l, m) \rangle = \frac{\sum_i I_i(l, m) v_{r_i}}{\sum_i I_i(l, m)} \,. \tag{10.91}$$

The intensity-weighted velocity dispersion

$$\sqrt{\frac{\sum_i I_i(l, m)(v_{r_i} - \langle v_r \rangle)^2}{\sum_i I_i(l, m)}} \tag{10.92}$$

is an indicator of random motions within the source. The summation in the velocity dimension is performed separately for each (l, m) pixel of the images. In each of the three quantities in expressions (10.90)–(10.92), the intensity values correspond to the specific line of interest, continuum features having been subtracted out. In obtaining the best estimates for these three quantities, it should be noted that including ranges of (l, m, v_r) that contain no discernable emission only adds noise to the results.

Exploring the relationships between three-dimensional images in (l, m, v_r) and the three-dimensional distribution of the radiating material is an astronomical concern. As a simple example, consider a spherical shell of radiating material. If the material is at rest, it will appear in (l, m, v_r) space as a circular disk in the plane of zero velocity, with brightening at the outer edge. If the shell is expanding with the same velocity in all directions, it will appear in (l, m, v_r) space as a hollow ellipsoidal shell. Interpretation of observations of rotating spiral galaxies is more complex. An example of a model galaxy is given by Roelfsema (1989), and a more extensive discussion can be found in Burton (1988).

10.6   Miscellaneous Considerations

10.6.1   Interpretation of Measured Intensity

The quantity measured in a synthesized image is the radio intensity, but $\mathcal{V}$ is usually calibrated in terms of the equivalent flux density of a point source, and the intensity unit in the resulting image is in units of flux density per beam area Ω_0, which is given by

$$\Omega_0 = \int \int_{\substack{\text{main} \\ \text{lobe}}} \frac{b_0(l, m) \, dl \, dm}{\sqrt{1 - l^2 - m^2}} .$$

(10.93)

The response to an extended source is the convolution of the sky intensity $I(l, m)$ with the synthesized beam $b_0(l, m)$. Note that since there is often no measured visibility value at the (u, v) origin, the integral of $b_0(l, m)$ over all angles is zero; that is to say, there is no response to a uniform level of intensity. At any point on the extended source where the intensity varies slowly compared with the width of the synthesized beam, the convolution with $b_0(l, m)$ results in a flux density that is approximately $I\Omega_0$. Thus, the scale of the image can also be interpreted as intensity measured in units of flux density per beam area Ω_0. For a discussion of imaging wide sources and measuring the intensity of extended components of low spatial frequency, see Sects. 11.5 and 10.4.

10.6.2   Ghost Images

Figure 10.14 illustrates how bandpass ripples are introduced into the visibility as a function of frequency, as a result of the sharp edges in the cross power spectrum. A related effect discussed by Bos (1984) is the introduction of "ghost" images into the image derived from the observations. The ghost structure appears at a position that, relative to the true structure, is diametrically opposite with respect to the field center. For each spectral channel, the amplitude of the ghost structure is proportional to the amplitude of the ripple component. Thus, it is most serious for the channels at the edges of the receiver passband, as can be seen from Fig. 10.14b.

The ghost phenomenon is most easily explained by considering a simple example. Suppose there is a point source of unit amplitude at position $(l, m) = (l_1, 0)$, where $(0, 0)$ is the field center, and it is observed over a range of baselines u. The fringe visibility of a point source is the Fourier transform[4] with respect to l of a delta function at l_1, which is

$$\mathcal{V}_1(u) = e^{-j2\pi u l_1} = \cos(2\pi u l_1) - j\sin(2\pi u l_1) . \tag{10.94}$$

Suppose that a multichannel spectral correlator is used and there is a visibility data set for each spectral channel. The ripples across the spectrum in Fig. 10.14 have the effect that the relative amplitudes of the sine and cosine components are no longer equal, as they are in Eq. (10.94), so we rewrite Eq. (10.94) as

$$\mathcal{V}_1(u) = \cos(2\pi u l_1) - j(1 + \Delta)\sin(2\pi u l_1) . \tag{10.95}$$

Here, a component of relative amplitude Δ has been added to the imaginary component, which has the most severe ripples. Δ is positive for a channel in which there is a peak in the imaginary-component ripple. To determine the effect of the term $-j\Delta\sin(2\pi u l_1)$ in the image, we take its Fourier transform with respect to u, which is $\Delta[\delta(u+l_1) - \delta(u-l_1)]/2$. Thus, the ripple adds to the image a delta function of amplitude $\Delta/2$ at $-l_1$, which is the ghost, and subtracts a delta function of the same amplitude from the true image[5] at l_1. For a source at the field center, the ghost and the true image combine, providing a correct measure of the source intensity.

Since the visibility data are usually not calibrated prior to the spectral filtering, the relative amplitudes of the real and imaginary components in Eq. (10.94) result from the instrumental phases introduced by the receiving system as well as from the structure of the source. If these instrumental phase data are lost after calibration of the visibility, precise removal of the ghost is not possible. However, the effect of the ripples can be reduced by use of smoothing functions on the spectral data before creating the image, as discussed earlier. If the spectral data are averaged to provide a continuum result before assigning (u, v) values, the effect of the frequency

---

[4] In the Fourier transformations used here, we follow Bracewell (2000), who, for the delta (impulse) function, defines a "transform in the limit" by considering two Gaussian functions, $|a|e^{-\pi a^2 l^2}$ and $e^{-\pi(u/a)^2}$, that are a Fourier pair. As $a \to \infty$, the first Gaussian tends toward a delta function at the l-origin and the second toward unity. For a delta function at l_1, we use the shift theorem and multiply by $e^{-2\pi u l_1}$.

[5] Bos (1984, 1985) refers to the ripple-induced component at the true image position as the "hidden component."

difference of the channels with high amplitude ripples at the two edges of the passband may be sufficient to separate the ghost into two components, as shown by Bos (1985). This separation will not occur if the (u, v) values are individually assigned for each spectral channel.

Bos (1984) points out that the ghost can be removed, or substantially attenuated, by $\pi/2$ switching of the relative phase between each signal pair before cross correlating, and restoring the phase before transformation of the visibility data to form an image. For the source considered in Eq. (10.94), the introduction of $\pi/2$ into the differential phase for an antenna pair results in the visibility

$$\mathcal{V}_2(u) = je^{-j2\pi ul_1} = j\cos(2\pi ul_1) + \sin(2\pi ul_1) \ . \tag{10.96}$$

The imaginary part consists of the cosine components, which are the real part in Eq. (10.94). Adding the visibility term resulting from the ripples in the imaginary part of the spectrum, as in Eq. (10.95), we have

$$\mathcal{V}_2(u) = je^{-j2\pi ul_1} = j(1 + \Delta)\cos(2\pi ul_1) + \sin(2\pi ul_1) \ . \tag{10.97}$$

To remove the effect of the quadrature phase switch, we multiply Eq. (10.96) by j. The visibility term introduced by the ripple then becomes $-\Delta\cos(2\pi ul_1)$, and taking the Fourier transform with respect to u, we find that the contribution of the ripple to the image is $-\Delta[\delta(u+l_1) + \delta(u-l_1)]/2$. Again, there are delta functions at $\pm l_1$, but in this case, they both have the same sign. Thus, the result of averaging the images with the two positions of the phase switch is to cancel the ghost but double the amplitude loss of the true image. Note that we have assumed that the quadrature phase shift introduced by the switch can be represented by the factor j in Eq. (10.96): If the sign of the phase shift is such that the factor is $-j$, then the sign of the right side of Eq. (10.96) must be reversed. If the sign is wrong, the effect is to double the amplitude of the ghost but restore the amplitude of the image.

10.6.3 Errors in Images

A very useful technique for investigating suspicious or unusual features in any synthesized image or continuum or spectral line is to compute an inverse Fourier transform (i.e., from intensity to visibility), including only the feature in question. A distribution in the (u, v) plane concentrated in a single baseline, or in a series of baselines with a common antenna, could indicate an instrumental problem. A distribution corresponding to a particular range of hour angle of the source could indicate the occurrence of sporadic interference.

An aid in identifying erroneous features is a familiarity with the behavior of functions under Fourier transformation; see, for example, Bracewell (2000) and the discussion by Ekers (1999). A persistent error in one antenna pair will, for an east–west spacing, be distributed along an elliptical ring centered on the (u, v)

origin, and in the (l, m) plane will give rise to an elliptical feature with a radial profile in the form of the zero-order Bessel function. An error of short duration on one baseline introduces two delta functions representing the measurement and its conjugate. In the image, these produce a sinusoidal corrugation over the (l, m) plane. The amplitude in the image plane may be only small, since in an $M \times N$ visibility matrix, the effect of the two erroneous points is diluted by a factor of $2(MN)^{-1}$, which is usually of order 10^{-3}–10^{-6}. Thus, a single short-duration error could be acceptable if, in the image plane, it is small compared with the noise.

Errors of an additive nature combine by addition with the true visibility values. In the image, the Fourier transform of the error distribution $\varepsilon_{\mathrm{add}}(u, v)$ is added to the intensity distribution, and we have

$$\mathcal{V}(u, v) + \varepsilon_{\mathrm{add}}(u, v) \longleftrightarrow I(l, m) + \overline{\varepsilon}_{\mathrm{add}}(l, m) . \tag{10.98}$$

Other types of additive errors result from interference, cross coupling of system noise between antennas, and correlator offset errors. The Sun is many orders of magnitude stronger than most radio sources and can produce interference of a different character from that of terrestrial sources because of its diurnal motion. The response to the Sun is governed mainly by the sidelobes of the primary beam, the difference in fringe frequencies for the Sun and the target source, and the bandwidth and visibility averaging effects. Solar interference is most severe for low-resolution arrays with narrow bandwidths. Cross coupling of noise (cross talk) occurs only between closely spaced antennas and is most severe for low elevation angles when shadowing of antennas may occur.

A second class of errors comprises those that combine with the visibility in a multiplicative manner, and for these, we can write

$$\mathcal{V}(u, v)\varepsilon_{\mathrm{mul}}(u, v) \longleftrightarrow I(l, m) * * \overline{\varepsilon}_{\mathrm{mul}}(l, m) . \tag{10.99}$$

The Fourier transform of the error distribution is convolved with the intensity distribution, and the resulting distortion produces erroneous structure connected with the main features in the image. In contrast, the distribution of errors of the additive type is unrelated to the true intensity pattern. Multiplicative errors mainly involve the gain constants of the antennas and result from calibration errors, including antenna pointing and, in the case of VLBI systems, radio interference (see Sect. 16.4).

Distortions that increase with distance from the center of the image constitute a third category of errors. These include the effects of noncoplanar baselines (Sect. 11.7), bandwidth (Sect. 6.3), and visibility averaging (Sect. 6.4), which are predictable and therefore somewhat different in nature from the other distortions mentioned above.

10.6.4   Hints on Planning and Reduction of Observations

Making the best use of synthesis arrays and similar instruments requires an empirical approach in some areas, and the best procedures for analyzing data are often gained by experience. Much helpful information exists in the handbooks on specific instruments, symposium proceedings, etc. [see, for example, Perley, Schwab, and Bridle (1989) and Taylor, Carilli, and Perley (1999)]. A few points are discussed below.

In choosing the observing bandwidth for continuum observations, the radial smearing effect should be considered, since the SNR for a point source near the edge of the field is not necessarily maximized by maximizing the bandwidth. Then in choosing the data-averaging time, the resulting circumferential smearing can be about equal to the radial effect. The required condition is obtained from Eqs. (6.75) and (6.80) and for high declinations is

$$\frac{\Delta \nu}{\nu_0} \simeq \omega_e \tau_a . \tag{10.100}$$

Here, ν_0 is the center frequency of the observing band, $\Delta \nu$ is the bandwidth, ω_e is the Earth's rotation velocity, and τ_a is the averaging time. When attempting to detect a weak source of measurable angular diameter, or an extended emission, it is important not to choose an angular resolution that is too high. The SNR for an extended source is approximately proportional to $I\Omega_0$, as discussed in the previous section. The observing time required to obtain a given SNR is proportional to Ω_0^{-2}, or to θ_b^{-4}, where θ_b is the synthesized beamwidth.

If the antenna beam contains a source that is much stronger than the features to be studied, the response to the strong source can be subtracted, provided it is a point source or one that can be accurately modeled. This is best done by subtracting the computed visibility before gridding the measurements for the FFT. The subtracted response will then accurately include the effect of the sidelobes of the synthesized beam. Nevertheless, the precision of the operation will be reduced if the source response is significantly affected by bandwidth, visibility averaging, and similar effects, so it may be best to place the source to be subtracted at the center of the field. When observing a very weak source, it may be advisable to place the source a few beamwidths away from the (l, m) origin to avoid confusion with residual errors from correlator offsets, etc.

As part of the procedure in making any image, it may be useful also to make a low-resolution image covering the entire area of the primary antenna beam. For this image, the data can be heavily tapered in the (u, v) plane to reduce the resolution and thus also the computation. Such an image will reveal any sources outside the field of the final image that may introduce aliased responses in the FFT. Aliasing of these sources can be suppressed by subtraction of their visibility or use of a suitable convolving function. The sidelobe or ringlobe responses to such a source are also eliminated by subtraction of the source but not by convolution in the (u, v) plane.

The low-resolution image will also emphasize any extended low-intensity features that might otherwise be overlooked.

10.7   Observations of Cosmological Fine Structure

10.7.1   Cosmic Microwave Background

The anisotropy of the cosmic microwave background (CMB), which is about 10^{-5} of the mean temperature of 2.7 K, was first detected by the COBE mission (Smoot et al. 1992), and its characteristics were explored in great detail by the WMAP mission (Bennett et al. 2003) and the Planck mission (Planck Collaboration 2016). The data from these missions were obtained using total-power beam-switching techniques, revealed a major peak in the angular spectrum of the background fluctuations at $\sim 1.6°$. Interferometry offers advantages for the study of the higher-resolution peaks that, like the major peak, are attributed to acoustic waves in the early photon-baryon plasma at the surface of last scattering. Since interferometers do not respond to uncorrelated signals such as those generated within the Earth's atmosphere, it is possible to use ground-based interferometers for investigation of the finer angular structure of the CMB. A number of special instruments have been developed specifically to cover structure of angular range $\sim 0.1°$ to $\sim 3°$. These include the Degree Angular Scale Interferometer (DASI) (Leitch et al. 2002b; Pryke et al. 2002), located at the South Pole; Cosmic Background Imager (CBI) (Padin et al. 2002; Readhead et al. 2004), located at Llano de Chajnantor, Chile; and the Very Small Array (VSA) (Watson et al. 2003; Scott et al. 2003), in Tenerife. Planar arrays, discussed in Sect. 5.6.5, were primarily used for this work.

In the study of the fluctuations in the CMB, it is the statistics of the temperature variations rather than images of specific fields on the sky that are of interest for comparison with theoretical models. Model power spectra are given in terms of spherical harmonics, that is, the amplitudes of multipole moments of the temperature variation. Measurements of the angular spectrum of the CMB in this form can be derived directly from the Fourier components measured by interferometry without forming images of the structure on the sky. It is assumed that the CMB spectrum can be expressed as a function with circular symmetry (rotational invariance), since there is no preferred direction in the structure on the sky. Thus, characteristics of the CMB lead to some design considerations that differ from those for general-purpose synthesis arrays. The individual antennas need to be large enough to allow accurate phase and amplitude calibration with observing times of a few minutes, using strong discrete sources. With regard to the antenna configuration, the main requirement is to obtain sampling in a radial coordinate, $q = \sqrt{u^2 + v^2}$, in the (u, v) plane, rather than uniform sampling in two dimensions, as required for imaging. To obtain sufficiently fine sampling in q, the antennas were usually configured so that, considered pairwise, the spacing between centers from the closest to the most widely spaced increases in increments that are smaller than the diameter of an antenna. This can be achieved, for example, by the curved arm configuration shown for the CBI in Fig. 5.24.

In CMB measurements, it is also essential to be able to separate out the effects of all foreground sources. These signals can be identified by their spectral characteristics, which, for synchrotron or optically thin thermal emissions, differ from the blackbody spectrum of the CMB. Another requirement for CMB interferometry is sufficient frequency coverage to allow the spectral characteristics of signals to be determined. All three of the systems mentioned above used 10 GHz-wide receiving bandwidths of 26–36 GHz, subdivided into channels. These frequencies were chosen to be high enough to take advantage of the increase of CMB flux density with frequency and also to avoid H_2O and O_2 atmospheric absorption lines.

DASI was designed to provide measurements over a range of multipole moments $\ell = 100$–900 and used 13 antenna of diameter 20 cm with baselines 0.25–1.21 m. For CBI, the range of ℓ is $400 - 4250$, and 13 antennas of diameter 90 cm with a range of baselines 1–5.51 m were used. Each array was small enough to allow the antennas to be mounted on a mechanically rigid faceplate that could be pointed in azimuth and altitude so that the normal would track the center of the field under observation. The faceplate could also be rotated about its axis, to control the parallactic angle of the interferometer fringe patterns on the sky. No delay system or fringe rotation was needed, but phase switching was included to remove instrumental offsets. In CBI and DASI, the antennas were arranged in patterns with threefold symmetry, and thus, a rotation of the faceplate through 120° caused the configuration of the antennas to repeat relative to the sky (see Fig. 5.24). This property was very useful since the response to the sky remains unchanged after such a rotation, and variations in the signals resulting from unwanted effects such as residual cross talk between antennas could be identified and removed.

A further problem at the high levels of sensitivity required to observe the CMB structure results from thermal radiation from the ground and nearby objects, incident through the antenna sidelobes. This can introduce a serious unwanted contribution in the responses of the more closely spaced antenna pairs, but the effect decreases with increasing antenna spacing. For analysis of the results of observations of this type, see Hobson et al. (1995) and White et al. (1999). Further details of observations can be found in Leitch et al. (2002a,b) and Padin et al. (2002).

10.7.2 Epoch of Reionization

At redshifts corresponding to the period prior to the Epoch of Reionization (EoR), it should be possible to detect radiation of the neutral hydrogen line (1420 MHz rest frequency). As stars were formed in the early Universe, much of the hydrogen became ionized, and this period is referred to as the EoR. This probably occurred at a redshift no higher than about 7 or 8 (Morales and Wyithe 2010). Radiation at the frequency of the neutral hydrogen line should, in principle, be detectable at a redshift corresponding to the beginning of the EoR or earlier and should be detectable in all directions over the sky. However, there is also the cosmic background and the foreground noise from our Galaxy, and the level of these exceeds the distant hydrogen line signal by an estimated factor of 10^4. For detection of a broad faint background of radiation, in contrast with detection of discrete sources, sensitivity can be increased by using a large number of small antennas, to maximize sensitivity to broad structural features. In the image domain, (l, m), the third variable added is the frequency, ν, and in the spatial frequency domain, (u, v), the corresponding conjugate variable, represents time delay. A basic concern is how redundancy in the array configuration can be chosen to maximize the sensitivity to different angular scales in the search for the reionization signal. Further discussion of the challenges associated with EoR imaging can be found in Parsons et al. (2010, 2012, 2014); Zheng et al. (2013), and Dillon et al. (2015).

Appendix 10.1   The Edge of the Moon as a Calibration Source

During the test phase of bringing an interferometer into operation, it is useful to observe sources that produce fringes with high SNR. At frequencies above ~ 100 GHz, there are not many such sources. The Sun, Moon, and planets, the disks of which are resolved by the interferometer fringes, can nevertheless provide significant correlated flux density because of their sharp edges. Consider the limb of the Moon and the case in which the primary beam of the interferometer elements is much smaller than $30'$, the lunar diameter. When the antenna beam tracks the Moon's limb, the apparent source distribution is the antenna pattern multiplied by a step function; it is assumed that the brightness temperature of the Moon is constant within the beam. Approximating the antenna pattern as a Gaussian function, assuming that the antennas track a fixed point on the west limb of the Moon, and ignoring the curvature of the lunar limb, we can express the effective source distribution as

$$
\begin{aligned}
I(x, y) &= I_0 e^{-4\,(\ln 2)(x^2 + y^2)/\theta_b^2} & x &\geq 0\,, \\
&= 0 & x &< 0\,,
\end{aligned}
\tag{A10.1}
$$

where x and y are angular coordinates centered on the beam axis, θ_b is the full width of the beam at the half-power level, and in the Rayleigh–Jeans regime, $I_0 = 2kT_m/\lambda^2$, where T_m is the temperature of the Moon. The visibility function is then

$$
\begin{aligned}
\mathcal{V}(u, v) = 2I_0 &\left[\int_0^\infty e^{-4\,(\ln 2)x^2/\theta_b^2} (\cos 2\pi ux - j \sin 2\pi ux)\, dx \right] \\
&\times \left[\int_0^\infty e^{-4\,(\ln 2)y^2/\theta_b^2} \cos 2\pi vy\, dy \right].
\end{aligned}
\tag{A10.2}
$$

The cosine integral is straightforward, and the sine integral can be written in terms of a degenerate hypergeometric function $_1F_1$ (see Gradshteyn and Ryzhik 1994, Eq. 3.896.3). The result is

$$\mathcal{V}(u, v) = I_0 S_0 e^{-\pi^2 \theta_b^2 (u^2+v^2)/4 \ln 2} \left[1 - j\sqrt{\frac{\pi}{\ln 2}} (\theta_b u) \, _1F_1 \left(\frac{1}{2}, \frac{3}{2}, \frac{\pi^2 \theta_b^2 u^2}{4 \ln 2} \right) \right] , \cdot$$

(A10.3)

where

$$S_0 = \frac{\pi k \tau_m \theta_b^2}{4 \lambda^2 \ln 2}$$

(A10.4)

is the flux density of the Moon in the half-Gaussian beam. In the limit $(u, v) \gg (0,0)$, the imaginary part of the visibility is zero, and $\mathcal{V}(u, v) = S_0$, as expected. For $T_m = 200$ K and $\theta_b = 1.2\lambda/d$, where d is the diameter of the interferometer antennas in meters, $S_0 \simeq 460,000/d^2$ Jy. The integral over x in Eq. (A10.2) can also be written in terms of the error function. For the limit where $u \gg d/\lambda$, the asymptotic expansion of the error function leads to the convenient approximation

$$\mathcal{V}(u, v = 0) = j\sqrt{\frac{4 \ln 2}{\pi^3}} \frac{S_0}{\theta_b u} \simeq j 0.41 \frac{kT_m}{dD} ,$$

(A10.5)

where D is the baseline length. Hence, we have the interesting situation that the visibility for a given baseline length increases as the antenna diameter decreases, as long as $\theta_b \ll 30'$. The approximation in Eq. (A10.5) is accurate to 2% for $D > 2d$. The full visibility function as a function of projected baseline length is shown in Fig. A10.1. Note that the visibility measured with an interferometer having an east–west baseline orientation and tracking the north or south limb of the Moon will be essentially zero. In the general case, the maximum fringe visibility is obtained by tracking the limb of the Moon that is perpendicular to the baseline.

Although the Moon may produce strong fringes, it is not an ideal calibration source. First, libration may make it difficult to track the exact edge of the Moon. Second, because the apparent source distribution is determined by the antennas, tracking errors introduce amplitude and phase fluctuations. Third, because the temperature of the Moon depends on solar illumination, variations around the mean temperature of ~ 200 K are significant, especially at short wavelengths. For accurate results, the lunar temperature variation should be incorporated into the brightness temperature model.

Appendix 10.2   Doppler Shift of Spectral Lines

The Doppler shift [e.g., Rybicki and Lightman (1979)] is given by the relation

$$\frac{\lambda}{\lambda_0} = \frac{\nu_0}{\nu} = \frac{1 + \frac{v}{c} \cos \theta}{\sqrt{1 - \left(\frac{v}{c} \right)^2}} ,$$

(A10.6)

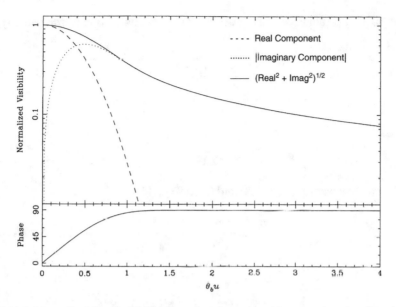

Fig. A10.1 Normalized fringe visibility for an interferometer with an east–west baseline observing the west limb of the Moon at transit ($v = 0$), vs. $\theta_b u$. $\theta_b \simeq 1.2\lambda/d$ is the half-power beamwidth of the antenna, d is the antenna diameter, and $u = D/\lambda$ is the baseline in wavelengths. On the horizontal axis, $\theta_b u$ is approximately equal to $1.2D/d$. The dotted line is the imaginary component of visibility, the dashed line is the real part, and the solid line is the magnitude. Since the portion of the curve for $D/d < 1$ is not accessible, the measured visibility is almost purely imaginary. For $d = 6$ m and $D/d = 3$, the zero-spacing flux density [see Eq. (A10.4)] is 12,700 Jy, and the visibility is about 1000 Jy [see Eq. (A10.5)]. Adapted from Gurwell (1998).

where λ_0 and ν_0 are the rest wavelength and frequency as measured in the reference frame of the source, the corresponding unsubscripted variables are the wavelength and frequency in the observer's frame, v is the magnitude of the relative velocity between the source and the observer, and θ is the angle between the velocity vector and the line-of-sight direction between source and observer in the observer's frame ($\theta < 90°$ for a receding source). The numerator in Eq. (A10.6) is the classical Doppler shift caused by the change in distance between the source and the observer. The denominator is the relativistic time dilation factor, which takes account of the difference between the period of the radiated wave as measured in the rest frame of the source and the rest frame of the observer.

Because of the time dilation effect, there will be a second-order Doppler shift even if the motion is transverse to the line of sight. For the rest of this discussion, we consider only radial velocities; that is, $\theta = 0$ or $180°$. In this case, the Doppler shift equation is

$$\frac{\lambda}{\lambda_0} = \frac{\nu_0}{\nu} = \sqrt{\frac{1 + \frac{v_r}{c}}{1 - \frac{v_r}{c}}}, \tag{A10.7}$$

where v_r is the radial velocity (positive for recession). Solving for velocity, we obtain

$$\frac{v_r}{c} = \frac{v_0^2 - v^2}{v_0^2 + v^2} , \tag{A10.8}$$

or

$$\frac{v_r}{c} = \frac{\lambda^2 - \lambda_0^2}{\lambda^2 + \lambda_0^2} . \tag{A10.9}$$

Taylor expansions of Eqs. (A10.8) and (A10.9) yield

$$\frac{v_r}{c} \simeq -\frac{\Delta v}{v_0} + \frac{1}{2} \frac{\Delta v^2}{v_0^2} \cdots \tag{A10.10}$$

and

$$\frac{v_r}{c} \simeq \frac{\Delta \lambda}{\lambda_0} - \frac{1}{2} \frac{\Delta \lambda^2}{\lambda_0^2} \cdots , \tag{A10.11}$$

where $\Delta v = v - v_0$ and $\Delta \lambda = \lambda - \lambda_0$. For negative Δv, the velocity is positive and the signal is "redshifted." Since $\Delta v / v_0 \simeq -\Delta \lambda / \lambda_0$, the second-order terms have approximately the same magnitude but opposite signs in Eqs. (A10.10) and (A10.11).

Devices for spectroscopy at radio and optical frequencies usually produce data that are uniformly spaced in frequency and wavelength, respectively. Hence, to first order, the velocity axis can be calculated as a linear transformation of the frequency or wavelength axes. Unfortunately, this has led to two different approximations of the velocity:

$$\frac{v_{r\text{radio}}}{c} = -\frac{\Delta v}{v_0} \tag{A10.12}$$

and

$$\frac{v_{r\text{optical}}}{c} = \frac{\Delta \lambda}{\lambda_0} . \tag{A10.13}$$

The difference between these two approximations can be appreciated by noting that $v_{r\text{radio}}/c = -\Delta \lambda / \lambda$. Each velocity scale produces a second-order error in its estimation of the true velocity; that is, the radio definition underestimates the velocity, and the optical definition overestimates the velocity by the same amount. The difference in velocity between the scales as a function of velocity is

$$\delta v_r = v_{r\text{optical}} - v_{r\text{radio}} \simeq \frac{v_r^2}{c} . \tag{A10.14}$$

Hence, the identification of the velocity scale used is very important for extragalactic sources. For example, if $v_r = 10,000$ km s^{-1}, $\delta v_r \simeq 330$ km s^{-1}. Failure to recognize the difference between the velocity conventions can cause considerable problems when observations are made with narrow bandwidth.

To interpret the velocities of spectral lines, it is necessary to refer them to an appropriate inertial frame. The rotation velocity of an observer at the equator about the Earth's center is about 0.5 km s^{-1}; the velocity of the Earth around the Sun is about 30 km s^{-1}; the velocity of the Sun with respect to the nearby stars is about 20 km s^{-1} [this defines the local standard of rest (LSR)]; the velocity of the LSR around the center of the Galaxy is about 220 km s^{-1}; the velocity of our Galaxy with respect to the local group is about 310 km s^{-1}; and the velocity of the local group with respect to the CMB radiation is about 630 km s^{-1}. The most accurate reference frame beyond the solar system is defined with respect to the CMB. The velocity of the Sun with respect to the CMB has been determined from measurements of the dipole anisotropy of the CMB ($v = cT_{dipole}/T_{CMB}$, where $T_{dipole} = 3364.3 \pm 1.5\,\mu$K and $T_{CMB} = 2.7255 \pm 0.0006$ K), which yields the remarkably precise result of 370.1 $\pm$ 0.1 km s^{-1} toward $\ell = 263.91° \pm 0.02°$ and $b = 48.265° \pm 0.002°$ (Planck Collaboration 2016). Information on these various reference frames is listed in Table A10.1. Most observations are reported with respect to either the solar

Table A10.1   Reference frames for spectroscopic observations

Name	Type of motion	Motion (km s^{-1})	Direction[a] ℓ (°)	b (°)
Topocentric	Rotation of Earth	0.5	–	–
Geocentric	Rotation of Earth around Earth/Moon barycenter	0.013	–	–
Heliocentric	Rotation of Earth around Sun	30	–	–
Barycentric	Rotation of Sun around solar system barycenter (planetary perturbations)	0.012	–	–
Local standard of rest (LSR)[b,c]	Sun with respect to local stars	20	57	23
Galactocentric[b]	LSR around center of the Galaxy	220	90	0
Local Galactic Standard of rest[d]	Sun with respect to Galaxies of the local group	308	105	−7
CMB[e]	Sun with respect to CMB	370	264	48

[a]Galactic longitude and latitude.
[b]Standard value adopted by the IAU in 1985 (see Kerr and Lynden-Bell 1986). See literature for more recent determinations.
[c]Converted from 20 km s^{-1} toward $\alpha = 18^h$, $\delta = 30°$ (1990). See literature for newer measurements.
[d]Cox (2000).
[e]Planck Collaboration (2016).

system barycenter or the LSR. Velocities of stars and galaxies are usually given in the former frame, and observations of nonstellar Galactic objects (e.g., molecular clouds) are usually given in the latter frame. Accurate determination of the rotation speed of the Galaxy and its structure depend on precise knowledge of the LSR. Velocity corrections at many radio observatories are based on a program called DOP [Ball (1969); see also Gordon (1976)], which has an accuracy of $\sim$0.01 km s^{-1} because it does not take planetary perturbations into account. Routines such as CVEL in AIPS are based on this code. Much higher accuracy can be obtained by more sophisticated programs such as the Planetary Ephemeris Program (Ash 1972) or the JPL Ephemeris (Standish and Newhall 1996). Precise comparison of velocity measurements at different observations requires comparison of their dynamical calculations. Interpretation of pulsar timing measurements also requires precise velocity correction.

There is sometimes confusion in the conversion of baseband frequency to true observed frequency. In the calculation of the spectrum in the baseband by Fourier transformation of either the data stream or the correlation function with the FFT algorithm, the first channel corresponds to zero frequency, and the channel increment is $\Delta \nu_{IF}/N$, where $\Delta \nu_{IF}$ is the bandwidth (half the Nyquist sampling rate) and N is the total number of frequency channels. The Nth channel corresponds to frequency $\Delta \nu_{IF}(1 - 1/N)$. If N is an even number (N is usually a power of two), channel $N/2$ corresponds to the center frequency of the baseband. For a system with only upper-sideband conversions, the sky frequency of the first channel (zero frequency in the baseband) is the sum of the local oscillator frequencies. Note that the velocity axes run in opposite directions ($v \propto -\nu$ and $v \propto \nu$) for systems with net upper- and lower-sideband conversion, respectively.

There are several velocity shifts of non-Doppler origin that sometimes need to be taken into account. For spectral lines originating in deep potential wells—for example, close to black holes—there is an additional time dilation term

$$\gamma_G = \frac{1}{\sqrt{1 - \frac{r_s}{r}}}, \qquad (A10.15)$$

where r is the distance from the center of the black hole and r_s is its Schwarzschild radius ($r_s = 2GM/c^2$), which is valid for $r \gg r_s$. The total frequency shift [obtained by generalizing Eq. (A10.6)] is therefore

$$\frac{\nu_0}{\nu} = \left(1 + \frac{v_r}{c} \cos \theta \right) \gamma_L \gamma_G, \qquad (A10.16)$$

where $\gamma_L = 1/\sqrt{1 - v_r^2/c^2}$ is known as the Lorentz factor. For example, the radiation from the water masers in NGC 4258 (see Fig. 1.23), which orbit a black hole at a radius of 40,000 r_s, undergoes a velocity shift of about 4 km s^{-1}.

The most important non-Doppler frequency shift for sources at cosmological distances is due to the expansion of the Universe. In the relatively nearby Universe, this velocity shift is

$$z = \frac{\lambda}{\lambda_0} - 1 \simeq \frac{H_0 d}{c} , \qquad (A10.17)$$

where H_0 is the Hubble constant and d is the distance. H_0 is about 70 km s^{-1} Mpc^{-1} (Mould et al. 2000). For greater distances ($z > 1$), the relations between z and the distance and look-back time depend on the cosmological model used [e.g., Peebles (1993)]. However, given the definition of z, the correct frequency will always be related to it by

$$\nu = \frac{\nu_0}{z+1} . \qquad (A10.18)$$

Other issues regarding observations of cosmologically distant spectral line sources are discussed by Gordon et al. (1992). An early example of spectroscopic interferometric observations of a molecular cloud at a cosmological distance ($z = 3.9$) can be found in Downes et al. (1999).

Appendix 10.3   Historical Notes

A10.3.1   *Images from One-Dimensional Profiles*

Early images of the Sun and a few other strong sources were made with linear arrays such as the grating array and compound interferometer shown in Fig. 1.13. The results were obtained in the form of fan-beam scans. With such an instrument, the visibility data sampled at any instant are located on a straight line through the origin in the (u, v) plane, as shown in Fig. 10.1. Fourier transformation of the visibility data sampled along such a line provides a corrugated surface with a profile given by the fan-beam scan, as shown in Fig. A10.2. This can be regarded as one component of a two-dimensional image. As the Earth rotates, the angle of the beam on the sky varies, so addition of these components builds up a two-dimensional image. However, in the fan-beam scans from such arrays, each pair of antennas contributes with equal weight to the profile, so an image built up from profiles in such a manner exhibits the undesirable characteristics of natural weighting. During the 1950s, before digital computers were generally available, the combination of such data to provide two-dimensional images with a desirable weighting was a laborious process. Christiansen and Warburton's (1955) solar image involved Fourier transformation, weighting, and retransformation of the data by manual calculation. A method of combining fan-beam scans without Fourier transformation was later devised by Bracewell and Riddle (1967) using convolution to adjust the visibility weighting. Basic relationships between one- and two-dimensional responses (Bracewell 1956a) are discussed in Sect. 2.4.

Fig. A10.2   A surface in the
(l, m) domain that is the
Fourier transform of visibility
data in the (u, v) plane
measured along a line making
an angle $\phi + \pi/2$ with the u
axis, as shown by the
broken line in Fig. 10.1.

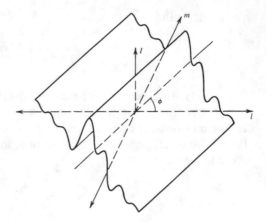

A10.3.2   Analog Fourier Transformation

An optical lens can be used as an analog device for Fourier transformation. Analog systems for data processing based on optical, acoustic, or electron-beam processes were investigated in the early years but generally have not proved successful for synthesis imaging. They lacked flexibility, and a further problem was limitation of the *dynamic range*, which is the ratio of the highest intensity levels to the noise in the image. Maintaining image quality in any iterative process that involves successive Fourier transformation and retransformation of the same data, as occurs in some deconvolution processes (see Chap. 11), requires high precision. Analog possibilities for Fourier transformation were discussed by Cole (1979) but became irrelevant as more powerful computers became available.

Further Reading

Perley, R.A., Schwab, F.R., and Bridle, A.H., Eds., *Synthesis Imaging in Radio Astronomy*, Astron. Soc. Pacific Conf. Ser., **6** (1989)

Sault, R.J., and Oosterloo, T.A., Imaging Algorithms in Radio Interferometry, in *Review of Radio Science 1993–1996*, Stone, W.R., Ed., Oxford Univ. Press, Oxford, UK (1996), pp. 883–912

Taylor, G.B., Carilli, C.L., and Perley, R.A., Eds., *Synthesis Imaging in Radio Astronomy II*, Astron. Soc. Pacific Conf. Ser., **180** (1999)

Thompson, A.R., and D'Addario, L.R., Eds., *Synthesis Mapping*, Proc. NRAO Workshop No. 5, National Radio Astronomy Observatory, Green Bank, WV (1982)

References

Akiyama, K., Lu, R.-S., Fish, V.L., Doeleman, S.S., Broderick, A.E., Dexter, J., Hada, K., Kino, M., Nagai, H., Honma, M., and 28 coauthors, 230 GHz VLBI Observations of M87: Event-Horizon-Scale Structure During an Enhanced Very-High-Energy γ-Ray State in 2012, *Astrophys. J.*, **807**:150 (11pp) (2015)

Ash, M.E., *Determination of Earth Satellite Orbits*, MIT Lincoln Laboratory Technical Note, 1972-5 (1972)

Baddour, N., Operational and Convolution Properties of Two-Dimensional Fourier Transforms in Polar Coordinates, *J. Opt. Soc. Am. A*, **26**, 1768–1778 (2009)

Ball, J.A., *Some Fortran Subprograms Used in Astronomy*, MIT Lincoln Laboratory Technical Note, 1969-42 (1969)

Bennett, C.L., Bay, M., Halpern, M., Hinshaw, G., Jackson, C., Jarosik, N., Kogut, A., Limon, M., Meyer, S.S., Page, L., and five coauthors, The Microwave Anisotropy Probe Mission, *Astrophys. J.*, **583**, 1–23 (2003)

Betzig, E., Trautman, J.K., Harris, T.D., Weiner, J.S., and Kostelak, R.L., Breaking the Diffraction Barrier: Optical Microscopy on a Nanometric Scale, *Science*, **251**, 1468–1470 (1991)

Bevington, P.R., and Robinson, D.K., *Data Reduction and Error Analysis for the Physical Sciences*, 2nd ed., McGraw-Hill, New York (1992)

Bos, A., On Ghost Source Mechanisms in Spectral Line Synthesis Observations with Digital Spectrometers, in *Indirect Imaging*, Roberts, J.A., Ed., Cambridge Univ. Press, Cambridge, UK (1984), pp. 239–243

Bos, A., "On Instrumental Effects in Spectral Line Synthesis Observations," Ph.D. thesis, Univ. of Groningen (1985), see section 10

Bower, G.C., Markoff, S., Brunthaler, A., Law, C., Falcke, H., Maitra, D., Clavel, M., Goldwurm, A., Morris, M.R., Witzel, G., Meyer, L., and Ghez, A.M., The Intrinsic Two-Dimensional Size of Sagittarius A*, *Astrophys. J.*, **790**:1 (10pp) (2014)

Bracewell, R.N., Strip Integration in Radio Astronomy, *Aust. J. Phys.*, **9**, 198–217 (1956a)

Bracewell, R.N., Two-Dimensional Aerial Smoothing in Radio Astronomy, *Aust. J. Phys.*, **9**, 297–314 (1956b)

Bracewell, R.N., *Two-Dimensional Imaging*, Prentice-Hall, Englewood Cliffs, NJ (1995)

Bracewell, R.N., *The Fourier Transform and Its Applications*, McGraw-Hill, New York (2000) (earlier eds. 1965, 1978)

Bracewell, R.N., and Riddle, A.C., Inversion of Fan-Beam Scans in Radio Astronomy, *Astrophys. J.*, **150**, 427–434 (1967)

Bracewell, R.N., and Roberts, J.A., Aerial Smoothing in Radio Astronomy, *Aust. J. Phys.*, **7**, 615–640 (1954)

Briggs, D.S., "High Fidelity Deconvolution of Moderately Resolved Sources," Ph.D. thesis, New Mexico Institute of Mining and Technology (1995). http://www.aoc.nrao.edu/dissertations/dbriggs

Briggs, D.S., Schwab, F.R., and Sramek, R.A., Imaging, in *Synthesis Imaging in Radio Astronomy II*, Taylor, G.B., Carilli, C.L., and Perley, R.A., Eds., Astron. Soc. Pacific Conf. Ser., **180**, 127–149 (1999)

Brouw, W.N., *Data Processing for the Westerbork Synthesis Radio Telescope*, Univ. of Leiden (1971)

Brouw, W.N., Aperture Synthesis, in *Methods in Computational Physics*, Vol. 14, Alder, B., Fernbach, S., and Rotenberg, M., Eds., Academic Press, New York (1975), pp. 131–175

Burton, W.B., The Structure of Our Galaxy Derived from Observations of Neutral Hydrogen, in *Galactic and Extragalactic Radio Astronomy*, Verschuur, G.L., and Kellermann, K.I., Eds., Springer-Verlag, Berlin (1988), pp. 295–358

Carlstrom, J.E., Joy, M., and Grego, L., Interferometric Imaging of the Sunyaev-Zeldovich Effect at 30 GHz, *Astrophys. J. Lett.*, **456**, L75–L78 (1996)

Christiansen, W.N., and Warburton, J.A., The Distribution of Radio Brightness over the Solar Disk at a Wavelength of 21 cm. III. The Quiet Sun–Two-Dimensional Observations, *Aust. J. Phys.*, **8**, 474–486 (1955)

Cole, T.W., Analog Processing Methods for Synthesis Observations, in *Image Formation from Coherence Functions in Astronomy*, van Schooneveld, C., Ed., Reidel, Dordrecht, the Netherlands (1979), pp. 123–141

Contreras, M.E., Rodríguez, L.F., and Arnal, E.M., New VLA Observations of WR 6 (= HR 50896): A Search for an Anisotropic Wind, *Rev. Mex. Astron. Astrof.*, **36**, 135–139 (2000)

Cox, A.N., Ed., *Allen's Astrophysical Quantities*, 4th ed., AIP Press, Springer, New York (2000)

Crane, P.C., and Napier, P.J., Sensitivity, in *Synthesis Imaging in Radio Astronomy*, Perley, R.A., Schwab, F.R., and Bridle, A.H., Eds., Astron. Soc. Pacific Conf. Ser., **6**, 139–165 (1989)

D'Addario, L.R., Cross Correlators, in *Synthesis Imaging in Radio Astronomy*, Perley, R.A., Schwab, F.R., and Bridle, A.H., Eds., Astron. Soc. Pacific Conf. Ser., **6**, 59–82 (1989)

Dillon, J.S., Tegmark, M., Lui, A., Ewall-Wice, A., Hewitt, J.N., Morales, M.F., Neben, A.R., Parsons, A.R., and Zheng, H., Mapmaking for Precision 2 km Cosmology, *Phys. Rev. D*, **91**, 023002 (26 pp.) (2015)

Doeleman, S.S., Weintroub, J., Rogers, A.E.E., Plambeck, R., Freund, R., Tilanus, R.P.J., Friberg, P., Ziurys, L.M., Moran, J.M., Corey, B., and 18 coauthors, Event-Horizon-Scale Structure in the Supermassive Black Hole Candidate at the Galactic Centre, *Nature*, **455**, 78–80 (2008)

Downes, D., Neri, R., Wiklind, T., Wilner, D.J., and Shaver, P.A., Detection of CO(4–3), CO(9–8), and Dust Emission in the Broad Absorption Line Quasar APM 08279+5255 at a Redshift of 3.9, *Astrophys. J. Lett.*, **513**, L1–L4 (1999)

Ekers, R.D., Error Recognition, in *Synthesis Imaging in Radio Astronomy II*, Taylor, G.B., Carilli, C.L., and Perley, R.A., Eds., Astron. Soc. Pacific Conf. Ser., **180**, 321–334 (1999)

Ekers, R.D., and van Gorkom, J.H., Spectral Line Imaging with Aperture Synthesis Radio Telescopes, in *Indirect Imaging*, Roberts, J.A., Ed., Cambridge Univ. Press, Cambridge, UK (1984), pp. 21–32

Escalante, V., Rodríguez, L.F., Moran, J.M., and Cantó, J., The Asymmetric Profile of the H76α Line Emission from MWC349, *Rev. Mex. Astron. Astrof.*, **17**, 11–14 (1989)

Fish, V.L., Johnson, M.D., Doeleman, S.S., Broderick, A.E., Psaltis, D., Lu, R.-S., Akiyama, K., Alef, W., Algaba, J.C., Asada, K., and 62 coauthors, Persistent Asymmetric Structure of Sagittarius A* on Event Horizon Scales, *Astrophys. J.*, **820**:90 (11pp) (2016)

Fomalont, E.B., The East–West Structure of Radio Sources at 1425 MHz, *Astrophys. J. Suppl.*, **15**, 203–274 (1968)

Fomalont, E.B., and Wright, M.C.H., Interferometry and Aperture Synthesis, in *Galactic and Extragalactic Radio Astronomy*, Verschuur, G.L., and Kellermann, K.I., Eds., Springer-Verlag, New York (1974), pp. 256–290

Gordon, M.A., Computer Programs for Radio Astronomy, in *Methods of Experimental Physics*, Vol. 12, Part C *(Astrophysics: Radio Observations)*, Meeks, M.L., Ed., Academic Press, New York (1976)

Gordon, M.A., Baars, J.W.M., and Cocke, W.J., Observations of Radio Lines from Unresolved Sources: Telescope Coupling, Doppler Effects, and Cosmological Corrections, *Astron. Astrophys.*, **264**, 337–344 (1992)

Gradshteyn, I.S., and Ryzhik, I.M., *Table of Integrals, Series, and Products*, 5th ed., Academic Press, New York (1994)

Gurwell, M., Lunar and Planetary Fluxes at 230 GHz: Models for the Haystack 15-m Baseline, SMA Technical Memo 127, Smithsonian Astrophysical Observatory (1998)

Hasler, N., Bulbul, E., Bonamente, M., Carlstrom, J.E., Culverhouse, T.L., Gralla, M., Greer, C., Hawkins, D., Hennessy, R., Joy, M., and 12 coauthors, Joint Analysis of X-Ray and Sunyaev-Zel'dovich Observations of Galaxy Clusters Using an Analytic Model of the Intracluster Medium, *Astrophys. J.*, **748**:113 (12pp) (2012)

Hobson, M.P., Lazenby, A.N., and Jones, M., A Bayesian Method for Analyzing Interferometer Observations of Cosmic Microwave Background Fluctuations, *Mon. Not. R. Astron. Soc.*, **275**, 863–873 (1995)

Jacquinot, P., and Roizen-Dossier, B., Apodisation, in *Progress in Optics*, Vol. 3, Wolf, E., Ed., North Holland, Amsterdam (1964), pp. 29–186

Jennison, R.C., A Phase Sensitive Interferometer Technique for the Measurement of the Fourier Transforms of Spatial Brightness Distributions of Small Angular Extent, *Mon. Not. R. Astron. Soc.*, **118**, 276–284 (1958)

Kazemi, S., Yatawatta, S., and Zaroubi, S., Clustered Calibration: An Improvement to Radio Interferometric Direction-Dependent Self-Calibration, *Mon. Not. R. Astron. Soc.*, **430**, 1457–1472 (2013)

Kerr, F.J., and Lynden-Bell, D., Review of Galactic Constants, *Mon. Not. R. Astron. Soc.*, **221**, 1023–1038 (1986)

Lannes, A., Phase and Amplitude Calibration in Aperture Synthesis: Algebraic Structures, *Inverse Problems*, **7**, 261–298 (1991)

Leitch, E.M., Kovac, J.M., Pryke, C., Carlstrom, J.E., Halverson, N.W., Holzapfel, W.L., Dragovan, M., Reddall, B., and Sandberg, E.S., Measurement of Polarization with the Degree Angular Scale Interferometer, *Nature*, **420**, 763–771 (2002a)

Leitch, E.M., Pryke, C., Halverson, N.W., Kovac, J., Davidson, G., LaRoque, S., Schartman, E., Yamasaki, J., Carlstrom, J.E., Holzapfel, W.L., and seven coauthors, Experiment Design and First Season Observations with the Degree Angular Scale Interferometer, *Astrophys. J.*, **568**, 28–37 (2002b)

Lobanov, A., Brightness Temperature Constraints from Interferometric Visibilities, *Astron. Astrophys.*, **574**, A84 (9pp) (2015)

Lochner, M., Natarajan, I., Zwart, J.T.L., Smirnov, L., Bassett, B.A., Oozeer, N., and Kunz, M., Bayesian Inference for Radio Observations, *Mon. Not. R. Astron. Soc.*, **450**, 1308–1319 (2015)

Lu, R.-S., Fish, V.L., Akiyama, K., Doeleman, S.S., Algaba, J.C., Bower, G.C., Brinkerink, C., Chamberlin, R., Crew, G., Cappallo, R.J., and 23 coauthors, Fine-Scale Structure of the Quasar 3C279 Measured with 1.3-mm Very Long Baseline Interferometry, *Astrophys. J.*, **772**:13 (10pp) (2013)

Maltby, P., and Moffet, A.T., Brightness Distribution in Discrete Radio Sources. III. The Structure of the Sources, *Astrophys. J. Suppl.*, **7**, 141–163 (1962)

Martí-Vidal, I., Pérez-Torres, M.A., and Lobanov, A.P., Over-Resolution of Compact Sources in Interferometric Observations, *Astron. Astrophys.*, **541**, A135 (4pp) (2012)

Martí-Vidal, I., Vlemmings, W.H.T., Mueller, S., and Casey, S., UVMULTIFIT: A Versatile Tool for Fitting Astronomical Radio Interferometric Data, *Astron. Astrophys.*, **563**, A136 (9pp) (2014)

Masson, C.R., Angular Expansion and Measurement with the VLA: The Distance to NGC 7027, *Astrophys. J. Lett.*, **302**, L27–L30 (1986)

Morales, M.F., and Wyithe, J.S.B., Reionization and Cosmology with 21-cm Fluctuations, *Ann. Rev. Astron. Astrophys.*, **48**, 127–171 (2010)

Mould, J.R., Huchra, J.P., Freedman, W.L., Kennicutt, R.C., Jr., Ferrarese, L., Ford, H.C., Gibson, B.K., Graham, J.A., Hughes, S.M.G., Illingworth, G.D., and seven coauthors, The Hubble Space Telescope Key Project on the Extragalactic Distance Scale. XXVIII. Combining the Constraints on the Hubble Constant, *Astrophys. J.*, **529**, 786–794 (2000)

Napier, P.J., and Crane, P.C., Signal-to-Noise Ratios, in *Synthesis Mapping*, Proc. NRAO Workshop No. 5, Thompson, A.R., and D'Addario, L.R., Eds., National Radio Astronomy Observatory, Green Bank, WV (1982), pp. 3-1–3-28

Ng, C.-Y., Gaensler, B.M., Staveley-Smith, L., Manchester, R.N., Kesteven, M.J., Ball, L., and Tzioumis, A.K., Fourier Modeling of the Radio Torus Surrounding SN 1987A, *Astrophys. J.*, **684**, 481–497 (2008)

Ortiz-León, G.N., Johnson, M.D., Doeleman, S.S., Blackburn, L., Fish, V.L., Loinard, L., Reid, M.J., Castillo, E., Chael, A.A., Hernández-Gómez, A., and 12 coauthors, The Intrinsic Shape of Sagittarius A* at 3.5-mm Wavelength, *Astrophys. J.*, **824**:40 (10pp) (2016)

Padin, S., Shepherd, M.C., Cartwright, J.K., Keeney, R.G., Mason, B.S., Pearson, T.J., Readhead, A.C.S., Schaal, W.A., Sievers, J., Udomprasert, P.S., and six coauthors, The Cosmic Background Imager, *Publ. Astron. Soc. Pacific*, **114**, 83–97 (2002)

Parsons, A., Backer, D.C., Foster, G.S., Wright, M.C.H., Bradley, R.F., Gugliucci, N.E., Parashare, C.R., Benoit, E.E., Aguirre, J.E., Jacobs, D.C., and five coauthors, The Precision Array for Probing the Epoch of Reionization: Eight Station Results, *Astron. J.*, **139**, 1468–1480 (2010)

Parsons, A., Pober, J., McQuinn, M., Jacobs, D., and Aguirre, J., A Sensitivity and Array Configuration Study for Measuring the Power Spectrum of 21-cm Emission from Reionization, *Astrophys. J.*, **753**:81 (16pp) (2012)

Parsons, A.R., Lui, A., Aguirre, J.E., Ali, Z.S., Bradley, R.F., Carilli, C.L., DeBoer, D.R., Dexter, M.R., Gugliucci, N.E., Jacobs, D.C., and seven coauthors, New Limits on 21 cm Epoch of Reionization from PAPER-32 Consistent with an X-Ray Heated Intergalactic Medium at $z = 7.7$, *Astrophys. J.*, **788**:106 (21pp) (2014)

Pearson, T.J., Non-Imaging Data Analysis, in *Synthesis Imaging in Radio Astronomy II*, Taylor, G.B., Carilli, C.L., and Perley, R.A., Eds., Astron. Soc. Pacific Conf. Ser., **180**, 335–355 (1999)

Peebles, P.J.E., *Principles of Physical Cosmology*, Princeton Univ. Press, Princeton, NJ (1993)

Perley, R.A., Schwab, F.R., and Bridle, A.H., Eds., *Synthesis Imaging in Radio Astronomy*, Astron. Soc. Pacific Conf. Ser., **6** (1989)

Planck Collaboration, *Planck* 2015 Results. XIII. Cosmological Parameters, *Astron. Astrophys.*, **594**, A13 (63pp) (2016)

Pryke, C., Halverson, N.W., Leitch, E.M., Kovac, J., Carlstrom, J.E., Holzapfel, W.L., and Dragovan, M., Cosmological Parameter Extraction from the First Season of Observations with the Degree Angular Scale Interferometer, *Astrophys. J.*, **568**, 46–51 (2002)

Rau, U., Bhatnagar, S., Voronkov, M.A., and Cornwell, T.J., Advances in Calibration and Imaging Techniques in Radio Interferometry, *Proc. IEEE*, **97**, 1472–1481 (2009)

Readhead, A.C.S., Myers, S.T., Pearson, T.J., Sievers, J.L., Mason, B.S., Contaldi, C.R., Bond, J.R., Bustos, R., Altamirano, P., Achermann, C., and 16 coauthors, Polarization Observations with Cosmic Background Imager, *Science*, **306**, 836–844 (2004)

Reid, M.J., Spectral-Line VLBI, in *Very Long Baseline Interferometry and the VLBA*, Zensus, J.A., Diamond, P.J., and Napier, P.J., Eds., Astron. Soc. Pacific Conf. Ser., **82**, 209–225 (1995)

Reid, M.J., Spectral-Line VLBI, in *Synthesis Imaging in Radio Astronomy II*, Taylor, G.B., Carilli, C.L., and Perley, R.A., Eds., Astron. Soc. Pacific Conf. Ser., **180**, 481–497 (1999)

Rhodes, D.R., On the Spheriodal Functions, *J. Res. Natl. Bureau of Standards B*, **74**, 187–209 (1970)

Roelfsema, P., Spectral Line Imaging I: Introduction, in *Synthesis Imaging in Radio Astronomy*, Perley, R.A., Schwab, F.R., and Bridle, A.H., Eds., Astron. Soc. Pacific Conf. Ser., **6**, 315–339 (1989)

Rybicki, G.B., and Lightman, A.P., *Radiative Processes in Astrophysics*, Wiley-Interscience, New York (1979) (reprinted 1985)

Schwab, F.R., Optimal Gridding of Visibility Data in Radio Interferometry, in *Indirect Imaging*, Roberts, J.A., Ed., Cambridge Univ. Press, Cambridge, UK (1984), pp. 333–346

Scott, P.F., Carreira, P., Cleary, K., Davies, R.D., Davis, R.J., Dickinson, C., Grainge, K., Gutiérrez, C.M., Hobson, M.P., Jones, M.E., and 16 coauthors, First Results from the Very Small Array. III. The Cosmic Microwave Background Power Spectrum, *Mon. Not. R. Astron. Soc.*, **341**, 1076–1083 (2003)

Sivia, D.S., with Skilling, J., *Data Analysis: A Bayesian Tutorial*, 2nd. ed., Oxford Univ. Press, Oxford, UK (2006)

Slepian, D., Analytic Solution of Two Apodization Problems, *J. Opt. Soc. Am.*, **55**, 1110–1115 (1965)

Slepian, D., and Pollak, H.O., Prolate Spheroidal Wave Functions, Fourier Analysis and Uncertainty. I. *Bell Syst. Tech. J.*, **40**, 43–63 (1961)

Smith, F.G., The Measurement of the Angular Diameter of Radio Stars, *Proc. Phys. Soc. B*, **65**, 971–980 (1952)

Smoot, G.F., Bennett, C.L., Kogut, A., Wright, E.L., Aymon, J., Boggess, N.W., Cheng, E.S., De Amici, G., Gulkis, S., Hauser, M.G., and 18 coauthors, Structure in the *COBE* Differential Microwave Radiometer First-Year Maps, *Astrophys. J. Lett.*, **396**, L1–L5 (1992)

Standish, E.M., and Newhall, X.X., New Accuracy Levels for Solar System Ephemerides, in *Dynamics, Ephemerides, and Astrometry of Solar System Bodies*, IAU Symp. 172, Kluwer, Dordrecht, the Netherlands, (1996), pp. 29–36

Taylor, G.B., Carilli, C.L., and Perley, R.A., Eds., *Synthesis Imaging in Radio Astronomy II*, Astron. Soc. Pacific Conf. Ser., **180** (1999)

Thompson, A.R., and Bracewell, R.N., Interpolation and Fourier Transformation of Fringe Visibilities, *Astron. J.*, **79**, 11–24 (1974)

Trott, C.M., Wayth, R.B., Macquart, J.-P.R., and Tingay, S.J., Source Detection in Interferometric Visibility Data. I. Fundamental Estimation Limits, *Astrophys. J.*, **731**:81 (14pp) (2011)

Trotter, A.S., Moran, J.M., and Rodríguez, L.F., Anisotropic Radio Scattering of NGC6334B, *Astrophys. J.*, **493**, 666–679 (1998)

Twiss, R.Q., Carter, A.W.L., and Little, A.G., Brightness Distribution Over Some Strong Radio Sources at 1427 Mc/s, *Observatory*, **80**, 153–159 (1960)

van Gorkom, J.H., and Ekers, R.D., Spectral Line Imaging. II. Calibration and Analysis, in *Synthesis Imaging In Radio Astronomy*, Perley, R.A., Schwab, F.R., and Bridle, A.H., Eds., Astron. Soc. Pacific Conf. Ser., **6**, 341–353 (1989)

Watson, R.A., Carreira, P., Cleary, K., Davies, R.D., Davis, R.J., Dickinson, C., Grainge, K., Gutiérrez, C.M., Hobson, M.P., Jones, M.E., and 16 coauthors, First Results from the Very Small Array. I. Observational Methods, *Mon. Not. R. Astron. Soc.*, **341**, 1057–1065 (2003)

White, M., Carlstrom, J.E., Dragovan, M., and Holzapfel, W.L., Interferometric Observations of Cosmic Microwave Background Anisotropies, *Astrophys. J.*, **514**, 12–24 (1999)

White, R.L., and Becker, R.H., The Resolution of P Cygni's Stellar Wind, *Astrophys. J.*, **262**, 657–662 (1982)

Wright, A.E., and Barlow, M.J., The Radio and Infrared Spectrum of Early-Type Stars Undergoing Mass Loss, *Mon. Not. R. Astron. Soc.*, **170**, 41–51 (1975)

Zheng, H., Tegmark, M., Buza, V., Dillon, J., Gharibyan, H., Hickish, J., Kunz, E., Liu, A., Losh, J., Lutomirski, A., and 28 coauthors, Mapping Our Universe in 3D with MITEoR, Proc. IEEE International Symposium on Phased Array Systems and Technology, Waltham, MA (2013), pp. 784–791

Chapter 11
Further Imaging Techniques

This chapter is concerned with techniques of processing that are largely nonlinear and include deconvolution, that is removing, to the extent possible, the limitations of the visibility measurements. There are two principal deficiencies in the visibility data that limit the accuracy of synthesis images. These are (1) the limited distribution of spatial frequencies in u and v and (2) errors in the visibility measurements. The limited spatial frequency coverage can be improved by deconvolution processes such as CLEAN that allow the unmeasured visibility to take nonzero values within some general constraints on the image. Calibration can be improved by adaptive techniques in which the antenna gains, as well as the required image, are derived from the visibility data. Wide-field imaging, multifrequency imaging, and compressed sensing are also discussed.

11.1 The CLEAN Deconvolution Algorithm

One of the most successful deconvolution procedures is the algorithm CLEAN devised by Högbom (1974). This is basically a numerical deconvolving process usually applied in the image (l, m) domain. It has become an essential tool in producing images from incomplete (u, v) data sets. The procedure is to break down the intensity distribution into point-source responses that correspond to the original imaging process, and then replace each one with the corresponding response to a beam that is free of sidelobes. CLEAN can be thought of as a type of compressed sensing (see Sect. 11.8.6).

© The Author(s) 2017   551
A.R. Thompson, J.M. Moran, and G.W. Swenson Jr., *Interferometry and Synthesis in Radio Astronomy*, Astronomy and Astrophysics Library,
DOI 10.1007/978-3-319-44431-4_11

11.1.1   CLEAN Algorithm

The principal steps are as follows.

1. Compute the image and the response to a point source by Fourier transformation of the visibility and the weighted transfer function. These functions, the synthesized intensity and the synthesized beam, are often referred to as the "dirty image" and the "dirty beam," respectively. The spacing of the sample points in the (l, m) plane should not exceed about one-third of the synthesized beamwidth.
2. Find the highest intensity point in the image and subtract the response to a point source, i.e., the dirty beam, including the full sidelobe pattern, centered on that position. The peak amplitude of the subtracted point-source response is equal to γ times the corresponding image amplitude. γ is called the loop gain, by analogy with negative feedback in electrical systems, and commonly has a value of a few tenths. Record the position and amplitude of the component removed by inserting a delta-function component into a model that will become the cleaned image.
3. Return to step 2 and repeat the procedure iteratively until all significant source structure has been removed from the image. There are several possible indicators of this condition. For example, one can compare the highest peak with the rms level of the residual intensity, look for the first time that the rms level fails to decrease when a subtraction is made, or note when significant numbers of negative components start to be removed.
4. Convolve the delta functions in the cleaned model with a clean-beam response, that is, replace each delta function with a clean-beam function of corresponding amplitude. The clean beam is often chosen to be a Gaussian with a half-amplitude width equal to that of the original synthesized (dirty) beam, or some similar function that is free from negative values.
5. Add the residuals (the residual intensity from step 3) into the clean-beam image, which is then the output of the process. (When the residuals are added, the Fourier transform of the image is equal to the measured visibilities.)

It is assumed that each dirty-beam response that is subtracted represents the response to a point source. As discussed in Sect. 4.4, the visibility function of a point source is a pair of real and imaginary sinusoidal corrugations that extend to infinity in the (u, v) plane. Any intensity feature for which the visibility function is the same within the (u, v) area sampled by the transfer function would produce a response in the image identical to the point-source response. Högbom (1974) has pointed out that much of the sky is a random distribution of point sources on an empty background, and CLEAN was initially developed for this situation. Nevertheless, experience shows that CLEAN also works on well-extended and complicated sources.

The result of the first three steps in the CLEAN procedure outlined above can be represented by a model intensity distribution that consists of a series of delta functions with magnitudes and positions representing the subtracted components. Since the modulus of the Fourier transform of each delta function extends uniformly to infinity in the (u, v) plane, the visibility is extrapolated as required beyond the cutoff of the transfer function.

The delta-function components do not constitute a satisfactory model for astronomical purposes. Groups of delta functions with separations no greater than the beamwidth may actually represent extended structure. Convolution of the delta-function model by the clean beam, which occurs in step 4, removes the danger of overinterpretation. Thus, CLEAN performs, in effect, an interpolation in the (u, v) plane. Desirable characteristics of a clean beam are that it should be free from sidelobes, particularly negative ones, and that its Fourier transform should be constant inside the sampled region of the (u, v) plane and rapidly fall to a low level outside it. These characteristics are essentially incompatible since a sharp cutoff in the (u, v) plane results in oscillations in the (l, m) plane. The usual compromise is a Gaussian beam, which introduces a Gaussian taper in the (u, v) plane. Since this function tapers the measured data and the unmeasured data generated by CLEAN, the resulting intensity distribution no longer agrees with the measured visibility data. However, the absence of large, near-in sidelobes improves the dynamic range of the image, that is, it increases the range of intensity over which the structure of the image can reliably be measured.

As discussed in Chap. 10, we cannot directly divide out the weighted spatial transfer function on the right side of Eq. (10.4) because it is truncated to zero outside the areas of measurement. In CLEAN, this problem is solved by analyzing the measured visibility into sinusoidal visibility components and then removing the truncation so that they extend over the full (u, v) plane. Selecting the highest peak in the (l, m) plane is equivalent to selecting the largest complex sinusoid in the (u, v) plane.

At the point that the component subtraction is stopped, it is generally assumed that the residual intensity distribution consists mainly of the noise. Retaining the residual distribution within the image is, like the convolution with the clean beam, a nonideal procedure that is necessary to prevent misinterpretation of the final result. Without the residuals added in step 5, there would be an amplitude cutoff in the structure corresponding to the lowest subtracted component. Also, the presence of the background fluctuations provides an indication of the level of uncertainty in the intensity values. An example of the effect of processing with the CLEAN algorithm is shown in Fig. 11.1.

11.1.2   Implementation and Performance of the CLEAN Algorithm

As a procedure for removing sidelobe responses, CLEAN is easy to understand. Being highly nonlinear, however, CLEAN does not yield readily to a complete mathematical analysis. Some conclusions have been derived by Schwarz (1978, 1979), who has shown that conditions for convergence of CLEAN are that the synthesized beam must be symmetrical and its Fourier transform, that is, the

(a)   (b)

(c)   (d)

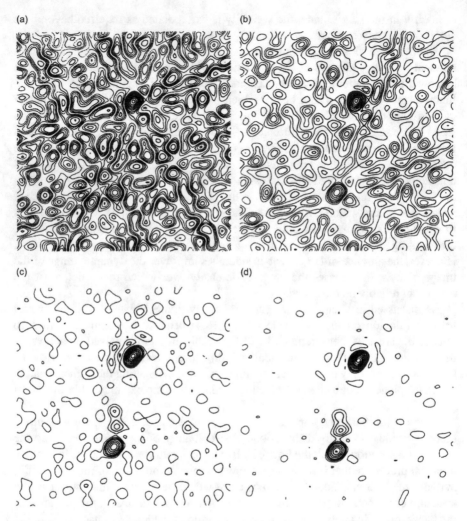

Fig. 11.1 Illustration of the CLEAN procedure using observations of 3C224.1 at 2695 MHz made with the interferometer at Green Bank, and rather sparse (u, v) coverage. (**a**) The synthesized "dirty" image; (**b**) the image after one iteration with the loop gain $\gamma = 1$; (**c**) after two iterations; (**d**) after six iterations. The components removed were restored with a clean beam in all cases. The contour levels are 5, 10, 15, 20, 30, etc., percent of the maximum value. From J. A. Högbom (1974), reproduced with permission. © ESO.

weighted transfer function, must be nonnegative. These conditions are fulfilled in the usual synthesis procedure. Schwarz's analysis also indicates that if the number of delta-function components in the CLEAN model does not exceed the number of independent visibility data, CLEAN converges to a solution that is the least-mean-squares fit of the Fourier transforms of the delta-function components to the measured visibility. In enumerating the visibility data, either the real and imaginary parts or the conjugate values (but not both) are counted independently. In images made using the fast Fourier transform (FFT) algorithm, there are equal numbers of grid points in the (u, v) and (l, m) planes, but not all (u, v) grid points contain visibility measurements. To maintain the condition for convergence, it is a common procedure to apply CLEAN only within a limited area, or "window," of the original image.

In order to clean an image of a given dimension, it is necessary to have a dirty beam pattern of twice the image dimensions so that a point source can be subtracted from any location in the image. However, it is often convenient for the image and beam to be the same size. In that case, only the central quarter of the image can be properly processed. Thus, it is commonly recommended that the image obtained from the initial Fourier transform should have twice the dimensions required for the final image. As mentioned above, the use of such a window also helps to ensure that the number of components removed does not exceed the number of visibility data and, in the absence of noise, allows the residuals within the window area to approach zero.

Several arbitrary choices influence the result of the CLEAN process. These include the parameter γ, the window area, and the criterion for termination. Note that a point-source component in the image can be removed in one step of CLEAN only if it is centered on an image cell. This is an important reason for choosing $\gamma \ll 1$. A value between 0.1 and 0.5 is usually assigned to γ, and it is a matter of general experience that CLEAN responds better to extended structure if the loop gain is in the lower part of this range. The computation time for CLEAN increases rapidly as γ is decreased, because of the increasing number of subtraction cycles required. If the signal-to-noise ratio is $\mathcal{R}_{sn}$, then the number of cycles required for one point source is $-\log \mathcal{R}_{sn} / \log(1 - \gamma)$. Thus, for example, with $\mathcal{R}_{sn} = 100$ and $\gamma = 0.2$, a point source requires 21 cycles.

A well-known problem of CLEAN is the generation of spurious structure in the form of spots or ridges as modulation on broad features. A heuristic explanation of this effect is given by Clark (1982). The algorithm locates the maximum in the broad feature and removes a point-source component, as shown in Fig. 11.2. The negative sidelobes of the beam add new maxima, which are selected in subsequent cycles, and thus, there is a tendency for the component subtraction points to be located at intervals equal to the spacing of the first sidelobe of the synthesized (dirty) beam. The resulting image contains a lumpy artifact introduced by CLEAN, but the image is consistent with the measured visibility data. Cornwell (1983) introduced a modification of the CLEAN algorithm that is intended to reduce this unwanted modulation. The original CLEAN algorithm minimizes

$$\sum_k w_k |\mathcal{V}_k^{\text{meas}} - \mathcal{V}_k^{\text{model}}|^2 , \tag{11.1}$$

where $\mathcal{V}_k^{\text{meas}}$ is the measured visibility at (u_k, v_k), w_k is the applied weighting, and $\mathcal{V}_k^{\text{model}}$ is the corresponding visibility of the CLEAN-derived model. The summation is taken over the points with nonzero data in the input transformation for the dirty

Fig. 11.2 Subtraction of the point-source response (broken line) at the maximum of a broad feature, as in the process CLEAN. Adapted from Clark (1982).

image. Cornwell's algorithm minimizes

$$\sum_k w_k |\mathcal{V}_k^{\text{meas}} - \mathcal{V}_k^{\text{model}}|^2 - \kappa s \,, \qquad (11.2)$$

where s is a measure of smoothness, and κ is an adjustable parameter. Cornwell found that the mean-squared intensity of the model, taken with a negative sign, is an effective implementation of s.

The effects of visibility tapering appear in both the original image and the beam, and thus the magnitudes and positions of the components subtracted in the CLEAN process should be largely independent of the taper. However, since tapering reduces the resolution, it is a common practice to use uniform visibility weighting for images that are processed using CLEAN. Alternately, in difficult cases such as those involving extended smooth structure, reduction of sidelobes by tapering may improve the performance of CLEAN.

An important reduction in the computation required for CLEAN was introduced by Clark (1980). This is based on subtraction of the point-source responses in the (u, v) plane and using the FFT for moving data between the (u, v) and (l, m) domains. The procedure consists of minor and major cycles. A series of minor cycles is used to locate the components to be removed by performing approximate subtractions using only a small patch of the synthesized dirty beam that includes the main beam and the major sidelobes. Then in a major cycle, the identified point-source responses are subtracted, without approximation, in the (u, v) plane. That is, the convolution of the delta functions with the dirty beam is performed by multiplying their Fourier transforms. The series of minor and major cycles is then repeated until the required stop condition is reached. Clark devised this technique for use with data from the VLA and found that it reduced the computation by a factor of two to ten compared with the original CLEAN algorithm.

Other variations on the CLEAN process have been devised; one of the more widely used is the Cotton–Schwab algorithm [Schwab (1984); see Sect. IV], which is a variation of the Clark algorithm. The subtractions in the major cycle are performed on the ungridded visibility data, which eliminates aliasing at this point. The algorithm is also designed to permit processing of adjacent fields, which are treated separately in the minor cycles but in the major cycles, components are jointly removed from all fields.

To summarize the characteristics of CLEAN, we note that it is simple to understand from a qualitative viewpoint and straightforward to implement and that its usefulness is well proven. On the other hand, a full analysis of its response is difficult. The response of CLEAN is not unique, and it can produce spurious artifacts. It is sometimes used in conjunction with model-fitting techniques; for example, a disk model can be removed from the image of a planet and the residual intensity processed by CLEAN. A more stable and efficient version of CLEAN called multiscale CLEAN has been developed for extended objects (Wakker and Schwarz 1988; Cornwell 2008). The basic idea is that broad emission components are identified first and removed. More sophisticated methods are being developed to handle extended emission [e.g., Junklewitz et al. (2016)]. CLEAN is also used as part of more complex image construction techniques. For more details, including hints on usage, see Cornwell et al. (1999), and for extended objects, Cornwell (2008).

11.2   Maximum Entropy Method

11.2.1   MEM Algorithm

An important class of image-restoration algorithms operates to produce an image that agrees with the measured visibility to within the noise level, while constraining the result to maximize some measure of image quality. Of these, the maximum entropy method (MEM) has received particular attention in radio astronomy. If $I'(l, m)$ is the intensity distribution derived by MEM, a function $F(I')$ is defined, which is referred to as the entropy of the distribution. $F(I')$ is determined entirely by the distribution of I' as a function of solid angle and takes no account of structural forms within the image. In constructing the image, $F(I')$ is maximized within the constraint that the Fourier transform of I' should fit the observed visibility values.

In astronomical image formation, an early application of MEM is that of Frieden (1972) to optical images. In radio astronomy, the earliest discussions are by Ponsonby (1973) and Ables (1974). The aim of the technique, as described by Ables, is to obtain an intensity distribution consistent with all relevant data but minimally committal with regard to missing data. Thus, $F(I')$ must be chosen so that maximization introduces legitimate a priori information but allows the visibility in the unmeasured areas to assume values that minimize the detail introduced.

Several forms of $F(I')$ have been used, which include the following:

$$F_1 = -\sum_i \frac{I'_i}{I'_s} \log\left(\frac{I'_i}{I'_s}\right) \tag{11.3a}$$

$$F_2 = -\sum_i \log I'_i \tag{11.3b}$$

$$F_3 = -\sum_i I'_i \ln\left(\frac{I'_i}{M_i}\right), \tag{11.3c}$$

where $I'_i = I'(l_i, m_i)$, $I'_s = \sum_i I'_i$, M_i represents an a priori model, and the sums are taken over all pixels, I_i, in the image. F_3 can be described as relative entropy, since the intensity values are specified relative to a model.

A number of papers discuss the derivation of the expressions for entropy from theoretical and philosophical considerations. Bayesian statistics are invoked: see Jaynes (1968, 1982). Gull and Daniell (1979) consider the distributions of intensity quanta scattered randomly on the sky, and they derive the form F_1, which is also used by Frieden (1972). The entropy form F_2 is obtained by Ables (1974) and Wernecke and D'Addario (1977). Other investigators take a pragmatic approach to MEM (Högbom 1979, Subrahmanya 1979, Nityananda and Narayan 1982). They view the method as an effective algorithm, even though there may be no underlying physical or information-theoretical basis for the choice of constraints. Högbom (1979) points out that both F_1 and F_2 contain the required mathematical characteristics: the first derivatives tend to infinity as I' approaches zero, so maximizing F_1 or F_2 produces positivity in the image. The second derivatives are everywhere negative, which favors uniformity in the intensity. Narayan and Nityananda (1984) consider a general class of functions F that have the properties $d^2F/dI'^2 < 0$ and $d^3F/dI'^3 > 0$. F_1 and F_2, discussed above, are members of this class.

In the maximization of the entropy expression $F(I')$, the constraint that the resulting intensity model should be consistent with the measured visibility data is implemented through a χ^2 statistic. Here, χ^2 is a measure of the mean-squared difference between the measured visibility values, $\mathcal{V}_k^{\mathrm{meas}} = \mathcal{V}(u_k, v_k)$, and the corresponding values for the model $\mathcal{V}_k^{\mathrm{model}}$:

$$\chi^2 = \sum_k \frac{|\mathcal{V}_k^{\mathrm{meas}} - \mathcal{V}_k^{\mathrm{model}}|^2}{\sigma_k^2} , \tag{11.4}$$

where σ_k^2 is the variance of the noise in $\mathcal{V}_k^{\mathrm{meas}}$, and the summation is taken over the visibility data set. Obtaining a solution involves an iterative procedure; see Wernecke and D'Addario (1977), Wernecke (1977), Gull and Daniell (1978), Skilling and Bryan (1984), and a review by Narayan and Nityananda (1984). As an example, Cornwell and Evans (1985) maximize a parameter J given by

$$J = F_3 - \alpha\chi^2 - \beta S_{\mathrm{model}} , \tag{11.5}$$

where F_3 is defined in Eq. (11.3c). S_{model} is the total flux density of the model and is included because in order for the process to converge to a satisfactory result, it was found necessary to include a constraint that the total flux density of the model be equal to the measured flux density. Lagrange multipliers α and β are included, the values of which are adjusted as the model fitting proceeds so that χ^2 and S_{model} are equal to the expected values. Through the use of F_3, a priori information can be introduced into the final image. The various algorithms that have been developed for implementing MEM generally use the gradients of the entropy and of χ^2 to determine the adjustment of the model in each iteration cycle.

A feature of images derived by MEM is that the point-source response varies with position, so the angular resolution is not constant over the image. Comparison of maximum entropy images with those obtained using direct Fourier transformation often shows higher angular resolution in the former. The extrapolation of the visibility values can provide some increase in resolution over more conventional imaging techniques.

11.2.2   Comparison of CLEAN and MEM

CLEAN is defined in terms of a procedure, so the implementation is straightforward, but because of the nonlinearity in the processing, a noise analysis of the result is very difficult. In contrast, MEM is defined in terms of an image that fits the data to within the noise and is also constrained to maximize some parameter of the image. The noise in MEM is taken into account through the χ^2 statistic, and the resulting effect on the noise is more easily analyzed for MEM; see, for example, Bryan and Skilling (1980). Some further points of comparison are as follows:

- Implementation of MEM requires an initial source model, which is not necessary in CLEAN.
- CLEAN is usually faster than MEM for small images, but MEM is faster for very large images. Cornwell et al. (1999) give the break-even point as about 10^6 pixels.
- CLEAN images tend to show a small-scale roughness, attributable to the basic approach of CLEAN, which models all images as ensembles of point sources. In MEM, the constraint in the solution emphasizes smoothness in the image.
- Broad, smooth features are better deconvolved using MEM, since CLEAN may introduce stripes and other erroneous detail. MEM does not perform well on point sources, particularly if they are superimposed on a smooth background that prevents negative sidelobes from appearing as negative intensity in the dirty image.

To illustrate the characteristics of the CLEAN and MEM procedures, Fig. 11.3 shows examples of processing of a model jet structure from Cornwell (1995) and Cornwell et al. (1999), using model calculations by Briggs. The jet model is based on similar structure in M87 and is virtually identical to the contour levels shown in part (e). The left end of the jet is a point source smoothed to the resolution of the simulated observation. Visibility values for the model corresponding to the (u, v) coverage of the VLBA (Napier et al. 1994) were calculated for a frequency of 1.66 GHz and a declination of 50° with essentially full tracking range. Thermal noise was added, but the calibration was assumed to be fully accurate. Fourier transformation of the visibility data and the spatial transfer function provided the dirty image and dirty beam. The image shows the basic structure, but fine details are swamped by sidelobes. Parts (a) to (c) of Fig. 11.3 show the effects of processing by CLEAN. In the CLEAN deconvolution, 20,000 components were subtracted with a

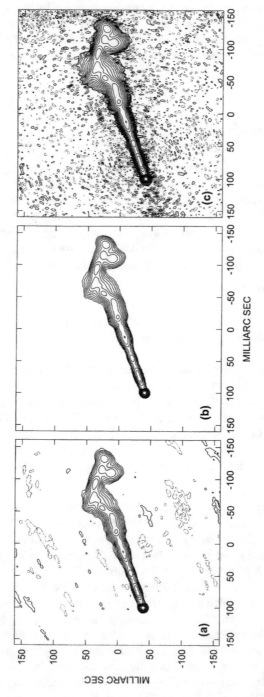

Fig. 11.3 Examples of deconvolution procedures applied to a model jet structure that includes a point source at the left end. Part (**a**) shows the result of application of CLEAN to the whole image, and (**b**) the result when components are taken only within a tight support region surrounding the source. Note the improvement obtained in (**b**). The contours approximately indicate the intensity increasing in powers of two from a low value of 0.05%. Part (**c**) shows the same image as (**b**) but with contours starting a factor of ten lower in intensity, and the roughness characteristic of CLEAN is visible in the low-level contours.

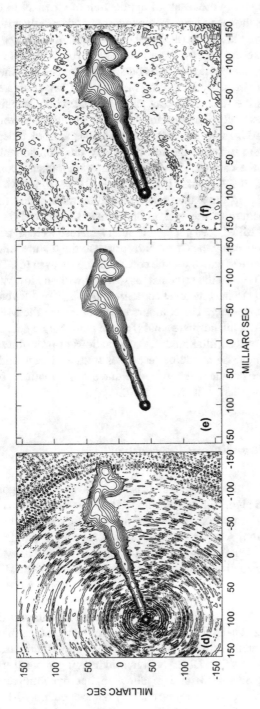

Fig. 11.3 Part (**d**) shows the result of MEM deconvolution using the same constraint region as in (**b**) and 80 iterations. The circular artifacts, centered on the point source, illustrate the well-known inability of MEM to handle sharp features well. In part (**e**), the point source was subtracted, using the CLEAN response to the feature, and then the MEM deconvolution was performed with the same constraint region as in (**d**). The source was then replaced. Part (**f**) shows the same response as (**e**) with the lowest contours at the same level as part (**c**). Note that the low-level contours are smoother in the MEM image than in the CLEAN one. The images in (**c**) and (**f**) show comparable fidelity to the model. All six parts are from Cornwell (1995), courtesy of and © the Astronomical Society of the Pacific.

loop gain of 0.1. Part (a) shows the result of application of CLEAN to the whole image, and part (b) shows the result when components are taken only within a tight support region surrounding the source (the technique sometimes referred to as use of a box or window). Note the improvement obtained in (b), which is a result of adding the information that there is no emission outside the box region. The contours approximately indicate the intensity increasing in powers of two from a low value of 0.05%. Part (c) shows the same image as panel (b) but with contours starting a factor of ten lower in intensity. The roughness visible in the low-level contours is characteristic of CLEAN, in which each component is treated independently and there is no mechanism to relate the result for any one component to those for its neighbors, unlike the case of MEM, in which a smoothness constraint is introduced. Parts (d) to (f) result from MEM processing. Part (d) shows the result of MEM deconvolution using the same constraint region as in panel (b) and 80 iterations. The circular pattern of the background artifacts, centered on the point source, clearly shows that MEM does not handle such a feature well. In part (e), the point source was subtracted, using the CLEAN response to the feature, and then the MEM deconvolution performed with the same constraint region as in (d). The source was then replaced. Part (f) shows the same response as (e) with the lowest contours at the same level as panel (c). The low-level contours show the structure contributed by the observation and processing. The contours are smoother in the MEM image than in the CLEAN one. The images in (c) and (f) have comparable *fidelity*, that is, accuracy of reproduction of the initial model. Combinations of procedures, such as the use of CLEAN to remove point-source responses from an image and then the use of MEM to process the broader background features can sometimes be used to advantage in complex images.

11.2.3   Further Deconvolution Procedures

Briggs (1995) has applied a nonnegative, least-squares (NNLS) algorithm for deconvolution. The NNLS algorithm was developed by Lawson and Hanson (1974) and provides a solution to a matrix equation of the form $\mathbf{AX} = \mathbf{B}$, where, in the radio astronomy application, $\mathbf{A}$ represents the dirty beam and $\mathbf{B}$ the dirty image. The algorithm provides a least-mean-squares solution for the intensity $\mathbf{X}$ that is constrained to contain no negative values. However, unlike the case for MEM, there is no smoothness criterion involved. The NNLS solution requires more computer capacity than CLEAN or MEM solutions, but Briggs's investigation indicated that it is capable of superior performance, particularly in cases of compact objects of width only a few synthesized beamwidths. NNLS was found to reduce the residuals to a level close to the system noise in the observations. In certain cases, it was found to work more effectively than CLEAN in hybrid imaging and self-calibration procedures (discussed in Sect. 11.3) and to allow higher dynamic range to be achieved. In MEM, the residuals may not be entirely random but may be correlated in the image plane, and this effect can introduce bias in the (u, v) data that limits the achievable dynamic range. CLEAN appears to behave somewhat similarly unless it is allowed to run long enough to work down into the noise. Some further discussion can be found in Briggs (1995) and Cornwell et al. (1999).

11.3   Adaptive Calibration and Imaging

Calibration of the visibility amplitude is usually accurate to a few percent or better, but phase errors expressed as a fraction of a radian may be much larger, sometimes as a result of variations in the ionosphere or troposphere. Nevertheless, the relative values of the uncalibrated visibility measured simultaneously on a number of baselines contain information about the intensity distribution that can be extracted through the closure relationships described in Chap. 10, Eqs. (10.34) and (10.44). Following Schwab (1980), we use the term *adaptive calibration* for both the hybrid imaging and self-calibration techniques that make use of this information. Imaging with amplitude data only has also been investigated and is briefly described

11.3.1   Hybrid Imaging

The rekindling of interest in closure techniques in the 1970s began with their rediscovery by Rogers et al. (1974), who used closure phases to derive model parameters for VLBI data. Fort and Yee (1976) and several later groups incorporated closure data into iterative imaging techniques, of which that by Readhead et al. (1980) is as follows:

1. Obtain an initial trial image based on inspection of visibility amplitudes and any a priori data such as an image at a different wavelength or epoch. If the trial image is inaccurate, the convergence will be slow, but if necessary, an arbitrary trial image such as a single point source will often suffice.
2. For each visibility integration period, determine a complete set of independent amplitude and/or phase closure equations. For each such set, compute a sufficient number of visibility values from the model such that when added to the closure relationships, the total number of independent equations is equal to the number of antenna spacings.
3. Solve for the complex visibility corresponding to each antenna spacing and make an image from the visibility data by Fourier transformation.
4. Process the image from step 3 using CLEAN but omitting the residuals.
5. Apply constraints for positivity and confinement (delete components having negative intensity or lying outside the area that is judged to contain the source).
6. Test for convergence and return to step 2 as necessary, using the image from step 5 as the new model.

Note that the solution improves with iteration because of the constraints of confinement and positivity introduced in step 5. These nonlinear processes can be envisioned as spreading the errors in the model-derived visibility values throughout

the visibility data, so that they are diluted when combined with the observed values
in the next iterative cycle.

In the process described, and most variants of it, the image is formed by using
some data from the model and some from direct measurements, and following
Baldwin and Warner (1978), the term *hybrid* imaging (or mapping) is sometimes
used as a generic description. With the use of phase closure, there is no absolute
position measurement, but there is no ambiguity with respect to the position angle
of the image. With the use of amplitude closure, only relative levels of intensity are
determined, but it is usually not difficult to calibrate enough of the data to establish
an intensity scale. In many cases, the amplitude data are sufficiently accurate as
observed, and only the phase closure relationships need be used; Readhead and
Wilkinson (1978) have described a version of the above program using phase
closure only. Other versions of this technique, which differ mainly in detail of
implementation from that described, have been developed by Cotton (1979) and
Rogers (1980). If there is some redundancy in the baselines, the number of free
parameters is reduced, which can be advantageous, as discussed by Rogers.

The number of antennas, n_a, is obviously an important factor in imaging
procedures that make use of the closure relationships, since it affects the efficiency
with which the data are used. We can quantify this efficiency by considering the
number of closure data as a fraction of the number of data that would be available
if full calibration were possible, as a function of n_a. The numbers of independent
closure data are given by Eqs. (10.42) and (10.45). The number of data with full
calibration is equal to the number of baselines, which, if we assume there is no
redundancy, is $\frac{1}{2}n_a(n_a - 1)$. For the phase data, the fraction is

$$\frac{\frac{1}{2}(n_a - 1)(n_a - 2)}{\frac{1}{2}n_a(n_a - 1)} = \frac{n_a - 2}{n_a}. \tag{11.6}$$

For the amplitude data, the fraction is

$$\frac{\frac{1}{2}n_a(n_a - 3)}{\frac{1}{2}n_a(n_a - 1)} = \frac{n_a - 3}{n_a - 1}. \tag{11.7}$$

These fractions are also equal to the ratios of observed data to observed plus
model-derived data in each iteration of the hybrid imaging procedure. Equations (11.6) and (11.7) are plotted in Fig. 11.4. For $n_a = 4$, the closure relationships
yield only 50% of the possible phase data and only 33% of the amplitude data. For
$n_a = 10$, however, the corresponding figures are 80% and 78%. Thus, in any array in
which the atmosphere or instrumental effects may limit the accuracy of calibration
by a reference source, it is desirable that the number of antennas should be at least
ten and preferably more. The number of iterations required to obtain a solution
with the hybrid technique depends on the complexity of the source, the number of
antennas, the accuracy of the initial model, and other factors including details of the
algorithm used.

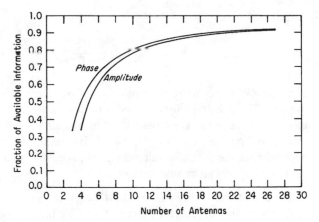

Fig. 11.4 Visibility data that can be obtained through adaptive calibration techniques expressed as a fraction of those available from a fully calibrated array. The curves correspond to Eqs. (11.6) and (11.7).

11.3.2   Self-Calibration

Hybrid imaging has largely been superseded by a more general approach called *self-calibration*. Here, the complex antenna gains are regarded as free parameters to be explicitly derived together with the intensity. In certain cases, the process is easily explained. For example, in imaging an extended source containing a compact component (as in many radio galaxies), the broad structure is resolved with the longer antenna spacings, leaving only the compact source. This can be used as a calibrator to provide the relative phases of the long-spacing antenna pairs, but not the absolute phase since the position is not known. Then, if there is a sufficient number of long spacings in the array, the relative gain factors of the antennas can be obtained using long spacings only. Such a special intensity distribution, however, is not essential to the method, and with an iterative technique, it is possible to use almost any source as its own calibrator. Programs of this type were developed by Schwab (1980) and by Cornwell and Wilkinson (1981). Reviews of the techniques are given by Pearson and Readhead (1984) and Cornwell (1989).

The procedure in self-calibration is to use a least-mean-squares method to minimize the square of the modulus of the difference between the observed visibilities, $\mathcal{V}_{mn}^{\mathrm{meas}}$, and the corresponding values for the derived model, $\mathcal{V}_{mn}^{\mathrm{model}}$. The expression that is minimized is

$$\sum_{\text{time}} \sum_{m<n} w_{mn} |\, \mathcal{V}_{mn}^{\mathrm{meas}} - g_m g_n^* \, \mathcal{V}_{mn}^{\mathrm{model}}|^2 \,, \tag{11.8}$$

where the weighting coefficient w_{mn} is usually chosen to be inversely proportional to the variance of $\mathcal{V}_{mn}^{\mathrm{meas}}$, and the quantities shown are all functions of time within the observing period. Expression (11.8) can be written

$$\sum_{\text{time}} \sum_{m<n} w_{mn} |\, \mathcal{V}_{mn}^{\mathrm{model}}|^2 |X_{mn} - g_m g_n^*|^2 \,, \tag{11.9}$$

where

$$X_{mn} = \frac{\mathcal{V}_{mn}^{\text{meas}}}{\mathcal{V}_{mn}^{\text{model}}} . \tag{11.10}$$

If the model is accurate, the ratio X_{mn} of the uncalibrated observed visibility to the visibility predicted by the model is independent of u and v but proportional to the antenna gains. Thus, the values of X_{mn} simulate the response to a calibrator and enable the gains to be determined. However, since the initial model is only approximate, the desired result must be approached by iteration.

The self-calibration procedure is:

1. Make an initial image as for hybrid imaging.
2. Compute the X_{mn} factors for each visibility integration period within the observation.
3. Determine the antenna gain factors for each integration period.
4. Use the gains to calibrate the observed visibility values and make an image.
5. Use CLEAN and select components to provide positivity and confinement of the image; Cornwell (1982) recommends omitting all features for which $|I(l, m)|$ is less than that for the most negative feature.
6. Test for convergence and return to step 2 as necessary.

The numbers of independent data used in the procedure above are, as in the case of hybrid imaging, equal to the numbers of independent closure relationships given in Eqs. (10.45) and (10.36), that is, $\frac{1}{2}n_a(n_a - 3)$ for amplitude and $\frac{1}{2}(n_a - 1)(n_a - 2)$ for phase. The two procedures, hybrid imaging and self-calibration, are basically equivalent but differ in details of approach and implementation. The efficiency as a function of the number of antennas (Fig. 11.4) applies to both. Examples of the performance of the self-calibration technique are shown in Figs. 11.5 and 11.6.

Treating the gain factors, which are the fundamental unknown quantities, as free parameters as in self-calibration is a rather more direct approach than that of hybrid imaging. A global estimate of the instrumental factors is obtained using the entire data set. Cornwell (1982) points out that it is easier to deal correctly with the noise when considering complex visibility as a vector quantity, as in self-calibration, than when considering amplitude and phase separately, as in hybrid imaging. The noise combines additively in the vector components, resulting in a Gaussian distribution, whereas in the amplitude and phase, the more complicated Rice distributions of Eqs. (6.63) result. Cornwell and Wilkinson (1981) have developed a form of adaptive calibration that takes account of the different probability distributions of the amplitude and phase fluctuations, including system noise, for the different antennas. It has been used with the MERLIN array of Jodrell Bank, UK, which incorporates antennas of different sizes and designs (Thomasson 1986). The probability distributions of the antenna-associated errors are legitimate a priori information, which can be empirically determined for an array.

Experience shows that adaptive calibration techniques in many cases converge to a satisfactory result using only a single point source as a starting model, although

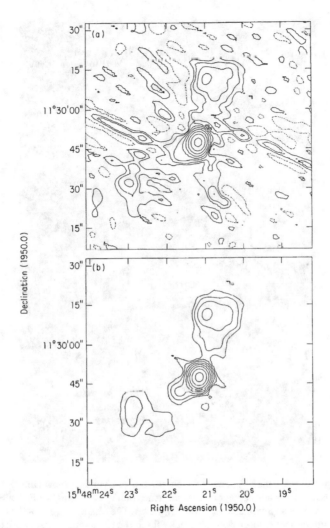

Fig. 11.5 Effect of self-calibration on a VLA radio image of the quasar 1548+115. (a) Image obtained by normal calibration techniques, which has spurious detail at the level of 1% of the peak intensity. (b) Image obtained by the self-calibration technique, in which the level of spurious detail is reduced below the 0.2% level. In both (a) and (b), the lowest contour level is 0.6%. © 1983 IEEE. Reprinted, with permission, from Napier et al. (1983).

inaccuracy in the initial model increases the number of iterative cycles required. A point source is a good model for the phase of a symmetrical intensity distribution but may be a poor model for the amplitude. It must also be remembered that the accuracy of the closure relationships depends on the accuracy of the matching of the frequency responses and polarization parameters from one antenna to another, as discussed in Sects. 7.3 and 7.4. In general, any effect that cannot be represented by a single gain factor for each antenna degrades the closure accuracy.

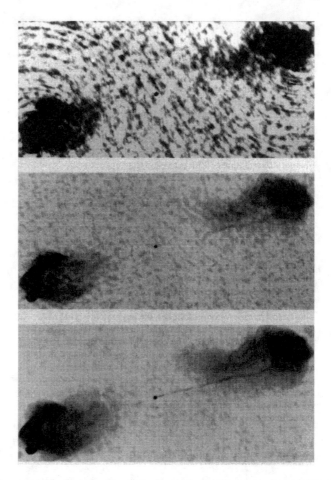

Fig. 11.6 Three stages in the reduction of the observation of Cygnus A shown in Fig. 1.18. The top image is the result of transformation of the calibrated visibility data using the FFT algorithm. The calibration source was approximately 3° from Cygnus A. The center image shows reduction using the MEM algorithm. This compensates principally for the undersampling in the spatial frequencies and thereby removes sidelobes from the synthesized beam. The result is similar to that obtainable using the CLEAN algorithm. The bottom image shows the effect of the self-calibration technique, in which the maximum entropy image is used as the initial model. The final step improves the dynamic range by a factor of 3. In observations in which the initial calibration is not as good as in this case, self-calibration usually provides a greater improvement. The long dimension of the field is 2.1′ and contains approximately 1000 pixels. Courtesy R. A. Perley, J. W. Dreher, and J. J. Cowan. Reproduced by permission of and © NRAO/AUI.

In using adaptive calibration techniques, the integration period of the data must not be longer than the coherence time of the phase variations; otherwise, the visibility amplitude may be reduced. The coherence time may be governed by the atmosphere, for which the timescale is of the order of minutes (see Sect. 13.4). In order for the imaging procedure to work, the field under observation must contain structure fine enough to provide a phase reference and bright enough to be detected with satisfactory signal-to-noise ratio within the coherence time. Thus, adaptive calibration does not solve all problems and cannot be used for the detection of a very weak source in an otherwise empty field.

11.3.3 Imaging with Visibility Amplitude Data Only

A number of early studies were made concerning the feasibility of producing images using only the amplitude values of the visibility. The Fourier transform of the squared modulus of the visibility is equal to the autocorrelation of the intensity distribution, $I \star \star I$:

$$|\mathcal{V}(u, v)|^2 = \mathcal{V}(u, v) \, \mathcal{V}^*(u, v) \longleftrightarrow I(l, m) \star \star I(l, m) . \tag{11.11}$$

The right side can also be written as a convolution: $I(l, m) * * I(-l, -m)$. The problem of imaging with $|\mathcal{V}|$ only is mainly one of interpreting an image of the autocorrelation of I. Without phase data, the position of the center of the field cannot be determined, and there is a $180°$ rotational ambiguity in the position angle of the image.

Examples of studies relevant to imaging without phase data are found in Bates (1969, 1984), Napier (1972), and Fienup (1978). Napier and Bates (1974) review some of the results. The positivity requirement is generally found to be insufficient to provide unique solutions for one-dimensional profiles, but for two-dimensional images, uniqueness is obtained in some cases (Bruck and Sodin 1979). Baldwin and Warner (1978, 1979) considered a two-dimensional distribution of sources, with some success in producing a source image from the autocorrelation function. Although these approaches showed some promise of providing useful interpretation of radio interferometer data, they have not been widely used. More importantly, the development of techniques that make use of closure relationships, such as hybrid imaging and self-calibration, has allowed visibility phases to contribute useful data even when not well calibrated.

11.4 Imaging with High Dynamic Range

The *dynamic range* of an image is usually defined as the ratio of the maximum intensity to the rms level at some part of the field where the background is mainly blank sky. This rms level is assumed to indicate the lowest measurable intensity. The term *image fidelity* is used to indicate the degree to which an image is an accurate representation of a source on the sky. Image fidelity is not directly measurable on an actual source, but simulation of an observation of a model source and reduction of the visibility data allow comparison of the resulting image with the model. This is a way of investigating antenna configurations, processing methods, and other details. The requirements and techniques are discussed in detail by Perley (1989, 1999a).

High dynamic range requires high accuracy in calibration, removal of any erroneous data, and careful deconvolution. That is, it requires high accuracy in the visibility measurements and very good (u, v) coverage. A phase error $\Delta\phi$ can be regarded as introducing an erroneous component of relative amplitude, $\sin\Delta\phi$, into the visibility data, in phase quadrature to the true visibility. An amplitude error of $\varepsilon_a\%$ can be regarded as introducing an error component of relative amplitude $\varepsilon_a\%$ into the visibility. Thus, for example, a phase error of $10°$ introduces as large an error component as does an amplitude error of 17%. An amplitude error of 17% would be considered unusually large in most cases, except in conditions of strong atmospheric attenuation. However, a $10°$ phase error would be more commonly encountered, especially at frequencies in which ionospheric or tropospheric irregularities are important. A phase error $\Delta\phi$ (rad) in a correlator output introduces an error component of rms relative amplitude $\Delta\phi/\sqrt{2}$ in the resulting image. With similar errors in $n_a(n_a - 1)/2$ baselines, the dynamic range of a snapshot is limited to $\sim n_a/\Delta\phi$.

Use of self-calibration is an essential step in minimizing gain errors. However, after calibration of the antenna-based gain factors, there remain small baseline-based terms that can also be calibrated. These result from variations, from one antenna to another, in the frequency passband or the polarization, and similar effects. Note that in arrays with very high sensitivity at the longer wavelengths, the requirement to observe down to the limit set by system noise, in the presence of background sources, places a lower limit on the required dynamic range. A large number of array elements is helpful in distinguishing individual sources (Lonsdale et al. 2000). Braun (2013) describes a detailed analysis of dynamic range in synthesis imaging and gives the results of application of this analysis to several large arrays.

Obtaining the highest possible dynamic range requires attention to details that are specific to particular instruments. For the VLA, the following figures were quoted as rough guidelines for a good observation. Basic calibration results in dynamic range of order $1,000 : 1$. After self-calibration, dynamic range up to $\sim 20,000 : 1$ is possible. After careful correction of baseline-based errors, it may be a few times higher. If a spectral correlator is used, which avoids errors in quadrature networks and also relaxes the requirement for delay accuracy, a dynamic range of $\sim 200,000 : 1$ is possible, with much care, assuming that the signal-to-noise ratio is adequate (Perley 1989).

11.5   Mosaicking

Mosaicking is a technique that allows imaging of an area of sky that is larger than the beam of the array elements. It becomes very important in the millimeter-wavelength range, where antenna beams are relatively narrow. Although radio astronomy antennas for millimeter wavelengths are generally smaller in diameter than are antennas for centimeter wavelengths, their beamwidths are often narrower because the wavelengths are so much shorter. For example, the Atacama Large Millimeter/submillimeter Array (ALMA) can operate at frequencies up to 950 GHz, at which the beamwidth of the 12-m-diameter antennas could be as small as $\sim 6''$.

Consider imaging a square field whose sides are n times the width of the antenna primary beam. One can divide the required area into n^2 subfields, each the size of a beam, and produce a separate image for each such area. The n^2 beam-area images can then be fitted together like mosaic pieces to cover the full field desired. One would anticipate that some difficulty might occur in obtaining uniform sensitivity, particularly near the joints of the mosaic pieces, but clearly the idea is feasible. From the sampling theorem described in Sect. 5.2, the number of visibility sample points in u and v required in an image covering n^2 beam areas is n^2 times as many as would be required in an image that covers just one beam area. In mosaicking, the increased data are obtained by using n^2 different pointing directions of the antennas. As a result, the sampling of the visibility in u and v must be at an interval $1/n$ of that for a field equal to the beam size, and this interval is usually less than the diameter of the antenna aperture. However, it is possible to determine how the visibility varies on a scale less than the diameter of an antenna, as discussed below.

Figure 5.9 shows two antennas that are tracking the position of a source. The antenna spacing projected normal to the direction of the source is u, and the antenna diameter is d_λ, both quantities being measured in wavelengths. In the u direction, the interferometer responds to spatial frequencies from $(u - d_\lambda)$ to $(u + d_\lambda)$, since spacings within this range can be found within the antenna apertures. Measurement of the variation of the visibility over this range of baselines can provide the fine sampling required in mosaicking. The difference in path lengths from the source to the two antenna apertures is w wavelengths, and as the antennas track, the variation in w gives rise to fringes at the correlator output. Since the apertures of the antennas remain normal to the direction of the source, the path difference w, and its rate of change, are the same for any pair of points of which one is in each aperture plane, regardless of their spacing. Thus, because of the tracking motion, the signals received at any two such points produce a component of the correlator output with the same fringe frequency. Such components cannot, therefore, be separated by Fourier analysis, and information on the variation of the visibility within the spatial frequency range $(u - d_\lambda)$ to $(u + d_\lambda)$ is lost. However, in mosaicking, the antenna beams are scanned across the field, either by moving periodically between different pointing centers or by continuously scanning, for example, in a raster pattern. The scanning is in addition to the usual tracking motion to follow the source across the sky. In Fig. 5.9, it can be seen that if the antennas are suddenly turned through a small angle $\Delta\theta$, then the position of the point B is changed by $\Delta u \, \Delta\theta$ wavelengths in a direction parallel to that of the source. This results in a phase change of approximately $2\pi \Delta u \, \Delta\theta$ in the fringe component corresponding to the spacing $(u + \Delta u)$, of which points A_1 and B are an example. Since this phase change is linearly proportional to Δu, the variation of the visibility within the range $(u - d_\lambda)$ to $(u + d_\lambda)$ can be obtained by Fourier transformation of the correlator output with respect to the pointing offset $\Delta\theta$. Thus, the changes in pointing induce variations in the fringe phase that are dependent on the spacing of the incoming rays within the antenna apertures, and this effect allows the information on the variation of the visibility to be retained.

The conclusion given above, that the scanning action of the antennas allows information on a range of visibility values to be retrieved, was first reached by Ekers and Rots (1979), using a mathematical analysis, as follows. Consider a pair of antennas with spacing (u_0, v_0) pointing in the direction (l_p, m_p). As the pointing angle is varied, the effective intensity distribution over the field of interest is represented by $I(l, m)$ convolved with the *normalized* antenna beam $A_N(l, m)$. The observed fringe visibility is the Fourier transform with respect to u and v of $I(l, m)$ multiplied by the antenna response for the particular pointing:

$$\mathcal{V}(u_0, v_0, l_p, m_p) = \int \int A_N(l - l_p, m - m_p)I(l, m)e^{-j2\pi(u_0 l + v_0 m)}dl\, dm . \quad (11.12)$$

Assuming that the antenna beam is symmetrical, we can write Eq. (11.12) as

$$\mathcal{V}(u_0, v_0, l_p, m_p) = \int \int A_N(l_p - l, m_p - m)I(l, m)e^{-j2\pi(u_0 l + v_0 m)}dl\, dm , \quad (11.13)$$

which has the form of a two-dimensional convolution:

$$\mathcal{V}(u_0, v_0, l_p, m_p) = [I(l, m)e^{-j2\pi(u_0 l + v_0 m)}] ** A_N(l, m) . \quad (11.14)$$

Now we take the Fourier transform of $\mathcal{V}$ with respect to u and v, which represents the full-field visibility data obtained by means of the ensemble of pointing angles used:

$$\mathcal{V}(u, v) = \int \int [I(l, m)e^{-j2\pi(u_0 l + v_0 m)}] ** A_N(l, m)e^{j2\pi(ul + vm)}dl\, dm$$

$$= [\mathcal{V}(u, v) ** {}^2\delta(u_0 - u, v_0 - v)]\overline{A}_N(u, v) . \quad (11.15)$$

Here, $\overline{A}_N(u, v)$ is the Fourier transform of $A_N(l, m)$, that is, the autocorrelation of the field distribution over the aperture of a single antenna, referred to as the transfer function or spatial sensitivity function of the antenna. The two-dimensional delta function ${}^2\delta(u_0 - u, v_0 - v)$ is the Fourier transform of $e^{-j2\pi(u_0 l + v_0 m)}$. As the final step, Eq. (11.15) becomes

$$\mathcal{V}(u, v) = \mathcal{V}[(u_0 - u), (v_0 - v)]\overline{A}_N(u, v) . \quad (11.16)$$

The conclusion from Eq. (11.16) is that if one observes a field of dimensions equal to several beamwidths, obtains the visibility for a number of pointing directions, and then for each antenna pair takes the Fourier transform of the visibility with respect to the pointing direction, the result will be values of the visibility extended over an area of the (u, v) plane as large as the support of the function $\overline{A}_N(u, v)$. For a circular reflector antenna of diameter d, $\overline{A}_N(u, v)$ is nonzero within a circle of diameter $2d$. Thus, if $\overline{A}_N(u, v)$ is known with sufficient accuracy, that is, the beam pattern is sufficiently well calibrated, the visibility can be obtained at the intermediate points required to provide the full-field image.

In the practical reduction of visibility data used in mosaicking, the Fourier transform with respect to pointing is usually not explicitly performed. The importance of the discussion above is that it shows that the information at the required spacing is present in the data if the antenna pointing is scanned with respect to the source, either as a continuous motion or as a series of discrete pointings. The reduction to obtain the intensity distribution is generally based on the use of nonlinear deconvolution algorithms.

Cornwell (1988) has pointed out that the angular spacing required between the pointing centers on the sky can be deduced from the sampling theorem of Fourier transforms (Sect. 5.2.1). A more general form of the theorem can be stated as follows: If a function $f(x)$ is nonzero only within an interval of width Δ in the x coordinate, then it is fully specified if its Fourier transform $F(s)$ is sampled at intervals no greater than Δ^{-1} in s. If the sampling is coarser than this, aliasing will occur, and the original function will not be reproducible from the samples. Here, we consider an antenna beam pointing toward a source that is wide enough to cover most of the reception pattern, that is, the main beam and major sidelobes. As we move the antenna beam to different pointing angles to cover the source, we are effectively sampling the convolution of the source and the antenna beam. The beam pattern is equal to the Fourier transform of the autocorrelation function of the field distribution over the antenna aperture. The field cuts off at the edges of the aperture, which is d_λ wavelengths wide. Thus, the autocorrelation function cuts off at a width $2d_\lambda$. The sampling theorem indicates that the interval between pointings Δl_p should not exceed $1/(2d_\lambda)$ in order to fully sample the source convolved with the beam. In practice, the antenna illumination function is likely to be tapered at the edge, so the autocorrelation function falls to low levels before it reaches the cutoff width $2d_\lambda$. Thus, if Δl_p slightly exceeds $1/(2d_\lambda)$, the error introduced may not be large.

11.5.1 Methods of Producing the Mosaic Image

The basic steps in the mosaicking method are:

1. Observe the visibility function for an appropriate series of pointing centers.
2. Reduce the data for each pointing center independently to produce a series of images, each covering approximately one antenna beam area.
3. Combine the beam-area images into the required full-field image.

In step 2, it is desirable also to deconvolve the synthesized beam response from each beam-area image to remove the effects of sidelobes in the response, and this can be done using, for example, CLEAN or MEM. Use of these nonlinear algorithms can fill in some of the frequency components of the intensity that were omitted from the coverage of the antenna array. Cornwell (1988) and Cornwell et al. (1993) describe two procedures for mosaic imaging. The first of these, which they refer to as linear mosaicking, is essentially the three steps above with a least-mean-squares procedure for combination of the individual pointing images in step 3. Although a nonlinear deconvolution is used individually on each beam-area image, the combination of the images is a linear process. The second procedure, which differs in that the deconvolution is performed jointly, is referred to as nonlinear mosaicking and involves a nonlinear algorithm such as MEM. Unmeasured visibility data can best be estimated in the deconvolution process if the full field that is covered by the ensemble of pointing angles contributes simultaneously to the deconvolution, rather than by treating each primary beam area separately. The benefit of a joint deconvolution of the combined beam-area images is illustrated by consideration of an unresolved component of the intensity distribution located at the edge of a beam area, where it occurs in two or more individual beam images. Being at the beam edge where the response is changing rapidly, the amplitude of the component is more likely to be inaccurately determined, but such errors will tend to average out in the combined data. In the application to mosaicking, maximum entropy can be envisaged as the formation of an image that is consistent with all the visibility data for the various pointings, within the uncertainty resulting from the noise.

Cornwell (1988) discusses use of the MEM algorithm of Cornwell and Evans (1985) in mosaicking. This algorithm is briefly described in Sect. 11.2.1 [see Eq. (11.5)]. The procedure is essentially the same as in the application to a single-pointing image, except for a few more steps in determining χ^2 and its gradient. As in Eq. (11.4), χ^2 is the statistic that indicates the deviation of the model from the measured visibility values and is here expressed as

$$\chi^2 = \sum_p \sum_k \frac{|\, \mathcal{V}_{kp}^{\text{meas}} - \mathcal{V}_{kp}^{\text{model}} \,|^2}{\sigma_{kp}^2} \, , \qquad (11.17)$$

where the subscripts k and p indicate the kth visibility value at the pth pointing position, and σ_{kp}^2 is the variance of the visibility. An initial model is required, and the procedure follows a series of steps described by Cornwell (1988):

1. For the first pointing center, multiply the current trial model with the antenna beam as pointed during the observation, and take the Fourier transform with respect to (l, m) to obtain the predicted visibility values.
2. Subtract the measured visibilities from the model visibilities to obtain a set of residual visibilities. Insert the residual visibilities into the accumulating χ^2 function of Eq. (11.17).
3. By Fourier transformation, convert the residual visibilities, weighted inversely as their variances, into an intensity distribution. Taper this distribution by multiplying it by the antenna beam pattern, and store in a data array of dimensions equal to the full MEM model.

4. Repeat steps 1–3 for each pointing. In step 2, add the value for χ^2 to those for the other pointings in this cycle. In step 3, add the residual intensity values into the data array. The accumulated values in this data array are used to obtain the gradient of χ^2 with respect to the MEM image.

The reason for the additional multiplication of the residual distribution by the beam function in step 3 is that it reduces unwanted responses from sidelobes of the primary beam that fall on adjacent pointing areas. It also weights the data with respect to the signal-to-noise ratio. Completion of the MEM procedure may require several tens of cycles through the steps given above to obtain convergence to the final image. To complete the process, smoothing with a two-dimensional Gaussian beam of width equal to the array resolution is recommended, to reduce the effects of variable resolution across the image.

A slightly different procedure for nonlinear mosaicking is described by Sault et al. (1996). In this case, the beam-area images are combined linearly without the individual deconvolution step, and then the final nonlinear deconvolution is applied to the combined image. In the linear combination, each pixel in the combined image is a weighted sum of the corresponding pixels in the individual beam-area images. As an example, Sault et al. also show results for a mosaic of the Small Magellanic Cloud made with the compact configuration of the Australia Telescope using 320 pointings. They demonstrate that the joint deconvolution used in nonlinear mosaicking is superior to the linear combination of the subfield images, even if these have been individually deconvolved. They also show the deconvolution using both their method and that described by Cornwell (1988) and conclude that the results are of comparable quality.

11.5.2   Short-Baseline Measurements

In imaging sources wider than the antenna beam, it is important to obtain visibility values at increments in u and v that are smaller than the diameter of an antenna. Data equivalent to an essentially continuous coverage in u and v can then be obtained by observing at various pointing positions as discussed above. The minimum spacing of two antennas is limited by mechanical considerations, and there is a gap or region of low sensitivity corresponding to a spacing of about half the minimum spacing between the centers of two antenna apertures. This is called the "short-spacing problem."

This minimum spacing depends on the antenna design, but in general, unless the range of zenith angles is restricted, two antennas of diameter d cannot be spaced much closer than about $1.4d$, or perhaps $1.25d$ with special design. Otherwise, there is danger of mechanical collision, especially if there is a possibility that the antennas may not always be pointing in the same direction. Total-power observations with a single antenna will, in principle, provide spacings from zero to d/λ, but with some antennas, measurements at spatial frequencies greater than $\sim 0.5d/\lambda$ are unreliable because the spatial sensitivity function of the antenna falls to low levels as a result of the tapered illumination of the reflector. Missing data at low (u, v) values result in broad negative sidelobes of the synthesized beam, such that the beam appears to be situated in a shallow bowl-shaped depression. This effect is most noticeable when the field to be imaged is wide enough that there are several empty (u, v) cells within the central area.

The transfer function $\overline{A}_N(u)$ is the autocorrelation function of the field distribution over the antenna aperture and depends on the particular design of the antenna, including the illumination pattern of the feed. The solid curve in Fig. 11.7 shows $\overline{A}_N$ for a uniformly illuminated circular aperture, which can be regarded as an ideal case. Since there is usually some tapering in the illumination of a reflector antenna, in practice, $\overline{A}_N$ will generally fall off somewhat more rapidly than the curves shown. The function $\overline{A}_N$ in Fig. 11.7 is proportional to the common area of two overlapping circles of diameter d, and the abscissa is the distance between their centers. In three dimensions, this function is sometimes referred to as the Chinese hat function, and its properties are discussed by Bracewell (1995). The dashed curves in Fig. 11.7 show the relative spatial sensitivity for an interferometer using two uniformly illuminated circular apertures of diameter d. Curve 1 is for a spacing of $1.4d$ between the centers of the apertures; curve 2 is for a spacing of $1.25d$. If both total-power and interferometer data are obtained, it can be seen that the minimum sensitivity occurs for spacings of approximately half of the antenna spacing.

One solution to increasing the minimum sensitivity in the spatial frequency coverage is the addition of total-power measurements from a larger antenna [see, for example, Bajaja and van Albada (1979), Welch and Thornton (1985), Stanimirović (2002)]. Stanimirović considered the requirements for the use of single-antenna measurements of fringe visibility and concluded that the diameter of the antenna

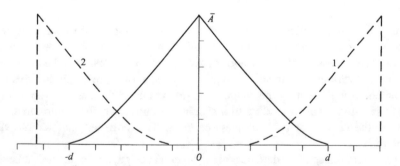

Fig. 11.7 The solid curve centered on the origin shows the spatial sensitivity function $\overline{A}_N$ for a single antenna of diameter d. The curve corresponds to the case of uniform excitation over the aperture. This curve indicates the relative sensitivity to spatial frequencies for total-power observations with a single antenna. The dashed curves show the spatial sensitivity for two antennas of diameter d, with uniform aperture excitation, working as an interferometer. Curve 1 is for a spacing of $1.4d$ between the centers of the antennas, and curve 2 is for a spacing of $1.25d$. If the aperture illumination is tapered, the curves will fall off to low values more rapidly than is shown.

should be at least 1.5 times the spacing for which the visibility value is required. Note, however, that since the cost of an antenna scales approximately as $d^{2.7}$ (see Sect. 5.7.2.2), the expected cost of an antenna of diameter $1.5d$ is roughly 4.4 times that of an antenna of diameter d. The process of merging total power and interferometric data is sometimes called "feathering."

Another possibility for covering the missing spatial frequencies is the use of one or more pairs of smaller antennas, say, $d/2$ in diameter, with spacing about $0.7d$. A pair of antennas of diameter $d/2$ have one-quarter of the area, and consequently one-quarter of the sensitivity to fine structure, of a pair of the standard antennas. Since the beam of the smaller antenna has four times the solid angle of a standard antenna, it will require one-quarter of the number of pointing directions, and the integration time for each one can be four times as long. Cornwell et al. (1993) present evidence that, for mosaicking, it is possible to obtain satisfactory performance with a homogeneous array, that is, one in which all antennas are the same size. This requires total-power observation as well as interferometry with some antennas spaced as closely as possible. The deconvolution steps in the data reduction help to fill in remaining (u, v) gaps.

At frequencies of several hundred gigahertz, where antenna beams are of minute-of-arc order, images of objects of order one degree in size require numbers of pointings in the range 10^2–10^4. Any given pointing cannot be quickly repeated, so dependence on Earth rotation to fill in small gaps in the (u, v) coverage may not be practicable. Thus, arrays designed for mosaicking of large objects require good instantaneous (u, v) coverage. At such high frequencies, it is also desirable to avoid high zenith angles to minimize atmospheric effects.

An alternative to tracking discrete pointing centers is to sweep the beams over the area of sky under investigation in a raster scan motion. This technique has been referred to as "on-the-fly" mosaicking. It has several advantages:

- The uniformity of the (u, v) coverage for all points in the field is maximized, which results in uniformity of the synthesized beam across the resulting image and thereby simplifies the image processing.
- Each point in the field is observed many times in as rapid succession as possible, so some advantage can be taken of Earth rotation to fill in the (u, v) coverage.
- If total-power measurements are made, the scanning motion of the beam can be used to remove atmospheric effects in a similar way to the use of beam switching in large single-dish telescopes.
- Waste of observing time during moves of the antennas from one pointing center to another is eliminated.

With on-the-fly observing, the real-time integration at the correlator output must be somewhat less than the time taken for the beam to scan over any point in the field, and thus a large number of visibility data, each with a separate pointing position, are generated.

11.6   Multifrequency Synthesis

Making observations at several different radio frequencies is an effective way of improving the sampling of the visibility in the (u, v) plane. This technique is referred to as multifrequency synthesis, or bandwidth synthesis. Generally, the range of frequencies is about $\pm 15\%$ of the midrange value. Such a range can be very effective in filling in gaps in the coverage, and since it is not too large, major changes in the source structure with frequency are avoided [see, e.g., Conway et al. (1990)]. However, the variation of structure with frequency may be large enough to limit the dynamic range unless some steps are taken to mitigate it, as discussed here. The principal continuum radio emission mechanisms produce radio spectra that vary smoothly in frequency (see Fig. 1.1), and the intensity usually follows a power-law variation with frequency:

$$I(v) = I(v_0)\left(\frac{v}{v_0}\right)^{\alpha} , \tag{11.18}$$

where α is the spectral index, which varies with (l, m). If the spectrum does not conform to a power law, then, in effect, we can write

$$\alpha = \frac{v}{I}\frac{\partial I}{\partial v} . \tag{11.19}$$

If the spectral index were a constant over the source, the spectral effects could be removed. Although this is generally not the case, the spectral effects are reduced by first correcting the data for a "mean" or "representative" spectral index for the overall structure to be imaged. Thus, from this point, α will represent the spectral index of the deviation of the intensity distribution from this first-order correction. Consider the case in which the intensity variation can be approximated by a linear term:

$$I(v) = I(v_0) + \frac{\partial I}{\partial v}(v - v_0)$$

$$= I(v_0) + \alpha I(v_0)\frac{(v - v_0)}{v}$$

$$\simeq I(v_0) + \alpha I(v_0)\frac{(v - v_0)}{v_0} , \tag{11.20}$$

where the reference frequency v_0 is near the center of the range of frequencies used. Equation (11.20) is the sum of a single-frequency term and a spectral term. To determine the synthesized beam of an array working in the multifrequency mode, consider the response to a point source with a spectrum given by Eq. (11.20). The response to the single-frequency term can be obtained by taking the Fourier transform of the spatial transfer function. The transfer function has a delta function of u and v for each visibility measurement. Each frequency used contributes a different set of delta functions. The response to the spectral term is obtained by multiplying the transfer function by $(v - v_0)/v_0$ and taking the Fourier transform. If we call the single-frequency and spectral responses b_0' and b_1', respectively, the synthesized beam is equal to

$$b_0(l, m) = b_0'(l, m) + \alpha(l, m)b_1'(l, m) . \qquad (11.21)$$

The first component is a conventional synthesized beam, and the second one is an unwanted artifact. The measured intensity distribution obtained as the Fourier transform of the measured visibilities is

$$I_0(l, m) = I(l, m) * * b_0'(l, m) + \alpha(l, m)I(l, m) * * b_1'(l, m) , \qquad (11.22)$$

where $I(l, m)$ is the true intensity on the sky. Conway et al. (1990) and Sault and Wieringa (1994) have both developed deconvolution processes based on the CLEAN algorithm that deconvolve both b_0' and b_1'. In the method used by the first of these groups, components representing each one of the two beams were removed alternately. In the method used by the second group, each component removed represented both beams. These methods provide the distribution of both the source intensity and the spectral index as functions of frequency. Conway et al. also consider a logarithmic rather than a linear form of the frequency offsets from v_0. These analyses show that for a frequency spread of approximately $\pm 15\%$, the magnitude of the response resulting from the b_1' component is typically 1% and can sometimes be ignored. Removing the b_1' component reduces the spectral effects to $\sim 0.1\%$.

For other approaches and extensions to multifrequency synthesis, see Rau and Cornwell (2011) and Junklewitz et al. (2015).

11.7 Noncoplanar Baselines

In Sect. 3.1, it was shown that, except in the case of east–west linear arrays, the baselines of a synthesis array do not remain in a plane as the Earth rotates. It was also shown that for fields of view of small angular size [as given approximately by Eq. (3.12)], the Fourier transform relationship between visibility and intensity can be expressed satisfactorily in two dimensions. However, particularly for frequencies less than a few hundred megahertz, the small-field assumption does not always apply. At meter wavelengths, the primary beams of the antennas are wide, for example, $\sim 6°$ for a 25-m-diameter antenna at a wavelength of 2 m. Also, the high density of strong sources on the sky at meter wavelengths requires that the full beam be imaged to avoid confusion. We now consider the case in which the condition in Eq. (3.12) ($\theta_f < \frac{1}{3}\sqrt{\theta_b}$) is not valid, so the two-dimensional solution should not be used. The following treatment follows those of Sramek and Schwab (1989) and others as noted. We start with the exact result in Eq. (3.7), which is

$$
\mathcal{V}(u, v, w) = \int_{-\infty}^{\infty} \int_{-\infty}^{\infty} \frac{A_N(l, m) I(l, m)}{\sqrt{1 - l^2 - m^2}}
$$
$$
\times \exp\left\{-j2\pi\left[ul + vm + w\left(\sqrt{1 - l^2 - m^2} - 1\right)\right]\right\} dl\,dm . \qquad (11.23)
$$

Here, $\mathcal{V}(u, v, w)$ is the visibility as a function of spatial frequency in three dimensions, $A_N(l, m)$ is the normalized primary beam pattern of an antenna, and $I(l, m)$ is the two-dimensional intensity distribution to be imaged.

The next step is to rewrite Eq. (11.23) in the form of a three-dimensional Fourier transform, which involves the third direction cosine n defined with respect to the w axis. The phase of the visibility $\mathcal{V}(u, v, w)$ is measured relative to the visibility of a (hypothetical) source at the phase reference position for the observation. This introduces a factor $e^{j2\pi w}$ within the exponential term on the right side of Eq. (11.23), as noted in the text following Eq. (3.7). The corresponding phase shift is inserted by the fringe rotation discussed in Sect. 6.1.6. As a result of this factor, we use $n' = n - 1$ as the conjugate variable of w in order to obtain the three-dimensional Fourier transform. Functions of n' will be indicated by a prime. Thus, we rewrite Eq. (11.23):

$$
\mathcal{V}(u, v, w) = \int_{-\infty}^{\infty} \int_{-\infty}^{\infty} \int_{-\infty}^{\infty} \frac{A_N(l, m) I(l, m)}{\sqrt{1 - l^2 - m^2}} \, \delta\left(\sqrt{1 - l^2 - m^2} - n' - 1\right)
$$
$$
\times \exp\{-j2\pi(ul + vm + wn')\} dl\,dm\,dn' . \qquad (11.24)
$$

The delta function $\delta(\sqrt{1 - l^2 - m^2} - n' - 1)$ is introduced to maintain the condition $n = \sqrt{1 - l^2 - m^2}$ and thereby to allow n' to be treated as an independent variable in the Fourier transformation. In a practical observation, $\mathcal{V}$ is measured only at points at which the sampling function $W(u, v, w)$ is nonzero. The Fourier transform of the sampled visibility defines a three-dimensional intensity function I'_3 as follows:

$$
I'_3(l, m, n') =
$$
$$
\int_{-\infty}^{\infty} \int_{-\infty}^{\infty} \int_{-\infty}^{\infty} W(u, v, w) \mathcal{V}(u, v, w) \, e^{j2\pi(ul + vm + wn')} du\,dv\,dw . \qquad (11.25)
$$

This is the Fourier transform of the product of the two functions $W(u, v, w)$ and $\mathcal{V}(u, v, w)$, which by the convolution theorem is equal to the convolution of the Fourier transforms of the two functions. Thus,

$$
I'_3(l, m, n') = \left\{ \frac{A_N(l, m) I(l, m) \, \delta\left(\sqrt{1 - l^2 - m^2} - n' - 1\right)}{\sqrt{1 - l^2 - m^2}} \right\} \overline{W}'(l, m, n') .
$$
$$
\qquad (11.26)
$$

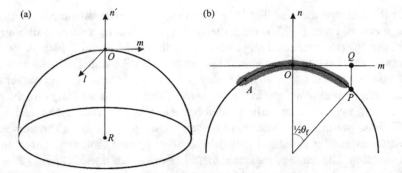

Fig. 11.8 (a) One hemisphere of the unit sphere in (l, m, n) coordinates. The point R is the origin of the (l, m, n) coordinates. O is the origin of the (l, m, n') coordinates, which is the phase reference point. (b) Section through the unit sphere in the (m, n) plane. The shaded area represents the extent of the function I_3. A source at point A would not appear, or would be greatly attenuated, in a two-dimensional analysis in the (l, m) plane. The width of the three-dimensional "beam" in the n direction should be comparable to that in l and m, since the range of the sampling function in w is comparable to that in u and v if the observations cover a large range in hour angle. (In the superficially similar case in Fig. 3.5, the intensity function is not confined to the surface of the sphere because the measurements are all made in the $w' = 0$ plane).

Here, $\overline{W}'(l, m, n')$ is the Fourier transform of the three-dimensional sampling function $W(u, v, w)$, and the triple asterisk denotes three-dimensional convolution. Having determined the result of the Fourier transformation, we can now replace n' by $(n - 1)$, and Eq. (11.26) becomes

$$I_3(l, m, n) = \left\{ \frac{A_N(l, m) I(l, m) \, \delta \left(\sqrt{1 - l^2 - m^2} - n \right)}{\sqrt{1 - l^2 - m^2}} \right\} \, \overline{W}(l, m, n) . \qquad (11.27)$$

The expression in the braces on the right side of Eq. (11.27) is confined to the surface of the unit sphere $n = \sqrt{1 - l^2 - m^2}$, since the delta function is nonzero only on the sphere. The function $\overline{W}$ with which it is convolved is the Fourier transform of the sampling function and is, in effect, a three-dimensional dirty beam. The convolution has the effect of spreading the expression so that I_3 has finite extent in the radial direction of the sphere. Figure 11.8a shows the unit sphere centered on the origin of (l, m, n) coordinates at R. The (l, m) plane in which the results of the conventional two-dimensional analysis lie is tangent to the unit sphere at O, at which point $n = 1$ and $n' = 0$. Note that since l, m, and n are direction cosines, the unit sphere in (l, m, n) is a mathematical concept, not a sphere in real space.

Several ways of obtaining an undistorted wide-field image are possible (Cornwell and Perley 1992).

1. *Three-Dimensional Transformation.* $I_3(l, m, n)$ can be deconvolved by means of a three-dimensional extension of the CLEAN algorithm. This is complicated

by the fact that the visibility is, in practice, not as well sampled in w as it is in u and v. From Fig. 3.4, the large values of w occur for large zenith angles of the target source. In Fig. 11.8b, the width of the angular field is θ_f. The transform must be computed over the range of (l, m) within this field, and over the range PQ in n. Cornwell and Perley (1992) suggest using a direct (rather than discrete) Fourier transform in the n to w transformation, since otherwise the poor sampling may result in serious sidelobes and aliasing. Thus, two-dimensional FFTs are performed in a series of planes normal to the n axis. The number of planes required is equal to PQ divided by the required sampling interval in the n direction. The range of measured visibility values has a width $2|w|_{max}$ in the w direction, so, by the sampling theorem, the intensity function is fully specified in the n coordinate if it is sampled at intervals of $(2|w|_{max})^{-1}$. The distance PQ is approximately equal to $\frac{1}{8}\theta_f^2 \approx \frac{1}{2}|l^2 + m^2|_{max}$ [note that the angle $POQ = \theta_f/4$, and $(\theta_f/2)^2 = |l^2 + m^2|_{max}$]. Thus, the number of planes in which the two-dimensional intensity must be calculated is $|l^2 + m^2|_{max} |w|_{max}$. [This result can also be obtained by taking the phase term in Eq. (3.8) that is omitted in going from three to two dimensions and sampling at the Nyquist interval of half a turn of phase.] The maximum possible value of w is D_{max}/λ, where D_{max} is the longest baseline in the array. If θ_f is limited by the beamwidth of antennas of diameter d, for which the angular distance from the beam center to the first null is $\sim \lambda/d$, the required number of planes is $\sim (\lambda/d)^2 \times D_{max}/\lambda = \lambda D_{max}/d^2$. Examples of images made using this method are given by Cornwell and Perley (1992).

2. *Polyhedron Imaging.* The area of the unit sphere for which the image is required can be divided into a number of subfields, which can be imaged individually using the small-field approximation. Each one is imaged in two dimensions onto a plane that is tangent to the unit sphere at a different point on the sphere. These tangent points are the phase centers for the individual subfields. For each subfield image, it is necessary to adjust both the visibility phases and the (u, v, w) coordinates of the whole database to the particular phase center. The subfields can be combined using methods similar to those used in mosaicking, including joint deconvolution. This approach has been referred to as *polyhedron imaging* because the various image planes form part of the surface of a polyhedron. Again, examples are given by Cornwell and Perley (1992).

3. *Combination of Snapshots.* In most synthesis arrays, the antennas are mounted on an area of approximately level ground and thus lie close to a plane at any given instant. In such cases, a long observation can be divided into a series of "snapshots," for each of which the planar baseline condition applies individually. It should therefore be possible to make an image by combining a series of snapshot responses. Each snapshot represents the true intensity distribution convolved with a different dirty beam, since the (u, v) coverage changes progressively as the source moves across the sky. Ideally, deconvolution would thus require optimization of the intensity distribution using the snapshot responses in a combined manner rather than individually. It should be noted that the plane in which the baselines lie for any snapshot is, in general, not normal to

the direction of the target source. As a result, the angle at which points on the unit sphere in Fig. 11.8a are projected onto the (l, m) plane is not parallel to the n axis and varies with the position of the source on the sky. Positions of sources in the snapshot images suffer an offset in (l, m) that is zero at the phase center but that increases with distance from the phase center. Images should be corrected for this effect before being combined. Since the required correction varies with the hour angle of the source, in long observations, the effect can cause smearing of source details in the outer part of the field. Perley (1999b) discusses this effect and its correction. Bracewell (1984) has discussed a method similar to the combination of snapshots described above.

4. *Deconvolution with Variable Point-Source Response.* In some cases, the effect of two-dimensional Fourier transformation is principally distortion of the point-source response in the outer parts of the field, without serious attenuation of the response. Then a possible procedure is deconvolution using a point-source response (the dirty beam) that is varied over the field to match the calculated response (McClean (1984)). This approach was used by Waldram and McGilchrist (1990) in analysis of a survey using the Cambridge Low-Frequency Synthesis Telescope, which operated at 151 MHz using Earth rotation and baselines that are offset from east–west by $3°$. Point-source responses were computed for a grid of positions within the field, and the response for any particular position could then be obtained by interpolation. The principal requirement was to obtain accurate positions and flux densities for sources identified in images obtained by two-dimensional transformation. Fitting the appropriate theoretical beam response for each source position allowed distortion of the beam, including any position offset, to be accounted for. The procedure was relatively inexpensive in computer time.

5. *W-Projection.* W-projection (Cornwell et al. 2008) is a more efficient method of handling the problem of noncoplanar baselines. This problem occurs when the width of the synthesized field is sufficiently large that the w term in the exact visibility equation [(3.7) and (11.23)] cannot be neglected. In w-projection, we start by rewriting the visibility equation, Eq. (11.23), as

$$\mathcal{V}(u, v, w) = \int_{-\infty}^{\infty} \int_{-\infty}^{\infty} \frac{A_N(l, m)I(l, m)}{\sqrt{1 - l^2 - m^2}} G(l, m, w) \, e^{[-j2\pi(ul+vm)]} \, dl \, dm \, ,$$

(11.28)

where

$$G(l, m, w) = e^{-j2\pi w(\sqrt{1 - l^2 - m^2} - 1)} \, ,$$

(11.29)

so that the w dependence is contained within $G(l, m, w)$, and the other parts of Eq. (11.28) represent $\mathcal{V}(u, v, w = 0)$. If $\mathcal{G}(u, v, w)$ is the Fourier transform of $G(l, m, w)$ with respect to (u, v) and (l, m), Eq. (11.28) can be written as a two-dimensional convolution in (u, v),

$$\mathcal{V}(u, v, w) = \mathcal{V}(u, v, w = 0) * * \mathcal{G}(u, v, w) \, .$$

(11.30)

Again, we can visualize (u, v, w) space with the u and v axes in a horizontal plane with w increasing vertically upward. The measured visibility values are located within a block of (u, v, w) space of dimensions limited by the longest antenna spacings and the geometry of the observations. Generally, the observations are designed to optimize the uniformity of sampling of the visibility in the u and v dimensions, but the sampling in w is usually relatively sparse. The procedure in w-projection is to project the three-dimensional visibility data onto the $(u, v, w = 0)$ plane, from which a two-dimensional Fourier transform provides an image in l and m. The $(u, v, w = 0)$ plane is parallel to the tangent plane on the celestial sphere at the field center and thus represents data for which the ray paths from a source at the field center to the corresponding pair of antennas are of equal length. Data for which w is nonzero are those for which the ray paths differ in length by w wavelengths. To use such data to obtain visibility in the $(u, v, w = 0)$ plane, it is necessary to account for the additional path length to one antenna of each pair. In propagating the extra distance in space, the radiation from a point is spread by diffraction, so a single (u, v, w) point is spread into a diffraction pattern at $w = 0$. This spread of the pattern results from the width of the convolution function $G(u, v, w)$ in Eq. (11.30) and is approximately proportional to $|w|$.

If we use the approximation $\sqrt{1 - l^2 - m^2} \approx 1 - (l^2 + m^2)/2$, Eq. (11.29) becomes

$$G(l, m, w) \approx e^{-j\pi w(l^2 + m^2)} . \tag{11.31}$$

Fourier transformation then gives

$$\mathcal{G}(u, v, w) \approx \frac{j}{w} e^{-j\pi(u^2 + v^2)/w} . \tag{11.32}$$

The visibility $\mathcal{V}(u, v, w)$ is entirely determined by $\mathcal{V}(u, v, w = 0)$ through convolution with $\mathcal{G}$. Thus, $\mathcal{V}(u, v, w = 0)$ contains all the data that are required to provide an accurate image, limited only by the synthesized (dirty) beam. Nothing essential to the image is lost in the transition from three dimensions to two. The same convolution function $\mathcal{G}$ applies to projection in both directions, i.e., from $(u, v, w = 0)$ to (u, v, w) and vice versa. Note that the convolving function is different for each (u, v, w) data point. Cornwell et al. (2008) point out that this convolutional relationship between a two-dimensional and a three-dimensional one is due to the fact that the original brightness is confined to the two-dimensional surface of the celestial sphere. They also discuss the result in terms of the diffraction of the electric field over the w-coordinate space.

The w-projection imaging procedure is as follows. First, the visibility data are gridded in (u, v, w) and then projected onto the $(u, v, w = 0)$ plane. In the projection, the data are spread in (u, v) space by the convolution,[1] and thus regridding in the $(u, v, w = 0)$ plane is required. A two-dimensional Fourier transform then provides the dirty image, from which the dirty beam must then be deconvolved by the CLEAN algorithm or some alternate procedure (see Sect. 11.1). CLEAN requires numerous transpositions of data between the visibility and image domains. In going from the model image to visibility, a two-dimensional transform provides $\mathcal{V}(u, v, w = 0)$, from which projection gives values at (u, v, w) points required for comparison with the observations. For the regridding steps, convolution with a spheroidal or other gridding function is required. Since convolution is commutative and associative, it can be computationally efficient to convolve the spheroidal function with the projection function G and thus store the combined convolution functions for use with each (u, v, w) grid point. Convolution of G with the spheroidal function has the additional benefit of damping the behavior of G as $w \to 0$.

Cornwell et al. (2008) also provide details of a simulated example of wide-field imaging using w-projection. They compare the results with the method of image-plane facets (similar to polyhedron imaging), and also uvw-space facets (similar to mosaicking), which projects the (u, v) space, rather than image space, onto tangent plane facets. Hitherto, the facets methods has been perhaps the most widely used procedure for wide-field imaging. Cornwell et al. conclude that, with regard to computing load, the facets method is roughly competitive with w-projection for images of low dynamic range but that w-projection is superior when high sensitivity and dynamic range are required.

A variation of w-projection imaging, which is computationally less expensive, is called w-stacking (Offringa et al. 2014).

11.8   Some Special Techniques of Image Analysis

11.8.1   Use of CLEAN and Self-Calibration with Spectral Line Data

A procedure that has been found to provide accurate separation of the continuum from the line features involves use of the deconvolving algorithm CLEAN (van Gorkom and Ekers 1989). However, if CLEAN is applied individually to the images for the different channels, errors in the CLEAN process appear as differences from channel to channel and may be confused with true spectral features. Such errors can be avoided by subtracting the continuum before applying CLEAN to the line data. First, CLEAN is applied to an average of the continuum-only channels,

---

[1]Cornwell et al. (2008) point out that, interestingly, this spreading in (u, v) space shows that, in general, a single antenna pair responds to a range of spatial frequencies, except when $w = 0$ (i.e., when the baseline is normal to the direction of incidence of the wavefront.)

and the visibility components removed from these channels are also removed from the visibility data for the channels containing line features. When the CLEAN process is terminated, the residuals are also removed from the line data. The resulting line-channel images, which should then contain only line data, can be deconvolved individually. Note that since absorption of the continuum may occur in the line frequency channels, images of line-minus-continuum may contain negative as well as positive intensity features. Thus, algorithms such as MEM that depend on positivity of the intensity may not be easily applicable in such cases.

In applying self-calibration to eliminate phase errors in spectral line data, it can generally be assumed that phase and amplitude differences between channels vary only very little with time and are removed by the bandpass calibration. This is true for both atmospheric and instrumental effects. Thus, the strongest spectral feature in the field under investigation can be used to determine the phase-calibration solution, which is then applied to all channels.

11.8.2   A-Projection

Observations of the most distant Universe, which require removal of the emissions from the foreground, require observations at the highest precision and correspondingly accurate calibration of instrumental effects. Calibration of the responses of the individual antennas includes correcting for DD gains in the deconvolution of images, as discussed by Bhatnagar et al. (2008), Smirnov (2011a,b), and others. DD gains[2] include instrumental and atmospheric effects that affect the pointing and polarization of the antenna responses. Correction for DD effects includes taking account of the rotation of the antennas relative to the sky that results from altazimuth tracking. The DD effects for each antenna can be represented in a 2×2 Jones matrix, a separate one for each pixel of the image. For each cross-correlated antenna pair, the signal product is represented by the outer product of the two Jones matrices, which provides a 4×4 Mueller matrix for each pixel. The diagonal elements of the Mueller matrix represent the four principal products of the two cross-polarization terms of each antenna, for either linear or circular polarization. The off-diagonal terms are small and result from errors in the cross-polarization adjustment and from leakage. These terms must be included if accuracy better than $\sim 1\%$ is required in the image. This procedure has been referred to as the narrowband A-projection algorithm, in which A refers to the elements of $A_{i,j}$ that are the complex convolution of the aperture illumination patterns of antennas i and j. The details of the cross products depend upon the details of the particular array, and Bhatnagar et al. (2008) consider the case for the VLA, in which shading by the feed legs is one of the factors represented by the off-axis terms. The derivation of the image from observations involves an

---

[2]Direction-independent gains include, for example, the gain of the receiver system and are generally much simpler to correct for.

iterative χ^2 minimalization in which χ represents the difference between observed visibility and the visibility of a model that is developed. Calculation of gradients of χ^2 can be used as an aid in the minimalization.

Bhatnagar et al. (2013) expand these concepts to cover a wider frequency bandwidth and develop an A-projection algorithm that includes variation of the model parameters as a function of frequency. A wide bandwidth ratio (e.g., 2 : 1) improves the sensitivity of the observations but requires careful consideration since in the outer parts of the beam, the response varies rapidly with frequency. On wideband, wide-field imaging, see also Rau and Cornwell (2011). An extension of the A-projection technique called fast holographic deconvolution, which is particularly useful for very wide field-of-view observations, has been developed by Sullivan et al. (2012).

11.8.3   Peeling

Synchrotron emission from radio sources usually becomes stronger as the frequency is reduced, and hence, the density of strong sources on the sky generally increases with decreasing frequency. At low frequencies, it is therefore often important to image the whole antenna beam to avoid source confusion resulting from aliasing. Also, the gain of the main beam of a reflector antenna decreases with decreasing frequency, and if phased arrays of dipoles are used, they have to be very large to maintain high gain. As a result, sources in the sidelobes may not be as effectively suppressed relative to a source in the main beam, as is possible at higher frequencies. In the data analysis, unwanted responses from strong sources with known positions can be subtracted. In this process, known as peeling (Noordam 2004; van der Tol et al. 2007), the response to such sources, down to the lowest calibrated level of the sidelobes, is removed. This usually starts with the strongest source in the field and then the second strongest, and so on. The removal can be done in the visibility domain. A procedure of this type is essential in the measurement of the weakest sources and the Epoch of Reionization signatures (see Sect. 10.7.2). Some further discussion of peeling can be found in Bhatnagar et al. (2008), Mitchell et al. (2008), and Bernardi et al. (2011).

11.8.4   Low-Frequency Imaging

In addition to source confusion, a complication of the wide-field imaging is the variation of ionospheric effects over the field of view (see Sect. 14.1). The excess path length in the ionosphere is proportional to ν^{-2}, so the resulting phase change is proportional to ν^{-1}. The term *isoplanatic patch* is used to denote an area of the sky over which the variation in the path length for an incoming wave is small compared with the observing wavelength. At centimeter and shorter wavelengths, the beams of reflector antennas used in synthesis arrays are generally smaller than the isoplanatic patch (see Table 14.1). Thus, the effect of an irregularity in the ionosphere (or troposphere) is constant over the beam and can be corrected by a single phase adjustment for each antenna, for example, by self-calibration. However, at meter wavelengths, the size of the antenna beam may be several times that of the ionospheric isoplanatic patch. In observations with the VLA in New Mexico, Erickson (1999) estimated that the size of the isoplanatic patch at 74 MHz is $\sim 3°-4°$, whereas the beamwidth of a 25-m-diameter antenna at the same frequency is $\sim 13°$. Later, low-frequency instruments have used phased arrays, which enable much smaller beams to be formed. These include LOFAR, covering 15–80 MHz and 110–240 MHz (de Vos et al. 2009); the Murchison Widefield Array, covering 80–300 MHz (Lonsdale et al. 2009); and the Long Wavelength Array, covering 10–88 MHz (Ellingson et al. 2009).

Although, at meter wavelengths, arrays of dipoles or similar elements are more generally used than parabolic reflectors, some early measurements using the 25-m-diameter antennas of the VLA by Kassim et al. (1993) are of interest. These include simultaneous measurements of a number of strong sources at 74 and 330 MHz, using a phase reference procedure to calibrate the phases at the lower frequency. At 74 MHz, the phase fluctuations are dominated by the ionosphere, and rates of phase change were found to be as high as one degree per second. These precluded calibration by the usual methods. However, at 330 MHz, the rates of phase change were slow enough to allow imaging of strong sources. The resulting 330-MHz phases were scaled to 74 MHz and used to remove the ionospheric component from the 74-MHz data that were recorded simultaneously. The procedure used for obtaining images at 74 MHz was essentially as follows:

1. Simultaneous observations of a strong source were made at 74 and 330 MHz, with periodic observations of a calibrator at 330 MHz.
2. An image of the target source was made at 330 MHz using the standard techniques (i.e., use of a calibrator as at centimeter wavelengths). This was used as a starting model for self-calibration of the 330-MHz data.
3. The self-calibration provided phase calibration for each antenna at 330 MHz. These values were then scaled to 74 MHz and used to remove the ionospheric variations from the 74 MHz data, the ionospheric phase changes being inversely proportional to frequency.
4. The instrumental phases at 330 and 74 MHz were different at each antenna as a result of dissimilar cable lengths, etc. To calibrate these differences, an unresolved calibrator was observed at both 330 and 74 MHz. The ionospheric variations could be removed from the 74-MHz calibrator phases using the scheme in step 3.
5. The 74-MHz image of the target source was made from the calibrated phase data. Self-calibration of the 74-MHz data was used to remove residual phase drifts, and for this, the 330-MHz image provided a suitable starting model.

For the strongest sources, for which it was possible to obtain a good signal-to-noise ratio in an averaging time of no more than 10 s, self-calibration at 74 MHz was sufficient in most cases. Although only eight VLA antennas were equipped for operation at 74 MHz, images with dynamic range of better than 20 dB were obtained for several sources. The problem of noncoplanar baselines did not arise in these measurements because the sources were compact enough for satisfactory two-dimensional imaging.

11.8.5   Lensclean

There are many cases in which the image of a quasar or radio galaxy is distorted by the gravitational field of a galaxy, following the discovery of this phenomenon by Walsh et al. (1979). The line of sight from the lens source intersects, or passes very close to, the galaxy. In some cases, the gravitational lensing results in multiple images of a single point-source quasar, and in other cases, extended structure is involved: see, for example, Narayan and Wallington (1992). In studies of gravitational lensing, the structure of the gravitational field is of major astrophysical importance. The term *lensclean* has been used to denote a method of analysis, including several variations of the original algorithm, that allows the lensing field to be determined by synthesis imaging. The basis of these methods is analogous to self-calibration, in which the image is sufficiently overdetermined by the visibility measurements that it is possible to determine also the complex gains of the antennas. In lensclean, it is the pattern of the gravitational field that is determined. An additional complication is that points in the source of the radiation can each contribute to more than one point in the synthesized image.

The original lensclean procedure (Kochanek and Narayan 1992) is based on an adaptation of the CLEAN algorithm. The basic principle can be described as follows. Consider the case in which the source that is imaged by the lens contains extended structure. An initial model for the lens is chosen. Each point in the source contributes to multiple points in the image, and this procedure from the source to the image is defined by the lens model. For any point in the source, the intensity in the image should ideally be the same at each point at which it appears, since the imaging involves only geometric bending of the radiation from the source, as in an optical system. Suppose that the jth source pixel contributes to n_j image pixels. In practice, the intensity of these pixels in the image is not equal because of defects in the lens model and noise in the image. The best estimate of the intensity of the pixel in the source is the mean intensity of the corresponding pixels in the image. Thus, one can subtract components from the image in the manner of CLEAN and build up an image of the source. For each source pixel for which $n_j > 1$, the mean-squared deviation of the intensity of the corresponding image pixels from the mean intensity of the $n_j > 1$ image pixels, σ_j^2, is calculated. For a good lens model, the mean value of σ_j^2 over the pixels in the source should be no greater than the variance of the noise in the image, $\sigma^2_{\mathrm{noise}}$. If the number of degrees of freedom in the source image is taken to be equal to the number of pixels, then the statistical measure of the quality of the lens model is $\chi^2 = \sum(\sigma^2_j/\sigma^2_{\mathrm{noise}})$, where the sum is taken over the j source pixels. The lens parameters can thus be varied to minimize χ^2. In practice, the procedure is more complicated than indicated by the description above. Modifications are included to take account of the finite resolution of the image, which has the effect of spreading the contribution of each source pixel over a number of image pixels. Also, for any unresolved structure in the source, the intensity of the corresponding structure in the image depends on the magnification of the lens.

Ellithorpe et al. (1996) introduced a *visibility lensclean* procedure in which the CLEAN components are removed from the ungridded visibility values under the constraints of a lens model. The squared deviations of the measured visibility from a model are used to determine a χ^2 statistic. The quality of the fit is judged from the variance of the measured visibility, and the number of degrees of freedom is $2N_{\mathrm{vis}} - 3N_{\mathrm{src}} - N_{\mathrm{lens}}$. Here, N_{vis} is the number of visibility measurements (which each have two degrees of freedom), N_{src} is the number of independent CLEAN components in the source model (three degrees of freedom, from position and amplitude), and N_{lens} is the number of parameters in the lens model. Ellithorpe et al. compared results of the original lensclean with visibility lensclean and found the best results from the latter, with a further improvement if a self-calibration step is added. The use of the MEM algorithm as an alternative to CLEAN has also been investigated (Wallington et al. 1994).

11.8.6 Compressed Sensing

Compressed sensing, also known as compressive sensing, compressive sampling, or sparse sampling, is a widely used signal processing technique generally employed to reduce the size of data sets, e.g., images, without loss of information. Sampling at the Nyquist interval provides the most general and complete representation of an image. However, if an image is sparse, i.e., it is mostly blank with isolated components or can be represented by a small number of basis functions such as wavelets, then it is possible to compress or reduce the image size far below that required for Nyquist sampling. The theory of compressed sensing has formal requirements such as sparsity and incoherent sampling. The latter requirement in interferometric imaging corresponds to random sampling in the (u, v) plane. Under such conditions, an image can be derived exactly with very high probability from a sparse set of visibility measurements. These conditions are not perfectly met in radio interferometry. Nonetheless, much can be learned from compressed sensing techniques [see Li et al. (2011a,b)].

In the application to interferometric imaging, the method is formulated in a way to obtain an accurate image from an incompletely sampled (u, v) plane data set. The degree of success of the method depends on the signal-to-noise ratio in the

(u, v) plane and the amount of information that can be supplied to constrain the image solution while being consistent with the (u, v) plane measurements. The simplest of these constraints are nonnegativity, compactness, and smoothness in the image plane. In other words, the method improves as the amount of a priori information available increases. For applications to radio interferometric data and specific algorithms, see, for example, Wiaux et al. (2009), Wenger et al. (2010), Li et al. (2011), Hardy (2013), Carrillo et al. (2012), Garsden et al. (2015), and Dabbech et al. (2015). For a general introduction to the field of compressed sensing as a signal-processing tool, see Candès and Wakin (2008), and for its application to image construction, see Candès et al. (2006a,b). Compressed sensing is widely used in medical imaging [e.g., Lustig et al. (2008)].

For a simplified overview of the method and some of its key concepts, consider a linear vector equation that can be expressed as

$$\mathcal{V} = \mathbf{AX} , \qquad (11.33)$$

where $\mathcal{V}$ represents visibility, $\mathbf{X}$ represents brightness in the image, and $\mathbf{A}$ is the operator that derives the visibility from the parameters of the image, i.e., the Fourier transform kernels.

Important quantities in the image-restoration process are the L_p norms defined as

$$L_p \equiv \|\mathbf{X}\|_p = \left[\sum_{n=1}^{N} |X_n|^p \right]^{1/p} , \qquad p > 0 , \qquad (11.34)$$

where X_n are the elements of $\mathbf{X}$. For $p = 0$, a pseudo norm L_0 can be written as

$$L_0 = \|\mathbf{X}\|_0 = \sum_{n=1}^{N} |X_n|^0 , \qquad (11.35)$$

with the understanding that $0^0 \equiv 0$. In our context, L_0 is the number of cells in the image with nonzero amplitude. Suppose that a point source is observed. In the absence of measurement noise, the normalized moduli of the visibilities will all be unity. Minimizing L_0 with the image constraint of Eq. (11.33) will lead to the recovery of a source distribution of a delta function. Note that the principal image solution is proportional to the dirty image. L_0 minimization tends to remove its high sidelobe response.

Unfortunately, exploration of the L_0 norm is computationally very intensive. In two of the foundational papers in compressed sensing, Candès and Tao (2006) and Donoho (2006) showed that under fairly general conditions, the L_1 norm, defined as

$$L_1 = \|\mathbf{X}\|_1 = \sum_{n=1}^{N} |X_n| , \qquad (11.36)$$

is a suitable proxy for the L_0 norm, and it is more easily computed. The L_1 norm is the total flux density for sources of positive brightness. Virtually all work in compressed sensing is based on L_1 minimization. In the interferometry case, the solution process can be described as

$$\text{minimize } \|\mathbf{X}\|_1 \, , \qquad \text{subject to } \|\boldsymbol{\mathcal{V}} - \mathbf{AX}\|_2^2 < \epsilon \, , \qquad (11.37)$$

where ϵ is the noise threshold based on the measurements, and $\|\boldsymbol{\mathcal{V}} - \mathbf{AX}\|_2^2$ is the squared sum of the visibility residuals or the goodness of fit. An equivalent optimization can be written as

$$\text{minimize } \left\{ \|\boldsymbol{\mathcal{V}} - \mathbf{AX}\|_2^2 + \Lambda \|\mathbf{X}\|_2 \right\} \, , \qquad (11.38)$$

where Λ is a regularization parameter that determines the relative importance of minimizing L_1 and the measurement residuals. In statistics, this approach is called the *least absolute shrinkage selection operator*, or LASSO, developed by Tibshirani (1996).

Another commonly used constraint is based on total variation (*TV*), often computed for two-dimensional images as

$$TV = \sum_{i,j} \left[\left(X_{i-1,j} - X_{i,j} \right)^2 + \left(X_{i,j+1} - X_{i,j} \right)^2 \right]^{1/2} \, . \qquad (11.39)$$

TV is also known as the L_1 norm for adjacent pixel differences. Minimizing *TV* minimizes the gradients and favors smoother images. *TV* minimization can be added to Eq. (11.38) with another Λ term. Note that MEM imposes a similar smoothness constraint (see Sect. 11.2.1). A nonnegative constraint can also be added. Collectively, the application of these constraints is known as regularization.

The potential for recovering source structure finer than the diffraction limit has been investigated by Honma et al. (2014). An example of this application is shown in Fig. 11.9. The possibilities of successful superresolution improve as the image plane becomes more sparse, i.e., a "near black" image [see Starck et al. (2002)].

Another variation on the above approach, more useful for extended source distributions, is to represent the image by a set of basis functions such as wavelets [see Starck and Murtagh (1994) and Starck et al. (1994)]. If the representation is sparse in such a basis space, L_1 minimization gives good results [e.g., Li et al. (2011a,b) and Garsden et al. (2015)].

The most efficient and reliable methods of producing images from visibility data are the subject of continuing development. As with the advent of MEM methods, it is often desirable for researchers to present multiple reconstructions of images that are consistent with their (u, v) plane measurements.

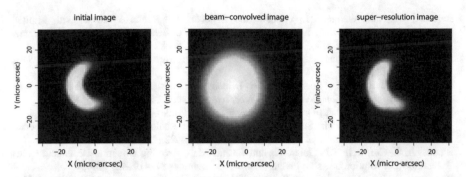

Fig. 11.9 Reconstructed images of a simulated black hole shadow source observed with a six-element Event Horizon Telescope Array at the position of M87. (**left**) Simulated image, (**middle**) CLEANed image with resolution equal to that of the dirty beam, and (**right**) image reconstructed with compressed sensing regularization methods. It may be difficult to apply techniques tested on simulated data. From M. Honma et al. (2014), by permission of and © Oxford University Press.

Further Reading

Roberts, J.A., Ed., *Indirect Imaging*, Cambridge Univ. Press, Cambridge, UK (1984)
Taylor, G.B., Carilli, C.L., and Perley, R.A., Eds., *Synthesis Imaging in Radio Astronomy II*, Astron. Soc. Pacific Conf. Ser., **180** (1999)
van Schooneveld, C., Ed., *Image Formation from Coherence Functions in Astronomy*, Reidel, Dordrecht (1979)

References

Ables, J.G., Maximum Entropy Spectral Analysis, *Astron. Astrophys. Suppl.*, **15**, 383–393 (1974)
Bajaja, E., and van Albada, G.D., Complementing Aperture Synthesis Radio Data by Short Spacing Components from Single Dish Observation, *Astron. Astrophys.*, **75**, 251–254 (1979)
Baldwin, J.E., and Warner, P.J., Phaseless Aperture Synthesis, *Mon. Not. R. Astron. Soc.*, **182**, 411–422 (1978)

Baldwin, J.E., and Warner, P.J., Fundamental Aspects of Aperture Synthesis with Limited or No Phase Information, in *Image Formation from Coherence Functions in Astronomy*, van Schooneveld, C., Ed., Reidel, Dordrecht (1979), pp. 67–82

Bates, R.H.T., Contributions to the Theory of Intensity Interferometry, *Mon. Not. R. Astron. Soc.*, **142**, 413–428 (1969)

Bates, R.H.T., Uniqueness of Solutions to Two-Dimensional Fourier Phase Problems for Localized and Positive Images, *Comp. Vision, Graphics, Image Process.*, **25**, 205–217 (1984)

Bernardi, G., Mitchell, D.A., Ord, S.M., Greenhill, L.J., Pindor, B., Wayth, R.B., and Wyithe, J.S.B., Subtraction of Point Sources from Interferometric Radio Images Through an Algebraic Modelling Scheme, *Mon. Not. R. Astron. Soc.*, **413**, 411–422 (2011)

Bhatnagar, S., Cornwell, T.J., Golap, K., and Uson, J.M., Correcting Direction-Dependent Gains in the Deconvolution of Radio Images, *Astron. Astrophys.*, **487**, 419–429 (2008)

Bhatnagar, S., Rau, U., and Golap, K., Wide-Field Wide-Band Interferometric Imaging: A WB A-Projection and Hybrid Algorithms, *Astrophys. J.*, **770**:91 (9pp) (2013)

Bracewell, R.N., Inversion of Nonplanar Visibilities, in *Indirect Imaging*, Roberts, J.A., Ed., Cambridge Univ. Press, Cambridge, UK (1984), pp. 177–183

Bracewell, R.N., *Two-Dimensional Imaging*, Prentice-Hall, Englewood Cliffs, NJ (1995)

Braun, R., Understanding Synthesis Imaging Dynamic Range, *Astron. Astrophys.*, **551**, A91 (26 pp.) (2013)

Briggs, D.S., "High Fidelity Deconvolution of Moderately Resolved Sources," Ph.D. thesis, New Mexico Institute of Mining and Technology (1995). http://www.aoc.nrao.edu/dissertations/dbriggs

Bruck, Y.M., and Sodin, L.G., On the Ambiguity of the Image Reconstruction Problem, *Opt. Commun.*, **30**, 304–308 (1979)

Bryan, R.K., and Skilling, J., Deconvolution by Maximum Entropy, as Illustrated by Application to the Jet of M87, *Mon. Not. R. Astron. Soc.*, **191**, 69–79 (1980)

Candès, E.M., Romberg, J., and Tao, T., Robust Uncertainty Principles: Exact Signal Reconstruction from Highly Incomplete Frequency Information, *IEEE Trans. Inf. Theory*, **52**, 489–509 (2006a)

Candès, E.M., Romberg, J., and Tao, T., Stable Signal Recovery from Incomplete and Inaccurate Measurements, *Comm. Pure Appl. Math.*, **59**, 1207–1223 (2006b)

Candès, E.M., and Tao, T., Near-Optimal Signal Recovery from Random Projections: Universal Encoding Strategies?, *IEEE Trans. Inf. Theory*, **52**, 5406–5425 (2006)

Candès, E.J., and Wakin, M.B., An Introduction to Compressive Sampling, *IEEE Signal Proc. Mag.*, **25**, 21–30 (2008)

Carrillo, R.E., McEwen, J.D., and Wiaux, Y., Sparsity Averaging Reweighted Analysis (SARA): A Novel Algorithm for Radio-Interferometric Imaging, *Mon. Not. R. Astron. Soc.*, **426**, 1223–1234 (2012)

Clark, B.G., An Efficient Implementation of the Algorithm "CLEAN," *Astron. Astrophys.*, **89**, 377–378 (1980)

Clark, B.G., Large Field Mapping, in *Synthesis Mapping*, Proc. NRAO Workshop No. 5, Thompson, A.R., and D'Addario, L.R., Eds., National Radio Astronomy Observatory, Green Bank, WV (1982)

Conway, J.E., Cornwell, T.J., and Wilkinson, P.N., Multi-Frequency Synthesis: A New Technique in Radio Interferometric Imaging, *Mon. Not. R. Astr. Soc.*, **246**, 490–509 (1990)

Cornwell, T.J., Self Calibration, in *Synthesis Mapping*, Proc. of NRAO Workshop No. 5, Thompson, A.R., and D'Addario, L.R., Eds., National Radio Astronomy Observatory, Green Bank, WV (1982)

Cornwell, T.J., A Method of Stabilizing the Clean Algorithm, *Astron. Astrophys.*, **121**, 281–285 (1983)

Cornwell, T.J., Radio-Interferometric Imaging of Very Large Objects, *Astron. Astrophys.*, **202**, 316–321 (1988)

Cornwell, T.J., The Applications of Closure Phase to Astronomical Imaging, *Science*, **245**, 263–269 (1989)

Cornwell, T.J., Imaging Concepts, in *Very Long Baseline Interferometry and the VLBA*, Zensus, J.A., Diamond, P.J., and Napier, P.J., Eds., Astron. Soc. Pacific Conf. Ser., **82**, 39–56 (1995)

Cornwell, T.J., Multiscale CLEAN Deconvolution of Radio Synthesis Images, *IEEE J. Selected Topics in Signal Proc.* **2**, 793–801 (2008)

Cornwell, T.J., Braun, R., and Briggs, D.S., Deconvolution, in *Synthesis Imaging in Radio Astronomy II*, Taylor, G.B., Carilli, C.L., and Perley, R.A., Eds., Astron. Soc. Pacific Conf. Ser., **180**, 151–170 (1999)

Cornwell, T.J., and Evans, K.F., A Simple Maximum Entropy Deconvolution Algorithm, *Astron. Astrophys.*, **143**, 77–83 (1985)

Cornwell, T.J., Golap, K., and Bhatnagar, S., The Non-Coplanar Baselines Effect in Radio Astronomy. The W Projection Algorithm, *IEEE J. Selected Topics Signal Proc.*, **2**, 647–657 (2008)

Cornwell, T.J., Holdaway, M.A., and Uson, J.M., Radio-Interferometric Imaging of Very Large Objects: Implications for Array Design, *Astron. Astrophys.*, **271**, 697–713 (1993)

Cornwell, T.J., and Perley, R.A., Radio-Interferometric Imaging of Very Large Fields, *Astron. Astrophys.*, **261**, 353–364 (1992)

Cornwell, T.J., and Wilkinson, P.N., A New Method for Making Maps with Unstable Radio Interferometers, *Mon. Not. R. Astron. Soc.*, **196**, 1067–1086 (1981)

Cotton, W.D., A Method of Mapping Compact Structure in Radio Sources Using VLBI Observations, *Astron. J.*, **84**, 1122–1128 (1979)

Dabbech, A., Ferrari, C., Mary, D., Slezak, E., Smirnov, O., and Kenyon, J.S., MORESANE: MOdel REconstruction by Synthesis-ANalysis Estimators: A Sparse Deconvolution Algorithm for Radio Interferometric Imaging, *Astron. Astrophys.*, **576**, A7 (16pp) (2015)

de Vos, M., Gunst, A.W., and Nijboer, R., The LOFAR Telescope: System Architecture and Signal Processing, *Proc. IEEE*, **97**, 1431–1437 (2009)

Donoho, D.L., Compressed Sensing, *IEEE Trans. Inf. Theory*, **52**, 1289–1306 (2006)

Ekers, R.D., and Rots, A.H., Short Spacing Synthesis from a Primary Beam Scanned Interferometer, in *Image Formation from Coherence Functions in Astronomy*, van Schoonveld, C., Ed., Reidel, Dordrecht (1979), pp. 61–63

Ellingson, S.W., Clarke, T.E., Cohen, A., Craig, J., Kassim, N.E., Pihlström, Y., Rickard, L.J., and Taylor, G.B., The Long Wavelength Array, *Proc. IEEE*, **97**, 1421–1430 (2009)

Ellithorpe, J.D., Kochanek, C.S., and Hewitt, J.N., Visibility Lensclean and the Reliability of Deconvolved Radio Images, *Astrophys. J.*, **464**, 556–567 (1996)

Erickson, W.C., Long Wavelength Interferometry, in *Synthesis Imaging in Radio Astronomy II*, Taylor, G.B., Carilli, C.L., and Perley, R.A., Eds., Astron. Soc. Pacific Conf. Ser., **180**, 601–612 (1999)

Fienup, J.R., Reconstruction of an Object from the Modulus of Its Fourier Transform, *Opt. Lett.*, **3**, 27–29 (1978)

Fort, D.N., and Yee, H.K.C., A Method of Obtaining Brightness Distributions from Long Baseline Interferometry, *Astron. Astrophys.*, **50**, 19–22 (1976)

Frieden, B.R., Restoring with Maximum Likelihood and Maximum Entropy, *J. Opt. Soc. Am.*, **62**, 511–518 (1972)

Garsden, H., Girard, J.N., Starck, J.L., Corbel, S., Tasse, C., Woiselle, A., McKean, J.P., van Amesfoort, A.S., Anderson, J., Avruch, I.M., and 71 coauthors, LOFAR Sparse Image Reconstruction, *Astron. Astrophys.*, **575**, A90 (18 pp) (2015)

Gull, S.F., and Daniell, G.J., Image Reconstruction from Incomplete and Noisy Data, *Nature*, **272**, 686–690 (1978)

Gull, S.F., and Daniell, G.J., The Maximum Entropy Method, in *Image Formation from Coherence Functions in Astronomy*, van Schoonveld, C., Ed., Reidel, Dordrecht (1979), pp. 219–225

Hardy, S.J., Direct Deconvolution of Radio Synthesis Images Using L_1 Minimization, *Astron. Astrophys.*, **557**, A134 (10pp) (2013)

Högbom, J.A., Aperture Synthesis with a Non-Regular Distribution of Interferometer Baselines, *Astron. Astrophys. Suppl.*, **15**, 417–426 (1974)

Högbom, J.A., The Introduction of A Priori Knowledge in Certain Processing Algorithms, in *Image Formation from Coherence Functions in Astronomy*, van Schooneveld, C., Ed., Reidel, Dordrecht (1979), pp. 237–239

Honma, M., Akiyama, K., Uemura, M., and Ikeda, S., Super-Resolution Imaging with Radio Interferometry Using Sparse Modeling, *Publ. Astron. Soc. Japan*, **66**, 95 (14pp) (2014)

Jaynes, E.T., Prior Probabilities, *IEEE Trans. Syst. Sci. Cyb.*, **SSC-4**, 227–241 (1968)

Jaynes, E.T., On the Rationale of Maximum-Entropy Methods, *Proc. IEEE*, **70**, 939–952 (1982)

Junklewitz, H., Bell, M.R., and Enßelin, T., A New Approach to Multifrequency Synthesis in Radio Interferometry, *Astron. Astrophys.*, **581**, A59 (11pp) (2015)

Junklewitz, H., Bell, M.R., Selig, M., and Enßelin, T., RESOLVE: A New Algorithm for Aperture Synthesis Imaging of Extended Emission in Radio Astronomy, *Astron. Astrophys.*, **586**, A76 (21pp) (2016)

Kassim, N.E., Perley, R.A., Erickson, W.C., and Dwarakanath, K.S., Subarcminute Resolution Imaging of Radio Sources at 74 MHz with the Very Large Array, *Astron. J.*, **106**, 2218–2228 (1993)

Kochanek, C.S., and Narayan, R., Lensclean: An Algorithm for Inverting Extended, Gravitationally Lensed Images with Application to the Radio Ring Lens PKS 1830-211, *Astrophys. J.*, **401**, 461–473 (1992)

Lawson, C.L., and Hanson, R.J., *Solving Least Squares Problems*, Prentice-Hall, Englewood Cliffs, NJ (1974)

Li, F., Brown, S., Cornwell, T.J., and de Hoog, F., The Application of Compressive Sampling to Radio Astronomy. 2. Faraday Rotation Measure Synthesis, *Astron. Astrophys.*, **531**, A126 (8pp) (2011a)

Li, F., Cornwell, T.J., and de Hoog, F., The Application of Compressive Sampling to Radio Astronomy. 1. Deconvolution, *Astron. Astrophys.*, **528**, A31 (10pp) (2011b)

Lonsdale, C.J., Cappallo, R.J., Morales, M.F., Briggs, F.H., Benkevitch, L., Bowman, J.D., Bunton, J.D., Burns, S., Corey, B.E., deSouza, L., and 38 coauthors, The Murchison Widefield Array: Design Overview, *Proc. IEEE*, **97**, 1497–1506 (2009)

Lonsdale, C.J., Doeleman, S.S., Capallo, R.J., Hewitt, J.N., and Whitney, A.R., Exploring the Performance of Large-*N* Radio Astronomical Arrays, in *Radio Telescopes*, Butcher, H.R., Ed., Proc. SPIE, **4015**, 126–134 (2000)

Lustig, M., Donoho, D.L., Santos, J.M., and Pauly, J.M., Compressed Sensing MRI, *IEEE Signal Proc. Mag.*, **25**, 72–82 (2008)

McClean, D.J., A Simple Expansion Method for Wide-Field Mapping, in *Indirect Imaging*, Roberts, J.A., Ed., Cambridge Univ. Press, Cambridge, UK (1984), pp. 185–191

Mitchell, D.A., Greenhill, L.J., Wayth, R.B., Sault, R.J., Lonsdale, C.J., Capallo, R.J., Morales, M.F., and Ord, S.M., Real-Time Calibration of the Murchison Widefield Array, *IEEE J. Selected Topics Signal Proc.*, **2**, 707–717 (2008)

Napier, P.J., The Brightness Temperature Distributions Defined by a Measured Intensity Interferogram, *NZ J. Sci.*, **15**, 342–355 (1972)

Napier, P.J., Bagri, D.S., Clark, B.G., Rogers, A.E.E., Romney, J.D., Thompson, A.R., and Walker, R.C., The Very Long Baseline Array, *Proc. IEEE*, **82**, 658–672 (1994)

Napier, P.J., and Bates, R.H.T., Inferring Phase Information from Modulus Information in Two-Dimensional Aperture Synthesis, *Astron. Astrophys. Suppl.*, **15**, 427–430 (1974)

Napier, P.J., Thompson, A.R., and Ekers, R.D., The Very Large Array: Design and Performance of a Modern Synthesis Radio Telescope, *Proc. IEEE*, **71**, 1295–1320 (1983)

Narayan, R., and Nityananda, R., Maximum Entropy—Flexibility vs. Fundamentalism, in *Indirect Imaging*, Roberts, J.A., Ed., Cambridge Univ. Press, Cambridge, UK (1984), pp. 281–290

Narayan, R., and Wallington, S., Introduction to Basic Concepts of Gravitational Lensing, in *Gravitational Lenses*, Kayser, R., Schramm, T., and Nieser, L., Eds., Springer-Verlag, Berlin (1992), pp. 12–26

Nityananda, R., and Narayan, R., Maximum Entropy Image Reconstruction–A Practical Non-Information-Theoretic Approach, *J. Astrophys. Astron.*, **3**, 419–450 (1982)

Noordam, J.E., LOFAR Calibration Challenges, in *Ground-Based Telescopes*, Oschmann, J.M., Jr., Ed., Proc. SPIE, **5489**, 817–825 (2004)

Offringa, A.R., McKinley, B., Hurley-Walker, N., Briggs, F.H., Wayth, R.B., Kaplan, D.L., Bell, M.E., Feng, L., Neben, A.R., Hughes, J.D., and 43 coauthors, WSCLEAN: An Implementation of a Fast, Generic Wide-Field Imager for Radio Astronomy, *Mon. Not. R. Astron. Soc.*, **444**, 606–619 (2014)

Pearson, T.J., and Readhead, A.C.S., Image Formation by Self-Calibration in Radio Astronomy, *Ann. Rev. Astron. Astrophys.*, **22**, 97–130 (1984)

Perley, R.A., High Dynamic Range Imaging, in *Synthesis Imaging in Radio Astronomy*, Perley, R.A., Schwab, F.R., and Bridle, A.H., Eds., Astron. Soc. Pacific Conf. Ser., **6**, 287–313 (1989)

Perley, R.A., High Dynamic Range Imaging, in *Synthesis Imaging in Radio Astronomy II*, Taylor, G.B., Carilli, C.L., and Perley, R.A., Eds., Astron. Soc. Pacific Conf. Ser., **180**, 275–299 (1999a)

Perley, R.A., Imaging with Non-Coplanar Arrays, in *Synthesis Imaging in Radio Astronomy II*, Taylor, G.B., Carilli, C.L., and Perley, R.A., Eds., Astron. Soc. Pacific Conf. Ser., **180**, 383–400 (1999b)

Ponsonby, J.E.B., An Entropy Measure for Partially Polarized Radiation and Its Application to Estimating Radio Sky Polarization Distributions from Incomplete "Aperture Synthesis" Data by the Maximum Entropy Method, *Mon. Not. R. Astron. Soc.*, **163**, 369–380 (1973)

Rau, U., and Cornwell, T.J., A Multi-Scale Multi-Frequency Deconvolution Algorithm for Synthesis Imaging in Radio Astronomy, *Astron. Astrophys.*, **532**, A71 (17pp) (2011)

Readhead, A.C.S., Walker, R.C., Pearson, T.J., and Cohen, M.H., Mapping Radio Sources with Uncalibrated Visibility Data, *Nature*, **285**, 137–140 (1980)

Readhead, A.C.S., and Wilkinson, P.N., The Mapping of Compact Radio Sources from VLBI Data, *Astrophys. J.*, **223**, 25–36 (1978)

Rogers, A.E.E., Methods of Using Closure Phases in Radio Aperture Synthesis, *Soc. Photo-Opt. Inst. Eng.*, **231**, 10–17 (1980)

Rogers, A.E.E., Hinteregger, H.F., Whitney, A.R., Counselman, C.C., Shapiro, I.I., Wittels, J.J., Klemperer, W.K., Warnock, W.W., Clark, T.A., Hutton, L.K., and four coauthors, The Structure of Radio Sources 3C273B and 3C84 Deduced from the "Closure" Phases and Visibility Amplitudes Observed with Three-Element Interferometers, *Astrophys. J.*, **193**, 293–301 (1974)

Sault, R.J., Stavely-Smith, L., and Brouw, W.N., An Approach to Interferometric Mosaicing, *Astron. Astrophys. Supp.*, **120**, 375–384 (1996)

Sault, R.J., and Wieringa, M.H., Multi-Frequency Synthesis Techniques in Radio Interferometric Imaging, *Astron. Astrophys. Suppl.*, **108**, 585–594 (1994)

Schwab, F.R., Adaptive Calibration of Radio Interferometer Data, *Soc. Photo-Opt. Inst. Eng.*, **231**, 18–24 (1980)

Schwab, F.R., Relaxing the Isoplanatism Assumption in Self Calibration: Applications to Low-Frequency Radio Astronomy, *Astron. J.*, **89**, 1076–1081 (1984)

Schwarz, U.J., Mathematical-Statistical Description of the Iterative Beam Removing Technique (Method CLEAN), *Astron. Astrophys.*, **65**, 345–356 (1978)

Schwarz, U.J., The Method "CLEAN"—Use, Misuse, and Variations, in *Image Formation from Coherence Functions in Astronomy*, van Schooneveld, C., Ed., Reidel, Dordrecht (1979), pp. 261–275

Skilling, J., and Bryan, R.K., Maximum Entropy Image Reconstruction: General Algorithm, *Mon. Not. R. Astron. Soc.*, **211**, 111–124 (1984)

Smirnov, O.M., Revisiting the Radio Interferometer Measurement Equation. I. A Full-Sky Jones Formalism, *Astron. Astrophys.*, **527**, A106 (2011a)

Smirnov, O.M., Revisiting the Radio Interferometer Measurement Equation. II. Calibration and Direction-Dependent Effects, *Astron. Astrophys.*, **527**, A107 (2011b)

Sramek, R.A., and Schwab, F.R., Imaging, in *Synthesis Imaging in Radio Astronomy*, Perley, R.A., Schwab, F.R., and Bridle, A.H., Eds., Astron. Soc. Pacific Conf. Ser., **6**, 117–138 (1989)

Stanimirović, S., Short Spacings Correction from the Single-Dish Perspective, in *Single-Dish Radio Astronomy: Techniques and Applications*, Stanimirović, S., et al., Eds., Astron. Soc. Pacific Conf. Ser., **278**, 375–396 (2002)

Starck, J.-L., Bijaoui, A., Lopez, B., and Perrier, C., Image Reconstruction by the Wavelet Transform Applied to Aperture Synthesis, *Astron. Astrophys*, **283**, 349–360 (1994)

Starck, J.-L., and Murtagh, F., Image Restoration with Noise Suppression Using the Wavelet Transform, *Astron. Astrophys*, **228**, 342–348 (1994)

Starck, J.-L., Pantin, E., and Murtagh, F., Deconvolution in Astronomy: A Review, *Publ. Astron. Soc. Pacific*, **114**, 1051–1069 (2002)

Subrahmanya, C.R., An Optimum Deconvolution Method, in *Image Formation from Coherence Functions in Astronomy*, van Schooneveld, C., Ed., Reidel, Dordrecht (1979), pp. 287–290

Sullivan, I.S., Morales, M.F., Hazelton, B.J., Arcus, W., Barnes, D., Bernardi, G., Briggs, F.H., Bowman, J.D., Bunton, J.D., Cappallo, R.J., and 41 coauthors, Fast Holographic Deconvolution: A New Technique for Precision Radio Interferometry, *Astrophys. J.*, **759**:17 (6pp) (2012)

Thomasson, P., MERLIN, *Quart. J. R. Astron. Soc.*, **27**, 413–431 (1986)

Tibshirani, R., Regression Shrinkage and Selection via the Lasso, *J. R. Statis. Soc. B*, **58**, 267–288 (1996)

van der Tol, S., Jeffs, B.D., and van der Veen, A.-J., Self-Calibration for the LOFAR Radio Astronomical Array, *IEEE Trans. Signal Proc.*, **55**, 4497–4510 (2007)

van Gorkom, J.H., and Ekers, R.D., Spectral Line Imaging II: Calibration and Analysis, in *Synthesis Imaging in Radio Astronomy*, Perley, R.A., Schwab, F.R., and Bridle, A.H., Eds., Astron. Soc. Pacific Conf. Ser., **6**, 341–353 (1989)

Wakker, B.P., and Schwarz, J.J., The Multi-Resolution Clean and Its Application to the Short-Spacing Problem in Interferometry, *Astron. Astrophys.*, **200**, 312–322 (1988)

Waldram, E.M., and McGilchrist, M.M., Beam-Sets: A New Approach to the Problem of Wide-Field Mapping with Non-Coplanar Baselines, *Mon. Not. R. Astron. Soc.*, **245**, 532–541 (1990)

Wallington, S., Narayan, R., and Kochanek, C.S., Gravitational Lens Inversion Using the Maximum Entropy Method, *Astrophys. J.*, **426**, 60–73 (1994)

Walsh, D., Carswell, R.F., and Weymann, R.J., 0957+561A,B: Twin Quasistellar Objects or Gravitational Lens?, *Nature*, **279**, 381–384 (1979)

Welch, W.J., and Thornton, D.D., An Introduction to Millimeter and Submillimeter Interferometry and a Summary of the Hat Creek System, in *Int. Symp. Millimeter and Submillimeter Wave Radio Astronomy*, International Scientific Radio Union, Institut de Radio Astronomie Millimétrique, Granada, Spain (1985), pp. 53–64

Wenger, S., Magnor, M., Pihlström, Y., Bhatnagar, S., and Rau, U., SparseRI: A Compressed Sensing Framework for Aperture Synthesis Imaging in Radio Astronomy, *Publ. Astron. Soc. Pacific*, **122**, 1367–1374 (2010)

Wernecke, S.J., Two-Dimensional Maximum Entropy Reconstruction of Radio Brightness, *Radio Sci.*, **12**, 831–844 (1977)

Wernecke, S.J., and D'Addario, L.R., Maximum Entropy Image Reconstruction, *IEEE Trans. Comput.*, **C-26**, 351–364 (1977)

Wiaux, Y., Jacques, L., Puy, G., Scaife, A.M.M., and Vandergheynst, P., Compressed Sensing Imaging Techniques for Radio Interferometry, *Mon. Not. R. Astron. Soc.*, **395**, 1733–1742 (2009)

Chapter 12
Interferometer Techniques for Astrometry and Geodesy

This chapter is concerned with the techniques by which angular positions of radio sources can be measured with the greatest possible accuracy, and with the design of interferometers for optimum determination of source-position, baseline, and geodetic[1] parameters.

The total fringe phase of an interferometer, where the effect of delay tracking is removed, can be expressed in terms of the scalar product of the baseline and source-position vectors $\mathbf{D}$ and $\mathbf{s}$, respectively, as

$$\phi = \frac{2\pi}{\lambda}\mathbf{D}\cdot\mathbf{s} = \frac{2\pi}{\lambda}D\cos\theta\,, \tag{12.1}$$

where θ is the angle between $\mathbf{D}$ and $\mathbf{s}$. Up to this point, we have assumed that these factors are describable by constants that can be specified with high accuracy. However, the measurement of source positions to an accuracy better than a milliarcsecond (mas) requires, for example, that variation in the Earth's rotation vector be taken into account. The baseline accuracy is comparable to that at which variation in the antenna positions resulting from crustal motions of the Earth can be detected. The calibration of the baseline and the measurement of source positions can be accomplished in a single observing period of one or more days. Geodetic data are obtained from repetition of this procedure over intervals of months or years, which reveals the variation in the baseline and Earth-rotation parameters.

---

[1] For simplicity, we use the term *geodetic* to include geodynamic and static phenomena regarding the shape and orientation of the Earth.

© The Author(s) 2017 599
A.R. Thompson, J.M. Moran, and G.W. Swenson Jr., *Interferometry and Synthesis in Radio Astronomy*, Astronomy and Astrophysics Library,
DOI 10.1007/978-3-319-44431-4_12

The redefinition of the meter from a fundamental to a derived quantity has an interesting implication for the units of baseline length derived from interferometric data. An interferometer measures the relative time of arrival of the signal wavefront at the two antennas, that is, the geometric delay. Baselines determined from interferometric data are therefore in units of light travel time. Conversion to meters formerly depended on the value chosen for c. However, in 1983, the Conférence Générale des Poids et Mesures adopted a new definition of the meter: "the meter is the length of the path traveled by light in vacuum during a time interval of 1/299,792,458 of a second." The second and the speed of light are now primary quantities, and the meter is a derived quantity. Thus, baseline lengths can be given unambiguously in meters. Issues related to fundamental units are discussed by Petley (1983).

12.1   Requirements for Astrometry

We begin with a heuristic discussion of how baseline and source-position parameters may be determined. A more formal discussion is given in Sect. 12.2.

The phase of the fringe pattern for a tracking interferometer [Eq. (12.1)] can be expressed in polar coordinates (see Fig. 4.2) as

$$\phi(H) = 2\pi D_\lambda [\sin d \sin \delta + \cos d \cos \delta \cos(H - h)] + \phi_{\text{in}} , \qquad (12.2)$$

where D_λ is the length of the baseline in wavelengths, H and δ are the hour angle and declination of the source, h and d are the hour angle and declination of the baseline, and ϕ_{in} is an instrumental phase term. We assume for the purpose of this discussion that ϕ_{in} is a fixed constant, unaffected by the atmosphere and electronic drifts. The hour angle is related to the right ascension α by

$$H = t_s - \alpha , \qquad (12.3)$$

where t_s is the sidereal time (in VLBI, t_s and H are referred to the Greenwich meridian, whereas in connected interferometry, they are often referred to the local meridian). Consider a short-baseline interferometer that is laid out in exactly the east–west direction, i.e., with $d = 0$, $h = \frac{\pi}{2}$ [see Ryle and Elsmore (1973)]. Then

$$\phi(H) = -2\pi D_\lambda \cos \delta \sin H + \phi_{\text{in}} , \qquad (12.4)$$

and the phase goes through one sinusoidal oscillation in a sidereal day. Suppose that the source is circumpolar, i.e., above the horizon for 24 h. From continuous observation of ϕ over a full day, the 2π crossings of ϕ can be tracked so that there are no phase ambiguities. The average value of the geometric term of Eq. (12.4) is zero, so that ϕ_{in} can be estimated and removed. When the source transits the local meridian, $H = 0$, the corrected phase will be zero, and the right ascension α can be determined for the time of this transit and Eq. (12.3).

The length of the baseline can be determined by observing sources close to the celestial equator, $|\delta| \sim 0$, where the dependence of phase on δ is small. With this calibration of D_λ and ϕ_{in}, the positions of other sources can be determined, i.e., their right ascensions from the transit times of the central fringe and their declinations from the diurnal amplitude of $\phi(H)$. The source declination can also be found from the rate of change of phase at $H = 0$. This rate of change of phase is

$$\left.\frac{d\phi}{dt}\right|_{H=0} = 2\pi D_\lambda \omega_e \cos \delta , \tag{12.5}$$

where $\omega_e = dII/dt$, the rotation rate of the Earth. From Eq. (12.5), it is clear that if the error in $(d\phi/dt)_{H=0} = \sigma_f$, then the error in position will be

$$\sigma_\delta \simeq \frac{1}{2\pi D_\lambda \omega_e \sin \delta} \sigma_f . \tag{12.6}$$

Note that accuracy of the derived declinations is poor for sources near the celestial equator. An informative review of the application of these techniques is given by Smith (1952).

In the determination of right ascension, interferometer observations provide relative measurements, that is, the differences in right ascension among different sources. The zero of right ascension is defined as the great circle through the pole and through the intersection of the celestial equator and the ecliptic at the vernal equinox at a specific epoch. The vernal equinox is the point at which the apparent position of the Sun moves from the Southern to the Northern celestial hemisphere. This direction can be located in terms of the motions of the planets, which are well-defined objects for optical observations. It has been related to the positions of bright stars that provide a reference system for optical measurements of celestial position. Relating the radio measurements to the zero of right ascension is less easy, since solar system objects are generally weak or do not contain sharp enough features in their radio structure. In the 1970s, results were obtained from the lunar occultation of the source 3C273B (Hazard et al. 1971) and from measurements of the weak radio emission from nearby stars such as Algol (β Persei) (Ryle and Elsmore 1973; Elsmore and Ryle 1976).

In the reduction of interferometer measurements in astrometry, the visibility data are interpreted basically in terms of the positions of point sources. The data processing is equivalent, in effect, to model fitting using delta-function intensity components, the visibility function for which has been discussed in Sect. 4.4. The essential position data are determined from the calibrated visibility phase or, in some VLBI observations, from the geometric delay as measured by maximization of the cross-correlation of the signals (i.e., the use of the bandwidth pattern) and from the fringe frequency. Because the position information is contained in the visibility phase, measurements of closure phase discussed in Sect. 10.3 are of use in astrometry and geodesy only insofar as they can provide a means of correcting for the effects of source structure. Uniformity of (u, v) coverage is less important than in imaging because high dynamic range is generally not needed. Determination of the position of an unresolved source depends on interferometry with precise phase calibration and a sufficient number of baselines to avoid ambiguity in the position.

12.1.1   Reference Frames

A reference frame based on the positions of distant extragalactic objects can be expected to show greater temporal stability than a frame based on stellar positions and to approach more closely the conditions of an inertial frame. An inertial frame is one that is at rest or in uniform motion with respect to absolute space and not in a state of acceleration or rotation [see, e.g., Mueller (1981)]. Newton's first law holds in such a frame. A detailed description of astronomical reference frames is given by Johnston and de Vegt (1999). The International Celestial Reference System (ICRS) adopted by the International Astronomical Union (IAU) specifies the zero points and directions of the axes of the coordinate system for celestial positions. The measured positions of a set of reference objects in the coordinates of the reference system provide the International Celestial Reference Frame (ICRF). Thus, the frame provides the reference points with respect to which positions of other objects are measured within the coordinate system.

The most accurate measurements of celestial positions are those of selected extragalactic sources observed by VLBI. Large databases of such high-resolution observations exist as a result of measurements made for purposes of geodesy and astrometry. These measurements have been made in a systematic way mainly since 1979, using VLBI systems with dual frequencies of 2.3 and 8.4 GHz to allow calibration of ionospheric effects. The positions are determined mainly by the 8.4 GHz data. The first catalog of source positions, now called ICRF1 (Ma et al. 1998), was adopted by the IAU in 1998. This frame supersedes earlier ones based on optical positions of stars, most recently those of the FK5 and Hipparcos catalogs. The ICRF1 was based on 1.6×10^6 measurements of group delay obtained between 1979 and 1995 of 608 sources. Criteria for exclusion of a source included inconsistency in the position measurements, evidence of motion, or presence of extended structure. In this study, 212 sources were found that passed all tests; 294 failed in one criterion; and 102 other sources, including 3C273, failed in several. The 212 sources in the best category were used to define the reference frame. Only 27% of these are in the Southern Hemisphere. A global solution provides the positions of the sources together with the antenna positions and various geodetic and atmospheric parameters. Position errors of the 212 defining sources are mostly less than 0.5 mas in both right ascension and declination and less than 1 mas in almost all cases.

In 2009, an updated reference frame catalog called ICRF2 was released (Fey et al. 2009, 2015) and was adopted by the IAU. It contains the positions of 3,414 sources derived from 6.5×10^6 measurements of group delay acquired over 30 years through

1998. About 28% of the data came from the VLBA. The core reference frame is based and maintained on data from 295 sources, whose distribution is much more uniform over the celestial sphere than those in ICRF1. The positional accuracy is about 40 μas, about five times better than achieved in ICRF1.

About 50% of sources in the ICRF have redshifts greater than 1.0. The use of such distant objects to define the reference frame provides a level of astrometric uncertainty at least an order of magnitude better than optical measurements of stars. The ultimate accuracy of this frame may depend on the structural stability of the radio sources involved [see, e.g., Fey and Charlot (1997), Fomalont et al. (2011)]. The level of uncertainty in the connection between the radio and optical frames is essentially the uncertainty in optical positions. Radio measurements of the positions of some of the nearer stars provide a comparison between the radio and optical frames. Lestrade (1991) and Lestrade et al. (1990, 1995) have measured the positions of about ten stars by VLBI, achieving accuracy in the range 0.5–1.5 mas. These results provide a link between the ICRF and the star positions in the Hipparcos catalog. The visual magnitudes of the known optical counterparts of the reference frame sources are mostly within the range 15–21, and precise positions of objects fainter than 18th magnitude are likely to be very difficult to obtain.

There are several methods of linking the extragalactic reference frame to the heliocentric reference frame. Pulsar positions can be derived from timing measurements and VLBI measurements (Bartel et al. 1985; Fomalont et al. 1992; and Madison et al. 2013). The timing analysis is inherently linked to the heliocentric frame. VLBI of space probes in orbit around solar system bodies can also help link the frames [see Jones et al. (2015)]. Radio observations of minor planets may be helpful (Johnston et al. 1982).

12.2 Solution for Baseline and Source-Position Vectors

We now discuss in a more formal way how interferometer baselines and source positions can be estimated simultaneously for phase, fringe rate, or group delay measurements. For discussion of early implementations of these techniques, see Elsmore and Mackay (1969), Wade (1970), and Brosche et al. (1973). An excellent tutorial is Fomalont (1995).

12.2.1 Phase Measurements

Consider an observation with a two-element tracking interferometer of arbitrary baseline in which the source is unresolved. Let $\mathbf{D}_\lambda$ be the assumed baseline vector, in units of the wavelength, and $(\mathbf{D}_\lambda - \Delta\mathbf{D}_\lambda)$ be the true vector. Similarly, let $\mathbf{s}$ be a unit vector indicating the assumed position of the source, and let $(\mathbf{s} - \Delta\mathbf{s})$ indicate the true position. Note that the convention used is Δ term = (approximate or assumed value) – (true value). The expected fringe phase, using the assumed positions, is $2\pi \mathbf{D}_\lambda \cdot \mathbf{s}$. The observed phase, measured relative to the expected phase, is a function of the hour angle H of the source given by

$$
\begin{aligned}
\Delta\phi(H) &= 2\pi \left[\mathbf{D}_\lambda \cdot \mathbf{s} - (\mathbf{D}_\lambda - \Delta\mathbf{D}_\lambda) \cdot (\mathbf{s} - \Delta\mathbf{s}) \right] + \phi_{\text{in}} \\
&= 2\pi \left(\Delta\mathbf{D}_\lambda \cdot \mathbf{s} + \mathbf{D}_\lambda \cdot \Delta\mathbf{s} \right) + \phi_{\text{in}} .
\end{aligned}
\tag{12.7}
$$

A second-order term involving $\Delta\mathbf{D}_\lambda \cdot \Delta\mathbf{s}$ has been neglected since we assume that fractional errors in $\mathbf{D}_\lambda$ and $\mathbf{s}$ are small.

The baseline vector can be written in terms of the coordinate system introduced in Sect. 4.1:

$$
\mathbf{D}_\lambda = \begin{bmatrix} X_\lambda \\ Y_\lambda \\ Z_\lambda \end{bmatrix} , \qquad \Delta\mathbf{D}_\lambda = \begin{bmatrix} \Delta X_\lambda \\ \Delta Y_\lambda \\ \Delta Z_\lambda \end{bmatrix} ,
\tag{12.8}
$$

where X_λ, Y_λ, and Z_λ form a right-handed coordinate system with Z_λ parallel to the Earth's spin axis and X_λ in the meridian plane of the interferometer. The source-position vector can be specified in the $(X_\lambda, Y_\lambda, Z_\lambda)$ system in terms of the hour angle H and declination δ of the source by using Eq. (4.2):

$$
\mathbf{s} = \begin{bmatrix} s_X \\ s_Y \\ s_Z \end{bmatrix} = \begin{bmatrix} \cos\delta \cos H \\ -\cos\delta \sin H \\ \sin\delta \end{bmatrix} .
\tag{12.9}
$$

Taking the differential of Eq. (12.9), we can write

$$
\Delta\mathbf{s} \simeq \begin{bmatrix} -\sin\delta \cos H \Delta\delta + \cos\delta \sin H \Delta\alpha \\ \sin\delta \sin H \Delta\delta + \cos\delta \cos H \Delta\alpha \\ \cos\delta \Delta\delta \end{bmatrix} ,
\tag{12.10}
$$

where $\Delta\alpha$ and $\Delta\delta$ are the angular errors in right ascension and declination. Note that $\Delta\alpha = -\Delta H$ [see Eq. (12.3)].

Consider the case in which there exists a catalog of sources whose positions are considered to be known perfectly. Most connected-element arrays, e.g., ALMA, the VLA, the SMA, and IRAM, have far more antenna pads than antennas, so that the arrays can be reconfigured for various resolutions. Each time an array is reconfigured, the baselines must be redetermined because of the mechanical imprecision of positioning the antennas on the pads. With only baseline errors, i.e., $\Delta\mathbf{s} = 0$, the residual phase [substitute Eqs. (12.8) and (12.9) into (12.7)] is

$$
\Delta\phi(H) = \phi_0 + \phi_1 \cos H + \phi_2 \sin H ,
\tag{12.11}
$$

where

$$\phi_0 = 2\pi \sin \delta \Delta Z_\lambda + \phi_{\text{in}} \,,$$

$$\phi_1 = 2\pi \cos \delta \Delta X_\lambda \,, \qquad\qquad (12.12)$$

$$\phi_2 = -2\pi \cos \delta \Delta Y_\lambda \,.$$

A long track on a single source can be fitted to a sinusoidal function in H with three free parameters, ϕ_0, ϕ_1, and ϕ_2. ΔX_λ and ΔY_λ can be found from ϕ_1 and ϕ_2, respectively. To separate the instrumental term from ΔZ_λ, it is necessary to observe several sources. A simple graphical analysis would be to plot ϕ_0 vs. $\sin \delta$ for these sources; ΔZ_λ is given by the slope, i.e., $d\phi_0/d(\sin \delta)$, and ϕ_{in} by the $\sin \delta = 0$ intercept.

In the general case, as encountered in geodetic applications of VLBI, it is necessary to determine both baseline and source positions. Here, the residual phase [substitute Eqs. (12.8)–(12.10) into (12.7)] is the same as Eq. (12.11) but with

$$\phi_0 = 2\pi \left(\sin \delta \Delta Z_\lambda + Z_\lambda \cos \delta \Delta \delta \right) + \phi_{\text{in}} \,,$$

$$\phi_1 = 2\pi \left(\cos \delta \Delta X_\lambda + Y_\lambda \cos \delta \Delta \alpha - X_\lambda \sin \delta \Delta \delta \right) \,, \qquad (12.13)$$

$$\phi_2 = 2\pi \left(-\cos \delta \Delta Y_\lambda + X_\lambda \cos \delta \Delta \alpha + Y_\lambda \sin \delta \Delta \delta \right) \,.$$

Interleaved observations need to be made of a set of sources over a period of ~ 12 h. Three parameters (ϕ_0, ϕ_1, and ϕ_2) can be derived for each source. If n_s sources are observed, $3n_s$ quantities are obtained. The number of unknown parameters required to specify the n_s positions, the baseline, and the instrumental phase (assumed to be constant) is $2n_s + 3$; the right ascension of one source is chosen arbitrarily. Thus, if $n_s \geq 3$, it is possible to solve for all the unknown quantities. Note that the sources should have as wide a range in declination as possible in order to distinguish ΔZ from ϕ_{in} in Eq. (12.12). Least-mean-squares analysis provides simultaneous solutions for the instrumental parameters and the source positions. Usually, many more than three sources are observed, so there is redundant information, and variation of the instrumental phase with time as well as other parameters can be included in the solution. A discussion of the method of least-mean-squares analysis can be found in Appendix 12.1.

Most astronomers are concerned with measuring the position of a source of interest with respect to a nearby calibrator taken from the ICRF or other catalog on an interferometer with well-calibrated baselines. In this case, the phase terms for Eq. (12.11) are

$$\phi_0 = 2\pi Z_\lambda \cos \delta \Delta \delta + \phi_{\text{in}} \,,$$

$$\phi_1 = 2\pi \left(Y_\lambda \cos \delta \Delta \alpha - X_\lambda \sin \delta \Delta \delta \right) \,, \qquad (12.14)$$

$$\phi_2 = 2\pi \left(X_\lambda \cos \delta \Delta \alpha + Y_\lambda \sin \delta \Delta \delta \right) \,.$$

However, the fringe visibility of a point source at position $l = \Delta\alpha\cos\delta$ and $m = \Delta\delta$ is

$$V = V_0 e^{-j2\pi\,(ul+vm)} = V_0 e^{-j\Delta\phi(H)}\,. \tag{12.15}$$

Thus, the source can be imaged by the usual interferometric techniques and its position determined by fitting a Gaussian (or similar) profile to the image plane. The accuracy of position determined in this manner will be limited by the thermal noise to approximately

$$\sigma_\theta \simeq \frac{1}{2}\frac{\theta_{\text{res}}}{\mathcal{R}_{\text{sn}}}\,, \tag{12.16}$$

where θ_{res} is the resolution of the interferometer, and $\mathcal{R}_{\text{sn}}$ is the signal-to-noise ratio (SNR) [Reid et al. 1988, Condon 1997; see also Eq. (10.68)]. It is shown in Appendix A12.1.3 that the Fourier transform used in imaging is equivalent to a grid parameter search with trial values of α and δ. However, to find baseline parameters or to analyze complex data sets, it is necessary to perform the data analysis in the (u, v) plane.

12.2.2   Measurements with VLBI Systems

The use of independent local oscillators at the antennas in VLBI systems does not easily permit calibration of absolute fringe phase. The earliest method used for obtaining positional information in VLBI was the analysis of the fringe frequency (fringe rate). The fringe frequency is the time rate of change of the interferometer phase. Thus, from Eq. (12.2), the fringe frequency is

$$\nu_f = \frac{1}{2\pi}\frac{d\phi}{dt} = -\omega_e D_\lambda \cos d\cos\delta\sin(H-h) + \nu_{\text{in}}\,, \tag{12.17}$$

where ω_e is the angular velocity of rotation of the Earth (dH/dt), and ν_{in} is an instrumental term equal to $d\phi_{\text{in}}/dt$. The component ν_{in} largely results from residual differences in the frequencies of the hydrogen masers, which provide the local oscillator references at the antennas.

The quantity $D_\lambda \cos d$ is the projection of the baseline in the equatorial plane, denoted D_E. Thus, Eq. (12.17) can be rewritten

$$\nu_f = -\omega_e D_E \cos\delta\sin(H-h) + \nu_{\text{in}}\,. \tag{12.18}$$

The polar component of the baseline (the projection of the baseline along the polar axis) does not appear in the equation for fringe frequency. An interferometer with a baseline parallel to the spin axis of the Earth has lines of constant phase parallel to the celestial equator, and the interferometer phase does not change with hour angle. Therefore, the polar component of the baseline cannot be determined from the analysis of fringe frequency.

The usual practice in VLBI is to refer hour angles to the Greenwich meridian. We follow this convention and use a right-handed coordinate system with X through the Greenwich meridian and with Z toward the north celestial pole. Thus, in terms of the Cartesian coordinates for the baseline, Eq. (12.18) becomes

$$\nu_f = -\omega_e \cos \delta \, (X_\lambda \sin H + Y_\lambda \cos H) + \nu_{in} \, . \tag{12.19}$$

The residual fringe frequency $\Delta \nu_f$, that is, the difference between the observed and expected fringe frequencies, can be calculated by taking the differentials of Eq. (12.19) with respect to δ, H, X_λ, and Y_λ and also including the unknown quantity ν_{in}. We thereby obtain

$$\Delta \nu_f = a_1 \cos H + a_2 \sin H + \nu_{in} \, , \tag{12.20}$$

where

$$a_1 = \omega_e (Y_\lambda \sin \delta \Delta \delta + X_\lambda \cos \delta \Delta \alpha - \cos \delta \Delta Y_\lambda) \tag{12.21}$$

and

$$a_2 = \omega_e (X_\lambda \sin \delta \Delta \delta - Y_\lambda \cos \delta \Delta \alpha - \cos \delta \Delta X_\lambda) \, . \tag{12.22}$$

Note that $\Delta \nu_f$ is a diurnal sinusoid and that the average value of $\Delta \nu_f$ is the instrumental term ν_{in}. Information about source positions and baselines must come from the two parameters a_1 and a_2. Therefore, unlike the case of fringe phase [Eq. (12.11)] where three parameters per source are available, it is not possible to solve for both source and baseline parameters with fringe-frequency data. For example, from observations of n_s sources, $2n_s + 1$ quantities are obtained. The total number of unknowns (two baseline parameters, $2n_s$ source parameters, and ν_{in}) is $2n_s + 3$. If the position of one source is known, the rest of the source positions and X_λ, Y_λ, and ν_{in} can be determined. Note that the accuracy of the measurements of source declinations is reduced for sources close to the celestial equator because of the $\sin \delta$ factor in Eqs. (12.21) and (12.22).

As an illustration of the order of magnitude of the parameters involved in fringe-frequency observations, consider two antennas with an equatorial component of spacing equal to 1000 km and an observing wavelength of 3 cm. Then $D_E \simeq 3 \times 10^7$ wavelengths, and the fringe frequency for a low-declination source is about 2 kHz. Assume that the coherence time of the independent frequency standards is about 10 min. In this period, 10^6 fringe cycles can be counted. If we suppose that the phase can be measured to 0.1 turn, ν_f will be obtained to a precision of 1 part in 10^7. The corresponding errors in D_E and angular position are 10 cm and $0.02''$, respectively.

To overcome the limitations of fringe-frequency analysis, techniques for the precise measurement of the relative group delay of the signals at the antennas were developed. The use of bandwidth synthesis to improve the accuracy of delay measurements has been discussed in Sect. 9.8. The group delay is equal to the geometric delay τ_g except that, as measured, it also includes unwanted components resulting from clock offsets at the antennas and atmospheric differences in the signal paths. The fringe phase measured with a connected-element interferometer observing at frequency ν is $2\pi\nu\tau_g$, modulo 2π. Except for the dispersive ionosphere, the group delay therefore contains the same type of information as the fringe phase, without the ambiguity resulting from the modulo 2π restriction. Thus, group delay measurements permit a solution for baselines and source positions similar to that discussed above for connected-element systems, except that clock offset terms also must be included.

It is interesting to compare the relative accuracies of group delay and the fringe frequency (or, equivalently, the rate of change of phase delay) measurements. The intrinsic precision with which each of these quantities can be measured is derived in Appendix 12.1 [Eqs. (A12.27) and (A12.34)] and can be written

$$\sigma_f = \sqrt{\frac{3}{2\pi^2}}\left(\frac{T_S}{T_A}\right)\frac{1}{\sqrt{\Delta\nu\tau^3}} \qquad (12.23)$$

and

$$\sigma_\tau = \frac{1}{\sqrt{8\pi^2}}\left(\frac{T_S}{T_A}\right)\frac{1}{\sqrt{\Delta\nu\tau}\,\Delta\nu_{rms}}\,, \qquad (12.24)$$

where σ_f and σ_τ are the rms errors in fringe frequency and delay, T_S and T_A are the system and antenna temperatures, $\Delta\nu$ is the IF bandwidth, τ is the integration time, and $\Delta\nu_{rms}$ is the rms bandwidth introduced in Sect. 9.8 [see also Eqs. (A12.32) and related text in Appendix 12.1]. $\Delta\nu_{rms}$ is typically 40% of the spanned bandwidth. For a single rectangular RF band, $\Delta\nu_{rms} = \Delta\nu/\sqrt{12}$. To express the measurement error as an angle, recall that the geometric delay is

$$\tau_g = \frac{D}{c}\cos\theta\,, \qquad (12.25)$$

where θ is the angle between the source vector and the baseline vector. Thus, the sensitivity of the delay to angular changes is

$$\frac{\Delta\tau_g}{\Delta\theta_\tau} = \frac{D}{c}\sin\theta\,, \qquad (12.26)$$

where $\Delta\theta_\tau$ is the increment in θ corresponding to an increment $\Delta\tau_g$ in τ_g. Similarly, the sensitivity of the fringe frequency to angular changes [since $v_f = v(d\tau_g/dt)$] is (for an east–west baseline)

$$\frac{\Delta v_f}{\Delta\theta_f} = D_\lambda\omega_e\sin\theta \ , \tag{12.27}$$

where $\Delta\theta_f$ is the increment in θ corresponding to an increment Δv_f in v_f. Thus, by setting $\Delta v_f = \sigma_f$ and $\Delta\tau_g = \sigma_\tau$ and ignoring geometric factors, we obtain the equation

$$\frac{\Delta\theta_\tau}{\Delta\theta_f} \simeq 2\pi\frac{\tau/t_e}{\Delta v/v} \ , \tag{12.28}$$

where $t_e = 2\pi/\omega_e$ is the period of the Earth's rotation. Equation (12.28) describes the relative precision of delay and fringe-frequency measurements. In practice, measurements of delay are generally more accurate because of the noise imposed by the atmosphere. Measurements of fringe frequency are sensitive to the time derivative of atmospheric path length, and in a turbulent atmosphere, this derivative can be large, while the average path length is relatively constant. Note that fringe-frequency and delay measurements are complementary. For example, with a VLBI system of known baseline and instrumental parameters, the position of a source can be found from a single observation using the delay and fringe frequency because these quantities constrain the source position in approximately orthogonal directions. The earliest analyses of fringe-frequency and delay measurements to determine source positions and baselines were made by Cohen and Shaffer (1971) and Hinteregger et al. (1972).

The accuracy with which group delay can be used to measure a source position is proportional to the reciprocal of the bandwidth $1/\Delta v$. Similarly, the accuracy with which phase can be used to measure a source position is proportional to the reciprocal of the observing frequency $1/v$. Since the proportionality constants are approximately the same, the relative accuracy of these techniques is $v/\Delta v$. This ratio of the observing frequency to the bandwidth, including effects of bandwidth synthesis, is commonly one to two orders of magnitude. On the other hand, the antenna spacings used in VLBI are one to two orders of magnitude greater than those used in connected-element systems. Thus, the accuracy of source positions estimated from group delay measurements with VLBI systems is comparable to the accuracy of those estimated from fringe phase measurements on connected-element systems having much shorter baselines. VLBI position measurements using phase referencing, as described below, are the most accurate of radio methods.

The ultimate limitations on ground-based interferometry are imposed by the atmosphere. Dual-frequency-band measurements effectively remove ionospheric phase noise (see Sect. 14.1.3). The rms phase noise of the troposphere increases about as $d^{5/6}$, where d is the projected baseline length, for baselines shorter than a few kilometers [see Eq. (13.101) and Table 13.3]. In this regime, measurement accuracies of angles improve only slowly with increasing baseline length. For baselines greater than ~ 100 km, the effects of the troposphere above the interferometer elements are uncorrelated, and the measurement accuracy might be expected to improve more rapidly with baseline length. However, for widely spaced elements, the zenith angle can be significantly different, and the atmospheric model becomes very important.

12.2.3 Phase Referencing (Position)

In VLBI measurements of relative positions of closely spaced sources, it is possible to measure the relative fringe phases and thus obtain positional accuracy corresponding to the very high angular resolution inherent in the long baselines. The most accurate measurements can be made when the sources are sufficiently close that both fall within the antenna beams [see, e.g., Marcaide and Shapiro (1983) and Rioja et al. (1997)] or when they are no more than a few degrees apart so that tropospheric and ionospheric effects are closely matched (Shapiro et al. 1979; Bartel et al. 1984; Ros et al. 1999). In such cases, one source can be used as a calibrator in a manner similar to that for phase calibration in connected-element arrays. In VLBI, this procedure is referred to as *phase referencing*. It allows imaging of sources for which the flux densities are too low to permit satisfactory self-calibration. The description here follows reviews of phase referencing procedures by Alef (1989) and Beasley and Conway (1995).

In phase referencing observations, measurements are made alternately on the target source and on a nearby calibrator, with periods on the order of a minute on each. (Note that the calibrator is also referred to as the phase reference source.) The rate of change of phase during these measurements must be slow enough that, from one calibrator measurement to the next, it is possible to interpolate the phase without ambiguity factors of 2π. It is therefore necessary to use careful modeling to remove geodetic and atmospheric effects, including tectonic plate motions, polar motion, Earth tides, and ocean loading, and to make precise corrections for precession and nutation on the source positions. More subtle effects may need to be taken into account; for example, gravitational distortions of antenna structures, which tend to cancel out in connected-element arrays, can affect VLBI baselines because of the difference in elevation angles at widely spaced locations. Phase referencing has become more useful as better models for these effects, together with increased sensitivity and phase stability of receiving systems, have been developed.

Consider the case in which we observe the calibration source at time t_1, then the target source at time t_2, and then the calibrator again at time t_3. For any one of these observations, the measured phase is

$$\phi_{\text{meas}} = \phi_{\text{vis}} + \phi_{\text{inst}} + \phi_{\text{pos}} + \phi_{\text{ant}} + \phi_{\text{atmos}} + \phi_{\text{ionos}} , \tag{12.29}$$

where the terms on the right side are, respectively, the components of the phase due to the source visibility, instrumental effects (cables, clock errors, etc.), the error in the assumed source position, errors in assumed antenna positions, the effect of the neutral atmosphere, and the effect of the ionosphere. To correct the phase of the target source, we need to interpolate the measurements on the calibrator at t_1 and t_3 to estimate what the calibrator phase would have been if measured at t_2, and then subtract the interpolated phase from the measured phase for the target source. If the positions of the target source and the calibrator are sufficiently close on the sky (not more than a few degrees apart), lines of sight from any antenna to the two sources will pass through the same isoplanatic patch, so the differences in the atmospheric and ionospheric terms can be neglected. We can assume that the instrumental terms do not differ significantly with small position changes, and if the calibrator is unresolved, then its visibility phase is zero. If the calibrator is partially resolved, it should be strong enough to allow imaging by self-calibration, and correction can be made for its phase. Thus, the corrected phase of the target source reduces to

$$\phi^t - \tilde{\phi}^c = \phi^t_{\text{vis}} + \left(\phi^t_{\text{pos}} - \tilde{\phi}^c_{\text{pos}} \right) , \tag{12.30}$$

where the superscripts t and c refer to the target source and calibrator, respectively, and the tilde indicates interpolated values. The right side of Eq. (12.30) depends only on the structure and position of the target source, and the position of the calibrator. Figure 12.1 shows an example of phase referencing in which fringe fitting was performed on the data for the reference source, that is, determination of baseline errors, offsets between time standards at the sites, and instrumental phases. The results for the phase reference source (calibrator) are shown as crosses, and the resulting phase and phase rate corrections were interpolated to the times of the data points for the target source, shown as open squares. The corrected phases for the target source are shown in the lower diagram. For fringe fitting, it is desirable to have a source that is unresolved and provides a strong signal, so a phase reference source should be chosen for these characteristics when the target source is weak or resolved.

   Of the various effects in Eq. (12.29) that are removed by phase referencing, those that vary most rapidly with time are the atmospheric ones, and at frequencies above a few gigahertz, they result primarily from the troposphere rather than the ionosphere. Thus, at centimeter wavelengths, the tropospheric variations limit the time that can be allowed for each cycle of observation of the target and calibrator sources. Variations resulting from a moving-screen model of the troposphere are described in Section 13.1.6; the characteristics of the screen are based on Kolmogorov turbulence theory (Tatarski 1961). The relative rms variation in phase for the target and calibrator sources, the rays from which pass through the atmosphere a distance d_{tc} apart, is proportional to $d_{tc}^{5/6}$:

$$\sigma = \sigma_0 \, d_{tc}^{5/6} , \tag{12.31}$$

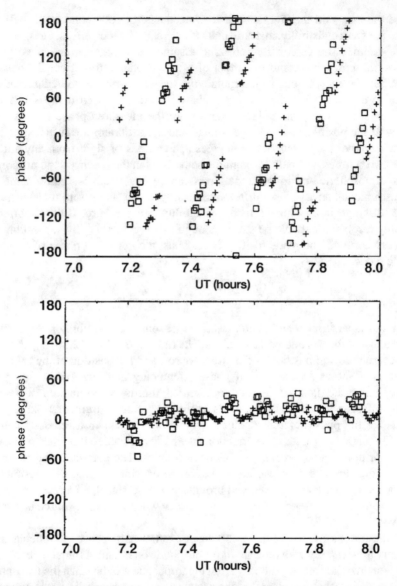

Fig. 12.1 An example of phase referencing with the VLBA. The data are from the Brewster–Pie Town baseline with an observing frequency of 8.4 GHz. The top figure shows the uncalibrated data for two sources: 1638+398 (the target source, open squares) and 1641+399 (the phase reference source, crosses). The bottom figure shows the data for 1641+399 after fringe fitting, and the data for 1638+398 after phase referencing, using 1641+399 as the reference source. From Beasley and Conway (1995), courtesy of and © the Astronomical Society of the Pacific.

where σ_0 is the phase variation for 1-km ray spacing. In order to be able to interpolate the VLBI phase reference values from one calibrator observation to the next without ambiguity in the number of turns, the rms path length should not change by more than $\sim \lambda/8$ between successive calibrator scans. Then if the scattering screen moves horizontally with velocity v_s, the criterion above results in a limit on the time for one cycle of the target source and calibrator, t_{cyc}. To determine this limit, we put $d_{tc} = v_s t_{cyc}$, and from Eq. (12.31) obtain

$$
t_{cyc} < \left(\frac{\pi}{8\sigma_0} \right)^{6/5} v_s^{-1} . \tag{12.32}
$$

This result can be used to illustrate the time limit on the switching cycle. The empirical data in Table 13.4 show that at $\lambda = 6\,\text{cm}$ (5 GHz frequency), the typical rms delay path is about 1 mm for $d_{tc} = 1\,\text{km}$, at the VLA site. The corresponding value of σ_0 for 6-cm wavelength is $6°$, and for $v_s = 0.01\,\text{km s}^{-1}$, $t_{cyc} < 19\,\text{min}$. This result is for typical conditions at the VLA site. For the same location and 1-km ray spacing, but under conditions described as "very turbulent," Sramek (1990) gives a value of 7.5 mm for the rms path deviation. The value of σ_0 for 6-cm wavelength is then $45°$, resulting in $t_{cyc} < 1.7\,\text{min}$. The elevation angle of the source was not less than $60°$ for this last observation, so even shorter switching times could apply at lower elevation angles. Specific recommendations for cycle times in VLBI applications are given by Ulvestad (1999).

At frequencies below ~ 1 GHz, the ionosphere becomes the limiting factor and medium-scale traveling ionospheric disturbances (MSTIDs), which have velocities of $100–300\,\text{m s}^{-1}$ and wavelengths up to several hundred kilometers, become important (Hocke and Schlegel 1996). Phase fluctuations resulting from the ionosphere or troposphere are minimized in the approximate range 5–15 GHz, in which good performance can be obtained by phase referencing in VLBI.

There are also limits on the angular range that should be used in switching to the phase reference source, since even with a static atmosphere phase, errors are introduced that increase with switching angle. Phase referencing over $3°$ with 50-μas precision has been demonstrated by Reid et al. (2009) and Reid and Honma (2014).

The offsets and uncertainties in the geometric parameters of the interferometer cause residual errors that scale in proportion to the angular separation between the target source and the reference or calibrator source. To first order, an offset in the position of the calibrator source is simply transferred to the estimate of the position of the target source. This is because the (u, v) coordinates of the target and calibrator sources can be considered to be the same for a separation of a few degrees or less. However, second-order corrections can be important. The total fringe phase [see Eq. (12.1)] is $2\pi D_\lambda \cos\theta$. For a target source in the direction θ_t and a calibrator source in direction θ_c, the difference in these interferometer phases will be

$$
\Delta\phi = \phi_t - \phi_c = 2\pi D_\lambda (\cos\theta_t - \cos\theta_c) . \tag{12.33}
$$

Since $\cos(\theta_c) \simeq \cos\theta_t - \sin\theta_t(\theta_c - \theta_t)$,

$$\Delta\phi \simeq 2\pi D_\lambda \sin\theta_t\,\theta_{\text{sep}}\,, \tag{12.34}$$

where $\theta_{\text{sep}} = \theta_t - \theta_c$.

Now we include the effect of a change in phase for an error in the baseline of ΔD_λ, giving a second-order phase term of

$$\Delta^2\phi \simeq 2\pi \Delta D_\lambda \sin\theta_t\,\theta_{\text{sep}}\,, \tag{12.35}$$

or, ignoring the geometric factor,

$$\Delta^2\phi \simeq 2\pi \Delta D_\lambda \theta_{\text{sep}}\,. \tag{12.36}$$

This is the phase that affects astrometric accuracy. Hence, the effect of the baseline error on the phase is reduced by the factor θ_{sep}. The phase shift has an equivalent position offset of

$$\Delta\theta \simeq \frac{\Delta^2\phi}{2\pi D_\lambda} \simeq \frac{\Delta D}{D}\theta_{\text{sep}}\,. \tag{12.37}$$

With $D = 8000\,\text{km}$, $\lambda = 4\,\text{cm}$, and $\theta_{\text{sep}} = 1$ degree (0.017 radians), $D_\lambda = 10^8$, and the resolution is $\theta_s = 1/D_\lambda = 1$ mas. An error in the baseline of 2 cm would cause a phase error of about 3 degrees, corresponding to an angle of $9\,\mu\text{as}$.

Equation (12.37) provides an excellent rule of thumb for astrometric accuracy, and $\Delta D/D$ can be interpreted as a rotational angular error in the baseline, or ΔD can be replaced by $c\Delta\tau$, where $\Delta\tau$ might characterize the atmospheric delay error.

Similarly, if there is an error in the calibrator position of $\Delta\theta_c$, then, from Eq. (12.34), there will be an error in the phase of

$$\Delta^2\phi \simeq 2\pi D_\lambda \cos\theta_t\,\Delta\theta_c\,\Delta\theta_{\text{sep}}\,, \tag{12.38}$$

where we assume $\sin\theta_t \simeq \sin\theta_c$. Again, ignoring the trigonometric factor, we can write

$$\Delta^2\phi \simeq 2\pi D_\lambda \Delta\theta_c\,\theta_{\text{sep}} \simeq 2\pi \frac{\Delta\theta_c\,\theta_{\text{sep}}}{\theta_{\text{res}}}\,. \tag{12.39}$$

If astrometry is done in the image plane with an array, then there will be a variety of phase errors of magnitude equal to Eq. (12.38) but differing by the trigonometric factors of the various array baselines and can be thought of as the rms phase noise. Then the image will be substantially degraded when $\Delta^2\phi \sim 1$. To meet this criterion with $D = 8000\,\text{km}$, $\lambda = 4$ cm, and $\theta_{\text{sep}} = 1$ degree, the calibrator position must be known to about 10 mas. With an rms phase error of 1 radian, the visibility of a point source is reduced by a factor of $\exp(-2\phi^2/2) \sim 0.6$. For 2 radians of phase error, the image would be destroyed. The equivalent angular offset for the phase shift given in Eq. (12.39) is

$$\Delta \vartheta \simeq \Delta \theta_c \, \theta_{\text{sep}} \, . \tag{12.40}$$

Note that $\Delta \theta_c$ plays the same role as $\Delta D/D$ in Eq. (12.37). Hence, an astrometric accuracy of 10 μas requires position error in the calibrator of 150 μas or less. An analysis in the (u, v) plane is given in Appendix 12.2.

12.2.4 Phase Referencing (Frequency)

At millimeter wavelengths, position phase referencing becomes more difficult because calibration sources are generally weaker and more sparsely distributed than at longer wavelengths. Coherence times are also shorter, e.g., a few tens of seconds above 100 GHz, requiring rapid antenna pointing changes for calibration by position switching. In this case, frequency switching on the target source itself is a valuable calibration technique. The goal is to remove effects in the fringe phase that scale as frequency, i.e., that can be characterized by a nondispersive or constant delay. ϕ_c, the phase at the lower frequency (ν_c), is used to calibrate the phase at the higher frequency (ν_t), ϕ_t by forming the quantity

$$\phi = \phi_t - R\phi_c \, , \tag{12.41}$$

where R is the ratio of frequencies ν_t/ν_c. This procedure will remove the effects of the atmosphere and the frequency standards but not the ionosphere or other dispersive processes. Note that at low frequencies, the goal of dual frequency calibration is to remove the ionospheric delay, and R in Eq. (12.41) is replaced by $1/R$ (e.g., see Sects. 12.6 and 14.1.3). It is usually convenient to choose R to be an exact integer to avoid the need to deal with phase wrap issues. To see this, focus on the term that describes the tropospheric excess path length L. For this exercise, $\phi_c = 2\pi \nu_c L/c + 2\pi n_c$ and $\phi_t = 2\pi \nu_t L/c + 2\pi n_t$, where n_c and n_t are the integers that characterize the phase wraps. The calibrated phase is thus

$$\phi = 2\pi \, (n_c - Rn_t) \, . \tag{12.42}$$

which will be an integral multiple of 2π if R is an integer. An early demonstration of this technique was carried out by Middelberg et al. (2005), who used phases at 14.375 GHz to calibrate those at 86.25 GHz $(R = 6)$. The residual error phase caused by the ionosphere and electronic drifts in local oscillator chains have a much longer timescale than tropospheric variations. These can be removed by adding a slower position switching cycle. The efficacy of this double switching technique has been demonstrated by Rioja and Dodson (2011), Rioja et al. (2014, 2015), and Jung et al. (2011). If the source structure consists of a compact core at both frequencies, as in many AGN, the shift in position with frequency caused by opacity effects in the source can be an important physical diagnostic. This shift can be accurately measured by application of frequency/position switching calibration.

12.3   Time and Motion of the Earth

We now consider the effect of changes in the magnitude and direction of the Earth's rotation vector on interferometric measurements. These changes cause variations in the apparent celestial coordinates of sources, the baseline vectors of the antennas, and universal time. The variations of the Earth's rotation can be divided into three categories:

1. There are variations in the direction of the rotation axis, resulting mainly from precession and nutation of the spinning body. Since the direction of the axis defines the location of the pole of the celestial coordinate system, the result is a variation in the right ascension and declination of celestial objects.
2. The axis of rotation varies slightly with respect to the Earth's surface; that is, the positions on the Earth at which this axis intersects the Earth's surface vary. This effect is known as polar motion. Since the (X, Y, Z) coordinate system of baseline specification introduced in Sect. 4.1 takes the direction of the Earth's axis as the Z axis, polar motion results in a variation of the measured baseline vectors (but not of the baseline length). It also results in a variation in universal time.
3. The rate of rotation varies as a result of atmospheric and crustal effects, and this again results in variation in universal time.

We briefly discuss these effects. Detailed discussions from a geophysical viewpoint can be found in Lambeck (1980).

12.3.1   Precession and Nutation

The gravitational effects of the Sun, Moon, and planets on the nonspherical Earth produce a variety of perturbations in its orbital and rotational motions. To take account of these effects, it is necessary to know the resulting variation of the ecliptic, which is defined by the plane of the Earth's orbit, as well as the variation of the celestial equator, which is defined by the rotational motion of the Earth. The gravitational effects of the Sun and Moon on the equatorial bulge (quadrupole moment) of the Earth result in a precessional motion of the Earth's axis around the pole of the ecliptic.

The Earth's rotation vector is inclined at about 23.5° to the pole of its orbital plane, the ecliptic. The period of the resulting precession is approximately 26,000 years, corresponding to a motion of the rotation vector of $20''$ per year

[$2\pi \sin(23.5°)/26,000$ radians per year]. The 23.5° obliquity is not constant but is currently decreasing at a rate of 47″ per century, due to the effect of the planets, which also cause a further component of precession. The lunisolar and planetary precessional effects, together with a smaller relativistic precession, are known as the general precession. Precession results in the motion of the line of intersection of the ecliptic and celestial equator. This line, called the line of nodes, defines the equinoxes and the zero of right ascension, which precess at a rate of 50″ per year. In addition, the time-varying lunisolar gravitation effects cause nutation of the Earth's axis with periods of up to 18.6 years and a total amplitude of about 9″. The principal variations of the ecliptic and equator are those just described, but other smaller effects also occur. The general accuracy within which positional variations can be calculated is better than 1 mas (Herring et al. 1985). Expressions for precession can be found in Lieske et al. (1977) and for nutation in Wahr (1981). The required procedures are discussed in texts on spherical astronomy, such as Woolard and Clemence (1966), Taff (1981), and Seidelmann (1992).

    Since precession and nutation result in variations in celestial coordinates that can be as large as 50″ per year for objects at low declinations, these effects must be taken into account in almost all observational work, whether astrometric or not. Positions of objects in astronomical catalogs are therefore reduced to the coordinates of standard epochs, B1900.0, B1950.0, or J2000.0. These dates denote the beginning of a Besselian year or Julian year, as indicated by the B or J. The positions correspond to the mean equator and equinox for the specified epoch, where "mean" indicates the positions of the equator and equinox resulting from the general precession, but not including nutation. For further explanation and a discussion of a method of conversion between standard epochs, see Seidelmann (1992). Correction is also required for aberration, that is, for the apparent shift in position resulting from the finite velocity of light and the motion of the observer. Two components are involved: annual aberration resulting from the Earth's orbital motions, which has a maximum value of about 20″; and diurnal aberration resulting from the rotational motion, which has a maximum value of 0.3″. The retarded baseline concept (Sect. 9.3) used in VLBI data reduction accounts for the diurnal aberration. For the nearer stars, corrections for proper motion (i.e., actual motion of the star through space) are required and in some cases also for the parallax resulting from the changing position of the Earth in its orbit (see Sect. 12.5). The impact of radio techniques, particularly VLBI, is resulting in refinement of the classical expressions and parameters. Effects such as the deflection of electromagnetic waves in the Sun's gravitational field must also be included in positional work of the highest accuracy (see Sect. 12.6).

12.3.2 Polar Motion

The term *polar motion* denotes the variation of the pole of rotation of the Earth (the geographic pole) with respect to the Earth's crust. This results in a component of motion of the celestial pole that is distinct from precessional and other motions.

Polar motion is largely, but not totally, of geophysical origin. The motion of the geographic pole around the pole of the Earth's figure is irregular, but over the last century, the distance between these two poles wandered by up to 0.5″, or 15 m on the Earth's surface. In a year's time, the excursion of the figure axis is typically 6 m or less. The motion can be analyzed into several components, some regular and some highly irregular, and not all are understood. The two major components have periods of 12 and 14 months. The 12-month component is a forced motion due to the annual redistribution of water and of atmospheric angular momentum and is far from any resonance. The 14-month component, known as the Chandler wobble (Chandler 1891), is the motion at a resonance frequency whose driving force is unknown. For a more detailed description, see Wahr (1996).

The motion of the pole of rotation is measured in angle or distance in the x and y directions, as shown in Fig. 12.2. The (x, y) origin is the mean pole of 1900–1905, which is referred to as the *conventional international origin* (CIO), and the x axis is in the plane of the Greenwich meridian (Markowitz and Guinot 1968). Since polar motion is a small angular effect, it can often be ignored in imaging observations, especially if the visibility is measured with respect to a calibrator that is only a few degrees from the center of the field being imaged.

Fig. 12.2 Coordinate system for the measurement of polar motion. The x coordinate is in the plane of the Greenwich meridian, and the y axis is 90° to the west. CIO is the conventional international origin.

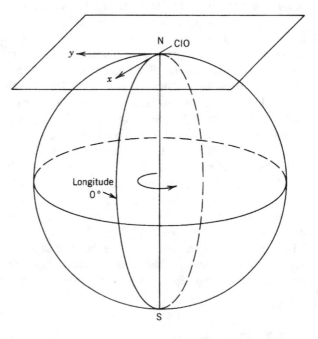

12.3.3 Universal Time

Like the motion of the Earth, the system of timekeeping based on Earth rotation is a complicated subject, and for a detailed discussion, one can refer to Smith (1972) or to the texts mentioned in the discussion of precession and nutation above. We shall briefly review some essentials. Solar time is defined in terms of the rotation of the Earth with respect to the Sun. In practice, the stars present more convenient objects for measurement, so solar time is derived from measurement of the sidereal rotation. The positions of stars or radio sources used for such measurements are adjusted for precession, nutation, and so on, and the resulting time measurements thus depend only on the angular velocity of the Earth and on polar motion. When converted to the solar timescale, these measurements provide a form of universal time (UT) known as UT0; this is not truly "universal" since the effects of polar motion, which can amount to about 35 ms, depend on the location of the observatory. When UT0 is corrected for polar motion, the result is known as UT1. Since it is a measure of the rotation of the Earth relative to fixed celestial objects, UT1 is the form of time required in astronomical observing, including the analysis of interferometric observations, navigation, and surveying. However, UT1 contains the effects of small variations in the Earth's rotation rate, attributable largely to geophysical effects such as the seasonal variations in the distribution of water between the surface and atmosphere. Fluctuations in the length of day over the period of a year are typically about 1 ms. To provide a more uniform measure of time, UT2 is derived from UT1 by attempting to remove seasonal variations. UT2 is rarely used. UT1 and UT2 include the effect of the gradual decrease of the rotation rate of the Earth. This causes the length of the UT1/UT2 day to increase slightly when compared with International Atomic Time (IAT), which is based on the frequency of the cesium line (see Sect. 9.5.4). The IAT second is the basis for another form of UT, Coordinated Universal Time (UTC), which is offset from IAT so that $|\text{UT1} - \text{UTC}| < 1$ s. This relationship is maintained by inserting one-second discontinuities (leap seconds) in UTC when required on specified days of the year.

The practice at many observatories is to maintain UTC or IAT using an atomic standard and then obtain UT1 from the published values of $\Delta\text{UT1} = \text{UT1} - \text{UTC}$. Since ΔUT1 is measured rather than computed, in principle it can be determined only after the fact. However, it is possible to predict it by extrapolation with satisfactory accuracy for periods of one or two weeks and thus to implement UT1 in real time. Values of ΔUT1 are available from the Bureau International de L'Heure (BLH), which was established in 1912 at the Paris Observatory to coordinate international timekeeping, and from the U.S. Naval Observatory. Rapid service data are available from these institutions with a timeliness suitable for extrapolation.

12.3.4   Measurement of Polar Motion and UT1

The classical optical methods of measuring polar motion and UT1 are by timing the meridian transits of stars of known positions. Observations at different longitudes, using stars at more than one declination, are required to determine all three parameters $(x, y, \Delta UT1)$. During the 1970s, it became evident that such astrometric tasks can also be performed by radio interferometry (McCarthy and Pilkington 1979).

To specify the baseline components of an interferometer for such measurements, we use the (X, Y, Z) system of Sect. 4.1, rotated so that the X axis lies in the Greenwich meridian instead of the local meridian. Let ΔX, ΔY, and ΔZ be the changes in the baseline components resulting from polar motion (x, y) (in radians) and a time variation (UT1 – UTC) corresponding to Θ radians. Then we may write

$$
\begin{bmatrix} \Delta X \\ \Delta Y \\ \Delta Z \end{bmatrix} = \begin{bmatrix} 0 & -\Theta & x \\ \Theta & 0 & -y \\ -x & y & 0 \end{bmatrix} \begin{bmatrix} X \\ Y \\ Z \end{bmatrix} , \tag{12.43}
$$

where the square matrix is a three-dimensional rotational matrix valid for small angles of rotation. Θ, x, and y are the rotation angles about the Z, Y, and X axes, respectively. From Eq. (12.43), we obtain

$$
\Delta X = - \Theta Y + xZ ,
$$
$$
\Delta Y = \Theta X - yZ , \tag{12.44}
$$
$$
\Delta Z = - xX + yY .
$$

Thus, if one observes a series of sources at periodic intervals and determines the variation in baseline parameters, Eqs. (12.44) can be used to determine UT1 and polar motion. For an interferometer with an east–west baseline $(Z = 0)$, one can determine Θ but cannot separate the effects of x and y. An east–west interferometer located on the Greenwich meridian $(X = Z = 0)$ would yield measures of Θ and y but not of x. If it had a north–south component of baseline $(Z \neq 0)$, one could still measure y but would not be able to separate the effects of x and Θ. In general, one cannot measure all three quantities with a single baseline, since a single direction is specified by two parameters only. Systems suitable for a complete solution might be, for example, two east–west interferometers separated by about 90° in longitude or a three-element noncollinear interferometer. An example of VLBI measurements of the pole position is shown in Fig. 12.3. The Global Positioning System provides a method of making pole-position measurements [see, e.g., Herring (1999)].

The methods just described are applicable to observations using connected-element interferometers in which the phase can be calibrated, and also to VLBI observations in which the bandwidth is sufficient to obtain accurate group delay measurements. An example of VLBI determination of the length of day is shown

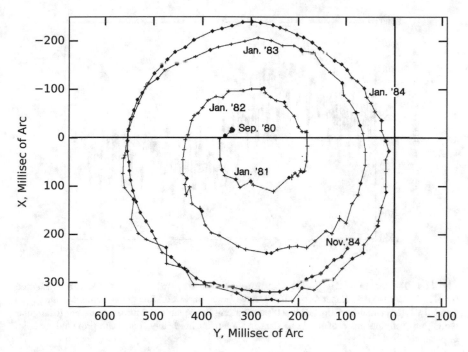

Fig. 12.3 VLBI determination of the position of the pole. The diamonds indicate measurements in which three or more stations particulated in the observations so that both pole coordinates could be determined from the observations. The crosses mark observations that employed only one baseline, for which only the x component of the pole was determined. The corresponding y components were obtained from the BLH. Note that 100 mas corresponds to 3.2 m. From Carter et al. (1985), © John Wiley & Sons.

in Fig. 12.4. The data show an annual variation of about 2 ms, which is caused by the angular momentum exchange between the Earth and the atmosphere due to the difference in land mass in the Northern and Southern Hemispheres [see, e.g., Paek and Huang (2012)]. The trend in the long-term variation is thought to be due to an exchange of angular momentum between the Earth's core and mantle. The effects of El Niño events can be seen in these data (Gipson and Ma 1999). A comparison of determinations of UT1 and polar motion by VLBI, satellite laser ranging, and BLH analyses of standard astrometric data is given in Robertson et al. (1983) and Carter et al. (1984).

VLBI is a unique tool for the study of many phenomena related to Earth dynamics. For example, the period and amplitude of the free-core nutation has been estimated (Krásná et al. 2013).

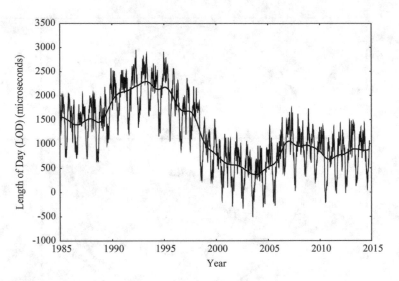

Fig. 12.4 The length of day (LOD) with respect to the standard day of 86,400 s, as measured by VLBI, 1980–2015. Lunar tidal effects have been removed. The measurement accuracy is typically 20 μs, which corresponds to a distance along the equator of 2 mm. A two-year (triangular weighting) running mean of the LOD data is shown by the heavy line. Data provided by John Gipson, NASA/GSFC.

12.4   Geodetic Measurements

Certain geophysical phenomena, for example, Earth tides (Melchior 1978) and movements of tectonic plates, can result in variations in the baseline vector of a VLBI system. Variations in the length of the baseline are clearly attributable to such phenomena, whereas variations in the direction can also result from polar motion and rotational variations. Magnitudes of the effects are of order 1–10 cm per year for plate motions and 30 cm (diurnal) for Earth tides. They are thus measurable using the techniques of VLBI. Solid-Earth tides were first detected by Shapiro et al. (1974), and refined measurements were reported by Herring et al. (1983). In addition to solid-Earth tides, displacement of land masses resulting from tidal shifts of water masses, called ocean loading, is measurable. The earliest evidence of contemporary motion of tectonic plates was found by Herring et al. (1986), who reported that the increase in the baseline between Westford, MA, and Onsala, Sweden, based on data from 1980 to 1984, was 17 ± 2 mm/yr. A plot of the extensive measurements of the Westford–Onsala baseline is shown in Fig. 12.5. For reviews of geodetic applications of VLBI, see Shapiro (1976), Counselman (1976), Clark et al. (1985), Carter and Robertson (1993), and Sovers et al. (1998).

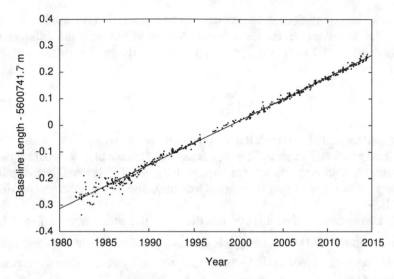

Fig. 12.5 The baseline length between Westford, MA, and Onsala, Sweden, determined by 499 VLB observations from 1981 to 2015. The formal error in recent measurements is typically less than 1 mm. The straight line fit to the data has a slope 16.34 ± 0.04 mm/yr. For an analysis of short-term systematic trends in baseline-length data, see Titov (2002). These data were taken from the website of the International VLBI Service for Geodesy and Astrometry (http://ivscc.gsfc.nasa. gov/products-data).

12.5   Proper Motion and Parallax Measurements

The position of a relatively nearby star or radio source changes with respect to the distant background due to the annual motion of the Earth around the Sun. This effect is called *annual parallax*. It can be used to measure distances by the classical technique of trigonometric triangulation, first demonstrated by Bessel (1838) from optical observations of the star 61 Cygni. The parallax angle, Π, is defined as one-half of the total excursion in apparent position over a year. The distance to the object, by simple trigonometry for small angles, is

$$D = \frac{1}{\Pi} . \tag{12.45}$$

By definition, an object with a parallactic angle of $1''$ has a distance of 1 parsec. Hence, a parsec is 206265 (the number of arcseconds in a radian) times the Sun–Earth distance [called the astronomical unit (AU)], or 3.1×10^{18} cm. The AU, which is determined by ranging measurements of the planets and spacecraft, is called the first rung on the cosmic distance ladder. Its value is $1.4959787070000 \times 10^{13}$ cm, to an accuracy of about 1 part in 50 billion (Pitjeva and Standish 2009). The intrinsic motion of nearby objects can also be measured. This is called *proper motion*. The precision of VLBI astrometric measurements has greatly extended the distances over which proper motions and parallaxes can be measured. If the parallax accuracy is σ_Π, the uncertainty in the distance can be determined from the differential of Eq. (12.45), i.e., $\Delta D = \Pi^{-2} \Delta \Pi$, to be $\sigma_D = D^2 \sigma_\Pi$. Hence, the fractional distance accuracy is

$$\left(\frac{\sigma_D}{D} \right) = D \sigma_\Pi \ . \tag{12.46}$$

The important result here is that the fractional distance accuracy grows with distance for a fixed positional accuracy. Hence, a fractional distance of 10% accuracy can be measured for objects to a distance of 10 pc with $\sigma_\Pi = 0.01''$ (ground-based optical), or 100 pc with $\sigma_\Pi = 1$ mas (Hipparcos satellite), and 10^4 pc with $\sigma_\Pi = 10$ μas (VLBI).

For measurements of parallax with better than 10% accuracy, $\sigma_\Pi / \Pi < 0.1$, the distance estimate of $D = \dfrac{1}{\Pi} \pm \dfrac{\sigma_\Pi}{\Pi^2}$ is essentially unbiased. The situation is more complex when the accuracy is lower. If the probability distribution for a parallax measurements is

$$p\left(\Pi \right) = \frac{1}{\sqrt{2\pi}\sigma_\Pi} e^{-\frac{(\Pi - \Pi_0)^2}{2\sigma_\Pi^2}} \ , \tag{12.47}$$

where Π_0 is the true, but unknown, parallax, then the probability distribution function of D, $p\left(D \right) = p\left(\Pi \right) \left| \dfrac{d\Pi}{dD} \right|$, is

$$p\left(D \right) = \frac{1}{\sqrt{2\pi}\sigma_\Pi} \frac{1}{D^2} e^{-\frac{\left(\frac{1}{D} - \Pi_0 \right)^2}{2\sigma_\Pi^2}} \ . \tag{12.48}$$

$p\left(D \right)$ becomes increasingly asymmetric with increasing σ_Π / Π and develops a long tail at large values of D. The expectation of D, i.e., $\frac{1}{\Pi}$, can be calculated from Eq. (12.47) by Taylor expansion of $D = \frac{1}{\Pi}$, which gives the result

$$\langle D \rangle \simeq \frac{1}{\Pi_0} \left[1 + \frac{\sigma_\Pi^2}{\Pi_0^2} \right] \ . \tag{12.49}$$

For the case of a single source, an accepted strategy is to perform a Markov chain Monte Carlo (MCMC) analysis of the position-vs.-time data with D as a parameter and apply an appropriate prior distribution of D to estimate the final distribution, $p\left(D \right)$. The difficulties of parallax analysis at low signal-to-noise ratio, including the Lutz–Kelker effect (Lutz and Kelker 1973), are discussed by Bailer-Jones (2015) and Verbiest and Lorimer (2014).

Parallaxes have been measured to many pulsars (Verbiest et al. 2010, 2012). These may be compared with indirect estimates based on dispersion measures and galactic models of electron density. Precision parallax distances may prove to be important in the use of pulsar timing measurements to detect gravitational radiation

Fig. 12.6 VLBI measurement of the position of the radio star IM Peg for 39 epochs from 1991 to 2005. The "cusps" show the annual parallax effect superimposed on a large proper motion of about 34 mas/yr (denoted by arrow). From Ratner et al. (2012). © AAS. Reproduced with permission.

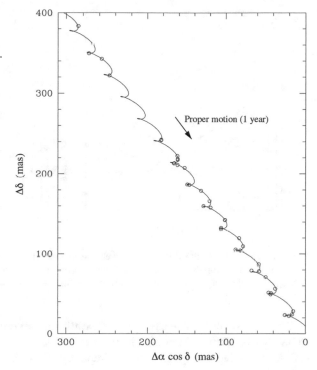

[see Madison et al. (2013)]. The distance to the pulsar PSR J2222-0137 has been determined to be $267.3^{+1.2}_{-0.9}$ pc, an accuracy of 0.4% (Deller et al. 2013).

The star IM Peg, which has detectable radio emission, provides an example of a precise measurement of parallax with VLBI (Bartel et al. 2015). Its position was precisely determined over 39 epochs spanning six years so that it could be used as a guide star for the physics experiment Gravity Probe B (Everitt et al. 2011). The position of the radio star is shown in Fig. 12.6. The position shift is dominated by the proper motion. The annual parallax can be readily seen when this proper motion, modeled as a constant velocity vector, is removed, as shown in Fig. 12.7.

An excellent example of the steady improvement in VLBI parallax measurements can be found in the work on the Orion Nebula, a galactic object of singular importance in astronomy. The results, shown in Table 12.1, made with a variety of continuum and spectral line sources over a considerable frequency range, have yielded a distance accurate to 1.5%. This corresponds to a parallactic accuracy [Eq. (12.46)] of ~ 30 μas.

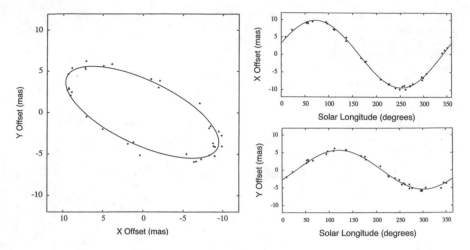

Fig. 12.7 (**left**) Motion of IM Peg with proper motion and orbital motion removed. The parallactic angle Π is the semimajor axis of the ellipse. (**right**) Relative position in Dec. (top) and RA (bottom) vs. solar longitude. Plots derived from the data in Table 1 of Ratner et al. (2012).

There are other notable examples of parallax measurements. The distance to the Pleiades Cluster was determined by VLBI to be 136.2 ± 1.2 pc by Melis et al. (2014), resolving a long-standing discrepancy in its distance estimates. VLBI has also been used to detect the apparent motion of Sgr A*, the radio source at the center of the Galaxy, against the extragalactic background caused by the rotation of the Galaxy. These results are shown in Fig. 12.8. A combination of these data and parallax measurements with the VLBA, VERA, and EVN of more than 100 masers

Table 12.1 VLBI parallax distance measurements to the Orion Nebula[a]

Parallax method[b]	Array	Source type[c]	No. of epochs	Freq. (GHz)	Distance (pc)	Reference
Expansion	VLBI[d]	H_2O	5	22	480 ± 80	Genzel et al. (1981)
Annual	VLBA	YSO	5	15	389 ± 24	Sandstrom et al. (2007)
Annual	VERA	H_2O	16	22	437 ± 19	Hirota et al. (2007)
Annual	VLBA	YSO	4	8	414 ± 7	Menten et al. (2007)
Annual	VERA	SiO	7	43	418 ± 6	Kim et al. (2008)
Annual	VLBA	YSO	5	5	383 ± 5	Kounkel et al. (2016)

[a]All of the sources in these studies were colocated within the Orion Nebula Cluster (ONC) to an angular distance of about $\pm 10'$ or a projected distance of ~ 2 pc. Distance measurements before 1980 were in the range 300–540 pc; the distance measured by Hipparcos to one star was 361^{+168}_{-87} pc (Bertout et al. 1999).
[b]Expansion parallax used a model of internal symmetrical expansion of 21 maser components.
[c]H_2O = water vapor masers; YSO = young stellar objects with nonthermal emission from GMR catalog (Garay et al. 1987); SiO = silicon monoxide masers.
[d]Ad hoc four-station VLBI array.

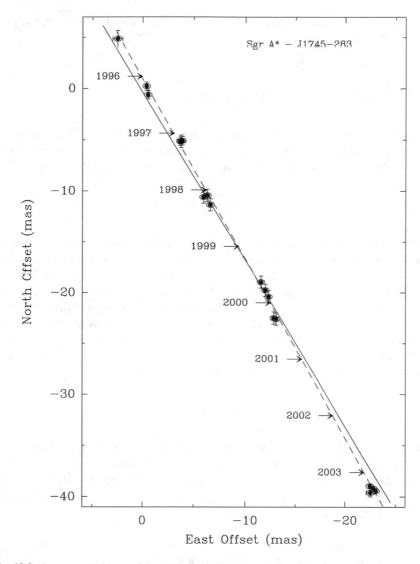

Fig. 12.8 Apparent positions of Sgr A*, the radio source in the Galactic center, relative to the extragalactic calibrator J1745-283, measured over an eight-year period at 43 GHz with the VLBA. The ellipses around the measurement points indicate the scatter-broadened size of Sgr A* (see Fig. 14.10). The one-sigma error bars in the measurements are also shown. The broken line is the variance-weighted least-mean-squares fit to the data, and the solid line indicates the orientation of the Galactic plane. The motion is almost entirely in galactic longitude, attributable to the solar motion around the center of the Galaxy of 241 ± 15 km s^{-1}, for a distance between the Sun and Galactic center of 8 ± 0.5 kpc. The limit on the residual motion of Sgr A* is nearly two orders of magnitude less than that of the motions of stars lying within a projected distance of about 0.02 pc of Sgr A*. These stellar motions suggest that about $4.1 \times 10^6\ M_\odot$ of matter are contained within 0.02 pc of Sgr A*, and the lack of detected motion of Sgr A* itself suggests that a mass of at least $10^3\ M_\odot$ must be associated with the radio source Sgr A*. For a comparison of measurements made with the VLA, see Backer and Sramek (1999). From Reid and Brunthaler (2004). © AAS. Reproduced with permission.

gives Galactic structure parameters of $R_0 = 8.34 \pm 0.16\,\mathrm{pc}$ and $\theta_0 = 240 \pm 8\,\mathrm{km\,s^{-1}}$ (Reid et al. 2014).

12.6   Solar Gravitational Deflection

The bending of electromagnetic radiation passing a massive body is described in the parametrized post-Newtonian formalism of general relativity (GR) by the parameter γ and is normally written as [see, e.g., Misner et al. (1973), or Will (1993)]

$$\Delta\epsilon = (1 + \gamma)\frac{GM}{pc^2}(1 + \cos\epsilon). \qquad (12.50)$$

G is the gravitational constant. M is the mass of the perturbing body, which we take to be the Sun in this discussion; p is the impact parameter (closest approach of the unperturbed ray to the Sun); and ϵ is the elongation angle (the angle between the direction to the source and the direction to the Sun as seen by the observer). Equation (12.50) holds for sources at infinite distance. This parametrization reflects the fact that the bending predicted by Newtonian physics is exactly half the value predicted by GR, i.e., $\gamma = 1$ for GR and 0 for Newtonian physics. GM/c^2 is known as the gravitational radius, which is 1.48 km for the Sun. For a ray path passing close to the Sun where $\epsilon \ll 1$, Eq. (12.50) can be approximated as

$$\Delta\epsilon = (1 + \gamma)\frac{2GM}{pc^2}. \qquad (12.51)$$

For a ray grazing the surface of the Sun, where $p = r_0$ (corresponding to $\epsilon = 0.267°$) and r_0 is the solar radius, the deflection angle is 1.75″.

Equation (12.50) can be rewritten so as to eliminate p, since $p = R_0 \sin\epsilon$, where R_0 is the distance from the Sun to the Earth. After some trigonometric manipulation, the deflection angle can be expressed as

$$\Delta\epsilon = (1 + \gamma)\frac{GM}{R_0 c^2}\sqrt{\frac{1 + \cos\epsilon}{1 - \cos\epsilon}}. \qquad (12.52)$$

$\Delta\epsilon$ declines monotonically with ϵ, as shown in Fig. 12.9, and for $\gamma = 1$ has a value of 4.07 mas at $\epsilon = 90°$, 1 mas at $\epsilon = 150°$, and 0 at $\epsilon = 180°$. Furthermore, two sources separated by 1° near $\epsilon = 90°$ will suffer a 70-μas shift in their relative positions.

Shapiro (1967) first suggested that GR could be tested by observing the deflection of radio waves passing in the vicinity of the Sun. This is just the radio version of the famous optical experiment first performed in 1919 by the Eddington expedition (Dyson et al. 1920). For a long time, the radio astronomical experiments were based on the two sources 3C279 and 3C273, which are separated by about 10 degrees and

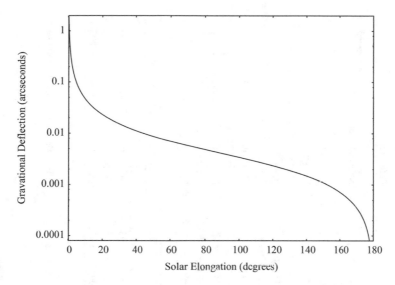

Fig. 12.9 The gravitational deflection, $\Delta\alpha$, of a radio wave passing the Sun at an elongation angle, α, calculated from Eq. (12.52) when $\gamma = 1$.

pass fortuitously close in angle to the Sun each October. In fact, 3C279 is occulted by the Sun on October 8. The measurement of the change in relative position of these sources can be used to estimate γ. The challenge of such measurements is to overcome the effects of the ionized plasmas surrounding the Sun, i.e., the corona and solar wind, whose effects diminish with distance from the Sun and as λ^2 (see Sect. 14.3.1). Note that the ray bending caused by the solar plasma has the opposite sign as caused by GR, i.e., plasma bending makes sources appear closer to the Sun in angle and GR makes them appear farther.

The first radio interferometry experiments were undertaken for the 1969 passage, one with two antennas forming an interferometer at the Owens Valley Radio Observatory at a frequency of 9.1 GHz and a baseline of 1.4 km, and the other with two antennas forming an ad hoc interferometer at the JPL Goldstone facility at a frequency of 2.4 GHz and a baseline of 21 km. The solar plasma was modeled by both groups as two power-law components with amplitude parameters estimated from the data (see Sect. 14.3.1). The results from both experiments confirmed GR to an accuracy of about 30%, with the JPL instrument's advantage of longer baseline and higher resolution compensating for the OVRO instrument's advantage of higher frequency. The experiment has been repeated many times with ever more sensitive equipment and more refined techniques, and the results are listed in Table 12.2. The first VLBI experiment was reported by Counselman et al. (1974) for an 845-km baseline between Haystack and NRAO. Each site employed two antennas so that two coherent interferometers were formed to track both sources simultaneously.

Table 12.2 Measurements of solar gravitational bending with radio interferometry

Instrument	Measurement[a]	Baseline (km)	Frequency (GHz)	Epoch	γ	Reference
				3C279/3C273		
OVRO	P	1.1	9.6	1969[b]	1.01 ± 0.24[c]	Seielstad et al. (1970)
JPL	P	21.5	2.4	1969	1.08 ± 0.30[c]	Muhleman et al. (1970)
Green Bank	P	2.7	2.7/8.1	1970	0.80 ± 0.1[c,d]	Sramek (1971)
Cambridge	P	1.4	2.7/5	1970	1.03 ± 0.034[c]	Hill (1971)
Green Bank	P	27	2.7/8.1	1971	0.94 ± 0.16[c]	Sramek (1974)
Cambridge	P	5	5	1972	1.02 ± 0.16[c]	Riley (1973)
WSRT[e]	P	1.4	5	1972	0.98 ± 0.10[c]	Weiler et al. (1974)
Haystack/Green Bank[f]	P	845	8.1	1972	0.98 ± 0.06	Counselman et al. (1974)
WSRT[e]	P	1.4	1.4/5	1973	1.02 ± 0.066	Weiler et al. (1975)
Green Bank	P	35	2.7/8.1	1974/5	1.003 ± 0.018	Fomalont and Sramek (1976, 1977)
Haystack/OVRO	D	3930	2.4/8.6/23	1987	0.9996 ± 0.0017	Lebach et al. (1995)
VLBA[g]	P	8610	15/23/43	2005	0.9998 ± 0.00030	Fomalont et al. (2009)
				0116+08/0111+02		
Green Bank	P	35	2.7/8.1	1974	1.030 ± 0.022	Fomalont and Sramek (1975)
			All-sky geodetic database			
IVS	D	10,000	2.4/8.6	1979–1990	1.000 ± 0.002	Robertson et al. (1991)
IVS	D	10,000	2.4/8.6	1979–1999	0.9998 ± 0.0004	Shapiro et al. (2004)
IVS	D	10,000	2.4/8.6	1979–2010	0.9992 ± 0.00012	Lambert and Le Poncin-Lafitte (2011)

[a]P = phase; D = group delay.

[b]Typically ten-day experiments, bracketing closest approaches of 3C279 to the Sun.

[c]The results from these experiments were reported as measurements of $\gamma' = $ (observed deflection)/(GR prediction) $= (\gamma + 1)/2$. For this table, all results are reported in terms of γ. Note that $\sigma_\gamma = 2\sigma_{\gamma'}$.

[d]With error updated by Sramek (1974).

[e]Array split to allow simultaneous tracking of 3C273 and 3C279.

[f]First VLBI measurement. Two antennas used at each site to allow simultaneous tracing of 3C273 and 3C279.

[g]3C279 and three other sources.

Another major step forward was the use of dual-frequency observations. This allowed a phase (or delay) observable to be formed from the two phases ϕ_1 and ϕ_2 measured at frequencies ν_1 and ν_2,

$$\phi_c = \phi_2 - \left(\frac{\nu_1}{\nu_2}\right)\phi_1 , \qquad (12.53)$$

from which the dispersive effects of the solar plasma (and ionosphere) are largely removed (see Sect. 14.1.3). The best result so far for the targeted 3C279/3C273 experiment is $\gamma = 0.9998 \pm 0.0003$ with the VLBA at 15, 23, and 43 GHz (Fomalont et al. 2009). It should be noted that this result was based largely on the 43-GHz results alone, where the plasma effects were greatly diminished. Various changes are expected to be made that will improve this experiment by a factor of four (i.e., to a fractional accuracy of better than 1 part in 10^4).

In addition, the huge geodetic VLBI database has been used to estimate γ, giving the results listed in Table 12.2. The analysis described by Lambert and Le Poncin-Lafitte (2011) is based on 5,055 observing sessions (1979–2010) of 3,706 sources and 7 million delay measurements. The postfit delay residual is 23 picoseconds, and $\gamma = 0.9992 \pm 0.0001$. The continual accrual of geodetic data will lead to better results in the future. The best measurement overall of γ to date, $\gamma = 1.000021 \pm 0.000023$, was made by analyzing the delay residuals from tracking the Cassini spacecraft as it passed the Sun in 2002 (Bertotti et al. 2003).

12.7 Imaging Astronomical Masers

In the envelopes of many newly formed stars, and also those of highly evolved stars and the accretion disks of AGN, radio emission from molecules such as H_2O and OH is caused by a maser process. The frequency spectrum of the emission is often complicated, containing many spectral features or components caused by clouds of gas moving at different line-of-sight velocities. Maps of strong maser sources reveal hundreds of compact components with brightness temperatures approaching 10^{15} K, angular sizes as small as 10^{-4} arcsec, and flux densities as high as 10^6 Jy. The components are typically distributed over an area of several arcseconds in diameter and a Doppler velocity range of 10–3000 km s^{-1} (0.7–200 MHz for the H_2O maser transition at 22 GHz). Individual features have line widths of about 1 km s^{-1} or less (74 kHz at 22 GHz). The physics and phenomenology of masers are discussed by Reid and Moran (1988); Elitzur (1992); and Gray (2012). The processing and analysis of maser data require large correlator systems because the ratio of required bandwidth to spectral resolution is large (10^2–10^4). They also require prodigious amounts of image processing because the ratio of the field of view to the spatial resolution is large (10^2–10^4). As an extreme example, the H_2O maser in W49 has hundreds of features distributed over 3 arcsec (Gwinn et al. 1992). The complete mapping of this source at a resolution of 10^{-3} arcsec with 3 pixels per resolution interval would require the production of 600 maps, each with at least 10^8 pixels. However, most of the map cells would contain no emission. Thus, the usual procedure is to measure the positions of the features crudely by fringe-frequency analysis and then to map small fields around these locations by Fourier synthesis techniques. Examples of maps made by fringe-frequency analysis can be found in Walker et al. (1982); by phase analysis in Genzel et al. (1981) and Norris and Booth (1981); and by Fourier synthesis in Reid et al. (1980), Norris et al. (1982), and Boboltz et al. (1997). We shall briefly discuss some of the techniques used in mapping masers and their accuracies. Note that geometric (group) delays cannot be measured accurately because of the narrow bandwidths of the maser lines.

In mapping masers, we must explicitly consider the frequency dependence of the fringe visibility. We assume that a maser source consists of a number of point sources. Furthermore, we assume that the measurements are made with a VLBI system and that the desired RF band is converted to a single baseband channel. Adapting Eq. (9.28), we can write the residual fringe phase of one maser component at frequency ν as

$$\Delta\phi(\nu) = 2\pi \left[\nu\Delta\tau_g(\nu) + (\nu - \nu_{LO})\tau_e + \nu\tau_{at} \right] + \phi_{in} + 2\pi n , \qquad (12.54)$$

where τ_e is the relative delay error due to clock offsets; τ_{at} is the differential atmospheric delay; $\Delta\tau_g(\nu)$ is the difference between the true geometric delay of the source $\tau_g(\nu)$ and the expected (reference) delay; ν_{LO} is the local oscillator frequency; ϕ_{in} is the instrumental phase, which includes the local oscillator frequency difference and can be a rapidly varying function of time; and $2\pi n$ represents the phase ambiguity. A frequency can usually be found that has only one unresolved maser component, and this component can then be used as a phase reference. The use of a phase reference feature is fundamental to all maser analysis procedures, and it allows maps of the relative positions of maser components to be made with high accuracy. The difference in residual fringe phase between a maser feature at frequency ν and the reference feature at frequency ν_R is

$$\Delta^2\phi(\nu) = \Delta\phi(\nu) - \Delta\phi(\nu_R) , \qquad (12.55)$$

which, with the use of Eq. (12.54), becomes

$$\Delta^2\phi(\nu) = 2\pi \left\{ \nu \left[\tau_g(\nu) - \tau_g(\nu_R) \right] \right.$$
$$\left. + (\nu - \nu_R) \left[\tau_g(\nu_R) - \tau_g'(\nu_R) \right] + (\nu - \nu_R) \left[\tau_e + \tau_{at} \right] \right\} , \qquad (12.56)$$

where $\tau_g'(\nu_R)$ is the expected delay of the reference feature, and $\tau_g(\nu_R)$ is the true delay. The frequency-independent terms ϕ_{in} and $2\pi n$ cancel in Eq. (12.56). However, there are residual terms in Eq. (12.56) that are proportional to the difference in frequency between the feature of interest and the reference feature. These terms arise because phases at different frequencies are differenced in Eq. (12.55). Following the notation of Eq. (12.7), which uses the convention Δ term = (assumed value) – (true value), we can write Eq. (12.56) as

$$\Delta^2\phi(\nu) - \frac{2\pi\nu}{c}\mathbf{D}\cdot\Delta\mathbf{s}_{\nu R} - \frac{2\pi\nu}{c}\Delta\mathbf{D}\cdot\Delta\mathbf{s}_{\nu R}$$

$$- \frac{2\pi}{c}[(\nu-\nu_R)(\Delta\mathbf{D}\cdot\mathbf{s}_R + \mathbf{D}\cdot\Delta\mathbf{s}_R)] + 2\pi(\nu-\nu_R)(\tau_e + \tau_{at}) ,$$

$$(12.57)$$

where $\mathbf{D}$ is the assumed baseline, $\Delta\mathbf{D}$ is the baseline error, $\mathbf{s}_R$ is the assumed position of the reference feature, and $\Delta\mathbf{s}_R$ is the corresponding position error. $\Delta\mathbf{s}_{\nu R}$ is the separation vector from the feature at frequency ν to the reference feature, and thus the true position of the feature at frequency ν is $\mathbf{s}_R - \Delta\mathbf{s}_R + \Delta\mathbf{s}_{\nu R}$.

The first term on the right side of Eq. (12.57) is the desired quantity from which the position of the feature relative to the reference feature can be determined, and the remaining terms describe the phase errors introduced by uncertainty in baseline, source position, clock offset, and atmospheric delay. These phase error terms can be converted approximately to angular errors by dividing them by $c/2\pi\nu D$. Thus, for example, an error of 0.3 m in a baseline component would cause a delay error of about 1 ns in the term $\Delta\mathbf{D}\cdot\mathbf{s}_R$ in Eq. (12.57) and a phase error of 10^{-3} turns for features separated by 1 MHz. This phase error corresponds to a nominal error of 10^{-6} arcsec on a baseline of 2500 km at 22 GHz, which provides a fringe spacing of 10^{-3} arcsec. Similarly, a clock or atmospheric error of 1 ns would cause the same positional error. The same baseline error also causes additional positional errors, through the $\Delta\mathbf{D}\cdot\Delta\mathbf{s}_{\nu R}$ term, of 10^{-7} arcsec per arcsecond separation of the features. A detailed discussion of mapping errors caused by this calibration method can be found in Genzel et al. (1981).

Another method of calibrating the fringe phase is to scale the phase of the reference feature to the frequency of the feature to be calibrated. That is,

$$\Delta^2\phi(\nu) = \Delta\phi(\nu) - \Delta\phi(\nu_R)\frac{\nu}{\nu_R} .$$

$$(12.58)$$

This method of calibration is more accurate than the method of Eq. (12.55) because error terms proportional to $\nu-\nu_R$ do not appear. However, there are additional terms involving the phase ambiguity and the instrumental phase. Thus, this calibration method is applicable only if the fringe phase can be followed carefully enough to avoid the introduction of phase ambiguities.

Maps of lower accuracy and sensitivity than those obtainable from phase data can be made with fringe-frequency data. Suppose that the interferometer is well calibrated. The differential fringe frequency, that is, the difference in fringe frequency between the feature at frequency ν and the reference feature, can then be written [using Eq. (12.20)]

$$\Delta^2\nu_f(\nu) \simeq \dot{u}\Delta\alpha'(\nu) + \dot{v}\Delta\delta(\nu) ,$$

$$(12.59)$$

where $\dot{u}$ and $\dot{v}$ are the time derivatives of the projected baseline components, $\Delta\alpha'(\nu)$ and $\Delta\delta(\nu)$ are the coordinate offsets from the reference feature, and $\Delta\alpha'(\nu) = \Delta\alpha(\nu)\cos\delta$. The relative positions of the maser feature can then be found by fitting Eq. (12.59) to a series of fringe-frequency measurements at various hour angles. This technique was first employed by Moran et al. (1968) for the mapping of an OH maser. The errors in fringe-frequency measurements decrease as $\tau^{3/2}$ [see Eq. (A12.27)], where τ is the length of an observation, but for large values of τ, the differential fringe frequency $\Delta^2 \nu_f$ is not constant, because $\ddot{u}$ and $\ddot{v}$ are not zero. Thus, there is a limited field of view available for accurate mapping with fringe-frequency measurements. This field of view can be estimated by equating the rms fringe-frequency error in Eq. (A12.27) with τ times the derivative of the differential fringe frequency with respect to time. Therefore, for an east–west baseline,

$$D_\lambda \omega_e^2 \Delta\theta\tau \, \cos\theta \simeq \sqrt{\frac{3}{2\pi^2}}\left(\frac{T_S}{T_A}\right)\frac{1}{\sqrt{\Delta\nu\tau^3}}\,, \qquad (12.60)$$

where $\Delta\theta$ is the field of view. For $\sqrt{2\pi^2/3}\cos\theta \simeq 1$, the field of view is

$$\Delta\theta \simeq \frac{T_S}{D_\lambda T_A \omega_e^2 \tau^2 \sqrt{\Delta\nu\tau}}\,, \qquad (12.61)$$

or

$$\Delta\theta \simeq \frac{1}{\mathcal{R}_{\mathrm{sn}}D_\lambda \omega_e^2 \tau^2}\,, \qquad (12.62)$$

where $\mathcal{R}_{\mathrm{sn}}$ is the signal-to-noise ratio. Let $\mathcal{R}_{\mathrm{sn}} = 10$ and $\tau = 100$ s. The field of view is then about equal to 2000 times the fringe spacing. This restriction is often important. Usually when a feature is found, the phase center of the field is moved to the estimated position of the feature, and the position is then redetermined. Only components that are detected in individual observations on each baseline can be mapped with the fringe-frequency mapping technique. Thus, fringe-frequency mapping is less sensitive than synthesis mapping, in which fully coherent sensitivity is achieved.

The fringe-frequency analysis procedure can be extended to handle the case in which there are many point components in one frequency channel. From each observation (i.e., a measurement on one baseline lasting for a few minutes), the fringe-frequency spectrum is calculated. Multiple components will appear as distinct fringe-frequency features, as shown in Fig. 12.10. The fringe frequency of each feature defines a line in $(\Delta\alpha', \Delta\delta)$ space on which a maser component lies. The slope of the line is $\tan^{-1}(\dot{v}/\dot{u})$. As the projected baseline changes, the slopes of the lines change. The intersections of the lines define the source positions (see Fig. 12.10). For this method to work, the components must be sufficiently separate to produce separate peaks in the fringe-frequency spectrum. The fringe-frequency

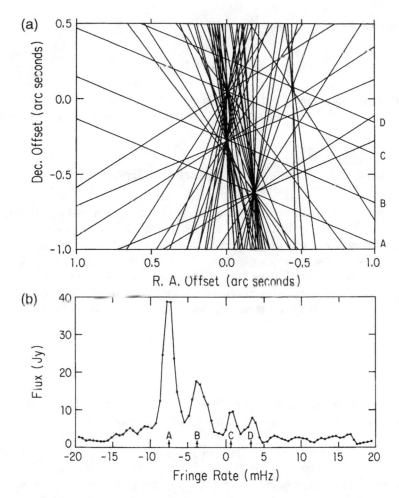

Fig. 12.10 Plot (**b**) is the fringe-frequency spectrum of the water vapor maser in W49N, at one particular hour angle and one frequency in the radio spectrum of the maser. The ordinate is flux density. There are four peaks, each corresponding to a separate feature on the sky. Plot (**a**) shows such lines from many scans. The peaks in the lower plot and their corresponding lines in the upper plot are labeled A–D. There are at least four separate features at the frequency of these data. Their positions are marked by the locations where many lines intersect. The feature corresponding to line D is sufficiently far from the phase center that its fringe frequency changes enough during the 20-min integration to degrade significantly the estimate of the feature position. The window in which accurate positions can be determined is 0.5″ in right ascension and 2″ in declination. The window can be moved by shifting the phase center of the data. From Walker (1981). © AAS. Reproduced with permission.

resolution is about τ^{-1}, which defines an effective beam of width

$$\Delta\theta_f = \frac{1}{D_\lambda \omega_e \tau \cos\theta} \, . \tag{12.63}$$

Fringe-frequency mapping is discussed in detail, for example, by Walker (1981). It remains a useful technique for arrays that involve instruments such as RadioAstron.

Appendix 12.1   Least-Mean-Squares Analysis

The principles of least-mean-squares analysis play a fundamental role in astrometry, where the goal is to extract a number of parameters from a set of noisy measurements. We briefly discuss these principles in an elementary way, ignoring mathematical subtleties, and apply them to the problems encountered in interferometry. Detailed discussions of the statistical analysis of data can be found in books such as Bevington and Robinson (1992) and Hamilton (1964). The exhaustive treatment of how to fit a straight line, by Hogg et al. (2010), is highly recommended.

A12.1.1   Linear Case

Suppose we wish to measure a quantity m. We make a set of measurements y_i that are the sum of the desired quantity m and a noise contribution n_i:

$$y_i = m + n_i \, , \tag{A12.1}$$

where n_i is a Gaussian random variable with zero mean and variance σ_i^2. The probability that the ith measurement will take any specific value of y_i is given by the probability (density) function

$$p\,(y_i) = \frac{1}{\sqrt{2\pi}\sigma_i} \, e^{-(y_i - m)^2 / 2\sigma_i^2} \, . \tag{A12.2}$$

If all the measurements are independent, then the probability that an experiment will yield a set of N measurements $y_1, y_2, \ldots, y_N$ is

$$L = \prod_{i=1}^{N} p\,(y_i) \, , \tag{A12.3}$$

where the $\prod$ denotes the product of the $p\,(y_i)$ terms. L, viewed as a function of m, is called the likelihood function. The method of maximum likelihood is based on the assumption that the best estimate of m is the one that maximizes L. Maximizing L is the same as maximizing $\ln L$, where

$$\ln L = \sum_{i=1}^{N} \ln \frac{1}{\sqrt{2\pi}\sigma_i} - \frac{1}{2} \sum_{i=1}^{N} \frac{(y_i - m)^2}{\sigma_i^2} . \tag{A12.4}$$

Since the first summation term on the right side of Eq. (A12.4) is a constant and the second summation term is multiplied by $-\frac{1}{2}$, the maximization of L is equivalent to the minimization of the second summation term in Eq. (A12.4) with respect to m. Thus, we wish to minimize the quantity χ^2 given by

$$\chi^2 = \sum_{i=1}^{N} \frac{(y_i - m)^2}{\sigma_i^2} . \tag{A12.5}$$

In the more general problem discussed later in this appendix, m is replaced by a function with one or more parameters describing the system model. With this generalization, Eq. (A12.5) becomes the fundamental equation of the method of weighted least-mean-squares. In this method, the parameters of the model are determined by minimizing the sum of the squared differences between the measurements and the model, weighted by the variances of the measurements. The quantity χ^2, which indicates the goodness of fit, is a random variable whose mean value equals the number of data points less the number of parameters when the model adequately describes the measurements. The method of least-mean-squares, appropriate when the noise is a Gaussian random process, is a special case of the more general method of maximum likelihood. Gauss invented the method of least-mean-squares, perhaps as early as 1795, using arguments similar to those given here, for the purpose of estimating the orbital parameters of planets and comets (Gauss 1809). The method was independently developed by Legendre in 1806 (Hall 1970).

Returning to Eq. (A12.5), we can estimate m by setting the derivative of χ^2 with respect to m equal to zero. The resulting estimate of m, denoted by m_e, is

$$m_e = \frac{\sum \dfrac{y_i}{\sigma_i^2}}{\sum \dfrac{1}{\sigma_i^2}} , \tag{A12.6}$$

where the summation goes from $i = 1$ to N. Using Eq. (A12.2), we note that $\langle y_i \rangle = m$ and $\langle y_i^2 \rangle = m^2 + \sigma_i^2$. Therefore, by calculating the expectation of Eq. (A12.6), it is clear that $\langle m_e \rangle = \langle y_i \rangle = m$, and it is easy to show that

$$\langle m_e^2 \rangle = m^2 + \left(\sum \frac{1}{\sigma_i^2} \right)^{-1} . \tag{A12.7}$$

Hence the variance of the estimate of m_e is

$$\sigma_m^2 = \langle m_e^2 \rangle - \langle m_e \rangle^2 = \left(\sum \frac{1}{\sigma_i^2} \right)^{-1} . \qquad \text{(A12.8)}$$

Equation (A12.8) shows that when poor quality or noisy data are added to better data, the value of σ_m may be reduced only slightly. If the statistical error σ_i of each of the measurements has the same value, σ, then Eq. (A12.8) reduces to the well-known result

$$\sigma_m = \frac{\sigma}{\sqrt{N}} , \qquad \text{(A12.9)}$$

and m_e is the average of the measurements. In many instances, σ is not known. An estimate of σ is

$$\sigma_e^2 = \frac{1}{N} \sum (y_i - m)^2 . \qquad \text{(A12.10)}$$

However, m is not known, only its estimate, m_e. If m_e were used in place of m in Eq. (A12.10), the value of σ_e^2 would be an underestimate of σ^2 because of the manner in which m_e was determined in minimizing χ^2. The unbiased estimate of σ^2 is

$$\sigma_e^2 = \frac{1}{N-1} \sum (y_i - m_e)^2 . \qquad \text{(A12.11)}$$

It is easy to show by substitution of Eq. (A12.6) into Eq. (A12.11) that $\langle \sigma_e^2 \rangle = \sigma^2$. The term $N - 1$, which is called the number of degrees of freedom, appears in Eq. (A12.11) because there are N data points and one free parameter.

Consider a model described by the function $f(x; p_1, \ldots, p_n)$, where x is the independent variable, which takes values x_i, where $i = 1$ to N, at the sample points, and $p_1, \ldots, p_n$ are a set of parameters. We assume that the values of the independent variable are exactly known. If the function f correctly models the measurement system, the measurement set is given by

$$y_i = f(x_i; p_1, \ldots, p_n) + n_i , \qquad \text{(A12.12)}$$

where n_i represents the measurement error. The general problem is to find the values of the parameters for which χ^2, given by the generalization of Eq. (A12.5),

$$\chi^2 = \sum \frac{[y_i - f(x_i)]^2}{\sigma_i^2} , \qquad \text{(A12.13)}$$

is a minimum.

A simple example of this problem is the fitting of a straight line to a data set. Let

$$f(x;a,b) = a + bx,$$

(A12.14)

where a and b are the parameters to be found. Minimizing χ^2 is accomplished by solving the equations

$$\frac{\partial \chi^2}{\partial a} = -\sum \frac{2(y_i - a - bx_i)}{\sigma_i^2} = 0,$$

(A12.15a)

and

$$\frac{\partial \chi^2}{\partial b} = -\sum \frac{2(y_i - a - bx_i)x_i}{\sigma_i^2} = 0.$$

(A12.15b)

In matrix notation, we have

$$
\begin{bmatrix} \sum \frac{y_i}{\sigma_i^2} \\ \sum \frac{x_i y_i}{\sigma_i^2} \end{bmatrix}
=
\begin{bmatrix} \sum \frac{1}{\sigma_i^2} & \sum \frac{x_i}{\sigma_i^2} \\ \sum \frac{x_i}{\sigma_i^2} & \sum \frac{x_i^2}{\sigma_i^2} \end{bmatrix}
\begin{bmatrix} a_e \\ b_e \end{bmatrix},
$$

(A12.16)

where we distinguish between the true values of the parameters and their estimates by the subscript e. The solution is

$$a_e = \frac{1}{\Delta}\left[\left(\sum \frac{x_i^2}{\sigma_i^2}\right)\left(\sum \frac{y_i}{\sigma_i^2}\right) - \left(\sum \frac{x_i}{\sigma_i^2}\right)\left(\sum \frac{x_i y_i}{\sigma_i^2}\right)\right]$$

(A12.17)

and

$$b_e = \frac{1}{\Delta}\left[\left(\sum \frac{1}{\sigma_i^2}\right)\left(\sum \frac{x_i y_i}{\sigma_i^2}\right) - \left(\sum \frac{x_i}{\sigma_i^2}\right)\left(\sum \frac{y_i}{\sigma_i^2}\right)\right],$$

(A12.18)

where Δ is the determinant of the square matrix in Eq. (A12.16), given by

$$\Delta = \left(\sum \frac{1}{\sigma_i^2}\right)\left(\sum \frac{x_i^2}{\sigma_i^2}\right) - \left(\sum \frac{x_i}{\sigma_i^2}\right)^2.$$

(A12.19)

Estimates of the errors in the parameters a_e and b_e can be calculated from Eqs. (A12.17) and (A12.18) and are given by

$$\sigma_a^2 = \langle a_e^2 \rangle - \langle a_e \rangle^2 = \frac{1}{\Delta}\sum \frac{x_i^2}{\sigma_i^2}$$

(A12.20)

and

$$\sigma_b^2 = \langle b_e^2 \rangle - \langle b_e \rangle^2 = \frac{1}{\Delta} \sum \frac{1}{\sigma_i^2} . \tag{A12.21}$$

Note that a_e and b_e are random variables, and in general $\langle a_e b_e \rangle$ is not zero, so the parameter estimates are correlated. The error estimates in Eqs. (A12.20) and (A12.21) include the deleterious effects of the correlation between parameters. In this particular example, the correlation can be made equal to zero by adjusting the origin of the x axis so that $\sum (x_i/\sigma_i^2) = 0$.

The above analysis can be used to estimate the accuracy of measurements of fringe frequency and delay made with an interferometer. Fringe frequency, the rate of change of fringe phase with time,

$$\nu_f = \frac{1}{2\pi} \frac{\partial \phi}{\partial t} , \tag{A12.22}$$

can be estimated by fitting a straight line to a sequence of uniformly spaced measurements of phase with respect to time. The fringe frequency is proportional to the slope of this line. Assume that N measurements of phase ϕ_i, each having the same rms error σ_ϕ, are made at times t_i, spaced by interval T, running from time $-NT/2$ to $NT/2$, such that the total time of the observation is $\tau = NT$. From Eq. (A12.21) and the above definitions, including Eq. (A12.22), the error in the fringe-frequency estimate is

$$\sigma_f^2 = \frac{\sigma_\phi^2}{(2\pi)^2 \sum t_i^2} , \tag{A12.23}$$

since $\sum t_i = 0$. The term $\sum t_i^2$ is approximately given by

$$\sum t_i^2 \simeq \frac{1}{T} \int_{-\tau/2}^{\tau/2} t^2 dt = \frac{1}{T} \frac{\tau^3}{12} = \frac{N\tau^2}{12} . \tag{A12.24}$$

$\tau/\sqrt{12}$ can be thought of as the rms time span of the data. Thus, Eq. (A12.23) becomes

$$\sigma_f^2 = \frac{12\sigma_\phi^2}{(2\pi)^2 N \tau^2} . \tag{A12.25}$$

The expression for σ_ϕ, given in Eq. (6.64) for the case when the source is unresolved and there are no processing losses, is

$$\sigma_\phi = \frac{T_S}{T_A \sqrt{2\Delta\nu T}} , \tag{A12.26}$$

where T_S is the system temperature, T_A is the antenna temperature due to the source, and Δv is the bandwidth. Substitution of Eq. (A12.26) into Eq. (A12.25) yields

$$\sigma_f = \sqrt{\frac{3}{2\pi^2}} \left(\frac{T_S}{T_A}\right) \frac{1}{\sqrt{\Delta v \tau^3}} \text{ (Hz) .} \qquad (A12.27)$$

Note that this result does not depend on the details of the analysis procedure, such as the choice of N. Equivalently, one can estimate the fringe frequency by finding the peak of the fringe-frequency spectrum, that is, the peak of the Fourier transform of $e^{j\phi_i}$.

The delay is the rate of change of phase with frequency,

$$\tau = \frac{1}{2\pi} \frac{\partial \phi}{\partial v} . \qquad (A12.28)$$

Thus, the delay can be estimated by finding the slope of a straight line fitted to a sequence of phase measurements as a function of frequency. For a single band, such data can be obtained from the cross power spectrum, the Fourier transform of the cross-correlation function. Assume that N measurements of phase are made at frequencies v_i, each with a bandwidth $\Delta v/N$ and with an error σ_ϕ. In this calculation, only the relative frequencies are important. It is convenient for the purpose of analysis to set the zero of the frequency axis such that $\sum v_i = 0$. The error in delay [from Eqs. (A12.19), (A12.21), and (A12.28)] is

$$\sigma_\tau^2 = \frac{\sigma_\phi^2}{(2\pi)^2 \sum v_i^2} . \qquad (A12.29)$$

Using a calculation for $\sum v_i^2$ analogous to the one in Eq. (A12.24), we can write Eq. (A12.29) as

$$\sigma_\tau^2 = \frac{12\sigma_\phi^2}{(2\pi)^2 N \Delta v^2} . \qquad (A12.30)$$

Thus, substitution of Eq. (A12.26) (with an integration time of τ and bandwidth $\Delta v/N$) into Eq. (A12.30) yields

$$\sigma_\tau = \sqrt{\frac{3}{2\pi^2}} \left(\frac{T_S}{T_A}\right) \frac{1}{\sqrt{\Delta v^3 \tau}} . \qquad (A12.31)$$

We can define the rms bandwidth as

$$\Delta v_{\text{rms}} = \sqrt{\frac{1}{N} \sum v_i^2} \qquad (A12.32)$$

and obtain from Eqs. (A12.26) and (A12.29) the result quoted in Sect. 9.8 [Eq. (9.179)],

$$\sigma_\tau = \frac{1}{\zeta} \left(\frac{T_S}{T_A} \right) \frac{1}{\sqrt{\Delta \nu_{rms}^3 \, \tau}} \, , \tag{A12.33}$$

where $\zeta = \pi (768)^{1/4}$. (Note that in Sect. 9.8, σ_ϕ applies to the full bandwidth $\Delta \nu$.) The expressions for σ_τ in Eqs. (A12.30), (A12.31), and (A12.33) incorporate the condition $\Delta \nu_{rms} = \Delta \nu / \sqrt{12}$ and apply to a continuous passband of width $\Delta \nu$.

In bandwidth synthesis, which is described in Sect. 9.8, the measurement system consists of N channels of width $\Delta \nu / N$, which are not in general contiguous. The rms delay error is obtained by substituting Eqs. (A12.26) and (A12.32) into Eq. (A12.29), yielding

$$\sigma_\tau = \frac{1}{\sqrt{8\pi^2}} \left(\frac{T_S}{T_A} \right) \frac{1}{\sqrt{\Delta \nu \tau} \, \Delta \nu_{rms}} \, , \tag{A12.34}$$

where $\Delta \nu_{rms}$ is given by Eq. (A12.32) and $\Delta \nu$ is the total bandwidth. $\Delta \nu_{rms}$ is generally equal to about 40% of the total frequency range spanned.

A general formulation of the linear least-mean-squares solution can be found when the model function f is a linear function of the parameters p_k, that is, when

$$f(x; p_1, \ldots, p_n) = \sum_{k=1}^{n} \frac{\partial f}{\partial p_k} p_k \, , \tag{A12.35}$$

where n is the number of parameters. For example, the model could be a cubic polynomial

$$f(x; p_0, p_1, p_2, p_3) = p_0 + p_1 x + p_2 x^2 + p_3 x^3 \, , \tag{A12.36}$$

in which case $\partial f / \partial p_k = x^k$ for $k = 0, 1, 2,$ and 3. If the parameters appear as linear multiplicative factors, then the minimization of Eq. (A12.13) leads to a set of n equations of the form

$$\frac{\partial \chi^2}{\partial p_k} = 0 \, , \qquad k = 1, 2, \ldots, n \, . \tag{A12.37}$$

Substitution of Eq. (A12.13) into Eq. (A12.37) and use of Eq. (A12.35) yield the set of n equations

$$D_k = \sum_{j=1}^{n} T_{jk} p_j, \qquad k = 1, 2, \ldots, n \, , \tag{A12.38}$$

where

$$D_k = \sum_{i=1}^{N} \frac{y_i}{\sigma_i^2} \frac{\partial f(x_i)}{\partial p_k} \qquad (A12.39)$$

and

$$T_{jk} = \sum_{i=1}^{N} \frac{1}{\sigma_i^2} \frac{\partial f(x_i)}{\partial p_j} \frac{\partial f(x_i)}{\partial p_k} , \qquad (A12.40)$$

and the summations are carried out over the set of N independent measurements. In matrix notation, the equation set (A12.38) is

$$[D] = [T][P_e] , \qquad (A12.41)$$

where $[D]$ is a column matrix with elements D_k, $[P_e]$ is a column matrix containing the estimates of the parameters p_{ek}, and $[T]$ is a symmetric square matrix with elements T_{jk}. For obvious reasons, $[T]$ is sometimes called the matrix of the normal equations. Note that Eq. (A12.41) is a generalization of Eq. (A12.16). The matrices $[T]$ and $[D]$ are sometimes written as the product of other matrices (Hamilton 1964, Ch. 4). Let $[M]$ be the variance matrix (size $N \times N$) whose diagonal elements are σ_i^2 and whose off-diagonal elements are zero; let $[F]$ be a column matrix containing the data y_i; and let $[A]$ be the partial derivative matrix (size $n \times N$) whose elements are $\partial f(x_i)/\partial p_k$. Then one can write $[T] = [A]^T[M]^{-1}[A]$ and $[D] = [A]^T[M]^{-1}[F]$, where $[A]^T$ is the transpose of $[A]$ and $[M]^{-1}$ is the inverse of $[M]$. The analysis can be generalized to include the situation in which the errors between measurements are correlated. In this case, $[M]$ is modified to include off-diagonal elements $\sigma_i\sigma_j\rho'_{ij}$, where ρ'_{ij} is the correlation coefficient for the ith and jth measurements.

The solution to Eq. (A12.41) is

$$[P_e] = [T]^{-1}[D] , \qquad (A12.42)$$

where $[T]^{-1}$ is the inverse matrix of $[T]$, and $[P_e]$ is the column matrix containing the parameter estimates. The elements of $[T]^{-1}$ are denoted T'_{jk}. It can be shown by direct calculation that the estimates of the errors of the parameters σ_{ek}^2 are the diagonal elements of $[T]^{-1}$, which is called the covariance matrix. Thus,

$$\sigma_{ek}^2 = T'_{kk} . \qquad (A12.43)$$

The probability that parameter p_k will be within $\pm\sigma_k$ of its true value is 0.68, which is the integral under the one-dimensional Gaussian probability distribution between $\pm\sigma_k$. The probability that all of the n parameters will be within $\pm\sigma$ of their true values (i.e., within the error "box" in the n-dimensional space) is approximately 0.68^n when the correlations are moderate.

Fig. A12.1 The error ellipse, or contour, defining the e^{-1} level of the joint probability function [Eq. (A12.45)] for the estimates of parameters p_k and p_j. The quantities $p_{ek} - p_k$ and $p_{ej} - p_j$ are the parameter estimates minus their true values. The angle ψ_{jk} is defined by Eq. (A12.46).

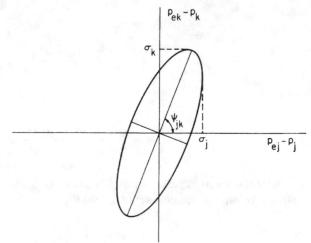

The normalized correlation coefficients between parameters are proportional to the off-diagonal elements of $[T]^{-1}$:

$$\rho_{jk} = \frac{\langle (p_{ej} - p_j)(p_{ek} - p_k) \rangle}{\sigma_{ek}\,\sigma_{ej}} = \frac{T'_{jk}}{\sqrt{T'_{jj}T'_{kk}}}\,. \tag{A12.44}$$

For any two parameters, there is a bivariate Gaussian probability distribution that describes the distribution of errors

$$p\left(\epsilon_j, \epsilon_k\right) = \frac{1}{2\pi\sigma_j\sigma_k\sqrt{1-\rho_{jk}^2}}\exp\left\{-\frac{1}{2(1-\rho_{jk}^2)}\left[\frac{\epsilon_j^2}{\sigma_j^2} + \frac{\epsilon_k^2}{\sigma_k^2} - \frac{2\rho_{jk}\epsilon_j\epsilon_k}{\sigma_j\sigma_k}\right]\right\}, \tag{A12.45}$$

where $\epsilon_k = p_{ek} - p_k$ and $\epsilon_j = p_{ej} - p_j$. The contour of $p\left(\epsilon_k, \epsilon_j\right) = p(0,0)e^{-1/2}$ defines an ellipse, shown in Fig. A12.1, which is known as the error ellipse. The probability that both parameters will lie within the error ellipse is the integral of Eq. (A12.45) over the area of the error ellipse, which equals 0.46. The orientation of the error ellipse is given by

$$\psi_{jk} = \frac{1}{2}\tan^{-1}\left(\frac{2\rho_{jk}\sigma_j\sigma_k}{\sigma_j^2 - \sigma_k^2}\right). \tag{A12.46}$$

The errors in the parameters p_k are completely determined by the matrix $[T]^{-1}$ through Eqs. (A12.43)–(A12.45). The elements of $[T]^{-1}$ depend only on the partial derivatives of the model function and the values of the measurement errors, which can usually be predicted in advance from the characteristics of the measurement apparatus. Therefore, once an experiment is planned, the errors in the parameters can be predicted from $[T]^{-1}$ without reference to the data. For this reason, $[T]$ is sometimes called the design matrix. Studies of the design matrix for a specific experiment might reveal a very high correlation between two parameters, leading to large errors in their estimated values. It is often possible to modify the experiment to obtain more data that will reduce the correlation. After the data are analyzed, the value of χ^2 can be computed. If the model is a good fit to the data, χ^2 should be approximately equal to $N - n$, the number of measurements minus the number of parameters. If it is not, the difficulty is often that the values of σ_i are estimated incorrectly or that the model does not describe adequately the measurement system, that is, the model has too few parameters or is not correct. Even if $\chi^2 \simeq N - n$, the derived errors in Eq. (A12.43) may not be realistic, and they are referred to as "formal errors." The formal errors describe the *precision* of the parameter estimates. The *accuracy* of the parameter measurements is the deviation between the estimates of the parameters and the true values of the parameters. The accuracy of the measurements is often difficult to determine. For example, an unknown effect that closely mimics the functional dependence of one of the model parameters may be present in an experiment. The model may appear to be a good one, but the accuracy of the particular model parameter in question will be much poorer than expected because of the systematic error introduced by the unmodeled effect.

We can envision how the principles of least-mean-squares analysis are applied to a large astrometric experiment. Consider a hypothetical VLBI experiment made on a three-station array. Suppose that ten recordings are made of each of 20 sources during observations made over one day (an epoch). The observations are repeated six times a year for five years. The data set would consist of 18,000 measurements (20 sources × 10 observations × 3 baselines × 30 epochs) of delay and fringe frequency, or 36,000 total measurements. The measurements of delay and fringe frequency can be combined in the analysis since, in the least-mean-squares method, the relevant quantities are the squares of the measurements divided by their variances, which are dimensionless, as in Eq. (A12.13). Now we can count the number of parameters in the analysis model: 39 source coordinates (1 right ascension fixed), 9 station coordinates, 90 atmospheric parameters (a zenith excess path length at each station at each epoch), 120 clock parameters (a clock error and clock rate error at two of the stations per epoch), and 90 polar motion and UT1–UTC parameters, as well as several other parameters to model precession, nutation, solid-Earth tides, gravitational deflection by the Sun, movement of stations, and other effects such as antenna axis offsets (see Sect. 4.6.1). The total number of parameters is about 360. The parameters within each observation epoch are linked because of the common clock and atmosphere parameters. Parameters among epochs are linked because of baseline, precession, and nutation parameters. Naturally, partial solutions from subsets of the data should be obtained before a grand global solution is attempted. Procedures are available for obtaining global solutions that do not require the inversion of matrices as large as the total number of parameters [see, e.g., Morrison (1969)]. Experiments of the scale described here, and larger ones, have been carried out [e.g., Fanselow et al. (1984), Herring et al. (1985), and Ma et al. (1998)].

A12.1.2   Nonlinear Case

The discussion of linear least-mean-squares analysis can be generalized to include nonlinear functions in a straightforward manner. Assume that $f(x;p)$ has one nonlinear parameter p_n. For the purpose of discussion, we can separate f into linear and nonlinear parts, $f_L(x;p_1,\ldots,p_{n-1})$ and $f_{NL}(x;p_n)$, and approximate the nonlinear function by the first two terms in a Taylor expansion

$$f_{NL}(x;p_n) \simeq f_{NL}(x,p_{0n}) + \frac{\partial f_{NL}}{\partial p_n} \Delta p_n \,, \tag{A12.47}$$

where p_{0n} is the initial guess of parameter p_n and $\Delta p_n = p_n - p_{0n}$. We assume that the initial parameter guesses are accurate enough for Eq. (A12.47) to be valid. We replace the data with $y_i - f_{NL}(x_i;p_{0n})$ and then compute the elements of the matrices $[D]$ and $[T]$ from the partial derivatives, including $\partial f_{NL}/\partial p_n$. The nth parameter in the matrix $[P_e]$ in Eq. (A12.42) will be the differential parameter Δp_n defined in Eq. (A12.47). The solution must be iterated with a new Taylor expansion centered on the parameter $p_{0n} + \Delta p_n$. Thus, nonlinear functions can be accommodated in the analysis through linearization, but initial guesses of the nonlinear parameters and solution iteration are required. In some cases, nonlinear estimation problems can cause difficulties [see, e.g., Lampton et al. (1976), Press et al. (1992)]. Recently, the use of the Markov chain Monte Carlo (MCMC) method has become almost universal (Sivia and Skilling 2006).

A12.1.3   (u, v) vs. Image Plane Fitting

One final topic concerns the estimation of the coordinates of a radio source with a well-calibrated interferometer, which has accurately known baselines and instrumental phases. In this case, the differential interferometer phase is, from Eq. (12.2),

$$\Delta\phi = 2\pi D_\lambda \left\{ [\sin d \cos \delta - \cos d \sin \delta \cos(H - h)] \,\Delta\delta \right.$$
$$\left. + \cos d \cos \delta \sin(H - h)\Delta\alpha \right\} \,. \tag{A12.48}$$

Expressing the geometric quantities in terms of projected baseline components, we can write Eq. (A12.48) as

$$\Delta\phi = 2\pi \left(u\Delta\alpha' + v\Delta\delta \right) , \tag{A12.49}$$

where $\Delta\alpha' = \Delta\alpha \cos \delta$. A set of phase measurements from one or more baselines can be analyzed by the method of least-mean-squares to determine $\Delta\alpha'$ and $\Delta\delta$. The partial derivatives are $\partial f/\partial p_1 = 2\pi u$ and $\partial f/\partial p_2 = 2\pi v$, where $p_1 = \Delta\alpha'$ and

$p_2 = \Delta\delta$. From Eqs. (A12.40) and (A12.49), the normal-equation matrix is

$$[T] = \frac{4\pi^2}{\sigma_\phi^2} \begin{bmatrix} \sum u_i^2 & \sum u_i v_i \\ \sum u_i v_i & \sum v_i^2 \end{bmatrix}, \tag{A12.50}$$

where all the measurements are assumed to have the same uncertainty σ_ϕ given by Eq. (A12.26). The inverse of $[T]$ is

$$[T]^{-1} = \frac{1}{\Delta} \begin{bmatrix} \sum v_i^2 & -\sum u_i v_i \\ -\sum u_i v_i & \sum u_i^2 \end{bmatrix}, \tag{A12.51}$$

where Δ is the determinant of the matrix in Eq. (A12.50),

$$\Delta = \frac{4\pi^2}{\sigma_\phi^2} \left[\sum u_i^2 \sum v_i^2 - \left(\sum u_i v_i \right)^2 \right]. \tag{A12.52}$$

The correlation coefficient defined by Eq. (A12.44) is

$$\rho_{12} = \frac{-\sum u_i v_i}{\sqrt{\sum u_i^2 \sum v_i^2}}. \tag{A12.53}$$

The variances of the estimates of the parameters are given by the diagonal elements of Eq. (A12.51),

$$\sigma_{\alpha'}^2 = \frac{\sigma_\phi^2 \sum v_i^2}{4\pi^2 \left[\sum v_i^2 \sum u_i^2 - \left(\sum u_i v_i \right)^2 \right]}, \tag{A12.54}$$

and

$$\sigma_\delta^2 = \frac{\sigma_\phi^2 \sum u_i^2}{4\pi^2 \left[\sum v_i^2 \sum u_i^2 - \left(\sum u_i v_i \right)^2 \right]}. \tag{A12.55}$$

If the (u, v) loci are long (that is, the observations extend over a large fraction of the day), then $\sum u_i v_i$ will be small compared to $\sum u_i^2$ and $\sum v_i^2$ so that

$$\sigma_{\alpha'} \simeq \frac{\sigma_\phi}{2\pi \sqrt{\sum u_i^2}}, \tag{A12.56}$$

and

$$\sigma_\delta \simeq \frac{\sigma_\phi}{2\pi \sqrt{\sum v_i^2}} \, . \tag{A12.57}$$

Furthermore, if only one baseline is used on a high-declination source, then $u_i \simeq v_i \simeq D_\lambda$, and both errors reduce to the intuitive result

$$\sigma_{\alpha'} \simeq \sigma_\delta \simeq \frac{\sigma_\phi}{2\pi \sqrt{N} D_\lambda} \, . \tag{A12.58}$$

Alternately, the source position can be found by Fourier transformation of the visibility data. This procedure can be thought of as image plane fitting or as multiplying the visibility data by the exponential factors $\exp[2\pi \, (u_i \Delta\alpha' + v_i \Delta\delta)]$ and summing over the data. The resulting "function" is maximized with respect to $\Delta\alpha'$ and $\Delta\delta$. In this latter view, it is easy to understand that (basic) image plane fitting (that is, no tapering or gridding of the data) is a maximum-likelihood procedure for finding the position of a point source and therefore formally equivalent to the method of least-mean-squares. The synthesized beam b_0 for N measurements is

$$b_0(\Delta\alpha', \Delta\delta) = \frac{1}{N} \sum \cos\left[2\pi \, (u_i \Delta\alpha' + v_i \Delta\delta)\right] \, . \tag{A12.59}$$

The shape of b_0 near its peak can be found by expanding Eq. (A12.59) to second order:

$$b_0(\Delta\alpha', \Delta\delta) \simeq 1 - \frac{2\pi^2}{N}\left(\Delta\alpha'^2 \sum u_i^2 + \Delta\delta^2 \sum v_i^2 - 2\Delta\alpha' \Delta\delta \sum u_i v_i\right) \, .$$
$$\tag{A12.60}$$

From Eq. (A12.60), it is easy to see that the contours of the synthesized beam are proportional to the error ellipse defined by Eqs. (A12.45), (A12.46), and (A12.53)–(A12.55). Note that the method of least-mean-squares can be applied only in the regime of high signal-to-noise ratio, where phase ambiguities can be resolved. However, the Fourier synthesis method can be applied in any case.

Appendix 12.2   Second-Order Effects in Phase Referencing

We present a more general analysis of how an error in the position of a calibrator source affects the determination of the position of a target source. Suppose the phase of an interferometer is referenced to its tracking center (corresponding to θ_c in Eq. (12.33). If the calibrator has coordinate errors $x_c = \Delta\alpha_c \cos\delta_c$ and $y_c = \Delta\delta_c$,

then the residual phase is

$$\Delta\phi_{c_1} - 2\pi\,(u_c x_c + v_t y_c)\,. \tag{A12.61}$$

This causes a shift in phase at the position of the target of

$$\Delta\phi_{c_2} = 2\pi\,(u_t x_c + v_t y_c)\,. \tag{A12.62}$$

Since the (u, v) coordinates are slightly different, there will be a second-order phase shift of $\Delta^2\phi = \Delta\phi_{c_2} - \Delta\phi_{c_1}$:

$$\Delta^2\phi = 2\pi\,[(u_t - u_c)x_c + (v_1 - v_2)y_c]$$
$$= 2\pi\,(\Delta u x_c + \Delta v y_c)\,. \tag{A12.63}$$

This leads to the same approximation given in Eq. (12.39). A complete expression for Eq. (A12.63) can be derived by calculating the differential quantities Δu and Δv from Eq. (4.3).

Further Reading

Enge, P., and Misra, P., Eds., *Proc. IEEE*, Special Issue on Global Positioning System, **87**, No. 1 (1999)

Jespersen, J., and Hanson, D.W., Eds., *Proc. IEEE*, Special Issue on Time and Frequency, **79**, No. 7 (1991)

Johnston, K.J., and de Vegt, C., Reference Frames in Astronomy, *Ann. Rev. Astron. Astrophys.*, **37**, 97–125 (1999)

NASA, *Radio Interferometry Techniques for Geodesy*, NASA Conf. Pub. 2115, National Aeronautics and Space Administration, Washington, DC (1980)

Petit, G., and Luzum, B., Eds., *IERS Conventions (2010)*, IERS Technical Note 36, International Earth Rotation and Reference Systems Service, Verlag des Bundesamts für Kartographie und Geodäsie, Frankfurt am Main (2010)

Reid, M.J., and Honma, M., Microarcsecond Radio Astrometry, *Ann. Rev. Astron. Astrophys.*, **52**, 339–372 (2014)

References

Alef, W., Introduction to Phase-Reference Mapping, in *Very Long Baseline Interferometry: Techniques and Applications*, Felli, M., and Spencer, R.E., Eds., Kluwer, Dordrecht (1989), pp. 261–274

Backer, D.C., and Sramek, R.A., Proper Motion of the Compact, Nonthermal Radio Source in the Galactic Center, Sagittarius A*, *Astrophys. J.*, **524**, 805–815 (1999)

Bailer-Jones, C.A.L., Estimating Distances from Parallaxes, *Publ. Astron. Soc. Pacific*, **127**, 994–1009 (2015)

Bartel, N., Bietenholz, M.F., Lebach, D.E., Ransom, R.R., Ratner, M.I., and Shapiro, I.I., VLBI for Gravity Probe B: The Guide Star, IM Pegasi, *Class. Quantum Grav.*, **32**, 224021 (21pp) (2015)

Bartel, N., Ratner, M.I., Shapiro, I.I., Cappallo, R.J., Rogers, A.E.E., and Whitney, A.R., Pulsar Astrometry via VLBI, *Astron. J.*, **90**, 318–325 (1985)

Bartel, N., Ratner, M.I., Shapiro, I.I., Herring, T.A., and Corey, B.E., Proper Motion of Components of the Quasar 3C345, in *VLBI and Compact Radio Sources*, Fanti, R., Kellermann, K., and Setti, G., Eds., IAU Symp. 110, Reidel, Dordrecht (1984), pp. 113–116

Beasley, A.J., and Conway, J.E., VLBI Phase-Referencing, in *Very Long Baseline Interferometry and the VLBA*, Zensus, J.A., Diamond, P.J., and Napier, P.J., Eds., Astron. Soc. Pacific Conf. Ser., **82**, 327–343 (1995)

Bertotti, B., Iess, L., and Tortora, P., A Test of General Relativity Using Radio Links with the Cassini Spacecraft, *Nature*, **425**, 372–376 (2003)

Bertout, C., Robichon, N., and Arenou, F., Revisiting Hipparcos Data for Pre-Main Sequence Stars, *Astron. Astrophys.*, **352**, 574–586 (1999)

Bessel, F.W., On the Parallax of 61 Cygni, *Mon. Not. R. Astron. Soc.*, **4**, 152–161 (1838)

Bevington, P.R., and Robinson, D.K., *Data Reduction and Error Analysis for the Physical Sciences*, 2nd ed., McGraw-Hill, New York (1992)

Boboltz, D.A., Diamond, P.J., and Kemball, A.J., R Aquarii: First Detection of Circumstellar SiO Maser Proper Motions, *Astrophys. J. Lett.*, **487**, L147–L150 (1997)

Brosche, P., Wade, C.M., and Hjellming, R.M., Precise Positions of Radio Sources. IV. Improved Solutions and Error Analysis for 59 Sources, *Astrophys. J.*, **183**, 805–818 (1973)

Carter, W.E., and Robertson, D.S., Very-Long-Baseline Interferometry Applied to Geophysics, in *Developments in Astrometry and Their Impact on Astrophysics and Geodynamics*, Mueller, I.I., and Kolaczek, B., Eds., Kluwer, Dordrecht (1993), pp. 133–144

Carter, W.E., Robertson, D.S., and MacKay, J.R., Geodetic Radio Interferometric Surveying: Applications and Results, *J. Geophys. Res.*, **90**, 4577–4587 (1985)

Carter, W.E., Robertson, D.S., Pettey, J.E., Tapley, B.D., Schutz, B.E., Eanes, R.J., and Lufeng, M., Variations in the Rotation of the Earth, *Science*, **224**, 957–961 (1984)

Chandler, S.C., On the Variation of Latitude, *Astron. J.*, **11**, 65–70 (1891)

Clark, T.A., Corey, B.E., Davis, J.L., Elgered, G., Herring, T.A., Hinteregger, H.F., Knight, C.A., Levine, J.I., Lundqvist, G., Ma, C., and 11 coauthors, Precise Geodesy Using the Mark-III Very-Long-Baseline Interferometer System, *IEEE Trans. Geosci. Remote Sensing*, **GE-23**, 438–449 (1985)

Cohen, M.H., and Shaffer, D.B., Positions of Radio Sources from Long-Baseline Interferometry, *Astron. J.*, **76**, 91–100 (1971)

Condon, J.J., Errors in Elliptical Gaussian Fits, *Publ. Astron. Soc. Pacific*, **109**, 166–172 (1997)

Counselman, C.C., III, Radio Astrometry, *Ann. Rev. Astron. Astrophys.*, **14**, 197–214 (1976)

Counselman, C.C., III, Kent, S.M., Knight, C.A., Shapiro, I.I., Clark, T.A., Hinteregger, H.F., Rogers, A.E.E., and Whitney, A.R., Solar Gravitational Deflection of Radio Waves Measured by Very Long Baseline Interferometry, *Phys. Rev. Lett.*, **33**, 1621–1623 (1974)

Deller, A.T., Boyles, J., Lorimer, D.R., Kaspi, V.M., McLaughlin, M.A., Ransom, S., Stairs, I.H., and Stovall, K., VLBI Astrometry of PSR J2222-0137: A Pulsar Distance Measured to 0.4% Accuracy, *Astrophys. J.*, **770**:145 (9pp) (2013)

Dyson, F.W., Eddington, A.S., and Davidson, C., A Determination of the Deflection of Light by the Sun's Gravitational Field, from Observations Made at the Total Eclipse of May 29, 1919, *Phil. Trans. R. Soc. Lond.* A, **220**, 291 333 (1920)

Elitzur, M., *Astronomical Masers*, Kluwer, Dordrecht (1992)

Elsmore, B., and Mackay, C.D., Observations of the Structure of Radio Sources in the 3C Catalogue. III. The Absolute Determination of Positions of 78 Compact Sources, *Mon. Not. R. Astron. Soc.*, **146**, 361–379 (1969)

Elsmore, B., and Ryle, M., Further Astrometric Observations with the 5-km Radio Telescope, *Mon. Not. R. Astron. Soc.*, **174**, 411–423 (1976)

Everitt, C.W.F., DeBra, D.B., Parkinson, B.W., Turneaure, J.P., Conklin, J.W., Heifetz, M.I., Keiser, G.M., Silbergleit, A.S., Holmes, T., Kolodziejczak, J., and 17 coauthors, Gravity Probe B: Final Results of a Space Experiment to Test General Relativity, *Phys. Rev. Lett.*, **106**, 221101-1– 221101-5 (2011)

Fanselow, J.L., Sovers, O.J., Thomas, J.B., Purcell, G.H., Jr., Cohen, E.J., Rogstad, D.H., Skjerve, L.J., and Spitzmesser, D.J., Radio Interferometric Determination of Source Positions Utilizing Deep Space Network Antennas—1971 to 1980, *Astron. J.*, **89**, 987–998 (1984)

Fey, A.L., and Charlot, P., VLBA Observations of Radio Reference Frame Structures. II. Astrometric Suitability Based on Observed Structure, *Astrophys. J. Suppl.*, **111**, 95–142 (1997)

Fey, A.L., Gordon, D., and Jacobs, C.S., Eds., *The Second Realization of the International Celestial Reference Frame by Very Long Baseline Interferometry*, IERS Technical Note 35, International Earth Rotation and Reference Systems Service, Verlag des Bundesamts für Kartographie und Geodäsie, Frankfurt am Main (2009)

Fey, A.L., Gordon, D., Jacobs, C.S., Ma, C., Gamme, R.A., Arias, E.F., Bianco, G., Boboltz, D.A., Böckmann, S., Bolotin, S., and 31 coauthors, The Second Realization of the International Celestial Reference Frame by Very Long Baseline Interferometry, *Astrophys. J.*, **150**:58 (16 pp) (2015)

Fomalont, E., Astrometry, in *Very Long Baseline Interferometry and the VLBA*, Zensus, J.A., Diamond, P.J., and Napier, P.J., Eds., Astron. Soc. Pacific Conf. Ser., **82** (1995), pp. 363–394

Fomalont, E.B., Goss, W.M., Lyne, A.G., Manchester, R.N., and Justtanont, K., Positions and Proper Motions of Pulsars, *Mon. Not. R. Astron. Soc.*, **258**, 479–510 (1992)

Fomalont, E., Johnston, K., Fey, A., Boboltz, D., Oyama, T., and Honma, M., The Position/Structure Stability of Four ICRF2 Sources, *Astron. J*, **141**, 91 (19pp) (2011)

Fomalont, E., Kopeikin, S., Lanyi, G., and Benson, J., Progress in Measurements of the Gravitational Bending of Radio Waves Using the VLBA, *Astrophys. J.*, **699**, 1395–1402 (2009)

Fomalont, E.B., and Sramek, R.A., A Confirmation of Einstein's General Theory of Relativity by Measuring the Bending of Microwave Radiation in the Gravitational Field of the Sun, *Astrophys. J.*, **199**, 749–755 (1975)

Fomalont, E.B. and Sramek, R.A., Measurements of the Solar Gravitational Deflection of Radio Waves in Agreement with General Relativity, *Phys. Rev. Lett.*, **36**, 1475–1478 (1976)

Fomalont, E.B., and Sramek, R.A., The Deflection of Radio Waves by the Sun, *Comments Astrophys.*, **7**, 19–33 (1977)

Garay, G., Moran, J.M., and Reid, M.J., Compact Continuum Radio Sources in the Orion Nebula, *Astrophys. J.*, **314**, 535–550 (1987)

Gauss, K.F., *Theoria Motus*, 1809; repr. in transl. as *Theory of the Motion of the Heavenly Bodies Moving about the Sun in Conic Sections*, Dover, New York (1963), p. 249

Genzel, R., Reid, M.J., Moran, J.M., and Downes, D., Proper Motions and Distances of H_2O Maser Sources. I. The Outflow in Orion-KL, *Astrophys. J.*, **244**, 884–902 (1981)

Gipson, J.M., and Ma, C., Signature of El Niño in Length of Day as Measured by VLBI, in *The Impact of El Niño and Other Low-Frequency Signals on Earth Rotation and Global Earth System Parameters*, Salstein, D.A., Kolaczek, B., and Gambis, D., Eds., IERS Technical Note 26, International Earth Rotation Service, Observatoire de Paris (1999), pp. 17–22

Gray, M., *Maser Sources in Astrophysics*, Cambridge Univ. Press, Cambridge, UK (2012)

Gwinn, C.R., Moran, J.M., and Reid, M.J., Distance and Kinematics of the W49N H_2O Maser Outflow, *Astrophys. J.*, **393**, 149–164 (1992)

Hall, T., *Karl Friedrich Gauss*, MIT Press, Cambridge, MA (1970), p. 74

Hamilton, W.C., *Statistics in Physical Science*, Ronald, New York (1964)

Hazard, C., Sutton, J., Argue, A.N., Kenworthy, C.M., Morrison, L.V., and Murray, C.A., Accurate Radio and Optical Positions of 3C273B, *Nature Phys. Sci.*, **233**, 89–91 (1971)

Herring, T.A., Geodetic Applications of GPS, *Proc. IEEE*, Special Issue on Global Positioning System, **87**, No. 1, 92–110 (1999)

Herring, T.A., Corey, B.E., Counselman, C.C., III, Shapiro, I.I., Rogers, A.E.E., Whitney, A.R., Clark, T.A., Knight, C.A., Ma, C., Ryan, J.W., and seven coauthors, Determination of Tidal Parameters from VLBI Observations, in *Proc. Ninth Int. Symp. Earth Tides*, Kuo, J., Ed., E. Schweizerbart'sche Verlagsbuchhandlung, Stuttgart (1983), pp. 205–211

Herring, T.A., Gwinn, C.R., and Shapiro, I.I., Geodesy by Radio Interferometry: Corrections to the IAU 1980 Nutation Series, in *Proc. MERIT/COTES Symp.*, Mueller, I.I., Ed., Ohio State Univ. Press, Columbus, OH (1985), pp. 307–325

Herring, T.A., Shapiro, I.I., Clark, T.A., Ma, C., Ryan, J.W., Schupler, B.R., Knight, C.A., Lundqvist, G., Shaffer, D.B., Vandenberg, N.R., and nine coauthors, Geodesy by Radio Interferometry: Evidence for Contemporary Plate Motion, *J. Geophys. Res.*, **91**, 8344–8347 (1986)

Hill, J.M., A Measurement of the Gravitational Deflection of Radio Waves by the Sun, *Mon. Not. R. Astron. Soc.*, **153**, 7p–11p (1971)

Hinteregger, H.F., Shapiro, I.I., Robertson, D.S., Knight, C.A., Ergas, R.A., Whitney, A.R., Rogers, A.E.E., Moran, J.M., Clark, T.A., and Burke, B.F., Precision Geodesy Via Radio Interferometry, *Science*, **178**, 396–398 (1972)

Hirota, T., Bushimata, T., Choi, Y.K., Honma, M., Imai, H., Iwadate, K., Jike, T., Kameno, S., Kameya, O., Kamohara, R., and 27 coauthors, Distance to Orion KL Measured with VERA, *Pub. Astron. Soc. Japan*, **59**, 897–903 (2007)

Hocke, K., and Schlegel, K. A Review of Atmospheric Gravity Waves and Travelling Ionospheric Disturbances: 1982–1995, *Ann. Geophysicae*, **14**, 917–940 (1996)

Hogg, D.W., Bovy, J., and Lang, D., Data Analysis Recipes: Fitting a Model to Data (2010), arXiv:1008.4686

Johnston, K.J., and de Vegt, C., Reference Frames in Astronomy, *Ann. Rev. Astron. Astrophys.*, **37**, 97–125 (1999)

Johnston, K.J., Seidelmann, P.K., and Wade, C.M., Observations of 1 Ceres and 2 Pallas at Centimeter Wavelengths, *Astron. J.*, **87**, 1593–1599 (1982)

Jones, D.L., Folkner, W.M., Jacobson, R.A., Jacobs, C.S., Dhawan, V., Romney, J., and Fomalont, E., Astrometry of *Cassini* with the VLBA to Improve the Saturn Ephemeris, *Astron. J.*, **149**, 28 (7pp) (2015)

Jung, T., Sohn, B.W., Kobayashi, H., Sasao, T., Hirota, T., Kameya, O., Choi, Y.K., and Chung, H.S., First Simultaneous Dual-Frequency Phase Referencing VLBI Observation with VERA, *Publ. Astron. Soc. Japan*, **63**, 375–385 (2011)

Kim, M.K., Hirota, T., Honma, M., Kobayashi, H., Bushimata, T., Choi, Y.K., Imai, H., Iwadate, K., Jike, T., Kameno, S., and 22 coauthors, SiO Maser Observations Toward Orion-KL with VERA, *Pub. Astron. Soc. Japan*, **60**, 991–999 (2008)

Kounkel, M., Hartmann, L., Loinard, L., Ortiz-León, G.N., Mioduszewski, A.J., Rodríguez, L.F., Dzib, S.A., Torres, R.M., Pech, G., Galli, P.A.B., and five coauthors, The Gould's Belt Distances Survey (Gobelins). III. Distances and Structure Towards the Orion Molecular Clouds, *Astrophys. J.*, accepted (2016). arXiv:1609.04041v2

Krásná, H., Böhm, J., and Schuh, H., Free Core Nutation Observed by VLBI, *Astron. Astrophys.*, **555**, A29 (5pp) (2013)

Lambeck, K., *The Earth's Variable Rotation: Geophysical Causes and Consequences*, Cambridge Univ. Press, Cambridge, UK (1980)

Lambert, S.B., and Le Poncin-Lafitte, C., Improved Determination of γ by VLBI, *Astron. Astrophys*, **529**, A70 (4 pp) (2011)

Lampton, M., Margon, B., and Bowyer, S., Parameter Estimation in X-Ray Astronomy, *Astrophys. J.*, **208**, 177–190 (1976)

Lebach, D.E., Corey, B.E., Shapiro, I.I., Ratner, M.I., Webber, J.C., Rogers, A.E.E., Davis, J.L., and Herring, T.A., Measurements of the Solar Deflection of Radio Waves Using Very Long Baseline Interferometry, *Phys. Rev. Lett.*, **75**, 1439–1442 (1995)

Lestrade, J.-F., VLBI Phase Referencing for Observations of Weak Radio Sources, *Radio Interferometry: Theory, Techniques and Applications*, Cornwell, T.J., and Perley, R.A., Eds., Astron. Soc. Pacific Conf. Ser., **19**, 289–297 (1991)

Lestrade, J.-F., Jones, D.L., Preston, R.A., Phillips, R.B., Titus, M.A., Kovalevsky, J., Lindegren, L., Hering, R., Froeschlé, M., Falin, J.-L., and five coauthors, Preliminary Link of the Hipparcos and VLBI Reference Frames, *Astron. Astrophys*, **304**, 182–188 (1995)

Lestrade, J.-F., Rogers, A.E.E., Whitney, A.R., Niell, A.E., Phillips, R.B., and Preston, R.A., Phase-Referenced VLBI Observations of Weak Radio Sources: Milliarcsecond Position of Algol, *Astron. J.*, **99**, 1663–1673 (1990)

Lieske, J.H., Lederle, T., Fricke, W., and Morando, B., Expressions for the Precession Quantities Based upon the IAU (1976) System of Astronomical Constants, *Astron. Astrophys.*, **58**, 1–16 (1977)

Lutz, T.E., and Kelker, D.H., On the Use of Trigonometric Parallaxes for the Calibration of Luminosity Systems: Theory, *Publ. Astron. Soc. Pacific*, **85**, 573–578 (1973)

Ma, C., Arias, E.F., Eubanks, T.M., Fey, A.L., Gontier, A.-M., Jacobs, C.S., Sovers, O.J., Archinal, B.A., and Charlot, P., The International Celestial Reference Frame as Realized by Very Long Baseline Interferometry, *Astron. J.*, **116**, 516–546 (1998)

Madison, D.R., Chatterjee, S., and Cordes, J.M., The Benefits of VLBI Astrometry to Pulsar Timing Array Searches for Gravitational Radiation, *Astrophys. J.*, **777**:104 (14pp) (2013)

Marcaide, J.M., and Shapiro, I.I., High Precision Astrometry via Very-Long-Baseline Radio Interferometry: Estimate of the Angular Separation between the Quasars 1038 + 528A and B, *Astron. J.*, **88**, 1133–1137 (1983)

Markowitz, W., and Guinot, B., Eds., *Continental Drift, Secular Motion of the Pole, and Rotation of the Earth*, IAU Symp. 32, Reidel, Dordrecht (1968), pp. 13–14

McCarthy, D.D., and Pilkington, J.D.H., Eds., *Time and the Earth's Rotation*, IAU Symp. 82, Reidel, Dordrecht (1979) (see papers on radio interferometry)

Melchior, P., *The Tides of the Planet Earth*, Pergamon Press, Oxford (1978)

Melis, C., Reid, M.J., Mioduszewski, A.J., Stauffer, J.R., and Bower, G.C., A VLBI Resolution of the Pleiades Distance Controversy, *Science*, **345**, 1029–1032 (2014)

Menten, K.M., Reid, M.J., Forbrich, J., and Brunthaler, A., The Distance to the Orion Nebula, *Astron. Astrophys.*, **474**, 515–520 (2007)

Middelberg, E., Roy, A.L., Walker, R.C., and Falcke, H., VLBI Observations of Weak Sources Using Fast Frequency Switching, *Astron. Astrophys.*, **433**, 897–909 (2005)

Misner, C.W., Thorne, K.S., and Wheeler, J.A., *Gravitation*, Freedman, San Francisco (1973), Sec. 40.3

Moran, J.M., Burke, B.F., Barrett, A.H., Rogers, A.E.E., Ball, J.A., Carter, J.C., and Cudaback, D.D., The Structure of the OH Source in W3, *Astrophys. J. Lett.*, **152**, L97–L101 (1968)

Morrison, N., *Introduction to Sequential Smoothing and Prediction*, McGraw-Hill, New York (1969), p. 645

Mueller, I.I., Reference Coordinate Systems for Earth Dynamics: A Preview, in *Reference Coordinate Systems for Earth Dynamics*, Gaposchkin, E.M., and Kołaczek, B., Eds., Reidel, Dordrecht (1981), pp. 1–22

Muhleman, D.O., Ekers, R.D., and Fomalont, E.B., Radio Interferometric Test of the General Relativistic Light Bending Near the Sun, *Phys. Rev. Lett.*, **24**, 1377–1380 (1970)

Norris, R.P., and Booth, R.S., Observations of OH Masers in W3OH, *Mon. Not. R. Astron. Soc.*, **195**, 213–226 (1981)

Norris, R.P., Booth, R.S., and Diamond, P.J., MERLIN Spectral Line Observations of W3OH, *Mon. Not. R. Astron. Soc.*, **201**, 209–222 (1982)

Paek, N., and Huang, H.-P., A Comparison of the Interannual Variability in Atmospheric Angular Momentum and Length-of-Day Using Multiple Reanalysis Data Sets, *J. Geophys. Res.*, **117**, D20102 (9pp) (2012)

Petley, B.W., New Definition of the Metre, *Nature*, **303**, 373–376 (1983)

Pitjeva, E.V., and Standish, E.M., Proposals for the Masses of the Three Largest Asteroids, the Moon–Earth Mass Ratio, and the Astronomical Unit, *Celest. Mech. Dyn. Astron.*, **103**, 365–372 (2009).

Press, W.H., Teukolsky, S.A., Vetterling, W.T., and Flannery, B.P., *Numerical Recipes*, 2nd ed., Cambridge Univ. Press, Cambridge, UK (1992)

Ratner, M.I., Bartel, N., Bietenholz, M.F., Lebach, D.E., Lestrade, J.-F., Ransom, R.R., and Shapiro, I.I., VLBI for Gravity Probe B. V. Proper Motion and Parallax of the Guide Star IM Pegasi, *Astrophys. J. Suppl.*, **201**:5 (16pp) (2012). doi: 10.1088/0067-0049/201/1/5

Reid, M.J., and Brunthaler, A., The Proper Motion of Sagittarius A*. II. The Mass of Sagittarius A*, *Astrophys. J.*, **616**, 872–884 (2004). doi: 10.1086/424960

Reid, M.J., Haschick, A.D., Burke, B.F., Moran, J.M., Johnston, K.J., and Swenson, G.W., Jr., The Structure of Interstellar Hydroxyl Masers: VLBI Synthesis Observations of W3(OH), *Astrophys. J.*, **239**, 89–111 (1980)

Reid, M.J., and Honma, M., Microarcsecond Radio Astrometry, *Ann. Rev. Astron. Astrophys.*, **52**, 339–372 (2014)

Reid, M.J., Menten, K.M., Brunthaler, A., Zheng, X.W., Dame, T.M., Xu, Y., Wu, Y., Zhang, B., Sanna, A., Sato, M., and six coauthors, Trigonometric Parallaxes of High-Mass Star-Forming Regions: The Structure and Kinematics of the Milky Way, *Astrophys. J.*, **783**:130 (14pp) (2014)

Reid, M.J., Menten, K.M., Brunthaler, A., Zheng, X.W., Moscadelli, L., and Xu, Y., Trigonometric Parallaxes of Massive Star-Forming Regions. I. S 252 and G232.6+1.0, *Astrophys. J.*, **693**, 397–405 (2009)

Reid, M.J., and Moran, J.M., Astronomical Masers, in *Galactic and Extragalactic Radio Astronomy*, 2nd ed., Verschuur, G.L., and Kellermann, K.I., Eds., Springer-Verlag, Berlin (1988), pp. 255–294

Reid, M.J., Schneps, M.H., Moran, J.M., Gwinn, C.R., Genzel, R., Downes, D., and Rönnäng, B., The Distance to the Center of the Galaxy: H_2O Maser Proper Motions in Sagittarius B2(N), *Astrophys. J.*, **330**, 809–816 (1988)

Riley, J.M., A Measurement of the Gravitational Deflection of Radio Waves by the Sun During 1972 October, *Mon. Not. R. Astron. Soc.*, **161**, 11p–14p (1973)

Rioja, M., and Dodson, R., High-Precision Astrometric Millimeter Very-Long-Baseline Interferometry Using a New Method for Atmospheric Calibration, *Astron. J.*, **141**, 114 (15pp) (2011)

Rioja, M.J., Dodson, R., Jung, T., and Sohn, B.W., The Power of Simultaneous Multi-Frequency Observations for mm-VLBI: Astrometry Up to 130 GHz with the KVN, *Astron. J.*, **150**:202 (14pp) (2015)

Rioja, M.J., Dodson, R., Jung, T., Sohn, B.W., Byun, D.-Y., Agudo, I., Cho, S.-H., Lee, S.-S., Kim, J., Kim, K.-T., and 16 coauthors, Verification of the Astrometric Performance of the Korean VLBI Network, Using Comparative SFPR Studies with the VLBA at 14/7 mm, *Astron. J.*, **148**, 84 (15pp) (2014)

Rioja, M.J., Marcaide, J.M., Elósegui, P., and Shapiro, I.I., Results from a Decade-Long VLBI Astrometric Monitoring of the Pair of Quasars 1038+528 A and B, *Astron. Astrophys.*, **325**, 383–390 (1997)

Robertson, D.S., Carter, W.E., and Dillinger, W.H., New Measurement of Solar Gravitational Deflection of Radio Signals Using VLBI, *Nature*, **349**, 768–770 (1991)

Robertson, D.S., Carter, W.E., Eanes, R.J., Schutz, B.E., Tapley, B.D., King, R.W., Langley, R.B., Morgan, P.J., and Shapiro, I.I., Comparison of Earth Rotation as Inferred from Radio Interferometric, Laser Ranging, and Astrometric Observations, *Nature*, **302**, 509–511 (1983)

Ros, E., Marcaide, J.M., Guirado, J.C., Ratner, M.I., Shapiro, I.I., Krichbaum, T.P., Witzel, A., and Preston, R.A., High Precision Difference Astrometry Applied to the Triplet of S5 Radio Sources B1803+784/Q1928+738/B2007+777, *Astron. Astrophys.*, **348**, 381–393 (1999)

Ryle, M., and Elsmore, B., Astrometry with the 5-km Telescope, *Mon. Not. R. Astron. Soc.*, **164**, 223–242 (1973)

Sandstrom, K.M., Peek, J.E.G., Bower, G.C., Bolatto, A.D., and Plambeck, R.L., A Parallactic Distance of 389^{+24}_{-21} Parsecs to the Orion Nebula Cluster from Very Long Baseline Array Observations, *Astrophys. J.*, **667**, 1161–1169 (2007)

Seidelmann, P.K., Ed., *Explanatory Supplement to the Astronomical Almanac*, University Science Books, Mill Valley, CA (1992)

Seielstad, G.A., Sramek, R.A., and Weiler, K.W., Measurement of the Deflection of 9.602-GHz Radiation from 3C279 in the Solar Gravitational Field, *Phys. Rev. Lett.*, **24**, 1373–1376 (1970)

Shapiro, I.I., New Method for the Detection of Light Deflection by Solar Gravity, *Science*, **157**, 806–808 (1967)

Shapiro, I.I., Estimation of Astrometric and Geodetic Parameters, in *Methods of Experimental Physics*, Vol. 12, Part C *(Astrophysics: Radio Observations)*, Meeks, M.L., Ed., Academic Press, New York (1976), pp. 261–276

Shapiro, I.I., Robertson, D.S., Knight, C.A., Counselman, C.C., III, Rogers, A.E.E., Hinteregger, H.F., Lippincott, S., Whitney, A.R., Clark, T.A., Niell, A.E., and Spitzmesser, D.J., Transcontinental Baselines and the Rotation of the Earth Measured by Radio Interferometry, *Science*, **186**, 920–922 (1974)

Shapiro, I.I., Wittels, J.J., Counselman, C.C., III, Robertson, D.S., Whitney, A.R., Hinteregger, H.F., Knight, C.A., Rogers, A.E.E., Clark, T.A., Hutton, L.K., and Niell, A.E., Submilliarcsecond Astrometry via VLBI. I. Relative Position of the Radio Sources 3C345 and NRAO512, *Astron. J.*, **84**, 1459–1469 (1979)

Shapiro, S.S., Davis, J.L., Lebach, D.E., and Gregory, J.S., Measurement of the Solar Gravitational Deflection of Radio Waves Using Geodetic Very-Long-Baseline Interferometry Data, 1979–1999, *Phys. Rev. Lett.*, **92**, 121101-1–121101-4 (2004)

Sivia, D.S. with Skilling, J., *Data Analysis: A Bayesian Tutorial*, 2nd ed., Oxford Univ. Press, Oxford, UK (2006)

Smith, F.G., The Determination of the Position of a Radio Star, *Mon. Not. R. Astron. Soc.*, **112**, 497–513 (1952)

Smith, H.M., International Time and Frequency Coordination, *Proc. IEEE*, **60**, 479–487 (1972)

Sovers, O.J., Fanselow, J.L., and Jacobs, C.S., Astrometry and Geodesy with Radio Interferometry: Experiments, Models, Results, *Rev. Mod. Phys.*, **70**, 1393–1454 (1998)

Sramek, R.A., A Measurement of the Gravitational Deflection of Microwave Radiation Near the Sun, 1970 October, *Astrophys. J. Lett.*, **167**, L55–L60 (1971)

Sramek, R., The Gravitational Deflection of Radio Waves, in *Experimental Gravitation*, Proc. International School of Physics "Enrico Fermi," Course 56, B. Bertotti, Ed., Academic Press, New York and London (1974)

Sramek, R.A., Atmospheric Phase Stability at the VLA, in *Radio Astronomical Seeing*, Baldwin, J.E., and Wang, S., Eds., International Academic Publishers and Pergamon Press, Oxford (1990), pp. 21–30

Taff, L.G., *Computational Spherical Astronomy*, Wiley, New York (1981)

Tatarski, V.I., *Wave Propagation in a Turbulent Medium*, transl. by Silverman, R.A., McGraw-Hill, New York (1961)

Titov, O., Spectral Analysis of the Baseline Length Time Series from VLBI Data, International VLBI Service for Geodesy and Astrometry: General Meeting Proceedings, Goddard Space Flight Center, Greenbelt, MD, National Technical Information Service (2002), pp. 315–319

Ulvestad, J., Phase-Referencing Cycle Times, VLBA Scientific Memo 20, National Radio Astronomy Observatory (1999)

Verbiest, J.P.W., and Lorimer, D.R., Why the Distance of PSR J0218+4232 Does Not Challenge Pulsar Emission Theories, *Mon. Not. R. Astron. Soc.*, **444**, 1859–1861 (2014)

Verbiest, J.P.W., Lorimer, D.R., and McLaughlin, M.A., Lutz–Kelker Bias in Pulsar Parallax Measurements, *Mon. Not. R. Astron. Soc.*, **405**, 564–572 (2010)

Verbiest, J.P.W., Weisberg, J.M., Chael, A.A., Lee, K.J., and Lorimer, D.R., On Pulsar Distance Measurements and Their Uncertainties, *Astrophys. J.*, **755**:39 (9pp) (2012)

Wade, C.M., Precise Positions of Radio Sources. I. Radio Measurements, *Astrophys. J.*, **162**, 381–390 (1970)

Wahr, J.M., The Forced Nutations of an Elliptical, Rotating, Elastic, and Oceanless Earth, *Geophys. J. R. Astron. Soc.*, **64**, 705–727 (1981)

Wahr, J.M., *Geodesy and Gravity*, Samezdot Press, Golden, CO (1996)

Walker, R.C., The Multiple-Point Fringe-Rate Method of Mapping Spectral-Line VLBI Sources with Application to H_2O Masers in W3-IRS5 and W3(OH), *Astron. J.*, **86**, 1323–1331 (1981). doi: 10.1086/113013

Walker, R.C., Matsakis, D.N., and Garcia-Barreto, J.A., H_2O Masers in W49N. I. Maps, *Astrophys. J.*, **255**, 128–142 (1982)

Weiler, K.W., Ekers, R.D., Raimond, E., and Wellington, K.J., A Measurement of Solar Gravitational Microwave Deflection with the Westerbork Synthesis Telescope, *Astron. Astrophys.*, **30**, 241–248 (1974)

Weiler, K.W., Ekers, R.D., Raimond, E., and Wellington, K.J., Dual-Frequency Measurement of the Solar Gravitational Microwave Deflection, *Phys. Rev. Lett.*, **35**, 134–137 (1975)

Will, C.M., *Theory and Experiment in Gravitational Physics*, Cambridge Univ. Press, Cambridge, UK (1993), ch. 7

Woolard, E.W., and Clemence, G.M., *Spherical Astronomy*, Academic Press, New York (1966)

Chapter 13
Propagation Effects: Neutral Medium

The neutral gas in the atmosphere has a significant effect on signals passing through it. We are concerned with three types of effects. First, the large-scale structures in the media give rise to refractive effects. These effects, which can be analyzed in terms of geometrical optics and Fermat's principle, are the deflection of the radio waves and the change of the propagation velocity. Second, radiation can be absorbed. Finally, radiation can be scattered by the turbulent structure of the media. The phenomenon of scattering results in scintillation, or seeing.

In the troposphere, water vapor plays a particularly important role in radio propagation. The refractivity of water vapor is about 20 times greater in the radio range than in the near-infrared or optical regimes. The phase fluctuations in radio interferometers at centimeter, millimeter, and submillimeter wavelengths are caused predominantly by fluctuations in the distribution of water vapor. Water vapor is poorly mixed in the troposphere, and the total column density of water vapor cannot be accurately sensed from surface meteorological measurements. Uncertainties in the water vapor content are a serious limitation to the accuracy of VLBI measurements. Small-scale (< 1 km) fluctuations in water vapor distribution limit the angular resolution of connected-element interferometers in the absence of wavefront correction techniques. Furthermore, spectral lines of water vapor cause substantial absorption at frequencies above 100 GHz and usually render the troposphere highly opaque at frequencies between 1 and 10 THz (300 and 30 μm). Thus, any discussion of the neutral atmosphere must be primarily concerned with the effects of water vapor. Propagation in the neutral atmosphere from the point of view of radio communications is discussed by Crane (1981) and Bohlander et al. (1985).

Our interest in the propagation media arises because the media degrade interferometric measurements of radio sources. Alternately, observations of radio sources can be used to probe the characteristics of the propagation media. Radio interferometric measurements have been used widely for this purpose.

© The Author(s) 2017
A.R. Thompson, J.M. Moran, and G.W. Swenson Jr., *Interferometry and Synthesis in Radio Astronomy*, Astronomy and Astrophysics Library,
DOI 10.1007/978-3-319-44431-4_13

Fig. 13.1 Vertical profiles of temperature (solid line) and the water vapor (H_2O) (dashed line) and ozone (O_3) (dotted line) volume-mixing ratios, averaged over northern and southern midlatitudes for the period 2005–2014, compiled from the NASA Program for Modern-Era Retrospective Analysis for Research and Application (MERRA) reanalysis (Rienecker et al. 2011). The averaging captures diurnal and annual variations.

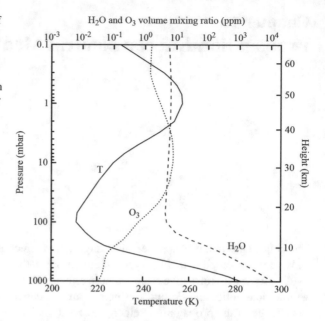

13.1   Theory

A temperature profile of the atmosphere is shown in Fig. 13.1. In the lowest part of the atmosphere, the temperature decreases monotonically from the surface at a rate of about 6.5 K km^{-1}, except for an occasional low-level inversion, until it reaches about 210 K at an altitude of approximately 12 km at midlatitudes. This lowermost layer is called the troposphere. Above 12 km, the temperature is relatively constant for a distance of about 10 km in the region called the tropopause. Above the tropopause, the temperature begins to rise with altitude in the stratosphere, due to the presence of ozone, reaching about 260 K at 45 km altitude. Above this level, the temperature drops with altitude through the mesosphere before rising again in the upper atmosphere, where the neutral atmosphere gives way to the ionosphere. Within the neutral atmosphere, the propagation of radio waves is most affected by the troposphere. Before discussing the refraction, absorption, and scattering of radio waves in the troposphere in detail, we introduce some basic physical concepts.

13.1.1   Basic Physics

Consider a plane wave propagating along the y direction in a uniform dissipative dielectric medium, as represented by the equation

$$\mathbf{E}(y, t) = \mathbf{E}_0 e^{j(kny - 2\pi \nu t)} \ , \tag{13.1}$$

where k is the propagation constant in free space and is equal to $2\pi v/c$, c is the velocity of light, and $\mathbf{E}_0$ is the electric field amplitude. n is the complex index of refraction, equal to $n_R + jn_I$. If the imaginary part of the index of refraction is positive, the wave will decay exponentially. The power absorption coefficient is defined as

$$\alpha = \frac{4\pi v}{c} n_I . \qquad (13.2)$$

Its units are m^{-1}. The propagation constant in the atmosphere is k multiplied by the real part of the index of refraction, which can be written

$$kn_R = \frac{2\pi n v}{c} = \frac{2\pi v}{v_p} , \qquad (13.3)$$

where $n = n_R$ is the index of refraction when absorption is neglected, and v_p is the phase velocity. The phase velocity of the wave, c/n, is less than c by about 0.03% in the lower atmosphere. The extra time required to traverse a medium with index of refraction $n(y)$ compared with the time necessary to traverse the same distance in free space is

$$\Delta t = \frac{1}{c} \int (n - 1) \, dy , \qquad (13.4)$$

where we assume that the effect of the difference in physical length between the actual ray path and the straight-line path is negligible. The *excess* path length is defined as $c \Delta t$, or

$$\mathcal{L} = 10^{-6} \int N(y) \, dy , \qquad (13.5)$$

where we have introduced the refractivity N, defined by $N = 10^6 (n - 1)$. Note that the concept of excess path length, which is used extensively in this chapter, does not represent an actual physical path.

A widely accepted expression for the radio refractivity is (Rüeger 2002)

$$N = 77.6898 \frac{p_D}{T} + 71.2952 \frac{p_V}{T} + 375463 \frac{p_V}{T^2} , \qquad (13.6)$$

where T is the temperature in kelvins, p_D is the partial pressure of the dry air, and p_V is the partial pressure of water vapor in millibars (1 mb = 100 newtons per square meter = 100 pascals = 1 hectopascal; 1 atmosphere = 1013 mb). The first two terms on the right side of Eq. (13.6) arise from the displacement polarizations of the gaseous constituents of the air (N_2, O_2, CO_2, and H_2O). The third term is due to the permanent dipole moment of water vapor. Equation (13.6) is formally known as the "zero-frequency" limit for the refractivity but is accurate to better than

1% for frequencies below 100 GHz. The contributions of dispersive components of refractivity associated with resonances below 100 GHz are very small. Between 100 and 1,000 GHz, the deviations from unity of the refractivity are more significant (see discussion in Sect. 13.1.4).

The refractivity can be expressed in terms of gas density, using the ideal gas law,

$$p = \frac{\rho R T}{M} , \tag{13.7}$$

where p and ρ are the partial pressure and density of any constituent gas; R is the universal gas constant, equal to $8.314 \, \text{J} \, \text{mol}^{-1} \, \text{K}^{-1}$; and M is the molecular weight, which for dry air in the troposphere is $M_D = 28.96 \, \text{g} \, \text{mol}^{-1}$ and for water vapor is $M_V = 18.02 \, \text{g} \, \text{mol}^{-1}$. Thus, $p_D = \rho_D R T / M_D$ and $p_V = \rho_V R T / M_V$, where ρ_D and ρ_V are the densities of dry air and water vapor, respectively. Since the total pressure P is the sum of the partial pressures, and the total density ρ_T is the sum of the constituent densities, Eq. (13.7) can be written $P = \rho_T R T / M_T$, where

$$M_T = \left(\frac{1}{M_D} \frac{\rho_D}{\rho_T} + \frac{1}{M_V} \frac{\rho_V}{\rho_T} \right)^{-1} . \tag{13.8}$$

Substitution of the appropriate forms of Eq. (13.7) and the equation $\rho_D = \rho_T - \rho_V$ into Eq. (13.6) yields

$$N = 0.2228\rho_T + 0.076\rho_V + 1742\frac{\rho_V}{T} , \tag{13.9}$$

where ρ_T and ρ_V are in $\text{g} \, \text{m}^{-3}$. Since the second term on the right side of Eq. (13.9) is small with respect to the third term, it can be combined with the third term to give, for $T = 280 \, \text{K}$,

$$N \simeq 0.2228\rho_T + 1763\frac{\rho_V}{T} = N_D + N_V . \tag{13.10}$$

Equation (13.10) defines the dry and wet refractivities, N_D and N_V, respectively. These definitions are not universally followed in the literature. Note that N_D is proportional to the total density and therefore has a contribution due to the induced dipole moment of water vapor. Mean values of the distribution of the column density of water vapor around the world are shown in Fig. 13.2. For a discussion of climatology of water vapor, see Peixoto and Oort (1996).

The atmosphere is in hydrostatic equilibrium to a high degree of accuracy (Andrews 2000). A parcel of gas in static equilibrium between pressure and gravity obeys the equation

$$\frac{dP}{dh} = -\rho_T g , \tag{13.11}$$

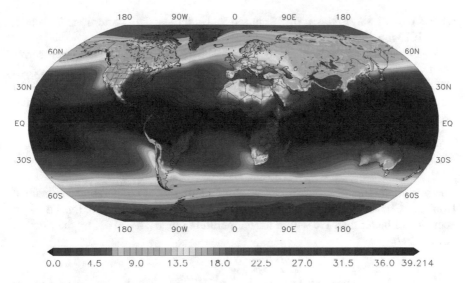

Fig. 13.2 Worldwide distribution of total water vapor content (w) based on satellite and ground-based observations over the ten-year period 2005–2014 in the framework of a global atmospheric model. The color scale denotes the column density in units of kg m^{-2} (equivalent to millimeters of precipitable water). Note that the resolution is not sufficient to show small localized areas of low water vapor, such as Mauna Kea. Data from the NASA MERRA Program. See Rienecker et al. (2011).

where g is the acceleration due to gravity, approximately equal to 980 cm s^{-2}, and h is the height above the Earth's surface. Using the ideal gas law, Eq. (13.7), we can integrate Eq. (13.11), assuming specific forms for the temperature profile and mixing ratio. If an isothermal atmosphere with constant mixing ratio is assumed, then ρ_T is an exponential function with a scale height of $RT/Mg \simeq 8.5$ km for 290 K, which is close to the observed scale height. Other models are described by Hess (1959). The excess path length caused by the dry component of refractivity does not depend on the height distribution of total density or temperature, but only on the surface pressure P_0, under conditions of hydrostatic equilibrium. If g is assumed to be constant with height, the surface pressure can be obtained by integrating Eq. (13.11),

$$P_0 = g \int_0^\infty \rho_T(h)\, dh .$$

(13.12)

From Eqs. (13.5), (13.10), and (13.12), the dry excess path length in the zenith direction is

$$\mathcal{L}_D = 10^{-6} \int_0^\infty N_D\, dh = AP_0 ,$$

(13.13)

where $A = 77.6\,R/gM_D = 0.228\,\text{cm mb}^{-1}$. Under standard conditions for which $P_0 = 1013\,\text{mb}$, the value of $\mathcal{L}_D$ is 231 cm.

Water vapor is not well mixed in the atmosphere and therefore is not well correlated with ground-based meteorological parameters (e.g., Reber and Swope 1972). On average, water vapor density has an approximately exponential distribution with a scale height of 2 km. This can be understood in the following way. The partial pressure and density of water vapor from Eq. (13.7) are related by

$$\rho_V = \frac{217 p_V}{T}\ (\text{g m}^{-3})\ .\tag{13.14}$$

The partial pressure of water vapor for saturated air, p_{VS}, at temperature T, obtained from the Clausius–Clapeyron equation (Hess 1959), can be approximated to an accuracy of better than 1% within the temperature range 240–310 K by the formula (Crane 1976)

$$p_{VS} = 6.11 \left(\frac{T}{273}\right)^{-5.3} e^{25.2(T-273)/T}\ (\text{mb})\ .\tag{13.15}$$

The relative humidity is p_V/p_{VS}. This approximation to the Clausius–Clapeyron equation is nearly an exponential function of temperature, dropping from 10.0 mb at 280 K to 3.7 (a factor of e^{-1}) at 266 K. For a lapse rate in temperature of 6 K km^{-1}, the profile of water vapor density is very close to an exponential function with a scale height of 2.5 km. For the purpose of this discussion, we adopt a simple model for the wet atmosphere as being isothermal with a scale height of 2.0 km, as is often observed.

The component of the path length resulting primarily from the permanent dipole moment of water vapor is, from Eq. (13.10),

$$\mathcal{L}_V = 1763 \times 10^{-6} \int_0^\infty \frac{\rho_V(h)}{T(h)}\, dh\ ,\tag{13.16}$$

where the units of $\mathcal{L}_V$ are the same as those of h. Hence, for the approximation above, we obtain

$$\mathcal{L}_V \simeq 350\, \frac{\rho_{V0}}{T}\ (\text{cm})\tag{13.17a}$$

or

$$\mathcal{L}_V = 7.6 \times 10^4 \frac{p_{V0}}{T^2}\ (\text{cm})\ ,\tag{13.17b}$$

where ρ_{V0} and p_{V0} are the density and partial pressure of water vapor at the surface of the Earth, respectively. Hence, for $T = 280\,\text{K}$, the path length is given by $\mathcal{L}_V = 1.26\rho_{V0} = 0.97 p_{V0}$.

The integrated water vapor density, or the height of the column of water condensed from the atmosphere, is given by

$$w = \frac{1}{\rho_w} \int_0^\infty \rho_V(h)\, dh \,, \tag{13.18}$$

where ρ_w is the density of water, 10^6 g m^{-3}. Hence, from Eq. (13.16), for an isothermal atmosphere at 280 K,

$$\mathcal{L}_V \simeq 6.3w \,. \tag{13.19}$$

This formula, which is widely used in the literature, is an excellent approximation for frequencies below 100 GHz. In the windows above 100 GHz, the ratio $\mathcal{L}_V/w$ can vary from 6.3 to about 8 (see Fig. 13.9 and associated discussion). The values of $\mathcal{L}_V$ under extreme conditions for a temperate, sea-level site can be calculated from the equations above. With $T = 303$ K ($30\,^\circ$C) and relative humidity $= 0.8$, we have $p_{V0} = 34$ mb, $\rho_{V0} = 24$ g m^{-3}, $w = 4.9$ cm, and $\mathcal{L}_V = 28$ cm. With $T = 258$ K ($-15\,^\circ$C) and relative humidity $= 0.5$, we have $p_{V0} = 1.0$ mb, $\rho_{V0} = 0.8$ g m^{-3}, $w = 0.15$ cm, and $\mathcal{L}_V = 1.1$ cm. The total zenith excess path length through the atmosphere is $\mathcal{L} \simeq \mathcal{L}_D + \mathcal{L}_V$, which, from Eqs. (13.13) and (13.19), is

$$\mathcal{L} \simeq 0.228 P_0 + 6.3w \quad \text{(cm)} \,, \tag{13.20}$$

where P_0 is in millibars, and w is in centimeters. Equation (13.20) is reasonably accurate for estimation purposes because the fractional variation in the temperature of the lower atmosphere, and in the scale height of water vapor, is usually less than 10%. However, it is usually not accurate enough to predict the path length to a small fraction of a wavelength at millimeter wavelengths.

13.1.2   Refraction and Propagation Delay

If the vertical distributions of temperature and water vapor pressure are known, then precise estimates of the angle of arrival and excess propagation time for a ray impinging on the atmosphere at an arbitrary angle can be computed by ray tracing. Here, we consider a few elementary cases in order to derive some simple analytic expressions. The simplest case is that of an interferometer in a uniform or plane-parallel atmosphere, as shown in Fig. 13.3. The refraction of the ray is governed by Snell's law, which is

$$n_0 \sin z_0 = \sin z \,, \tag{13.21}$$

where z is the zenith angle at the top of the atmosphere (where $n = 1$), and z_0 is the zenith angle at the surface (where $n = n_0$). The geometric delay for an

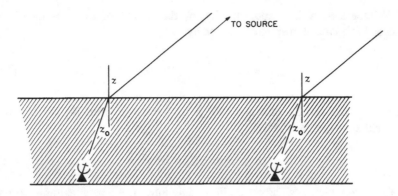

Fig. 13.3 Two-element interferometer with the atmosphere modeled as a uniform flat slab. The geometric delay is the same as it would be if the interferometer were in free space.

interferometer, as defined in Chap. 2, is

$$\tau_g = \frac{n_0 D}{c} \sin z_0 = \frac{D}{c} \sin z \, . \tag{13.22}$$

τ_g can be calculated from the angle of arrival z_0 and the velocity of light at the Earth's surface c/n_0, or from z and the velocity of light in free space. Thus, if Earth curvature is neglected and the atmosphere is uniform, the resulting geometric delay is the same as the free-space value. The angle of refraction need only be calculated to ensure that the antennas track the source properly. The angle of refraction, $\Delta z = z - z_0$, can be written, using Eq. (13.21), as

$$\Delta z = z - \sin^{-1} \left(\frac{1}{n_0} \sin z \right) \, . \tag{13.23}$$

This equation can be expanded in a Taylor series in $n_0 - 1$, which to first order gives

$$\Delta z \simeq (n_0 - 1) \tan z \, . \tag{13.24}$$

Since $n_0 - 1 \simeq 3 \times 10^{-4}$ at the surface of the Earth, Eq. (13.24) can be written

$$\Delta z \, (\text{arcmin}) \simeq \tan z \, . \tag{13.25}$$

The angle of refraction can also be calculated for more realistic cases. Ignore the curvature of the Earth, and consider the atmosphere to consist of a large number of plane-parallel layers numbered 0 through m, as shown in Fig. 13.4. Let the index of refraction at the surface be n_0, and at the top layer, $n_m = 1$. Applying Snell's law to

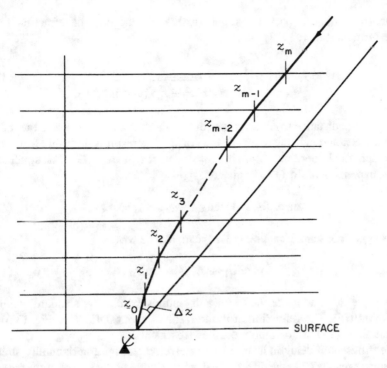

Fig. 13.4 The atmosphere modeled as a set of thin, uniform slabs. The angle of incidence on the topmost slab is z_m, which is equal to the free-space zenith angle z, and the angle of incidence at the surface is z_0. The total bending is $\Delta z = z - z_0$.

the various layers gives the following set of equations:

$$n_0 \sin z_0 = n_1 \sin z_1$$

$$n_1 \sin z_1 = n_2 \sin z_2$$

$$\vdots \qquad \vdots$$

$$n_{m-1} \sin z_{m-1} = \sin z \,, \tag{13.26}$$

where $z = z_m$. From these equations, we see that $n_0 \sin z_0 = \sin z$. This result is identical to that for the homogenous case. Thus, regardless of the vertical distribution of the index of refraction, the angle of refraction is given by Eq. (13.21), where n_0 is the surface value of the index of refraction. This result can also be obtained by an elementary application of Fermat's principle. An interesting application of this result is that if $n_0 = 1$, as would be the case if the measuring device were in a vacuum chamber at the surface of the Earth, then there would be no net refraction; that is, $z_0 = z$.

For an atmosphere consisting of spherical layers, the angle of refraction is given by the formula (Smart 1977)

$$\Delta z = r_0 n_0 \sin z_0 \int_1^{n_0} \frac{dn}{n\sqrt{r^2 n^2 - r_0^2 n_0^2 \sin^2 z_0}} , \tag{13.27}$$

where r is the distance from the center of the Earth to the layer where the index of refraction is n and r_0 is the radius of the Earth. This result is derivable from Snell's law in spherical coordinates: $nr \sin z = $ constant (Smart 1977). For small zenith angles, expansion of Eq. (13.27) gives

$$\Delta z \simeq (n_0 - 1) \tan z_0 - a_2 \tan z_0 \sec^2 z_0 , \tag{13.28}$$

where a_2 is a constant. Equation (13.28) can also be written

$$\Delta z \simeq a_1 \tan z_0 - a_2 \tan^3 z_0 , \tag{13.29}$$

where $a_1 \simeq 56''$ and $a_2 \simeq 0.07''$ for a dry atmosphere under standard conditions (COESA 1976). The refraction at the horizon is about $0.46°$ (see Fig. 13.6). See Saastamoinen (1972a) for a more detailed treatment.

The differential delay induced in an interferometer by a horizontally stratified troposphere results from the difference in zenith angle of the source at the antennas. Consider two closely spaced antennas. If the excess path in the zenith direction is $\mathcal{L}_0$, then the excess path in other directions is approximately $\mathcal{L}_0 \sec z$. This approximation becomes inaccurate at large zenith angles. The difference in excess paths, $\Delta\mathcal{L}$, by first-order expansion, is

$$\Delta\mathcal{L} \simeq \mathcal{L}_0 \Delta z \frac{\sin z}{\cos^2 z} , \tag{13.30}$$

where Δz is the difference in zenith angles at the two antennas.

If the antennas are on the equator and the source has a declination of zero, then Δz is equal to the difference in longitudes, or approximately D/r_0, where D is the separation between antennas. For this case,

$$\Delta\mathcal{L} \simeq \frac{\mathcal{L}_0 D}{r_0} \frac{\sin z}{\cos^2 z} . \tag{13.31}$$

If $D = 10$ km, $\mathcal{L}_0 = 230$ cm, $r_0 = 6370$ km, and $z = 80°$, then $\Delta\mathcal{L}$ is 12 cm. The calculation of the difference in excess paths can be easily generalized as follows. Let $\mathbf{r}_1$ and $\mathbf{r}_2$ be vectors from the center of the Earth to each antenna. The geometric delay is $(\mathbf{r}_1 \cdot \mathbf{s} - \mathbf{r}_2 \cdot \mathbf{s})/c$, where $\mathbf{s}$ is the unit vector in the direction of the source.

Since $\cos z_1 = (\mathbf{r}_1 \cdot \mathbf{s})/r_0$ and $\cos z_2 = (\mathbf{r}_2 \cdot \mathbf{s})/r_0$, where z_1 and z_2 are the zenith angles at the two antennas, the geometric delay can be written

$$\tau_{\mathrm{g}} = \frac{r_0}{c}(\cos z_1 - \cos z_2) \simeq \frac{r_0}{c}\Delta z \sin z . \tag{13.32}$$

Substitution of Δz from Eq. (13.32) into Eq. (13.30) yields an expression for the difference in excess path lengths, valid for short-baseline interferometers and moderate values of zenith angle:

$$\Delta \mathcal{L} \simeq \frac{c\tau_{\mathrm{g}}\mathcal{L}_0}{r_0} \sec^2 z . \tag{13.33}$$

For very-long-baseline interferometers, the expression in Eq. (13.30) is not appropriate. The difference in excess path lengths is approximately $\Delta \mathcal{L} = \mathcal{L}_1 \sec z_1 - \mathcal{L}_2 \sec z_2$, where $\mathcal{L}_1$, $\mathcal{L}_2$, z_1, and z_2 are the excess zenith path lengths and the zenith angles at the two antennas. We now derive a more accurate expression for the excess path length to each antenna. The geometry is shown in Fig. 13.5. Assume the index of refraction to be exponentially distributed with a scale height h_0. The

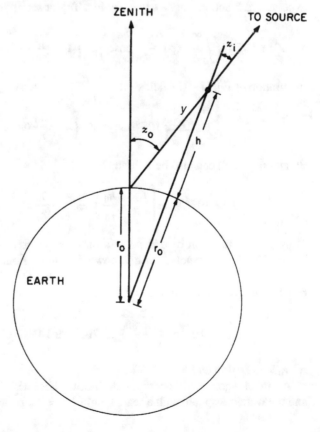

Fig. 13.5 Geometry for calculating the propagation delay, taking into account the sphericity of the Earth. The ray path along the y coordinate is assumed to be straight. The angle z_i is the zenith angle of the ray at height h. This angle is needed in the calculation of the excess path length through the ionosphere.

ZENITH

TO SOURCE

z_i

y

z_0

h

r_0

r_0

EARTH

excess path length is

$$\mathcal{L} = 10^{-6} N_0 \int_0^\infty \exp\left(-\frac{h}{h_0}\right) dy , \tag{13.34}$$

where N_0 is the refractivity at the Earth's surface, h is the height above the surface, and dy is the differential length along the ray path. Bending of the ray is neglected. From the geometry of Fig. 13.5, $(h+r_0)^2 = r_0^2 + y^2 + 2r_0 y \cos z$. Using the quadratic formula and the second-order expansion $(1 + \Delta)^{1/2} \simeq 1 + \Delta/2 - \Delta^2/8$, where $\Delta = (y^2 + 2yr_0 \cos z)/r_0^2$, one can show that

$$h \simeq y\cos z + \frac{y^2}{2r_0} \sin^2 z . \tag{13.35}$$

Therefore

$$\mathcal{L} \simeq 10^{-6} N_0 \int_0^\infty \exp\left(-\frac{y}{h_0}\cos z\right) \exp\left(-\frac{y^2}{2r_0 h_0}\sin^2 z\right) dy . \tag{13.36}$$

The argument of the rightmost exponential function in Eq. (13.36) is small, and this exponential function can be expanded in a Taylor series so that

$$\mathcal{L} \simeq 10^{-6} N_0 \int_0^\infty \exp\left(-\frac{y}{h_0}\cos z\right) \times \left(1 - \frac{y^2}{2r_0 h_0}\sin^2 z \cdots\right) dy . \tag{13.37}$$

Integration of Eq. (13.37) yields

$$\mathcal{L} \simeq 10^{-6} N_0 h_0 \sec z \left(1 - \frac{h_0}{r_0}\tan^2 z\right) . \tag{13.38}$$

Equation (13.38) can also be written

$$\mathcal{L} \simeq 10^{-6} N_0 h_0 \left[\left(1 + \frac{h_0}{r_0}\right)\sec z - \frac{h_0}{r_0}\sec^3 z\right] . \tag{13.39}$$

Thus, $\mathcal{L}$ is a function of odd powers of $\sec z$, whereas the bending angle, given in Eq. (13.29), is a function of odd powers of $\tan z$. Equations (13.38) and (13.39) both diverge as z approaches $90°$. For $z = 90°$, Eq. (13.35) shows that $h \simeq y^2/2r_0$. Hence, for direct integration of Eq. (13.34), the excess path at the horizon is

$$\mathcal{L} \simeq 10^{-6} N_0 \sqrt{\frac{\pi r_0 h_0}{2}} \simeq 70\mathcal{L}_0 \simeq 14 N_0 \ \text{(cm)} \tag{13.40}$$

for $r_0 = 6370\,\text{km}$ and $h_0 = 2\,\text{km}$.

A model incorporating both the dry atmosphere with a scale height $h_D = 8$ km and the wet atmosphere with a scale height $h_V = 2$ km can be obtained by applying

Eq. (13.38) to both the dry and wet components using Eqs. (13.13) and (13.17). This result is

$$\mathcal{L} \simeq 0.228 P_0 \sec z \, (1 - 0.0013 \tan^2 z)$$

$$+ \frac{7.5 \times 10^4 p_{V0} \sec z}{T^2} (1 - 0.0003 \tan^2 z) \, . \qquad (13.41)$$

More sophisticated models have been derived by Marini (1972), Saastamoinen (1972b), Davis et al. (1985), Niell (1996), and others. A comparison of the approximate formula of Eq. (13.41), a simple $\sec z$ model, and a ray-tracing solution is given in Fig. 13.6.

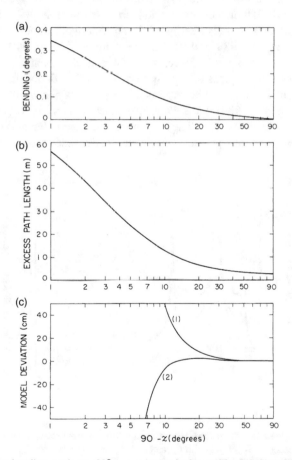

Fig. 13.6 (a) The bending angle vs. $90° - z$, where z is the zenith angle that the ray would have in the absence of refraction, calculated by a ray-tracing algorithm for a standard dry atmosphere (COESA 1976). (b) The excess path length vs. $90° - z$ calculated by a ray-tracing algorithm. The zenith excess path is 2.31 m. (c) Deviation between the excess path length and (1) the $\mathcal{L}_0 \sec z$ model and (2) the model of Eq. (13.41); in both cases, $p_{V0} = 0$, and the zenith excess path is the same as in (b).

13.1.3   Absorption

When the sky is clear, the principal sources of atmospheric attenuation are the molecular resonances of water vapor, oxygen, and ozone. The resonances of water vapor and oxygen are strongly pressure broadened in the troposphere and cause attenuation far from the resonance frequencies. A plot of the absorption vs. frequency is shown in Fig. 13.7. Below 30 GHz, absorption is dominated by the weak 6_{16}–5_{23} transition of H_2O at 22.2 GHz (Liebe 1969). Absorption by this line rarely exceeds 20% in the zenith direction. (See Appendix 13.1 for the history of research on this line.)

The oxygen lines in the band 50–70 GHz are considerably stronger, and no astronomical observations can be made from the ground in this band. An isolated oxygen line at 118 GHz makes observations impossible in the band 116–120 GHz. At higher frequencies, there is a series of strong water vapor lines at 183, 325, 380, 448, 475, 557, 621, 752, 988, and 1097 GHz and higher Liebe (1981). Observations can be made in the windows between these lines at dry locations, usually found at high altitudes. The physics of atmospheric absorption is discussed in detail by Waters (1976), and a model of absorption at frequencies below 1000 GHz is given by Liebe (1981, 1985, 1989). We are concerned here only with the phenomenology of absorption and its calibration. The absorption coefficient depends on the temperature, gas density, and total pressure. For example, the absorption coefficient for the

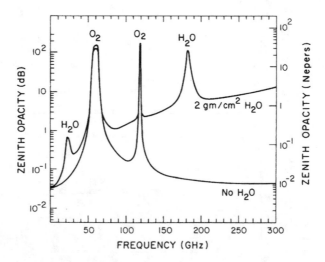

Fig. 13.7 Atmospheric zenith opacity. The absorption from narrow ozone lines has been omitted. Adapted from Waters (1976). For zenith opacity at frequencies above 300 GHz, see Liebe (1981, 1989). Note that 2 g cm^{-2} of H_2O corresponds to $w = 2$ cm.

22 GHz H_2O line can be written (Staelin 1966)

$$\alpha = \left(3.24 \times 10^{-4} e^{-644/T}\right) \frac{\nu^2 P \rho_v}{T^{3.125}} \left(1 + 0.0147 \frac{\rho_v T}{P}\right)$$

$$\times \left[\frac{1}{(\nu - 22.235)^2 + \Delta\nu^2} + \frac{1}{(\nu + 22.235)^2 + \Delta\nu^2}\right]$$

$$+ 2.55 \times 10^{-8} \rho_v \nu^2 \frac{\Delta\nu}{T^{3/2}} \quad (cm^{-1}) . \tag{13.42}$$

Here, $\Delta\nu$ is approximately the half-width at half-maximum of the line in gigahertz, given by the equation

$$\Delta\nu = 2.58 \times 10^{-3} \left(1 + 0.0147 \frac{\rho_v T}{P}\right) \frac{P}{(T/318)^{0.625}} , \tag{13.43}$$

where ν is the frequency in gigahertz, T is the temperature in kelvins, P is the total pressure in millibars, and ρ_v is the water vapor density in grams per cubic meter. The lineshape specified by Eq. (13.42), the Van Vleck–Weisskopf profile, appears to fit the empirical data better than other theoretical profiles (Hill 1986). Other line parametrizations of the line profile are available, for example, Pol et al. (1998).

The intensity of a ray passing through an absorbing medium obeys the radiative transfer equation. We assume that the medium is in local thermodynamic equilibrium at temperature T and that scattering is negligible. In the domain where the Rayleigh–Jeans approximation to the Planck function is valid, so that the intensity is proportional to the brightness temperature, the equation of radiative transfer can be written (Rybicki and Lightman 1979)

$$\frac{dT_B}{dy} = -\alpha(T_B - T) , \tag{13.44}$$

where T_B is the brightness temperature and α is the absorption coefficient defined in Eqs. (13.2) and (13.42). The solution to Eq. (13.44) for radiation propagating along the y axis is

$$T_B(\nu) = T_{B0}(\nu)e^{-\tau_\nu} + \int_0^\infty \alpha(\nu, y)T(y)e^{-\tau_\nu'} dy , \tag{13.45}$$

where T_{B0} is the brightness temperature in the absence of absorption, including the cosmic background component,

$$\tau_\nu' = \int_0^y \alpha(\nu, y') dy' , \tag{13.46}$$

and

$$\tau_\nu = \int_0^\infty \alpha(\nu, y')\, dy' . \tag{13.47}$$

Here, y is the distance measured from the observer. τ_ν is called the *optical depth*, or *opacity*. The first term on the right side of Eq. (13.45) describes the absorption of the signal, and the second describes the emission contribution of the atmosphere. Equation (13.45) illustrates the fundamental law that an absorbing medium must also radiate. If $T(y)$ is constant throughout the medium, then Eq. (13.45) can be written

$$T_B(\nu) = T_{B0}(\nu)e^{-\tau_\nu} + T(1 - e^{-\tau_\nu}) . \tag{13.48}$$

The presence of absorption can have a very significant effect on system performance. If the receiver temperature is T_R, then the system temperature, which is the sum of T_R and the atmospheric brightness temperature (the effects of ground radiation being neglected), is

$$T_S = T_R + T_{at}(1 - e^{-\tau_\nu}) , \tag{13.49}$$

where T_{at} is the temperature of the atmosphere. In the absence of a source, the antenna temperature is taken as equal to the brightness temperature of the sky. Furthermore, if the brightness temperature scale is referenced to a point outside the atmosphere by multiplying the measurements of brightness temperature [see Eq. (13.48)] by e^{τ_ν}, then the effective system temperature is $T_S e^{\tau_\nu}$, or

$$T_S' = T_R e^{\tau_\nu} + T_{at}(e^{\tau_\nu} - 1) . \tag{13.50}$$

In effect, the atmospheric loss is modeled by an equivalent attenuator at the receiver input. Suppose that $T_R = 30\,\text{K}$, $T_{at} = 290\,\text{K}$, and $\tau_\nu = 0.2$; then the effective system temperature is 100 K. In such a situation, the atmosphere would degrade the system sensitivity by a factor of more than three. Note that the loss in sensitivity results primarily from the increase in system temperature rather than from the attenuation of the signal, which is only 20%. The emission from the atmosphere induces signals in spaced antennas that are uncorrelated and thus contributes only to the noise in the output of an interferometer.

The absorption can be estimated directly from measurements made with a radio telescope. In one technique introduced by Dicke et al. (1946), called the tipping-scan method, the opacity is determined from the atmospheric emission. If the antenna is scanned from the zenith to the horizon, the observed brightness temperature, in the absence of background sources, will depend on the zenith angle, since the opacity is proportional to the path length through the atmosphere, which varies approximately as $\sec z$. Thus, the atmospheric brightness temperature is

$$T_B = T_{at}(1 - e^{-\tau_0 \sec z}) , \tag{13.51}$$

where τ_0 is the zenith opacity. When $\tau_0 \sec z \ll 1$,

$$T_B \simeq T_{at} \tau_0 \sec z . \tag{13.52}$$

For narrow-beamed antennas, the antenna temperature is equal to the brightness temperature. For broad-beamed antennas, the antenna temperature is a zenith angle weighted version using Eq. (13.51). The opacity can be found from the slope of T_B plotted vs. $\sec z$, assuming that T_{at} is the surface temperature. The accuracy of this method is affected by ground pickup through the sidelobes, which varies as a function of zenith angle.

The opacity can also be estimated from measurements of the absorption suffered by a radio source over a range of zenith angles. The observed antenna temperature on-source minus the antenna temperature off-source at the same zenith angle to remove the emission [see Eq. (13.48)] is

$$\Delta T_A = T_{S0} e^{-\tau_0 \sec z} , \tag{13.53}$$

where T_{S0} is the component of antenna temperature due to the source in the absence of the atmosphere. From Eq. (13.53),

$$\ln \Delta T_A = \ln T_{S0} - \tau_0 \sec z . \tag{13.54}$$

Thus, τ_0 can be found without knowledge of T_A if a sufficient range in $\sec z$ is covered. This method is affected by changes in antenna gain as a function of zenith angle.

Another technique, called the chopper-wheel method, is commonly used at millimeter wavelengths. A wheel consisting of alternate open and absorbing sections is placed in front of the feed horn. As the wheel rotates, the radiometer alternately views the sky and the absorbing sections and synchronously measures the difference in antenna temperature between the sky and the chopper wheel at temperature T_0. Thus, the on-source and off-source antenna temperatures are

$$\Delta T_{on} = T_{S0} e^{-\tau_\nu} + T_{at}(1 - e^{-\tau_\nu}) - T_0 \tag{13.55}$$

and

$$\Delta T_{off} = T_{at}(1 - e^{-\tau_\nu}) - T_0 . \tag{13.56}$$

Table 13.1 Empirical coefficients for estimating opacity from surface absolute humidity[a]

ν	α_0	α_1
(GHz)	(nepers)	(nepers m^3 g^{-1})
15	0.013	0.0009
22.2	0.026	0.011
35	0.039	0.0030
90	0.039	0.0090

Source: Waters (1976).
[a]From the equation $\tau_0 = \alpha_0 + \alpha_1 \rho_{V0}$ fitted to opacity data derived from radiosonde measurements and measurements of surface absolute humidity, ρ_{V0} g m^{-3}.

These measurements can be combined to obtain T_0 and thereby eliminate the effect of atmospheric absorption. In the case in which $T_0 = T_{\mathrm{at}}$,

$$T_{S0} = \left(\frac{\Delta T_{\mathrm{off}} - \Delta T_{\mathrm{on}}}{\Delta T_{\mathrm{off}}} \right) T_0 . \tag{13.57}$$

When sensitivity is critical, the chopper wheel is used only to calibrate the output in the off-source position. $\Delta T_{\mathrm{off}} - \Delta T_{\mathrm{on}}$ in the numerator of Eq. (13.57) is then replaced by $T_{\mathrm{off}} - T_{\mathrm{on}}$. Measurement of T_{S0} provides the flux density of the source, which determines the visibility at the origin of the (u, v) plane.

The opacity can be estimated also from surface meteorological measurements when other data are not available. This method is not as accurate as the direct radiometric measurement techniques described above but has the advantage of not expending observing time. Waters (1976) has analyzed data on absorption vs. surface water vapor density for a sea-level site at various frequencies by fitting them to an equation of the form $\tau_0 = \alpha_0 + \alpha_1 \rho_{V0}$. The coefficients α_0 and α_1 are listed in Table 13.1.

13.1.4 Origin of Refraction

For practical reasons, we have discussed separately the effects of the propagation delay and the absorption in the neutral atmosphere. However, the delay and the absorption are intimately related because they are derived from the real and imaginary parts of the dielectric constant of the gas in the atmosphere. The real and imaginary parts of the dielectric constant are not independent but are related by the Kramers–Kronig relation, which is similar to the mathematical relation known as the Hilbert transform (Van Vleck et al. 1951; Toll 1956). We now discuss this relationship from the physical viewpoint of the classical theory of dispersion. From this analysis, it will become clear why the atmospherically induced delay is essentially independent of frequency, even in the vicinity of spectral lines that cause significant absorption.

A dilute gas of molecules can be modeled as bound oscillators. In each molecule, an electron with mass m and charge $-e$ is harmonically bound to the nucleus, and the electron's motion is characterized by a resonance frequency v_0 and damping constant $2\pi\Gamma$. The equation of motion with a harmonic driving force $-eE_0e^{-j2\pi vt}$ caused by the electric field of an electromagnetic wave can be approximated as

$$m\ddot{x} + 2\pi m\Gamma\dot{x} + 4\pi^2 mv_0^2 x = -eE_0e^{-j2\pi vt} , \tag{13.58}$$

where x is the displacement of the bound electron, E_0 and v are the amplitude and frequency of the applied electric field, and the dots denote time derivatives. The steady-state solution has the form $x = x_0e^{-j2\pi vt}$, where

$$x_0 = \frac{eE_0/4\pi^2 m}{v^2 - v_0^2 + jv\Gamma} . \tag{13.59}$$

The magnitude of the dipole moment per unit volume, $\mathbf{P}$, is equal to $-n_m ex_0$, where n_m is the density of gas molecules. The dielectric constant[1] ε is equal to $1 + \mathbf{P}/(\epsilon_0\mathbf{E})$, so that

$$\varepsilon = 1 - \frac{n_m e^2/4\pi^2 m\epsilon_0}{v^2 - v_0^2 + jv\Gamma} . \tag{13.60}$$

This classical model predicts neither the resonance frequency nor the absolute amplitude of the oscillation. A full treatment of the problem requires the application of quantum mechanics. The proper quantum-mechanical calculation for a system with many resonances yields a result that closely resembles Eq. (13.60) [e.g., Loudon (1983)]:

$$\varepsilon = 1 - \frac{n_m e^2}{4\pi^2 m\epsilon_0} \sum_i \frac{f_i}{v^2 - v_{0i}^2 + jv\Gamma_i} , \tag{13.61}$$

where f_i is the so-called oscillator strength of the ith resonance. The f_i values obey the sum rule, $\sum f_i = 1$.

---

[1] In this section and in Sect. 13.3, we use SI (System International) units, also known as rationalized MKS units. In this system, the constitutive relation between the displacement vector $\mathbf{D}$, the electric field vector $\mathbf{E}$, and the polarization vector $\mathbf{P}$ is $\mathbf{D} = \epsilon_0\mathbf{E} + \mathbf{P} = \epsilon\mathbf{E}$, where ϵ_0 is the permittivity of free space, and ϵ is the permittivity of the medium. The dielectric constant ε is ϵ/ϵ_0. A comparison of various systems of units and equations in electricity and magnetism can be found in Jackson (1999).

The dielectric constant ($\varepsilon = \varepsilon_R + j\varepsilon_I$) and index of refraction ($n = n_R + jn_I$) are connected by Maxwell's relation,

$$n^2 = \varepsilon .$$

(13.62)

Thus, $\varepsilon_R = n_R^2 - n_I^2$ and $\varepsilon_I = 2n_I n_R$. Since for a dilute gas $n_R \simeq 1$ and $n_I \ll 1$, we have $n_R \simeq \sqrt{\varepsilon_R}$ and $n_I \simeq \varepsilon_I/2$. Therefore, for a gas with a single resonance,

$$n_R \simeq 1 - \frac{n_m e^2 (v^2 - v_0^2)/8\pi^2 m\epsilon_0}{(v^2 - v_0^2)^2 + v^2\Gamma^2}$$

(13.63)

and

$$n_I \simeq \frac{n_m e^2 v\Gamma/8\pi^2 m\epsilon_0}{(v^2 - v_0^2)^2 + v^2\Gamma^2} .$$

(13.64)

The resonance is usually sharp, that is, $\Gamma \ll v_0$, and the expressions for n_R and n_I can be simplified by considering their behavior in the vicinity of the resonance frequency v_0, in which case

$$v^2 - v_0^2 = (v + v_0)(v - v_0) \simeq 2v_0(v - v_0) .$$

(13.65)

Thus

$$n_R \simeq 1 - \frac{2b(v - v_0)}{(v - v_0)^2 + \Gamma^2/4} ,$$

(13.66)

and

$$n_I \simeq \frac{b\Gamma}{(v - v_0)^2 + \Gamma^2/4} ,$$

(13.67)

where $b = n_m e^2/32\pi^2 m\epsilon_0 v_0$.

Equation (13.67) defines an unnormalized Lorentzian profile for n_I that is symmetric about frequency v_0 and has a full width at half-maximum of Γ and a peak amplitude of $4b/\Gamma$. The function $n_R - 1$ is antisymmetric about frequency v_0 and has extreme values of $\pm 2b/\Gamma$ at frequencies $v_0 \pm \Gamma/2$, respectively. The functions n_R and n_I are plotted in Fig. 13.8. Note that the peak deviation from unity in the real part of the index of refraction, Δn, is equal to one-half the peak value of n_I, denoted n_{Imax}. Thus, from Eq. (13.2), we see that the peak absorption coefficient, $\alpha_m = 4\pi n_{Imax} v_0/c$, is related to Δn by the formula

$$\Delta n = \frac{\alpha_m \lambda_0}{8\pi} ,$$

(13.68)

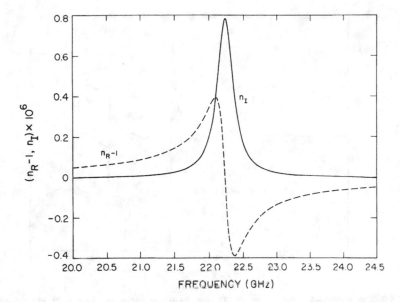

Fig. 13.8 Real and imaginary parts of the index of refraction vs. frequency for a single resonance, given by Eqs. (13.63) and (13.64). The case shown is for the 6_{16}–5_{23} transition in *pure* water vapor with $\rho_V = 7.5\,\mathrm{g\,m^{-3}}$. In the atmosphere at the standard sea-level pressure of 1013 mb, the line is broadened to about 2.6 GHz (Liebe 1969). For the curve $n_R - 1$, the peak deviation is Δn [see Eq. (13.68)], and the change in level passing through the line is δn [see Eq. (13.69)].

where λ_0 is the wavelength of the resonance, c/ν_0. The magnitude of the real part of the index of refraction is equal to the peak absorption over a distance of $\lambda_0/8\pi$. In addition, Eq. (13.66) shows that the real part of the index of refraction is not exactly symmetric about ν_0; that is, n_R tends to unity as ν tends to ∞, and n_R tends to $1 + 2b/\nu_0 = 1 + \Delta n\Gamma/\nu_0 = 1 + (\lambda_0\alpha_m/8\pi)(\Gamma/\nu_0)$ as ν tends to zero. Hence, the change δn in the asymptotic value of the index of refraction on passing through a resonance is given by

$$\delta n = \frac{\alpha_m \Gamma \lambda_0^2}{8\pi c}. \tag{13.69}$$

Thus, $\delta n/\Delta n = \gamma/\nu_0$, but unless the resonance is extremely strong, Δn and δn are both negligible. Consider the 22-GHz water vapor line. The attenuation in the atmosphere when $\rho_V = 7.5\,\mathrm{g\,m^{-3}}$ is 0.15 dB $\mathrm{km^{-1}}$, so $\alpha_m = 3.5 \times 10^{-7}\,\mathrm{cm^{-1}}$. Equation (13.68) then predicts that $\Delta n = 1.9 \times 10^{-8}$, or $\Delta N = 0.019$, which agrees with the value measured in the laboratory (Liebe 1969). For the same value of ρ_V, the contribution of all transitions of water vapor to the value of the index of refraction at low frequencies ($10^{-6}N_V$), from Eq. (13.10), is equal to 4.4×10^{-5}. Thus, the fractional change in refractivity near the 22-GHz line is only 1 part in 2500. The change in asymptotic level is even smaller. At sea level, $\Gamma = 2.6\,\mathrm{GHz}$ and $\delta n = 2.2 \times 10^{-8}$. The water vapor line at 557 GHz (the 1_{10}–1_{01} transition)

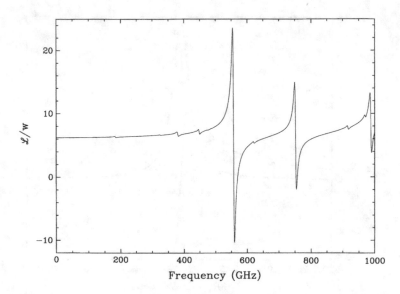

Fig. 13.9 The predicted excess path length due to water vapor per unit column density vs. frequency, from formulas by Liebe (1989) for $T = 270\,\text{K}$ and $P = 750\,\text{mb}$. From Sutton and Hueckstaedt 1996, reproduced with permission. © ESO.

has an absorption coefficient of $29,000\,\text{dB}\,\text{km}^{-1}$, or $0.069\,\text{cm}^{-1}$. The values of Δn and δn are 1.44×10^{-6} and 0.7×10^{-6}, respectively. In the atmospheric windows above $400\,\text{GHz}$, where radio astronomical observations are possible only from very dry sites, the refractive index can be noticeably different from the value at lower frequencies. The normalized refractivity is shown in Fig. 13.9.

Equation (13.68) is an important result of very general validity. We derived it from a specific model [Eq. (13.58)] that led to an approximately Lorentzian profile for the absorption spectrum. In practice, line profiles are found to differ slightly from the Lorentzian form, and more sophisticated models are needed to fit them exactly. However, Eqs. (13.68) and (13.69) could be derived from the Kramers–Kronig relation.

The low-frequency value of the index of refraction, as given by Eq. (13.9), results from the contributions of all transitions at higher frequencies. Summing the contributions [see Eq. (13.69)] of many lines, each characterized by parameters Δn_1, Γ_i, α_{mi}, and ν_{0i}, we obtain the low-frequency value of the index of refraction:

$$n_S = 1 + \sum_i \frac{\alpha_{mi}\lambda_{0i}^2\Gamma_i}{8\pi c} = 1 + \sum_i \frac{\Delta n_i\Gamma_i}{\nu_{0i}} \,. \qquad (13.70)$$

The water vapor molecule has a large number of strong rotational transitions in the band from $10\,\mu\text{m}$ to $0.3\,\text{mm}$ (from $30\,\text{THz}$ to $1000\,\text{GHz}$). The atmosphere is opaque through most of this region because of these lines, which contribute about 98% of the low-frequency refractivity. The remainder comes from the 557-GHz line.

Grischkowsky et al. (2013) show that the full theoretical calculation behind Eq. (13.70), based on Van Vleck–Weisskopf line shapes and incorporating all water lines from 22.2 GHz through 30 THz, gives agreement with the empirical expression for refractivity without any ad hoc corrections. Complete computer codes for the atmospheric absorption and refraction have been developed by Pardo et al. (2001a) and Paine (2016).

13.1.5 Radio Refractivity

A detailed discussion of the radio refractivity equation can be found in the report of a working group of the International Association of Geodesy (Rüeger 2002). Previous work on combining laboratory measurements includes Bean and Dutton (1966), Thayer (1974), Hill et al. (1982), and Bevis et al. (1994). From the classic work of Debye (1929), it can be shown that the refractivity of molecules with induced dipole transitions varies as pressure and T^{-1}, and the refractivity of molecules with permanent dipole moments varies as pressure and T^{-2}. The principal constituents of the atmosphere—oxygen molecules, O_2, and nitrogen molecules, N_2,—being homonuclear, have no permanent electric dipole moments. However, molecules such as H_2O and other minor trace constituents have permanent dipole moments. Thus, the general form of the refractivity equation is

$$N = \frac{K_1 p_D}{T \mathcal{Z}_D} + \frac{K_2 p_V}{T \mathcal{Z}_V} + \frac{K_3 p_V}{T^2 \mathcal{Z}_V} , \qquad (13.71)$$

where p_D and p_V are the partial pressures of the dry air and water vapor; K_1, K_2, and K_3 are constants; and $\mathcal{Z}_D$ and $\mathcal{Z}_V$ are compressibility factors for dry-air gases and water vapor, which correct for nonideal gas behavior and deviate from unity in atmospheric conditions by less than 1 part in 10^3. These compressibility factors are given by Owens (1967) but are usually assumed to be equal to unity and their effects absorbed into the K coefficients.

The first and second terms in Eq. (13.71) are due to ultraviolet electronic transitions of the induced dipole type for dry-air molecules and water vapor, respectively, and the third term is due to the permanent dipole infrared rotational transitions of water vapor. The best values of the parameters are $K_1 = 77.6898$, $K_2 = 71.2952$, and $K_3 = 375463$, based on a weighted average of all available experimentally derived values before 2002, as presented by Rüeger (2002). These values were the result of working groups of the IUGG and the IAG. Thus, as in Eq. (13.6),

$$N = 77.6898 \frac{p_D}{T} + 71.2952 \frac{p_V}{T} + 375463 \frac{p_V}{T^2} . \qquad (13.72)$$

The accuracy of this expression at the zero-frequency limit is conservatively estimated to be 0.02% for the p_D term and 0.2% for the p_V terms. We can rewrite Eq. (13.72) in terms of the total pressure ($P = p_D + p_V$) as

$$N = 77.7\frac{P}{T} - 6.4\frac{p_V}{T} + 375463\frac{p_V}{T^2} . \tag{13.73}$$

For temperatures around 280 K, the last two terms on the right side of Eq. (13.73) can be combined to give the well-known two-term *Smith–Weintraub equation* (Smith and Weintraub 1953) that has been widely used in the radio science community. Using the best available parameters in 1953, this equation is

$$N \simeq \frac{77.6}{T}\left(P + 4810\frac{p_V}{T}\right) . \tag{13.74}$$

The accuracy of Eqs. (13.73) and (13.74) at frequencies above zero can be improved by adding a small term that increases monotonically with frequency to account for the effect of the wings of the infrared transitions (see Fig. 13.9). Hill and Clifford (1981) show that because of this effect, the wet refractivity increases by about 0.5% at 100 GHz, and 2% at 200 GHz, over its value at low frequencies.

It is interesting to compare the refractivities at radio and optical wavelengths. The term proportional to T^{-2} is due to the infrared resonances of H_2O, because of its permanent dipole moment, and does not affect the optical refractivity. On the other hand, the terms proportional to T^{-1} arise from the induced dipole moments associated with resonances of oxygen and nitrogen and also water vapor in the ultraviolet. Hence, to a first approximation, we estimate the optical refractivity by omitting the permanent dipole term from Eq. (13.72) and obtain

$$N_{\text{opt}} \simeq 77.7\frac{p_D}{T} + 71.3\frac{p_V}{T} . \tag{13.75}$$

For precise work, Cox (2000) and Rüeger (2002) provide more accurate values for N_{opt} that include small terms having wavelength dependence to account for the effects of the wings of ultraviolet transitions that cause it to increase about 3% going from 1 to 0.3 μm. The ratio of the wet refractivity in the radio and optical regions is obtained by omitting the dry-air terms from Eqs. (13.72) and (13.75): $N_{V\text{rad}}/N_{V\text{opt}} \simeq 1 + 5830/T$. For $T \simeq 280$ K, this ratio is about equal to 22. Hence, water vapor plays a much more prominent role in propagation issues in radio than in optical astronomy.

13.1.6   Phase Fluctuations

In the radio region, the most important nonuniformly distributed quantity in the troposphere is the water vapor density. Variations in water vapor distribution in

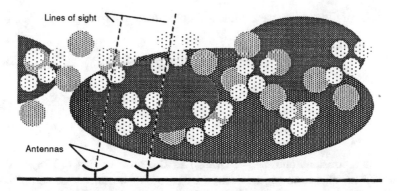

Lines of sight

Antennas

Fig. 13.10 A cartoon of a two-element interferometer beneath a tropospheric screen of water vapor irregularities of various sizes. The screen moves over the interferometer at a velocity component v_s parallel to the baseline. The distribution of these irregularities is important in designing the phase compensation schemes discussed in Sect. 13.2. Note that fluctuations with scale sizes larger than the baseline cover both antennas and do not affect the interferometer phase significantly. From Masson (1994a), courtesy of and © the Astronomical Society of the Pacific.

the troposphere that move across an interferometer cause phase fluctuations that degrade the measurements. In the optical region, variations in temperature, rather than in water vapor content, are the principal cause of phase fluctuations. The situation is depicted in Fig. 13.10. A critical dimension is the size of the first Fresnel zone, $\sqrt{\lambda h}$, where h is the distance between the observer and the screen. For $\lambda = 1\,\mathrm{cm}$ and $h = 1\,\mathrm{km}$, the Fresnel scale is about 3 m. The atmospherically induced phase fluctuations on this scale are very small ($\ll 1\,\mathrm{rad}$). In this case, the phase fluctuation can cause image distortion but not amplitude fluctuation (i.e., scintillation). This is known as the regime of weak scattering. Plasma scattering in the interstellar medium belongs to the regime of strong scattering, where the phenomena are considerably more complex (see Sect. 14.4).

The fluctuations along an initially plane wavefront that has traversed the atmosphere can be characterized by a so-called structure function of the phase. This function is defined as

$$D_\phi(d) = \langle [\Phi(x) - \Phi(x-d)]^2 \rangle , \qquad (13.76)$$

where $\Phi(x)$ is the phase at point x, $\Phi(x - d)$ is the phase at point $x - d$, and the angle brackets indicate an ensemble average. In practical applications, the ensemble average must be approximated by a time average of suitable duration. We assume that D_ϕ depends only on the magnitude of the separation between the measurement points, that is, the projected baseline length d of the interferometer. The rms deviation in the interferometer phase is

$$\sigma_\phi = \sqrt{D_\phi(d)} . \qquad (13.77)$$

For the sake of illustration, we assume a simple functional form for σ_ϕ given by

$$\sigma_\phi = \frac{2\pi a d^\beta}{\lambda}, \qquad d \le d_m, \qquad (13.78a)$$

and

$$\sigma_\phi = \sigma_m, \qquad d > d_m, \qquad (13.78b)$$

where a is a constant, and $\sigma_m = 2\pi a d_m^\beta / \lambda$. The form of Eqs. (13.78) is shown in Fig. 13.11a. This form can be derived by assuming a multiple-scale power-law model for the spectrum of the phase fluctuations. There is a limiting distance d_m beyond which fluctuations do not increase noticeably, a few kilometers, roughly the size of clouds. This limit is called the outer scale length of the fluctuations. Beyond this dimension, the fluctuations in the path lengths become uncorrelated.

First, consider an interferometer that operates in the domain of baselines shorter than d_m. The measured visibilities $\mathcal{V}_m$ are related to the true visibilities $\mathcal{V}$ by the

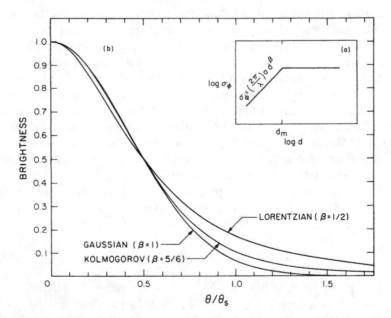

Fig. 13.11 (a) Simple model for the rms phase fluctuation induced by the troposphere in an interferometer of baseline length d given by Eqs. (13.78). (b) The point-source response function $\overline{w}_a(\theta)$ for various power-law models is obtained by taking the Fourier transform of the visibility in the regime $d < d_m$. The values of θ_s, the full width at half-maximum of $\overline{w}_a(\theta)$, for each model are: Gaussian ($\beta = 1$), $\sqrt{8 \ln 2}\, a$; modified Lorentzian ($\beta = \frac{1}{2}$), $1.53\pi\lambda^{-1}a^2$; and Kolmogorov ($\beta = \frac{5}{6}$), $2.75\lambda^{-1/5}a^{6/5}$. λ is the wavelength and a is the constant defined in Eq. (13.78a).

equation

$$\mathcal{V}_m = \mathcal{V}e^{j\phi} , \tag{13.79}$$

where $\phi = \Phi(x) - \Phi(x - d)$ is a random variable describing the phase fluctuations introduced by the atmosphere. If we assume ϕ is a Gaussian random variable with zero mean, then the expectation of the visibility is

$$\langle \mathcal{V}_m \rangle = \mathcal{V}\langle e^{j\phi} \rangle = \mathcal{V}e^{-\sigma_\phi^2/2} = \mathcal{V}e^{-D_\phi/2} . \tag{13.80}$$

Consider the conceptually useful case in which $\beta = 1$. It would arise in an atmosphere consisting of inhomogeneous wedges of scale size larger than the baseline. In this case, σ_ϕ is proportional to d, and the constant a is dimensionless. Substituting Eq. (13.78a) into Eq. (13.80) yields

$$\langle \mathcal{V}_m \rangle = \mathcal{V}e^{-2\pi^2 a^2 q^2} , \tag{13.81}$$

where $q = \sqrt{u^2 + v^2} = d/\lambda$. On average, therefore, the measured visibility is the true visibility multiplied by an atmospheric weighting function $w_a(q)$ given by

$$w_a(q) = e^{-2\pi^2 a^2 q^2} . \tag{13.82}$$

In the image plane, the derived map is the convolution of the true source distribution and the Fourier transform of $w_a(q)$, which is

$$\overline{w}_a(\theta) \propto e^{-\theta^2/2a^2} , \tag{13.83}$$

where θ is here the conjugate variable of q. The full width at half-maximum of $\overline{w}_a(\theta)$ is θ_s, given by

$$\theta_s = \sqrt{8 \ln 2}\, a . \tag{13.84}$$

Thus, the resolution is degraded because the derived map is convolved with a Gaussian beam of width θ_s (in addition to the effects of any other weighting functions, as described in Sect. 10.2.2). θ_s is the seeing angle. Images with finer resolution than θ_s can often be obtained by use of adaptive calibration procedures described in Sect. 11.3. Now, from Eq. (13.78a), we obtain

$$a = \frac{\sigma_\phi \lambda}{2\pi d} = \frac{\sigma_d}{d} , \tag{13.85}$$

where $\sigma_d = \sigma_\phi \lambda/2\pi$ is the rms uncertainty in path length. Thus, we obtain

$$\theta_s = 2.35 \frac{\sigma_d}{d} \quad \text{(radians)} . \tag{13.86}$$

Since σ_d/d is constant, θ_s is independent of wavelength. This independence results from the condition $\beta = 1$ in Eq. (13.78a). For the radio regime, σ_d is about 1 mm on a baseline of 1 km, so $a \simeq 10^{-6}$ and $\theta_s \simeq 0.5''$. Let d_0 be the baseline length for which $\sigma_\phi = 1$ rad. From Eq. (13.85), we see that Eq. (13.84) can be written in the form

$$\theta_s = \frac{\sqrt{2\ln 2}}{\pi}\frac{\lambda}{d_0} \simeq 0.37\frac{\lambda}{d_0} \; . \tag{13.87}$$

For the case in which β is arbitrary, we find $\overline{w}_a(\theta)$ by substituting Eq. (13.78a) into Eq. (13.80) and writing the two-dimensional Fourier transform as a Hankel transform (Bracewell 2000). Thus

$$\overline{w}_a(\theta) \propto \int_0^\infty \exp\left[-2\pi^2 a^2 \lambda^{2(\beta-1)} q^{2\beta}\right] J_0(2\pi q\theta)\, q\, dq \; , \tag{13.88}$$

where J_0 is the Bessel function of order zero and a has dimensions cm$^{(1-\beta)}$. In general, $\overline{w}_a(\theta)$ cannot be evaluated analytically. However, by making appropriate substitutions in Eq. (13.88), it is easy to show that $\theta_s \propto a^{1/\beta}\lambda^{(\beta-1)/\beta}$. A case that can be treated analytically is the one for which $\beta = \frac{1}{2}$. In this case, we obtain (Bracewell 2000, p. 338)

$$\overline{w}_a(\theta) \propto \frac{1}{[\theta^2 + (\pi a^2/\lambda)^2]^{3/2}} \; , \tag{13.89}$$

which represents a Lorentzian profile raised to the 3/2 power and has very broad skirts. The full width at half-maximum of $\overline{w}_a(\theta)$ is

$$\theta_s = \frac{1.53\pi a^2}{\lambda} \; , \tag{13.90}$$

or

$$\theta_s = \frac{0.77}{2\pi}\frac{\lambda}{d_0} \simeq 0.12\frac{\lambda}{d_0} \; . \tag{13.91}$$

In the case of Kolmogorov turbulence, which is discussed later in this section, $\beta = 5/6$. Numerical integration of Eq. (13.88) yields

$$\theta_s \simeq 2.75 a^{6/5}\lambda^{-1/5} \simeq 0.30\frac{\lambda}{d_0} \; . \tag{13.92}$$

Plots of $\overline{w}_a(\theta)$ for various power-law models of phase fluctuations are shown in Fig. 13.11b.

Now consider the case of an interferometer operating in the domain of baselines greater than d_m, where σ_ϕ is a constant equal to σ_m. This case is most applicable to VLBI arrays or to large connected-element arrays. If the timescale of the fluctuation is short with respect to the measurement time, then, on average, all the visibility measurements are reduced by a constant factor $e^{-\sigma_m^2/2}$. Thus, this type of atmospheric fluctuation does not reduce the resolution. However, on average, the measured flux density is reduced from the true value by the factor $e^{-\sigma_m^2/2}$. If the timescale of the fluctuations is long with respect to the measurement time, then each visibility measurement suffers a phase error $e^{j\phi}$. Assume that K visibility measurements are made of a point source of flux density S. The image of the source, considering only one dimension for simplicity, is

$$\overline{w}_a(\theta) = \frac{S}{K} \sum_{i=1}^{K} e^{j\phi_i} e^{j2\pi u_i \theta} . \tag{13.93}$$

The expectation of $\overline{w}_a(\theta)$ at $\theta = 0$ is

$$\langle \overline{w}_a(0) \rangle = S e^{-\sigma_m^2/2} . \tag{13.94}$$

The measured flux density is less than S. (Note: $\langle \overline{w}_a(0) \rangle /S$ is sometimes called the coherence factor of the interferometer.) The missing flux density is scattered around the map. This is immediately evident from Parseval's theorem:

$$\sum_i \left|\overline{w}_a(\theta_i)\right|^2 = \frac{1}{K} \sum_i \left|\mathcal{V}(u_i)\right|^2 = S^2 . \tag{13.95}$$

Thus, the total flux density could be obtained by integrating the square of the image-plane response. The rms deviation in the flux density, measured at the peak response for a source at $\theta = 0$, is $\sqrt{\langle \overline{w}_a^2(\theta) \rangle - \langle \overline{w}_a(\theta) \rangle^2}$, which we call σ_S. This quantity can be calculated from Eq. (13.93) and is given by

$$\sigma_S = \frac{S}{\sqrt{K}} \sqrt{1 - e^{-\sigma_m^2}} , \tag{13.96}$$

which reduces to $\sigma_S \simeq S\sigma_m/\sqrt{K}$ when $\sigma_m \ll 1$.

13.1.7 Kolmogorov Turbulence

The theory of propagation through a turbulent neutral atmosphere has been treated in detail in the seminal publications of Tatarski (1961, 1971). This theory has been developed and applied extensively to problems of optical seeing [e.g., Roddier (1981), Woolf (1982), Coulman (1985)] and to infrared interferometry (Sutton et al. 1982). We confine the discussion here to a few central ideas concerning the structure

function of phase and indicate how it is related to other functions that are used to
characterize atmospheric turbulence.

When the Reynolds number (a dimensionless parameter that involves the vis-
cosity, a characteristic scale size, and the velocity of a flow) exceeds a critical
value, the flow becomes turbulent. In the atmosphere, the Reynolds number is nearly
always high enough that turbulence is fully developed. In the Kolmogorov model
for turbulence, the kinetic energy associated with large-scale turbulent motions
is transferred to smaller and smaller scale sizes of turbulence until it is finally
dissipated into heat by viscous friction. If the turbulence is fully developed and
isotropic, then the two-dimensional power spectrum of the phase fluctuations (or the
refractive index) varies as $q_s^{-11/3}$, where q_s (cycles per meter) is the spatial frequency
(q_s, the conjugate variable of d, is analogous to q, the conjugate variable of θ). The
structure function for the refractive index $D_n(d)$ is defined in a fashion similar to
the structure function of phase in Eq. (13.76); that is, $D_n(d)$ is the mean-squared
deviation of the difference in the refractive index at two points a distance d apart, or
$D_n(d) = \langle [n(x) - n(x-d)]^2 \rangle$. Note that only the scalar separation d is important for
isotropic turbulence. For the conditions stated above, D_n can be shown to be given
by the equation

$$D_n(d) = C_n^2 d^{2/3} , \qquad d_{\text{in}} \ll d \ll d_{\text{out}} , \qquad (13.97)$$

where d_{in} and d_{out} are called the inner and outer scales of turbulence, which may
be less than a centimeter and a few kilometers, respectively. The parameter C_n^2
characterizes the strength of the turbulence. Note that water vapor, which is the
dominant cause of fluctuation in the index of refraction, is poorly mixed in the
troposphere and therefore may be only an approximate tracer of the mechanical
turbulence.

The details of the derivation structure function of phase from the structure
function of the index of refraction given in Eq. (13.97) are given in Appendix 13.2.
The result is that $D_\phi(d)$ for a uniform layer of turbulence of thickness L has several
important power-law segments:

$$\begin{aligned}
D_\phi(d) &\sim d^{5/3} , & d_r < d < d_2 , \\
&\sim d^{2/3} , & d_2 < d < d_{\text{out}} , \\
&\sim d^0 , & d_{\text{out}} < d .
\end{aligned} \qquad (13.98)$$

d_r is the limit where diffractive effects become important. $d_r \simeq \sqrt{L\lambda}$, so for
an atmospheric layer of $L = 2$ km, d_r varies from 1.4 to 40 m for λ ranging
from 1 mm to 1 m. This inner turbulence scale d_{in} is considerably smaller and of
interest only at optical wavelengths. d_2 marks the transition from 3-D turbulence
and 2-D turbulence caused by the thickness of the layer. Stotskii (1973 and 1976)
was the first to recognize the importance of this break for radio arrays (see also
Dravskikh and Finkelstein 1979). d_{out} is the distance beyond which the fluctuations

are uncorrelated, as described in Sect. 13.1.6. d_{out} is nominally the scale size of clouds, a few kilometers. However, some correlation remains out to the scale size of weather systems and beyond.

The structure function is formally an ensemble average. For practical purposes, the turbulent eddies are assumed to remain fixed as the atmospheric layer moves across an array. This is the frozen-screen hypothesis, sometimes attributed to Taylor (1938). Practically, the rms fluctuations in phase increase with time up to the cross time $t_c = d/v_s$, where v_s is the wind speed parallel to the baseline direction corresponding to d. t_c is called the corner time, beyond which the rms fluctuations flatten out and $D_\phi(d)$ can be estimated. Atmospheric fluctuations on scales larger than d cover both receiving elements and do not contribute to the structure function. An example of the structure function as a function of time measured at the ALMA site by the 300-m satellite site-testing interferometer is shown in Fig. 13.12. $t_c \sim 20$ s, implying a wind speed of about 15 m s^{-1}.

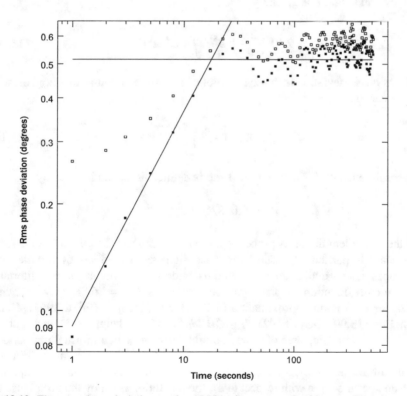

Fig. 13.12 The rms phase deviation at the ALMA site, measured with a satellite site-testing interferometer with a 300-m baseline at 11 GHz. The open symbols represent actual measurements; the solid symbols have the instrument noise removed. The line through the data has a slope of 0.6, as approximately expected by Kolmogorov theory [see Eq. (13.108)]. The break in the slope of the data occurs at the instrument crossing time, t_c. An estimate of the ensemble average of the structure function is reached for $t > t_c$. From Holdaway et al. (1995a).

We continue this section with a discussion of the primary case of 3-D turbulence in which $D_\phi \sim d^{5/3}$ and then generalize the results for other power-law indices. As derived in Appendix 13.2, for a uniform turbulence layer,

$$D_\phi(d) = 2.91 \left(\frac{2\pi}{\lambda} \right)^2 C_n^2 L \, d^{5/3} , \qquad (13.99)$$

which is valid in the range $\sqrt{L\lambda} \ll d \ll L$. The lower limit on d is equivalent to the requirement that diffraction effects be negligible. Note that the factor 2.91 is a dimensionless constant, and C_n^2 has units of length$^{-2/3}$. This factor appears in calculations based on D_n as defined in Eq. (13.97). (It is sometimes absorbed into C_n^2.)

We can generalize Eq. (13.99) for a stratified turbulent layer. The structure function of phase for an atmosphere in which C_n^2 varies with height from the surface to an overall height L is given by

$$D_\phi(d) = 2.91 \left(\frac{2\pi}{\lambda} \right)^2 d^{5/3} \int_0^L C_n^2(h) \, dh . \qquad (13.100)$$

The rms phase deviation is the square root of the phase structure function, or, when C_n^2 is a constant,

$$\sigma_\phi = 1.71 \left(\frac{2\pi}{\lambda} \right) \sqrt{C_n^2 L} \, d^{5/6} . \qquad (13.101)$$

The baseline length for which $\sigma_\phi = 1$ rad is defined as d_0 and is given by

$$d_0 = 0.058 \lambda^{6/5} (C_n^2 L)^{-3/5} . \qquad (13.102)$$

Another scale length that is proportional to d_0 is the Fried length, d_f (Fried 1966). This scale is particularly useful for discussions of the effects of turbulence in telescopes with circular apertures and is widely used in the optical literature. The structure function of phase can be written as $D_\phi = 6.88(d/d_f)^{5/3}$, where the factor 6.88 is an approximation of $2[(24/5)\Gamma(6/5)]^{5/6}$ (Fried 1967). Hence, from Eqs. (13.99) and (13.102), $d_f = 3.18d_0$. The Fried length is defined such that the effective collecting area of a large circular aperture with uniform illumination in the presence of Kolmogorov turbulence is $\pi d_f^2/4$. Hence, for an aperture of diameter small with respect to d_f, the resolution is dominated by diffraction at the aperture. With an aperture large with respect to d_f, the resolution is set by the turbulence and is approximately λ/d_f. The exact resolution in this latter case can be derived from Eq. (13.92), with the result $\theta_s = 0.97\lambda/d_f$. In addition, the rms phase error over an aperture of diameter d_f is 1.01 rad. The reason that d_f is larger than d_0 is related to the downweighting of long baselines in two-dimensional apertures [see Eq. (15.13) and related discussion]. For an aperture of diameter d_f, the ratio of the collecting area to the geometric area, which is called the Strehl ratio in the optical literature, is equal to 0.45 (Fried 1965).

Equation (13.102) shows that d_0 is proportional to $\lambda^{6/5}$, and thus the angular resolution or seeing limit ($\sim \lambda/d_0$) is proportional to $\lambda^{-1/5}$ [see Fig. 13.11 and Eq. (13.92)]. This relationship may hold over broad wavelength ranges when C_n^2 is constant. In the optical range, C_n^2 is related to temperature fluctuations, whereas in the radio range, C_n^2 is dominated by turbulence in the water vapor. It is an interesting coincidence that the seeing angle is about $1''$ at both optical and radio wavelengths, for good sites. The important difference is the timescale of fluctuations, τ_{cr}. If the critical level of fluctuation is 1 radian, then $\tau_{cr} \simeq d_0/v_s$, where v_s is the velocity component of the screen parallel to the baseline. Any adaptive optics compensation must operate on a timescale short with respect to τ_{cr}. From Eq. (13.92), τ_{cr} can be expressed as

$$\tau_{cr} \simeq 0.3 \frac{\lambda}{\theta_s v_s} \ . \tag{13.103}$$

For $v_s = 10 \text{ m s}^{-1}$ and $\theta_s = 1''$, $\tau_{cr} = 3 \text{ ms}$ at $0.5 \ \mu\text{m}$ wavelength and 60 s at 1 cm wavelength.

The two-dimensional power spectrum of phase, $S_2(q_x, q_y)$, is the Fourier transform of the two-dimensional autocorrelation function of phase, $R_\phi(d_x, d_y)$. If R_ϕ is a function only of d, where $d^2 = d_x^2 + d_y^2$, then S_2 is a function of q_s, where $q_s^2 = q_x^2 + q_y^2$, and $S_2(q_s)$ and $R_\phi(d)$ form a Hankel transform pair. Since $D_\phi(d) = 2[R_\phi(0) - R_\phi(d)]$, we can write

$$D_\phi(d) = 4\pi \int_0^\infty [1 - J_0(2\pi q_s d)] S_2(q_s) q_s \, dq_s \ , \tag{13.104}$$

where J_0 is the Bessel function of order zero. When $D_\phi(d)$ is given by Eq. (13.100), $S_2(q_s)$ is

$$S_2(q_s) = 0.0097 \left(\frac{2\pi}{\lambda} \right)^2 C_n^2 L q_s^{-11/3} \ . \tag{13.105}$$

It is often useful to study temporal variations caused by atmospheric turbulence. In order to relate the temporal and spatial variations, we invoke the frozen-screen hypothesis. The one-dimensional temporal spectrum of the phase fluctuations $S_\phi'(f)$ (the two-sided spectrum) can be calculated from $S_2(q_s)$ by

$$S_\phi'(f) = \frac{1}{v_s} \int_{-\infty}^\infty S_2 \left(q_x = \frac{f}{v_s}, q_y \right) dq_y \ , \tag{13.106}$$

where v_s is in meters per second. Substitution of Eq. (13.105) into Eq. (13.106) yields

$$S'_\phi(f) = 0.016 \left(\frac{2\pi}{\lambda}\right)^2 C_n^2 L v_s^{5/3} f^{-8/3} \quad (\text{rad}^2 \text{ Hz}^{-1}) . \tag{13.107}$$

Examples of the temporal spectra of water vapor fluctuations can be found in Hogg et al. (1981) and Masson (1994a) (see Fig. 13.17). The temporal structure function $D_\tau(\tau) = \langle [\phi(t) - \phi(t - \tau)]^2 \rangle$ is related to the spatial structure function by $D_\tau(\tau) = D_\phi(d = v_s\tau)$. Hence, for Kolmogorov turbulence, we obtain from Eq. (13.99)

$$D_\tau(\tau) = 2.91 \left(\frac{2\pi}{\lambda}\right)^2 C_n^2 L v_s^{5/3} \tau^{5/3} . \tag{13.108}$$

$D_\tau(\tau)$ and $S'_\phi(f)$ are related by a transformation similar to Eq. (13.104). The use of temporal structure functions to estimate the effects of fluctuations on interferometers is discussed by Treuhaft and Lanyi (1987) and Lay (1997a).

The Allan variance $\sigma_y^2(\tau)$, or fractional frequency stability for time interval τ, associated with $S'_\phi(f)$ has been defined in Sect. 9.5.1. It can be calculated by substituting Eq. (9.119) into Eq. (9.131), which gives

$$\sigma_y^2(\tau) = \left(\frac{2}{\pi v_0 \tau}\right)^2 \int_0^\infty S'_\phi(f) \sin^4(\pi\tau f) \, df . \tag{13.109}$$

By substituting Eq. (13.107) into Eq. (13.109), and noting that

$$\int_0^\infty [\sin^4(\pi x)]/x^{8/3} \, dx = 4.61 ,$$

we obtain

$$\sigma_y^2(\tau) = 1.3 \times 10^{-17} C_n^2 L v_s^{5/3} \tau^{-1/3} . \tag{13.110}$$

Armstrong and Sramek (1982) give general expressions for the relations among S_2, S'_ϕ, D_ϕ, and σ_y for an arbitrary power-law index. If $S_2 \propto q^{-\alpha}$, then $D_\phi(d) \propto d^{\alpha-2}$, $S'_\phi \propto f^{1-\alpha}$, and $\sigma_y^2 \propto \tau^{\alpha-4}$. These relations are summarized in Table 13.2.

The actual behavior of the atmosphere is more complex than described above, but the theory developed provides a general guide. An example of a structure function of phase from the VLA is shown in Fig. 13.13 (see also Fig. 13.22 for a similar plot for ALMA). It clearly shows the three power-law regions, with power-law exponents close to their expected values. The effects of phase noise on VLBI observations are discussed by Rogers and Moran (1981) and Rogers et al. (1984). The plot of Allan variance by Rogers and Moran is shown in Fig. 9.17.

Table 13.2 Power law relations for turbulence

Quantity			Exponent 3-D turbulence ($\alpha = 11/3$)	2-D turbulence ($\alpha = 8/3$)
2-D, 3-D power spectrum	$S_2(q_s), S(q_s)$	$-\alpha$	$-11/3$	$-8/3$
Structure function	$D_\phi(d)$	$\alpha - 2$	5/3	2/3
Temporal phase spectrum	$S'_\phi(f)$	$1 - \alpha$	$-8/3$	$-5/3$
Allan variance	$\sigma_y^2(\tau)$	$\alpha - 4$	$-1/3$	$-4/3$
Temporal structure function	$D_\tau(\tau)$	$\alpha - 2$	5/3	2/3

Adapted from Wright (1996, p. 526).

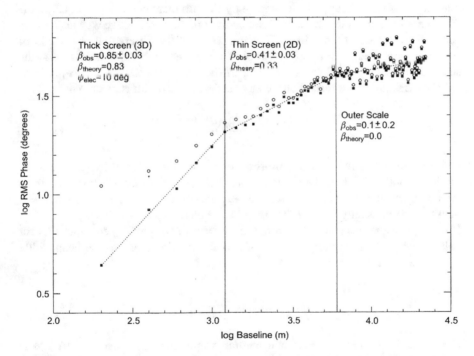

Fig. 13.13 The root phase structure function (rms phase) from observations with the VLA at 22 GHz. The open circles show the rms phase variation vs. baseline length measured on the source 0748+240 over a period of 90 min. The filled squares show the data after removal of a constant receiver-induced noise component of rms amplitude $10°$. The three regimes of the phase structure function are indicated by vertical lines (at 1.2 and 6 km). Note that $\beta = \alpha/2$. From Carilli and Holdaway (1999). © 1999 by the American Geophys. Union.

13.1.8   Anomalous Refraction

The beamwidths of many millimeter radio telescopes are sufficiently small that the effect of atmospheric phase fluctuations can be detected. This effect was first noticed with the 30-m-diameter millimeter-wavelength telescope on Pico de Veleta, where the apparent positions of unresolved sources were observed to wander by about 5″ on timescales of a few seconds under certain meteorological conditions [see, e.g., Altenhoff et al. (1987), Downes and Altenhoff (1990), and Coulman (1991)]. This motion is due to the flow of the turbulent layer of water vapor across the telescope aperture, which is distinct from the refraction caused by the quasi-static atmosphere, and hence the term "anomalous refraction." This effect can be understood by a simple application of the theory developed in Sect. 13.1.7. The magnitude of the refraction is dominated by the turbulent cells of size equal to the diameter of the antenna. These cells can be thought of as refractive wedges moving across the aperture of the antenna. The rms value of the differential phase shift of such a wedge is equal to the square root of the structure function evaluated at the separation distance equal to the diameter of the antenna, $\sqrt{D_\phi(d)}$. Hence, the rms value of the anomalous refraction for an observation at zenith is given by

$$
\epsilon = \frac{\sqrt{2D_\phi(d)}}{d} \, ,
\tag{13.111}
$$

where the structure function is in units of length and the factor of two accounts for motion in both azimuth and elevation. Note that fluctuations on larger scales than d are unimportant as long as the power-law exponent on the structure function is less than two, as is usually the case. In the 3-D turbulence case, ϵ will vary as $\sqrt{\sec z}$. If we express the rms phase fluctuations as $\sigma = \sigma_0 (d/100m)^{5/6}$ (see Table 13.4 for values of σ_0), then the ratio of the anomalous refraction angle to the beamwidth, $\theta_b \sim 1.2\lambda/d$, is

$$
\epsilon/\theta_b \simeq 1.2 \frac{\sigma_0}{\lambda} \left(\frac{d}{100m} \right)^{5/6} .
\tag{13.112}
$$

For example, the range of seasonal median values for σ_0 for the ALMA site is 0.045–0.17 mm. Since the diameter of the ALMA antennas is 12 m, the range of ϵ is 0.2–0.6″ from Eq. (13.111), which is independent of wavelength. The timescale of this effect is d/v_s, where v_s is the wind speed. At a wavelength of 1 mm, the beamwidth is about 20″, so the ratio ϵ/θ_b has the range of 1% to 3%. There is no effect on the amplitude of the incident electric field because the phase fluctuations arise in a layer close to ground. However, the fractional changes in antenna gain at the half-beamwidth point would range from 1.5% to 5%, which could have an effect on the quality of mosaic images derived from array observations under some conditions. For further details, see Holdaway and Woody (1998). Methods of real-time correction of anomalous refraction have been proposed by Lamb and Woody (1998).

13.2   Site Evaluation and Data Calibration

13.2.1   Opacity Measurements

At millimeter and submillimeter wavelengths, absorption and path length fluctuations in the atmosphere limit performance in synthesis imaging. This section is
concerned with monitoring of atmospheric parameters for optimum choice of sites
and with methods of calibrating the atmosphere to reduce phase errors. This subject
has received much attention as a result of the development of major instruments at
millimeter and submillimeter wavelengths.

For given atmospheric parameters, the zenith opacity (optical depth) τ_0 can be
calculated as a function of frequency using the propagation models of Liebe (1989),
Pardo et al. (2001a), or Paine (2016). Figure 13.14 shows curves of transmission,
$\exp(-\tau_0)$, for 4 mm of precipitable water at an elevation of 2124 m and 1 mm at
5000 m, corresponding to the VLA and ALMA sites, respectively. For the purpose of
choosing a suitable observatory site, detailed monitoring of the atmosphere covering
both diurnal and annual variation is necessary. We assume that the zenith opacity has
the form

$$\tau_\nu = A_\nu + B_\nu w \,, \tag{13.113}$$

where A_ν and B_ν are empirical constants that depend on frequency, site elevation,
and meteorological conditions. Selected measurements of these constants are given
in Table 13.3.

The opacity can be monitored by measuring the total noise power received in a
small antenna as a function of zenith angle (i.e., the tipping-scan method described
in Sect. 13.1.3). A commonly used frequency for opacity monitoring is 225 GHz,
which lies within the 200–310 GHz atmospheric window (see Figs. 13.7 and 13.14)
in the vicinity of the important CO 2–1 rotational transition at 230 GHz.

A typical site-test radiometer designed for opacity measurements uses a small
parabolic primary reflector with a beamwidth of $\sim 3°$ at 225 MHz. A wheel with
blades that act as plane reflectors is inserted at the beam waist between the primary
and secondary reflectors and sequentially directs the input of the receiver to the
output of the antenna, a reference load at 45 °C, and a calibration load at 65 °C. The
amplified signals go to a power-linear detector and then to a synchronous detector
that produces voltages proportional to the difference between the antenna and the
45 °C load, which is the required output, and the difference between the 45 ° and
65 °C loads, which provides a calibration. Measurements of the antenna temperature
are made at a range of different zenith angles. When connected to the antenna, the
measured noise temperature of such a system, T_{meas}, consists of three components:

$$T_{\mathrm{meas}} = T_{\mathrm{const}} + T_{\mathrm{at}}(1 - e^{-\tau_0 \sec z}) + T_{\mathrm{cmb}} e^{-\tau_0 \sec z} \,. \tag{13.114}$$

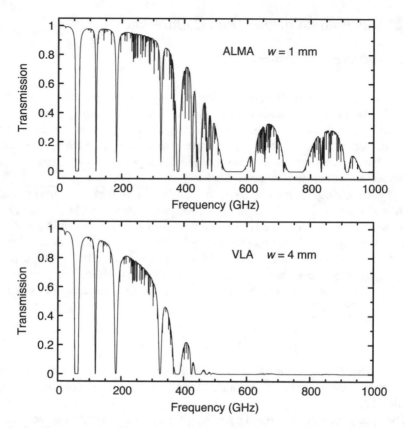

Fig. 13.14 (**top**) The zenith atmospheric transmission (equal to e^τ) at the ALMA site (5000 m, with 1 mm of precipitable water vapor). There are additional windows with transmissions of about 0.3% near 1100, 1300, and 1500 GHz. There are no additional windows with higher transmission up to 7500 GHz (40 μm). (**bottom**) The zenith transmission at the VLA site (2124 m, with 4 mm of precipitable water). Note that the transmission depends on the altitude because of the pressure broadening of the absorption lines. Because of this effect, for a fixed value of w, the transmission at any frequency in an atmospheric window will be lower for lower altitude sites. The many narrow absorption features (line widths of $\sim$ 100 MHz) are caused by stratospheric ozone lines [for a catalog of these lines, see Lichtenstein and Gallagher (1971)]. The effects of these lines in astronomical observations can be removed by careful bandpass calibration. These transmission plots were calculated with the *am* code (Paine 2016) with profiles for mean midlatitude atmospheric conditions.

Here T_{const} represents the sum of noise components that remain constant as the antenna elevation is varied, that is, the receiver noise, thermal noise resulting from losses between the antenna and the receiver input, any offset in the radiometer detector, and so on. The second term in Eq. (13.114) represents the component of noise from the atmosphere: T_{at} is the temperature of the atmosphere, and z is the zenith angle. $T_{cmb} \simeq 2.7$ K represents the cosmic microwave background radiation. It will be assumed that T_{at} and T_{cmb} represent brightness temperatures that are related

Table 13.3 Zenith opacity as a function of column height of water vapor

ν (GHz)	Location[a]	Altitude (m)	A_ν (nepers)	B_ν (nepers mm^{-1})	Method[b]	Ref.[c]
15	Sea level	0	0.013	0.002	1	1
22.2	Sea level	0	0.026	0.02	1	1
35	Sea level	0	0.039	0.006	1	1
90	Sea level	0	0.039	0.018	1	1
225	South Pole	2835	0.030	0.069	2	2
225	Mauna Kea	4070	0.01	0.04	2	3
225	Chajnantor	5000	0.006	0.033	2	4
225	Chajnantor	5000	0.007	0.041	2	5
493	South Pole	2835	0.33	1.49	2	6

[a]Locations: South Pole = Amundsen–Scott station; Mauna Kea = site of submillimeter telescopes on Mauna Kea; Chajnantor = Llano de Chajnantor, Atacama Desert, Chile.
[b]Methods: (1) opacity derived from radiosonde data, water vapor estimated from surface humidity and scale height of 2 km; (2) opacity derived from tipping radiometer, water vapor column height derived from radiosonde data.
[c]References: (1) Waters (1976); (2) Chamberlin and Bally (1995); (3) Masson (1994a); (4) Holdaway et al. (1996); (5) Delgado et al. (1998); (6) Chamberlin et al. (1997).

to the physical temperatures by the Planck or Callen and Welton formulas (see Sect. 7.1.2). If T_{at} is known, it is straightforward to determine τ_0 from T_{meas} as a function of z. The temperature of the atmosphere is assumed to fall off from the ambient temperature at the Earth's surface T_{amb}, with a lapse rate l considered to be constant with height. Thus, at height h, the temperature is $T_{amb} - lh$. We require the mean temperature weighted in proportion to the density of water vapor, which is exponentially distributed with scale height h_0:

$$T_{at} = T_{amb} - \frac{l \int_0^\infty h e^{h/h_0} \, dh}{\int_0^\infty e^{h/h_0} \, dh} = T_{amb} - lh_0 . \qquad (13.115)$$

The lapse rate resulting from adiabatic expansion of rising air, 9.8 K km^{-1}, can be used as an approximate value, but as indicated earlier, a typical measured value is ~ 6.5 K km^{-1}. The scale height of water vapor is approximately 2 km. Thus, T_{at} is typically less than T_{amb} by ~ 13–20 K.

Figure 13.15 displays examples of data taken on Mauna Kea, which show the diurnal and seasonal effects at this site. The cumulative distribution of zenith opacity at 225 and 850 GHz as measured at Cerro Chajnantor, Llano de Chajnantor in Chile; Mauna Kea; and the South Pole are shown in Fig. 13.16. Measurements of mean opacity provide a basis for calculating the loss in sensitivity due to absorption of the signal and the addition of noise from the atmosphere [see Eq. (13.50)]. The opacity varies both diurnally and annually, so measurements at hourly intervals

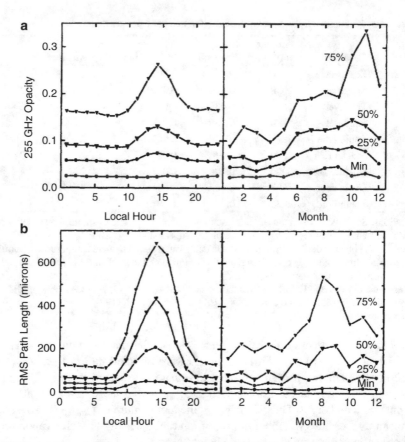

Fig. 13.15 (a) Diurnal and seasonal zenith opacity at 225 GHz measured at the CSO site on Mauna Kea (4070-m elevation) for a three-year period (August 1989–July 1992) computed from 14,900 measurements. The minimum value and the 25th, 50th, and 75th percentiles are shown. The increase in opacity during the day is caused by an inversion layer that rises above the mountain in the afternoons. (b) Diurnal and seasonal variation of the rms path length on Mauna Kea on a 100-m baseline, determined from observations of a geostationary satellite at 11 GHz. From Masson (1994a), courtesy of and © the Astronomical Society of the Pacific.

over a year or more are required for reliable comparison of different sites. Long-term variability due to climatic effects (e.g., El Niño) can be significant. Table 13.3 shows the effect of site altitude on opacity. Comparison of the measurements of A_ν and B_ν show that both parameters decrease with altitude because of the effects of pressure broadening. Comparisons of opacities at various frequencies can be made with broadband Fourier transform spectrometers (Hills et al. 1978; Matsushita et al. 1999; Paine et al. 2000; Pardo et al. 2001b).

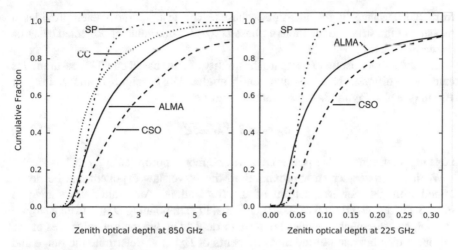

Fig. 13.16 Cumulative distributions of the zenith optical depth at 850 GHz (**left panel**) and at 225 GHz (**right panel**) at Cerro Chajnantor, Chile (5612-m elevation); the ALMA site on Llano de Chajnantor, Chile (5060-m elevation); the CSO on Mauna Kea, Hawaii (4100-m elevation); and the South Pole (2835-m elevation) for the periods April 1995–April 1999, Jan. 1997–July 1999, and Jan. 1992–Dec. 1992, respectively. Note that the median opacity at 225 GHz at the VLBA site on Mauna Kea (3720-m elevation) for the same time interval at the CSO site was 0.13. The median opacity for the VLA site (2124-m elevation) for the period 1990–1998 was 0.3 (Butler 1998). Conditions at lower elevation sites are correspondingly worse. For example, at a sea-level site in Cambridge, MA, the 225-GHz opacity, inferred from measurements at 115 GHz, was 0.5 for the six-month winter observing seasons spanning 1994–1997. Conditions on Dome C, Antarctica (3260-m elevation), are significantly better than at the South Pole (Calisse et al. 2004), and Ridge A on the Antarctic high plateau (4050-m elevation) may have the lowest water vapor on the planet (Sims et al. 2012). Adapted from Radford and Peterson (2016).

13.2.2 Site Testing by Direct Measurement of Phase Stability

Interferometer observations provide a direct method of determining atmospheric phase fluctuations. Signals from a geostationary satellite are usually used, since strong signals can be obtained using small, nontracking antennas. This technique, called satellite-tracking interferometry (STI), was developed by Ishiguro et al. (1990); Masson (1994a); and Radford et al. (1996). It was used in site testing for the SAO Submillimeter Array on Mauna Kea, Hawaii, Atacama Large Millimeter/submillimeter Array at Llano de Chajnantor, and potential SKA sites. Several suitable geostationary-orbit satellites operate in bands allocated to the fixed and broadcasting services near 11 GHz. Two commercial satellite TV antennas of diameter 1.8 m provide signal-to-noise ratios close to 60 dB. For measurements of atmospheric phase, baselines of 100–300 m have been used. The residual motion of the satellite, as well as any temperature variations, can cause unwanted phase drifts. These are generally slow compared with the atmospheric effects and can be removed by subtracting a mean and slope from the output data. The variance of the fluctuations resulting from the system noise can also be determined and subtracted from the variance of the measured phases. The test interferometer provides a measure of the structure function of phase $D_\phi(d)$ for one value of projected baseline d (see Fig. 13.15b).

It is sometimes useful to compare the quality of sites based on STI measurements made with different baselines and zenith angles. For baselines in the vicinity of 100 m (see also Fig. A13.4), a reasonable scaling is

$$\sigma_\phi \sim \sigma_0 d^{5/6} \sqrt{\sec z} \, . \tag{13.116}$$

For longer baselines, other power laws will be more appropriate.

With the frozen-screen approximation, the power-law exponent can be determined from the power spectrum of the fluctuations. An example is shown in Fig. 13.17 (see also Bolton et al. 2011). Thus, in extrapolating $D_\phi(d)$ from a single-spacing measurement, one does not have to depend on the theoretical values of the exponent of d but can use the measurements of $D_\phi(\tau)$ to determine the range and variation [see Eq. (13.108) and Table 13.2]. For the example shown in Fig. 13.17, the power-law slope for frequencies above 0.01 Hz is 2.5, slightly below the value of 8/3 or 2.67 predicted for Kolmogorov turbulence. The spectrum flattens at frequencies below 0.01 Hz because of the filtering effect of the interferometer. Fluctuations larger than the baseline, 100 m in this case, cause little phase effect. For the corner frequency $f_c = v_s/d$, the wind speed along the baseline direction can be inferred to be about 1 m s^{-1}.

Table 13.4 shows a compilation of the measurements of the structure function referred to a baseline of 100 m. The range of values reported for a fiducial baseline

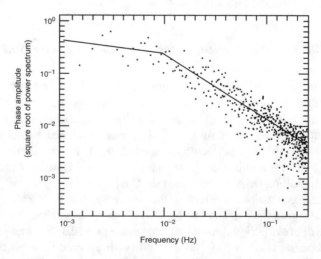

Fig. 13.17 The square root of the temporal power spectrum [i.e., Eq. (13.107)] measured on a 100-m baseline on Mauna Kea (CSO site). The tropospheric wind speed along the baseline can be computed from the break in the spectrum. From Masson (1994a), courtesy of and © the Astronomical Society of the Pacific.

Table 13.4 Site phase stability characteristics

Location	ID	Altitude (m)	Baseline (m)	Frequency (GHz)	Type[a]	No. of days	Date of data	σ[b] (mm)	β[c]	Ref.[d]
Cambridge, UK		17	1000	5	A	50	Jan 1969–Sep 1969	0.12–0.5	0.6	1
Murchison, Australia	MRO	370	200	11.7	S	180	Jun 2011–Sep 2011	0.15–0.38	—	2
Green Bank, USA		840	2400	2.7	A	25	Mar 1965–Aug 1965	0.2–0.8	—	3
Goldstone (A), USA	DSN	952	190	12.5	S	700	Jan 2011–Dec 2012	0.12–0.31	0.7	4
Hat Creek, USA		1043	6–850	86	A	10	Nov 1993–Feb 1995	0.11–0.3	0.4–0.7	5
Goldstone (V), USA		1070	256	20.2	S	700	Jan 2011–Dec 2012	0.083–0.22	0.6	4
Karoo, South Africa	RSA	1081	200	11.7	S	180	Mar 2011–Oct 2011	0.12–0.25	—	6
Nobeyama, Japan		1350	50–500	22	A	2	Feb 1985–May 1985	0.15–0.4	0.8	7
VLA, USA		2124	50–8000	5.15	A	109	Dec 1983–Dec 1985	0.09–0.22	0.1–0.6	8
VLA, USA	VLA	2124	300	11.3	S	350	Sep 1998–Aug 1999	0.065–0.26	—	9
Plateau de Bure, France	PdB	2552	24–290	86	A	200	Feb 1990–Aug 1990	0.10–0.6	0.7	10
Mauna Kea, USA		4070	33–260	12	S	60	Nov 2011–Dec 2011	0.095	0.6	11
Mauna Kea, USA	SMA	4070	100	11.7	S	600	Dec 1990–Sep 1992	0.065–0.17	0.75	12
Pampa la Bola, Chile	PlaB	4800	300	11.7	S	1000	Jul 1996–Mar 1999	0.056–0.22	—	13
ALMA, Chile	ALMA	5000	300	11.7	S	1000	Jul 1996–Mar 1999	0.045–0.17	—	13

[a] A = astronomical data; S = satellite-tracking interferometry (STI) data.

[b] The rms phase deviation range referred to a baseline of 100 m usually represents the span from the median condition at nighttime during winter to daytime during summer.

[c] Power-law exponent for baseline length dependence. For astronomical data, β is derived from rms phase vs. baseline. For satellite-tracking interferometry data, β is derived from temporal power spectrum. β = 5/6 or 0.833 for 3-D turbulence and 2/3 or 0.667 for 2-D turbulence.

[d] References: (1) Hinder (1970), Hinder and Ryle (1971), Hinder (1972); (2) Millenaar (2011a); (3) Baars (1967); (4) Morabito et al. (2013); (5) Wright (1996), see also Wright and Welch (1990), Bieging et al. (1984); (6) Millenaar (2011b); (7) Kasuga et al. (1986), see also Ishiguro et al. (1990); (8) Sramek (1990), see also Sramek (1983), Carilli and Holdaway (1999), Armstrong and Sramek (1982); (9) Butler and Desai (1999); (10) Olmi and Downes (1992); (11) Kimberk et al. (2012); (12) Masson (1994a); (13) Butler et al. (2001).

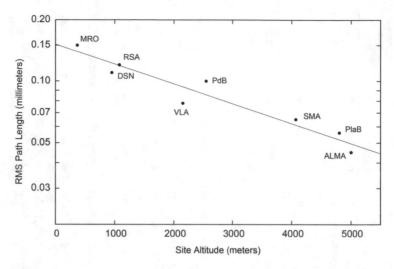

Fig. 13.18 Rms path length vs. site elevation referred to zenith and baseline of 100 m. Data taken in good weather conditions, i.e., winter nighttime (see Table 13.4 for station identifiers). Straight line is best-fit exponential with scale height of 2200 m and sea level intercept of 0.15.

of 100 m are meant to reflect the median conditions at night during the winter and daytime during the summer. The measurements were obtained either by satellite interferometry or astronomical measurements. A plot of the rms phase noise vs. site altitude for "best conditions" is presented in Fig. 13.18. The decrease of rms phase noise vs. altitude is evident. With the assumption that the turbulence, i.e., C_n^2, is proportional to water vapor density and that the water vapor is distributed exponentially with a scale height of h_0, we obtain from Eq. (13.100) the result

$$\sigma = \sigma_0 \, e^{-h/2h_0} \ . \tag{13.117}$$

(The factor of 2 arises from the fact that $\sigma_\phi = \sqrt{D_\phi}$. The line shown is a fit to this equation.) The value of h_0 is 2.2 km, close to the nominal scale height of 2 km, and $\sigma_0 = 0.05$ mm. Because the distillation of this information from disparate sources is difficult, the results are meant to show the importance of altitude rather than make small distinctions among observatories. Local conditions can also be important. See Masson (1994b) for further discussion.

A wide range of power-law indices has been observed (see Table 13.4). Much of the variation between 0.33 and 0.833 is due to the effects of thin scattering layers in the troposphere, which effectively moves or blurs the crossover from 2-D to 3-D turbulence (see Bolton et al. 2011). Beaupuits et al. (2005) explored this problem by pointing two 183-GHz water-vapor radiometers so that their beam intersected at an altitude of about 1500 m. By analyzing the delay between the radiometer signals, they identified a significant turbulent layer near 600 m.

The atmospheric phase noise, if left uncorrected, causes a coherence loss in an interferometer. For the model in Fig. 13.18, the baseline for which the coherence factor C, equal to measured/true visibility defined in Eq. (13.80), can be derived from Eq. (13.117), giving

$$d_c = 100 \left[-\frac{\ln C}{2\pi^2} \left(\frac{\lambda}{\sigma_0} \right)^2 e^{h/h_0} \right]^{3/5}. \tag{13.118}$$

For example, with $\sigma_0 = 0.10$ mm, $h_0 = 2200$ m, $h = 5000$ m, $\lambda = 1.3$ mm, and $C = 0.9$, $d_c = 80$ m.

13.3 Calibration via Atmospheric Emission

A practical method of estimating phase fluctuations is to measure the integrated water vapor in the direction of each antenna beam. This usually requires an auxiliary radiometer at each antenna to measure the sky brightness temperature. Water vapor is the main cause of opacity at radio frequencies (except for the oxygen bands at 50–70 and 118 GHz), even at frequencies well away from the centers of water vapor lines, as can be seen in Fig. 13.7. Away from the centers of spectral lines, the opacity is due to the far line wings of infrared transitions. There is also an important *continuum* component of the absorption caused by water vapor, which varies as ν^2 (Rosenkranz 1998). This component includes various quantum mechanical effects involving water molecules such as dimers (Chylek and Geldart 1997). It is usually necessary to model this component with an empirical coefficient. In addition, as described in Sect. 13.3.2, the water droplets in the form of clouds and fog, as well as ice crystals, contribute absorption that varies as ν^2. Hence, there are two distinct methods of calibration: those based on measurement of sky brightness in the bands between the lines (continuum) and those based on measurements near a spectral line (see Welch 1999). The sensitivities of the brightness temperature to the propagation delay are listed in Table 13.5 for selected frequencies.

13.3.1 Continuum Calibration

The method of measuring the continuum sky brightness at, say, 90 or 225 GHz has several advantages, as first described by Zivanovic et al. (1995). The same radiometers used for the astronomical measurements can be used for the sky brightness measurements. At 225 GHz, if phase calibration to an accuracy of a twentieth of a wavelength is required, then, from the sensitivity listed in Table 13.5, the brightness temperature accuracy required is 0.1 K. For a system temperature of 200 K, this accuracy requires a gain stability of 5×10^{-4}. Such stability usually requires special attention to the temperature stabilization of the receiver cryogenics. In addition, the gain scales must be accurately calibrated. Changes in ground pickup

Table 13.5 Brightness temperature sensitivity dT_b/dw (K/mm) for various frequencies at site elevations of 0 and 5 km for various values of precipitable water vapor[a]

v (GHz)	Origin of opacity	0-km elevation		5-km elevation	
		$w = 0$ mm	15 mm	$w = 0$ mm	15 mm
22.2	Line center ($6_{16} - 5_{23}$)	1.9	1.7	2.8	2.8
90.0	Continuum	1.8	2.1[b]	1.2	1.2
183.3	Line center ($3_{13} - 2_{20}$)	294	0.0	527	51.4
185.0	Line wing ($3_{13} - 2_{20}$)	222	0.1	280	91.2
230.0	Continuum	15.9	7.3	11.4	9.8
690.0	Continuum	380	0.0	297	82.5

[a]Entries in this table were calculated with the *am* model (Paine 2016) for median midlatitude atmospheric profiles. Note that $w = 15$ and 0 mm are close to the measured values for midlatitudes for altitudes of 0 and 5 km, respectively. The effects of pressure broadening are clearly evident. For example, at $w = 0$, dT_b/dw is less at sea level than at 5 km for the 22.2- and 183.3-GHz spectral lines, while the opposite is true for the continuum bands. More detailed information about the brightness temperature sensitivity near the 183-GHz transition can be found in Fig. 13.21.
[b]More sensitive than for $w = 0$ because of the effect of H_2O line self-broadening.

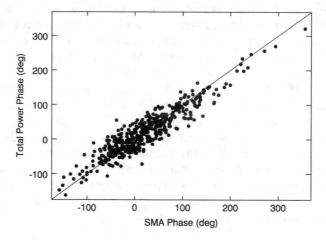

Fig. 13.19 Correlation between interferometric phase predicted by total power measurement at 230 GHz vs. interferometric phase. The data were taken over a period of 20 min on a 140-m baseline of the Submillimeter Array (SMA) on Mauna Kea. The total powers (i.e., antenna temperatures) at each antenna were used to estimate phase with a linear model having free parameters. The straight line shown has unity slope and zero intercept. The rms phase error is improved from 72 to 27°, corresponding to path length residuals of 260 to 98 μm, respectively. From Battat et al. (2004). © AAS. Reproduced with permission.

can be misinterpreted as sky brightness temperatures change. The presence of clouds defeats this method, because of the contribution of liquid water to the opacity. An example of viability of this type of calibration is shown in Fig. 13.19. The application of this method for the Plateau de Bure interferometer is described by Bremer (2002). For further discussion, see Matsushita et al. (2002).

13.3.2   22-GHz Water-Vapor Radiometry

The idea of determining the vertical distribution of water vapor in the atmosphere from brightness temperature measurement at frequencies near the 22-GHz line was first investigated theoretically by Barrett and Chung (1962). It was further developed into a technique for determining the excess propagation path by Westwater (1967) and Schaper et al. (1970). To appreciate the degree of correlation between wet path length and brightness temperature, we need to examine the dependence of these quantities on pressure, water vapor density, and temperature. We consider here the interpretation of measurements near the 22.2-GHz resonance. The absorption coefficient given by Eq. (13.42) is complicated, but at line center it can be approximated by

$$\alpha_m \simeq 0.36 \frac{\rho_V}{PT^{1.875}} e^{-644/T} , \tag{13.119}$$

where T is in kelvins, and we have neglected all except the leading terms in Eq. (13.42). We assume that the opacity given by Eq. (13.47) is small, so that the brightness temperature defined by Eq. (13.45) can be written

$$T_B \simeq 17.8 \int_0^\infty \frac{\rho_V}{PT^{0.875}} e^{-644/T} \, dh , \tag{13.120}$$

when we neglect the background temperature T_{B0} and any contributions from clouds. Recall that Eq. (13.16) shows that

$$\mathcal{L}_V = 1763 \times 10^{-6} \int_0^\infty \frac{\rho_V(h)}{T(h)} \, dh . \tag{13.121}$$

Thus, if P and T were constant with height and equal to 1013 mb and 280 K, respectively, we could use Eq. (13.19), $\mathcal{L}_V \simeq 6.3w$, to obtain from Eq. (13.120) the relation $T_B \simeq 12.7w$, where w is the column height of water vapor [see Eq. (13.18)]. Hence, to the degree of approximation used above, we obtain

$$T_B \, (22.2 \text{ GHz}) \, (K) \simeq 2.1 \mathcal{L}_V \, (cm) . \tag{13.122}$$

Note that this approximation is valid at sea level. Since, because of pressure broadening, the brightness temperature scales inversely with total pressure [see Eq. (13.120)], the coefficient in Eq. (13.122) is increased to 3.9 for a site at 5000-m elevation, where the pressure is approximately 540 mb. Measurements of brightness temperature and path length estimated from radiosonde profiles show that Eq. (13.122) is a good approximation (see, e.g., Moran and Rosen 1981). Recall that ρ_V is approximately exponentially distributed with a scale height of 2 km. The temperature, on average, decreases by about 2% per kilometer. This change affects the proportionality between T_B and $\mathcal{L}_V$ only through the exponential factor in Eq. (13.120) and the slight difference in the power law for temperature. Thus, temperature has a small effect. The pressure decreases by 10% per kilometer, so water vapor at higher altitudes contributes more heavily to T_B than is desirable for estimation by radiometry. The sensitivity of T_B to pressure is decreased by moving off the resonance frequency to a frequency near the half-power point of the transition. The reason for this is that as pressure increases, the line profile broadens while the integrated line profile is constant. Therefore, the absorption at line center decreases and the absorption in the line wings increases. Westwater (1967) showed that at 20.6 GHz, the absorption is nearly invariant with pressure. This particular frequency is called the *hinge point*. The opacity at this frequency is less than at the line center, so the nonlinear relationship between T_B and opacity is less important.

The foregoing discussion assumes that measurements of T_B are made in clear-sky conditions. The water droplet content in clouds or fog causes substantial absorption but small change in the index of refraction compared with that of water vapor. Fortunately, the effect of clouds can be eliminated by combining measurements at two frequencies. In nonprecipitating clouds, the sizes of the water droplets are generally less than $100\,\mu$m, and at wavelengths greater than a few millimeters, the scattering is small and the attenuation is due primarily to absorption. The absorption coefficient is given by the empirical formula (Staelin 1966)

$$\alpha_{\text{clouds}} \simeq \frac{\rho_L 10^{0.0122(291-T)}}{\lambda^2} \quad (\text{m}^{-1})\,, \tag{13.123}$$

where ρ_L is the density of liquid water droplets in grams per cubic meter, λ is the wavelength in meters, and T is in kelvins. This formula is valid for λ greater than ~ 3 mm where the droplet sizes are small compared with $\lambda/(2\pi)$. For shorter wavelengths, the absorption is less than predicted by Eq. (13.123) (Freeman 1987; Ray 1972). A very wet cumulus cloud with a water density of $1\,$g m^{-3} and a size of 1 km will have an absorption coefficient of 7×10^{-5} m^{-1} and will therefore have a brightness temperature of about 20 K at 22 GHz. The index of refraction of liquid water is about 5 at 22 GHz for $T = 280\,$K (Goldstein 1951). The actual excess propagation path through the cloud due to liquid water would be about 4 mm, but the predicted excess path from Eq. (13.122) is 10 cm. Thus, the brightness temperature at a single frequency cannot be used reliably to estimate the excess path length when clouds are present. In order to eliminate the brightness temperature contribution of clouds, measurements must be made at two frequencies, ν_1 and ν_2, one near the water line and one well off the water line, respectively. The brightness temperature is

$$T_{Bi} = T_{BVi} + T_{BCi}\,, \tag{13.124}$$

where T_{BVi} and T_{BCi} are the brightness temperatures due to water vapor and clouds at frequency i. Here we neglect the effects of atmospheric O_2. Since, from

Eq. (13.123), $T_{BC} \propto \nu^2$, we can form the observable

$$T_{B1} - T_{B2}\frac{\nu_1^2}{\nu_2^2} = T_{BV1} - T_{BV2}\frac{\nu_1^2}{\nu_2^2} \,, \tag{13.125}$$

which eliminates the effect of clouds. The correlation between $T_{BV1} - T_{BV2} \times \nu_1^2/\nu_2^2$ and $\mathcal{L}_V$ can be estimated from model calculations based on Eqs. (13.45) and (13.16). The off-resonance frequency ν_2 is generally chosen to be about 31 GHz. The problem of finding the two best frequencies and the appropriate correlation coefficients to use in predicting $\mathcal{L}_V$ has been widely discussed (Westwater 1978; Wu 1979; Westwater and Guiraud 1980). The liquid content of clouds can also be measured by dual-frequency techniques [see, e.g., Snider et al. (1980)].

The application of multifrequency microwave radiometry to the calibration of wet path length has been described by Guiraud et al. (1979), Elgered et al. (1982), Resch (1984), Elgered et al. (1991), and Tahmoush and Rogers (2000). A high-performance receiver design is discussed by Tanner and Riley (2003). The results show that $\mathcal{L}_V$ can be estimated to an accuracy better than a few millimeters. This is useful for calibrating VLBI delay measurements and extending coherence times. Measurements of T_B at the antennas of short-baseline interferometers can be useful in correcting the interferometer phase. More accurate predictions of $\mathcal{L}_V$, or interferometer phase, can be obtained by including measurements in other bands. For example, measurements of the wings of the terrestrial oxygen line near 50 GHz can be used to probe the vertical temperature structure of the troposphere [see, e.g., Miner et al. (1972), Snider (1972)]. The accuracy of these schemes has been analyzed by Solheim et al. (1998).

The observation of the 22-GHz line provides a calibration technique that is not sensitive to gain variations and ground pickup. Multiple frequencies can be monitored to correct for clouds and the variable distribution of water vapor with height [see Eq. (13.125)]. For millimeter observations at moderately dry sites, the 22-GHz line may be the best choice [see Bremer (2002) for a description of the system operating at Plateau de Bure]. An example of phase correction based on this line is shown in Fig. 13.20.

13.3.3   183-GHz Water-Vapor Radiometry

For observations at very dry sites, the 183-GHz line may give better results (Lay 1998; Wiedner and Hills 2000). The 183-GHz line is intrinsically about 30 times more sensitive than the 22-GHz line. However, the 183-GHz line is much more easily saturated (i.e., its opacity exceeds unity) than is the 22-GHz line.

A phase-correcting system utilizing the 183-GHz lines was developed for ALMA, and its application is described by Nikolic et al. (2013). Each telescope of

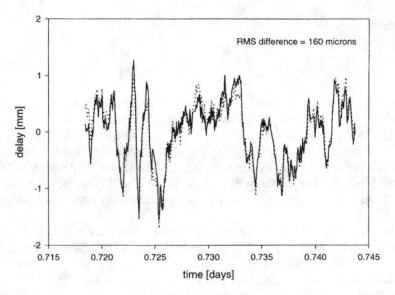

Fig. 13.20 The interferometric phase (in units of delay) measured at 3-mm wavelength on one baseline of the interferometer at Owens Valley Radio Observatory (solid line), and the delay predicted by 22-GHz water-vapor radiometer measurements (dotted line), vs. time. The rms deviation of the difference is $160\,\mu$m. The source is 3C273. From Welch (1999), with kind permission from and © URSI; see also Woody et al. (2000).

the array is equipped with a boresighted radiometer having four channels sampling parts of the 183-GHz line profile. The radiometers are double-sideband, and the four channels are symmetrically placed around line center at offsets of 0.5, 3.1, 5.2, and 8.3 GHz. Theoretical line profiles are shown in Fig. 13.21 for various values of precipitable water vapor, w [see Eq. (13.18)]. For low values of w, e.g., 0.3 mm, the maximum sensitivity, dT_B/dw, is obtained at line center. This sensitivity decreases to zero as the line saturates. The channels away from line center become more important as w increases. Combining the measurements with appropriate statistical weights allows accurate estimates of the propagation path over a wide range of conditions. The actual sensitivity coefficients are derived empirically. An example of the reduction in phase noise is shown in Fig. 13.22.

The system does not work well in the presence of clouds, which add a brightness temperature contribution $\sim \nu^2$ [see Eq. (13.123)]. Separate measurements on either side of the line could allow the estimation of the cloud contribution. At low levels of w, an unmodeled contribution to the fluctuations is detected that is attributed to fluctuation in the dry-air component of refractivity. Ancillary measurements of total pressure at each antenna may allow the effects of these fluctuations to be corrected.

The 183-GHz line has been used to estimate w by the atmospheric remote sensing community [e.g., Racette et al. (2005)].

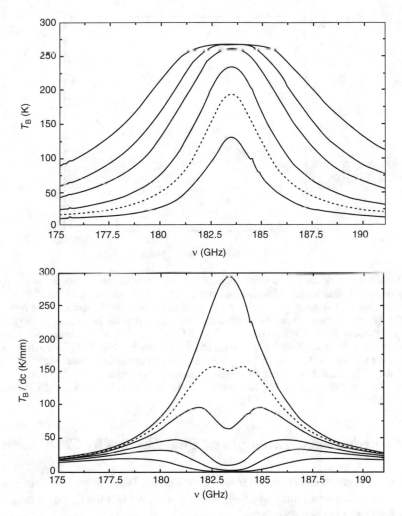

Fig. 13.21 (**top**) Theoretical brightness temperature profiles for the water vapor transition centered at 183.3 GHz appropriate for a site at 5000-m altitude for six values of w, the water vapor column density: (from bottom to top) 0.3, 0.6, 1, 2, 3, and 5 mm. The small blip noticeable at 184.4 GHz is the $10_{0\,10}$–$9_{1\,9}$ ozone transition originating in the upper atmosphere, where pressure broadening is small. The brightness temperature profiles become increasingly saturated at the atmospheric temperature as w increases [see Eq. (13.48)]. (**bottom**) The change of brightness temperature with water vapor column density, dT_B/dw, for the same values of w (from top to bottom). At $w = 5$ mm, the brightness temperature sensitivity to a change in water vapor column density is essentially zero at line center and reaches a broad maximum about 5 GHz from line center. From B. Nikolic et al. (2013), reproduced with permission. © ESO.

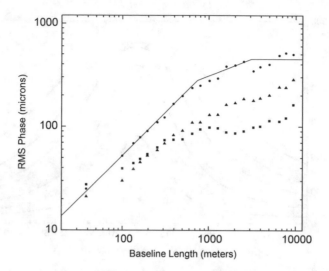

Fig. 13.22 The rms phase (in microns) deviation (square root of the phase structure function, D_ϕ) vs. projected baseline length. Measurements were made on the source 3C138 at 230 GHz in a period of 15 min. The water vapor column density was 1.4 mm, and the surface wind speed was 7 m s^{-1}. The circles show the uncalibrated results. The three-part power law is a suggestive fit to the data. The break at 670 m marks the transition from 3-D to 2-D turbulence and indicates a thickness to the turbulent layer of about 2 km [see Eq. (A13.17)]. The break at 3 km indicates the outer scale of the turbulence. The triangles show the rms phase deviation after water-vapor radiometer corrections. The squares show the rms phase deviations after phase referencing to a calibrator source offset by 1.3° (target/calibrator cycle time was 20 s). Adapted from ALMA Partnership et al. (2015).

13.4   Reduction of Atmospheric Phase Errors by Calibration

Atmospheric phase errors can be treated like antenna-based phase errors in considering their effect on an image. In Sect. 11.4, it is shown that the dynamic range of a snapshot image is approximately

$$\frac{\sqrt{n_a(n_a - 1)}}{\phi_{\text{rms}}},\qquad(13.126)$$

where ϕ_{rms} is the rms of the phase error in radians measured with pairs of antennas, and n_a is the number of antennas. For example, if ϕ_{rms} is 1 rad and $n_a = 30$, the dynamic range is ~ 30. As a rough guide, the range of ϕ_{rms} from 0.5 to 1 rad represents array performance from fair to marginal. The improvement in the image with longer integration depends on the spectrum of the phase fluctuations.

For phase calibration at centimeter wavelengths, it is common to observe a phase calibrator at intervals of $\sim$ 20–30 min. At millimeter wavelengths, this is generally not satisfactory, because of the much greater phase fluctuations resulting from the atmosphere. Procedures that can be used at millimeter and submillimeter wavelengths to reduce the effect of atmospheric phase fluctuations are described below. These methods are analogous to those of adaptive optics at optical wavelengths.

Self-Calibration. The simplest way to remove the effects of atmospheric phase fluctuations is to use self-calibration, as described in Sects. 10.3 and 11.3. This method depends on phase closure relationships in groups of three or more antennas. In applying this method, it is necessary to integrate the correlator output data for a long enough time that the source can be detected; that is, the measured visibility phase must result mainly from the source, not the instrumental noise. However, the integration time is limited by the fluctuation rate, so self-calibration is not useful for sources that require long integration times to detect.

Frequent Calibration (Fast Switching).   Frequent phase calibration using an unresolved source close to the target source (the source under study) can greatly reduce atmospheric phase errors (Holdaway et al. 1995b; Lay 1997b). To ensure that the atmospheric phase measured on the calibrator is close to that for the target source, the angular distance between the two sources must be no more than a few degrees. The time difference must be less than ~ 1 min, so fast position switching between the target source and the calibrator is required. In the layer in which most of the water vapor occurs, the lines of sight from the antennas to the target source and the calibrator pass within a distance d_{tc} of one another. For a nominal screen height of 1 km, $d_{tc} \simeq 17\theta$, where θ is the angular separation in degrees and d_{tc} is in meters. For one antenna, the rms phase difference between the two paths is $\sqrt{D_\phi(d_{tc})}$ at any instant. If t_{cyc} is the time to complete one observing cycle of the target source and the calibrator, then the mean time difference between the measurements on these two sources is $t_{cyc}/2$. In time $t_{cyc}/2$, the atmosphere will have moved $v_s t_{cyc}/2$. Thus, the phase difference between the measurements on the two paths is effectively $D_\phi(d_{tc} + v_s t_{cyc}/2)$. This is a worst-case estimate, since we have taken the scalar sum of vector quantities corresponding to d_{tc} and v_s. For the difference in the paths to the two antennas as measured by the interferometer, the rms value will be $\sqrt{2}$ times that for one antenna, so the residual atmospheric phase error in the measured visibility is

$$\phi_{rms} = \sqrt{2D_\phi(d_{tc} + v_s t_{cyc}/2)} \ . \tag{13.127}$$

Note that ϕ_{rms} is independent of the baseline, so the phase errors should not increase with baseline length. The total time for one cycle of observation of the two sources is the sum of the observing times on the target source and the calibrator, plus twice the antenna slew time between the sources and twice the setup time between ending the slew motion and starting to record data. The observing times required on each of the sources depend on the flux densities and the sensitivity of the instrument. For the calibrator, there may be a choice between a weak source nearby and a stronger one that requires less observing time but more antenna slew time. In order to use calibration sources as a general solution to the atmospheric phase problem, suitable calibrators must be available within a few degrees of any point on the sky. Since calibrator flux densities generally decrease with increasing frequency, it may

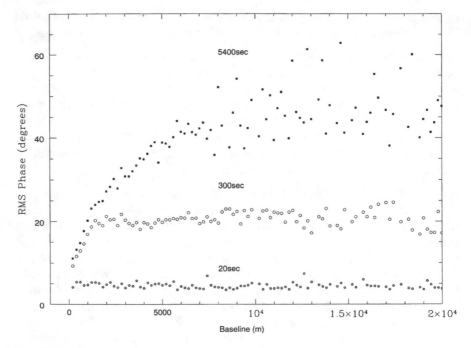

Fig. 13.23 The square root of the phase structure, that is, the rms phase deviation vs. baseline length, for data taken at the VLA at 22 GHz for various averaging times. These data show the effectiveness of fast switching. In these measurements, the target source and calibrator source were the same, 0748+240. The solid squares (labeled 5400 sec) show the rms phase fluctuations with no switching (same data as in Fig. 13.13). The circles and the stars show the rms phase deviation for cycle times 300 s and 20 s, respectively. From Carilli and Holdaway (1999). © 1999 by the American Geophys. Union.

be necessary to observe the calibrator at a lower frequency than is used for the target source. The measured phase for the calibrator must then be multiplied by $\nu_{\text{source}}/\nu_{\text{cal}}$ (since the troposphere is essentially nondispersive) before subtraction from the target source phases, so the accuracy required for the calibrator phase is increased. Thus, the observing frequency for the calibrator should not be too low; a frequency near 90 GHz may be a practical choice with observations of the target source up to a few hundred gigahertz. The effectiveness of the fast-switching technique is demonstrated by the data in Fig. 13.23. Note that the break in the curve for the 300-s averaging time at antenna spacing 1500 m indicates that the wind speed was about $2 \times 1500/300 = 10\,\text{m s}^{-1}$ (Carilli and Holdaway 1999). The effectiveness of fast switching for ALMA is described by Asaki et al. (2014).

Paired or Clustered Antennas.   Location of antennas in closely spaced pairs is an alternative to fast movement between the target source and the calibrator. One antenna of each pair continuously observes the target source and the other observes the calibrator. With this scheme, t_{cyc} is zero in Eq. (13.127), but the spacing of the paired antennas, d_p, should be included. The rms residual atmospheric error in the visibility phase becomes

$$\phi_{\text{rms}} = \sqrt{2D_\phi(d_{\text{tc}} + d_p)} \, . \tag{13.128}$$

As in Eq. (13.127), ϕ_{rms} is a worst-case estimate, since we have taken a scalar sum of vector quantities corresponding to d_{tc} and d_p. For a 2° position difference between the target source and the calibrator, and an effective height of 1 km for the water vapor, $d_{\text{tc}} = 35$ m. For antennas of diameter ~ 10 m, which is typical for antennas operating up to 300 GHz, d_p should be about 15 m to avoid serious shadowing, and this is smaller than $v_s t_{\text{cyc}}/2$ for the fast-switching scheme, since v_s is typically 6–12 m s^{-1} and t_{cyc} is 10 s or more. Thus, with paired antennas, the residual phase errors are somewhat less than with fast switching. Also, observing time is not wasted during antenna slewing and setup. However, with fast switching, about half of the *time* is devoted to the target source, whereas with paired antennas, half of the *antennas* are devoted to the target source, so in the latter case, the sensitivity is less by a factor $\sim \sqrt{2}$. In some cases, the paired antennas are available for use in an array. If the "science array" and the "reference array" are separate, there is no loss of capability in the "science array." Demonstration of the technique for the VLA is described by Carilli and Holdaway (1999) and for the Nobeyama Radio Observatory by Asaki et al. (1996). Another example is the CARMA array of 6-m- and 10-m-diameter antennas. The reference array is comprised of 3.5-m-diameter antennas. This system is described by Peréz et al. (2010) and Zauderer et al. (2016).

Appendix 13.1   Importance of the 22-GHz Line in WWII Radar Development

The history of the 22-GHz transition of water vapor is quite interesting. The water vapor molecule has different moments of inertia about its three axes of rotation, and hence its rotational spectrum is complex, as shown in Fig. A13.1. The rotational energy levels were first determined through measurements of the infrared spectra by Randall et al. (1937). Van Vleck noted in an MIT Radiation Laboratory report (Van Vleck 1942) that these energy levels indicated the existence of an allowable microwave line in the range of 1.2–1.5 cm (20–25 GHz), due to a chance near-coincidence of two energy levels in adjacent rotational ladders. Lying at an energy level of 447 cm^{-1} above the ground state, corresponding to a temperature of 640 K, the line has a Boltzmann population factor at atmospheric temperature that is about 0.1. Van Vleck calculated that atmospheric opacity along horizontal path lengths due to the absorption of H_2O and absorption in the wing of the O_2 lines near 60 GHz would cause problems for radars operating at short centimeter wavelengths. However, there was little empirical data about the pressure-broadening constants for the line widths [see Eq. (13.43) and Fig. 13.7] of these lines, and the estimates Van Vleck used were almost three times larger than than their actual values. Therefore, he substantially overestimated the absorption of O_2 and underestimated the absorption of H_2O at 1.25-cm wavelength. Nonetheless, he raised an important

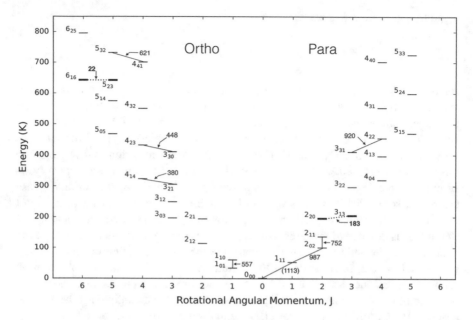

Fig. A13.1 Energy levels of the ground vibrational state of H_2O, an axisymmetric rotating molecule. The quantum numbers are denoted J_{K-1K+1}. $K - 1K + 1$ are even/odd and odd/even for ortho states and even/even and odd/odd for para states. The seven most important transitions responsible for making the atmosphere opaque at ALMA for a water vapor pressure of 1 mm and frequencies less than 1 THz (see Fig. 13.14) are marked with their frequencies (380, 448, 557, 621, 752, 920, and 987 GHz) along with the ground state transition at 1113 GHz. The diagnostic lines at 22 and 183 GHz used in water-vapor radiometry (Sects. 13.3.2 and 13.3.3) are shown with dotted lines. Other molecular lines causing high opacity are O_2 at 60 GHz and O_3 at 118 GHz. Data from Splatalogue (2016).

concern, which was to go unheeded. His later absorption estimates (Van Vleck 1945, 1947) were more accurate.

Late in World War II, the 3-cm airborne radar had proved to be highly successful. To obtain even higher resolution for antennas of similar size, a new system at 1.25 cm was planned as more powerful microwave signal sources became available. Van Vleck and Townes warned that the new system would have difficulties because of water vapor absorption (see Townes 1952, 1999; Buderi 1996; and Sullivan 2009), but development proceeded nonetheless. The range of the new system, looking along horizontal paths, was found to be only typically 20 km or less, a tremendous disappointment. The cause was quickly traced to atmospheric water vapor absorption. Dicke et al. (1946) traced out the line shape from atmospheric brightness temperature measurements in Florida in 1945 and established the wavelength to be 1.34 cm ($\nu = 22.2\,\text{GHz}$) and also determined the line profile and absorption coefficient. Planned deployment of the system to the moist South Pacific war zone was canceled. Townes and Merritt (1946) measured the transition at low pressure in the lab to high accuracy ($\nu = 22237 \pm 5\,\text{MHz}$, 1.349 cm). The modern standard frequency of the transition, weighted over its hyperfine transitions, is 22235.080 MHz (Kukolich 1969).

Appendix 13.2   Derivation of the Tropospheric Phase Structure Function

The purpose of this appendix is to derive the phase structure function for the troposphere from the structure function for the index of refraction in a turbulent medium. We follow the derivation of Tatarski (1961).

The structure function of phase is defined as

$$D_\phi = \langle [\phi_1(x_1) - \phi_2(x_2)]^2 \rangle , \qquad (A13.1)$$

where x_1 and x_2 are two measurement points as shown in Fig. A13.2, which for our purposes form a baseline interferometer normal to the incoming propagation direction as viewed outside the homogeneous scattering layer of thickness L. The turbulence is considered to be "frozen" as it moves along the x axis. The ensemble average is usually approximated as a time average of duration T, where T is much longer than the crossing time of the turbulent cells, i.e., $T \gg d/v_s$, where v_s is the wind speed component along the baseline. The initially flat phase front is distorted by the turbulent medium, as shown in the right side of Fig. A13.2. The phase structure function depends only on the separation $d = |x_1 - x_2|$. The structure function of the index of refraction for Kolmogorov turbulence has the general form

$$D_n = C_n^2 \mathbf{r}^{2/3} , \qquad (A13.2)$$

where $\mathbf{r}$ is the vector separation between two points in the turbulent medium. With the assumption that the medium is isotropic and homogeneous, the structure function becomes a function of only the scalar separation, r,

$$D_n = C_n^2 r^{2/3} . \qquad (A13.3)$$

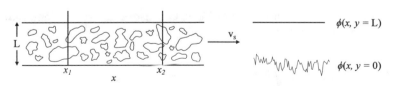

Fig. A13.2 (**left**) Cartoon of a frozen turbulent layer of the troposphere moving along the x axis at velocity v_s. The structure function is measured at two points, x_1 and x_2. (**right**) The incoming signal phase from a point source at the bottom and top of the scattering layer.

From a strict ray-tracing calculation, the phases at some instant will be given by

$$
\begin{aligned}
\phi_1 &= \frac{2\pi}{\lambda} \int_0^L n(y, x_1)\, dy \\
\phi_2 &= \frac{2\pi}{\lambda} \int_0^L n(y, x_2)\, dy\,,
\end{aligned}
\tag{A13.4}
$$

where n is the index of refraction along the y axis perpendicular to the baseline, and λ is the wavelength. Except at very dry sites, the fluctuations in refraction are primarily caused by variation in the water vapor density. The difference in phase at points x_1 and x_2 is therefore

$$
\phi_1 - \phi_2 = \frac{2\pi}{\lambda} \int_0^L [n(y, x_1) - n(y, x_2)]\, dy\,,
\tag{A13.5}
$$

and the squared difference of phase is

$$
(\phi_1 - \phi_2)^2 = \left(\frac{2\pi}{\lambda}\right)^2 \int_0^L [n(y_a, x_1) - n(y_a, x_2)]\, dy_a
$$
$$
\times \int_0^L [n(y_b, x_1) - n(y_b, x_2)]\, dy_b\,,
\tag{A13.6}
$$

or

$$
(\phi_1 - \phi_2)^2 = \left(\frac{2\pi}{\lambda}\right)^2 \int_0^L \int_0^L [n(y_a, x_1) - n(y_a, x_2)]
$$
$$
\times [n(y_b, x_1) - n(y_b, x_2)]\, dy_a\, dy_b\,.
\tag{A13.7}
$$

We could expand the integrand in Eq. (A13.7) into cross products of n at different positions. However, we prefer to express the final result in terms of structure functions rather than correlation functions. To proceed, we use the algebraic identity

$$
(a - b)(c - d) = \frac{1}{2}[(a - d)^2 + (b - c)^2 - (a - c)^2 - (b - d)^2]\,.
\tag{A13.8}
$$

Substituting Eq. (A13.7) into Eq. (A13.1), making use of Eq. (A13.8), and taking the expectation term by term, we obtain

$$
\begin{aligned}
D_\phi(d) = \frac{1}{2}\left(\frac{2\pi}{\lambda}\right)^2 \int_0^L \int_0^L \Big\{ &\langle [n(y_a, x_1) - n(y_b, x_2)]^2\rangle \\
&+ \langle [n(y_a, x_2) - n(y_b, x_1)]^2\rangle \\
&- \langle [n(y_a, x_1) - n(y_b, x_1)]^2\rangle \\
&- \langle [n(y_a, x_2) - n(y_b, x_2)]^2\rangle \Big\}\, dy_a\, dy_b\,.
\end{aligned}
\tag{A13.9}
$$

The four terms in Eq. (A13.9) are structure functions of the index of refraction for various separations, as defined in Eq. (A13.3). Note that the separation in the first two terms is $[(y_a - y_b)^2 + (x_1 - x_2)^2]^{1/2}$, while the separation in the second two terms is $|y_a - y_h|$. Hence, the structure function of phase can now be written as

$$D_\phi(d) = \left(\frac{2\pi}{\lambda}\right)^2 \int_0^L \int_0^L \left[D_n \left(\sqrt{(y_a - y_b)^2 + (x_1 - x_2)^2} \right) \right.$$

$$\left. - D_n(|y_a - y_b|) \right] dy_a \, dy_b \, . \qquad (A13.10)$$

The integral in Eq. (A13.10) can be simplified because the arguments of D_n are a function of only $y_a - y_b$. Note that an integral of the form

$$I = \int_0^L \int_0^L f(y_a - y_b) \, dy_a \, dy_b \qquad (A13.11)$$

can be simplified (see Fig. A13.3) by a change in variables to $y = y_a - y_b$ and y_b. For the case in which $f(y_a - y_b)$ is an even function, Eq. (A13.11) becomes

$$I = 2 \int_0^L (L - y) f(y) \, dy \, . \qquad (A13.12)$$

By use of this relation, the structure function of phase becomes

$$D_\phi(d) = 2 \left(\frac{2\pi}{\lambda}\right)^2 \int_0^L (L - y) \left[D_n \left(\sqrt{y^2 + d^2} \right) - D_n(y) \right] dy \, . \qquad (A13.13)$$

Substitution of Eq. (A13.3) into Eq. (A13.13) gives

$$D_\phi(d) = 2 \left(\frac{2\pi}{\lambda}\right)^2 C_n^2 \int_0^L (L - y)[(y^2 + d^2)^{1/3} - y^{2/3}] \, dy \, . \qquad (A13.14)$$

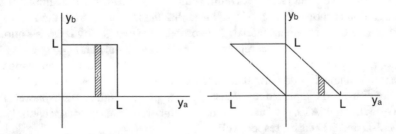

Fig. A13.3 The change in integration variables from y_a, y_b to y, y_b, where $y = y_a - y_b$ for the derivation of Eq. (A13.12).

This equation is the general starting point for most discussions (see Tatarski 1961, eq. 6.27). The distinction of two major regimes, $d \ll L$ and $d \gg L$, was first made in the context of radio interferometry by Stotskii (1973, 1976) and further discussed by Dravskikh and Finkelstein (1979) and Coulman (1990).

The case $d \ll L$ is called the "three-dimensional," or 3-D, turbulence solution. The integral in Eq. (A13.14) is maximum at $y = 0$, where it equals $L d^{2/3}$ and decreases monotonically to zero as y increases. It declines slowly for $y < d$, and for larger y, it decreases as $y^{-4/3}$. Hence, the integrand is approximately constant in the range of y from 0 to d, and most of the contribution to the integral is within this range. Thus, from Eq. (A13.14), $D_\phi \sim L d^{2/3} \times d \sim L d^{5/3}$. The proportionality constant, as reported by Tatarski (1961, eq. 6.65), based on analytic approximation, is about 2.91. Hence,

$$D_\phi(d) = 2.91 \left(\frac{2\pi}{\lambda}\right)^2 C_n^2 L \, d^{5/3} \,, \qquad d_f, d_{\text{in}} < d \ll L \,. \qquad \text{(A13.15)}$$

The case $d \gg L$ is called the "two-dimensional," or 2-D, turbulence case. Stotskii (1973) and Coulman (1990) presented the reasons why Eq. (A13.14), strictly valid for isotropic turbulence, can be used for this case. When $d \gg L$, the argument in brackets in Eq. (A13.14) becomes $\sim d^{2/3} - y^{2/3}$, and Eq. (A13.14) can be integrated directly. The leading term in the integral is $\frac{1}{2} L^2 d^{2/3}$, which gives

$$D_\phi(d) \simeq \left(\frac{2\pi}{\lambda}\right)^2 C_n^2 L^2 d^{2/3} \,, \qquad L \ll d < L_{\text{out}} \,. \qquad \text{(A13.16)}$$

For $d > L_{\text{out}}$, D_n becomes independent of distance, and D_ϕ becomes flat.

It is interesting to note that the two structure functions given in Eqs. (A13.15) and (A13.16) intersect at a distance

$$d_2 = L/2.9 \,, \qquad \text{(A13.17)}$$

which can be taken to be the nominal transition point from 3-D to 2-D turbulence. For a scale height of 2 km, this would be about 700 m. Numerical integrations of Eq. (A13.14) have been done by Treuhaft and Lanyi (1987). An example of such an integration is shown in Fig. A13.4. Note that the transition from the 2-D to 3-D structure functions is rather gradual. This probably explains why large variations in the power-law index have been reported from observational data.

The results described above can be generalized for the case in which the propagation angle is not normal to the baseline but rather is at an angle y. In this situation, L is replaced by $L \sec y$ for a plane-parallel atmosphere. Thus, Eqs. (A13.15) and (A13.16) show that the structure functions vary as $\sec y$ and $\sec^2 y$ for the 3-D and 2-D cases, respectively.

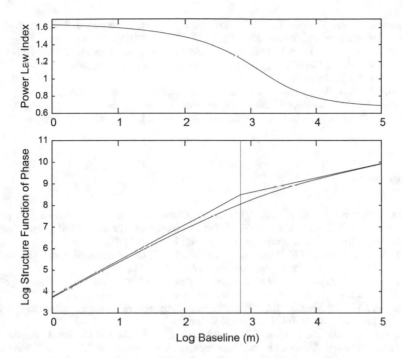

Fig. A13.4 (**bottom**) Structure function of phase vs. baseline length (d) and its power-law approximations for a layer thickness of 2 km and turbulence parameter $C_n^2 = 1$. The intersection of the two power-law components, which occurs at $d = L/2.9 = d_2$, or about 700 m, is marked by the thin vertical line. (**top**) Power-law index as a function of baseline for the structure function of phase.

Further Reading

Andrews, D.G., *An Introduction to Atmospheric Physics*, Cambridge Univ. Press, Cambridge, UK (2000)

Baldwin, J.E. and Wang, S., Eds., *Radio Astronomical Seeing*, International Academic Publishers and Pergamon Press, Oxford, UK (1990)

Janssen, M.A., *Atmospheric Remote Sensing by Microwave Radiometry*, Wiley, New York (1993)

Mangum, J.G., and Wallace, P., Atmospheric Refractive Electromagnetic Wave Bending and Propagation Delay, *Pub. Astron. Soc. Pacific*, **127**, 74–91, (2015)

Proc. RadioNet Workshop on Measurement of Atmospheric Water Vapour: Theory, Techniques, Astronomical, and Geodetic Applications, Wettzell/Hoellenstein, Germany, Oct. 9–11 (2006). http://bit.ly/1Knb11E

Tatarski, V.I., *Wave Propagation in a Turbulent Medium*, Dover, New York (1961)

Westwater, R., Ed., *Specialist Meeting on Microwave Radiometry and Remote Sensing Applications*, National Oceanic and Atmospheric Administration, U.S. Dept. Commerce (1992)

References

ALMA Partnership, Fomalont, E.B., Vlahakis, C., Corder, S., Remijan, A., Barkats, D., Lucas, R., Hunter, T.R., Brogan, C.L., Asaki, Y., and 239 coauthors, The 2014 ALMA Long Baseline Campaign: An Overview, *Astrophys. J. Lett.*, **808**, L1 (11 pp) (2015)

Altenhoff, W.J., Baars, J.W.M., Downes, D., and Wink, J.E., Observations of Anomalous Refraction at Radio Wavelengths, *Astron. Astrophys.*, **184**, 381–385 (1987)

Andrews, D.G., *An Introduction to Atmospheric Physics*, Cambridge Univ. Press, Cambridge, UK (2000), p. 24.

Armstrong, J.W., and Sramek, R.A., Observations of Tropospheric Phase Scintillations at 5 GHz on Vertical Paths, *Radio Sci.*, **17**, 1579–1586 (1982)

Asaki, Y., Matsushita, S., Kawabe, R., Fomalont, E., Barkats, D., and Corder, S., ALMA Fast Switching Phase Calibration on Long Baselines, in *Ground-Based and Airborne Telescopes V*, L. M. Stepp, R. Gilmozzi, and H. J. Hall, Eds., Proc. SPIE, **9145**, 91454K-1 (2014)

Asaki, Y., Saito, M., Kawabe, R., Morita, K.-I., and Sasao, T., Phase Compensation Experiments with the Paired Antennas Method, *Radio Sci.*, **31**, 1615–1625 (1996)

Baars, J.W.M., Meteorological Influences on Radio Interferometer Phase Fluctuations, *IEEE Trans. Antennas Propag.*, **AP-15**, 582–584 (1967)

Barrett, A.H., and Chung, V.K., A Method for the Determination of High-Altitude Water-Vapor Abundance from Ground-Based Microwave Observations, *J. Geophys. Res.*, **67**, 4259–4266 (1962)

Battat, J.B., Blundell, R., Moran, J.M., and Paine, S., Atmospheric Phase Correction Using Total Power Radiometry at the Submillimeter Array, *Astrophys. J. Lett.*, **616**, L71–L74 (2004)

Bean, B.R., and Dutton, E.J., *Radio Meteorology*, National Bureau of Standards Monograph 92, U.S. Government Printing Office, Washington, DC (1966)

Beaupuits, J.P.P., Rivera, R.C., and Nyman, L.-A., Height and Velocity of the Turbulence Layer at Chajnantor Estimated from Radiometric Measurements, ALMA Memo 542 (2005)

Bevis, M., Businger, S., Chiswell, S., Herring, T.A., Anthes, R.A., Rocken, C., and Ware, R.H., GPS Meteorology: Mapping Zenith Wet Delays onto Precipitable Water, *J. Appl. Meteor.*, **33**, 379–386 (1994)

Bieging, J.H., Morgan, J., Welch, W.J., Vogel, S.N., Wright, M.C.H., Interferometer Measurements of Atmospheric Phase Noise at 86 GHz, *Radio Sci.*, **19**, 1505–1509 (1984)

Bohlander, R.A., McMillan, R.W., and Gallagher, J.J., Atmospheric Effects on Near-Millimeter-Wave Propagation, *Proc. IEEE*, **73**, 49–60 (1985)

Bolton, R., Nikolic, B., and Richer, J., The Power Spectrum of Atmospheric Path Fluctuations at the ALMA Site from Water Vapour Radiometer Observations, ALMA Memo 592 (2011)

Bracewell, R.N., *The Fourier Transform and Its Applications*, 3rd ed., McGraw-Hill, New York (2000) (earlier eds. 1965, 2000).

Bremer, M., Atmospheric Phase Correction for Connected-Element Interferometry and for VLBI, in *Astronomical Site Evaluation in the Visible and Radio Range*, J. Vernin, Z. Benkhaldoun, and C. Muñoz-Tuñón, Eds., Astron. Soc. Pacific Conf. Ser., **266**, 238–245 (2002)

Buderi, R., *The Invention That Changed the World*, Simon and Schuster, New York (1996), pp. 261 and 340.

Butler, B., Precipitable Water at the VLA—1990–1998, MMA Memo 237, National Radio Astronomy Observatory (1998)

Butler, B., and Desai, K., Phase Fluctuations at the VLA Derived from One Year of Site Testing Interferometer Data, VLA Test Memo 222, National Radio Astronomy Observatory (1999)

Butler, B.J., Radford, S.J.E., Sakamoto, S., Kohno, K., Atmospheric Phase Stability at Chajnantor and Pampa la Bola, ALMA Memo 365 (2001)

Calisse, P.G., Ashley, M.C.B., Burton, M.G., Phillips, M.A., Storey, J.W.V., Radford, S.J.E., and Peterson, J.B., Submillimeter Site Testing at Dome C, Antarctica, *Pub. Astron. Soc. Austr.*, **21**, 256–263 (2004)

Carilli, C.L., and Holdaway, M.A., Tropospheric Phase Calibration in Millimeter Interferometry, *Radio Sci.*, **34**, 817–840 (1999)

Chamberlin, R.A., and Bally, J., The Observed Relationship Between the South Pole 225-GHz Atmospheric Opacity and the Water Vapor Column Density, *Int. J. Infrared and Millimeter Waves*, **16**, 907–920 (1995)

Chamberlin, R.A., Lane, A.P., and Stark, A.A., The 492 GHz Atmospheric Opacity at the Geographic South Pole, *Astrophys. J.*, **476**, 428–433 (1997)

Chylek, P., and Geldart, D.J.W., Water Vapor Dimers and Atmospheric Absorption of Electromagnetic Radiation, *Geophys. Res. Lett.*, **24**, 2015–2018 (1997)

COESA, *U.S. Standard Atmosphere, 1976*, NOAA-S/T 76-1562, U.S. Government Printing Office, Washington, DC (1976)

Coulman, C.E., Fundamental and Applied Aspects of Astronomical "Seeing," *Ann. Rev. Astron. Astrophys.*, **23**, 19–57 (1985)

Coulman, C.E., Atmospheric Structure, Turbulence, and Radioastronomical "Seeing," in *Radio Astronomical Seeing*, J. E. Baldwin and S. Wang, Eds., International Academic Publishers and Pergamon Press, Oxford, UK (1990), pp. 11–20.

Coulman, C.E., Tropospheric Phenomena Responsible for Anomalous Refraction at Radio Wavelengths, *Astron. Astrophys.*, **251**, 743–750 (1991)

Cox, A.N., Ed., *Allen's Astrophysical Quantities*, 4th ed., AIP Press, Springer, New York (2000), Sec. 11.20, p. 262.

Crane, R.K., Refraction Effects in the Neutral Atmosphere, in *Methods of Experimental Physics*, Vol. 12, Part B *(Astrophysics: Radio Telescopes)*, Meeks, M.L., Ed., Academic Press, New York (1976), pp. 186–200.

Crane, R.K., Fundamental Limitations Caused by RF Propagation, *Proc. IEEE*, **69**, 196–209 (1981)

Davis, J.L., Herring, T.A., Shapiro, I.I., Rogers, A.E.E., and Elgered, G., Geodesy by Radio Interferometry: Effects of Atmospheric Modeling Errors on Estimates of Baseline Length, *Radio Sci.*, **20**, 1593–1607 (1985)

Debye, P., *Polar Molecules*, Dover, New York (1929)

Delgado, G., Otárola, A., Belitsky, V., and Urbain, D., The Determination of Precipitable Water Vapour at Llano de Chajnantor from Observations of the 183 GHz Water Line, ALMA Memo 271 (1998)

Dicke, R.H., Beringer, R., Kyhl, R.L., and Vane, A.B., Atmospheric Absorption Measurements with a Microwave Radiometer, *Phys. Rev.*, **70**, 340–348 (1946)

Downes, D., and Altenhoff, W.J., Anomalous Refraction at Radio Wavelengths, in *Radio Astronomical Seeing*, Baldwin, J.E., and Wang, S., Eds., International Academic Publishers and Pergamon Press, Oxford, UK (1990), pp. 31–40.

Dravskikh, A.F., and Finkelstein, A.M., Tropospheric Limitations in Phase and Frequency Coordinate Measurements in Astronomy, *Astrophys. Space Sci.*, **60**, 251–265 (1979)

Elgered, G., Davis, J.L., Herring, T.A., and Shapiro, I.I., Geodesy by Radio Interferometry: Water Vapor Radiometry for Estimation of the Wet Delay, *J. Geophys. Res.*, **96**, 6541–6555 (1991)

Elgered, G., Rönnäng, B.O., and Askne, J.I.H., Measurements of Atmospheric Water Vapor with Microwave Radiometry, *Radio Sci.*, **17**, 1258–1264 (1982)

Freeman, R.L., *Radio System Design for Telecommunications (1–100 GHz)*, Wiley, New York (1987)

Fried, D.L., Statistics of a Geometric Representation of Wavefront Distortion, *J. Opt. Soc. Am.*, **55**, 1427–1435 (1965)

Fried, D.L., Optical Resolution Through a Randomly Inhomogeneous Medium for Very Long and Very Short Exposures, *J. Opt. Soc. Am.*, **56**, 1372–1379 (1966)

Fried, D.L., Optical Heterodyne Detection of an Atmospherically Distorted Signal Wave Front, *Proc. IEEE*, **55**, 57–67 (1967)

Goldstein, H., Attenuation by Condensed Water, in *Propagation of Short Radio Waves*, MIT Radiation Laboratory Ser., Vol. 13, Kerr, D.E., Ed., McGraw-Hill, New York (1951), pp. 671–692.

Grischkowsky, D., Yang, Y., and Mandehgar, M., Zero-Frequency Refractivity of Water Vapor: Comparison of Debye and Van Vleck–Weisskopf Theory, *Optics Express*, **21**, 18899–18908 (2013)

Guiraud, F.O., Howard, J., and Hogg, D.C., A Dual-Channel Microwave Radiometer for Measurement of Preciptable Water Vapor and Liquid, *IEEE Trans. Geosci. Electron.*, **GE-17**, 129–136 (1979)

Hess, S.L., *Introduction to Theoretical Meteorology*, Holt, Rinehart, Winston, New York (1959)

Hill, R.J., Water Vapor-Absorption Line Shape Comparison Using the 22-GHz Line: The Van Vleck–Weisskopf Shape Affirmed, *Radio Sci.*, **21**, 447–451 (1986)

Hill, R.J., and Clifford, S.F., Contribution of Water Vapor Monomer Resonances to Fluctuations of Refraction and Absorption for Submillimeter through Centimeter Wavelengths, *Radio Sci.*, **16**, 77–82 (1981)

Hill, R.J., Lawrence, R.S., and Priestley, J.T., Theoretical and Calculational Aspects of the Radio Refractive Index of Water Vapor, *Radio Sci.*, **17**, 1251–1257 (1982)

Hills, R.E., Webster, A.S., Alston, D.A., Morse, P.L.R., Zammit, C.C., Martin, D.H., Rice, D.P., and Robson, E.I., Absolute Measurements of Atmospheric Emission and Absorption in the Range 100–1000 GHz, *Infrared Phys.*, **18**, 819–825 (1978)

Hinder, R.A., Observations of Atmospheric Turbulence with a Radio Telescope at 5 GHz, *Nature*, **225**, 614–617 (1970)

Hinder, R.A., Fluctuations of Water Vapour Content in the Troposphere as Derived from Interferometric Observations of Celestial Radio Sources, *J. Atmos. Terr. Phys.*, **34**, 1171–1186 (1972)

Hinder, R.A., and Ryle, M., Atmospheric Limitations to the Angular Resolution of Aperture Synthesis Radio Telescopes, *Mon. Not. R. Astron. Soc.*, **154**, 229–253 (1971)

Hogg, D.C., Guiraud, F.O., and Sweezy, W.B., The Short-Term Temporal Spectrum of Precipitable Water Vapor, *Science*, **213**, 1112–1113 (1981)

Holdaway, M.A., Ishiguro, M., Foster, S.M., Kawabe, R., Kohno, K., Owen, F.N., Radford, S.J.E., and Saito, M., Comparison of Rio Frio and Chajnantor Site Testing Data, MMA Memo 152, National Radio Astronomy Observatory (1996)

Holdaway, M.A., Radford, S.J.E., Owen, F.N., and Foster, S.M., Data Processing for Site Test Interferometers, ALMA Memo 129 (1995a)

Holdaway, M.A., Radford, S.J.E., Owen, F.N., and Foster, S.M., Fast Switching Phase Calibration: Effectiveness at Mauna Kea and Chajnantor, MMA Memo 139, National Radio Astronomy Observatory (1995b)

Holdaway, M.A., and Woody, D., Yet Another Look at Anomalous Refraction, MMA Memo 223, National Radio Astronomy Observatory (1998)

Ishiguro, M., Kanzawa, T., and Kasuga, T., Monitoring of Atmospheric Phase Fluctuations Using Geostationary Satellite Signals, in *Radio Astronomical Seeing*, Baldwin, J.E., and Wang, S., Eds., International Academic Publishers and Pergamon Press, Oxford, UK (1990), pp. 60–63

Jackson, J.D., *Classical Electrodynamics*, 3rd ed., Wiley, New York (1999), pp. 775–784

Kasuga, T., Ishiguro, M., and Kawabe, R., Interferometric Measurement of Tropospheric Phase Fluctuations at 22 GHz on Antenna Spacings of 27 to 540 m, *IEEE Trans. Antennas Propag.*, **AP-34**, 797–803 (1986)

Kimberk, R.S., Hunter, T.R., Leiker, P.S., Blundell, R., Nystrom, G.U., Petitpas, G.R., Test, J., Wilson, R.W., Yamaguchi, P., and Young, K.H., A Multi-Baseline 12 GHz Atmospheric Phase Interferometer with One Micron Path Length Sensitivity, *J. Astron. Inst,*, **1**, 1250002 (2012)

Kukolich, S.G., Measurement of the Molecular g Values in H_2O and D_2O and Hyperfine Structure in H_2O, *J. Chem. Phys.*, **50**, 3751–3755 (1969)

Lamb, J.W., and Woody, D., Radiometric Correction of Anomalous Refraction, MMA Memo 224, National Radio Astronomy Observatory (1998)

Lay, O.P., The Temporal Power Spectrum of Atmospheric Fluctuations Due to Water Vapor, *Astron. Astrophys. Suppl.*, **122**, 535–545 (1997a)

Lay, O.P., Phase Calibration and Water Vapor Radiometry for Millimeter-Wave Arrays, *Astron. Astrophys. Suppl.*, **122**, 547–557 (1997b)

Lay, O.P., 183 GHz Radiometric Phase Correction for the Millimeter Array, MMA Memo 209, National Radio Astronomy Observatory (1998)

Lichtenstein, M., and Gallagher, J.J., Millimeter Wave Spectrum of Ozone, *J. Molecular Spectroscopy*, **40**, 10–26 (1971)

Liebe, H.J., Calculated Tropospheric Dispersion and Absorption Due to the 22-GHz Water Vapor Line, *IEEE Trans. Antennas Propag.*, **AP-17**, 621–627 (1969)

Liebe, H.J., Modeling Attenuation and Phase of Radio Waves in Air at Frequencies below 1000 GHz, *Radio Sci.*, **16**, 1183–1199 (1981)

Liebe, H.J., An Updated Model for Millimeter Wave Propagation in Moist Air, *Radio Sci.*, **20**, 1069–1089 (1985)

Liebe, H.J., MPM: An Atmospheric Millimeter-Wave Propagation Model, *Int. J. Infrared and Millimeter Waves*, **10**, 631–650 (1989)

Loudon, R., *The Quantum Theory of Light*, 2nd ed., Oxford Univ. Press, London (1983)

Marini, J.W., Correction of Satellite Tracking Data for an Arbitrary Tropospheric Profile, *Radio Sci.*, **7**, 223–231 (1972)

Masson, C.R., Atmospheric Effects and Calibrations, in *Astronomy with Millimeter and Submillimeter Wave Interferometry*, Ishiguro, M., and Welch, W.J., Eds., Astron. Soc. Pacific Conf. Ser., **59**, 87–95 (1994a)

Masson, C.R., Seeing, in *Very High Angular Resolution Imaging*, IAU Symp. 158, Robertson, J.G., and Tango, W.J., Eds., Kluwer, Dordrecht, the Netherlands (1994b), pp. 1–10

Matsushita, S., Matsuo, H., Pardo, J.R., and Radford, S.J.E., FTS Measurements of Submillimeter-Wave Atmospheric Opacity at Pampa la Bola II: Supra-Terahertz Windows and Model Fitting, *Pub. Ast. Soc. Japan*, **51**, 603–610 (1999)

Matsushita, S., Matsuo, H., Wiedner, M.C., and Pardo, J.R., Phase Correction Using Submillimeter Atmospheric Continuum Emission, ALMA Memo 415 (2002)

Millenaar, R.P., Tropospheric Stability at Candidate SKA Sites: Australia Edition, SKA Doc. WP3-040.020.001-TR-003 (2011a)

Millenaar, R.P., Tropospheric Stability at Candidate SKA Sites: South Africa Edition, SKA Doc. WP3-040.020.001-TR-002 (2011b)

Miner, G.F., Thornton, D.D., and Welch, W.J., The Inference of Atmospheric Temperature Profiles from Ground-Based Measurements of Microwave Emission from Atmospheric Oxygen, *J. Geophys. Res.*, **77**, 975–991 (1972)

Morabito, D.D., D'Addario, L.R., Acosta, R.J., and Nessel, J.A., Tropospheric Delay Statistics Measured by Two Site Test Interferometers at Goldstone, California, *Radio Sci.*, **48**, 1–10 (2013)

Moran, J.M., and Rosen, B.R., Estimation of the Propagation Delay through the Troposphere from Microwave Radiometer Data, *Radio Sci.*, **16**, 235–244 (1981)

Niell, A.E., Global Mapping Functions for the Atmospheric Delay at Radio Wavelengths, *J. Geophys Res.*, **101**, 3227–3246 (1996)

Nikolic, B., Bolton, R.C., Graves, S.F., Hills, R.E., and Richter, J.S., Phase Collection for ALMA with 183 GHz Water Vapour Radiometers, *Astron. Astrophys.*, **552**, A104 (11pp) (2013)

Olmi, L., and Downes, D., Interferometric Measurement of Tropospheric Phase Fluctuations at 86 GHz on Antenna Spacings of 24 m to 288 m, *Astron. Astrophys.*, **262**, 634–643 (1992)

Owens, J.C., Optical Refractive Index of Air: Dependence on Pressure, Temperature, and Composition, *Appl. Opt.*, **6**, 51–58 (1967)

Paine, S., The *am* Atmospheric Model, SMA Technical Memo 152, Smithsonian Astrophysical Observatory, Cambridge, MA (2016)

Paine, S., Blundell, R., Papa, D.C., Barrett, J.W., and Radford, S.J.E., A Fourier Transform Spectrometer for Measurement of Atmospheric Transmission at Submillimeter Wavelengths, *Publ. Astron. Soc. Pacific*, **112**, 108–118 (2000)

Pardo, J.R., Cernicharo, J., and Serabyn, E., Atmospheric Transmission at Microwaves (ATM): An Improved Model for Millimeter/Submillimeter Applications, *IEEE Trans. Antennas Propag.*, **49**, 1683–1694 (2001a)

Pardo, J.R., Serabyn, E., and Cernicharo, J., Submillimeter Atmospheric Transmission Measurements on Mauna Kea During Extremely Dry El Niño Conditions, *J. Quant. Spect. Rad. Trans.*, **68**, 419–433 (2001b)

Peixoto, J.P., and Oort, A.H., The Climatology of Relative Humidity in the Atmosphere, *J. Climate*, **9**, 3443–3463 (1996)

Peréz, L.M., Lamb, J.W., Woody, D.P., Carpenter, J.M., Zauderer, B.A., Isella, A., Bock, D.C., Bolatto, A.D., Carlstrom, J., Culverhouse, T.L., and nine coauthors, Atmospheric Phase Correction Using CARMA-PACs: High-Angular-Resolution Observations of the Fu Orionis Star PP 13S*, *Astrophys. J.*, **724**, 493–501 (2010)

Pol, S.L.C., Ruf, C.S., and Keihm, S.J., Improved 20- to 32-GHz Atmospheric Absorption Model, *Radio Sci.*, **33**, 1319–1333 (1998)

Racette, P.E., Westwater, E.R., Han, Y., Gasiewski, A.J., Klein, M., Cimini, D., Jones, D.C., Manning, W., Kim, E.J., Wang, J.R., Leuski, V., and Kiedron, P., Measurement of Low Amounts of Precipitable Water Vapor Using Ground-Based Millimeter-Wave Radiometry, *J. Atmos. Oceanic Tech.*, **22**, 317–337 (2005)

Radford, S.J.E., and Peterson, J.B., Submillimeter Atmospheric Transparency at Maunakea, at the South Pole, and at Chajnantor, *Publ. Astron. Soc. Pacific*, **128**:075001 (13pp) (2016)

Radford, S.J.E., Reiland, G., and Shillue, B., Site Test Interferometer, *Publ. Astron. Soc. Pacific*, **108**, 441–445 (1996)

Randall, H.M., Dennison, D.M., Ginsburg, N., and Weber, L.R., The Far Infrared Spectrum of Water Vapor, *Phys. Rev.*, **52**, 160–174 (1937)

Ray, P.S., Broadband Complex Refractive Indices of Ice and Water, *Applied Optics*, **11**, 1836–1843 (1972)

Reber, E.E., and Swope, J.R., On the Correlation of Total Precipitable Water in a Vertical Column and Absolute Humidity, *J. Appl. Meteor.*, **11**, 1322–1325 (1972)

Resch, G.M., Water Vapor Radiometry in Geodetic Applications, in *Geodetic Refraction: Effects of Electromagnetic Wave Propagation Through the Atmosphere*, Brunner, F.K., Ed., Springer-Verlag, Berlin (1984), pp. 53–84

Rienecker, M.M., Suarez, M.J., Gelaro, R., Todling, R., Bacmeister, J., Liu, E., Bosilovich, M.G., Schubert, S.D., Takacs, L., Kim, G.-K., and 19 coauthors, MERRA: NASA's Modern Era Retrospective Analysis for Research and Applications, *J. Climate*, **24**, 3624–3648 (2011)

Roddier, F., The Effects of Atmospheric Turbulence in Optical Astronomy, in *Progress in Optics XIX*, E. Wolf, Ed., North-Holland, Amsterdam (1981), pp. 281–376

Rogers, A.E.E., and Moran, J.M., Coherence Limits for Very-Long-Baseline Interferometry, *IEEE Trans. Instrum. Meas.*, **IM-30**, 283–286 (1981)

Rogers, A.E.E., Moffet, A.T., Backer, D.C., and Moran, J.M., Coherence Limits in VLBI Observations at 3-MillimeterWavelength, *Radio Sci.*, **19**, 1552–1560 (1984)

Rosenkranz, P.W., Water Vapor Microwave Continuum Absorption: A Comparison of Measurements and Models, *Radio Sci.*, **33**, 919–928 (1998)

Rüeger, J.M., Refractive Indices of Light, Infrared, and Radio Waves in the Atmosphere, Unisurv Report S-68, School of Surveying and Spatial Information Systems, University of New South Wales, Sydney, Australia (2002)

Rybicki, G.B., and Lightman, A.P., *Radiative Processes in Astrophysics*, Wiley-Interscience, New York (1979) (reprinted 1985)

Saastamoinen, J., Introduction to Practical Computation of Astronomical Refraction, *Bull. Géodésique*, **106**, 383–397 (1972a)

Saastamoinen, J., Atmospheric Correction for the Troposphere and Stratosphere in Radio Ranging of Satellites, in *The Use of Artificial Satellites for Geodesy*, Geophys. Monograph 15, American Geophysical Union, Washington, DC (1972b), pp. 247–251

Schaper, Jr. L.W., Staelin, D.H., and Waters, J.W., The Estimation of Tropospheric Electrical Path Length by Microwave Radiometry, *Proc. IEEE*, **58**, 272–273 (1970)

Sims, G., Kulesa, C., Ashley, M.C.B., Lawrence, J.S., Saunders, W., and Storey, J.W.V., Where is Ridge A?, in *Ground-Based and Airborne Telescopes IV*, Proc. SPIE, **8444**, 84445H-1–84445H-9 (2012)

Smart, W.M., *Textbook on Spherical Astronomy*, 6th ed., revised by R. M. Green, Cambridge Univ. Press, Cambridge, UK (1977)

Smith, Jr. E.K., and Weintraub, S., The Constants in the Equation for Atmospheric Refractive Index at Radio Frequencies, *Proc. IRE*, **41**, 1035–1037 (1953)

Snider, J.B., Ground-Based Sensing of Temperature Profiles from Angular and Multi-Spectral Microwave Emission Measurements, *J. Appl. Meteor.*, **11**, 958–967 (1972)

Snider, J.B., Burdick, H.M., and Hogg, D.C., Cloud Liquid Measurement with a Ground-Based Microwave Instrument, *Radio Sci.*, **15**, 683–693 (1980)

Solheim, F., Godwin, J.R., Westwater, E.R., Han, Y., Keihm, S.J., Marsh, K., and Ware, R., Radiometric Profiling of Temperature, Water Vapor, and Cloud Liquid Water Using Various Inversion Methods, *Radio Sci.*, **33**, 393–404 (1998)

Sramek, R., VLA Phase Stability at 22 GHz on Baselines of 100 m to 3 km, VLA Test Memo 143, National Radio Astronomy Observatory (1983)

Sramek, R.A., Atmospheric Phase Stability at the VLA, in *Radio Astronomical Seeing*, Baldwin, J.E., and Wang, S., Eds., International Academic Publishers and Pergamon Press, Oxford, UK (1990), pp. 21–30

Staelin, D.H., Measurements and Interpretation of the Microwave Spectrum of the Terrestrial Atmosphere near 1-Centimeter Wavelength, *J. Geophys. Res.*, **71**, 2875–2881 (1966)

Stotskii, A.A., Concerning the Fluctuation Characteristics of the Earth's Troposphere, *Radiophys. and Quantum Elect.*, **16**, 620–622 (1973)

Stotskii, A.A., Tropospheric Limitations of the Measurement Accuracy on Coordinates of Cosmic Radio Source, *Radiophys. and Quantum Elect.*, **19**, 1167–1169 (1976)

Sullivan, W.T., III, *Cosmic Noise: A History of Early Radio Astronomy*, Cambridge Univ. Press, Cambridge, UK (2009)

Sutton, E.C., and Hueckstaedt, R.M., Radiometric Monitoring of Atmospheric Water Vapor as It Pertains to Phase Correction in Millimeter Interferometry, *Astron. Astrophys. Suppl.*, **119**, 559–567 (1996)

Sutton, E.C., Subramanian, S., and Townes, C.H., Interferometric Measurements of Stellar Positions in the Infrared, *Astron. Astrophys.* **110**, 324–331 (1982)

Tahmoush, D.A., and Rogers, A.E.E., Correcting Atmospheric Variations in Millimeter Wavelength Very Long Baseline Interferometry Using a Scanning Water Vapor Radiometer, *Radio Sci.*, **35**, 1241–1251 (2000)

Tanner, A.B., and Riley, A.L., Design and Performance of a High-Stability Water Vapor Radiometer, *Radio Sci.*, **38**, 8050 (17pp) (2003)

Tatarski, V.I., *Wave Propagation in a Turbulent Medium*, Dover, New York (1961)

Tatarski, V.I., *The Effects of the Turbulent Atmosphere on Wave Propagation*, National Technical Information Service, Springfield, VA (1971)

Taylor, G.I., Spectrum of Turbulence, *Proc. R. Soc. London A*, **164**, 476–490 (1938)

Thayer, G.D., An Improved Equation for the Radio Refractive Index of Air, *Radio Sci.*, **9**, 803–807 (1974)

Toll, J.S., Causality and the Dispersion Relation: Logical Foundations, *Phys. Rev.*, **104**, 1760–1770 (1956)

Townes, C.H., Microwave Spectroscopy, *Am. Scientist*, **40**, 270–290 (1952)

Townes, C.H., *How the Laser Happened*, Oxford Univ. Press, Oxford, UK (1999), p. 40

Townes, C.H., and Merritt, F.R., Water Spectrum Near One-Centimeter Wave-Length, *Phys. Rev.*, **70**, 558–559 (1946)

Treuhaft, R.N., and Lanyi, G.E., The Effect of the Dynamic Wet Troposphere on Radio Interferometric Measurements, *Radio Sci.*, **22**, 251–265 (1987)

Van Vleck, J.H., *Atmospheric Absorption of Microwaves*, MIT Radiation Laboratory Report 43-2 (1942)

Van Vleck, J.H., Further Theoretical Investigations of the Atmospheric Absorption of Microwaves, MIT Radiation Laboratory Report 664 (1945)

Van Vleck, J.H., The Absorption of Microwaves by Uncondensed Water Vapor, *Phys. Rev.*, **71**, 425–433 (1947)

Van Vleck, J.H., Purcell, E.M., and Goldstein, H., Atmospheric Attenuation, in *Propagation of Short Radio Waves*, MIT Radiation Laboratory Ser., Vol. 13, D. E. Kerr, Ed., McGraw-Hill, New York (1951), pp. 641–692

Waters, J.W., Absorption and Emission by Atmospheric Gases, in *Methods of Experimental Physics*, Vol. 12, Part B *(Astrophysics: Radio Telescopes)*, Meeks, M.L., Ed., Academic Press, New York (1976), pp. 142–176

Welch, W.J., Correcting Atmospheric Phase Fluctuations by Means of Water-Vapor Radiometry, in *The Review of Radio Science, 1996–1999*, Stone, W.R., Ed., Oxford Univ. Press, Oxford, UK (1999), pp. 787–808

Westwater, E.R., *An Analysis of the Correction of Range Errors Due to Atmospheric Refraction by Microwave Radiometric Techniques*, ESSA Technical Report IER 30-ITSA 30, Institute for Telecommunication Sciences and Aeronomy, Boulder, CO (1967)

Westwater, E.R., The Accuracy of Water Vapor and Cloud Liquid Determination by Dual-Frequency Ground-Based Microwave Radiometry, *Radio Sci.*, **13**, 677–685 (1978)

Westwater, E.R., and Guiraud, F.O., Ground-Based Microwave Radiometric Retrieval of Precipitable Water Vapor in the Presence of Clouds with High Liquid Content, *Radio Sci.*, **15**, 947–957 (1980)

Wiedner, M.C., and Hills, R.E., Phase Correction on Mauna Kea Using 183 GHz Water Vapor Monitors, in *Imaging at Radio through Submillimeter Wavelengths*, Mangum, J.G., and Radford, S.J.E., Eds., Astron. Soc. Pacific Conf. Ser., **217**, 327–335 (2000)

Woody, D., Carpenter, J., and Scoville, N., Phase Correction at OVRO Using 22 GHz Water Line, in *Imaging at Radio through Submillimeter Wavelengths*, Mangum, J.G., and Radford, S.J.E., Eds., Astron. Soc. Pacific Conf. Ser., **217**, 317–326 (2000)

Woolf, N.J., High Resolution Imaging from the Ground, *Ann. Rev. Astron. Astrophys.*, **20**, 367–398 (1982)

Wright, M.C.H., Atmospheric Phase Noise and Aperture-Synthesis Imaging at Millimeter Wavelengths, *Publ. Astron. Soc. Pacific*, **108**, 520–534 (1996)

Wright, M.C.H., and Welch, W.J., Interferometer Measurements of Atmospheric Phase Noise at 3 mm, in *Radio Astronomical Seeing*, Baldwin, J.E., and Wang, S., Eds., International Academic Publishers and Pergamon Press, Oxford, UK (1990), pp. 71–74

Wu, S.C., Optimum Frequencies of a Passive Microwave Radiometer for Tropospheric Path-Length Correction, *IEEE Trans. Antennas Propag.*, **AP-27**, 233–239 (1979)

Zauderer, B.A., Bolatto, A.D., Vogel, S.N., Carpenter, J.M., Peréz, L.M., Lamb, J.W., Woody, D.P., Bock, D.C.-J., Carlstrom, J.E., Culverhouse, T.L., and 12 coauthors, The CARMA Paired Antenna Calibration System: Atmospheric Phase Correction for Millimeter-Wave Interferometry and Its Application to Mapping the Ultraluminous Galaxy Arp 193, *Astron. J.*, **151**, 18 (19pp) (2016)

Zivanovic, S.S., Forster, J.R., and Welch, W.J., A New Method for Improving the Interferometric Resolution by Compensating for the Atmospherically Induced Phase Shift, *Radio Sci.*, **30**, 877–884 (1995)

Chapter 14
Propagation Effects: Ionized Media

Three distinct ionized media, or plasmas, affect the propagation of radio signals passing through them: the Earth's ionosphere; the interplanetary medium, also known as the solar wind; and the interstellar medium of our Galaxy. The effects of scattering in other galaxies or in the media between galaxies are not usually important. There are several essential differences between neutral and ionized media with regard to propagation. For neutral media, the index of refraction is greater than unity and is unaffected by magnetic fields. In ionized media, the index of refraction is less than unity and is strongly affected by magnetic fields. Most plasma phenomena scale as ν^{-2}, and their effects can be avoided or mitigated, if desired, by observations at high frequency. Absorption plays an important role in neutral media but very little in ionized media since most radio astronomical observations occur at frequencies far above the plasma frequency. Descriptions of scattering phenomena in both types of media are based on Kolmogorov theory. However, the situation in the neutral troposphere is greatly simplified because the turbulent layer lies close to the observer, and only phase fluctuations develop. The ionized media lie far from the observer, and both phase and amplitude fluctuations are often present in the wavefront when it reaches the observer.

14.1 Ionosphere

The ionosphere has been studied extensively since the pioneering experiments of Appleton and Barnett (1925) and Breit and Tuve (1926). The literature on the subject is vast. Magneto-ionic propagation theory relevant to the ionosphere is treated in depth by Ratcliffe (1962) and Budden (1961); the general physics and chemistry of the ionosphere is described by Schunk and Nagy (2009); and an excellent general treatment of ionospheric propagation is given by Davies (1965). Reviews of particular relevance to radio astronomy can be found in Evans and Hagfors

© The Author(s) 2017
A.R. Thompson, J.M. Moran, and G.W. Swenson Jr., *Interferometry and Synthesis in Radio Astronomy*, Astronomy and Astrophysics Library,
DOI 10.1007/978-3-319-44431-4_14

Table 14.1 Maximum likely values of ionospheric effects at 100 MHz for a zenith angle[a] of 60°

Effect	Maximum[b] (Daytime)	Minimum[c] (Night)	Frequency dependence
Faraday rotation	15 rotations	1.5 rotations	ν^{-2}
Group delay	12 μs	1.2 μs	ν^{-2}
Excess (phase) path length	3500 m	350 m	ν^{-2}
Phase change	7500 rad	750 rad	ν^{-1}
Phase stability			
(peak to peak)	$\pm$150 rad	$\pm$15 rad	ν^{-1}
Frequency stability (rms)	$\pm$0.04 Hz	$\pm$0.004 Hz	ν^{-1}
Absorption			
(in D and F regions)	0.1 dB[d]	0.01 dB	ν^{-2}
Refraction (ambient)	0.05°	0.005°	ν^{-2}
Isoplanatic patch	–	$\sim$5°	ν

Adapted from Evans and Hagfors (1968).

[a]For values of parameters at the zenith, divide numbers (except refraction) by $\sec z_i$, which is approximately 1.7 [see Eq. (14.14)]. For typical (rather than maximum) parameters, divide numbers by 2.

[b]Total electron content = 5×10^{17} m^{-2}.

[c]Total electron content = 5×10^{16} m^{-2}.

[d]1 dB = 0.230 nepers.

(1968) and Hagfors (1976). Beynon (1975) gives interesting historical anecdotes on the early development of ionospheric research. In this section, we treat only those aspects of the ionosphere that have a deleterious effect on interferometric observations. Table 14.1 gives the magnitude of various propagation effects for the daytime and nighttime ionosphere. Most of these effects scale as ν^{-2}, and they can be minimized by observing at higher frequencies. For small zenith angles, the magnitude of the ionospheric excess path typically equals that of the neutral atmosphere at approximately 2 GHz, but the frequency of this equality can vary from about 1 to 5 GHz. Thus, at 20 GHz and small zenith angles, the ionospheric excess path length is typically only 1% of the tropospheric excess path length. However, at very large zenith angles, i.e., near the horizon, the effects are equal at about 300 MHz.

14.1.1   Basic Physics

The ionization of the upper atmosphere is caused by the ultraviolet radiation from the Sun. Typical daytime and nighttime vertical profiles of the electron density are shown in Fig. 14.1. The electron distribution and the total electron content vary also with geomagnetic latitude, time of year, and sunspot cycle. There are also substantial winds, traveling disturbances, and irregularities in the ionosphere. The ionosphere is

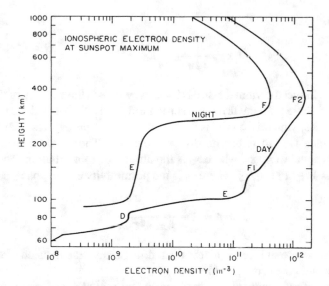

Fig. 14.1 Idealized electron density distribution in the Earth's ionosphere. The curves indicate the densities to be expected at sunspot maximum in temperate latitudes. Peak sunspot activity occurs at 11-year intervals, most recently peaking in 2001 (cycle 23) and $\sim$ 2012 (cycle 24). Cycle 24 was about a year late and rather weak [see Janardhan et al. (2015)]. From J. V. Evans and T. Hagfors, *Radar Astronomy*, 1968. © McGraw-Hill Education.

permeated by the quasi-dipole magnetic field of the Earth. Propagation is governed by the theory of waves in a magnetized plasma with collisions.

We derive some of the fundamental properties of the ionosphere related to the propagation of electromagnetic waves by considering elementary cases. First, consider a plane monochromatic wave of linear polarization that propagates through a uniform plasma of electron density n_e, where the magnetic field and collisions between particles can be neglected. The electrons oscillate with the electric field, but the protons, because of their greater mass, remain relatively unperturbed. The index of refraction can be found by calculating either the induced current or the dipole moment. Either method yields the same result. We use the latter method, as we did when considering the index of refraction of water vapor using the bound oscillator model in Sect. 13.1.4. The equation of motion of a free electron in the plasma is

$$m\ddot{\mathbf{x}} = -e\mathbf{E}_0 e^{-j2\pi \nu t} , \tag{14.1}$$

where m, e, and $\mathbf{x}$ are the mass, charge *magnitude*, and displacement of the electron, and $\mathbf{E}_0$ and ν are the amplitude and frequency of the electric field $\mathbf{E}$ of the incident wave. The magnetic field of the plane wave has negligible influence on the electrons as long as the electron velocity is much less than c, and the electric field has negligible influence on the motion of the protons. The steady-state solution to

Eq. (14.1) is

$$\mathbf{x} = \frac{e}{(2\pi\nu)^2 m} \mathbf{E}_0 e^{-j2\pi\nu t} . \tag{14.2}$$

Note that the induced current density is $\mathbf{i} = n_e e \dot{\mathbf{x}}$, where $\dot{\mathbf{x}}$, the velocity of the particle, is 90° out of phase with the driving electric field. Thus, the work done by the wave on the particles, which is $\langle \mathbf{i} \cdot \mathbf{E} \rangle$, is zero, and the wave propagates without loss, as expected, since Eq. (14.1) has no dissipative terms. The dipole moment per unit volume $\mathbf{P}$ is equal to $n_e e \mathbf{x}_0$, where $\mathbf{x}_0$ is the amplitude of oscillation. The dielectric constant ε is $1 + (\mathbf{P}/\mathbf{E}_0)/\epsilon_0$, where ϵ_0 is the permittivity of free space, so that

$$\varepsilon = 1 - \frac{n_e e^2}{4\pi^2 \nu^2 \epsilon_0 m} . \tag{14.3}$$

The dielectric constant is real and less than unity because the induced dipole is 180° out of phase with the driving field.

The index of refraction n is equal to the square root of ε, and in this case is real, so

$$n = \sqrt{1 - \frac{\nu_p^2}{\nu^2}} , \tag{14.4}$$

where

$$\nu_p = \frac{e}{2\pi} \sqrt{\frac{n_e}{\epsilon_0 m}} \simeq 9\sqrt{n_e} \ (\text{Hz}) , \tag{14.5}$$

and n_e is in m^{-3}. ν_p is known as the *plasma frequency*, which is also the natural frequency of mechanical oscillations in the plasma [see, e.g., Holt and Haskell (1965)]. The plasma frequency of the ionosphere (see Fig. 14.1) is usually less than 12 MHz. Waves normally incident on a plasma with frequencies below ν_p are perfectly reflected. The phase velocity of a wave with $\nu > \nu_p$ in the plasma is c/n, which is greater than c, and the group velocity of a wave packet is cn, which is less than c.

Now consider a plasma with a static magnetic field $\mathbf{B}$ in the direction of propagation of the plane wave. The vector equation of motion of an electron, called the Lorentz equation, is

$$m\dot{\mathbf{v}} = -e \left[\mathbf{E} + \mathbf{v} \times \mathbf{B} \right] , \tag{14.6}$$

where $\mathbf{v}$ is the vector velocity. Let the incident field be a circularly polarized wave. If $\mathbf{B}$ is zero, the particle will follow the tip of the electric field vector and move in a circular orbit. If $\mathbf{B}$ is nonzero, the sum of the $\mathbf{v} \times \mathbf{B}$ force term, which will be in the radial direction, and the electric force term must be balanced by centripetal acceleration. Thus, there is a basic anisotropy in the plasma depending on whether the wave is right or left circularly polarized, since the sign of the $\mathbf{v} \times \mathbf{B}$ term changes between the two cases. The radius R_e of the circular orbit of the electron is derived from the balance-of-forces equation $eE_0 \pm evB = mv^2/R_e$, where the scalar velocity $v = 2\pi v R_e$, B is the magnitude of the magnetic field, and the upper and lower signs refer to left and right circular polarization, respectively. Thus, we obtain

$$R_e = \frac{eE_0}{4\pi^2 m v^2 \mp 2\pi v e B} .$$ (14.7)

Following the same procedure as the one described below Eq. (14.2), we find that the index of refraction is given by the equation

$$n^2 = 1 - \frac{v_p^2}{v(v \mp v_B)} ,$$ (14.8)

where v_B is the gyrofrequency, or cyclotron frequency, given by

$$v_B = \frac{eB}{2\pi m} .$$ (14.9)

The gyrofrequency is the frequency at which an electron would spiral around a magnetic field line in the absence of any electromagnetic radiation. In the absence of damping, R_e would go to infinity if the applied electric field frequency were v_B. The gyrofrequency of the Earth's magnetic field in the ionosphere ($\sim 0.5 \times 10^{-4}$ tesla) is about 1.4 MHz.

Equation (14.8) gives the index of refraction for the case of a longitudinal magnetic field, that is, where the field is parallel to the direction of wave propagation. The solution for the transverse case is different. The solution for the quasi-longitudinal case is obtained by replacing B with $B\cos\theta$, where θ is the angle between the propagation vector and the direction of the magnetic field. The quasi-longitudinal solution is applicable when the angle θ is less than that specified by the inequality (Ratcliffe 1962)

$$\frac{1}{2} \sin\theta \tan\theta < \frac{v^2 - v_p^2}{v v_B} .$$ (14.10)

When $v > 100$ MHz, $v_p \simeq 10$ MHz, and $v_B \simeq 1.4$ MHz, the quasi-longitudinal solution is valid for $|\theta| < 89°$, or virtually all cases of interest. Therefore, to a high accuracy, when $v \gg (v_p$ and $v_B)$, we can expand Eq. (14.8) to obtain

$$n \simeq 1 - \frac{1}{2}\frac{v_p^2}{v^2} \mp \frac{1}{2}\frac{v_p^2 v_B}{v^3}\cos\theta ,$$ (14.11)

where we neglect terms in ν^4 and higher order. For propagation in the direction of **B**, the index of refraction is lower for a left circularly polarized wave than for a right circularly polarized wave.

The difference in the index of refraction for right and left circularly polarized waves leads to the important phenomenon of Faraday rotation, whereby a linearly polarized wave has its plane of polarization rotated as it propagates through the plasma. A linearly polarized wave with position angle ψ can be decomposed into right and left circularly polarized waves of equal amplitude and phase difference 2ψ. The phase of the two circular waves as they propagate in the y direction through a plasma is $2\pi\nu n_r y/c$ and $2\pi\nu n_\ell y/c$, where n_r and n_ℓ are the indices of refraction for the right circular and left circular modes, respectively. The phase difference between the waves is $2\pi\nu(n_r - n_\ell)y/c$. From Eq. (14.11), $n_r - n_\ell = \nu_p^2\nu_B\nu^{-3}\cos\theta$, so it is clear that the plane of polarization is rotated by the angle

$$\Delta\psi = \frac{\pi}{c\nu^2} \int \nu_p^2\nu_B \cos\theta \, dy , \qquad (14.12)$$

where ν_p, ν_B, and θ may be functions of y.

For constant magnetic field and electron density, Eq. (14.12) can be written

$$\Delta\psi = 2.6 \times 10^{-13} n_e B \lambda^2 L \cos\theta , \qquad (14.13)$$

where $\Delta\psi$ is in radians, n_e is in m^{-3}, B is in tesla and is positive when the field is pointed toward the observer, λ is the wavelength in meters, and L is the path length in meters. A magnetic field pointed toward the observer causes the position angle to increase (i.e., a counterclockwise rotation of the plane of polarization of incident radiation as viewed from the surface of the Earth).

14.1.2   Refraction and Propagation Delay

The situation with ionospheric refraction is different from that of tropospheric refraction. The latter occurs in a layer within about 10 km from the ground, and most effects can be understood, at least to first order, with a model of a plane-parallel medium. Since the index of refraction is slightly larger than unity, incoming rays are bent toward the zenith. In contrast, the layers responsible for ionospheric refraction occur several hundred kilometers above the surface, as shown in Fig. 14.2. If the ionosphere were modeled as a plane-parallel layer, then an incoming ray at a certain zenith angle would be bent away from the normal upon entering the layer and then bent back by an equal amount when exiting below. In this case, there would be no net change in the zenith angle. However, the Earth's curvature, in combination with the index of refraction of less than unity, results in a net deflection toward the zenith, the same sense as for tropospheric refraction. A concept that is especially important

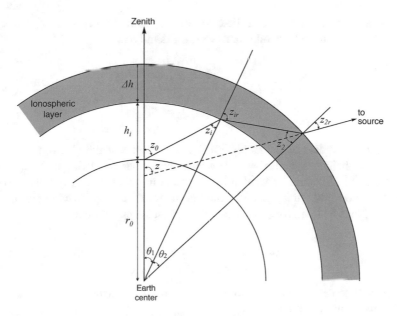

Fig. 14.2 Diagram of a ray passing through a homogeneous but exaggerated ionospheric layer from height h_i to $h_i + \Delta h$. Because of the curvature of the Earth, $z_{ir} = z_2 + \theta_2 \neq z_2$, the net bending angle $\Delta z = z - z_0$ is positive, as it is for the troposphere (see Appendix 14.1 for a derivation of Δz for this single-layer model). Note that if $z_2 > 90$, there will be total internal reflection, and the ray will not reach the Earth's surface. The effect of this internal reflection condition on the effective horizon is also discussed in Appendix 14.1. For the exaggerated case shown, $z_0 = 60°$, $n = 0.8$, and $z = 63°$. For a case with more realistic parameters, see Appendix 14.1.

in the context of all-sky or very-wide-field imaging is that the static ionosphere acts like an achromatic spherical lens that bends incoming rays toward the zenith (Vedantham et al. 2014).

To understand ionospheric refraction, consider a ray passing through a simple ionized layer as shown in Fig. 14.2. Note that the zenith angle of the ray at the bottom of the ionosphere is rather different than the zenith angle at the observer. That is, with the law of sines,

$$z_i = \sin^{-1}\left[\left(\frac{r_0}{r_0 + h_i}\right) \sin z_0\right] . \tag{14.14}$$

We can apply the law of sines to the triangle involving the upper boundary of the layer, as well as Snell's law, to obtain the bending angle of interest, $\Delta z = z - z_0$, where $z = z_{2r} + \theta_1 + \theta_2$ and z_{2r}, θ_1, and θ_2 are defined in Fig. 14.2. Δz is always ≥ 0. The details of the calculation of Δz are given in Appendix 14.1.

The case of a radially stratified ionosphere can be handled by rewriting Eq. (13.27) in the form (see Sukumar 1987, for details) as

$$\Delta z = \frac{A \sin z_0}{r_0} \frac{1}{v^2} \int_0^\infty \frac{\left[1 + \frac{h}{r_0}\right] n_e(h) dh}{\left[\left(1 + \frac{h}{r_0}\right)^2 - \sin^2 z_0\right]^{3/2}} , \tag{14.15}$$

where r_0 is the radius of the Earth, $n_e(h)$ is the profile of the electron density as a function of height, h, and $A = e^2/8\pi^2 m\varepsilon_0$. Note that $v_p^2 = An_e$ [see Eq. (14.5)], so that Eq. (14.15) could be written as a vertical distribution of v_p. Note also that $\Delta z = 0$ in the zenith direction and would go to zero for r_0 approaching infinity, as expected. Since $h \ll r_0$, for $z_0 \ll 1$, the deflection is approximately

$$\Delta z = \frac{A \sin z_0}{r_0} \frac{1}{v^2} \int_0^h n_e(h) \, dh . \tag{14.16}$$

The ionosphere can be modeled to reasonable accuracy with a parabolic electron density distribution of the form

$$n_e(h) = n_{e0} \left[1 - \frac{2(h - h_m)^2}{\Delta h^2}\right] , \tag{14.17}$$

as described by Bailey (1948) for $|h - h_m| \leq 1/\sqrt{2}$, and where h_m is the height of the peak of the electron density, n_{e0}, and Δh is the thickness of the layer. In this case, the bending is approximately given by

$$\Delta z = \frac{\Delta h \sin z}{3r_0} \left(\frac{v_p}{v}\right)^2 \left(1 + \frac{h_m}{r_0}\right) \left(\cos^2 z + \frac{2h_m}{r_0}\right)^{-3/2} . \tag{14.18}$$

The excess path length (see Eqs. 13.4 and 13.5) in the zenith direction can be calculated using Eqs. (14.5) and (14.11) with the assumption that $v \gg (v_p$ and $v_B)$. The result is

$$\mathcal{L}_0 \simeq -\frac{1}{2} \int_0^\infty \left[\frac{v_p(h)}{v}\right]^2 dh \simeq -\frac{40.3}{v^2} \int_0^\infty n_e(h) \, dh , \tag{14.19}$$

where v is in hertz and $n_e(h)$ and $v_p(h)$ are the electron density (m^{-3}) and plasma frequency as a function of height. The integral of electron density over height in Eq. (14.19) is called the *total electron content* (TEC) or *column density*. The excess path corresponds to a phase delay and is negative for the ionosphere. If we approximate the ionosphere by a thin layer at height h_i, then the excess path length will vary as the secant of the zenith angle of the ray as it passes through the layer. Thus,

$$\mathcal{L} \simeq \mathcal{L}_0 \sec z_i , \tag{14.20}$$

Fig. 14.3 (a) Ionospheric bending angle vs. zenith angle at 1000 MHz from a ray-tracing calculation for the daytime electron density profile in Fig. 14.1. The bending predicted by Eq. (14.18), with parameters $v_p = 12$ MHz, $h_i = 350$ km, $\Delta h = 225$ km, and $r_0 = 6370$ km, differs from the curve shown by no more than 5%. (b) Normalized ionospheric excess path length vs. zenith angle for the same electron density profile from a ray-tracing calculation (solid curve) and from Eq. (14.20) (dashed curve). The total electron content is 6.03×10^{17} m^{-2}, and the excess path length at the zenith is 24.3 m. The bending and excess path length scale as v^{-2}. The function forms are rather different from those of the troposphere shown in Fig. 13.6.

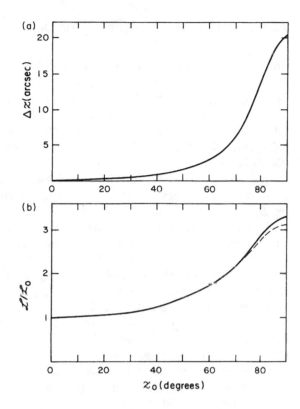

where z_i (see Fig. 14.2) is given by Eq. (14.14). Because of the diurnal variation in n_e, it may be important to use the ionospheric penetration coordinates (defined by θ_1 and θ_2 in Fig. 14.2) to calculate $\mathcal{L}_0$ in Eq. (14.19) for a particular site.

When $z = 90°$ and $h_i = 400$ km, sec z_i is only ~ 3. The secant law provides a reasonable model for estimating the excess ionospheric path length. A more complex model can be found in Spoelstra (1983)). Plots of Δz from ray tracing and $\mathcal{L}$ obtained from Eqs. (14.18) and (14.20) as well as from actual ray-tracing calculations are shown in Fig. 14.3.

In some applications, it is necessary to correct the measurements of fringe frequency for the effects of ionospheric delay. The ionospherically induced frequency shift at an antenna is $(v/c)d\mathcal{L}/dt$. The time rate of change in excess path length $d\mathcal{L}/dt$ has two components: one caused by the time rate of change of zenith angle dz/dt, and the other caused by the time rate of change of $\mathcal{L}_0$, $d\mathcal{L}_0/dt$. At many times, especially near sunrise and sunset, the latter term may dominate (Mathur et al. 1970; Hagfors 1976).

14.1.3   Calibration of Ionospheric Delay

The excess ionospheric path length must be calibrated as accurately as possible in experiments involving precise determination of source positions or baselines. Three approaches are possible. In the first approach, models of the ionosphere can be constructed that depend on parameters such as geomagnetic latitude, solar time, season, and solar activity. Two such models are the International Reference Ionosphere (IRI) (Bilitza 1997) and the Parametrized Ionosphere Model (PIM) (Daniell et al. 1995).

In the second approach, estimates of the total electron content can be obtained from measurements of the dual-frequency transmissions from the Global Positioning System (GPS) (Ho et al. 1997; Mannucci et al. 1998). GPS has replaced the more traditional methods such as ionosondes, Faraday rotation of satellite signals, and incoherent backscatter radar (Evans 1969). The usefulness of GPS for phase correction of array data has been tested at the VLA (Erickson et al. 2001). Four GPS receivers were installed, one at the array center, and one at the end of each arm. The GPS receiver measured the TEC along lines of sight to the GPS satellite. When compared with interferometric phases at 330 MHz, the GPS system was effective in predicting wavefront slopes from large-scale structures ($>$ 1,000 km) in the ionosphere. GPS methods have also been used in the calibration of VLBI observations (e.g., Ros et al. 2000).

In the third approach, the differential path length effects can be virtually eliminated for unresolved sources by making astronomical observations simultaneously at two widely separated frequencies, ν_1 and ν_2. If the interferometer phases are ϕ_1 and ϕ_2 at the two frequencies, then the quantity

$$\phi_c = \phi_2 - \left(\frac{\nu_1}{\nu_2}\right)\phi_1 \qquad (14.21)$$

will preserve source position information and be substantially free of ionospheric delay effects. This technique will correct for the effects of all plasmas along the line of sight, not only the ionosphere. A small residual error remains because of higher-order frequency terms in the index of refraction and because the rays at the two frequencies traverse slightly different paths through the ionosphere. Dual-frequency observations are widely used in astrometric radio interferometry where source structure can be neglected [see, e.g., Sect. 12.6; Fomalont and Sramek (1975); Kaplan et al. (1982); Shapiro (1976)]. Note that the difference in TEC along the ray paths to the interferometer elements can be estimated from measurement of $\phi_2 - (\nu_2/\nu_1)\phi_1$. Similar dual-frequency systems can be employed for the transfer of a local oscillator reference to a space-based VLBI station [see, for example, Moran (1989) and Sect. 9.10].

14.1.4 Absorption

Absorption in the ionosphere is caused by collisions of electrons with ions and neutral particles. At frequencies much greater than ν_p, the power absorption coefficient is

$$\alpha = 2.68 \times 10^{-7} \frac{n_e \nu_c}{\nu^2} \quad (\text{m}^{-1}) , \qquad (14.22)$$

where ν_c is the collision frequency and n_e is in m^{-3}. The collision frequency in hertz is approximately

$$\nu_c \simeq 6.1 \times 10^{-9} \left(\frac{T}{300} \right)^{-3/2} n_i + 1.8 \times 10^{-14} \left(\frac{T}{300} \right)^{1/2} n_n , \qquad (14.23)$$

where n_i is the ion density and n_n is the neutral particle density, both in m^{-3} (Evans and Hagfors 1968). Numerical values of absorption are listed in Table 14.1. Radiometric measurements of both electron temperature and opacity have been made by Rogers et al. (2015).

14.1.5 Small- and Large-Scale Irregularities

The small-scale irregularities in the electron density distribution introduce random changes in the wavefront of a passing electromagnetic wave. As a consequence, fluctuations in fringe amplitude and phase can be readily observed with an interferometer at frequencies below a few hundred megahertz. In the early days of radio astronomy, signals from Cygnus A and other compact sources were observed to fluctuate on timescales of 0.1–1 min. At first, these fluctuations were thought to be intrinsic to the sources (Hey et al. 1946), but later observations with spaced receivers showed that the fluctuations were uncorrelated for receiver separations of more than a few kilometers (Smith et al. 1950). This result led to the conclusion that irregularities in the ionosphere were perturbing the cosmic signals. The predominant scale sizes in the ionization irregularities were found to be a few kilometers or less. The timescale of the fluctuations indicates that ionospheric wind speeds are in the range of 50–300 m s^{-1}. The effects of these fluctuations have been studied extensively at frequencies between about 20 and 200 MHz and have been observed at frequencies as high as 7 GHz (Aarons et al. 1983). An early example of the fluctuations seen in interferometer measurements is given in Fig. 14.4. Hewish (1952), Booker (1958), and Lawrence et al. (1964) reviewed the early results and techniques. A comprehensive review of theory and observations of ionospheric fluctuations can be found in Crane (1977), Fejer and Kelley (1980), and Yeh and Liu (1982), and summaries of global morphology can be found in Aarons (1982) and

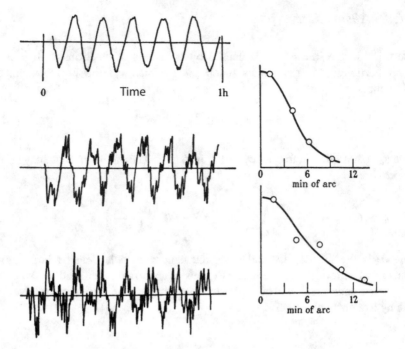

Fig. 14.4 (**left**) Typical records of the correlator output on three occasions from a phase-switching interferometer at Cambridge, England, having a 1-km baseline and operating at a wavelength of 8 m. The irregular responses are caused by disturbances in the ionosphere. (**right**) Probability distributions of the angle of arrival deduced from the zero crossings of the correlator response. Reprinted with permission of and © the Royal Society, conveyed through Copyright Clearance Center Inc. From Hewish (1952).

Aarons et al. (1999). Measurements with the GPS can be very useful in monitoring ionospheric fluctuations [e.g., Ho et al. (1996), Pi et al. (1997)]. The effects of ionospheric scintillation on a synthesis telescope have been described by Spoelstra and Kelder (1984). Loi et al. (2015a) report extensive measurements of ionospheric-induced position wander in sources at 150 MHz, and Loi et al. (2015b) demonstrated a parallax technique for determining the height of the perturbing layer. In Sect. 14.2, we discuss a theory of scintillation, which can be applied to the ionosphere as well as to the interplanetary and interstellar media.

Large-scale variations in the electron density integrated along the line of sight are caused by traveling ionospheric disturbances (TIDs). TIDs, which are manifestations of acoustic-gravity waves in the upper atmosphere, are quasi-periodic, large-scale perturbations in electron density. The atmosphere has a natural buoyancy, so that a parcel of gas displaced vertically and released will oscillate at a frequency known as the Brunt–Väisälä, or buoyancy, frequency. This frequency is about 0.5–2 mHz (periods of 10–20 min) at ionospheric heights. For waves with frequencies above the buoyancy frequency, the restoring force is pressure (acoustic wave), and for waves with frequencies below the buoyancy frequency, the restoring force is gravity (gravity wave). Hunsucker (1982) and Hocke and Schlegel (1996) have reviewed the literature on acoustic-gravity waves. There are many potential sources of TIDs, including auroral heating, severe weather fronts, earthquakes, and volcanic eruptions. Medium-scale TIDs have scale lengths of 100–200 km and timescales of 10–20 min and cause a variation in TEC of 0.5–5%. Such TIDs are present for a substantial fraction of the time. Large-scale TIDs, which are relatively uncommon, have scale lengths of 1000 km and timescales of hours and can cause variations in TEC of up to 8%. One such disturbance, excited by a volcano, was observed by VLBI (Roberts et al. 1982). A variety of ionospheric disturbances have been studied by observations of compact sources with the VLA [e.g., Helmboldt et al. (2012), Helmboldt (2014)].

14.2 Scattering Caused by Plasma Irregularities

Understanding the propagation of radiation in a random medium is an important problem in many fields. The signals from cosmic radio sources propagate through several ionized random media, including the ionized interstellar gas of our Galaxy, the solar wind, and the ionosphere. In the observer's plane, there are two effects. First, the amplitude varies with the position of the observer, which leads to temporal amplitude variations if there are relative motions among the source, scattering medium, and observer. Second, the image of the source is also distorted in a frequency-dependent manner. Much of the research in this area has been motivated by the attempt to understand the observational characteristics of pulsars [see, e.g., Gupta (2000)]. Propagation effects in the turbulent troposphere are described in Chap. 13.

14.2.1 Gaussian Screen Model

We begin the discussion by considering a simple model that serves to illustrate many features of the problem. This model was first developed by Booker et al. (1950)) to explain ionospheric scintillation and was refined by Ratcliffe (1956). Scheuer (1968) applied it to pulsar observations. The model assumes that the irregular medium is confined to a thin screen and that the irregularities (blobs) have one characteristic scale size a. Diffraction effects are neglected within the irregular medium; only the phase change imposed by the medium is considered. Diffraction is taken into account in the free-space region between the irregular medium and the receivers.

The geometric situation is shown in Fig. 14.5. The thin-screen assumption is not particularly restrictive. However, the assumption that the screen is filled with plasma blobs having one characteristic size is restrictive and distinguishes this model from the power-law model, in which a range of scale sizes is present. From Eqs. (14.5)

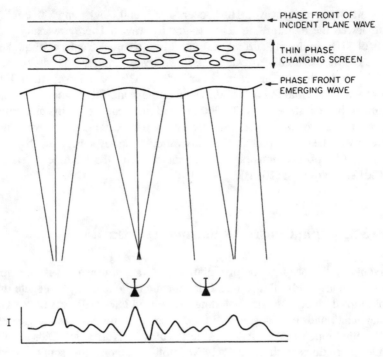

Fig. 14.5 Geometry of a thin-screen scintillation model. An initially plane wave is incident on a thin phase-changing screen. The emerging wavefront is irregular. As the wave propagates to the observer, amplitude fluctuations develop, as suggested by the crossing rays. Below the antenna is a plot of intensity vs. position along the wavefront. If there is motion between the screen and the observer, the spatial fluctuations will be observed as temporal fluctuations in the power received or the fringe visibility.

and (14.11), the index of refraction of the plasma can be written

$$n \simeq 1 - \frac{r_e n_e \lambda^2}{2\pi} , \tag{14.24}$$

where r_e is the classical electron radius, equal to $e^2/4\pi\epsilon_0 mc^2$ or 2.82×10^{-15} m, and the term in ν_B is neglected. Thus, the excess phase shift (a phase advance in this situation) across one blob is

$$\Delta\phi_1 = r_e \lambda a \, \Delta n_e , \tag{14.25}$$

where Δn_e is the excess electron density in the blob over the ambient level. If the thickness of the screen is L, then the wave will encounter about L/a blobs, and the rms phase deviation $\Delta\phi$ will be $\Delta\phi_1 \sqrt{L/a}$, or

$$\Delta\phi = r_e \lambda \, \Delta n_e \sqrt{La} . \tag{14.26}$$

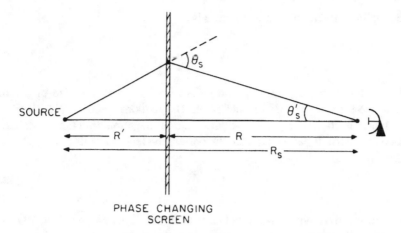

Fig. 14.6 Path of a refracted ray in the thin-screen model. The rms scattering angle, θ_s, is given by Eq. (14.27).

The wave emerging from the screen is crinkled; that is, the amplitude is unchanged, but the phase is no longer constant and has random fluctuations with rms deviation $\Delta\phi$. The wave can therefore be decomposed into an angular spectrum of waves propagating with a variety of angles. The full width of the angular spectrum, θ_s, can be estimated by imagining that the random medium consists of refracting wedges that tilt the wavefront by the amount $\pm\Delta\phi\lambda/2\pi$ over a distance a. Thus,

$$\theta_s = \frac{1}{\pi}r_e\lambda^2\Delta n_e\sqrt{\frac{L}{a}}\ . \tag{14.27}$$

If the source is not infinitely distant, then the incident wave will not be plane. In that case, the observed scattering angle θ_s' depends on the location of the screen with respect to the source and the observer. Since θ_s and θ_s' are small angles, it follows from the geometry in Fig. 14.6 that

$$\theta_s' = \frac{R'}{R+R'}\theta_s\ , \tag{14.28}$$

where R and R' are defined in Fig. 14.6. Therefore, the effectiveness of the scattering screen is diminished if the screen is moved toward the source. This lever effect is very important in astrophysical situations. It can be used to distinguish galactic and extragalactic sources whose radiation passes through the same scattering screen (Lazio and Cordes 1998).

Amplitude fluctuations build up as the wave propagates away from the screen. If the phase fluctuations are large, $\Delta\phi > 1$, then significant amplitude fluctuations occur when rays cross (see Fig. 14.5). The critical distance beyond which large-amplitude fluctuations are observed is

$$R_f \simeq \frac{a}{\theta'_s} \, . \tag{14.29}$$

Note that if $\Delta\phi = 2\pi$, then R_f is the distance for which the size of a blob is equal to the size of the first Fresnel zone. The random electric field distribution at the Earth, in the plane perpendicular to the propagation direction, is called the diffraction pattern. It has a characteristic correlation length d_c, given by

$$d_c \simeq \frac{\lambda}{\theta'_s} \, . \tag{14.30}$$

If the screen moves with relative velocity v_s in the direction perpendicular to the propagation direction, so that the diffraction pattern sweeps across the observer, then the timescale of variability is

$$\tau_d \simeq \frac{d_c}{v_s} \frac{R'}{R + R'} \simeq \frac{\lambda}{\theta_s v_s} \, . \tag{14.31}$$

The signal reaching the observer by traveling along the scattered ray path is delayed by an amount

$$\tau_c \simeq \frac{RR'\theta_s^2}{2c(R + R')} \tag{14.32}$$

with respect to the unscattered signal. The phase of the scattered wave is $2\pi \nu \tau_c$ with respect to the direct (unscattered) wave, and interference between these two waves causes scintillation. The bandwidth over which the relative phase changes by 2π is called the correlation bandwidth, $\Delta\nu_c$. The correlation bandwidth is the reciprocal of τ_c, and for the case $R = R'$ is

$$\Delta\nu_c \simeq \frac{8c}{R_s\theta_s^2} \, , \tag{14.33}$$

where R_s is the distance between the source and the observer. If the observations are made with a receiver of bandwidth greater than $\Delta\nu_c$, the amplitude fluctuations will be greatly reduced. Note from Eqs. (14.33) and (14.27) that $\Delta\nu_c$ varies as λ^{-4}.

Finally, if the source has two equal components separated by distance ℓ, then each component will produce the same diffraction pattern, but these patterns will be displaced at the Earth by distance $\ell R/R'$. If this distance is greater than d_c, then the diffraction pattern will be smoothed and the amplitude fluctuations reduced. Thus, if the source size is greater than a critical size θ_c, amplitude fluctuations will be sharply reduced because the diffraction patterns from the component parts overlap and are smoothed out. From Eqs. (14.28) and (14.30), θ_c can be written as

$$\theta_c = \frac{\lambda}{R\theta_s} . \tag{14.34}$$

Hence, only sources of small angular diameter scintillate. In the optical regime, the analogous phenomenon is that stars twinkle, but usually planets do not. An elegant application of Eq. (14.34) was made by Frail et al. (1997) to determine the angular size of the expanding radio source associated with a gamma-ray burst. They determined that the amplitude fluctuations in the radio emission, assumed to be caused by interstellar scattering, ceased during the first weeks after the burst, indicating that the source diameter had increased beyond the critical size of 3 μas at that time.

A useful quantity is the ensemble average fringe visibility, $\mathcal{V}_m$, measured by an interferometer in the presence of scintillation. Assume that the phases ϕ_1 and ϕ_2 at two points along the phase screen, separated by distance d, are random variables with a joint Gaussian distribution with variance $\Delta\phi^2$ and normalized correlation $\rho(d)$. $\rho(d)$ is the correlation function of the phase or of the variable component of the index of refraction. The joint probability density function of the phase along the wavefront is

$$p(\phi_1, \phi_2) = \frac{1}{2\pi\Delta\phi^2\sqrt{1 - \rho(d)^2}} \exp\left[-\frac{\phi_1^2 + \phi_2^2 - 2\rho(d)\phi_1\phi_2}{2\Delta\phi^2[1 - \rho(d)^2]}\right] , \tag{14.35}$$

where $\rho(d) = \langle\phi_1\phi_2\rangle/\Delta\phi^2$, the correlation function of the phase fluctuations. The expectation of $e^{j(\phi_1 - \phi_2)}$ is

$$\langle e^{j(\phi_1 - \phi_2)}\rangle = \iint e^{j(\phi_1 - \phi_2)} p(\phi_1, \phi_2)\, d\phi_1\, d\phi_2 , \tag{14.36}$$

which can be evaluated directly from Eq. (14.35) with the result

$$\langle e^{j(\phi_1 - \phi_2)}\rangle = e^{-\Delta\phi^2[1 - \rho(d)]} . \tag{14.37}$$

For a point source of flux density S, the ensemble average of the fringe visibility is

$$\langle\mathcal{V}_m\rangle = S\langle e^{j\phi_1}e^{-j\phi_2}\rangle , \tag{14.38}$$

or

$$\langle\mathcal{V}_m\rangle = Se^{-\Delta\phi^2[1 - \rho(d)]} . \tag{14.39}$$

If the source has an intrinsic visibility $\mathcal{V}_0$, the ensemble average is

$$\langle\mathcal{V}_m\rangle = \mathcal{V}_0 e^{-\Delta\phi^2[1 - \rho(d)]} . \tag{14.40}$$

This result was first derived by Ratcliffe (1956) and Mercier (1962). Note that the structure function of phase is $D_\phi(d) = 2\Delta\phi^2[1 - \rho(d)]$, so that Eq. (14.40) is equivalent to Eq. (13.80). In much of the early radio astronomical literature, $\rho(d)$ is assumed to be a Gaussian function

$$\rho(d) = e^{-d^2/2a^2} ,\qquad(14.41)$$

where the characteristic scale length a corresponds to the blob size in the discussion above. This model, called the Gaussian screen model, is probably unrealistically restrictive because there are undoubtedly many scale sizes present. In the case in which $\Delta\phi \gg 1$, $\mathcal{V}_m$ decreases rapidly as d increases, and we need consider only the situation of $d \ll a$. Then, substitution of Eq. (14.41) into Eq. (14.40) yields

$$\langle \mathcal{V}_m \rangle \simeq \mathcal{V}_0 e^{-\Delta\phi^2 d^2/2a^2} .\qquad(14.42)$$

Thus, the intensity distribution of a point source observed through a Gaussian screen is a Gaussian distribution with a diameter (full width at half-maximum) of

$$\theta_s \simeq \sqrt{2\ln 2}\frac{\Delta\phi\lambda}{\pi a} = \frac{\sqrt{2\ln 2}}{\pi}r_e\lambda^2\Delta n_e\sqrt{\frac{L}{a}} .\qquad(14.43)$$

This formula for θ_s is essentially equivalent to the one given in Eq. (14.27). In the case in which $\Delta\phi \ll 1$, the normalized visibility function drops from unity to $e^{-\Delta\phi^2}$ when $d \gg a$. Therefore, the resulting intensity distribution for a point source is an unresolved core surrounded by a halo. The ratio of the flux density in the halo to the flux density in the core is $e^{\Delta\phi^2} - 1$.

14.2.2   Power-Law Model

The spectrum of fluctuations in the electron density in ionized astrophysical plasmas is normally modeled as a power law,

$$P_{ne} = C_{ne}^2 q^{-\alpha} ,\qquad(14.44)$$

where q is the three-dimensional spatial frequency (cycles per meter), $q^2 = q_x^2 + q_y^2 + q_z^2$, and C_{ne}^2 characterizes the strength of the turbulence. The definition of C_{ne}^2 varies in the literature, depending on whether it is used as a constant in the spectrum or in the structure function. The two-dimensional power spectrum of phase [see Eq. (14.22) for the relation between $\Delta\phi$ and Δn_e] is

$$P_\phi(q) = 2\pi r_e^2\lambda^2 LP_{ne} .\qquad(14.45)$$

Hence, from Eq. (13.104), the structure function of phase is

$$D_\phi(d) = 8\pi^2 r_e^2 \lambda^2 L \int_0^\infty [1 - J_0(qd)]\Gamma_{ne}(q)q\,dq \ . \tag{14.46}$$

For a power-law spectrum of the form of Eq. (14.44), the structure function is

$$D_\phi(d) = 8\pi^2 r_e^2 \lambda^2 C_{ne}^2 L f(\alpha)d^{\alpha-2} \ , \tag{14.47}$$

where $f(\alpha)$ is of order unity. The index α is often taken to be 11/3, which is its value for Kolmogorov turbulence, for which $f(\alpha) = 1.45$ [see Cordes et al. (1986)) for other values of $f(\alpha)$]. The ensemble average of the interferometric visibility [see Eq. (13.80)] is

$$\langle \mathcal{V} \rangle = \mathcal{V}_0 e^{-D_\phi/2} \ , \tag{14.48}$$

or

$$\langle \mathcal{V} \rangle = \mathcal{V}_0 e^{-4\pi^2 r_e^2 \lambda^2 C_{ne}^2 L f(\alpha)d^{\alpha-2}} \ . \tag{14.49}$$

The observed intensity distribution, the Fourier transform of Eq. (14.49), differs slightly from a Gaussian distribution, as can be seen in Fig. 13.11b. The scattering angle (full width at half-maximum) obtained from the width of the intensity distribution is

$$\theta_s \simeq 4.1 \times 10^{-13} (C_{ne}^2 L)^{3/5} \lambda^{11/5} \ \text{(arcsec)} \ , \tag{14.50}$$

where λ is in units of meters and $C_{ne}^2 L$ is in $\text{m}^{-17/3}$. Thus, a difference between the power-law model and the Gaussian screen model is that θ_s, measured by Fourier transformation of visibility data over a range of baselines, is proportional to $\lambda^{2.2}$ in the former model and to λ^2 in the latter. Note that if $\langle \mathcal{V} \rangle$ were measured on a single baseline, that is, with d fixed, and if θ_s were estimated from comparison of the measured visibility with the visibility expected for a Gaussian intensity distribution, θ_s would appear to vary as λ^2 in both models.

Measurements of visibility must be made over sufficiently long integration times to achieve an ensemble average if Eqs. (14.48), (14.49), and (14.50) are to be valid (Cohen and Cronyn 1974). A detailed discussion of the averaging time necessary to achieve an ensemble average is given by Narayan (1992) (see also Scct. 14.4.3).

For plasmas, we can expect that the power law will hold from an inner scale q_0 to an outer scale q_1; that is, there are no fluctuations on length scales smaller than $\ell_{inner} = 1/q_1$ or larger than $\ell_{outer} = 1/q_0$. For the case in which $qd \ll 1$, that is, when the baseline is shorter than the inner length scale, the Bessel function in

Eq. (14.46) becomes $1 - q^2 r^2 / 4$, and the integration is straightforward, yielding

$$D_\phi(d) = \frac{2\pi^2 r_e^2 \lambda^2 L C_{ne}^2}{4 - \alpha} (q_1^{4-\alpha} - q_0^{4-\alpha}) \, d^2 \, . \qquad (14.51)$$

This is a very important result that has two interesting consequences. First, the structure function varies as d^2 regardless of α. Second, for $\alpha < 4$, the structure function is dominated by the effect of the smallest irregularities, whereas for $\alpha > 4$, it is dominated by the effect of the largest-scale irregularities. This result also suggests an important demarcation in phenomena between plasmas with $\alpha < 4$ and those with $\alpha > 4$. The case in which $\alpha < 4$ is called Type A (shallow spectrum), and the case in which $\alpha > 4$ is called Type B (steep spectrum) (Narayan 1988).

Consider the situation in which the spectrum has three regimes:

$$
\begin{aligned}
P_{ne} &= C_{ne}^2 q_0^{-\alpha} \, , & q < q_0 \\
&= C_{ne}^2 q^{-\alpha} \, , & q_0 < q < q_1 \\
&= 0 \, , & q > q_1 \, .
\end{aligned}
\qquad (14.52)
$$

Substitution of Eq. (14.52) into Eq. (14.46) gives

$$
\begin{aligned}
D_\phi(d) &\simeq c_1 d^2 \, , & d < 1/q_1 = \ell_{\text{inner}} \\
&\simeq \left(\frac{d}{d_0} \right)^{\alpha - 2} \, , & 1/q_1 < d < 1/q_0 \\
&\simeq c_2 \, , & d > 1/q_0 = \ell_{\text{outer}} \, ,
\end{aligned}
\qquad (14.53)
$$

where c_1 and c_2 are constants, and we have introduced the normalization factor d_0, such that $D_\phi(d_0) = 1$, as in the discussion of the troposphere in Sect. 13.1.7. We have also assumed that $1/q_1 < d_0 < 1/q_0$. The constants needed to join the power-law segments are $c_1 = q_1^{4-\alpha} d_0^{-2}$ and $c_2 = (q_0 d_0)^{1-\alpha}$. This spectrum and structure function for the model are shown in Fig. 14.7.

14.3 Interplanetary Medium

14.3.1 Refraction

Radio waves passing near the Sun are bent by the ionization of the solar corona and the solar wind. The general characteristics of the solar corona and the solar wind can be found in Winterhalter et al. (1996). Calculation of the refraction in the extended solar atmosphere is important for the understanding of solar radio emission at low frequencies, where the bending angles are large (Kundu 1965), and for tests of the

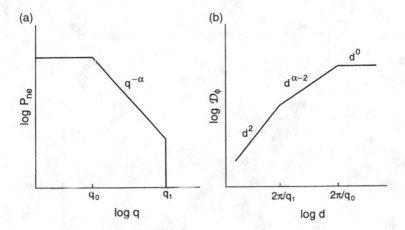

Fig. 14.7 (a) A model spectrum of the electron density fluctuations with inner and outer scales of spatial frequency q_0 and q_1. (b) The corresponding structure function of phase: see Eqs. (14.52) and (14.53). Note that $\ell_{inner} = 2\pi q_1$ and $\ell_{outer} = 2\pi/q_0$. From Moran (1989), © Kluwer Academic Publishers. With kind permission from Springer Science and Business Media.

general relativistic bending of electromagnetic radiation passing near the Sun (see Sect. 12.6).

The electron density as a function of distance from the Sun can be measured in a variety of ways. Optical observations of Thomson scattering during solar eclipses have been analyzed to give an electron density model

$$n_e = \left(1.55r^{-6} + 2.99r^{-16}\right) \times 10^{14} \ (\text{m}^{-3}),\tag{14.54}$$

where r, the radial distance from the Sun in units of the solar radius, is less than ~ 4. Equation (14.54) is the well-known Allen–Baumbach formula (Allen 1947).

The electron density profile over a broad range of radii can be determined from satellites that can track the plasma frequency measured during solar radio bursts. For example, observations with the Wind spacecraft with observations from 14 MHz to a few kHz could be reasonably represented by the model

$$n_e = 3.3 \times 10^{11} r^{-2} + 4.1 \times 10^{11} r^{-4} + 8.0 \times 10^{13} r^{-6} \ (\text{m}^{-3})\tag{14.55}$$

for $1.2 < r < 215$ (Leblanc et al. 1998). The value of n_e at $r = 217$ (1 AU) is 7.2×10^6 m^{-3}. This model is based on data taken near sunspot minimum, and the range of conditions is shown in Fig. 14.8. Ground-based measurements of radio sources during solar occultations (e.g., scintillations of the Crab Nebula) (Erickson 1964; Evans and Hagfors 1968) and dispersion measurements of pulsars (Counselman and Rankin 1972; Counselman et al. 1974) give about the same result as Eq. (14.55) for $r > 10$.

The angle of refraction of a ray passing near the Sun can be calculated readily for the case in which this angle is small. A ray obeys Snell's law in spherical

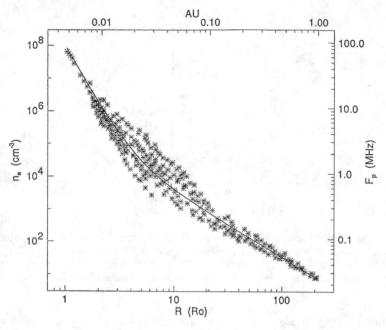

Fig. 14.8 The electron density vs. radial distance from the Sun, measured by the Wind satellite (orbiting at 1 AU) from observations of solar radio bursts. Scatter in data derived from observations at 11 epochs indicates the range of condition in the solar wind. For other data and in situ measurements, see Bougeret et al. (1984). From Leblanc et al. (1998), © *Solar Phys.* With kind permission from Springer Science and Business Media.

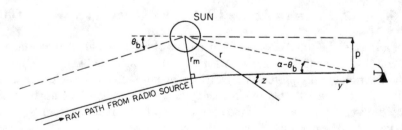

Fig. 14.9 Path of a ray passing through the ionized gas surrounding the Sun. p is the impact parameter, and α is the solar elongation angle, that is, the angle between the Sun and the source in the absence of solar bending.

coordinates, $nr\sin z = $ constant (Smart 1977), where n is the index of refraction and z is the angle between the ray and a line from the center of the Sun, as shown in Fig. 14.9. From this relation, the bending angle is found to be

$$\theta_b = \pi - 2 \int_{r_m}^{\infty} \frac{dr}{r\sqrt{(nr/p)^2 - 1}} \, , \qquad (14.56)$$

where r_m is the distance of closest approach of the ray to the Sun, and p is the impact parameter (see Fig. 14.9). Assume that the electron density has a single power-law distribution given by

$$n_e = n_{e0} r^{-\beta} , \qquad (14.57)$$

where n_{e0} is the electron density in m^{-3} at one solar radius, and β is a constant. For a fully ionized solar wind, characterized by a constant mass loss rate and velocity, β is equal to 2. This case is applicable for $r \gtrsim 10$ (see Fig. 14.8). The index of refraction is obtained by substituting Eqs. (14.57) and (14.5) into Eq. (14.11) and neglecting the term in ν_B. Graphical solutions of Eq. (14.56) for large bending angles are given by Jaeger and Westfold (1950). For small bending angles, an approximate solution to Eq. (14.56) can be obtained by the use of the substitution $nr/p = \sec\theta$,

$$\theta_b \simeq 80.6\sqrt{\pi}\, \frac{n_{e0}}{\nu^2} \frac{\Gamma\left(\frac{\beta+1}{2}\right)}{\Gamma\left(\frac{\beta}{2}\right)} p^{-\beta} , \qquad (14.58)$$

where p is in units of the solar radius, and Γ is the gamma function. Note that the rays are bent away from the Sun. The bending angle associated with the model in Eq. (14.55) (using only the quadratic term) is

$$\theta_b \simeq 2.4\lambda^2 p^{-2}\ \text{(arcmin)} , \qquad (14.59)$$

where λ is the wavelength in meters. For a multiple power-law model of electron density such as given in Eqs. (14.54) and (14.55), the bending angles for each component can be summed when the bending angles are small.

There is another pedagogically interesting way to determine the bending angle from the change in excess propagation path with impact parameters. The excess (phase) path for a ray passing through the corona, for the case in which the effect of ray bending can be neglected, is, from Eq. (14.19),

$$\mathcal{L} \simeq -\frac{40.3}{\nu^2} \int_{-\infty}^{\infty} n_e\, dy , \qquad (14.60)$$

where y is measured along the ray path as shown in Fig. 14.9. For a power-law model given by Eq. (14.57), the excess path is

$$\mathcal{L} \simeq -\frac{40.3 n_{e0}}{\nu^2} \int_{-\infty}^{\infty} \frac{dy}{(p^2 + y^2)^{\beta/2}} , \qquad (14.61)$$

which can be integrated to give

$$\mathcal{L} \simeq -\frac{40.3\sqrt{\pi}}{\nu^2} \frac{\Gamma\left(\frac{\beta-1}{2}\right)}{\Gamma\left(\frac{\beta}{2}\right)} n_{e0} p^{1-\beta} . \qquad (14.62)$$

The change in $\mathcal{L}$ with p describes the tilting of the wavefront and is the bending angle; hence $\theta_b \simeq d\mathcal{L}/dp$ (Bracewell et al. 1969). Differentiation of Eq. (14.62) with respect to p gives Eq. (14.58).

We mention the effect of the general relativistic bending of waves passing close to the Sun here because the phenomenon can be described classically by an effective index of refraction given by $1 + 2GM_\odot/rc^2$, where G is the gravitational constant, and $M_\odot$ is the mass of the Sun. The bending angle, for small values of p, is (Weinberg 1972)

$$\theta_{GR} \simeq -1.75p^{-1} \ (\text{arcsec}) . \tag{14.63}$$

The negative sign indicates that the bending is toward the Sun, which is the opposite sense of bending by interplanetary medium. Measurements of the solar general relativistic bending are discussed in more detail in Sect. 12.6.

14.3.2   Interplanetary Scintillation (IPS)

Scintillation of extragalactic radio sources due to irregularities in the solar wind was first observed by Clarke (1964) and reported by Hewish et al. (1964). Clarke was studying 88 of the 3C sources at 178 MHz with the Cambridge one-mile interferometer. She noticed that three of the sources, the only ones smaller than $2''$, showed anomalous rapid (< 1 s) scintillation, which could not be attributed to the ionosphere. They were all within $30°$ in angle from the Sun. Interplanetary scintillation is readily distinguishable from ionospheric scintillation, since the timescale [Eq. (14.31)] and critical source size [Eq. (14.34)] are approximately 1 s and $0.5''$ for interplanetary scintillation and 30 s and 10 arcmin for ionospheric scintillation. Further observations of interplanetary scintillation by Cohen et al. (1967a) showed that the angular size of the radio source 3C273B is smaller than $0.02''$, based on the application of Eq. (14.34). This result, and the long-baseline interferometric results, stimulated the development of VLBI. Interplanetary scintillation can be studied with the modern generation of low-frequency arrays [e.g., Kaplan et al. (2015)].

A comprehensive discussion of the interpretation of interplanetary scintillation can be found in Salpeter (1967), Young (1971), and Scott et al. (1983). For rough calculations, the scattering angle due to the interplanetary medium may be approximated by (Erickson 1964)

$$\theta_s \simeq 50 \left(\frac{\lambda}{p}\right)^2 \ (\text{arcmin}) , \tag{14.64}$$

where λ is in meters, and p, the impact parameter, is in solar radii. This relationship is based on measurements taken in 1960–61 at 11-m wavelength for impact parameters between 5 and 50 solar radii. Analysis of VLBI observations at 3.6

and 6 cm obtained in 1991 for a range of impact parameters of 10–50 solar radii led to a model for C_{ne}^2 of the form $C_{ne}^2 = 1.5 \times 10^{14}(r/R_{Sun})^{-3.7}$ (Spangler and Sakurai 1995). Note that the power-law exponent is expected to be about -4 from the elementary consideration that C_{ne}^2 is proportional to the variance of the electron density, which is proportional to the square of the density. For a constant wind speed, the density is proportional to r^{-2}, and hence C_{ne}^2 is proportional to r^{-4}. Deviations from 4 are caused by the radial dependence of the magnetic field strength, which plays a role in driving the turbulence. Integrating C_{ne}^2 along the line of sight, and using Eq. (14.50), we derive an estimate for the scattering angle of $\theta_s = 3100(p/\lambda)^{-2.2}$ arcsec, which is comparable to the result in Eq. (14.64).

The concept that extended sources do not scintillate as much as point sources [see Eq. (14.34)] can be generalized to obtain more information about source structure. We assume that the scintillation is caused by a screen at a distance R from the Earth, as shown in Fig. 14.6, where $R \ll R_s$, and that the intensity at the Earth is $I(x, y)$, where x and y are coordinates in a plane parallel to the screen in Fig. 14.5. The function $\Delta I(x, y)$ is equal to $I(x, y) - \langle I(x, y)\rangle$, where $\langle I(x, y)\rangle$ is the mean intensity. It has a power spectrum $S_{I0}(q_x, q_y)$ for a point source and $S_I(q_x, q_y)$ for an extended source, where q_x and q_y are the spatial frequencies (cycles per meter). If the visibility of the source is $\mathcal{V}(q_x R, q_y R)$, then it can be shown (Cohen 1969) that

$$S_I(q_x, q_y) = S_{I0}(q_x, q_y)|\mathcal{V}(q_x R, q_y R)|^2 , \qquad (14.65)$$

where $q_x R$ and $q_y R$ correspond to the projected baseline coordinates u and v. The scintillation index of the source m_s is defined by

$$m_s^2 = \frac{\langle \Delta I(x, y)^2\rangle}{\langle I(x, y)\rangle^2} = \frac{1}{\langle I(x, y)\rangle^2} \int_{-\infty}^{\infty} \int_{-\infty}^{\infty} S_I(q_x, q_y) \, dq_x \, dq_y . \qquad (14.66)$$

In principle, $S_I(q_x, q_y)$ could be computed from the simultaneous measurements of $\Delta I(x, y)$ with a large number of spaced receivers. In practice, the motion of the solar wind sweeps the diffraction pattern across a single telescope so that, from measurements of $\Delta I(t)$, the temporal power spectrum $S(f)$ can be calculated. If the diffraction pattern moves with velocity v_s in the x direction, then $S(f)$ can be related to the spatial spectrum since $q_x = f/v_s$:

$$S(f) = \frac{1}{v_s} \int_{-\infty}^{\infty} S_I\left(q_x = \frac{f}{v_s}, q_y\right) dq_y . \qquad (14.67)$$

In principle, $|\mathcal{V}|^2$ can be recovered from Eq. (14.65) by observing a source over a range of different orientations with respect to the solar wind vector. The situation is entirely analogous to that of lunar occultation observations (Sect. 17.2) except that with lunar occultation observations, the visibility phase can also be obtained. An estimate of the source diameter can be deduced from the width of the temporal power spectrum (Cohen et al. 1967b) or from the scintillation index [Eq. (14.66)] (Little and Hewish 1966).

Table 14.2 Typical values of the effects of the interstellar medium on radiation at 100 MHz

Effect	Equation number	Magnitude[a]	Frequency dependence[b]
Angular broadening[c]	14.43	$0.3''$	ν^{-2}
Pulse broadening[c]	14.32	10^{-4} s	ν^{-4}
Scintillation bandwidth[c]	14.33	10^4 Hz	ν^4
Spectral broadening[c]	—	1 Hz	ν^{-1}
Scintillation timescale[c]	14.31	10 s	ν
Scintillation timescale[d]	—	10^6 s	ν^{-2}
Free-free optical depth	14.22	0.01	ν^{-2}
Faraday rotation	14.71	10 rad	ν^{-2}

Adapted from Cordes (2000).
[a]For a source in the Galactic plane at a distance of 1 kpc. Actual values can differ by an order of magnitude.
[b]Valid for the Gaussian screen model or the power-law turbulence model when $D_\phi(d) \sim d^2$ [see Eq. (14.46)].
[c]Diffractive scattering.
[d]Refractive scattering (see Sect. 14.4.3).

Interplanetary scattering is generally weak, except in directions close to the Sun. An interesting phenomenon is that the scintillation index, m_s, increases monotonically with decreasing impact parameter, reaching $m_s \sim 1$ for small diameter sources around $p \sim 0.1$ and then decreasing for smaller values of p [e.g., Armstrong and Coles (1978), Gapper et al. (1982), Manoharan et al. (1995)]. The effects of refractive scattering (discussed in the next section and in Sect. 15.3), which can be important in the strong scattering regime, have been studied by Narayan et al. (1989).

A substantial effort has been made to study the 3-D characteristics of the interplanetary medium by monitoring the scintillation of radio sources over the past decades. See Manoharan (2012) for results from the Ooty Radio Telescope and Asai et al. (1998) and Tokumaru et al. (2012) for results from the Solar-Terrestrial Environment Laboratory of Nagoya University. Long-term trends are discussed by Janardhan et al. (2015).

14.4　Interstellar Medium

Table 14.2 lists the typical magnitudes and scale sizes of various effects caused by the interstellar medium. These are discussed individually in the following sections.[1]

---

[1]In this section, we follow the commonly used symbols *DM*, *RM*, and *SM* for dispersion measure, rotation measure, and scintillation measure.

14.4.1   Dispersion and Faraday Rotation

The smooth, ionized component of the Interstellar medium of our Galaxy affects propagation by introducing delay and Faraday rotation. The time of arrival of a pulse of radiation, such as that from a pulsar, is

$$t_p = \int_0^L \frac{dy}{v_g} \; , \qquad (14.68)$$

where L is the propagation path, $v_g = cn$ is the group velocity, and n is given by Eq. (14.11), where we neglect the effect of the magnetic field. Differentiation of Eq. (14.68) gives

$$\frac{dt_p}{dv} \simeq -\frac{e^2}{4\pi\epsilon_0 mcv^3} \int_0^L n_e \, dy \; . \qquad (14.69)$$

The integral of n_e over the path length is called the *dispersion measure*,

$$DM = \int_0^L n_e \, dy \; , \qquad (14.70)$$

which is the same quantity as the total electron content. dt_p/dv can be estimated by measuring the time of arrival of pulsar pulses at different frequencies, and the dispersion measure can then be found from Eq. (14.69). If the distance to the pulsar is known, then the average electron density can be calculated. A typical value of $\langle n_e \rangle$ in the plane of our Galaxy is 0.03 cm^{-3} (Weisberg et al. 1980). Alternately, if a pulsar's distance is unknown, it can be estimated from Eq. (14.69) using an estimated average value of n_e.

The magnetic field of the Galaxy causes Faraday rotation of the polarization plane of radiation from extragalactic radio sources. Equation (14.12) can be rewritten

$$\Delta\psi = \lambda^2 RM \; , \qquad (14.71)$$

where *RM* is the *rotation measure* given by

$$RM = 8.1 \times 10^5 \int_0^L n_e B_\parallel \, dy \; . \qquad (14.72)$$

Here, *RM* is in radians per square meter, λ is in meters, $B_\parallel$ is the longitudinal component of magnetic field in gauss (1 gauss = 10^{-4} tesla), n_e is in cm^{-3}, and dy is in parsecs (pc) (1 pc = 3.1×10^{16} m). The interstellar magnetic field can be estimated by dividing the rotation measure by the dispersion measure. Typical values of the magnetic field obtained in this way are 2 μG (Heiles 1976). This procedure underestimates the magnetic field if the field reverses direction along the line of sight. A formula for roughly estimating the rotation measure due to the galactic magnetic field is (Spitzer 1978)

$$RM \simeq -18|\cot b| \cos(\ell - 94°) , \qquad (14.73)$$

where ℓ and b are the galactic longitude and latitude. Extensive measurements of rotation measure as a function of direction can be found in Oppermann et al. (2012).

Faraday rotation that occurs within a radio source depolarizes the emergent radiation. This depolarization happens because radiation emitted from different depths in the source suffers different amounts of Faraday rotation. Such a source might be a relativistic gas emitting polarized synchrotron radiation immersed in a thermal plasma that causes the Faraday rotation. The degree of polarization of the observed radiation can be succinctly described in a Fourier transform relationship when self-absorption is negligible. We first introduce the function M, the complex degree of linear polarization, defined by

$$M = m_\ell \, e^{j2\psi} = \frac{Q + jU}{I} , \qquad (14.74)$$

where m_ℓ is the degree of linear polarization, ψ is the position angle of the electric field, and Q, U, and I are the Stokes parameters as defined in Sect. 4.7. If y is the linear distance into the source, $\psi(y)$ is the intrinsic position angle of the radiation at depth y, $j_v(y)$ is the volume emissivity of the source, and $\lambda^2 \beta(y)$ is the Faraday rotation suffered by radiation emitted at depth y, then the degree of polarization of the observed radiation can be written

$$M(\lambda^2) = \frac{\int_0^\infty m_\ell(y) \, j_v(y) \, e^{j2[\psi(y) + \lambda^2 \beta(y)]} \, dy}{\int_0^\infty j_v(y) \, dy} . \qquad (14.75)$$

The denominator in Eq. (14.75) is the total intensity. $\beta(y)$ is the Faraday depth, which increases monotonically into the source as long as the sign of the longitudinal magnetic field direction does not change. In any case, we can superpose all the radiation from the same Faraday depth and write the integrals in Eq. (14.75) as a function of β instead of y, yielding

$$M(\lambda^2) = \int_{-\infty}^\infty F(\beta) e^{j2\lambda^2 \beta} \, d\beta , \qquad (14.76)$$

where

$$F(\beta) = \frac{m_\ell(y) j_v(y) e^{j2\psi(y)}}{\int_0^\infty j_v(y) \, dy} . \qquad (14.77)$$

Thus, $M(\lambda^2)$ and $F(\beta)$ form a Fourier transform pair. $F(\beta)$ is sometimes called the Faraday dispersion function. Unfortunately, $F(\beta)$, in general, cannot be found since M cannot be measured for negative values of λ^2. Because of this difficulty with the Fourier transform, $F(\beta)$ is usually estimated by model fitting. However, if $\psi(y)$ is constant, then $M(-\lambda^2) = M^*(\lambda^2)$, and $F(\beta)$ can be obtained by Fourier transformation.

Consider the result for a simple source model for which m_ℓ, ψ, and j_ν are constant. From Eq. (14.76), we have

$$M(\lambda^2) = M(0) \left[\frac{\sin \lambda^2 RM}{\lambda^2 RM} \right] e^{j\lambda^2 RM} , \qquad (14.78)$$

where RM is the Faraday rotation measure through the whole source. If the Faraday rotation originates in front of the radiation source, the complex degree of polarization is

$$M(\lambda^2) = M(0)\, e^{j2\lambda^2 RM} . \qquad (14.79)$$

In this case, there is no depolarization, and the Faraday rotation is twice that of Eq. (14.78), in which the source is uniformly distributed throughout the rotation medium. For detailed treatment of intrinsic Faraday rotation, see Burn (1966), Gardner and Whiteoak (1966), and Brentjens and de Bruyn (2005).

14.4.2   Diffractive Scattering

Diffractive interstellar scattering has been extensively investigated by observation of pulsars and compact extragalactic radio sources. For pulsars, the temporal broadening of the pulses [Eq. (14.32)], the decorrelation bandwidth [Eq. (14.33)], and the angular broadening [Eq. (14.27)] can be measured. Interpretation of the measurements in terms of a thin-screen model suggests that $\Delta n_e / n_e \simeq 10^{-3}$ and that the scale size responsible for the scintillation is on the order of 10^{11} cm. The temporal variations or scintillation of the signal from a pulsar are caused by the motions of the observer and the pulsar relative to the quasi-stationary interstellar medium. A measurement of the decorrelation bandwidth can be used to estimate the scattering angle [Eq. (14.33)]. This estimate of the scattering angle and the measurement of the timescale of fading (10^2–10^3 s at 408 MHz) can be used to estimate the relative velocity of the scattering screen by Eq. (14.31). From the relative velocity of the screen, the transverse velocity of the pulsar can be found. Velocities, and thus proper motions, of pulsars estimated in this way (Lyne and Smith 1982) agree with those measured directly with interferometers [see, e.g., Campbell et al. (1996)]. The transverse component of the orbital velocity of a binary pulsar has also been measured (Lyne 1984).

Observations show that the fluctuations in electron density can be described by a power-law spectrum with a power-law exponent of about 3.7 ± 0.3, which is similar to the value of 11/3 for Kolmogorov turbulence (Rickett 1990; Cordes et al. 1986). The power-law spectrum appears to extend over a range of scale sizes from less than 10^{10} cm to more than 10^{15} cm. The inner scale may be set by the proton gyrofrequency ($\sim 10^7$ cm) and the outer scale by the scale height of the Galaxy ($\sim 10^{20}$ cm). Observational evidence for the inner scale is given by Spangler and Gwinn (1990).

Extensive measurements of the angular sizes of extragalactic radio sources have been used to derive an approximate formula for θ_s [see Eq. (14.27)] based on the Gaussian screen model, by Harris et al. (1970), Readhead and Hewish (1972), Cohen and Cronyn (1974), Duffett-Smith and Readhead (1976), and others. This formula is

$$\theta_s \simeq \frac{15}{\sqrt{|\sin b|}} \lambda^2 \ (\text{mas}) \ , \qquad |b| > 15° \qquad (14.80)$$

where b is the Galactic latitude and λ is the wavelength in meters. The pulsar data have been interpreted by Cordes (1984) in terms of the power-law model to arrive at approximate formulas for θ_s:

$$\theta_s \simeq 7.5 \lambda^{11/5} \ (\text{arcsec}) \ , \qquad\qquad |b| \leq 0°\!.6$$

$$\simeq 0.5 |\sin b|^{-3/5} \lambda^{11/5} \ (\text{arcsec}) \ , \qquad 0°\!.6 < |b| < 3°\!-\!5°$$

$$\simeq 13 |\sin b|^{-3/5} \lambda^{11/5} \ (\text{mas}) \ , \qquad |b| \geq 3°\!-\!5° \ . \qquad (14.81)$$

The accuracy of the representations in Eqs. (14.81) decreases with decreasing $|b|$. In particular, the scattering angle at low latitudes, $|b| < 1°$, can take on a wide range of values (Cordes et al. 1984). A much more detailed model with 23 parameters characterizing the electron distribution in the Galaxy was constructed by Taylor and Cordes (1993). This model has been superseded by another model called NE2001 (Cordes and Lazio 2002, 2003). They define a scattering measure to characterize the strength of turbulence given by

$$SM = \int_0^L C_n^2 \, dy \ , \qquad (14.82)$$

where C_n^2 is defined in Eq. (14.44). With this definition, the angular broadening of an extragalactic radio source is given by

$$\theta_s \simeq 71 \nu^{-11/5} SM^{3/5} \ (\text{mas}) \ , \qquad (14.83)$$

where ν is in GHz. There are several regions in the Galaxy of anomalously high scattering (Cordes and Lazio 2001). The most highly scattered source among them is a quasar along the line of sight to a galactic HII region known as NGC6334B, which has an angular size of $3''$ at 1.5 GHz (Trotter et al. 1998). The apparent sizes of interstellar masers, which are mostly found in the Galaxy at low galactic latitudes, are sometimes set by interstellar scattering (Gwinn et al. 1988).

An example of a compact radio source that suffers a high degree of interstellar scattering is Sagittarius A* at the dynamical center of our Galaxy. This source has an angular size of about $1.0''$ at a wavelength of 30 cm (1.5 GHz) [compared with $0.5''$ predicted by Eq. (14.81)]. The angular size varies approximately as the wavelength squared over the entire measuring range ~ 0.3–30 cm, as shown in Fig. 14.10. The measurements by Doeleman et al. (2008) show that the intrinsic source size exceeds the scattering size at 1.3 mm. If the scattering can be modeled accurately, as in the case of Sgr A*, then its effects on the image can in principle be removed. The observed visibility $\mathcal{V}_m$ is the true visibility times $\mathcal{V}_s = e^{-D_\phi^2/2}$ [see Eq. (14.48)]. If, for example, $D_\phi = a\lambda^2 d^2$, appropriate if the baseline is less than the inner scale of turbulence, then $\mathcal{V}_s$ is a simple Gaussian function, and the true visibility can be

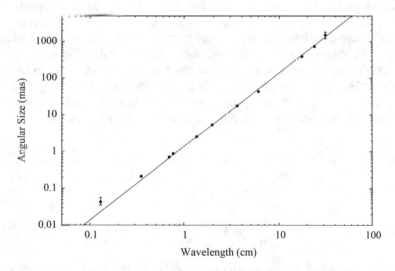

Fig. 14.10 A clear example of interstellar scattering demonstrated by the observed angular size of the compact source in the center of our Galaxy (Sgr A*). The measurements were made with interferometric arrays (Jodrell Bank at the longest two wavelengths, the Event Horizon Telescope at the shortest wavelength, and the VLBA at the intermediate wavelengths). In all cases, the visibility or image data were fitted with Gaussian profiles to determine the major axis (full width at half-maximum). Error bars not visible are smaller than the symbol size. The line is an approximate fit to the data at wavelengths longer than 6 cm and has the form λ^2. The λ-squared dependence suggests that if the scattering is caused by a turbulent medium following the Kolmogorov prescription, it has an inner scale that is longer than the size of the measurement arrays [see Eqs. (14.51) and (14.53)]. Angular sizes for this plot were taken from Davies et al. (1976), Bower et al. (2004, 2006), Shen et al. (2005), and Doeleman et al. (2008). At 0.13 cm, the intrinsic source size exceeds the scattering size. Interstellar scattering was first identified as an image broadening agent in Sgr A* by Davies et al. (1976).

recovered as

$$\mathcal{V} = \mathcal{V}_m/\mathcal{V}_s = \mathcal{V}_m e^{a\lambda^2 d^2/2} . \tag{14.84}$$

The success of this inversion clearly depends on the signal-to-noise ratio. Further discussion of "deblurring" techniques can be found in Fish et al. (2014).

14.4.3 Refractive Scattering

The realization by Sieber (1982) that the characteristic periods of amplitude scintillations of pulsars, on timescales of days to months, were correlated with their dispersion measures led Rickett et al. (1984) to the identification of another important scale length in the turbulent interstellar medium, the refractive scale d_{ref}. Refractive scattering is important in the strong scattering regime ($d_0 < d_{\text{Fresnel}} = \sqrt{\lambda R}$), where d_0 is the diffractive scale size defined by $D_\phi(d_0) = 1$. The refractive scale is the size of the diffractive scattering disk, which is the projection of the cone of scattered radiation on the scattering screen, located a distance R from the observer. The diameter of the diffractive scattering disk is $R\theta_s$. The scattering disk represents the maximum extent on the screen from which radiation can reach the observer. With a power-law distribution of irregularities, it is the irregularities at the maximum allowed scale that have the largest amplitude and are the most influential. Thus, the refractive scale is $d_{\text{ref}} \simeq R\theta_s$. Since $\theta_s \simeq \lambda/d_0$, we can write

$$d_{\text{ref}} = \frac{\lambda R}{d_0} , \tag{14.85}$$

or

$$d_{\text{ref}} = \frac{d_{\text{Fresnel}}^2}{d_0} . \tag{14.86}$$

The scale lengths d_{ref} and d_0 are widely separated. Hence, the timescale associated with scintillation scattering for a screen velocity of v_s, $t_{\text{ref}} = d_{\text{ref}}/v_s$, is much longer than that associated with diffractive scattering, $t_{\text{dif}} = d_0/v_s$. Suppose that a source is observed through a scattering screen located at a distance of 1 kpc, at $b \simeq 20°$, and a wavelength of 0.5 m. For this case, the diffractive scale length is 2×10^9 cm, the Fresnel scale is 4×10^{11} cm, and the refractive scale is 8×10^{13} cm. The typical velocity associated with the interstellar medium is 50 km s^{-1} (the sum of the Earth's orbital motion and the motion of the Sun with respect to the local standard of rest; see Table A10.1). For this velocity, the diffractive and refractive timescales for amplitude scintillation are 6 min and 6 months, respectively. Sgr A*, in addition to its diffractive scattering, also shows the effect of refractive scattering in its visibility function (see Fig. 14.11).

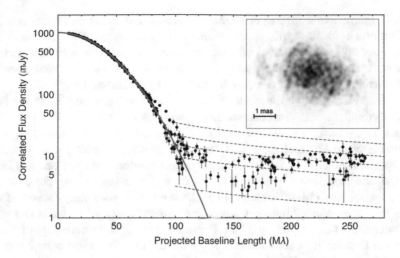

Fig. 14.11 The effect of refractive interstellar scattering on Sgr A* seen on a plot of fringe visibility (correlated flux density) vs. projected baseline at 23.8 GHz. The array consisted of the VLBA augmented with the phased VLA and the GBT. Note the logarithmic flux density scale. The projected baseline has been calculated so as to remove the source elongation. Errors are 1σ values. The solid line through the data shows the visibility model for a Gaussian diffraction-scattered disk with 735-mas diameter (see Fig. 14.10), while the dashed lines above 100 Mλ show the percentage of time the expected visibility for the refractive scattering component should be below the levels indicated for 97, 75, 50, 25, and 3% of the time, respectively. The inset shows a simulated image of Sgr A*, which shows the refractive substructure ($t_{ref} > t_{int} > t_{dif}$) calculated from the algorithm described by Johnson and Gwinn (2015) and smoothed to 0.3 mas. Data from Gwinn et al. (2014).

Refractive scattering is thought to be responsible for the slow amplitude variations observed in some pulsars and quasars at meter and decimeter wavelengths. This realization solved the long-standing problem of understanding the behavior of "long-wavelength variables," which could not be explained by intrinsic variability models based on synchrotron emission. The identification of two scales in the interstellar medium provides strong support for the power-law model. The two scales provide a way of estimating the power-law index, because the relative importance of refractive scattering increases as the power spectrum steepens. It is interesting to note that these two scales arise from a power-law phenomenon, which has no intrinsic scale. The scales are related to the propagation and depend on the wavelength and distance of the screen.

In addition to amplitude scintillation, refractive scattering causes the apparent position of the source to wander with time. The amplitude and timescale are about equal to θ_s and t_{ref}, respectively. The character of this wander depends on the power-law index of the fluctuations. Limits on the power-law index have been established from the limits on the amplitude of image wander in the relative positions among clusters of masers (Gwinn et al. 1988).

Rare sudden changes in the intensity of several extragalactic sources, called *Fiedler events*, or *extreme scattering events* (Fiedler et al. 1987), are probably caused by refractive scattering in the interstellar medium. In the archetypal example, the flux density of the extragalactic source 0954+658 increased by 30% and then dropped by 50% over a period of a month, after which it recovered in symmetric fashion. A large-scale plasma cloud presumably drifted between the source and the Earth, creating flux density changes due to focusing and refraction.

Because there are two timescales associated with strong scattering in the interstellar medium, three distinct data-averaging regimes are important for constructing images from interferometry data obtained on a timescale t_{int}. These are: $t_{int} > t_{ref}$ (ensemble average image), $t_{ref} > t_{int} > t_{dif}$ (average image), and $t_{int} < t_{dif}$ (snapshot image). The characteristics of these image regimes are described by Narayan (1992), Narayan and Goodman (1989), and Goodman and Narayan (1989). For ensemble averaging [see Eqs. (14.48) through (14.50)], the image is essentially convolved with the appropriate "seeing" function. An example of an "average" image is shown for the simulation of Sgr A* in Fig. 14.11. For more analysis and simulations of images in various time regimes, see Johnson and Gwinn (2015). The snapshot regime offers intriguing possibilities for image restoration. In this regime, it should be possible to image the source with a resolution of λ/d_{ref}, which can be very much smaller than that achievable with terrestrial interferometry. In this case, the scattering screen functions as the aperture of the interferometer. Because of the multipath propagation provided by refractive scattering, which brings radiation from widely separated parts of the scattering screen to the observer, the effective baselines can be very large. See Sect. 15.3 for further discussion, including an observation by Wolszczan and Cordes (1987).

Appendix 14.1   Refractive Bending in the Ionosphere

In this appendix, we show that a ray incident on an ionospheric layer will be bent so that it has a smaller zenith angle upon arrival at the Earth's surface, as shown in Fig. 14.2. Application of the law of sines to the two triangles with opening angles θ_1 and θ_2 gives

$$\frac{\sin z_i}{r_0} = \frac{\sin z_0}{r_0 + h_i} \tag{A14.1}$$

and

$$\frac{\sin z_{ir}}{r_0 + h_i + \Delta h} = \frac{\sin z_2}{r_0 + h_i} . \tag{A14.2}$$

Snell's law gives the relations

$$n \sin z_{ir} = \sin z_i \tag{A14.3}$$

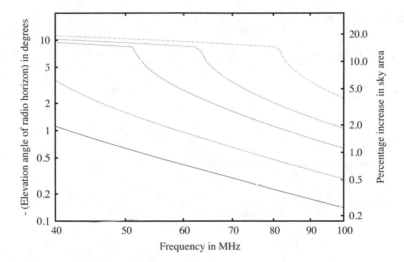

Fig. A14.1 The elevation angle of the radio horizon vs. frequency for various values of electron density in a uniform layer from 200 to 400 km, which approximates the F layer. The densities from the bottom to the top curves are $(5, 3, 2, 1,$ and $0.5) \times 10^{12}$ m^{-3}, corresponding to plasma frequencies of 20.1, 15.6, 12.7, 9.0, and 6.4 MHz. The knees of the curves (shown only for the higher densities) occur at $\nu \sim 4\nu_p$ [see Eq. (A14.12)]. For frequencies above the knees, the radio horizon is given by Eq. (A14.8) for $z_0 = 90°$. Below the knee, the radio horizon is limited by internal reflection and is given by Eq. (A14.11). From Vedantham et al. (2014). © Royal Astronomical Society, used with permission.

and

$$\sin z_{2r} = n \sin z_2 . \tag{A14.4}$$

Combining relations Eqs. (A14.1)–(A14.4) gives

$$
\begin{aligned}
\sin z_{2r} &= n \sin z_2 \\[4pt]
&= \frac{r_0 + h_i}{r_0 + h_i + \Delta h}\, n \sin z_{ir} \\[4pt]
&= \frac{r_0 + h_i}{r_0 + h_i + \Delta h}\, \sin z_i \\[4pt]
&= \frac{r_0}{r_0 + h_i + \Delta h}\, \sin z_0 .
\end{aligned}
\tag{A14.5}
$$

Note that z_0 is related to z_{2r} without reference to n. Since $z = z_{2r} + \theta_1 + \theta_2$, the net bending angle is

$$\Delta z = z - z_0 = z_{2r} + \theta_1 + \theta_2 - z_0 . \tag{A14.6}$$

Since $\theta_2 = z_{ir} - z_2$,

$$\theta_2 = \sin^{-1}\left\{\frac{1}{n}\frac{r_0}{r_0 + h_i}\sin z_0\right\} - \sin^{-1}\left\{\frac{1}{n}\frac{r_0}{r_0 + h_i + \Delta h}\sin z_0\right\}. \tag{A14.7}$$

Since $\theta_1 = z_0 - z_i$,

$$\theta_1 = z_0 - \sin^{-1}\left\{\frac{r_0}{r_0 + h_i}\sin z_0\right\}. \tag{A14.8}$$

The final result of Δz in terms of $\sin z_0$ is

$$\Delta z = \sin^{-1}\left\{\frac{r_0}{r_0 + h_i + \Delta h}\sin z_0\right\} - \sin^{-1}\left\{\frac{r_0}{r_0 + h_i}\sin z_0\right\}$$
$$+ \sin^{-1}\left\{\frac{1}{n}\frac{r_0}{r_0 + h_i}\sin z_0\right\} - \sin^{-1}\left\{\frac{1}{n}\frac{r_0}{r_0 + h_i + \Delta h}\sin z_0\right\}. \tag{A14.9}$$

Note that

$$\Delta z = (z_{2r} - z_2) + (z_{ir} - z_i). \tag{A14.10}$$

As an example, let us assume that $h_i = 300\,\text{km}$, $\Delta h = 200\,\text{km}$, $r_0 = 6370\,\text{km}$, $n_e = 3 \times 10^{11}\,\text{m}^{-3}$, and $\nu = 50\,\text{MHz}$. From Eqs. (14.4) and (14.5), we find that $\nu_p = 4.9\,\text{MHz}$ and $n = 0.9951$. For $z_0 = 75°$, the other angles are $z_i = 67.29°$, $z_{ir} = 67.98°, z_2 = 64.16°, z_{2r} = 63.59°, \theta_1 = 7.71°, \theta_2 = 3.81°, z = 75.11°$, and $\Delta z = 0.11°$. Equation (14.15) gives the same result. This demonstrates the counterintuitive result that the sign of the change in zenith angle is the same for the ionosphere and for the troposphere.

For the case $z_0 = 90°$, the result is $z = 90.22°$, so that the radiation from $0.22°$ below the horizon can in principle be received.

The phenomenon of internal reflection will occur when $z_{ir} = 90°$. This gives a critical zenith angle, z_c, below which incoming rays will not reach the observer, given by

$$\sin z_c = \frac{r_0 + h_i}{r_0}n. \tag{A14.11}$$

For $z_c = 90°$, the frequency at which this effect limits the incoming zenith angle is

$$\nu \simeq \sqrt{\frac{r_0}{z_{h_i}}}\nu_p \simeq 4\nu_p. \tag{A14.12}$$

The combination of the normal refraction and the critical angle defines the radio horizon. An example of the radio horizon is shown in Fig. A14.1. The radio horizon may affect studies of the Epoch of Reionization, as described by Vedantham et al. (2014).

Further Reading

Cordes, J.M., Rickett, B.J., and Backer, D.C., Eds., *Radio Wave Scattering in the Interstellar Medium*, Am. Inst. Physics Conf. Proc., **174**, New York (1988)

Lazio, T.J.W., Cordes, J.M., de Bruyn, A.G., and Macquart, J.-P., The Microarcsecond Sky and Cosmic Turbulence, *New Astron. Rev.*, **48**, 1439–1457 (2004)

Narayan, R., The Physics of Pulsar Scintillation, *Phil. Tran. R. Soc. Lond. A*, **341**, 151–165 (1992)

Schunk, R., and Nagy, A., *Ionospheres: Physics, Plasma Physics, and Chemistry*, 2nd ed., Cambridge Univ. Press, Cambridge, UK (2009)

References

Aarons, J., Global Morphology of Ionospheric Scintillations, *Proc. IEEE*, **70**, 360–378 (1982)

Aarons, J., Klobuchar, J.A., Whitney, H.E., Austen, J., Johnson, A.L., and Rino, C.L., Gigahertz Scintillations Associated with Equatorial Patches, *Radio Sci.*, **18**, 421–434 (1983)

Aarons, J., Mendillo, M., Lin, B., Colerico, M., Beach, T., Kintner, P., Scali, J., Reinisch, B., Sales, G., and Kudeki, E., Equatorial F-Region Irregularity Morphology during an Equinoctial Month at Solar Minimum, *Space Science Reviews*, **87**, 357–386 (1999)

Allen, C.W., Interpretation of Electron Densities from Corona Brightness, *Mon. Not. R. Astron. Soc.*, **107**, 426–432 (1947)

Appleton, E.V., and Barnett, M.A.F., On Some Direct Evidence for Downward Atmospheric Reflection of Electric Rays, *Proc. R. Soc. Lond. A*, **109**, 621–641 (1925)

Armstrong, J.W., and Coles, W.A., Interplanetary Scintillations of PSR 0531+21 at 74 MHz, *Astrophys. J.*, **220**, 346–352 (1978)

Asai, K., Kojima, M., Tokumaru, M., Yokobe, A., Jackson, B.V., Hick, P.L., and Manoharan, P.K., Heliospheric Tomography Using Interplanetary Scintillation Observations. 3. Correlation Between Speed and Electron Density Fluctuations in the Solar Wind, *J. Geophys. Res.*, **103**, 1991–2001 (1998)

Bailey, D.K., On a New Method of Exploring the Upper Atmosphere, *J. Terr. Mag. Atmos. Elec.*, **53**, 41–50 (1948)

Beynon, W.J.G., Marconi, Radio Waves, and the Ionosphere, *Radio Sci.*, **10**, 657–664 (1975)

Bilitza, D., International Reference Ionosphere—Status 1995/96, *Adv. Space Res.*, **20**, 1751–1754 (1997)

Booker, H.G., The Use of Radio Stars to Study Irregular Refraction of Radio Waves in the Ionosphere, *Proc. IRE*, **46**, 298–314 (1958)

Booker, H.G., Ratcliffe, J.A., and Shinn, D.H., Diffraction from an Irregular Screen with Applications to Ionospheric Problems, *Philos. Tran. R. Soc. Lond. A*, **242**, 579–607 (1950)

Bougeret, J.-L., King, J.H., and Schwenn, R., Solar Radio Burst and In Situ Determination of Interplanetary Electron Density, *Solar Phys.*, **90**, 401–412 (1984)

Bower, G.C., Falcke, H., Herrnstein, R.M., Zhao, J.-H., Goss, W.M., and Backer, D.C., Detection of the Intrinsic Size of Sagittarius A* Through Closure Amplitude Imaging, *Science*, **304**, 704–708 (2004)

Bower, G.C., Goss, W.M., Falcke, H., Backer, D.C., and Lithwick, Y., The Intrinsic Size of Sagittarius A* from 0.35 to 6 cm, *Astrophys. J. Lett.*, **648**, L127–L130 (2006)

Bracewell, R.N., Eshleman, V.R., and Hollweg, J.V., The Occulting Disk of the Sun at Radio Wavelengths, *Astrophys. J.*, **155**, 367–368 (1969)

Breit, G., and Tuve, M.A., A Test of the Existence of the Conducting Layer, *Phys. Rev.*, **28**, 554–575 (1926)

Brentjens, M.A., and de Bruyn, A.G., Faraday Rotation Measure Synthesis, *Astron. Astrophys.*, **441**, 1217–1228 (2005)

Budden, K.G., *Radio Waves in the Ionosphere*, Cambridge Univ. Press, Cambridge, UK (1961)

Burn, B.J., On the Depolarization of Discrete Radio Sources by Faraday Dispersion, *Mon. Not. R. Astron. Soc.*, **133**, 67–83 (1966)

Campbell, R.M., Bartel, N., Shapiro, I.I., Ratner, M.I., Cappallo, R.J., Whitney, A.R., and Putnam, N., VLBI-Derived Trigonometric Parallax and Proper Motion of PSR B2021+51, *Astrophys. J. Lett.*, **461**, L95–L98 (1996)

Clarke, M., "Two Topics in Radiophysics," Ph.D. thesis, Cambridge Univ. (1964) (see App. II)

Cohen, M.H., High-Resolution Observations of Radio Sources, *Ann. Rev. Astron. Astrophys.*, **7**, 619–664 (1969)

Cohen, M.H., and Cronyn, W.M., Scintillation and Apparent Angular Diameter, *Astrophys. J.*, **192**, 193–197 (1974)

Cohen, M.H., Gundermann, E.J., Hardebeck, H.E., and Sharp, L.E., Interplanetary Scintillations. II. Observations, *Astrophys. J.*, **147**, 449–466 (1967a)

Cohen, M.H., Gundermann, E.J., and Harris, D.E., New Limits on the Diameters of Radio Sources, *Astrophys. J.*, **150**, 767–782 (1967b)

Cordes, J.M., Interstellar Scattering, in *VLBI and Compact Radio Sources*, IAU Symp. 110, Fanti, R., Kellermann, K., and Setti, G., Eds., Reidel, Dordrecht, the Netherlands, (1984), pp. 303–307

Cordes, J.M., Interstellar Scattering: Radio Sensing of Deep Space Through the Turbulent Interstellar Medium, in *Radio Astronomy at Long Wavelengths*, Stone, R.G., Weiler, K.W., Goldstein, M.L., and Bougeret, J.-L., Eds., Geophysical Monograph 119, Am. Geophys. Union, pp. 105–114 (2000)

Cordes, J.M., Ananthakrishnan, S., and Dennison, B., Radio Wave Scattering in the Galactic Disk, *Nature*, **309**, 689–691 (1984)

Cordes, J.M., and Lazio, T.J.W., Anomalous Radio-Wave Scattering from Interstellar Plasma Structures, *Astrophys. J.*, **549**, 997–1010 (2001)

Cordes, J.M., and Lazio, T.J.W., NE2001. I. A New Model for the Galactic Distribution of Free Electrons and Its Fluctuations (2002), astro-ph/0207156

Cordes, J.M., and Lazio, T.J.W., NE2001. II. Using Radio Propagation Data to Construct a Model for the Galactic Distribution of Free Electrons (2003), astro-ph/0301598

Cordes, J.M., Pidwerbetsky, A., and Lovelace, R.V.E., Refractive and Diffractive Scattering in the Interstellar Medium, *Astrophys. J.*, **310**, 737–767 (1986)

Counselman, C.C., III, Kent, S.M., Knight, C.A., Shapiro, I.I., Clark, T.A., Hinteregger, H.F., Rogers, A.E.E., and Whitney, A.R., Solar Gravitational Deflection of Radio Waves Measured by Very-Long-Baseline Interferometry, *Phys. Rev. Lett.*, **33**, 1621–1623 (1974)

Counselman, C.C., III, and Rankin, J.M., Density of the Solar Corona from Occultations of NP0532, *Astrophys. J.*, **175**, 843–856 (1972)

Crane, R.K., Ionospheric Scintillation, *Proc. IEEE*, **65**, 180–199 (1977)

Daniell, R.E., Brown, L.D., Anderson, D.N., Fox, M.W., Doherty, P.H., Decker, D.T., Sojka, J.J., and Schunk, R.W., Parameterized Ionospheric Model: A Global Ionospheric Parameterization Based on First Principles Models, *Radio Sci.*, **30**, 1499–1510 (1995)

Davies, K., *Ionospheric Radio Propagation*, National Bureau of Standards Monograph 80, U.S. Government Printing Office, Washington, DC (1965)

Davies, R.D., Walsh, D., and Booth, R.S., The Radio Source at the Galactic Nucleus, *Mon. Not. R. Astron. Soc.*, **177**, 319–333 (1976)

Doeleman, S.S., Weintroub, J., Rogers, A.E.E., Plambeck, R., Freund, R., Tilanus, R.P.J., Friberg, P., Ziurys, L.M., Moran, J.M., Corey, B., and 18 coauthors, Event-Horizon-Scale Structure in the Supermassive Black Hole Candidate at the Galactic Center, *Nature*, **455**, 78–80 (2008)

Duffett-Smith, P.J., and Readhead, A.C.S., The Angular Broadening of Radio Sources by Scattering in the Interstellar Medium, *Mon. Not. R. Astron. Soc.*, **174**, 7–17 (1976)

Erickson, W.C., The Radio-Wave Scattering Properties of the Solar Corona, *Astrophys. J.*, **139**, 1290–1311 (1964)

Erickson, W.C., Perley, R.A., Flatters, C., and Kassim, N.E., Ionospheric Corrections for VLA Observations Using Local GPS Data, *Astron. Astrophys.*, **366**, 1071–1080 (2001)

Evans, J.V., Theory and Practice of Ionosphere Study by Thomson Scatter Radar, *Proc. IEEE*, **57**, 496–530 (1969)

Evans, J.V., and Hagfors, T., *Radar Astronomy*, McGraw-Hill, New York (1968)

Fejer, B.G., and Kelley, M.C., Ionospheric Irregularities, *Rev. Geophys. Space Sci.*, **18**, 401–454 (1980)

Fiedler, R.L., Dennison, B., Johnston, K.J., and Hewish, A., Extreme Scattering Events Caused by Compact Structures in the Interstellar Medium, *Nature*, **326**, 675–678 (1987)

Fish, V.L., Johnson, M.D., Lu, R.-S., Doeleman, S.S., Bouman, K.L., Zoran, D., Freeman, W.T., Psaltis, D., Narayan, R., Pankratius, V., Broderick, A.E., Gwinn, C.R., and Vertatschitch, L.E., Imaging an Event Horizon: Mitigation of Scattering Toward Sagittarius A*, *Astrophys. J.*, **795**:134 (7pp) (2014)

Fomalont, E.B., and Sramek, R.A., A Confirmation of Einstein's General Theory of Relativity by Measuring the Bending of Microwave Radiation in the Gravitational Field of the Sun, *Astrophys. J.*, **199**, 749–755 (1975)

Frail, D.A., Kulkarni, S.R., Nicastro, L., Feroci, M., and Taylor, G.B., The Radio Afterglow from the γ-Ray Burst of 8 May 1997, *Nature*, **389**, 261–263 (1997)

Gapper, G.R., Hewish, A., Purvis, A., and Duffett-Smith, P.J., Observing Interplanetary Disturbances from the Ground, *Nature*, **296**, 633–636 (1982)

Gardner, F.F., and Whiteoak, J.B., The Polarization of Cosmic Radio Waves, *Ann. Rev. Astron. Astrophys.*, **4**, 245–292 (1966)

Goodman, J., and Narayan, R., The Shape of a Scatter-Broadened Image: II. Interferometric Visibilities, *Mon. Not. R. Astron. Soc.*, **238**, 995–1028 (1989)

Gupta, Y., Pulsars and Interstellar Scintillations, in *Pulsar Astrometry—2000 and Beyond*, M. Kramer, N. Wex, and R. Wielebinski, Eds., Astron. Soc. Pacific Conf. Ser., **202**, 539–544 (2000)

Gwinn, C.R., Kovalev, Y.Y., Johnson, M.D., and Soglasnov, V.A., Discovery of Substructure in the Scatter-Broadened Image of Sgr A*, *Astrophys. J. Lett.*, **794**:L14 (5pp) (2014)

Gwinn, C.R., Moran, J.M., Reid, M.J., and Schneps, M.H., Limits on Refractive Interstellar Scattering Toward Sagittarius B2, *Astrophys. J.*, **330**, 817–827 (1988)

Hagfors, T., The Ionosphere, in *Methods of Experimental Physics*, Vol. 12, Part B (*Astrophysics: Radio Telescopes*), M. L. Meeks, Ed., Academic Press, New York (1976), pp. 119–135

Harris, D.E., Zeissig, G.A., and Lovelace, R.V., The Minimum Observable Diameter of Radio Sources, *Astron. Astrophys.*, **8**, 98–104 (1970)

Heiles, C., The Interstellar Magnetic Field, *Ann. Rev. Astron. Astrophys.*, **14**, 1–22 (1976)

Helmboldt, J.F., Drift-Scan Imaging of Traveling Ionospheric Disturbances with the Very Large Array, *Geophys. Res. Lett.*, **41**, 4835–4843 (2014)

Helmboldt, J.F., Lane, W.M., and Cotton, W.D., Climatology of Midlatitude Ionospheric Disturbances from the Very Large Array Low-Frequency Sky Survey, *Radio Sci.*, **37**, RS5008 (19pp) (2012)

Hewish, A., The Diffraction of Galactic Radio Waves as a Method of Investigating the Irregular Structure of the Ionosphere, *Proc. R. Soc. Lond. A*, **214**, 494–514 (1952)

Hewish, A., Scott, P.F., and Wills, D., Interplanetary Scintillation of Small Diameter Radio Sources, *Nature*, **203**, 1214–1217 (1964)

Hey, J.S., Parsons, S.J., and Phillips, J.W., Fluctuations in Cosmic Radiation at Radio Frequencies, *Nature*, **158**, 234 (1946)

Ho, C.M., Mannucci, A.J., Lindqwister, U.J., Pi, X., and Tsurutani, B.T., Global Ionospheric Perturbations Monitored by the Worldwide GPS Network, *Geophys. Res. Lett.*, **23**, 3219–3222 (1996)

Ho, C.M., Wilson, B.D., Mannucci, A.J., Lindqwister, U.J., and Yuan, D.N., A Comparative Study of Ionospheric Total Electron Content Measurements Using Global Ionospheric Maps of GPS, TOPEX Radar, and the Bent Model, *Radio Sci.*, **32**, 1499–1512 (1997)

Hocke, K., and Schlegel, K., A Review of Atmospheric Gravity Waves and Travelling Ionospheric Disturbances: 1982–1995, *Ann. Geophysicae*, **14**, 917–940 (1996)

Holt, E.H., and Haskell, R.E., *Foundations of Plasma Dynamics*, Macmillan, New York (1965), p. 254

Hunsucker, R.D., Atmospheric Gravity Waves Generated in the High-Latitude Ionosphere: A Review, *Rev. Geophys. Space Phys.*, **20**, 293–315 (1982)

Jaeger, J.C., and Westfold, K.C., Equivalent Path and Absorption for Electromagnetic Radiation in the Solar Corona, *Aust. J. Phys.*, **3**, 376–386 (1950)

Janardhan, P., Bisoi, S.K., Ananthakrishnan, S., Tokumaru, M., Fujiki, K., Jose, L., and Sridharan, R., A Twenty-Year Decline in Solar Photospheric Magnetic Fields: Inner-Heliospheric Signatures and Possible Implications, *J. Geophys. Res.: Space Phys.*, **120**, 5306–5317 (2015)

Johnson, M.D., and Gwinn, C.R., Theory and Simulations of Refractive Substructure in Resolved Scatter-Broadened Images, *Astrophys. J.*, **805**:180 (15pp) (2015)

Kaplan, D.L., Tingay, S.J., Manoharan, P.K., Macquart, J.-P., Hancock, P., Morgan, J., Mitchell, D.A., Ekers, R.D., Wayth, R.B., Trott, C., and 27 coauthors, Murchison Widefield Array Observations of Anomalous Variability: A Serendipitous Nighttime Detection of Interplanetary Scintillation, *Astrophys. J. Lett.*, **809**:L12 (7pp) (2015)

Kaplan, G.H., Josties, F.J., Angerhofer, P.E., Johnston, K.J., and Spencer, J.H., Precise Radio Source Positions from Interferometric Observations, *Astron. J.*, **87**, 570–576 (1982)

Kundu, M.R., *Solar Radio Astronomy*, Wiley-Interscience, New York (1965), p. 104

Lawrence, R.S., Little, C.G., and Chivers, H.J.A., A Survey of Ionospheric Effects upon Earth-Space Radio Propagation, *Proc. IEEE*, **52**, 4–27 (1964)

Lazio, T.J.W., and Cordes, J.M., Hyperstrong Radio-Wave Scattering in the Galactic Center. I. A Survey for Extragalactic Sources Seen through the Galactic Center, *Astrophys. J. Suppl.*, **118**, 201–216 (1998)

Leblanc, Y., Dulk, G.A., and Bougeret, J.-L., Tracing the Electron Density from the Corona to 1 AU, *Solar Phys.*, **183**, 165–180 (1998)

Little, L.T., and Hewish, A., Interplanetary Scintillation and Relation to the Angular Structure of Radio Sources, *Mon. Not. R. Astron. Soc.*, **134**, 221–237 (1966)

Loi, S.T., Murphy, T., Bell, M.E., Kaplan, D.L., Lenc, E., Offinga, A.R., Hurley-Walker, N., Bernardi, G., Bowman, J.D., Briggs, F., and 32 coauthors, Quantifying Ionospheric Effects on Time-Domain Astrophysics with the Murchison Widefield Array, *Mon. Not. R. Astron. Soc.*, **453**, 2731–2746 (2015a)

Loi, S.T., Murphy, T., Cairns, I.H., Menk, F.W., Waters, C.L., Erickson, P.J., Trott, C.M., Hurley-Walker, N., Mortan, J., Lenc, E., and 31 coauthors, Real-Time Imaging of Density Ducts Between the Plasmasphere and Ionosphere, *Geophys. Res. Lett.*, **42**, 3707–3714 (2015b)

Lyne, A.G., Orbital Inclination and Mass of the Binary Pulsar PSR0655+64, *Nature*, **310**, 300–302 (1984)

Lyne, A.G., and Smith, F.G., Interstellar Scintillation and Pulsar Velocities, *Nature*, **298**, 825–827 (1982)

Mannucci, A.J., Wilson, B.D., Yuan, D.N., Ho, C.H., Lindqwister, U.J., and Runge, T.F., A Global Mapping Technique for GPS-Derived Ionospheric Total Electron Content Measurements, *Radio Sci.*, **33**, 565–582 (1998)

Manoharan, P.K., Three-Dimensional Evolution of Solar Wind During Solar Cycles 22–24, *Astrophy. J.*, **751**:128 (13pp) (2012)

Manoharan, P.K., Ananthakrishnan, S., Dryer, M., Detman, T.R., Leinbach, H., Kojima, M., Watanabe, T., and Kahn, J., Solar Wind Velocity and Normalized Scintillation Index from Single-Station IPS Observations, *Solar Phys.*, **156**, 377–393 (1995)

Mathur, N.C., Grossi, M.D., and Pearlman, M.R., Atmospheric Effects in Very Long Baseline Interferometry, *Radio Sci.*, **5**, 1253–1261 (1970)

Mercier, R.P., Diffraction by a Screen Causing Large Random Phase Fluctuations, *Proc. R. Soc. Lond. A*, **58**, 382–400 (1962)

Moran, J.M., The Effects of Propagation on VLBI Observations, in *Very Long Baseline Interferometry: Techniques and Applications*, Felli, M., and Spencer, R.E., Eds., Kluwer, Dordrecht, the Netherlands, (1989), pp. 47–59

Narayan, R., From Scintillation Observations to a Model of the ISM–The Inverse Problem, in *Radio Wave Scattering in the Interstellar Medium*, Cordes, J.M., Rickett, B.J., and Backer, D.C., Eds., Am. Inst. Physics Conf. Proc., **174**, New York (1988), pp. 17–31

Narayan, R., The Physics of Pulsar Scintillation, *Phil. Tran. R. Soc. Lond. A*, **341**, 151–165 (1992)

Narayan, R., Anantharamaiah, K.R., and Cornwell, T.J., Refractive Radio Scintillation in the Solar Wind, *Mon. Not. R. Astron. Soc.*, **241**, 403–413 (1989)

Narayan, R., and Goodman, J., The Shape of a Scatter-Broadened Image: I. Numerical Simulations and Physical Principles, *Mon. Not. R. Astron. Soc.*, **238**, 963–994 (1989)

Oppermann, N., Junklewitz, H., Robbers, G., Bell, M.R., Enßlin, T.A., Bonafede, A., Braun, R., Brown, J.C., Clarke, T.E., Feain, I.J., and 21 coauthors, An Improved Map of the Galactic Faraday Sky, *Astron. Astrophys.*, **542**, A93 (14pp) (2012)

Pi, X., Mannucci, A.J., Lindqwister, U.J., and Ho, C.M., Monitoring of Global Ionospheric Irregularities Using the Worldwide GPS Network, *Geophys. Res. Lett.*, **24**, 2283–2286 (1997)

Ratcliffe, J.A., Some Aspects of Diffraction Theory and Their Application to the Ionosphere, *Rep. Prog. Phys.*, **19**, 188–267 (1956)

Ratcliffe, J.A., *The Magneto-Ionic Theory and Its Application to the Ionosphere*, Cambridge Univ. Press, Cambridge, UK (1962)

Readhead, A.C.S., and Hewish, A., Galactic Structure and the Apparent Size of Radio Sources, *Nature*, **236**, 440–443 (1972)

Rickett, B.J., Radio Propagation Through the Turbulent Interstellar Medium, *Ann. Rev. Astron. Astrophys.*, **28**, 561–605 (1990)

Rickett, B.J., Coles, W.A., and Bourgois, G., Slow Scintillation in the Interstellar Medium, *Astron. Astrophys.*, **134**, 390–395 (1984)

Roberts, D.H., Rogers, A.E.E., Allen, B.R., Bennet, C.L., Burke, B.F., Greenfield, P.E., Lawrence, C.R., and Clark, T.A., Radio Interferometric Detection of a Traveling Ionospheric Disturbance Excited by the Explosion of Mt. St. Helens, *J. Geophys. Res.*, **87**, 6302–6306 (1982)

Rogers, A.E.E., Bowman, J.D., Vierinen, J., Monsalve, R., and Mozdzen, T., Radiometric Measurements of Electron Temperature and Opacity of Ionospheric Perturbations, *Radio Sci.*, **50**, 130–137 (2015)

Ros, E., Marcaide, J.M., Guirado, J.C., Sardón, E., and Shapiro, I.I., A GPS-Based Method to Model the Plasma Effects in VLBI Observations, *Astron. Astrophys.*, **356**, 357–362 (2000)

Salpeter, E.E., Interplanetary Scintillations. I. Theory, *Astrophys. J.*, **147**, 433–448 (1967)

Scheuer, P.A.G., Amplitude Variations in Pulsed Radio Sources, *Nature*, **218**, 920–922 (1968)

Schunk, R., and Nagy, A., *Ionospheres: Physics, Plasma Physics, and Chemistry*, 2nd ed., Cambridge Atmospheric and Space Science Series, Cambridge Univ. Press, Cambridge, UK, (2009)

Scott, S.L., Coles, W.A., and Bourgois, G., Solar Wind Observations Near the Sun Using Interplanetary Scintillation, *Astron. Astrophys.*, **123**, 207–215 (1983)

Shapiro, I.I., Estimation of Astrometric and Geodetic Parameters, in *Methods of Experimental Physics*, Vol. 12, Part C *(Astrophysics: Radio Observations)*, Meeks, M.L., Ed., Academic Press, New York (1976), pp. 261–276

Shen, Z.-Q., Lo, K.Y., Liang, M.-C., Ho, P.T.P. and Zhao, J.-H., A Size of ∼ 1 AU for the Radio Source Sgr A* at the Center of the Milky Way, *Nature*, **438**, 62–64 (2005)

Sieber, W., Causal Relationship Between Pulsar Long-Term Intensity Variations and the Interstellar Medium, *Astron. Astrophys.*, **113**, 311–313 (1982)

Smart, W.M., *Textbook on Spherical Astronomy*, 6th ed., revised by Green, R.M., Cambridge Univ. Press, Cambridge, UK (1977)

Smith, F.G., Little, C.G., and Lovell, A.C.B., Origin of the Fluctuations in the Intensity of Radio Waves from Galactic Sources, *Nature*, **165**, 422–424 (1950)

Spangler, S.R., and Gwinn, C.R., Evidence for an Inner Scale to the Density Turbulence in the Interstellar Medium, *Astrophys. J. Lett.*, **353**, L29–L32 (1990)

Spangler, S.R., and Sakurai, T., Radio Interferometry of Solar Wind Turbulence from the Orbit of Helios to the Solar Corona, *Astrophys. J.*, **445**, 999–1061 (1995)

Spitzer, L., *Physical Processes in the Interstellar Medium*, Wiley-Interscience, New York (1978), p. 65

Spoelstra, T.A.T., The Influence of Ionospheric Refraction on Radio Astronomy Interferometry, *Astron. Astrophys.*, **120**, 313–321 (1983)

Spoelstra, T.A.T., and Kelder, H., Effects Produced by the Ionosphere on Radio Interferometry, *Radio Sci.*, **19**, 779–788 (1984)

Sukumar, S., Ionospheric Refraction Effects on Radio Interferometer Phase, *J. Astrophys. Astr.*, **8**, 281–294 (1987)

Taylor, J.H., and Cordes, J.M., Pulsar Distances and the Galactic Distribution of Free Electrons, *Astrophys. J.*, **411**, 674–684 (1993)

Tokumaru, M., Kojima, M., and Fujiki, K., Long-Term Evolution in the Global Distribution of Solar Wind Speed and Density Fluctuations During 1997–2009, *J. Geophys. Res.*, **117**, A06108 (14 pp) (2012)

Trotter, A.S., Moran, J.M., and Rodríguez, L.F., Anisotropic Radio Scattering of NGC6334B, *Astrophys. J.*, **493**, 666–679 (1998)

Vedantham, H.K., Koopmans, L.V.E., de Bruyn, A.G., Wijnholds, A.J., Ciardi, B., and Brentjens, M.A., Chromatic Effects in the 21-cm Global Signal from the Cosmic Dawn, *Mon. Not. R. Astron. Soc.*, **437**, 1056–1059 (2014)

Weinberg, S., *Gravitation and Cosmology: Principles and Applications of the General Theory of Relativity*, Wiley, New York (1972), p. 188

Weisberg, J.M., Rankin, J., and Boriakoff, V., HI Absorption Measurements of Seven Low Latitude Pulsars, *Astron. Astrophys.*, **88**, 84–93 (1980)

Winterhalter, D., Gosling, J.T., Habbal, S.R., Kurth, W.S., and Neugebauer, M., Eds., *Solar Wind Eight*, Proc. 8th Int. Solar Wind Conf., Vol. 382, Am. Inst. Physics, New York (1996)

Wolszczan, A., and Cordes, J.M., Interstellar Interferometry of the Pulsar PSR 1237+25, *Astrophys. J. Lett.*, **320**, L35–L39 (1987)

Yeh, K.C., and Liu, C.H., Radio Wave Scintillations in the Ionosphere, *Proc. IEEE*, **70**, 324–360 (1982)

Young, A.T., Interpretation of Interplanetary Scintillations, *Astrophys. J.*, **168**, 543–562 (1971)

Chapter 15
Van Cittert–Zernike Theorem, Spatial Coherence, and Scattering

This chapter is concerned with the van Cittert–Zernike theorem, including an examination of the assumptions involved in its derivation, the requirement of spatial incoherence of a source, and the interferometer response to a coherent source. Some optical terminology is used, for example, *mutual coherence*, which includes complex visibility. There is also a brief discussion of some aspects of scattering by irregularities in the propagation medium. Much of the development of the theory of coherence and similar concepts of electromagnetic radiation is to be found in the literature of optics. The terminology is sometimes different from that which has evolved in radio interferometry, but many of the physical situations are similar or identical. However, in spite of the similarity, the literature shows that in the early development of radio astronomy, the optical experience was hardly ever mentioned, an exception being the reference by Bracewell (1958) to Zernike (1938) for the concept of the complex degree of coherence. The van Cittert–Zernike theorem contains a simple formalism that includes the basic principles of correlation in electromagnetic fields.

15.1 Van Cittert–Zernike Theorem

We showed in Chaps. 2 and 3 that the cross-correlation of the signals received in spaced antennas can be used to form an image of the intensity distribution of a distant cosmic source through a Fourier transform relationship. This result is a form of the van Cittert–Zernike theorem, which originated in optics. The basis for the theorem is a study published by van Cittert in 1934 and followed a few years later by a simpler derivation by Zernike. A description of the result established by van Cittert and Zernike is given by Born and Wolf (1999, Chap. 10). The original form of the result does not specifically refer to the Fourier transform relationship between intensity and mutual coherence but is essentially as follows.

© The Author(s) 2017
A.R. Thompson, J.M. Moran, and G.W. Swenson Jr., *Interferometry and Synthesis in Radio Astronomy*, Astronomy and Astrophysics Library,
DOI 10.1007/978-3-319-44431-4_15

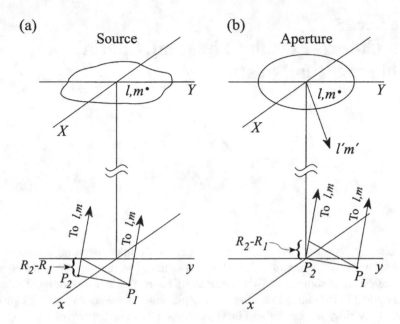

Fig. 15.1 (**a**) Geometry of a distant spatially incoherent source and the points P_1 and P_2 at which the mutual coherence of the radiation is measured. The source plane (X, Y) is parallel to the measurement plane (x, y) but at a large distance from it. (**b**) Similar geometry for measurement of the radiation field from an aperture in the (X, Y) plane that is illuminated from above by a coherent wavefront. The radiated field has a maximum at the point P_2. Direction cosines (l, m) are defined with respect to the (x, y) axes in the measurement plane, and direction cosines (l', m') are defined with respect to the (X, Y) axes in the plane of the aperture.

Consider an extended, quasi-monochromatic, incoherent source, and let the mutual coherence of the radiation be measured at two points P_1 and P_2 in a plane normal to the direction of the source, as in Fig. 15.1a. Then suppose that the source is replaced by an aperture of identical shape and size and illuminated from behind by a spatially coherent wavefront. The distribution of the electric field amplitude over the aperture is proportional to the intensity distribution over the source. The Fraunhofer diffraction pattern of the aperture is observable in the plane containing P_1 and P_2. The relative positions of the points P_1 and P_2 are the same in the two cases, but for the aperture, the geometric configuration is such that P_2 lies on the maximum of the diffraction pattern. Then the mutual coherence measured for the incoherent source, normalized to unity for zero spacing between P_1 and P_2, is equal to the complex amplitude of the field of the aperture diffraction pattern at the position P_1, normalized to the maximum value at P_2.

In this form, the theorem results from the fact that the behavior of both the mutual coherence and the Fraunhofer diffraction can be represented by similar Fourier transform relationships. Derivation of the theorem provides an opportunity to examine the assumptions involved and is given below. The analysis is similar to that given by Born and Wolf but with some modifications to take advantage of the simplified geometry when the source is at an astronomical distance. First, we note that in optics, the *mutual coherence function* for a field $E(t)$, measured at points 1 and 2, is represented by

$$\Gamma_{12}(u, v, \tau) = \lim_{T \to \infty} \frac{1}{2T} \int_{-T}^{T} E_1(t) E_2^*(t - \tau)\, dt \, , \tag{15.1}$$

where u and v are the coordinates of the spacing between the two measurement points, expressed in units of wavelength. $\Gamma_{12}(u, v, 0)$, for zero time offset, is equivalent to the complex visibility $\mathcal{V}(u, v)$ used in the radio case.

15.1.1   Mutual Coherence of an Incoherent Source

The geometric situation for the incoherent source is shown in Fig. 15.1a. Consider the source located in a distant plane, indicated by (X, Y). The radiated field is measured at two points, P_1 and P_2, in the (x, y) plane that is parallel to the source plane. In the radio case, these points are the locations of the interferometer antennas. It is convenient to specify the position of a point in the (X, Y) plane by the direction cosines (l, m) measured with respect to the (x, y) axes. The source is sufficiently distant that the direction of any point within it measured from P_1 is the same as that measured from P_2. The fields at P_1 and P_2 resulting from a single element of the source at the point (l, m) are given by

$$E_1(l, m, t) = \mathcal{E}\left(l, m, t - \frac{R_1}{c}\right) \frac{\exp\left[-j2\pi \nu(t - R_1/c)\right]}{R_1} \, , \tag{15.2}$$

and

$$E_2(l, m, t) = \mathcal{E}\left(l, m, t - \frac{R_2}{c}\right) \frac{\exp\left[-j2\pi \nu(t - R_2/c)\right]}{R_2} \, , \tag{15.3}$$

where $\mathcal{E}(l, m, t)$ is a phasor representation of the complex amplitude of the electric field at the source for an element at position (l, m). R_1 and R_2 are the distances from this element to P_1 and P_2, respectively, and c is the velocity of light. The exponential terms in Eqs. (15.2) and (15.3) represent the phase change in traversing the paths from the source to P_1 and P_2.

The complex cross-correlation of the field voltages at P_1 and P_2 due to the radiation from the element at (l, m) is, for zero time offset,

$$\left\langle E_1(l, m, t)\, E_2^*(l, m, t) \right\rangle$$

$$= \left\langle \mathcal{E}\left(l, m, t - \frac{R_1}{c}\right) \mathcal{E}^*\left(l, m, t - \frac{R_2}{c}\right) \right\rangle$$

$$\times \frac{\exp[-j2\pi\nu(t - R_1/c)] \exp[j2\pi\nu(t - R_2/c)]}{R_1 R_2}$$

$$= \left\langle \mathcal{E}(l, m, t)\, \mathcal{E}^*\left(l, m, t - \frac{R_2 - R_1}{c}\right) \right\rangle \frac{\exp\left[j2\pi\nu(R_1 - R_2)/c\right]}{R_1 R_2},$$

$$(15.4)$$

where the superscript asterisk denotes the complex conjugate, and the angle brackets $\langle\ \rangle$ represent a time average. Note that the source is assumed to be spatially incoherent, which means that terms of the form $\langle E_1(l_p, m_p, t)E_2^*(l_q, m_q, t)\rangle$, where p and q denote different elements of the source, are zero. If the quantity $(R_2 - R_1)/c$ is small compared with the reciprocal receiver bandwidth, we can neglect it within the angle brackets of Eq. (15.4), where it occurs in the amplitude term for $\mathcal{E}$. Equation (15.4) then becomes

$$\langle E_1(l, m, t)\, E_2^*(l, m, t) \rangle = \frac{\langle \mathcal{E}(l, m, t)\mathcal{E}^*(l, m, t) \rangle \exp\left[j2\pi\nu(R_1 - R_2)/c\right]}{R_1 R_2}.$$

$$(15.5)$$

The quantity $\langle \mathcal{E}(l, m, t)\, \mathcal{E}^*(l, m, t) \rangle$ is a measure of the time-averaged intensity, $I(l, m)$, of the source. To obtain the mutual coherence function of the fields at points P_1 and P_2, we integrate over the source, using ds to represent an element of area within the (X, Y) plane:

$$\Gamma_{12}(u, v, 0) = \int_{\text{source}} \frac{I(l, m) \exp\left[j2\pi\nu(R_1 - R_2)/c\right]}{R_1 R_2}\, ds, \qquad (15.6)$$

where u and v are the x and y components of the spacing between the points P_1 and P_2 measured in wavelengths. Note that $(R_1 - R_2)$ is the differential distance in the path lengths from (l, m) in the source to P_1 and P_2. The points P_1 and P_2 have coordinates (x_1, y_1) and (x_2, y_2) respectively, so $u = (x_1 - x_2)\nu/c$ and $v = (y_1 - y_2)\nu/c$, where c/ν is the wavelength. Thus, we obtain $(R_2 - R_1) = (ul + vm)c/\nu$. Because the distance of the source is very much greater than the distance between P_1 and P_2, for the remaining R terms, we can put $R_1 \simeq R_2 \simeq R$, where R is the distance between the (X, Y) and (x, y) origins. Then $ds = R^2 dl\, dm$, and from Eq. (15.6),

$$\Gamma_{12}(u, v, 0) = \int\!\!\int_{\text{source}} I(l, m)\, e^{-j2\pi(ul + vm)} dl\, dm. \qquad (15.7)$$

Since the integrand in Eq. (15.7) is zero outside the source boundary, the limits of the integral effectively extend to infinity, and the mutual coherence $\Gamma_{12}(u, v, 0)$, which is equivalent to the complex visibility $\mathcal{V}(u, v)$, is the Fourier transform of the intensity distribution $I(l, m)$ of the source. This result is generally referred to as the van Cittert–Zernike theorem. However, it is instructive to examine the definition of the theorem in terms of the diffraction pattern of an aperture, given at the beginning of this section.

15.1.2   *Diffraction at an Aperture and the Response of an Antenna*

The Fraunhofer diffraction field of an aperture, as a function of angle, can be analyzed using the geometry shown in Fig. 15.1b. Here, an aperture is illuminated by an electromagnetic field of amplitude $\mathcal{E}(l, m, t)$, where again we use direction cosines with respect to the x and y axes to indicate points within the aperture as seen from P_1 and P_2. The (x, y) plane is in the far field of a wavefront from any point in the aperture, so such a wavefront can be considered plane over the distance $P_1 P_2$. The aperture is centered on the (X, Y) origin and is normal to the line from the (X, Y) origin to P_2. The phase over the aperture is assumed to be uniform, and components of the field therefore combine in phase at P_2. Thus, in the (x, y) plane, the maximum field strength occurs at P_2. Now consider the field at the point P_1, which has coordinates (x, y). The component of the field at P_1 due to radiation from an element of the aperture at position (l, m) is given by Eq. (15.2). The path lengths from the point (l, m) at the source to P_1 and P_2 are R_1 and R_2, respectively, and $R_2 - R_1 = lx + my$. Thus, from Eq. (15.2), we can write

$$
E_1(l, m, t) = \frac{e^{-j2\pi v(t-R_2/c)}}{R_1} \, \mathcal{E}\left(l, m, t - \frac{R_1}{c}\right) e^{-j2\pi v(xl+ym)/c} \, . \tag{15.8}
$$

Again, for the remaining R terms, we put $R_1 \simeq R_2 \simeq R$. Integration over the aperture then gives the total field at P_1,

$$
E(x, y) = \frac{e^{-j2\pi v(t-R/c)}}{R} \int_{\text{aperture}} \mathcal{E}\left(l, m, t - \frac{R}{c}\right) e^{-j2\pi[(x/\lambda)l+(y/\lambda)m]} ds \, , \tag{15.9}
$$

where λ is the wavelength, and the element of area ds is proportional to $dl\,dm$. The term on the right side that is outside the integral is a propagation factor that represents the variation in amplitude and phase over the path from the source to P_2 in Fig. 15.1b. In applying the result to the radiation pattern of an aperture, we replace the time-dependent functions E and $\mathcal{E}$ by the corresponding rms field amplitudes, which will be denoted by $\overline{E}$ and $\overline{\mathcal{E}}$, respectively:

$$\overline{E}(x, y) \propto \int\!\!\int_{\text{aperture}} \overline{\mathcal{E}}(l, m)\, e^{-j2\pi[(x/\lambda)l+(y/\lambda)m]} dl\, dm\ , \tag{15.10}$$

where the propagation factor in Eq. (15.9) has been omitted. A comparison of Eqs. (15.7) and (15.10) explains the van Cittert–Zernike theorem as described at the beginning of this section. With the specified proportionality between the incoherent intensity and the coherent field amplitude, it will be found that

$$\frac{\Gamma_{12}(u, v, 0)}{\Gamma_{12}(0, 0, 0)} = \frac{\overline{E}(x, y)}{\overline{E}(0, 0)}\ . \tag{15.11}$$

In Eqs. (15.7) and (15.10), the integrand is zero outside the source or aperture. Thus, in each case, the limits of integration can be extended to $\pm\infty$, and the equations are seen to be Fourier transforms. The calculations of the mutual coherence of the source and the radiation pattern of the aperture yield similar results because the geometry and the mathematical approximations are the same in each case. It should be emphasized, however, that the physical situations are different. In the first case considered, the source is spatially incoherent over its surface, whereas in the second case, the field across the aperture is fully coherent.

The result in Eq. (15.10) also gives the angular radiation pattern for an antenna that has the form of an excited aperture. The application to an antenna is more useful if the radiation pattern is specified in terms of an angular representation (l', m') of the direction of radiation from the antenna aperture instead of the position of the point P_1, and if the field distribution over the aperture is specified in terms of units of length rather than angle. (l', m') are direction cosines with respect to the (X, Y) axes. Since the angles concerned are small, we can substitute into Eq. (15.10) $x = Rl'$, $y = Rm'$, $l = X/R$, $m = Y/R$, $dl = dX/R$, and $dm = dY/R$, and obtain

$$\overline{E}'(l', m') \propto \int\!\!\int_{\text{aperture}} \overline{\mathcal{E}}_{XY}(X, Y)e^{-j2\pi[(X/\lambda)l'+(Y/\lambda)m']} dX\, dY\ . \tag{15.12}$$

This is the expression for the field distribution resulting from Fraunhofer diffraction at an aperture [see, e.g., Silver (1949)]. It includes the case of a transmitting antenna in which the aperture of a parabolic reflector is illuminated by a radiator at the focus. If such an antenna is used in reception, the received voltage from a source in direction (l', m') is proportional to the right side of Eq. (15.12). Thus, the voltage reception pattern $V_A(l', m')$, introduced in Sect. 3.3.1, is proportional to the right side of Eq. (15.12).

To obtain the power radiation pattern for an antenna, we need the response in terms of $|\overline{E}'(l', m')|^2$. From an autocorrelation theorem of Fourier transforms, the squared amplitude of $\overline{E}'(l', m')$ is equal to the autocorrelation of the Fourier transform of $\overline{E}'(l', m')$ [see, e.g., Bracewell (2000), and note that this relationship is also a generalization of the Wiener–Khinchin relationship derived in Sect. 3.2]. Thus, the power radiated as a function of angle is given by

$$|\overline{E}'(l',m')|^2 \propto$$

$$\int\int_{\text{aperture}} [\overline{\mathcal{E}}_{XY}(X, Y) \star \star \overline{\mathcal{E}}_{XY}(X, Y)] e^{-j2\pi[(X/\lambda)l'+(Y/\lambda)m']} \, dX \, dY \ ,$$

$$(15.13)$$

where $\overline{\mathcal{E}}(X, Y) \star \star \overline{\mathcal{E}}(X, Y)$ is the two-dimensional autocorrelation function of the field distribution over the aperture. To obtain absolute values of the radiated field, the required constant of proportionality can be determined by integrating Eq. (15.13) over 4π steradians to obtain the total radiated power and equating this to the power applied to the antenna terminals. In reception, the power collected by an antenna is proportional to the power radiated in transmission, so the form of the beam is identical in the two cases. To illustrate the physical interpretation of Eq. (15.13), consider the simple case of a rectangular aperture with uniform excitation of the electric field. The function $\overline{\mathcal{E}}_{XY}(X, Y)$ is then the product of two one-dimensional functions of X and Y. If d is the aperture width in the X direction, the autocorrelation function in X is triangular with a width $2d$, and Fourier transformation gives

$$|\overline{E}_X(l')|^2 \propto \left[\frac{\sin(\pi dl'/\lambda)}{\pi dl'/\lambda}\right]^2 . \qquad (15.14)$$

In the l' dimension, the full width of this beam at the half-power level is $0.886\lambda/d$, for example, $1°$ for $d/\lambda = 50.8$ wavelengths. For a uniformly illuminated circular aperture of diameter d, the response pattern is circularly symmetrical and is given by

$$|\overline{E}_r(l'_r)|^2 \propto \left[\frac{2J_1(\pi dl'_r/\lambda)}{\pi dl'_r/\lambda}\right]^2 , \qquad (15.15)$$

where the subscript r indicates a radial profile in which l'_r is measured from the center of the beam, and J_1 is the first-order Bessel function. The full width of the beam at the half-power level is $\sim 1.03\lambda/d$.

A more direct way of obtaining the Fraunhofer radiation pattern of an aperture antenna is to start by considering the field strength of the radiated wavefront as a function of direction, rather than the field strength at a single point P_1, as above. However, the method used was chosen to provide a more direct comparison with the interferometer response to a spatially incoherent source. For a more detailed analysis of the response of an antenna, see, for example, Booker and Clemmow (1950), Bracewell (1962), or the textbooks on antennas in the Further Reading of Chapter 5.

15.1.3 Assumptions in the Derivation and Application of the van Cittert–Zernike Theorem

At this point, it is convenient to collect and review the assumptions and limitations that are involved in the theory of the interferometer response.

1. *Polarization of the electric field.* Although the electric fields are vector quantities with directions that depend on the polarization of the radiation, the components received by antennas from different elements of the source can be combined in the manner of scalar quantities. The fields are measured by antennas at P_1 and P_2, and each antenna responds to the component of the radiation for which the polarization matches that of the antenna. If the fields are randomly polarized and the antennas are identically polarized, then the signal product in Eq. (15.4) represents half the total power at each antenna. However, the antenna polarizations do not have to be identical since, in general, the interferometer system will respond to some combination of components of the source intensity determined by the antenna polarizations. The ways in which the antenna polarizations can be chosen to examine all polarizations of the incident radiation are described in Sect. 4.7.2. Thus, the scalar treatment of the field involves no loss of generality.

2. *Spatial incoherence of the source.* The radiation from any point on the source is statistically independent from that from any other point. This applies almost universally to astronomical sources and permits the integration in Eq. (15.6) by allowing cross products representing different elements of the source to be omitted. The Fourier transform relationship provided by the van Cittert–Zernike theorem requires the source to be spatially incoherent. Spatial coherence and incoherence are discussed in Sect. 15.2. Note that an incoherent source gives rise to a coherent or partially coherent wavefront as its radiation propagates through space. If this were not the case, the mutual coherence (or visibility) of an incoherent source, measured by spaced antennas, would always be zero.

3. *Bandwidth pattern.* The assumption required in going from Eqs. (15.4) to (15.5), that $(R_2 - R_1)/c$ is less than the reciprocal bandwidth $(\Delta \nu)^{-1}$, can be written

$$\frac{\Delta \nu}{\nu} < \frac{1}{l_d u} , \qquad \frac{\Delta \nu}{\nu} < \frac{1}{m_d v} , \qquad (15.16)$$

where l_d and m_d are the maximum angular dimensions of the source. This is the requirement that the source be within the limits imposed by the bandwidth pattern of the interferometer, which is discussed in Sect. 2.2. Conversely, the required field of view limits the maximum bandwidth that can be used in a single receiving channel. The distortion caused by the bandwidth effect is discussed further in Sect. 6.3.1 and, if not severe, can often be corrected.

4. *Distance of the source.* For an array with maximum baseline D, the departure of the wavefront from a plane, for a source of distance R, is $\sim D^2/R$. Thus, the *far-field* distance R_{ff}, defined as that for which the divergence is small compared

with the wavelength λ, is given by

$$R_{ff} \gg D^2/\lambda \,. \tag{15.17}$$

The far-field condition implies that the antenna spacing subtends a small angle as seen from the source and results in the approximation for Fraunhofer diffraction. If the source is at a known distance closer than the far-field distance, then the phase term can be compensated. This may sometimes be necessary in solar system studies. For example, for an antenna spacing of 35 km and a wavelength of 1 cm, the far-field distance is greater than 1.2×10^{11} m, or approximately the distance to the Sun. On the other hand, the distances to sources in the near field such as Earth-orbiting satellites can be determined from measurements of the wavefront curvature (e.g., Sect. 9.11). When the source is in the far-field distance, no information concerning its structure in the line-of-sight direction is possible, only the intensity distribution as projected onto the celestial sphere. (Line-of-sight structure can be determined by modeling velocity structure.)

5. *Use of direction cosines.* In going from Eqs. (15.6) to (15.7), the path difference $(R_2 - R_1)$ is specified in terms of the baseline coordinates (u, v) and angular coordinates (l, m). The expression for the path difference is precise if l and m are specified as direction cosines. In integration over the source, the element of area bounded by increments $dl\,dm$ is equal to $dl\,dm/n$, where n is the third direction cosine and is equal to $\sqrt{1 - l^2 - m^2}$. In optics, derivation of the van Cittert–Zernike theorem usually involves the assumption that the source subtends only small angles at the measurement plane. Then l and m can be approximated by the corresponding small angles, and n can be approximated by unity. As a result, the relationship between $\mathcal{V}$ and I becomes a two-dimensional Fourier transform, as in the approximation for limited field size discussed in Sect. 3.1.1. In the radio case, the less restrictive result in Eq. (3.7) is sometimes required.

6. *Three-dimensional distribution of the visibility measurements.* As antennas track a source, the antenna-spacing vectors, designated above by (u, v) components, may not lie in a plane, and three coordinates, (u, v, w), are then required to specify them. The Fourier transform relationship is then more complicated, but a simplifying approximation can be made if the field of view to be imaged is small. These effects are discussed in Sect. 11.7.

7. *Refraction in space.* It has been implicitly assumed in the analysis above that the space between the source and the antennas is empty, or at least that any medium within it has a uniform refractive index, so that there is no distortion of the incoming wavefront from the source. However, the interstellar and interplanetary media, and the Earth's atmosphere and ionosphere, can introduce effects including rotation of the position angle of a linearly polarized component, as discussed in Chaps. 13 and 14.

15.2   Spatial Coherence

In the derivation of the interferometer response in Chaps. 2 and 3, and in Eq. (15.5), it is assumed that the source under discussion is spatially incoherent. This means that the waveforms received from different spatial elements of the source are not correlated, which enables us to add the correlator output from the different angular increments in the integration over the source. We now examine this requirement in more detail. To illustrate the principles involved, it is sufficient to work in one dimension on the sky, for which the position is given by the direction cosine l.

15.2.1   Incident Field

Consider the electric field $E(l, t)$ at the Earth's surface resulting from a wavefront incident from the direction l at time t. Figure 15.2 shows the geometry of the situation, in which $l = 0$ in the direction OS of the center, or nominal position, of the source under observation. l is a direction cosine measured from OB, the normal to OS. A path OS' is shown that indicates the direction of another part of the source. Radiation from the direction OS' produces a wavefront parallel to OB'. The wavefronts from points on the source are plane because we are considering a source in the far field of the interferometer. The line OA represents the projection of the baseline normal to the direction of the source, and the distance OA measured in wavelengths is equal to u. Now consider wavefronts from the directions S and S' that arrive at the same time at O. To reach the point A, the wavefront from S' has to travel a farther distance AA'. With the usual small-angle approximation, we find that the distance AA' is equal to ulc/v, that is, ul wavelengths. Thus, the wave from direction S' is delayed at A by a time interval $\tau = ul/v$, relative to the wave from S. If we represent the wave from direction S' by $E(l, t)$ at O, at A it is $E(l, t - \tau)$.

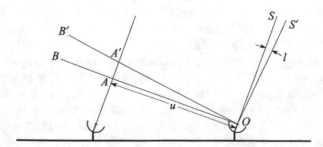

Fig. 15.2 Diagram to illustrate the variation of phase along a line OB that is perpendicular to the direction of a source OS, where l is the direction cosine that specifies the direction OS' and is defined with respect to OB. The angle SOS' is small and is thus approximately equal to l, as indicated. The line OS' points toward another part of the same source, and OB' is perpendicular to it.

Now because the incident wavefronts are plane, the amplitude of the wave does not change over the distance AA'. However, the phase changes by $\nu\tau = ul$, so for the wave from S' at A, we have

$$E(l, t - \tau) = E(l, t)\, e^{-j2\pi ul}\,. \tag{15.18}$$

If $e(u, t)$ is the field at A resulting from radiation from all parts of the source, then

$$e(u, t) = \int_{-\infty}^{\infty} E(l, t)\, e^{-j2\pi ul}\, dl\,. \tag{15.19}$$

It will be assumed that the angular dimensions of the source are not large, so also we have

$$E(l, t) = 0, \qquad\qquad |l| \geq 1\,. \tag{15.20}$$

The condition specified in Eq. (15.20) allows us to write the limits of the integral in Eq. (15.19) as $\pm\infty$. Note that Eq. (15.19) has the form of a Fourier transform, and the inverse transform gives $E(l, t)$ from $e(u, t)$. Equation (15.19) will be required in the following subsection.

15.2.2   Source Coherence

We now return to the spatial coherence of the source and follow part of a more extensive analysis by Swenson and Mathur (1968). As a measure of the spatial coherence, we introduce the *source coherence function* γ. This is defined in terms of the cross-correlation of signals received from two different directions, l_1 and l_2, at two different times:

$$
\begin{aligned}
\gamma(l_1, l_2, \tau) &= \lim_{T\to\infty} \frac{1}{2T} \int_{-T}^{T} E(l_1, t)E^*(l_2, t - \tau)\, dt \\
&= \langle E(l_1, t)E^*(l_2, t - \tau)\rangle\,.
\end{aligned}
\tag{15.21}
$$

Finite limits are used in the integral to ensure convergence. $\gamma(l_1, l_2, \tau)$ is similar to the coherence function of a source or object discussed by Drane and Parent (1962) and Beran and Parent (1964).

The *complex degree of coherence* of an extended source is the normalized source coherence function

$$\gamma_N(l_1, l_2, \tau) = \frac{\gamma(l_1, l_2, \tau)}{\sqrt{\gamma(l_1, 0)\gamma(l_2, 0)}}\,, \tag{15.22}$$

where $\gamma(l_1, \tau)$ is defined by putting $l_1 = l_2$ in Eq. (15.21), that is, $\gamma(l_1, \tau) = \gamma(l_1, l_2, \tau)$. It can be shown by using the Schwarz inequality that $0 \leq |\gamma_N(l_1, l_2, \tau)| \leq 1$. The extreme values of 0 and 1 correspond to the cases of complete incoherence and complete coherence, respectively. When dealing with extended sources of arbitrary spectral width, it is possible that, for a given pair of points l_1 and l_2, $|\gamma_N(l_1, l_2, \tau)|$ is zero for one value of τ and nonzero for another value. Therefore, more stringent definitions of complete coherence and incoherence are necessary. The following definitions are adapted from Parrent (1959):

1. The emissions from the directions l_1 and l_2 are completely coherent (incoherent) if $|\gamma_N(l_1, l_2, \tau)| = 1$ (0) for all values of τ.
2. An extended source is coherent (incoherent) if the emissions from all pairs of directions l_1, l_2 within the source are coherent (incoherent).

In all other cases, the extended source is described as partially coherent.

Consider now the coherence function of the field $e(x_\lambda, t)$ of a distant source measured, say, at the Earth's surface, x_λ being a linear coordinate measured in wavelengths in a direction normal to $l = 0$:

$$\Gamma(x_{\lambda 1}, x_{\lambda 2}, \tau) = \lim_{T \to \infty} \frac{1}{2T} \int_{-T}^{T} e(x_{\lambda 1}, t) e^*(x_{\lambda 2}, t - \tau)\, dt$$

$$= \langle e(x_{\lambda 1}, t) e^*(x_{\lambda 2}, t - \tau) \rangle .$$ (15.23)

This is a variation of the mutual coherence function Γ_{12} in Eq. (15.1), in which the positions of the measurement points defined by $\dot{x}_{\lambda 1}$ and $x_{\lambda 2}$ are retained, rather than just the relative positions given by the baseline components. By using the Fourier transform relationship between $E(l, t)$ and $e(u, t)$ derived in Eq. (15.19), and replacing u by x_λ, we obtain

$$\Gamma(x_{\lambda 1}, x_{\lambda 2}, \tau) = \int_{-\infty}^{\infty} \int_{-\infty}^{\infty} \gamma(l_1, l_2, \tau) e^{-j2\pi(x_{\lambda 1}l_1 - x_{\lambda 2}l_2)}\, dl_1\, dl_2 ,$$ (15.24)

and the inverse transform, which is

$$\gamma(l_1, l_2, \tau) = \int_{-\infty}^{\infty} \int_{-\infty}^{\infty} \Gamma(x_{\lambda 1}, x_{\lambda 2}, \tau)\, e^{j2\pi(x_{\lambda 1}l_1 - x_{\lambda 2}l_2)}\, dx_{\lambda 1}\, dx_{\lambda 2} .$$ (15.25)

The relationships in Eqs. (15.24) and (15.25) do not provide a means of measuring the intensity distribution of a source, except in the case of complete incoherence. For complete incoherence, the coherence function can be expressed as

$$\gamma(l_1, l_2, \tau) = \gamma(l_1, \tau)\, \delta(l_1 - l_2) ,$$ (15.26)

where δ is the delta function. Using the relation in Eq. (15.26) in conjunction with Eqs. (15.24) and (15.25), we find that the self-coherence function of a completely incoherent source and its spatial frequency spectrum are Fourier transforms of each other:

$$\Gamma(u, \tau) = \int_{-\infty}^{\infty} \gamma(l, \tau)\, e^{-j2\pi u l}\, dl \tag{15.27}$$

$$\gamma(l, \tau) = \int_{-\infty}^{\infty} \Gamma(u, \tau)\, e^{j2\pi u l}\, du\,, \tag{15.28}$$

where $u = x_{\lambda 1} - x_{\lambda 2}$. It is clear that $\Gamma(u, \tau)$ is independent of $x_{\lambda 1}$ and $x_{\lambda 2}$ and depends only on their difference. Thus, u can be interpreted as the spacing of two sample points between which the coherence of the field is measured, and also as the spatial frequency of the visibility measured over the same baseline. For $\tau = 0$, from Eqs. (15.21) and (15.26), we obtain

$$\gamma(l, 0) = \langle |E(l)|^2 \rangle\,, \tag{15.29}$$

which is the one-dimensional intensity distribution of the source, I_1, introduced in Eq. (1.10). Then from Eqs. (15.27) and (15.29),

$$\Gamma(u, 0) = \int_{-\infty}^{\infty} \langle |E(l)|^2 \rangle\, e^{-j2\pi u l}\, dl\,. \tag{15.30}$$

$\Gamma(u, 0)$ is measured between points along a line normal to the direction $l = 0$. As measured with an interferometer, it is also the complex visibility $\mathcal{V}$. Eq. (15.30) is the Fourier transform relationship between mutual coherence (visibility) and intensity.

When the incoherence condition in Eq. (15.26) is introduced into Eqs. (15.24) and (15.25), two results appear: the van Cittert–Zernike relation between mutual coherence and intensity, and the stationarity of the mutual coherence with respect to u. The physical reason underlying these results is seen in Fig. 15.2. When the wavefronts incident at different angles combine at any point, the relative phases of their (Fourier) frequency components vary linearly with the position of the point (e.g., the position of A along the line OB in Fig. 15.2), and for small l, they also vary linearly with the angle on the sky. As a result, the phase differences of the Fourier components at two points depend only on the relative positions of the points, not their absolute positions. Interferometer measurements of mutual coherence incorporate the phase differences for a range of angles of incidence governed by the angular dimensions of the source and the width of the antenna beams. The linear relationship between phase and position angle allows us to recover the angular distribution of the incident wave intensity from the variation of the mutual coherence as a function of u, by Fourier analysis. If the angular width of the source is small enough that the distance AA' in Fig. 15.2 is always much less than the wavelength, then the form of the electric field remains constant along the line OA, and the source is not resolved.

15.2.3   Completely Coherent Source

Parrent (1959) has shown that an extended source can be completely coherent only if it is monochromatic. As examples of such a source, one may visualize the aperture of a distant, large antenna, or an ensemble of radiating elements all driven by the same monochromatic signal. The aperture considered in Sect. 15.1.2 is a conceptual example of a coherent source. The difference between the responses of an interferometer to a fully coherent source and to a fully incoherent one can be explained by the following physical picture. The source can be envisioned as an ensemble of radiators distributed over a solid angle on the sky. In the case of a coherent source, the signals from the radiators are monochromatic and coherent. The radiation in any direction combines into a single monochromatic wavefront and produces a monochromatic signal in each antenna of an interferometer. The output of the correlator is directly proportional to the product of the two (complex) signal amplitudes from the antennas. Thus, if a coherent source is observed with n_a antennas, the $n_a(n_a - 1)/2$ pairwise cross-correlations of the signals that are measured can be factored into n_a values of complex signal amplitude.

In contrast, for an incoherent source, the outputs from radiating elements are uncorrelated and must be considered independently. Each one produces a component of the fringe pattern in the correlator output. But since the phases of these fringe components depend on the positions of the radiators within the source, the combined response is proportional not only to the signal amplitudes at the antennas but also to a factor that depends on the angular distribution of the radiators. This factor, of magnitude ≤ 1, is equal to the modulus of the visibility normalized to unity for an unresolved (point) source of flux density equal to that of the source under observation. Unless the source is unresolved, it is not possible to factor the measured cross-correlations into signal amplitude values at the antennas. Because the emissions of the radiating elements of a source are uncorrelated, the information on the source distribution is preserved in the ensemble of wavefronts they produce at the antennas.

As shown by the derivation of the angular dependence of the radiation from a coherently illuminated aperture [Eq. (15.12)], and suggested by the analogy with a large antenna, the radiation from a coherent source is highly directional. Thus, the signal strengths observed depend on the absolute positions of the two antennas of an interferometer, as in Eqs. (15.24) and (15.25), not only on their relative positions, as is the case for an incoherent source. The ability to factor the signal outputs from a series of baselines, and the nonstationarity of the correlator output measurements with the absolute positions of the antennas, are two characteristics that could allow a coherent source to be recognized (MacPhie 1964). From the analysis in Sect. 15.1, it is clear that a similar range of antenna spacings is required to resolve an incoherent source or to explore the radiation pattern of a coherent source of the same angular size.

15.3   Scattering and the Propagation of Coherence

It is well known that optical telescope images of single stars made with exposure times short compared with the timescale of atmospheric scintillation exhibit multiple stellar images (see Sect. 17.6.4). These images result from the scattering of light from the star by irregularities in the Earth's atmosphere. Something closely analogous to this occurs in the case of imaging of an unresolved radio source through a medium with strong irregular scattering, such as the interplanetary medium within a few degrees of the Sun, as described in Sect. 14.3. Since each scattered image results from the emission of the same source, one is led to expect that such a situation would simulate the effect of a distribution of coherent point sources. In this section, we examine the effects of scattering by considering the propagation of coherence in space, following in part a discussion by Cornwell et al. (1989). This formalism suggests methods for the recovery of the unscattered image from the observed image.

Given a radiating surface, we wish to know the mutual coherence function on another (possibly virtual) surface in space. In the typical radio astronomy situation, a number of simplifying assumptions can be made about the geometry of the problem. Consider the situation illustrated in Fig. 15.3, in which narrowband radio waves propagate from surface S to surface Q. The mutual coherence of two points in space is the expectation of the product of the (copolarized) electric fields at the two points.

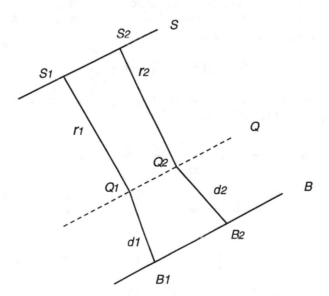

Fig. 15.3 Simplified geometry for examining the propagation of coherence. S represents an extended source, Q is the location of a scattering screen, and B is the measurement plane. Surfaces S, Q, and B are plane and parallel, and r_1, r_2, d_1, and d_2 are much greater than the wavelength. All rays are nearly (but not necessarily exactly) perpendicular to the surfaces.

For signals correlated with arbitrary time delay, the mutual coherence is

$$\Gamma(Q_1, Q_2, \tau) = \langle E(Q_1, t)E^*(Q_2, t - \tau)\rangle . \tag{15.31}$$

The mutual coherence function Γ is a function of the field at two points and the time difference τ. We consider the propagation of *mutual intensity*, that is, the mutual coherence evaluated for $\tau = 0$. Following common practice, we represent the mutual intensity by $J(Q_1, Q_2) \equiv \Gamma(Q_1, Q_2, 0)$. J will be subscripted by S, Q, or B to indicate the corresponding plane (Fig. 15.3) of the mutual intensity value. We assume that the emitting surface is completely incoherent, as is usually the case for astronomical objects, and that the observed radiation is restricted to a narrow band of frequencies, as dictated by the characteristics of the receiving system. From Eq. (15.31) and the Huygens–Fresnel formulation of radiation, it can be shown (Born and Wolf 1999, Goodman 1985), by a calculation similar to the one used in deriving Eq. (15.6), that the mutual intensity for points Q_1 and Q_2 is

$$J_Q(Q_1, Q_2) = \lambda^{-2} \int \int_S J_S(S_1, S_2) \frac{\exp[-j2\pi(r_1 - r_2)/\lambda]}{r_1 r_2} dS_1 \, dS_2 , \tag{15.32}$$

where $dS_1 \, dS_2$ is a surface element of S, and λ is the wavelength at the center of the observed frequency band.

The condition of incoherence can be represented by the use of a delta function (Beran and Parrent 1964), as in Eq. (15.26). Here, the mutual intensity is represented by a delta function, and thus, the intensity distribution on the surface Q is found by allowing points Q_1 and Q_2 to merge:

$$J_S(S_1, S_2) = \lambda^2 I(S_1) \, \delta(S_1 - S_2) , \tag{15.33}$$

where the factor λ^2 has been included to preserve the physical dimension of intensity. Equation (15.32) then becomes

$$J_Q(Q_1, Q_2) = \int_S I(S_1) \frac{\exp[-j2\pi(r_1 - r_2)/\lambda]}{r_1 r_2} dS . \tag{15.34}$$

When the angular dimension of the source is infinitesimal, that is, when the source is unresolved, the integration over the source becomes trivial, and the mutual intensity can be factored into terms depending, respectively, on r_1 and r_2:

$$J_Q(Q_1, Q_2) = I(S) \left(\frac{\exp(-j2\pi r_1/\lambda)}{r_1} \right) \left(\frac{\exp(j2\pi r_2/\lambda)}{r_2} \right) , \tag{15.35}$$

where r_1 and r_2 now originate at a single point S. In the more general case of a resolved source, Eq. (15.34) cannot be factored. Equations (15.34) and (15.35) describe for their respective cases the propagation of mutual coherence in situations subject to the constraints of Fig. 15.3 and thus can be used to determine the mutual intensity on surface Q resulting from incoherent radiation from surface S. Examination of Eq. (15.31) reveals that, for the extended source S, the mutual intensity on Q depends on both r_1 and r_2 for all pairs of points on Q. Thus, the field at Q is at least partially coherent for all sources, including those of finite extent. This is intuitively reasonable, as all points on Q are illuminated by all points on S. In fact, it can be demonstrated rigorously that an incoherent field cannot exist in free space (Parrent 1959).

Suppose now that we have a situation in which the surface Q is actually a screen of irregularities in the transmission medium, such as plasma or dust, which scatters the radiation from S. The mutual intensity incident on the screen is modified by a complex transmission factor $T(Q)$ to produce the transmitted mutual intensity

$$J_{Qt}(Q_1, Q_2) = T(Q_1)T^*(Q_2)J_{Qi}(Q_1, Q_2) \,, \tag{15.36}$$

where subscripts i and t indicate the incident and transmitted mutual intensity, respectively. From Eq. (15.34), we now define a "propagator" (Cornwell et al. 1989) for mutual intensity:

$$W(S, B) = \int_S \frac{T(Q)\exp[-j2\pi(r+d)/\lambda]}{rd}\, dS \,, \tag{15.37}$$

where r and d are defined in Fig. 15.3. Then the mutual intensity on surface B is given, in terms of the mutual intensity of an extended source S, by

$$J_B(B_1, B_2) = \lambda^{-4} \int\int_S J_S(S_1, S_2)W(S_1, B_1)W^*(S_2, B_2)\, dS_1\, dS_2 \,. \tag{15.38}$$

For an incoherent extended source,

$$J_B(B_1, B_2) = \lambda^{-2} \int_S I(S)W(S, B_1)W^*(S, B_2)\, dS \,, \tag{15.39}$$

and for a point source of flux density F, the mutual intensity on B becomes

$$J_B(B_1, B_2) = F\lambda^{-2}W(S, B_1)W^*(S, B_2) \,. \tag{15.40}$$

Again, for the unresolved source, the mutual intensity on B consists of two factors, each depending only on one position on B. However, for an extended incoherent source distribution on S, the mutual intensity depends on *differences* in position and therefore cannot be factored.

The existence of a scattering screen between a source and an observer, with an instrument of limited aperture, raises the possibility of greatly increased angular resolution resulting from the much larger extent of the scattering screen. The partial coherence of radiation from the screen requires that the intensity be measured at all points on the measurement plane B, spaced as dictated by the Nyquist criterion, rather than at all points in the spatial frequency spectrum as allowed by the van Cittert–Zernike theorem. The former observing mode results in very much more data than does the latter. In two spatial dimensions, a large redundancy of data results, so that in principle, not only can the scattering screen be characterized, but the source as well. In this respect, the problem is similar to that of self-calibration (Sect. 11.3.2). Unfortunately, in the case of the scattering screen, the practical difficulties of such observations are enormous, and few significant attempts have been made to apply the principle. Cornwell and Narayan (1993) discuss the possibilities of statistical image synthesis using scattering to obtain ultrafine resolution in a manner somewhat analogous to speckle imaging (see Sect. 17.6.4).

Emission from a radio source that undergoes strong scattering during propagation through space has been investigated by Anantharamaiah et al. (1989), and Cornwell et al. (1989). To demonstrate the response of a radio telescope to such a spatially coherent source distribution, they observed the strong and essentially pointlike source 3C279, which passes close to the Sun each year. Under these conditions, the scattering is strong enough to cause amplitude scintillation of the received signals. Anantharamaiah and colleagues used the VLA in its most extended configuration, for which the longest baselines are approximately 35 km. The velocity of the solar wind, of order 100–400 km s^{-1}, causes irregularities to sweep across the array in $\sim$100 ms, so it was necessary to make snapshot observations of duration 10–40 ms to avoid smearing of the image by the movement of the scattering screen. Observations were made at wavelengths of 20, 6, and 2 cm, with the source at angular distances of 0.9° to 5° from the Sun. It was found that the correlator output values could be factored as expected for a coherent source. When correlated signals were averaged for about 6 s, an enlarged image of the source was obtained, and the enlargement increased as the distance from the Sun decreased. It was also demonstrated that it would be possible to determine the characteristics of the scattering screen by measuring the mutual intensity function on the ground, provided that the latter is measured completely in the two-dimensional spatial frequency domain. It is not possible to distinguish between a spatially coherent extended source and a scattering screen illuminated by a point source.

A significant observation was made by Wolszczan and Cordes (1987), who were able to infer the dimensions of structure within pulsar PSR 1237+25 from an occurrence of interstellar scattering. The pulsar was observed with a single antenna, the 308-m-diameter spherical reflector at Arecibo, at a frequency of 430 MHz. Dynamic spectra of the received signal (i.e., the received power displayed as a function of both time and frequency) showed prominent band structure with maxima separated by $\sim$ 300–700 kHz in frequency. This was interpreted in terms of a thin-screen model of the interstellar medium, in which refraction of rays from the pulsar occurred at two separated points in the screen. The analysis of such a model is complicated by the occurrence of both diffractive and refractive scattering, resulting from structure smaller and larger than the Fresnel scale, respectively (Cordes et al. 1986). The refraction gave rise to two images of the source at the radio telescope, resulting in fringes in the intensity of the received signal. The distance of the pulsar (0.33 kpc) and its transverse velocity (178 km s^{-1}) were known from other observations, and the distance of the screen was taken to be half the distance of the pulsar. It was deduced that the angular separation of the images was ~ 3.3 mas, corresponding to a spacing of ~1 AU (astronomical unit) between the refracting structures. In effect, the refracting structures constitute a two-element interferometer, with fringe spacing ~ 1 μas. For comparison, the angular resolution of a baseline equal to the diameter of the Earth at 430 MHz would be 44 mas. The particular conditions that resulted in this observation lasted for at least 19 days, and during that period, observations of other pulsars did not show similar scattering. This strongly suggests that the observed phenomenon resulted from a fortuitous configuration of the interstellar medium in the direction of the pulsar.

Apart from cases of scattering such as that described, there are essentially no clear cases of spatially coherent astronomical sources, although coherent mechanisms may occur in pulsars and masers (Verschuur and Kellermann 1988). Fully coherent sources are not amenable to synthesis imaging using the van Cittert–Zernike principle and thus do not fall within the area of principal concern of this book. Further material on coherence and partial coherence can be found, for example, in Beran and Parrent (1964), Born and Wolf (1999), Drane and Parrent (1962), Mandel and Wolf (1965, 1995), MacPhie (1964), and Goodman (1985).

References

Anantharamaiah, K.R., Cornwell, T.J., and Narayan, R., Synthesis Imaging of Spatially Coherent Objects, in *Synthesis Imaging in Radio Astronomy*, Perley, R.A., Schwab, F.R., and Bridle, A.H., Eds., Astron. Soc. Pacific Conf. Ser., **6**, 415–430 (1989)

Beran, M.J., and Parrent Jr., G.B., *Theory of Partial Coherence*, Prentice-Hall, Englewood Cliffs, NJ, 1964; repr. by Society of Photo-Optical Instrumentation Engineers, Bellingham, WA (1974)

Booker, H.G., and Clemmow, P.C., The Concept of an Angular Spectrum of Plane Waves, and Its Relation to That of Polar Diagram and Aperture Distribution, *Proc. IEE*, **97**, 11–17 (1950)

Born, M., and Wolf, E., *Principles of Optics*, 7th ed., Cambridge Univ. Press, Cambridge, UK (1999)

Bracewell R.N., Radio Interferometry of Discrete Sources, *Proc. IEEE*, **46**, 97–105 (1958)

Bracewell, R.N., Radio Astronomy Techniques, in *Handbuch der Physik*, Vol. 54, Flugge, S., Ed., Springer-Verlag, Berlin (1962), pp. 42–129.

Bracewell, R.N., *The Fourier Transform and Its Applications*, McGraw-Hill, New York, 2000 (earlier eds. 1965, 1978)

Cordes, J.M., Pidwerbetsky, A., and Lovelace, R.V.E., Refractive and Diffractive Scattering in the Interstellar Medium, *Astrophys. J.*, **310**, 737–767 (1986)

Cornwell, T.J., Anantharamaiah, K.R., and Narayan, R., Propagation of Coherence in Scattering: An Experiment Using Interplanetary Scintillation, *J. Opt. Soc. Am.*, **6A**, 977–986 (1989)

Cornwell, T.J., and Narayan, R., Imaging with Ultra-Resolution in the Presence of Strong Scattering, *Astrophys. J. Lett.*, **408**, L69–L72 (1993)

Drane, C.J., and Parrent Jr., G.B., On the Mapping of Extended Sources with Nonlinear Correlation Antennas, *IRE Trans. Antennas Propag.*, **AP-10**, 126–130 (1962)

Goodman, J.W., *Statistical Optics*, Wiley, New York (1985)

MacPhie, R.H., On the Mapping by a Cross Correlation Antenna System of Partially Coherent Radio Sources, *IEEE Trans. Antennas Propag.*, **AP-12**, 118–124 (1964)

Mandel, L., and Wolf, E., Coherence Properties of Optical Fields, *Rev. Mod. Phys.*, **37**, 231–287 (1965)

Mandel, L., and Wolf, E., *Optical Coherence and Quantum Optics*, Cambridge Univ. Press, Cambridge, UK (1995)

Parrent, Jr. G.B., Studies in the Theory of Partial Coherence, *Opt. Acta*, **6**, 285–296 (1959)

Silver, S., *Microwave Antenna Theory and Design*, Radiation Laboratory Series, Vol. 12, McGraw-Hill, New York (1949), p. 174

Swenson, Jr. G.W., and Mathur, N.C., The Interferometer in Radio Astronomy, *Proc. IEEE*, **56**, 2114–2130 (1968)

Verschuur, G.L., and Kellermann, K.I., Eds., *Galactic and Extragalactic Astronomy*, Springer-Verlag, New York (1988)

Wolszczan, A., and Cordes, J.M., Interstellar Interferometry of the Pulsar PSR 1237+25, *Astrophys. J. Lett.*, **320**, L35–L39 (1987)

Zernike, F., Concept of Degree of Coherence and Its Application to Optical Problems, *Physica*, **5**, 785–795 (1938)

Chapter 16
Radio Frequency Interference

A basic requirement of radio astronomy is access to a spectrum in which observations can be made without detrimental interference from transmissions by other services. In the early years of radio astronomy, when most of the radio astronomy bands below a few GHz were allocated, bandwidths of radio astronomy systems were generally no greater than a few MHz, and the comparable allocated bandwidths largely sufficed. Some allocations were made for radio lines, most importantly the hydrogen (H1) line, for which 1420–1427 MHz was reserved. In the following decades, as radio astronomy at frequencies in the range of tens of GHz developed, bandwidths of order 1 GHz were allocated, and later, a substantial fraction of the spectrum above ~ 100 GHz was allocated to radio astronomy. However, spillover of radiation from transmitting services into radio astronomy bands occurs, and generally it has been necessary to choose observatory sites in radio-quiet areas of low population density and to take advantage of terrain shielding where possible. These considerations have led to the choice of sites in South Africa and Western Australia for international development of several of the largest arrays. Also, with the increase in computing capacity at observatories, detection and removal of interfering signals in astronomical observations have become important parts of data analysis. In particular, digital analysis allows the received bandwidths to be divided into as many as 10^6 spectral channels, which allows those containing interference to be identified and removed. A general discussion of interference in radio astronomy is given by Baan (2010).

The most serious interference usually results from intentional radio radiation such as those used for transmission of information in many forms, or for radio location, etc. Interference can also occur as a result of unintentional emissions from electrical machinery and industrial processes such as welding. Such emissions often take the form of trains of short electromagnetic pulses. For example, a rectangular

© The Author(s) 2017
A.R. Thompson, J.M. Moran, and G.W. Swenson Jr., *Interferometry and Synthesis in Radio Astronomy*, Astronomy and Astrophysics Library,
DOI 10.1007/978-3-319-44431-4_16

pulse of width δt has a power spectrum

$$P(v) \sim \left[\frac{\sin \pi v \delta t}{\pi v \delta t}\right]^2 . \tag{16.1}$$

Most of the power is contained in the frequency range between DC and $1/\delta t$. However, the envelope of the power spectrum decreases only as v^{-2}, and hence, such interference can be a problem at frequencies much higher than $1/\delta t$. Electromagnetic interference (EMI) of this type is usually most serious at frequencies below a few GHz. A detailed analysis of radiation produced by sparks in power lines was developed by Beasley (1970). To avoid such interference, sites for radio astronomy observatories are located in relatively undeveloped areas, and proximity to industries and major highways is avoided. Necessary vehicles and machinery on observatory sites are generally fitted with filtering components that strongly reduce unwanted emissions. Shielding of electronic equipment is very important.

16.1   Detection of Interference

A basic problem is the identification of the contaminated data. In the simplest case, this is a matter of examining the output of a correlator or detector and deleting data in which the signal amplitude is larger than expected or does not vary with time or antenna pointing in the manner expected for astronomical sources. In earlier times, interference removal sometimes meant losing the whole bandwidth received, but as mentioned above, use of multichannel spectral processing permits deletion of only the contaminated channels. The greatest difficulty is the detection of weak interference. Use of channel bandwidths comparable to the bandwidth of an interfering transmitter also has the advantage of maximizing the interference-to-noise ratio, thus improving the detectability.

When inspecting data for variations that indicate the presence of interference, data averaging times of seconds or minutes are often appropriate when the interference varies on similar timescales. However, the astronomical measurements may require averaging of data over periods of many hours to obtain the required sensitivity, so interference at levels that can introduce errors in the data may be too weak relative to the noise to be easily detected. With the high data output rates produced by large synthesis arrays, it is impractical to examine all of the data manually, and algorithms by which contaminated data can be flagged by computer are important. Methods of dealing with radio interference include (1) those that simply delete receiver output data that are believed to be contaminated by interference; (2) those that cancel or reduce the interference without removing the astronomical data that occur at the same time and frequency; and (3) those that involve spatial filtering, in which a null is generated in the antenna reception pattern in the direction of an interferer.

The common characteristic of interfering signals is that, in general, their sources do not move relative to the antennas at the sidereal rates of astronomical objects.

When the effects of the sidereal motion of the astronomical target are removed from the correlated data, the fringe frequency variations are transferred to any extraneous signals, and thus the unwanted signals can be identified by the fringe rate variations in their phases. For long baselines where the fringe rates are high, the interference will be attenuated by the subsequent time averaging of the data (e.g., Perley 2002). However, in many cases, some further analysis is required to remove the effects of the interference, as considered in Sect. 16.2. Athreya (2009) describes the application to the Giant Metrewave Radio Telescope array in India.

Some examples of techniques to detect the presence of interference are as follows. For a general overview, see also Fridman and Baan (2001), Briggs and Kocz (2005), and Baan (2010).

1. Use of monitoring receivers with antennas pointed toward likely sources of interference, such as toward the horizon for terrestrial transmitters (e.g., Rogers et al. 2005).
2. Comparison of data taken simultaneously at two observatories that are sufficiently widely separated that interference from any transmitter is unlikely to be received at both. This has been used in the search for pulses and other transient astronomical emissions (e.g., Bhat et al. 2005).
3. Detection of transmissions with cyclostationary characteristics, that is, transmission in which some characteristic repeats at intervals τ_c in time. Examples are the frame cycles in TV signals and repeated data cycles in GPS (global positioning system) transmissions. Values of τ_c can be determined for the expected signal environment. The occurrence of components of the data with cyclostationary characteristics can be investigated by performing an autocorrelation and looking for features that repeat at intervals τ_c. Bretteil and Weber (2005) find that searching for a Fourier component at frequency $1/\tau_c$ in the data is an advantageous method.
4. Use of the closure amplitude relationships [Eq. (10.44)] can provide an indication of interference. If observations are made of a point source in the absence of interference, the visibility ratios on the right side of the equation are equal to unity, and the values of r_{ij} on the left side are proportional to the corresponding main-beam gains. A signal from an interfering source will add a component to the output, the magnitude of which will depend upon the sidelobe gains of the antennas, which will vary with time as the antennas track. Thus, variation of the closure gain values is an indication of possible interference. Note, however, that the correction for the fringe frequency of the target source will cause variation of the same frequency in the response to interference from a stationary transmitter, so the effect of the interference will decrease (i.e., the interference threshold will increase) with an increase of the baseline component u, as discussed in Sect. 16.3.2.
5. In the case of interference from radar pulses, the individual pulses are sometimes strong enough to be seen by looking at the output of a detector, especially if the transmitter is close enough that a direct signal path exists. It may be possible to determine the timing pattern of the pulses and generate blanking pulses for the radio astronomy receivers. The situation may be such that it is necessary to

extend the blanking to include reflections from nearby aircraft, etc. See, e.g., the discussion by Dong et al. (2005). A buffer memory for the data allows blanking to begin just before the pulse is detected, to ensure effective removal.

6. An interesting object lesson is the identification of solitary radio bursts of millisecond duration with swept frequency vs. time characteristics similar to those of pulsars. These events were known as perytons. The radiation entered multiple channels of the telescope's multibeam receivers, showed peculiar kinks in the time frequency dependence, and preferentially occurred at midmorning. These characteristics led to the source: prematurely opened doors of microwave ovens at the observatory (Petroff et al. 2015).

7. Examination of the statistics of the receiver output data. Kurtosis is defined as μ_4/μ_2^2, where μ_2 and μ_4 are the mean values of the second and fourth powers of the data with respect to the mean value. For Gaussian noise, kurtosis has a value of 3.0, and other values are an indication of non-Gaussian data (Nita et al. 2007; Nita and Gary 2010).

8. With high-resolution multichannel receivers, interference can be detected and excised from deviant channels (e.g., Leshem et al. 2000).

16.1.1   Low-Frequency Radio Environment

The LOw Frequency ARray (LOFAR array; see Sect. 5.7.1), which is located in the Netherlands with long baseline extensions in other countries, covers the frequency ranges 10–80 and 110–240 MHz, thus avoiding the FM broadcast band. Discussions of the problem of radio interference in these frequency ranges are given by Boonstra and van der Tol (2005) and Offringa et al. (2013). The latter provides a detailed examination of the radio environment in the 30–78 MHz and 115–163 MHz ranges. For these measurements, the received signals were split into 512 sub-bands of 195 kHz width. The spectral resolution of the data was 0.76 kHz. Offringa et al. (2013) found that the interference occupancy was,1.8% for the lower band and 3.2% for the higher one. They concluded that these levels of narrowband interference should not significantly restrict astronomical observations, but that it is important that the frequency range of LOFAR remain free of broadband interference. A similar analysis has been carried out for the Murchison Widefield Array in Western Australia (Offringa et al. 2015).

16.2   Removal of Interference

When possible, mitigation of interference by cancellation, leaving the astronomical data intact, is clearly preferable to total deletion of the corrupted data. Cancellation requires not only detection but also an accurate estimation of the interfering signal in order to remove it. In *adaptive* cancellation [see, e.g., Barnbaum and Bradley (1998)], a separate antenna (usually smaller than the astronomy antenna) is pointed toward the interferer. The received signal from this antenna is digitized, passed through an adaptive filter, and combined with the signal from the astronomy antenna. The combined outputs are processed by an algorithm that provides a control of the adaptive filter in such a way as to cause the interfering signal voltages from the two antennas to cancel each other. Of various algorithms that could be used to control the adaptive filter, Barnbaum and Bradley used a least-mean-squares algorithm, which is computationally simple and thus easily adapted to follow the relative variation of the astronomical and interfering signals as the astronomy antenna tracks. All of this takes place before the signals reach the correlator or detector. Briggs and Kocz (2005) give an example of an interference cancellation scheme in which the outputs of the astronomy and interference antennas are cross-correlated to provide a control for the adaptive filter. For a detailed discussion of methods involving cross-correlation of astronomy antenna outputs with those of axillary antennas directed toward the source of interference, see Briggs et al. (2000). In some cases in which the structure of the interfering signal is known in detail, as in the case of the GLONASS (Global Navigation Satellite System) navigational satellite signals, it is possible to recreate the interfering signal from the interference received in the astronomy antenna with sufficient accuracy for cancellation. In an example discussed by Ellingson et al. (2001), interference from GLONASS was reduced by 20 dB.

16.2.1   Nulling for Attenuation of Interfering Signals

Spatial nulling involves using a group of antennas in which a null in the combined spatial response is formed in the direction of the source of interference. In low-frequency arrays in which the individual receiving elements are dipoles with beams covering much of the sky, such nulling may also result in loss of astronomical sky coverage.

In *deterministic* nulling, the direction of the interferer is known, and a null is formed in that direction by weighting the signals received. Weighting factors (in amplitude and phase) can be applied to the signals from individual antennas if they are being combined, as in a phased array, or to the correlated products from antenna pairs before they are combined to form an image. It is not necessary to be able to identify the interference within the received signal, but if the angular responses in the direction of the null differ from one antenna to another, it is necessary to calibrate the antenna responses, which may not be practicable for the far sidelobes. Deterministic nulling can be applied to a synthesis array in two ways. First, the nulls can be formed by adjusting the weights with which the cross products of the outputs of pairs of antennas (the visibility values) are combined. In this case, the nulls are formed in the synthesized beam pattern, i.e., most likely in the sidelobes of the synthesized beam unless the direction of the interferer is within the synthesized field. Second, in the case of synthesis arrays in which the elements between which cross-correlations are formed are themselves phased subarrays of antennas, the nulls can be formed in the subarray beams. In this case, the weighting is applied directly to the signals from the individual antennas.

16.2.2   Further Considerations of Deterministic Nulling

Consider an array of n nominally identical antennas, each of which is connected through a phase shifter to an n-to-1 power combiner. Each picks up a power level p from an interfering signal. In the power combiner, the power is divided n ways between the other antennas and the output. Thus, each antenna contributes a power level p/n to the combiner output. The voltage contributions of the antennas can be represented by vectors of amplitude $\sqrt{p/n}$. If the phase shifters are adjusted so that the contributions combine in phase, the vectors are aligned and the output voltage is $\sqrt{np}$. The output power is np, as expected, since the total collecting area is n times that of a single antenna. Now suppose that the phase shifters are set so that the signal vectors combine with random phase angles. The combined voltage has an expectation of $\sqrt{n}$ times that of a single antenna. Thus, the expectation of the combined power received is equal to that from a single antenna, p (see Sect. 9.9 for a related discussion of phasing of arrays). Finally, consider the case in which the phases are adjusted so that the vectors form a closed loop with zero resultant, thus producing a null in the direction of incidence of the signal. If each signal vector has a random error in amplitude and phase of relative rms amplitude ϵ, then the vector sum will fail to close by an amount equal to the sum of the errors, i.e., $\sim \epsilon \sqrt{p}$, resulting in a power level of $\epsilon^2 p$. Thus, a null of depth x dB below the response of a single antenna requires $\epsilon = 10^{-x/20}$, e.g., $\epsilon = 0.03$ for a null depth of 30 dB. These requirements on the accuracy of the voltage responses apply to the interference components identified in adaptive nulling and to the accuracy of the antenna responses in deterministic nulling. In closing the vector loop for a null, the shape of the loop is not constrained, so free parameters remain for forming beams or nulls in other directions.

In forming a null in a given direction, one can start by determining the complex gain factors required to close the vector loop on the assumption that the antennas are all ideal isotropic radiators. Then, to take account of the actual gain of the antennas, each signal vector has to be multiplied by a further complex gain factor. If this second gain factor is the same for each antenna, the size and orientation of the vector loop may be changed, but it will remain closed. Thus, the response factor in the direction of the interferer need not be known so long as it is identical for all antennas. If the gain factor differs from one antenna to another, as is likely to be the case for signals received through far sidelobes, the loop will not close unless the individual gain factors are known and taken into account. The need to calibrate the far sidelobes of a high-gain reflector antenna over a large fraction of 4π steradians, and perhaps also as a function of frequency across receiver bandwidth, limits the usefulness of deterministic nulling in such cases. Deterministic nulling is discussed by Smolders and Hampson (2002); Ellingson and Hampson (2002); Ellingson and Cazemier (2003); Raza et al. (2002); and van der Tol and van der Veen (2005).

16.2.3   *Adaptive Nulling in the Synthesized Beam*

A way of removing the effects of an interfering signal is to place a null in the reception pattern of an array of antennas in the direction of incidence of the interference. This can be done in software and is referred to as adaptive nulling. In general, the direction of incidence of the interference is not known and must be deduced from the observations. Adaptive nulling, in which the system reacts automatically to an interfering signal, generally requires that the interference is not too strong and originates from a single source. It can also result in a significant computing load. Details can be found in Leshem and van der Veen (2000); Ellingson and Hampson (2002); Raza et al. (2002); and van der Tol and van der Veen (2005).

16.3   Estimation of Harmful Thresholds

In the efforts to obtain protection for radio astronomy observations in the systems of frequency regulation within the International Telecommunication Union (ITU), and also within the regulatory systems of individual nations, it has been essential for radio astronomers to provide quantitative estimates of the threshold levels of signal power that are harmful to astronomical observations. These vary with frequency and also with the type of radio telescopes involved. This section is concerned with the estimation of these harmful thresholds, particularly for the frequency bands allocated to radio astronomy.

The ultimate limit on the sensitivity of a radio telescope is set by the system noise, and an interfering signal can generally be tolerated if its contribution to the output image is small compared with the noise fluctuations. A response to interference of one-tenth of the rms level of the noise in the measurements is a useful criterion in interference threshold calculations. The corresponding flux density of such a signal can be calculated if the effective collecting area of the antenna, in the direction of the interference, is known. Except at the longer wavelengths, radio astronomy antennas usually have narrow beams, and the probability of the interfering signal being received in the main beam or nearby sidelobes is low, especially if the interfering transmitter is ground based. Thus, we assume here that interference usually enters the far sidelobes of the antenna. Figure 16.1 shows an empirical model curve for the maximum sidelobe gain as a function of angle from the main-beam axis. This curve is derived from the measured response patterns of a number of large reflector antennas. For the present estimate, it is appropriate to use a gain of 0 dBi (i.e., 0 dB with respect to an isotropic radiator), which occurs at about 19° from the main beam. Zero dBi is also the mean gain of an antenna over 4π steradians, and the effective collecting area for this gain is equal to $\lambda^2/4\pi$, where λ is the wavelength. If F_h (W m^{-2}) is the flux density of an interfering signal within the receiver passband, the

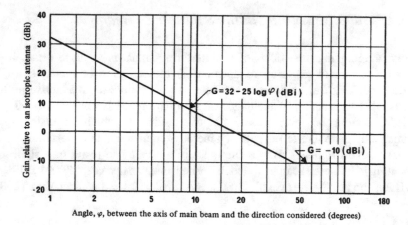

Fig. 16.1 Empirical sidelobe-envelope model for reflector antennas of diameter greater than 100 wavelengths. Measurements on antennas show that 90% of sidelobe peaks lie below the curve. Sidelobe levels can be reduced by 3 dB or more in designs in which aperture blockage by feed structure is eliminated or minimized. The model shown is representative of large antennas with tripod or quadrupod feed supports of the type commonly used in radio astronomy. From ITU-R Recommendation SA.509-1 (1997).

interference-to-noise power ratio in the receiver is

$$\frac{F_h \lambda^2}{4\pi k T_S \Delta v} , \tag{16.2}$$

where k is Boltzmann's constant, T_S is the system noise temperature, and Δv is the receiver bandwidth. In this expression, it is assumed that the polarization of the interfering signal matches that of the antenna. Since radio astronomy antennas commonly receive two polarizations, crossed linear or opposite circular, choice of antenna polarization is of little help in avoiding interference. In practice, the received level of the interfering signal varies with time because of propagation effects and the tracking motion of the radio telescope, which sweeps the sidelobe pattern across the direction of the transmitter.

For comparison with correlator systems, we first consider the simpler case of a receiver that measures the total power at the output of a single antenna. The interference-to-noise ratio of the output, after square-law detection and averaging for a time τ_a, is expression (16.2) multiplied by $\sqrt{\Delta v \tau_a}$. This result follows from considerations similar to those discussed in Sect. 6.2.1. Then for an output interference-to-noise ratio of 0.1, which we use as the criterion for the threshold of harmful interference,

$$F_h = \frac{0.4\pi k T_S v^2 \sqrt{\Delta v}}{c^2 \sqrt{\tau_a}} . \tag{16.3}$$

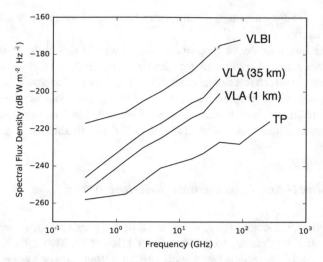

Fig. 16.2 Curves of the estimated harmful threshold of interference in dBW m^{-2} Hz^{-1}. The lowest curve is for total power (TP) measurement on a single antenna, and the topmost curve is for VLBI. The data in these are from ITU-R Recommendation RA.769-2 (2003) and essentially represent the considerations in Sects. 16.3 and 16.5. The smaller variations result from the characteristics of individual instruments. The two middle curves in the figure represent synthesis arrays and are based on the VLA at the most compact and most extended configurations, for which the corresponding spacings vary by a factor of about 35. The major feature in all of the curves is the increase with frequency, which results mainly from the effective collecting area of the receiving sidelobes, which varies as ν^{-2}. Because of various simplifying assumptions, the results in this figure are only approximate but also provide an indication of the relative vulnerability of different types of observations.

Note that the harmful threshold increases with frequency as ν^2 as a result of the dependence of the sidelobe collecting area. With increasing frequency, the system temperature and the usable bandwidth also generally increase. Expressed in spectral power flux density, the corresponding threshold level, S_h (W m^{-2} Hz^{-1}), is

$$S_h = \frac{F_h}{\Delta \nu} = \frac{0.4\pi k T_S \nu^2}{c^2 \sqrt{\tau_a \, \Delta \nu}} \; . \tag{16.4}$$

To determine the harmful interference level for continuum observations within a band allocated to radio astronomy, $\Delta \nu$ is usually taken to be the width of the allocated band. The total-power type of radio telescope is the most sensitive to interference. Thus, the results in Eqs. (16.3) and (16.4) provide a worst-case specification for the harmful thresholds of interference for radio astronomy. Values of F_h and S_h computed for total-power systems using typical parameters for the various radio astronomy bands are given in ITU-R[1] documentation (ITU-R 2013). For S_h, the values are plotted as the bottom curve in Fig. 16.2. Since much of the

---

[1] ITU-R denotes the Radiocommunication Sector of the International Telecommunication Union.

interference to radio astronomy results from broadband spurious emissions, S_h is particularly useful.

Low-level interference, of amplitude comparable to the noise in the receiver output, degrades the sensitivity and impedes the ability to detect weak sources. Thus, in observations in which interference has occurred, it is often necessary to delete any data that appear to be corrupted. The analysis that follows considers the response to interference resulting from basic methods of observation and data reduction and does not include procedures designed specifically for mitigation of interference.

16.3.1   Short- and Intermediate-Baseline Arrays

We now consider the interference response of a correlator array with antenna spacings from a few meters to a few tens of kilometers. Two effects reduce the response to interference compared with that of a total-power system. First, the source of interference does not move across the sky with the sidereal motion of the object under observation, and thus it produces fringe oscillations of a different frequency from those of the wanted signal. Second, the instrumental delays are adjusted to equalize the signal paths for radiation incident from the direction of observation, and signals from another direction, if they are broadband, are to some extent decorrelated. The following analysis is based on Thompson (1982).

16.3.2   Fringe-Frequency Averaging

Consider first the fringe-frequency effect. Suppose that instrumental phase shifts are introduced, as described in Sect. 6.1.6, to slow the fringe oscillations of the wanted signal to zero frequency. The removal of the fringe-frequency phase shifts from the cosmic signals introduces corresponding shifts into the interfering signals. If the source of interference is stationary with respect to the antennas, the interference at the correlator output has the form of oscillations at the natural fringe frequency for the source under observation, which from Eq. (4.9) (omitting the sign of dw/dt) is

$$\nu_f = \omega_e u \cos \delta \ . \tag{16.5}$$

Here ω_e is the angular rotation velocity of the Earth, u is a component of antenna spacing, and δ is the declination of the source under observation. Averaging of such a fringe-frequency waveform for a period τ_a is equivalent to convolution with a rectangular function of width τ_a. The amplitude is thus decreased by a factor that follows from the Fourier transform of the convolving function. This factor is

$$f_1 = \frac{\sin(\pi \nu_f \tau_a)}{\pi \nu_f \tau_a} \ . \tag{16.6}$$

In order to estimate a harmful threshold for interference (F_h), we compute the ratio of the rms level of interference to the rms level of noise in a radio image and, as before, equate the result to 0.1. The first step is to determine the mean squared value of the modulus of the interference component in the visibility data. Figure 6.7b, which depicts the spectral components at the correlator output, shows that the output from the correlated signal component, in this case the interference, is represented by a delta function. Assuming, as before, that the interference enters sidelobes of gain 0 dBi and that the polarization is matched, we substitute in the magnitude of the delta function $kT_A \, \Delta v = F_h c^2/4\pi v^2$. Thus, the sum of the squared modulus of the interference over n_r grid points in the (u, v) plane is

$$\sum_{n_r} \langle |r_i|^2 \rangle = \left(\frac{H_0^2 F_h c^2}{4\pi v^2} \right) n_r \langle f_1^2 \rangle \, . \tag{16.7}$$

Here r_i is the correlator response to the interference, H_0 is a voltage gain factor, and $\langle f_1^2 \rangle$ is the mean squared value of f_1, as given in Eq. (16.6), which represents the effect of the visibility averaging on the fringe-frequency oscillations. To determine the mean squared value of f_1, a simple approach is to consider the variation of this factor in the (u', v') plane in which the antenna-spacing vector rotates with constant angular velocity ω_e and sweeps out a circular locus, as described in Sect. 4.2. Also, suppose that to interpolate the values of visibility at the rectangular grid points in the (u, v) plane, the measured values are averaged with uniform weight within rectangular cells centered on the grid points (see the description of cell averaging in Sect. 5.2.2). Then the effective averaging time τ for the interference is equal to the time taken by the baseline vector to cross a cell, as shown in Fig. 16.3. Note from Eq. (16.5) that the fringe frequency goes through zero at the v' axis, and f_1 is then unity. For small values of ψ, as defined in Fig. 16.3, the path length through a cell is closely equal to Δu, and the cell crossing time is $\tau = \Delta u/\omega_e q'$, where $q' = \sqrt{X_\lambda^2 + Y_\lambda^2}$, and where X_λ and Y_λ are the components of antenna spacing measured in wavelengths and projected onto the equatorial plane, as defined in Sect. 4.1. Also, $v_f \tau = \Delta u \sin \psi \cos \delta$. Now Δu is equal to the reciprocal of the width of the synthesized field, which, except at long wavelengths, is unlikely to be more than $\sim 0.5°$. We therefore assume that Δu is of order 100 or greater, which permits the following simplification. For $\Delta u = 100$ and $\delta < 70°, f_1^2$ goes from 1 to 10^{-3} as ψ goes from 0 to $< 17°$. Thus, most of the contribution to f_1^2 occurs for small ψ, and we can substitute $v_f \tau = \psi \Delta u \cos \delta$ in Eq. (16.6) and obtain

$$\langle f_1^2 \rangle = \frac{2}{\pi} \int_0^{\pi/2} \frac{\sin^2(\pi \psi \Delta u \cos \delta)}{(\pi \psi \Delta u \cos \delta)^2} d\psi \simeq \frac{1}{\pi \Delta u \cos \delta} \, . \tag{16.8}$$

Since Δu is large, we have used an upper limit of ∞ in evaluating the integral.

For the noise, we again refer to Fig. 6.7b. The power spectral density of the noise near zero frequency is $H_0^4 k^2 T_S^2 \Delta v$, and an equivalent bandwidth τ^{-1}, including

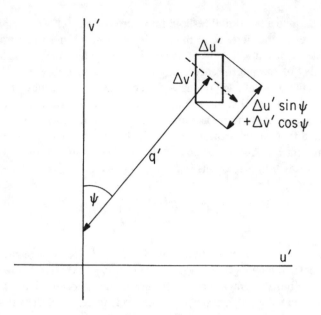

Fig. 16.3 Derivation of the mean cell crossing time for the spatial frequency locus indicated by the broken line. The velocity of the spatial frequency vector in the (u', v') plane is $\omega_e q'$. The mean path length through the cell in the direction of the broken line is the cell area $\Delta u' \Delta v'$ divided by the cell width projected normal to that direction.

negative frequencies, is passed by the averaging process; see Eq. (6.44). The mean-squared component of the noise over the n_r grid points is thus

$$\sum_{n_r} \langle |r_n| \rangle^2 = H_0^4 k^2 T_S^2 \Delta v n_r \langle \tau^{-1} \rangle \,, \tag{16.9}$$

where $\langle \tau^{-1} \rangle$ is the mean value of τ^{-1}. From Fig. 16.3, the mean cell crossing time is

$$\tau = \frac{\Delta u \, |\mathrm{cosec} \, \delta|}{q' \omega_e \, (| \sin \psi | + |\mathrm{cosec} \, \delta| \, | \cos \psi |)} \,. \tag{16.10}$$

We have assumed that $\Delta u' = \Delta v' \sin \delta$ (i.e., $\Delta u = \Delta v$) and that for all except a small number of cells, the path of the spatial frequency locus through a cell can be approximated by a straight line. The mean value of τ^{-1} around a locus in the (u', v') plane (see Sect. 4.2) is, from Eq. (16.10),

$$\frac{2}{\pi} \int_0^{\pi/2} \tau^{-1} d\psi = \frac{2\omega_e q'}{\pi \Delta u} (1 + | \sin \delta |) \,, \tag{16.11}$$

and the mean for the n_r points in the (u, v) plane is

$$\langle \tau^{-1} \rangle = \frac{2\omega_e}{\pi \Delta u} (1 + |\sin \delta|) \frac{1}{n_r} \sum_{n_r} q' . \tag{16.12}$$

From Eqs. (16.7)–(16.9) and Eq. (16.12), the interference-to-noise ratio is

$$\frac{(|r_i|)_{\mathrm{rms}}}{(|r_n|)_{\mathrm{rms}}} = \frac{F_h c^2}{4\pi k T_S v^2 \sqrt{2\Delta v \omega_e} \cos \delta (1 + |\sin \delta}} \frac{1}{\sqrt{\frac{1}{n_r} \sum_{n_r} q'}} . \tag{16.13}$$

By Parseval's theorem, the ratio of the rms values of the interference and noise in the image is equal to the same ratio in the visibility domain, which is given by Eq. (16.13). To evaluate the harmful threshold F_h, we equate the right side to 0.1 and obtain

$$F_h = \frac{0.4\pi k T_S v^2 \sqrt{2\Delta v \omega_e}}{c^2} \sqrt{\frac{1}{n_r} \sum_{n_r} q'} . \tag{16.14}$$

The factor $\sqrt{\cos \delta (1 + |\sin \delta}$ has been replaced by unity, the resulting error being less than 1 dB for $0 < |\delta| < 71°$, and 2.3 dB for $\delta = 80°$. The number of points in the (u', v') plane to which an antenna pair contributes is proportional to q', so in evaluating Eq. (16.14), it is convenient to write

$$\frac{1}{n_r} \sum_{n_r} q' = \frac{\sum_{n_p} q'^2}{\sum_{n_p} q'} , \tag{16.15}$$

where n_p is the number of correlated antenna pairs in the array.

The interference threshold S_h, in units of dBW m^{-2} Hz^{-1}, is given by

$$S_h = \frac{F_h}{\Delta v} = \frac{0.4\pi k T_S v^2 \sqrt{2\omega_e}}{c^2 \sqrt{\Delta v}} \sqrt{\frac{1}{n_r} \sum_{n_r} q'} . \tag{16.16}$$

Note that q' is proportional to v, so S_h is proportional to $v^{2.5}$. Values of S_h for the VLA are shown by two middle curves in Fig. 16.2, which correspond to configurations in which the maximum baselines are 35 and 1 km, respectively (see Fig. 5.17b).

Since the averaging is ineffective in reducing the interference when u goes through zero, visibility values containing the greatest contributions from interference cluster around the v axis. Some degree of randomness in the occurrence of high values is to be expected, as a result of the varying sidelobe levels through which the interference enters. Because of the (u, v) distribution, the interference in the image domain takes the form of quasi-random structure that is elongated in the east–west direction; for an example, see Thompson (1982). The clustering also suggests the possibility of reducing the interference response by deleting any questionable visibility data near the v axis. The resulting degradation of the (u, v) coverage would increase the sidelobes of the synthesized beam.

The discussion above applies to cases in which the observation is of sufficiently long duration that the (u, v) plane is well sampled, and in which the strength of the interfering signal remains approximately constant during this time. If only a fraction α of the (u, v) loci cross the v axis, then a factor of $\sqrt{\alpha}$ should be introduced into the denominators of Eqs. (16.14) and (16.16). Strong, sporadic interference can produce different responses from that considered above.

16.3.3   Decorrelation of Broadband Signals

Since interfering signals are usually incident from directions other than that of the desired radiation, their time delays to the correlator inputs are generally not equal. Broadband interfering signals are thereby decorrelated to some extent, which further reduces their response. The reduction is not amenable to a general-case analysis like that resulting from averaging of the fringe frequency, but it can be computed for each particular antenna configuration and position of the interfering source. For this reason, and the fact that only broadband signals are reduced, the effect has not been included in the threshold equations (16.14) and (16.16).

At any instant during an observation, let θ_s be the angle between a plane normal to the baseline for a pair of antennas and the direction of the source under observation. θ_s defines a circle on the celestial sphere for which the delays are equalized. Similarly, let θ_i be the corresponding angle for the source of interference. The delay difference for the interfering signals at the correlator is

$$\tau_d = \frac{D \, |\sin \theta_s - \sin \theta_i|}{c} \, , \tag{16.17}$$

where D is the baseline length. Expressions for θ_s and θ_i can be derived from Eq. (4.3), since $\sin \theta_s = w\lambda/D$, where w is the third spacing coordinate as shown in Fig. 3.2, and λ is the wavelength. Suppose that the received interfering signal has an effectively rectangular spectrum of width Δv and center frequency v_0, defined either by the signal itself or by the receiving passband. By the Wiener–Khinchin relation, the autocorrelation function of the signal is equal to

$$\frac{\sin(\pi \Delta v \tau_d)}{\pi \Delta v \tau_d} \cos(2\pi v_0 \tau_d) \, . \tag{16.18}$$

Expression (16.18) represents the real output of a complex correlator as a function of the differential delay τ_d. The imaginary output is represented by a similar expression in which the cosine function is replaced by a sine. Thus, the decorrelation of the modulus of the complex output for a delay τ_d is given by the factor

$$f_2 = \frac{\sin(\pi \Delta \nu \tau_d)}{\pi \Delta \nu \tau_d} \; . \tag{16.19}$$

For a fixed transmitter location, θ_i remains constant, but θ_s varies as the antennas track. Thus, τ_d may go through zero, causing f_2 to peak, but unlike f_1 in Eq. (16.6), a peak in f_2 can occur at any point on the (u, v) plane. Those antenna pairs for which the f_1 and f_2 peaks overlap contribute most strongly to the interference in the image, and those for which the peaks are well separated contribute less. Therefore, for broadband signals, the fringe-frequency and decorrelation effects should be considered in combination. For example, in calculations for the response of the VLA to a geostationary satellite on the meridian, a factor

$$\sqrt{\frac{\sum q' f_1^2 f_2^2}{\sum q' f_1^2}} \tag{16.20}$$

was computed that represents the additional decrease in the rms interference resulting from decorrelation (Thompson 1982). The summations in (16.20) were taken over all antenna pairs for equal increments in hour angle, and the q' factors were inserted to compensate for the uneven density of the sampled points in the (u, v) plane. The antenna spacings of the VLA for both the most compact and most extended configurations were considered, with observing frequencies from 1.4 to 23 GHz and bandwidths of 25 and 50 MHz. The results indicate that suppression of broadband interference by decorrelation varies from 4 to 34 dB, with strong dependence on the observing declination. The interference was assumed to extend uniformly across the bandwidth, which would tend to overestimate the suppression in a practical situation.

16.4   Very-Long-Baseline Systems

In VLBI arrays, in which the antenna spacings are hundreds or thousands of kilometers, the output resulting from correlated components of an interfering signal at the correlator inputs is usually negligible. This is because the natural fringe frequencies are higher than those in arrays with baselines up to a few tens of kilometers, and the delay inequalities for signals that do not come from the direction of observation are also much greater. Furthermore, unless the interfering signal originates in a satellite or spacecraft, it is unlikely to be present at two widely separated locations.

Consider an interfering signal entering one antenna of a correlated pair. The interference reduces the measured correlation, and the overall effect is similar to

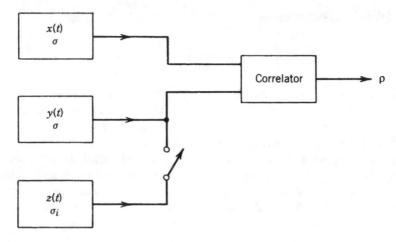

Fig. 16.4 Components of the correlator input signals used in the discussion of the effects of interference on VLBI observations.

an increase in the system noise for the antenna. In Fig. 16.4, $x(t)$ and $y(t)$ represent the signals plus system noise from two antennas in the absence of interference, and $z(t)$ represents an interfering signal at one antenna. The three waveforms have zero means, and the standard deviations are σ for x and y and σ_i for z. In the absence of interference, the measured correlation coefficient is

$$\rho_1 = \frac{\langle xy \rangle}{\sqrt{\langle x^2 \rangle \langle y^2 \rangle}} = \frac{\langle xy \rangle}{\sigma^2} . \tag{16.21}$$

When the interference is present, the correlation becomes

$$\rho_2 = \frac{\langle xy \rangle + \langle xz \rangle}{\sqrt{\langle x^2 \rangle (\langle y^2 \rangle + 2\langle yz \rangle + \langle z^2 \rangle)}} . \tag{16.22}$$

The interference is uncorrelated with x and y, so $\langle xz \rangle = \langle yz \rangle = 0$. Also, at the harmful threshold, $\sigma_i^2 \ll \sigma^2$. Thus, from Eqs. (16.21) and (16.22),

$$\rho_2 \simeq \rho_1 \left[1 - \frac{1}{2} \left(\frac{\sigma_i}{\sigma} \right)^2 \right] . \tag{16.23}$$

The interference reduces the measured correlation. In a system with automatic level control (ALC), the reduction in correlation can be envisaged as resulting from a reduction in the system gain in response to the added power of the interference. The error introduced in the correlation measurement therefore takes the form of a multiplicative factor, rather than an additive error component. Interference causes additive errors in single antennas or arrays that have short enough baselines that the detector or correlator responds directly to the interfering signal. The different effects of these two types of error have been discussed in Sect. 10.6.3. In principle, the change in the effective gain can be monitored by using a calibration signal, as discussed in Sect. 7.6. However, such a calibration process could be difficult if the strength of the interference varies rapidly. The harmful interference threshold should therefore be specified so it is just small enough that the errors introduced do not significantly increase the level of uncertainty in the measurements. In general, a value of 1% for variations in the visibility amplitude resulting from interference is a reasonable choice. If we include the possibility of simultaneous but uncorrelated interference in both antennas, the resulting condition is

$$\left(\frac{\sigma_i}{\sigma}\right)^2 \leq 0.01 \ . \tag{16.24}$$

It follows from Parseval's theorem, that a 1% rms error in the visibility introduces into the intensity an error of which the rms over the image is 1% of the corresponding rms of the true intensity distribution. The effect on the dynamic range of intensity within the image depends on the form of the intensity distribution and of the error distribution. For an image of a single point source, the rms intensity error would be about $10^{-2}\sqrt{f/n_r}$ times the peak intensity, where f is the fraction of the n_r gridded visibility data that contain interference. Here it is assumed that the fluctuations in the received interfering signal are sufficiently fast that the values of the interference level are essentially independent for each gridded visibility point. If this is not the case, the resulting error will be greater.

To comply with the criterion in Eq. (16.24), the ratio of the powers of the interference to system noise as given by Eq. (16.2) must not exceed 0.01. Thus, for the harmful threshold, we have

$$F_h = \frac{0.04\pi k T_S \nu^2 \Delta\nu}{c^2} \ . \tag{16.25}$$

The interference threshold in units of W m^{-2} Hz^{-1} is

$$S_h = \frac{F_h}{\Delta\nu} = \frac{0.04\pi k T_S \nu^2}{c^2} \ . \tag{16.26}$$

Note that the interference-to-noise ratio of 0.01 here refers to the levels at the correlator input. In the case of total-power systems (single antennas) and the arrays considered in Sect. 16.3, for which the errors are additive, the criterion of an interference-to-noise ratio of 0.1 applies to the time-averaged output of the correlator or detector. This therefore results in lower (i.e., more stringent) thresholds than those for VLBI in Eqs. (16.25) and (16.26). A curve for VLBI is shown in Fig. 16.2, using typical values for T_S. The harmful thresholds are approximately 40–50 dB less stringent than those for total-power systems.

16.5 Interference from Airborne and Space Transmitters

In application of the F_h and S_h values obtained above, it was assumed that the angular distance between the pointing direction of the antenna beams and the direction of the source of interference is large enough that the interference enters through sidelobes of gain ~ 1 dB or less, i.e., that the angular distance is $\sim 19°$ or more. Thus, airborne and satellite transmitters present a special problem. Radio astronomy cannot share bands with space-to-Earth (downlink) transmissions of satellites. However, because of the pressure for more spectrum for communications, allocations have been made in bands adjacent or close to those allocated to radio astronomy. Spurious emissions from satellite transmitters that fall outside the allocated band of the satellite pose a very serious threat to radio astronomy. Motion of the transmitter across the sky is most likely to increase the fringe frequency at the correlator outputs of a synthesis array and thereby reduce the response to interference. On the other hand, these signals may be received in high-level sidelobes near the main beam.

Examples of spurious emissions that extend far outside the allocated band of the satellite system are described by Galt (1990) and Combrinck et al. (1994). In these cases, the spurious emission resulted largely from the use of simple phase shift keying for the modulation, and newer techniques (e.g., Gaussian-filtered minimum shift keying) provide much sharper reduction in spectral sidebands (Murota and Hirade 1981; Otter 1994). However, intermodulation products resulting from the nonlinearity of amplifiers carrying many communication channels can remain a problem.

In some cases, operating requirements and limitations associated with space tend to make reduction of spurious emissions difficult. Some satellites use a large number of narrow beams to cover their area of operation, so that the same frequency channels can be used a corresponding number of times to accommodate a large number of customers. This requires phased-array antennas with many (of order 100 or more) small radiating elements, each with its own power amplifier [see, e.g., Schuss et al. (1999)]. Because of power limitations from the solar cells, these amplifiers are operated at levels that maximize power efficiency but could compromise linearity, resulting in spurious emissions from intermodulation products.

The recommended limits on spurious emissions (ITU-R 2012) in effect require that, for space services, the power in spurious emissions measured in a 4-kHz band at the transmitter output should be no more than –43 dBW. Thus, for example, spurious emission at this level from a low-Earth-orbit satellite at 800 km height and radiated from a sidelobe of 0 dBi gain would produce a spurious spectral power flux density of –208 dBW m^{-2} Hz^{-1} at the Earth's surface. This figure may be compared with the harmful interference thresholds for radio astronomy of –239 and –255 dBW m^{-2} Hz^{-1} for spectral line and continuum measurements, respectively, at 1.4 GHz. Although this very simple calculation considers only the worst-case situation, the differences of several tens of decibels show that such limits do not effectively protect radio astronomy.

16.6 Regulation of the Radio Spectrum

Regulation of the usage of the radio spectrum is organized through ITU, based in Geneva, which is a specialized agency of the United Nations. Radio astronomy was first officially recognized as a radiocommunication service by the ITU in 1959. The ITU-R was created in March 1993 and replaced the International Radio Consultative Committee (CCIR), an earlier entity within the ITU. A system of study groups within the ITU-R is responsible for technical matters. Study Group 7, entitled Science Services, includes radio astronomy, various aspects of space research, environmental monitoring, and standards for time and frequency. Study groups are subdivided into working parties that deal with specific areas. Their primary function is to study problems of current importance in frequency coordination, for example, specific cases of sharing of frequency bands between different services, and to produce documented Recommendations on the solutions. Decisions within the ITU are made largely by consensus. Recommendations must be approved by all of the radiocommunication study groups and then effectively become part of the ITU Radio Regulations. Recommendations in the RA series are specific to radio astronomy.

The ITU-R organizes meetings of study groups, working parties, and other groups required from time to time to deal with specific problems. It also organizes World Radiocommunication Conferences (WRCs) at intervals of two to three years, at which new spectrum allocations are made and the ITU Radio Regulations are revised as necessary. Administrations of many countries send delegations to WRCs, and the results of these conferences have the status of treaties. Participating countries can take exceptions to the international regulations so long as these do not affect spectrum usage in other countries. As a result, many administrations have their own system of radio regulations, based largely on the ITU Radio Regulations but with exceptions to accommodate their particular requirements: see Pankonin and Price (1981) and Thompson et al. (1991). See also ITU-R *Handbook on Radio Astronomy* (2013) and ITU-R Recommendation RA.769-2 (2003).

Further Reading

Crawford, D.L., Ed., *Light Pollution, Radio Interference, and Space Debris*, Astron. Soc. Pacific Conf. Ser., **17** (1991)

Ellingson, S., Introduction to Special Section on Mitigation of Radio Frequency Interference in Radio Astronomy, *Radio Sci.*, **40**, No. 5 (2005) (*Radio Sci.*, **40**, No. 5, contains 17 articles of interest.)

ITU-R, *Handbook on Radio Astronomy*, International Telecommunication Union, Geneva (2013)

Kahlmann, H.C., Interference: The Limits of Radio Astronomy, in *Review of Radio Science 1996–1999*, Stone, W.R., Ed., Oxford Univ. Press, Oxford, UK (1999), pp. 751–785

National Research Council, *Handbook of Frequency Allocations and Spectrum Protection for Scientific Uses,* 2nd ed., National Academies Press, Washington, DC (2015)

National Research Council, *Spectrum Management for Science in the 21st Century*, National Academies Press, Washington, DC (2010)

Swenson, G.W., Jr., and Thompson, A.R., Radio Noise and Interference, in *Reference Data for Engineers: Radio, Electronics, Computer, and Communications*, 8th ed., Sams, Indianapolis, IN (1993)

References

Athreya, R., A New Approach to Mitigation of Radio Frequency Interference in Interferometric Data, *Astrophys. J.*, **696**, 885–890 (2009)

Baan, W.A., RFI Mitigation in Radio Astronomy, in *Proc. of RFI Mitigation Workshop*, Proc. Science, PoS(RFI2010)011 (2010)

Barnbaum, C., and Bradley, R.F., A New Approach to Interference Excision in Radio Astronomy: Real-Time Adaptive Cancellation, *Astron. J.*, **116**, 2598–2614 (1998)

Beasley, W.L., "An Investigation of the Radiated Signals Produced by Small Sparks on Power Lines," Ph.D. thesis, Texas A&M Univ. (1970)

Bhat, N.D.R., Cordes, J.M., Chatterjee, S., and Lazio, T.J.W., Radio Frequency Interference Identification and Mitigation, Using Simultaneous Dual-Station Observations, *Radio Sci.*, **40**, RS5S14 (2005)

Boonstra, A.J., and van der Tol, S., Spatial Filtering of Interfering Signals at the Initial Low-Frequency Array (LOFAR) Phased Array Test Station, *Radio Sci.*, **40**, RS5S09 (2005)

Bretteil, S., and Weber, R., Comparison of Two Cyclostationary Detectors for Radio Frequency Interference Mitigation in Radio Astronomy, *Radio Sci.*, **40**, RS5S15 (2005)

Briggs, F.H., Bell, J.F., and Kesteven, M.J., Removing Radio Interference from Contaminated Astronomical Spectra Using an Independent Reference Signal and Closure Relations, *Astron. J.*, **120**, 3351–3361 (2000)

Briggs, F.H., and Kocz, J., Overview of Approaches to Radio Frequency Interference Mitigation, *Radio Sci.*, **40**, RS5S02 (2005)

Combrinck, W.L., West, M.E., and Gaylord, M.J., Coexisting with GLONASS: Observing the 1612-MHz Hydroxyl Line, *Publ. Astron. Soc. Pacific*, **106**, 807–812 (1994)

Dong, W., Jeffs., B.D., and Fisher, J.R., Radar Interference Blanking in Radio Astronomy Using a Kalman Tracker, *Radio Sci.*, **40**, RS5S04 (2005)

Ellingson, S.W., Bunton, J.D., and Bell, J.F., Removal of the GLONASS C/A Signal from OH Spectral Line Observations Using a Parametric Modeling Technique, *Astron. J. Suppl.*, **135**, 87–93 (2001)

Ellingson, S.W., and Cazemier, W., Efficient Multibeam Synthesis with Interference Nulling for Large Arrays, *IEEE Trans. Antennas Propag.*, **51**, 503–511 (2003)

Ellingson, S.W., and Hampson, G.A., A Subspace Tracking Approach to Interference Nulling for Phased Array Based Radio Telescopes, *IEEE Trans. Antennas Propag.*, **50**, 25–30 (2002)

Fridman, P.A., and Baan, W.A., RFI Mitigation in Radio Astronomy, *Astron. Astrophys.*, **378**, 327–344 (2001)

Galt, J., Contamination from Satellites, *Nature*, **345**, 483 (1990)

ITU-R, *Handbook on Radio Astronomy*, International Telecommunication Union, Geneva (2013)

ITU-R Recommendation SA.509-1, Generalized Space Research Earth Station Antenna Radiation Pattern for Use in Interference Calculations, Including Coordination Procedures, *ITU-R Recommendations, SA Series*, International Telecommunication Union, Geneva (1997) (see also updated Recommendation SA.509-3, 2013)

ITU-R Recommendation RA.769-2, Protection Criteria Used for Radio Astronomical Measurements, *ITU-R Recommendations, RA Series*, International Telecommunication Union, Geneva (2003) (or current revision)

ITU-R Recommendation SM.329-12, Unwanted Emissions in the Spurious Domain, *ITU-R Recommendations, SA Series*, International Telcommunication Union, Geneva (2012) (or current revision)

Leshem, A., and van der Veen, A.-J., Radio-Astronomical Imaging in the Presence of Strong Radio Interference, *IEEE Trans. Inform. Theory*, **46**, 1730–1747 (2000)

Leshem, A., van der Veen, A.-J., and Boonstra, A.-J., Multichannel Interference Mitigation Techniques in Radio Astronomy, *Astrophys. J. Suppl.*, **131**, 355–373 (2000)

Murota, K., and Hirade, K., GMSK Modulation for Digital Mobile Radio Telephony, *IEEE Trans. Commun.*, **COM-29**, 1044–1050 (1981)

Nita, G.M., and Gary, D.E., Statistics of the Spectral Kurtosis Estimator, *Publ. Astron. Soc. Pacific*, **122**, 595–607 (2010)

Nita, G.M., Gary, D.E., Lui, Z., Hurford, G.H., and White, S.M., Radio Frequency Interference Excision Using Spectral Domain Statistics, *Publ. Astron. Soc. Pacific*, **119**, 805–827 (2007)

Offringa, A.R., de Bruyn, A.G., Zaroubi, S., van Diepen, G., Martinez-Ruby, O., Labropoulos, P., Brentjens, M.A., Ciardi, B., Daiboo, S., Harker, G., and 86 coauthors, The LOFAR Radio Environment, *Astron. Astrophys.*, **549**, A11 (15pp) (2013)

Offringa, A.R., Wayth, R.B., Hurley-Walker, N., Kaplan, D.L., Barry, N., Beardsley, A.P., Bell, M.E., Bernardi, G., Bowman, J.D., Briggs, F., and 55 coauthors, The Low-Frequency Environment of the Murchison Widefield Array: Radio-Frequency Interference Analysis and Mitigation, *Publ. Astron. Soc. Aust.*, **32**, e008 (13pp) (2015)

Otter, M., *A Comparison of QPSK, OQPSK, BPSK, and GMSK Modulation Schemes*, Report of the European Space Agency, European Space Operations Center, Darmstadt, Germany (1994)

Pankonin, V., and Price, R.M., Radio Astronomy and Spectrum Management: The Impact of WARC-79, *IEEE Trans. Electromag. Compat.*, **EMC-23**, 308–317 (1981)

Perley, R., Attenuation of Radio Frequency Interference by Interferometric Fringe Rotation, EVLA Memo 49, National Radio Astronomy Observatory (2002)

Petroff, E., Keane, E.F., Barr, E.D., Reynolds, J.E., Sarkissian, J., Edwards, P.G., Stevens, J., Brem, C., Jameson, A., Burke-Spolaor, S., and four coauthors, Identifying the Source of Perytons at the Parkes Radio Telescope, *Mon. Not. R. Astron. Soc.*, **451**, 3933–3940 (2015)

Raza, J., Boonstra, A.-J., and van der Veen, A.-J., Spatial Filtering of RF Interference in Radio Astronomy, *IEEE Signal Proc. Lett.*, **9**, 64–67 (2002)

Rogers, A.E.E., Pratap, P., Carter, J.C., and Diaz, M.A., Radio Frequency Interference Shielding and Mitigation Techniques for a Sensitive Search for the 327-MHz Line of Deuterium, *Radio Sci.*, **40**, RS5S17 (2005)

Schuss, J.J., Upton, J., Myers, B., Sikina, T., Rohwer, A., Makridakas, P., Francois, R., Wardle, L., and Smith, R., The IRIDIUM Main Mission Antenna Concept, *IEEE Trans. Antennas Propag.*, **AP-47**, 416–424 (1999)

Smolders, B., and Hampson, G., Deterministic RF Nulling in Phased Arrays for the Next Generation of Radio Telescopes, *IEEE Antennas Propag. Mag.*, **44**, 13–22 (2002)

Thompson, A.R., The Response of a Radio-Astronomy Synthesis Array to Interfering Signals, *IEEE Trans. Antennas Propag.*, **AP-30**, 450–456 (1982)

Thompson, A.R., Gergely, T.E., and Vanden Bout, P., Interference and Radioastronomy, *Physics Today*, **44**, 41–49 (1991)

van der Tol, S., and van der Veen, A.-J., Performance Analysis of Spatial Filtering of RF Interference in Radio Astronomy, *IEEE Trans. Signal Proc.*, **53**, 896–910 (2005)

Chapter 17
Related Techniques

Concepts and techniques similar to those used in radio interferometry and synthesis imaging occur in various areas of astronomy, Earth remote sensing, and space science. Here we introduce a few of them, including optical techniques, to leave the reader with a broader view. All of these subjects are described in detail elsewhere, so here the aim is mainly to outline the principles involved and to make connections between them and the material developed in earlier chapters.

17.1 Intensity Interferometer

In long-baseline interferometry, the intensity interferometer offers some technical simplifications that were mainly of importance in radio astronomy during the early development of the subject. As mentioned in Sect. 1.3.7, its practical applications in radio astronomy have been limited (Jennison and Das Gupta 1956; Carr et al. 1970; Dulk 1970) because, in comparison with a conventional interferometer, an intensity interferometer requires a much higher signal-to-noise ratio (SNR) in the receiving system, and only the modulus of the visibility function is measured. This type of interferometer was devised by Hanbury Brown, who has described its development and application (Hanbury Brown 1974).

In the intensity interferometer, the signals from the antennas are amplified and then passed through square-law (power-linear) detectors before being applied to a correlator, as shown in Fig. 17.1. As a result, the rms signal voltages at the correlator inputs are proportional to the powers delivered by the antennas and therefore also proportional to the intensity of the signal. No fringes are formed because the phase of the radio frequency (RF) signals is lost in the detection, but the correlator output indicates the degree of correlation of the detected waveforms. Let the voltages at the detector inputs be V_1 and V_2. The outputs of the detectors are V_1^2 and V_2^2 and

© The Author(s) 2017   809
A.R. Thompson, J.M. Moran, and G.W. Swenson Jr., *Interferometry and Synthesis in Radio Astronomy*, Astronomy and Astrophysics Library,
DOI 10.1007/978-3-319-44431-4_17

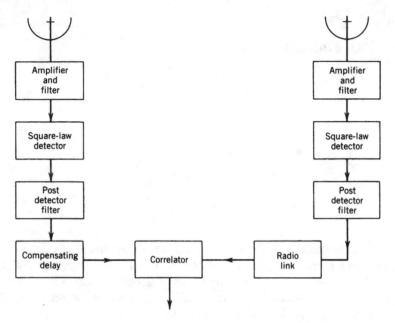

Fig. 17.1 The intensity interferometer. The amplifier and filter block may also incorporate a local oscillator and mixer. The compensating delay equalizes the time delays of signals from the source to the correlator inputs. The post-detector filters remove dc and radio frequency components.

each consists of a dc component, which is removed by a filter, and a time-varying component, which goes to an input of the correlator. From the fourth-order moment relation [Eq. (6.36)], the correlator output is

$$\langle (V_1^2 - \langle V_1^2 \rangle)(V_2^2 - \langle V_2^2 \rangle) \rangle = \langle V_1^2 V_2^2 \rangle - \langle V_1^2 \rangle \langle V_2^2 \rangle$$
$$= 2\langle V_1 V_2 \rangle^2 . \tag{17.1}$$

The correlator output is proportional to the square of the correlator output for a conventional interferometer and measures the squared modulus of the visibility of a source under observation.

We now give an alternative derivation of the response, which provides a physical picture of how the signals from different parts of the source combine within the instrument. The source is represented as a one-dimensional intensity distribution in Fig. 17.2. We suppose that it can be considered as a linear distribution of many small regions, each of which is large enough to emit a signal with the characteristics of stationary random noise, but of angular width small compared with $1/u$, which defines the angular resolution of the interferometer. The source is assumed to be spatially incoherent so the signals from different regions are uncorrelated. Consider two regions of the source, k and ℓ, at angular positions θ_k and θ_ℓ and subtending

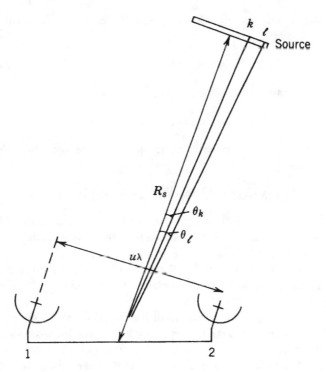

Fig. 17.2 Distances and angles used in the discussion of the intensity interferometer. u is the projected antenna spacing in wavelengths.

angles $d\theta_k$ and $d\theta_\ell$, as in Fig. 17.2. Each radiates a broad spectrum, but we first consider only the output resulting from a Fourier component at frequency ν_k from region k and similarly a component at ν_ℓ from region ℓ. Let $A_1(\theta)$ be the power reception pattern of the two antennas and $I_1(\theta)$ the intensity distribution of the source, these two functions being one-dimensional representations. Then the detector output of the first receiver is equal to

$$[V_k \cos 2\pi \nu_k t + V_\ell \cos(2\pi \nu_\ell t + \phi_1)]^2 \,, \tag{17.2}$$

where ϕ_1 is a phase term resulting from path-length differences, and the signal voltages V_k and V_ℓ are given by

$$V_k^2 = A_1(\theta_k)I_1(\theta_k)\, d\theta_k\, d\nu_k \tag{17.3}$$

and

$$V_\ell^2 = A_1(\theta_\ell)I_1(\theta_\ell)\, d\theta_\ell\, d\nu_\ell \,. \tag{17.4}$$

After expanding (17.2) and removing the dc and RF terms, we obtain for the detector output from receiver 1:

$$V_k V_\ell \cos\left[2\pi(\nu_k - \nu_\ell)t - \phi_1\right] . \tag{17.5}$$

Similarly, the detector output from receiver 2 is

$$V_k V_\ell \cos\left[2\pi(\nu_k - \nu_\ell)t - \phi_2\right] . \tag{17.6}$$

The correlator output is proportional to the time-averaged product of (17.5) and (17.6), that is, to

$$\langle A_1(\theta_k)A_1(\theta_\ell)I_1(\theta_k)I_1(\theta_\ell) \, d\theta_k \, d\theta_\ell \, d\nu_k \, d\nu_\ell \, \cos(\phi_1 - \phi_2) \rangle . \tag{17.7}$$

The change in the phase term with respect to frequency is small so long as the fractional bandwidth is much less than the ratio of the resolution to the field of view [see Eq. (6.69) and related discussion]. With this restriction, expression (17.7) is effectively independent of the frequencies ν_k and ν_ℓ, so that if we integrate it with respect to ν_k and ν_ℓ over a rectangular receiving passband of width $\Delta\nu$, $d\nu_k \, d\nu_\ell$ is replaced by $\Delta\nu^2$.

The phase angles ϕ_1 and ϕ_2 result from the path differences kk' and $\ell\ell'$ shown in Fig. 17.3. Note that ϕ_1 and ϕ_2 have opposite signs since the excess path length to antenna 1 is from point ℓ and that to antenna 2 is from point k. If R_s is the distance of the sources from the antennas, the distance $k\ell$ in the source is approximately equal to $R_s(\theta_k - \theta_\ell)$. The angle $\alpha_k + \alpha_\ell$ is approximately equal to $u\lambda/R_s$, since u represents the antenna spacing projected normal to the source and measured in wavelengths. The preceding approximations are accurate if α_k, α_ℓ, and the angle subtended by the source are all small. Thus, the difference of the phase angles is

$$\phi_1 - \phi_2 = 2\pi R_s(\theta_k - \theta_\ell)\frac{(\sin\alpha_k + \sin\alpha_\ell)}{\lambda}$$

$$\simeq 2\pi u(\theta_k - \theta_\ell) . \tag{17.8}$$

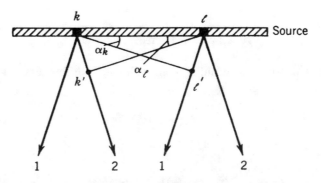

Fig. 17.3 Relative delay paths kk' and $\ell\ell'$ from regions k and ℓ of the source for rays traveling in the directions of antennas 1 and 2.

From (17.7), the output of the correlator now becomes

$$\langle A_1(\theta_k)A_1(\theta_\ell)I_1(\theta_k)I_1(\theta_\ell)\Delta v^2 \cos\left[2\pi u(\theta_k - \theta_\ell)\right] d\theta_k\, d\theta_\ell\rangle \ . \tag{17.9}$$

To obtain the output from all pairs of regions within the source, expression (17.9) can, with the assumption of spatial incoherence, be integrated with respect to θ_k and θ_ℓ over the source, giving

$$\left\langle \left[\Delta v \int A_1(\theta_k)I_1(\theta_k)\cos(2\pi u\theta_k)d\theta_k\right]\left[\Delta v \int A_1(\theta_\ell)I_1(\theta_\ell)\cos(2\pi u\theta_\ell)d\theta_\ell\right]\right.$$
$$\left. + \left[\Delta v \int A_1(\theta_k)I_1(\theta_k)\sin(2\pi u\theta_k)d\theta_k\right]\left[\Delta v \int A_1(\theta_\ell)I_1(\theta_\ell)\sin(2\pi u\theta_\ell)d\theta_\ell\right]\right\rangle$$
$$= A_0^2\Delta v^2\left[\mathcal{V}_R^2 + \mathcal{V}_I^2\right] = A_0^2\Delta v^2\,|\mathcal{V}|^2\ , \tag{17.10}$$

where we assume that the antenna response $A_1(\theta)$ has a constant value A_0 over the source, and the subscripts R and I denote the real and imaginary parts of the visibility. This result follows from the definition of visibility that is given for a two-dimensional source in Sect. 3.1.1. Thus, the correlator output is proportional to the square of the modulus of the complex visibility. For a more detailed discussion following the same approach, see Hanbury Brown and Twiss (1954). An analysis based on the mutual coherence of the radiation field is given by Bracewell (1958).

Some characteristics of the intensity interferometer offer advantages over the conventional interferometer. The intensity interferometer is much less sensitive to atmospheric phase fluctuations, because each signal component at the correlator input is generated as the difference between two radio frequency components that have followed almost the same path through the atmosphere. The phase fluctuations in the difference-frequency components at the detectors are less than those in the radio frequency signals by the ratio of the difference frequency to the radio frequency, which may be of order 10^{-5}. In the conventional interferometer, such phase fluctuations can make the amplitude, as well as the phase, of the visibility difficult to measure. Similarly, fluctuations in the phases of the local oscillators in the two receivers do not contribute to the phases of the difference-frequency components. Thus, it is not necessary to synchronize the local oscillators or even to use high-stability frequency standards, as in VLBI. These advantages were helpful, although by no means essential, in the early radio implementation of the intensity interferometer. Had the diameters of the sources under investigation then been of order of arcseconds, rather than arcminutes, the characteristics of the intensity interferometer would have played a more essential role.

The serious disadvantage of the intensity interferometer is its relative lack of sensitivity. Because of the action of the detectors in the receivers, the ratio of the signal power to the noise power at the correlator inputs is proportional to the square of the corresponding ratio in the RF (predetector) stages, the exact value being dependent on the bandwidths of these and the post-detector stages (Hanbury Brown and Twiss 1954). In a conventional interferometer, it is possible to detect signals that are $\sim$ 60 dB below the noise at the correlator inputs. In the intensity interferometer, a similar SNR at the correlator output would require SNRs greater by $\sim$ 30 dB in the RF stages. This effect, together with the lack of sensitivity to the visibility phase, has greatly restricted the radio usage of the intensity interferometer. Intensity interferometry played a similar role in the early days of optical interferometry (see Sect. 17.6.3) before the development of the modern Michelson interferometer.

17.2   Lunar Occultation Observations

Measurement of the light intensity from a star as a function of time during occultation by the Moon was suggested by MacMahon (1909) as a means of determining the star's size and position. His analysis, which was based on a simple consideration of geometric optics, was criticized by Eddington (1909), who stated that diffraction effects would mask the detail at the angular scale of the star. Eddington's paper probably discouraged observations of lunar occultations for some time. The first occultation measurements were reported 30 years later by Whitford (1939), who observed the stars β Capricorni and ν Aquarii and obtained clear diffraction patterns.

What was not realized by Eddington and others at the time was that although the temporal response to an occultation is not a simple step function, as it would be for the case of geometrical optics and a point source, the Fourier transform of the point-source response, which represents the sensitivity to spatial frequency on the sky, has the same amplitude as that of a step function and differs only in the phase. Hence, the lunar occultation is sensitive to all Fourier components, and there is no intrinsic limit to the resolution that can be obtained, except for that imposed by the finite SNR. This equality of the amplitudes was recognized by Scheuer (1962), who devised a method of deriving the one-dimensional intensity distribution I_1 from the occultation curve. By that time, the concept of spatial frequency had become widely understood through application to radio interferometry. Since, in lunar occultations, the diffraction occurs outside the Earth's atmosphere, the high angular resolution is not corrupted significantly by atmospheric effects, as it is in the case of ground-based interferometry. Furthermore, the only dependence of the obtainable resolution on the telescope size results from the SNR. An early radio application of the technique was the measurement of the position and size of 3C273 by Hazard et al. (1963), which led to the identification of quasars. As mentioned in Sect. 12.1, this position measurement was used for many years as the right ascension reference for VLBI position catalogs. Radio occultation measurements have been most important at meter wavelengths, since at shorter wavelengths, the high thermal flux density from the Moon presents a difficulty. At radio frequencies, lunar occultations have been largely superseded by interferometry, but lunar occultations are still useful at optical and infrared wavelengths.

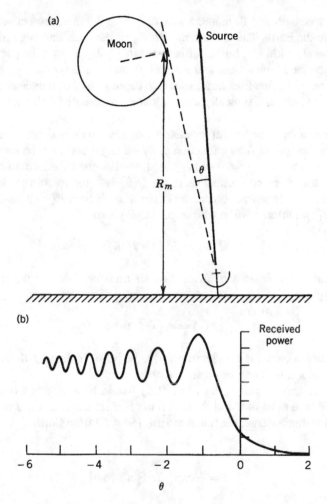

Fig. 17.4 Occultation of a radio source by the Moon: (**a**) the geometrical situation, in which θ is measured clockwise from the direction of the source, and is negative as shown; (**b**) the occultation curve for a point source, which is proportional to $\mathcal{P}(\theta)$. The units of θ on the abscissa are equal to $\sqrt{\lambda/2R_m}$, where λ is the wavelength and R_m the Moon's distance.

Figure 17.4 shows the geometrical situation and the form of an occultation record. The departure of the Moon's limb from a straight edge, as a result of curvature and roughness, is small compared with the size of the first Fresnel zone at radio frequencies. Thus, the point-source response is the well-known diffraction pattern of a straight edge, which is derived in most texts on physical optics. The main change in the received power in Fig. 17.4b corresponds to the covering or uncovering of the first Fresnel zone by the Moon, and the oscillations result from higher-order zones. The critical scale is the size of the first Fresnel zone, $\sqrt{(\lambda R_m/2)}$, where $R_m \simeq 3.84 \times 10^5$ km is the Earth–Moon distance. This corresponds to 4400 m at 10-cm wavelength and 10 m at 0.5 μm, or 2.3″ and 5 mas, respectively, in angle as seen from the Earth. The maximum velocity of the occulting edge of the Moon is approximately 1 km s^{-1}, but the effective velocity depends on the position of the occultation on the Moon's limb, and we use 0.6 km s^{-1} as a typical figure. Thus, the coverage time of the first Fresnel zone, which determines the characteristic fall time and oscillation period, is typically about 7 s at a wavelength of 10 cm and 16 ms at 0.5 μm.

In the case of the hypothetical geometrical-optics occultation, the observed curve would be the integral of I_1 as a function of θ, the angle between the source and the Moon's limb as in Fig. 17.4a. Then I_1 could be obtained by differentiation. In the actual case, the observed occultation curve $G(\theta)$ is equal to convolution of $I_1(\theta)$ with the point-source diffraction pattern of the Moon's limb $P(\theta)$. This convolution is $I_1(\theta) * P(\theta)$. Differentiation with respect to θ yields

$$G'(\theta) = I_1(\theta) * P'(\theta) , \qquad (17.11)$$

where the primes indicate derivatives. Fourier transformation of the two sides of Eq. (17.11) gives

$$\overline{G}'(u) = \overline{I}_1(u)\,\overline{P}'(u) , \qquad (17.12)$$

where the bar indicates the Fourier transform, the prime indicates a derivative in the θ domain, and u is the conjugate variable of θ.

Now in the geometrical-optics case, $P(\theta)$ would be a step function, and thus $P'(\theta)$ would be a delta function for which the Fourier transform is a constant. For the diffraction-limited case, the function $\overline{P}(u)$ [adapted from Cohen (1969)] is given by

$$\overline{P}(u) = \frac{j}{u}\exp\left[j2\pi\theta_F^2 u^2\mathrm{sgn}\,u\right] , \qquad (17.13)$$

where θ_F is the angular size of the first Fresnel zone, $\sqrt{\lambda/2R_m}$, and sgn is the sign function, which takes values ± 1 to indicate the sign of u. It follows from the derivative theorem of Fourier transforms that $\overline{P}'(u) = j2\pi\overline{P}(u)$, which has a constant amplitude with no zeros and can be divided out from Eq. (17.12). Thus, $I_1(\theta)$ is equal to $G'(\theta)$ convolved with a function whose Fourier transform is $1/\overline{P}'(u)$. Scheuer (1962) shows that this last function is proportional to $P'(-\theta)$, which can be used as a restoring function as follows:

$$I_1(\theta) = G'(\theta) * P'(-\theta)$$
$$= G(\theta) * P''(-\theta) . \qquad (17.14)$$

The second form on the right side is more useful since it avoids the practical difficulty of differentiating a noisy occultation curve. In principle, this restoration provides I_1 without limit on the angular resolution, in contrast with the performance of an array. Remember, however, that the amplitude of the spatial frequency sensitivity of the occultation curve, which is given by Eq. (17.13), is proportional to $1/u$. Thus, in the restoration in Eq. (17.14), the amplitudes of the Fourier components, which also include the noise, are increased in proportion to u. The increase of the noise sets a limit to the useful resolution. This limit can be conveniently introduced by replacing $\mathcal{P}''(\theta)$ in Eq. (17.14) by $\mathcal{P}''(\theta)$ convolved with a Gaussian function of θ θ with a resolution $\Delta\theta$. One then derives I_1 as it would be observed with a beam of the same Gaussian shape. In practice, the introduction of the Gaussian function is essential to the method, since it ensures the convergence of the convolution integral in Eq. (17.14). The optimal choice of $\Delta\theta$ depends on the SNR. Examples of restoring functions for various resolutions can be found in von Hoerner (1964).

The discussion above follows the classical approach to reduction of Moon-occultation observations, which developed from the geometrical optics analogy. One can envisage the reduction more succinctly as taking the Fourier transform of the occultation curve, dividing by $\overline{\mathcal{P}}(u)$ (with suitable weighting to control the increase of the noise), and retransforming to the θ domain. This process is mathematically equivalent to that in Eq. (17.14).

An estimate of the noise-imposed limit on the angular resolution can be obtained using the geometrical optics model, since the SNRs of the Fourier components are the same as for the actual point-source response. Consider the region of an occultation curve (see Fig. 17.4b) in which the main change in the received power occurs, and let τ be a time interval in which the change in the record level is equal to the rms noise. Then if v_m is the rate of angular motion of the Moon's limb over the radio source, the obtainable angular resolution is approximately

$$\Delta\theta = v_m \tau .$$
(17.15)

During the interval τ, the flux density at the antenna changes by ΔS. Let θ_s be the width of the main structure of the source in a direction normal to the Moon's limb, and let S be the total flux density of the source. Then for a source of approximately similar dimension in any direction, the average intensity is approximately S/θ_s^2. The change in solid angle of the covered part of the source in time τ is $\theta_s \Delta\theta$, and

$$\frac{\Delta\theta}{\theta_s} \simeq \frac{\Delta S}{S} .$$
(17.16)

The SNR at the receiver output for a component of flux density ΔS is

$$\mathcal{R}_{sn} = \frac{A \Delta S \sqrt{\Delta\nu\tau}}{2kT_S} ,$$
(17.17)

where A is the collecting area of the antenna, $\Delta\nu$ and T_S are the bandwidth and system temperature of the receiving system, and k is Boltzmann's constant. Note that the thermal contribution from the Moon can contribute substantially to T_S. The conditions that we are considering correspond to $\mathcal{R}_{sn} \simeq 1$, and from Eqs. (17.15)–(17.17), we obtain

$$\Delta\theta = \left(\frac{2kT_S\theta_s}{AS}\right)^{2/3}\left(\frac{v_m}{\Delta v}\right)^{1/3}. \qquad (17.18)$$

Note that frequency (or wavelength) does not enter directly into Eq. (17.18), but the values of several parameters, for example, S, Δv, and T_S, depend upon the observing frequency. As an example, consider an observation at a frequency in the 100–300-MHz range for which we use $A = 2000$ m^2, $T_S = 200$ K, and $\Delta v = 2$ MHz. For an example of a radio source, we take $S = 10^{-26}$ W m^{-2} Hz^{-1} (1 Jy) and $\theta_s = 5''$. v_m is typically $0.3''$ s^{-1}. With these values, Eq. (17.18) gives $\Delta\theta = 0.7''$. Although Eq. (17.18) is derived using a geometrical optics approach, this does not limit its applicability. For an observed occultation curve, the equivalent curve for geometrical optics can be obtained by adjustment of the phases of the Fourier components.

The bandwidth of the receiving system has the effect of smearing out angular detail in an occultation observation. Thus, since the SNR increases with bandwidth, for any observation there exists a bandwidth with which the sensitivity to fine angular structure is maximized. This bandwidth is approximately $v^2\Delta\theta^2 R_m/c$, which can be derived from the requirement that the phase term in Eq. (17.13) not change significantly over the bandwidth. This result can be compared to the bandwidth limitation for an array [given by Eq. (6.70)] by noting that a measurement by lunar occultation with resolution $\Delta\theta$ involves examination of the wavefront, at the distance of the Moon, on a linear scale of $\lambda/\Delta\theta$. Such an interval subtends an angle $\lambda/\Delta\theta R_m$ at the Earth. Further discussion of such details, and of the practical implementation of Scheuer's restoration technique, is given by von Hoerner (1964), Cohen (1969), and Hazard (1976). Note that a source may undergo a number of occultations within a period of a few months, with the Moon's limb traversing the source at different position angles. If a sufficient range of position angles is observed, the one-dimensional intensity distributions can be combined to obtain a two-dimensional image of the source [see, e.g., Taylor and De Jong (1968)]. In radio astronomy, the use of lunar occultations has become less important since the development of very-long-baseline interferometry.

The method of lunar occultation has been widely used in optical and infrared astronomy to measure the size and the limb darkening of stars, and the separation of close binary stars. Consistency of the results with those of optical interferometry proves that the lunar occultation method is not corrupted by variations in the lunar topography, which can be expected to become important when the size of the variations is comparable to the Fresnel scale. Angular sizes have been routinely measured down to about 1 mas. The analysis of stellar occultation curves is usually done by fitting parameterized models, rather than the reconstruction methods used in radio observations described above. A review of special considerations for lunar occultation observations at optical and infrared wavelengths can be found in Richichi (1994). Extensive measurements of stellar diameters [see, e.g., White and

Feierman (1987)] and binary star separations [see, e.g., Evans et al. (1985)] have been made. Other applications include the measurement of subarcsecond dust shells surrounding Wolf–Rayet stars [see, e.g., Ragland and Richichi (1999)].

17.3   Measurements on Antennas

Measurement of the electric field distribution over the aperture of an antenna is an important step in optimizing the aperture efficiency, especially in the case of a reflector antenna for which such results indicate the accuracy of the surface adjustment. The Fourier transform relationship between the voltage response pattern of an antenna and the field distribution in the aperture has been derived in Sect. 15.1.2. If x and y are axes in the aperture plane, the field distribution $\mathcal{E}(x_\lambda, y_\lambda)$ is the Fourier transform of the far-field voltage radiation (reception) pattern $V_A(l, m)$ (see Sect. 3.3.1), where l and m are here the direction cosines measured with respect to the x and y axes and the subscript λ indicates measurement in wavelengths. Thus,

$$V_A(l, m) \propto \int\int_{-\infty}^{\infty} \mathcal{E}(x_\lambda, y_\lambda) \, e^{j2\pi(x_\lambda l + y_\lambda m)} dx_\lambda \, dy_\lambda \ . \tag{17.19}$$

Direct measurement of $\mathcal{E}$ can be made by moving a probe across the aperture plane, but care must be taken to avoid disturbing the field. Such a technique is useful for characterizing horn antennas for millimeter wavelengths (Chen et al. 1998). However, in many applications, especially for large antennas on fully steerable mounts, it is easier to measure V_A. It is necessary to measure both the amplitude and phase of $V_A(l, m)$ in order to perform the Fourier transform for $\mathcal{E}(x_\lambda, y_\lambda)$. To accomplish this, the beam of the antenna under test can be scanned over the direction of a distant transmitter, and a second, nonscanning, antenna can be used to receive a phase reference signal. The function $V_A(l, m)$ is obtained from the product of the signals from the two antennas. This technique resembles the use of a reference beam in optical holography, and antenna measurements of this type have been described as holographic (Napier and Bates 1973; Bennett et al. 1976).

The holographic technique is readily implemented for measurements of antennas in interferometers and synthesis arrays. If the instrumental parameters (baselines, etc.) and the source position are accurately known, and the phase fluctuations introduced by the atmosphere are negligible, then for an unresolved source, calibrated visibility values will have a real part corresponding to the flux density of the source and an imaginary part equal to zero (except for the noise). If one antenna of a correlated pair is scanned over the source, while the other antenna continues to track the source, the corresponding visibility values will be proportional to the amplitude and phase of $V_A(l, m)$ for the scanning antenna. Measurement of synthesis array antennas as outlined above was first described by Scott and Ryle (1977), whose analysis, and that of D'Addario (1982), we largely follow below.

It is convenient to visualize the data in the aperture plane $\mathcal{E}(x_\lambda, y_\lambda)$ and in the sky plane $V_A(l, m)$ as discrete measurements at grid points in two $N \times N$ arrays to be used in the discrete Fourier transformation. For simplicity, consider a square antenna aperture with dimensions $d_\lambda \times d_\lambda$. Since $\mathcal{E}(x_\lambda, y_\lambda)$ is zero outside a range $\pm d_\lambda/2$, the sampling theorem of Fourier transforms indicates that the response must be sampled at intervals in (l, m) no greater than $1/d_\lambda$. [This interval is twice the sampling interval for the power beam because the power beam is the Fourier transform of the autocorrelation function of $\mathcal{E}(x_\lambda, y_\lambda)$.] If the $V_A(l, m)$ samples are spaced at $1/d_\lambda$, the aperture data just fill the $\mathcal{E}(x_\lambda, y_\lambda)$ array. The spacing of the measurements in the aperture is d_λ/N. Therefore, N is usually chosen so that the sample interval provides several measurements on each surface panel. In the (l, m) plane, the range of angles over which the scanning takes place is N times the pointing interval, that is, N/d_λ. This scan range is approximately N beamwidths. The procedure is to scan with the antenna under test in N^2 discrete pointing steps and thereby obtain the $V_A(l, m)$ data.

As a measure of the strength of the signal, let $\mathcal{R}_{sn}$ be the SNR obtained in time τ_a with the beams of both antennas pointed directly at the source. Now suppose that the (x_λ, y_λ) aperture plane is divided into square cells (as in Fig. 5.3) with sides d_λ/N centered on the measurement points. Consider the contribution to the correlator output of the signal from one such aperture cell, of area $(d_\lambda/N)^2$, in the antenna under test. The effective beamwidth of such an aperture cell is N times the antenna beamwidth, that is, approximately the total scan width required. Such an area contributes a fraction $1/N^2$ to the signal at the correlator output, so relative to the noise at the correlator output, the component resulting from one aperture cell is $\mathcal{R}_{sn}/N^2$ in time τ_a, or $\mathcal{R}_{sn}/N$ in time $N^2\tau_a$, which is the total measurement time. The accuracy of the phase measurement for the signal component from one aperture cell, $\delta\phi$, is the reciprocal of $\sqrt{2}$ times the corresponding SNR, that is, $N/(\sqrt{2}\mathcal{R}_{sn})$. The factor $\sqrt{2}$ is introduced because only the component of the system noise that is normal to the signal (visibility) vector introduces error in the phase measurement; see Fig. 6.8. Now a displacement ϵ in the surface of the aperture cell causes a change of phase $4\pi\epsilon/\lambda$ in the reflected signal. Thus, an uncertainty $\delta\phi$ in the phase of this signal component results in an uncertainty in ϵ of $\delta\epsilon = \lambda\delta\phi/(4\pi) = \lambda N/(4\sqrt{2}\pi\mathcal{R}_{sn})$. From the accuracy $\delta\epsilon$ desired for the surface measurement, we determine that the signal strength should be such that the SNR in time τ_a, with both beams on source, is

$$\mathcal{R}_{sn} = \frac{N\lambda}{4\sqrt{2}\pi\delta\epsilon} . \tag{17.20}$$

Having determined $\mathcal{R}_{sn}$, we can use Eqs. (6.48) and (6.49) to obtain values of antenna temperature or flux density (W m^{-2} Hz^{-1}) for the signal. If the two antennas used are not of the same size, then in Eqs. (6.48) and (6.49), A, T_A, and T_S are replaced by the geometric means of the corresponding quantities. Several simplifying approximations have been made. The statement that one aperture cell contributes $1/N^2$ of the antenna output implies the assumption that the field strength is uniform over the aperture. If the aperture illumination is tapered, a higher value of $\mathcal{R}_{sn}$ will be required to maintain the accuracy at the outer edges. Consideration of a square antenna overestimates $\mathcal{R}_{sn}$ for a circular aperture of diameter d_λ by $4/\pi$. The situation can be significantly different when the signal used in the holography measurement is a cw (continuous wave) tone, for example, from a satellite. The received signal power P can be large compared with the receiver noise $kT_R \Delta v$ (D'Addario 1982). In that case, the noise in the correlator output is dominated by the cross products formed by the signal and the receiver noise voltages. The resulting SNR in time τ is $\sqrt{P \Delta v\, \tau/(kT_R \Delta v)}$, which is independent of the receiver bandwidth.

An example of holographic measurements on an antenna of a submillimeter-wavelength synthesis array is shown in Fig. 17.5. Some practical points are listed below

- The source used in a holographic measurement is ideally strong enough to allow a high SNR to be obtained. Usually, either a signal from a transmitter on a satellite or a cosmic maser source is used. Morris et al. (1988) describe measurements on the 30-m-diameter antenna at Pico de Veleta in which a measurement accuracy (repeatability) of 25 μm was achieved using the 22.235-GHz water maser in Orion. For holography with interferometer elements, sources that are partially resolved can be used (Serabyn et al. 1991).
- If the test antenna is on an altazimuth mount, the beam will rotate relative to the sky as the observation proceeds. In determining the pointing directions, the (l, m) axes of the sky plane should remain aligned with the local horizontal and vertical directions. If the antenna is on an equatorial mount, the (l, m) axes should be the directions of east and north on the sky plane [i.e., the usual (l, m) definition].
- If the source is strongly linearly polarized and the antennas are on altazimuth mounts, it may be necessary to compensate for rotation of the beam. This is possible if the antennas receive on two orthogonal polarizations.
- When using two separate antennas, differences in the signal paths resulting from tropospheric irregularities can cause phase errors. It may be necessary to make periodic recordings with both beams centered on the source to determine the magnitude of such effects. In the case of measurement on a single large antenna, a small antenna mounted on the feed support structure of the large one, and pointing in the same direction as the large antenna's beam, is sometimes used to provide the on-source reference signal. Tropospheric effects on the phase should then cancel.

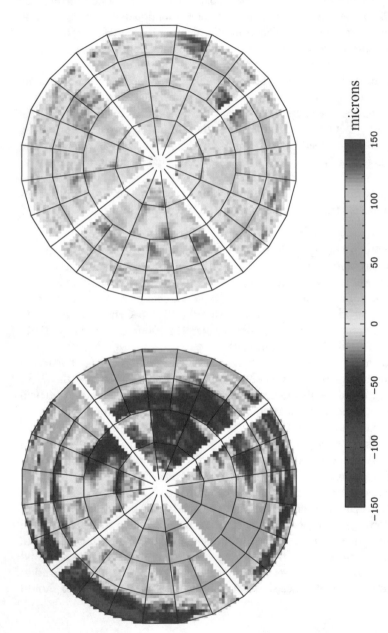

microns

Fig. 17.5 Surface deviations of one of the 6-m-diameter antennas of the Submillimeter Array measured with holography. There are 64 panels in the reflector, which are shown by the black lines. (**left**) The surface error map when the antenna's reflector was first deployed, with an rms deviation of 65 microns. (**right**) The error map after four rounds of adjustments were completed, which achieved an accuracy of 12 microns. The expected aperture efficiency with uniform illumination at 0.5 mm [see Eq. (5.4)] is 0.91. The near-field holography measurements were carried out using a 232.4-GHz cw tone from a beacon at a distance of 219 m and standard SMA science receivers. The maps have 128 × 128 pixels and a resolution of 6 cm. Adapted from Sridharan et al. (2004).

- An antenna may be rotated (through a limited angular range) about any axis
  through its phase center without varying the phase of the received signal. The
  phase center of a parabolic reflector lies on the axis of the paraboloid and is
  roughly near the midpoint between the vertex and the aperture plane.[1] In the
  scanning, the maximum angle through which the antenna is turned from the on-
  source direction is $N/(2d_\lambda)$. If the axis about which it is turned is a distance r
  from the phase center, the phase path length to the antenna will be increased by
  $r[1 - \cos(N/2d_\lambda)]$. If this distance is a significant fraction of a wavelength, a
  phase correction must be applied to the signal at the correlator output.
- For an antenna in a radome, structural members of which can cause scattering
  of the incident radiation, corrections are necessary. Rogers et al. (1993) describe
  such corrections for measurements on the Haystack 37-m-diameter antenna.
- In measurements on the antennas of a correlator array in which the number of
  antennas n_a is large, a possible procedure would be to use one antenna to track
  the source and provide the reference signal and to scan all the others over the
  source. However, a better procedure would be to use $n_a/2$ antennas to track the
  source while the other $n_a/2$ antennas are scanned. The averaging time would be
  half that of the first procedure to allow the roles of the two sets of antennas
  to be interchanged at the midpoint of the observation. However, there would
  be $n_a/2$ different measurements for each antenna, so compared with the first
  procedure, the sensitivity would be increased by a factor $\sqrt{n_a/4}$. Also, cross-
  correlation of the signals from the tracking antennas would provide information
  about the phase stability of the atmosphere, which would be useful in interpreting
  the measurements.

A method that requires only measurement of the amplitude of the far-field pattern
has been developed by Morris (1985). In such a procedure the reference antenna
is not required. The method is based on the Misell algorithm (Misell 1973), and
the procedure can be outlined as follows. Input requirements are an initial "first
guess" model of the amplitude and phase of the field distribution across the antenna
aperture, and two measurements of the far-field amplitude pattern, one with the
antenna correctly focused and the other with the antenna defocused sufficiently
to produce phase errors of a few radians at the antenna edge. The model aperture
distribution is used to calculate the in-focus far-field pattern in amplitude and phase,
and the calculated in-focus amplitude is replaced by the measured amplitude. The
measured in-focus amplitude and the calculated phase are then used to calculate

---

[1]Consider transmission from an antenna in which the parabolic surface is formed by rotation of the
parabola $x = ay^2$ around the x axis. Radiation from a ring-shaped element of the surface between
the planes $x = x'$ and $x = x' + dx$ has an effective phase center on the x axis at x'. The area of
such an element projected onto the aperture plane (i.e., normal to the x axis) is independent of x'. If
the aperture illumination is uniform, each surface element between planes normal to the x axis and
separated by the same increment makes an equal contribution to the electric vector in the far field.
Thus, the effective phase center of the total radiation should be on the x axis, midway between
the vertex and the aperture plane. Note that this is an approximate analysis based on geometrical
optics.

the corresponding aperture amplitude and phase, which then become the new aperture model. This new model is then used to calculate the defocused far-field pattern. In calculating the defocused pattern, it is assumed that in the aperture, the defocusing affects only the phase and that it introduces a component that varies in the aperture as the radius squared. The calculated defocused amplitude pattern is then replaced by the measured defocused pattern, and the corresponding in-focus aperture distribution is calculated and becomes the new model. In the continuing iterations, the in-phase and defocused amplitudes are calculated alternately. After each calculation, the amplitude pattern is replaced by the corresponding measured pattern, and the result is used to upgrade the model. The required solution to which the procedure should converge is a model that fits both the in-focus and defocused responses. This technique requires a higher SNR than when phase measurements are made. For measurements near nulls in the beam, the required SNR is approximately equal to the square of that when the phase is measured (Morris 1985).

A holographic method involving only one antenna, suitable for a large submillimeter-wavelength telescope, is described by Serabyn et al. (1991). Measurements are made in the focal plane using a shearing interferometer, an adaptation of a technique used for optical instruments.

17.4   Detection and Tracking of Space Debris

Tracking of satellites and space debris by reception of scattered broadcast signals (called "noncooperative transmitters") is known as passive radar. The technique generally requires a large separation between transmitter and receiver to avoid RFI generated by receipt of the direct transmission from the transmitter. The scattering cross section for a sphere of radius a is approximately

$$\sigma = \pi a^2 \qquad\qquad \lambda \ll 2\pi a \,,$$

$$\sigma \simeq \beta \pi a^2 \left(\frac{a}{\lambda}\right)^4 \qquad\qquad \lambda \gg 2\pi a \,, \qquad\qquad (17.21)$$

where $\beta \sim 10^4$. The short wavelength limit is called geometrical scattering, and the long wavelength limit is called Rayleigh scattering. These two limits are part of the general theory of Mie scattering [see, e.g., Jackson (1998)]. The cross section of dielectric spheres scales with a and λ in the same way. Equation (17.21) shows that $\sigma/(\pi a^2)$ decreases as $(a/\lambda)^4$, so there is a sharp decrease in sensitivity for scatters smaller than $\sim \lambda$. The tracking of satellites and space debris is an important part of space situational awareness.

The use of radio arrays to passively track space objects has been demonstrated by Tingay et al. (2013) with the Murchison Widefield Array (MWA). The MWA, which operates in the 80–300 MHz frequency range, is located in Western Australia, a region of low population density and radio-quiet environment. The antennas are dipoles mounted at ground level, which is helpful in shielding from direct reception of broadcast signals in the 87.5–108 MHz FM band. The FM signals originate from the area several hundred kilometers distant from Perth in southwestern Australia and, after scattering from objects in space, have been detected by the MWA. The directions of the incoming signals can be measured by the array, and as a test of the ability to track individual objects in space, reflections from the International Space Station were detected.

For this exercise, the astronomical interferometric delay model was adapted in an ad hoc fashion. Calculations based on Eq. (17.21) indicate that for a radius greater than 0.5 m, an object could plausibly be expected to be detected up to altitudes of approximately 1,000 km. The large collecting area of the MWA is helpful for such observations. At the FM-band frequencies, the field of view of the MWA is $\sim 2,400$ sq. deg., and the beamwidth is ~ 6 arcmin. It is estimated that on average, ~ 50 meter-size pieces of debris will be present within the MWA field of view at any time. Most will be at distances between the near-field and far-field distances for meter-wavelength observation.

A related application of radio interferometry is the near-field three-dimensional positioning of active satellites with VLBI arrays, which is described in Sect. 9.11.

17.5   Earth Remote Sensing by Interferometry

Global radio measurements have been made of the Earth since the beginning of the satellite era. For these measurements, the basic principle is that, in the absence of radiative transfer effects, the brightness temperature is related to the physical temperature of the surface through the emissivity e,

$$T_B = eT . \tag{17.22}$$

Since the emissivity of a material is related to its dielectric constant, the properties of the Earth's surface—e.g., moisture content of soil, salinity of sea water, and the structure of ice in the polar regions—can be deduced from maps of T_B. To obtain sufficient resolution at radio wavelengths, relatively large apertures are needed. In 2009, the European Space Agency launched the Soil Moisture and Ocean Salinity (SMOS) mission (McMullan et al. 2008; Kerr et al. 2010). This instrument strongly resembles a miniature version of the VLA. It has 69 antennas in a Y configuration (see Fig. 17.6). The system operates in the protected band of 1420–1427 MHz, which turns out to be an excellent frequency range to determine soil parameters and ocean salinity. The arm lengths are about 4 m, and the satellite is in a circular orbit with a height of 758 km and a period of 1.7 h. The maximum resolution is $\sim 2.6°$, corresponding to a linear resolution of about 35 km. The instantaneous field of view is about 1100 km. The (u, v) plane is sampled every 1.2 s. Most points on the Earth are revisited every three days. The theory of image formulation is a modified version of that presented in this book (Anterrieu 2004; Corbella et al. 2004).

Fig. 17.6 Artist's conception of the SMOS satellite, a downward-looking interferometric array operating at 21-cm wavelength and imaging the Earth at a resolution of 35 km. The length of each of the three arms of the array is 4 m, and the array is tilted by 32° to the tangent plane of the Earth below it. Image courtesy of and © European Space Agency.

The recovery of soil properties from the brightness temperature is a complex endeavor. The first step in the recovery process is based on a dielectric mixing model (Dobson et al. 1985) and the Fresnel reflection laws giving the relation between the emissivity and the dielectric constant of a surface material. A plane wave in free space incident on a flat surface with a dielectric constant ε at an incidence angle α will have power reflection coefficients of

$$r_{\parallel} = \left[\frac{\varepsilon \cos \alpha - \sqrt{\varepsilon - \sin^2 \alpha}}{\varepsilon \cos \alpha + \sqrt{\varepsilon - \sin^2 \alpha}} \right]^2$$

$$(17.23)$$

$$r_{\perp} = \left[\frac{\cos \alpha - \sqrt{\varepsilon - \sin^2 \alpha}}{\cos \alpha + \sqrt{\varepsilon - \sin^2 \alpha}} \right]^2 ,$$

where $r_{\parallel}$ is for the electric vector component in the plane of propagation and $r_{\perp}$ is for the component normal to the plane. These are the Fresnel reflection coefficients (note that the index of refraction is $\sqrt{\varepsilon}$). The emissivity, as a function of incidence angle, is

$$e(\alpha) = 1 - r(\alpha) .$$

$$(17.24)$$

The emissivity goes to unity for $r_\parallel$ at the Brewster angle α_B,[2] given by

$$\tan \alpha_B = \sqrt{\varepsilon} .$$ (17.25)

In the case of normal incidence ($\alpha = 0$), the emissivity is

$$e_n = 1 - \left[\frac{\sqrt{\varepsilon} - 1}{\sqrt{\varepsilon} + 1} \right]^2 .$$ (17.26)

The values of ε for various types of soils and water saturation vary from about 2 to 50, corresponding to a range of e_n from 0.5 to 0.97 and a brightness temperature range, for a nominal surface temperature of 280 K, of 140 to 270 K.[3]

The actual retrieval of soil moisture requires careful modeling of the surface temperature, subsurface temperature gradient, surface roughness, and radiative transfer through the vegetation layer using a physical based algorithm (Kerr et al. 2012) or statistical methods such as neural networks (Rodríguez-Fernández et al. 2015). An example of a soil moisture map is shown in Fig. 17.7. The nominal accuracy for the technique is 4% in volumetric moisture content. The dielectric constant of sea water is ~ 80, so the ocean brightness temperature is usually below 100 K. Retrieval of ocean salinity is challenging, and even the reflection of the galactic emission on the ocean surface needs to be taken into account for accurate determinations (Font et al. 2012).

17.6   Optical Interferometry

The principles of optical interferometry are essentially identical to those at radio frequencies, but accurate measurements are more difficult to make at optical wavelengths. One difficulty arises because irregularities in the atmosphere introduce variations in the effective path length that are large compared with the wavelength and thus cause the phase to vary irregularly by many rotations. Also, obtaining the mechanical stability of an instrument required to obtain fringes at a wavelength of order $0.5\,\mu$m presents a formidable problem. However, the practicality of synthesis imaging in the optical spectrum has been demonstrated using phase closure techniques, see, e.g., Haniff et al. (1987) and Baldwin et al. (1996). In the absence of visibility phase, the amplitude data can be interpreted in terms of the autocorrelation of the intensity distribution, as explained in Sect. 11.3.3, or

---

[2]Clark and Kuz'min (1965) used the Owens Valley interferometer to make the first passive measurement of the dielectric constant of the surface of Venus ($\varepsilon = 2.2 \pm 0.2$) by effectively measuring the Brewster angle.

[3]The dielectric constant of sea water is ~ 80, so the ocean brightness temperature is usually below 100 K.

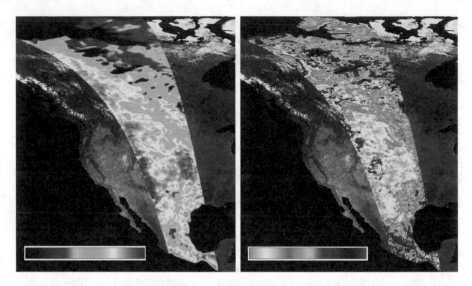

Fig. 17.7 Observations of the Earth made with the European Space Agency's SMOS orbiting synthesis array from data taken on 1 July 2015, overlaid on a visual image. (**left**) A "swath" map constructed from many snapshot images of brightness temperature in one polarization at a fixed incidence angle of 42.5°. The range of the color bar is 180–290 K. (**right**) Reconstructed map of the soil moisture based on complex retrieval algorithms and observations of each location with multiple angles of incidence and two polarizations. The range of the color bar is 0–0.5 m^3/m^3 (fractional volume). The brown shading indicates areas where the soil moisture value was not accurately retrieved. Images courtesy of Nemesio Rodríguez-Fernández and Arnaud Mialon.

in terms of models of the intensity distribution. Techniques for two-dimensional reconstruction without phase data [see, e.g., Bates (1984)] are also applicable. Optical interferometry is an active and growing field, and here we attempt only to give an overview of some basic principles. See Further Reading at the end of this chapter for a collection of important publications in optical interferometry.

Before discussing instruments, we briefly review some relevant atmospheric parameters. The irregularities in the atmosphere give rise to random variations in the refractive index over a large range of linear scales. For any particular wavelength, there exists a scale size over which portions of a wavefront remain substantially plane compared with the wavelength, that is, atmospheric phase variations are small compared with 2π. This scale size is represented by a parameter, the Fried length d_f (Fried 1966); see the discussion following Eq. (13.102). The Fried length is equal to $3.2d_0$, where d_0 is the spacing between paths through the atmosphere for which the rms phase difference is one radian; see Eq. (13.102). Regions for which the uniformity of the phase path lies within this range are sometimes referred to as seeing cells. The scale size d_f and the height at which the dominant irregularities occur define an isoplanatic angle (or isoplanatic patch) size, that is, an angular range of the sky within which the incoming wavefronts from different points encounter similar phase shifts. Within an isoplanatic patch, the point-spread function remains

Table 17.1 Atmospheric and instrumental parameters at visible and infrared wavelengths

Wavelength (μm)	d_f (m)	Isoplanatic Angle at Zenith	Resolution of 1-m-diameter aperture	Atmospheric resolution (λ/d_f)
0.5 (visible)	0.14	5.5″	0.13″	0.70″
2.2 (near infrared)	0.83	33″	0.55″	0.55″
20 (far infrared)	11.7	8′	5.0″	0.35″

Updated from Woolf (1982).

constant, so the convolution relationship between source and image holds. Typical figures for the 50th percentile value of d_f, which scales as $\lambda^{6/5}$ [see Eq. (13.102)], and the isoplanatic angle are given in Table 17.1. Also included for comparison are the corresponding values of the diffraction-limited resolution of a telescope of 1-m-diameter aperture. Optical interferometers provide a powerful means of studying the structure functions of the atmosphere at infrared and optical wavelengths; see, for example, Bester et al. (1992) and Davis et al. (1995). Note that techniques involving correction of atmospheric distortion of the wavefront by means of the telescope hardware are referred to as adaptive optics [see, e.g., Roggemann et al. (1997) and Milonni (1999)]. Most large telescopes have adaptive optics systems. Such systems are strongly analogous to the techniques of self-calibration and phase referencing in radio astronomy.

17.6.1   Instruments and Their Usage

The use of interferometry for measurement of the angular sizes of stars was suggested by Fizeau (1868), and the earliest attempted measurements of this type are those of Stéphan (1874), using a mask with two apertures on the objective lens of a telescope. Unfortunately, Stéphan's telescope was not large enough to resolve any of the stars he observed. The first successful measurement of the diameter of a star was made by Michelson and Pease (1921) on the supergiant star Betelgeuse, as described in Sect. 1.3.2. For this measurement, four plane mirrors were mounted on a beam attached to the telescope, so that signals received with a baseline spacing of 6 m were reflected into the telescope objective to form fringes. In this type of measurement, the whole instrument was carried on the mounting of the telescope, which simplified the pointing. However, attempts to use a similar system with an increased spacing between the mirrors were generally unsuccessful, because of the extreme stability required to maintain the relative positions of the mirrors to an accuracy of a few tenths of the optical wavelength. Thus, little progress was made in optical interferometry until the development of modern electronics and computer control for positioning of instrumental components. This allows longer baselines to be used but has become possible only in recent decades.

Figure 17.8 illustrates some of the basic features of a modern optical interferometer. The two mirrors S are mounted as siderostats and track the optical source under

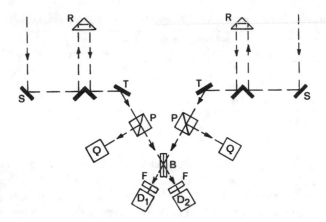

Fig. 17.8 Basic features of an optical interferometer. The broken line represents the light path from a star. From Davis and Tango (1985). With permission from and © W. J. Tango.

study. The positions of the retroreflectors R are continuously adjusted to equalize the lengths of the paths from the source to the combination point B. This delay compensation is usually implemented in evacuated tubes because the geometric delay of the interferometer largely occurs above the atmosphere. If air delay lines were used, a separate mechanism would be needed to compensate for the dispersive component of the delay, which is difficult to implement in wide bandwidth systems [see, e.g., Benson et al. (1997)]. The siderostats are mounted on stable foundations, and the rest of the system is usually mounted on optical benches within a controlled environment. The apertures of the interferometer, determined by the mirrors S, are made no larger than the Fried length d_f. Thus, the wavefront across the mirror remains essentially plane, and the effect of the irregularities is to produce a variation in the angle of arrival of the wavefront. The variation cannot be tolerated since the angles of the beams at the combination point B must be correct to within $1''$. To mitigate this effect, the polarizing beamsplitter cubes P reflect light to quadrant detectors Q, which produce a voltage proportional to any displacement of the angle of the light beam. These voltages are then used to control the tilt angles of the mirrors T, to compensate for the wavefront variation. A servo loop with bandwidth ~ 1 kHz is required to follow the fastest atmospheric effects. The filters F define the operating wavelength. The two detectors D_1 and D_2 respond to points on the fringe pattern spaced by one-quarter of a fringe cycle, and their outputs provide a measure of the instantaneous amplitude and phase of the fringes. This method is described, for example, by Rogstad (1968), who has also pointed out that with a multielement system, the phase information can be utilized by means of closure relationships, as introduced in Sect. 10.3. The system in Fig. 17.8 is shown to illustrate some of the important features used in modern optical interferometers. In practice, the siderostat mirrors may be replaced by large-aperture telescopes, and the paths of the light to the point at which the fringes are formed may be considerably more complicated.

Optical interferometers can be built with very wide bandwidths, that is, $\Delta\lambda/\lambda \simeq$ 0.1 or possibly more, so the central, or white light, fringe is readily identifiable. If such a system is made to operate at two such wide wavelength bands simultaneously, the effects of the atmosphere, which is slightly dispersive, can be removed. Ground-based optical astrometry with dual-wavelength phase-tracking interferometers can yield accurate positions of stars (Colavita et al. 1987, 1999). As examples of earlier interferometry, Currie et al. (1974)) made measurements using two apertures on a single large telescope, and Labeyrie (1975) obtained the first successful measurements using two telescopes. For descriptions of later, more complex instruments, see, for example, Davis and Tango (1985); Shao et al. (1988); Baldwin et al. (1994); Mourard et al. (1994); Armstrong et al. (1998, 2013); Davis et al. (1999a,b); ten Brummelaar et al. (2005); and Jankov (2010).

For use in space where the Earth's atmosphere is avoided, optical interferometry holds great promise. The Space Interferometry Mission (SIM) (Shao 1998; Allen and Böker 1998; Böker and Allen 1999) was a space-based interferometer for the wavelength band 0.4–1.0 μm with variable baseline up to 10 m, intended to provide synthesis imaging with a resolution of 10 mas, and to measure fringe phases with sufficient accuracy to provide positions of stars to within 4 μas. It was never launched. An application of space interferometry to the detection of planets around distant stars is discussed by Bracewell and MacPhie (1979). The ratio of the signal from the planet to that from the star is maximized by choosing an infrared wavelength on the long-wavelength side of 20 μm and by placing a fringe-pattern null in the direction of the star. A demonstration of the nulling technique using ground-based telescopes is described by Hinz et al. (1998).

Rogstad (1968) describes a technique for measurement of the visibility phase using an interferometer in the presence of an atmospheric component of seeing (refraction). Consider a linear arrangement of mirrors (i.e., the optical receivers) in which a unit spacing occurs twice and all integral multiples of the unit spacing, up to a maximum value, occur at least once. The receivers are designed to measure both the amplitude and the phase of the visibility function. The phase of the visibility for each of these spacings can be derived from the measured phases plus a unit-spacing phase component. This unit-spacing phase contains a component due to the atmosphere, although the longer spacing values are free from the atmospheric effect. However, the unit-spacing phase affects only the position of the resulting image, i.e., the coordinates on the sky, and it can be set to zero without affecting the form of the image. This method has been implemented on several interferometers [e.g., Jorgensen et al. (2012)], where it is referred to as baseline bootstrapping.

Several optical interferometers have been constructed from large telescopes that are used independently part of the time. For example, the Keck Observatory on Mauna Kea has two 10-m-diameter telescopes with a spacing of 85 m. As an interferometer, these antennas can provide an effective angular resolution of 5 mas at 2.2-μm wavelength and 24 mas at 10-μm wavelength. The European Southern Observatory in Chile has constructed the Very Large Telescope Interferometer (VLTI), which consists of four 8.2-m-diameter telescopes and four auxiliary 1.8-m-diameter telescopes. With current instrumentation [Petrov et al. (2007) and

Le Bouquin et al. (2011)], up to six baselines can be correlated at once, providing multiple phase closures and imaging capability. Operating in the bands between 1.5-and 2.4-μm, resolution as fine as 2 mas can be achieved on baselines up to 130 m. Spectral line capability is also provided with $\lambda/\Delta\lambda = 12000$ (velocity resolution of 25 km s^{-1}).

In the systems mentioned above, the fringes are formed by combining the incoming radiation at the same wavelength as it is received, as in the classical Michelson stellar interferometer. They are therefore also referred to as *direct detection systems*. A disadvantage of arrays built in this way is that the light cannot be divided, with loss in SNR. An alternative to the direct detection system is the *heterodyne system*, in which the light from each aperture is mixed with coherent light from a central laser to produce an intermediate frequency (IF). The IF waveforms are then amplified and correlated in an electronic system, in a manner basically identical to that used in radio interferometry. In comparison with a direct detection system, the sensitivity is greatly limited by the quantum effects mentioned in Sect. 1.4. It is also limited by the bandwidth that can be handled by the electronic amplifiers, unless the mixer outputs are split into many frequency channels, each of which is processed in parallel. A large bandwidth can then be processed using a correspondingly large number of amplifiers and correlators. The bandwidth division also has the effect of increasing the path length difference over which the signals remain coherent. The heterodyne technique has been used in infrared interferometry; see, for example, Johnson et al. (1974), Assus et al. (1979), and Bester et al. (1990). Possible application to large multielement telescopes with multiband processing in the infrared and visible ranges has been discussed by Swenson et al. (1986).

From the submillimeter radio range to the optical is a factor of $\sim 10^3$ in wavelength, and a further factor of $\sim 10^3$ takes one to the X-ray region. X-ray astronomy could benefit greatly by the potentially high angular resolution obtainable through interferometry. The viability of X-ray interferometry, suitable for astronomical imaging, has been demonstrated in the laboratory by Cash et al. (2000). It holds the promise of providing extremely high angular resolution in observations above the atmosphere. At a wavelength of 2 nm, a baseline of 1 m provides a fringe spacing of 40 μas. In the laboratory instrument, the apertures are defined by flat reflecting surfaces, which are used at grazing incidence to minimize the requirement for surface accuracy. Direct detection is the only available technique, and if the fringes are formed by simply allowing the reflected beams to converge on a detector surface, a long distance is required to obtain sufficient fringe spacing. With 400-μas angular spacing of the fringes, adjacent maxima would be separated by only 1 μm at 500-m distance. Astronomical interferometry at X-ray wavelengths will be a challenging enterprise.

17.6.2 Sensitivity of Direct Detection and Heterodyne Systems

Factors that determine the sensitivity of optical systems, such as losses due to scattering, partial reflection, and absorption, are different from corresponding effects at radio wavelengths. However, in heterodyne systems, the most important difference is the role of quantum effects. The energy of optical photons is five or more orders of magnitude greater than that of radio photons, and quantum effects are largely negligible in the radio domain at frequencies lower than ~ 100 GHz. In the optical range (wavelength ~ 500 μm), the frequency is of order 600 THz, and the bandwidth could be as high as 100 THz. In a typical heterodyne system in the infrared, the wavelength of 10 μm corresponds to 30 THz, and the bandwidth used is ~ 3 GHz [see, e.g., Townes et al. (1998)].

In direct detection systems, the detector or photon counter does not preserve the phase of the signal, and thus the noise resulting from the uncertainty principle, discussed in Sect. 1.4, does not occur. The noise is principally shot noise resulting from the random arrival times of the signal photons. The number of photons received from a source of intensity I is

$$N = \frac{I \Omega_s A \Delta \nu}{h\nu} \text{ (photons s}^{-1}\text{)} , \tag{17.27}$$

where Ω_s is the solid angle of the source (with no atmospheric blurring), A is the collecting area of the telescope, $\Delta \nu$ is the bandwidth, ν is the frequency, and h is Planck's constant. If the source is a blackbody at temperature T, the Planck formula gives

$$I = \frac{2h\nu^3}{c^2} \frac{1}{e^{h\nu/kT} - 1} . \tag{17.28}$$

Note that for direct detection, we are considering the signal in both polarizations. Thus, we have

$$N = \frac{2\Omega_s A \Delta \nu}{\lambda^2} \frac{1}{e^{h\nu/kT} - 1} \text{ (photons s}^{-1}\text{)} . \tag{17.29}$$

The received power is

$$P = h\nu N . \tag{17.30}$$

The fluctuations in the power, ΔP_D, are caused by photon shot noise and therefore are proportional to $\sqrt{N}$. Thus,

$$\Delta P_D = h\nu \sqrt{N} . \tag{17.31}$$

ΔP_D is known as the noise equivalent power. The SNR in one second is $P/\Delta P_D = \sqrt{N}$, and therefore for an integration time τ_a, the SNR for direct detection is

$$\mathcal{R}_{\text{snD}} = \left[\left(\frac{2\Omega_s A}{\lambda^2}\right) \frac{\Delta v \tau_a}{e^{hv/kT} - 1}\right]^{1/2}, \tag{17.32}$$

where the subscript D indicates direct detection. Note that $\mathcal{R}_{\text{snD}}$ is proportional to $\sqrt{A}$, because of the shot noise, rather than to A, as in the radio case.

In a heterodyne system, the noise is determined by the uncertainty principle, since the mixer is a linear device that preserves phase. The minimum noise is one photon per mode (one photon per hertz per second), as noted in the discussion following Eq. (1.15). This is equivalent to saying that the system temperature is hv/k [see, e.g., Heffner (1962), Caves (1982)]. Hence, in a period of one second, the uncertainty in power is

$$\Delta P_H = hv \sqrt{\Delta v}. \tag{17.33}$$

The heterodyne detector responds only to the component of the radiation to which its polarization is matched, and the received power is half of that in Eq. (17.30). The SNR for a heterodyne system (indicated by subscript H) is therefore $P/(2\Delta P_H)$ in one second, and in time τ_a, it is

$$\mathcal{R}_{\text{snH}} = \left(\frac{\Omega_s A}{\lambda^2}\right) \frac{\sqrt{\Delta v \tau_a}}{e^{hv/kT} - 1}. \tag{17.34}$$

Note that Eq. (17.34) reduces to the usual radio form in Eq. (1.8) when $hv/kT \ll 1$. In that case, $T_A = T\Omega_s A/\lambda^2$ and the minimum value of hv/k can be used for system temperature. The ratio of SNRs for the direct detection and heterodyne systems, when parameters other than the bandwidth are the same, is

$$\frac{\mathcal{R}_{\text{snH}}}{\mathcal{R}_{\text{snD}}} \simeq \sqrt{\left(\frac{\Omega_s A}{2\lambda^2}\right) \frac{1}{e^{hv/kT} - 1} \left(\frac{\Delta v_H}{\Delta v_D}\right)}. \tag{17.35}$$

As indicated earlier, $\sqrt{\Delta v_H/\Delta v_D}$ could be as low as $\sim 4 \times 10^{-3}$. However, for direct detection, the propagation delays through the different siderostats to the fringe-forming point must be maintained constant to $\sim 1/10$ of the reciprocal bandwidth. This requirement restricts the bandwidths that can practicably be used, especially with baselines of hundreds of meters. The heterodyne system offers simpler hardware that provides useful sensitivity at $10\ \mu$m wavelength and possibly to the next atmospheric window at $5\ \mu$m. It also allows the amplified IF signals to be split without loss in sensitivity, to provide multiple simultaneous correlations in multielement arrays. Relative advantages of the heterodyne and direct detection systems are discussed by Townes and Sutton (1981) and de Graauw and van de Stadt (1981).

17.6.3   Optical Intensity Interferometer

The use of the intensity interferometer for optical measurements on stars was demonstrated by Hanbury Brown and Twiss (1956a), shortly after the success of the radio intensity interferometer described in Sects. 1.3.7 and 17.1. At that time, the possibility of coherence between photons in different light rays from the same source was questioned, and the physical basis and consistency with quantum mechanics is explained by Hanbury Brown and Twiss (1956c) and Purcell (1956). The laboratory demonstration of the correlation of intensity fluctuations of light by Hanbury Brown and Twiss (1956b) led to the appreciation of the phenomenon of photon bunching and to the broader development of quantum statistical studies and to their application to particle beams as well as electromagnetic radiation (Henny et al. 1999).

In the optical intensity interferometer, a photomultiplier tube at the focus of each telescope mirror replaces the RF and IF stages and the detectors of the radio instrument. The photomultiplier outputs are amplified and fed to the inputs of the correlator. The optical intensity interferometer is largely insensitive to atmospheric phase fluctuations, as explained for the radio case in Sect. 17.1. The size of the light-gathering apertures is therefore unrestricted by the scale size of the irregularities. Also, it is not necessary that the reflecting mirrors produce a diffraction-limited image, and their accuracy need only be sufficient to deliver all the light to the photomultiplier cathodes. This is fortunate since the low sensitivity mentioned earlier for the radio case necessitates the use of large light-gathering areas. Hanbury Brown (1974) gave an analysis of the response of the optical instrument and showed that it is proportional to the square of the visibility modulus as in the radio case. Either a correlator or a photon coincidence counter can be used to combine the photomultiplier outputs.

The intensity interferometer constructed at Narrabri, Australia (Hanbury Brown et al. 1967; Hanbury Brown 1974), used two 6.5-m-diameter reflectors and a bandwidth of 60 MHz for the signals at the correlator inputs. The resulting limiting magnitude of +2.5 enabled measurements of 32 stars to be made. Davis (1976) has discussed the relative merits of the intensity interferometer and modern implementations of the Michelson interferometer for development of more sensitive instruments.

17.6.4   Speckle Imaging

The image of an unresolved point source observed with a telescope of which the width of the aperture is large compared with the Fried length d_f depends on the exposure time over which the image is averaged. An exposure no longer than 10 ms shows a group of bright speckles, each of which is the approximate size of the Airy disk (i.e., the diffraction-limited point-source image) of the telescope. If the exposure is much longer, the pattern is blurred into a single patch (the "seeing" disk) of typical diameter $1''$, determined by the atmosphere. The characteristic fluctuation time of 10 ms in the optical range corresponds to the time taken for an atmospheric cell of size $d_f \simeq 0.14$ m to move past any point in the telescope aperture at a typical wind speed of 10–20 m s^{-1}. The use of sequences of short-exposure images to obtain information at the diffraction limit of a large telescope is known as speckle imaging. Speckle patterns reflect the random distribution of atmospheric irregularities over the aperture and differ from one exposure to the next on the 10-ms timescale. Reduction of many exposures is required to observe faint objects by this technique.

For the theory of the speckle response, see, for example, Dainty (1973), Bates (1982), or Goodman (1985). Here we note that the high-resolution image represented by a single speckle can be understood if one considers each speckle as resulting from several seeing cells of the wavefront, located at points distributed across the telescope aperture. These cells are the ones that present approximately equal phase shifts in the ray paths from the wavefront to the speckle image (Worden 1977). Then, by analogy with an array of antennas, the resolution corresponds to the maximum spacing of the cells, that is, it is of the order λ/d, where d is the telescope aperture. Aberrations in the reflector do not significantly degrade the speckle pattern as long as the dominant phase irregularities are those of the atmosphere. The area of the image over which the speckles are spread corresponds to λ/d_f on the sky and becomes the seeing disk in a long exposure. The seeing cells can be regarded as subapertures within the main telescope aperture, the responses of which combine with random phases in the image. The number of speckles is of the order of the number of subapertures, that is, $(d/d_f)^2$. With a large telescope ($d \sim 1$ m), this number is of the order of 50 at optical wavelengths. Also, the size of the seeing cells increases with wavelength, and in the infrared, only a few speckles appear in the image.

A rather simple image restoration technique called the "shift-and-add" algorithm can be applied to speckle images (Christou 1991). It works best when there is a point source in the field, and at infrared wavelengths where there are relatively few speckles per frame and the isoplanatic patch is relatively large (see Table 17.1). The short exposure speckle frames are aligned on their brightest speckles and summed. The point-spread function ("dirty beam"), which can be obtained from the image of a point source within the field, will have a diffraction-limited component and a much broader component composed of the fainter speckles. This step can be followed by other restoration algorithms such as CLEAN (see Sect. 11.1) to improve the image quality further [see, e.g., Eckart et al. (1994)].

When the shift-and-add algorithm is not applicable, the modulus of the visibility can be obtained by the technique of speckle interferometry, which originated with Labeyrie (1970). This procedure can be understood from the following simplified discussion. On a single image of short exposure, a number of approximately diffraction-limited speckles appear at random locations within the seeing disk. The speckle image $I_s(l, m)$ can be described as the convolution of the actual intensity distribution $I(l, m)$ with the speckle point-spread function $\mathcal{P}(l, m)$. Thus,

$$I_s(l, m) = I(l, m) * * \mathcal{P}(l, m) . \tag{17.36}$$

The function $\mathcal{P}(l, m)$ is a random function that cannot be specified exactly. As a first approximation, we will assume that $\mathcal{P}(l, m)$ is the point-spread function of the telescope in the absence of atmospheric effects, $b_0(l, m)$, replicated at the position of each speckle. Thus, we can write

$$\mathcal{P}(l, m) = \sum b_0(l - l_i, m - m_i) , \tag{17.37}$$

where l_i and m_i are the locations of the speckles, all of which are assumed to have the same intensity. From Eqs. (17.36) and (17.37), we obtain

$$I_s(l, m) = \sum I(l, m) * * b_0(l - l_i, m - m_i) \tag{17.38}$$

If the Fourier transform of $b_0(l, m)$ is $\overline{b}_0(u, v)$, then the Fourier transform of $b_0(l - l_i, m - m_i)$ is $\overline{b}_0(u, v) \exp[j2\pi(ul_i + vm_i)]$. Hence, the Fourier transform of Eq. (17.38) can be written as

$$\overline{I}_s(u, v) = \sum \mathcal{V}(u, v)\, \overline{b}_0(u, v)\, e^{j2\pi(ul_i + vm_i)} , \tag{17.39}$$

where $\mathcal{V}$ and $\overline{I}$ are the Fourier transforms of I and I_s, respectively. The speckle transforms $\overline{I}_s$ cannot be summed directly because of random phase factors in Eq. (17.39). To eliminate these phase factors, we calculate $|\overline{I}_s|^2$ (i.e., $\overline{I}_s \overline{I}_s^*$), which is

$$|\overline{I}_s(u, v)|^2 = \sum_i \sum_k |\mathcal{V}(u, v)|^2 |\overline{b}_0(u, v)|^2\, e^{j2\pi[u(l_i - l_k) + v(m_i - m_k)]}$$

$$= |\mathcal{V}(u, v)|^2 |\overline{b}_0(u, v)|^2 \left[N + \sum_{i \neq k} e^{j2\pi[u(l_i - l_k) + v(m_i - m_k)]} \right] , \tag{17.40}$$

where N is the number of speckles. Since the expectation of the summation term in the second line of Eq. (17.40) is zero, the expectation of Eq. (17.40) is

$$\langle |\overline{I}_s(u, v)|^2 \rangle = N_0 |\mathcal{V}(u, v)|^2 |\overline{b}_0(u, v)|^2 , \tag{17.41}$$

where N_0 is the average number of speckles. Hence, the average of a series of measurements of $|I_s(u, v)|^2$, estimated from short exposures, is proportional to the squared modulus of $\mathcal{V}(u, v)$ times the squared modulus of $\overline{b}_0(u, v)$. Since $\overline{b}_0(u, v)$ is nonzero for $|u|$ and $|v| < D/\lambda$, the function $|\mathcal{V}(u, v)|^2$ can be determined over the same range of u and v, if $\overline{b}_0(u, v)$ is known. In practice, the speckles cannot be accurately modeled by Eq. (17.37). However, we can write

$$\langle |\bar{I}_s(u, v)|^2 \rangle = |\mathcal{V}(u, v)|^2 \langle |\overline{\mathcal{P}}(u, v)|^2 \rangle , \qquad (17.42)$$

where $\overline{\mathcal{P}}(u, v)$ is the Fourier transform of $\mathcal{P}(l, m)$. From Eqs. (17.41) and (17.42), $\langle |\overline{\mathcal{P}}(u, v)|^2 \rangle$ should be approximately proportional to $|\bar{b}_0(u, v)|^2$. It can be estimated by observing a point source under the same conditions as those for the source under study.

The phase information can be extracted from the speckle frames but with considerably more computational effort. Most phase-retrieval algorithms are variations of two basic methods: the Knox–Thompson, or cross-spectral, method (Knox and Thompson 1974; Knox 1976) and the bispectrum method (Lohmann et al. 1983). These methods are described in detail by Roggemann et al. (1997).

Further Reading in Optical Interferometry

Labeyrie, A., Lipson, S.G., and Nisenson, P., *An Introduction to Optical Stellar Interferometry*, Cambridge Univ. Press, Cambridge, UK (2006)

Lawson, P.R., Ed., *Selected Papers on Long Baseline Stellar Interferometry*, SPIE Milestone Ser., MS139, SPIE, Bellingham, WA (1997)

Lawson, P.R., Ed., *Principles of Long Baseline Stellar Interferometry*, Course Notes from the 1999 Michelson Summer School, Jet Propulsion Laboratory, Pasadena, CA (2000)

Léna, P.J., and Quirrenbach, A., Eds., *Interferometry in Optical Astronomy*, Proc. SPIE, **4006**, SPIE, Bellingham, WA (2000)

Reasenberg, R.D., Ed., *Astronomical Interferometry*, Proc. SPIE, **3350**, SPIE, Bellingham, WA (1998)

Robertson, J.G., and Tango, W.J., Eds., *Very High Angular Resolution Imaging*, IAU Symp. 158, Kluwer, Dordrecht, the Netherlands (1994)

Saha, S.K., *Aperture Synthesis*, Springer, New York (2011)

Shao, M., and Colavita, M.M., Long-Baseline Optical and Stellar Interferometry, *Ann. Rev. Astron. Astrophys.*, **30**, 457–498 (1992)

ten Brummelaar, T., Tuthill, P., and van Belle, G., *J. Astron. Instrum.*, Special Issue on Optical and Infrared Interferometry, **2** (2013)

References

Allen, R.J., and Böker, T., Optical Interferometry and Aperture Synthesis in Space with the Space Interferometry Mission, in *Astronomical Interferometry*, Reasenberg, R.D., Ed., Proc. SPIE, **3350**, 561–570 (1998)

Anterrieu, E., A Resolving Matrix Approach for Synthetic Aperture Imaging Radiometers, *IEEE Trans. Geosci. Remote Sensing*, **42**, 1649–1656 (2004)

Armstrong, J.T., Hutter, D.J., Baines, E.K., Benson, J.A., Bevilacqua, R.M., Buschmann, T., Clark III J.H., Ghasempour, A., Hall, J.C., Hindsley, R.B., and ten coauthors, The Navy Precision Optical Interferometer (NPOI): An Update, *J. Astron. Instrum.*, **2**, 1340002 (8pp) (2013)

Armstrong, J.T., Mozurkewich, D., Rickard, L.J., Hutter, D.J., Benson, J.A., Bowers, P.F., Elias II N.M., Hummel, C.A., Johnston, K.J., Buscher, D.F., and five coauthors, The Navy Prototype Optical Interferometer, *Astrophys. J.*, **496**, 550–571 (1998)

Assus, P., Choplin, H., Corteggiani, J.P., Cuot, E., Gay, J., Journet, A., Merlin, G., and Rabbia, Y., L'Interféromètre Infrarouge du C.E.R.G.A., *J. Opt. (Paris)*, **10**, 345–350 (1979)

Baldwin, J.E., Beckett, M.G., Boysen, R.C., Burns, D., Buscher, D.F., Cox, G.C., Haniff, C.A., Mackay, C.D., Nightingale, N.S., Rogers, J., and six coauthors, The First Images from an Optical Aperture Synthesis Array: Mapping of Capella with COAST at Two Epochs, *Astron. Astrophys.*, **306**, L13–L16 (1996)

Baldwin, J.E., Boysen, R.C., Cox, G.C., Haniff, C.A., Rogers, J., Warner, P.J., Wilson, D.M.A., and Mackay, C.D., Design and Performance of COAST, *Amplitude and Intensity Spatial Interferometry. II*, Breckinridge, J.B., Ed., Proc. SPIE, **2200**, 118–128 (1994)

Bates, R.H.T., Astronomical Speckle Imaging, *Phys. Rep.*, **90**, 203–297 (1982)

Bates, R.H.T., Uniqueness of Solutions to Two-Dimensional Fourier Phase Problems for Localized and Positive Images, *Comp. Vision, Graphics, Image Process.*, **25**, 205–217 (1984)

Bennett, J. C., Anderson, A.P., and McInnes, P.A., Microwave Holographic Metrology of Large Reflector Antennas, *IEEE Trans. Antennas Propag.*, **AP-24**, 295–303 (1976)

Benson, J.A., Hutter, D.J., Elias, N.M., Bowers, P.F., Johnston, K.J., Haijian, A.R., Armstrong, J.T., Mozurkewich, D., Pauls, T.A., Rickard, L.J., and four coauthors, Multichannel Optical Aperture Synthesis Imaging of Eta 1 Ursae Majoris with the Navy Optical Prototype Interferometer, *Astron. J.*, **114**, 1221–1226 (1997)

Bester, M., Danchi, W.C., Degiacomi, C.G., Greenhill, L.J., and Townes, C.H., Atmospheric Fluctuations: Empirical Structure Functions and Projected Performance of Future Instruments, *Astrophys. J.*, **392**, 357–374 (1992)

Bester, M., Danchi, W.C., and Townes, C.H., Long Baseline Interferometer for the Mid-Infrared, *Amplitude and Intensity Spatial Interferometry*, Breckinridge, J.B., Ed., Proc. SPIE, **1237**, 40–48 (1990)

Böker, T., and Allen, R.J., Imaging and Nulling with the Space Interferometer Mission, *Astrophys. J. Suppl.*, **125**, 123–142 (1999)

Bracewell, R.N., Radio Interferometry of Discrete Sources, *Proc. IRE*, **46**, 97–105 (1958)

Bracewell, R.N., and MacPhie, R.H., Searching for Nonsolar Planets, *Icarus*, **38**, 136–147 (1979)

Carr, T.D., Lynch, M.A., Paul, M.P., Brown, G.W., May, J., Six, N.F., Robinson, V.M., and Block, W.F., Very Long Baseline Interferometry of Jupiter at 18 MHz, *Radio Sci.*, **5**, 1223–1226 (1970)

Cash, W., Shipley, A., Osterman, S., and Joy, M., Laboratory Detection of X-Ray Fringes with a Grazing-Incidence Interferometer, *Nature*, **407**, 160–162 (2000)

Caves, C.M., Quantum Limits on Noise in Linear Amplifiers, *Phys. Rev.*, **26D**, 1817–1839 (1982)

Chen, M.T., Tong, C.-Y.E., Blundell, R., Papa, D.C., and Paine, S., Receiver Beam Characterization for the SMA, in *Advanced Technology MMW, Radio, and Terahertz Telescopes*, Phillips, T.G., Ed., Proc. SPIE, **3357**, 106–113 (1998)

Christou, J.C., Infrared Speckle Imaging: Data Reduction with Application to Binary Stars, *Experimental Astron.*, **2**, 27–56 (1991)

Clark, B.G., and Kuz'min, A.D., The Measurement of the Polarization and Brightness Distribution of Venus at 10.6-cm Wavelength, *Astrophys. J.*, **142**, 23–44 (1965)

Cohen, M.H., High Resolution Observations of Radio Sources, *Ann. Rev. Astron. Astrophys.*, **7**, 619–664 (1969)

Colavita, M.M., Shao, M., and Staelin, D.H., Two-Color Method for Optical Astrometry: Theory and Preliminary Measurements with the Mark III Stellar Interferometer, *Appl. Opt.*, **26**, 4113–4122 (1987)

Colavita, M.M., Wallace, J.K., Hines, B.E., Gursel, Y., Malbet, F., Palmer, D.L., Pan, X.P., Shao, M., Yu, J.W., Boden, A.F., and seven coauthors, The Palomar Testbed Interferometer, *Astrophys. J.*, **510**, 505–521 (1999)

Corbella, I., Duffo, N., Vall-Ilossera, M., Camps, A., and Torres, F., The Visibility Function in Interferometric Aperture Synthesis Radiometry, *IEEE Trans. Geosci. Remote Sensing*, **42**, 1677–1682 (2004)

Currie, D.G., Knapp, S.L., and Liewer, K.M., Four Stellar-Diameter Measurements by a New Technique: Amplitude Interferometry, *Astrophys. J.*, **187**, 131–134 (1974)

D'Addario, L.R., Holographic Antenna Measurements: Further Technical Considerations, 12-Meter Millimeter Wave Telescope Memo 202, National Radio Astronomy Observatory (1982)

Dainty, J.C., Diffraction-Limited Imaging of Stellar Objects Using Telescopes of Low Optical Quality, *Opt. Commun.*, **7**, 129–134 (1973)

Davis, J., High-Angular-Resolution Stellar Interferometry, *Proc. Astron. Soc. Aust.*, **3**, 26–32 (1976)

Davis, J., Lawson, P.R., Booth, A.J., Tango, W.J., and Thorvaldson, E.D., Atmospheric Path Variations for Baselines Up to 80 m Measured with the Sydney University Stellar Interferometer, *Mon. Not. R. Astron. Soc.*, **273**, L53–L58 (1995)

Davis, J., and Tango, W.J., The Sydney University 11.4 m Prototype Stellar Interferometer, *Proc. Astron. Soc. Aust.*, **6**, 34–38 (1985)

Davis, J., Tango, W.J., Booth, A.J., ten Brummelaar, T.A., Minard, R.A., and Owens, S.M., The Sydney University Stellar Interferometer—I. The Instrument, *Mon. Not. R. Astron. Soc.*, **303**, 773–782 (1999a)

Davis, J., Tango, W.J., Booth, A.J., Thorvaldson, E.D., and Giovannis, J., The Sydney University Stellar Interferometer—II. Commissioning Observations and Results, *Mon. Not. R. Astron. Soc.*, **303**, 783–791 (1999b)

de Graauw, T., and van de Stadt, H., Coherent Versus Incoherent Detection for Interferometry at Infrared Wavelengths, *Proc. ESO Conf. Scientific Importance of High Angular Resolution at Infrared and Optical Wavelengths*, Ulrich, M.H., and Kjär, K., Eds., European Southern Observatory, Garching (1981)

Dobson, M.C., Ulaby, F.T., Hallikainen, M.T., and El-Rayes, M.A., Microwave Dielectric Behavior of Wet Soil—Part II: Dielectric Mixing Models, *IEEE Trans. Geosci. Remote Sensing*, **GE-23**, 35–46 (1985)

Dulk, G.A., Characteristics of Jupiter's Decametric Radio Source Measured with Arc-Second Resolution, *Astrophys. J.*, **159**, 671–684 (1970)

Eckart, A., Genzel, R., Hofmann, R., Sams, B.J., Tacconi-Garman, L.E., and Cruzalebes, P., Diffraction-Limited Near-Infrared Imaging of the Galactic Center, in *The Nuclei of Normal Galaxies*, Genzel, R., and Harris, A., Eds., Kluwer, Dordrecht, the Netherlands (1994), pp. 305–315

Eddington, A.S., Note on Major MacMahon's Paper "On the Determination of the Apparent Diameter of a Fixed Star," *Mon. Not. R. Astron. Soc.*, **69**, 178–180 (1909)

Evans, D.S., Edwards, D.A., Frueh, M., McWilliam, A., and Sandmann, W., Photoelectric Observations of Lunar Occultations. XV, *Astron. J.*, **90**, 2360–2371 (1985)

Fizeau, H., Prix Bordin: Rapport sur le concours de l'année 1867, *Comptes Rendus des Séances de L'Académie des Sciences*, **66**, 932–934 (1868)

Font, J., Boutin, J., Reul, N., Spurgeon, P., Ballabrera-Poy, J., Chuprin, A., Gabarró, C., Gourrion, J., Guimbard, S., Hénocq, C., and 17 coauthors, SMOS First Data Analysis for Sea Surface Salinity Determination, *Int. J. Remote Sensing*, **34**, 3654–3670 (2012)

Fried, D.L., Optical Resolution Through a Randomly Inhomogenious Medium for Very Long and Very Short Exposures, *J. Opt. Soc. Am.*, **56**, 1372–1379 (1966)

Goodman, J.W., *Statistical Optics*, Wiley, New York (1985), pp. 441–459

Hanbury Brown, R., *The Intensity Interferometer*, Taylor and Francis, London (1974)

Hanbury Brown, R., Davis, J., and Allen, L.R., The Stellar Interferometer at Narrabri Observatory. I, *Mon. Not. R. Astron. Soc.*, **137**, 375–392 (1967)

Hanbury Brown, R., and Twiss, R.Q., A New Type of Interferometer for Use in Radio Astronomy, *Philos. Mag.*, Ser. 7, **45**, 663–682 (1954)

Hanbury Brown, R., and Twiss, R.Q., A Test of a New Type of Stellar Interferometer on Sirius, *Nature*, **178**, 1046–1048 (1956a)

Hanbury Brown, R., and Twiss, R.Q., Correlation Between Photons in Two Coherent Light Beams, *Nature*, **177**, 27–29 (1956b)

Hanbury Brown, R., and Twiss, R.Q., A Question of Correlation Between Photons in Coherent Light Rays, *Nature*, **178**, 1447–1448 (1956c)

Haniff, C.A., Mackay, C.D., Titterington, D.J., Sivia, D., Baldwin, J.E., and Warner, P.J., The First Images from Optical Aperture Synthesis, *Nature*, **328**, 694–696 (1987)

Hazard, C., Lunar Occultation Measurements, in *Methods of Experimental Physics*, Vol. 12, Part C *(Astrophysics: Radio Observations)*, Meeks, M.L., Ed., Academic Press, New York (1976), pp. 92–117

Hazard, C., Mackey, M.D., and Shimmins, A.J., Investigation of the Radio Source 3C273 by the Method of Lunar Occultations, *Nature*, **197**, 1037–1039 (1963)

Heffner, H., The Fundamental Noise Limit of Linear Amplifiers, *Proc. IRE*, **50**, 1604–1608 (1962)

Henny, M., Oberholzer, S., Strunk, C., Heinzel, T., Ensslin, K., Holland, M., and Schönenberger, C., The Fermionic Hanbury Brown and Twiss Experiment, *Science*, **284**, 296–298 (1999)

Hinz, P.M., Angel, J.R.P., Hoffmann, W.F., McCarthy Jr. D.W., McGuire, P.C., Cheselka, M., Hora, J.L., and Woolf, N.J., Imaging Circumstellar Environments with a Nulling Interferometer, *Nature*, **395**, 251–253 (1998)

Jackson, J.D., *Classical Electrodynamics*, Wiley, New York (1998)

Jankov, S., Astronomical Optical Interferometry. 1. Methods and Instrumentation, *Serb. Astron. J.*, **181**, 1–17 (2010)

Jennison, R.C., and Das Gupta, M.K., The Measurement of the Angular Diameter of Two Intense Radio Sources, Parts I and II, *Philos. Mag.*, Ser. 8, **1**, 55–75 (1956)

Johnson, M.A., Betz, A.L., and Townes, C.H., 10-μm Heterodyne Stellar Interferometer, *Phys. Rev. Lett.*, **33**, 1617–1620 (1974)

Jorgensen, A.M., Schmitt, H.R., van Belle, G.T., Mozurkewich, D., Hutter, D., Armstrong, J.T., Baines, E.K., Restaino, S., and Hall, T., Coherent Integration in Optical Interferometry, in *Optical and Infrared Interferometry III*, Proc. SPIE, **8445**, 844519 (2012)

Kerr, Y.H., Waldteufel, P., Wigneron, J.-P., Delwart, S., Cabot, F., Boutin, J., Escorihuela, M.-J., Font, J., Reul, N., Gruhier, C., and five coauthors, The SMOS Mission: New Tool for Monitoring Key Elements of the Global Water Cycle, *Proc. IEEE*, **98**, 666–687 (2010)

Kerr, Y.H., Waldteufel, P., Richaume, P., Wigneron, J.P., Ferrazzoli, P., Mahmoodi, A., Al Bitar, A., Cabot, F., Gruhier, C., Enache Juglea, S., and three coauthors, The SMOS Soil Moisture Retrieval Algorithm, *IEEE Trans. Geosci. Remote Sensing*, **50**, 1384–1403 (2012)

Knox, K.T., Image Retrieval from Astronomical Speckle Patterns, *J. Opt. Soc. Am.*, **66**, 1236–1239 (1976)

Knox, K.T., and Thompson, B.J., Recovery of Images from Atmospherically Degraded Short-Exposure Photographs, *Astrophys. J. Lett.*, **193**, L45–L48 (1974)

Labeyrie, A., Attainment of Diffraction-Limited Resolution in Large Telescopes by Fourier Analysing Speckle Patterns in Star Images, *Astron. Astrophys.*, **6**, 85–87 (1970)

Labeyrie, A., Interference Fringes Obtained on Vega with Two Optical Telescopes, *Astrophys. J. Lett.*, **196**, L71–L75 (1975)

Le Bouquin, J.-B., Berger, J.-P., Lazareff, B., Zins, G., Haguenauer, P., Jocou, L., Kern, P., Millan-Gabet, R., Traub, W., Absil, O., and 36 coauthors, PIONIER: A Four-Telescope Visitor Instrument at VLTI, *Astron. Astrophys.*, **535**, A67 (14pp) (2011)

Lohmann, A.W., Weigelt, G., and Wirnitzer, B., Speckle Masking in Astronomy: Triple Correlation Theory and Applications, *Appl. Optics*, **22**, 4028–4037 (1983)

MacMahon, P.A., On the Determination of the Apparent Diameter of a Fixed Star, *Mon. Not. R. Astron. Soc.*, **69**, 126–127 (1909)

McMullan, K.D., Brown, M.A., Martín-Neira, M., Rits, W., Ekholm, S., Martí, J., and Lemanczyk, J., SMOS: The Payload, *IEEE Trans. Geosci. Remote Sensing*, **46**, 594–605 (2008)

Michelson, A.A., and Pease, F.G., Measurement of the Diameter of α Orionis with the Interferometer, *Astrophys. J.*, **53**, 249–259 (1921)

Milonni, P.W., Resource Letter: Orionis with the Interferometer AOA-1: Adaptive Optics in Astronomy, *Am. J. Phys.*, **67**, 476–485 (1999)

Misell, D.L., A Method for the Solution of the Phase Problem in Electron Microscopy, *J. Phys. D.*, **6**, L6–L9 (1973)

Morris, D., Phase Retrieval in the Radio Holography of Reflector Antennas and Radio Telescopes, *IEEE Trans. Antennas Propag.*, **AP-33**, 749–755 (1985)

Morris, D., Baars, J.W.M., Hein, H., Steppe, H., Thum, C., and Wohlleben, R., Radio-Holographic Reflector Measurement of the 30-m Millimeter Radio Telescope at 22 GHz with a Cosmic Signal Source, *Astron. Astrophys.*, **203**, 399–406 (1988)

Mourard, D., Tallon-Bosc, I., Blazit, A., Bonneau, D., Merlin, G., Morand, F., Vakili, F., and Labeyrie, A., The G12T Interferometer on Plateau de Calern, *Astron. Astrophys.*, **283**, 705–713 (1994)

Napier, P.J., and Bates, R.H.T., Antenna-Aperture Distributions from Holographic Type of Radiation-Pattern Measurements, *Proc. IEEE*, **120**, 30–34 (1973)

Petrov, R.G., Malbet, F., Weigelt, G., Antonelli, P., Beckmann, U., Bresson, Y., Chelli, A., Dugué, M., Duvert, G., Gennari, S., and 88 coauthors, AMBER, The Near-Infrared Spectro-Interferometric Three-Telescope VLTI Instrument, *Astron. Astrophys.*, **464**, 1–12 (2007)

Purcell, E.M., A Question of Correlation between Photons in Coherent Light Rays, *Nature*, **178**, 1449–1450 (1956)

Ragland, S., and Richichi, A., Detection of a Sub-Arcsecond Dust Shell around the Wolf–Rayet Star WR112, *Mon. Not. R. Astron. Soc.*, **302**, L13–L16 (1999)

Richichi, A., Lunar Occultations, in *Very High Angular Resolution Imaging*, IAU Symp. 158, Robertson, J.G., and Tango, W.J., Eds., Kluwer, Dordrecht, the Netherlands (1994), pp. 71–81

Rodríguez-Fernández, N.J., Aires, F., Richaume, P., Kerr, Y.H., Prigent, C., Kolassa, J., Cabot, F., Jiménez, C., Mahmoodi, A., and Drusch, M., Soil Moisture Retrieval Using Neural Networks: Application to SMOS, *IEEE Trans. Geosci. Remote Sensing*, **53**, 5991–6006 (2015)

Rogers, A.E.E., Barvainis, R., Charpentier, P.J., and Corey, B.E., Corrections for the Effect of a Radome on Antenna Surface Measurements Made by Microwave Holography, *IEEE Trans. Antennas Propag.*, **AP-41**, 77–84 (1993)

Roggemann, M.C., Welch, B.M., and Fugate, R.Q., Improving the Resolution of Ground-Based Telescopes, *Rev. Mod. Phys.*, **69**, 437–505 (1997)

Rogstad, D.H., A Technique for Measuring Visibility Phase with an Optical Interferometer in the Presence of Atmospheric Seeing, *Appl. Opt.*, **7**, 585–588 (1968)

Scheuer, P.A.G., On the Use of Lunar Occultations for Investigating the Angular Structure of Radio Sources, *Aust. J. Phys.*, **15**, 333–343 (1962)

Scott, P.F., and Ryle, M., A Rapid Method for Measuring the Figure of a Radio Telescope Reflector, *Mon. Not. R. Astron. Soc.*, **178**, 539–545 (1977)

Serabyn, E., Phillips, T.G., and Masson, C.R., Surface Figure Measurements of Radio Telescopes with a Shearing Interferometer, *Appl. Optics*, **30**, 1227–1241 (1991)

Shao, M., SIM: The Space Interferometry Mission, in *Astronomical Interferometry*, Reasenberg, R.D., Ed., Proc. SPIE, **3350**, 536–540 (1998)

Shao, M., Colavita, M., Hines, B.E., Staelin, D.H., Hutter, H.J., Johnston, K.J., Mozurkewich, D., Simon, R.S., Hershey, J.L., Hughes, J.A., and Kaplan, G.H., The Mark III Stellar Interferometer, *Astron. Astrophys.*, **193**, 357–371 (1988)

Sridharan, T.K., Saito, M., Patel, N.A., and Christensen, R.D., Holographic Surface Setting of the Submillimeter Array Antennas, in *Astronomical Structures and Mechanisms Technology*, Antebi, J., and Lemke, D., Eds., Proc. SPIE, 5495, 441–446 (2004)

Stéphan, E., Sur l'extrême petitesse du diamètre apparent des étoiles fixes, *Comptes Rendus des Séances de L'Académie des Sciences*, 78, No. 15 (meeting of April 13, 1874), 1008–1012 (1874)

Swenson, Jr. G.W., Gardner, C.S., and Bates, R.H.T., Optical Synthesis Telescopes, in *Infrared, Adaptive, and Synthetic Aperture Optical Systems*, Proc. SPIE, 643, 129–140 (1986)

Taylor, J.H., and De Jong, M.L., Models of Nine Radio Sources from Lunar Occultation Observations, *Astrophys. J.*, 151, 33–42 (1968)

ten Brummelaar, T.A., McAlister, H.A., Ridgway, S.T., Bagnuolo Jr. W.G., Turner, N.H., Sturmann, L., Sturmann, J., Berger, D.H., Ogden, C.E., Cadman, R., and three coauthors, First Results from the Chara Array. II. A Description of the Instrument, *Astrophys. J.*, 628, 453–465 (2005)

Tingay, S.J., Kaplan, D.L., McKinley, B., Briggs, F., Wayth, R.B., Hurley-Walker, N., Kennewell, J., Smith, C., Zhang, K., Arcus, W., and 53 coauthors, On the Detection and Tracking of Space Debris Using the Murchison Widefield Array: I. Simulations and Test Observations Demonstrate Feasibility, *Astron. J.*, 146, 103–111 (2013)

Townes, C.H., Bester, M., Danchi, W.C., Hale, D.D.S., Monnier, J.D., Lipman, E.A., Tuthill, P.G., Johnson, M.A., and Walters, D., Infrared Spatial Interferometer, in *Astronomical Interferometry*, Reasonberg, R.D., Ed., Proc. SPIE, 3350, 908–932 (1998)

Townes, C.H., and Sutton, E.C., Multiple Telescope Infrared Interferometry, *Proc. ESO Conf. on Scientific Importance of High Angular Resolution at Infrared and Optical Wavelengths*, Ulrich, M.H., and Kjär, K., Eds., European Southern Observatory, Garching (1981), pp. 199–223

von Hoerner, S., Lunar Occultations of Radio Sources, *Astrophys. J.*, 140, 65–79 (1964)

White, N.M., and Feierman, B.H., A Catalog of Stellar Angular Diameters Measured by Lunar Occultation, *Astrophys. J.*, 94, 751–770 (1987)

Whitford, A.E., Photoelectric Observation of Diffraction at the Moon's Limb, *Astrophys. J.*, 89, 472–481 (1939)

Woolf, N.J., High Resolution Imaging from the Ground, *Ann. Rev. Astron. Astrophys.*, 20, 367–398 (1982)

Worden, S.P., Astronomical Image Reconstruction, *Vistas in Astronomy*, 20, 301–318 (1977)

Author Index

A

Aarons, J., 735, 736
Ables, J.G., 557, 558
Abramowitz, M., 233, 341
Adatia, N.A., 156
Adgie, R.L., 26
Agrawal, G.P., 262
Akiyama, K., 513
Alef, W., 610
Allan, D.W., 428, 447, 472
Allen, C.W., 745
Allen, L.R., 264
Allen, R.J., 831
ALMA Partnership, 708
Altenhoff, W.J., 692
Anantharamaiah, K.R., 227, 784
Anderson, S.B., 316, 336
Andrews, D.G., 660
Apostol, T.M., 93
Appleton, E.V., 18, 725
Archer, J.W., 261, 301
Armstrong, E.H., 46
Armstrong, J.T., 831
Armstrong, J.W., 690, 699, 750
Arsac, J., 174
Asai, K.M., 750
Asaki, Y., 710, 711
Ash, M.E., 542
Ashby, N., 447, 472
Assus, P., 832
Athreya, R., 789
Audoin, C., 429

B

Baade, W., 23
Baan, W.A., 787, 789
Baars, J.W.M., 5, 35, 153, 175, 195, 255, 699
Backer, D.C., 40, 627
Baddour, N., 518
Bagri, D.S., 282
Bailer-Jones, C.A.L., 624
Bailey, D.K., 732
Bajaja, E., 576
Bakker, I., 198
Baldwin, J.E., 564, 569, 827, 831
Ball, J.A., 359, 542
Bally, J., 695
Bare, C., 38, 451
Barlow, M.J., 520
Barnbaum, C., 790
Barnes, J.A., 426, 427, 433, 438
Barnett, M.A.F., 725
Barrett, A.H., 703
Bartel, N., 603, 610, 625
Bates, R.H.T., 569, 819, 828, 836
Battat, J.B., 702
Batty, M.J., 255
Baudry, A., 470
Baver, K.D., 394
Bean, B.R., 679
Beasley, A.J., 610, 612
Beasley, W.L., 788
Beauchamp, K.G., 292
Beaupuits, J.P.P., 700
Becker, R.H., 522

© The Author(s) 2017

A.R. Thompson, J.M. Moran, and G.W. Swenson Jr., *Interferometry and Synthesis in Radio Astronomy*, Astronomy and Astrophysics Library,
DOI 10.1007/978-3-319-44431-4

845

Beevers, C.A., 1
Behrend, D., 394
Bellanger, M.G., 369
Benkevitch, L.V., 343
Bennett, A.S., 27
Bennett, C.L., 535
Bennett, J.C., 819
Benson, J.M, 360
Benson, J.A., 830
Beran, M.J., 777, 782, 785
Berkeland, D.J., 440
Bertotti, B., 631
Bertout, C., 626
Bessel, F.W., 623
Bester, M., 829, 832
Betzig, E., 518
Bevington, P.R., 514, 636
Bevis, M., 679
Beynon, W.J.G., 726
Bhat, N.D.R., 789
Bhatnagar, S., 586, 587
Bieging, J.H., 699
Bignell, R.C., 130, 140
Bilitza, D., 734
Blackman, R.B., 356
Blair, B.E., 426
Blake, G.A., 8
Blum, E.J., 30, 227
Blythe, J.H., 28, 169
Boboltz, D.A., 632
Boccardi, B., 36
Bohlander, R.A., 657
Böker, T., 831
Bolton, J.G., 20, 21, 23
Bolton, R., 698, 700
Booker, H.G., 735, 737, 773
Boonstra, A-J., 143, 149, 790
Booth, R.S., 632
Borella, M.S., 262
Born, M., 1, 91, 104, 105, 121, 767, 782,
      785
Bos, A., 523, 531, 532
Bougeret, J.-L., 746
Bower, G.C., 510, 755
Bowers, F.K., 316, 325, 326, 347
Braccesi, A., 28
Bracewell, R.N., 29, 70, 74–76, 80, 91, 104,
      107, 157, 158, 161, 165, 167, 174,
      175, 177, 178, 242, 255, 312, 352,
      378, 490, 491, 493, 494, 496, 497,
      518, 519, 522, 531, 532, 543, 576,
      583, 684, 748, 767, 772, 773, 813,
      831
Bradley, R.F., 790

Braude, S.Ya., 186
Braun, R., 570
Bregman, J.D., 250, 356
Breit, G., 725
Bremer, M., 702
Brentjens, M.A., 753
Bretteil, S., 789
Bridle, A.H., 11
Briggs, D.S., 495, 496, 501, 562, 563
Briggs, F.H., 789, 791
Brigham, E.O., 160
Brooks, J.W., 139
Brosche, P., 603
Broten, N.W., 30, 37, 451
Brouw, W.N., 97, 175, 499
Brown, G.W., 38
Bruck, Y.M., 569
Brunthaler, A., 40, 627
Bryan, R.K., 558, 559
Budden, K.G., 725
Buderi, R., 712
Bunton, J.D., 366, 369, 370
Burke, B.F., 38, 44, 188
Burn, B.J., 753
Burns, W.R., 316
Burton, W.B., 530
Butler, B.J., 697, 699

C
Calisse, P.G., 697
Callen, H.B., 258
Campbell, R.M., 753
Candès, E.J., 591
Cane, H.V., 196
Cannon, A.J., 10
Cannon, W.H., 407, 451
Cao, H.-M., 304
Carilli, C.L., 36, 187, 691, 699, 710, 711
Carlson, B.R., 287, 300, 368, 473
Carlstrom, J.E., 522
Carr, T.D., 809
Carrillo, R.E., 591
Carter, W.E., 621, 622
Carver, K.R., 121
Cash, W., 832
Casse, J.L., 257
Caves, C.M., 834
Cazemier, W., 792
Chamberlin, R.A., 695
Champeney, D.C., 76
Chandler, S.C., 618
Charlot, P., 603
Chen, M.T., 819

Chennamangalam, J., 369
Chesalın, L.S., 452
Chikada, Y., 360
Chio, T.H., 197
Chow, Y.L., 181
Christiansen, W.N., 29, 33, 543
Christou, J.C., 836
Chu, T.-S., 156
Chung, V.K., 703
Chylek, P., 701
Clark, B.G., 31, 40, 417, 419, 448, 451, 455, 555, 556, 827
Clark, T.A., 40, 442, 448, 451, 622
Clarke, M., 748
Clemence, G.M., 617
Clemmow, P.C., 773
Clifford, S.F., 680
Coe, J.R., 288
COESA, 666
Cohen, M.H., 37, 38, 405, 462, 609, 743, 748, 749, 754, 816, 818
Colavita, M.M., 831
Cole, T.W., 316, 338, 544
Coles, W.A., 750
Colvin, R.S., 221, 227
Combrinck, W.L., 804
Condon, J.J., 5, 11, 28, 606
Contreras, M.E., 522
Conway, J.E., 578, 579, 610, 612
Conway, R.G., 128, 129, 140
Cooley, J.W., 310
Cooper, B.F.C., 316, 321, 325, 326
Corbella, I., 825
Cordes, J.M., 739, 743, 750, 754, 758, 784
Cornwell, T.J., 183, 556–559, 561, 563, 565, 566, 573–575, 577, 579–585, 587, 781, 783, 784
Cotton, W.D., 140, 420–422, 564
Coulman, C.E., 685, 692, 716
Counselman, C.C., 622, 629, 630, 745
Covington, A.E., 30
Cox, A.N., 541, 680
Crane, P.C., 502
Crane, R.K., 657, 662, 735
Crochiere, R.E., 369
Cronyn, W.M., 743, 754
Currie, D.G., 831
Cutler, L.S., 445

D
D'Addario, L.R., 253, 279, 282, 286, 288, 292, 344, 348, 349, 351, 366, 449, 450, 471, 524, 558, 820, 821

Dainty, J.C., 836
Daniell, G.J., 558
Daniell, R.E., 734
Das Gupta, M.K., 24, 26, 809
Davenport, W.B., 48
Davies, J.G., 274
Davies, K., 725
Davies, R.D., 755
Davis, J., 829–831, 835
Davis, J.L., 669
Davis, M.M., 439, 448
Davis, W.F., 348
de Bruijne, J.H.J., 2
de Bruyn, A.G., 753
de Graauw, T., 834
De Jong, M.L., 818
de Pater, I., 76
de Vegt, C., 602
de Vos, M., 115, 187, 195, 255, 369, 588
DeBoer, D.R., 121, 187, 199
Debye, P., 679
Delgado, G., 695
Deller, A.T., 375, 394, 398, 454, 625
Desai, K., 699
Dewdney, P.E., 199, 287, 300, 368
Dewey, R.J., 468
Dhawan, V., 405
Dicke, R.H., 13, 48, 672, 712
Dillon, J.S., 537
Dobson, M.C., 826
Dodson, R., 615
Doeleman, S.S., 394, 439, 440, 510, 755
Doi, A., 394
Dong, W., 790
Douglas, J.N., 240
Downes, D., 543, 692, 699
Drane, C.J., 777, 785
Dravskikh, A.F., 686, 716
Dreyer, J.L.E, 10
Drullinger, R.E., 440
Duffett-Smith, P.J., 754
Dulk, G.A., 196, 251, 809
Dutta, P., 433
Dutton, E.J., 679
Dyson, F.W., 628

E
Eckart, A., 836
Eddington, A.S., 814
Edge, D.O., 10, 27
Edson, W.A., 426, 445
Edwards, P., 394

Ekers, R.D., 523, 524, 532, 572, 585
Elgaroy, O., 26
Elgered, G., 705
Elitzur, M., 6, 631
Ellingson, S.W., 196, 369, 588, 791–793
Ellithorpe, J.D., 590
Elmer, M., 155
Elsmore, B., 255, 600, 601, 603
Emerson, D.T., 295–297
Erickson, W.C., 186, 195, 255, 588, 734, 745,
     748
Escalante, V., 522
Escoffier, R.P., 358, 362, 367, 368
Evans, D.S., 819
Evans, J.V., 725–727, 734, 735, 745
Evans, K.F., 558, 574
Everitt, C.W.F., 625

F
Fanselow, J.L., 645
Farley, T., 316, 321, 322, 325, 326
Feierman, B.H., 819
Fejer, B.G., 735
Feldman, M.J., 221, 260
Fey, A.L., 11, 40, 602, 603
Fiebig, D., 140
Fiedler, R.L., 757
Fienup, J.R., 569
Finkelstein, A.M., 686, 716
Fish, V.L., 513, 756
Fisher, J.R., 197
Fizeau, H., 829
Fomalont, E.B., 31, 32, 133, 511, 603, 630,
     631, 734
Ford, H.A., 474
Forman, P., 442
Fort, D.N., 563
Frail, D.A., 465, 741
Frank, R.L., 447
Frater, R.H., 139, 186, 264
Freeman, R.L., 704
Fridman, P.A., 789
Fried, D.L., 688, 689, 828
Frieden, B.R., 557, 558

G
Gabor, D., 104
Gallagher, J.J., 694
Galt, J., 804
Gapper, G.R., 750
Garay, G., 626

Gardner, F.F., 753
Gardner, F.M., 274
Garsden, H., 591, 592
Gary, D.E., 374, 790
Gauss, K.F., 637
Geldart, D.J.W., 701
Genee, R., 175
Genzel, R., 38, 626, 632, 633
Gilbert, S.W., 447
Gipson, J.M., 621
Gold, B., 160
Gold, T., 40
Goldstein, H., 704
Goldstein, R.M., 310
Goldstein, S.J., 240
Golomb, S.W., 174
Goodman, J., 758
Goodman, J.W., 104, 782, 785, 836
Gordon, M.A., 542, 543
Gower, J.F.R., 28
Gradshteyn, I.S., 538
Granlund, J., 298, 299, 364
Gray, M., 6
Grischkowsky, D., 679
Guilloteau, S.J., 37, 367
Guinot, B., 618
Guiraud, F.O., 705
Gull, S.F., 558
Gupta, Y., 737
Gurwell, M.A., 5, 539
Güsten, R., 140
Gwinn, C.R., 316, 338, 631, 754, 755, 757, 758

H
Hagen, J.B., 316, 321, 322, 325, 326
Hagfors, T., 44, 725–727, 733, 735, 745
Hall, P.J., 199
Hall, T., 637
Hamaker, J.P., 126, 134, 142, 143, 174
Hamilton, W.C., 643
Hampson, G.A., 792, 793
Hanbury Brown, R., 18, 24, 25, 809, 813, 835
Haniff, C.A., 827
Hanson, R.J., 562
Harmuth, H.F., 292
Harris, D.E., 754
Harris, F.C., 369
Harris, F.J., 356
Harvey, P.M., 5
Haskell, R.E., 728
Hasler, N., 522
Hay, S.G., 197, 198

Hazard, C., 28, 601, 814, 818
Heffner, H., 834
Heiles, C., 344, 751
Hellwig, H., 439, 443
Helmboldt, J.F., 737
Henny, M., 835
Herbst, E., 6
Herring, T.A., 40, 454, 459, 617, 620, 622, 645
Herrnstein, J.R., 40
Hess, S.L., 661, 662
Hewish, A., 26, 28, 169, 735, 736, 748, 749, 754
Hey, J.S., 735
Hill, J.M., 630
Hill, R.J., 671, 679, 680, 696
Hills, R.E., 705
Hinder, R.A., 699
Hinkley, N., 440
Hinteregger, H.F., 451, 609
Hinz, P.M., 831
Hirabayashi, H., 43, 188, 474
Hirade, K., 804
Hirota, T., 626
Hjellming, R.M., 180, 407
Ho, C.M., 734, 736
Ho, P.T.P., 37, 187, 367
Hobbs, G., 439
Hobson, M.P., 536
Hocke, K., 613, 737
Högbom, J.A., 175, 551, 552, 554, 558
Hogg, D.C., 690
Hogg, D.E., 34
Hogg, D.W., 636
Holdaway, M.A., 687, 691, 692, 695, 699, 709–711
Holt, E.H., 728
Honma, M., 41, 592, 593, 613
Hooghoudt, H.G., 175
Horn, P.M., 433
Hotan, A.W., 198
Huang, H.-P., 621
Hueckstaedt, R.M., 678
Hughes, M.P., 31
Humphreys, E.M.L., 42
Hunsucker, R.D., 737

I
Ingalls, R.P., 157
International Astronomical Union, 10, 11, 124, 126, 602
Ishiguro, M., 174, 697, 699
ITU-R, 795, 804, 805
Ivashina, M.V., 198

J
Jackson, J.D., 675, 824
Jacobs, I.M., 224
Jacquinot, P., 494
Jaeger, J.C., 747
James, G.L., 142
Jan, M.T., 367
Janardhan, P., 727, 750
Jankov, S., 831
Jansky, K.G., 18
Jaynes, E.T., 558
Jenet, F.A., 316, 336
Jenkins, C.R., 337
Jennison, R.C., 24–26, 505, 809
Johnson, M.A., 832
Johnson, M.D., 336, 338, 339, 418, 757, 758
Johnston, K.J., 602, 603
Jonas, J.L., 199
Jones, D.L., 603
Jones, R.C., 134
Jorgensen, A.M., 831
Jung, T., 615
Junklewitz, H., 579

K
Kalachev, D., 28
Kaplan, D.L., 734, 748
Kardashev, N.S., 43, 188, 191, 474
Kartashoff, P., 438
Kassim, N.E., 588
Kasuga, T., 699
Kaula, W., 40
Kawaguchi, N., 451
Kazemi, S., 489
Kelder, H., 736
Kelker, D.H., 624
Kellermann, K.I., 2, 5, 37, 187, 785
Kelley, M.C., 735
Kemball, A.J., 140
Kerr, A.R., 45, 259, 260
Kerr, F.J., 541
Kerr, Y.H., 825
Keshner, M.S., 433
Kesteven, M.J.L., 11
Keto, E., 172, 183–185, 295
Kim, M.K., 626
Kimberk, R.S., 699
Klemperer, W.K., 393
Kleppner, D., 442, 444–446
Klingler, R.J., 316, 325, 326
Knowles, S.H., 38
Knox, K.T., 838
Ko, H.C., 123

Kobayashi, H., 394
Kochanek, C.S., 589
Kocz, J., 789, 791
Kogan, L.R., 164, 452
Kokkeler, A.B.J., 470
Koles, W.A., 154
König, A., 93
Koyama, Y., 452
Krásná, H., 621
Kraus, J.D., 48, 121, 198, 258
Krishnan, T., 30
Kronberg, P.P., 128, 129, 140, 141, 418
Kukolich, S.G., 713
Kulkarni, S.R., 344, 424
Kundu, M.R., 744
Kuz'min, A.D., 827

L
Labeyrie, A., 831, 836
Labrum, N.R., 30
Lal, D.V., 161
Lamb, J.W., 692
Lambeck, K., 616
Lambert, S.B., 630, 631
Lampton, M., 646
Lannes, A., 509
Lanyi, G.E., 690, 716
Latham, V., 25
Lawrence, C.R., 157
Lawrence, R.S., 735
Lawson, C.L., 562
Lawson, J.L., 224, 318
Lay, O.P., 690, 705, 709
Lazio, T.J., 739, 754
Le Bouquin, J.-B., 832
Le Poncin-Lafitte, C., 630, 631
Lebach, D.E., 630
Leblanc, Y., 745, 746
Lee, S.-S., 394
Leech, J., 175
Leick, A., 447
Leitch, E.M., 535, 536
Lequeux, J., 31
Lesage, P., 429
Leshem, A., 143, 146, 793
Lestrade, J.F., 603
Levine, M.W., 446
Levy, G.S., 42, 473
Lewandowski, W., 447
Lewis, L.L., 436
Li, F., 591
Lichtenstein, M., 694
Liebe, H.J., 670, 677, 678, 693

Lieske, J.H., 617
Lightman, A.P., 4, 5, 9, 538, 671
Linfield, R.P., 35, 43, 473
Lipson, H., 1
Little, A.G., 28, 273
Little, L.T., 749
Liu, C.H., 735
Lo, W.F., 215, 352
Lobanov, A.P., 161, 517
Lochner, M., 511
Lohmann, A.W., 838
Loi, S.T., 736
Longair, M.S., 4
Lonsdale, C.J., 154, 187, 196, 570, 588
Lorimer, D.R., 624
Loudon, R., 44, 675
Lu, R.-S., 513
Lutz, T.E., 624
Lynden-Bell, D., 541
Lyne, A.G., 753

M
Ma, C., 11, 602, 621, 645
Mackay, C.D., 603, 621
MacMahon, P.A., 814
MacPhie, R.H., 780, 785, 831
Madison, D.R., 603, 625
Maltby, P., 31, 511
Mandel, L., 785
Mangum, J.G., 157, 195
Mannucci, A.J., 734
Manoharan, P.K., 750
Marcaide, J.M., 610
Marini, J.W., 669
Markowitz, W., 618
Martin, D.H., 251
Martí-Vidal, I., 516, 517
Masson, C.R., 511, 681, 690, 695–700
Mathur, N.C., 102, 180, 181, 187, 240, 733,
    777
Matsushita, S., 696, 702
Matveenko, L.I., 37
Mayer, C.E., 157
McCarthy, D.D., 620
McClean, D.J., 583
McCready, I.I., 20
McGilchrist, M.M., 583
Meinel, A.B., 198, 466
Melchior, P., 622
Melis, C., 626
Menten, K.M., 626
Menu, J.R., 5
Mercier, R.P., 742

Merritt, F.R., 712
Messier, C., 10
Michelson, A.A., 14
Mickelson, R.L., 239
Middelberg, E., 615
Middleton, D., 224, 318
Millenaar, R.P., 699
Mills, B.Y., 23, 24, 26–29
Milonni, P.W., 829
Miner, G.F., 705
Minkowski, R., 23
Misell, D.L., 823
Misner, C.W., 628
Mitchell, D.A., 587
Miyoshi, M., 42
Moffet, A.T., 31, 174, 175, 511
Morabito, D.D., 699
Morales, M.F., 536
Moran, J.M., 6, 37, 38, 234, 274, 361, 405,
        407, 434, 439, 451, 468, 469, 631,
        634, 690, 703, 734, 745
Morgan, J.S., 398
Morita, K.-I., 37
Morris, D., 26, 126, 821, 823, 824
Morrison, N., 645
Mould, J.R., 543
Mourard, D., 831
Mueller, I.I., 602
Muhleman, D.O., 630
Mullaly, R.F., 29
Murota, K., 804

N
Nagy, A., 725
Napier, P.J., 35, 40, 180, 181, 188, 190, 255,
        351, 394, 395, 502, 559, 567, 569,
        819
Narayan, R., 558, 589, 743, 744, 750, 758, 784
NASA/IPAC, 5, 11
National Radio Astronomy Observatory, 272
Neville, A.C., 34
Newhall, X.X., 542
Ng, C.-V., 517
Niell, A.E., 669
Nikolic, B., 705, 707
Nita, G.M., 790
Nityananda, R., 45, 558
Noordam, J.E., 587
Norris, R.P., 632
Nyquist, H., 12, 257, 312

O
O'Brien, P.A., 33
Offringa, A.R., 585, 790
Okumura, S.K., 359
Oliver, B.M., 45
Olmi, L., 699
O'Neill, E.L., 136
Oort, A.H., 660
Oppenheim, A.V., 312, 378
Oppermann, N., 752
Ortiz-León, G.N., 510
O'Sullivan, J.D., 362
Otter, M., 804
Owens, J.C., 679

P
Padin, S.A., 192, 264, 535, 536
Paek, N., 621
Paine, S., 679, 693, 694, 696, 702
Paley, R.E., 294
Pankonin, V., 805
Papoulis, A., 76, 160, 234
Pardo, J.R., 679, 693, 696
Parkinson, B.W., 447
Parrent, G.B., 777, 778, 780, 782, 783, 785
Parsons, A.D., 365, 537
Pauliny-Toth, I.I.K., 5
Pawsey, J.L., 27–28
Payne, J.M., 251, 257, 274
Pearson, T.J., 39, 565
Pease, F.G., 14, 18, 829
Peebles, P.J.E., 543
Peixoto, J.P., 660
Percival, B.D., 362
Peréz, L.M., 711
Perley, R.A., 35, 36, 133, 180, 255, 256, 261,
        262, 358, 359, 368, 534, 568, 570,
        581–583, 789
Perryman, M.A.C., 2
Petley, B.W., 600
Petroff, E., 790
Petrov, R.G., 831
Phillips, T.G., 257
Pi, X., 736
Picken, J.S., 30
Pickering, E.C., 10
Pierce, J.A., 447
Pilkington, J.D.H., 620
Planck Collaboration, 541
Pol, S.L.C., 671

Pollak, H.O., 501
Ponsonby, J.E.B., 557
Porcas, R.W., 394
Pospieszalski, M.W., 257
Prabu, T., 255, 289
Press, W.H., 433, 646
Preston, J.B., 697
Preston, R.A., 188, 191, 476
Price, R., 322
Price, R.M., 805
Pryke, C., 535
Puplett, E., 251
Purcell, E.M., 835

R
Rabiner, L.R., 160, 369
Racette, P.E., 706
Rademacher, H., 183
Radford, S.J.E., 697
Radhakrishnan, V., 45, 49
Ragland, S., 819
Raimond, E., 127, 175
Räisänen, A.V., 257
Randall, H.M., 711
Rankin, J., 745
Ratcliffe, J.A., 725, 729, 737, 742
Ratner, M.I., 625, 626
Rau, U., 142, 143, 504, 579, 587
Rawlings, S., 187
Ray, P.S., 704
Raza, J., 792, 793
Read, R.B., 30, 221, 255
Readhead, A.C.S., 563–565, 754
Reber, E.E., 662
Reber, G., 18
Reid, M.J., 6, 40, 405, 525, 606, 613, 627, 628, 631, 632
Reid, M.S., 257
Resch, G.M., 705
Rhodes, D.R., 501
Rice, S.O., 234
Richichi, A., 818, 819
Rickett, B.J., 754, 756
Riddle, A.C., 75, 543
Rienecker, M.M., 658, 661
Riley, A.L., 705
Riley, J.M., 630
Rioja, M.J., 440, 610, 615
Roberts, D.H., 140, 737
Roberts, J.A., 157, 490, 494
Robertson, D.S., 621, 622, 630
Robinson, D.K., 514, 636
Roddier, F., 685

Rodríguez-Fernández, N.J., 828
Roelfsema, P., 529, 530
Roger, R.S., 31
Rogers, A.E.E., 40, 236, 403, 422, 423, 425, 434, 439, 451, 452, 462–464, 563, 564, 705, 735, 789, 823
Roggemann, M.C., 829, 838
Rogstad, D.H., 830, 831
Roizen-Dossier, B., 494
Romney, J.D., 360, 362, 366
Root, W.I., 48
Ros, E., 610, 734
Rosen, B.R., 703
Rosenkranz, P.W., 701
Roshi, D.A., 197
Rots, A.H., 572
Rowson, B., 34, 111, 112
Rudge, A.W., 156
Rüeger, J.M., 659, 679, 680
Rutman, J., 426
Ruze, J., 156
Rybicki, G.B., 4, 5, 9, 538, 671
Ryle, M., 13, 18, 19, 22, 23, 26, 28, 31, 33–35, 169, 175, 600, 601, 699, 820
Ryzhik, I.M., 538

S
Saastamoinen, J., 666, 669
Sakurai, T., 141, 749
Salpeter, E.E., 748
Sandstrom, K.M., 626
Sault, R.J., 76, 130, 131, 134, 138–140, 147, 575, 579
Schafer, R.W., 312, 378
Schaper, L.W., 703
Schaubert, D.H., 197
Scheuer, P.A.G., 28, 737, 814, 816
Schilke, P., 8
Schlegel, K., 613, 737
Schunk, R., 725
Schuss, J.J., 804
Schwab, F.R., 377, 420–422, 497, 501, 556, 563, 565, 579
Schwarz, U.J., 553
Scott, P.F., 535, 820
Scott, S.L., 748
Scoville, N.J., 37
Searle, C.L., 445
Seidelmann, P.K., 115, 617
Seielstad, G.A., 187, 630
Sekido, M., 452
Serabyn, E., 821, 824
Shaffer, D.B., 405, 609

Shakeshaft, J.R., 27
Shannon, C.E., 312
Shao, M., 831
Shapiro, I.I., 407, 610, 622, 628, 630, 734
Shen, Z.-Q., 755
Shimoda, K., 442
Sieber, W., 756
Siegman, A.E., 444
Silver, S., 772
Sims, G., 697
Sinclair, M.W., 255
Sivia, D.S., 407, 513, 646
Skilling, J., 558, 559, 646
Slee, O.B., 20
Slepian, D., 494, 501
Smart, W.M., 666, 746
Smegal, R.J., 140
Smirnov, O.M., 142, 143, 586
Smith, E.K., 680
Smith, F.G., 23, 24, 509, 601, 735, 753
Smith, H.M., 619
Smolders, B., 792
Smoot, G.F., 13
Snel, R.C., 157, 195
Snider, J.B., 705
Sodin, L.G., 569
Solheim, F., 705
Southworth, G.C., 18
Sovers, O.J., 407, 622
Spangler, S.R., 141, 749, 754
Spitzer, L., 752
Splatalogue, 712
Spoelstra, T.A., 733, 736
Sramek, R.A., 40, 579, 613, 627, 630, 690, 699, 734
Staelin, D.H., 671, 704
Standish, E.M., 542
Stanimirović, S., 576
Stanley, G.J., 20, 21, 23
Stegun, I.A., 233, 341
Stéphan, E., 829
Stewart, K.P., 196
Stotskii, A.A., 686, 716
Subrahmanya, C.R., 558
Sukumar, S., 732
Sullivan, I.S., 587
Sutton, E.C., 678, 685, 834
Swarup, G., 29, 30, 35, 186, 265, 266
Swenson, G.W., 102, 187, 239, 240, 777, 832
Swope, J.R., 662

T
Taff, L.G., 617
Talmoush, D.A., 705
Tango, W.J., 830, 831
Tanner, A.B., 705
Tatarski, V.I., 611, 685, 713, 716
Taylor, G.B., 534
Taylor, G.I., 687
Taylor, J.H., 818
Tegmark, M., 200
ten Brummelaar, T.A., 831
Thayer, G.D., 679
Thomas, J.B., 454
Thomasson, P., 26, 187, 566
Thompson, A.R., 5, 30, 35, 37, 161, 177, 178, 180, 181, 187, 195, 253, 255, 265, 274, 282, 288, 290, 296, 328, 332, 346, 369, 447, 491, 497, 796, 801, 805
Thompson, D.J., 838
Thornton, D.D., 576
Tingay, S.J., 824
Titov, O., 623
Tiuri, M.E., 48, 227, 257
Toeplitz, O., 183
Tokumaru, M., 750
Toll, J.S., 674
Townes, C.H., 712, 833, 834
Treuhaft, R.N., 690, 716
Trott, C.M., 511
Trotter, A.S., 510, 755
Tucker, J.R., 221, 260
Tukey, J.W., 310, 356
Turrin, R.H., 156
Tuve, M.A., 725
Twiss, R.Q., 24, 509, 813, 835

U
Uhlenbeck, G.E., 224, 318
Ulvestad, J., 613
Unser, M., 159

V
Vaidyanathan, P.P., 369
Valley, S.L., 264
van Albada, G.D., 576
van Ardenne, A., 197
van Cappellen, W.A., 198

van de Stadt, H., 834
van der Tol, S., 587, 790, 792, 793
van der Veen, A.-J., 143, 146, 792, 793
van Dishoeck, E.F., 6
van Gorkom, J.H., 523, 524, 585
van Haarlem, M.P., 195, 394
Van Vleck, J.H., 318, 674, 711, 712
Vander Vorst, A. S., 221
Vanier, J., 437
Vedantham, H.K., 731, 759, 761
VERA, 394
Verbiest, J.P.W., 439, 624
Verschuur, G.L., 785
Vertatschitsch, L., 452
Vessot, R.F.C., 432, 437, 438, 442, 443, 446,
    472
Vinokur, M., 234, 408
Vitkevich, V.V., 28
von Hoerner, S., 818
Vonberg, D., 13, 18

W
Wade, C.M., 119, 603
Wahr, J.M., 617, 618
Wakin, M.B., 591
Walden, A.T., 362
Waldram, E.M., 583
Walker, R.C., 188, 189, 422, 632, 635, 636
Wall, J.V., 337
Wallington, S., 589, 590
Walsh, D., 28, 589
Walsh, J.L., 292
Warburton, J.A., 29, 33, 543
Wardle, J.F.C., 141, 418
Warner, P.J., 564, 569
Waters, J.W., 670, 674, 695
Watson, R.A., 535
Webber, J.C., 257
Weber, R., 789
Weiler, K.W., 126–128, 140, 630
Weinberg, S., 748
Weinreb, S., 257, 261, 310, 316, 366
Weintraub, S., 680
Weisberg, J.M., 751
Welch, P.D., 362
Welch, W.J., 37, 157, 196, 255, 576, 699, 701,
    706
Welton, T.A., 258
Wenger, S., 591
Wengler, M.J., 260
Wernecke, S.J., 558
Westerhout, G., 10

Westfold, K.C., 747
Westwater, E.R., 703–705
White, M., 536
White, N.M., 818
White, R.L., 522
Whiteoak, J.B., 753
Whitford, A.E., 814
Whitney, A.R., 38, 40, 405, 447, 450, 452
Wiaux, Y., 591
Wiedner, M.C., 705
Wieringa, M.H., 579
Wietfeldt, R.D., 450, 451, 465
Wild, J.P., 187
Wilkinson, P.N., 564–566
Will, C.M., 628
Williams, W.F., 155
Willis, A.G., 356
Wilson, T.L., 121
Winterhalter, D., 744
Wolf, E., 1, 91, 104, 105, 121, 767, 782,
    785
Wolszczan, A., 758, 784
Woody, D.P., 257, 260, 692, 706
Woolard, E.W., 617
Woolf, N.J., 685, 829
Wootten, A., 37, 187, 195, 255, 296, 369
Worden, S.P., 836
Wozencraft, J.M., 224
Wright, A.E., 520
Wright, M.C.H., 220, 255, 511, 691, 699
Wu, S.C., 705
Wyithe, J.S.B., 536

Y
Yang, K.S., 265, 266
Yao, S.S., 316
Yee, H.K.C., 563
Yeh, K.C., 735
Yen, J.L., 38, 360, 391
Young, A., 470
Young, A.C., 255
Young, A.T., 748

Z
Zaldarriaga, M., 200
Zauderer, B.A., 711
Zernike, F., 767
Zhang, X., 394
Zheng, H., 200, 537
Zivanovic, S.S., 701
Zorin, A.B., 260

Subject Index

3C sources. *See* Radio source

A-projection, 586
Aberration
   chromatic, 527
   diurnal, 10, 406, 617
Absorption, 726, 735
   bound oscillator model, 675–679
   by water vapor, 703–705
   chopper-wheel calibration, 673
   coefficient, 659, 671, 678, 703
   in clouds (liquid water), 704
   ionospheric, 726, 735
   spectra, 31
   tipping-scan calibration, 672
   tropospheric, 670–673
Academia Sinica, 187
Adaptive calibration, 563, 569. *See also* Hybrid
      mapping; Self-calibration
   comparison of methods, 566
   limitations of, 568
Adaptive optics, 829
Airy disk (diffraction pattern of circular
      aperture), 492, 773, 835
Aliasing, 158, 232, 312, 534, 587. *See also*
      Ringlobe
   suppression in maps, 497–504
Allan variance, 428–430, 438
   atmosphere, 690
   frequency standards, 428, 440–446
Allen–Baumbach formula, 745
ALMA, 14, 37, 195, 693, 697. *See also*
      Chajnantor, Chile

Altazimuth mount, 118–120, 139, 141
Amplifiers, at antennas, 164
Amplitude closure, 508, 564. *See also* Adaptive
      calibration
Amplitude scintillation, 756
Analog processing
   comparison with digital, 309, 312
   Fourier transformation, 544
Analytic signal, 104–107
Antenna spacing coordinates, 109–113
Antenna(s), 100, 153–157, 772
   angular resolution (beamwidth), 1, 772–773
   aperture efficiency, 11, 155, 156
   aperture illumination, 165, 166, 819
   asymmetric feed geometry, 155
   axis offsets, 119–120, 645
   collecting area, 11
   design, 153–157
   feed displacement, 141
   feeds, bandwidth of, 141
   in space, 42, 44, 188–191, 470–476
   measurements, holographic, 200, 819–824
      practical considerations, 821–824
      required sensitivity, 823
   minimum number, 564
   mounts, 118–120
   Naysmith focus, 154
   offset-Cassegrain focus, 154
   polarization
      oppositely polarized. *See* Polarization,
         cross-polarized
      polarization method, 123–126
   prime focus, 155
   received power, 11, 66

© The Author(s) 2017
A.R. Thompson, J.M. Moran, and G.W. Swenson Jr., *Interferometry and Synthesis
in Radio Astronomy*, Astronomy and Astrophysics Library,
DOI 10.1007/978-3-319-44431-4

reception pattern, 67, 68, 90, 100, 166, 788
reflections in structure, 357
response (reception) pattern, 67, 166, 772,
        819
shadowing, 486
shaping of reflector, 155
sidelobe model, 794
surface accuracy, 156
surface measurements, 819–824
temperature, 11
tracking effect on fringe frequency,
        169–172
unblocked aperture, 155
voltage reception (response) pattern,
        100–101, 166, 772, 819
Yagi, 24
APERTIF, 198
Aperture efficiency, 11, 155
Aperture synthesis, 169, 172
Aperture, radiation pattern of, 492, 771–773
Apodization, 494
Area density, 491
Arecibo, Puerto Rico, 784
Arrays
    Arsac, 174
    Bracewell, 174
    circular (ring), 183–187
    closed configurations, 183–187
    collecting area, 100, 156
    correlator, 162–164
    cross-shaped, 26–28, 168, 180. See also
        Mills cross
    grating, 29–30
    linear, 173–178
    minimum redundancy, 174
    mixed, 120
    mosaicking, 570–577
    nontracking, 153, 165
    one-dimensional. See Arrays, linear
    open-ended configurations, 180–182
    phased. See Phased array
    planar, 192–193, 535–536
    Reuleaux triangle, 153, 168, 183–187
    tracking, transfer function of, 169–172
    T-shaped, 168, 180
    two-dimensional, 179–193
    VLBI, 187–191
    Y-shaped, 180–182
ASKAP, 198–199
Astrometry, 2, 407, 600–649, 734
    accuracy, 609
    nutation, 617
    polar motion, 617–621
    precession, 616–617

reference frames, 602–603
    VLBI, 606–610
Atacama Large Millimeter/submillimeter
        Array. See ALMA
Atmosphere, neutral, 657–702
    absorption, 670–679
    brightness temperature, 703–705
    calibration. See Calibration, of atmospheric
    effects
        of clouds, 704
        on astrometry, 609
        on visibility, 680–692
        on VLBI, 685–690
    excess path length, 667–669
    Fried length in, 688, 828
    opacity, 693–695. See also Absorption
    oxygen, 670, 704
    phase fluctuations, 680–692
    refraction, 663–669
    water vapor. See Water vapor
Attenuation in cable, 271
Attenuation in optical fiber, 261
Australia Telescope, 139, 186, 575
Autocorrelation function, 63, 164–167, 224,
        312, 426, 800
    definition, 63, 164–167
    of intensity distribution (image), 569, 827
Autocorrelator, 310
Automatic level control (ALC), 299, 348, 487,
        802
Automatic phase correction, 272, 273
Azimuth, 112, 146

Bandpass calibration, 523–524
Bandwidth
    correlation, 740
    effect in maps, 240–246, 278
    Gaussian, 64
    output (postcorrelator), 225
    pattern, 64, 774
    rectangular, 64, 278
    rms, 463, 608, 641
    synthesis, 462–464, 608
Bartlett weighting (smoothing), 356
Baseband response, 312, 459
Baseline
    calibration, 117, 603–606, 645
    coordinates, 109–113
    definition, 18, 59
    equatorial component, 607
    reference point, 119
    retarded, 405–407
    surveying, 3, 111

Baseline bootstrapping, 831
Baselines, noncoplanar, 97–98, 579–585
Bayesian statistics, 513, 558
Beam
    CLEAN, 552–553
    fan, 26, 176, 543
    fringe frequency, 636
    pencil, 27, 177
    synthesized (dirty), 180, 241–242,
        490–495, 530, 648, 800, 836
Beam-shape effects, 120–121, 491–494
Beamwidth effects, 120–121, 492–495
Besselian year, 617
Bias
    in MEM, 562
    in polarization measurements (Rice
        distribution), 141
    in variance of mean, 638
    in visibility measurements, 412
Bispectrum, 423–424
Bivariate (joint) Gaussian probability
        distribution, 311–312, 644, 741
Blackbody. See Planck formula
Blackman weighting (smoothing), 356, 373
Blackman-Harris weighting (smoothing), 356
Bologna, Italy, 28
Borrego Springs, California, 186
Brewster angle, 121, 827
Brightness, 9, 10
Brunt–Väisälä frequency, 736
Bureau International de L'Heure (BIH), 619
Burst mode, VLBI, 465
Burst radiation, Jupiter, 37, 391

C_n^2, 686
C_{ne}^2, 742
Cables
    attenuation, 270
    reflections, 267, 270, 279
    velocity dispersion, 267
Calibration, 485–489
    bandpass, 523–524
    baseline. See Baseline, calibration
    cables, 101, 588
    gain, 488
    of atmospheric
        delay, 734
        phase, by fast-switching, 709–711
        phase, by paired antennas, 710
        phase, in VLBI, 447
    polarization, 137–142
    sources, 395, 487–489
    spectral-line, 522–530

Calibration sources, 23
Calibrator (source), 487. See also Calibration
        sources
Callen and Welton formula, 258–260
Cambridge, England, 26
    Five-Kilometre Radio Telescope, 35
    fourth survey, 28
    Low-Frequency Synthesis Telescope, 583
    One-Mile Radio Telescope, 33, 34, 175
    third survey, 10, 27
Canadian VLBI array, 38, 453
Cassegrain focus, 154, 155
Causal function, 107
Cavity pulling, 444, 445
CCIR, 805
Celestial
    coordinates, 146
    equator, 179
    sphere, 90
Cell
    averaging, 161, 289, 502, 797
    crossing time, 798
Chajnantor, Chile, 37, 693, 695, 697
Chandler wobble, 618
Chinese hat function, 167, 576
Chopper-wheel method, 673
Circular array, 183–187
Circular polarization, 140
    degree, 122
    IEEE definition, 124
Circulator, 266
Classical electron radius, 738
Clausius–Clapeyron equation, 662
CLEAN algorithm, 72, 178, 551–557
    application, 553–557
    Clark's algorithm, 555–556
    comparison with MEM, 559–562
    Cotton–Schwab algorithm, 556
    extended sources, 556
    hybrid mapping, 563
    loop gain, 552, 554
    self-calibration, 566
    speckle imaging, 835–838
    spectral line data, 585, 586
Clipping (clipped noise), 317, 318. See also
        Quantization
Closure relationships, 25, 505–510, 563,
        827
    amplitude, 508
    phase, 25, 423, 505
Clouds, atmospheric
    absorption, 704
    index of refraction of, 704
CMB. See Cosmic microwave background

COESA (Committee on the Extension to the
        Standard Atmosphere), 666, 669
Coherence
    complex degree of, 777
    factor, 685
    function, source, 90, 777–780
    function, temporal, 434–436. *See also*
        Autocorrelation function
    mutual, 767–771, 813
    of hydrogen maser, 438
    of oscillator, 434–437
    partial, 778
    propagation of, 781–785
    pulsars and masers, 785
    time, 392, 396, 412, 415, 434–437
Coherency matrix (polarization), 123
Coherent source, 780
Collecting area, 11, 100, 156
Comb spectrum, 275
Compensating delay. *See* Delay, instrumental
Complex correlator. *See* Correlator, complex
Complex degree of coherence, 777
Complex visibility, 30, 70, 91
    definition, 91
Compound interferometer, 29
Compressibility factors, 679
Compton loss, 393
Confusion of radio sources, 27, 172, 579
Continuum radiation, 4
Conventional International Origin (CIO), 618.
        See also Polar motion
Conversion
    frequency. *See* Frequency, conversion
    serial-to-parallel, 352. *See also*
        Demultiplexing
Convolution, 67
    theorem, 69, 82
Convolving functions, 497–501. *See also*
        Smoothing functions
    Gaussian, 498
    Gaussian-sinc, 499, 500
    rectangular, 498
    spheroidal, 500–501
Coordinate
    conversion, 146
    systems, 73, 109–114
Cornell University, 38
Correlator, 102n
    analog, 264
    comparison, lag and FX, 360–366
    complex, 214–215, 228
    digital, 353–360, 375
    FX, 360–361
    hybrid, 366–367

    lag (XF), 358
    multiplexing in, 367
    output in the complex plane, 217, 229
    recirculating, 358
    simple (single-multiplier), 214–215, 238
    software, 373
    systems, 102n
    voltage offset in, 290, 347, 533, 534
Cosmic Background Explorer, 13, 535
Cosmic Background Imager, 192
Cosmic distance ladder, 623
Cosmic microwave background (CMB), 192,
        535–536, 541, 671
    anisotropy of, 192
Covariance matrix, 643
Cross. *See* Mills cross
Cross power spectrum, 99, 353–358, 457, 641.
        See also Spectral line(s)
Cross talk (cross coupling), 193, 295, 533
Cross-correlation, 98–99, 102
    coefficient, 313
Cryogenic cooling, 221, 256
Crystal-controlled (quartz) oscillator, 274, 425,
        436, 439, 442
Culgoora array (Australia), 187
Cycle time, 613, 709
Cyclotron
    frequency, 729
    radiation, 121

Data cube, 528
dBi, 793
Deblurring, 756
Declination, 10
    coordinate conversion, 646
    measurement of, 600–601
Deconvolution, 551–557
    comparison of CLEAN and MEM, 559–562
Delay
    adjustment, 64, 283
    analog, 264
    circuits, digital, 351
    compensating. *See* Delay, instrumental
    delay pattern. *See* Bandwidth, pattern
    errors, 282–283
    fractional sample correction, 364
    geometric, 60, 90, 209, 399, 453, 664, 830
    group, 396, 403, 462, 608, 620, 726
    instrumental, 62, 115, 211, 282–283
    measurement error, 463
    reference, 213, 283
    subsystem, 264, 351
    tracking, 213

Delay resolution function, 463. *See also*
    Bandwidth, synthesis
Delay-setting tolerances, 215, 283
Delta function
    CLEAN components, 552–553
    Fourier transform of series of, 176–177
    LO frequency, 446
    point source, 116, 171, 247, 601
    shah function, 157
    visibility sampling, 230
Demultiplexing, 367
Depolarization, 752–753
Derivative property, 81
Detector
    power-linear (square-law), 19, 794, 809
    synchronous, 22, 266, 290, 673
Diameter, stellar, 18, 835
Dielectric constant, 675, 676, 825–827
    of plasma, 728
Diffraction at an aperture, 771
Diffraction pattern
    lunar occultation, 814
    scintillation, 740, 750
Digital filters, 368
Digital processing, 309–385
    multipication, 353
    sampling, 312–351
        accuracy, 347–351
    spectral measurements, 353–375
Diode, 208–209, 265, 275, 445, 447
Direct detection, optical, 45, 832–834
Direct Fourier Transform Telescope, 199–200
Direction cosine, 73, 92, 775
Directional coupler, 266
Dirty beam. *See* Beam, synthesized (dirty)
Dirty image, 552
Discrete Fourier transform. *See* Fourier
    transform, discrete
Dispersion in optical fiber, 263, 264, 302–303
Dispersion measure, 751
Dispersion, classical theory, 674–679
Diurnal aberration. *See* Aberration, diurnal
Doppler effect, 61, 170, 358, 444, 446, 631
    analysis and formulas, 538–543
    reference frames, 541, 543
Double-sideband system, 215–236
Dynamic range, 544, 569–570, 602, 803

Earth. *See also* Geodetic measurements; Polar
        motion
    atmosphere, 657–663
    ionosphere, 725–737
    magnetic field, 729

    radius vector, 406, 666
    tectonic plate motion, 3, 622
    tides, 622
Earth remote sensing, 825–827
Earth rotation. *See also* Universal time
Earth rotation synthesis imaging. *See* Synthesis
        imaging
East–west array, 96, 173–178, 491–493
East–west baseline, 59, 247
Ecliptic, 616–617
Editing of data, 485
    for interference, 796
Efficiency
    aperture, 11, 156, 819
    quantization, 228, 335, 461
El Niño, 621, 696
Electron density
    Galaxy, 751
    interplanetary medium, 745
    interstellar medium, 750
Electronics
    historical development, 255
    subsystems, 255–256, 261–264
Elevation, 120, 146
Emissivity, 825–827
Entropy, 557–559
Epoch of Reionization, 200, 536–537, 587
Equatorial mount, 118–120
Equinox, 601, 617
Ergodic waveform, 4, 104
Error function (erf), 233, 323, 330
Errors
    additive, 533, 802
    clock (VLBI), 399–405
    in images, 532–533
    (l, m) origin, 534
    multiplicative, 533, 802
    phase, 277, 570
    pointing, 486, 533
Evolution of synthesis techniques, 13–14
Excess path length, 659
    interplanetary medium, 747
    ionosphere, 726, 732–733
    troposphere, 667–669, 704–705
Extended (broad) sources
    deconvolution, 552, 555
    mosaicking, 570, 577
    response, 530

Fan beam, 26, 176
Far-field assumption, 59, 90, 775
Faraday depth, 752
Faraday dispersion function, 753

Faraday rotation, 4, 121, 141, 730
    dispersion function, 753
    interstellar, 751, 753
    ionospheric, 726, 730
Fast Fourier transform (FFT), 160–161,
            495–497
    Telescope, 199
Fast Fourier transform telescope. *See* Direct
            Fourier Transform Telescope
Fast holographic deconvolution, 587
Feathering, 577
Feeds, bandwidth, 141
Fermat's principle, 665
Fiber optics. *See* Optical fiber
Fiedler events, 757
Field
    far, requirement, 775
    near, observations in, 775
Field of view
    bandwidth effect, 240–246
    fringe-frequency mapping, 634
    restrictions, 774, 775
    visibility averaging effect, 246–251
Field of view problem, 396–398
Filters, 101, 208
    baseband, 262
    Butterworth, 461
    effect on signal-to-noise ratio, 279
    narrowband, 276
    number of poles, 276–277
    phase stability, 276–277
    phase-locked oscillator, use of, 276
    Q-factor, 276, 444–446
    spectral line, 360–367
First point of Aries, 10
Fleurs, Australia, 26, 29
Flux density, 6
Fort Davis, Texas, 38, 190
Fourier series, 86
Fourier synthesis, 20
Fourier transform, 76–87
    derivative property, 81, 427, 816
    direct, 490
    discrete, 159–161, 378–386, 495–497
    fast, 160–161, 360
    integral theorem, 225
    moment property, 81
    projection-slice theorem, 74–75
    relationships, imaging, 166
    sign of exponent, 91
    similarity theorem, 80, 242
    three-dimensional, 98, 580–585
    two-dimensional, 85

Fourth-order moment relation, 224, 314, 810
Fractional bit shift loss. *See* VLBI, discrete
            delay step loss
Fractional frequency deviation, 426
Fraunhofer diffraction, 768, 771. *See also*
            Field, far, requirement
Frequency
    channels, 354
    conversion, 207–209
        multiple, 213, 218
    demultiplexing, 367
    multiplication, 275, 446
    regulation, 805
    response, 277–282
        optimum, 277–279
        tolerances, 279–282
Frequency standards, 425–442
    cesium beam, 441
    cryogenic sapphire, 439
    crystal oscillator, 274, 425, 436, 439, 442
    hydrogen maser, 442–446
    phase noise processes, 431–434
    rubidium vapor, 440–442
Fresnel zone, 740, 815
Fried length, 688, 828
Fringe
    envelope, 61, 64
    finding, 395, 412
    fitting, 236
        global, 419–425
    function (pattern), 20, 61, 68
    rotation (stopping), 214, 220, 261, 298
    rotation, digital, 364, 454–457
    search. *See* Signal search (VLBI)
    visibility, 16
    white light, 64, 395, 831
Fringe frequency, 115–117, 212
    averaging, 796
    baseline solution, 606
    beam, effective, 636
    definition, 115
    effect of tracking, 115, 170
    in astrometry, 606–610
    in VLBI, 411, 412
    interference suppression effect, 796–800
    ionospheric effect on, 733
    mapping with, 633–636
    measurement accuracy, 640
    natural, 212
    spectrum, 416, 634, 641
Fringe rate. *See* Fringe frequency
Fringes, first radio record, 20
Front end, 257. *See also* Receiver

Frozen-screen approximation, 687, 689, 698, 713
FX spectral line processor, 360

Gaia mission, 2
Gain
  calibration, 299
  errors, 281–283, 350
  factor, 212
Gamma function, 747
Gamma-ray burst, 741
Gaussian convolving function, 498
Gaussian probability distribution, 311–312
  piecewise linear approximation, 328
Gaussian random noise, 4
Gaussian random variable, 4, 224, 407–412, 434. See also Bivariate (joint) Gaussian probability distribution
Gaussian taper, 169, 172, 492–495, 553
Gaussian-sinc function, 499, 500
Geodetic measurements, 40, 599, 622
  length of day, 621
Geometric delay, 60, 90, 209, 399, 453, 664, 830
Gibbs phenomenon, 357, 523
Global fringe fitting, 419–422
GLONASS, 791
GMRT, 35, 186
Golomb ruler, 174n
GPS (global positioning system), 3, 447, 620, 734, 736
Granlund system, 272, 273
Grating array, 29–30
Gravitational deflection. See Relativistic effect
Gravity Probe B, 625
Green Bank, West Virginia, 34, 38, 554, 631
Greenwich meridian, 109, 406, 607, 618, 620
Gridding (convolutional), 496–497. See also Cell, averaging
Group delay, 396, 403, 462, 608, 620, 632, 726
Group velocity, 728, 751
Gyrofrequency, 729

Hadamard matrices, 293
HALCA satellite, 190, 473, 475. See also VLBI, VSOP project
Half-order derivative, 177
Hamming weighting (smoothing), 356, 373
Hankel transform, 86, 684
Hann weighting (smoothing), 355, 373, 524
Hanning weighting, 356. See also Hann weighting (smoothing)

Harmful thresholds, 795
Hat Creek Observatory, California, 37, 38
Haystack Observatory, Westford, Massachusetts, 38, 415, 823
Heaviside step function, 79
Heterodyne (superheterodyne receiver), 3, 46
Heterodyne conversion. See Frequency, conversion
Hilbert transform, 104, 106, 215, 229, 352
Hinge point, 704
Hipparcos satellite, 2
  star catalog, 602, 603
Historical development, 13–44
  analog Fourier transformation, 544
  imaging from one-dimensional profiles, 543
  receivers, 221, 256
  VLBI, 37–42, 391–394
Holes in spatial frequency coverage, 173
Holography. See Antenna(s), measurements, holographic
Hour angle, 109–113, 146
Hubble constant, 543
Hybrid correlator, 366
Hybrid mapping, 40, 563
Hydrogen line, 6, 31, 442–446
Hydrostatic equilibrium, 661

IAU
  polarization standard, 124, 125
  radio source nomenclature, 10
ICRF, 11, 40, 602
ICRS, 602
IEEE
  committee on frequency stability, 426
  polarization standard, 124
  power flux density, 6
IF. See Intermediate frequency
Illumination, aperture. See Antenna(s), aperture illumination
Image defects. See also Phase, noise
  correlator offset, 533
  distortion, 533
  errors in visibility data, 533
  ghost, 530–532
Incoherence assumption (spatial), 90, 774, 810
Incoherent averaging, 415–419, 425
Incoherent source, response to, 777–779
Inertial reference frame, 602
Infrared interferometry
  detection of planets, 831
  heterodyne detection, 834

Instrumental (compensation) delay. *See* Delay,
        instrumental
Instrumental polarization, 129–133,
        137–142
    degrees of freedom, 139
Intensity, 9, 530
    derivation, 490, 504
    interpretation, 530
    scale, 530, 564
Intensity interferometer, 24–25, 809–814
    optical, 835
    sensitivity of, 419
Interference, radio
    satellites, 804–805
    (u, v) plane distribution, 797–800
    VLBI, 801–803
Interferometer
    adding (simple), 19, 20
    basic components, 99–102
    compound, 30
    correlator, 22
    infrared. *See* Infrared interferometry
    intensity. *See* Intensity interferometer
    Michelson, 14–18
    optical (modern Michelson), 827–835
    sea, 20
    spectral-line, 31
Intermediate frequency (IF), 208
    amplifier, 262
    subsystem, 261–262
International Astronomical Union. *See* IAU
International Atomic Time (IAT), 619
International Celestial Reference Frame, 11,
        40, 602
International Reference Ionosphere (IRI),
        734
International Telecommunication Union (ITU),
        793
    Radiocommunication Sector, 795
International VLBI Service (IVS), 40
Interplanetary medium
    excess path length, 747
    refraction, 744–748
    scintillation, 748–750
    scintillation index, 749
Interpolation, 159, 496–497. *See also* Gridding
        (convolutional)
Interstellar masers, 755
Interstellar medium, 750–758
    dispersion measure, 751
    electron density, 751
    Faraday rotation, 751–753

pulsar signals, effects on, 751
scattering
    diffractive, 753–758
    Fiedler, 757
    refractive, 756–758
Invisible distribution, 490
Ionosphere
    absorption, 725–735
    achromatic spherical lens, 731
    acoustic-gravity waves, 736
    effects of irregularities, 735–737
    Faraday rotation, 726, 730
    Gaussian screen model, 737–742
    index of refraction, 728–730
    phase stability, 726
    power-law model, 742–744
    propagation delay, 730–742
    refraction, 730–733
    scintillation, 736, 737
    total electron content, 726, 732, 737
    traveling ionospheric disturbances (TIDs),
        736
Isoplanatic
    angle, neutral atmosphere, 828
    patch
        ionosphere, 509, 587, 726
        neutral atmosphere, 828
ITU, 805
ITU-R. *See* International Telecommunication
        Union, Radiocommunication Sector

J^2 synthesis (J-squared synthesis), 187
Jansky (unit), 6
Jinc function, 492
Jodrell Bank Observatory, England, 24, 26, 34,
        187, 391
Johnson noise, 4
Jones matrix, 134
Julian year, 617
Jupiter, 391

Kolmogorov turbulence, 685, 690, 743–744,
        754
Kramers–Kronig relation, 674, 678
Kronecker delta function, 380

Leakage (polarization), 130, 146–147
Leakage (sampling), 158

Leap second, 619
Least-mean-squares analysis, 636–648
    accuracy, 645
    correlated measurements, 643
    covariance matrix, 643
    design matrix, 645
    error ellipse, 644, 648
    estimation of delay, 641
    estimation of fringe frequency, 641
    likelihood function, 636
    matrix formulation, 643
    nonlinear case, 646
    normal equation matrix, 643, 644, 647
    partial derivative matrix, 643
    precision, 645
    self-calibration, 566
    sinusoid fitting, 236
    variance matrix, 643
    weighted, 637
Length of day, 620
Lensclean, 589
Light, speed (velocity) of, 60, 600
Likelihood function, 636
Line of nodes, 617. *See also* Equinox
Linear arrays, 173–178
Lines, radio. *See* Spectral line(s)
Lloyd's mirror, 20
LO. *See* Local oscillator
Local oscillator, 3, 46, 207–209, 261. *See also*
        Frequency standards
    independent, 37. *See also* VLBI
    laser, 832
    multipication, stability, 446
    nonsynchronized, 813
    phase switching of, 298
    synchronization of, 264–277
Local standard of rest, 541
Long-baseline interferometer, 391
Loran, 447
Lorentz equation, 728
Lorentz factor, 542
Lorentzian profile, 676, 678, 684
Low-frequency imaging, 587–589
Low-noise input stage, 221, 256
Lunar occultation
    optical, 814, 818
    radio, 24, 601, 749, 814–819
Lutz–Kelker effect, 624

Magellanic Cloud, Small, 575
Magnetic fields
    in frequency standards, 442, 443,
        446

interstellar, 750
    terrestrial, 729
Magnetic tape recording, 37, 450
Magnitude of visibility, $72n$
Mapping
    two-dimensional, 73
    visibility amplitude only, 569
    wide-field, 95–98, 245–246, 570–577
Markov chain Monte Carlo (MCMC)
        algorithm, 513, 624, 646
Maryland Point Observatory, Maryland, 417
Maser frequency standard, 442–446
Maser radio sources, 6, 38, 391, 393, 417
    mapping procedures, 631–636
    scattering, 755
    spatial coherence, 785
Master oscillator, 261
Mauna Kea, Hawaii, 37, 187, 693, 695
Mauritius Radio Telescope, 186
Maximum entropy method (MEM), 557–559,
        590
Maximum-likelihood method, 464, 636,
        648
Maxwell's relation, 676
MeerKAT, 199
Meridian, 109
    Greenwich, 109, 406, 607, 618, 620
    local, 109, 600, 620
    plane, 109, 604
    transit (crossing), 620
MERLIN, 26, 187, 566
Meter, definition, 600
Michelson interferometer, 13–18
Microwave link. *See* Radio link
Mie scattering, 824
Millibar, 659
Millimeter-wavelength arrays, 37
Mills cross, 26–28, 168–172
Minimum redundancy. *See* Arrays, minimum
        redundancy; Bandwidth, synthesis
Mirror-image reception pattern, 67
Misell algorithm, 823
Mixer, 208–209. *See also* Frequency,
        conversion
    sideband-separating, 301
MKSA units, $675n$
Model
    adaptive calibration, 566
    circular disk, 17
    Cygnus A, 25
    delta function (CLEAN), 552
    fitting, 510–520
    Gaussian, 17, 32
    rectangular, 17

Modulated reflector, 265
Molonglo, Australia, 28
Moment property, 81
Moon, 44. *See also* Lunar occultation;
    Precession
Mosaicking (mosaic imaging), 570–577, 692
    arrays for, 570–577
    linear, 574
    nonlinear, 574–575
    on-the-fly, 577
Mueller matrix, 136
Mullard Radio Astronomy Observatory. *See*
    Cambridge, England
Multifrequency synthesis, 578
Multiplier (voltage), 22, 210. *See also*
    Correlator
Mutual coherence function, 767–771

Nançay Observatory, 30, 31
Narrabri, Australia, 835
National Aeronautics and Space
    Administration (NASA), 3,
    40, 453
    Extragalactic Database (NED), 5, 11
National Geodetic Survey (NGS), 3
National Radio Astronomy Observatory
    (NRAO), 34, 38, 180, 392, 453, 568.
    See also ALMA; Green Bank, West
    Virginia; Very Large Array; Very
    Long Baseline Array
Natural weighting, 232, 491, 494
Naval Observatory, U.S. (USNO), 3, 619
Naval Research Laboratory (NRL), 3, 417
NAVSTAR. *See* GPS
Near-field observations, 473, 775
Negative frequencies, 64, 70, 105, 106
Network Users Group (US), 38–41
Neutral atmosphere
    opacity, 693–696
    phase stability, 697–701
Nobel lecture, Ryle, 35
Nobeyama Radio Observatory (NRO), Japan,
    37
Noise. *See also* Signal-to-noise ratio
    amplitude and phase, 233–235
    equivalent power (NEP), 834
    in complex visibility, 228–230, 236
    in image, 230–233
    in oscillators
        flicker-frequency, 433–434
        flicker-phase, 433–434
        random-walk-of-frequency, 433–434

white-frequency, 433–434
    white-phase, 433–434
    in VLBI, 407–412
photon shot noise, 441, 445, 834
power, 11, 257
quantum effect, 44–45
response to, 223–235
temperature measurement, 257–260
Noncoplanar baselines, 97–98, 579–585
    3D Fourier transform, 581–582
    polyhedron mapping, 582
    snapshot combination, 582
    variable point-source response, 583
Nonnegative, least-squares, 562
North Liberty, Iowa, 38
NRAO. *See* National Radio Astronomy
    Observatory
NRAO VLA Sky Survey (NVSS), 11
Nuffield Radio Astronomy Laboratories. *See*
    Jodrell Bank Observatory, England
Nutation, 2, 10, 616–617
NVSS. *See* NRAO VLA Sky Survey
Nyquist power theorem, 12
Nyquist rate (frequency), 312–313. *See also*
    Sampling theorem
Nyquist sampling theorem, 46

Observation, planning, and reduction, 534–535
Occultation observations. *See* Lunar
    occultation; Precession
On-the-fly mosaicking, 577
Opacity, 670–672
    measurement of, 672–673
Optical depth. *See* Opacity
Optical fiber, 262–264, 273–274
    dispersion, 263, 302–303
    high stability, 274
Optical interferometry, 45, 827–835
    direct and heterodyne detection, 832–834
Orbiting VLBI. *See* OVLBI
Oscillator coherence time, 434–436
Oscillator strength, 675
Outer product, 135, 293
Overlap processing, 360
Overlapping segment averaging, 360
Oversampling, 313, 315–316, 325, 328, 347
OVLBI, 42–44, 188–191, 470–476
    data link, 471–472
    round-trip phase, 471
    timing link, 471–472
Owens Valley Radio Observatory, California,
    30, 32, 37, 38, 706

Parabolic main reflector, 155
Parabolic-cylinder reflector, 153
Parallactic angle, 121, 128, 139, 141
Parallax, 617, 623
Parametric amplifier, degenerate, 221
Parametrized Ionospheric Model (PIM), 734
Parseval's theorem, 84, 376, 381, 685, 799, 803
Partial coherence, 778
Passband
    Gaussian, 64, 244
    rectangular, 64, 243, 278
    tolerances, 279–282
passive radar, 824
Peeling, 587
Pencil beam, 27, 177
Permittivity, 675n
Peryton, 790
Phase
    errors, effect on sensitivity, 277
    noise
        effects on maps, 570, 633
        in frequency multipliers, 446
        in frequency standards, 415, 425–434
        in receivers, 233–235, 407–408
        neutral atmospheric, 680–692, 813
Phase center shifting, 398
Phase closure, 25–26, 40, 393, 489, 505–510
Phase data
    imaging without, 569
    uncalibrated, 563–569
Phase fluctuations, 680–690
Phase reference
    feature, 632–636
    position, 73, 89, 93, 110
Phase referencing
    atmospheric effects, 709–711
    for masers, 632
    in VLBI, 610–616
Phase stability
    analysis of, 425–436
    in reference distribution, 264–274
    of filters, 276–277
    of frequency standards, 436–440, 446
    of local oscillators, 446
Phase switching, 30, 290–298, 348, 349
    in Mills cross, 27
    in simple interferometers, 21–23, 29
    interaction with fringe rotation and delay,
        298
Phase tracking center. See also Phase reference,
    position
Phase-locked oscillator, 267, 274–276, 436
    loop natural frequency, 274
Phase-tracking center, 89

Phased array, 162–164, 187, 365n,
    466–470
    as VLBI element, 466–470
    correlator array, comparison with,
        162–164
    randomly phased, 466
Photo bunching noise, 49
Pico de Veleta, Spain, 821
Planar arrays, 192–193
Plancherel's theorem, 84n
Planck formula, 9, 12, 259–264, 833
Planck mission, 535, 539
Planetary nebula, 5, 511
Planets, 557, 601, 616. See also Burst
        radiation, Jupiter
    as calibration sources, 537–538
Plasma. See also Interplanetary medium;
            Interstellar medium; Ionosphere
    absorption in, 735
    frequency, 728
    index of refraction, 729–730
    oscillations, 121
    propagation in, 727–758
    RF discharge, 440, 442
    turbulence, 742–744
Plateau de Bure, France, 37
Pleiades, 626
Point-source response, 65, 165, 171, 551.
        See also Beam, synthesized (dirty)
Point-spread function, 836. See also
        Point-source response
Pointing correction, 486
Poisson distribution, 45
Polar motion, 2–3, 617, 618
    measurement of, 620–621
Polarimetry, 121–142
Polarization
    calibration, 137–142
    circular, 122, 123, 128, 129, 140, 141
    complex degree of, 752–753
    cross-polarized, 127–129
    degree of, 122
    design considerations, 141, 142
    ellipse, 123–125
    emission processes, 4, 121
    identically polarized, 126–127
    instrumental, 129–133
    linear, 122, 127–129, 141
    matrix formulation, 134–137
    mismatch tolerance, 289
    parallactic angle effect, 128
    position angle, 122. See also Faraday
        rotation
Polyphase filter banks, 369–373

Position measurements
    early, 24
    methods. *See* Astrometry
Power (density) spectrum, 63, 98
    atmospheric phase, 686–690
    correlator output, 227
    interplanetary scintillation, 748–750
    phase and frequency fluctuations,
        425–434
Power combiner, 162, 164
Power flux density, 6
Power reception pattern. *See* Antenna(s),
        reception pattern
Power-law antenna spacing, 180, 182
Power-law turbulence relations, 690
Poynting vector, 6
Precession, 2, 10, 616–617
Price's theorem, 322, 327
Principal response, 494, 591
Principal solution. *See* Principal response
Probability
    of error, 412–415
    of misidentification, 415
Probability density function, 741
Probability distribution
    bivariate Gaussian, 311–312, 644, 741
    Gaussian, 156, 224, 311, 332, 407, 411,
        434
    Rayleigh, 234, 407, 410, 414
    Rice, 234, 408
Projection-slice theorem, 74–75
Prolate spheroidal wave functions, 501
Propagation
    constant, 405, 659
    interplanetary, 744–750
    interstellar, 750–758
    ionospheric, 725–737
    neutral atmosphere, 658–705
Proper motion, 10, 617, 623–627
Pulsar, 465
    astrometry, 602
    correlator gating, 365
    determination of vernal equinox, 601
    dispersion measure, 745, 751
    proper motions, 753
    scintillation, 756
    spatial coherence, 784
    timing accuracy, 439, 448
Pulse calibration (VLBI), 447

Q-factor of
    cavity, 444–446
    filter, 276

Quadrature
    network, 215, 348, 355
    phase shift, 222, 229, 298, 352
Quadruple moment theorem. *See* Fourth-order
        moment relation
Quadrupod, 157, 822
Quantization
    comparison of schemes, 346
    correction, 365, 377–378
    efficiency factor, 228, 328–336, 453, 461
    four-level, 320–326, 377–378
    in VLBI systems, 453, 460
    indecision regions, 350–351
    noise, 309
    repeated (requantization), 368, 470
    three-level, 332–336
    thresholds, 319–320, 327
    two-level, 316–320
Quantum noise, 44, 833
Quantum paradox, 44
Quasar, 4, 38, 41, 393, 567. *See also* Radio
        source

Rademacher functions, 291*n*
Radial smearing. *See* Bandwidth, effect in
        maps
Radiative transfer equation, 671
Radio interference
    adaptive cancellation, 790
    adaptive nulling, 793
    airborne and space transmitters,
        804–805
    decorrelation, 800–801
    deterministic nulling, 791, 792
    fringe-frequency averaging, 796–800
    threshold pfd and spfd
        short and intermediate baselines,
            796–801
        total power systems, 794
        VLBI, 801–803
Radio lines. *See* Spectral line(s)
Radio link, 24, 26, 261
Radio source
    0134+329, 11
    0748+240, 691
    1548+115, 567
    1622+633, 190
    1638+398, 612
    1641+399, 612
    3C138, 140
    3C147, 489
    3C224.1, 554
    3C273, 39, 601, 628, 630, 631, 706, 748

3C279, 628–631, 784
3C286, 140, 489
3C295, 489
3C33.1, 32
3C48, 4, 5, 10, 489
Algol, 601
Cassiopeia A, 4, 5, 19, 23–25, 34
Centaurus A. See NGC5128
Crab Nebula, 24, 745
Cygnus A, 4, 5, 10, 11, 19, 20, 24–26, 34,
    35, 37, 735
  central component (VLBI), 36
  fringe pattern, 19, 21
  map or image, 26, 34–36, 568
IM Peg, 625
J1745-283, 627
Jupiter, 37
M82, 4, 5
M87. See NGC4486
MG J0751+2716, 41
MWC349A, 5, 6
NGC4258, 41, 542
NGC4486, 24, 559, 593
NGC5128, 24
NGC6334B, 755
NGC7027, 5, 10, 489
Orion Nebula, 6, 8, 625, 626
Orion Nebula Cluster (ONC), 626
Orion water-line maser, 821
Pleiades, 626
PSR 1237+25, 784
Sagittarius A*, 40, 626–627, 755, 757
Sun, 20, 28, 30, 187, 543
Taurus A. See Crab Nebula
TW Hydrae, 4, 5
Venus, 5, 6, 310, 827
Virgo A. See NGC4486
W3 (OH), 417
W49, 631
Radio source nomenclature, 10
Radio spectrum, regulation of, 805
Radiometer equation, 46–49
Radiosonde data, 674, 693, 703
Raised cosine weighting. See Hann weighting
    (smoothing)
Rayleigh distribution, 234, 407
Rayleigh–Jeans approximation, 9, 12, 257,
    258, 260. See also Planck formula
Rayleigh scattering, 824
Rayleigh theorem, 84n
Receiver
  electronics, 255–264
  phase switching, 21–23, 27, 29, 290–298.
    See also Phase switching

temperature, 12, 46, 257
  for cascaded components, 258
Reception pattern. See Antenna(s), reception
    pattern; Voltage reception (response)
    pattern
Recording systems (VLBI), 448–450
Rectangular function, 78
Redundancy measure, 174
Reference frames. See ICRF; ICRS
Reflections
  in cable, 267, 270, 279
  in optical fiber, 263
Reflector, modulated, 265
Refraction
  anomalous, 692
  in neutral atmosphere, 658, 663–669
  in plane-parallel atmosphere, 663
  index of, 659
  interplanetary, 744–748
  ionospheric, 728–733
  optical, 657
  origin of, 674–679
  spherically symmetric, 666,
    744–748
Refractivity. See also Refraction, index of
  optical, 680
Relative sensitivity of systems, 235–239
Relativistic effect
  general relativistic bending, 628, 748
  Lorentz factor, 542
  time transfer effects, 447
Resolution
  atmospheric limitation of, 680–692
  MEM, 559
Restoration from samples. See Sampling
    theorem
Retarded baseline, 405–407
Reuleaux triangle, 183–187
Reynolds number, 686
Rice distribution, 234, 408, 566
Right ascension, 10
  measurement of, 601–606, 645
  zero of, 601
Ringlobe, 176–178
RMS bandwidth, 463, 608, 641
Robust weighting, 494
Rotation measure, 751, 753
Round-trip phase, 264–271, 471, 487
Ruze formula, 156

Sampling, 312–351. See also Quantization
  digital, accuracy of, 347–351
  of bandpass spectrum, 312

Sampling theorem, 157–159, 175, 177, 312,
　　　573, 820
Satellite
　　data link, 38, 471–472
　　interference from, 801, 804
　　signals, Faraday rotation, 734
　　time transfer, 447
　　tracking, 475, 824
Scalloping, 361
Scattering, 750, 753–758, 783. *See also*
　　　Scintillation
Schwarzschild radius, 542
Scintillation
　　angular spectrum of, 739, 743
　　correlation bandwidth, 740
　　correlation length, 740
　　critical source size, 740
　　Gaussian screen model, 737–742
　　index, 749
　　interplanetary, 748–750
　　interstellar, 188, 753–758
　　ionospheric, 736, 737
　　neutral atmosphere, 685–690
　　power-law model, 742–744
　　scattering angle, 739, 743
　　thin screen, 737–742
Sea interferometer, 20
Second, definition of, 441, 600. *See also* Time
Seeing, 657. *See also* Scintillation
　　cell, 828
　　disk, 836
Self-absorption, 4
Self-calibration, 565–569
Self-coherence, 778
Sequency, 292
Serial-to-parallel conversion, 352
Serpukhov, Russian Federation, 28
Sgn function, 79, 105
Shah function, 157, 496–497
Shift theorem, 80, 526
Shift-and-add algorithm, 836
Short-spacing data, 177, 576–577
Shot noise, photon, 441, 445, 834
Sideband(s), 208
　　double, 215–223, 238–239
　　fringe-frequency dependence, 115–117
　　partial rejection of, 251–253
　　relative advantages of single, double,
　　　221
　　separation, 221–223
　　sideband-separating (image-rejection)
　　　mixer, 301
　　single (upper, lower), 209, 210, 212
　　unequal responses, 251–253

Sidelobe. *See also* Ringlobe; Beam,
　　　synthesized (dirty)
　　bandwidth smearing of, 243
　　envelope model, 794
Sidereal rate (Earth rotation), 119
Signal search (VLBI), 412–419
Signal transmission subsystems, 261–264
Signal-to-noise ratio, 231. *See also* Noise
　　aliasing effect, 501–503
　　basic analysis, 223–240
　　coherent averaging, 415–419
　　　of frequency standard, 436–440
　　frequency response, effect of, 277–282
　　fringe-frequency mapping, 634
　　in images, 230–233
　　in interference calculations, 793–803
　　in lunar occultations, 818
　　in phased arrays, 466–470
　　incoherent averaging, 415–419
　　intensity interferometer, 419, 814
　　loss factors, VLBI, 454–461
　　optical, 833–834
　　quantization, effect of, 316–347
　　quantum effect, 834
　　receiving system, 12
　　systems, relative, 235–239
Signals
　　cosmic, 4–10
　　ergodic, 4, 104
　　spurious, 290, 298. *See also* Errors
Sinc function, 61
　　definition, 61
Single sideband mixer. *See* Sideband(s),
　　　sideband-separating mixer
Site testing
　　opacity, 693–696
　　phase stability, 697–706
SMA (Submillimeter Array). *See also*
　　　Submillimeter Array
Smearing, circumferential. *See* Visibility,
　　　averaging
　　radial. *See* Bandwidth, effect in maps
Smith–Weintraub equation, 680
Smithsonian Astrophysical Observatory
　　　(SAO), 14, 187
Smoothing functions, 355
SMOS (Soil Moisture and Ocean Salinity), 825
Snapshot, 182, 185
Snell's law, 663, 745
　　spherical coordinates, 666, 745
Soil temperature, 264
Solar imaging, 28–30, 543
Solar system studies, 28–30, 775
Solar wind, 744–748

Source. *See* Radio source
  calibration, 395, 485–489
  coherence, 777–780
  completely coherent, 780
  extended. *See* Extended (broad) sources
  far-field condition, 59, 90, 775
  incoherence requirement, 90, 769–771, 774
  model. *See* Model
  radio. *See* Radio source
  subtraction, 534. *See also* CLEAN
      algorithm
Source catalog
  3C, 10, 27
  ICRF, 11
  Messier, 10
  NED, 5, 11
  NGC, 10
  NVSS, 11
South Pole, 693, 697
Space debris, tracking, 824
Space Interferometry Mission (NASA),
    831
Space VLBI. *See* OVLBI
Spatial frequency, 66, 70, 164–166
  coverage, 165–166
  filter, 165
Spatial incoherence. *See* Source, incoherence
    requirement
Spatial sensitivity
  of aperture antenna, 576–577
  of correlator array, 164–167
  support of, 165
Spatial transfer function. *See* Transfer function
Specific intensity. *See* Intensity
Speckle imaging, 835–838
  phase information, 838
  shift-and-add, 836
Spectral
  flux density, 6
  power flux density, 6
Spectral line(s)
  absorption, 31
  accuracy, 528
  adaptive calibration, 585
  analog correlator, 264
  atmospheric absorption, 670–673, 694
  bandpass calibration, 523–524
  bandpass ripple, 523–524
  baseline ripple, 357
  calibration procedure, 523–530
  chromatic aberration, 527
  CLEAN procedures, 585
  correlators, 353–375
  digital correlators, 353–375

Doppler shifts, 538–543
  reference frames, 541–543
double-sideband observation, 221
examples of
  CO, 10, 693
  H_2, 701–702
  H_2O, 6, 38, 417, 631, 635
  hydrogen, 6, 31, 442
  OH, 6, 391, 631
  SiO, 6
presentation, 528–530
radiation. *See* Spectral line(s); Maser radio
    sources
systems, 31, 353–375
table of important, 6
velocity reference frames, 541
VLBI procedures, 41–42, 405, 524–527,
    631–636
Spheroidal wave functions, 500–501
Splatalogue, 6
Stanford, California, 29
Stars
  observation of, 14–18, 600, 814, 819, 831,
      835
  proper motion, 617
Statially coherent source, 780
Step-recovery diode, 447
Stokes parameters, 121–123
Stokes visibilities, 126–129
Strehl ratio, 156
Structure function
  phase (spatial), 681, 686–690, 743–744
  phase (temporal), 435, 689
  refractive index (spatial), 686
Submillimeter Array (SMA), 14, 187, 470, 822
Sun
  coronal refraction, 744–748
  gravitational effects, 616
  interference from, 533
  ionosphere, 726
  observation of, 20, 28–30, 187, 543
  relativistic deflection, 748
  solar time, 619
  solar wind, 744–748
Sunyaev–Zeldovich effect, 522
Superluminal motion, 38
Support of a function, 165
Survey interferometers, 26–28
Swarup and Yang system, 265–266
Symmetry, *n*-fold, 180
Synchronous detector, 266, 290, 673
Synchrotron radiation, 4, 121, 393, 752
Synthesis imaging
  evolution of techniques, 13, 31, 34

Synthesized beam. *See* Beam, synthesized
    (dirty)
System equivalent flux density (SEFD), 12,
    489, 526
System temperature, 12, 225–228, 239, 486
    correction for atmospheric absorption,
        672–673
    measurement of, 300

Tangent plane, 93, 96, 98, 581
Taper. *See* Gaussian taper; Weighting
Target source, 487
TDRSS experiment, 14, 42
Tectonic plates, 3, 42, 622
Telephone signal transmission, 395
Temperature
    antenna, 12
    receiver, 12, 257–260
    system. *See* System temperature
Temperature coefficient of length, 264
Thomson scattering (incoherent backscatter),
        734, 745
Time
    averaging of visibility, 246–249
    definition of second, 441
    demultiplexing, 367
    International Atomic (IAT), 441, 619
    multiplexing, 274
    solar, 619
    time synchronization, 447
    transfer methods, 447
    universal time, 447, 619–621
Timing accuracy, 117, 448
Tipping-scan method, 672
Tolerances in
    bandpass (frequency response), 281–282
    delay-setting, 283
    polarization, 290
    three-level sampling, 348–351
Total electron content. *See* Ionosphere, total
    electron content
Transfer function, 164–166, 169–172. *See also*
        Spatial sensitivity
    OVLBI, 188–191
    VLA, 185
    VLBA, 189
Transmission lines. *See* Cables; Local
        oscillator, synchronization of;
        Optical fiber; Waveguide
Traveling ionospheric disturbances (TIDs), 736
Tripod, 157
Troposphere. *See* Atmosphere, neutral
Truncated function, 107

T-shaped array, 28, 168, 180
Turbulence
    Allan variance of, 445
    in neutral atmosphere, 685–690
    inner and outer scales of, 686–690
    Kolmogorov, 685–690
    power-law relations, 690
    spectrum of phase fluctuations, 689
    structure function of phase, 686
Two-dimensional array, 73–74, 179–193
Two-dimensional synthesis, 73

(u, v) plane (spatial frequency plane), 73, 92
    coordinates, 73, 92
    coverage. *See* Spatial frequency, coverage
    holes in coverage, 173
    in CLEAN algorithm, 556
    in interference susceptibility, 797–800
    interpolation in, 159, 161, 497–501
(u', v') plane, 96–98, 113–114
(u, v, w) component, 91–93, 580–585
    in fringe-frequency averaging, 797
    in visibility (time) averaging, 246–249
Uncertainty principle, 44, 81, 440, 834
Undersampling, 313, 316
Uniform weighting, 492–495
Unit rectangle function, 78, 279–280, 498,
        517
Universal time, 619–621
Usuda, Japan, 188, 475
UTR-2, 186

Van Cittert–Zernike theorem, 1, 94, 767–775
    assumptions, 774
    derivation, 767–771
Van Vleck relationship, 318
Van Vleck–Weisskopf profile, 671, 679
Varactor diode, 275, 445
Variance matrix, 643
Velocity standard. *See* Spectral line(s)
Vermilion River Observatory, Illinois, 38
Very Large Array (VLA), 11, 36, 261, 282,
        588, 784
    antenna configuration, 180–182
    atmospheric phase noise, 613, 691
    delay increments, 283
    dynamic range, 570
    images from, 35, 567, 568
    interference thresholds, 795, 799
    opacity at site, 693
    phase switching, 299
    phased-array mode (VLBI), 466

self-calibration, 567
(*u, v*) spacing loci, 183
Very Long Baseline Array (VLBA), 40, 188,
    453
  phase referencing, 612
Very-long-baseline interferometry. *See* VLBI
Visibility
  at low spatial frequencies, 571–577
  averaging, 246–249
  complex, 30, 70, 91
  fringe (Michelson), 16
  lensclean, 590
  model fitting, 510–522
  reduction due to phase noise, 277, 680–685,
    741–742
Visibility frequencies, 115–117
Visibility–intensity relationship, 89–91,
    767–775
VLA. *See* Very Large Array
VLBA. *See* Very Long Baseline Array
VLBI
  antenna polarization (parallactic) angle),
    141
  antennas
    in space, 188–191, 470–476
    nonidentical, 120–121
  arrays, 37–42
  astrometry, 606–616, 645
  atmospheric limitations, 609, 685
  bandwidth synthesis, 462–464
  burst mode, 465
  calibration sources, 395, 489. *See also*
    Phase referencing
  clock errors, 399–405
  closure phase, 421–422, 563–569
  coherence time, 392, 434–436
  coherent and incoherent averaging,
    415–419
  data encoding, 448–450
  data storage systems, 448–450
  development of, 37–42, 391–394, 748
  discrete delay step loss, 459–461
  double-sideband system, 223, 236
  fractional bit shift loss. *See* VLBI, discrete
    delay step loss
  frequency standards, precise, 436–440,
    446
  fringe detection, 422
  fringe fitting
    global (multielement), 419–425
    two-element, 412–419
  fringe rotation, 454–457, 461
  fringe rotation loss, 454–457
  fringe sideband rejection loss, 457–459

  geodesy, 40
  group delay, 403, 462
  hybrid mapping, 563–564
  in geodesy, 622
  interference sensitivity, 801–803
  K-4 system, 453
  local oscillator stability, 446
  Mark I system, 392, 417, 450, 453
  Mark II, III, and IV systems, 450, 453
  masers, mapping, 631–636
  networks, 38–41
  noise in, 407–412
  orbiting. *See* OVLBI
  phase calibration system, 447
  phase closure, 40
  phase noise, 233–235. *See also* VLBI,
    atmospheric limitations
  phase referencing, 610–616
  phase stability, oscillators, 425–436
  phased-array elements, 466–470
  polar motion observations, 620–621
  probability distributions, 407–415
  pulse calibration system, 447
  quantization loss, 392, 461
  RadioAstron, 473, 474, 636
  recording systems, 448–450
  relativistic bending measurements, 748
  retarded baseline, 405–407
  S2 system, 453
  satellite link, 38
  satellite positioning, 473–476
  sideband separation, 223, 236
  signal-to-noise ratio, 392–393, 418,
    454–461
  spectral line, 405, 460, 524–527
  TIDs, observation of, 736
  time synchronization, 447, 448
  triple product, 423–424
  VSOP project, 473, 475
  water-vapor radiometry, 700, 703–705
VLBI Space Observatory Programme (VSOP),
    14, 43
Voltage reception (response) pattern, 100–101,
    166, 772
  measurement of, 819
VSOP. *See* VLBI Space Observatory
    Programme

w component, 92–94, 115, 580–585, 800
Walsh functions, 292–295
  natural order, 294
  orthogonality, period of, 291
  sequency, 292

Water vapor
    22-GHz line, 677
    absorption, 670–673
    effect on phase, 680–685
    maser, 43, 417, 631–636
    resonance model, 675–679
    turbulence, 685–690
    worldwide distribution, 661
Water-vapor radiometry, 703–705
Water-vapor refractivity, 657–663, 680
Waveguide, 258, 261
Weighting
    antenna excitation, 169, 576–577
    function
        atmospheric, 683
        spectral, 355–357, 524
    natural, 491, 494
    of visibility, 231–232, 491–495
Westerbork Synthesis Radio Telescope, 35,
        128, 175, 176, 466, 499
Westford, Massachusetts. *See* Haystack
        Observatory

White light fringe, 64, 395, 831
WIDAR, 359
Wide-field imaging, 95–98, 245, 487, 503–504,
        570–577, 581–585
Wiener–Khinchin relation, 63, 87, 98–99, 225,
        312, 354, 800
Wind spacecraft, 745
WMAP mission, 535

X-ray interferometry, 832

Young's two-slit interferometer, 44
Y-shaped array, 180–182

Zeeman effect, 121, 140, 442, 446
Zenith opacity, 670, 693–696
Zero padding, 360, 384–385
Zero spacing problem. *See* Short-spacing data

Printed in the United States
By Bookmasters